physics

for scientists and engineers

Paul A. Tipler

physics for scientists and engineers

Fourth Edition

Volume 2

Electricity and Magnetism
Light

W.H. FREEMAN AND COMPANY/WORTH PUBLISHERS

Physics for Scientists and Engineers
Fourth Edition, Volume 2
Paul A. Tipler

Manufactured in the United States of America
Library of Congress Catalog Card Number: 98-60168
Volume 1 (Chapters 1–21) paperback ISBN: 1-57259-491-8
Volume 1 (Chapters 1–21) hardcover ISBN: 1-57259-812-3
Volume 2 (Chapters 22–35) paperback ISBN: 1-57259-492-6
Volume 2 (Chapters 22–35) hardcover ISBN: 1-57259-813-1
Volume 3 (Chapters 36–41) paperback ISBN: 1-57259-490-X
Volume 3 (Chapters 36–41) hardcover ISBN: 1-57259-814-X
Volumes 1 and 2, ISBN: 1-57259-614-7
Volumes 1, 2, and 3, ISBN: 1-57259-615-5

Printing: 1 2 3 4 5 02 01 00 99 98

Executive Editors: Anne C. Duffy and Susan Finnemore Brennan
Development Editors: Steven Tenney and Morgan Ryan, with Richard Mickey
Marketing Managers: Kimberly Manzi and John Britch
Design: Malcolm Grear Designers
Art Director: George Touloumes
Production Editor: Margaret Comaskey
Production Manager: Patricia Lawson
Layout: Fernando Quiñones and Lee Mahler
Picture Editor: Elyse Rieder
Graphic Arts Manager: Demetrios Zangos
Three-dimensional art by DreamLight Incorporated
Illustrations: DreamLight Incorporated and Mel Erikson Art Services
Composition: Compset, Inc.
Separations: Creative Graphic Services
Printing and Binding: R. R. Donnelley and Sons
Cover Image: Sand atop a vertically driven shaker table spontaneously
forms a roughly sinusoidal outline. Image by Max Aguilera-Hellweg.

Illustration credits begin on page IC-1 and constitute
an extension of the copyright page.

W.H. Freeman and Company
41 Madison Avenue
New York, NY 10010 U.S.A.

For Claudia

preface

In this fourth edition I have worked toward four goals:

1. To help students increase their experience and ability in problem solving
2. To make the reading of the text easier and more fun for students
3. To bring the presentation of physics up to date to reflect the importance of the role of quantum theory
4. To make the text more flexible for the instructor in a wide variety of course formats

Enhanced Problem Solving

To help students learn how to solve problems, the number of worked *Examples* that correspond to intermediate-level problems has been greatly increased. Especially notable is a new two-column side-by-side example format that has been developed to better display the text and equations in worked examples. Care has been taken to show the students a logical method of solving problems. Examples begin with strategies, and often diagrams, in a *Picture the Problem* prologue. When possible, the first step gives an equation relating the quantity asked for to other quantities. This is usually followed by a statement of the general physical principle that applies. For example, this step may be "Apply Newton's second law" or "Use conservation of energy." Examples usually conclude with *Remarks* that discuss the problem and solution, and in many cases there are additional *Check the Result* sections that teach the student how to check the answer, as well as *Exercises* that present additional related problems, which students can solve on their own.

Also new are innovative, interactive types of examples, each labeled *Try it yourself.* In these, students are told in the left column how to proceed with each step of the problem-solving process, but in the right column are given only the answer. Thus, students are guided through the problem, but must independently work through the actual derivations and calculations.

A *Problem-Solving Guide* appears at the end of each chapter in the form of a summary of the worked examples in the chapter. The Problem-Solving Guide is designed to help students recognize types of problems and find the right conceptual strategy for solving them. Here again, general principles such as applying Newton's second law or the conservation of energy are emphasized.

Concluding each chapter is a selection of approximately one hundred *Problems*. The problems are grouped by type, which may or may not coincide with the section titles in the chapter. Each problem is designated easy, intermediate, or challenging. Qualitative questions and problems are integrated with quantitative problems within each group, in the hope that this organization will elevate the stature of qualitative problems in the minds of students (and instructors). At the back of the book, *Answers* are given to the odd-numbered problems. Preceding the answers for each chapter is a *Problem Map* that charts which odd-numbered intermediate-level problems correspond with worked examples in the text. Complete solutions to every other odd-numbered problem, worked out in the two-column example format, are available in the *Solutions Manual for Students.*

I do not believe that students can be given too much help in solving problems. Students learn best when they are successful at the tasks they are given. The hierarchy of worked examples, "Try it yourself" examples, Problem-Solving Guide, and Problem Map gives the student and the instructor maximum flexibility by leading the student through progressive levels of independence. "Try it yourself" problems take students step by step through a problem without doing the math for them. The Problem-Solving Guide gives an overview of the techniques that have been demonstrated in the chapter. The Problem Map shows students who are having difficulty where help may lie in the chapter but gives no other assistance.

Student Interest

Much effort has gone into making the written text more lively and informal. Students build their understanding of physics on the physics they've already learned, each concept serving as a building block that will provide the foundation for further inquiry. Over one hundred enthusiastic student reviews indicate that the changes in the fourth edition will successfully reach the widest range of students and will help them to enjoy learning and doing physics rather than focusing on the difficulty of the subject. To further stimulate the interest of students, supplemental, brief *"Exploring ..."* sections offer essays on various topics of interest to science and engineering undergraduates.

Modern Physics in the Introductory Course

Although quantum theory revolutionized the way we describe the physical world more than 70 years ago, we have been slow to integrate it into our introductory physics courses. To make physics more relevant to today's students, the mass–energy relationship and energy quantization sections are included in the conservation of energy chapter, and the quantization of angular momentum is discussed in the chapter on the conservation of angular momentum. These ideas are then used throughout the text, for example, in Chapter 19 to explain the failure of the equipartition theorem.

In addition, two optional chapters, "Wave–Particle Duality and Quantum Physics" (Chapter 17) and "The Microscopic Theory of Electrical Conduction" (Chapter 27), have been written so that instructors who choose to do so can integrate them into a two-semester course along with the usual topics in classical physics. These chapters offer something completely new—support for professors who choose to introduce quantum physics earlier in the course. Chapter 17 on the wave–particle duality of nature is the concluding chapter in Part II, immediately following the chapter on superposition and standing waves. This chapter introduces the idea of the wave–particle duality of light and matter and uses the frequency quantization of standing waves, just studied in the previous chapter, to introduce energy quantization of confined systems. Many students have heard of quantum theory and are curious about it. Having just studied frequency quantization that arises in standing waves, students can easily grasp energy quantization from standing electron waves, once they have seen from diffraction and interference patterns that electrons have wave properties. Because there is little time to cover even the usual material in the introductory course, some instructors are reluctant to consider adding even one more chapter such as Chapter 17. I would argue that quantum physics is at least as important as many of the other topics we teach.

Chapter 27 on the quantum explanation for electrical conduction is positioned so that it can be covered immediately after the discussion of electric current and dc circuits. The classical model of conduction is developed, concluding with the relation between resistivity and the average speed v_{av} and mean free path λ of electrons. The classical and quantum interpretations of v_{av} and λ are then discussed using the particle-in-a-box problem, discussed in the optional Chapter 17, to introduce the Fermi energy. Simple band theory is discussed to show why materials are conductors, insulators, or semiconductors. My hope in offering these optional chapters is that, given the choice, instructors will take advantage of the means to incorporate simple quantum theory into their elementary physics course.

Flexibility

To accommodate professors in a wide variety of course formats and to respond to the preferences of previous users of this text, there has been some revision in the order of material. With this new edition, instructors can give their students a brief exposure to modern physics integrated with the classical topics, or they can choose to skip the optional chapters on quantum physics entirely, perhaps returning to them in the final part of the course when this material is traditionally taught. To make room for these optional quantum chapters, some traditional material may be deleted from the course. To aid the instructor, material that can be skipped without jeopardizing coverage in other sections has been placed in optional sections. There are also two optional chapters in addition to Chapters 17 and 27. Chapter 12, "Static Equilibrium and Elasticity," and Chapter 21, "Thermal Properties and Processes," gather material that instructors sometimes choose to skip over or offer as added reading. The "optional" labeling of sections and chapters enables the instructor to pick and choose among topics with confidence that no material in nonoptional sections depends on previous coverage of an optional topic. Optional sections and chapters are clearly marked by gray borders down the side of the page. Some optional material, such as numerical methods and the use of complex numbers to solve the driven oscillator equation, is presented in "Exploring ..." essays.

Acknowledgments

Many people have contributed to this edition. I would like to thank everyone who used the earlier editions and offered comments and suggestions.

Gene Mosca, James Garland, Robert Lieberman, and Murray Scureman provided detailed reviews of nearly every chapter. Gene Mosca also wrote the student study guide along with Ron Gautreau. Robert Leiberman and Brooke Pridmore class-tested parts of the book, and assisted in obtaining student reviews and feedback. Howard McAllister was instrumental in the development of a standard approach to problem solving in the examples.

Many new problems were provided by Frank Blatt and Boris Korsunsky. Frank Blatt wrote the solutions manuals and offered many helpful suggestions. Jeff Culbert helped to enliven the problem sets with his story problems. Several of the graphs at the ends of the examples were provided by Robert Hollebeek.

I received invaluable help in manuscript checking from Murray Scureman, Thor Stromberg and Howard Miles, and in checking problems and solutions from Thor Stromberg, Howard Miles, Robert Detenbeck, Daniel G. Tekleab, Jeannette Myers, Scott Sinawi, John Pratte, Yuriy Zhestkov, Huidong Guo, Fred Watts, Ilon Joseph, Monwhea Jeng, Harry Chu, and Roy Wood. Any errors remaining are of course my responsibility.

I would particularly like to thank the more than one hundred students who read and studied from various chapters and provided detailed and valuable comments. Many instructors have provided extensive and invaluable reviews of one or more chapters. They have all made fundamental contributions to the quality of this revision. I would therefore like to thank:

Michael Arnett, *Iowa State University*
William Bassichis, *Texas A&M*
Joel C. Berlinghieri, *The Citadel*
Frank Blatt, *Retired*
John E. Byrne, *Gonzaga University*
Wayne Carr, *Stevens Institute of Technology*
George Cassidy, *University of Utah*
I. V. Chivets, *Trinity College, University of Dublin*
Harry T. Chu, *University of Akron*
Jeff Culbert, *London, Ontario*
Paul Debevec, *University of Illinois*
Robert W. Detenbeck, *University of Vermont*
Bruce Doak, *Arizona State University*
John Elliott, *University of Manchester, England*
James Garland, *Retired*
Ian Gatland, *Georgia Institute of Technology*
Ron Gautreau, *New Jersey Institute of Technology*
David Gavenda, *University of Texas at Austin*
Newton Greenburg, *SUNY Binghamton*
Huidong Guo, *Columbia University*
Richard Haracz, *Drexel University*
Michael Harris, *University of Washington*
Randy Harris, *University of California at Davis*
Dieter Hartmann, *Clemson University*
Robert Hollebeek, *University of Pennsylvania*
Madya Jalil, *University of Malaya*
Monwhea Jeng, *University of California, Santa Barbara*
Ilon Joseph, *Columbia University*
David Kaplan, *University of California, Santa Barbara*
John Kidder, *Dartmouth College*
Boris Korsunsky, *Northfield Mt. Hermon School*
Andrew Lang (graduate student), *University of Missouri*
David Lange, *University of California, Santa Barbara*
Isaac Leichter, *Jerusalem College of Technology*
William Lichten, *Yale University*
Robert Lieberman, *Cornell University*
Fred Lipschultz, *University of Connecticut*
Graeme Luke, *Columbia University*
Howard McAllister, *University of Hawaii*
M. Howard Miles, *Washington State University*
Matthew Moelter, *University of Puget Sound*
Eugene Mosca, *United States Naval Academy*
Aileen O'Donughue, *St. Lawrence University*
Jack Ord, *University of Waterloo*
Richard Packard, *University of California*
George W. Parker, *North Carolina State University*
Edward Pollack, *University of Connecticut*
John M. Pratte, *Clayton College & State University*
Brooke Pridmore, *Clayton State College*
David Roberts, *Brandeis University*
Lyle D. Roelofs, *Haverford College*
Larry Rowan, *University of North Carolina at Chapel Hill*
Lewis H. Ryder, *University of Kent, Canterbury*
Bernd Schuttler, *University of Georgia*

Cindy Schwarz, *Vassar College*
Murray Scureman, *Amdahl Corporation*
Scott Sinawi, *Columbia University*
Wesley H. Smith, *University of Wisconsin*
Kevork Spartalian, *University of Vermont*
Kaare Stegavik, *University of Trondheim, Norway*
Jay D. Strieb, *Villanova University*
Martin Tiersten, *City College of New York*
Oscar Vilches, *University of Washington*
Fred Watts, *College of Charleston*
John Weinstein, *University of Mississippi*
David Gordon Wilson, *MIT*
David Winter, *Columbia University*
Frank L. H. Wolfe, *University of Rochester*
Roy C. Wood, *New Mexico State University*
Yuriy Zhestkov, *Columbia University*

Focus Group Participants

Cherry Hill, New Jersey, July 15, 1997

John DiNardo, *Drexel University*
Eduardo Flores, *Rowan College*
Jeff Martoff, *Temple University*
Anthony Novaco, *Lafayette College*
Jay Strieb, *Villanova University*
Edward Whittaker, *Stevens Institute of Technology*

Denver, Colorado, August 15, 1997

Edward Adelson, *Ohio State University*
David Bartlett, *University of Colorado at Boulder*
David Elmore, *Purdue University*
Colonel Rolf Enger, *United States Air Force Academy*
Kendal Mallory, *University of Northern Colorado*
Samuel Milazzo, *University of Colorado at Colorado Springs*
Anders Schenstrom, *Milwaukee School of Engineering*
Daniel Schroeder, *Weber State University*
Ashley Schultz, *Fort Lewis College*

Student Reviewers

For this edition we invited the input of student reviewers at all stages of manuscript development. A number of the student reviews were blind submissions. The reviews of the following students were especially helpful:

Jesper Anderson, *Haverford College*
Anthony Bak, *Haverford College*
Luke Benes, *Cornell University*
Deborah Brown, *Northwestern University*
Andrew Burgess, *University of Kent, Canterbury*
Sarah Burnett, *Cornell University*
Sara Ellison, *University of Kent, Canterbury*
Ilana Greenstein, *Haverford College*
Sharon Hovey, *Northwestern University*
Samuel LaRoque, *Cornell University*
Valerie Larson, *Northwestern University*
Jonathan McCoy, *Haverford College*
Aaron Todd, *Cornell University*
Katalin Varju, *University of Kent, Canterbury*
Ryan Walker, *Haverford College*
Matthew Wolpert, *Haverford College*
Julie Zachiariadis, *Haverford College*

I would also like to thank the reviewers of previous editions, whose contributions are part of the foundation of this edition:

Walter Borst, *Texas Technological University*
Edward Brown, *Manhattan College*
James Brown, *The Colorado School of Mines*
Christopher Cameron, *University of Southern Mississippi*
Roger Clapp, *University of South Florida*
Bob Coakley, *University of Southern Maine*
Andrew Coates, *University College, London*
Miles Dresser, *Washington State University*
Manuel Gómez-Rodríguez, *University of Puerto Rice, Río Piedras*
Allin Gould, *John Abbott College C.E.G.E.P., Canada*

Dennis Hall, *University of Rochester*
Grant Hart, *Brigham Young University*
Jerold Izatt, *University of Alabama*
Alvin Jenkins, *North Carolina State University*
Lorella Jones, *University of Illinois, Urbana-Champaign*
Michael Kambour, *Miami-Dade Junior College*
Patrick Kenealy, *California State University at Long Beach*
Doug Kurtze, *Clarkson University*
Lui Lam, *San Jose State University*
Chelcie Liu, *City College of San Francisco*
Robert Luke, *Boise State University*
Stefan Machlup, *Case Western Reserve University*
Eric Matthews, *Wake Forest University*
Konrad Mauersberger, *University of Minnesota, Minneapolis*
Duncan Moore, *University of Rochester*
Elizabeth Nickles, *Albany College of Pharmacy*
Harry Otteson, *Utah State University*
Jack Overley, *University of Oregon*
Larry Panek, *Widener University*
Malcolm Perry, *Cambridge University, United Kingdom*
Arthur Quinton, *University of Massachusetts, Amherst*
John Risley, *North Carolina State University*
Robert Rundel, *Mississippi State University*
John Russell, *Southeastern Massachusetts University*
Michael Simon, *Housatonic Community College*
Jim Smith, *University of Illinois, Urbana-Champaign*
Richard Smith, *Montana State University*
Larry Sorenson, *University of Washington*
Thor Stromberg, *New Mexico State University*
Edward Thomas, *Georgia Institute of Technology*
Colin Thomson, *Queens University, Canada*
Gianfranco Vidali, *Syracuse University*
Brian Watson, *St. Lawrence University*
Robert Weidman, *Michigan Technological University*
Stan Williams, *Iowa State University*
Thad Zaleskiewicz, *University of Pittsburgh, Greensburg*
George Zimmerman, *Boston University*

Finally, I would like to thank everyone at Worth and W. H. Freeman Publishers for their help and encouragement. I was fortunate to work with two talented developmental editors. Steve Tenney worked on the beginning phases of the book and is responsible for many of the innovative ideas, such as the example format, summary format, problem-solving guide, and problem map. Morgan Ryan worked on the final stages, including the entire art program, and made significant improvements in the entire book. I am grateful also for the contributions of Kerry Baruth, Anne Duffy, Margaret Comaskey, Elizabeth Geller, Yuna Lee, Sarah Segal, Patricia Lawson, and George Touloumes.

Berkeley, California
December 1997

Paul Tipler

supplements

For Students

Study Guide

Volume 1 (Chapters 1–21) ISBN: 1-57259-511-6
Volumes 2 and 3 (Chapters 22–41) ISBN: 1-57259-512-4

Each chapter contains a description of key ideas, potential pitfalls, true-false questions that test essential definitions and relations, questions and answers that require qualitative reasoning, and problems and solutions.

Solutions Manual for Students

Volume 1 (Chapters 1–21) ISBN: 1-57259-513-2
Volumes 2 and 3 (Chapters 22–41) ISBN: 1-57259-524-8

The *Solutions Manual for Students* provides answers to every other odd end-of-chapter problem, presented in the same format and with the same level of detail as the *Instructor's Solutions Manual* (see below).

Tipler PLUS⊕ CD-ROM

The *Tipler PLUS⊕* CD-ROM is specifically designed to complement the learning process started in the text. On the CD-ROM, students will find a wealth of features to enhance the learning process. Interactive solution-builder exercises based on additional example problems build problem-solving skills. Video clips of lab demonstrations and applied physics bring main objectives to life. Animated quizzes based on the 3D graphics in the text test concepts from each chapter. And Web links lead the student to the sprawling world of physics on the Web. The student version of *Tipler PLUS⊕*, like the instructor's version, can be updated via the Web.

For Instructors

Instructor's Solutions Manual

Volume 1 (Chapters 1–21) ISBN: 1-57259-514-0
Volumes 2 and 3 (Chapters 22–41) ISBN: 1-57259-515-9

Complete solutions to all problems in the text are worked out in the same two-column format as the examples.

Test Bank

Approximately 3500 multiple-choice questions span all sections of the text. Each question is identified by topic and noted as factual, conceptual, or numerical. ISBN: 1-57259-517-5

Computerized Test-Generation System

A database comprises the questions in the *Test Bank.* Instructors can custom design their tests with the *Computerized Test Bank.* For Windows, ISBN: 1-57259-519-1; for Macintosh, ISBN: 1-57259-520-5

Instructor's Resource Manual

Demonstrations and a film and video cassette guide are included. ISBN: 1-57259-516-7

Transparencies

Approximately 150 full-color acetates of figures and tables from the text are included, with type enlarged for projection. Volume 1, ISBN: 1-57259-521-3; Volumes 2 and 3, ISBN: 1-57259-674-0

Instructor's Tipler PLUS⊕ CD-ROM

The instructor's version of the *Tipler PLUS⊕* CD-ROM includes everything on the student CD as well as syllabus-making software in one easy-to-navigate environment. Just indicate what part of the book you are teaching and when and *Tipler PLUS⊕* will link your syllabus to a wealth of CD-ROM and Web content. You can click the update button for new Web links, exercises, and updated content or create your own annotated study and lecture aids. *Tipler PLUS⊕* can even create an e-mail list of your students and fellow instructors. In addition to the syllabus-maker software and the material on the student CD-ROM, the instructor's CD-ROM also features selected items from the *Study Guide* and the *Instructor's Resource Manual.*

about the author

Paul Tipler was born in the small farming town of Antigo, Wisconsin, in 1933. He graduated from high school in Oshkosh, Wisconsin, where his father was superintendent of the Public Schools. He received his B.S. from Purdue University in 1955 and his Ph.D. at the University of Illinois in 1962, where he studied the structure of nuclei. He taught for one year at Wesleyan University in Connecticut while writing his thesis, then moved to Oakland University in Michigan, where he was one of the original members of the physics department, playing a major role in developing the physics curriculum. During the next 20 years, he taught nearly all the physics courses and wrote the first and second editions of his widely used textbooks *Modern Physics* (1969, 1978) and *Physics* (1976, 1982). In 1982, he moved to Berkeley, California, where he now resides, and where he wrote *College Physics* (1987) and the third edition of *Physics* (1991). In addition to physics, his interests include music, hiking, and camping, and he is an accomplished jazz pianist and poker player.

The author as a student, 1954

For over 20 years, the formula has been

Tipler = Quality

Tipler *Physics for Scientists and Engineers, 4/* continues to be the best resource a student ca[n] have for learning physics. Dynamic features like these guide the student to mastery . . .

EXAMPLES

- Text includes a greater number of intermediate-level worked examples.
- Each example has a **"Picture the Problem"** section that teaches students how to solve the problem conceptually before solving it mathematically. By learning how to find and organize the relevant information in a problem, students learn to think like a physicist.
- A major innovation is the potent **two-column side-by-side format** for the solutions to examples. Concepts are explained on the left, and the math is presented on the right. This format allows students to make the connections between the equation and what it means.
- Most examples conclude with a **"Remark"** that supplies additional information, discussion of common errors, and advice on solving problems as a physicist would.

Example 6-8

You ski downhill on waxed skis that are nearly frictionless. (*a*) What work is done on you as you ski a distance *s* down the hill? (*b*) What is your speed on reaching the bottom of the run? Assume the length of the ski run is *s*, its angle of incline is θ, and your mass is *m*. The height of the hill is then $h = s \sin\theta$.

Picture the Problem We assume that you are a particle. Two forces act on you: gravity, $m\vec{g}$, and the normal force exerted by the hill, $\vec{F}_n$ (Figure 6-15*a*). Only gravity does work on you, because the normal force is perpendicular to the hill, and hence has no component in the direction of your motion. The work–kinetic energy theorem with $v_i = 0$ gives the final speed v.

Figure 6-15*b* shows a free-body diagram for you on skies. The net force is $mg \sin\theta$, which is the component of the weight in the direction of the displacement Δs.

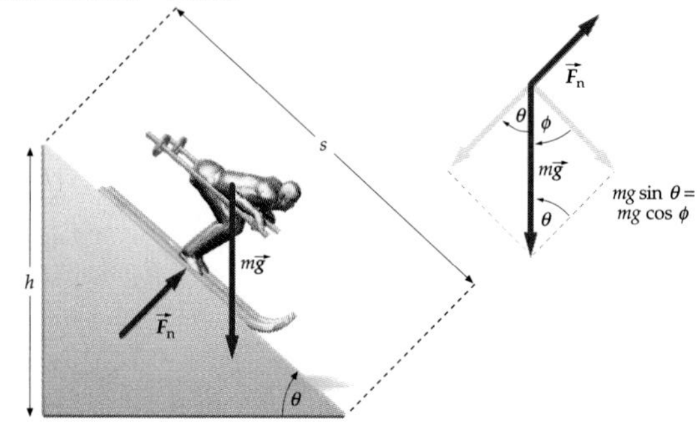

(*a*) 1. The work done by gravity as you traverse the slope is $m\vec{g}\cdot\vec{s}$:	$W = m\vec{g}\cdot\vec{s} = mgs\cos\phi = mgs\sin\theta$
2. From Figure 6-15*a*, the angle θ is related to *h* and *s*:	$\sin\theta = \dfrac{h}{s}$
3. Substitute *h* for $s \sin\theta$:	$W = mgh$
(*b*) Apply the work–kinetic energy theorem to find the final speed v:	$W = mgh = \frac{1}{2}mv^2 - 0$ or $v = \sqrt{2gh}$

Remarks $mg\sin\theta = mg\cos\phi$ is the component of the weight in the direction of the displacement. This is the component that does work on you. The final speed is independent of the angle θ, and the same as if the skier had dropped vertically a height *h* with acceleration *g*. If θ were smaller, the skier would travel a greater distance to drop the same vertical distance *h*, but the component of the force of gravity in the direction of motion would be less. The two effects cancel, and the work done by gravity is *mgh* independent of the angle of the slope. Figure 6-16 shows that for a hill of arbitrary shape, the work done by the earth on the skier is *mgh*.

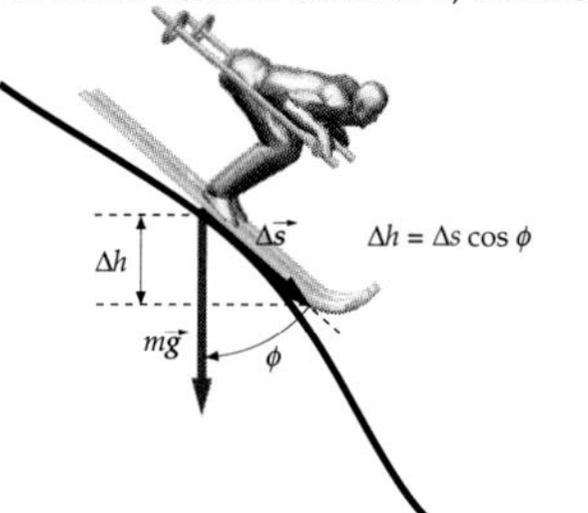

Figure 6-16 Skier skiing down a hill of arbitrary shape. The work done by the earth during a displacement $\Delta\vec{s}$ is $m\vec{g}\cdot\Delta\vec{s} = mg\,\Delta s\cos\phi = mg\,\Delta h$, where Δh is the vertical distance dropped. The total work done by the earth when the skier skis down a vertical distance *h* is $W = \int_0^s m\vec{g}\cdot d\vec{s} = mg\int_0^s \cos\phi\, ds = mg\int_0^h dh = mgh$, independent of the shape of the hill.

- When appropriate, **"Check the Result"** sections teach students how to check their own work.
- Many examples are followed by one or more related **exercises.** Answers are given, but it is up to the student to relate the exercise to the worked-out example.

Check the Result The component of *A* along *B* is $A\cos\phi = (\sqrt{13}\text{ m})\cos 70.6° = 1.2$ m.

Exercise (*a*) Find $\vec{A}\cdot\vec{B}$ for $\vec{A} = (3\text{ m})\hat{i} + (4\text{ m})\hat{j}$ and $\vec{B} = (2\text{ m})\hat{i} + (8\text{ m})\hat{j}$. (*b*) Find *A*, *B*, and the angle between $\vec{A}$ and $\vec{B}$ for these vectors. (*Answers* (*a*) 38 m^2, (*b*) $A = 5$ m, $B = 8.25$ m, $\phi = 23°$)

- Instructive illustrations accompany nearly all examples.
- Students are involved in every step of the complete solution for each example.

Example 2-15

A car is speeding at 25 m/s (≈ 90 km/h ≈ 56 mi/h) in a school zone. A police car starts from rest just as the speeder passes and accelerates at a constant rate of 5 m/s². (*a*) When does the police car catch the speeding car? (*b*) How fast is the police car traveling when it catches up with the speeder?

Picture the Problem To determine when the two cars will be at the same position, we write the positions x_s of the speeder and x_p of the police car as functions of time and solve for the time t when $x_s = x_p$.

(*a*) 1. Write the position functions for the speeder and the police car: $x_s = v_s t$ and $x_p = \frac{1}{2}a_p t^2$

2. Set $x_s = x_p$ and solve for the time t: $v_s t = \frac{1}{2}a_p t^2; \quad t = 0$ (initial condition)

$$t = \frac{2v_s}{a_p} = \frac{2(25\text{ m/s})}{5\text{ m/s}^2} = 10\text{ s}$$

(*b*) The velocity of the police car is given by $v = v_0 + at$ with $v_0 = 0$: $v_p = a_p t = (5\text{ m/s}^2)(10\text{ s}) = 50\text{ m/s}$

Remark The final speed of the police car in (*b*) is exactly twice that of the speeder. Since the two cars covered the same distance in the same time, they must have had the same average speed. The speeder's average speed, of course, is 25 m/s. For the police car to start from rest and have an average speed of 25 m/s, it must reach a final speed of 50 m/s.

Exercise How far have the cars traveled when the police car catches the speeder? (*Answer* 250 m)

Remark In Figure 2-13 the solid lines depict the speeder and the police car in this example. The dashed lines are variations on the example. The smaller acceleration depicted by the lower dashed line means the police car takes longer to reach the speeder. In the higher dashed line, the acceleration is the same as in the example, but the police car does not start accelerating until 4 s after the speeder passes by.

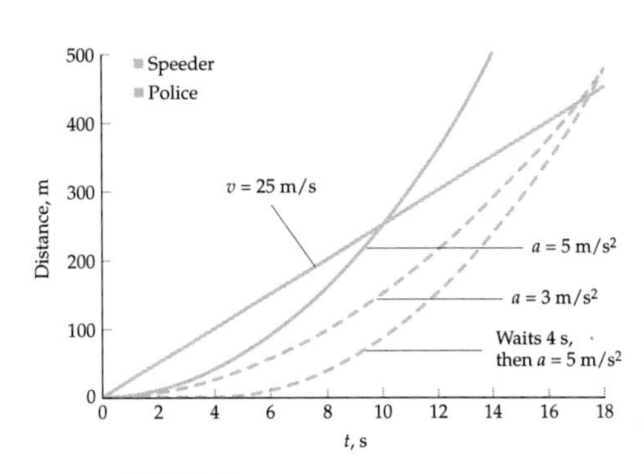

Figure 2-13

"TRY IT YOURSELF" EXAMPLES

- **"Try it yourself"** examples follow the same format as regular examples, except that the student is given only brief instructions on the left side, and only the relevant formulas and the answer on the right side, without being shown the mathematical steps in between.
- **"Try it yourself"** examples test understanding, identify weaknesses, and reinforce proper problem-solving procedures.

Example 6-7 *try it yourself*

A particle is given a displacement $\Delta\vec{s} = 2\text{ m}\,\hat{i} - 5\text{ m}\,\hat{j}$ along a straight line. During the displacement, a constant force $\vec{F} = 3\text{ N}\,\hat{i} + 4\text{ N}\,\hat{j}$ acts on the particle (Figure 6-14). Find (*a*) the work done by the force, and (*b*) the component of the force in the direction of the displacement.

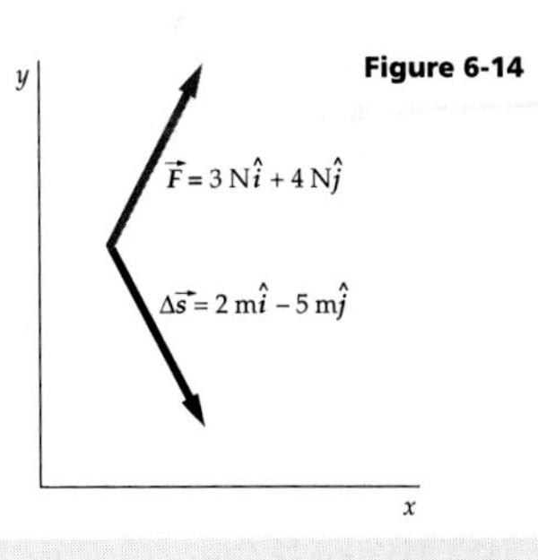

Figure 6-14

Picture the Problem The work W is found by computing $W = \vec{F}\cdot\Delta\vec{s} = F_x\Delta x + F_y\Delta y + F_z\Delta z$. Since $\vec{F}\cdot\Delta\vec{s} = F\cos\phi|\Delta\vec{s}|$, we can find the component of $\vec{F}$ in the direction of the displacement from

$$F\cos\phi = \frac{(\vec{F}\cdot\Delta\vec{s})}{|\Delta\vec{s}|} = \frac{W}{|\Delta\vec{s}|}$$

Cover the column to the right and try these on your own before looking at the answers.

Steps	***Answers***
(*a*) Compute the work done W.	$W = -14\text{ N}\cdot\text{m}$
(*b*) 1. Compute $\Delta\vec{s}\cdot\Delta\vec{s}$ and use your result to find the distance $\lvert\Delta\vec{s}\rvert$.	$\lvert\Delta\vec{s}\rvert = \sqrt{29}\text{ m}$
2. Compute $F\cos\phi = W/\lvert\Delta\vec{s}\rvert$.	$F\cos\phi = -2.60\text{ N}$

Remark The component of the force in the direction of the displacement is negative, so the work done is negative.

Exercise Find the magnitude of $\vec{F}$, and the angle ϕ between $\vec{F}$ and $\Delta\vec{s}$. (*Answer* $F = 5$ N, $\phi = 121°$)

"EXPLORING" ESSAYS

- "Exploring" essays enrich students by covering topics that are outside the core part of the course—but are too interesting to pass over completely.

Numerical Methods: Euler's Method

If a particle moves under the influence of a *constant* force, its acceleration is constant and we can find its velocity and position from the constant-acceleration formulas in Chapter 2. But consider a particle moving through space where the force on it, and therefore its acceleration, depends on its position and velocity. The velocity and acceleration of the particle at one instant determine its position and velocity at the next instant, which then determines its acceleration at that instant. The actual position, velocity, and acceleration of an object all change continuously with time. We can approximate this by replacing the continuous time variations with small time steps of duration Δt. The simplest approximation is to assume constant acceleration during each step. This approximation is called **Euler's method**. If the time interval is sufficiently short, the change in acceleration during the interval will be small and can be neglected.

Let x_0, v_0, and a_0 be the known position, veloc-

$$x_2 = x_1 + v_1 \Delta t$$

In general, the connection between the position and velocity at time t_n and time $t_{n+1} = t_n + \Delta t$ is given by

$$v_{n+1} = v_n + a_n \Delta t \qquad 3$$

and

$$x_{n+1} = x_n + v_n \Delta t \qquad 4$$

To find the velocity and position at some time t, we therefore divide the time interval $t - t_0$ into a large number of smaller intervals Δt and apply Equations 3 and 4, beginning at the initial time t_0. This involves a large number of simple, repetitive calculations that are easily done on a computer. The technique of breaking the time interval into small steps and computing the acceleration, velocity, and position at each step using the values from the previous step is called numerical integration.

Drag Forces

To illustrate the use of numerical methods, let us consider a problem in which a sky diver is dropped from rest at some height under the influences of gravity and a drag force that is proportional to the square of the speed. We will find the velocity v and the distance traveled x as functions of time.

The equation describing the motion of an object of mass m dropped from rest is Equation 5-7 with $n = 2$:

$$\sum F_y = mg - bv^n = ma_y$$

SUMMARY

- Detailed **Summary** sections reinforce the lessons of the chapter. The summary is not a substitute for attentive study of the text, but a skeletal outline to help students organize what they have learned.

Summary

1. Work, kinetic energy, potential energy, and power are important derived dynamic quantities.
2. The work–kinetic energy theorem is an important relation derived from Newton's laws applied to a particle.
3. The dot product of vectors is a mathematical definition that is useful throughout physics.

Topic	Remarks and Relevant Equations	
1. Work		
Constant force	The work done by a constant force is the product of the component of the force in the direction of motion and the displacement of the force: $W = F\cos\theta\,\Delta x = F_x\,\Delta x$	6-1
Variable force	$W = \int_{x_1}^{x_2} F_x\,dx$ = area under the F_x-versus-x curve	6-9
Force in three dimensions	$W = \int_1^2 \vec{F}\cdot d\vec{s}$	6-14
Units	The SI unit of work and energy is the joule (J): $1\text{ J} = 1\text{ N}\cdot\text{m}$	6-2
2. Kinetic Energy	$K = \frac{1}{2}mv^2$	6-6

PROBLEMS

- Types of problems are denoted by color swatches: **yellow** denotes conceptual problems and a **gray** band indicates optional or exploring sections.
- The difficulty level is denoted by bullets.
- Qualitative problems are included in context with related quantitative problems.

Problems

Conceptual Problems

Problems from Optional and Exploring sections

• Single-concept, single-step, relatively easy

•• Intermediate-level, may require synthesis of concepts

••• Challenging, for advanced students

In a few problems, you are given more data than you actually need; in a few other problems, you are required to supply data from your general knowledge, outside sources, or informed estimates.

Conditions for Equilibrium

1 • True or false:

(*a*) $\Sigma\vec{F} = 0$ is sufficient for static equilibrium to exist.
(*b*) $\Sigma\vec{F} = 0$ is necessary for static equilibrium to exist.
(*c*) In static equilibrium, the net torque about any point is zero.
(*d*) An object is in equilibrium only when there are no forces acting on it.

2 • A seesaw consists of a 4-m board pivoted at the center. A 28-kg child sits on one end of the board. Where should a 40-kg child sit to balance the seesaw?

3 • In Figure 12-23, Misako is about to do a push-up. Her center of gravity lies directly above point P on the floor, which is 0.9 m from her feet and 0.6 m from her hands. If her mass is 54 kg, what is the force exerted by the floor on her hands?

Figure 12-23 Problem 3

Center of gravity

0.9 m 0.6 m

P

4 • Juan and Bettina are carrying a 60-kg block on a 4-m board as shown in Figure 12-24. The mass of the board is 10 kg. Since Juan spends most of his time reading cookbooks, whereas Bettina regularly does push-ups, they place the block 2.5 m from Juan and 1.5 m from Bettina. Find the force in newtons exerted by each to carry the block.

Figure 12-24 Problem 4

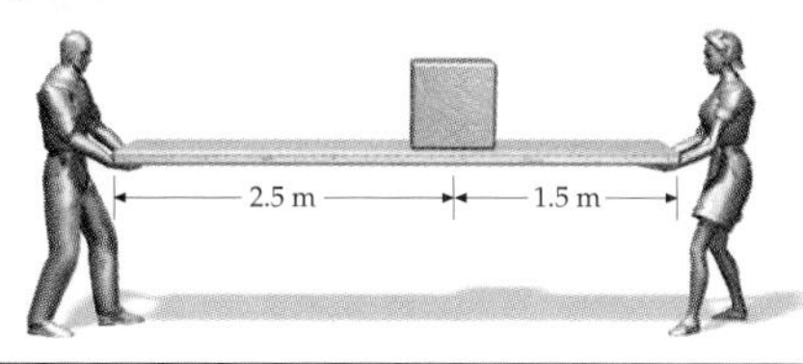

Figure 12-25 Problem 5

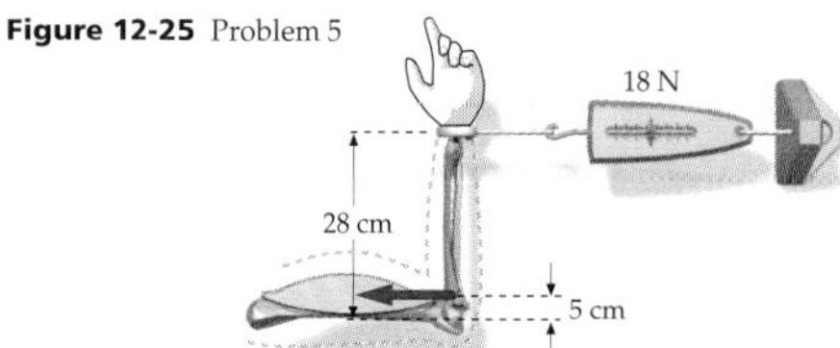

the pivot point. If the scale reads 18 N when she exerts her maximum force, what force is exerted by the biceps muscle?

6 • A crutch is pressed against the sidewalk with a force $\vec{F}_c$ along its own direction as in Figure 12-26. This force is balanced by a normal force $\vec{F}_n$ and a frictional force $\vec{f}_s$. (*a*) Show that when the force of friction is at its maximum value, the coefficient of friction is related to the angle θ by $\mu_s = \tan\theta$. (*b*) Explain how this result applies to the forces on your foot when you are not using a crutch. (*c*) Why is it advantageous to take short steps when walking on ice?

Figure 12-26 Problem 6

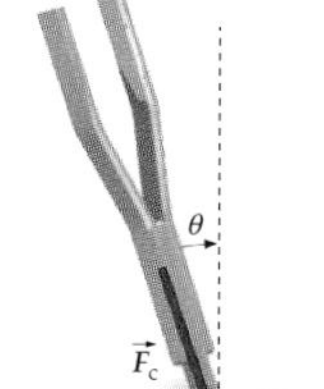

The Center of Gravity

7 • True or false: The center of gravity is always at the geometric center of a body.

8 • Must there be any material at the center of gravity of an object?

9 • If the acceleration of gravity is not constant over an object, is it the center of mass or the center of gravity that is the pivot point when the object is balanced?

10 • Two spheres of radius R rest on a horizontal table with their centers a distance $4R$ apart. One sphere has twice the weight of the other sphere. Where is the center of gravity of this system?

11 • An automobile has 58% of its weight on the front wheels. The front and back wheels are separated by 2 m.

General Problems

66 • If the net torque about some point is zero, must it be zero about any other point? Explain.

67 • The horizontal bar in Figure 12-52 will remain horizontal if

(*a*) $L_1 = L_2$ and $R_1 = R_2$.
(*b*) $L_1 = L_2$ and $M_1 = M_2$.
(*c*) $R_1 = R_2$ and $M_1 = M_2$.
(*d*) $L_1M_1 = L_2M_2$.
(*e*) $R_1L_1 = R_2L_2$.

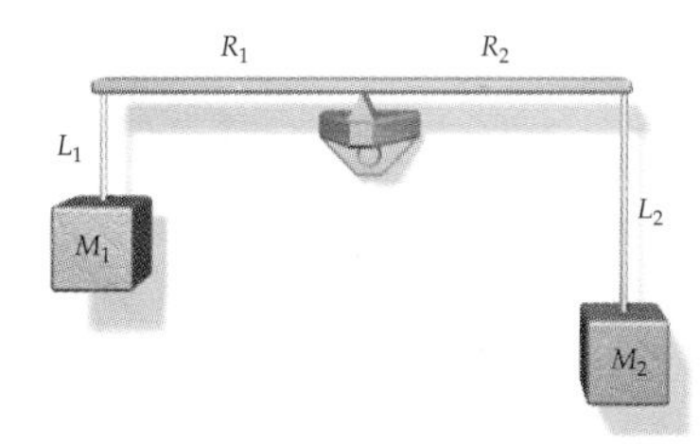

68 • Which of the following could not have units of N/m^2?

(*a*) Young's modulus
(*b*) Shear modulus
(*c*) Stress
(*d*) Strain

69 •• Sit in a chair with your back straight. Now try to stand up without leaning forward. Explain why you cannot do it.

70 • A 90-N board 12 m long rests on two supports, each 1 m from the end of the board. A 360-N block is placed on the board 3 m from one end as shown in Figure 12-53. Find the force exerted by each support on the board.

Figure 12-53 Problem 70

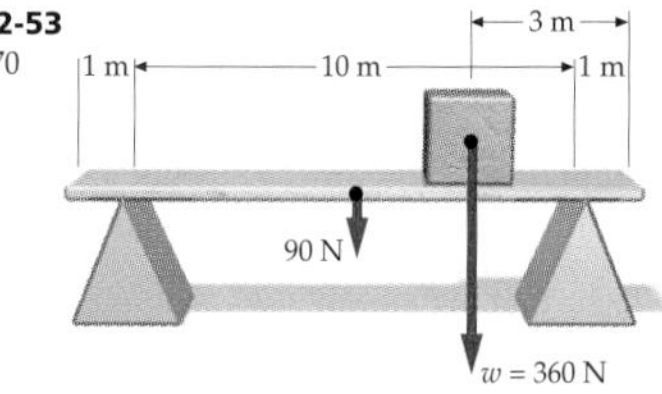

contents in brief

Volume 1

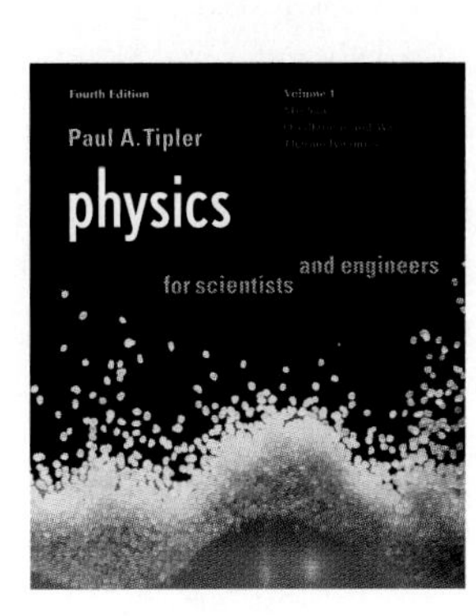

Volume 2

PART IV electricity and magnetism

PART V light

PART VI modern physics: quantum mechanics, relativity, and the structure of matter

Volume 3

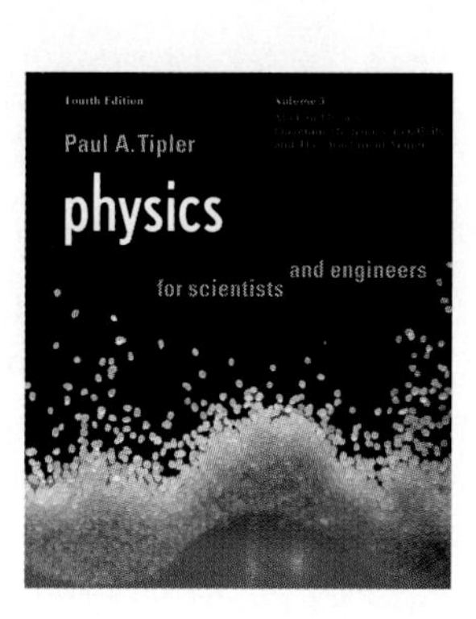

OPTIONAL SECTIONS: These sections contain material that can be omitted without jeopardizing coverage of other sections.

contents

PART IV electricity and magnetism

PART V light

physics

for scientists and engineers

PART IV

electricity and magnetism

A small, permanent magnet levitates above a disk of the superconductor yttrium–barium copper oxide cooled to 77 K. The magnetic field of the cube sets up circulating electric currents in the superconducting disk, such that the resultant magnetic field in the superconductor is zero. These currents produce a magnetic field that repels the cube.

CHAPTER 22

The Electric Field I: Discrete Charge Distributions

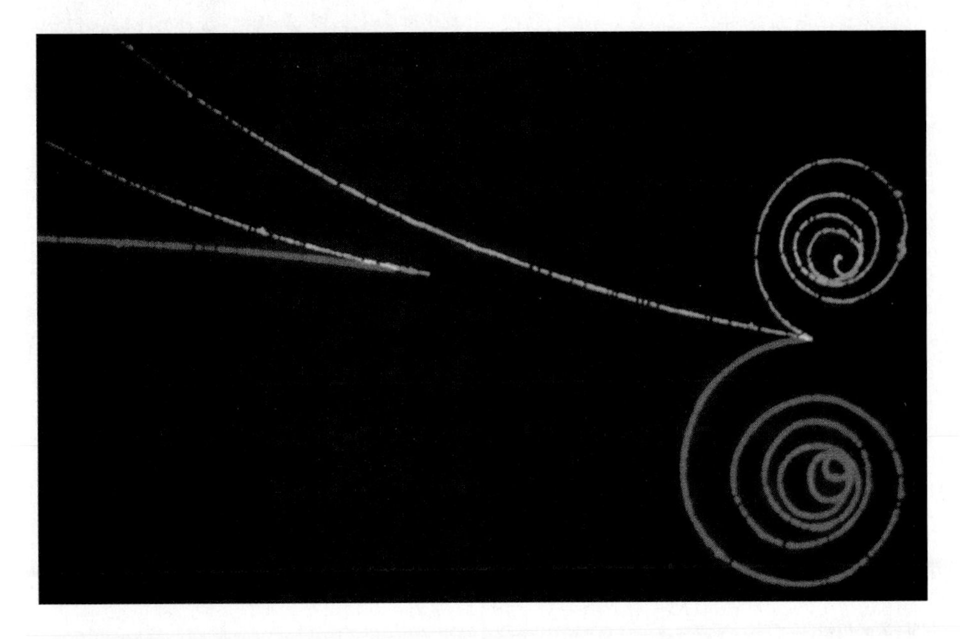

Pair production. An electron of charge $-e$ and a positron of charge $+e$ are created by the interaction of electromagnetic radiation with matter. The paths of the oppositely charged particles, made visible in a bubble chamber, are bent in opposite directions by a magnetic field.

We are extremely dependent on electricity in our daily lives, while just a century ago we had nothing more than a few electric lights. Yet the study of electricity has a history reaching long before the first electric lamp glowed. Observations of electrical attraction can be traced back to the ancient Greeks, who noticed that after amber has been rubbed, it attracts small objects such as straw or feathers. Indeed, the word "electric" comes from the Greek word for amber, *elektron.*

We begin our examination of electricity with **electrostatics**, the study of electrical charges at rest. After introducing the concept of electric charge, we briefly look at conductors and insulators and how conductors can be given a net charge. We then study Coulomb's law, which gives the force exerted by one electric charge on another. Next, we introduce the electric field and show how it can be described by electric field lines that indicate the magnitude and direction of the field. Finally, we discuss the behavior of point charges and electric dipoles in electric fields.

22-1 Electric Charge

Suppose we rub a plastic rod with fur and suspend it from a string so that it is free to rotate. Now we bring a second similarly rubbed plastic rod near it. The rods repel each other (Figure 22-1). We get the same results if we use two glass rods that have been rubbed with silk. But a plastic rod rubbed with fur and a glass rod rubbed with silk attract each other.

Rubbing a rod causes it to become electrically charged. If we repeat the experiment with various materials, we find that all charged objects fall into one of just two groups—those like the plastic rod rubbed with fur and those like the glass rod rubbed with silk. Benjamin Franklin explained this by a model in which every object has a "normal" amount of electricity that can be transferred from one object to the other when two objects are in close contact, as when they are rubbed together. This leaves one with an excess charge and the other with a deficiency of charge in the same amount as the excess. Franklin described the resulting charges with plus and minus signs, choosing positive to be the charge acquired by a glass rod when it is rubbed with a piece of silk. The piece of silk acquires a negative charge of equal magnitude during the procedure. Based on Franklin's convention, plastic rubbed with fur acquires a negative charge and the fur acquires a positive charge. Two objects that carry the same type of charge repel each other, and two objects that carry opposite charges attract each other (Figure 22-2).

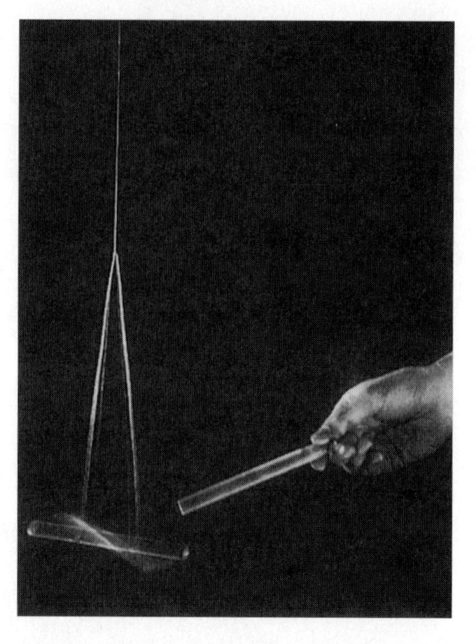

Figure 22-1 Two plastic rods that have been rubbed with fur repel each other.

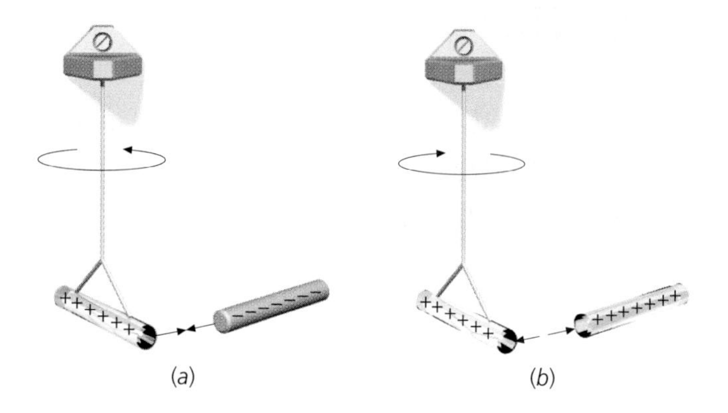

Figure 22-2 (*a*) Objects carrying charges of opposite sign attract each other. (*b*) Objects carrying charges of the same sign repel each other.

Today, we know that when glass is rubbed with silk, electrons are transferred from the glass to the silk. Since the silk is negatively charged (according to Franklin's classification, which we still use), electrons are said to carry a negative charge.

Charge Quantization

Matter consists of atoms that are electrically neutral. Each atom has a tiny but massive nucleus that contains protons and neutrons. Protons are positively charged, whereas neutrons are uncharged. The number of protons in the nucleus is the atomic number Z of the element. Surrounding the nucleus is an equal number of negatively charged electrons, leaving the atom with zero net charge. The electron is about 2000 times less massive than the proton, yet the charges of these two particles are exactly equal in magnitude. The charge of the proton is e and that of the electron is $-e$, where e is called the **funda-**

mental unit of charge. The charge of an electron or proton is an intrinsic property of the particle, just as mass and spin are intrinsic properties of these particles.

All observable charges occur in integral amounts of the fundamental unit of charge e. That is, *charge is quantized.* Any charge Q occurring in nature can be written $Q = \pm Ne$, where N is an integer.* For large systems, however, N is usually very large and charge appears to be continuous, just as air appears to be continuous even though air consists of many discrete molecules. To give an everyday example of N, charging a plastic rod by rubbing it with a piece of fur typically transfers 10^{10} or more electrons to the rod.

Charge Conservation

When objects are rubbed together, one object is left with an excess number of electrons and is therefore negatively charged; the other object is left lacking electrons and is therefore positively charged. The net charge of the two objects remains constant; that is, *charge is conserved.* The **law of conservation of charge** is a fundamental law of nature. In certain interactions among elementary particles, charged particles such as electrons are created or annihilated. However, in all these processes, equal amounts of positive and negative charge are produced or destroyed, so the net charge of the universe is unchanged.

The SI unit of charge is the coulomb, which is defined in terms of the unit of electric current, the ampere.† The **coulomb** (C) is the amount of charge flowing through a wire in one second when the current in the wire is one ampere. The fundamental unit of electric charge e is related to the coulomb by

$$e = 1.602\,177 \times 10^{-19}\,\text{C} \qquad \text{22-1}$$

Fundamental unit of charge

Exercise A charge of magnitude 50 nC (1 nC = 10^{-9} C) can be produced in the laboratory by simply rubbing two objects together. How many electrons must be transferred to produce this charge? (*Answer* $N = Q/e = (50 \times 10^{-9}\,\text{C})/(1.6 \times 10^{-19}\,\text{C}) = 3.12 \times 10^{11}$. Charge quantization cannot be detected in a charge of this size; even adding or subtracting a million electrons produces a negligibly small effect.)

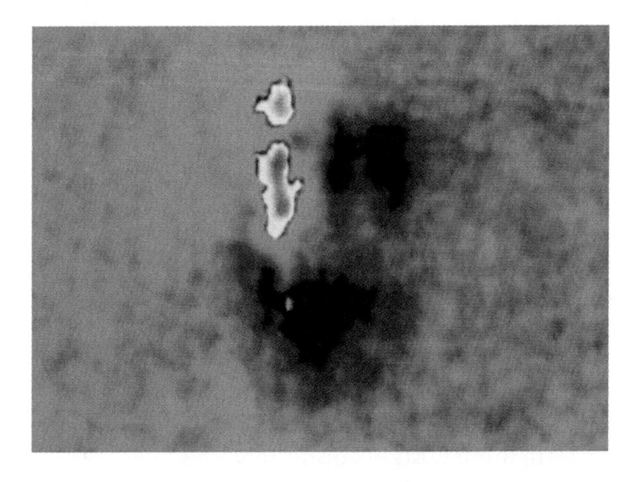

Charging by contact. A sample of plastic about 0.02 mm wide that was charged by contact with a piece of nickel. Although the plastic carries a net positive charge, regions of negative charge (dark) as well as regions of positive charge (yellow) are indicated. The photograph was taken by sweeping a charged needle of width 10^{-7} m over the sample and recording the electrostatic force on the needle.

* In the Standard Model of elementary particles, protons, neutrons, and some other elementary particles are made up of more fundamental particles called quarks that carry charges of $\pm\frac{1}{3}e$ or $\pm\frac{2}{3}e$. Only combinations that result in a net charge of $\pm Ne$ or 0 are known.

† The ampere (A) is the unit of current used in everyday electrical work. It will be defined later.

Example 22-1

A copper penny ($Z = 29$) has a mass of 3 g. What is the total charge of all the electrons in the penny?

Picture the Problem The electrons have a total charge given by the number of electrons in the penny, N_e, times the charge of an electron, $-e$. The number of electrons is 29 times the number of copper atoms N. To find N, we use the fact that one mole of any substance has Avogadro's number ($N_A = 6.02 \times 10^{23}$) of molecules, and the number of grams in a mole is the molecular mass M, which is 63.5 for copper. Since each molecule of copper is just one copper atom, we find the number of atoms per gram by dividing N_A atoms/mole by M grams/mole.

1. The total charge is the number of electrons times the electronic charge:

$$Q = N_e(-e)$$

2. The number of electrons is Z times the number of copper atoms N_a:

$$N_e = ZN_a$$

3. Compute the number of copper atoms in 3 g of copper:

$$N_a = (3\text{ g})\frac{6.02 \times 10^{23}\text{ atoms/mol}}{63.5\text{ g/mol}} = 2.84 \times 10^{22}\text{ atoms}$$

4. Compute the number of electrons N_e:

$$N_e = ZN_a = (29\text{ electrons/atom})(2.84 \times 10^{22}\text{ atoms})$$
$$= 8.24 \times 10^{23}\text{ electrons}$$

5. Use this value of N_e to find the total charge:

$$Q = N_e(-e)$$
$$= (8.24 \times 10^{23}\text{ electrons})(-1.6 \times 10^{-19}\text{ C/electron})$$
$$= -1.32 \times 10^{5}\text{ C}$$

Exercise If one million electrons are given to each man, woman, and child in the United States (about 250,000,000 people), what percentage of the number of electrons in a penny would this represent? (*Answer* About thirty billionths of a percent.)

22-2 Conductors and Insulators

In many materials, such as copper and other metals, some of the electrons are free to move about the entire material. Such materials are called **conductors.** In other materials, such as wood or glass, all the electrons are bound to nearby atoms and none can move freely. These materials are called **insulators.**

In a single atom of copper, 29 electrons are bound to the nucleus by the electrostatic attraction between the negatively charged electrons and the positively charged nucleus. The outer electrons are more weakly bound than the inner electrons because of their greater distance from the nucleus and because of the repulsive force exerted by the inner electrons. When a large number of copper atoms are combined in a piece of metallic copper, the binding of the electrons of each individual atom is reduced by interactions with neighboring atoms. One or more of the outer electrons in each atom is no longer bound but is free to move throughout the whole piece of metal, much as a gas molecule is free to move about in a box. The number of free electrons depends on the particular metal, but it is typically about one per atom. An atom with an electron removed or added, resulting in a net charge on the atom, is called an **ion**. In metallic copper, the copper ions are arranged in a regular array called a *lattice*. Normally, a conductor is electrically neutral because there is a lattice ion carrying a positive charge $+e$ for each free electron carrying a negative charge $-e$. A conductor can be given a net charge by adding or removing free electrons.

The Electroscope

Figure 22-3 shows a device for detecting electric charge called an **electroscope**. Two gold leaves are attached to a conducting post that has a conducting ball on top. The leaves are otherwise insulated from the container. When uncharged, the leaves hang together vertically. When the ball is touched by a negatively charged plastic rod, some of the negative charge from the rod is transferred to the ball and moves to the gold leaves, which then spread apart because of electrical repulsion between their negative charges. Touching the ball with a positively charged glass rod also causes the leaves to spread apart. In this case, the positively charged glass rod attracts electrons from the metal ball, leaving a net positive charge on the leaves.

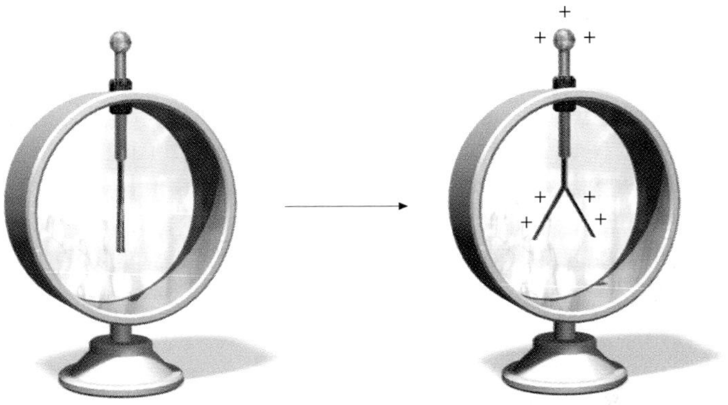

Figure 22-3 An electroscope. The two gold leaves are attached to a metal rod that has a metal ball on top. When a charge is placed on the metal ball, the charge is conducted to the leaves and they repel each other.

Charging by Induction

The conservation of charge is illustrated by a simple method of charging a conductor called **charging by induction,** as shown in Figure 22-4. Two uncharged metal spheres are in contact. When a charged rod is brought near one of the spheres, free electrons flow from one sphere to the other, toward a positively charged rod or away from a negatively charged rod. The positively charged rod in Figure 22-4*a* attracts the negatively charged electrons, and the sphere nearest the rod acquires electrons from the sphere farther away. This leaves the near sphere with a net negative charge and the far sphere with an equal net positive charge. A conductor that has separated equal and opposite charges is said to be **polarized**. If the spheres are separated before the rod is removed, they will be left with equal amounts of opposite charges (Figure 22-4*b*). A similar result would be obtained with a negatively charged rod, which would drive electrons from the near sphere to the far sphere.

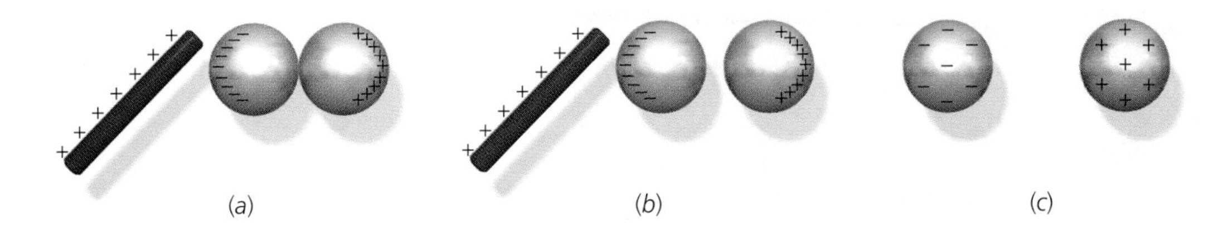

Figure 22-4 Charging by induction. (*a*) Conductors in contact become oppositely charged when a charged rod attracts electrons to the left sphere. (*b*) If the spheres are separated before the rod is removed, they will retain their equal and opposite charges. (*c*) When the rod is removed and the spheres are far apart, the spheres are uniformly charged with equal and opposite charges.

Exercise Two identical conducting spheres, one with an initial charge Q, the other initially uncharged, are brought into contact. (*a*) What is the new charge on each sphere? (*b*) While the spheres are in contact, a negatively charged rod is moved close to one sphere, causing it to have a charge of $2Q$. What is the charge on the other sphere? (*Answers* (*a*) $\frac{1}{2}Q$. Since the spheres are identical, they must share the total charge equally. (*b*) $-Q$, which is necessary to satisfy the conservation of charge)

Exercise Two identical spheres are charged by induction and then separated; sphere 1 has charge Q and sphere 2 has charge $-Q$. A third identical sphere is initially uncharged. If sphere 3 is touched to sphere 1 and separated, then touched to sphere 2 and separated, what is the final charge on each of the three spheres? (*Answer* $Q_1 = Q/2, Q_2 = -Q/4, Q_3 = -Q/4$)

For many purposes, the earth itself can be considered to be an infinitely large conductor with an abundant supply of free charge. When a conductor is connected to the earth, it is said to be **grounded** (indicated schematically in Figure 22-5*b* by a connecting wire ending in parallel horizontal lines). Figure 22-5 demonstrates how we can induce a charge in a single conductor by transferring charge from the earth through the ground wire and then breaking the connection to the ground.

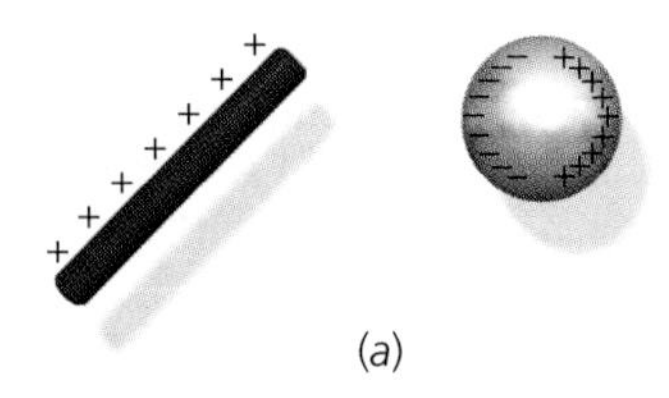

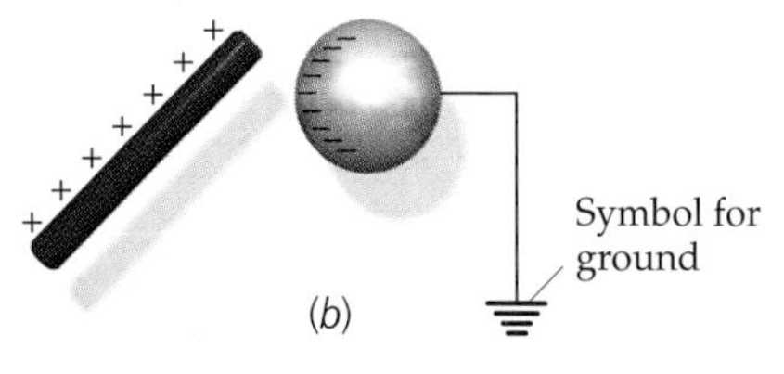

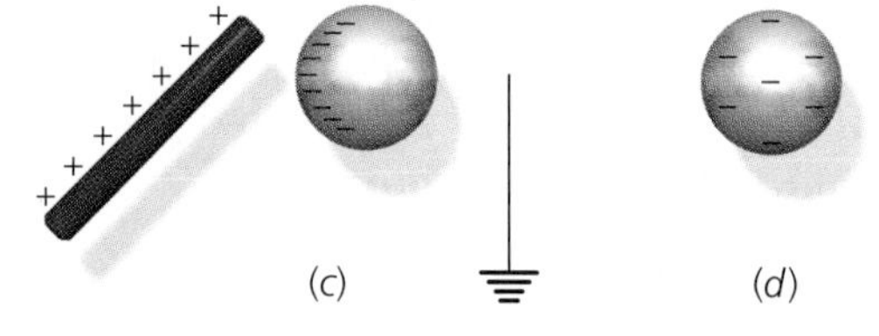

Figure 22-5 Induction via grounding. (*a*) The free charge on the single conducting sphere is polarized by the positively charged rod, which attracts negative charges on the sphere. (*b*) When the conductor is grounded by connecting it with a wire to a very large conductor, such as the earth, electrons from the ground neutralize the positive charge on the far face. The conductor is then negatively charged. (*c*) The negative charge remains if the connection to the ground is broken before the rod is removed. (*d*) When the rod is removed, the sphere has a uniform negative charge.

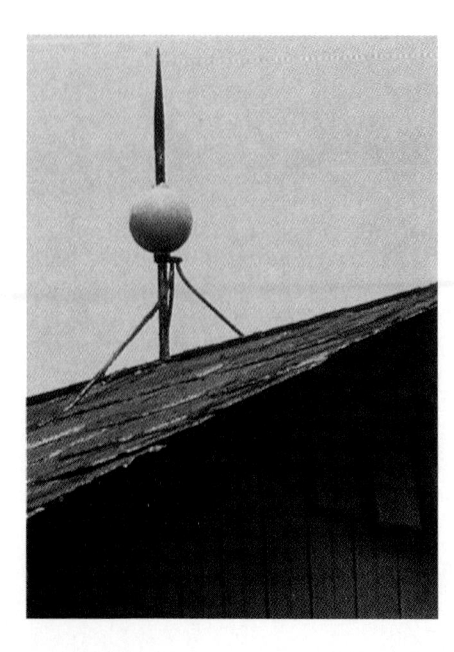

(*Left*) The lightning rod on this building is grounded so that it can conduct electrons from the ground to the positively charged clouds, thus neutralizing them. (*Right*) These fashionable ladies are wearing hats with metal chains that drag along the ground, which were supposed to protect them from lightning.

22-3 Coulomb's Law

The force exerted by one charge on another was studied by Charles Coulomb (1736–1806) using a torsion balance of his own invention.* In Coulomb's experiment, the charged spheres were much smaller than the distance between them so that the charges could be treated as point charges. Coulomb used the method of charging by induction to produce equally charged spheres and to vary the amount of charge on the spheres. For example, beginning with charge q_0 on each sphere, he could reduce the charge to $\frac{1}{2}q_0$ by temporarily grounding one sphere to discharge it and then placing the two spheres in contact. The results of the experiments of Coulomb and others are summarized in **Coulomb's law**:

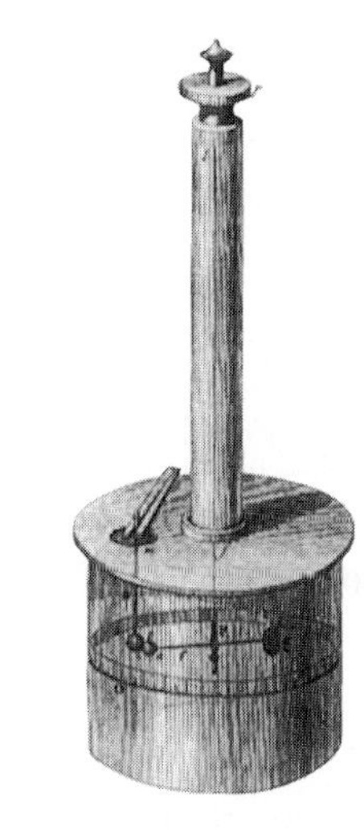

Coulomb's torsion balance.

> The force exerted by one point charge on another acts along the line between the charges. It varies inversely as the square of the distance separating the charges and is proportional to the product of the charges. The force is repulsive if the charges have the same sign and attractive if the charges have opposite signs.

Coulomb's law

The *magnitude* of the electric force exerted by a charge q_1 on another charge q_2 a distance r away is thus given by

$$F = \frac{k|q_1q_2|}{r^2} \qquad \text{22-2}$$

where k is an experimentally determined constant called the **Coulomb constant,** which has the value

$$k = 8.99 \times 10^9\ \text{N}\cdot\text{m}^2/\text{C}^2 \qquad \text{22-3}$$

If q_1 is at position $\vec{r}_1$ and q_2 is at $\vec{r}_2$ (Figure 22-6), the force $\vec{F}_{1,2}$ exerted by q_1 on q_2 is

$$\vec{F}_{1,2} = \frac{kq_1q_2}{r_{1,2}^2}\hat{r}_{1,2} \qquad \text{22-4}$$

Coulomb's law for the force exerted by q_1 on q_2

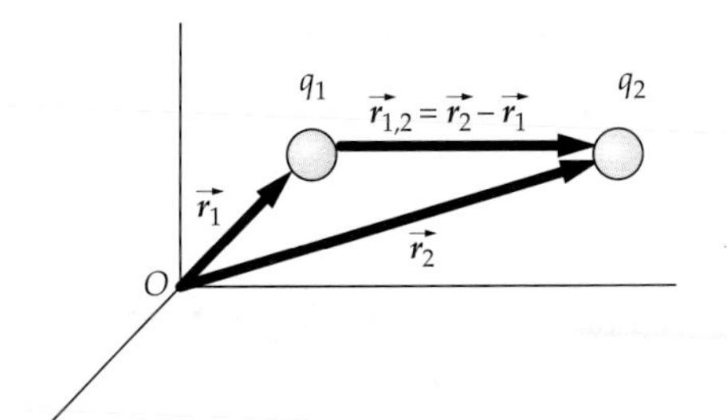

Figure 22-6 Charge q_1 at position $\vec{r}_1$ and charge q_2 at $\vec{r}_2$ relative to the origin O. The force exerted by q_1 on q_2 is in the direction of the vector $\vec{r}_{1,2} = \vec{r}_2 - \vec{r}_1$ if both charges have the same sign, and in the opposite direction if they have opposite signs.

where $\vec{r}_{1,2} = \vec{r}_2 - \vec{r}_1$ is the vector pointing from q_1 to q_2, and $\hat{r}_{1,2} = \vec{r}_{1,2}/r_{1,2}$ is a unit vector pointing from q_1 to q_2.

By Newton's third law, the force $\vec{F}_{2,1}$ exerted by q_2 on q_1 is the negative of $\vec{F}_{1,2}$. Note the similarity between Coulomb's law and Newton's law of gravity (Equation 11-3). Both are inverse-square laws. But the gravitational force between two particles is proportional to the masses of the particles and is always attractive, whereas the electric force is proportional to the charges of the particles and is repulsive if both charges have the same sign and attractive if they have opposite signs.

* Coulomb's experimental apparatus was essentially the same as that described for the Cavendish experiment in Chapter 11, with the masses replaced by small charged spheres. For the magnitudes of charges easily transferred by rubbing, the gravitational attraction of the spheres is completely negligible compared with their electric attraction or repulsion.

Example 22-2

In a hydrogen atom, the electron is separated from the proton by an average distance of about 5.3×10^{-11} m. Calculate the magnitude of the electrostatic force of attraction exerted by the proton on the electron.

Substitute the given values into Coulomb's law:

$$F = \frac{k|q_1 q_2|}{r^2} = \frac{ke^2}{r^2} = \frac{(8.99 \times 10^9\ \mathrm{N \cdot m^2/C^2})(1.6 \times 10^{-19}\ \mathrm{C})^2}{(5.3 \times 10^{-11}\ \mathrm{m})^2}$$

$$= 8.19 \times 10^{-8}\ \mathrm{N}$$

Remarks Compared with macroscopic interactions, this is a very small force. However, since the mass of the electron is only about 10^{-30} kg, this force produces an enormous acceleration of $F/m = 8 \times 10^{22}\ \mathrm{m/s^2}$.

Exercise Two point charges of 0.05 μC each are separated by 10 cm. Find the magnitude of the force exerted by one point charge on the other. (*Answer* 2.25×10^{-3} N)

Since the electrical force and the gravitational force between any two particles both vary inversely with the square of the separation between the particles, the ratio of these forces is independent of separation. We can therefore compare the relative strengths of the electrical and gravitational forces for elementary particles such as the electron and proton.

Example 22-3

Compute the ratio of the electric force to the gravitational force exerted by a proton on an electron in a hydrogen atom.

Picture the Problem We use Coulomb's law with $q_1 = e$ and $q_2 = -e$ to find the electric force, and Newton's law of gravity with the mass of the proton, $m_p = 1.67 \times 10^{-27}$ kg, and the mass of the electron, $m_e = 9.11 \times 10^{-31}$ kg.

1. Express the magnitudes of the electric force F_e and the gravitational force F_g in terms of the charges, masses, separation distance r, and electrical and gravitational constants:

$$F_e = \frac{ke^2}{r^2}, \quad F_g = \frac{Gm_p m_e}{r^2}$$

2. Take the ratio. Note that the separation distance r cancels:

$$\frac{F_e}{F_g} = \frac{ke^2}{Gm_p m_e}$$

3. Substitute numerical values:

$$\frac{F_e}{F_g} = \frac{(8.99 \times 10^9\ \mathrm{N \cdot m^2/C^2})(1.6 \times 10^{-19}\ \mathrm{C})^2}{(6.67 \times 10^{-11}\ \mathrm{N \cdot m^2/kg^2})(1.67 \times 10^{-27}\ \mathrm{kg})(9.11 \times 10^{-31}\ \mathrm{kg})}$$

$$= 2.27 \times 10^{39}$$

Remark This result shows why the effects of gravity are not considered when discussing atomic or molecular interactions.

Though the gravitational force is incredibly weak compared with the electric force and plays essentially no role on the atomic level, gravity is the dominant force between large objects such as planets and stars, because large objects contain almost equal numbers of positive and negative charges, and hence the attractive and repulsive electrical forces cancel. The net force between astronomical objects is therefore essentially the force of gravitational attraction alone.

Force Exerted by a System of Charges

In a system of charges, each charge exerts a force given by Equation 22-4 on every other charge. The net force on any charge is the vector sum of the individual forces exerted on that charge by all the other charges in the system. This follows from the principle of superposition of forces.

Example 22-4 *try it yourself*

Three point charges lie on the x axis; $q_1 = 25$ nC is at the origin, $q_2 = -10$ nC is at $x = 2$ m, and $q_0 = 20$ nC is at $x = 3.5$ m (Figure 22-7). Find the net force on q_0 due to q_1 and q_2.

Picture the Problem The net force on q_0 is the vector sum of the force $\vec{F}_{1,0}$ exerted by q_1, and the force $\vec{F}_{2,0}$ exerted by q_2. The individual forces are found using Coulomb's law. Note that $\hat{r}_{1,0} = \hat{r}_{2,0} = \hat{i}$ because q_0 is to the right of both q_1 and q_2.

Figure 22-7

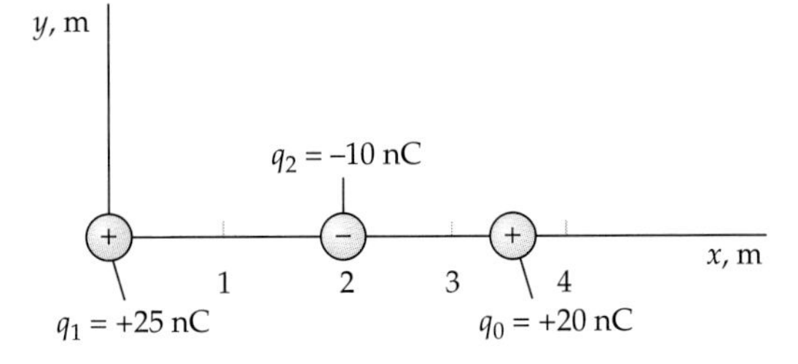

Cover the column to the right and try these on your own before looking at the answers.

Steps	*Answers*
1. Find the force $\vec{F}_{1,0}$ due to q_1.	$\vec{F}_{1,0} = (0.367\ \mu\text{N})\hat{i}$
2. Find the force $\vec{F}_{2,0}$ due to q_2.	$\vec{F}_{2,0} = (-0.799\ \mu\text{N})\hat{i}$
3. Combine your results to obtain the net force.	$\vec{F}_{\text{net}} = \vec{F}_{1,0} + \vec{F}_{2,0} = -(0.432\ \mu\text{N})\hat{i}$

Remark Figure 22-8 shows the force F_x on q_0 as a function of its position x. Near either of the other charges, the force on q_0 is essentially due to the nearest charge alone. Note that the force is undefined at the position of the charges q_1 at $x = 0$, and q_2 at $x = 2$.

Exercise If q_0 is at $x = 1$ m, find (*a*) $\hat{r}_{1,0}$, (*b*) $\hat{r}_{2,0}$, and (*c*) the net force acting on q_0. (*Answers* (*a*) $\hat{i}$, (*b*) $-\hat{i}$, (*c*) $(6.29\ \mu\text{N})\hat{i}$)

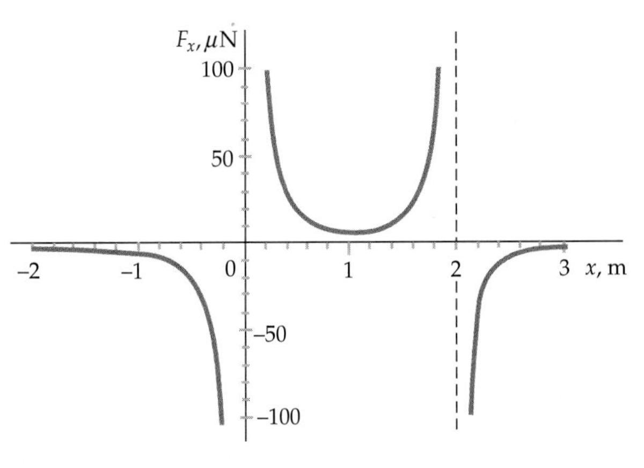

Figure 22-8

If a system of charges is to remain stationary, there must be other forces acting on the charges so that the net force from all sources acting on each charge is zero. In the preceding example and those that follow, we assume that there are such forces so that all the charges remain stationary.

Example 22-5

Charge $q_1 = +25$ nC is at the origin, charge $q_2 = -15$ nC is on the x axis at $x =$ 2 m, and charge $q_0 = +20$ nC is at the point $x = 2$ m, $y = 2$ m as shown in Figure 22-9. Find the resultant force ΣF on q_0.

Picture the Problem The resultant force is the vector sum of the individual forces exerted by each charge on q_0. We compute each force from Coulomb's law and write it in terms of its rectangular components. Figure 22-9*a* shows the resultant force on charge q_0 as the vector sum of the forces $\vec{F}_{1,0}$ due to q_1 and $\vec{F}_{2,0}$ due to q_2. Figure 22-9*b* shows the net force in Figure 22-9*a* and its x and y components.

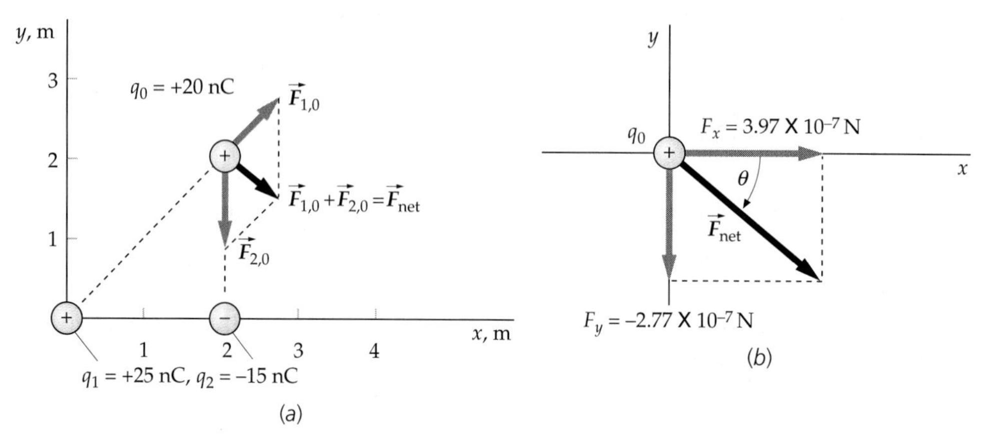

Figure 22-9

1. The resultant force $\Sigma\vec{F}$ on q_0 is the sum of the individual forces:
$$\sum \vec{F} = \vec{F}_{1,0} + \vec{F}_{2,0}$$
$$\sum F_x = F_{1,0x} + F_{2,0x}$$
$$\sum F_y = F_{1,0y} + F_{2,0y}$$

2. The force $\vec{F}_{1,0}$ is directed along the line from q_1 to q_0. Use $r_{1,0} = 2\sqrt{2}$ m for the distance between q_1 and q_0 to calculate its magnitude:
$$F_{1,0} = \frac{(8.99 \times 10^9\ \text{N}\cdot\text{m}^2/\text{C}^2)(25 \times 10^{-9}\ \text{C})(20 \times 10^{-9}\ \text{C})}{(2\sqrt{2}\ \text{m})^2}$$
$$= 5.62 \times 10^{-7}\ \text{N}$$

3. Since $\vec{F}_{1,0}$ makes an angle of 45° with the x and y axes, its x and y components are equal to each other:
$$F_{1,0x} = F_{1,0y} = \frac{F_{1,0}}{\sqrt{2}} = \frac{5.62 \times 10^{-7}\ \text{N}}{\sqrt{2}} = 3.97 \times 10^{-7}$$

4. The force $\vec{F}_{2,0}$ exerted by q_2 on q_0 is attractive and in the negative y direction as shown in Figure 22-9*a*:
$$\vec{F}_{2,0} = \frac{kq_2q_0}{r_{2,0}^2}\hat{r}_{2,0}$$
$$= \frac{(8.99 \times 10^9\ \text{N}\cdot\text{m}^2/\text{C}^2)(-15 \times 10^{-9}\ \text{C})(20 \times 10^{-9}\ \text{C})}{(2\ \text{m})^2}\hat{j}$$
$$= (-6.74 \times 10^{-7}\ \text{N})\hat{j}$$

5. Calculate the components of the resultant force:
$$\sum F_x = F_{1,0x} + F_{2,0x} = (3.97 \times 10^{-7}\ \text{N}) + 0 = 3.97 \times 10^{-7}\ \text{N}$$
$$\sum F_y = F_{1,0y} + F_{2,0y} = (3.97 \times 10^{-7}\ \text{N}) + (-6.74 \times 10^{-7}\ \text{N})$$
$$= -2.77 \times 10^{-7}\ \text{N}$$

6. The magnitude of the resultant force is found from its components:
$$F = \sqrt{F_x^2 + F_y^2} = \sqrt{(3.97 \times 10^{-7}\ \text{N})^2 + (-2.77 \times 10^{-7}\ \text{N})^2}$$
$$= 4.84 \times 10^{-7}\ \text{N}$$

7. The resultant force points to the right and downward as shown in Figure 22-9*b*, making an angle θ with the x axis given by:
$$\tan\theta = \frac{F_y}{F_x} = \frac{-2.77}{3.97} = -0.698$$
$$\theta = -34.9°$$

22-4 The Electric Field

The electric force exerted by one charge on another is an example of an action-at-a-distance force, similar to the gravitational force exerted by one mass on another. The idea of action at a distance presents a difficult conceptual problem. What is the mechanism by which one particle can exert a force on another across the empty space between the particles? Suppose that a charged particle at some point is suddenly moved. Does the force exerted on the second particle some distance r away change instantaneously? To avoid the problem of action at a distance, the concept of the **electric field** is introduced. One charge produces an electric field $\vec{E}$ everywhere in space, and this field exerts the force on the other charge. The force is thus exerted *by the field* at the position of the second charge, rather than by the first charge itself, which is some distance away. The field propagates through space with the speed of light, c. Thus, if a charge is suddenly moved, the force it exerts on another charge a distance r away does not change until a time r/c later.

Figure 22-10 shows a set of point charges, q_1, q_2, and q_3, arbitrarily arranged in space. These charges produce an electric field $\vec{E}$ everywhere in space. If we place a small **test charge** q_0 at some point nearby, there will be a force exerted on q_0 due to the other charges.* The net force on q_0 is the vector sum of the individual forces exerted on q_0 by each of the other charges in the system. Since each of these forces is proportional to q_0, the net force will be proportional to q_0. The electric field $\vec{E}$ at a point is this force divided by q_0†:

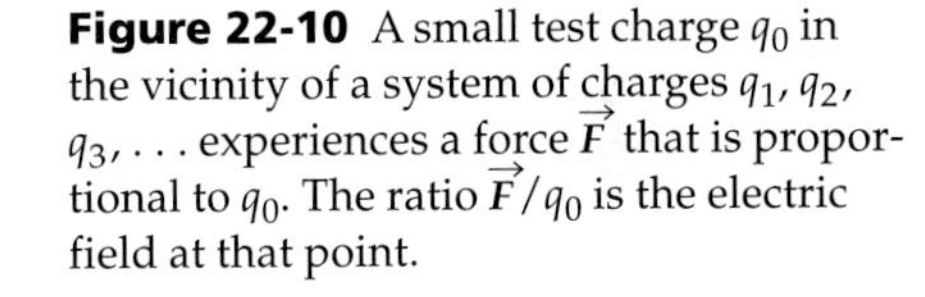

Figure 22-10 A small test charge q_0 in the vicinity of a system of charges q_1, q_2, q_3, . . . experiences a force $\vec{F}$ that is proportional to q_0. The ratio $\vec{F}/q_0$ is the electric field at that point.

$$\vec{E} = \frac{\vec{F}}{q_0} \qquad (q_0 \text{ small}) \qquad \text{22-5}$$

Definition—Electric field

The SI unit of the electric field is the newton per coulomb (N/C). Table 22-1 lists the magnitudes of some of the electric fields found in nature.

The electric field describes the condition in space set up by the system of point charges. By moving a test charge q_0 from point to point, we can find $\vec{E}$ at all points in space (except at any point occupied by a charge q). The electric field $\vec{E}$ is thus a vector function of position. The force exerted on a test charge q_0 at any point is related to the electric field at that point by

$$\vec{F} = q_0\vec{E} \qquad \text{22-6}$$

Exercise When a 5-nC test charge is placed at a certain point, it experiences a force of 2×10^{-4} N in the x direction. What is the electric field $\vec{E}$ at that point? (*Answer* $\vec{E} = \vec{F}/q_0 = [(2 \times 10^{-4}\,\text{N})\hat{i}]/(5 \times 10^{-9}\,\text{C}) = (4 \times 10^4\,\text{N/C})\hat{i}$)

Exercise What is the force on an electron placed at a point where the electric field is $\vec{E} = (4 \times 10^4\,\text{N/C})\hat{i}$? (*Answer* $(-6.4 \times 10^{-15}\,\text{N})\hat{i}$)

Table 22-1

Some Electric Fields in Nature

	E, N/C
In household wires	10^{-2}
In radio waves	10^{-1}
In the atmosphere	10^{2}
In sunlight	10^{3}
Under a thundercloud	10^{4}
In a lightning bolt	10^{4}
In an X-ray tube	10^{6}
At the electron in a hydrogen atom	6×10^{11}
At the surface of a uranium nucleus	2×10^{21}

* The presence of the charge q_0 will generally change the original distribution of the other charges, particularly if the charges are on conductors. However, we may choose q_0 to be small enough so that its effect on the original charge distribution is negligible.

† This definition is similar to that for the gravitational field of the earth, which was defined in Section 4-3 as the force per unit mass exerted by the earth on an object.

The electric field due to a single point charge can be calculated from Coulomb's law. Consider a small, positive test charge q_0 at some point P a distance r_{i0} away from a charge q_i. The force on it is

$$\vec{F}_{i,0} = \frac{kq_i q_0}{r_{i,0}^2}\hat{r}_{i,0}$$

The electric field at point P due to charge q_i is thus

$$\vec{E}_i = \frac{kq_i}{r_{i,0}^2}\hat{r}_{i,0} \qquad \text{22-7}$$

Coulomb's law for $\vec{E}$ due to a point charge

where $\hat{r}_{i0}$ is a unit vector pointing from the charge to the **field point** P. The net electric field due to a distribution of point charges is found by summing the fields due to each charge separately:

$$\vec{E} = \sum_i \vec{E}_i = \sum_i \frac{kq_i}{r_{i,0}^2}\hat{r}_{i,0} \qquad \text{22-8}$$

Electric field due to a system of point charges

Example 22-6

A positive charge $q_1 = +8$ nC is at the origin, and a second positive charge $q_2 = +12$ nC is on the x axis at $a = 4$ m (Figure 22-11). Find the net electric field (a) at point P_1 on the x axis at $x = 7$ m, and (b) at point P_2 on the x axis at $x = 3$ m.

Picture the Problem Because point P_1 is to the right of both charges, each charge produces a field to the right at that point. At point P_2, which is between the charges, the 5-nC charge gives a field to the right and the 12-nC charge gives a field to the left. We calculate each field using

$$\vec{E} = \sum_i \frac{kq_i}{r_{i,0}^2}\hat{r}_{i,0}$$

At point P_1, both unit vectors point along the x axis in the positive direction, so $\hat{r}_{1,0} = \hat{r}_{2,0} = \hat{i}$. At point P_2, $\hat{r}_{1,0} = \hat{i}$, but the unit vector from the 12-nC charge points along the negative x direction, so $\hat{r}_{2,0} = -\hat{i}$.

Figure 22-11

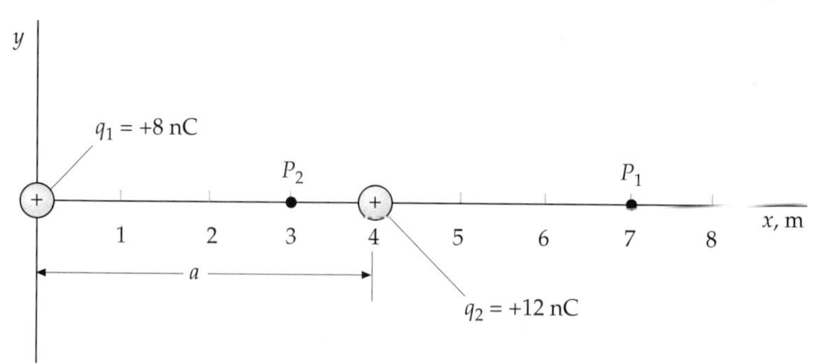

(a) Calculate $\vec{E}$ at point P_1, using $r_{1,0} = x = 7$ m and $r_{2,0} = (x - a) = 7\text{ m} - 4\text{ m} = 3$ m:

$$\vec{E} = \frac{kq_1}{r_{1,0}^2}\hat{r}_{1,0} + \frac{kq_2}{r_{2,0}^2}\hat{r}_{2,0} = \frac{kq_1}{x^2}\hat{i} + \frac{kq_2}{(x-a)^2}\hat{i}$$

$$= \frac{(8.99 \times 10^9\text{ N}\cdot\text{m}^2/\text{C}^2)(8 \times 10^{-9}\text{ C})}{(7\text{ m})^2}\hat{i}$$

$$+ \frac{(8.99 \times 10^9\text{ N}\cdot\text{m}^2/\text{C}^2)(12 \times 10^{-9}\text{ C})}{(3\text{ m})^2}\hat{i}$$

$$= (1.47\text{ N/C})\hat{i} + (12.0\text{ N/C})\hat{i} = (13.5\text{ N/C})\hat{i}$$

(*b*) Calculate $\vec{E}$ at point P_2, where $r_{1,0} = x = 3$ m and $r_{2,0} = a - x = 4\text{ m} - 3\text{ m} = 1$ m:

$$\vec{E} = \frac{kq_1}{r_{1,0}^2}\hat{r}_{1,0} + \frac{kq_2}{r_{2,0}^2}\hat{r}_{2,0} = \frac{kq_1}{x^2}\hat{i} + \frac{kq_2}{(a-x)^2}(-\hat{i})$$

$$= \frac{(8.99 \times 10^9\ \text{N}\cdot\text{m}^2/\text{C}^2)(8 \times 10^{-9}\ \text{C})}{(3\ \text{m})^2}\hat{i}$$

$$+ \frac{(8.99 \times 10^9\ \text{N}\cdot\text{m}^2/\text{C}^2)(12 \times 10^{-9}\ \text{C})}{(1\ \text{m})^2}(-\hat{i})$$

$$= (7.99\ \text{N/C})\hat{i} - (108\ \text{N/C})\hat{i} = (-100\ \text{N/C})\hat{i}$$

Remarks The electric field at point P_2 is in the negative x direction because the field due to the +12-nC charge, which is 1 m away, is larger than that due to the +8-nC charge, which is 3 m away. As we move toward the +8-nC charge, the magnitude of its field increases and that due to the +12-nC charge decreases. There is one point between the charges where the net electric field is zero. At this point, a test charge experiences no net force. A sketch of E_x versus x for this system is shown in Figure 22-12.

Exercise Find the point on the x axis where the electric field is zero. (*Answer* $x = 1.80$ m)

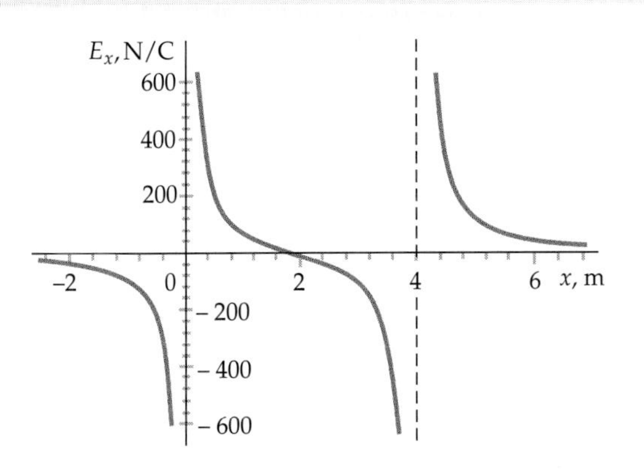

Figure 22-12

Example 22-7 *try it yourself*

Find the electric field on the y axis at $y = 3$ m for the charges in Example 22-6.

Picture the Problem On the y axis, the electric field $\vec{E}_1$ due to charge q_1 is directed along the y axis, and the field $\vec{E}_2$ due to charge q_2 makes an angle θ with the y axis (Figure 22-13*a*). To find the resultant field, we first find the x and y components of these fields, as shown in Figure 22-13*b*.

Figure 22-13

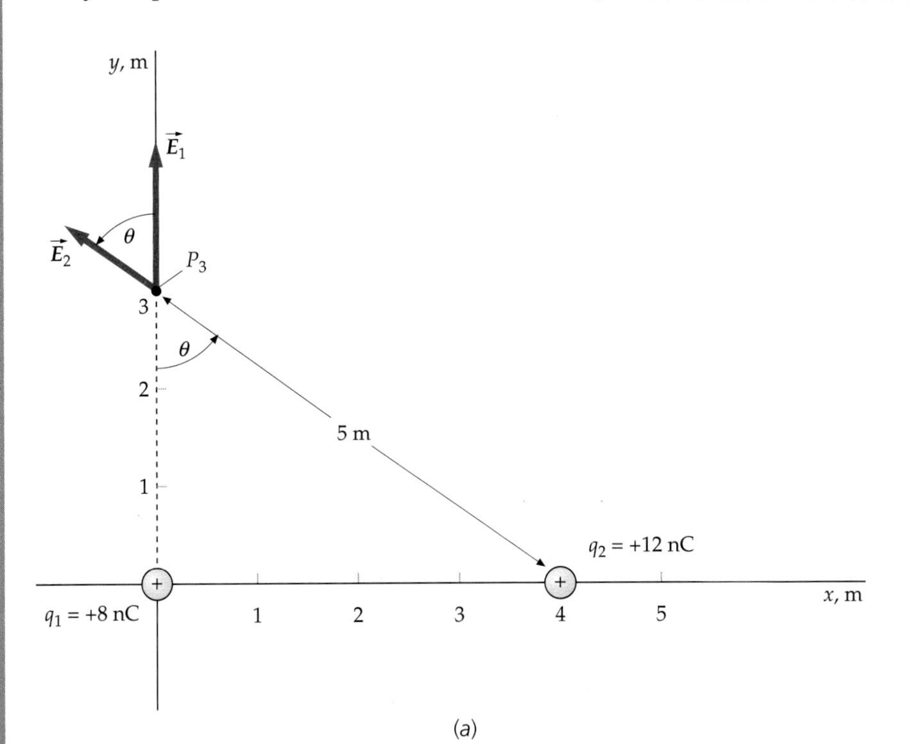

(*a*)

(*b*)

Cover the column to the right and try these on your own before looking at the answers.

Steps	***Answers***
1. Calculate the magnitude of the field $\vec{E}_1$ due to q_1. Find the x and y components of $\vec{E}_1$.	$E_1 = kq_1/y^2 = 7.99\text{ N/C}, \quad E_{1x} = 0, \quad E_{1y} = 7.99\text{ N/C}$
2. Calculate the magnitude of the field $\vec{E}_2$ due to q_2.	$E_2 = 4.32\text{ N/C}$
3. Write the x and y components of $\vec{E}_2$ in terms of the angle θ.	$E_x = -E_2 \sin\theta, \quad E_y = E_2 \cos\theta$
4. Compute $\sin\theta$ and $\cos\theta$.	$\sin\theta = 0.8, \quad \cos\theta = 0.6$
5. Calculate E_{2x} and E_{2y}.	$E_{2x} = -3.46\text{ N/C}, \quad E_{2y} = 2.59\text{ N/C}$
6. Find the x and y components of the resultant field $\vec{E}$.	$E_x = -3.46\text{ N/C}, \quad E_y = 10.6\text{ N/C}$
7. Calculate the magnitude of $\vec{E}$ from its components.	$E = \sqrt{E_x^2 + E_y^2} = 11.2\text{ N/C}$
8. Find the angle θ_1 made by $\vec{E}$ with the x axis.	$\theta_1 = \tan^{-1}\left(\frac{E_y}{E_x}\right) = 108°$

Example 22-8

A charge $+q$ is at $x = a$ and a second charge $-q$ is at $x = -a$ (Figure 22-14). (*a*) Find the electric field on the x axis at an arbitrary point $x > a$. (*b*) Find the limiting form of the electric field for $x >> a$.

Picture the Problem We calculate the electric field using

$$\vec{E} = \sum_i \frac{kq_i}{r_{i,0}^2}\hat{r}_{i,0}$$

(Equation 22-8). For $x > a$, the unit vector for each charge is $\hat{i}$. The distances are $x - a$ to the plus charge and $x - (-a) = x + a$ to the minus charge.

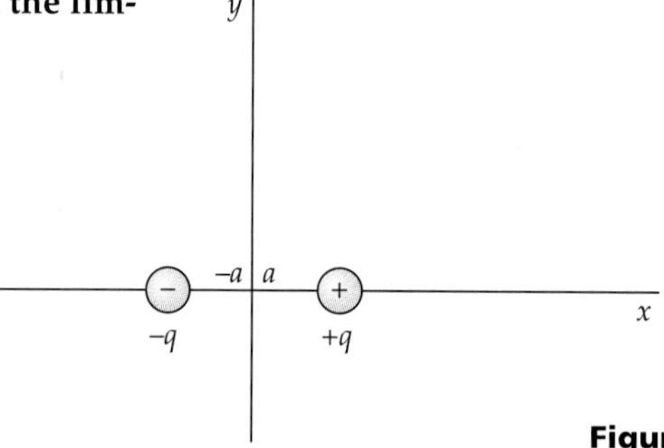

Figure 22-14

(*a*)1. Calculate $\vec{E}$ due to the two charges for $x > a$: [*Note:* The equation on the right holds only for $x > a$. For $x < a$, the signs of the two terms are reversed. For $-a < x < a$, both terms have negative signs.]

$$\vec{E} = \frac{kq}{(x-a)^2}\hat{i} + \frac{k(-q)}{(x+a)^2}\hat{i}$$

$$= kq\left[\frac{1}{(x-a)^2} - \frac{1}{(x+a)^2}\right]\hat{i}$$

2. Put the terms in square brackets under a common denominator and simplify:

$$\vec{E} = kq\left[\frac{(x+a)^2 - (x-a)^2}{(x+a)^2(x-a)^2}\right]\hat{i} = kq\frac{4ax}{(x^2-a^2)^2}\hat{i}$$

(*b*) In the limit $x >> a$, we can neglect a^2 compared with x^2 in the denominator:

$$\vec{E} = kq\frac{4ax}{(x^2-a^2)^2}\hat{i} \approx kq\frac{4ax}{x^4}\hat{i} = \frac{4kqa}{x^3}\hat{i}$$

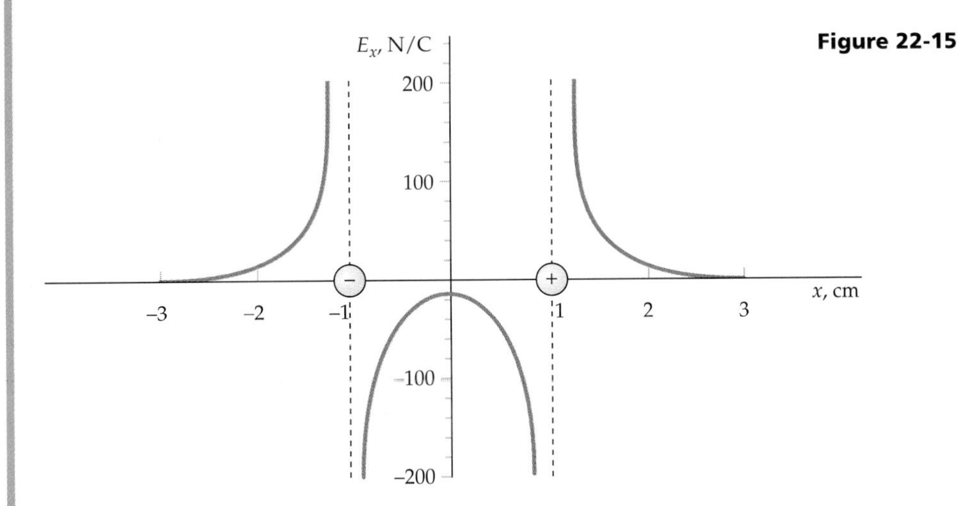

Figure 22-15

Remarks Figure 22-15 shows E_x versus x for all x, for $q = 1$ nC and $a = 1$ cm. Far from the charges, the field is given by

$$\vec{E} = \frac{4kqa}{|x^3|}\hat{i}$$

Between the charges, the contribution from each charge is in the negative direction. An expression that holds for all x is

$$\vec{E} = \frac{kq}{(x-a)^2}\left[\frac{(x-a)\hat{i}}{|x-a|}\right] + \frac{k(-q)}{(x+a)^2}\left[\frac{(x+a)\hat{i}}{|x+a|}\right]$$

Note that the unit vectors (quantities in square brackets) in this expression point in the proper direction for all x.

Electric Dipoles

A system of two equal and opposite charges q separated by a small distance L is called an **electric dipole.** Its strength and orientation are described by the **electric dipole moment** $\vec{p}$, which is a vector that points from the negative charge to the positive charge and has the magnitude qL (Figure 22-16).

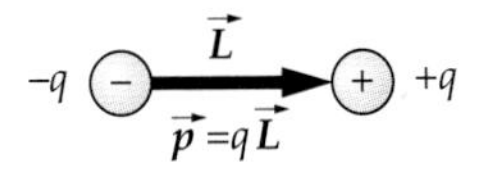

Figure 22-16 An electric dipole consists of two equal and opposite charges separated by some distance L. The dipole moment points from the negative charge to the positive charge and has the magnitude $p = qL$.

$$\vec{p} = q\vec{L} \qquad \text{22-9}$$

Definition—Electric dipole moment

For the configuration in Figure 22-14, $\vec{L} = 2a\hat{i}$ and the electric dipole moment is

$$\vec{p} = 2aq\hat{i}$$

In terms of the dipole moment, the electric field on the axis of the dipole at a point a great distance x away is in the direction of the dipole moment and has the magnitude

$$E = \frac{2kp}{|x^3|} \qquad \text{22-10}$$

(See Example 22-8.) At any point far from a dipole, the electric field is proportional to the dipole moment and decreases with the cube of the distance. When a system has a net charge, the electric field decreases as $1/r^2$ at large distances. In a system with zero net charge, the electric field falls off more rapidly with distance. In the case of an electric dipole, the field falls off as $1/r^3$.

22-5 Electric Field Lines

We can picture the electric field by drawing lines to indicate its direction. At any given point, the field vector $\vec{E}$ is tangent to the lines. Electric field lines are also called **lines of force** because they show the direction of the force exerted on a positive test charge. At any point near a positive charge, the electric field points radially away from the charge. Similarly, the electric field lines converge toward a point occupied by a negative charge.

Figure 22-17 shows the electric field lines of a single positive point charge. The spacing of the lines is related to the strength of the electric field. As we move away from the charge, the field becomes weaker and the lines become farther apart. Consider a spherical surface of radius r with its center at the charge. Its area is $4\pi r^2$. If N lines diverge from the point charge, the number of lines per unit area on a spherical surface a distance r away is $N/4\pi r^2$. Thus, as the distance increases, the density of the field lines (the number of lines per unit area) decreases as $1/r^2$, the same rate of decrease as E. So, if we adopt the convention of drawing a fixed number of lines from a point charge, the number being proportional to the charge q, and if we draw the lines symmetrically about the point charge, the field strength is indicated by the density of the lines. The more closely spaced the lines, the stronger the electric field.

Figure 22-18 shows the electric field lines for two equal positive point charges q separated by a small distance. Near each charge, the field is approximately due to that charge alone because the other charge is far away. Consequently, the field lines near either charge are radial and equally spaced. Since the charges are equal, we draw an equal number of lines originating from each charge. At very large distances, the details of the system are not important and the system looks like a point charge of magnitude $2q$. (For example, if the two charges were 1 mm apart and we were looking at them from a point 100 km away, they would look like a single charge.) So far from the charges, the field is approximately the same as that due to a point charge $2q$ and the lines are approximately equally spaced. Looking at the figure, we see that the electric field in the space between the charges is weak because there are few lines in this region compared with the region just to the right or left of the charges, where the lines are more closely spaced. This information can, of course, also be obtained by direct calculation of the field at points in these regions.

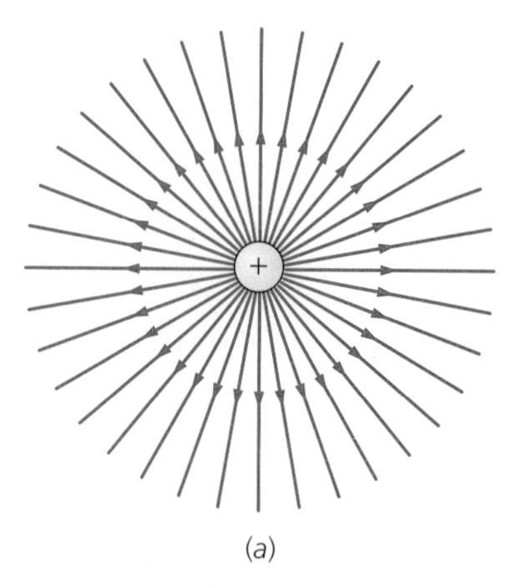

(a)

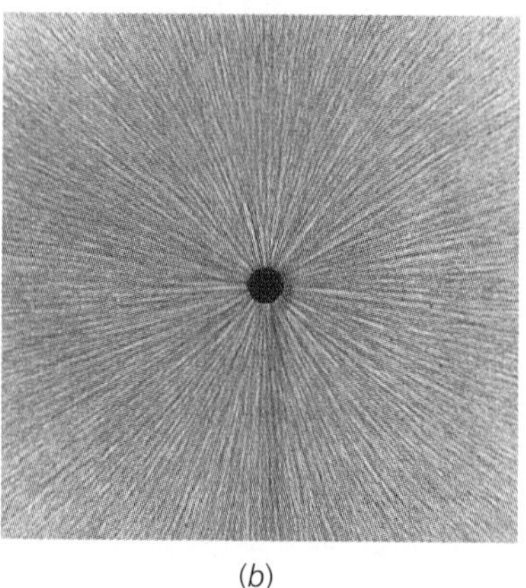

(b)

Figure 22-17 (*a*) Electric field lines of a single positive point charge. If the charge were negative, the arrows would be reversed. (*b*) The same electric field lines shown by bits of thread suspended in oil. The electric field of the charged object in the center induces opposite charges on the ends of each bit of thread, causing the threads to align themselves parallel to the field.

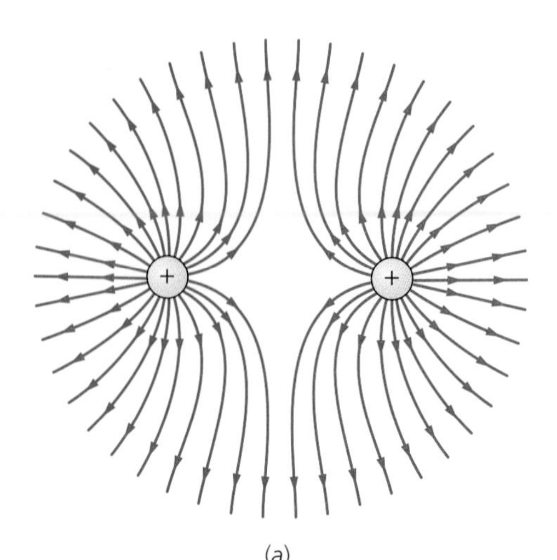

(a)

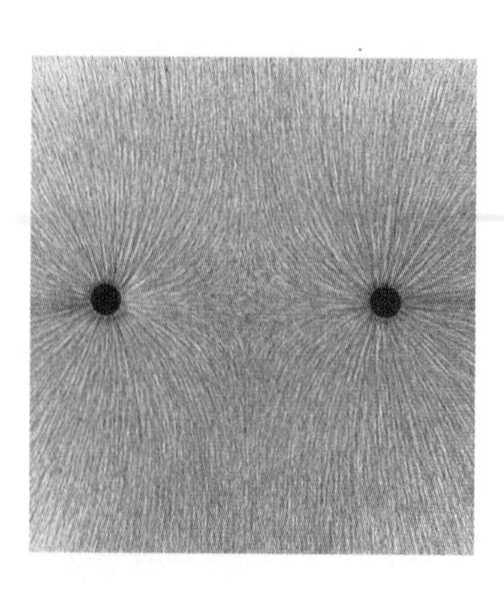

(b)

Figure 22-18 (*a*) Electric field lines due to two positive point charges. The arrows would be reversed if both charges were negative. (*b*) The same electric field lines shown by bits of thread in oil.

We can apply this reasoning to draw the electric field lines for any system of point charges. Very near each charge, the field lines are equally spaced and leave or enter the charge radially, depending on the sign of the charge. Very far from all the charges, the detailed structure of the system is not important so the field lines are just like those of a single point charge carrying the net charge of the system. For future reference, the rules for drawing electric field lines can be summarized as follows:

1. Electric field lines begin on positive charges (or at infinity) and end on negative charges (or at infinity).
2. The lines are drawn symmetrically entering or leaving an isolated charge.
3. The number of lines leaving a positive charge or entering a negative charge is proportional to the magnitude of the charge.
4. The density of the lines (the number of lines per unit area perpendicular to the lines) at any point is proportional to the magnitude of the field at that point.
5. At large distances from a system of charges, the field lines are equally spaced and radial, as if they came from a single point charge equal to the net charge of the system.
6. Field lines do not cross. (If two field lines crossed, that would indicate two directions for $\vec{E}$ at the point of intersection.)

Rules for drawing electric field lines

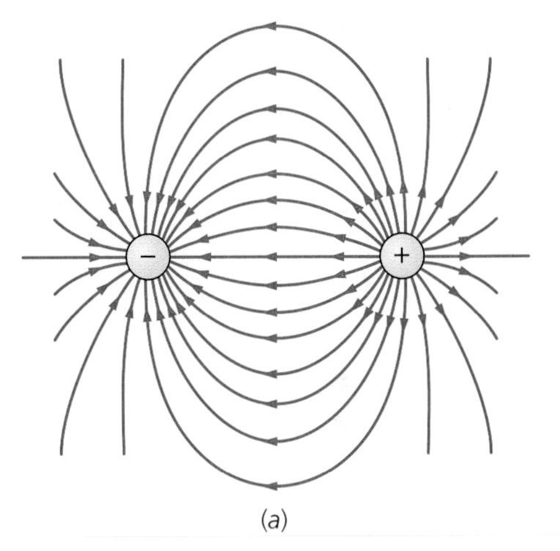

(*a*)

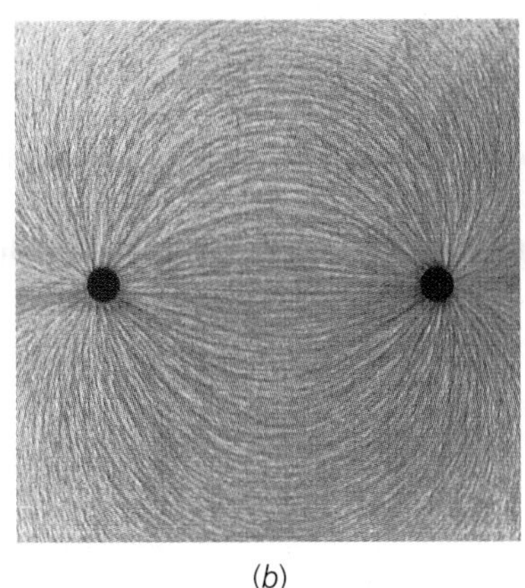

(*b*)

Figure 22-19 (*a*) Electric field lines for an electric dipole. (*b*) The same field lines shown by bits of thread in oil.

Figure 22-19 shows the electric field lines due to an electric dipole. Very near the positive charge, the lines are directed radially outward. Very near the negative charge, the lines are directed radially inward. Since the charges have equal magnitudes, the number of lines that begin at the positive charge equals the number that end at the negative charge. In this case, the field is strong in the region between the charges, as indicated by the high density of field lines in this region in the figure.

Figure 22-20 shows the electric field lines for a negative charge $-q$ at a small distance from a positive charge $+2q$. Twice as many lines leave the positive charge as enter the negative charge. Thus, half the lines beginning on the positive charge $+2q$ enter the negative charge $-q$; the rest leave the system. Very far from the charges, the lines leaving the system are approximately symmetrically spaced and point radially outward, just as they would for a single positive point charge $+q$.

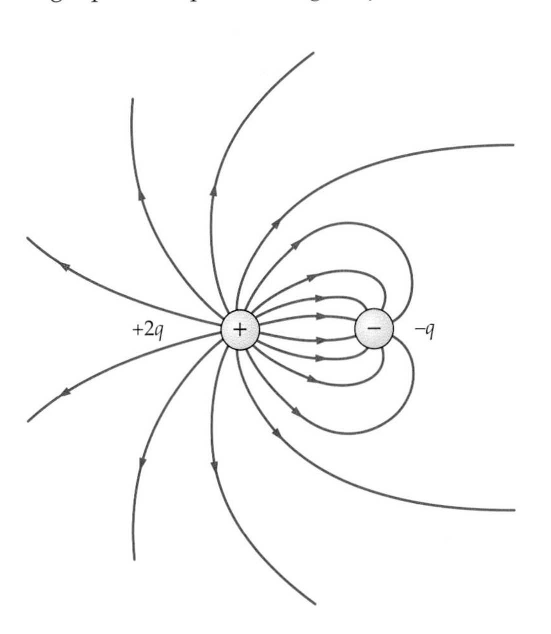

Figure 22-20 Electric field lines for a point charge $+2q$ and a second point charge $-q$. At great distances from the charges, the lines are the same as those for a single charge $+q$.

Example 22-9

The electric field lines for two conducting spheres are shown in Figure 22-21. What is the relative sign and magnitude of the charges on the two spheres?

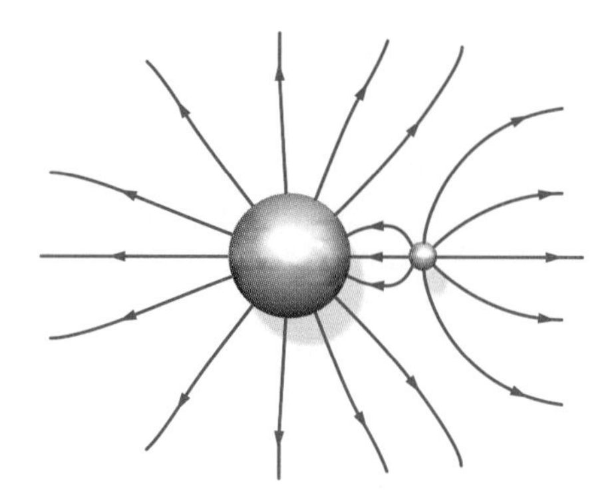

Figure 22-21

Picture the Problem The charge on a sphere is positive if more lines leave than enter and negative if more enter than leave. The ratio of the magnitudes of the charges equals the ratio of the net number of lines entering or leaving.

Since 11 electric field lines leave the large sphere on the left and 3 enter, the net number leaving is 8, so the charge on the large sphere is positive. For the small sphere on the right, 8 lines leave and none enter, so its charge is also positive. Since the net number of lines leaving each sphere is 8, the spheres carry equal positive charges. The charge on the small sphere creates an intense field at the nearby surface of the large sphere that causes a local accumulation of negative charge on the large sphere—indicated by the three entering field lines. Most of the large sphere's surface has positive charge, however, so its total charge is positive.

The convention relating the electric field strength to the electric field lines works because the electric field varies inversely as the square of the distance from a point charge. Since the gravitational field of a point mass also varies inversely as the square of the distance, field-line drawings are also useful for picturing the gravitational field. Near a point mass, the gravitational field lines converge toward the mass just as electric field lines converge toward a negative charge. However, there are no points in space where gravitational field lines diverge like electric field lines near a positive charge, because the gravitational force is always attractive, never repulsive.

22-6 Motion of Point Charges in Electric Fields

When a particle with a charge q is placed in an electric field $\vec{E}$, it experiences a force $q\vec{E}$. If the electric force is the only significant force acting on the particle, the particle has an acceleration

$$\vec{a} = \frac{\Sigma\vec{F}}{m} = \frac{q}{m}\vec{E}$$

where m is the mass of the particle.* If the electric field is known, the charge-to-mass ratio of the particle can be determined from the measured acceleration. The deflection of electrons in a uniform electric field was used by J. J. Thomson in 1897 to demonstrate the existence of electrons and to measure their charge-to-mass ratio. Familiar examples of devices that rely on the motion of electrons in electric fields are oscilloscopes, computer monitors, and television picture tubes.

* If the particle is an electron, its speed in an electric field is often a significant fraction of the speed of light. In such cases, Newton's laws of motion must be modified by Einstein's special theory of relativity.

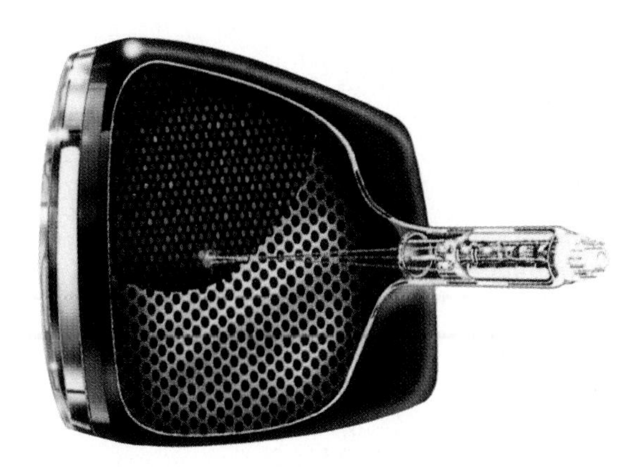

Schematic drawing of a cathode-ray tube used for color television. The beams of electrons from the electron gun on the right activate phosphors on the screen at the left, giving rise to bright spots whose colors depend on the relative intensity of each beam. Electric fields between deflection plates in the gun (or magnetic fields from coils within the gun) deflect the beams. The beams sweep across the screen in a horizontal line, are deflected downward, then sweep across again. The entire screen is covered in this way 30 times per second.

Example 22-10

An electron is projected into a uniform electric field $\vec{E} = (1000\ \mathrm{N/C})\hat{i}$ with an initial velocity $\vec{v}_0 = (2 \times 10^6\ \mathrm{m/s})\hat{i}$ in the direction of the field (Figure 22-22). How far does the electron travel before it is brought momentarily to rest?

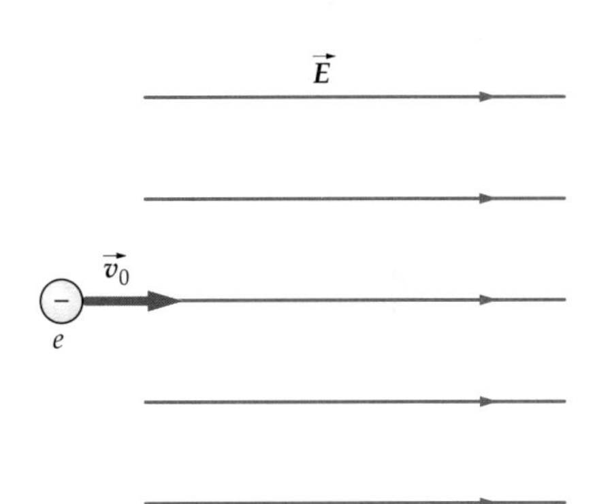

Figure 22-22

Picture the Problem Since the charge of the electron is negative, the force $-e\vec{E}$ acting on the electron is in the direction opposite that of the field. Since $\vec{E}$ is constant, the force is constant and we can use constant acceleration formulas from Chapter 2. We choose the field to be in the positive x direction.

1. The displacement Δx is related to the initial and final velocities: $v^2 = v_0^2 + 2a\,\Delta x$

2. The acceleration is obtained from Newton's second law: $a = \dfrac{F}{m} = \dfrac{-eE}{m}$

3. When $v = 0$, the displacement is:

$$\Delta x = \frac{v^2 - v_0^2}{2a} = \frac{0 - v_0^2}{2(-eE/m)} = \frac{mv_0^2}{2eE}$$

$$= \frac{(9.11 \times 10^{-31}\ \mathrm{kg})(2 \times 10^6\ \mathrm{m/s})^2}{2(1.6 \times 10^{-19}\ \mathrm{C})(1000\ \mathrm{N/C})} = 1.14 \times 10^{-2}\ \mathrm{m}$$

Example 22-11

An electron enters a uniform electric field $\vec{E} = (-2000\ \mathrm{N/C})\hat{j}$ with an initial velocity $\vec{v}_0 = (10^6\ \mathrm{m/s})\hat{i}$ perpendicular to the field (Figure 22-23). (*a*) Compare the gravitational force acting on the electron to the electric force acting on it. (*b*) By how much has the electron been deflected after it has traveled 1 cm in the x direction?

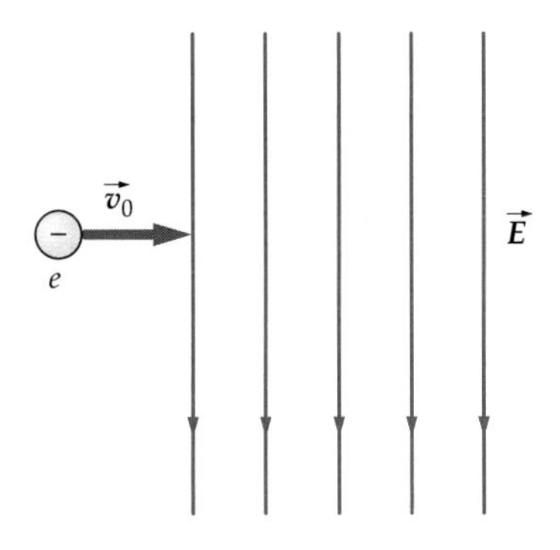

Figure 22-23

Picture the Problem (*a*) Calculate the ratio of the electric force $qE = -eE$ to the gravitational force mg. (*b*) Since mg is negligible, the force on the electron is $-eE$ vertically upward. The electron thus moves with constant horizontal velocity v_x and is deflected upward by an amount $y = \frac{1}{2}at^2$, where t is the time to travel 1 cm in the x direction.

(*a*) Calculate the ratio of the magnitude of the electric force, F_e, to the magnitude of the gravitational force, F_g:

$$\frac{F_e}{F_g} = \frac{eE}{mg} = \frac{(1.6 \times 10^{-19}\ \text{C})(2000\ \text{N/C})}{(9.11 \times 10^{-31}\ \text{kg})(9.81\ \text{N/kg})} = 3.6 \times 10^{13}$$

(*b*)1. Express the vertical deflection in terms of the acceleration a and time t:

$$y = \frac{1}{2}at^2$$

2. The time required for the electron to travel a distance x with constant horizontal velocity v_0 is:

$$t = \frac{x}{v_0} = \frac{10^{-2}\ \text{m}}{10^6\ \text{m/s}} = 10^{-8}\ \text{s}$$

3. Use this result for t and eE/m for a to calculate y:

$$y = \frac{1}{2}\frac{eE}{m}t^2 = \frac{1}{2}\frac{(1.6 \times 10^{-19}\ \text{C})(2000\ \text{N/C})}{9.11 \times 10^{-31}\ \text{kg}}(10^{-8}\ \text{s})^2$$
$$= 1.76\ \text{cm}$$

Remarks (*a*) As is usually the case, the electric force is huge compared with the gravitational force. Thus, it is not necessary to consider gravity when designing a cathode-ray tube, for example, or when calculating the deflection in the problem above. In fact, a television picture tube works equally well upside down and right side up, as if gravity were not even present. (*b*) The path of an electron moving in a uniform electric field is a parabola, the same as the path of a mass moving in a uniform gravitational field.

22-7 Electric Dipoles in Electric Fields

In Example 22-6 we found the electric field produced by a dipole, a system of two equal and opposite point charges that are close together. Here we consider the behavior of an electric dipole in an external electric field. Some molecules have permanent electric dipole moments due to a nonuniform distribution of charge within the molecule. Such molecules are called **polar molecules**. An example is HCl, which is essentially a positive hydrogen ion of charge $+e$ combined with a negative chlorine ion of charge $-e$. The center of charge of the positive ion does not coincide with the center of charge for the negative ion, so the molecule has a permanent dipole moment. Another example is water (Figure 22-24). A uniform external electric field exerts no net force on a dipole, but it does exert a torque that tends to rotate the dipole into the direction of the field. We see in Figure 22-25 that the torque about the negative charge* has the magnitude $F_1L \sin\theta = qEL \sin\theta = pE \sin\theta$. The direction of the torque is into the paper such that it rotates the dipole moment $\vec{p}$ into the direction of $\vec{E}$. The torque can be conveniently written as the cross product of the dipole moment $\vec{p}$ and the electric field $\vec{E}$:

$$\vec{\tau} = \vec{p} \times \vec{E} \qquad 22\text{-}11$$

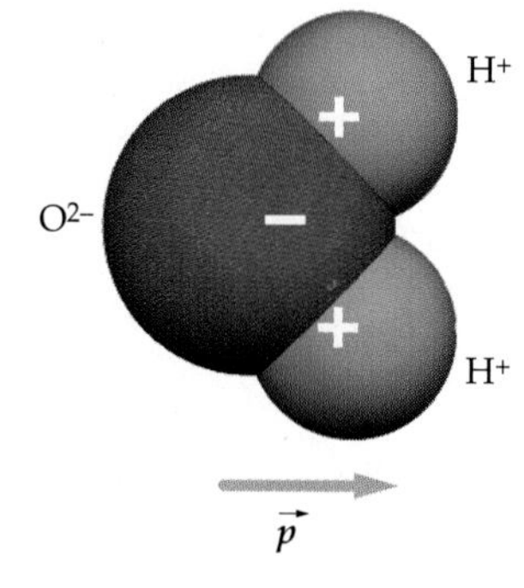

Figure 22-24 An H_2O molecule has a permanent electric dipole moment that points in the direction from the center of negative charge to the center of positive charge.

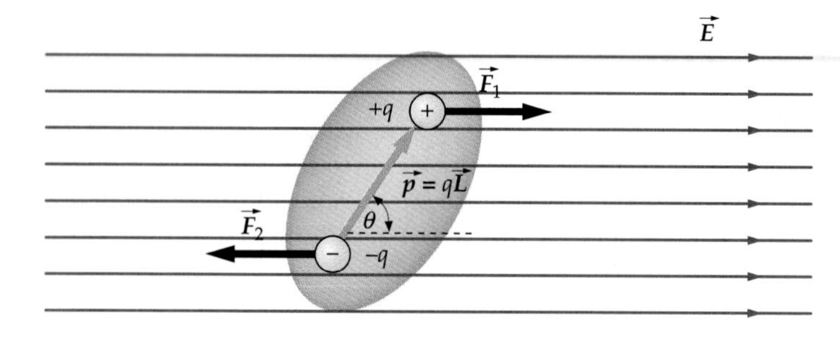

Figure 22-25 A dipole in a uniform electric field experiences equal and opposite forces that tend to rotate the dipole so that its dipole moment is aligned with the electric field.

* The torque produced by two equal and opposite forces (an arrangement called a couple) is the same about any point in space.

When the dipole rotates through $d\theta$, the electric field does work:

$$dW = -\tau\, d\theta = -pE \sin\theta\, d\theta$$

(The minus sign arises because the torque tends to decrease θ.) Setting this work equal to the decrease in potential energy, we have

$$dU = -dW = +pE \sin\theta\, d\theta$$

Integrating, we obtain

$$U = -pE \cos\theta + U_0$$

If we choose the potential energy U_0 to be zero when $\theta = 90°$, then the potential energy of the dipole is

$$U = -pE \cos\theta = -\vec{p}\cdot\vec{E} \qquad \text{22-12}$$

Potential energy of a dipole in an electric field

Microwave ovens take advantage of the electric dipole moment of water molecules to cook food. Like all electromagnetic waves, microwaves have an oscillating electric field that can cause electric dipoles to vibrate. The microwaves in home ovens are tuned to the natural frequency of vibration of water molecules. The water molecules in food resonate with the oscillating electric field and absorb large amounts of energy, accounting for the rapid cooking times that make microwave ovens so convenient.

Nonpolar molecules have no permanent electric dipole moment. However, all neutral molecules contain equal amounts of positive and negative charge. In the presence of an external electric field $\vec{E}$, the charges become separated in space. The positive charges are pushed in the direction of $\vec{E}$ and the negative charges are pushed in the opposite direction. The molecule thus acquires an induced dipole moment parallel to the external electric field and is said to be **polarized**.

In a nonuniform electric field, an electric dipole experiences a net force because the electric field has different magnitudes at the positive and negative poles. Figure 22-26 shows how a positive point charge polarizes a nonpolar molecule and then attracts it. A familiar example is the attraction that holds an electrostatically charged balloon against a wall. The nonuniform field produced by the charge on the balloon polarizes molecules in the wall and attracts them. An equal and opposite force is exerted by the wall molecules on the balloon.

The diameter of an atom or molecule is of the order of 10^{-10} m = 0.1 nm. A convenient unit for the electric dipole moment of atoms and molecules is the fundamental electronic charge e times the distance 1 nm. For example, the dipole moment of H_2O in these units has a magnitude of about 0.04 e·nm.

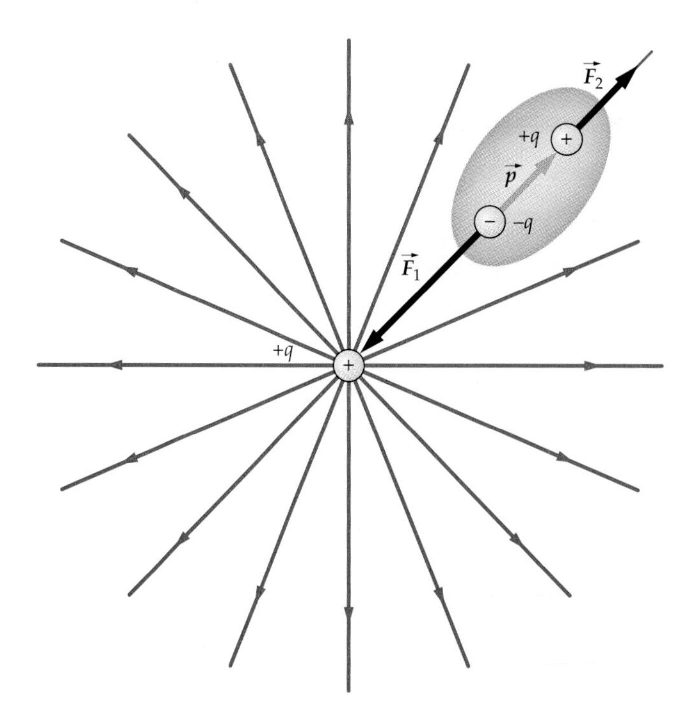

Figure 22-26 A nonpolar molecule in the nonuniform electric field of a positive point charge. The induced electric dipole moment $\vec{p}$ is parallel to the field of the point charge. Since the point charge is closer to the center of negative charge than to the center of positive charge, there is a net force of attraction between the dipole and the point charge. If the point charge were negative, the induced dipole moment would be reversed, and the molecule would again be attracted to the point charge.

Example 22-12

A dipole with a moment of magnitude 0.02 e·nm makes an angle of 20° with a uniform electric field of magnitude 3×10^3 N/C (Figure 22-27). Find (*a*) the magnitude of the torque on the dipole and (*b*) the potential energy of the system.

Figure 22-27

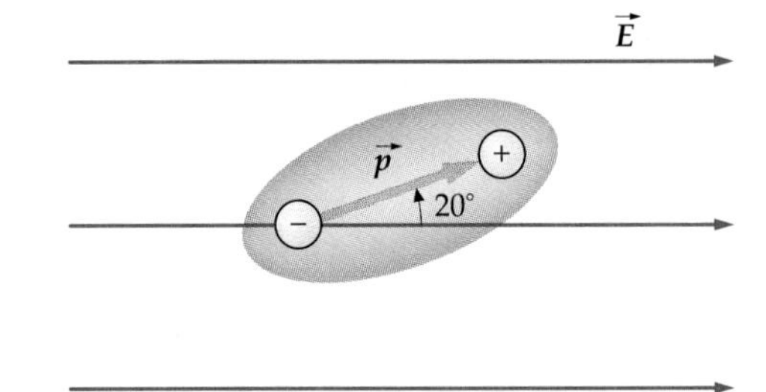

Picture the Problem The torque is found from $\vec{\tau} = \vec{p} \times \vec{E}$ and the potential energy is found from $U = -\vec{p} \cdot \vec{E}$.

(*a*) Calculate the magnitude of the torque:

$$\tau = |\vec{p} \times \vec{E}| = pE \sin\theta = (0.02\ \text{e}\cdot\text{nm})(3 \times 10^3\ \text{N/C})(\sin 20°)$$
$$= (0.02)(1.6 \times 10^{-19}\ \text{C})(10^{-9}\ \text{m})(3 \times 10^3\ \text{N/C})(\sin 20°)$$
$$= 3.28 \times 10^{-27}\ \text{N}\cdot\text{m}$$

(*b*) Calculate the potential energy:

$$U = -\vec{p} \cdot \vec{E} = -pE \cos\theta$$
$$= -(0.02)(1.6 \times 10^{-19}\ \text{C})(10^{-9}\ \text{m})(3 \times 10^3\ \text{N/C})\cos 20°$$
$$= -9.02 \times 10^{-27}\ \text{J}$$

Summary

1. Quantization and conservation are fundamental properties of electric charge.
2. Coulomb's law is the fundamental law of interaction between charges at rest.
3. The electric field describes the condition in space set up by a charge distribution.

Topic	Remarks and Relevant Equations	
1. Electric Charge	There are two kinds of electric charge, positive and negative.	
Quantization	Electric charge is quantized—it always occurs in integral multiples of the fundamental unit of charge e. The charge of the electron is $-e$ and that of the proton is $+e$.	
Magnitude	$e = 1.60 \times 10^{-19}$ C	**22-1**
Conservation	Charge is conserved. It is neither created nor destroyed in any process, but is merely transferred.	

Topic	Relevant Equations and Remarks	
2. Conductors and Insulators	In conductors, about one electron per atom is free to move about the entire material. In insulators, all the electrons are bound to nearby atoms.	
Ground	A very large conductor that can supply an unlimited amount of charge (such as the earth) is called a ground.	
3. Charging by Induction	A conductor can be charged by holding a charge near it to attract or repel the free electrons and then grounding the conductor to drain off the far away charges.	
4. Coulomb's Law	The force exerted by a charge q_1 on q_2 is given by $\vec{F}_{1,2} = \frac{kq_1q_2}{r_{1,2}^2}\hat{r}_{1,2}$ where $\hat{r}_{1,2}$ is a unit vector that points from q_1 to q_2	22-2
Coulomb constant	$k = 8.99 \times 10^9\ \mathrm{N{\cdot}m^2/C^2}$	
5. Electric Field	The electric field due to a system of charges at a point is defined as the net force exerted by those charges on a very small positive test charge q_0 divided by q_0: $\vec{E} = \frac{\vec{F}}{q_0}$	22-5
Due to a point charge	$\vec{E}_i = \frac{kq_i}{r_{i,0}^2}\hat{r}_{i,0}$	22-7
Due to a system of point charges	The electric field due to several charges is the vector sum of the fields due to the individual charges: $\vec{E} = \sum_i \vec{E}_i = \sum_i \frac{kq_i}{r_{i,0}^2}\hat{r}_{i,0}$	22-8
6. Electric Field Lines	The electric field can be represented by electric field lines that originate on positive charges and end on negative charges. The strength of the electric field is indicated by the density of the electric field lines.	
7. Electric Dipole	An electric dipole is a system of two equal but opposite charges separated by a small distance.	
Dipole moment	$\vec{p} = q\vec{L}$ where $\vec{L}$ points from the negative charge to the positive charge.	
Field due to dipole	The electric field far from a dipole is proportional to the dipole moment and decreases with the cube of the distance.	
Torque on a dipole	In a uniform electric field, the net force on a dipole is zero, but there is a torque $\vec{\tau}$ that tends to align the dipole in the direction of the field. $\vec{\tau} = \vec{p} \times \vec{E}$	
Potential energy of a dipole	$U = -\vec{p}\cdot\vec{E}$	
8. Polar and Nonpolar Molecules	Polar molecules, such as H_2O, have permanent dipole moments because their centers of positive and negative charge do not coincide. They behave like simple dipoles in an electric field. Nonpolar molecules do not have permanent dipole moments, but they acquire induced dipole moments in the presence of an electric field.	

Problem-Solving Guide

Begin by drawing a neat diagram that includes the important features of the problem. Include the location, sign and magnitude of the charges, as well as the individual force vectors or electric field vectors in your sketch. Show the vector sum of these force or field vectors when appropriate.

Summary of Worked Examples

Type of Calculation	Procedure and Relevant Examples	
1. Charges and Coulomb's Law		
Find the number of electrons in a given charge.	The number of electrons in a charge Q is $N = Q/e$.	**Example 22-1**
Find the electric force between two point charges.	Use Coulomb's law.	**Examples 22-2, 22-3**
Find the electric force on one point charge due to two or more other charges.	Use Coulomb's law to find the force due to each charge, then add the force vectors.	**Examples 22-4, 22-5**
2. The Electric Field		
Find the electric field due to two or more point charges.	Find the electric field due to each charge and add the electric field vectors.	**Examples 22-6, 22-7, 22-8**
3. Electric Field Lines		
Using electric field lines to determine the relative signs and magnitudes of charges.	The lines begin on positive charges and end on negative charges with the number of lines being proportional to the magnitude of the charge.	**Example 22-9**
4. Point Charges and Dipoles in Electric Fields		
Motion of a charged particle in a constant electric field.	Find the acceleration of the particle using $\vec{a} = q\vec{E}/m$, then use constant-acceleration kinematics.	**Examples 22-10 and 22-11**
Find the torque acting on an electric dipole.	Use $\vec{\tau} = \vec{p} \times \vec{E}$.	**Example 22-12**
Find the potential energy of a dipole in an electric field.	Use $U = -pE \cos\theta = -\vec{p}\cdot\vec{E}$.	**Example 22-12**

Problems

In a few problems, you are given more data than you actually need; in a few other problems, you are required to supply data from your general knowledge, outside sources, or informed estimates.

Conceptual Problems

Problems from Optional and Exploring sections

- • Single-concept, single-step, relatively easy
- •• Intermediate-level, may require synthesis of concepts
- ••• Challenging, for advanced students

Electric Charge

1 • If the sign convention for charge were changed so that the charge on the electron were positive and the charge on the proton were negative, would Coulomb's law still be written the same?

2 •• Discuss the similarities and differences in the properties of electric charge and gravitational mass.

3 • A plastic rod is rubbed against a wool shirt, thereby acquiring a charge of $-0.8\ \mu C$. How many electrons are transferred from the wool shirt to the plastic rod?

4 • A charge equal to the charge of Avogadro's number of protons ($N_A = 6.02 \times 10^{23}$) is called a *faraday*. Calculate the number of coulombs in a faraday.

5 • How many coulombs of positive charge are there in 1 kg of carbon? Twelve grams of carbon contain Avogadro's number of atoms, with each atom having six protons and six electrons.

Conductors, Insulators, and Charging by Induction

6 • Can insulators be charged by induction?

7 •• A metal rectangle B is connected to ground through a switch S that is initially closed (Figure 22-28). While the charge $+Q$ is near B, switch S is opened. The charge $+Q$ is then removed. Afterward, what is the charge state of the metal rectangle B?

(*a*) It is positively charged.
(*b*) It is uncharged.
(*c*) It is negatively charged.
(*d*) It may be any of the above depending on the charge on B before the charge $+Q$ was placed nearby.

Figure 22-28 Problem 7

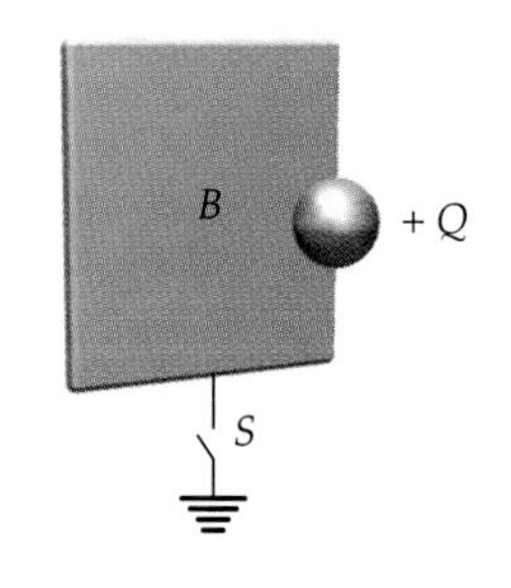

8 •• Explain, giving each step, how a positively charged insulating rod can be used to give a metal sphere (*a*) a negative charge, and (*b*) a positive charge. (*c*) Can the same rod be used to simultaneously give one sphere a positive charge and another sphere a negative charge without the rod having to be recharged?

9 •• Two uncharged conducting spheres with their conducting surfaces in contact are supported on a large wooden table by insulated stands. A positively charged rod is brought up close to the surface of one of the spheres on the side opposite its point of contact with the other sphere. (*a*) Describe the induced charges on the two conducting spheres, and sketch the charge distributions on them. (*b*) The two spheres are separated far apart and the charged rod is removed. Sketch the charge distributions on the separated spheres.

Coulomb's Law

10 • Three charges, $+q$, $+Q$, and $-Q$, are placed at the corners of an equilateral triangle as shown in Figure 22-29. The net force on charge $+q$ due to the other two charges is

(*a*) vertically up.
(*b*) vertically down.
(*c*) zero.
(*d*) horizontal to the left.
(*e*) horizontal to the right.

Figure 22-29 Problem 10

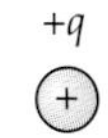

11 • A charge $q_1 = 4.0\ \mu C$ is at the origin, and a charge $q_2 = 6.0\ \mu C$ is on the x axis at $x = 3.0$ m. (*a*) Find the force on charge q_2. (*b*) Find the force on q_1. (*c*) How would your answers for parts (*a*) and (*b*) differ if q_2 were $-6.0\ \mu C$?

12 • Three point charges are on the x axis: $q_1 = -6.0\ \mu C$ is at $x = -3.0$ m, $q_2 = 4.0\ \mu C$ is at the origin, and $q_3 = -6.0\ \mu C$ is at $x = 3.0$ m. Find the force on q_1.

13 •• Two equal charges of $3.0\ \mu C$ are on the y axis, one at the origin and the other at $y = 6$ m. A third charge $q_3 = 2\ \mu C$ is on the x axis at $x = 8$ m. Find the force on q_3.

14 •• Three charges, each of magnitude 3 nC, are at separate corners of a square of side 5 cm. The two charges at opposite corners are positive, and the other charge is negative. Find the force exerted by these charges on a fourth charge $q = +3$ nC at the remaining corner.

15 •• A charge of $5\ \mu C$ is on the y axis at $y = 3$ cm, and a second charge of $-5\ \mu C$ is on the y axis at $y = -3$ cm. Find the force on a charge of $2\ \mu C$ on the x axis at $x = 8$ cm.

16 •• A point charge of $-2.5\ \mu C$ is located at the origin. A second point charge of $6\ \mu C$ is at $x = 1$ m, $y = 0.5$ m. Find the x and y coordinates of the position at which an electron would be in equilibrium.

17 •• A charge of $-1.0\ \mu C$ is located at the origin, a second charge of $2.0\ \mu C$ is located at $x = 0$, $y = 0.1$ m, and a third charge of $4.0\ \mu C$ is located at $x = 0.2$ m, $y = 0$. Find the forces that act on each of the three charges.

18 •• A charge of $5.0\ \mu C$ is located at $x = 0$, $y = 0$ and a charge Q_2 is located at $x = 4.0$ cm, $y = 0$. The force on a $2\text{-}\mu C$ charge at $x = 8.0$ cm, $y = 0$ is 19.7 N, pointing in the negative x direction. When this $2\text{-}\mu C$ charge is positioned at $x = 17.75$ cm, $y = 0$, the force on it is zero. Determine the charge Q_2.

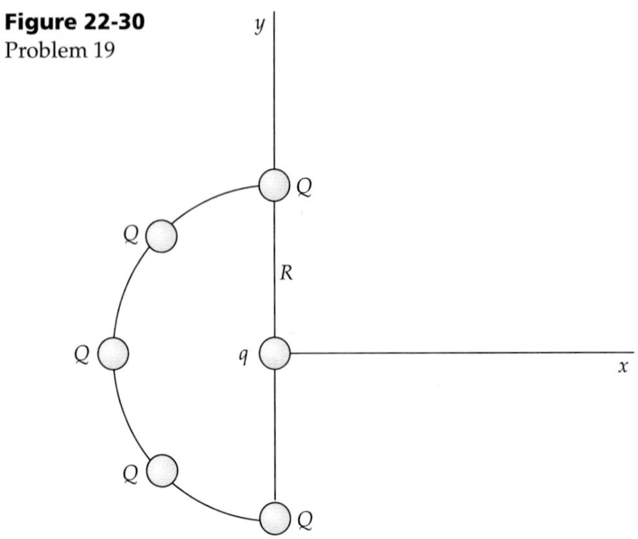

Figure 22-30 Problem 19

19 •• Five equal charges Q are equally spaced on a semicircle of radius R as shown in Figure 22-30. Find the force on a charge q located at the center of the semicircle.

20 ••• The configuration of the NH_3 molecule is approximately that of a regular tetrahedron, with three H^+ ions forming the base and an N^{3-} ion at the apex of the tetrahedron. The length of each side is 1.64×10^{-10} m. Calculate the force that acts on each ion.

The Electric Field

21 • A positive charge that is free to move but is at rest in an electric field $\vec{E}$ will

(*a*) accelerate in the direction perpendicular to $\vec{E}$.
(*b*) remain at rest.
(*c*) accelerate in the direction opposite to $\vec{E}$.
(*d*) accelerate in the same direction as $\vec{E}$.
(*e*) do none of the above.

22 • If four charges are placed at the corners of a square as shown in Figure 22-31, the field $\vec{E}$ is zero at

Figure 22-31 Problem 22

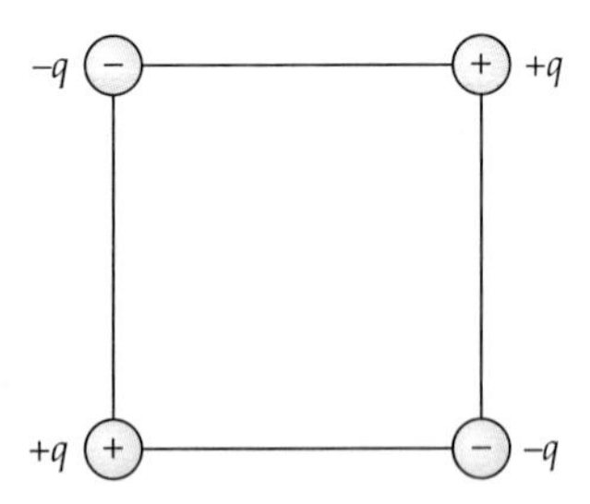

(*a*) all points along the sides of the square midway between two charges.
(*b*) the midpoint of the square.
(*c*) midway between the top two charges and midway between the bottom two charges.
(*d*) none of the above.

23 •• At a particular point in space, a charge Q experiences no force. It follows that

(*a*) there are no charges nearby.
(*b*) if charges are nearby, they have the opposite sign of Q.
(*c*) if charges are nearby, the total positive charge must equal the total negative charge.
(*d*) none of the above need be true.

24 • A charge of 4.0 μC is at the origin. What is the magnitude and direction of the electric field on the x axis at (*a*) $x = 6$ m and (*b*) $x = -10$ m? (*c*) Sketch the function E_x versus x for both positive and negative values of x. (Remember that E_x is negative when $\vec{E}$ points in the negative x direction.)

25 • Two charges, each $+4\ \mu$C, are on the x axis, one at the origin and the other at $x = 8$ m. Find the electric field on the x axis at (*a*) $x = -2$ m, (*b*) $x = 2$ m, (*c*) $x = 6$ m, and (*d*) $x = 10$ m. (*e*) At what point on the x axis is the electric field zero? (*f*) Sketch E_x versus x.

26 • When a test charge $q_0 = 2$ nC is placed at the origin, it experiences a force of 8.0×10^{-4} N in the positive y direction. (*a*) What is the electric field at the origin? (*b*) What would be the force on a charge of -4 nC placed at the origin? (*c*) If this force is due to a charge on the y axis at $y = 3$ cm, what is the value of that charge?

27 • An oil drop has a mass of 4×10^{-14} kg and a net charge of 4.8×10^{-19} C. An upward electric force just balances the downward force of gravity so that the oil drop is stationary. What is the direction and magnitude of the electric field?

28 • The electric field near the surface of the earth points downward and has a magnitude of 150 N/C. (*a*) Compare the upward electric force on an electron with the downward gravitational force. (*b*) What charge should be placed on a penny of mass 3 g so that the electric force balances the weight of the penny near the earth's surface?

29 •• Two equal positive charges of magnitude $q_1 = q_2 = 6.0$ nC are on the y axis at $y_1 = +3$ cm and $y_2 = -3$ cm. (*a*) What is the magnitude and direction of the electric field on the x axis at $x = 4$ cm? (*b*) What is the force exerted on a third charge $q_0 = 2$ nC when it is placed on the x axis at $x = 4$ cm?

30 •• A point charge of $+5.0\ \mu$C is located at $x = -3.0$ cm, and a second point charge of $-8.0\ \mu$C is located at $x = +4.0$ cm. Where should a third charge of $+6.0\ \mu$C be placed so that the electric field at $x = 0$ is zero?

31 •• A point charge of $-5\ \mu$C is located at $x = 4$ m, $y = -2$ m. A second point charge of 12 μC is located at $x = 1$ m, $y = 2$ m. (*a*) Find the magnitude and direction of the electric field at $x = -1$ m, $y = 0$. (*b*) Calculate the magnitude and direction of the force on an electron at $x = -1$ m, $y = 0$.

32 •• Two equal positive charges q are on the y axis, one at $y = a$ and the other at $y = -a$. (*a*) Show that the electric field on the x axis is along the x axis with $E_x = 2kqx(x^2 + a^2)^{-3/2}$. (*b*) Show that near the origin, when x is much smaller than a, E_x is approximately $2kqx/a^3$. (*c*) Show that for values of x much larger than a, E_x is approximately $2kq/x^2$. Explain why you would expect this result even before calculating it.

33 •• A 5-μC point charge is located at $x = 1$ m, $y = 3$ m, and a -4-μC point charge is located at $x = 2$ m, $y = -2$ m. (*a*) Find the magnitude and direction of the electric field at $x = -3$ m, $y = 1$ m. (*b*) Find the magnitude and direction of the force on a proton at $x = -3$ m, $y = 1$ m.

34 •• (*a*) Show that the electric field for the charge distribution in Problem 32 has its greatest magnitude at the points $x = a/\sqrt{2}$ and $x = -a/\sqrt{2}$ by computing dE_x/dx and setting the derivative equal to zero. (*b*) Sketch the function E_x versus x using your results for part (*a*) of this problem and parts (*b*) and (*c*) of Problem 32.

35 ••• For the charge distribution in Problem 32, the electric field at the origin is zero. A test charge q_0 placed at the origin will therefore be in equilibrium. (*a*) Discuss the stability of the equilibrium for a positive test charge by considering small displacements from equilibrium along the x axis and small displacements along the y axis. (*b*) Repeat part (*a*) for a negative test charge. (*c*) Find the magnitude and sign of a charge q_0 that when placed at the origin results in a net force of zero on each of the three charges. (*d*) What will happen if any of the charges is displaced slightly from equilibrium?

36 ••• Two positive point charges $+q$ are on the y axis at $y = +a$ and $y = -a$ as in Problem 32. A bead of mass m carrying a negative charge $-q$ slides without friction along a thread that runs along the x axis. (*a*) Show that for small displacements of $x \ll a$, the bead experiences a restoring force that is proportional to x and therefore undergoes simple harmonic motion. (*b*) Find the period of the motion.

Electric Field Lines

37 • Which of the following statements about electric field lines is (are) *not* true?
(*a*) The number of lines leaving a positive charge or entering a negative charge is proportional to the charge.
(*b*) The lines begin on positive charges and end on negative charges.
(*c*) The density of lines (the number per unit area perpendicular to the lines) is proportional to the magnitude of the field.
(*d*) Electric field lines cross midway between charges that have equal magnitude and sign.

38 • Figure 22-32 shows the electric field lines for a system of two point charges. (*a*) What are the relative magnitudes of the charges? (*b*) What are the signs of the charges? (*c*) In what regions of space is the electric field strong? In what regions is it weak?

Figure 22-32
Problem 38

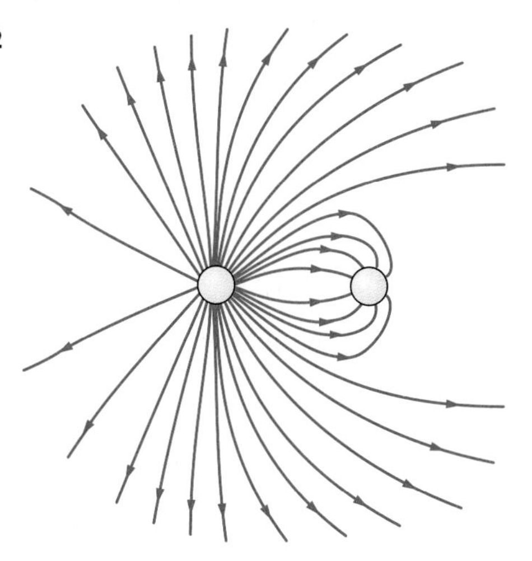

39 • Two charges $+4q$ and $-3q$ are separated by a small distance. Draw the electric field lines for this system.

40 • Two charges $+q$ and $-3q$ are separated by a small distance. Draw the electric field lines for this system.

41 • Three equal positive point charges are situated at the corners of an equilateral triangle. Sketch the electric field lines in the plane of the triangle.

Motion of Point Charges in Electric Fields

42 • The acceleration of a particle in an electric field depends on the ratio of the charge to the mass of the particle. (*a*) Compute e/m for an electron. (*b*) What is the magnitude and direction of the acceleration of an electron in a uniform electric field with a magnitude of 100 N/C? (*c*) When the speed of an electron approaches the speed of light c, relativistic mechanics must be used to calculate its motion, but at speeds significantly less than c, Newtonian mechanics applies. Using Newtonian mechanics, compute the time it takes for an electron placed at rest in an electric field with a magnitude of 100 N/C to reach a speed of $0.01c$. (*d*) How far does the electron travel in that time?

43 • (*a*) Compute e/m for a proton, and find its acceleration in a uniform electric field with a magnitude of 100 N/C. (*b*) Find the time it takes for a proton initially at rest in such a field to reach a speed of $0.01c$ (where c is the speed of light).

44 • An electron has an initial velocity of 2×10^6 m/s in the x direction. It enters a uniform electric field $\vec{E} = (400\ \text{N/C})\hat{j}$, which is in the y direction. (*a*) Find the acceleration of the electron. (*b*) How long does it take for the electron to travel 10 cm in the x direction in the field? (*c*) By how much and in what direction is the electron deflected after traveling 10 cm in the x direction in the field?

45 •• An electron, starting from rest, is accelerated by a uniform electric field of 8×10^4 N/C that extends over a distance of 5.0 cm. Find the speed of the electron after it leaves the region of uniform electric field.

46 •• An electron moves in a circular orbit about a stationary proton. The centripetal force is provided by the electrostatic force of attraction between the proton and the electron. The electron has a kinetic energy of 2.18×10^{-18} J. (*a*) What is the speed of the electron? (*b*) What is the radius of the orbit of the electron?

47 •• A mass of 2 g located in a region of uniform electric field $\vec{E} = (300\ \text{N/C})\hat{i}$ carries a charge Q. The mass, released from rest at $x = 0$, has a kinetic energy of 0.12 J at $x = 0.50$ m. Determine the charge Q.

48 •• A particle leaves the origin with a speed of 3×10^6 m/s at 35° to the x axis. It moves in a constant electric field $\vec{E} = E_y\hat{j}$. Find E_y such that the particle will cross the x axis at $x = 1.5$ cm if the particle is (*a*) an electron and (*b*) a proton.

49 •• An electron starts at the position shown in Figure 22-33 with an initial speed $v_0 = 5 \times 10^6$ m/s at 45° to the x axis. The electric field is in the positive y direction and has a magnitude of 3.5×10^3 N/C. On which plate and at what location will the electron strike?

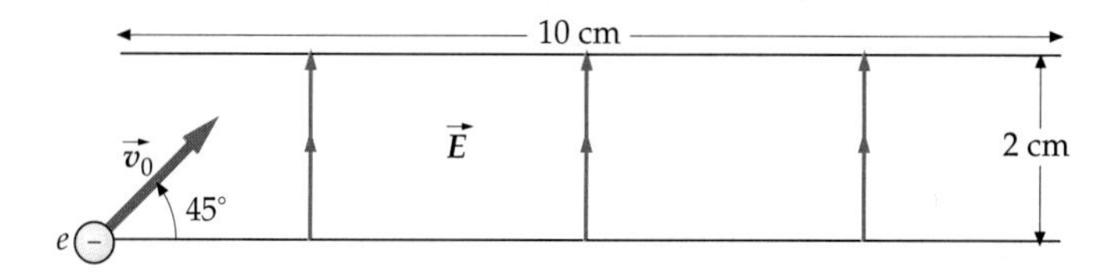

Figure 22-33 Problem 49

Figure 22-34 Problem 50

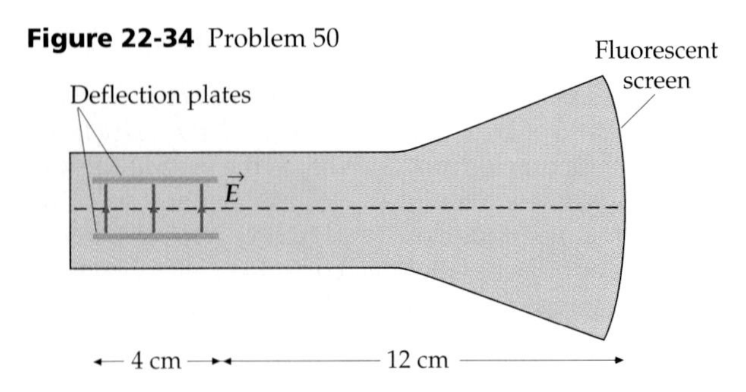

50 •• An electron with kinetic energy of 2×10^{-16} J is moving to the right along the axis of a cathode-ray tube as shown in Figure 22-34. There is an electric field $\vec{E} = (2 \times 10^4 \text{ N/C})\hat{j}$ in the region between the deflection plates. Everywhere else, $\vec{E} = 0$. (*a*) How far is the electron from the axis of the tube when it reaches the end of the plates? (*b*) At what angle is the electron moving with respect to the axis? (*c*) At what distance from the axis will the electron strike the fluorescent screen?

Electric Dipoles

51 • Two point charges, $q_1 = 2.0$ pC and $q_2 = -2.0$ pC, are separated by 4 μm. (*a*) What is the dipole moment of this pair of charges? (*b*) Sketch the pair, and show the direction of the dipole moment.

52 • A dipole of moment 0.5 e·nm is placed in a uniform electric field with a magnitude of 4.0×10^4 N/C. What is the magnitude of the torque on the dipole when (*a*) the dipole is parallel to the electric field, (*b*) the dipole is perpendicular to the electric field, and (*c*) the dipole makes an angle of 30° with the electric field? (*d*) Find the potential energy of the dipole in the electric field for each case.

53 •• For a dipole oriented along the x axis, the electric field falls off as $1/x^3$ in the x direction and $1/y^3$ in the y direction. Use dimensional analysis to prove that, in any direction, the field far from the dipole falls off as $1/r^3$.

54 •• A water molecule has its oxygen atom at the origin, one hydrogen nucleus at $x = 0.077$ nm, $y = 0.058$ nm, and the other hydrogen nucleus at $x = -0.077$ nm, $y = 0.058$ nm. If the hydrogen electrons are transferred completely to the oxygen atom so that it has a charge of $-2e$, what is the dipole moment of the water molecule? (Note that this characterization of the chemical bonds of water as totally ionic is simply an approximation that overestimates the dipole moment of a water molecule.)

55 •• An electric dipole consists of two charges $+q$ and $-q$ separated by a very small distance $2a$. Its center is on the x axis at $x = x_1$, and it points along the x axis in the positive x direction. The dipole is in a nonuniform electric field, which is also in the x direction, given by $\vec{E} = Cx\hat{i}$, where C is a constant. (*a*) Find the force on the positive charge and that on the negative charge, and show that the net force on the dipole is $Cp\hat{i}$. (*b*) Show that, in general, if a dipole of moment $\vec{p}$ lies along the x axis in an electric field in the x direction, the net force on the dipole is given approximately by $(dE_x/dx)p\hat{i}$.

56 ••• A positive point charge $+Q$ is at the origin, and a dipole of moment $\vec{p}$ is a distance r away and in the radial direction as in Figure 22-26. (*a*) Show that the force exerted by the electric field of the point charge on the dipole is attractive and has a magnitude of approximately $2kQp/r^3$ (see Problem 55). (*b*) Now assume that the dipole is centered at the origin and that a point charge Q is a distance r away along the line of the dipole. From your result for part (*a*) and Newton's third law, show that the magnitude of the electric field of the dipole along the line of the dipole a distance r away is approximately $2kp/r^3$.

57 ••• A quadrupole consists of two dipoles that are close together, as shown in Figure 22-35. The effective charge at the origin is $-2q$ and the other charges on the y axis at $y = a$ and $y = -a$ are each $+q$. (*a*) Find the electric field at a point on the x axis far away so that $x >> a$. (*b*) Find the electric field on the y axis far away so that $y >> a$.

Figure 22-35 Problem 57

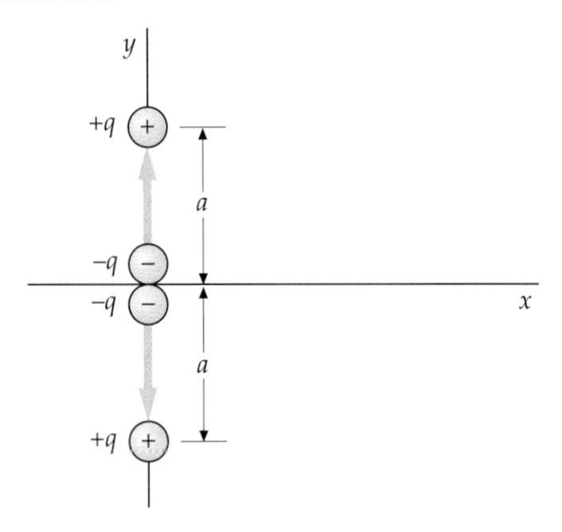

General Problems

58 • A charged insulator and an uncharged metal
(*a*) always repel one another.
(*b*) exert no electrostatic force on one another.
(*c*) always attract one another.
(*d*) may attract or repel, depending on the sign of the charge on the insulator.

59 • Which of the following statements are true?
(*a*) A positive charge experiences an attractive electrostatic force toward a nearby neutral conductor.
(*b*) A positive charge experiences no electrostatic force near a neutral conductor.
(*c*) A positive charge experiences a repulsive force, away from a nearby conductor.
(*d*) Whatever the force on a positive charge near a neutral conductor, the force on a negative charge is then oppositely directed.
(*e*) None of the above is correct.

60 • The electric field lines around an electrical dipole are best represented by which, if any, of the diagrams in Figure 22-36?

61 •• A molecule with electric dipole moment $\vec{p}$ is oriented so that $\vec{p}$ makes an angle θ with a uniform electric field $\vec{E}$. The dipole is free to move in response to the force from the field. Describe the motion of the dipole. Suppose the electric field is nonuniform and is larger in the x direction. How will the motion be changed?

Figure 22-36 Problem 60

(a)

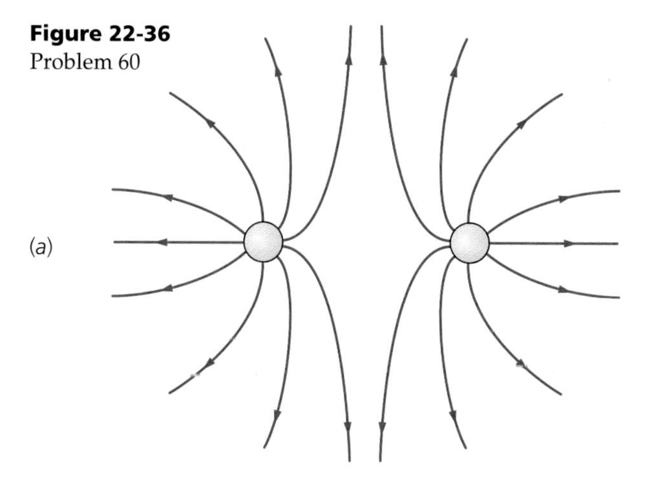

(b)

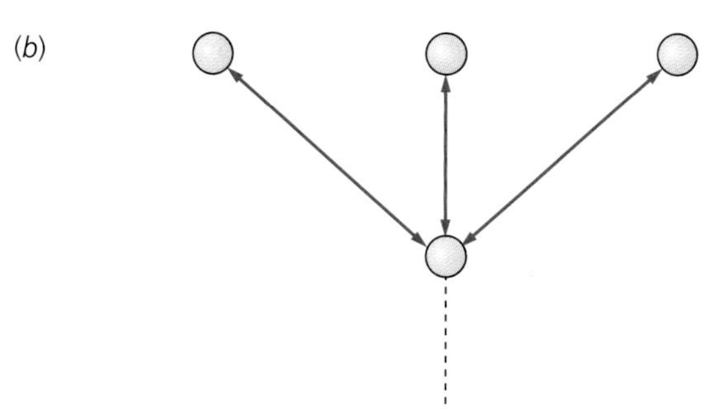

(c)

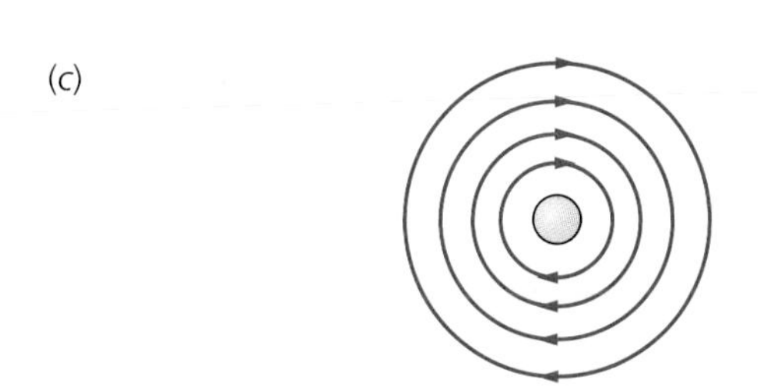

(d)

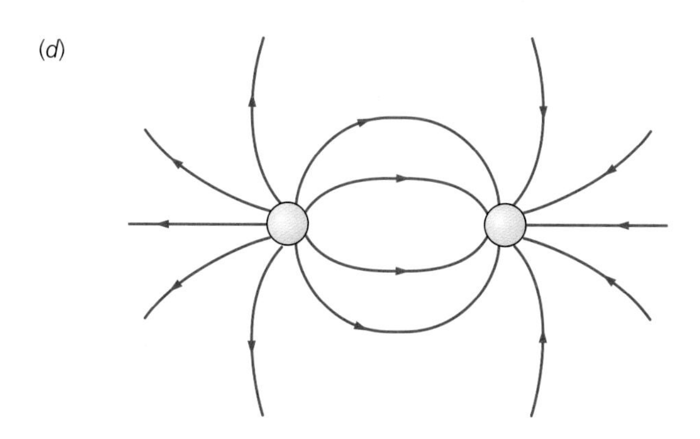

62 •• True or false:

(*a*) The electric field of a point charge always points away from the charge.
(*b*) All macroscopic charges Q can be written as $Q = \pm Ne$, where N is an integer and e is the charge of the electron.
(*c*) Electric field lines never diverge from a point in space.
(*d*) Electric field lines never cross at a point in space.
(*e*) All molecules have electric dipole moments in the presence of an external electric field.

63 •• A small, nonconducting ball with no net charge is suspended from a thread. When a positive charge is brought near the ball, the ball is attracted toward the charge. How does this come about? How would the situation be different if the charge brought near the ball were negative instead of positive?

64 •• Two metal balls have charges $+q$ and $-q$. How will the force on one of them change if (*a*) the balls are placed in water, the distance between them being unchanged, and (*b*) a third uncharged metal ball is placed between the first two? Explain.

65 •• A metal ball is positively charged. Is it possible for it to attract another positively charged ball? Explain.

66 • In interstellar space, two charged point-like objects, each of mass m and charge q, are separated by a distance d and released. They remain motionless at that separation. Find an expression for q in terms of m, G, and k.

67 •• Point charges of $-5.0\ \mu$C, $+3.0\ \mu$C, and $+5.0\ \mu$C are located along the x axis at $x = -1.0$ cm, $x = 0$, and $x = +1.0$ cm, respectively. Calculate the electric field at $x = 3.0$ cm and at $x = 15.0$ cm. Is there some point on the x axis where the magnitude of the electric field is zero? Locate that point.

68 •• For the charge distribution of Problem 67, find the electric field at $x = 15.0$ cm as the vector sum of the electric field due to a dipole formed by the two 5.0-μC charges and a point charge of 3.0 μC, both located at the origin. Compare your result with the result obtained in Problem 67 and explain any difference between these two.

69 •• In copper, about one electron per atom is free to move about. A copper penny has a mass of 3 g. (*a*) What percentage of the free charge would have to be removed to give the penny a charge of 15 μC? (*b*) What would be the force of repulsion between two pennies carrying this charge if they were 25 cm apart? Assume that the pennies are point charges.

70 •• Two charges q_1 and q_2 have a total charge of 6 μC. When they are separated by 3 m, the force exerted by one charge on the other has a magnitude of 8 mN. Find q_1 and q_2 if (*a*) both are positive so that they repel each other, and (*b*) one is positive and the other is negative so that they attract each other.

71 •• Three charges, $+q$, $+2q$, and $+4q$, are connected by strings as shown in Figure 22-37. Find the tensions T_1 and T_2.

Figure 22-37 Problem 71

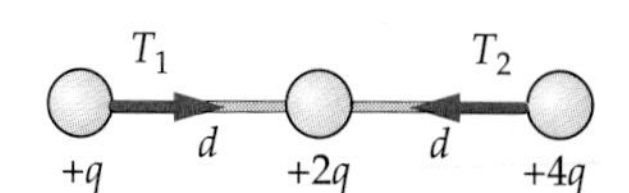

72 •• A positive charge Q is to be divided into two positive charges q_1 and q_2. Show that, for a given separation D, the force exerted by one charge on the other is greatest if $q_1 = q_2 = \frac{1}{2}Q$.

73 •• A charge Q is located at $x = 0$ and a charge $4Q$ is at $x = 12.0$ cm. The force on a charge of $-2\ \mu$C is zero if that charge is placed at $x = 4.0$ cm and is 126.4 N in the positive x direction if placed at $x = 8.0$ cm. Determine the charge Q.

74 •• Two small spheres (point charges) separated by 0.60 m carry a total charge of 200 μC. (*a*) If the two spheres repel each other with a force of 80 N, what are the charges on each of the two spheres? (*b*) If the two spheres attract each other with a force of 80 N, what are the charges on the two spheres?

75 •• A ball of known charge q and unknown mass m, initially at rest, falls freely from a height h in a uniform electric field $\vec{E}$ that is directed vertically downward. The ball hits the ground at a speed $v = 2\sqrt{gh}$. Find m in terms of E, q, and g.

76 •• Charges of 3.0 μC are located at $x = 0$, $y = 2.0$ m and at $x = 0$, $y = -2.0$ m. Charges Q are located at $x = 4.0$ m, $y = 2.0$ m and at $x = 4.0$ m, $y = -2.0$ m (Figure 22-38). The electric field at $x = 0$, $y = 0$ is $(4.0 \times 10^3\ \text{N/C})\hat{\imath}$. Determine Q.

Figure 22-38 Problem 76

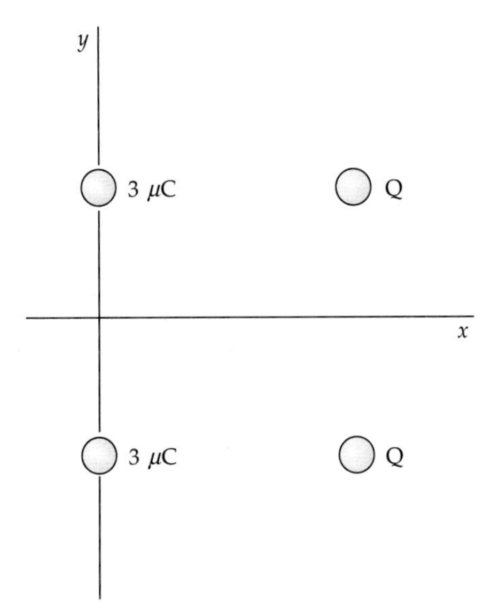

77 •• Two identical small spherical conductors (point charges), separated by 0.60 m, carry a total charge of 200 μC. They repel one another with a force of 120 N. (*a*) Find the charge on each sphere. (*b*) The two spheres are placed in electrical contact and then separated so that each carries 100 μC. Determine the force exerted by one sphere on the other when they are 0.60 m apart.

78 •• Repeat Problem 77 if the two spheres initially attract one another with a force of 120 N.

79 •• A charge of -3.0 μC is located at the origin; a charge of 4.0 μC is located at $x = 0.2$ m, $y = 0$; a third charge Q is located at $x = 0.32$ m, $y = 0$. The force on the 4.0-μC charge is 240 N, directed in the positive x direction. (*a*) Determine the charge Q. (*b*) With this configuration of three charges, where, along the x direction, is the electric field zero?

80 •• Two small spheres of mass m are suspended from a common point by threads of length L. When each sphere carries a charge q, each thread makes an angle θ with the vertical, as shown in Figure 22-39. (*a*) Show that the charge q is given by

$$q = 2L \sin\theta \sqrt{\frac{mg \tan\theta}{k}}$$

where k is the Coulomb constant. (*b*) Find q if $m = 10$ g, $L = 50$ cm, and $\theta = 10°$.

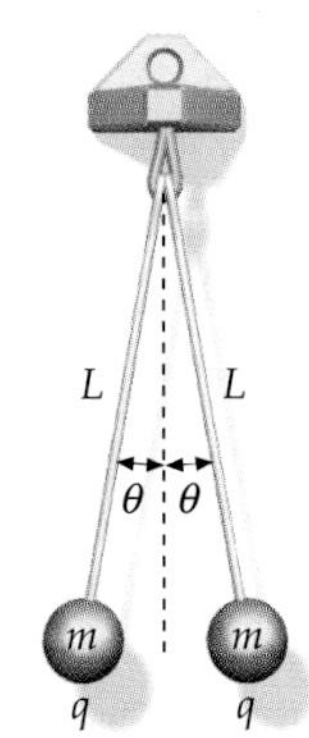

Figure 22-39 Problems 80 and 90

81 •• (*a*) Suppose that in Problem 80, $L = 1.5$ m, $m = 0.01$ kg, and $q = 0.75$ μC. What is the angle that each string makes with the vertical? (*b*) Find the angle that each string makes with the vertical if one mass carries a charge of 0.50 μC, the other a charge of 1.0 μC.

82 •• Four charges of equal magnitude are arranged at the corners of a square of side L as shown in Figure 22-40. (*a*) Find the magnitude and direction of the force exerted on the charge in the lower left corner by the other charges. (*b*) Show that the electric field at the midpoint of one of the sides of the square is directed along that side toward the negative charge and has a magnitude E given by

$$E = k\frac{8q}{L^2}\left(1 - \frac{\sqrt{5}}{25}\right)$$

Figure 22-40 Problem 82

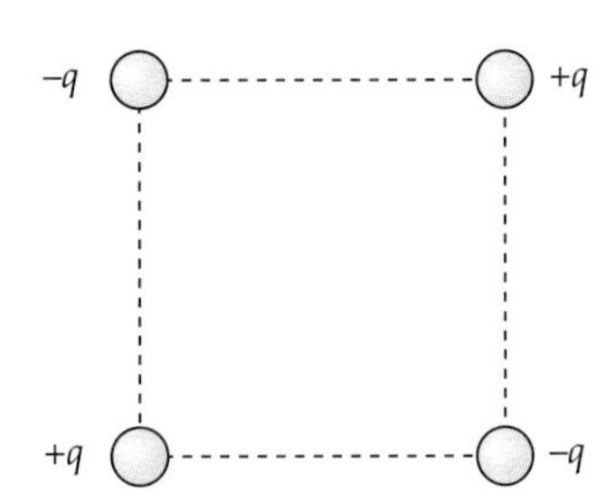

83 •• Figure 22-41 shows a dumbbell consisting of two identical masses m attached to the ends of a thin (massless) rod of length a that is pivoted at its center. The masses carry charges of $+q$ and $-q$ and the system is located in a uniform electric field $\vec{E}$. Show that for small values of the angle θ between the direction of the dipole and the electric field, the system displays simple harmonic motion, and obtain an expression for the period of that motion.

Figure 22-41 Problems 83 and 84

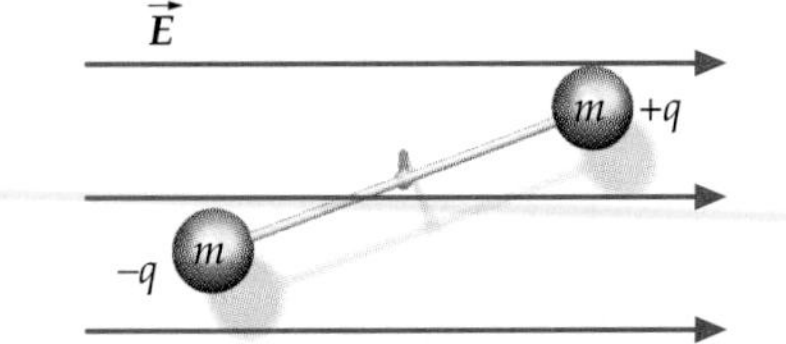

84 •• For the dumbell in Figure 22-41, let $m = 0.02$ kg, $a = 0.3$ m, and $\vec{E} = (600\ \text{N/C})\hat{\imath}$. Initially the dumbbell is at rest and makes an angle of 60° with the x axis. The dumbbell is then released, and when it is momentarily aligned with the electric field, its kinetic energy is 5×10^{-3} J. Determine the magnitude of q.

85 •• An electron (charge $-e$, mass m) and a positron (charge $+e$, mass m) revolve around their common center of mass under the influence of their attractive coulomb force. Find the speed of each particle v in terms of e, m, k, and their separation r.

86 •• The equilibrium separation between the nuclei of the ionic molecule KBr is 0.282 nm. The masses of the two ions, K^+ and Br^-, are very nearly the same, 1.4×10^{-25} kg, and each of the two ions carries a charge of magnitude e. Use the result of Problem 83 to determine the frequency of oscillation of a KBr molecule in a uniform electric field of 1000 N/C.

87 ••• A small (point) mass m, which carries a charge q, is constrained to move vertically inside a narrow, frictionless cylinder (Figure 22-42). At the bottom of the cylinder is a point mass of charge Q having the same sign as q. (*a*) Show that the mass m will be in equilibrium at a height $y_0 = (kqQ/mg)^{1/2}$. (*b*) Show that if the mass m is displaced by a small amount from its equilibrium position and released, it will exhibit simple harmonic motion with angular frequency $\omega = (2g/y_0)^{1/2}$.

Figure 22-42 Problem 87

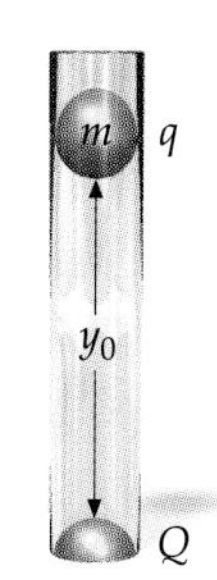

88 ••• A small bead of mass m and carrying a negative charge $-q$ is constrained to move along a thin frictionless rod (Figure 22-43). A distance L from this rod is a positive charge Q. Show that if the bead is displaced a distance x, where $x \ll L$, and released, it will exhibit simple harmonic motion. Obtain an expression for the period of this motion in terms of the parameters L, Q, q, and m.

Figure 22-43 Problem 88

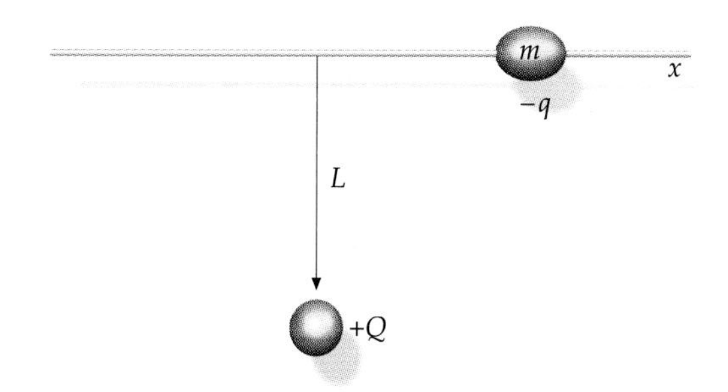

89 ••• Repeat Problem 81 with the system located in a uniform electric field of 1.0×10^5 N/C that points vertically downward.

90 ••• Suppose that the two masses in Problem 80 are not equal. One mass is 0.01 kg, the other is 0.02 kg. The charges on the two masses are 2.0 and 1.0 μC, respectively. Determine the angle that each of the strings supporting the masses makes with the vertical.

91 ••• A simple pendulum of length $L = 1.0$ m and mass $M = 5.0 \times 10^{-3}$ kg is placed in a uniform, vertically directed electric field $\vec{E}$. The bob carries a charge of -8.0 μC. The period of the pendulum is 1.2 s. What is the magnitude and direction of $\vec{E}$?

92 ••• Two neutral polar molecules attract each other. Suppose that each molecule has a dipole moment $\vec{p}$ and that these dipoles are aligned along the x axis and separated by a distance d. Derive an expression for the force of attraction in terms of p and d.

93 ••• A small bead of mass m, carrying a charge q, is constrained to slide along a thin rod of length L. Charges Q are fixed at each end of the rod (Figure 22-44). (*a*) Obtain an expression for the electric field due to the two charges Q as a function of x, where x is the distance from the midpoint of the rod. (*b*) Show that for $x \ll L$, the magnitude of the field is proportional to x. (*c*) Show that if q is of the same sign as Q, the force that acts on the object of mass m is always directed toward the center of the rod and is proportional to x. (*d*) Find the period of oscillation of the mass m if it is displaced by a small distance from the center of the rod and then released.

Figure 22-44 Problem 93

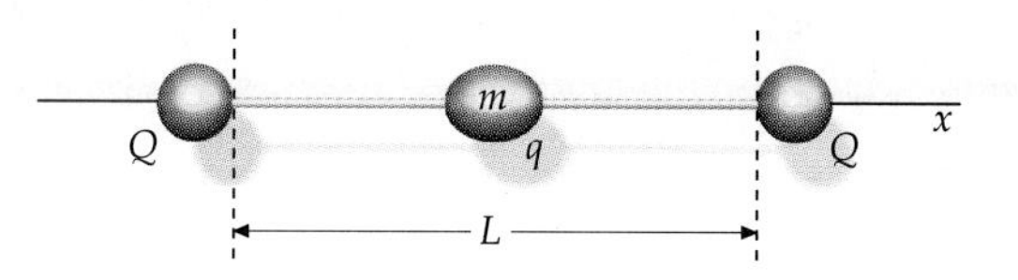

94 ••• Two equal positive charges Q are on the x axis at $x = \frac{1}{2}L$ and $x = -\frac{1}{2}L$. (*a*) Obtain an expression for the electric field as a function of y on the y axis. (*b*) A ring of mass m, which carries a charge q, moves on a thin frictionless rod along the y axis. Find the force that acts on the charge q as a function of y; determine the sign of q such that this force always points toward $y = 0$. (*c*) Show that for small values of y the ring exhibits simple harmonic motion. (*d*) If $Q = 5$ μC, $|q| = 2$ μC, $L = 24$ cm, and $m = 0.03$ kg, what is the frequency of the oscillation for small amplitudes?

CHAPTER 23

The Electric Field II: Continuous Charge Distributions

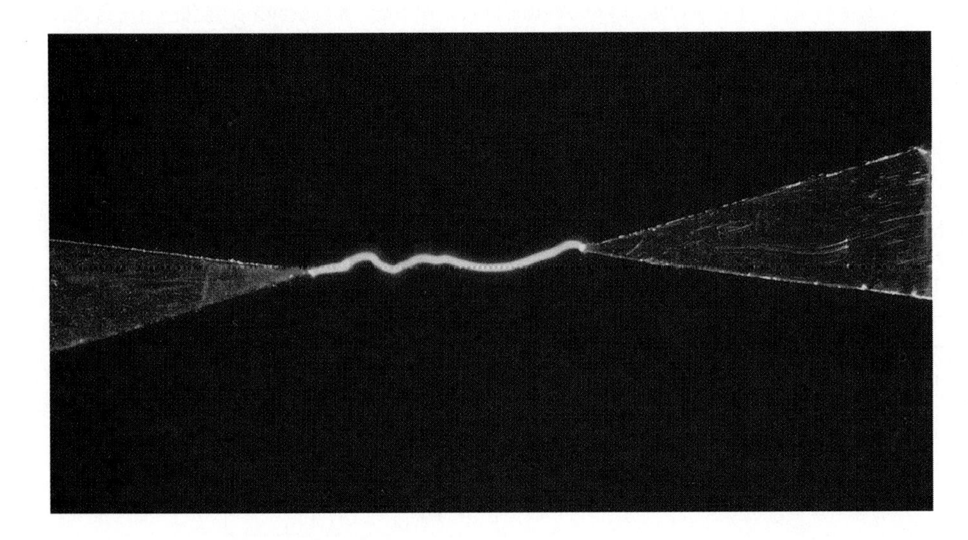

Electrical discharge between two charged conductors. The electric field near the sharp points is strong enough to strip the electrons from nearby air molecules, thus ionizing them and causing the air to conduct.

On a microscopic scale, electric charge is quantized. However, there are often situations in which many charges are so close together that they can be considered to be continuously distributed. The use of a continuous charge density to describe a large number of discrete charges is similar to the use of a continuous mass density to describe air, which actually consists of a large number of discrete molecules. In either case, it is usually easy to find a volume element ΔV that is large enough to contain a multitude of individual charges or molecules and yet is small enough that replacing ΔV by a differential dV and using calculus introduces negligible error.

We describe the charge per unit volume by the **volume charge density** ρ:

$$\rho = \frac{\Delta Q}{\Delta V} \tag{23-1}$$

Often charge is distributed in a thin layer on the surface of an object. We define the **surface charge density** σ as the charge per unit area:

$$\sigma = \frac{\Delta Q}{\Delta A} \tag{23-2}$$

Similarly, we sometimes encounter charge distributed along a line in space.

We define the **linear charge density** λ as the charge per unit length:

$$\lambda = \frac{\Delta Q}{\Delta L} \tag{23-3}$$

In this chapter, we show how Coulomb's law is used to calculate the electric field due to various types of continuous charge distributions. We then introduce Gauss's law, which relates the electric field on a closed surface to the net charge within the surface, and we use this relation to calculate the electric field for symmetric charge distributions.

23-1 Calculating $\vec{E}$ From Coulomb's Law

Figure 23-1 shows an element of charge $dq = \rho\, dV$ that is small enough to be considered a point charge. The electric field $d\vec{E}$ at a field point P due to this charge element is given by Coulomb's law:

$$d\vec{E} = \frac{k\,dq}{r^2}\hat{r}$$

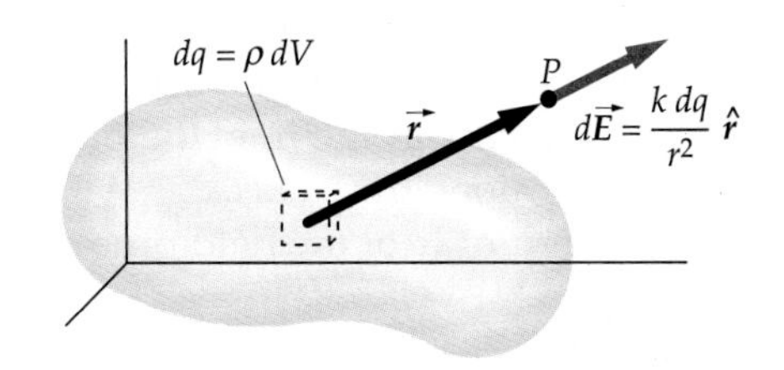

Figure 23-1 An element of charge dq produces a field $d\vec{E} = (k\,dq/r^2)\hat{r}$ at point P. The field at P due to the total charge is found by integrating over the entire charge distribution.

where $\hat{r}$ is a unit vector that points from the charge element to the field point P. The total field at P is found by integrating this expression over the entire charge distribution. Thus,

$$\vec{E} = \int_V \frac{k\,dq}{r^2}\hat{r} \tag{23-4}$$

Electric field due to a continuous charge distribution

where $dq = \rho\, dV$. If the charge is distributed on a surface or line, we use $dq = \sigma\, dA$ or $dq = \lambda\, dL$ and integrate over the surface or line.

$\vec{E}$ on the Axis of a Finite Line Charge

A uniform charge Q lies along the x axis from $x = 0$ to $x = L$ as shown in Figure 23-2. The linear charge density for this charge is $\lambda = Q/L$. We wish to find the electric field produced by this line charge at some point P on the x axis at $x = x_0$, for $x_0 > L$. In the figure, we have chosen a small differential element dx at a distance x from the origin. The field point P is at a distance $r = x_0 - x$ from this charge element. The electric field due to this element of charge is given by Coulomb's law for a point charge. It is directed along the x axis and is given by

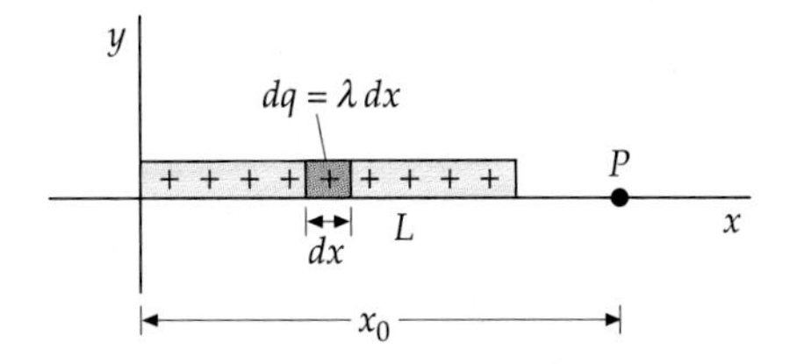

Figure 23-2 Geometry for the calculation of the electric field on the axis of a uniform line charge of length L, charge Q, and linear charge density $\lambda = Q/L$. An element $dq = \lambda\, dx$ is treated as a point charge.

$$dE_x = \frac{k\,dq}{(x_0 - x)^2} = \frac{k\lambda\, dx}{(x_0 - x)^2}$$

We find the total field by integrating over the entire line charge from $x = 0$ to $x = L$:

$$E_x = k\lambda \int_0^L \frac{dx}{(x_0 - x)^2} = k\lambda \left[\frac{1}{x_0 - x}\right]_0^L$$

$$= k\lambda\left\{\frac{1}{x_0 - L} - \frac{1}{x_0}\right\} = k\lambda\left\{\frac{L}{x_0(x_0 - L)}\right\}$$

Using $\lambda = Q/L$, we obtain

$$E_x = \frac{kQ}{x_0(x_0 - L)} \tag{23-5}$$

We can see that if x_0 is much larger than L, the electric field at x_0 is approximately kQ/x_0^2. That is, if we are sufficiently far away from the line charge, it looks like a point charge.

$\vec{E}$ off the Axis of a Finite Line Charge

We first consider the simple case of the field at a point P on the perpendicular bisector of the line charge. The geometry is shown in Figure 23-3. We chose a coordinate system such that the charge is on the x axis with the origin at its center, and the field point P is on the y axis. A typical charge element $dq = \lambda\, dx$ that produces a field $d\vec{E}$ is shown in the figure.

The field has a component parallel to the line charge and a component perpendicular to it. We can see from symmetry that when we sum over all charge elements in the line, the parallel components sum to zero. Thus, the field $\vec{E}$ lies along the y axis.

The magnitude of the field produced by an element of charge $dq = \lambda\, dx$ is

$$|d\vec{E}| = \frac{k\,dq}{r^2} = \frac{k\lambda\, dx}{r^2}$$

The y component is

$$dE_y = \frac{k\lambda\, dx}{r^2}\cos\theta = \frac{k\lambda y\, dx}{r^3} \tag{23-6}$$

where $\cos\theta = y/r$ and $r = \sqrt{x^2 + y^2}$. The total field E_y is computed by integrating from $x = -\frac{1}{2}L$ to $x = +\frac{1}{2}L$. Because of symmetry, the contribution of each half of the line charge to the total field is the same, so we can integrate from $x = 0$ to $x = \frac{1}{2}L$ and multiply by 2:

$$E_y = \int_{x=-\frac{1}{2}L}^{x=+\frac{1}{2}L} dE_y = 2\int_{x=0}^{x=\frac{1}{2}L} dE_y = 2k\lambda y \int_{x=0}^{x=\frac{1}{2}L} \frac{dx}{r^3} \tag{23-7}$$

This integral is of a standard form found in tables:

$$\int \frac{dx}{r^3} = \frac{1}{y^2}\frac{x}{r} = \frac{1}{y^2}\sin\theta$$

We see from the figure that at $x = 0$, $\theta = 0$, so $\sin\theta = 0$ at the lower limit. At the upper limit at $x = L/2$, $\theta = \theta_0$, as shown in Figure 23-3. The field is thus

$$E_y = \frac{2k\lambda y}{y^2}\sin\theta_0$$

or

$$E_y = \frac{2k\lambda}{y}\sin\theta_0 = \frac{2k\lambda}{y}\frac{\frac{1}{2}L}{\sqrt{(\frac{1}{2}L)^2 + y^2}} \tag{23-8}$$

$\vec{E}$ on the perpendicular bisector of a finite line charge

where (from Figure 23-3) $\sin\theta_0$ is related to L and y by

$$\sin\theta_0 = \frac{\frac{1}{2}L}{\sqrt{(\frac{1}{2}L)^2 + y^2}}$$

Exercise Show that when y is much greater than L, a finite line charge looks like a point charge, that is, Equation 23-8 reduces to $E_y \approx kQ/y^2$.

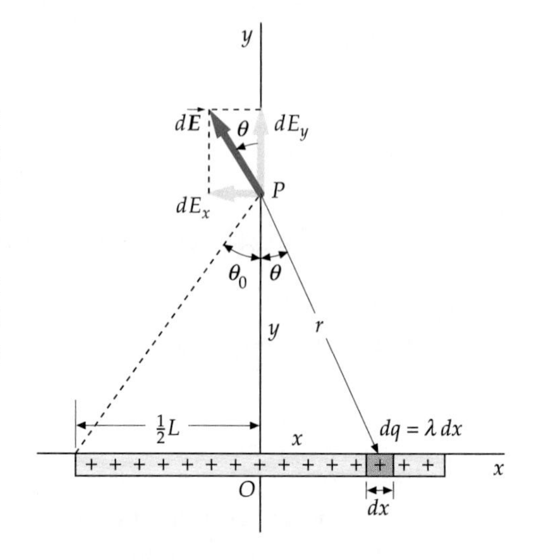

Figure 23-3 Geometry for the calculation of the electric field at a point on the perpendicular bisector of a uniform finite line charge. By symmetry, the net electric field is perpendicular to the line charge.

At a point a distance y from the line charge, but not on the perpendicular bisector, the field has both x and y components. Calculation of these components is left to the Problems section at the end of the chapter (see Problems 23-12 and 23-13).

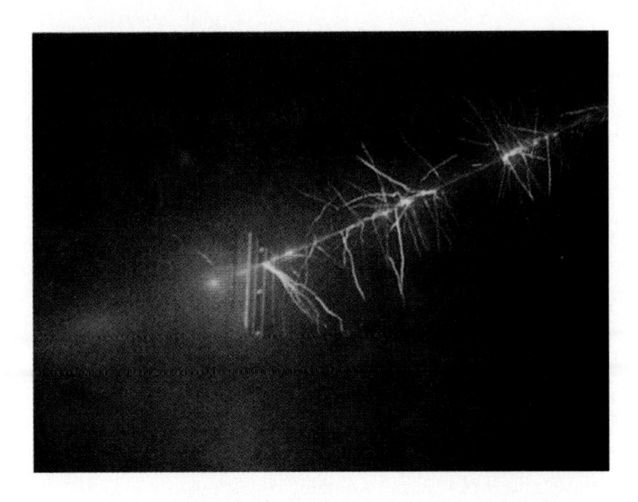

Electric field lines near a long wire. The electric field near a high-voltage power line can be large enough to strip the electrons from air molecules, thus ionizing them and making the air a conductor. The glow resulting from the recombination of free electrons with the ions is called corona discharge.

$\vec{E}$ due to an Infinite Line Charge

When we are very close to a line charge (when $y \ll L$) or when the line charge is very long, the angle θ_0 in Figure 23-3 is approximately 90°, and $\sin\theta_0 \approx 1$. Then

$$E_y = \frac{2k\lambda}{y} \qquad \text{23-9}$$

$\vec{E}$ at a distance y from an infinite line charge

Thus, $\vec{E}$ is directed away from the line (for a positive line charge) and its magnitude decreases as $1/y$.

Exercise Show that Equation 23-9 has the correct units for the electric field.

Example 23-1

A line charge of linear density $\lambda = 4.5$ nC/m lies on the x axis and extends from $x = -5$ cm to $x = 5$ cm. Calculate the electric field on the y axis using the exact expression

$$E_y = \frac{2k\lambda}{y}\sin\theta_0 = \frac{2k\lambda}{y}\frac{\frac{1}{2}L}{\sqrt{(\frac{1}{2}L)^2 + y^2}}$$

(Equation 23-8) at (*a*) $y = 1$ cm, (*b*) $y = 4$ cm, and (*c*) $y = 40$ cm. (*d*) Calculate the electric field on the y axis at $y = 1.0$ cm, assuming the line charge to be infinite. (*e*) Find the total charge and calculate the field at $y = 40$ cm, assuming the line charge to be a point charge.

Picture the Problem In the expression for $\sin\theta_0$, we can express L and y in centimeters because the units cancel. (*d*) To find the field very near the line charge, we use $E_y = 2k\lambda/y$. (*e*) To find the field very far from the charge, we use $E_y = kQ/y^2$ with $Q = \lambda L$.

(*a*) Calculate E_y at $y = 1$ cm for $\lambda = 4.5$ nC/m and $L = 0.5$ cm:

$$E_y = \frac{2k\lambda}{y}\frac{\frac{1}{2}L}{\sqrt{(\frac{1}{2}L)^2 + y^2}}$$

$$= \frac{2(8.99\times10^9\ \text{N}\cdot\text{m}^2/\text{C}^2)(4.5\times10^{-9}\ \text{C/m})}{0.01\ \text{m}}\frac{5\ \text{cm}}{\sqrt{(5\ \text{cm})^2 + (1\ \text{cm})^2}}$$

$$= \frac{80.9\ \text{N}\cdot\text{m/C}}{0.01\ \text{m}}\frac{5\ \text{cm}}{\sqrt{26}\ \text{cm}}$$

$$= 7.93\times10^3\ \text{N/C} = 7.93\ \text{kN/C}$$

(*b*) Repeat the calculation for $y = 4$ cm $= 0.04$ m using the result $2k\lambda = 80.9$ N·m/C to simplify the notation:

$$E_y = \frac{2k\lambda}{y}\frac{\frac{1}{2}L}{\sqrt{(\frac{1}{2}L)^2 + y^2}} = \frac{80.9\ \text{N}\cdot\text{m/C}}{0.04\ \text{m}}\frac{5\ \text{cm}}{\sqrt{(5\ \text{cm})^2 + (4\ \text{cm})^2}}$$

$$= 1.58\times10^3\ \text{N/C} = 1.58\ \text{kN/C}$$

(*c*) Repeat the calculation for y = 40 cm:

$$E_y = \frac{2k\lambda}{y}\frac{\frac{1}{2}L}{\sqrt{(\frac{1}{2}L)^2 + y^2}} = \frac{80.9\text{ N}\cdot\text{m/C}}{0.4\text{ m}}\frac{5\text{ cm}}{\sqrt{(5\text{ cm})^2 + (40\text{ cm})^2}} = 25.1\text{ N/C}$$

(*d*) Calculate the field at y = 1 cm = 0.01 m due to an infinite line charge:

$$E_y \approx \frac{2k\lambda}{y} = \frac{80.9\text{ N}\cdot\text{m/C}}{0.01\text{ m}} = 8.09\text{ kN/m}$$

(*e*) Calculate the total charge λL for L = 0.1 m and use it to find the field of a point charge at y = 4 m:

$$Q = \lambda L = (4.5\text{ nC/m})(0.1\text{ m}) = 0.45\text{ nC}$$

$$E_y \approx \frac{k\lambda L}{y^2} = \frac{kQ}{y^2} = \frac{(8.99 \times 10^9\text{ N}\cdot\text{m}^2/\text{C}^2)(0.45 \times 10^{-9}\text{ C})}{(0.40\text{ m})^2} = 25.3\text{ N/C}$$

Remarks At 1 cm from the 10-cm-long line charge, the appoximate value of 8.09 kN/C obtained by assuming an infinite line charge differs from the exact value of 7.93 calculated in (*a*) by about 2%. At 40 cm from the line charge, the approximate value of 25.3 N/C obtained by assuming the line charge to be a point charge differs from the exact value of 25.1 N/C obtained in (*c*) by about 1%. Figure 23-4 shows the exact result for this line segment of length 10 cm and charge density 4.5 nC/m, and for the limiting cases of an infinite line charge of the same charge density, and a point charge $Q = \lambda L$.

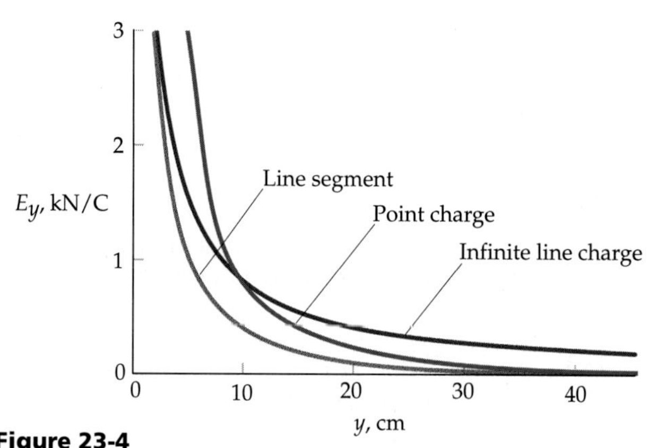

Figure 23-4

Example 23-2 try it yourself

An infinite line charge of linear charge density λ = 0.6 μC/m lies along the z axis, and a point charge q = 8 μC lies on the y axis at y = 3 m. Find the electric field at the point P on the x axis at x = 4 m.

Picture the Problem The electric field for this system is the superposition of the fields due to the infinite line charge and the point charge. The field of the line charge, $\vec{E}_L$, points radially away from the z axis (Figure 23-5). Thus, at point P on the x axis, $\vec{E}_L$ is in the positive x direction. The point charge produces a field $\vec{E}_p$ along the line connecting q and the point P. The distance from q to P is $r = \sqrt{(3\text{ m})^2 + (4\text{ m})^2} = 5$ m.

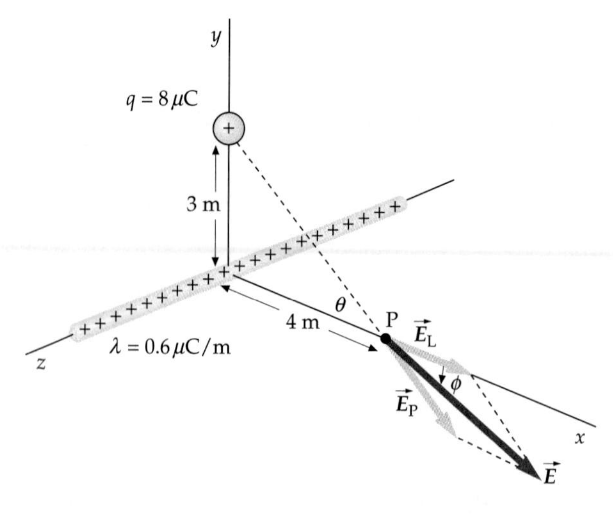

Figure 23-5

Cover the column to the right and try these on your own before looking at the answers.

Steps	***Answers***
1. Calculate the field $\vec{E}_L$ at point P due to the infinite line charge.	$\vec{E}_L = 2.70\text{ kN/C}\,\hat{i}$
2. Find the field $\vec{E}_p$ at point P due to the point charge. Express $\vec{E}_p$ in terms of the unit vector $\hat{r}$ that points from q to P.	$\vec{E}_p = 2.88\text{ kN/C}\,\hat{r}$
3. Find the x and y components of $\vec{E}_p$.	$E_{px} = E_p\,(0.8) = 2.30\text{ kN/C}$ $E_{py} = E_p(-0.6) = -1.73\text{ kN/C}$
4. Find the x and y components of the total field at point P.	$E_x = 5.00\text{ kN/C}, \quad E_y = -1.73\text{ kN/C}$
5. Use your result in step 4 to calculate the magnitude of the total field.	$E = \sqrt{E_x^2 + E_y^2} = 5.29\text{ kN/C}$
6. Use your results in step 4 to find the angle ϕ between the field and the x axis.	$\phi = -19.1°$

$\vec{E}$ on the Axis of a Ring Charge

Figure 23-6 shows a uniform ring charge of radius a and total charge Q. The field $d\vec{E}$ at point P on the axis due to the charge element dq is shown in the figure. This field has a component dE_x directed along the axis of the ring and a component $dE_\perp$ directed perpendicular to the axis. From the symmetry of the charge distribution, we can see that the net field due to the entire ring must lie along the axis of the ring; that is, the perpendicular components sum to zero. The axial component of the field due to the charge element shown is

$$dE_x = \frac{k\,dq}{r^2}\cos\theta = \frac{k\,dq}{r^2}\frac{x}{r} = \frac{k\,dq\,x}{(x^2 + a^2)^{3/2}}$$

where

$$r^2 = x^2 + a^2$$

and

$$\cos\theta = \frac{x}{r} = \frac{x}{\sqrt{x^2 + a^2}}$$

The field due to the entire ring of charge is

$$E_x = \int \frac{kx\,dq}{(x^2 + a^2)^{3/2}}$$

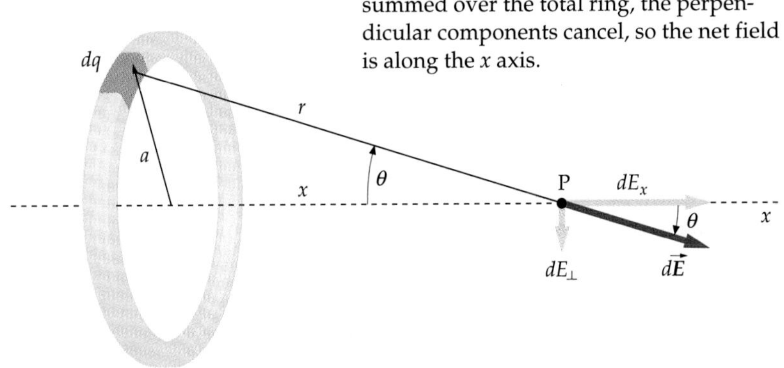

Figure 23-6 A ring charge of radius a. The electric field at point P on the x axis due to the charge element dq shown has one component along the x axis and one perpendicular to the x axis. When summed over the total ring, the perpendicular components cancel, so the net field is along the x axis.

Since x does not vary as we integrate over the elements of charge, we can remove it from the integral. Then

$$E_x = \frac{kx}{(x^2 + a^2)^{3/2}} \int dq$$

or

$$E_x = \frac{kQx}{(x^2 + a^2)^{3/2}} \tag{23-10}$$

$\vec{E}$ on the axis of a ring charge

Exercise Find the field on the axis of a ring charge when (*a*) $x = 0$ and (*b*) x is much greater than a. (*Answers* (*a*) $E_x = 0$ at $x = 0$, (*b*) $E_x \approx kQ/x^2$ for $x >> a$)

$\vec{E}$ on the Axis of a Uniformly Charged Disk

Figure 23-7 shows a uniformly charged disk of radius R and total charge Q. We can calculate the field on the axis of the disk by treating the disk as a set of concentric ring charges. By symmetry, $\vec{E}$ on the axis of the disk is along the axis. A ring of radius a and width da is shown in the figure. The area of this ring is $dA = 2\pi a\, da$, and its charge is $dq = \sigma\, dA = 2\pi\sigma a\, da$, where $\sigma = Q/\pi R^2$ is the charge per unit area. The field produced by this ring is given by Equation 23-10 if we replace Q with $dq = 2\pi\sigma a\, da$.

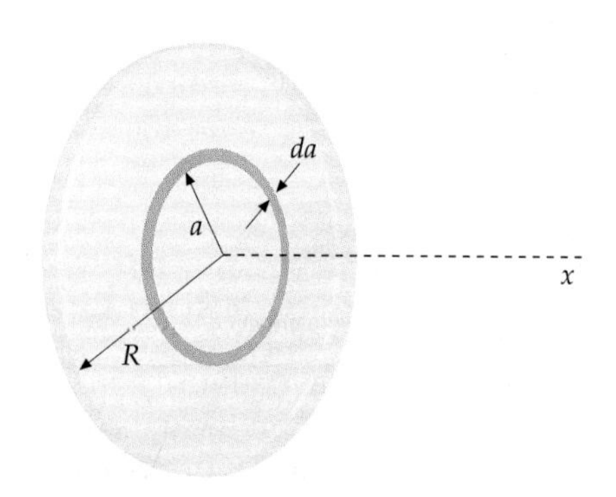

Figure 23-7 A uniform disk of charge can be treated as a set of ring charges, each of radius a and thickness da and carrying a charge $dq = \sigma\, dA = (Q/\pi R^2) 2\pi a\, da$.

$$dE_x = \frac{kx 2\pi\sigma a\, da}{(x^2 + a^2)^{3/2}}$$

The total field is found by integrating from $a = 0$ to $a = R$:

$$E_x = \int_0^R \frac{kx 2\pi\sigma a\, da}{(x^2 + a^2)^{3/2}} = kx\pi\sigma \int_0^R (x^2 + a^2)^{-3/2}\, 2a\, da$$

This integral is of the form $\int u^n\, du$, with $u = x^2 + a^2$ and $n = -\frac{3}{2}$. The integration thus gives

$$E_x = kx\pi\sigma \left[\frac{(x^2 + a^2)^{-1/2}}{-\frac{1}{2}}\right]_0^R = -2kx\pi\sigma\left(\frac{1}{\sqrt{x^2 + R^2}} - \frac{1}{x}\right)$$

or

$$E_x = 2\pi k\sigma\left(1 - \frac{x}{\sqrt{x^2 + R^2}}\right) \tag{23-11}$$

$\vec{E}$ on the axis of a disk charge

When we are very far from the disk, we expect it to look like a point charge. If we merely replace $x^2 + R^2$ with x^2 for $x >> R$, we get $E_x \to 0$. Although this is correct, it does not tell us anything about how E_x depends on x for large x. We can find this dependence by using the binomial expansion, $(1 + \epsilon)^n \approx 1 + n\epsilon$, for $\epsilon \ll 1$. Using this expansion on the second term in Equation 23-11, we obtain

$$\frac{x}{\sqrt{x^2 + R^2}} = \frac{x}{x(1 + R^2/x^2)^{1/2}} = \left(1 + \frac{R^2}{x^2}\right)^{-1/2} \approx 1 - \frac{R^2}{2x^2} + \cdots$$

If we use just the first term, we get $E_x \approx 0$ as before. But if we use the first two terms, Equation 23-11 becomes

$$E_x \approx 2\pi k\sigma\left(1 - 1 + \frac{R^2}{2x^2} + \cdots\right) = \frac{k\pi R^2\sigma}{x^2} = \frac{kQ}{x^2}$$

where $Q = \sigma\pi R^2$ is the total charge on the disk.

$\vec{E}$ due to an Infinite Plane of Charge

The field of an infinite plane of charge can be obtained from Equation 23-11 by either letting R go to infinity or letting x go to zero. Then

$$E_x = 2\pi k\sigma, \qquad x > 0 \qquad \text{23-12a}$$

$\vec{E}$ near an infinite plane of charge

Thus, the field due to an infinite-plane charge distribution is uniform; that is, the field does not depend on x. On the other side of the infinite plane, for negative values of x, the field points in the negative x direction, so

$$E_x = -2\pi k\sigma, \qquad x < 0 \qquad \text{23-12b}$$

As we move along the x axis, the electric field jumps from $-2\pi k\sigma\hat{i}$ to $+2\pi k\sigma\hat{i}$ when we pass through an infinite plane of charge (Figure 23-8). There is thus a discontinuity in E_x in the amount $4\pi k\sigma$.

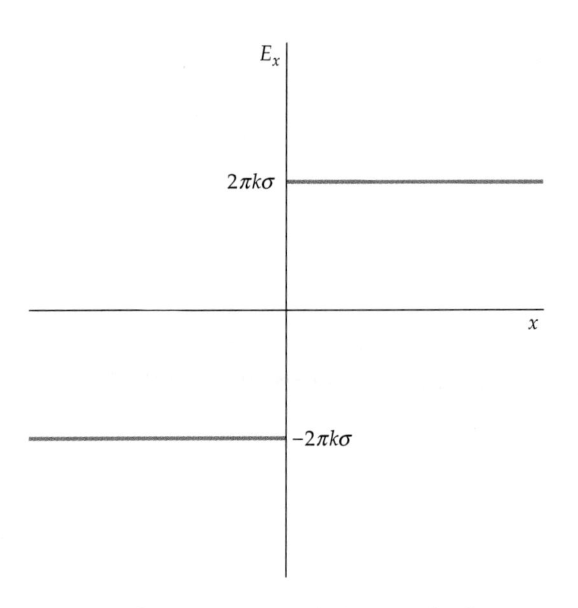

Figure 23-8 Graph showing the discontinuity of $\vec{E}$ at a plane charge.

Example 23-3

A disk of radius 5 cm carries a uniform surface charge density of 4 μC/m². Using reasonable approximations, find the electric field on the axis of the disk at distances of (*a*) 0.01 cm, (*b*) 0.03 cm, (*c*) 6 m, and (*d*) 6 cm.

Picture the Problem For (*a*) and (*b*), the field point is very near the disk compared with its radius, so we can approximate the disk as an infinite plane. For (*c*), the field point is far enough from the disk that we can approximate the disk as a point charge. (*d*) Since 6 cm is neither much less than nor much greater than the radius of 5 cm, we use the exact expression.

(*a*) The electric field near the disk is that due to an infinite plane charge:

$$E_x = 2\pi k\sigma = 2\pi(8.99 \times 10^9\ \text{N}\cdot\text{m}^2/\text{C}^2)(4 \times 10^{-6}\ \text{C/m}^2) = 226\ \text{kN/C}$$

(*b*) Since 0.03 cm is still very near the disk, the disk still looks like an infinite plane charge:

$$E_x = 2\pi k\sigma = 226\ \text{kN/C}$$

(*c*) 1. Far from the disk, the field is approximately that due to a point charge:

$$E_x = \frac{kQ}{x^2}$$

2. The total charge Q is:

$$Q = \sigma\pi r^2 = (4\ \mu\text{C/m}^2)\pi(0.05\ \text{m})^2 = 31.4\ \text{nC}$$

3. Substitute this value into step 1:

$$E_x = \frac{(8.99 \times 10^9\ \text{N}\cdot\text{m}^2/\text{C}^2)(31.4 \times 10^{-9}\ \text{C})}{(6\ \text{m})^2} = 7.84\ \text{N/C}$$

(*d*) For $x = 6$ cm, we use the exact expression for E_x:

$$E_x = 2\pi k\sigma\left[1 - \frac{x}{\sqrt{(x^2 + R^2)}}\right] = (226\ \text{kN/C})\left[1 - \frac{6\ \text{cm}}{\sqrt{(6\ \text{cm})^2 + (5\ \text{cm})^2}}\right] = (226\ \text{kN/C})(1 - 0.768) = 52.4\ \text{kN/C}$$

Remarks Note that in (*d*) we did not need to convert from centimeters to meters to find $x/\sqrt{x^2 + R^2}$ because the units cancel. Figure 23-9 shows E_x versus x for the disk charge in this example, for an infinite plane with the same charge density, and for a point charge.

Exercise Calculate E_x for parts (*a*) and (*b*) to five significant figures using the infinite plane approximation and then using the exact expression of Equation 23-11 and compare your results. (*Answers* (*a*) $E_x \approx 225.94$ kN/C and $E_x = 225.49$ kN/C, a difference of about 0.2%; (*b*) $E_x \approx 225.94$ kN/C and $E_x = 224.58$ kN/C, a difference of about 0.6%.)

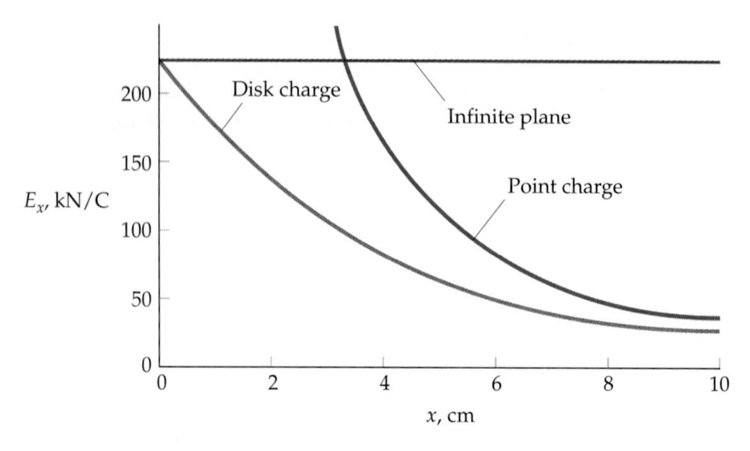

Figure 23-9

23-2 Gauss's Law

The qualitative description of the electric field using electric field lines, discussed in Chapter 22, is related to a mathematical equation known as Gauss's law, which relates the electric field on a closed surface to the net charge within the surface. Gauss's law is the basis for one of Maxwell's equations, the fundamental equations of electromagnetism, which is discussed in Chapter 32. For static charges, Gauss's law and Coulomb's law are equivalent. Electric fields arising from some symmetrical charge distributions, such as a spherical shell of charge or an infinite line of charge, can be easily calculated using Gauss's law. In this section, we give an argument for the validity of Gauss's law based on the properties of electric field lines. A rigorous derivation of Gauss's law is given in Section 23-6.

Figure 23-10 shows a surface of arbitrary shape enclosing a dipole. The number of electric field lines emanating from the positive charge and crossing the surface to the outside depends on where the surface is drawn, but it is exactly equal to the number of lines entering the surface and ending on the negative charge. If we count the number of lines leaving as positive and the number entering as negative, the net number leaving and entering is zero. For surfaces enclosing other types of charge distributions, such as that shown in Figure 23-11, *the net number of lines leaving any surface enclosing the charges is proportional to the net charge enclosed by the surface.* This rule is a qualitative statement of Gauss's law.

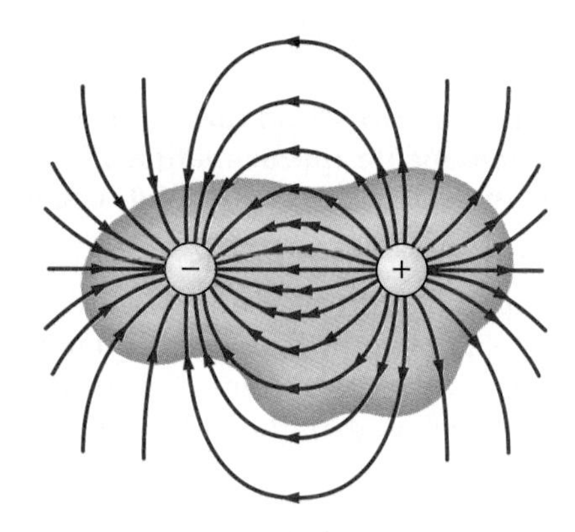

Figure 23-10 A surface of arbitrary shape enclosing an electric dipole. As long as the surface encloses both charges, the number of lines leaving the surface is exactly equal to the number of lines entering the surface no matter where the surface is drawn.

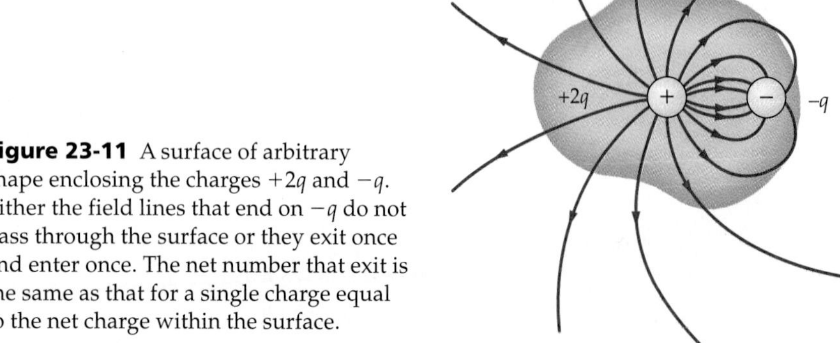

Figure 23-11 A surface of arbitrary shape enclosing the charges $+2q$ and $-q$. Either the field lines that end on $-q$ do not pass through the surface or they exit once and enter once. The net number that exit is the same as that for a single charge equal to the net charge within the surface.

Electric Flux

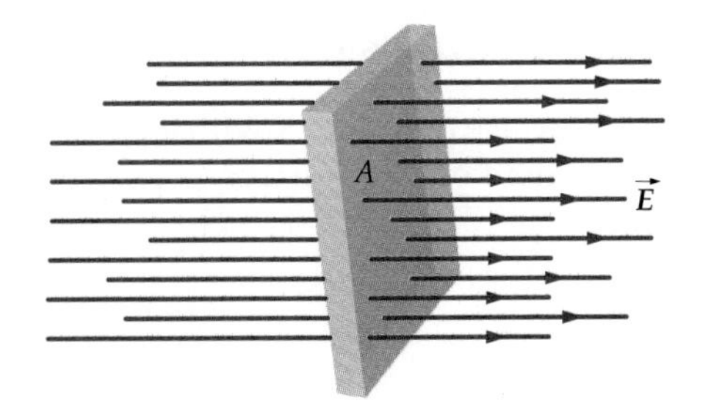

Figure 23-12 Electric field lines of a uniform field crossing an area A that is perpendicular to the field. The product EA is the electric flux through the area.

The mathematical quantity that corresponds to the number of field lines crossing a surface is called the **electric flux** ϕ, which for a surface perpendicular to $\vec{E}$ (Figure 23-12) is defined as the product of the magnitude of the field E and the area A:

$$\phi = EA$$

The units of flux are $N{\cdot}m^2/C$. Since E is proportional to the number of lines per unit area, the flux is proportional to the number of field lines through the area.

In Figure 23-13, the surface of area A_2 is not perpendicular to the electric field $\vec{E}$. However, the number of lines that cross area A_2 is the same as the number that cross area A_1, which is perpendicular to $\vec{E}$. These areas are related by

$$A_2 \cos\theta = A_1 \qquad 23\text{-}13$$

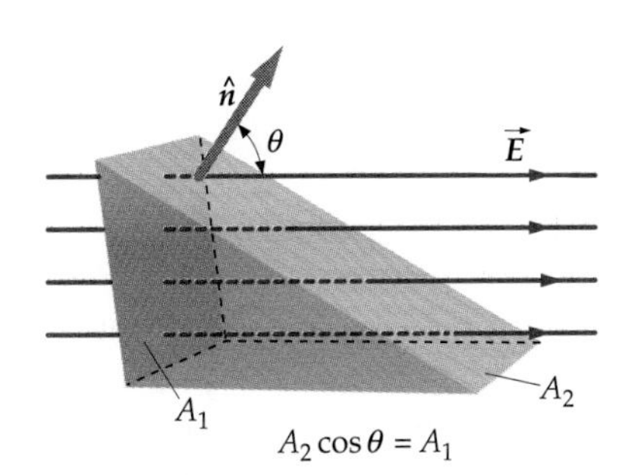

Figure 23-13 Electric field lines of a uniform electric field that is perpendicular to the area A_1 but makes an angle θ with the unit vector $\hat{n}$ that is normal to the area A_2. When $\vec{E}$ is not perpendicular to the area, the flux is E_nA, where $E_n = E\cos\theta$ is the component of $\vec{E}$ that is perpendicular to the area. The flux through A_2 is the same as the flux through A_1.

where θ is the angle between $\vec{E}$ and the unit vector $\hat{n}$ that is perpendicular to the surface A_2, as shown in the figure. The flux through a surface that is not perpendicular to $\vec{E}$ is defined to be

$$\phi = E{\cdot}\hat{n}A = EA\cos\theta = E_nA$$

where $E_n = \vec{E}{\cdot}\hat{n}$ is the component of $\vec{E}$ that is perpendicular, or normal, to the surface.

Figure 23-14 shows an arbitrary surface over which $\vec{E}$ may vary. If the area element ΔA_i that we choose is small enough, it can be considered to be a plane, and the variation of the electric field across the element can be neglected. The flux of the electric field through this element is

$$\Delta\phi_i = \vec{E}{\cdot}\hat{n}_i\,\Delta A_i$$

where $\hat{n}_i$ is the unit vector perpendicular to the element. If the surface is curved, the unit vectors for different elements will have different directions. The total flux through the surface is the sum of $\Delta\phi_i$ over all the elements. In the limit as the number of elements approaches infinity and the area of each element approaches zero, this sum becomes an integral. The general definition of electric flux is thus

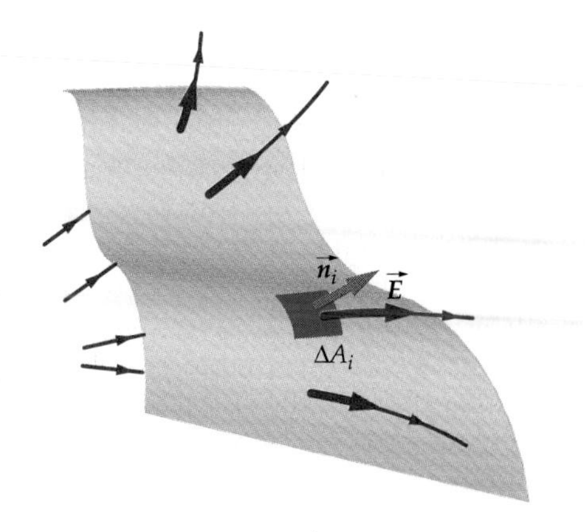

Figure 23-14 When $\vec{E}$ varies in either magnitude or direction, the area of the surface is divided into small elements ΔA_i. The flux through the area is computed by summing $\vec{E}{\cdot}\hat{n}\,\Delta A_i$ over all the area elements.

$$\phi = \lim_{\Delta A_i \to 0} \sum_i \hat{n}_i\,\Delta A_i = \int_S \vec{E}{\cdot}\hat{n}\,dA \qquad 23\text{-}14$$

Definition—Electric flux

where the S reminds us that we are integrating over a surface.

On a *closed* surface, the normal unit vector $\hat{n}$ is defined to be outward at each point. The integral over a closed surface is indicated by the symbol $\oint$. The total or net flux through a closed surface is therefore written

$$\phi_{net} = \oint_S \vec{E}{\cdot}\hat{n}\,dA = \oint_S E_n\,dA \qquad 23\text{-}15$$

The net flux ϕ_{net} through the closed surface is positive or negative, depending on whether $\vec{E}$ is predominantly outward or inward at the surface.

Quantitative Statement of Gauss's Law

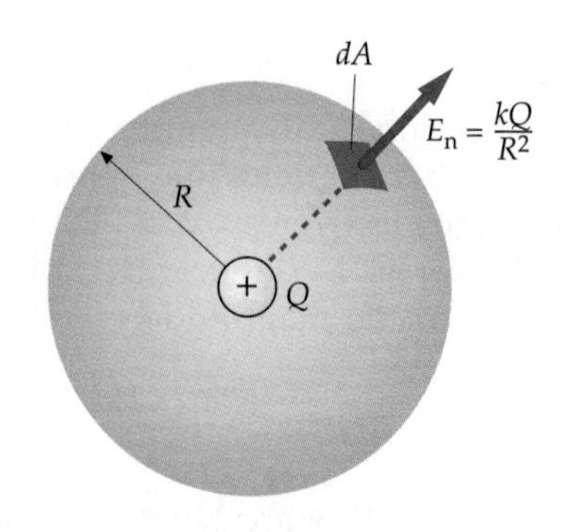

Figure 23-15 A spherical surface enclosing a point charge Q. The same number of electric field lines that pass through this surface will pass through any surface that encloses Q. The flux is easily calculated for a spherical surface. It equals E_n times the surface area, or $E_n \cdot 4\pi R^2$.

Figure 23-15 shows a spherical surface of radius R with a point charge Q at its center. The electric field everywhere on this surface is perpendicular to the surface and has the magnitude

$$E_n = \frac{kQ}{R^2}$$

The net flux through this spherical surface is

$$\phi_{net} = \oint_S E_n \, dA = E_n \oint_S dA$$

where we have taken E_n out of the integral because it is constant everywhere on the surface. The integral of dA over the surface is just the total area of the surface, which equals $4\pi R^2$. Using this and substituting kQ/R^2 for E_n, we obtain

$$\phi_{net} = \frac{kQ}{R^2} 4\pi R^2 = 4\pi kQ \qquad \text{23-16}$$

Thus, the net flux through a spherical surface with a point charge at its center is independent of the radius of the sphere and is equal to $4\pi k$ times the magnitude of the point charge. This is consistent with our previous observation that the net number of lines going out of a surface is proportional to the net charge inside the surface. *This number of lines is the same for all surfaces surrounding the charge, independent of the shape of the surface.* Thus, the net flux through *any surface* surrounding a point charge Q equals $4\pi kQ$.

We can extend this result to systems containing multiple charges. In Figure 23-16, the surface encloses two point charges q_1 and q_2, and there is a third point charge q_3 outside the surface. Since the electric field at any point on the surface is the vector sum of the electric fields produced by each of the three charges, the net flux $\phi_{net} = \oint \vec{E} \cdot \hat{n} \, dA$ through the surface is just the sum of the fluxes due to the individual charges. The flux due to charge q_3, which is outside the surface, is zero because every field line from q_3 that enters the surface at one point leaves the surface at some other point. The flux through the surface due to charge q_1 is $4\pi kq_1$ and that due to charge q_2 is $4\pi kq_2$. The net flux through the surface therefore equals $4\pi k(q_1 + q_2)$, which may be positive, negative, or zero, depending on the signs and magnitudes of the two charges.

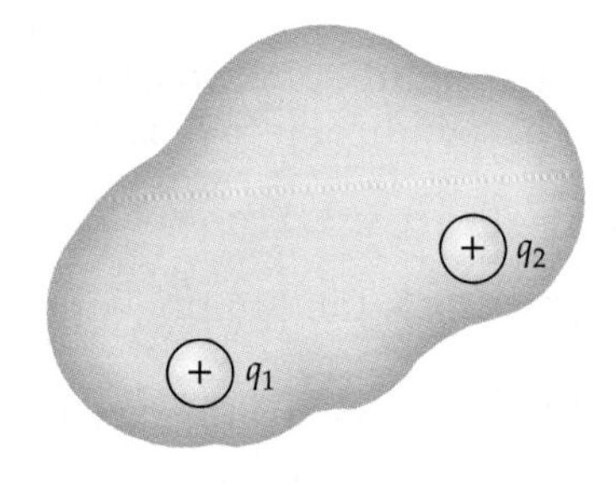

Figure 23-16 A surface enclosing point charges q_1 and q_2 but not q_3. The net flux through this surface is $4\pi k(q_1 + q_2)$.

The net flux through any surface equals $4\pi k$ times the net charge inside the surface:

$$\phi_{net} = \oint_S E_n \, dA = 4\pi kQ_{inside} \qquad \text{23-17}$$

Gauss's law

This is **Gauss's law.** Its validity depends on the fact that the electric field due to a single point charge varies inversely with the square of the distance from the charge. It was this property of the electric field that made it possible to draw a fixed number of electric field lines from a charge and have the density of lines be proportional to the field strength.

It is customary to write the Coulomb constant k in terms of another constant ϵ_0, which is called the **permittivity of free space:**

$$k = \frac{1}{4\pi\epsilon_0} \qquad \text{23-18}$$

Using this notation, Coulomb's law for $\vec{E}$ is written

$$\vec{E} = \frac{1}{4\pi\epsilon_0}\frac{q}{r^2}\hat{r} \qquad \text{23-19}$$

Coulomb's law in terms of ϵ_0

and Gauss's law is written

$$\phi_{\text{net}} = \oint_S E_{\text{n}}\, dA = \frac{1}{\epsilon_0} Q_{\text{inside}} \qquad \text{23-20}$$

Gauss's law in terms of ϵ_0

The value of ϵ_0 in SI units is

$$\epsilon_0 = \frac{1}{4\pi k} = \frac{1}{4\pi(8.99 \times 10^9\ \text{N}\cdot\text{m}^2/\text{C}^2)} = 8.85 \times 10^{-12}\ \text{C}^2/\text{N}\cdot\text{m}^2$$

Gauss's law is valid for all surfaces and all charge distributions. For charge distributions that have high degrees of symmetry, it can be used to calculate the electric field, as we illustrate in the next section. For static charge distributions, Gauss's law and Coulomb's law are equivalent. However, Gauss's law is more general in that it is always valid whether or not the charges are static.

Example 23-4

An electric field is $\vec{E}$ = (200 N/C) $\hat{i}$ for $x > 0$ and $\vec{E}$ = (−200 N/C) $\hat{i}$ for $x < 0$. A cylinder of length 20 cm and radius R = 5 cm has its center at the origin and its axis along the x axis, so that one end is at x = +10 cm and the other is at x = −10 cm (Figure 23-17). (*a*) What is the flux through each end? (*b*) What is the flux through the curved surface of the cylinder? (*c*) What is the net outward flux through the entire closed surface? (*d*) What is the net charge inside the cylinder?

Figure 23-17

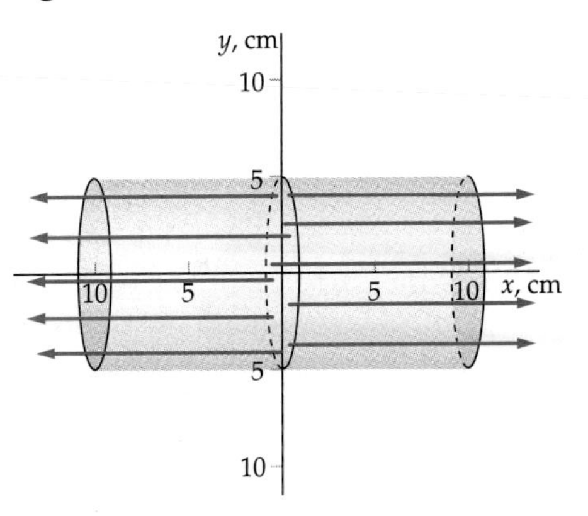

Picture the Problem The field at both circular faces of the cylinder is parallel to the outward vector normal to the surface, so the flux is just EA. There is no flux through the curved surface because the normal to that surface is perpendicular to $\vec{E}$. The net flux through the closed surface is related to the net charge inside by Gauss's law.

(*a*) 1. Calculate the flux through the right circular surface:

$$\phi_{\text{right}} = \vec{E}_{\text{right}}\cdot\hat{n}_{\text{right}}\, A = (200\ \text{N/C})\hat{i}\cdot\hat{i}\ (\pi)(0.05\ \text{m})^2 = 1.57\ \text{N}\cdot\text{m}^2/\text{C}$$

2. Calculate the flux through the left circular surface:

$$\phi_{\text{left}} = \vec{E}_{\text{left}}\cdot\hat{n}_{\text{left}}\, A = (-200\ \text{N/C})\hat{i}\cdot(-\hat{i})(\pi)(0.05\ \text{m})^2 = +1.57\ \text{N}\cdot\text{m}^2/\text{C}$$

(*b*) The flux through the curved surface is zero because $\vec{E}$ is perpendicular to $\hat{n}$:

$$\phi_{\text{curved}} = \vec{E}_{\text{curved}}\cdot\hat{n}_{\text{curved}}\, A = 0$$

(*c*) The total flux is the sum through all surfaces:

$$\phi_{\text{net}} = \phi_{\text{right}} + \phi_{\text{left}} + \phi_{\text{curved}} = 1.57\ \text{N}\cdot\text{m}^2/\text{C} + 1.57\ \text{N}\cdot\text{m}^2/\text{C} + 0 = 3.14\ \text{N}\cdot\text{m}^2/\text{C}$$

(*d*) Gauss's law relates the charge inside to the net flux:

$$Q_{\text{inside}} = \epsilon_0\phi_{\text{net}} = (8.85 \times 10^{-12}\ \text{C}^2/\text{N}\cdot\text{m}^2)(3.14\ \text{N}\cdot\text{m}^2/\text{C}) = 2.78 \times 10^{-11}\ \text{C}$$

23-3 Calculating $\vec{E}$ From Gauss's Law

The electric field due to a highly symmetrical charge distribution can often be easily calculated using Gauss's law. We first find a surface, called a **Gaussian surface**, on which the magnitude of the field E is constant. The flux through this surface will be proportional to the field E on the surface. Gauss's law then relates this field to the charge inside the surface.

Plane Geometry

$\vec{E}$ due to an Infinite Plane of Charge Figure 23-18 shows an infinite plane of charge of surface charge density σ. By symmetry, $\vec{E}$ must be perpendicular to the plane and can depend only on the distance from it. Also, $\vec{E}$ must have the same magnitude but the opposite direction at points the same distance above and below the plane. For our Gaussian surface, we choose a pillbox-shaped cylinder as shown. Let each face of the cylinder have an area A. Since $\vec{E}$ is parallel to the curved cylindrical surface, there is no flux through this surface. The flux through each face is $E_n A$, so the total flux is $2E_n A$. The net charge inside the surface is σA. Gauss's law then gives

$$\phi_{\text{net}} = \oint E_n\, dA = \frac{1}{\epsilon_0} Q_{\text{inside}}$$

$$2E_n A = \frac{1}{\epsilon_0}\sigma A$$

or

$$E_n = \frac{\sigma}{2\epsilon_0} = 2\pi k\sigma \qquad 23\text{-}21$$

$\vec{E}$ near an infinite plane of charge

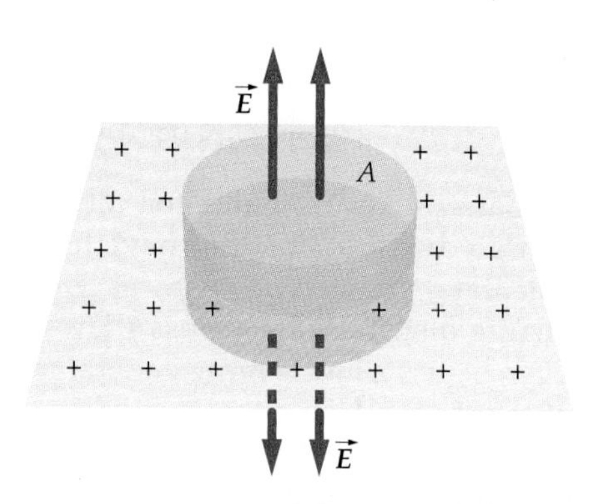

Figure 23-18 Gaussian surface for the calculation of $\vec{E}$ due to an infinite plane of charge. On the upper and lower faces of this pillbox, $\vec{E}$ is perpendicular to the surface and constant in magnitude.

This is the same result that we obtained with much more difficulty using Coulomb's law (Equation 23-12*a*). Note that the field is discontinuous at the plane. If the charge is in the xy plane, the field is $E_z = \sigma/2\epsilon_0$ just above the plane and $E_z = -\sigma/2\epsilon_0$ just below the plane. Thus, the field is discontinuous by $\Delta E_z = \sigma/2\epsilon_0 - (-\sigma/2\epsilon_0) = \sigma/\epsilon_0$.

Example 23-5

In Figure 23-19, an infinite plane of surface charge density $\sigma = +4.5\ \text{nC/m}^2$ lies in the yz plane, and a second infinite plane of surface charge density $\sigma = -4.5\ \text{nC/m}^2$ lies in a plane parallel to the yz plane at $x = 2$ m. Find the electric field at (*a*) $x = 1.8$ m and (*b*) $x = 5$ m.

Picture the Problem Each plane produces a uniform electric field of magnitude $E = \sigma/2\epsilon_0$. We use superposition to find the resultant field. Between the planes the fields add, producing a net field of magnitude σ/ϵ_0 in the positive x direction. For $x > 2$ m or $x < 0$, the fields point in opposite directions and cancel.

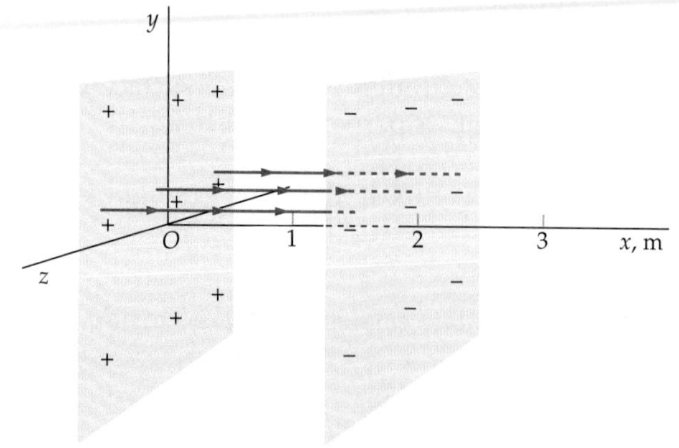

Figure 23-19

(*a*) 1. Calculate the magnitude of the field E produced by each plane:	$E = \frac{\sigma}{2\epsilon_0} = \frac{4.5 \times 10^{-9}\ \text{C/m}^2}{2(8.85 \times 10^{-12}\ \text{C}^2/\text{N}\cdot\text{m}^2)} = 254\ \text{N/C}$
2. At $x = 1.8$ m, between the planes, the field due to each plane points in the positive x direction:	$E_{x,\text{net}} = E_1 + E_2 = 254\ \text{N/C} + 254\ \text{N/C} = 508\ \text{N/C}$
(*b*) At $x = 5$ m, the fields due to the two planes are in opposite directions:	$E_{x,\text{net}} = E_1 - E_2 = 0$

Remarks Because the two planes carry equal and opposite charge densities, the electric field lines begin on the positive plane and end on the negative plane. $\vec{E}$ is zero except between the planes. Note that $E_{x,\text{net}} = 508$ N/C not just at $x = 1.8$ m but at any point between the planes.

Spherical Geometry

To calculate the electric field due to spherically symmetric charge distributions, we use a spherical surface for our Gaussian surface. We illustrate this by first finding the electric field at a distance r from a point charge q. We choose a spherical surface of radius r centered at the charge for our Gaussian surface. By symmetry, $\vec{E}$ must be radial, and its magnitude can depend only on the distance from the charge. The normal component of $\vec{E}$, $E_n = \vec{E}\cdot\hat{n} = E_r$, has the same value everywhere on the spherical surface. The net flux through this surface is thus

$$\phi_{\text{net}} = \oint \vec{E}\cdot\hat{n}\, dA = \oint E_r\, dA = E_r \oint dA$$

But $\oint dA$ is equal to $4\pi r^2$, the total area of the spherical surface. Since the total charge inside the surface is just the point charge q, Gauss's law gives

$$E_r 4\pi r^2 = \frac{q}{\epsilon_0}$$

or

$$E_r = \frac{1}{4\pi\epsilon_0}\frac{q}{r^2}$$

which is Coulomb's law. We have thus derived Coulomb's law from Gauss's law. Since Gauss's law can also be obtained from Coulomb's law (see Section 23-6), we have shown that the two laws are equivalent for static charges.

$\vec{E}$ due to a Spherical Shell of Charge

Consider a uniformly charged spherical shell of radius R and total charge Q. By symmetry, $\vec{E}$ must be radial, and its magnitude can depend only on the distance r from the center of the sphere. In Figure 23-20, we have chosen a spherical Gaussian surface of radius $r > R$. Since $\vec{E}$ is perpendicular to this surface and constant in magnitude everywhere on it, the flux through the surface is

$$\phi_{\text{net}} = \oint E_r\, dA = E_r 4\pi r^2$$

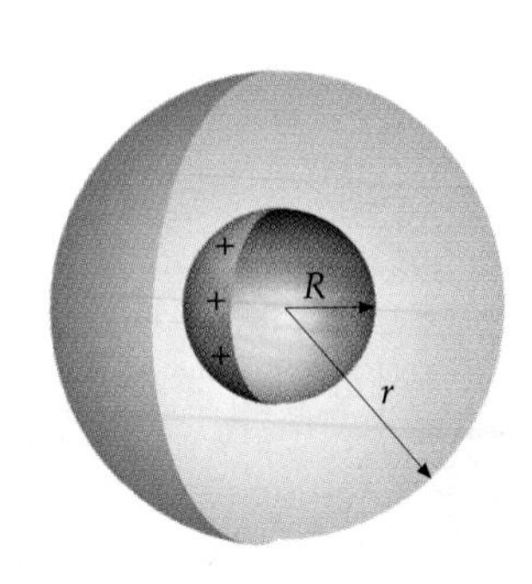

Figure 23-20 Spherical Gaussian surface of radius $r > R$ for the calculation of the electric field outside a uniformly charged spherical shell of radius R.

Since the total charge inside the Gaussian surface is the total charge on the shell Q, Gauss's law gives

$$E_r 4\pi r^2 = \frac{Q}{\epsilon_0}$$

or

$$E_r = \frac{1}{4\pi\epsilon_0}\frac{Q}{r^2}, \qquad r > R \qquad \text{23-22a}$$

$\vec{E}$ outside a spherical shell of charge

Thus, the electric field outside a uniformly charged spherical shell is the same as if all the charge were at the center of the shell.

If we choose a spherical Gaussian surface inside the shell, where $r < R$, the net flux is again $E_r 4\pi r^2$, but the total charge inside the surface is zero. Therefore, for $r < R$, Gauss's law gives

$$\phi_{\text{net}} = E_r 4\pi r^2 = 0$$

and

$$E_r = 0, \qquad r < R \qquad \text{23-22b}$$

$\vec{E}$ inside a spherical shell of charge

These results can also be obtained by direct integration of Coulomb's law, but that calculation is much more difficult.

Figure 23-21 shows E_r versus r for a spherical-shell charge distribution. Again, note that the electric field is discontinuous at $r = R$, where the surface charge density is σ. Just outside the shell at $r \approx R$, the electric field is $E_r = Q/4\pi\epsilon_0 R^2 = \sigma/\epsilon_0$ since $\sigma = Q/4\pi R^2$. Since the field just inside the shell is zero, the electric field is discontinuous by the amount σ/ϵ_0 as we pass through the shell.

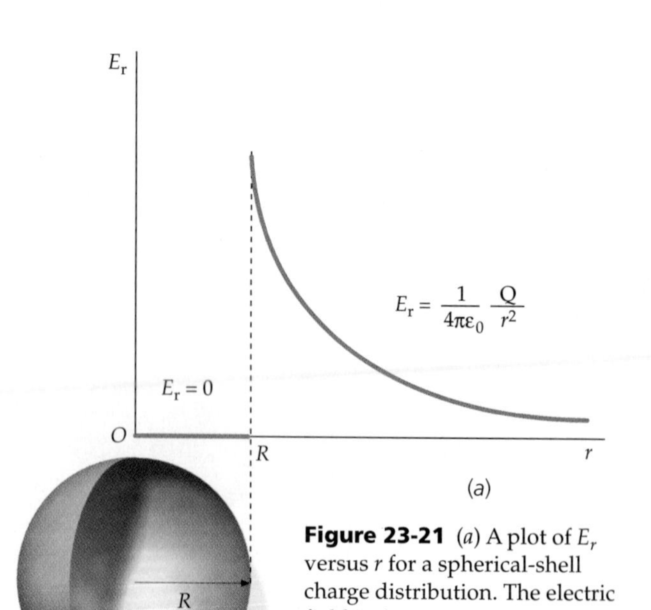

(a)

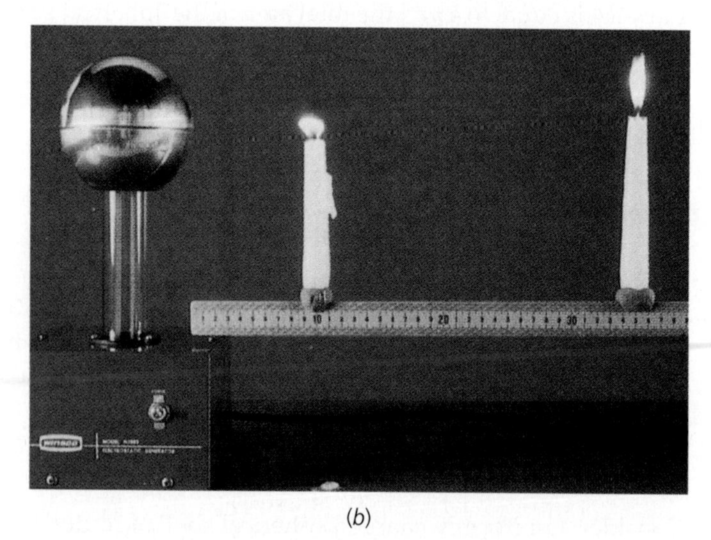

(b)

Figure 23-21 (*a*) A plot of E_r versus r for a spherical-shell charge distribution. The electric field is discontinuous at $r = R$, where there is a surface charge of density σ. (*b*) The decrease in E_r over distance due to a charged spherical shell is evident by the effect of the field on the flames of these two candles. The spherical shell at the left (part of a Van de Graaff generator, a device that is discussed in Chapter 24) carries a large negative charge that attracts the positive ions in the nearby candle flame. The flame at right, which is much farther away, is not affected.

Example 23-6

A spherical shell of radius $R = 3$ m has its center at the origin and carries a surface charge density of $\sigma = 3$ nC/m². A point charge $q = 250$ nC is on the y axis at $y = 2$ m. Find the electric field on the x axis at (*a*) $x = 2$ m and (*b*) $x = 4$ m.

Picture the Problem We find the field due to the point charge and that due to the spherical shell and sum the field vectors. For (*a*), the field point is inside the shell, so the field is due only to the point charge (Figure 23-22*a*). For (*b*), the field point is outside the shell, so the shell can be considered to be a point charge at the origin. We then find the field due to two point charges (Figure 23-22*b*).

Figure 23-22

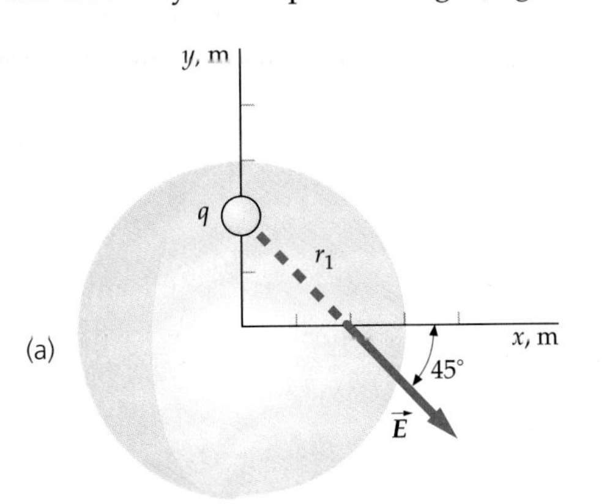

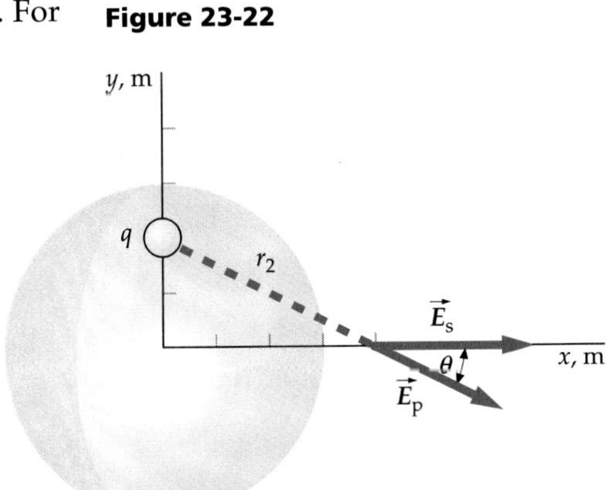

(*a*) 1. Inside the shell, $\vec{E}_1$ is due only to the point charge:

$$\vec{E}_1 = \frac{kq}{r_1^2}\hat{r}$$

2. Calculate the distance r_1:

$$r_1 = \sqrt{(2\text{ m})^2 + (2\text{ m})^2} = \sqrt{8}\text{ m}$$

3. Use r_1 to calculate the magnitude of the field:

$$E_1 = \frac{kq}{r_1^2} = \frac{(8.99 \times 10^9\text{ N}\cdot\text{m}^2/\text{C}^2)(250 \times 10^{-9}\text{ C})}{(\sqrt{8}\text{ m})^2} = 281\text{ N/C}$$

4. From Figure 23-22*a*, we can see that the field makes an angle of $-45°$ with the x axis:

$$\theta = -45°$$

(*b*) 1. Outside of its perimeter, the shell can be treated as a point charge at the origin, and the field due to the shell $\vec{E}_s$ is therefore along the x axis:

$$\vec{E}_s = \frac{kQ}{x_2^2}\hat{i}$$

2. Calculate the total charge Q on the shell:

$$Q = \sigma 4\pi R^2 = (3\text{ nC/m}^2)4\pi(3\text{ m})^2 = 339\text{ nC}$$

3. Use Q to calculate the field due to the shell:

$$E_s = \frac{(8.99 \times 10^9\text{ N}\cdot\text{m}^2/\text{C}^2)(339 \times 10^{-9}\text{ C})}{(4\text{ m})^2} = 190\text{ N/C}$$

4. The field due to the point charge is:

$$\vec{E}_p = \frac{kq}{r_2^2}\hat{r}_2$$

5. Calculate the distance from the point charge q on the y axis to the field point at $x = 4$ m:

$$r_2 = \sqrt{(2\text{ m})^2 + (4\text{ m})^2} = \sqrt{20}\text{ m}$$

6. Calculate the magnitude of the field due to the point charge:

$$E_p = \frac{kq}{r_2^2} = \frac{(8.99 \times 10^9\text{ N}\cdot\text{m}^2/\text{C}^2)(250 \times 10^{-9}\text{ C})}{(\sqrt{20}\text{ m})^2} = 112\text{ N/C}$$

7. This field makes an angle θ with the x axis, where:

$$\cos\theta = \frac{4}{\sqrt{20}}$$

$$\sin\theta = \frac{-2}{\sqrt{20}}$$

8. The x and y components of this field are thus:

$$E_{px} = E_p \cos\theta = (112\ \text{N/C})\left(\frac{4}{\sqrt{20}}\right) = 100\ \text{N/C}$$

$$E_{py} = E_p \sin\theta = (112\ \text{N/C})\left(-\frac{2}{\sqrt{20}}\right) = -50\ \text{N/C}$$

9. Find the x and y components of the net electric field:

$$E_x = E_{sx} + E_{px} = 190\ \text{N/C} + 100\ \text{N/C} = 290\ \text{N/C}$$

$$E_y = E_{sy} + E_{py} = 0 - 50\ \text{N/C} = -50\ \text{N/C}$$

Remark Giving the x and y components of a field completely specifies the field. The magnitude and direction of the net field can be found from $E = \sqrt{E_x^2 + E_y^2}$ and $\tan\theta' = E_y/E_x$.

$\vec{E}$ due to a Uniformly Charged Sphere

Example 23-7

Find the electric field (*a*) outside and (*b*) inside a uniformly charged sphere of radius R carrying a total charge Q that is uniformly distributed throughout the volume of the sphere with charge density $\rho = Q/V$, where $V = \frac{4}{3}\pi R^3$ is the volume of the sphere.

Picture the Problem By symmetry, the electric field must be radial. (*a*) To find E_r outside the charged sphere, we choose a spherical Gaussian surface of radius $r > R$ (Figure 23-23a). (*b*) To find E_r inside the charge we choose a spherical Gaussian surface of radius $r < R$ (Figure 23-23b). On each of these surfaces, E_r is constant. Gauss's law then relates E_r to the total charge inside the surface.

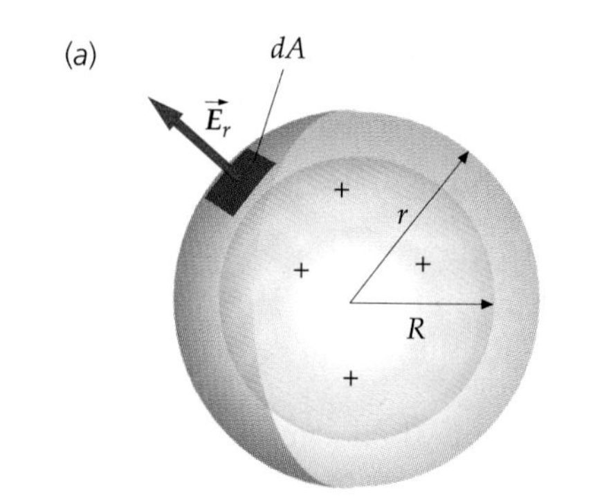

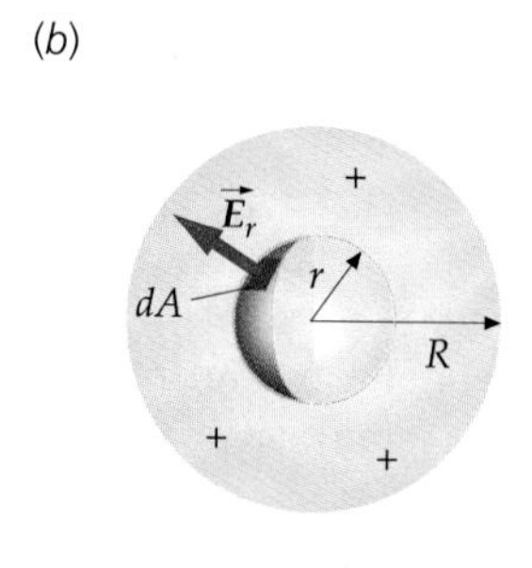

Figure 23-23

(*a*) 1. (Outside) Relate the flux through the Gaussian surface to the electric field E_r on the Gaussian surface at $r > R$:

$$\phi_{\text{net}} = E_r 4\pi r^2$$

2. Apply Gauss's law to relate the field to the total charge inside the surface, which is Q:

$$E_r 4\pi r^2 = \frac{Q}{\epsilon_0}$$

3. Solve for E_r:

$$E_r = \frac{1}{4\pi\epsilon_0}\frac{Q}{r^2}, \qquad r > R$$

(*b*) 1. (Inside) Relate the flux through the Gaussian surface to the electric field E_r on the Gaussian surface at $r < R$:

$$\phi_{\text{net}} = E_r 4\pi r^2$$

2. Apply Gauss's law to relate the field to the total charge inside the surface Q_{inside}:

$$E_r 4\pi r^2 = \frac{Q_{\text{inside}}}{\epsilon_0}$$

3. The total charge inside the surface is $\rho V'$, where $V' = \frac{4}{3}\pi r^3$ is the volume inside the Gaussian surface:

$$Q_{\text{inside}} = \rho V' = \left(\frac{Q}{V}\right)V' = \left(\frac{Q}{\frac{4}{3}\pi R^3}\right)\left(\frac{4}{3}\pi r^3\right) = Q\frac{r^3}{R^3}$$

4. Substitute this value for Q_{inside} and solve for E_r:

$$E_r 4\pi r^2 = \frac{Q_{\text{inside}}}{\epsilon_0} = \frac{1}{\epsilon_0} Q \frac{r^3}{R^3}$$

$$E_r = \frac{1}{4\pi\epsilon_0} \frac{Q}{R^3} r, \qquad r \leq R$$

Remarks Figure 23-24 shows E_r versus r for the charge distribution in this example. Inside a sphere of charge, E_r increases with r. Note that E_r is continuous at $r = R$. A uniformly charged sphere is sometimes used as a model to describe the electric field of an atomic nucleus.

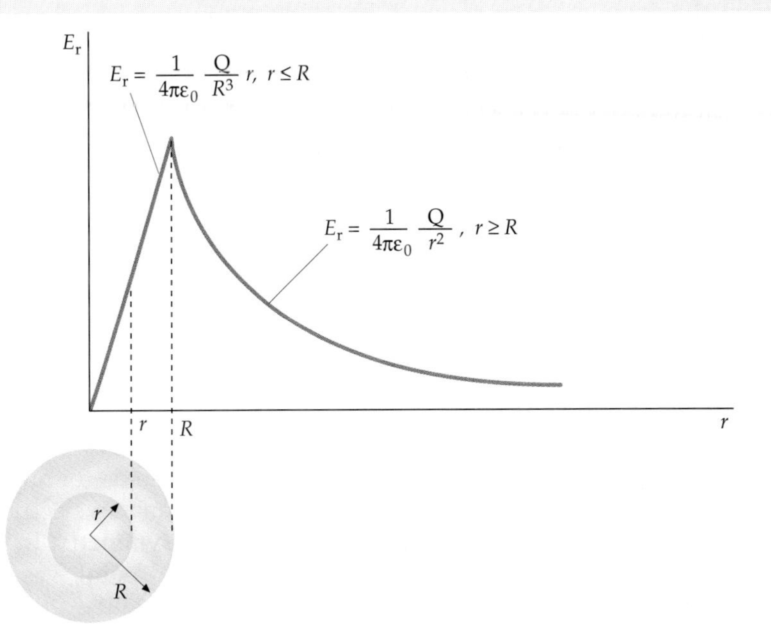

Figure 23-24

We see from Example 23-7 that the electric field at a distance r from the center of a uniformly charged sphere of radius R is given by

$$E_r = \frac{1}{4\pi\epsilon_0} \frac{Q}{r^2}, \qquad r \geq R \qquad \text{23-23a}$$

$$E_r = \frac{1}{4\pi\epsilon_0} \frac{Q}{R^3} r, \qquad r \leq R \qquad \text{23-23b}$$

where Q is the total charge of the sphere.

Cylindrical Geometry

To calculate the electric field due to cylindrically symmetric charge distributions, we use a cylindrical Gaussian surface. We illustrate this by calculating the electric field due to an infinitely long line charge of uniform linear charge density, a problem we have already solved using Coulomb's law.

Example 23-8

Use Gauss's law to find the electric field at a distance r from an infinitely long line charge of uniform charge density λ.

Picture the Problem By symmetry, the electric field lines radiate uniformly from the line, outward if λ is positive and inward if λ is negative. We therefore choose a cylindrical Gaussian surface of length L and radius r (Figure 23-25). The electric field is perpendicular to the cylindrical surface and has the same value E_r everywhere on the surface. The electric flux is then just the product of the electric field and the area of the cylindrical surface, which is $2\pi rL$. There is no flux through the flat surfaces at the ends of the cylinder because $\vec{E} \cdot \hat{n} = 0$ on these surfaces.

Figure 23-25

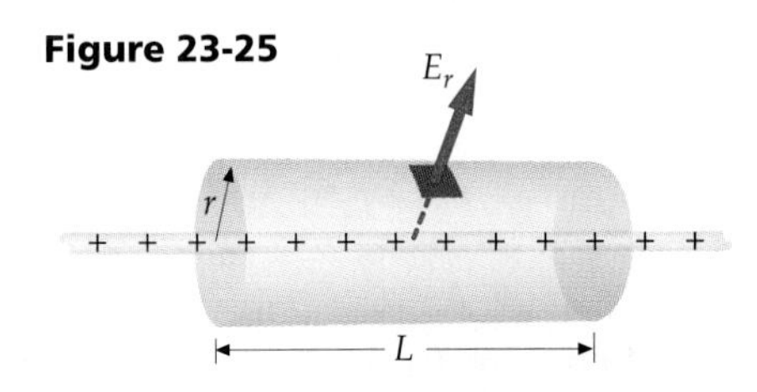

1. Relate the flux through the Gaussian surface to the electric field E_r on the Gaussian surface at $r > R$: $\phi_{\text{net}} = E_r \oint dA = E_r 2\pi rL$

2. Apply Gauss's law to relate the field to the total charge inside the surface Q_{inside}: $E_r 2\pi rL = \dfrac{Q_{\text{inside}}}{\epsilon_0}$

3. The charge inside is the charge on a length L of the line: $Q_{\text{inside}} = \lambda L$

4. Substitute this value for Q_{inside} and solve for E_r: $E_r 2\pi rL = \dfrac{\lambda L}{\epsilon_0}$

$$E_r = \frac{1}{2\pi\epsilon_0}\frac{\lambda}{r}$$

Remark Since $1/(2\pi\epsilon_0) = 2k$, the field is $2k\lambda/r$, the same as Equation 23-9 with $r = y$.

It is important to realize that although Gauss's law holds for any surface surrounding any charge distribution, its use in calculating an electric field is limited to charge distributions with a high degree of symmetry. In the preceding calculation, we needed to assume that the field point was very far from the ends of the line charge so that E_n would be constant everywhere on the cylindrical Gaussian surface. This is equivalent to assuming that, at the distance r from the line, the line charge appears to be infinitely long. If we are near the end of a finite line charge, we cannot assume that $\vec{E}$ is perpendicular to the cylindrical surface, or that E_n is constant everywhere on it, so we cannot use Gauss's law to calculate the electric field.

23-4 Discontinuity of E_n

We have seen that the electric field for an infinite plane of charge and a spherical shell of charge is discontinuous by the amount σ/ϵ_0 at a point where there is a surface charge density σ. We now show that this is a general result for the component of the electric field that is perpendicular to a surface carrying a charge density of σ.

Figure 23-26 shows a pill box-shaped Gaussian surface with faces of area A on each side of an arbitrary surface carrying a surface charge density σ. Let the normal component of the electric field be E_{n2} on one side of the surface and E_{n1} on the other side, as shown in the figure. If we make the height of the cylinder very small compared with the radius of the faces, we can neglect the flux through the sides of the cylinder compared with the flux through the faces of area πR^2. The net flux through the Gaussian surface is then $E_{n2}A - E_{n1}A$, and the charge inside the surface is σA. Gauss's law gives

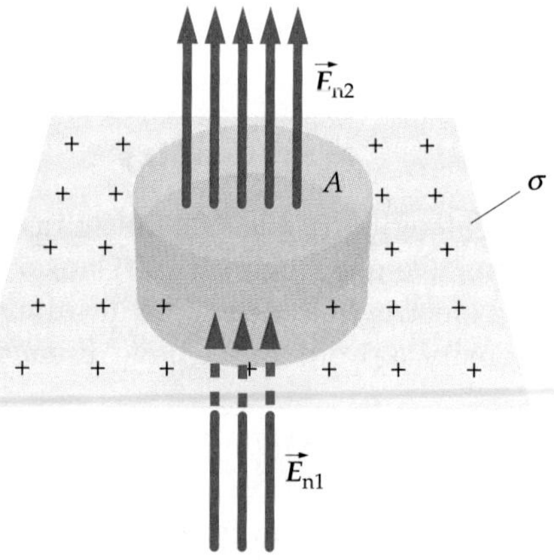

$$E_{n2}A - E_{n1}A = \frac{\sigma A}{\epsilon_0}$$

or

$$E_{n2} - E_{n1} = \frac{\sigma}{\epsilon_0} \qquad 23\text{-}24$$

Discontinuity of E_n at a surface charge

which is the result we wished to prove.

Figure 23-26 A Gaussian, pillbox-shaped surface with faces of area A on each side of a surface charge of charge density of σ. The net flux through the pillbox is $(E_{n2} - E_{n1})A$. The electric field E_{n2} on one side is greater than the electric field E_{n1} on the other side by the amount σ/ϵ_0.

Note that there is no discontinuity in the electric field at the surface of a uniform ball of charge. The field just inside the surface of a charged sphere is the same as the field just outside such a charge distribution, as can be seen from Figure 23-24.

23-5 Charge and Field at Conductor Surfaces

A conductor contains charge that is free to move about the volume of the conductor. If there is an electric field inside the conductor, there will be a net force on this charge causing a momentary electric current (electric currents are discussed in Chapter 26). However, unless there is a source of energy to maintain this current, the free charge in a conductor will merely redistribute itself to create an electric field that cancels the external field within the conductor. The conductor is then said to be in **electrostatic equilibrium.** Thus, in equilibrium, the electric field inside a conductor must be zero. The time to reach equilibrium depends on the conductor. For copper and other good conductors, the time is so small that, for all practical purposes, electrostatic equilibrium is reached instantaneously.

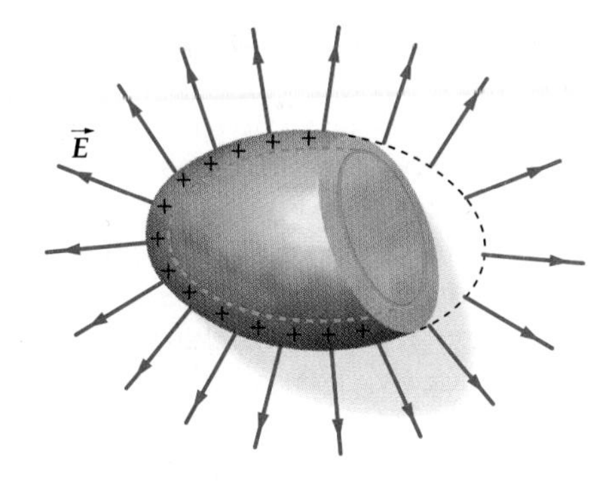

Figure 23-27 A Gaussian surface (the blue line) just inside the surface of a conductor. Since the electric field is zero inside a conductor in electrostatic equilibrium, the net flux through this surface must also be zero. Therefore, the net charge inside this surface must be zero.

We can use Gauss's law to show that any net electric charge on a conductor resides on the surface of the conductor. Consider a Gaussian surface just inside the actual conductor surface in electrostatic equilibrium (Figure 23-27). The electric field is zero everywhere on the Gaussian surface because the surface is completely within the conductor. The net flux through the surface must therefore be zero, and by Gauss's law, the net charge inside the surface must be zero. Thus there can be no net charge inside any surface lying completely within the conductor. If there is any net charge on the conductor, it must be on the conductor's surface. At the surface of a conductor in equilibrium, $\vec{E}$ must be perpendicular to the surface. If the electric field had a tangential component, the free charge would be accelerated tangential to the surface until equilibrium was established.

Since E_n is discontinuous at any surface by the amount σ/ϵ_0, and since $\vec{E}$ is zero inside a conductor, the field just outside a conductor is given by

$$E_n = \frac{\sigma}{\epsilon_0} \qquad 23\text{-}25$$

E_n just outside the surface of a conductor

This result is exactly twice the field produced by an infinite plane of charge. We can understand this result from Figure 23-28. The charge on the conductor consists of two parts: (1) the charge near point P, and (2) all the rest of the charge. The charge near point P looks like an infinite plane and produces a

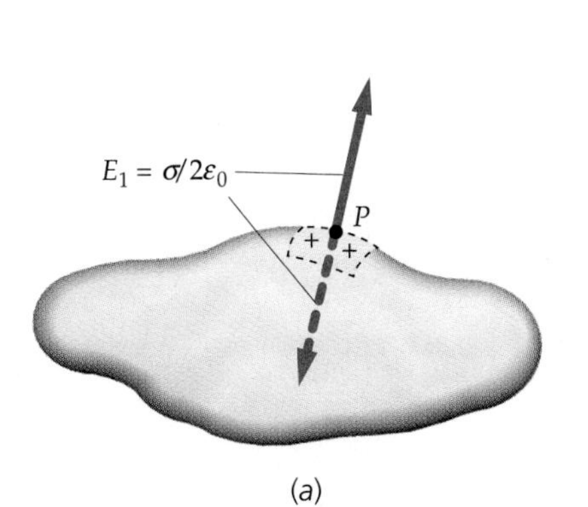

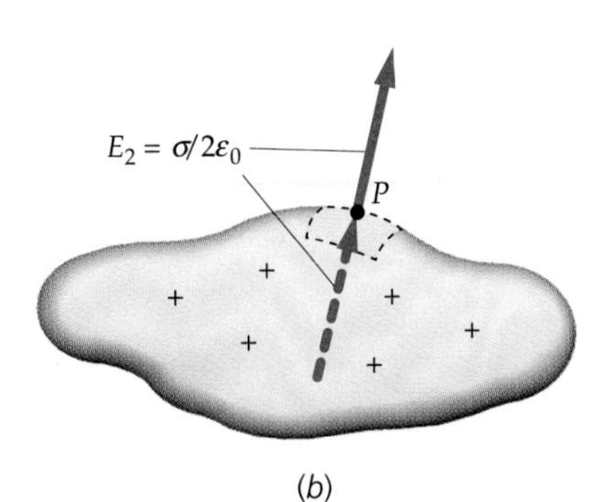

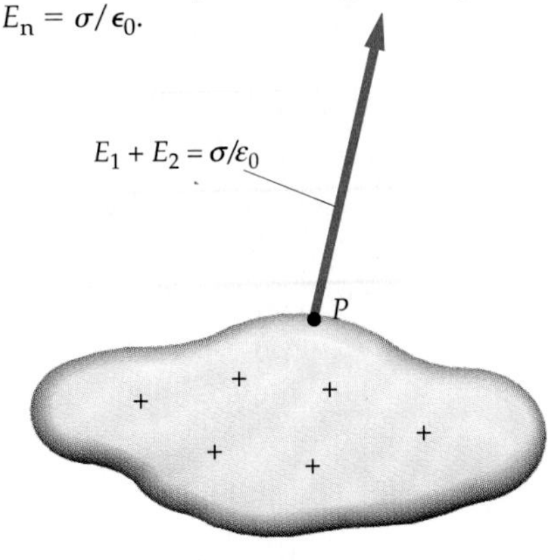

Figure 23-28 An arbitrarily shaped conductor carrying a charge on its surface. (*a*) The charge in the vicinity of point P near the surface looks like an infinite plane of charge, giving an electric field of magnitude $\sigma/2\epsilon_0$ pointing away from the surface both inside and outside the surface. Inside the conductor this field points down from point P. (*b*) Since the net field inside the conductor is zero, the rest of the charge must produce a field of magnitude $\sigma/2\epsilon_0$ in the upward direction. The field due to this charge is the same inside the surface as outside the surface. (*c*) Inside the surface, the fields shown in (*a*) and (*b*) cancel, but outside at point P they add to give $E_n = \sigma/\epsilon_0$.

field of magnitude $\sigma/2\epsilon_0$ just inside and just outside the conductor. The rest of the charge must produce a field of magnitude $\sigma/2\epsilon_0$ that cancels the field inside the conductor. This field due to the rest of the charge adds to the field due to the local charge just outside the conductor to give a total field of σ/ϵ_0.

Figure 23-29 shows a positive point charge q at the center of a spherical cavity inside a spherical conductor. Since the net charge must be zero within any surface drawn within the conductor, there must be a negative charge $-q$ induced in the inside surface. In Figure 23-30, the point charge has been moved so that it is no longer at the center of the cavity. The field lines in the cavity are altered, and the surface charge density of the induced negative charge on the inner surface is no longer uniform. However, the positive surface charge density on the outside surface is not disturbed—it is still uniform—because it is shielded from the cavity by the conductor.

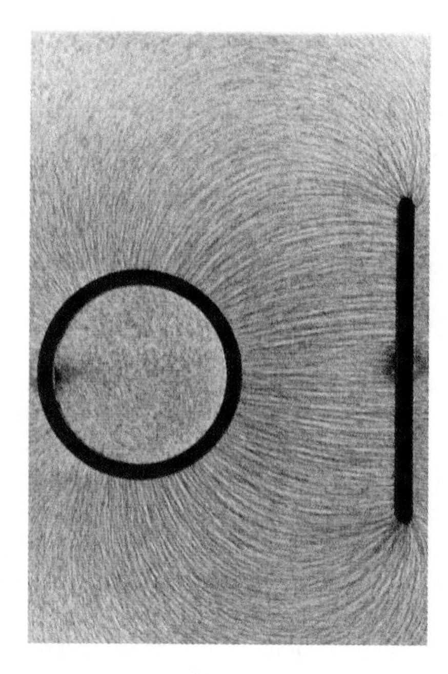

Electric field lines for an oppositely charged cylinder and plate, shown by bits of fine thread suspended in oil. Note that the field lines are perpendicular to the conductors and that there are no lines inside the cylinder.

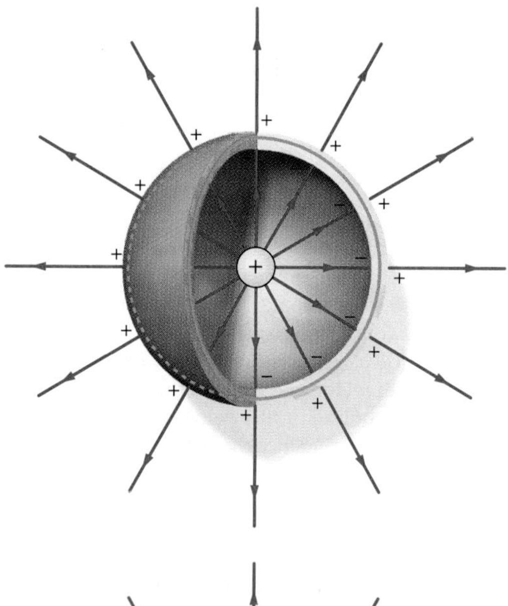

Figure 23-29 A point charge q in the cavity at the center of a thick spherical conducting shell. Since the net charge within the Gaussian surface (indicated in blue) must be zero, a surface charge $-q$ is induced on the inner surface of the shell, and since the conductor is neutral, an equal but opposite charge $+q$ is induced on the outer surface. Electric field lines begin on the point charge and end on the inner surface. The field lines begin again on the outer surface.

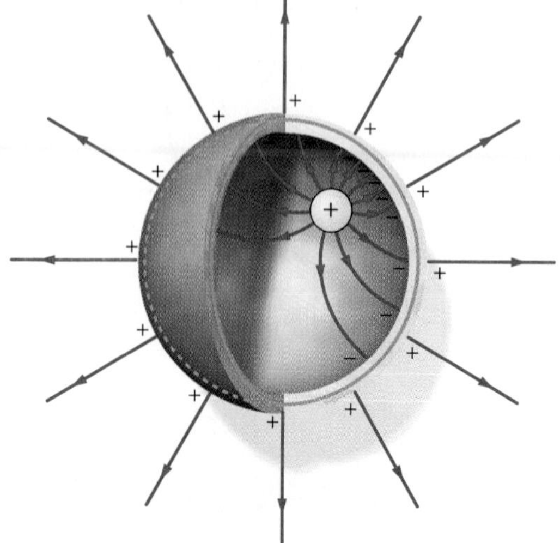

Figure 23-30 The same conductor as in Figure 23-29 with the point charge moved away from the center of the sphere. The charge on the outer surface and the electric field lines outside the sphere are not affected.

Example 23-9

An uncharged square conducting slab of negligible thickness and with 4-m sides is placed in an external uniform field $\vec{E} = (450\ \text{kN/C})\ \hat{i}$ that is perpendicular to the faces of the slab (Figure 23-31). (*a*) Find the charge density on each face of the slab. (*b*) A net charge of 96 μC is placed on the slab. Find the new charge density on each face and the electric field near each face but far from the edges of the slab.

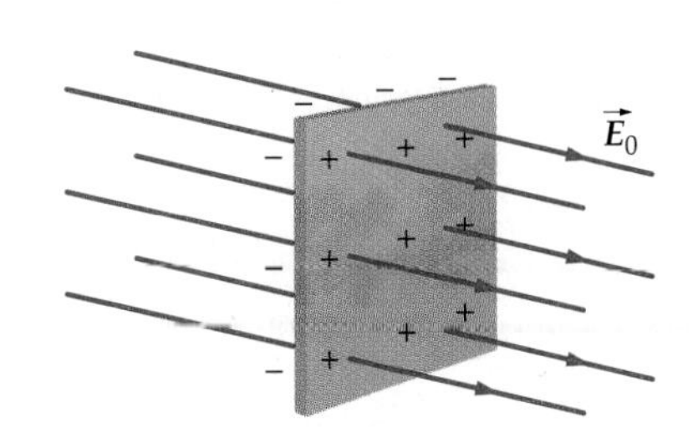

Figure 23-31

Picture the Problem (*a*) We find the charge density by using the value of E_n near each surface. (*b*) The additional charge of 96 μC must be distributed uniformly on each face of the slab so that the electric field inside the slab remains zero. The net charge density is the sum of the original density plus the added density.

(*a*)1. The charge density is related to the field just outside the face:

$$\sigma = \epsilon_0 E_n$$

2. At the right face, $\vec{E}$ points away from the slab so E_n = 450 kN/C. Use the value of E_n to calculate σ_R:

$$\sigma_R = \epsilon_0 E_n$$
$$= (8.85 \times 10^{-12}\ \text{C}^2/\text{N}\cdot\text{m}^2)(450\ \text{kN/C})$$
$$= 3.98 \times 10^{-6}\ \text{C/m}^2 = 3.98\ \mu\text{C/m}^2$$

3. At the left face, the electric field points toward the slab, so $E_n = -450$ kN/C. Use this to calculate σ_L:

$$\sigma_L = \epsilon_0 E_n$$
$$= (8.85 \times 10^{-12}\ \text{C}^2/\text{N}\cdot\text{m}^2)(-450\ \text{kN/C})$$
$$= -3.98\ \mu\text{C/m}^2$$

(*b*)1. The new charge density on a face equals the old charge density plus the additional charge density:

$$\sigma' = \sigma + \sigma_a$$

2. The additional charge density on each face equals the total charge on the face divided by its area:

$$\sigma_a = \frac{Q}{A}$$

3. The additional charge on each face is half the total additional charge (or 48 μC). Use this fact and the given dimensions of the slab to calculate σ_a:

$$\sigma_a = \frac{Q}{A} = \frac{48\ \mu\text{C}}{(4\ \text{m})^2} = 3.0\ \mu\text{C/m}^2$$

4. Add this additional charge density to the original charge density to find the new charge density:

$$\sigma_R = 3.98\ \mu\text{C/m}^2 + 3.0\ \mu\text{C/m}^2$$
$$= 6.98\ \mu\text{C/m}^2$$

$$\sigma_L = -3.98\ \mu\text{C/m}^2 + 3.0\ \mu\text{C/m}^2$$
$$= -0.98\ \mu\text{C/m}^2$$

5. Use these charge densities to calculate the normal component of the electric field just outside each face:

$$E_{nR} = \frac{\sigma_R}{\epsilon_0}$$
$$= \frac{6.98\ \mu\text{C/m}^2}{8.85 \times 10^{-12}\ \text{C}^2/\text{N}\cdot\text{m}^2}$$
$$= 789\ \text{kN/C}$$

$$E_{nL} = \frac{\sigma_R}{\epsilon_0}$$
$$= \frac{-0.98\ \mu\text{C/m}^2}{8.85 \times 10^{-12}\ \text{C}^2/\text{N}\cdot\text{m}^2}$$
$$= -111\ \text{kN/m}^2$$

Exercise Adding the positive charge density of 3.0 $\mu C/m^2$ to each face in this example is equivalent to adding two planes of positive charge. (*a*) Find the electric field due to these planes outside the slabs. (*b*) Combine this field with the original field to find the resultant electric fields on the right and left of the slab. (*Answers* (*a*) $E = \frac{1}{2}\sigma/\epsilon_0 + \frac{1}{2}\sigma/\epsilon_0 = \sigma/\epsilon_0 = 339$ kN/C (*b*) $E_R =$ 450 kN/C + 339 kN/C = 789 kN/C; E_L = 450 kN/C − 339 kN/C = 111 kN/C)

Exercise The electric field just outside the surface of a certain conductor points away from the conductor and has a magnitude of 2000 N/C. What is the surface charge density on the surface of the conductor? (*Answer* 17.7 nC/m^2)

optional

23-6 Derivation of Gauss's Law from Coulomb's Law

Gauss's law can be derived mathematically using the concept of the **solid angle**. Consider an area element ΔA on a spherical surface. The solid angle $\Delta\Omega$ subtended by ΔA at the center of the sphere is defined to be

$$\Delta\Omega = \frac{\Delta A}{r^2}$$

where r is the radius of the sphere. Since ΔA and r^2 both have dimensions of length squared, the solid angle is dimensionless. The unit of the solid angle is the **steradian** (sr). Since the total area of a sphere is $4\pi r^2$, the total solid angle subtended by a sphere is

$$\frac{4\pi r^2}{r^2} = 4\pi \text{ steradians}$$

There is a close analogy between the solid angle and the ordinary plane angle $\Delta\theta$, which is defined to be the ratio of an element of arc length of a circle Δs to the radius of the circle:

$$\Delta\theta = \frac{\Delta s}{r} \text{ radians}$$

The total plane angle subtended by a circle is 2π radians.

In Figure 23-32, the area element ΔA is not perpendicular to the radial lines from point O. The unit vector $\hat{n}$ normal to the area element makes an angle θ with the radial unit vector $\hat{r}$. In this case, the solid angle subtended by ΔA at point O is defined to be

$$\Delta\Omega = \frac{\Delta A\hat{n}\cdot\hat{r}}{r^2} = \frac{\Delta A\cos\theta}{r^2} \qquad 23\text{-}26$$

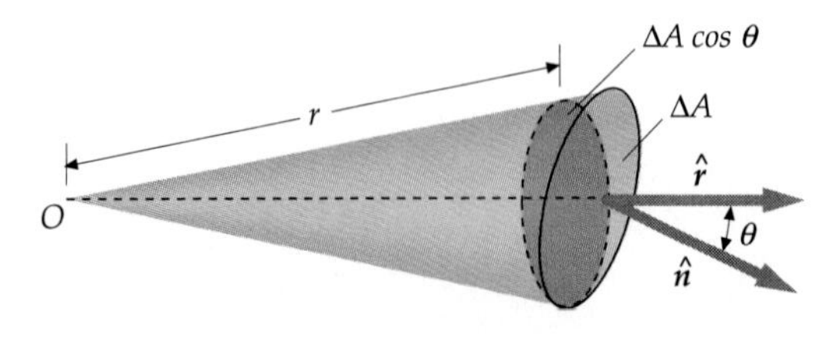

Figure 23-32 An area element ΔA whose normal is not parallel to the radial line from O to the center of the element. The solid angle subtended by this element at O is defined to be $(\Delta A\cos\theta)/r^2$.

Figure 23-33 shows a point charge q surrounded by a surface S of arbitrary shape. To calcuate the flux through this surface, we want to find $\vec{E}\cdot\hat{n}\,\Delta A$ for each element of area on the surface and sum over the entire surface. The flux through the area element shown is

$$\Delta\phi = \vec{E}\cdot\hat{n}\,\Delta A = \frac{kq}{r^2}\,\hat{r}\cdot\hat{n}\,\Delta A = kq\,\Delta\Omega$$

The solid angle $\Delta\Omega$ is the same as that subtended by the corresponding area element of a spherical surface of any radius. The sum of the flux through the entire surface is kq times the total solid angle subtended by the closed surface, which is 4π steradians:

$$\phi_{\text{net}} = \oint_S \vec{E}\cdot\hat{n}\,dA = kq\oint d\Omega = 4\pi kq = \frac{q}{\epsilon_0} \qquad \text{23-27}$$

which is Gauss's law.

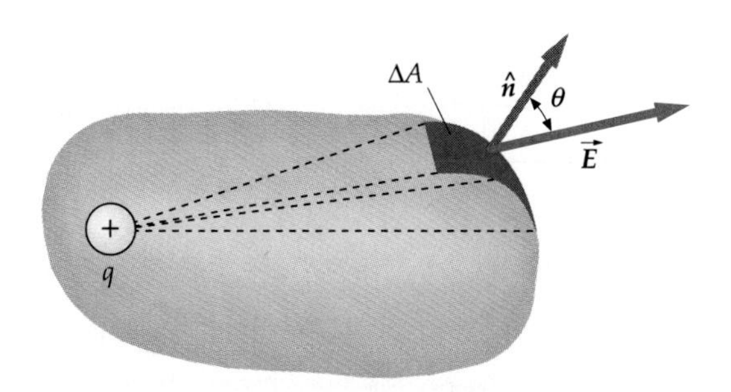

Figure 23-33 A point charge enclosed by an arbitrary surface S. The flux through an area element ΔA is proportional to the solid angle subtended by the area element at the charge. The net flux through the surface, found by summing over all the area elements, is proportional to the total solid angle 4π at the charge, which is independent of the shape of the surface.

Summary

1. Gauss's law is a fundamental law of physics that is equivalent to Coulomb's law for static charges.
2. For highly symmetric charge distributions, Gauss's law can be used to calculate the electric field.

Topic	Remarks and Relevant Equations	
Definitions and General Relations		
1. Electric field for a continuous charge distribution	$\vec{E} = \int_V \frac{k\,dq}{r^2}\,\hat{r}$ (Coulomb's law) where $dq = \rho\,dV$ for a charge distributed throughout a volume, $dq = \sigma\,dA$ for a charge distributed on a surface, and $dq = \lambda\,dL$ for a charge distributed along a line.	23-4
2. Electric flux	$\phi = \lim_{\Delta A_i\to 0}\sum_i \hat{n}_i\,\Delta A_i = \int_S \vec{E}\cdot\hat{n}\,dA$	23-14
3. Gauss's law	$\phi_{\text{net}} = \oint_S E_n\,dA = 4\pi k Q_{\text{inside}}$ The net flux through a closed surface equals $4\pi k$ times the net charge within the surface.	23-17
4. Coulomb constant k and permittivity of free space ϵ_0	$k = \frac{1}{4\pi\epsilon_0} = 8.99\times10^9\ \text{N·m}^2/\text{C}^2$	23-18
	$\epsilon_0 = 8.85\times10^{-12}\ \text{C}^2/\text{N·m}^2$	
5. Coulomb's law and Gauss's law in terms of ϵ_0	$\vec{E} = \frac{1}{4\pi\epsilon_0}\frac{q}{r^2}\,\hat{r}$	23-19
	$\phi_{\text{net}} = \oint_S E_n\,dA = \frac{1}{\epsilon_0}Q_{\text{inside}}$	23-20

6. Discontinuity of E_n	At a surface carrying a surface charge density σ, the component of the electric field perpendicular to the surface is discontinuous by σ/ϵ_0: $$E_{n2} - E_{n1} = \frac{\sigma}{\epsilon_0}$$	23-24
7. Charge on a conductor	In electrostatic equilibrium, the net electric charge on a conductor resides on the surface of the conductor.	
8. $\vec{E}$ just outside a conductor	The electric field just outside the surface of a conductor is perpendicular to the surface and has the magnitude σ/ϵ_0, where σ is the local surface charge density at that point on the conductor: $$E_n = \frac{\sigma}{\epsilon_0}$$ The force per unit area exerted on the charge on the surface of a conductor by all the other charges is called the electrostatic stress.	23-25

Electric Fields for Various Charge Distributions

At the bisector of a finite line charge	$E_r = \frac{2k\lambda}{r}\sin\theta_0$	23-8
At the bisector of an infinite line charge	$E_r = \frac{1}{2\pi\epsilon_0}\frac{\lambda}{r} = 2k\frac{\lambda}{r}$	23-9
On the axis of a ring charge	$E_x = \frac{kQx}{(x^2 + a^2)^{3/2}}$	23-10
On the axis of a disk charge	$E_x = 2\pi k\sigma\left(1 - \frac{x}{\sqrt{x^2 + R^2}}\right)$	23-11
Of an infinite plane of charge	$E_n = \frac{\sigma}{2\epsilon_0} = 2\pi k\sigma$	23-12, 23-21
Of a spherical shell of charge	$E_r = \frac{1}{4\pi\epsilon_0}\frac{Q}{r^2}, \quad r > R$	23-22a
	$E_r = 0, \quad r < R$	23-22b
Of a solid sphere of charge	$E_r = \frac{1}{4\pi\epsilon_0}\frac{Q}{r^2}, \quad r \geq R$	23-23a
	$E_r = \frac{1}{4\pi\epsilon_0}\frac{Q}{R^3}r, \quad r \leq R$	23-23b

Problem-Solving Guide

1. Begin by drawing a neat diagram that includes the important features of the problem. When using Coulomb's law, your sketch should show dq, the unit vector $\hat{r}$ from dq to the field point P, and the field element $d\vec{E}$. Resolve $d\vec{E}$ into components and use symmetry when possible. When finding $\vec{E}$ using superposition, show the individual $\vec{E}$ vectors in your sketch, along with an appropriate coordinate system.
2. For symmetrical charge distributions, $\vec{E}$ is most easily found using Gauss's law. Show the Gaussian surface in your sketch.

Summary of Worked Examples

Type of Calculation	Procedure and Relevant Examples
1. Coulomb's Law	
Calculate $\vec{E}$ using Coulomb's law.	Find $d\vec{E}$ for a given charge element dq, then integrate over the entire charge distribution. **(A sample calculation appears in the text of Section 23-1.)**
2. Gauss's Law	
Find the charge enclosed by a surface, given $\vec{E}$ on the surface.	Calculate the flux through the surface, then use Gauss's law. **Example 23-4**
Calculate $\vec{E}$ using Gauss's law.	Choose an appropriate Gaussian surface, calculate ϕ_{net} (in terms of E), calculate Q_{inside}, then use Gauss's law to solve for E. **Examples 23-7, 23-8**
3. Find $\vec{E}$ due to two or more charge distributions.	Determine $\vec{E}$ for each distribution separately, then do a vector sum to find the total field. **Examples 23-2, 23-3, 23-5, 23-6**

Problems

Conceptual Problems

Problems from Optional and Exploring sections

In a few problems, you are given more data than you actually need; in a few other problems, you are required to supply data from your general knowledge, outside sources, or informed estimates.

- • Single-concept, single-step, relatively easy
- •• Intermediate-level, may require synthesis of concepts
- ••• Challenging, for advanced students

Calculate $\vec{E}$ from Coulomb's Law

1 • A uniform line charge of linear charge density $\lambda = 3.5$ nC/m extends from $x = 0$ to $x = 5$ m. (*a*) What is the total charge? Find the electric field on the x axis at (*b*) $x = 6$ m, (*c*) $x = 9$ m, and (*d*) $x = 250$ m. (*e*) Find the field at $x = 250$ m, using the approximation that the charge is a point charge at the origin, and compare your result with that for the exact calculation in part (*d*).

2 • Two infinite vertical planes of charge are parallel to each other and are separated by a distance $d = 4$ m. Find the electric field to the left of the planes, to the right of the planes, and between the planes (*a*) when each plane has a uniform surface charge density $\sigma = +3\ \mu\text{C/m}^2$ and (*b*) when the left plane has a uniform surface charge density $\sigma = +3\ \mu\text{C/m}^2$ and that of the right plane is $\sigma = -3\ \mu\text{C/m}^2$. Draw the electric field lines for each case.

3 • A 2.75-μC charge is uniformly distributed on a ring of radius 8.5 cm. Find the electric field on the axis at (*a*) 1.2 cm, (*b*) 3.6 cm, and (*c*) 4.0 m from the center of the ring. (*d*) Find the field at 4.0 m using the approximation that the ring is a point charge at the origin, and compare your results with that for part (*c*).

4 • A disk of radius 2.5 cm carries a uniform surface charge density of 3.6 $\mu\text{C/m}^2$. Using reasonable approximations, find the electric field on the axis at distances of (*a*) 0.01 cm, (*b*) 0.04 cm, (*c*) 5 m, and (*d*) 5 cm.

5 • For the disk charge of Problem 4, calculate exactly the electric field on the axis at distances of (*a*) 0.04 cm and (*b*) 5 m, and compare your results with those for parts (*b*) and (*c*) of Problem 4.

6 • A uniform line charge extends from $x = -2.5$ cm to $x = +2.5$ cm and has a linear charge density of $\lambda = 6.0$ nC/m. (*a*) Find the total charge. Find the electric field on the y axis at (*b*) $y = 4$ cm, (*c*) $y = 12$ cm, and (*d*) $y = 4.5$ m. (*e*) Find the field at $y = 4.5$ m, assuming the charge to be a point charge, and compare your result with that for part (*d*).

7 • A disk of radius a lies in the yz plane with its axis along the x axis and carries a uniform surface charge density σ. Find the value of x for which $E_x = \frac{1}{2}\sigma/2\epsilon_0$.

8 • A ring of radius a with its center at the origin and its axis along the x axis carries a total charge Q. Find E_x at (*a*) $x = 0.2a$, (*b*) $x = 0.5a$, (*c*) $x = 0.7a$, (*d*) $x = a$, and (*e*) $x = 2a$. (*f*) Use your results to plot E_x versus x for both positive and negative values of x.

9 • Repeat Problem 8 for a disk of uniform surface charge density σ.

10 •• A disk of radius 30 cm carries a uniform charge density σ. (*a*) Compare the approximation $E = 2\pi k\sigma$ with the exact expression (Equation 23-11) for the electric field on the axis of the disk by computing the fractional difference $\Delta E/E \approx x/\sqrt{x^2 + R^2}$ for the distances $x = 0.1$, $x = 0.2$, and $x = 3$ cm. (*b*) At what distance is the neglected term 1% of $2\pi k\sigma$?

11 •• Show that E_x on the axis of a ring charge of radius a has its maximum and minimum values at $x = +a/\sqrt{2}$ and $x = -a/\sqrt{2}$. Sketch E_x versus x for both positive and negative values of x.

12 •• A line charge of uniform linear charge density λ lies along the x axis from $x = 0$ to $x = a$. (*a*) Show that the x component of the electric field at a point on the y axis is given by

$$E_x = -\frac{k\lambda}{y} + \frac{k\lambda}{\sqrt{y^2 + a^2}}$$

(*b*) Show that if the line charge extends from $x = -b$ to $x = a$, the x component of the electric field at a point on the y axis is given by

$$E_x = \frac{k\lambda}{\sqrt{y^2 + a^2}} - \frac{k\lambda}{\sqrt{y^2 + b^2}}$$

13 •• (*a*) A finite line charge of uniform linear charge density λ lies on the x axis from $x = 0$ to $x = a$. Show that the y component of the electric field at a point on the y axis is given by

$$E = \frac{k\lambda}{y}\sin\theta_1 = \frac{k\lambda}{y}\frac{a}{\sqrt{y^2 + a^2}}$$

where θ_1 is the angle subtended by the line charge at the field point. (*b*) Show that if the line charge extends from $x = -b$ to $x = a$, the y component of the electric field at a point on the y axis is given by

$$E_y = \frac{k\lambda}{y}(\sin\theta_1 + \sin\theta_2)$$

where $\sin\theta_2 = b/\sqrt{y^2 + b^2}$.

14 •• A semicircular ring of radius R carries a uniform line charge of λ. Find the electric field at the center of the semicircle.

15 ••• A hemispherical thin shell of radius R carries a uniform surface charge σ. Find the electric field at the center of the hemispherical shell ($r = 0$).

16 ••• A line charge of linear charge density λ with the shape of a square of side L lies in the yz plane with its center at the origin. Find the electric field on the x axis at an arbitrary distance x, and compare your result to that for the field on the axis of a charged ring of radius $r = \frac{1}{2}L$ with its center at the origin and carrying the same total charge. (*Hint:* Use Equation 23-8 for the field due to each segment of the square).

Gauss's Law

17 •• True or false:

(*a*) Gauss's law holds only for symmetric charge distributions.
(*b*) The result that $E = 0$ inside a conductor can be derived from Gauss's law.

18 •• What information in addition to the total charge inside a surface is needed to use Gauss's law to find the electric field?

19 ••• Is the electric field E in Gauss's law only that part of the electric field due to the charge inside a surface, or is it the total electric field due to all charges both inside and outside the surface?

20 • Consider a uniform electric field $\vec{E} = 2\text{ kN/C}\,\hat{i}$. (*a*) What is the flux of this field through a square of side 10 cm in a plane parallel to the yz plane? (*b*) What is the flux through the same square if the normal to its plane makes a 30° angle with the x axis?

21 • A single point charge $q = +2\ \mu\text{C}$ is at the origin. A spherical surface of radius 3.0 m has its center on the x axis at $x = 5$ m. (*a*) Sketch electric field lines for the point charge. Do any lines enter the spherical surface? (*b*) What is the net number of lines that cross the spherical surface, counting those that enter as negative? (*c*) What is the net flux of the electric field due to the point charge through the spherical surface?

22 • An electric field is $\vec{E} = 300\text{ N/C}\,\hat{i}$ for $x > 0$ and $\vec{E} = -300\text{ N/C}\,\hat{i}$ for $x < 0$. A cylinder of length 20 cm and radius 4 cm has its center at the origin and its axis along the x axis such that one end is at $x = +10$ cm and the other is at $x = -10$ cm. (*a*) What is the flux through each end? (*b*) What is the flux through the curved surface of the cylinder? (*c*) What is the net outward flux through the entire cylindrical surface? (*d*) What is the net charge inside the cylinder?

23 • A positive point charge q is at the center of a cube of side L. A large number N of electric field lines are drawn from the point charge. (*a*) How many of the field lines pass through the surface of the cube? (*b*) How many lines pass through each face, assuming that none pass through the edges or corners? (*c*) What is the net outward flux of the electric field through the cubic surface? (*d*) Use symmetry arguments to find the flux of the electric field through one face of the cube. (*e*) Which, if any, of your answers would change if the charge were inside the cube but not at its center?

24 • Careful measurement of the electric field at the surface of a black box indicates that the net outward flux through the surface of the box is 6.0 kN·m^2/C. (*a*) What is the net charge inside the box? (*b*) If the net outward flux through the surface of the box were zero, could you conclude that there were no charges inside the box? Why or why not?

25 • A point charge $q = +2\ \mu\text{C}$ is at the center of a sphere of radius 0.5 m. (*a*) Find the surface area of the sphere. (*b*) Find the magnitude of the electric field at points on the surface of the sphere. (*c*) What is the flux of the electric field due to the point charge through the surface of the sphere? (*d*) Would your answer to part (*c*) change if the point charge were moved so that it was inside the sphere but not at its center? (*e*) What is the net flux through a cube of side 1 m that encloses the sphere?

26 • Since Newton's law of gravity and Coulomb's law have the same inverse-square dependence on distance, an expression analogous in form to Gauss's law can be found for gravity. The gravitational field $\vec{g}$ is the force per unit mass on a test mass m_0. Then for a point mass m at the origin, the gravitational field g at some position r is

$$\vec{g} = -\frac{Gm}{r^2}\hat{r}$$

Compute the flux of the gravitational field through a spherical surface of radius r centered at the origin, and show that the gravitational analog of Gauss's law is $\phi_{net} = -4\pi G m_{inside}$.

27 •• A charge of 2 μC is 20 cm above the center of a square of side length 40 cm. Find the flux through the square. (*Hint:* Don't integrate.)

28 •• In a particular region of the earth's atmosphere, the electric field above the earth's surface has been measured to be 150 N/C downward at an altitude of 250 m and 170 N/C downward at an altitude of 400 m. Calculate the volume charge density of the atmosphere, assuming it to be uniform between 250 and 400 m. (You may neglect the curvature of the earth. Why?)

Spherical Charge Distributions

29 •• Explain why the electric field increases with r rather than decreasing as $1/r^2$ as one moves out from the center inside a spherical charge distribution of constant volume charge density.

30 • A spherical shell of radius R_1 carries a total charge q_1 that is uniformly distributed on its surface. A second, larger spherical shell of radius R_2 that is concentric with the first carries a charge q_2 that is uniformly distributed on its surface. (*a*) Use Gauss's law to find the electric field in the regions $r < R_1$, $R_1 < r < R_2$, and $r > R_2$. (*b*) What should the ratio of the charges q_1/q_2 and their relative signs be for the electric field to be zero for $r > R_2$? (*c*) Sketch the electric field lines for the situation in part (*b*) when q_1 is positive.

31 • A spherical shell of radius 6 cm carries a uniform surface charge density $\sigma = 9\ \text{nC/m}^2$. (*a*) What is the total charge on the shell? Find the electric field at (*b*) $r = 2$ cm, (*c*) $r = 5.9$ cm, (*d*) $r = 6.1$ cm, and (*e*) $r = 10$ cm.

32 •• A sphere of radius 6 cm carries a uniform volume charge density $\rho = 450\ \text{nC/m}^3$. (*a*) What is the total charge of the sphere? Find the electric field at (*b*) $r = 2$ cm, (*c*) $r = 5.9$ cm, (*d*) $r = 6.1$ cm, and (*e*) $r = 10$ cm. Compare your answers with Problem 31.

33 •• Consider two concentric conducting spheres (Figure 23-34). The outer sphere is hollow and initially has a charge $-7Q$ deposited on it. The inner sphere is solid and has a charge $+2Q$ on it. (*a*) How is the charge distributed on the outer sphere? That is, how much charge is on the outer surface and how much charge is on the inner surface? (*b*) Suppose a wire is connected between the inner and outer spheres. After electrostatic equilibrium is established, how much total charge is on the outside sphere? How much charge is on the outer surface of the outside sphere and how much is on the inner surface? Does the electric field at the surface of the inside sphere change when the wire is connected? If so, how? (*c*) Suppose we return to the original conditions in (*a*), with $+2Q$ on the inner sphere and $-7Q$ on the outer. We now connect the outer sphere to ground with a wire and then disconnect it. How much total charge will be on the outer sphere? How much charge will be on the inner surface of the outer sphere and how much will be on the outer surface?

Figure 23-34 Problem 33

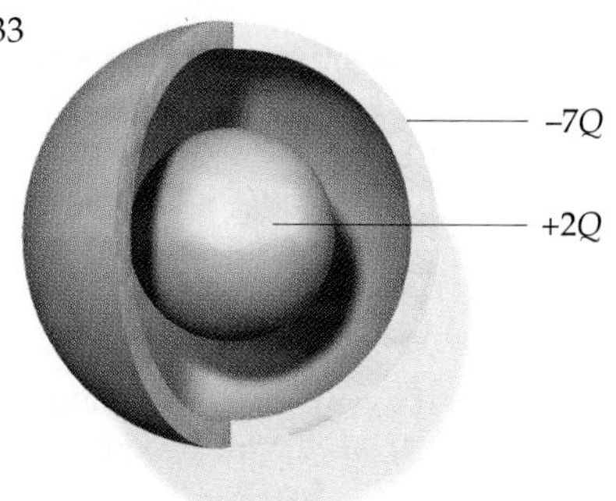

34 •• A nonconducting sphere of radius $R = 0.1$ m carries a uniform volume charge of charge density $\rho = 2.0\ \text{nC/m}^3$. The magnitude of the electric field at $r = 2R$ is 1883 N/C. Find the magnitude of the electric field at $r = 0.5R$.

35 •• A nonconducting sphere of radius R carries a volume charge density that is proportional to the distance from the center: $\rho = Ar$ for $r \leq R$, where A is a constant; $\rho = 0$ for $r > R$. (*a*) Find the total charge on the sphere by summing the charges on shells of thickness dr and volume $4\pi r^2\,dr$. (*b*) Find the electric field E_r both inside and outside the charge distribution, and sketch E_r versus r.

36 •• Repeat Problem 35 for a sphere with volume charge density $\rho = B/r$ for $r < R$; $\rho = 0$ for $r > R$.

37 •• Repeat Problem 35 for a sphere with volume charge density $\rho = C/r^2$ for $r < R$; $\rho = 0$ for $r > R$.

38 ••• The charge density in a region of space is spherically symmetric and is given by $\rho(r) = Ce^{-r/a}$ when $r < R$ and $\rho = 0$ when $r > R$. Find the electric field as a function of r.

39 ••• A thick, nonconducting spherical shell of inner radius a and outer radius b has a uniform volume charge density ρ. Find (*a*) the total charge and (*b*) the electric field everywhere.

40 ••• A point charge of +5 nC is located at the origin. This charge is surrounded by a spherically symmetric negative charge distribution with volume density $\rho(r) = Ce^{-r/a}$. (*a*) Find the constant C in terms of a if the total charge of the system is zero? (*b*) What is the electric field at $r = a$?

41 ••• A nonconducting solid sphere of radius a with its center at the origin has a spherical cavity of radius b with its center at the point $x = b$, $y = 0$ as shown in Figure 23-35. The sphere has a uniform volume charge density ρ. Show that the electric field in the cavity is uniform and is given by $E_y = 0$, $E_x = \rho b/3\epsilon_0$. (*Hint:* Replace the cavity with spheres of equal positive and negative charge densities.)

Figure 23-35 Problem 41

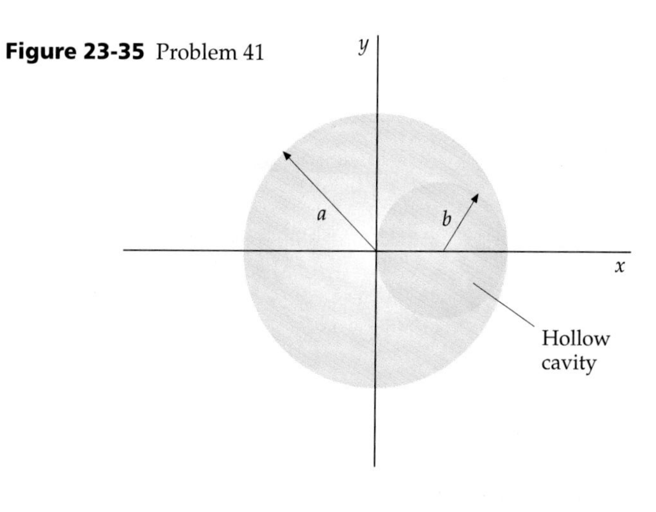

Cylindrical Charge Distributions

42 •• Show that the electric field due to an infinitely long, uniformly charged cylindrical shell of radius R carrying a surface charge density σ is given by

$$E_r = 0, \qquad r < R \qquad \text{23-28}a$$

$$E_r = \frac{\sigma R}{\epsilon_0 r} = \frac{\lambda}{2\pi\epsilon_0 r}, \qquad r > R \qquad \text{23-28}b$$

where $\lambda = 2\pi R\sigma$ is the charge per unit length on the shell.

43 •• A cylindrical shell of length 200 m and radius 6 cm carries a uniform surface charge density of $\sigma = 9\ \text{nC/m}^2$. (*a*) What is the total charge on the shell? Find the electric field at (*b*) $r = 2$ cm, (*c*) $r = 5.9$ cm, (*d*) $r = 6.1$ cm, and (*e*) $r = 10$ cm. (Use the results of Problem 42.)

44 •• An infinitely long nonconducting cylinder of radius R carries a uniform volume charge density of $\rho(r) = \rho_0$. Show that the electric field is given by

$$E_r = \frac{\rho R^2}{2\epsilon_0 r} = \frac{1}{2\pi\epsilon_0}\frac{\lambda}{r}, \qquad r > R \qquad \text{23-29}a$$

$$E_r = \frac{\rho}{2\epsilon_0} r = \frac{\lambda}{2\pi\epsilon_0 R^2} r, \qquad r < R \qquad \text{23-29}b$$

where $\lambda = \rho\pi R^2$ is the charge per unit length.

45 •• A cylinder of length 200 m and radius 6 cm carries a uniform volume charge density of $\rho = 300\ \text{nC/m}^3$. (*a*) What is the total charge of the cylinder? Use the formulas given in Problem 44 to calculate the electric field at a point equidistant from the ends at (*b*) $r = 2$ cm, (*c*) $r = 5.9$ cm, (*d*) $r = 6.1$ cm, and (*e*) $r = 10$ cm. Compare your results with those in Problem 43.

46 •• Consider two infinitely long, concentric cylindrical shells. The inner shell has a radius R_1 and carries a uniform surface charge density of σ_1, and the outer shell has a radius R_2 and carries a uniform surface charge density of σ_2. (*a*) Use Gauss's law to find the electric field in the regions $r < R_1$, $R_1 < r < R_2$, and $r > R_2$. (*b*) What is the ratio of the surface charge densities σ_2/σ_1 and their relative signs if the electric field is zero at $r > R_2$? What would the electric field between the shells be in this case? (*c*) Sketch the electric field lines for the situation in (*b*) if σ_1 is positive.

47 •• Figure 23-36 shows a portion of an infinitely long, concentric cable in cross section. The inner conductor carries a charge of 6 nC/m; the outer conductor is uncharged. (*a*) Find the electric field for all values of r, where r is the distance from the axis of the cylindrical system. (*b*) What are the surface charge densities on the inside and the outside surfaces of the outer conductor?

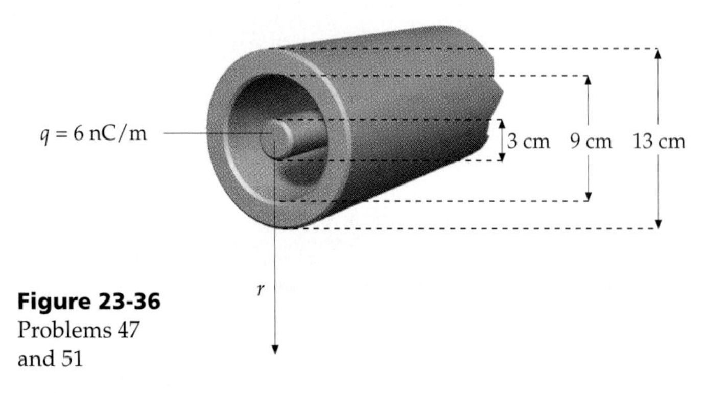

Figure 23-36
Problems 47 and 51

48 •• Repeat Problem 44 for a cylinder with volume charge density (a) $\rho(r) = ar$ and (b) $\rho = Cr^2$.

49 •• Repeat Problem 44 with $\rho = C/r$.

50 ••• An infinitely long, thick, nonconducting cylindrical shell of inner radius a and outer radius b has a uniform volume charge density ρ. Find the electric field everywhere.

51 ••• Suppose that the inner cylinder of Figure 23-36 is made of nonconducting material and carries a volume charge distribution given by $\rho(r) = C/r$, where $C = 200\ \text{nC/m}^2$. The outer cylinder is metallic. (*a*) Find the charge per meter carried by the inner cylinder. (*b*) Calculate the electric field for all values of r.

Charge and Field at Conductor Surfaces

52 • A penny is in an external electric field of magnitude 1.6 kN/C directed perpendicular to its faces. (*a*) Find the charge density on each face of the penny, assuming the faces are planes. (*b*) If the radius of the penny is 1 cm, find the total charge on one face.

53 • An uncharged metal slab has square faces with 12-cm sides. It is placed in an external electric field that is perpendicular to its faces. The total charge induced on one of the faces is 1.2 nC. What is the magnitude of the electric field?

54 • A charge of 6 nC is placed uniformly on a square sheet of nonconducting material of side 20 cm in the yz plane. (*a*) What is the surface charge density σ? (*b*) What is the magnitude of the electric field just to the right and just to the left of the sheet? (*c*) The same charge is placed on a square conducting slab of side 20 cm and thickness 1 mm. What is the surface charge density σ? (Assume that the charge distributes itself uniformly on the large square surfaces.) (*d*) What is the magnitude of the electric field just to the right and just to the left of each face of the slab?

55 • A spherical conducting shell with zero net charge has an inner radius a and an outer radius b. A point charge q is placed at the center of the shell. (*a*) Use Gauss's law and the properties of conductors in equilibrium to find the electric field in the regions $r < a$, $a < r < b$, and $b < r$. (*b*) Draw the electric field lines for this situation. (*c*) Find the charge density on the inner surface ($r = a$) and on the outer surface ($r = b$) of the shell.

56 •• The electric field just above the surface of the earth has been measured to be 150 N/C downward. What total charge on the earth is implied by this measurement?

57 •• A positive point charge of magnitude 2.5 μC is at the center of an uncharged spherical conducting shell of inner radius 60 cm and outer radius 90 cm. (*a*) Find the charge densities on the inner and outer surfaces of the shell and the total charge on each surface. (*b*) Find the electric field everywhere. (*c*) Repeat (*a*) and (*b*) with a net charge of +3.5 μC placed on the shell.

58 •• If the magnitude of an electric field in air is as great as 3×10^6 N/C, the air becomes ionized and begins to conduct electricity. This phenomenon is called dieletric breakdown. A charge of 18 μC is to be placed on a conducting sphere. What is the minimum radius of a sphere that can hold this charge without breakdown?

59 •• A square conducting slab with 5-m sides carries a net charge of 80 μC. (*a*) Find the charge density on each face of the slab and the electric field just outside one face of the slab. (*b*) The slab is placed to the right of an infinite charged nonconducting plane with charge density 2.0 $\mu C/m^2$ so that the faces of the slab are parallel to the plane. Find the electric field on each side of the slab far from its edges and the charge density on each face.

60 •• Imagine that a small hole has been punched through the wall of a thin, uniformly charged spherical shell whose surface charge density is σ. Find the electric field near the center of the hole.

General Problems

61 • True or false:

(*a*) If there is no charge in a region of space, the electric field on a surface surrounding the region must be zero everywhere.
(*b*) The electric field inside a uniformly charged spherical shell is zero.
(*c*) In electrostatic equilibrium, the electric field inside a conductor is zero.
(*d*) If the net charge on a conductor is zero, the charge density must be zero at every point on the surface of the conductor.

62 • If the electric field E is zero everywhere on a closed surface, is the net flux through the surface necessarily zero? What, then, is the net charge inside the surface?

63 • A point charge $-Q$ is at the center of a spherical conducting shell of inner radius R_1 and outer radius R_2 as shown in Figure 23-37. The charge on the inner surface of the shell is

(*a*) $+Q$.
(*b*) zero.
(*c*) $-Q$.
(*d*) dependent on the total charge carried by the shell.

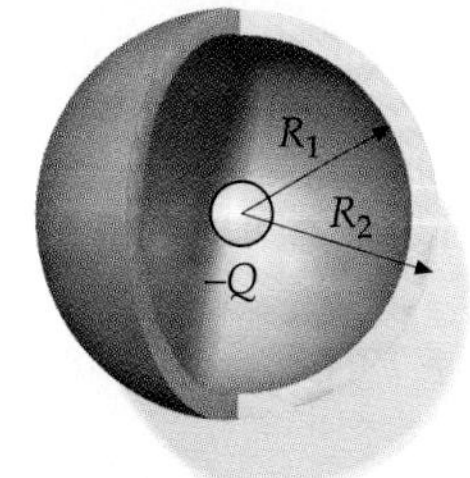

Figure 23-37 Problems 63–67

64 • For the configuration of Figure 23-37, the charge on the outer surface of the shell is

(*a*) $+Q$.
(*b*) zero.
(*c*) $-Q$.
(*d*) dependent on the total charge carried by the shell.

65 •• Suppose that the total charge on the conducting shell of Figure 23-37 is zero. It follows that the electric field for $r < R_1$ and $r > R_2$ points

(*a*) away from the center of the shell in both regions.
(*b*) toward the center of the shell in both regions.
(*c*) toward the center of the shell for $r < R_1$ and is zero for $r > R_2$.
(*d*) away from the center of the shell for $r < R_1$ and is zero for $> R_2$.

66 •• If the conducting shell in Figure 23-37 is grounded, which of the following statements is then correct?

(*a*) The charge on the inner surface of the shell is $+Q$ and that on the outer surface is $-Q$.
(*b*) The charge on the inner surface of the shell is $+Q$ and that on the outer surface is zero.
(*c*) The charge on both surfaces of the shell is $+Q$.
(*d*) The charge on both surfaces of the shell is zero.

67 •• For the configuration described in Problem 66, in which the conducting shell is grounded, the electric field for $r < R_1$ and $r > R_2$ points

(*a*) away from the center of the shell in both regions.
(*b*) toward the center of the shell in both regions.
(*c*) toward the center of the shell for $r < R_1$ and is zero for $r > R_2$.
(*d*) toward the center of the shell for $r < R_1$ and is zero for $r > R_1$.

68 •• If the net flux through a closed surface is zero, does it follow that the electric field E is zero everywhere on the surface? Does it follow that the net charge inside the surface is zero?

69 •• Equation 23-8 for the electric field on the perpendicular bisector of a finite line charge is different from Equation 23-9 for the electric field near an infinite line charge, yet Gauss's law would seem to give the same result for these two cases. Explain.

70 •• True or false: The electric field is discontinuous at all points at which the charge density is discontinuous.

71 •• Consider the three concentric metal spheres shown in Figure 23-38. Sphere I is solid, with radius R_1. Sphere II is hollow, with inner radius R_2 and outer radius R_3. Sphere III is hollow, with inner radius R_4 and outer radius R_5. Initially, all three spheres have zero excess charge. Then a negative charge $-Q_0$ is placed on sphere I and a positive charge $+Q_0$ is placed on sphere III. (*a*) After the charges have reached equilibrium, will the electric field in the space between spheres I and II point *toward* the center, *away* from the center, or neither? (*b*) How much charge will be on the inner surface of sphere II? Give the correct sign. (*c*) How much charge will be on the outer surface of sphere II? (*d*) How much charge will be on the inner surface of sphere III? (*e*) How much charge will be on the outer surface of sphere III? (*g*) Plot E versus r.

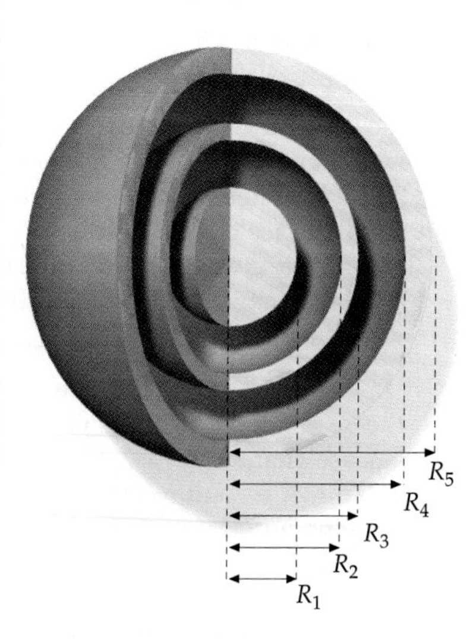

Figure 23-38 Problem 71

72 •• An early model of the hydrogen atom considered the atom to consist of a proton, which is a uniform charged sphere of radius R, with an electron in an orbit of radius r_0 *inside* the proton as shown in Figure 23-39. (*a*) Use Gauss's law to obtain the magnitude of E (the field due to the proton) at the position of the electron. Give your answer in terms of e (the charge on a proton), r_0, and R. (*b*) Find the frequency of revolution f in terms of r_0 and the velocity of the electron v. (*c*) What is the force on the electron in terms of m,v and r_0? (*d*) What is the frequency f in terms of m, e, R, ϵ_0, and r_0? (Each of your answers need not include all of the specified quantities.)

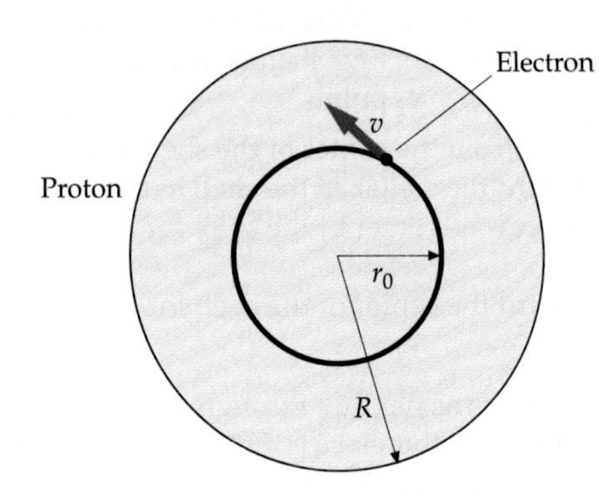

Figure 23-39 Problem 72

73 •• A nonuniform surface charge lies in the yz plane. At the origin, the surface charge density is $\sigma = 3.10\ \mu\text{C/m}^2$. Other charged objects are present as well. Just to the right of the origin, the x component of the electric field is $E_x = 4.65 \times 10^5$ N/C. What is E_x just to the left of the origin?

74 •• An infinite line charge of uniform linear charge density $\lambda = -1.5\ \mu\text{C/m}$ lies parallel to the y axis at $x = -2$ m. A point charge of 1.3 μC is located at $x = 1$ m, $y = 2$ m. Find the electric field at $x = 2$ m, $y = 1.5$ m.

75 •• Two infinite planes of charge lie parallel to each other and to the yz plane. One is at $x = -2$ m and has a surface charge density of $\sigma = -3.5\ \mu\text{C/m}^2$. The other is at $x = 2$ m and has a surface charge density of $\sigma = 6.0\ \mu\text{C/m}^2$. Find the electric field for (*a*) $x < -2$ m; (*b*) $-2 \text{ m} < x < 2$ m; and (*c*) $x > 2$ m.

76 •• An infinitely long cylindrical shell is coaxial with the y axis and has a radius of 15 cm. It carries a uniform surface charge density $\sigma = 6\ \mu\text{C/m}^2$. A spherical shell of radius 25 cm is centered on the x axis at $x = 50$ cm and carries a uniform surface charge density $\sigma = -12\ \mu\text{C/m}^2$. Calculate the magnitude and direction of the electric field at (*a*) the origin; (*b*) $x = 20$ cm, $y = 10$ cm; and (*c*) $x = 50$ cm, $y = 20$ cm. (See Problem 42.)

77 •• An infinite plane in the xz plane carries a uniform surface charge density $\sigma_1 = 65\ \text{nC/m}^2$. A second infinite plane carrying a uniform charge density $\sigma_2 = 45\ \text{nC/m}^2$ intersects the xz plane at the z axis and makes an angle of 30° with the xz plane as shown in Figure 23-40. Find the electric field in the xy plane at (*a*) $x = 6$ m, $y = 2$ m and (*b*) $x = 6$ m, $y = 5$ m.

Figure 23-40 Problem 77

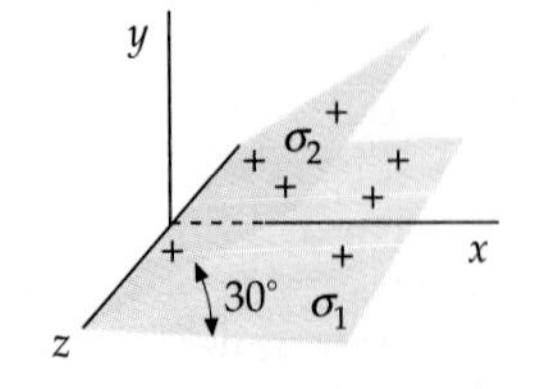

78 •• A ring of radius R carries a uniform, positive, linear charge density λ. Figure 23-41 shows a point P in the plane of the ring but not at the center. Consider the two elements of the ring of lengths s_1 and s_2 shown in the figure at distances r_1 and r_2, from point P. (*a*) What is the ratio of the charges of these elements? Which produces the greater field at point P? (*b*) What is the direction of the field at point P due to each element? What is the direction of the total electric field at point P? (*c*) Suppose that the electric field due to a point charge varied as $1/r$ rather than $1/r^2$. What would the electric field be at point P due to the elements shown? (*d*) How would your answers to parts (*a*), (*b*), and (*c*) differ if point P were inside a spherical shell of uniform charge and the elements were of areas s_1 and s_2?

Figure 23-41 Problem 78

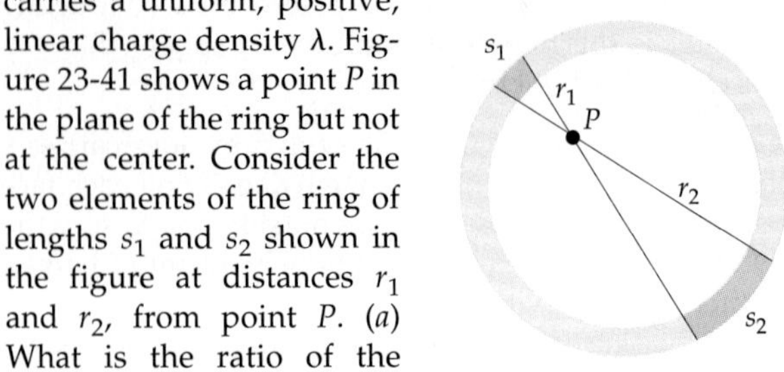

79 •• A ring of radius R that lies in the horizontal (xy) plane carries a charge Q uniformly distributed over its length. A mass m carries a charge q whose sign is opposite that of Q. (*a*) What is the minimum value of $|q|/m$ such that the mass will be in equilibrium under the action of gravity and the electrostatic force on the charge q? (*b*) If $|q|/m$ is twice what is calculated in (*a*), where will the mass be when it is in equilibrium?

80 •• A long, thin, nonconducting plastic rod is bent into a loop with radius R. Between the ends of the rods, a small gap of length l ($l \ll R$) remains. A charge Q is equally distributed on the rod. (*a*) Indicate the direction of the electric field at the center of the loop. (*b*) Find the magnitude of the electric field at the center of the loop.

81 •• A rod of length L lies perpendicular to an infinitely long, uniform line charge of charge density λ C/m (Figure 23-42). The near end of the rod is a distance d above the line charge. The rod carries a total charge Q uniformly distributed along its length. Find the force that the infinitely long line charge exerts on the rod.

Figure 23-42 Problem 81

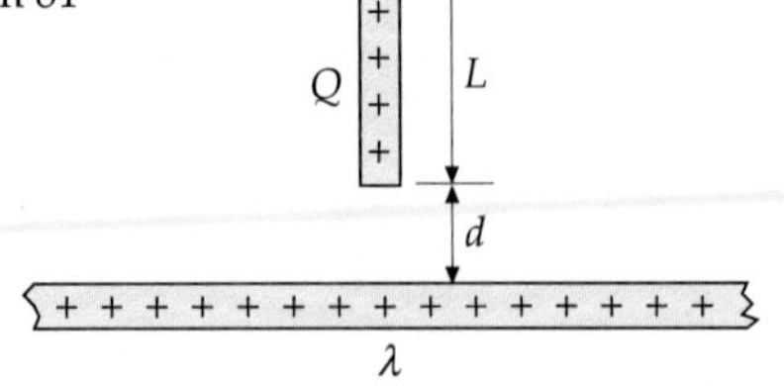

82 •• A nonconducting sphere 1.2 m in diameter with its center on the x axis at $x = 4$ m carries a uniform volume charge of density $\rho = 5\ \mu\text{C/m}^3$. Surrounding the sphere is a spherical shell with a diameter of 2.4 m and a uniform surface charge density $\sigma = -1.5\ \mu\text{C/m}^2$. Calculate the magnitude and direction of the electric field at (*a*) $x = 4.5$ m, $y = 0$; (*b*) $x = 4.0$ m, $y = 1.1$ m; and (*c*) $x = 2.0$ m, $y = 3.0$ m.

83 •• An infinite plane of charge with surface charge density $\sigma_1 = 3\ \mu\text{C/m}^2$ is parallel to the xz plane at $y = -0.6$

m. A second infinite plane of charge with surface charge density $\sigma_2 = -2\ \mu C/m^2$ is parallel to the yz plane at $x = 1$ m. A sphere of radius 1 m with its center in the xy plane at the intersection of the two charged planes ($x = 1$ m, $y = -0.6$ m) has a surface charge density $\sigma_3 = -3\ \mu C/m^2$. Find the magnitude and direction of the electric field on the x axis at (*a*) $x = 0.4$ m and (*b*) $x = 2.5$ m.

84 •• An infinite plane lies parallel to the yz plane at $x =$ 2 m and carries a uniform surface charge density $\sigma = 2\ \mu C/m^2$. An infinite line charge of uniform linear charge density $\lambda = 4\ \mu C/m$ passes through the origin at an angle of 45° with the x axis in the xy plane. A sphere of volume charge density $\rho = -6\ \mu C/m^3$ and radius 0.8 m is centered on the x axis at $x = 1$ m. Calculate the magnitude and direction of the electric field in the xy plane at $x = 1.5$ m, $y = 0.5$ m.

85 •• An infinite line charge λ is located along the z axis. A mass m that carries a charge q whose sign is opposite to that of λ is in a circular orbit in the xy plane about the line charge. Obtain an expression for the period of the orbit in terms of m, q, R, and λ, where R is the radius of the orbit.

86 •• A ring of radius R that lies in the yz plane carries a positive charge Q uniformly distributed over its length. A particle of mass m that carries a negative charge of magnitude q is at the center of the ring. (*a*) Show that if $x \ll R$, the electric field along the axis of the ring is proportional to x. (*b*) Find the force on the mass m as a function of x. (*c*) Show that if m is given a small displacement in the x direction, it will perform simple harmonic motion. Calculate the period of that motion.

87 •• When the charges Q and q of Problem 86 are 5 μC and $-5\ \mu$C, respectively, and the radius of the ring is 8.0 cm, the mass m oscillates about its equilibrium position with an angular frequency of 21 rad/s. Find the angular frequency of oscillation of the mass if the radius of the ring is doubled to 16 cm and all other parameters remain unchanged.

88 •• Given the initial conditions of Problem 87, find the angular frequency of oscillation of the mass if the radius of the ring is doubled to 16 cm while keeping the linear charge density on the ring constant.

89 •• A nonconducting cylinder of radius 1.2 m and length 2.0 m carries a charge of 50 μC uniformly distributed throughout the cylinder. Find the electric field *on the cylinder axis* at a distance of (*a*) 0.5 m, (*b*) 2.0 m, and (*c*) 20 m from the center of the cylinder.

90 •• A uniform line charge of density λ lies on the x axis between $x = 0$ and $x = L$. Its total charge is $Q = 8$ nC. The electric field at $x = 2L$ is 600 N/C $\hat{i}$. Find the electric field at $x = 3L$.

91 •• Find the linear charge density λ (in C/m) of the line charge of Problem 90.

92 ••• A uniformly charged sphere of radius R is centered at the origin with a charge of Q. Find the force on a uniformly charged line oriented radially having a total charge q with its ends at $r = R$ and $r = R + d$.

93 ••• Two equal uniform line charges of length L lie on the x axis a distance d apart as shown in Figure 23-43. (*a*) What is the force that one line charge exerts on the other line charge? (*b*) Show that when $d \gg L$, the force tends toward the expected result of $k(\lambda L)^2/d^2$.

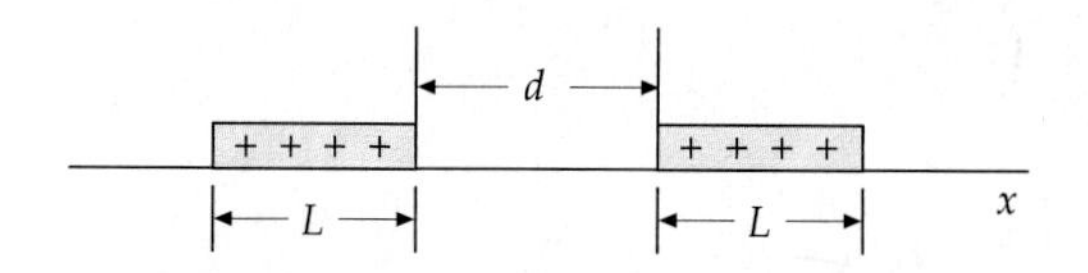

Figure 23-43 Problem 93

94 ••• A dipole $\vec{p}$ is located at a distance r from an infinitely long line charge with a uniform linear charge density λ. Assume that the dipole is aligned with the field due to the line charge. Determine the force that acts on the dipole.

95 ••• Suppose that the charge on the rod in Problem 81 is given by $\lambda(y) = ay^2$, where y is the distance from the midpoint of the rod, and that the total charge on the rod is Q. (*a*) Determine the constant a. (*b*) Find the force $d\vec{F}$ that acts on an element of charge $\lambda(y)\ dy$. (*c*) Integrate the force obtained in part (*b*) between $-L/2$ and $L/2$ to obtain the total force that acts on the rod.

96 ••• Repeat Problem 95 with the charge on the rod being $\lambda(y) = by$, where y is measured from the midpoint of the rod with the positive y direction up.

CHAPTER 24

Electric Potential

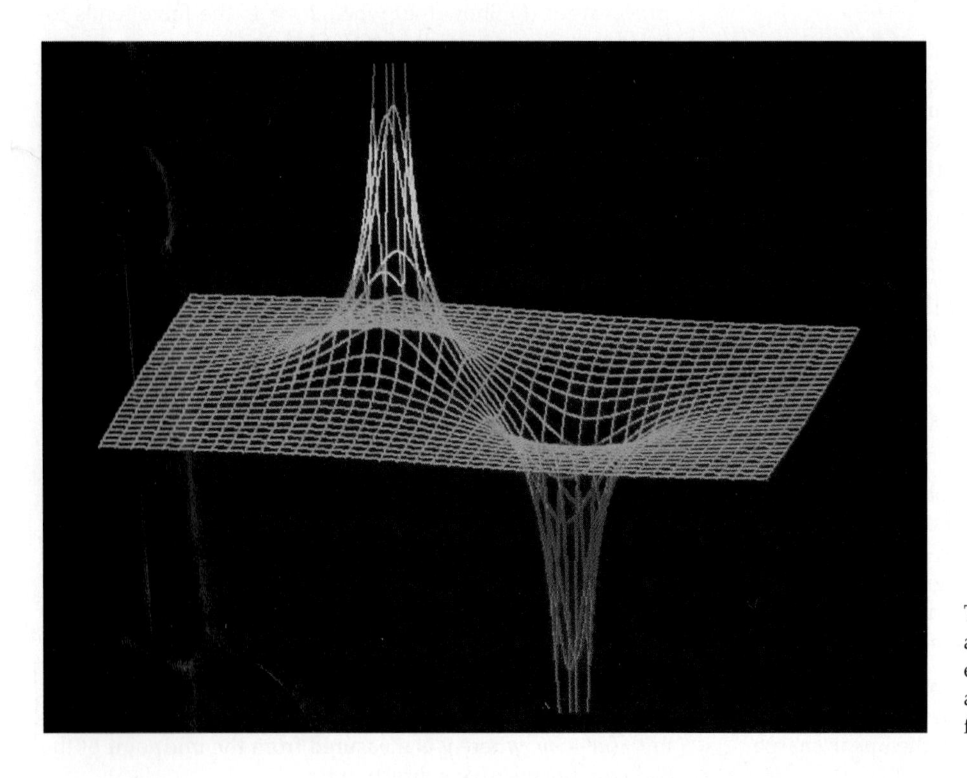

The electrostatic potential in the plane of an electric dipole. The potential due to each charge is proportional to the charge and inversely proportional to the distance from the charge.

The electric force between two charges is directed along the line of the charges and depends on the inverse square of their separation, the same as the gravitational force between two masses. Like the gravitational force, the electric force is conservative, so there is a potential-energy function U associated with it. If we place a test charge q_0 in an electric field, its potential energy is proportional to q_0. The potential energy per unit charge is a function of the position in space of the charge and is called the electric potential.

24-1 Potential Difference

In general, when a conservative force $\vec{F}$ undergoes a displacement $d\vec{\ell}$, the change in the potential-energy function dU is given by (Equation 6-21*b*)

$$dU = -\vec{F}\cdot d\vec{\ell}$$

The force exerted by an electric field $\vec{E}$ on a point charge q_0 is

$$\vec{F} = q_0\vec{E}$$

Thus, when a charge undergoes a displacement $d\vec{\ell}$ in an electric field $\vec{E}$, the change in the electrostatic potential energy is

$$dU = -q_0\vec{E}\cdot d\vec{\ell} \qquad 24\text{-}1$$

The potential-energy change is proportional to the charge q_0. The potential-energy change *per unit charge* is called the **potential difference** dV:

$$dV = \frac{dU}{q_0} = -\vec{E}\cdot d\vec{\ell} \qquad 24\text{-}2a$$

Definition—Potential difference

For a finite displacement from point a to point b, the change in potential is

$$\Delta V = V_b - V_a = \frac{\Delta U}{q_0} = -\int_a^b \vec{E}\cdot d\vec{\ell} \qquad 24\text{-}2b$$

Definition—Finite potential difference

The potential difference $V_b - V_a$ is the negative of the work done per unit charge by the electric field on a positive test charge when it moves from point a to point b. ΔV is also the positive work per charge that *you* must do against the electric field to move the charge from a to b.

The function V is called the **electric potential** or often just the **potential**. Like the electric field, the potential V is a function of position. Unlike the electric field, V is a scalar function, whereas $\vec{E}$ is a vector function. As with potential energy U, only *changes* in the potential V are important. We are free to choose the potential to be zero at any convenient point, just as we are when dealing with potential energy. If the electric potential and potential energy of a test charge are chosen to be zero at the same point, they are related by

$$U = q_0V \qquad 24\text{-}3$$

Relation between potential energy and potential

Continuity of *V*

In Chapter 23, we saw that the electric field is discontinuous by σ/ϵ_0 at a point where there is a surface charge density σ. The potential function, on the other hand, is continuous everywhere in space. We can see this from its definition. Consider an electric field in the x direction $\vec{E} = E_x\hat{i}$. The change in potential is given by Equation 24-2a.

$$\begin{aligned} dV &= -\vec{E}\cdot d\vec{\ell} = -E_x\hat{i}\cdot(dx\,\hat{i} + dy\,\hat{j} + dz\,\hat{k}) \\ &= -E_x\,dx \end{aligned}$$

Consider two nearby points x_1 and x_2. If V_1 is the potential at x_1 and V_2 is the potential at x_2, the potential difference can be written

$$\Delta V = (E_x)_{\text{av}}\,\Delta x = (E_x)_{\text{av}}\,(x_2 - x_1)$$

where $(E_x)_{\text{av}}$ is the average value of the electric field between the points. As x_2 approaches x_1, the potential difference ΔV approaches zero as long as $(E_x)_{\text{av}}$ is not infinite. The potential function V is thus continuous at any point not occupied by a point charge. Physically, if a test charge is moved a distance Δx, the work done by the field approaches zero as Δx approaches zero, as long as the electric field is not infinite.

Units

Since electric potential is the potential energy per unit charge, the SI unit for potential and potential difference is the joule per coulomb, called the **volt** (V):

$$1\ \text{V} = 1\ \text{J/C} \qquad \text{24-4}$$

The potential difference between two points (measured in volts) is sometimes called the **voltage**. In a 12-V car battery, the positive terminal has a potential 12 V higher than the negative terminal. If we attach an external circuit to the battery and one coulomb of charge is transferred from the positive terminal through the circuit to the negative terminal, the potential energy of the charge decreases by $Q\,\Delta V = (1\ \text{C})(12\ \text{V}) = 12\ \text{J}$.

We can see from Equation 24-2 that the dimensions of potential are also those of electric field times distance. Thus, the unit of the electric field is equal to one volt per meter:

$$1\ \text{N/C} = 1\ \text{V/m} \qquad \text{24-5}$$

In atomic and nuclear physics, we often have elementary particles with charges of magnitude e, such as electrons and protons, moving through potential differences of several to thousands or even millions of volts. Since energy has dimensions of electric charge times electric potential, a convenient unit of energy is the product of the electron charge e times a volt. This unit is called an **electron volt** (eV). The conversion between electron volts and joules is obtained by expressing the electronic charge in coulombs:

$$1\ \text{eV} = 1.6 \times 10^{-19}\ \text{C}\cdot\text{V} = 1.6 \times 10^{-19}\ \text{J} \qquad \text{24-6}$$

The electron volt

For example, an electron moving from the negative terminal to the positive terminal of a 12-V car battery gains potential energy of 12 eV.

Potential and Electric Field Lines

If we place a positive test charge q_0 in an electric field $\vec{E}$ and release it, it accelerates in the direction of $\vec{E}$. As the kinetic energy of the charge increases, its potential energy decreases. The charge therefore moves toward a region of lower potential energy, just as a mass falls toward a region of lower gravitational potential energy (Figure 24-1). Thus, as illustrated in Figure 24-2,

> Electric field lines point in the direction of decreasing electric potential.

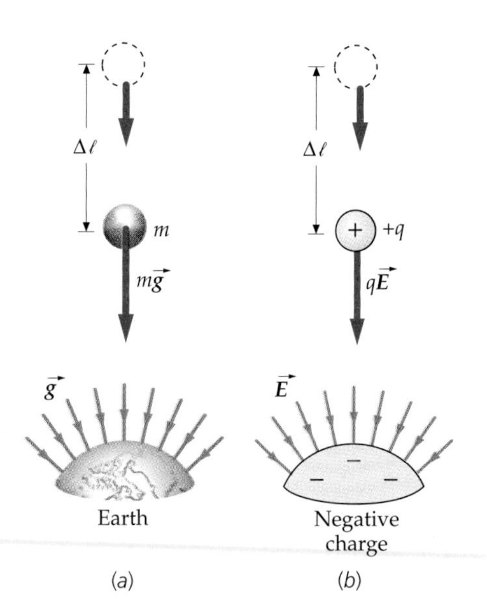

Figure 24-1 (*a*) The work done by the gravitational field on a mass decreases the gravitational potential energy. (*b*) The work done by the electric field on a positive charge $+q$ decreases the electrostatic potential energy.

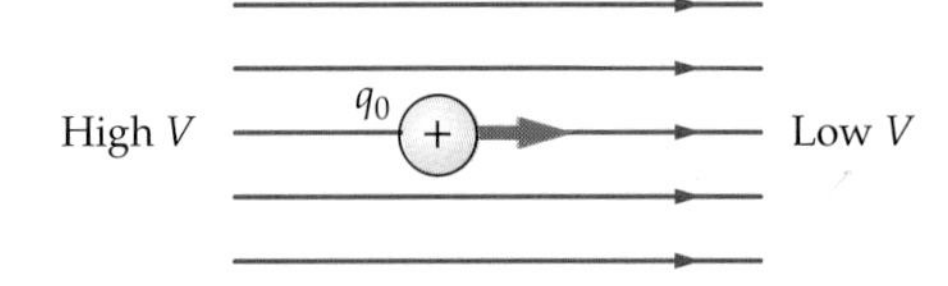

Figure 24-2 Electric field lines point in the direction of decreasing potential. When a positive test charge q_0 is placed in an electric field, it accelerates in the direction of the field. Its kinetic energy increases and its potential energy decreases.

Example 24-1

An electric field points in the positive x direction and has a constant magnitude of 10 N/C = 10 V/m. Find the potential as a function of x, assuming that $V = 0$ at $x = 0$.

1. By definition, the change in potential dV is related to the displacement $d\vec{\ell}$ and the electric field $\vec{E}$:

$$dV = -\vec{E}\cdot d\vec{\ell} = -(10\text{ V/m})\hat{i}\cdot(dx\,\hat{i} + dy\,\hat{j} + dz\,\hat{k})$$
$$= -(10\text{ V/m})\,dx$$

2. Integrate dV:

$$V = \int dV = \int -(10\text{ V/m})\,dx = -(10\text{ V/m})x + V_0$$

3. The constant of integration V_0 is found by setting $V = 0$ at $x = 0$:

$$V(0) = V_0 = 0$$

4. The potential is then:

$$V = -(10\text{ V/m})\,x$$

Remark The potential is zero at $x = 0$, and decreases by 10 V/m in the positive x direction.

Exercise Repeat this example for the electric field $\vec{E} = (10\text{ V/m}^2)x\hat{i}$. [*Answer* $V(x) = -(5\text{ V/m}^2)x^2$]

24-2 Potential due to a System of Point Charges

The electric potential at a distance r from a point charge q at the origin can be calculated from the electric field:

$$\vec{E} = \frac{kq}{r^2}\hat{r}$$

For an infinitesimal radial displacement $d\vec{\ell} = dr\,\hat{r}$, the change in potential is

$$dV = -\vec{E}\cdot d\vec{\ell} = -\frac{kq}{r^2}\hat{r}\cdot dr\hat{r} = -\frac{kq}{r^2}dr$$

Integrating, we obtain

$$V = +\frac{kq}{r} + V_0 \qquad \text{24-7}$$

Potential due to a point charge

where V_0 is a constant of integration.

If we define the potential to be zero at an infinite distance from the point charge ($r = \infty$), the constant V_0 is zero, and the potential at a distance r from the point charge is

$$V = \frac{kq}{r}, \qquad V = 0 \text{ at } r = \infty \qquad \text{24-8}$$

Coulomb potential

The potential given by Equation 24-8 is called the **Coulomb potential**. It is positive or negative depending on the sign of the charge q.

The potential energy U of a test charge q_0 placed a distance r from the point charge q is

$$U = q_0 V = \frac{kq_0 q}{r}, \qquad U = 0 \text{ at } r = \infty \tag{24-9}$$

Electrostatic potential energy of a two-charge system

This is the electrostatic potential energy of the two-charge system relative to $U = 0$ at infinite separation. If we release a test charge q_0 from rest at a distance r from q (and hold q fixed at the origin), the test charge will be accelerated outward (assuming that q has the same sign as q_0). Its kinetic energy at a great distance from q will be kq_0q/r. Alternatively, the work we must do against the electric field to bring a test charge q_0 from a great distance to a distance r from q is kq_0q/r (Figure 24-3).

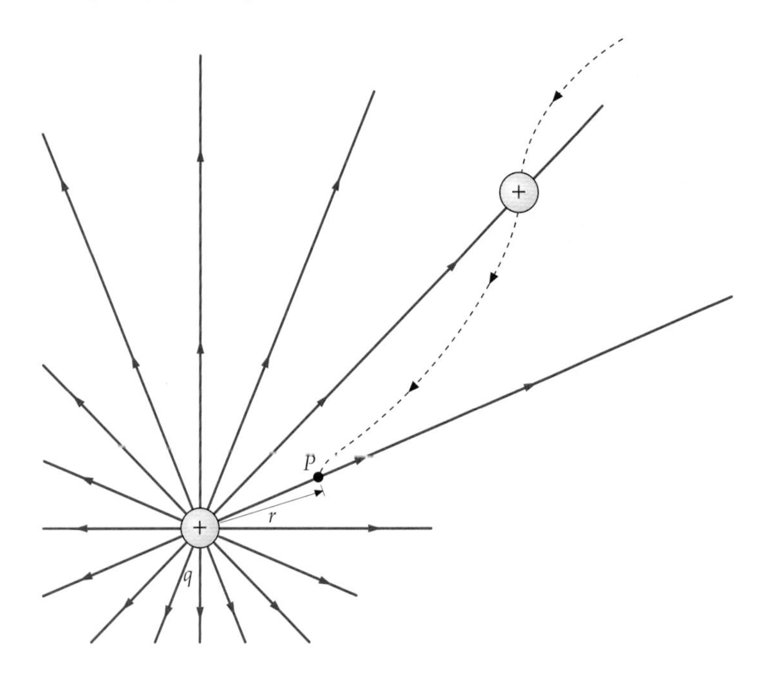

Figure 24-3 The work required to bring a test charge q_0 from infinity to a point P is kq_0q/r, where r is the distance from P to a charge q at the origin. The work per unit charge is kq/r, the electric potential at point P relative to zero potential at infinity. If the test charge is released from point P, the electric field does work kq_0q/r on the charge as the charge moves out to infinity.

Choosing the electrostatic potential energy of two charges to be zero at an infinite separation is analogous to the choice we made in Chapter 11 when we chose the gravitational potential energy of two point masses to be zero if the masses were very far apart.

Example 24-2

(*a*) What is the electric potential at a distance $r = 0.529 \times 10^{-10}$ m from a proton? (This is the average distance between the proton and electron in a hydrogen atom.) (*b*) What is the potential energy of the electron and the proton at this separation?

(*a*) Use $V = kq/r$ to calculate the potential V due to the proton:

$$V = \frac{kq}{r} = \frac{ke}{r} = \frac{(8.99 \times 10^9\ \text{N}\cdot\text{m}^2/\text{C}^2)(1.6 \times 10^{-19}\ \text{C})}{0.529 \times 10^{-10}\ \text{m}}$$

$$= 27.2\ \text{J/C} = 27.2\ \text{V}$$

(*b*) Use $U = q_0V$, with $q_0 = -e$ to calculate the electrostatic potential energy:

$$U = q_0V = (-e)(27.2\ \text{V}) = -27.2\ \text{eV}$$

Remarks If the electron were at rest at this distance from the proton, it would take 27.2 eV to remove it from the atom. However, the electron has kinetic energy equal to 13.6 eV, so its total energy in the atom is 13.6 eV − 27.2 eV = −13.6 eV. The energy needed to remove the electron from the atom is thus 13.6 eV. This energy is called the ionization energy.

Exercise What is the potential energy of the electron and proton in SI units? (*Answer* -4.35×10^{-18} J)

Example 24-3

In nuclear fission, a uranium-235 nucleus captures a neutron and splits apart into two lighter nuclei. Sometimes the two fission products are a barium nucleus (charge 56*e*) and a krypton nucleus (charge 36*e*). Assume that these nuclei are positive point charges separated by $r = 14.6 \times 10^{-15}$ m. Calculate the potential energy of this two-charge system in electron volts.

Picture the Problem The potential energy for two point charges separated by a distance r is $U = kq_1q_2/r$. To find this energy in electron volts we calculate the potential due to one of the charges kq_1/r in volts and multiply by the other charge.

1. Equation 24-9 gives the potential energy of the two charges: $$U = \frac{kq_1q_2}{r} = \frac{k(56e)(36e)}{r}$$

2. Factor out e and substitute the given values: $$U = \frac{k(56e)(36e)}{r} = e\frac{ke(56)(36)}{r}$$ $$= e\frac{(8.99 \times 10^9\ \text{N}\cdot\text{m}^2/\text{C}^2)(1.6 \times 10^{-19}\ \text{C})(56)(36)}{14.6 \times 10^{-15}\ \text{m}}$$ $$= e(1.99 \times 10^8\ \text{V}) = 199\ \text{MeV}$$

Remarks The separation distance r was chosen to be the sum of the radii of the two nuclei. After the fission, the two nuclei fly off because of their electrostatic repulsion. Their original potential energy of 199 MeV is converted into kinetic energy and thermal energy. Two or three neutrons are also released in the fission process. In a chain reaction, one or more of these neutrons produces a fission of another uranium nucleus. The average energy given off in chain reactions of this type is about 200 MeV per nucleus, as calculated in this example.

The potential at some point due to several point charges is the sum of the potentials due to each charge separately. (This follows from the superposition principle for the electric field. The work done by the net electric field is the sum of the work done by the electric fields due to each of the charges separately.) The potential due to a system of point charges q_i is thus given by

$$V = \sum_i \frac{kq_i}{r_i} \qquad \text{24-10}$$

Potential due to a system of point charges

where the sum is over all the charges, and r_i is the distance from the ith charge to the point P at which the potential is to be found.

Example 24-4

Two equal positive point charges of magnitude +5 nC are on the x axis. One is at the origin and the other is at x = 8 cm. Find the potential at (*a*) point P_1 on the x axis at x = 4 cm and (*b*) point P_2 on the y axis at y = 6 cm.

Picture the Problem The two positive point charges on the x axis are shown in Figure 24-4, and the potential is to be found at points P_1 and P_2.

Figure 24-4

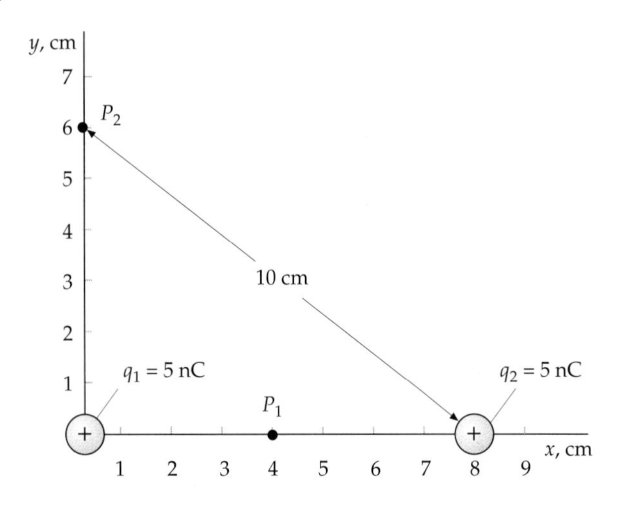

(*a*) 1. Use Equation 24-10 to write V as a function of the distances r_1 and r_2 to the charges:

$$V = \sum_i \frac{kq_i}{r_i} = \frac{kq_1}{r_1} + \frac{kq_2}{r_2}$$

2. Point P_1 is 4 cm from each charge, and the charges are equal.

$$r_1 = r_2 = r = 0.04\text{ m}$$

$$q_1 = q_2 = q = 5 \times 10^{-9}\text{ C}$$

3. Use these to find the potential at point P_1:

$$V = \frac{kq_1}{r_1} + \frac{kq_2}{r_2} = \frac{2kq}{r}$$

$$= \frac{2 \times (8.99 \times 10^9\text{ N}\cdot\text{m}^2/\text{C}^2)(5 \times 10^{-9}\text{ C})}{0.04\text{ m}}$$

$$= 2250\text{ V}$$

(*b*) Point P_2 is 6 cm from one charge and 10 cm from the other. Use these to find the potential at point P_2:

$$V = \frac{(8.99 \times 10^9\text{ N}\cdot\text{m}^2/\text{C}^2)(5 \times 10^{-9}\text{ C})}{0.06\text{ m}}$$

$$+ \frac{(8.99 \times 10^9\text{ N}\cdot\text{m}^2/\text{C}^2)(5 \times 10^{-9}\text{ C})}{0.10\text{ m}}$$

$$= 749\text{ V} + 450\text{ V} \approx 1200\text{ V}$$

Remarks Note that in (*a*), the electric field is zero at the point midway between the charges but the potential is not. It takes work to bring a test charge to this point from a long distance away, because the electric field is zero only at the final position.

Example 24-5

In Figure 24-5, a point charge q_1 is at the origin, and a second point charge q_2 is on the x axis at $x = a$. Find the potential everywhere on the x axis.

Picture the Problem The total potential is the sum of the potential due to each charge separately. The distance from q_1 to any point is $r_1 = |x|$ and the distance from q_2 to any point is $r_2 = |x - a|$.

Figure 24-5

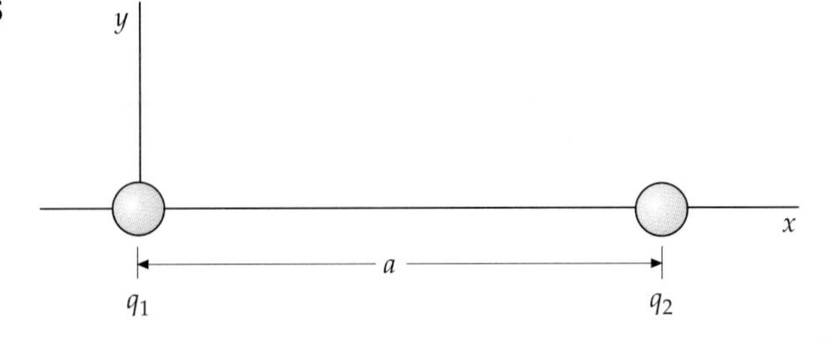

1. Write the potential as a function of the distances to the two charges: $$V = \frac{kq_1}{r_1} + \frac{kq_2}{r_2} = \frac{kq_1}{|x|} + \frac{kq_2}{|x - a|}$$

2. To the right of both charges, $|x| = x$, and $|x - a| = x - a$: $$V = \frac{kq_1}{x} + \frac{kq_2}{x - a}, \quad x > a$$

   Between the charges, $|x| = x$ and $|x - a| = a - x$: $$V = \frac{kq_1}{x} + \frac{kq_2}{a - x}, \quad 0 < x < a$$

   To the left of both charges, $|x| = -x$ and $|x - a| = a - x$: $$V = -\frac{kq_1}{x} + \frac{kq_2}{a - x}, \quad x < 0$$

Remark Figure 24-6 shows V versus x for equal charges. The potential is always positive, and becomes infinite at each charge.

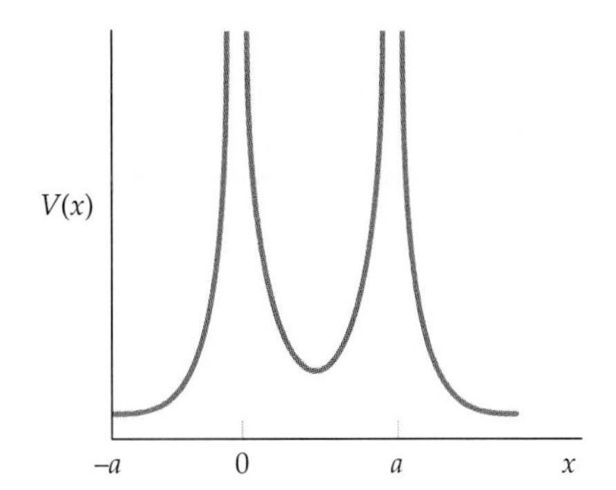

Figure 24-6

Example 24-6

An electric dipole consists of a positive charge $+q$ on the x axis at $x = +a$ and a negative charge $-q$ on the x axis at $x = -a$, as shown in Figure 24-7. Find the potential on the x axis for $x >> a$ in terms if the dipole moment $p = 2qa$.

Figure 24-7

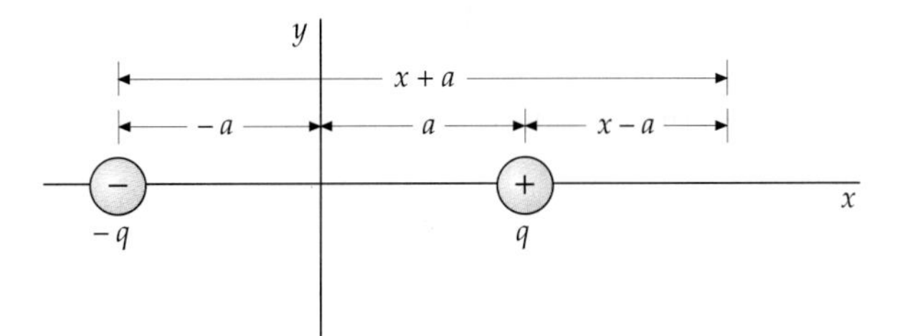

Picture the Problem The potential is the sum of the potential for each charge separately. For $x > a$, the distance to the positive charge is $x - a$ and the distance to the negative charge is $x + a$.

1. For $x > a$, the potential due to the two charges is:

$$V = \frac{kq}{x - a} + \frac{k(-q)}{x + a} = \frac{2kqa}{x^2 - a^2}$$

2. For $x >> a$, we can neglect a^2 compared with x^2 in the denominator. We then have:

$$V \approx \frac{2kqa}{x^2} = \frac{kp}{x^2}, \quad x >> a \qquad \text{24-11}$$

Remarks Far from the dipole, the potential decreases as $1/r^2$, compared to $1/r$ for the potential of a point charge. A three-dimensional plot of the potential near the charges is shown on the opening page of this chapter.

24-3 Finding the Electric Field From the Potential

If we know the potential, we can use it to calculate the electric field. Consider a small displacement $d\vec{\ell}$ in an arbitrary electric field $\vec{E}$. The change in potential is

$$dV = \vec{E}\cdot d\vec{\ell} = E_\ell \, d\ell \qquad \text{24-12}$$

where E_ℓ is the component of $\vec{E}$ parallel to the displacement. Then

$$E_\ell = -\frac{dV}{d\ell} \qquad \text{24-13}$$

If the displacement $d\vec{\ell}$ is perpendicular to the electric field, the potential does not change. The greatest change in V occurs when the displacement $d\vec{\ell}$ is along $\vec{E}$. A vector that points in the direction of the greatest change in a scalar function and has a magnitude equal to the derivative of that function with respect to the distance in that direction is called the **gradient** of the function. The electric field $\vec{E}$ is the negative gradient of the potential V. The field lines point in the direction of the greatest decrease in the potential function.

If the potential depends only on x, there will be no change for displacements in the y or z direction, so $\vec{E}$ must be in the x direction. For a displacement in the x direction, $d\vec{\ell} = dx\hat{i}$, and Equation 24-12 becomes

$$dV(x) = -\vec{E}\cdot d\vec{\ell} = -\vec{E}\cdot dx\,\hat{i} = -E_x\,dx$$

Then

$$E_x = -\frac{dV(x)}{dx} \qquad \text{24-14}$$

Similarly, for a spherically symmetric charge distribution, the potential can be a function only of the radial distance r. Displacements perpendicular to the radial direction give no change in $V(r)$, so the electric field must be radial. A displacement in the radial direction is written $d\vec{\ell} = dr\,\hat{r}$. Equation 24-12 is then

$$dV\,(r) = -\vec{E}\cdot d\vec{\ell} = -\vec{E}\cdot dr\,\hat{r} = -E_r\,dr$$

and

$$E_r = -\frac{dV(r)}{dr} \qquad \text{24-15}$$

If we know either the potential or the electric field over some region of space, we can use one to calculate the other. The potential is often easier to calculate because it is a scalar function, whereas the electric field is a vector function. Note that we cannot calculate $\vec{E}$ if we know the potential V at just a single point—we must know V over a region of space to calculate $\vec{E}$.

Example 24-7

Find the electric field for the electric potential function $V(x)$ given by $V(x) = 100\text{ V} - (25\text{ V/m})x$.

This potential function depends only on x. The electric field is found from Equation 24-14:

$$\vec{E} = -\frac{dV}{dx}\hat{\imath} = +(25\text{ V/m})\hat{\imath}$$

Remarks This electric field is uniform and in the x direction. Note that the constant 100 V in the expression for $V(x)$ has no effect on the electric field. The electric field does not depend on the choice of zero for the potential function.

Exercise (*a*) At what point does V equal zero in this example? (*b*) Write the potential function corresponding to the same electric field with $V = 0$ at $x = 0$. [*Answers* (*a*) $x = 4$ m, (*b*) $V = -(25\text{ V/m})\,x$]

General Relation Between $\vec{E}$ and V

In vector notation, the gradient of V is written $\overrightarrow{grad}\,V$. Then

$$\vec{E} = -\overrightarrow{grad}\,V \qquad \text{24-16}$$

In general, the potential function can depend on x, y, and z. The rectangular components of the electric field are related to the partial derivatives of the potential with respect to x, y, or z, while the other variables are held constant. For example, the x component of the electric field is given by

$$E_x = -\frac{\partial V}{\partial x} \qquad \text{24-17a}$$

Similarly, the y and z components of the electric field are related to the potential by

$$E_y = -\frac{\partial V}{\partial y} \qquad \text{24-17b}$$

and

$$E_y = -\frac{\partial V}{\partial z} \qquad \text{24-17c}$$

Thus, Equation 24-16 in rectangular coordinates is

$$\vec{E} = -\overrightarrow{grad}\,V = -\left(\frac{\partial V}{\partial x}\hat{\imath} + \frac{\partial V}{\partial y}\hat{\jmath} + \frac{\partial V}{\partial z}\hat{k}\right) \qquad \text{24-18}$$

optional

24-4 Calculation of *V* for Continuous Charge Distributions

The potential due to a continuous distribution of charge can be calculated by choosing an element of charge dq, which we treat as a point charge, and changing the sum in Equation 24-10 to an integral:

$$V = \int \frac{k\,dq}{r} \tag{24-19}$$

Potential due to a continuous charge distribution

This equation assumes that $V = 0$ at an infinite distance from the charges, so we cannot use it when there is charge at infinity, as is the case for artificial charge distributions like an infinite line charge or an infinite plane charge.

V on the Axis of a Charged Ring

Figure 24-8 shows a uniformly charged ring of radius a and charge Q. The distance from an element of charge dq to the field point P on the axis of the ring is $r = \sqrt{x^2 + a^2}$. Since this distance is the same for all elements of charge on the ring, we can remove this term from the integral in Equation 24-19. The potential at point P due to the ring is thus

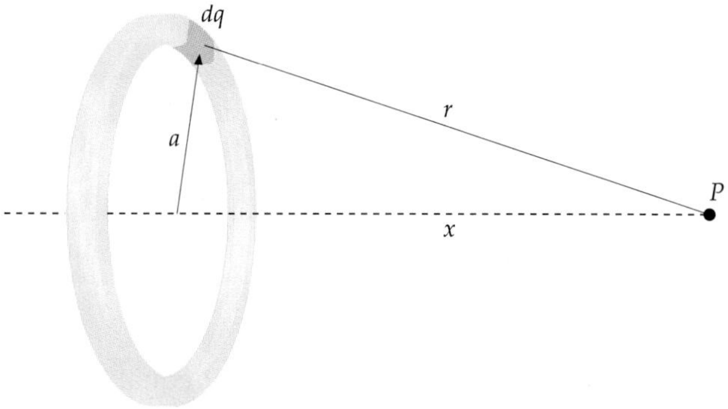

$$V = \int \frac{k\,dq}{r} = \int \frac{k\,dq}{\sqrt{x^2 + a^2}} = \frac{k}{\sqrt{x^2 + a^2}} \int dq$$

or

$$V = \frac{kQ}{\sqrt{x^2 + a^2}} \tag{24-20}$$

Potential on the axis of a uniformly charged ring

dq
r
a
P
x

Figure 24-8 Geometry for the calculation of the electric potential at a point on the axis of a uniformly charged ring of radius a.

Note that when x is much greater than a, the potential approaches kQ/x, the same as for a point charge at the origin.

Example 24-8 *try it yourself*

A ring of radius 4 cm is in the yz plane with its center at the origin. The ring carries a uniform charge of 8 nC. A small particle of mass m = 6 mg = 6 × 10^{-6} kg and charge q_0 = 5 nC is placed at x = 3 cm and released. Find the speed of the particle when it is a great distance from the ring.

Picture the Problem As the particle moves along the x axis, its potential energy decreases and its kinetic energy increases. When the particle is very far away, its potential energy is zero, and by conservation of energy, its kinetic energy equals its original potential energy. The final speed is found from the final kinetic energy.

Cover the column to the right and try these on your own before looking at the answers.

Steps	***Answers***
1. Use $U = q_0 V$ with V given by Equation 24-20 to calculate the initial potential energy of the point charge q_0 at a distance 3 cm from the center of the ring.	$U = q_0 V = \dfrac{kQq_0}{\sqrt{x^2 + a^2}} = 7.19 \times 10^{-6}\text{ J}$

2. Use conservation of energy to write an equation for the kinetic energy of the particle when it is far from the ring.

$$U_f + K_f = U_i + K_i$$
$$0 + \frac{1}{2}mv^2 = 7.19 \times 10^6\ \text{J} + 0$$

3. Solve for the speed v.

$$v = 1.55\ \text{m/s}$$

Exercise What is the potential energy of the particle when it is at $x =$ 9 cm? (*Answer* 3.65×10^{-6} J)

V on the Axis of a Uniformly Charged Disk

We can use our result for the potential on the axis of a ring charge to calculate the potential on the axis of a uniformly charged disk.

Example 24-9

Find the potential on the axis of a disk of radius R that carries a total charge Q distributed uniformly on its surface.

Picture the Problem We take the axis of the disk to be the x axis and we treat the disk as a set of ring charges. The ring of radius a and thickness da in Figure 24-9 has an area of $2\pi a\, da$ and its charge is $dq = \sigma\, dA = \sigma 2\pi a\, da$ where $\sigma = Q/\pi R^2$ is the surface charge density. The potential at point P is given by Equation 24-20. We then integrate from $a = 0$ to $a = R$ to find the total potential due to the disk.

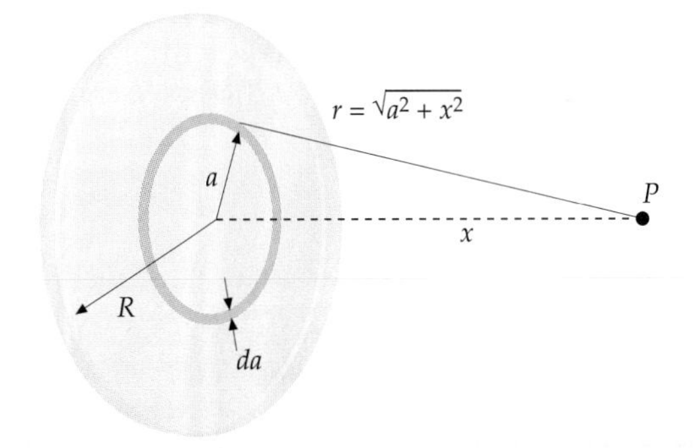

Figure 24-9

1. Write the potential dV at point P due to the charged ring of radius a:

$$dV = \frac{k\, dq}{(x^2 + a^2)^{1/2}} = \frac{k\sigma 2\pi a\, da}{(x^2 + a^2)^{1/2}}$$

2. Integrate from $a = 0$ to $a = R$:

$$V = \int_0^R \frac{k\sigma 2\pi a\, da}{(x^2 + a^2)^{1/2}} = k\sigma\pi \int_0^R (x^2 + a^2)^{-1/2} 2a\, da$$

3. The integral is of the form $\int u^n\, du$, with $u = x^2 + a^2$ and $n = -\frac{1}{2}$:

$$\int_0^R (x^2 + a^2)^{-1/2}\, 2a\, da = \left.\frac{(x^2 + a^2)^{+1/2}}{\frac{1}{2}}\right|_{a=0}^{a=R} = 2[(x^2 + R^2)^{1/2} - x]$$

4. Use this result to find V:

$$V = 2\pi k\sigma[(x^2 + R^2)^{1/2} - x]$$

Check the Result To find how the potential varies with x for $x >> R$, we write $V = 2\pi k\sigma x[(1 + R^2/x^2)^{1/2} - 1]$ and use the binomial expansion

$$\left(1 + \frac{R^2}{x^2}\right)^{1/2} \approx 1 + \frac{1}{2}\frac{R^2}{x^2} + \cdots$$

Then

$$V \approx 2\pi k\sigma x\left[1 + \frac{1}{2}\frac{R^2}{x^2} + \cdots - 1\right] = \frac{k(\sigma\pi R^2)}{x} = \frac{kQ}{x}$$

From Example 24-9 we see that the potential on the axis of a uniformly charged disk is

$$V = 2\pi k\sigma[(x^2 + R^2)^{1/2} - x] \qquad \text{24-21}$$

Potential on the axis of a disk charge

Example 24-10

Calculate the electric field on the axis of (*a*) a uniformly charged ring and (*b*) a uniformly charged disk using the potential functions given above for these charge distributions.

Picture the Problem Since V depends only on x, the electric field has only an x component, given by $E_x = -dV/dx$.

(*a*) 1. Write Equation 24-20 for the potential on the axis of a uniformly charged ring:

$$V = \frac{kQ}{\sqrt{x^2 + a^2}} = kQ(x^2 + a^2)^{-1/2}$$

2. Compute $-dV/dx$ to find E_x:

$$E_x = -\frac{dV}{dx} = +\frac{1}{2}kQ(x^2 + a^2)^{-3/2}(2x) = \frac{kQx}{(x^2 + a^2)^{3/2}}$$

(*b*) 1. Write Equation 24-21 for the potential on the axis of a uniformly charged disk:

$$V = 2\pi k\sigma[(x^2 + R^2)^{1/2} - x]$$

2. Compute $-dV/dx$ to find E_x:

$$E_x = -\frac{dV}{dx} = -2\pi k\sigma\left[\frac{1}{2}(x^2 + a^2)^{-1/2}2x - 1\right]$$

$$= 2\pi k\sigma\left(1 - \frac{x}{\sqrt{x^2 + R^2}}\right)$$

Remarks The results for (*a*) and (*b*) are the same as Equations 23-10 and 23-11, which were calculated directly from Coulomb's law.

V due to an Infinite Plane of Charge

If we let R become very large, our disk approaches an infinite plane. As R approaches infinity the potential function (Equation 24-21) approaches infinity. However, we obtained Equation 24-21 from Equation 24-19, which assumes that $V = 0$ at infinity, so Equation 24-21 can't be used. For infinite charge distributions, we must choose $V = 0$ at some finite point rather than at infinity. For such cases, we first find the electric field $\vec{E}$ (by direct integration or from Gauss's law) and then calculate the potential from its definition $dV = -\vec{E}\cdot d\vec{\ell}$. For an infinite plane of charge of density σ in the yz plane, the electric field for positive x is given by

$$\vec{E} = \frac{\sigma}{2\epsilon_0}\hat{i} = 2\pi k\sigma\hat{i}$$

The potential is then

$$dV = -\vec{E}\cdot d\vec{\ell} = -(2\pi k\sigma\hat{i})\cdot(dx\hat{i} + dy\hat{j} + dz\hat{k}) = -2\pi k\sigma\, dx$$

Integrating, we obtain

$$V = V_0 - 2\pi k\sigma x$$

where the arbitrary constant V_0 is the potential at $x = 0$. Note that the potential decreases with distance from the plane and approaches $-\infty$ as x approaches $-\infty$ Therefore, we cannot choose the potential to be zero at $x = \infty$. For negative x, the electric field is

$$\vec{E} = -2\pi k\sigma\hat{i}$$

so

$$dV = -\vec{E}\cdot d\vec{\ell} = +2\pi k\sigma\, dx$$

and the potential is

$$V = V_0 + 2\pi k\sigma x$$

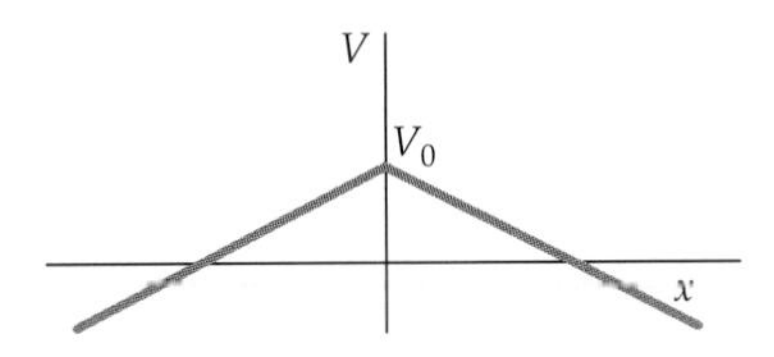

Figure 24-10 Plot of V versus x for an infinite plane of charge in the yz plane. Note that the potential is continuous at $x = 0$ even though the electric field is not.

Since x is negative, the potential again decreases with distance from the plane and approaches $-\infty$ as x approaches $-\infty$. For either positive or negative x, the potential can be written

$$V = V_0 - 2\pi k\sigma|x| \qquad \text{24-22}$$

Potential near an infinite plane of charge

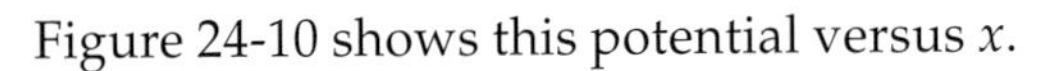

Figure 24-10 shows this potential versus x.

Example 24-11

An infinite plane of charge density σ is in the yz plane at $x = 0$, and a point charge q is on the x axis at $x = a$ (Figure 24-11). Find the potential at some point P a distance r from the point charge for $x > 0$ (that is, to the right of the plane charge).

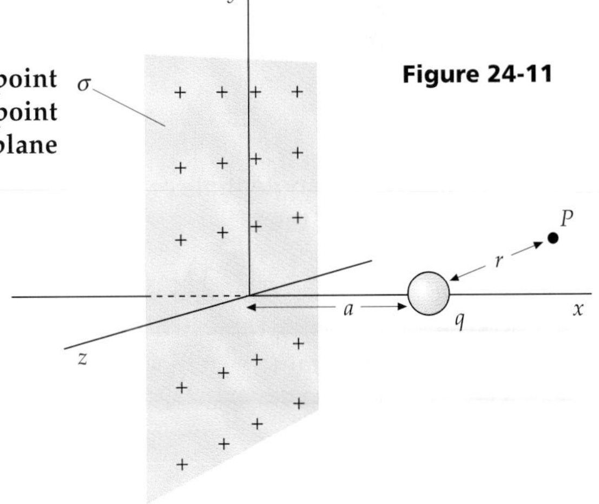

Figure 24-11

Picture the Problem The total potential is the sum of the potential due to the plane and the potential due to the point charge. Since we cannot choose V to be zero at $r = \infty$, we must include an arbitrary constant in our expression for $V(r)$ for a point charge. Since we have two constants, we will call one V_{01} and the other V_{02}. We are free to choose the zero of potential to be at any convenient finite point. For this example, we choose $V = 0$ at the origin.

1. The potential due to the plane is given by Equation 24-22 with $|x| = x$:

$$V_{\text{plane}} = V_{01} - 2\pi k\sigma x, \qquad V_{01} = \text{constant}$$

2. Equation 24-7 gives the potential due to a point charge:

$$V_{\text{point}} = \frac{kq}{r} + V_{02}, \qquad V_{02} = \text{constant}$$

3. Sum the above results to find the total potential V:

$$V = V_{\text{plane}} + V_{\text{point}} = V_{01} - 2\pi kx + \frac{kq}{r} + V_{02}$$

4. We choose $V = 0$ at the origin; that is, at $x = 0$ and $r = a$. This choice determines $V_{01} + V_{02}$:

$$V = 0 = V_{01} - 2\pi k\sigma(0) + \frac{kq}{a} + V_{02}$$

or

$$V_{01} + V_{02} = -\frac{kq}{a}$$

5. Substitute this for $V_{01} + V_{02}$ in the general expression for V:

$$V = -2\pi k\sigma x + \frac{kq}{r} - \frac{kq}{a}$$

Remarks In rectangular coordinates, $r = [(x-a)^2 + y^2 + z^2]^{1/2}$, thus

$$V = -2\pi k\sigma x + \frac{kq}{[(x-a)^2 + y^2 + z^2]^{1/2}} - \frac{kq}{a}$$

Exercise Find V for $x < 0$.

Answer

$$V = 2\pi k\sigma x + \frac{kq}{[(x-a)^2 + y^2 + z^2]^{1/2}} - \frac{kq}{a}$$

V Inside and Outside a Spherical Shell of Charge

We next find the potential due to a spherical shell of radius R with charge Q uniformly distributed on its surface. We are interested in the potential at all points inside and outside the shell. Since this shell is of finite extent, we could calculate the potential by direct integration of Equation 24-19, but this integration is somewhat difficult. Since the electric field for this charge distribution is easily obtained from Gauss's law, it is easiest to find the potential from the known electric field using $dV = -\vec{E}\cdot d\vec{\ell}$.

Outside the spherical shell, the electric field is radial and is the same as if all the charge were a point charge at the origin:

$$\vec{E} = \frac{kQ}{r^2}\hat{r}$$

The change in the potential for some displacement $d\vec{\ell} = dr\,\hat{r}$ outside the shell is then

$$dV = -\vec{E}\cdot d\vec{\ell} = -\frac{kQ}{r^2}\hat{r}\cdot dr\,\hat{r} = -\frac{kQ}{r^2}dr$$

Integrating, we obtain

$$V = \frac{kQ}{r} + V_0$$

where V_0 is the potential at $r = \infty$. Choosing the potential to be zero at $r = \infty$ gives

$$V = \frac{kQ}{r}, \qquad r > R$$

Inside the spherical shell, the electric field is zero. The change in potential for any displacement inside the shell is therefore also zero. Thus, the potential inside the shell must be constant. As r approaches R from outside the shell, the potential approaches kQ/R. Hence, the constant value of V inside must be kQ/R to make V continuous. Thus,

$$V = \begin{cases} \dfrac{kQ}{r}, & r \geq R \\[2ex] \dfrac{kQ}{R}, & r \leq R \end{cases} \qquad \text{24-23}$$

Potential due to a spherical shell

This potential function is plotted in Figure 24-12.

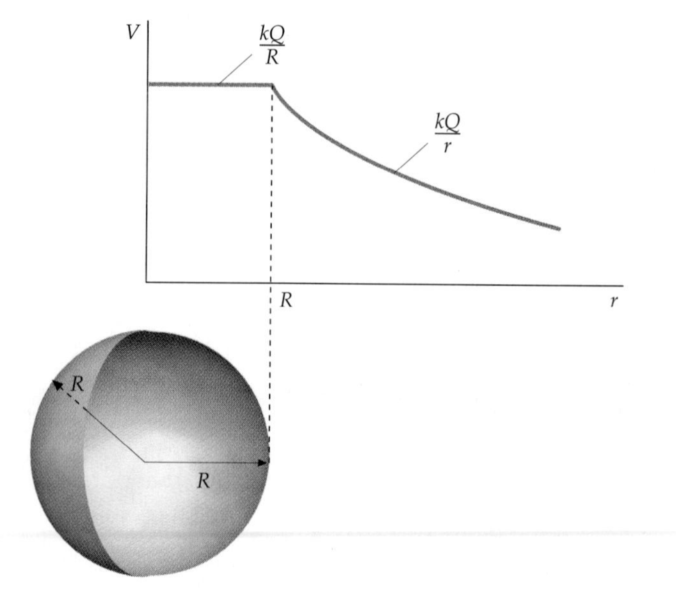

Figure 24-12 Electric potential of a uniformly charged spherical shell of radius R as a function of the distance r from the center of the shell. Inside the shell, the potential has the constant value kQ/R. Outside the shell, the potential is the same as that due to a point charge at the center of the sphere.

A common mistake is to think that the potential must be zero inside a spherical shell because the electric field is zero. But zero electric field merely implies that the potential does not change. Consider a spherical shell with a small hole so that we can move a test charge in and out of the shell. If we move the test charge from an infinite distance to the shell, the work per charge we must do is kQ/R. Inside the shell, there is no electric field, so it takes no work to move the test charge around inside the shell. The total amount of work per charge it takes to bring the test charge from infinity to any point inside the shell is just the work per charge it takes to bring it up to the shell radius R, which is kQ/R. The potential is therefore kQ/R everywhere inside the shell.

Exercise What is the potential of a spherical shell of radius 10 cm carrying a charge of 6 μC? (*Answer* 5.39×10^5 V = 539 kV)

Example 24-12 *try it yourself*

In one model, a proton is considered to be a spherical ball of charge of uniform volume charge density with radius R and total charge Q. The electric field inside the sphere is given by Equation 23-23*b*,

$$E_r = k\frac{Q}{R^3}\,r$$

Find the potential V both inside and outside the sphere.

Picture the Problem Outside the sphere, the charge looks like a point charge, so the potential is $V = kQ/r$. Inside the sphere, V can be found by integrating $dV = -\vec{E}\cdot d\vec{\ell}$. The constant of integration is found by requiring that V is continuous at $r = R$.

Cover the column to the right and try these on your own before looking at the answers.

Steps	***Answers***
1. Write down the potential V for $r \geq R$.	$V = \dfrac{kQ}{r}, \quad r \geq R$
2. For $r \leq R$, find dV from $dV = -\vec{E}\cdot d\vec{\ell}$	$dV = -\dfrac{kQ}{R^3}\,r\,dr, \quad r \leq R$
3. Find the indefinite integral of your expression in step 2. Be sure to include a constant of integration V_0.	$V = -\dfrac{kQ}{2R^3}\,r^2 + V_0$
4. Set your expression from step 3 equal to that from step 1 at $r = R$ and solve for the constant of integration V_0.	$V_0 = \dfrac{3kQ}{2R}$
5. Substitute your value for V_0 into your expression in step 3 to obtain $V(r)$ for $r \leq R$.	$V(r) = \dfrac{kQ}{2R}\left(3 - \dfrac{r^2}{R^2}\right), \quad r \leq R$

Check the Result Substituting $r = R$ in the result gives $V(R) = kQ/R$ as required. At $r = 0$, $V(0) = 3kQ/2R = 1.5\,kQ/R$, which is greater than $V(R)$, as it should be, because the electric field is in the positive radial direction for $r < R$, so work must be done to move a test charge from $r = R$ to $r = 0$.

Remarks Figure 24-13 shows $V(r)$ as a function of r. Note that both $V(r)$ and $E_r = -dV/dr$ are continuous everywhere.

Exercise What is $V(r)$ if we choose $V(R) = 0$? (*Answer* $V(r) = kQ/r - kQ/R$ for $r \geq R$; $V(r) = \frac{1}{2}(kQ/R)(1 - r^2/R^2)$ for $r \leq R$)

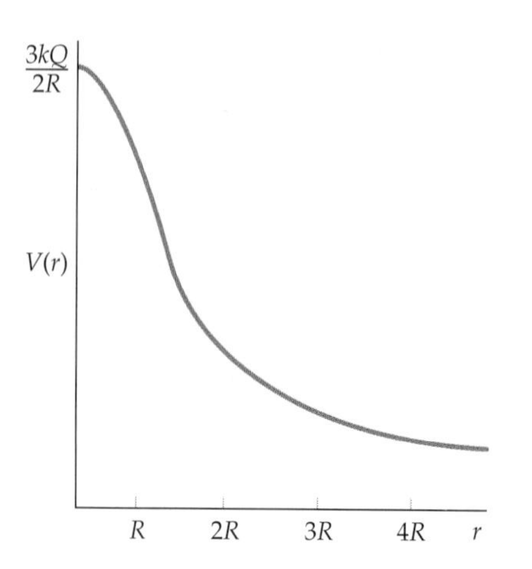

Figure 24-13

V due to an Infinite Line Charge

We will now calculate the potential due to a uniform infinite line charge. Let the charge per unit length be λ. Since this charge distribution extends to infinity, we find the potential from the electric field. In Chapter 23, we found that the electric field produced by an infinite line charge points away from the line (assuming λ to be positive) and is given by $E_r = 2k\lambda/r$. The change in potential for a displacement $d\vec{\ell}$ is then

$$dV = -\vec{E}\cdot d\vec{\ell} = -E_r\,dr = -\frac{2k\lambda}{r}\,dr$$

Integrating, we obtain

$$V = V_0 - 2k\lambda \ln r \qquad \text{24-24}$$

For a positive line charge, the electric field lines point away from the line and the potential decreases with increasing distance from the line charge. At large values of r, the potential decreases without limit. We therefore cannot choose for the potential to be zero at $r = \infty$. (Neither can we choose the potential to be zero at $r = 0$, because $\ln r$ approaches $-\infty$ as r approaches zero.) Instead, we choose V to be zero at some distance $r = a$. Substituting $r = a$ into Equation 24-24 and setting $V = 0$, we obtain

$$V = 0 = V_0 - 2k\lambda \ln a$$

or

$$V_0 = 2k\lambda \ln a$$

Then Equation 24-24 is

$$V = 2k\lambda \ln a - 2k\lambda \ln r$$

or

$$V = -2k\lambda \ln \frac{r}{a} \qquad \text{24-25}$$

Potential due to a line charge

24-5 Equipotential Surfaces

Since there is no electric field inside a conductor that is in static equilibrium, the change in potential as we move about inside the conductor is zero. The electric potential is thus the same throughout the conductor, that is, the conductor is an **equipotential volume** and its surface is an **equipotential surface**. Because the potential is constant on an equipotential surface, the change in V when a test charge is given a displacement $d\vec{\ell}$ parallel to the surface is $dV = -\vec{E}\cdot d\vec{\ell} = 0$. Then, since $\vec{E}\cdot d\vec{\ell}$ is zero, the electric field lines emanating from the equipotential surface must be perpendicular to the surface. Figures 24-14 and 24-15 show equipotential surfaces near a spherical conductor and a nonspherical conductor. Note that the field lines are everywhere perpendicular to the equipotential surfaces. If we move a short distance $d\ell$ along a field line from one equipotential surface to another, the potential changes by $dV = -\vec{E}\cdot d\vec{\ell} = -E\,d\ell$. Equipotential surfaces that have a fixed potential difference between them are more closely spaced where the electric field E is greater.

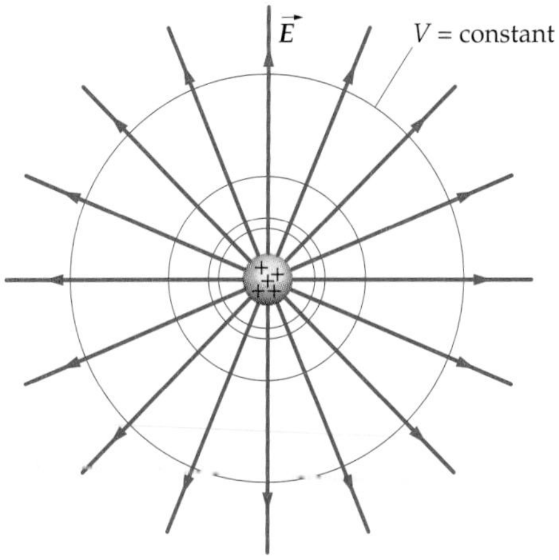

Figure 24-14 Equipotential surfaces and electric field lines outside a uniformly charged spherical conductor. The equipotential surfaces are spherical and the field lines are radial and perpendicular to the equipotential surfaces.

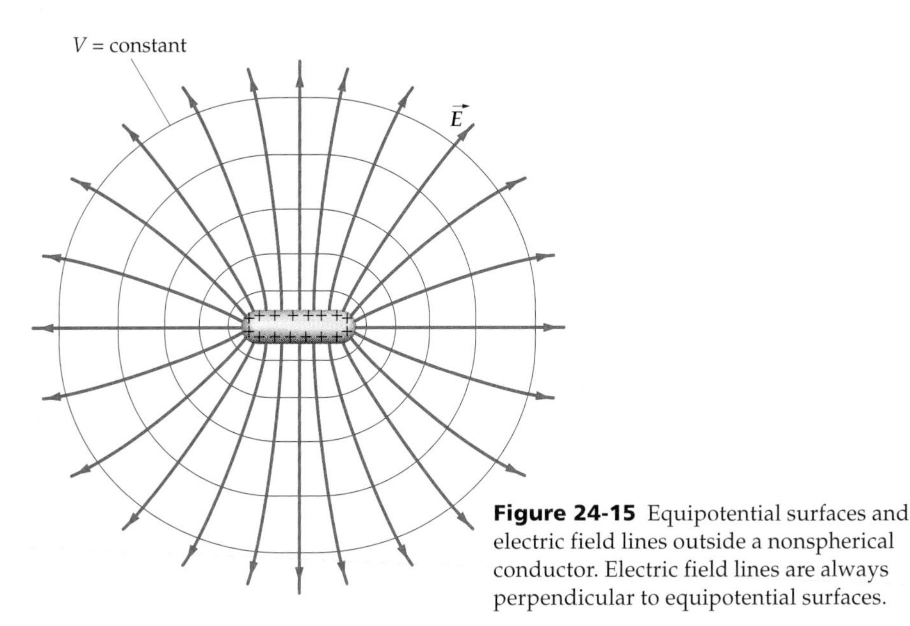

Figure 24-15 Equipotential surfaces and electric field lines outside a nonspherical conductor. Electric field lines are always perpendicular to equipotential surfaces.

Example 24-13

A hollow, uncharged spherical conductor has inner radius a and outer radius b. A positive point charge $+q$ is in the cavity at the center of the sphere. Find the potential $V(r)$ everywhere, assuming that $V = 0$ at $r = \infty$.

Picture the Problem We obtain the potential from $dV = -\vec{E}\cdot d\vec{\ell} = -E_r\,dr$. Inside the cavity, $E_r = kq/r^2$, so V is of the form $V = kq/r + V_0$. The conductor is an equipotential volume, so V is constant for $a \le r \le b$. The field lines inside the cavity must end on the inner surface of the cavity, so this surface has an induced charge of $-q$. Since the shell is uncharged, a positive charge $+q$ is on the outer surface. The three charges q at the center, $-q$ on the inner surface, and $+q$ on the outer surface produce a field $E_r = kq/r^2$ for $r > b$, so the potential for $r > b$ is $V = kq/r$.

1. Outside the shell, $V(r)$ is the same as that due to a point charge q at the origin. Choosing $V = 0$ at $r = \infty$, we have:

$$V(r) = \frac{kq}{r}, \qquad r \ge b$$

2. At $r = b$, the potential is kq/b. V remains at this constant value throughout the spherical shell:

$$V(r) = \frac{kq}{b}, \qquad a \le r \le b$$

3. Inside the cavity, V is the same as that due to a point charge q at the origin, but the arbitrary constant cannot be set equal to zero, because V has already been chosen to be zero at $r = \infty$:

$$V(r) = \frac{kq}{r} + V_0, \qquad r \le a$$

4. The constant V_0 is determined by the condition that V is continuous at $r = a$; that is, V must be kq/b at $r = a$:

$$V(a) = \frac{kq}{a} + V_0 = \frac{kq}{b}$$

or

$$V_0 = \frac{kq}{b} - \frac{kq}{a}$$

5. Use this value of V_0 to find $V(r)$ for $r \leq a$:

$$V(r) = \frac{kq}{r} + \frac{kq}{b} - \frac{kq}{a}, \qquad r \leq a$$

Remarks Figure 24-16 shows the electric potential as a function of the distance from the center of the cavity. Inside the conducting material, where $a \leq r \leq b$, the potential has the constant value kq/b. Outside the shell, the potential is the same as that of a point charge. Note that $V(r)$ is continuous everywhere. The electric field is discontinuous at the conductor surfaces, as reflected in the discontinuous slope of $V(r)$ at $r = a$ and $r = b$.

Figure 24-16

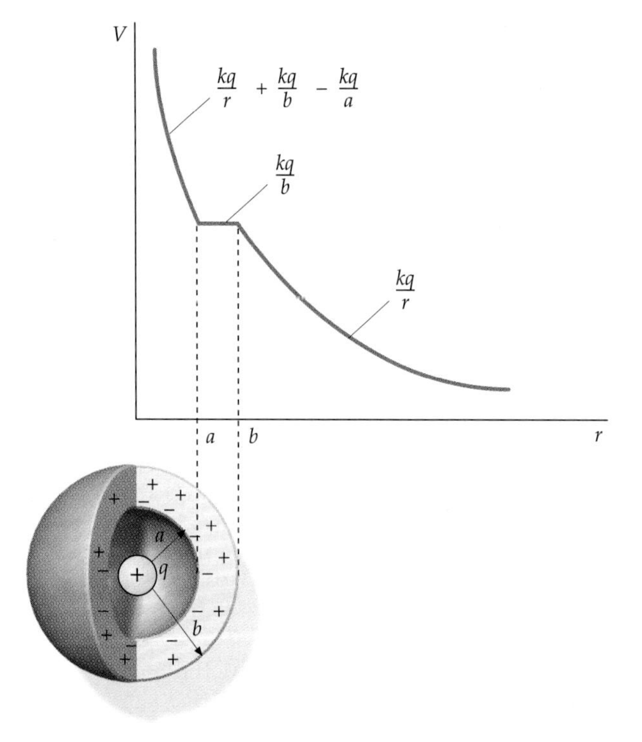

In general, two conductors that are separated in space will not be at the same potential. The potential difference between the conductors depends on their geometrical shapes, their separation, and the net charge on each. When two conductors are brought into contact, the charge on the conductors distributes itself so that electrostatic equilibrium is established and the electric field is zero inside both conductors. While in contact, the two conductors may be considered to be a single conductor with a single equipotential surface. If we put a spherical charged conductor in contact with a second spherical conductor that is uncharged, charge will flow to the neutral conductor until both conductors are at the same potential. If the conductors are identical, they share the original charge equally. If the conductors are now separated, each carries half the original charge, and both are at the same potential.

The Van de Graaff Generator

In Figure 24-17, a small conductor carrying a positive charge q is inside the cavity of a larger conductor. In equilibrium, the electric field is zero inside the conducting material of both conductors. The electric field lines that leave the positive charge q must end on the inner surface of the large conductor. This must occur no matter what the charge may be on the outside surface of the large conductor. Regardless of the charge on the large conductor, the small conductor in the cavity is at a greater potential because the electric field lines go from this conductor to the larger conductor. If the conductors are now connected, say, with a fine conducting wire, *all* the charge originally on the smaller conductor will flow to the larger one. When the connection is broken, there is no charge on the small conductor in the cavity, and there are no field lines between the conductors. The positive charge transferred from the smaller conductor resides completely on the outside surface of the larger conductor. If we put more positive charge on the small conductor in the cavity and again connect the conductors with a fine wire, all of the charge on the inner conductor will again flow to the outer conductor. This procedure can be repeated indefinitely. This method is used to produce large potentials in a device called the Van de Graaff generator, in which the charge is brought to the inner surface of a larger spherical conductor by a continuous charged belt (Figure 24-18). Work must be done by the motor driving the belt to bring the charge from the bottom to the top of the belt where the potential is very high. The greater the net charge on the outer conductor, the greater its potential, and the greater the electric field outside the conductor. A Van de Graaff accelerator is a device that uses the intense electric field produced by a Van de Graaff generator to accelerate positive particles, such as protons.

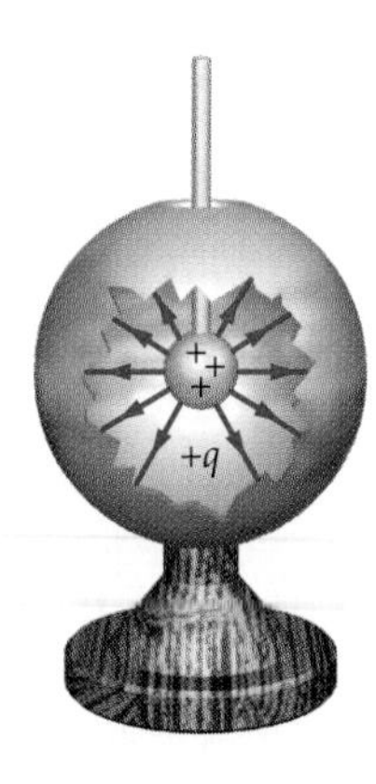

Figure 24-17 Small conductor carrying a positive charge inside a larger conductor.

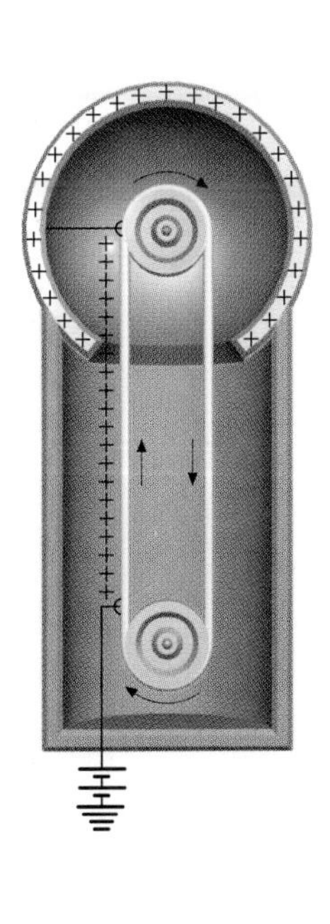

(*a*)

(*b*)

(*c*)

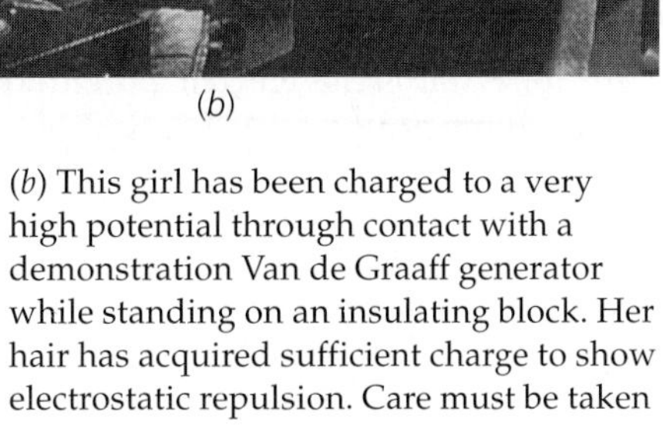

Figure 24-18 (*a*) Schematic diagram of a Van de Graaff generator. Charge leaks off the pointed conductor near the bottom onto the belt. Near the top, the charge leaks off of the belt onto the pointed conductors attached to the large spherical conductor. (*b*) This girl has been charged to a very high potential through contact with a demonstration Van de Graaff generator while standing on an insulating block. Her hair has acquired sufficient charge to show electrostatic repulsion. Care must be taken to acquire the charge gradually and to avoid rapid discharge to prevent a painful shock. (*c*) These large demonstration Van de Graaff generators in the Boston science museum are discharging to the grounded wire cage housing the operator.

Dielectric Breakdown

Many nonconducting materials become ionized in very high electric fields and become conductors. This phenomenon, called **dielectric breakdown,** occurs in air at an electric field strength of $E_{max} \approx 3 \times 10^6$ V/m = 3 MN/C. This limits the maximum potential that can be obtained in a Van de Graaff generator. The magnitude of the electric field for which dielectric breakdown occurs in a material is called the **dielectric strength** of that material. The dielectric strength of air is thus about 3 MV/m. The discharge through the conducting air resulting from dielectric breakdown is called **arc discharge.** The electric shock you receive when you touch a metal door knob after walking across a rug on a dry day is a familiar example of arc discharge. This occurs more often on dry days because moist air can conduct away charge as you acquire it. Lightning is an example of arc discharge on a large scale.

Example 24-14

A spherical conductor has a radius of 2 m. (*a*) What is the maximum charge that can be placed on the sphere before dielectric breakdown of the surrounding air occurs? (*b*) What is the maximum potential of the sphere?

Picture the Problem (*a*) We find the maximum charge by relating the charge to the electric field and setting the field equal to the dielectric strength of air, E_{max}. (*b*) The maximum potential is then found from the maximum charge calculated in (*a*).

(*a*) 1. The surface charge density on the conductor σ is related to the electric field just outside the conductor:

$$E = \frac{\sigma}{\epsilon_0}$$

2. Set this field equal to E_{max}:

$$E_{max} = 3 \times 10^6 \text{ N/C} = \frac{\sigma_{max}}{\epsilon_0}$$

3. The maximum charge Q is found from σ_{max}:

$$Q = 4\pi R^2 \sigma_{max} = 4\pi R^2(\epsilon_0 E_{max})$$
$$= 4\pi(2 \text{ m})^2(8.85 \times 10^{-12} \text{ C}^2/\text{N}\cdot\text{m}^2)(3 \times 10^6 \text{ N/C})$$
$$= 1.33 \times 10^{-3} \text{ C}$$

(*b*) Use this maximum charge to calculate the maximum potential of the sphere:

$$V_{max} = \frac{kQ_{max}}{R} = \frac{(8.99 \times 10^9 \text{ N}\cdot\text{m}^2/\text{C}^2)(1.33 \times 10^{-3} \text{ C})}{2 \text{ m}}$$
$$= 5.98 \times 10^6 \text{ V}$$

Example 24-15

Two charged spherical conductors of radius $R_1 = 6$ cm and $R_2 = 2$ cm are separated by a distance much greater than 6 cm and are connected by a conducting wire. A total charge $Q = +80$ nC is placed on one of the spheres. (*a*) What is the electric field near the surface of each sphere? (*b*) What is the electric potential of each sphere? (Assume that the charge on the connecting wire is negligible.)

Picture the Problem The total charge will be distributed with q_1 on sphere 1 and q_2 on sphere 2 such that the spheres will be at the same potential. We can use $V = kq/r$ for the potential of each sphere because they are far apart.

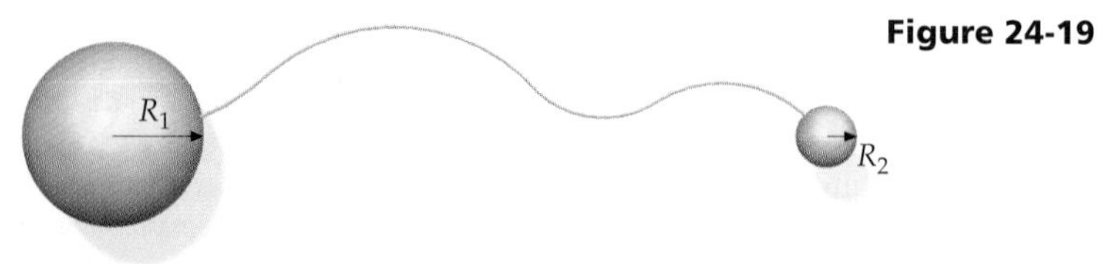

Figure 24-19

(*a*) 1. The electric field at the surface of each sphere is related to the charge on the sphere and its radius:

$$E_1 = \frac{kq_1}{R_1^2} \quad \text{and} \quad E_2 = \frac{kq_2}{R_2^2}$$

2. Conservation of charge gives us one relation between the charges q_1 and q_2:

$$q_1 + q_2 = Q = 80 \text{ nC}$$

3. Equating the potential of the spheres gives us a second relation for the charges q_1 and q_2:

$$\frac{kq_1}{R_1} = \frac{kq_2}{R_2}$$

$$q_1 = \frac{R_1}{R_2} q_2 = \frac{6 \text{ cm}}{2 \text{ cm}} q_2 = 3q_2$$

4. Combine these results and solve for q_1 and q_2:

$$q_1 = 60 \text{ nC}$$

$$q_2 = 20 \text{ nC}$$

5. Use these results to calculate the electric fields at the surface of the spheres:

$$E_1 = \frac{kq_1}{R_1^2} = \frac{(8.99 \times 10^9 \text{ N}\cdot\text{m}^2/\text{C}^2)(60 \times 10^{-9} \text{ C})}{(0.06 \text{ m})^2} = 150 \text{ kN/C}$$

$$E_2 = \frac{kq_2}{R_2^2} = \frac{(8.99 \times 10^9 \text{ N}\cdot\text{m}^2/\text{C}^2)(20 \times 10^{-9} \text{ C})}{(0.02 \text{ m})^2} = 450 \text{ kN/C}$$

(*b*) Calculate the common potential from kq/R for either sphere:

$$V_1 = \frac{kq_1}{R_1} = \frac{(8.99 \times 10^9 \text{ N}\cdot\text{m}^2/\text{C}^2)(60 \times 10^{-9} \text{ C})}{0.06 \text{ m}} = 8.99 \text{ kV}$$

Check the Result If we use sphere 2 to calculate V, we obtain $V_2 = kq_2/R_2 = (8.99 \times 10^9 \text{ N}\cdot\text{m}^2/\text{C}^2)(20 \times 10^{-9} \text{ C})/0.02 \text{ m} = 8.99 \times 10^3 \text{ V}$. An additional check is available, since the electric field at the surface of each sphere is proportional to its charge density. Since the radius of sphere 1 is three times that of sphere 2, its surface area is 9 times that of sphere 2. And since it carries 3 times the charge, its charge density is $\frac{1}{3}$ that of sphere 2. Therefore, the field of sphere 1 should be $\frac{1}{3}$ that of sphere 2, which is what we found above.

When a charge is placed on a conductor of nonspherical shape, like that in Figure 24-20*a*, the surface of the conductor will be an equipotential surface, but the surface charge density and the electric field just outside the conductor will vary from point to point. Near a point where the radius of curvature is small, such as point A in the figure, the surface charge density and electric field will be large, whereas near a point where the radius of curvature is large, such as point B in the figure, the field and surface charge density will be small. We can understand this qualitatively by considering the ends of the conductor to be spheres of different radii. Let σ be the surface charge density.

A
r_1
r_2
B

(*a*)

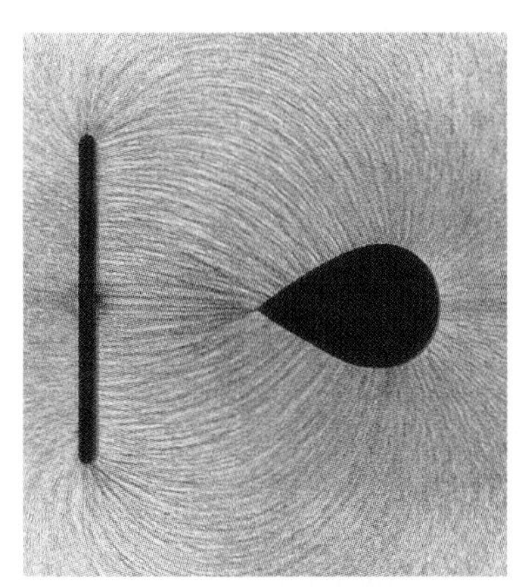

(*b*)

Figure 24-20 (*a*) A nonspherical conductor. If a charge is placed on such a conductor, it will produce an electric field that is stronger near point A, where the radius of curvature is small, than near point B, where the radius of curvature is large. (*b*) Electric field lines near a nonspherical conductor and plate carrying equal and opposite charges. The lines are shown by small bits of thread suspended in oil. Note that the electric field is strongest near points of small radius of curvature, such as at the ends of the plate and at the pointed left side of the conductor.

The potential of a sphere of radius r is

$$V = \frac{kq}{r} = \frac{1}{4\pi\epsilon_0}\frac{q}{r} \qquad \text{24-26}$$

Since the area of a sphere is $4\pi r^2$, the charge on a sphere is related to the charge density by $q = 4\pi r^2\sigma$. Substituting this expression for q into Equation 24-26 we have

$$V = \frac{1}{4\pi\epsilon_0}\frac{4\pi r^2\sigma}{r} = \frac{r\sigma}{\epsilon_0}$$

Solving for σ, we obtain

$$\sigma = \frac{\epsilon_0 V}{r} \qquad \text{24-27}$$

Since both "spheres" are at the same potential, the one with the smaller radius must have the greater surface charge density. And since $E_n = \sigma/\epsilon_0$, the electric field is greatest at points on the conductor where the radius of curvature is least.

For an arbitrarily shaped conductor, the potential at which dielectric breakdown occurs depends on the smallest radius of curvature of any part of the conductor. If the conductor has sharp points of very small radius of curvature, dielectric breakdown will occur at relatively low potentials. In the Van de Graaff generator (Figure 24-18*a*), the charge is transferred onto the belt by sharp-edged conductors near the bottom of the belt. The charge is removed from the belt by sharp-edged conductors near the top of the belt. Lightning rods at the top of a tall building draw the charge off a nearby cloud before the potential of the cloud can build up to a destructively large value.

Summary

1. Electric potential, which is defined as the electrostic potential energy per charge, is an important derived physical concept that is related to the electric field.
2. Because potential is a scalar quantity, it is often easier to calculate than the vector electric field. Once V is known, $\vec{E}$ can be calculated from V.

Topic	Remarks and Relevant Equations	
1. Potential Difference	The potential difference $V_b - V_a$ is defined as the negative of the work per unit charge done by the electric field when a test charge moves from point a to point b: $\Delta V = V_b - V_a = \frac{\Delta U}{q_0} = -\int_a^b \vec{E}\cdot d\vec{\ell}$	**24-2*b***
Potential difference for infinitesimal displacements	$dV = -\vec{E}\cdot d\vec{\ell}$	**24-2*a***
2. Electric Potential		
Potential due to a point charge	$V = \frac{kq}{r} + V_0 \quad (V = V_0 \text{ at } r = \infty)$	**24-7**
Coulomb potential	$V = \frac{kq}{r} \quad (V = 0 \text{ at } r = \infty)$	**24-8**

	Potential due to a system of point charges	$V = \sum_i \frac{kq_i}{r_i} \quad (V = 0 \text{ at } r = \infty)$	24-10
	Potential due to continuous charge distributions	$V = \int \frac{k\,dq}{r} \quad (V = 0 \text{ at } r = \infty)$ This expression can be used only if the charge distribution is contained in a finite volume so that the potential can be chosen to be zero at infinity.	24-19
	Potential and electric field lines	Electric field lines point in the direction of decreasing electric potential.	
	Continuity of electric potential	The potential function V is continuous everywhere in space.	
3.	**Finding the Electric Field from the Potential**	The electric field points in the direction of the greatest decrease in the potential. $E_\ell = -\frac{dV}{d\ell}$	24-13
	Gradient	A vector that points in the direction of the greatest change in a scalar function and has a magnitude equal to the derivative of that function with respect to the distance in that direction is called the gradient of the function. $\vec{E}$ is the negative gradient of V.	
	$\vec{E}$ in the x direction	$E_x = -\frac{dV(x)}{dx}$	24-14
	Radial electric field	$E_r = -\frac{dV(r)}{dr}$	24-15
4.	**General Relation Between $\vec{E}$ and V** (optional)	$\vec{E} = -\overrightarrow{grad}\, V = -\left(\frac{\partial V}{\partial x}\hat{i} + \frac{\partial V}{\partial y}\hat{j} + \frac{\partial V}{\partial z}\hat{k}\right)$	24-18
5.	**Units**		
	V and ΔV	The SI unit of potential and potential difference is the volt (V): $1\text{ V} = 1\text{ J/C}$	24-4
	Electric field	$1\text{ N/C} = 1\text{ V/m}$	24-5
	eV	The electron volt (eV) is the potential energy of a particle of charge e at a point where the potential is 1 volt: $1\text{ eV} = 1.6 \times 10^{-19}\text{ J}$	24-6
6.	**Potential Energy of Two Point Charges**	$U = q_0 V = \frac{kq_0 q}{r} \quad (U = 0 \quad \text{at} \quad r = \infty)$	24-9
7.	**Potential Functions**		
	On the axis of a uniformly charged ring	$V = \frac{kQ}{\sqrt{x^2 + a^2}}$	24-20
	On the axis of a uniformly charged disk	$V = 2\pi k\sigma[(x^2 + R^2)^{1/2} - x]$	24-21
	Near an infinite plane of charge	$V = V_0 - \frac{\sigma}{2\epsilon_0}\lvert x\rvert = V_0 - 2\pi k\sigma\lvert x\rvert$	24-22

	For a spherical shell of charge	$$V = \begin{cases} \dfrac{kQ}{r}, & r \geq R \\ \dfrac{kQ}{R}, & r \leq R \end{cases}$$	**24-23**
	For an infinite line charge	$V = -2k\lambda \ln \dfrac{r}{a}, \quad V = 0 \text{ at } r = a$	**24-25**
8.	**Charge on a Nonspherical Conductor**	On a conductor of arbitrary shape, the surface charge density σ is greatest at points where the radius of curvature is smallest.	
9.	**Dielectric Breakdown**	The amount of charge that can be placed on a conductor is limited by the fact that molecules of the surrounding medium become ionized in very high electric fields, causing the medium to become a conductor.	
	Dielectric strength	The dielectric strength is the magnitude of the electric field at which dielectric breakdown occurs. The dielectric strength of air is $E_{max} \approx 3 \times 10^6 \text{ V/m} = 3 \text{ MV/m}$	

Problem-Solving Guide

1. Begin by drawing a neat diagram that includes the important features of the problem, including the location and value of charges in the system. It is often helpful to indicate equipotential surfaces and electric field lines.
2. The potential due to a system of point charges can be calculated from $V = \Sigma\, kq_i/r_i$.
3. The potential due to a continuous, finite system of charges can be calculated from $V = \int kdq/r$, or if the electric field is known, the potential can be found from $dV = -\vec{E}\cdot d\vec{\ell}$.

Summary of Worked Examples

Type of Calculation	Procedure and Relevant Examples	
1. Potential and Electric Field		
Calculate the change in electric potential V given the electric field $\vec{E}$.	Use $dV = -\vec{E}\cdot d\vec{\ell}$ and integrate.	**Examples 24-1, 24-12, and 24-13**
Determine $\vec{E}$ given the potential V.	Use $E_x = -dV(x)/dx$ for a potential that depends only on x or $E_r = -dV(r)/dr$ for a spherically symmetric charge distribution.	**Examples 24-7, 24-10**
Find the potential V for discrete charges:		
Point charge	Use $V = kq/r$.	**Example 24-2**
System of point charges	Use $V = \Sigma_i\, kq_i/r_i$.	**Examples 24-4, 24-5, 24-6**
Find V for a continuous charge distribution.	Integrate over the charge distribution ($V = \int kdq/r$), or calculate V from the electric field $V = -\int \vec{E}\cdot d\vec{\ell}$.	**Examples 24-9, 24-12, 24-13**
Find V for a system that includes point charges and continuous charge distributions.	Use $V_{point} = (kq/r) + V_i$ for each point charge where V_i is a different constant for each point charge that is determined by the choice of zero for V. Sum the potential due to the point charges and the potential due to the continuous charge distributions.	**Example 24-11**

2. Potential and Energy

Find the potential energy of a two-charge system.	Use $U = kq_1q_2/r$.	Examples 24-2, 24-3
Find the change in kinetic energy ΔK as a charge moves through a region of electric potential.	Use conservation of energy, $\Delta K + \Delta U = 0$, where $\Delta U = q\,\Delta V$.	Example 24-8

3. Conductors

Find the potential in problems involving conductors.	Use the fact that the surface of a conductor is an equipotential surface, and that V is constant everywhere within a conductor.	Example 24-12
Find the maximum charge on a conductor before dielectric breakdown.	Use $\sigma_{\text{max}} = \epsilon_0 E_{\text{max}}$, where E_{max} for air is equal to 3 MN/C.	Example 24-14
Charge sharing between conductors.	If two conductors are in contact, they must be at the same potential. This condition determines the ratio of charge on the conductors.	Example 24-15

Problems

Conceptual Problems

Problems from Optional and Exploring sections

In a few problems, you are given more data than you actually need; in a few other problems, you are required to supply data from your general knowledge, outside sources, or informed estimates.

• Single-concept, single-step, relatively easy

•• Intermediate-level, may require synthesis of concepts

••• Challenging, for advanced students

Potential and Potential Difference

1 • A uniform electric field of 2 kN/C is in the x direction. A positive point charge $Q = 3\ \mu$C is released from rest at the origin. (*a*) What is the potential difference $V(4\text{ m}) - V(0)$? (*b*) What is the change in the potential energy of the charge from $x = 0$ to $x = 4$ m? (*c*) What is the kinetic energy of the charge when it is at $x = 4$ m? (*d*) Find the potential $V(x)$ if $V(x)$ is chosen to be (*d*) zero at $x = 0$, (*e*) 4 kV at $x = 0$, and (*f*) zero at $x = 1$ m.

2 • An infinite plane of surface charge density $\sigma = +2.5\ \mu\text{C/m}^2$ is in the yz plane. (*a*) What is the magnitude of the electric field in newtons per coulomb? In volts per meter? What is the direction of $\vec{E}$ for positive values of x? (*b*) What is the potential difference $V_b - V_a$ when point b is at $x = 20$ cm and point a is at $x = 50$ cm? (*c*) How much work is required by an outside agent to move a test charge $q_0 = +1.5$ nC from point a to point b?

3 • Two large parallel conducting plates separated by 10 cm carry equal and opposite surface charge densities such that the electric field between them is uniform. The difference in potential between the plates is 500 V. An electron is released from rest at the negative plate. (*a*) What is the magnitude of the electric field between the plates? Is the positive or negative plate at the higher potential? (*b*) Find the work done by the electric field on the electron as the electron moves from the negative plate to the positive plate. Express your answer in both electron volts and joules. (*c*) What is the change in potential energy of the electron when it moves from the negative plate to the positive plate? What is its kinetic energy when it reaches the positive plate?

Potential and Potential Energy

4 • Explain the distinction between electric potential and electrostatic potential energy.

5 • A positive charge is released from rest in an electric field. Will it move toward a region of greater or smaller electric potential?

6 •• A lithium nucleus and an α particle are at rest. The lithium nucleus has a charge of $+3e$ and a mass of 7 u; the α particle has a charge of $+2e$ and a mass of 4 u. Which of the methods below would accelerate them both to the same kinetic energy?

(*a*) Accelerate them through the same electrical potential difference.

(*b*) Accelerate the α particle through potential V_1 and the lithium nucleus through $\frac{2}{3}V_1$.

(*c*) Accelerate the α particle through potential V_1 and the lithium nucleus through $\frac{7}{4}V_1$.

(*d*) Accelerate the α particle through potential V_1 and the lithium nucleus through $(2 \times 7)/(3 \times 4)V$.

(*e*) None of the above.

7 • A positive charge of magnitude 2 μC is at the origin. (*a*) What is the electric potential V at a point 4 m from the origin relative to $V = 0$ at infinity? (*b*) How much work must be done by an outside agent to bring a 3-μC charge from infinity to $r = 4$ m, assuming that the 2-μC charge is held fixed at the origin? (*c*) How much work must be done by an outside agent to bring the 2-μC charge from infinity to the origin if the 3-μC charge is first placed at $r = 4$ m and is then held fixed?

8 •• The distance between the K^+ and Cl^- ions in KCl is 2.80×10^{-10} m. Calculate the energy required to separate the two ions to an infinite distance apart, assuming them to be point charges initially at rest. Express your answer in eV.

9 •• Two identical masses m that carry equal charges q are separated by a distance d. Show that if both are released simultaneously their speeds when they are separated a great distance are $v/\sqrt{2}$, where v is the speed that one mass would have at a great distance from the other if it were released and the other held fixed.

10 •• Protons from a Van de Graaff accelerator are released from rest at a potential of 5 MV and travel through a vacuum to a region at zero potential. (*a*) Find the final speed of the 5-MeV protons. (*b*) Find the accelerating electric field if the same potential change occurred *uniformly* over a distance of 2.0 m.

11 •• An electron gun fires electrons at the screen of a television tube. The electrons start from rest and are accelerated through a potential difference of 30,000 V. What is the energy of the electrons when they hit the screen (*a*) in electron volts and (*b*) in joules? (*c*) What is the speed of impact of electrons with the screen of the picture tube?

12 •• (*a*) Derive an expression for the distance of closest approach of an α particle with kinetic energy E to a massive nucleus of charge Ze. Assume that the nucleus is fixed in space. (*b*) Find the distance of closest approach of a 5.0- and a 9.0-MeV α particle to a gold nucleus; the charge of the gold nucleus is $79e$. Neglect the recoil of the gold nucleus.

Systems of Point Charges

13 • Four 2-μC point charges are at the corners of a square of side 4 m. Find the potential at the center of the square (relative to zero potential at infinity) if (*a*) all the charges are positive, (*b*) three of the charges are positive and one is negative, and (*c*) two are positive and two are negative.

14 • Three point charges are on the x axis: q_1 is at the origin, q_2 is at $x = 3$ m, and q_3 is at $x = 6$ m. Find the potential at the point $x = 0$, $y = 3$ m if (*a*) $q_1 = q_2 = q_3 = 2$ μC, (*b*) $q_1 = q_2 = 2$ μC and $q_3 = -2$ μC, and (*c*) $q_1 = q_3 = 2$ μC and $q_2 = -2$ μC.

15 • Points A, B, and C are at the corners of an equilateral triangle of side 3 m. Equal positive charges of 2 μC are at A and B. (*a*) What is the potential at point C? (*b*) How much work is required to bring a positive charge of 5 μC from infinity to point C if the other charges are held fixed? (*c*) Answer parts (*a*) and (*b*) if the charge at B is replaced by a charge of -2 μC.

16 • A sphere with radius 60 cm has its center at the origin. Equal charges of 3 μC are placed at 60° intervals along the equator of the sphere. (*a*) What is the electric potential at the origin? (*b*) What is the electric potential at the north pole?

17 • Two point charges q and q' are separated by a distance a. At a point $a/3$ from q and along the line joining the two charges the potential is zero. Find the ratio q/q'.

18 •• Two positive charges $+q$ are on the x axis at $x = +a$ and $x = -a$. (*a*) Find the potential $V(x)$ as a function of x for points on the x axis. (*b*) Sketch $V(x)$ versus x. (*c*) What is the significance of the minimum on your curve?

19 •• A point charge of $+3e$ is at the origin and a second point charge of $-2e$ is on the x axis at $x = a$. (*a*) Sketch the potential function $V(x)$ versus x for all x. (*b*) At what point or points is $V(x)$ zero? (*c*) How much work is needed to bring a third charge $+e$ to the point $x = \frac{1}{2}a$ on the x axis?

Finding the Electric Field From the Potential

20 • If the electric potential is constant throughout a region of space, what can you say about the electric field in that region?

21 • If E is known at just one point, can V be found at that point?

22 • In what direction can you move relative to an electric field so that the electric potential does not change?

23 • A uniform electric field is in the negative x direction. Points a and b are on the x axis, a at $x = 2$ m and b at $x = 6$ m. (*a*) Is the potential difference $V_b - V_a$ positive or negative? (*b*) If the magnitude of $V_b - V_a$ is 10^5 V, what is the magnitude E of the electric field?

24 • The potential due to a particular charge distribution is measured at several points along the x axis as shown in Figure 24-21. For what value(s) in the range $0 < x < 10$ m is $E_x = 0$?

Figure 24-21 Problem 24

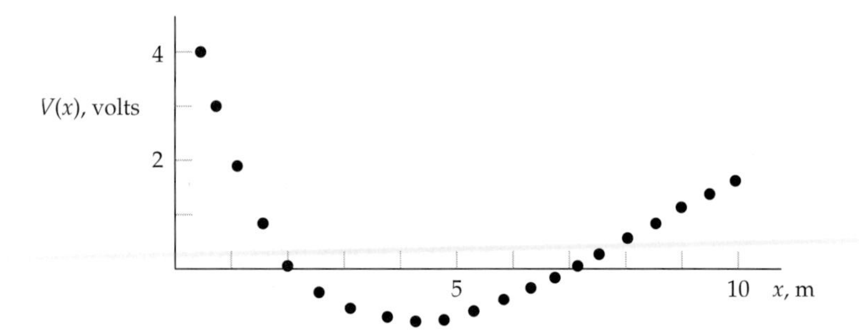

25 • A point charge $q = 3.00$ μC is at the origin. (*a*) Find the potential V on the x axis at $x = 3.00$ m and at $x = 3.01$ m. (*b*) Does the potential increase or decrease as x increases? Compute $-\Delta V/\Delta x$, where ΔV is the change in potential from $x = 3.00$ m to $x = 3.01$ m and $\Delta x = 0.01$ m. (*c*) Find the electric field at $x = 3.00$ m, and compare its magnitude with $-\Delta V/\Delta x$ found in part (*b*). (*d*) Find the potential (to three significant

figures) at the point $x = 3.00$ m, $y = 0.01$ m, and compare your result with the potential on the x axis at $x = 3.00$ m. Discuss the significance of this result.

26 • A charge of $+3.00\ \mu$C is at the origin, and a charge of $-3.00\ \mu$C is on the x axis at $x = 6.00$ m. (*a*) Find the potential on the x axis at $x = 3.00$ m. (*b*) Find the electric field on the x axis at $x = 3.00$ m. (*c*) Find the potential on the x axis at $x = 3.01$ m, and compute $-\Delta V/\Delta x$, where ΔV is the change in potential from $x = 3.00$ m to $x = 3.01$ m and $\Delta x = 0.01$ m. Compare your result with your answer to part (*b*).

27 • A uniform electric field is in the negative x direction. Points a and b are on the x axis, a at $x = 2$ m and b at $x = 6$ m. (*a*) Is the potential difference $V_b - V_a$ positive or negative? (*b*) If the magnitude of $V_b - V_a$ is 10^5 V, what is the magnitude E of the electric field?

28 • In the following, V is in volts and x is in meters. Find E_x when (*a*) $V(x) = 2000 + 3000x$; (*b*) $V(x) = 4000 + 3000x$; (*c*) $V(x) = 2000 - 3000x$; and (*d*) $V(x) = -2000$, independent of x.

29 • The electric potential in some region of space is given by $V(x) = C_1 + C_2x^2$, where V is in volts, x is in meters, and C_1 and C_2 are positive constants. Find the electric field E in this region. In what direction is E?

30 •• A charge q is at $x = 0$ and a charge $-3q$ is at $x = 1$ m. (*a*) Find $V(x)$ for a general point on the x axis. (*b*) Find the points on the x axis where the potential is zero. (*c*) What is the electric field at these points? (*d*) Sketch $V(x)$ versus x.

31 •• An electric field is given by $E_x = 2.0x^3$ kN/C. Find the potential difference between the points on the x axis at $x = 1$ m and $x = 2$ m.

32 •• Three equal charges lie in the xy plane. Two are on the y axis at $y = -a$ and $y = +a$, and the third is on the x axis at $x = a$. (*a*) What is the potential $V(x)$ due to these charges at a point on the x axis? (*b*) Find E_x along the x axis from the potential function $V(x)$. Evaluate your answers to (*a*) and (*b*) at the origin and at $x = \infty$ to see if they yield the expected results.

General Relation Between $\vec{E}$ and V (optional)

33 ••• The electric potential in a region of space is given by $V = (2\ \text{V/m}^2)x^2 + (1\ \text{V/m}^2)yz$. Find the electric field at the point $x = 2$ m, $y = 1$ m, $z = 2$ m.

34 ••• A potential is given by

$$V(x, y, z) = \frac{kQ}{\sqrt{(x-a)^2 + y^2 + z^2}}$$

(*a*) Find the components E_x, E_y, and E_z of the electric field by differentiating this potential function. (*b*) What simple charge distribution might be responsible for this potential?

Calculating V for Continuous Charge Distributions

35 •• In the calculation of V at a point x on the axis of a ring of charge, does it matter whether the charge Q is uniformly distributed around the ring? Would either V or E_x be different if it were not?

36 •• (*a*) Sketch $V(x)$ versus x for the uniformly charged ring in the yz plane given by Equation 24-20. (*b*) At what point is $V(x)$ a maximum? (*c*) What is E_x at this point?

37 • A charge of $q = +10^{-8}$ C is uniformly distributed on a spherical shell of radius 12 cm. (*a*) What is the magnitude of the electric field just outside and just inside the shell? (*b*) What is the magnitude of the electric potential just outside and just inside the shell? (*c*) What is the electric potential at the center of the shell? What is the electric field at that point?

38 • A disk of radius 6.25 cm carries a uniform surface charge density $\sigma = 7.5$ nC/m^2. Find the potential on the axis of the disk at a distance from the disk of (*a*) 0.5 cm, (*b*) 3.0 cm, and (*c*) 6.25 cm.

39 • An infinite line charge of linear charge density $\lambda = 1.5\ \mu$C/m lies on the z axis. Find the potential at distances from the line charge of (*a*) 2.0 m, (*b*) 4.0 m, and (*c*) 12 m, assuming that $V = 0$ at 2.5 m.

40 •• Derive Equation 24-21 by integrating the electric field E_x along the axis of the disk. (See Equation 23-11.)

41 •• A rod of length L carries a charge Q uniformly distributed along its length. The rod lies along the y axis with its center at the origin. (*a*) Find the potential as a function of position along the x axis. (*b*) Show that the result obtained in (*a*) reduces to $V = kQ/x$ for $x >> L$.

42 •• A disk of radius R carries a surface charge distribution of $\sigma = \sigma_0 R/r$. (*a*) Find the total charge on the disk. (*b*) Find the potential on the axis of the disk a distance x from its center.

43 •• Repeat Problem 42 if the surface charge density is $\sigma = \sigma_0 r^2/R^2$.

44 •• A rod of length L carries a charge Q uniformly distributed along its length. The rod lies along the y axis with one end at the origin. Find the potential as a function of position along the x axis.

45 •• A disk of radius R carries a charge density $+\sigma_0$ for $r < a$ and an equal but opposite charge density $-\sigma_0$ for $a < r < R$. The total charge carried by the disk is zero. (*a*) Find the potential a distance x along the axis of the disk. (*b*) Obtain an approximate expression for $V(x)$ when $x >> R$.

46 •• Use the result obtained in Problem 45(*a*) to calculate the electric field along the axis of the disk. Then calculate the electric field by direct integration using Coulomb's law.

47 •• A rod of length L has a charge Q uniformly distributed along its length. The rod lies along the x axis with its center at the origin. (*a*) What is the electric potential as a function of position along the x axis for $x > L/2$? (*b*) Show that for $x >> L/2$, your result reduces to that due to a point charge Q.

48 •• A conducting spherical shell of inner radius b and outer radius c is concentric with a small metal sphere of radius $a < b$. The metal sphere has a positive charge Q. The total charge on the conducting spherical shell is $-Q$. (*a*) What is the potential of the spherical shell? (*b*) What is the potential of the metal sphere?

49 •• Two very long, coaxial cylindrical shell conductors carry equal and opposite charges. The inner shell has radius a and charge $+q$; the other shell has radius b and charge $-q$. The length of each cylindrical shell is L. Find the potential difference between the shells.

50 •• A uniformly charged sphere has a potential on its surface of 450 V. At a radial distance of 20 cm from this surface, the potential is 150 V. What is the radius of the sphere, and what is the charge of the sphere?

51 •• Consider two infinite parallel planes of charge, one in the yz plane and the other at distance $x = a$. (*a*) Find the potential everywhere in space when $V = 0$ at $x = 0$ if the planes carry equal positive charge densities $+\sigma$. (*b*) Repeat the problem with charge densities equal and opposite, and the charge in the yz plane positive.

52 •• Show that for $x \gg R$ the potential on the axis of a disk charge approaches kQ/x, where $Q = \sigma\pi R^2$ is the total charge on the disk. (*Hint:* Write $(x^2 + R^2)^{1/2} = x(1 + R^2/x^2)^{1/2}$ and use the binomial expression.)

53 •• In Example 24-12 you derived the expression

$$V(r) = \frac{kQ}{2R}\left(3 - \frac{r^2}{R^2}\right)$$

for the potential inside a solid sphere of constant charge density by first finding the electric field. In this problem you derive the same expression by direct integration. Consider a sphere of radius R containing a charge Q uniformly distributed. You wish to find V at some point $r < R$. (*a*) Find the charge q' inside a sphere of radius r and the potential V_1 at r due to this part of the charge. (*b*) Find the potential dV_2 at r due to the charge in a shell of radius r' and thickness dr' at $r' > r$. (*c*) Integrate your expression in (*b*) from $r' = r$ to $r' = R$ to find V_2. (*d*) Find the total potential V at r from $V = V_1 + V_2$.

54 ••• A nonconducting sphere of radius R has a volume charge density $\rho = \rho_0 r/R$, where ρ_0 is a constant. (*a*) Show that the total charge is $Q = \pi R^3 \rho_0$. (*b*) Show that the total charge inside a sphere of radius $r < R$ is $q = Qr^4/R^4$. (*c*) Use Gauss's law to find the electric field E_r everywhere. (*d*) Use $dV = -E_r\, dr$ to find the potential V everywhere, assuming that $V = 0$ at $r = \infty$. (Remember that V is continuous at $r = R$.)

Equipotential Surfaces and Dielectric Breakdown

55 • Two charged metal spheres are connected by a wire, and sphere A is larger than sphere B (Figure 24-22). The magnitude of the electric potential of sphere A is

(*a*) greater than that at the surface of sphere B.
(*b*) less than that at the surface of sphere B.
(*c*) the same as that at the surface of sphere B.
(*d*) greater than or less than that at the surface of sphere B, depending on the radii of the spheres.
(*e*) greater than or less than that at the surface of sphere B, depending on the charge on the spheres.

Figure 24-22 Problem 55

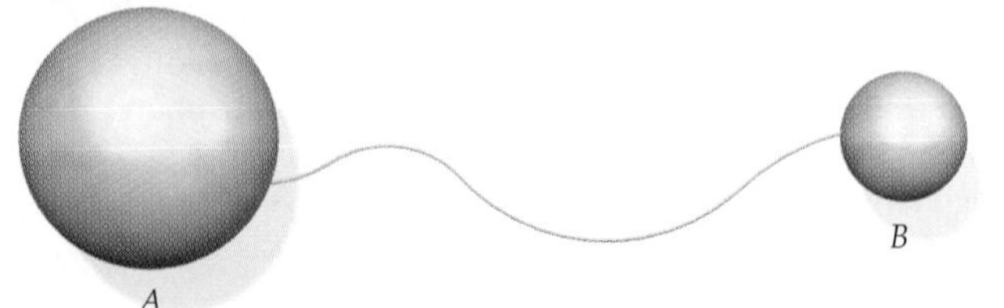

56 •• Figure 24-23 shows two parallel metal plates maintained at potentials of 0 and 60 V. Midway between the plates is a copper sphere. Sketch the equipotential surfaces and the electric field lines between the two plates.

Figure 24-23 Problem 56

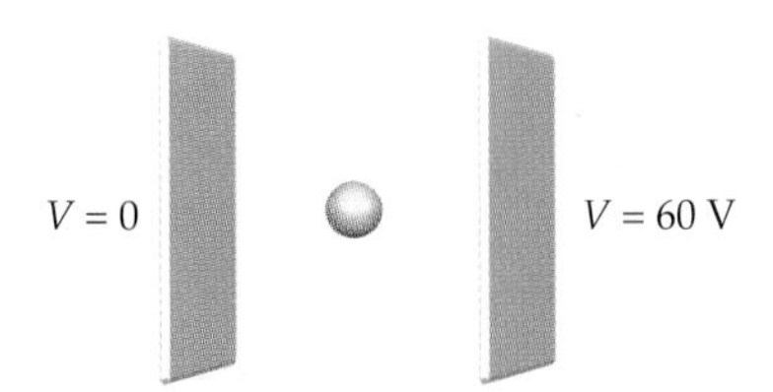

57 •• Figure 24-24 shows a metal sphere carrying a charge $-Q$ and a point charge $+Q$. Sketch the electric field lines and equipotential surfaces in the vicinity of this charge system.

Figure 24-24 Problems 57 and 58

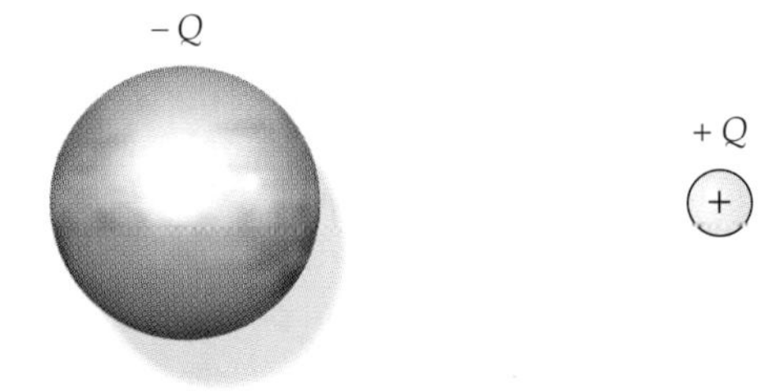

58 •• Repeat Problem 57 with the charge on the metal sphere changed to $+Q$.

59 •• Sketch the electric field lines and the equipotential surfaces both near and far from the conductor shown in Figure 24-20*a*, assuming that the conductor carries some charge Q.

60 •• Two equal positive charges are separated by a small distance. Sketch the electric field lines and the equipotential surfaces for this system.

61 • An infinite plane of charge has surface charge density 3.5 μC/m^2. How far apart are the equipotential surfaces whose potentials differ by 100 V?

62 • A point charge $q = +\frac{1}{9} \times 10^{-8}$ C is at the origin. Taking the potential to be zero at $r = \infty$, locate the equipotential surfaces at 20-V intervals from 20 to 100 V, and sketch them to scale. Are these surfaces equally spaced?

63 • (*a*) Find the maximum net charge that can be placed on a spherical conductor of radius 16 cm before dielectric breakdown of the air occurs. (*b*) What is the potential of the sphere when it carries this maximum charge?

64 • Find the greatest surface charge density σ_{max} that can exist on a conductor before dielectric breakdown of the air occurs.

65 •• Charge is placed on two conducting spheres that are very far apart and connected by a long thin wire (Figure 24-25). The larger sphere has a diameter twice that of the smaller. Which sphere has the largest electric field near its surface? By what factor is it larger than that at the surface of the other sphere?

Figure 24-25 Problem 65

66 •• Charge is placed on two conducting spheres that are very far apart and connected by a long thin wire. The radius of the smaller sphere is 5 cm and that of the larger sphere is 12 cm. The electric field at the surface of the larger sphere is 200 kV/m. Find the surface charge density on each sphere.

67 •• Two concentric spherical shell conductors carry equal and opposite charges. The inner shell has radius a and charge $+q$; the outer shell has radius b and charge $-q$. Find the potential difference between the shells, $V_a - V_b$.

68 •• Two identical uncharged metal spheres connected by a wire are placed close by two similar conducting spheres with equal and opposite charges as shown in Figure 24-26. (*a*) Sketch the electric field lines between spheres 1 and 3 and between spheres 2 and 4. (*b*) What can be said about the potentials V_1, V_2, V_3, and V_4 of the spheres? (*c*) If spheres 3 and 4 are connected by a wire, prove that the final charge on each must be zero.

Figure 24-26 Problem 68

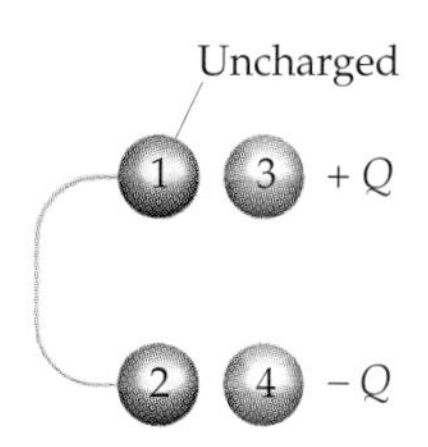

General Problems

69 • Two equal positive point charges $+Q$ are on the x axis. One is at $x = -a$ and the other is at $x = +a$. At the origin,

(*a*) $E = 0$ and $V = 0$.
(*b*) $E = 0$ and $V = 2kQ/a$.
(*c*) $\vec{E} = (2kQ^2/a^2)\,\hat{i}$ and $V = 0$.
(*d*) $\vec{E} = (2kQ^2/a^2)\,\hat{i}$ and $V = 2kQ/a$.
(*e*) none of the above is correct.

70 • The electrostatic potential is measured to be $V(x,y,z) = 4|x| + V_0$, where V_0 is a constant. The charge distribution responsible for this potential is

(*a*) a uniformly charged thread in the xy plane.
(*b*) a point charge at the origin.
(*c*) a uniformly charged sheet in the yz plane.
(*d*) a uniformly charged sphere of radius $1/\pi$ at the origin.

71 • Two point charges of equal magnitude but opposite sign are on the x axis; $+Q$ is at $x = -a$ and $-Q$ is at $x = +a$. At the origin,

(*a*) $E = 0$ and $V = 0$.
(*b*) $E = 0$ and $V = 2kQ/a$.
(*c*) $E = (2kQ^2/a^2)\,\hat{i}$ and $V = 0$.
(*d*) $E = (2kQ^2/a^2)\,\hat{i}$ and $V = 2kQ/a$.
(*e*) none of the above is correct.

72 •• True or false:

(*a*) If the electric field is zero in some region of space, the electric potential must also be zero in that region.
(*b*) If the electric potential is zero in some region of space, the electric field must also be zero in that region.
(*c*) If the electric potential is zero at a point, the electric field must also be zero at that point.
(*d*) Electric field lines always point toward regions of lower potential.
(*e*) The value of the electric potential can be chosen to be zero at any convenient point.
(*f*) In electrostatics, the surface of a conductor is an equipotential surface.
(*g*) Dielectric breakdown occurs in air when the potential is 3×10^6 V.

73 •• (*a*) V is constant on a conductor surface. Does this mean that σ is constant? (*b*) If E is constant on a conductor surface, does this mean that σ is constant? Does it mean that V is constant?

74 • An electric dipole has a positive charge of 4.8×10^{-19} C separated from a negative charge of the same magnitude by 6.4×10^{-10} m. What is the electric potential at a point 9.2×10^{-10} m from each of the two charges?

(*a*) 9.4 V
(*b*) Zero
(*c*) 4.2 V
(*d*) 5.1×10^9 V
(*e*) 1.7 V

75 • An electric field is given by $\vec{E} = ax\hat{i}$, where $\vec{E}$ is in newtons per coulomb, x is in meters, and a is a positive constant. (*a*) What are the SI units of a? (*b*) How much work is done by this field on a positive point charge q_0 when the charge moves from the origin to some point x? (*c*) Find the potential function $V(x)$ such that $V = 0$ at $x = 0$.

76 • Two positive charges $+q$ are on the y axis at $y = +a$ and $y = -a$. (*a*) Find the potential V for any point on the x axis. (*b*) Use your result in (*a*) to find the electric field at any point on the x axis.

77 • If a conducting sphere is to be charged to a potential of 10,000 V, what is the smallest possible radius of the sphere such that the electric field will not exceed the dielectric strength of air?

78 • An isolated aluminum sphere of radius 5.0 cm is at a potential of 400 V. How many electrons have been removed from the sphere to raise it to this potential?

79 • A point charge Q resides at the origin. A particle of mass $m = 0.002$ kg carries a charge of 4.0 μC. The particle is

released from rest at $x = 1.5$ m. Its kinetic energy as it passes $x = 1.0$ m is 0.24 J. Find the charge Q.

80 • A conducting wedge is charged to a potential V with respect to a large conducting sheet (Figure 24-27). (*a*) Sketch the electric field lines and the equipotentials for this configuration. Where along the x axis is $|E|$ greatest? (*b*) An electron of mass m_e leaves the sheet with zero velocity. What is its speed v when it arrives at the wedge? (Ignore the effect of gravity.)

Figure 24-27 Problem 80

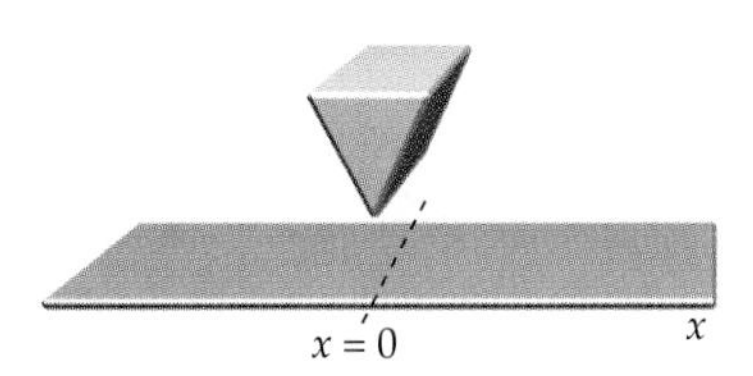

81 •• A Van de Graaff generator has a potential difference of 1.25 MV between the belt and the outer shell. Charge is supplied at the rate of 200 μC/s. What minimum power is needed to drive the moving belt?

82 •• A positive point charge $+Q$ is located at $x = -a$. (*a*) How much work is required to bring a second equal positive point charge $+Q$ from infinity to $x = +a$? (*b*) With the two equal positive point charges at $x = -a$ and $x = +a$, how much work is required to bring a third charge $-Q$ from infinity to the origin? (*c*) How much work is required to move the charge $-Q$ from the origin to the point $x = 2a$ along the semicircular path shown (Figure 24-28)?

Figure 24-28 Problem 82

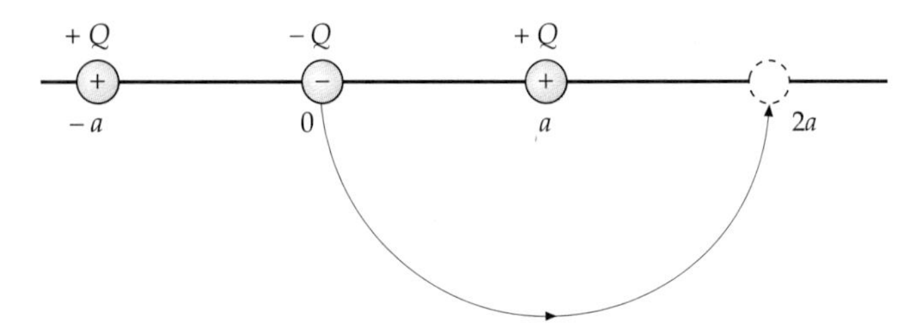

83 •• A charge of 2 nC is uniformly distributed around a ring of radius 10 cm that has its center at the origin and its axis along the x axis. A point charge of 1 nC is located at $x =$ 50 cm. Find the work required to move the point charge to the origin. Give your answer in both joules and electron volts.

84 •• The centers of two metal spheres of radius 10 cm are 50 cm apart on the x axis. The spheres are initially neutral, but a charge Q is transferred from one sphere to the other, creating a potential difference between the spheres of 100 V. A proton is released from rest at the surface of the positively charged sphere and travels to the negatively charged sphere. At what speed does it strike the negatively charged sphere?

85 •• A spherical conductor of radius R_1 is charged to 20 kV. When it is connected by a long, fine wire to a second conducting sphere far away, its potential drops to 12 kV. What is the radius of the second sphere?

86 •• A uniformly charged ring of radius a and charge Q lies in the yz plane with its axis along the x axis. A point charge Q' is placed on the x axis at $x = 2a$. (*a*) Find the potential at any point on the x axis due to the total charge $Q + Q'$. (*b*) Find the electric field for any point on the x axis.

87 •• A metal sphere centered at the origin carries a surface charge of charge density $\sigma = 24.6$ nC/m^2. At $r = 2.0$ m, the potential is 500 V and the magnitude of the electric field is 250 V/m. Determine the radius of the metal sphere.

88 •• Along the axis of a uniformly charged disk, at a point 0.6 m from the center of the disk, the potential is 80 V and the magnitude of the electric field is 80 V/m; at a distance of 1.5 m, the potential is 40 V and the magnitude of the electric field is 23.5 V/m. Find the total charge residing on the disk.

89 •• When you touch a friend after walking across a rug on a dry day, you typically draw a spark of about 2 mm. Estimate the potential difference between you and your friend before the spark.

90 •• When ^{235}U captures a neutron, it fissions (splits) into two nuclei, in the process emitting several neutrons that can cause other uranium nuclei to fission. Assume that the fission products are two nuclei of equal charges of $+46e$ and that these nuclei are at rest just after fission and are separated by twice their radius, $2R \approx 1.3 \times 10^{-14}$ m. (*a*) Calculate the electrostatic potential energy of the fission fragments. This is approximately the energy released per fission. (*b*) About how many fissions per second are needed to produce 1 MW of power in a reactor?

91 •• A radioactive ^{210}Po nucleus emits an α particle of charge $+2e$ and energy 5.30 MeV. Assume that just after the α particle is formed and escapes from the nucleus, it is a distance R from the center of the daughter nucleus ^{206}Pb, which has a charge $+82e$. Calculate R by setting the electrostatic potential energy of the two particles at this separation equal to 5.30 MeV. (Neglect the size of the α particle.)

92 •• Two large, parallel, nonconducting planes carry equal and opposite charge densities of magnitude σ. The planes have area A and are separated by a distance d. (*a*) Find the potential difference between the planes. (*b*) A conducting slab having thickness a and area A, the same area as the planes, is inserted between the original two planes. The slab carries no net charge. Find the potential difference between the original two planes and sketch the electric field lines in the region between the original two planes.

93 •• A uniformly charged ring with a total charge of 100 μC and a radius of 0.1 m lies in the yz plane with its center at the origin. A meterstick has a point charge of 10 μC on the end marked 0 and a point charge of 20 μC on the end marked 100 cm. How much work does it take to bring the meterstick from a long distance away to a position along the x axis with the end marked 0 at $x = 0.2$ m and the other end at $x = 1.2$ m.

94 •• Three large conducting plates are parallel to one another with the outer plates connected by a wire. The inner plate is isolated and carries a charge density σ_t on the upper surface and σ_b on the lower surface, where $\sigma_t + \sigma_b = 12\ \mu C/m^2$. The inner plate is 1 mm from the top plate and 3 mm from the bottom plate. Find the surface charge densities σ_t and σ_b.

95 ••• A point charge q_1 is at the origin and a second point charge q_2 is on the x axis at $x = a$ as in Example 24-5. (*a*) Calculate the electric field everywhere on the x axis from the potential function given in that example. (*b*) Find the potential at a general point on the y axis. (*c*) Use your result from (*b*) to calculate the y component of the electric field on the y axis. Compare your result with that obtained directly from Coulomb's law.

96 ••• A particle of mass m carrying a positive charge q is constrained to move along the x axis. At $x = -L$ and $x = L$ are two ring charges of radius L (Figure 24-29). Each ring is centered on the x axis and lies in a plane perpendicular to it. Each carries a positive charge Q. (*a*) Obtain an expression for the potential due to the ring charges as a function of x for $-L < x < L$. (*b*) Show that in this region, $V(x)$ is a minimum at $x = 0$. (*c*) Show that for $x \ll L$, the potential is of the form $V(x) = V(0) + \alpha x^2$. (*d*) Derive an expression for the angular frequency of oscillation of the mass m if it is displaced slightly from the origin and released.

Figure 24-29 Problem 96

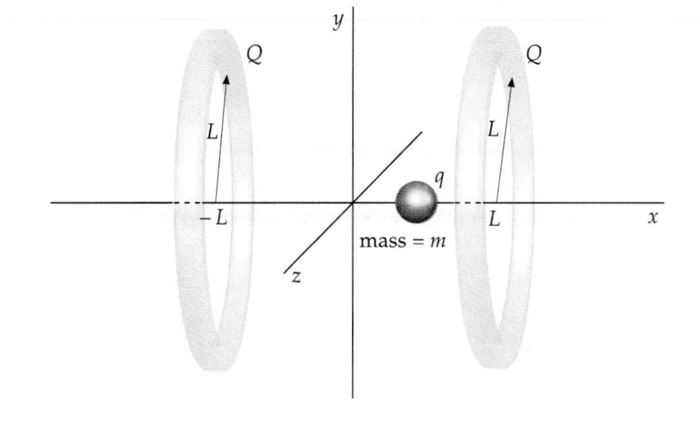

97 ••• Three concentric conducting spherical shells have radii a, b, and c such that $a < b < c$. Initially, the inner shell is uncharged, the middle shell has a positive charge Q, and the outer shell has a negative charge $-Q$. (*a*) Find the electric potential of the three shells. (*b*) If the inner and outer shells are now connected by a wire that is insulated as it passes through the middle shell, what is the electric potential of each of the three shells, and what is the final charge on each shell?

98 ••• Consider two concentric spherical metal shells of radii a and b, where $b > a$. The outer shell has a charge Q, but the inner shell is grounded. This means that the inner shell is at zero potential and that electric field lines leave the outer shell and go to infinity but other electric field lines leave the outer shell and end on the inner shell. Find the charge on the inner shell.

CHAPTER 25

Electrostatic Energy and Capacitance

A Leyden jar capacitor.

When we bring a point charge q from far away to a region where other charges are present, we must do work qV, where V is the potential at the final position due to the other charges in the vicinity. The work done is stored as electrostatic potential energy. The electrostatic potential energy of a system of charges is the total work needed to assemble the system.

When charge is placed on an isolated conductor, the potential of the conductor increases. The ratio of the charge to the potential is called the **capacitance** of the conductor. A useful device for storing charge and energy is the capacitor, which consists of two conductors, closely spaced but insulated from each other. When attached to a source of potential difference such as a battery, the conductors carry equal and opposite charges. The ratio of the magnitude of the charge on either conductor to the potential difference between the conductors is the capacitance of the capacitor. Capacitors have many uses. The flash attachment for your camera uses a capacitor to store the energy needed to provide the sudden flash of light. Capacitors are also used in the tuning circuits of devices such as radios, televisions, and cellular phones, allowing them to operate at specific frequencies.

The first capacitor was the Leyden jar, a glass container lined inside and out with gold foil. It was invented at the University of Leyden in the Netherlands by eighteenth-century experimenters who, while studying the effects of electric charges on people and animals, got the idea of trying to store a large amount of charge in a bottle of water. An experimenter held up a jar of water in one hand while charge was conducted to the water by a chain from a static electric generator. When he reached over to lift the chain out of the water with his other hand, he was knocked unconscious. Benjamin Franklin realized that the device for storing charge did not have to be jar-shaped and used foil-covered window glass, called Franklin panes. With several of these connected in parallel, he stored a large charge and attempted to kill a turkey with it. Instead, he knocked himself out. He later wrote, "I tried to kill a turkey but nearly succeeded in killing a goose."

25-1 Electrostatic Potential Energy

If we have a point charge q_1, the potential at a distance $r_{1,2}$ away is given by

$$V = \frac{kq_1}{r_{1,2}}$$

To bring up a second point charge q_2 from an infinite distance away to a distance $r_{1,2}$, we must do work:

$$W_2 = q_2 V = \frac{kq_2q_1}{r_{1,2}}$$

To bring up a third charge, work must be done against the electric field produced by both q_1 and q_2. The work required to bring up a third charge q_3 to a distance $r_{1,3}$ from q_1 and a distance $r_{2,3}$ from q_2 is

$$W_3 = \frac{kq_3q_1}{r_{1,3}} + \frac{kq_3q_2}{r_{2,3}}$$

The total work required to assemble the three charges is the **electrostatic potential energy** U of the system of three point charges:

$$U = \frac{kq_2q_1}{r_{1,2}} + \frac{kq_3q_1}{r_{1,3}} + \frac{kq_3q_2}{r_{2,3}} \qquad \text{25-1}$$

This quantity of work is independent of the order in which the charges are brought to their final positions. In general,

> The electrostatic potential energy of a system of point charges is the work needed to bring the charges from an infinite separation to their final positions.

Electrostatic potential energy of a system

The first two terms on the right-hand side of Equation 25-1 can be written

$$\frac{kq_2q_1}{r_{1,2}} + \frac{kq_3q_1}{r_{1,3}} = q_1\left(\frac{kq_2}{r_{1,2}} + \frac{kq_3}{r_{1,3}}\right) = q_1V_1$$

where V_1 is the potential due to charges q_2 and q_3. Similarly, the second and third terms represent the charge q_3 times the potential due to charges q_1 and q_2, and the first and third terms equal the charge q_2 times the potential due to charges q_1 and q_2. We can thus rewrite Equation 25-1 as

$$\begin{aligned} U &= \frac{kq_2q_1}{r_{1,2}} + \frac{kq_3q_1}{r_{1,3}} + \frac{kq_3q_2}{r_{2,3}} \\ &= \frac{1}{2}\left(\frac{kq_2q_1}{r_{1,2}} + \frac{kq_3q_1}{r_{1,3}} + \frac{kq_3q_2}{r_{2,3}} + \frac{kq_2q_1}{r_{1,2}} + \frac{kq_3q_1}{r_{1,3}} + \frac{kq_3q_2}{r_{2,3}}\right) \\ &= \frac{1}{2}\left[q_1\left(\frac{kq_2}{r_{1,2}} + \frac{kq_3}{r_{1,3}}\right) + q_2\left(\frac{kq_3}{r_{2,3}} + \frac{kq_1}{r_{1,2}}\right) + q_3\left(\frac{kq_1}{r_{1,3}} + \frac{kq_2}{r_{2,3}}\right)\right] \end{aligned}$$

The electrostatic potential energy U of a system of n point charges is thus

$$U = \frac{1}{2}\sum_{i=1}^{n} q_iV_i \qquad \text{25-2}$$

Electrostatic potential energy of a system of point charges

where V_i is the potential at the location of the ith charge due to all of the other charges.

Equation 25-2 also describes the electrostatic potential energy of a continuous charge distribution. Consider a spherical conductor of radius R. When the sphere carries a charge q, its potential relative to $V = 0$ at infinity is

$$V = \frac{kq}{R}$$

The work needed to bring an additional amount of charge dq from infinity to the conductor is $V\,dq$. This work equals the increase in the potential energy of the conductor:

$$dU = V\,dq = \frac{kq}{R}\,dq$$

The total potential energy U is the integral of dU as q increases from zero to its final value Q. Integrating from $q = 0$ to $q = Q$, we obtain

$$U = \frac{kQ^2}{2R} = \frac{1}{2}QV \qquad 25\text{-}3$$

where $V = kQ/R$ is the potential on the surface of the charged sphere. Although we derived Equation 25-3 for a spherical conductor, it holds for any conductor. The potential of any conductor is proportional to its charge q, so we can write $V = \alpha q$, where α is some constant. Then the work needed to bring an additional charge dq from infinity to the conductor is $V\,dq = \alpha q\,dq$, and the total work needed to put a charge Q on the conductor is $\frac{1}{2}\alpha Q^2 = \frac{1}{2}QV$. If we have a set of n conductors with the ith conductor at potential V_i and carrying a charge Q_i, the electrostatic potential energy is

$$U = \frac{1}{2}\sum_{i=1}^{n} Q_i V_i \qquad 25\text{-}4$$

Electrostatic potential energy of a system of conductors

Example 25-1

Points A, B, C, and D are at the corners of a square of side a, as shown in Figure 25-1. (*a*) Calculate the work required to place a positive charge q at each corner of the square by separately calculating the work required to bring each charge to its final position. (*b*) Show that Equation 25-2 gives the total work.

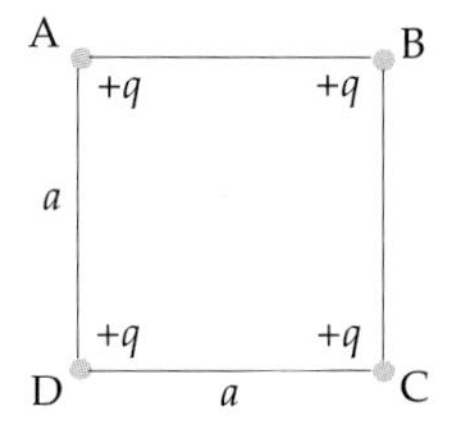

Figure 25-1

Picture the Problem No work is needed to place the first charge at point A since the potential there is zero when the other three charges are at infinity. As each additional charge is brought into place, work must be done because of the presence of the previous charges.

(*a*) 1. Place the first charge at A. To accomplish this step, the work W_A that is needed is zero:

$$W_A = 0$$

2. Bring the second charge to point B. The work required is $W_B = qV_A$, where V_A is the potential at B due to the first charge at A a distance a away:

$$W_B = qV_A = q\left(\frac{kq}{a}\right) = \frac{kq^2}{a}$$

3. $W_C = qV_C$, where V_C is the potential at C due to q at A a distance $\sqrt{2}\,a$ away, and q at B a distance a away:

$$W_C = qV_C = q\left(\frac{kq}{\sqrt{2}a} + \frac{kq}{a}\right) = \frac{kq^2}{\sqrt{2}a} + \frac{kq^2}{a}$$

4. Similar considerations give W_D, the work needed to bring the fourth charge to point D:

$$W_D = qV_D = q\left(\frac{kq}{a} + \frac{kq}{\sqrt{2}a} + \frac{kq}{a}\right) = \frac{2kq^2}{a} + \frac{kq^2}{\sqrt{2}a}$$

5. Summing the individual contributions gives the total work required to assemble the four charges:

$$W_{\text{total}} = W_A + W_B + W_C + W_D = \frac{4kq^2}{a} + \frac{2kq^2}{\sqrt{2}a}$$

$$= (4 + \sqrt{2})\frac{kq^2}{a}$$

(*b*) Calculate W_{total} from Equation 25-2. There are four identical terms, one from each charge:

$$W_{\text{total}} = U = \frac{1}{2}\sum_{i=1}^{4} q_iV_i = \frac{1}{2}\left[4 \times q\left(\frac{kq}{a} + \frac{kq}{a} + \frac{kq}{\sqrt{2}a}\right)\right]$$

$$= \frac{4kq^2}{a} + \frac{2kq^2}{\sqrt{2}a} = (4 + \sqrt{2})\frac{kq^2}{a}$$

Remark W_{total} is the total electrostatic energy of the charge distribution.

Exercise (*a*) How much work is required to bring a fifth positive charge q from infinity to the center of the square? (*b*) What is the total work required to assemble the five-charge system? (*Answers* (*a*) $4\sqrt{2}\,kq^2/a$, (*b*) $(4 + 5\sqrt{2})kq^2/a$)

25-2 Capacitance

The potential (relative to zero potential at infinity) of a single isolated conductor carrying a charge Q is proportional to the charge Q, and depends on the size and shape of the conductor. In general, the larger the conductor, the greater the amount of charge it can carry for a given potential. For example, the potential of a spherical conductor of radius R carrying a charge Q is

$$V = \frac{kQ}{R}$$

The ratio of charge Q to the potential V of an isolated conductor is called its capacitance C:

$$C = \frac{Q}{V} \qquad \text{25-5}$$

Definition—Capacitance

Capacitance is a measure of the capacity to store charge for a given potential difference. Since the potential is always proportional to the charge, this ratio does not depend on either Q or V, but only on the size and shape of the conductor. The capacitance of a spherical conductor is

$$C = \frac{Q}{V} = \frac{Q}{kQ/R} = \frac{R}{k} = 4\pi\epsilon_0 R \qquad \text{25-6}$$

The SI unit of capacitance is the coulomb per volt, which is called a **farad** (F) after the great English experimentalist Michael Faraday:

$$1\text{ F} = 1\text{ C/V} \qquad \text{25-7}$$

Since the farad is a rather large unit, submultiples such as the microfarad ($1\ \mu\text{F} = 10^{-6}$ F) or the picofarad ($1\text{ pF} = 10^{-12}$ F) are often used. Since capacitance is in farads and R is in meters, we can see from Equation 25-6 that the SI unit for the permittivity of free space, ϵ_0, can also be written as a farad per meter:

$$\epsilon_0 = 8.85 \times 10^{-12}\text{ F/m} = 8.85\text{ pF/m} \qquad \text{25-8}$$

Exercise Find the radius of a spherical conductor that has a capacitance of 1 farad. (*Answer* 8.99×10^9 m, which is about 1400 times the radius of the earth)

We see from the above exercise that the farad is indeed a very large unit.

Exercise A sphere of capacitance C_1 carries a charge of 20 μC. If the charge is increased to 60 μC, what is the new capacitance C_2? (*Answer* $C_2 = C_1$. The capacitance does not depend on the charge. If the charge is tripled, the potential of the sphere will be tripled and the ratio Q/V, which depends only on the radius of the sphere, remains unchanged.)

Capacitors

A system of two conductors carrying equal but opposite charges is called a **capacitor.** A capacitor is usually charged by transferring a charge Q from one conductor to the other leaving one of the conductors with a charge $+Q$ and the other with a charge $-Q$. The capacitance of the device is defined to be Q/V, where Q is the magnitude of the charge on either conductor and V is the magnitude of the potential difference between the conductors.* To calculate the capacitance, we place equal and opposite charges on the conductors and then find the potential difference V by first finding the electric field $\vec{E}$ between them.

Parallel-Plate Capacitors A common capacitor is the **parallel-plate capacitor,** which utilizes two parallel conducting plates. In practice, the plates may be thin metallic foils that are separated and insulated from one another by a thin plastic film. This "sandwich" is then rolled up, which allows for a large surface area in a relatively small space. Let A be the area of each plate and let d be the separation distance, which is small compared to the length and width of the plates. We place a charge $+Q$ on one plate and $-Q$ on the other. These charges attract each other and become uniformly distributed on the inside surfaces of the plates. Since the plates are close together, the electric field between them is approximately the same as the field between two equal and opposite infinite planes of charge. Each plate contributes a uniform field of magnitude $E = \sigma/2\epsilon_0$ (Equation 23-21) giving a total field $E = \sigma/\epsilon_0$, where $\sigma = Q/A$ is the magnitude of the charge per unit area on either plate. Since E is uniform between the plates (Figure 25-2), the potential difference between the plates equals the field times the plate separation d:

$$V = Ed = \frac{\sigma}{\epsilon_0}d = \frac{Qd}{\epsilon_0 A} \qquad \text{25-9}$$

The capacitance of the parallel-plate capacitor is thus

$$C = \frac{Q}{V} = \frac{\epsilon_0 A}{d} \qquad \text{25-10}$$

Capacitance of a parallel-plate capacitor

Note that since V is proportional to Q, the capacitance does not depend on either Q or V. For a parallel-plate capacitor, the capacitance is proportional to the area of the plates and is inversely proportional to the separation distance. In general, capacitance depends on the size, shape, and geometrical arrangement of the conductors and on the insulating medium between them.

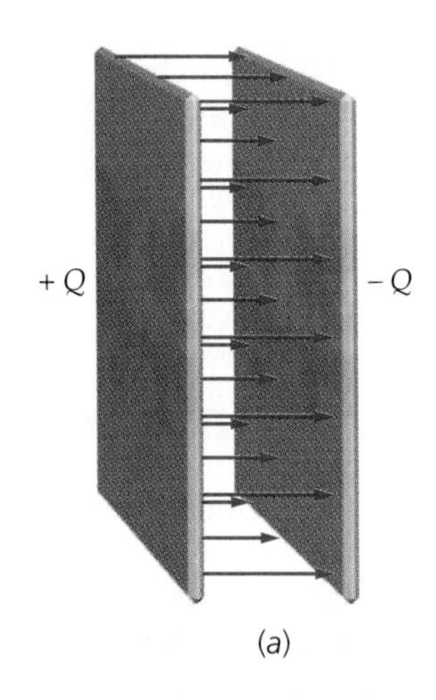

(*a*)

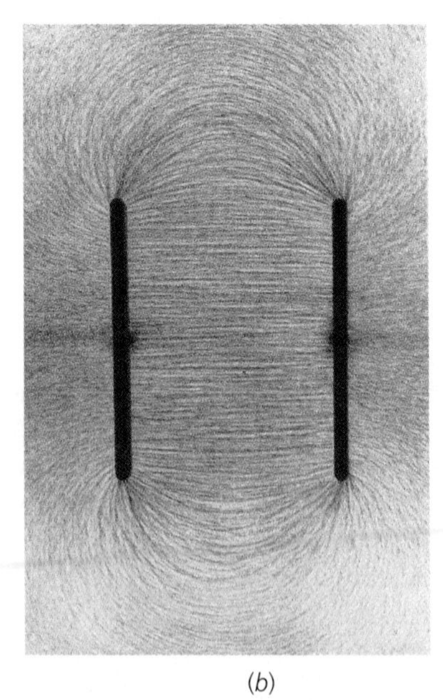

(*b*)

Figure 25-2 (*a*) Electric field lines between the plates of a parallel-plate capacitor. The lines are equally spaced between the plates, indicating that the field is uniform. (*b*) Electric field lines in a parallel-plate capacitor shown by small bits of thread suspended in oil.

* When we speak of the charge on a capacitor, we mean the magnitude of the charge on either conductor. The use of V rather than ΔV for the magnitude of the potential difference between the plates is standard and simplifies many of the equations relating to capacitance.

When a capacitor is connected to a battery,* as in Figure 25-3, charge is transferred from one conductor to the other until the potential difference between the conductors equals the potential difference across the battery terminals. The amount of charge transferred is $Q = CV$.

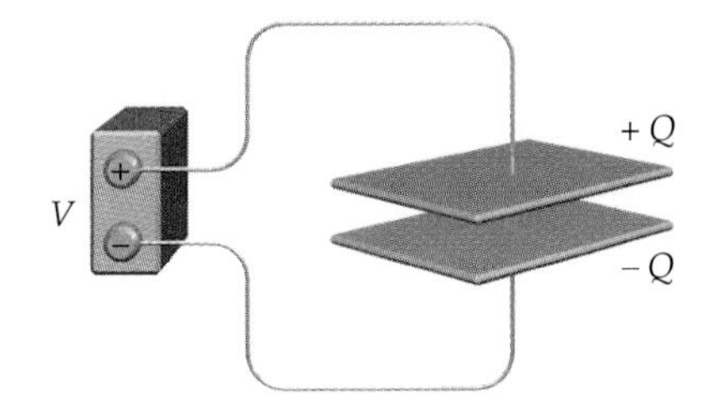

Figure 25-3 When the conductors in a capacitor are connected to the terminals of a battery, the battery transfers charge from one conductor to the other until the potential difference between the conductors equals that between the battery terminals. The amount of charge transferred is proportional to the potential difference.

*We will discuss batteries more fully in Chapter 26. Here, all we need to know is that a battery is a device that stores and supplies electrical energy and maintains a constant potential difference V between its terminals.

Example 25-2

A parallel-plate capacitor has square plates of side 10 cm separated by 1 mm. (*a*) Calculate the capacitance of this device. (*b*) If this capacitor is charged to 12 V, how much charge is transferred from one plate to another?

Picture the Problem The capacitance C is determined by the area and the separation of the plates. Once C is found, the charge for a given voltage V is found from the definition of capacitance $C = Q/V$.

(*a*) We find the capacitance using Equation 25-10:
$$C = \frac{\epsilon_0 A}{d} = \frac{(8.85\ \text{pF/m})(0.1\ \text{m})^2}{0.001\ \text{m}} = 88.5\ \text{pF}$$

(*b*) The charge transferred is found from the definition of capacitance:
$$Q = CV = (88.5\ \text{pF})(12\ \text{V}) = 1.06 \times 10^{-9}\ \text{C} = 1.06\ \text{nC}$$

Remarks Q is the magnitude of the charge on each plate of the capacitor. In this case, Q corresponds to roughly 6.6×10^9 electrons.

Exercise How large would the plates have to be for the capacitance to be 1 F? (*Answer* $A = 1.13 \times 10^8\ \text{m}^2$, which corresponds to a square 10.6 km on a side)

Cylindrical Capacitors A cylindrical capacitor consists of a small conducting cylinder or wire of radius r_1 and a larger, concentric cylindrical conducting shell of radius r_2. A coaxial cable, such as that used for cable television, can be thought of as a cylindrical capacitor. The capacitance per unit length of a coaxial cable is important in determining the transmission characteristics of the cable.

Example 25-3

Find an expression of the capacitance of a cylindrical capacitor consisting of two conductors both of length L. One cylinder has radius r_1 and the other is a coaxial cylindrical shell of inner radius r_2, with $r_1 < r_2 \ll L$ as shown in Figure 25-4.

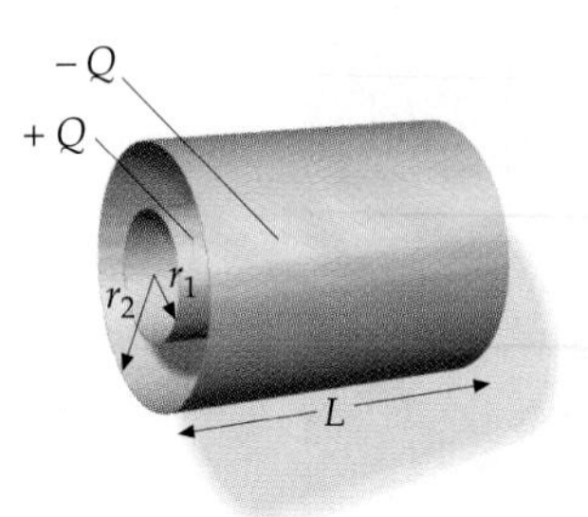

Figure 25-4

Picture the Problem We place charge $+Q$ on the inner conductor and $-Q$ on the outer conductor and calculate the potential difference $V = V_1 - V_2$ from the electric field between the conductors, which is found from Gauss's law. Since the electric field depends on r, we must integrate to find the potential difference.

1. The capacitance is defined as the ratio Q/V: $C = Q/V$

2. V is related to the electric field between the shells: $dV = -\vec{E}\cdot d\vec{\ell} = -E_r\,dr$

3. To find E_r we choose a cylindrical Gaussian surface of radius r between the conductors ($r_1 < r < r_2$). The area of the Gaussian surface is then $2\pi rL$. Gauss's law gives:

$$\oint_S E_n\,dA = \frac{1}{\epsilon_0}Q_{\text{inside}}$$

$$E_r 2\pi rL = \frac{Q}{\epsilon_0}$$

4. Solve for E_r:

$$E_r = \frac{1}{2\pi L\epsilon_0}\frac{Q}{r}$$

5. Integrate to find $V = V_1 - V_2$:

$$V = V_1 - V_2 = \int_{r_2}^{r_1} dV = \int_{r_2}^{r_1} -E_r\,dr$$

$$= \int_{r_1}^{r_2} +E_r\,dr = \frac{Q}{2\pi\epsilon_0 L}\int_{r_1}^{r_2}\frac{dr}{r} = \frac{Q}{2\pi\epsilon_0 L}\ln\frac{r_2}{r_1}$$

6. Substitute this result to find C:

$$C = \frac{Q}{V} = \frac{2\pi\epsilon_0 L}{\ln(r_2/r_1)}$$

Remarks The capacitance of a cylindrical capacitor is proportional to the length of the conductors.

Exercise How is the capacitance affected if the potential across a cylindrical capacitor is increased from 20 to 80 V? (*Answer* The capacitance of any capacitor does not depend on the potential. To increase V, you must increase the charge Q. The ratio Q/V depends only on the geometry of the capacitor.)

From Example 25-3 we see that the capacitance of a cylindrical capacitor is given by

$$C = \frac{2\pi\epsilon_0 L}{\ln(r_2/r_1)} \qquad \text{25-11}$$

Capacitance of a cylindrical capacitor

A coaxial cable is a long cylindrical capacitor with a solid wire for the inner conductor and a braided-wire shield for the outer conductor. The outer rubber coating has been pealed back here to show the conductors and the white plastic insulator that separates the conductors.

Cutaway of a 200-μF capacitor used in an electronic strobe light.

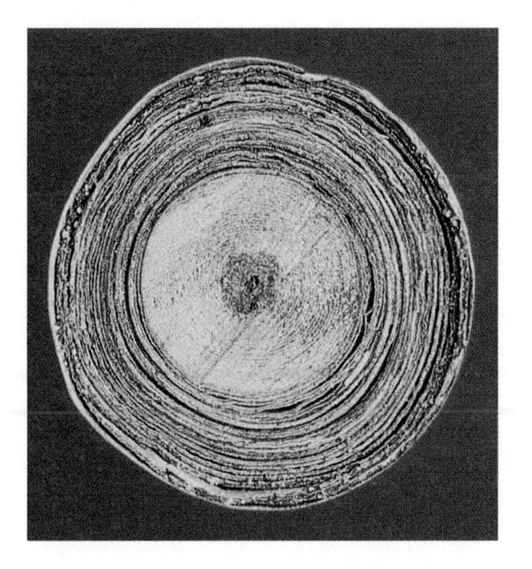

Cross section of a foil-wound capacitor.

A variable air-gap capacitor like those that were used in the tuning circuits of old radios. The semicircular plates rotate through the fixed plates, changing the amount of surface area between them, and hence the capacitance.

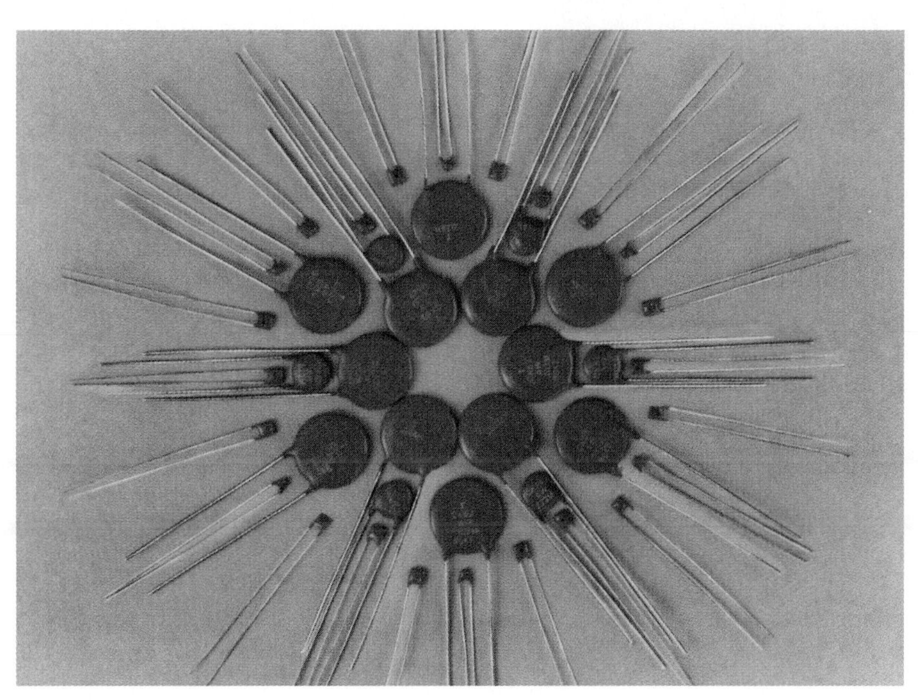

Ceramic capacitors for use in electronic circuits.

25-3 The Storage of Electrical Energy

When a capacitor is being charged, positive charge is transferred from the negatively charged conductor to the positively charged conductor. Work must therefore be done to charge a capacitor. Some of this work is stored as electrostatic potential energy.

Let q be the charge that has been transferred at some time during the charging process. The potential difference is then $V = q/C$. If a small amount of additional charge dq is now transferred from the negative conductor to the positive conductor through a potential increase of V (Figure 25-5), the potential energy of the charge is increased by

$$dU = V\,dq = \frac{q}{C}\,dq$$

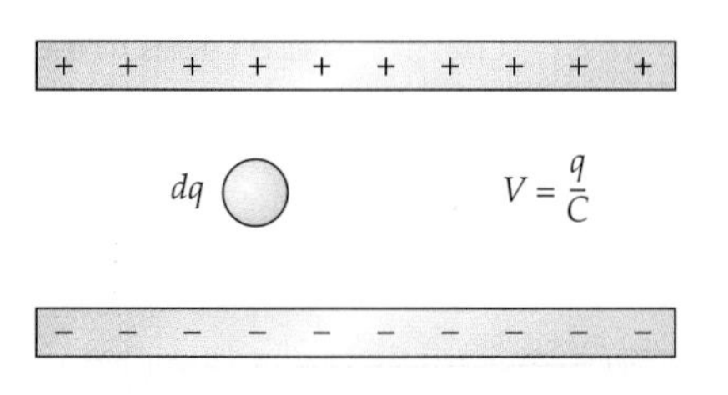

Figure 25-5 When a small amount of positive charge dq is moved from the negative conductor to the positive conductor, its potential energy is increased by $dU = V\,dq$, where V is the potential difference between the conductors.

The total increase in potential energy U is the integral of dU as q increases from zero to its final value Q (Figure 25-6):

$$U = \int dU = \int_0^Q \frac{q}{C}\,dq = \frac{1}{2}\frac{Q^2}{C}$$

This potential energy is the energy stored in the capacitor. Using $C = Q/V$, we can express this energy in a variety of ways:

$$U = \frac{1}{2}\frac{Q^2}{C} = \frac{1}{2}QV = \frac{1}{2}CV^2 \qquad 25\text{-}12$$

Energy stored in a capacitor

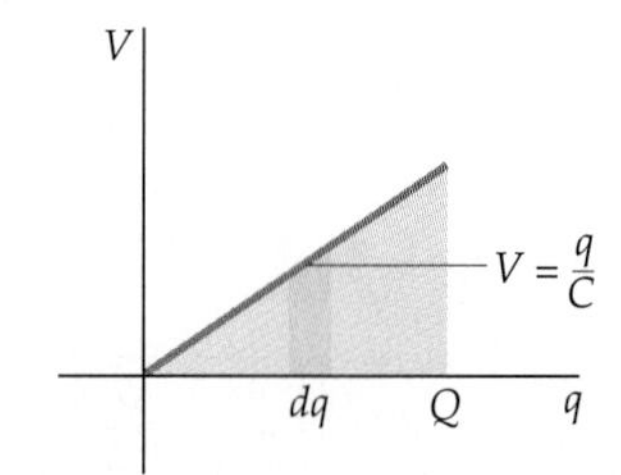

Figure 25-6 The work needed to charge a capacitor is the integral of $V\,dq$ from the original charge of $q = 0$ to the final charge of $q = Q$. This work is the area under the curve $\frac{1}{2}Q(Q/C)$.

Exercise A 15-μF capacitor is charged to 60 V. How much energy is stored in the capacitor? (*Answer* 0.027 J)

Exercise Obtain the expression for the electrostatic energy stored in a capacitor (Equation 25-12) from Equation 25-4 using $Q_1 = +Q$, $Q_2 = -Q$, and $V = V_1 - V_2$.

A capacitor is usually charged by connecting it to a battery, which maintains a potential difference V between its terminals. The work done by the battery in delivering a charge Q to the capacitor is QV, which is twice the energy stored in the capacitor. The additional work done by the battery is dissipated in heat in the connecting wires or radiation.*

*We will show in Section 26-6 that if the capacitor is connected to a battery by wires of some resistance R, half the energy supplied by the battery in charging the capacitor is dissipated as thermal energy in the wires.

Example 25-4

A parallel-plate capacitor with square plates 14 cm on a side and separated by 2.0 mm is connected to a battery and charged to 12 V. The battery is then disconnected from the capacitor and the plate separation is increased to 3.5 mm. (*a*) What is the charge on the capacitor? (*b*) How much energy was originally stored in the capacitor? (*c*) By how much is the energy increased when the plate separation is changed?

Picture the Problem (*a*) The charge on the capacitor can be calculated from the capacitance and then used to calculate the energy in (*b*). (*c*) Since the capacitor is removed from the battery, the charge remains constant as the plates are separated. The energy increase is found by using the charge and new potential to calculate the new energy, from which we subtract the original energy.

(*a*) 1. The charge on the capacitor equals the product of C and V: $\quad Q = CV$

2. Calculate the capacitance of the parallel-plate capacitor: $\quad C = \frac{\epsilon_0 A}{d} = \frac{(8.85\text{ pF/m})(0.14\text{ m})^2}{0.002\text{ m}} = 86.7\text{ pF}$

3. Substitute to calculate Q: $\quad Q = CV = (86.7\text{ pF})(12\text{ V}) = 1.04\text{ nC}$

(*b*) Calculate the original energy stored: $\quad U = \frac{1}{2}QV = \frac{1}{2}(1.04\text{ nC})(12\text{ V}) = 6.24\text{ nJ}$

(*c*) 1. After the capacitor is removed from the battery, the charge on the plates remains constant. The potential between the plates is the field E times the plate separation d: $\quad V = Ed$

2. Since E does not change, and the potential difference is 12 V when the plate separation is 2.0 mm, the potential difference when the separation is 3.5 mm is: $V' = (12\text{ V})\dfrac{3.5\text{ mm}}{2.0\text{ mm}} = 21\text{ V}$

3. At $d = 3.5$ mm, with $V' = 21$ V, the energy stored is: $U = \frac{1}{2}QV = \frac{1}{2}(1.04\text{ nC})(21\text{ V}) = 10.92\text{ nJ}$

4. Subtract the original energy to find the increase: $\Delta U = 10.92\text{ nJ} - 6.24\text{ nJ} = 4.68\text{ nJ}$

Remarks The additional energy calculated in part (*c*) comes from work done by the agent responsible for increasing the separation between the plates, which attract each other.

Figure 25-7 shows how this effect is used in a switch.

Exercise Find the original energy in the capacitor of this example directly from $U = \frac{1}{2}CV^2$. (*Answer* 6.24 nJ)

Exercise (*a*) Find the new capacitance C' in this example when separation of the plates is 3.5 mm. (*b*) Use your result to calculate the new energy from $U = Q^2/2C'$. (*Answers* (*a*) $C' = 49.6$ pF (*b*) $U = 10.92$ nJ)

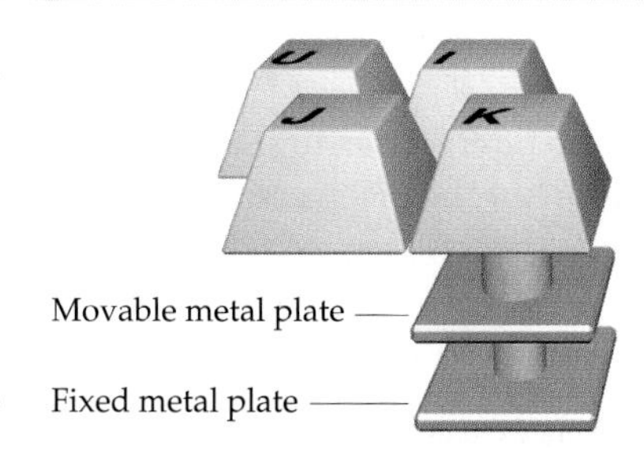

Figure 25-7 Capacitance switching in computer keyboards. A metal plate attached to each key acts as the top plate of a capacitor. Depressing the key decreases the separation between the top and bottom keys and increases the capacitance, which triggers the electronic circuitry of the computer to acknowledge the keystroke.

It is instructive to work part *c* of Example 25-4 in another way. The oppositely charged plates of a capacitor exert attractive forces on one another. Work must be done against these forces to increase the plate separation. Assume that the lower plate is held fixed and the upper plate is moved. The force on the upper plate is the charge on the plate Q times the electric field *due to the lower plate.* This field is half the total field between the plates, because the charge on the upper plate also contributes equally to the field. When the potential difference is 12 V and the separation is 2 mm, the total field between the plates is

$$E = \frac{V}{d} = \frac{12\text{ V}}{2\text{ mm}} = 6\text{ V/mm} = 6\text{ kV/m}$$

The force exerted on the upper plate by the bottom plate is thus

$$F = QE' = Q(\tfrac{1}{2}E) = (1.04\text{ nC})(3\text{ kV/m}) = 3.12\ \mu\text{N}$$

The work that must be done to move the upper plate a distance of $\Delta d =$ 1.5 mm is then

$$W = F\,\Delta d = (3.12\ \mu\text{N})(1.5\text{ mm}) = 4.68\text{ nJ}$$

This work equals the increase in the energy stored.

Electrostatic Field Energy

In the process of charging a capacitor, an electric field is produced between the plates. The work required to charge the capacitor can be thought of as the work required to create the electric field. That is, we can think of the energy stored in a capacitor as energy stored in the electric field, called **electrostatic field energy**.

Consider a parallel-plate capacitor. We can relate the energy stored in the capacitor to the electric field E between the plates. The potential difference between the plates is related to the electric field by $V = Ed$, where d is the plate separation. The capacitance is given by $C = \epsilon_0 A/d$ (Equation 25-10). The energy stored is

$$U = \frac{1}{2}CV^2 = \frac{1}{2}\left(\frac{\epsilon_0 A}{d}\right)(Ed)^2 = \frac{1}{2}\epsilon_0 E^2(Ad)$$

The quantity Ad is the volume of the space between the plates of the capacitor containing the electric field. The energy per unit volume is called the **energy density** u_e. The energy density in an electric field E is thus

$$u_e = \frac{\text{energy}}{\text{volume}} = \frac{1}{2}\epsilon_0 E^2 \qquad \text{25-13}$$

Energy density of an electrostatic field

Thus, the energy per unit volume of the electrostatic field is proportional to the square of the electric field. *Although we obtained Equation 25-13 by considering the electric field between the plates of a parallel-plate capacitor, the result applies to any electric field.* Whenever there is an electric field in space, the electrostatic energy per unit volume is given by Equation 25-13.

Exercise (*a*) Calculate the energy density u_e for Example 25-4 when the plate separation is 2.0 mm. (*b*) Show that the increase in energy in Example 25-4 is equal to u_e times the increase in volume (Δ vol) between the plates. (*Answers* (*a*) $u_e = \frac{1}{2}\epsilon_0 E^2 = 159.3\ \mu\text{J/m}^3$, (*b*) $\Delta\text{vol} = A\,\Delta d = 2.94 \times 10^{-5}\ \text{m}^3$, $u_e\,\Delta\text{vol} = 4.68$ nJ, in agreement with Example 25-4)

We can illustrate the generality of Equation 25-13 by calculating the electrostatic field energy of a spherical conductor of radius R that carries a charge Q. The electrostatic potential energy in terms of the charge Q and potential V is given by Equation 25-12:

$$U = \frac{kQ^2}{2R} = \frac{1}{2}QV \qquad \text{25-14}$$

We now obtain the same result by considering the energy density of an electric field given by Equation 25-13. When the conductor carries a charge Q, the electric field is radial and is given by

$$E_r = 0, \qquad r < R \text{ (inside the conductor)}$$

$$E_r = \frac{kQ}{r^2}, \qquad r > R \text{ (outside the conductor)}$$

Since the electric field is spherically symmetric, we choose a spherical shell for our volume element. If the radius of the shell is r and its thickness is dr, the volume is $d\mathcal{V} = 4\pi r^2\,dr$ (Figure 25-8). The energy dU in this volume element is

$$dU = u_e\,d\mathcal{V} = \frac{1}{2}(\epsilon_0 E^2)4\pi r^2 dr$$

$$= \frac{1}{2}\epsilon_0\left(\frac{kQ}{r^2}\right)^2(4\pi r^2\,dr) = \frac{1}{2}(4\pi\epsilon_0 k^2)Q^2\frac{dr}{r^2} = \frac{1}{2}kQ^2\frac{dr}{r^2}$$

where we have used $4\pi\epsilon_0 = 1/k$. Since the electric field is zero for $r < R$, we obtain the total energy in the electric field by integrating from $r = R$ to $r = \infty$:

$$U = \int u_e\,d\mathcal{V} = \frac{1}{2}kQ^2\int_R^\infty \frac{dr}{r^2} = \frac{1}{2}k\frac{Q^2}{R} = \frac{1}{2}Q\left(\frac{kQ}{R}\right) = \frac{1}{2}QV \qquad \text{25-15}$$

which is the same as Equation 25-12.

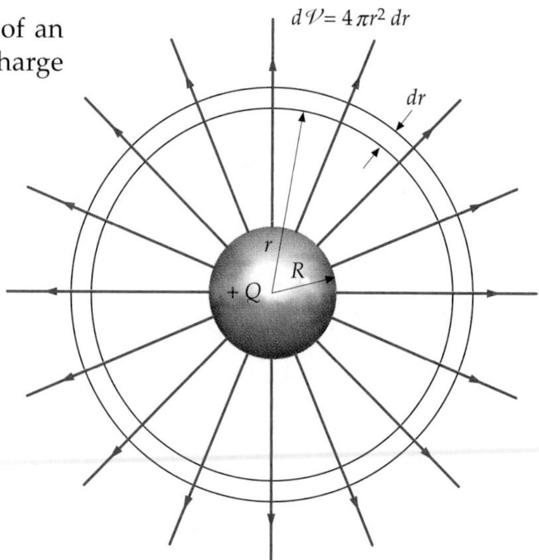

Figure 25-8 Geometry for the calculation of the electrostatic energy of a spherical conductor carrying a charge Q. The volume of the space between r and $r + dr$ is $d\mathcal{V} = 4\pi r^2\,dr$. The electrostatic field energy in this volume element is $u_e d\mathcal{V}$, where $u_e = \frac{1}{2}\epsilon_0 E^2$ is the energy density.

25-4 Combinations of Capacitors

Two or more capacitors are often used in combination. When two capacitors are connected as shown in Figure 25-9 such that the upper plates of the two capacitors are connected by a conducting wire and are therefore at a common potential V_a, and the lower plates are also connected together and are at a common potential V_b, the capacitors are said to be connected in **parallel.** (In electric circuits, a capacitor is indicated by the symbol ⫞⫟.) When capacitors are connected in parallel, the potential difference is the same across each capacitor.

In Figure 25-10, two capacitors are connected so that the magnitude of the charge on the two capacitors must be the same. These capacitors are said to be connected in **series.** The potential difference across the series combination is the sum of the potential differences across the individual capacitors.

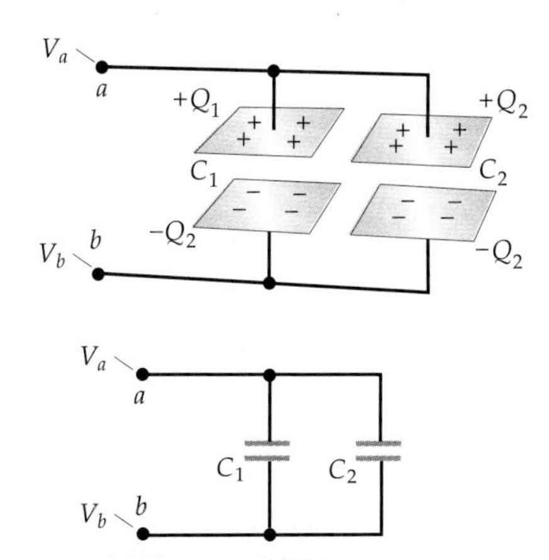

Figure 25-9 Two capacitors in parallel. The upper plates are connected together and are therefore at a common potential V_a; the lower plates are similarly connected together and therefore at a common potential V_b.

Parallel Capacitors

In Figure 25-9, assume that points a and b are connected to a battery or some other device that maintains a potential difference $V = V_a - V_b$ between the plates of each capacitor. If the capacitances are C_1 and C_2, the charges Q_1 and Q_2 stored on the plates are given by

$$Q_1 = C_1V$$

and

$$Q_2 = C_2V$$

The total charge stored is

$$Q = Q_1 + Q_2 = C_1V + C_2V = (C_1 + C_2)V$$

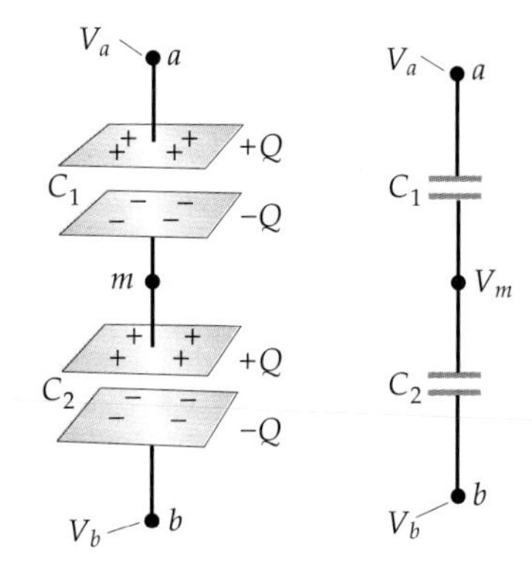

Figure 25-10 Two capacitors in series. The charge is the same on each capacitor.

A combination of capacitors in a circuit can sometimes be replaced by a single capacitor that stores the same amount of charge for a given potential difference. The substitute capacitor is said to have an **equivalent capacitance**. The equivalent capacitance of two capacitors in parallel is the ratio of the total charge stored to the potential difference:

$$C_{eq} = \frac{Q}{V} = C_1 + C_2 \qquad \text{25-16}$$

Thus, for two capacitors in parallel, C_{eq} is the sum of the individual capacitances. When we add a second capacitor in parallel, we increase the capacitance. The conductor area is essentially increased, allowing more charge to be stored for the same potential difference.

The same reasoning can be extended to three or more capacitors connected in parallel, as in Figure 25-11:

$$C_{eq} = C_1 + C_2 + C_3 + \cdots \qquad \text{25-17}$$

Equivalent capacitance for capacitors in parallel

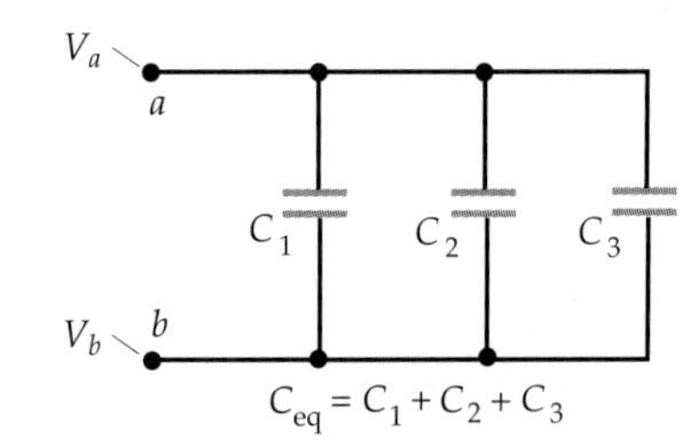

Figure 25-11 Three capacitors in parallel. The effect of adding a parallel capacitor to a circuit is an increase in the equivalent capacitance.

Series Capacitors

The capacitors in Figure 25-10 are connected in series. When points a and b are connected to the terminals of a battery, there is a potential difference $V = V_a - V_b$ across the two capacitors. If a charge $+Q$ is placed on the upper plate of the first capacitor, the electric field produced by that charge will induce an equal negative charge $-Q$ on its lower plate. This charge comes from electrons drawn from the upper plate of the second capacitor. Thus, there will be an equal charge $+Q$ on the upper plate of the second capacitor and a corresponding charge $-Q$ on its lower plate. The potential difference across the first capacitor is

$$V_1 = V_a - V_m = \frac{Q}{C_1}$$

where V_m is the potential of the adjacent plates and connecting wire. Similarly, the potential difference across the second capacitor is

$$V_2 = V_m - V_b = \frac{Q}{C_2}$$

The potential difference across the two capacitors in series is the sum of these potential differences:

$$V = V_a - V_b = V_1 + V_2 = \frac{Q}{C_1} + \frac{Q}{C_2} = Q\left(\frac{1}{C_1} + \frac{1}{C_2}\right) \tag{25-18}$$

The equivalent capacitance of two capacitors in series is defined as

$$C_{eq} = \frac{Q}{V} \tag{25-19}$$

Solving Equations 25-18 and 25-19 for V/Q gives

$$\frac{1}{C_{eq}} = \frac{1}{C_1} + \frac{1}{C_2} \tag{25-20}$$

Equation 25-20 can be generalized to three or more capacitors connected in series:

$$\frac{1}{C_{eq}} = \frac{1}{C_1} + \frac{1}{C_2} + \frac{1}{C_3} + \cdots \tag{25-21}$$

Equivalent capacitance for capacitors in series

Capacitor bank for storing energy to be used by the pulsed Nova laser at Lawrence Livermore Laboratories. The laser is used in fusion studies.

Exercise Two capacitors have capacitances of 20 and 30 μF. Find the equivalent capacitance if the capacitors are connected (*a*) in parallel and (*b*) in series. (*Answers* (*a*) 50 μF, (*b*) 12 μF)

Note that in the preceding exercise, the equivalent capacitance of the two capacitors in series is less than the capacitance of either capacitor. Adding a capacitor in series increases $1/C_{eq}$, which means the equivalent capacitance C_{eq} decreases.

Example 25-5

A 2-μF capacitor and a 4-μF capacitor are connected in series across an 18-V battery. Find the charge on the capacitors and the potential difference across each.

Picture the Problem Figure 25-12*a* shows the circuit in this Example and Figure 25-12*b* shows an equivalent capacitor that carries the same charge $Q = C_{eq}V$. After finding the charge, we can find the potential drop across each capacitor.

Figure 25-12

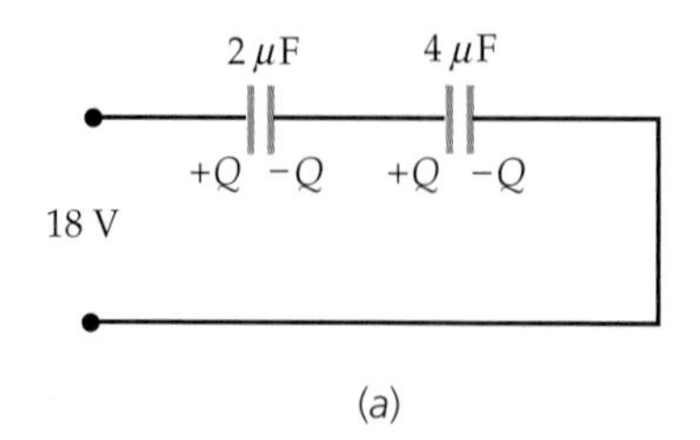

(*a*)

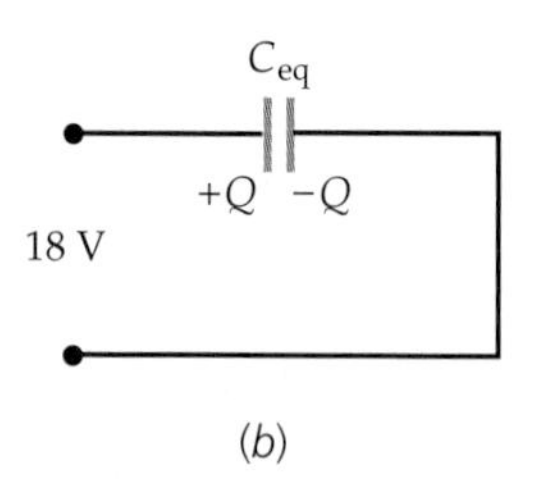

(*b*)

1. The charge on each capacitor equals the charge on the equivalent capacitor: $Q = C_{eq}V$

2. The equivalent capacitance of the series capacitors is found from: $\frac{1}{C_{eq}} = \frac{1}{C_1} + \frac{1}{C_2} = \frac{1}{2\mu F} + \frac{1}{4\mu F} = \frac{3}{4\mu F}$

   $C_{eq} = \frac{4}{3}\ \mu F$

3. Use this value to find the charge Q: $Q = C_{eq}V = \left(\frac{4}{3}\mu F\right)(18\ V) = 24\ \mu C$

4. Use the result for Q to find the potential across the 2-μF capacitor: $V_1 = \frac{Q}{C_1} = \frac{24\ \mu C}{2\ \mu F} = 12\ V$

5. Again use the result for Q to find the potential across the 4-μF capacitor: $V_2 = \frac{Q}{C_2} = \frac{24\ \mu C}{4\ \mu F} = 6\ V$

Check the Result The sum of these potential differences is 18 V, as required.

Example 25-6 *try it yourself*

The two capacitors in Example 25-5 are removed from the battery and carefully disconnected from each other so that the charge on the plates is not disturbed (Figure 25-13*a*). They are then reconnected in a circuit containing open switches, positive plate to positive plate and negative plate to negative plate (Figure 25-13*b*). Find the potential difference across the capacitors and the charge on each capacitor when the switches are closed.

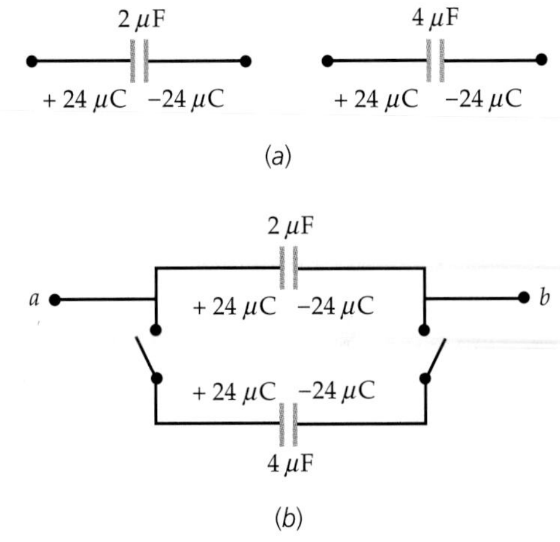

Figure 25-13

Picture the Problem Just after the two capacitors are disconnected from the battery, they carry equal charges of 24 μC. After switches S_1 and S_2 in the new circuit are closed, the capacitors are in parallel between points a and b. The potential across each of them is the same, and the equivalent capacitance of the system is $C_{eq} = C_1 + C_2$. The two positive plates form a single conductor with charge $Q = 48\ \mu C$, and the negative plates form a conductor with charge $-Q = -48\ \mu C$. Therefore, the potential difference is $V = Q/C_{eq}$, and the charges on the two capacitors are $Q_1 = C_1V$ and $Q_2 = C_2V$.

Cover the column to the right and try these on your own before looking at the answers.

Steps	*Answers*
1. Write V in terms of C_{eq} and the total charge Q.	$V = \frac{Q}{C_{eq}}$
2. Evaluate the equivalent capacitance.	$C_{eq} = 6\ \mu F$
3. Calculate V from C_{eq} and the charge $Q = 48\ \mu C$.	$V = 8\ V$
4. Use V to find the charge on each capacitor.	$Q_1 = 16\ \mu C, \quad Q_2 = 32\ \mu C$

Check the Result Note that $Q_1 + Q_2 = Q = 48\ \mu C$, as required.

Exercise Find the energy stored in the capacitors before and after they are connected. (*Answer* $U_i = q^2/2C_1 + q^2/2C_2$, where $q = 24\ \mu C$. Thus, $U_i = 216\ \mu J$. $U_f = Q_1^2/2C_1 + Q_2^2/2C_2 = 192\ \mu J$. Note that 24 μJ is "lost" to thermal energy in the wires or radiated away.)

Example 25-7

(*a*) Find the equivalent capacitance of the network of three capacitors in Figure 25-14. (*b*) Find the charge on each capacitor and the voltage drop across it when the system is connected to a 6-V battery.

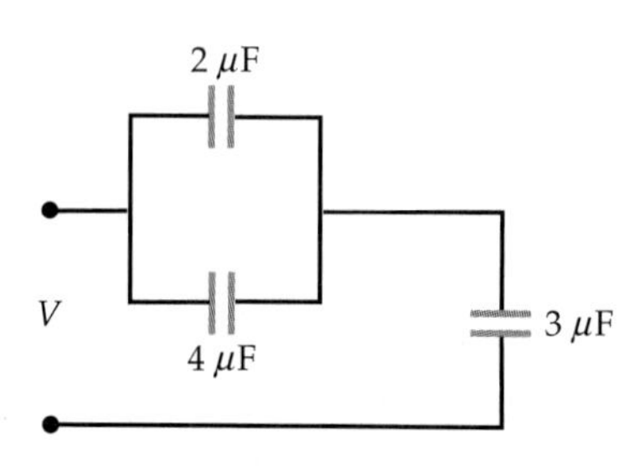

Figure 25-14

Picture the Problem (*a*) The 2- and 4-μF capacitors are connected in parallel, and the parallel combination is connected in series with the 3-μF capacitor. We first find the equivalent capacitance of the 2- and 4-μF capacitors (Figure 25-15*a*), then combine this equivalent capacitance with the 3-μF capacitor to reach a final equivalent capacitance (25-15*b*). (*b*) The charge on the 3-μF capacitor is the charge delivered by the battery $Q = C_{eq} V$ as shown in Figure 25-15*a*.

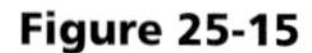
Figure 25-15

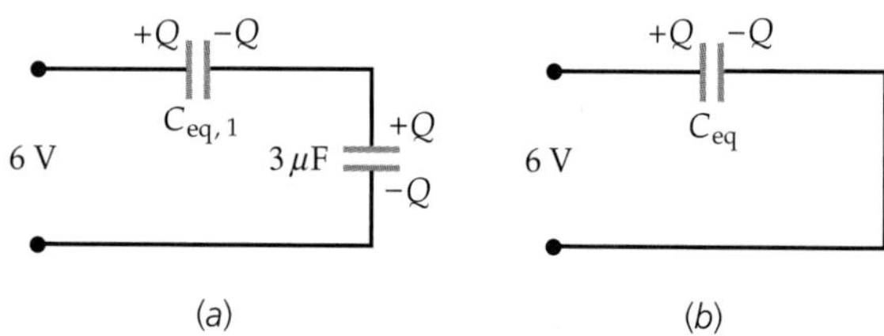

(*a*) 1. The equivalent capacitance of the two capacitors in parallel is the sum of the capacitances:

$$C_{eq,1} = C_1 + C_2 = 2\ \mu F + 4\ \mu F = 6\ \mu F$$

2. Find the equivalent capacitance of a 6-μF capacitor in series with a 3-μF capacitor:

$$\frac{1}{C_{eq}} = \frac{1}{C_1} + \frac{1}{C_2} = \frac{1}{6\ \mu F} + \frac{1}{3\ \mu F} = \frac{1}{2\ \mu F}$$

$$C_{eq} = 2\ \mu F$$

(*b*) 1. Calculate the charge Q delivered by the battery. This is also the charge on the 3-μF capacitor:

$$Q = C_{eq}V = (2\ \mu F)(6\ V) = 12\ \mu C$$

2. The potential drop across the 3-μF capacitor is Q/C:

$$V_3 = \frac{Q}{C_3} = \frac{12\ \mu C}{3\ \mu F} = 4\ V$$

3. The potential drop across the parallel combination $V_{2,4}$ is $Q/C_{eq,1}$:

$$V_{2,4} = \frac{Q}{C_{eq,1}} = \frac{12\ \mu C}{6\ \mu F} = 2\ V$$

4. The charge on each of the parallel capacitors is found from $Q_i = C_i V_{2,4}$, where $V_{2,4} = 2$ V:

$$Q_2 = C_2 V_{2,4} = (2\ \mu F)(2\ V) = 4\ \mu C$$

$$Q_4 = C_4 V_{2,4} = (4\ \mu F)(2\ V) = 8\ \mu C$$

Check the Result The voltage drop across the parallel combination (2 V) plus that across the 3-μF capacitor (4 V) equals the voltage of the battery. Also, the sum of the charges on the parallel capacitors (4 μC + 8 μC) equals the total charge (12 μC) on the 3-μF capacitor.

Exercise Find the energy stored in each capacitor. (*Answer* $U_2 = 4\ \mu J$, $U_3 = 24\ \mu J$, $U_4 = 8\ \mu J$. Note that $U_2 + U_3 + U_4 = 36\ \mu J = \frac{1}{2}QV = \frac{1}{2}C_{eq}V^2$.)

25-5 Dielectrics

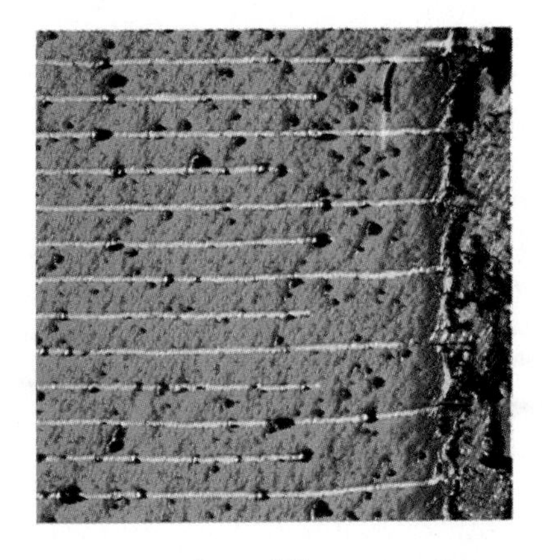

A cut section of a multilayer capacitor with a ceramic dielectric. The white lines are the edges of the conducting plates.

A nonconducting material, such as air, glass, paper, or wood, is called a **dielectric**. When the space between the two conductors of a capacitor is occupied by a dielectric, the capacitance is increased by a factor κ that is characteristic of the dielectric, a fact discovered experimentally by Michael Faraday. The reason for this increase is that the electric field between the plates of a capacitor is weakened by the dielectric. Thus, for a given charge on the plates, the potential difference is reduced and the capacitance (Q/V) is increased.

If the original electric field between the plates of a capacitor without a dielectric is E_0, the field in a dielectric slab inserted between the plates is

$$E = \frac{E_0}{\kappa} \tag{25-22}$$

Electric field inside a dielectric

where κ is called the **dielectric constant.** For a parallel-plate capacitor of separation d, the potential difference between the plates is

$$V = Ed = \frac{E_0 d}{\kappa} = \frac{V_0}{\kappa}$$

where V is the potential difference with the dielectric and $V_0 = E_0 d$ is the original potential difference without the dielectric. The new capacitance is

$$C = \frac{Q}{V} = \frac{Q}{V_0/\kappa} = \kappa\frac{Q}{V_0}$$

or

$$C = \kappa C_0 \tag{25-23}$$

Effect of a dielectric on capacitance

where $C_0 = Q/V_0$ is the capacitance without the dielectric. The capacitance of a parallel-plate capacitor filled with a dielectric of constant κ is thus

$$C = \frac{\kappa\epsilon_0 A}{d} = \frac{\epsilon A}{d} \tag{25-24}$$

where

$$\epsilon = \kappa\epsilon_0 \tag{25-25}$$

is called the **permittivity** of the dielectric.

In the preceding discussion, we assumed that the charge on the plates of the capacitor did not change when the dielectric was inserted. This is true if the capacitor is charged and then removed from the charging source (the battery) before the insertion of the dielectric. If the dielectric is inserted while the battery is still connected, the battery supplies more charge to maintain the original potential difference. The total charge on the plates is then $Q = \kappa Q_0$. In either case, the capacitance (Q/V) is increased by the factor κ.

Exercise The 88.5-pF capacitor of Example 25-2 is filled with a dielectric of constant $\kappa = 2$. (*a*) Find the new capacitance. (*b*) Find the charge on the capacitor with the dielectric in place if the capacitor is attached to a 12-V battery. (*Answers* (*a*) 177 pF, (*b*) 2.12 nC)

Exercise The capacitor in the previous exercise is charged to 12 V without the dielectric and is then disconnected from the battery. The dielectric of constant $\kappa = 2$ is then inserted. Find the new values for (*a*) the charge Q, (*b*) the voltage V, and (*c*) the capacitance C. (*Answers* (*a*) $Q = 1.06$ nC, which is unchanged; (*b*) $V = 6$ V; (*c*) $C = 177$ pF)

Dielectrics not only increase the capacitance of a capacitor, they also provide a means for keeping parallel conducting plates apart, and they raise the potential difference at which dielectric breakdown occurs.* Consider a parallel-plate capacitor made from two sheets of metal foil that are separated by a thin plastic sheet. The plastic sheet allows the metal sheets to be very close together without actually being in electrical contact, and because the dielectric strength of plastic is greater than that of air, a greater potential difference can be attained before dielectric breakdown occurs. Table 25-1 lists the dielectric constants and dielectric strengths of some dielectrics. Note that for air, $\kappa \approx 1$, so for most situations we do not need to distinguish between air and a vacuum.

Table 25-1

Dielectric Constants and Strengths of Various Materials

Material	Dielectric Constant κ	Dielectric Strength, kV/mm
Air	1.00059	3
Bakelite	4.9	24
Glass (Pyrex)	5.6	14
Mica	5.4	10–100
Neoprene	6.9	12
Paper	3.7	16
Paraffin	2.1–2.5	10
Plexiglas	3.4	40
Polystyrene	2.55	24
Porcelain	7	5.7
Transformer oil	2.24	12

* Recall from Chapter 24 that for electric fields greater than about 3×10^6 V/m, air breaks down, that is, it becomes ionized and begins to conduct.

Example 25-8

A parallel-plate capacitor has square plates of side 10 cm and a separation of $d = 4$ mm. A dielectric slab of constant $\kappa = 2$ has the same area as the plates. (*a*) What is the capacitance without the dielectric? (*b*) What is the capacitance if the dielectric slab fills the space between the plates? (*c*) What is the capacitance if a dielectric slab of thickness 3 mm is inserted into the 4-mm gap?

Picture the Problem The capacitance without the dielectric, C_0, is found from the area and spacing of the plates (Figure 25-16*a*). When the capacitor is filled with a dielectric κ, (Figure 25-16*b*), the capacitance is $C = \kappa C_0$ (Equation (25-25). If the dielectric only partially fills the capacitor (Figure 25-16*c*), we calculate the potential difference V for a given charge Q, then apply the definition of capacitance, $C = Q/V$.

Figure 25-16

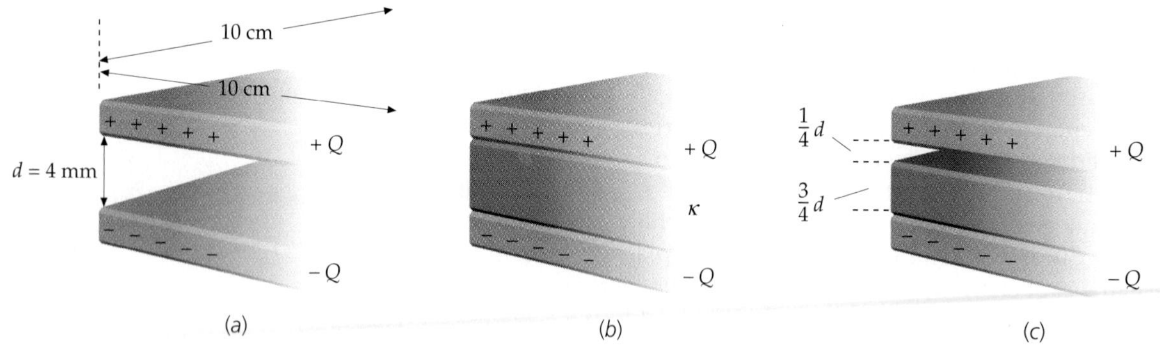

(*a*) If there is no dielectric, the capacitance C_0 is given by Equation 25-10:

$$C_0 = \frac{\epsilon_0 A}{d} = \frac{(8.85\ \text{pF/m})(0.1\ \text{m})^2}{0.004\ \text{m}} = 22.1\ \text{pF}$$

(*b*) When the capacitor is filled with a dielectric κ, its capacitance C is increased by the factor κ:

$$C = \kappa C_0 = (2)(22.1\ \text{pF}) = 44.2\ \text{pF}$$

(*c*) 1. The new capacitance is related to the original charge Q and the new potential difference V:

$$C = \frac{Q}{V}$$

2. The potential difference V between the plates is the sum of the potential difference for the gap plus the potential difference for the dielectric slab:

$$V = V_{\text{gap}} + V_{\text{slab}} = E_{\text{gap}}\left(\tfrac{1}{4}d\right) + E_{\text{slab}}\left(\tfrac{3}{4}d\right)$$

3. The field in the gap just outside the conductor is the original field E_0:

$$E_{\text{gap}} = E_0 = \frac{Q}{\epsilon_0 A}$$

4. The field in the dielectric slab is reduced by the factor κ:

$$E_{\text{slab}} = \frac{E_0}{\kappa}$$

5. Combining the previous two results yields V in terms of κ. Note that the original potential difference is $V_0 = E_0 d$:

$$V = E_0\left(\frac{1}{4}d\right) + \frac{E_0}{\kappa}\left(\frac{3}{4}d\right) = E_0 d\left(\frac{1}{4} + \frac{3}{4\kappa}\right) = V_0\left(\frac{\kappa + 3}{4\kappa}\right)$$

6. Using $C = Q/V$, we find the new capacitance in terms of the original capacitance, $C_0 = Q/V_0$:

$$C = \frac{Q}{V} = \frac{Q}{V_0(\kappa + 3)/4\kappa} = \frac{Q}{V_0}\left(\frac{4\kappa}{\kappa + 3}\right)$$

$$= C_0\left(\frac{4\kappa}{\kappa + 3}\right) = (22.1\text{ pF})\left(\frac{8}{5}\right) = 35.4\text{ pF}$$

Check the Result The absence of a dielectric corresponds to $\kappa = 1$. In this case, our result for the final step in (*c*) would reduce to $C = C_0$ as expected. Suppose that the dielectric slab were a conducting slab. In a conductor, $E = 0$, so according to Equation 25-22, κ for a conductor would equal infinity. For very large κ, the quantity $4\kappa/(\kappa + 3)$ is approximately 4, so the result for the final step in (*c*) approaches $4C_0$. A conducting slab simply extends the capacitor plate, hence the plate separation with the conducting dielectric in place would be $\frac{1}{4}d$. This means that C should be $4C_0$, as it is for very large κ.

Remarks Note that the results of this example are independent of the vertical position of the dielectric or conducting slab between the plates.

Energy Stored in the Presence of a Dielectric

The energy stored in a parallel-plate capacitor with dielectric is

$$U = \tfrac{1}{2}QV = \tfrac{1}{2}CV^2$$

We can express the capacitance C in terms of the area and separation of the plates, and the voltage difference V in terms of the electric field and plate separation, to obtain

$$U = \frac{1}{2}\left(\frac{\epsilon A}{d}\right)(Ed)^2 = \frac{1}{2}\epsilon E^2(Ad)$$

The quantity Ad is the volume between the plates containing the electric field. The energy per unit volume is thus

$$u_e = \tfrac{1}{2}\epsilon E^2 = \tfrac{1}{2}\kappa\epsilon_0 E^2 \qquad \text{25-26}$$

Example 25-9

Two parallel-plate capacitors, each having a capacitance of $C_1 = C_2 = 2\ \mu F$, are connected in parallel across a 12-V battery. Find (*a*) the charge on each capacitor and (*b*) the total energy stored in the capacitors.

The capacitors are then disconnected from the battery and a dielectric of constant $\kappa = 2.5$ is inserted between the plates of the capacitor C_2. After the dielectric is inserted, find (*c*) the potential difference across each capacitor, (*d*) the charge on each capacitor, and (*e*) the total energy stored in the capacitors.

Picture the Problem (*a* and *b*) The charge Q and total energy U can be found for each capacitor from its capacitance C and voltage V. (*c*) After the capacitors are removed from the battery, the total charge on them must remain the same. When the dielectric is inserted into one of the capacitors, its capacitance C_2 changes. The potential across the parallel combination can be found from the total charge and the equivalent capacitance.

(*a*) The charge on each capacitor is found from its capacitance C and voltage V:

$$Q = CV = (2\ \mu\text{F})(12\ \text{V}) = 24\ \mu\text{C}$$

(*b*) 1. The energy stored in each capacitor is found from its capacitance C and voltage V:

$$U = \tfrac{1}{2}CV^2 = \tfrac{1}{2}(2\ \mu\text{F})(12\ \text{V})^2 = 144\ \mu\text{J}$$

2. The total energy is twice that stored in each capacitor:

$$U_{\text{total}} = 2U = 288\ \mu\text{J}$$

(*c*) 1. The potential across the parallel combination is related to the total charge Q_{total} and the equivalent capacitance C_{eq}:

$$V = \frac{Q_{\text{total}}}{C_{\text{eq}}}$$

2. The capacitance C_2 of the capacitor with the dielectric is increased by the factor κ. The equivalent capacitance is the sum of the capacitances:

$$C_{\text{eq}} = C_1 + C_2 = C_1 + \kappa C_1 = (2\ \mu\text{F}) + (2.5)(2\ \mu\text{F})$$
$$= 2\ \mu\text{F} + 5\ \mu\text{F} = 7\ \mu\text{F}$$

3. The total charge remains $48\ \mu$C. Substitute for Q_{total} and C_{eq} to calculate V:

$$V = \frac{Q_{\text{total}}}{C_{\text{eq}}} = \frac{48\ \mu\text{C}}{7\ \mu\text{F}} = 6.86\ \text{V}$$

(*d*) The charge on each capacitor is again derived from its capacitance and the voltage V:

$$Q_1 = C_1V = (2\ \mu\text{F})(6.86\ \text{V}) = 13.7\ \mu\text{C}$$
$$Q_2 = C_2V = (5\ \mu\text{F})(6.86\ \text{V}) = 34.3\ \mu\text{C}$$

(*e*) 1. The energy stored in each capacitor is found from its new capacitance C and new voltage V:

$$U_1 = \tfrac{1}{2}C_1V^2 = \tfrac{1}{2}(2\ \mu\text{F})(6.86\ \text{V})^2 = 47.1\ \mu\text{J}$$
$$U_2 = \tfrac{1}{2}C_2V^2 = \tfrac{1}{2}(5\ \mu\text{F})(6.86\ \text{V})^2 = 118\ \mu\text{J}$$

2. Add these energies to get the total energy stored:

$$U = U_1 + U_2 = 47.1\ \mu\text{J} + 118\ \mu\text{J} = 165\ \mu\text{J}$$

Check the Result When the dielectric is inserted into one of the capacitors, the field is weakened and the potential difference is lowered. Since the two capacitors are connected in parallel, charge must flow from the other capacitor so that the potential difference is the same across both capacitors. Note that the capacitor with the dielectric has the greater charge, and that when the charges calculated for each capacitor in (*d*) are added, $Q_1 + Q_2 = 13.7\ \mu\text{C} + 34.3\ \mu\text{C} = 48\ \mu\text{C}$, the result is the same as the original sum.

Remarks The total energy of 165 μJ is less than the original energy of 288 μJ. When the dielectric is inserted, it is pulled in and work is done on whatever was holding it. To remove the dielectric, work $W = 288\ \mu\text{J} - 165\ \mu\text{J} = 123\ \mu\text{J}$ must be done, and this work is stored as electrostatic potential energy.

Example 25-10 *try it yourself*

Find (*a*) the charge on each capacitor, and (*b*) the total energy stored in the capacitors of Example 25-9, if the dielectric is inserted into one of the capacitors while the battery is still connected.

Picture the Problem Since the battery is still connected, the potential difference across the capacitors remains 12 V. This condition determines the charge and energy stored in each capacitor. Let subscript 1 refer to the capacitor without the dielectric, and subscript 2 refer to the capacitor with the dielectric.

Cover the column to the right and try these on your own before looking at the answers.

Steps	***Answers***
(*a*) Calculate the charge on each capacitor from $Q = CV$ using the result that $C_1 = 2\ \mu$F and $C_2 = 5\ \mu$F as found in Example 25-9.	$Q_1 = C_1V = 24\ \mu\text{C}$ $Q_2 = C_2V = 60\ \mu\text{C}$
(*b*) 1. Calculate the energy stored in each capacitor from $U = \frac{1}{2}CV^2$. Check your results by using $U = \frac{1}{2}QV$.	$U_1 = 144\ \mu\text{J}$ $U_2 = 360\ \mu\text{J}$
2. Add your results for U_1 and U_2 to obtain the final energy.	$U_{\text{total}} = 504\ \mu\text{J}$

Remarks Note that Q_2 is two and a half times its value before the dielectric was inserted (since $\kappa = 2.5$). The battery supplies this additional charge in order to maintain a fixed potential difference. Because of the work done by the battery to supply this charge, the total energy of the system is higher with the dielectric in place (504 μJ) than without the dielectric (288 μJ).

25-6 Molecular View of a Dielectric

A dielectric weakens the electric field between the plates of a capacitor because the molecules in the dielectric produce an electric field in a direction opposite to the field produced by the plates. This electric field is due to the electric dipole moments of the molecules in the dielectric.

Although atoms and molecules are electrically neutral, they are affected by electric fields because they contain positive and negative charges that can respond to external fields. We can think of an atom as a very small, positively charged nucleus surrounded by a negatively charged electron cloud. In some atoms and molecules, the electron cloud is spherically symmetric, so its "center of charge" is at the center of the atom or molecule, coinciding with the positive charge. An atom or molecule like this has zero dipole moment and is said to be nonpolar. But in the presence of an external electric field, the positive and negative charge experience forces in opposite directions. The positive and negative charges then separate until the attractive force they exert on each other balances the forces due to the external electric field (Figure 25-17). The molecule is then said to be polarized and it behaves like an electric dipole.

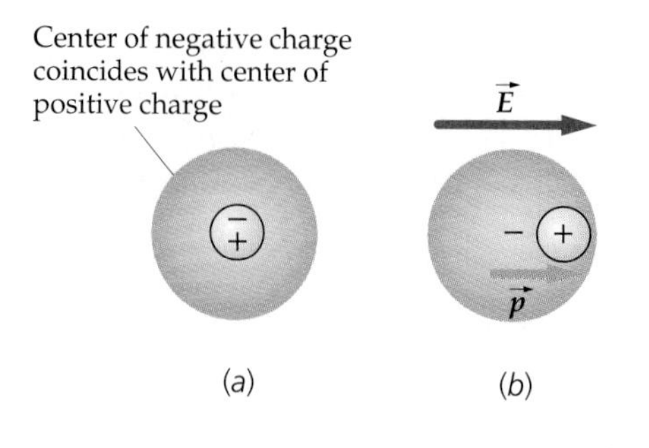

Figure 25-17 Schematic diagrams of the charge distributions of an atom or nonpolar molecule. (*a*) In the absence of an external electric field, the center of positive charge coincides with the center of negative charge. (*b*) In the presence of an external electric field, the centers of positive and negative charge are displaced, producing an induced dipole moment in the direction of the external field.

In some molecules (for example, HCl and H_2O), the centers of positive and negative charge do not coincide even in the absence of an external electric field. As we noted in Chapter 22, these polar molecules have a permanent electric dipole moment.

When a dielectric is placed in the field of a capacitor, its molecules are polarized such that there is a net dipole moment parallel to the field. If the molecules are polar, their dipole moments, originally oriented at random, tend to become aligned due to the torque exerted by the field.* If the molecules are nonpolar, the field induces dipole moments that are parallel to the field. In either case, the molecules in the dielectric are polarized in the direction of the external field (Figure 25-18).

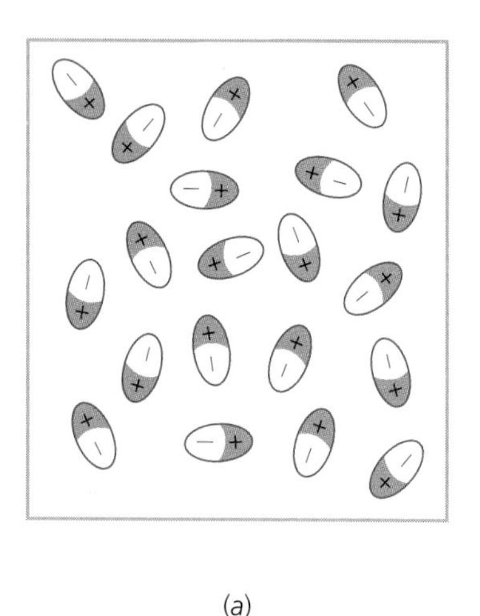

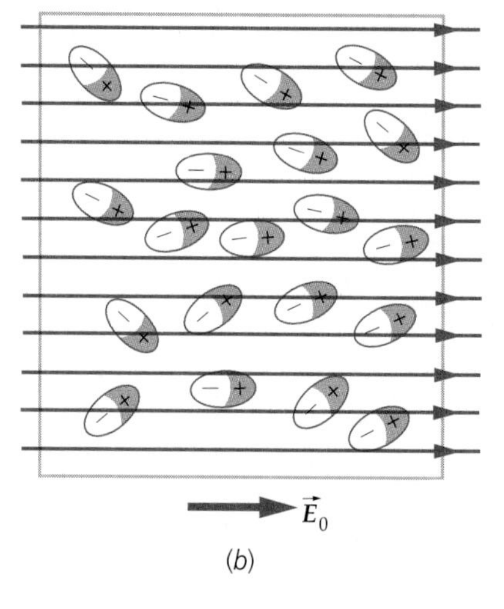

Figure 25-18 (*a*) The randomly oriented electric dipoles of a polar dielectric in the absence of an external electric field. (*b*) In the presence of an external electric field, the dipoles are partially aligned parallel to the field.

The net effect of the polarization of a homogeneous dielectric in a parallel-plate capacitor is the creation of a surface charge on the dielectric faces near the plates, as shown in Figure 25-19. The surface charge on the dielectric is called a **bound charge** because it is bound to the molecules of the dielectric and cannot move about like the free charge on the conducting capacitor plates. This bound charge produces an electric field opposite in direction to that produced by the free charge on the conductors. Thus, the net electric field between the plates is reduced, as illustrated in Figure 25-20.

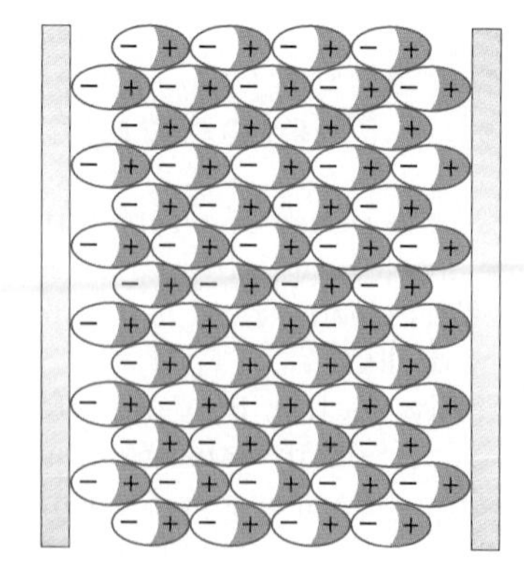

Figure 25-19 When a dielectric is placed between the plates of a capacitor, the electric field of the capacitor polarizes the molecules of the dielectric. The result is a bound charge on the surface of the dielectric that produces its own electric field; this field opposes the external field. The electric field between the plates is thus weakened by the dielectric.

* The degree of alignment depends on the external field and on the temperature. It is approximately proportional to pE/kT, where pE is the maximum energy of a dipole in a field E, and kT is the characteristic thermal energy.

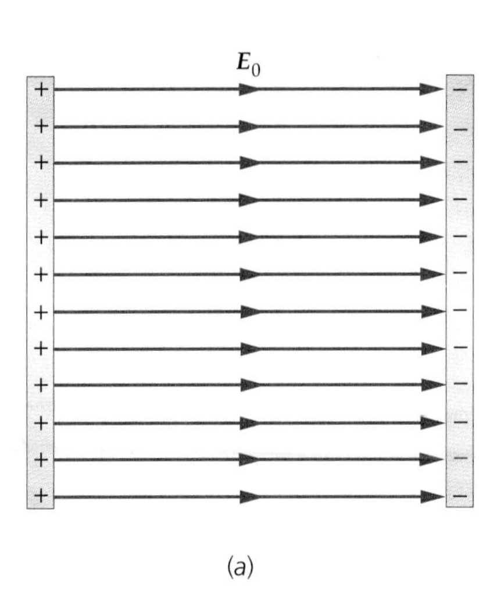

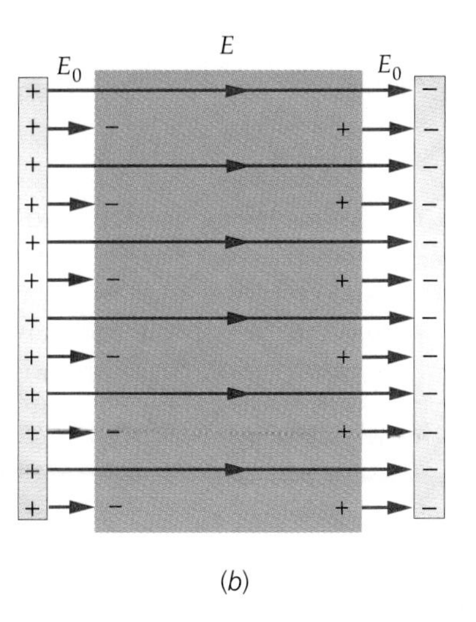

Figure 25-20 The electric field between the plates of a capacitor (*a*) with no dielectric and (*b*) with a dielectric. The surface charge on the dielectric weakens the original field between the plates.

Example 25-11

A hydrogen atom consists of a proton nucleus of charge $+e$ and an electron of charge $-e$. The charge distribution of the atom is spherically symmetric so the atom is nonpolar. Consider a model in which the hydrogen atom consists of a positive point charge $+e$ at the center of a uniformly charged spherical cloud of radius R and total charge $-e$. Show that when such an atom is placed in a uniform external electric field $\vec{E}$, the induced dipole moment is proportional to E, that is, $p = \alpha E$, where α is called the polarizability.

Picture the Problem In the external field, the positive charge is displaced from the center of the cloud by an amount L such that the force exerted by the field $e\vec{E}$ is balanced by the force exerted by the negative cloud $e\vec{E}'$, where $\vec{E}'$ is the field due to the cloud (Figure 25-21). We use Gauss's law to find E', and then we calculate the induced dipole moment $\vec{p} = e\vec{L}$.

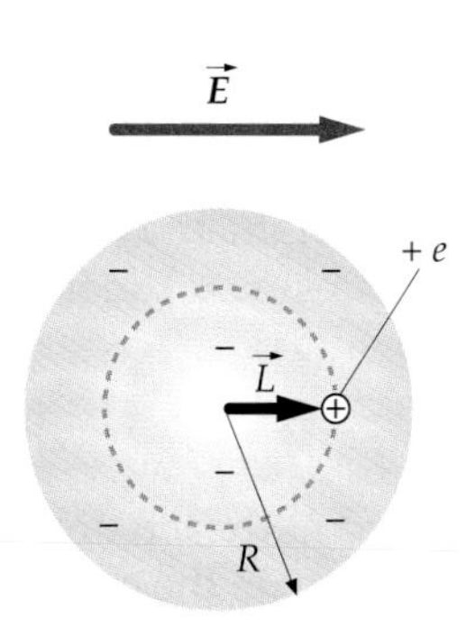

Figure 25-21

1. Write the magnitude of the induced dipole moment in terms of e and L:

$$p = eL$$

2. We can find L by calculating the field E' due to the negatively charged cloud at a distance L from the center. We use Gauss's law to compute E'. Choose a spherical Gaussian surface of radius L concentric with the cloud. Then E' is constant on this surface:

$$\phi_{\text{net}} = \oint E_n\, dA = \frac{1}{\epsilon_0} Q_{\text{inside}}$$

$$E'(4\pi L^2) = \frac{1}{\epsilon_0} Q_{\text{inside}}$$

$$E' = \frac{1}{4\pi\epsilon_0 L^2} Q_{\text{inside}}$$

3. The charge inside the sphere of radius L equals the charge density times the volume:

$$Q_{\text{inside}} = \rho \frac{4}{3}\pi L^3 = \frac{-e}{\frac{4}{3}\pi R^3}\frac{4}{3}\pi L^3 = -e\frac{L^3}{R^3}$$

4. Substitute this value of Q_{inside} to calculate E':

$$E' = \frac{1}{4\pi\epsilon_0 L^2} Q_{\text{inside}} = \frac{1}{4\pi\epsilon_0 L^2}\left(-e\frac{L^3}{R^3}\right) = -e\frac{L}{4\pi\epsilon_0 R^3}$$

5. Solve for L:

$$L = \frac{-4\pi\epsilon_0 R^3 E'}{e}$$

6. E' is negative because it points inward on the Gaussian surface. At the positive charge, E' points to the left, so $E' = -E$:

$$E' = -E$$

7. Substitute these results for L and E' to express p in terms of the external field E:

$$p = eL = -4\pi\epsilon_0 R^3 E' = 4\pi\epsilon_0 R^3 E$$

Remarks The charge distribution of the negative charge in a hydrogen atom, obtained from quantum theory, is spherically symmetric, but the charge density decreases exponentially with distance rather than being uniform. Nevertheless, the above calculation shows that the dipole moment is indeed proportional to the external field $p = \alpha E$, and the polarizability α is of the order of $4\pi\epsilon_0 R^3$ where R is the radius of the atom or molecule. The dielectric constant κ can be related to the polarizability and to the number of molecules per unit volume.

Magnitude of the Bound Charge

The bound charge density σ_b on the surfaces of the dielectric is related to the dielectric constant κ and to the free charge density σ_f on the plates. Consider a dielectric slab between the plates of a parallel-plate capacitor as shown in Figure 25-22. If the dielectric is a very thin slab between plates that are close together, the electric field inside the dielectric slab due to the bound charge densities $+\sigma_b$ on the right and $-\sigma_b$ on the left is just the field due to two infinite-plane charge densities. The field E_b thus has the magnitude

$$E_b = \frac{\sigma_b}{\epsilon_0}$$

This field is directed to the left and subtracts from the electric field E_0 due to the free charge density on the capacitor plates, which has the magnitude

$$E_0 = \frac{\sigma_f}{\epsilon_0}$$

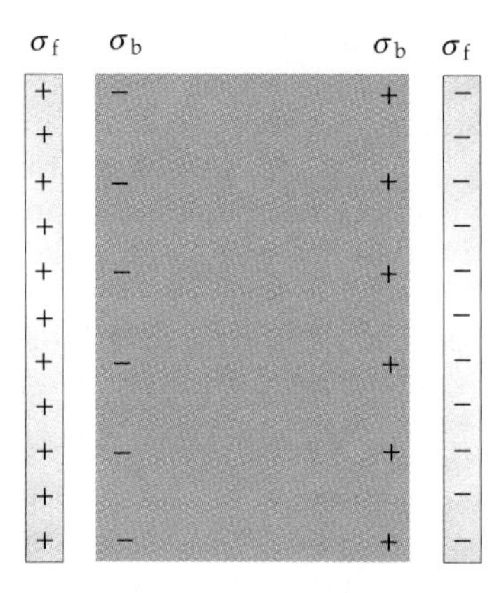

Figure 25-22 Parallel-plate capacitor with dielectric slab between the plates. If the plates are closely spaced, each of the surface charges can be considered an infinite plane charge. The electric field due to the free charge on the plates is directed to the right and has a magnitude $E_0 = \sigma_f/\epsilon_0$. That due to the bound charge is directed to the left and has a magnitude $E_b = \sigma_b/\epsilon_0$.

The magnitude of the net field $E = E_0/\kappa$ is the difference between these magnitudes:

$$E = E_0 - E_b = \frac{E_0}{\kappa}$$

or

$$E_b = E_0\left(1 - \frac{1}{\kappa}\right) = \frac{\kappa - 1}{\kappa}E_0$$

Writing σ_b/ϵ_0 for E_b and σ_f/ϵ_0 for E_0, we obtain

$$\sigma_b = \frac{\kappa - 1}{\kappa}\sigma_f = \left(1 - \frac{1}{\kappa}\right)\sigma_f \qquad \text{25-27}$$

The bound charge density σ_b is always less than the free charge density σ_f on the capacitor plates, and it is zero if $\kappa = 1$, which is the case when there is no dielectric. For a conducting slab, $\kappa = \infty$ and $\sigma_b = \sigma_f$.

The Piezoelectric Effect

optional

In certain crystals that contain polar molecules, such as quartz, tourmaline, and topaz, a mechanical stress applied to the crystal produces polarization of the molecules. This is known as the **piezoelectric effect**. The polarization of the stressed crystal causes a potential difference across the crystal, which can be used to produce an electric current. Piezoelectric crystals are used in transducers such as microphones, phonograph pickups, and vibration-sensing devices for converting mechanical strain into electrical signals. The converse piezoelectric effect, in which a voltage applied to such a crystal induces mechanical strain (deformation), is used in headphones and many other devices.

Because the natural frequency of vibration of quartz is in the range of radio frequencies, and because its resonance curve is very sharp,* it is used extensively to stabilize radio-frequency oscillators and to make accurate clocks.

* Resonance in ac circuits, which will be discussed in Chapter 31, is analogous to mechanical resonance, which was discussed in Chapter 14.

optional

Summary

Topic	Remarks and Relevant Equations	
1. Electrostatic Potential Energy	The electrostatic potential energy of a system of charges is the work needed to bring the charges from an infinite separation to their final positions.	
Of point charges	$U = \frac{1}{2}\sum_{i=1}^{n} q_i V_i$	25-2
Of a conductor with charge Q at potential V	$U = \frac{1}{2}QV$	25-3
Of a system of conductors	$U = \frac{1}{2}\sum_{i=1}^{n} Q_i V_i$	25-4
Energy stored in a capacitor	$U = \frac{1}{2}\frac{Q^2}{C} = \frac{1}{2}QV = \frac{1}{2}CV^2$	25-12
Energy density in an electric field	$u_e = \frac{\text{energy}}{\text{volume}} = \frac{1}{2}\epsilon E^2 = \frac{1}{2}\kappa\epsilon_0 E^2$	25-26
2. Capacitor	A capacitor is a device for storing charge and energy. It consists of two conductors insulated from each other that carry equal and opposite charges.	
3. Capacitance	$C = \frac{Q}{V}$	25-5
Isolated conductor	Q is its total charge, V is its potential.	
Capacitor	Q is the magnitude of charge on either conductor, V is the potential difference.	
Isolated spherical conductor	$C = 4\pi\epsilon_0 R$	25-6

Parallel-plate capacitor	$C = \frac{\epsilon_0 A}{d}$	**25-10**
Cylindrical capacitor	$C = \frac{2\pi\epsilon_0 L}{\ln(r_2/r_1)}$	**25-11**

4. Equivalent Capacitance

Parallel capacitors	When capacitors are in parallel, the voltage is the same across each one, and the capacitances add: $C_{eq} = C_1 + C_2 + C_3 + \cdots$	**25-17**
Series capacitors	When capacitors are in series, the charge stored is the same on each one, and the voltage drops add: $\frac{1}{C_{eq}} = \frac{1}{C_1} + \frac{1}{C_2} + \frac{1}{C_3} + \cdots$	**25-21**

5. Dielectric

Macroscopic behavior	A nonconducting material is called a dielectric. When a dielectric is inserted between the plates of a capacitor, the electric field within the dielectric is weakened and the capacitance is thereby increased by the factor κ, the dielectric constant.	
Microscopic picture	The field in the dielectric of a capacitor is weakened because the dipole moments of the molecules (either preexisting or induced) tend to align with the field and thereby produce an electric field that opposes the external field. The aligned dipole moment of the dielectric is proportional to the external field.	
Electric field	$E = \frac{E_0}{\kappa}$	**25-22**
Capacitance	$C = \kappa C_0$	**25-23**
Permittivity ϵ	$\epsilon = \kappa\epsilon_0$	**25-25**
Uses of a dielectric	1. Increases capacitance 2. Increases dielectric strength 3. Physically separates conductors	

6. Piezoelectric Effect (optional)

In certain crystals containing polar molecules, a mechanical stress polarizes the molecules, inducing a voltage across the crystal. Conversely, an applied voltage induces mechanical strain (deformation) in the crystal.

Problem-Solving Guide

1. Begin by drawing a neat diagram that includes the important features of the problem. Be sure to show the correct parallel or series connections between capacitors. When appropriate, indicate the charge on the plates of a capacitor and any dielectrics between the plates.
2. To find the capacitance of a capacitor, place a charge $+Q$ on one conductor and $-Q$ on the other. Then compute the electric field between the conductors and use it to find the potential difference V between the conductors. The capacitance is then $C = Q/V$.

Summary of Worked Examples

Type of Calculation	Procedure and Relevant Examples	
1. Energy		
Find the electrostatic potential energy of a system of point charges.	Use $U = \frac{1}{2}\Sigma\, q_i V_i$.	**Example 25-1**
2. Single Capacitors		
Find the capacitance of a parallel-plate capacitor.	Use $C = \epsilon_0 A/d$.	**Example 25-2**
Find the capacitance of any capacitor.	Place charges $+Q$ and $-Q$ on the two conductors and find the electric field E between them. Then calculate the potential difference by integrating $dV = -\vec{E}\, d\ell$ from one conductor to the other.	**Example 25-3**
Find the charge on a capacitor.	The charge is given by $Q = CV$.	**Example 25-2**
Find the energy stored in a capacitor.	Use $U = \frac{1}{2}Q^2/C = \frac{1}{2}QV = \frac{1}{2}CV^2$.	**Examples 25-4, 25-8, 25-6**
3. Combinations of Capacitors		
Calculate the equivalent capacitance.	Start by combining the smallest parallel or series units in the network, using $C_{eq} = C_1 + C_2 + \cdots$ (parallel) and $1/C_{eq} = 1/C_1 + 1/C_2 + \cdots$ (series). Keep replacing parallel and series units until the network is reduced to a single capacitor.	**Examples 25-5, 25-6, 25-7**
4. Capacitors with Dielectrics		
Determine the capacitance of a capacitor filled with a dielectric.	The capacitance is $C = \kappa C_0$, where C_0 is the capacitance without the dielectric.	**Examples 25-8, 25-9**
Calculate the capacitance of a capacitor partially filled a dielectric, or filled with more than one type of dielectric.	Find the potential difference V for a given charge Q, then use $C = Q/V$.	**Example 25-8**

Problems

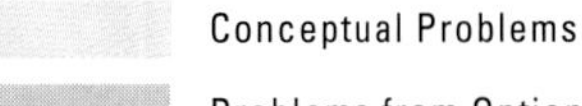

Conceptual Problems

Problems from Optional and Exploring sections

In a few problems, you are given more data than you actually need; in a few other problems, you are required to supply data from your general knowledge, outside sources, or informed estimates.

- • Single-concept, single-step, relatively easy
- •• Intermediate-level, may require synthesis of concepts
- ••• Challenging, for advanced students

Electrostatic Potential Energy

1 • Three point charges are on the x axis: q_1 at the origin, q_2 at $x = 3$ m, and q_3 at $x = 6$ m. Find the electrostatic potential energy for (*a*) $q_1 = q_2 = q_3 = 2\ \mu$C, (*b*) $q_1 = q_2 = 2\ \mu$C and $q_3 = -2\ \mu$C, and (*c*) $q_1 = q_3 = 2\ \mu$C and $q_2 = -2\ \mu$C.

2 • Point charges q_1, q_2, and q_3 are at the corners of an equilateral triangle of side 2.5 m. Find the electrostatic potential energy of this charge distribution if (*a*) $q_1 = q_2 = q_3 = 4.2\ \mu$C, (*b*) $q_1 = q_2 = 4.2\ \mu$C and $q_3 = -4.2\ \mu$C, (*c*) $q_1 = q_2 = -4.2\ \mu$C and $q_3 = +4.2\ \mu$C.

3 • What is the electrostatic potential energy of an isolated spherical conductor of radius 10 cm that is charged to 2 kV?

4 •• Four point charges of magnitude 2 μC are at the corners of a square of side 4 m. Find the electrostatic potential energy if (*a*) all the charges are negative, (*b*) three of the charges are positive and one is negative, and (*c*) two are positive and two are negative.

5 •• Four charges are at the corners of a square centered at the origin as follows: q at $(-a, +a)$; $2q$ at (a, a); $-3q$ at

$(a, -a)$; and $6q$ at $(-a, -a)$. A fifth charge $+q$ is placed at the origin and released from rest. Find its speed when it is a great distance from the origin.

6 •• Four identical particles each with charge Q are at the corners of a square of side L. The particles are released one at a time proceeding clockwise around the square. Each particle is allowed to reach its final speed a long distance from the square before the next particle is released. What is the final kinetic energy of (*a*) the first particle released, (*b*) the second particle released, (*c*) the third particle released, and (*d*) the fourth particle released? (*e*) What is the final kinetic energy of each particle if the four particles are released simultaneously? Does this depend on whether or not the particles have identical masses?

Capacitance

7 • If the voltage across a parallel-plate capacitor is doubled, its capacitance

(*a*) doubles.
(*b*) drops by half.
(*c*) remains the same.

8 • If the charge on an isolated spherical conductor is doubled, its capacitance

(*a*) doubles.
(*b*) drops by half.
(*c*) remains the same.

9 • An isolated spherical conductor of radius 10 cm is charged to 2 kV. (*a*) How much charge is on the conductor? (*b*) What is the capacitance of the sphere? (*c*) How does the capacitance change if the sphere is charged to 6 kV?

10 • A capacitor has a charge of 30 μC. The potential difference between the conductors is 400 V. What is the capacitance?

11 • (*a*) If a parallel-plate capacitor has a 0.15-mm separation, what must its area be for it to have a capacitance of 1 F? (*b*) If the plates are square, what is the length of their sides?

Storage of Electrical Energy

12 • True or false: The electrostatic energy per unit volume at some point is proportional to the square of the electric field at that point.

13 • If the potential difference of a parallel-plate capacitor is doubled by changing the plate separation without changing the charge, by what factor does its stored electric energy change?

14 •• Half the charge is removed from a capacitor without changing its capacitance. What fraction of its stored energy is removed along with the charge?

15 •• A parallel-plate air capacitor is connected to a constant-voltage battery. If the separation between the capacitor plates is doubled while the capacitor remains connected to the battery, the energy stored in the capacitor

(*a*) quadruples.
(*b*) doubles.
(*c*) remains unchanged.
(*d*) drops to half its previous value.
(*e*) drops to one-fourth its previous value.

16 •• If the capacitor of Problem 15 is disconnected from the battery before the separation between the plates is doubled, the energy stored in the capacitor upon separation of the plates

(*a*) quadruples.
(*b*) doubles.
(*c*) remains unchanged.
(*d*) drops to half its previous value.
(*e*) drops to one-fourth its previous value.

17 • (*a*) A 3-μF capacitor is charged to 100 V. How much energy is stored in the capacitor? (*b*) How much additional energy is required to charge the capacitor from 100 to 200 V?

18 • A 10-μF capacitor is charged to $Q = 4\ \mu$C. (*a*) How much energy is stored in the capacitor? (*b*) If half the charge is removed, how much energy remains?

19 • (*a*) Find the energy stored in a 20-pF capacitor when it is charged to 5 μC. (*b*) How much additional energy is required to increase the charge from 5 to 10 μC?

20 • Find the energy per unit volume in an electric field that is equal to 3 MV/m, the dielectric strength of air.

21 • A parallel-plate capacitor with a plate area of 2 m^2 and a separation of 1.0 mm is charged to 100 V. (*a*) What is the electric field between the plates? (*b*) What is the energy per unit volume in the space between the plates? (*c*) Find the total energy by multiplying your answer to part (*b*) by the total volume between the plates. (*d*) Find the capacitance C. (*e*) Calculate the total energy from $U = \frac{1}{2}CV^2$, and compare your answer with your result for part (*c*).

22 •• Energy prospectors from a distant planet are inspecting the earth to decide if its electrical energy resources are worth stealing. Measurements reveal that earth's electric field extends upward for 1000 m and has an average magnitude of 200 V/m. Estimate the electrical energy stored in the atmosphere. (*Hint:* You may treat the atmosphere as a flat slab with an area equal to the surface area of the earth. Why?)

23 •• A parallel-plate capacitor with plates of area 500 cm^2 is charged to a potential difference V and is then disconnected from the voltage source. When the plates are moved 0.4 cm farther apart, the voltage between the plates increases by 100 V. (*a*) What is the charge Q on the positive plate of the capacitor? (*b*) How much does the energy stored in the capacitor increase due to the movement of the plates?

24 ••• A ball of charge of radius R has a uniform charge density ρ and a total charge $Q = \frac{4}{3}\pi R^3\rho$. (*a*) Find the electrostatic energy density at a distance r from the center of the ball for $r < R$ and for $r > R$. (*b*) Find the energy in a spherical shell of volume $4\pi r^2\,dr$ for both $r < R$ and $r > R$. (*c*) Compute the total electrostatic energy by integrating your expressions from part (*b*), and show that your result can be written $U = \frac{3}{5}kQ^2/R$. Explain why this result is greater than that for a spherical conductor of radius R carrying a total charge Q.

Combinations of Capacitors

25 • True or false:

(*a*) The equivalent capacitance of two capacitors in parallel equals the sum of the individual capacitances.
(*b*) The equivalent capacitance of two capacitors in series is less than the capacitance of either capacitor alone.

26 •• Two initially uncharged capacitors of capacitance C_0 and $2C_0$, respectively, are connected in series across a battery. Which of the following is true?

(*a*) The capacitor $2C_0$ carries twice the charge of the other capacitor.
(*b*) The voltage across each capacitor is the same.
(*c*) The energy stored by each capacitor is the same.
(*d*) None of the above statements is correct.

27 • (*a*) How many 1.0-μF capacitors connected in parallel would it take to store a total charge of 1 mC with a potential difference of 10 V across each capacitor? (*b*) What would be the potential difference across the combination? (*c*) If the number of 1.0-μF capacitors found in part (*a*) is connected in series and the potential difference across each is 10 V, find the charge on each and the potential difference across the combination.

28 • A 3.0-μF capacitor and a 6.0-μF capacitor are connected in series, and the combination is connected in parallel with an 8.0-μF capacitor. What is the equivalent capacitance of this combination?

29 • Three capacitors are connected in a triangular network as shown in Figure 25-23. Find the equivalent capacitance across terminals *a* and *c*.

Figure 25-23 Problem 29

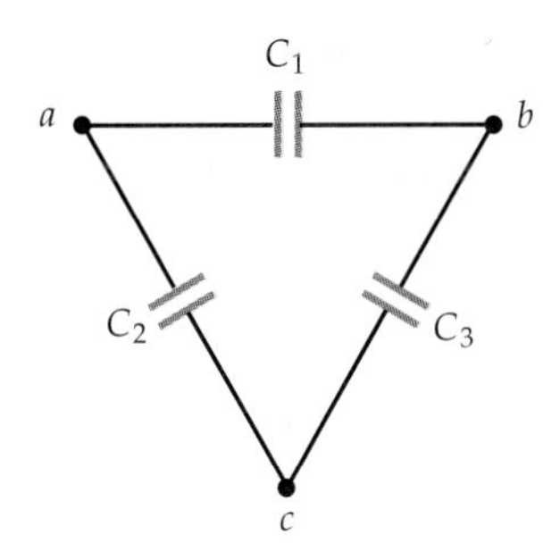

30 • A 10.0-μF capacitor and a 20.0-μF capacitor are connected in parallel across a 6.0-V battery. (*a*) What is the equivalent capacitance of this combination? (*b*) What is the potential difference across each capacitor? (*c*) Find the charge on each capacitor.

31 •• A 10.0-μF capacitor is connected in series with a 20.0-μF capacitor across a 6.0-V battery. (*a*) Find the charge on each capacitor. (*b*) Find the potential difference across each capacitor.

32 •• Three identical capacitors are connected so that their maximum equivalent capacitance is 15 μF. (*a*) Describe how the capacitors are combined. (*b*) There are three other ways to combine all three capacitors in a circuit. What are the equivalent capacitances for each arrangement?

33 •• For the circuit shown in Figure 25-24, find (*a*) the total equivalent capacitance between the terminals, (*b*) the charge stored on each capacitor, and (*c*) the total stored energy.

Figure 25-24
Problem 33

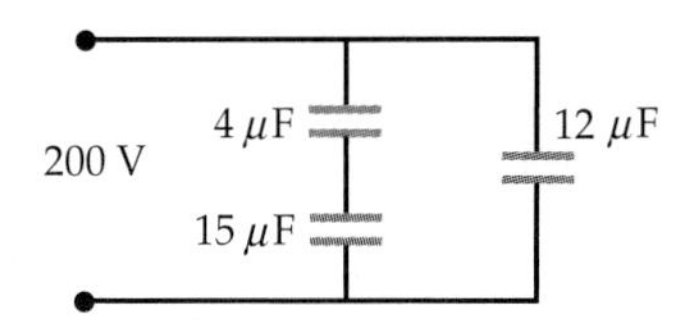

34 •• (*a*) Show that the equivalent capacitance of two capacitors in series can be written

$$C_{eq} = \frac{C_1C_2}{C_1 + C_2}$$

(*b*) Use this expression to show that $C_{eq} < C_1$ and $C_{eq} < C_2$. (*c*) Show that the correct expression for the equivalent capacitance of three capacitors in series is

$$C_{eq} = \frac{C_1C_2C_3}{C_1C_2 + C_2C_3 + C_1C_3}$$

35 •• For the circuit shown in Figure 25-25, find (*a*) the total equivalent capacitance between the terminals, (*b*) the charge stored on each capacitor, and (*c*) the total stored energy.

Figure 25-25
Problem 35

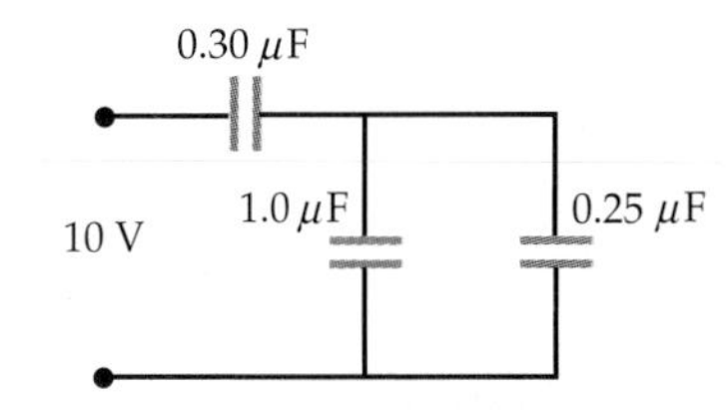

36 •• Five identical capacitors of capacitance C_0 are connected in a bridge network as shown in Figure 25-26. (*a*) What is the equivalent capacitance between points *a* and *b*? (*b*) Find the equivalent capacitance if the capacitance between *a* and *b* is changed to $10C_0$.

Figure 25-26 Problem 36

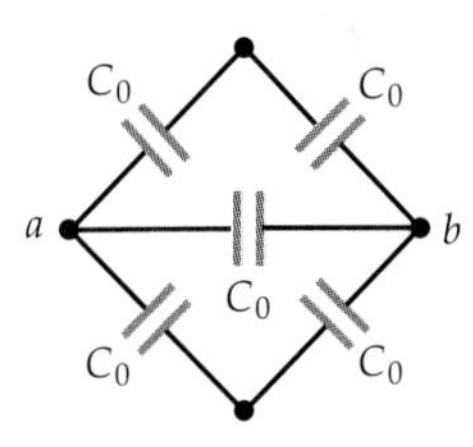

37 •• In Figure 25-27, $C_1 = 2\ \mu$F, $C_2 = 6\ \mu$F, and $C_3 = 3.5\ \mu$F. (*a*) Find the equivalent capacitance of this combination. (*b*) If the breakdown voltages of the individual capacitors are $V_1 = 100$ V, $V_2 = 50$ V, and $V_3 = 400$ V, what maximum voltage can be placed across points *a* and *b*?

Figure 25-27 Problem 37

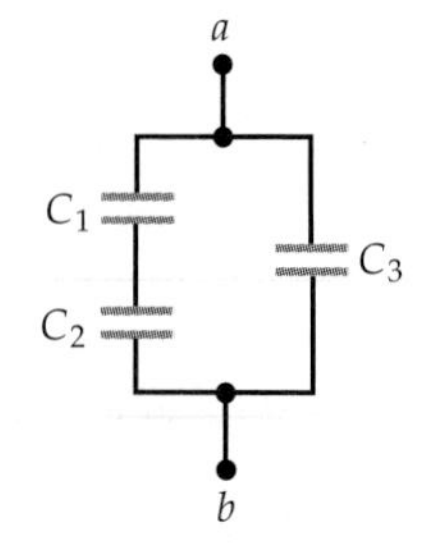

38 •• Design a network of capacitors that has a capacitance of 2 μF and breakdown voltage of 400 V using only

2-μF capacitors that have individual breakdown voltages of 100 V.

39 •• Find all the different possible equivalent capacitances that can be obtained using a 1.0-, a 2.0-, and a 4.0-μF capacitor in any combination that includes all three or any two of the capacitors.

Parallel-Plate Capacitors

40 • A parallel-plate capacitor has a capacitance of 2.0 μF and a plate separation of 1.6 mm. (*a*) What is the maximum potential difference between the plates such that dielectric breakdown of the air between the plates does not occur? (Use $E_{max} = 3$ MV/m.) (*b*) How much charge is stored at this maximum potential difference?

41 • An electric field of 2×10^4 V/m exists between the plates of a circular parallel-plate capacitor that has a plate separation of 2 mm. (*a*) What is the voltage across the capacitor? (*b*) What plate radius is required if the stored charge is 10 μC?

42 •• A parallel-plate, air-gap capacitor has a capacitance of 0.14 μF. The plates are 0.5 mm apart. (*a*) What is the area of each plate? (*b*) What is the potential difference if the capacitor is charged to 3.2 μC? (*c*) What is the stored energy? (*d*) How much charge can the capacitor carry before dielectric breakdown of the air between the plates occurs?

43 •• Design a 0.1-μF parallel-plate capacitor with air between the plates that can be charged to a maximum potential difference of 1000 V. (*a*) What is the minimum possible separation between the plates? (*b*) What minimum area must the plates of the capacitor have?

Cylindrical Capacitors

44 • A coaxial communications cable connecting two cities has an inner radius of 0.8 mm and an outer radius of 6 mm. Its length is 8×10^5 m (about 500 mi). Treat this cable as a cylindrical capacitor and calculate its capacitance.

45 • A Geiger tube consists of a wire of radius 0.2 mm and length 12 cm and a coaxial cylindrical shell conductor of the same length and a radius of 1.5 cm. (*a*) Find the capacitance, assuming that the gas in the tube has a dielectric constant of 1. (*b*) Find the charge per unit length on the wire when the potential difference between the wire and shell is 1.2 kV.

46 •• A cylindrical capacitor consists of a long wire of radius R_1 and length L with a charge $+Q$ and a concentric outer cylindrical shell of radius R_2, length L, and charge $-Q$. (*a*) Find the electric field and energy density at any point in space. (*b*) How much energy resides in a cylindrical shell between the conductors of radius r, thickness dr, and volume $2\pi rL\,dr$? (*c*) Integrate your expression from part (*b*) to find the total energy stored in the capacitor, and compare your result with that obtained using $U = \frac{1}{2}CV^2$.

47 ••• Three concentric thin conducting cylindrical shells have radii of 0.2, 0.5, and 0.8 cm. The space between the shells is filled with air. The innermost and outermost cylinders are connected together. Find the capacitance per unit length of this system.

Spherical Capacitors

48 •• A spherical capacitor consists of two thin concentric spherical shells of radii R_1 and R_2. (*a*) Show that the capacitance is given by $C = 4\pi\epsilon_0 R_1R_2/(R_2 - R_1)$. (*b*) Show that when the radii of the shells are nearly equal, the capacitance is given approximately by the expression for the capacitance of a parallel-plate capacitor, $C = \epsilon_0 A/d$, where A is the area of the sphere and $d = R_2 - R_1$.

49 •• A spherical capacitor has an inner sphere of radius R_1 with a charge of $+Q$ and an outer concentric spherical shell of radius R_2 with a charge of $-Q$. (*a*) Find the electric field and the energy density at any point in space. (*b*) Calculate the energy in the electrostatic field in a spherical shell of radius r, thickness dr, and volume $4\pi r^2\,dr$ between the conductors? (*c*) Integrate your expression from part (*b*) to find the total energy stored in the capacitor, and compare your result with that obtained using $U = \frac{1}{2}QV$.

50 ••• A spherical shell of radius R carries a charge Q distributed uniformly over its surface. Find the radius r of the sphere such that half the total electrostatic field energy of the system is contained within that sphere.

51 ••• Repeat Problem 50 if the charge Q resides not on a spherical shell but is distributed uniformly throughout a spherical volume of radius R. (See Problem 24.)

Disconnected and Reconnected Capacitors

52 •• A 2.0-μF capacitor is charged to a potential difference of 12.0 V. The wires connecting the capacitor to the battery are then disconnected from the battery and connected across a second, initially uncharged, capacitor. The potential difference across the 2.0-μF capacitor then drops to 4 V. What is the capacitance of the second capacitor?

53 •• A 100-pF capacitor and a 400-pF capacitor are both charged to 2.0 kV. They are then disconnected from the voltage source and are connected together, positive plate to positive plate and negative plate to negative plate. (*a*) Find the resulting potential difference across each capacitor. (*b*) Find the energy lost when the connections are made.

54 •• Two capacitors $C_1 = 4$ μF and $C_2 = 12$ μF are connected in series across a 12-V battery. They are carefully disconnected so that they are not discharged and are reconnected to each other with positive plate to positive plate and negative plate to negative plate. (*a*) Find the potential difference across each capacitor after they are connected. (*b*) Find the initial and final energy stored in the capacitors.

55 •• A 1.2-μF capacitor is charged to 30 V. After charging, the capacitor is disconnected from the voltage source and is connected to another uncharged capacitor. The final voltage is 10 V. (*a*) What is the capacitance of the other capacitor? (*b*) How much energy was lost when the connection was made?

56 •• Work Problem 53 if the capacitors are connected positive plate to negative plate after they have been charged to 2.0 kV.

57 •• Work Problem 54 if the two capacitors are first connected in parallel across the 12-V battery and are then connected, with the positive plate of each capacitor connected to the negative plate of the other.

58 •• A 20-pF capacitor is charged to 3.0 kV and then removed from the battery and connected to an uncharged 50-pF capacitor. (*a*) What is the new charge on each capacitor? (*b*) Find the initial energy stored in the 20-pF capacitor and the final energy stored in the two capacitors. Is electrostatic potential energy gained or lost when the two capacitors are connected?

59 ••• A parallel combination of three capacitors, $C_1 = 2\ \mu F$, $C_2 = 4\ \mu F$, and $C_3 = 6\ \mu F$, is charged with a 200-V source. The capacitors are then disconnected from both the voltage source and each other and are reconnected positive plates to negative plates as shown in Figure 25-28. (*a*) What is the voltage across each capacitor with switches S_1 and S_2 closed but switch S_3 open? (*b*) After switch S_3 is closed, what is the final charge on each capacitor? (*c*) Give the voltage across each capacitor after switch S_3 is closed.

Figure 25-28 Problem 59

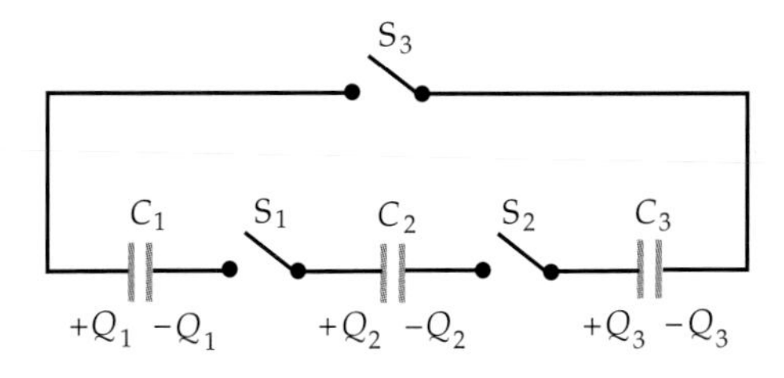

Dielectrics

60 • True or false: A dielectric inserted into a capacitor increases the capacitance.

61 • A parallel-plate capacitor is made by placing polyethylene ($\kappa = 2.3$) between two sheets of aluminum foil. The area of each sheet is 400 cm^2, and the thickness of the polyethylene is 0.3 mm. Find the capacitance.

62 •• Suppose the Geiger tube of Problem 45 is filled with a gas of dielectric constant $\kappa = 1.8$ and breakdown field of 2×10^6 V/m. (*a*) What is the maximum potential difference that can be maintained between the wire and shell? (*b*) What is the charge per unit length on the wire?

63 •• Repeat Problem 49 with the space between the two spherical shells filled with a dielectric of dielectric constant κ.

64 •• A certain dielectric with a dielectric constant $\kappa = 24$ can withstand an electric field of 4×10^7 V/m. Suppose we want to use this dielectric to construct a 0.1-μF capacitor that can withstand a potential difference of 2000 V. (*a*) What is the minimum plate separation? (*b*) What must the area of the plates be?

65 •• A parallel-plate capacitor has plates separated by a distance *s*. The space between the plates is filled with two dielectrics, one of thickness $\frac{1}{4}s$ and dielectric constant κ_1, the other with thickness $\frac{3}{4}s$ and dielectric constant κ_2. Find the capacitance of this capacitor in terms of C_0, the capacitance with no dielectrics.

66 •• A parallel-plate capacitor with no dielectric has a capacitance C_0. If the separation between the plates is *d*, and a slab with dielectric constant κ and thickness $t < d$ is placed in the capacitor, find the new capacitance.

67 •• The membrane of the axon of a nerve cell is a thin cylindrical shell of radius $r = 10^{-5}$ m, length $L = 0.1$ m, and thickness $d = 10^{-8}$ m. The membrane has a positive charge on one side and a negative charge on the other, and acts as a parallel-plate capacitor of area $A = 2\pi rL$ and separation *d*. Its dielectric constant is about $\kappa = 3$. (*a*) Find the capacitance of the membrane. If the potential difference across the membrane is 70 mV, find (*b*) the charge on each side of the membrane, and (*c*) the electric field through the membrane.

68 •• What is the dielectric constant of a dielectric on which the induced bound charge density is (*a*) 80% of the free charge density on the plates of a capacitor filled by the dielectric, (*b*) 20% of the free charge density, and (*c*) 98% of the free charge density?

69 •• Two parallel plates have charges *Q* and $-Q$. When the space between the plates is devoid of matter, the electric field is 2.5×10^5 V/m. When the space is filled with a certain dielectric, the field is reduced to 1.2×10^5 V/m. (*a*) What is the dielectric constant of the dielectric? (*b*) If $Q = 10$ nC, what is the area of the plates? (*c*) What is the total induced charge on either face of the dielectric?

70 •• Find the capacitance of the parallel-plate capacitor shown in Figure 25-29.

Figure 25-29 Problem 70

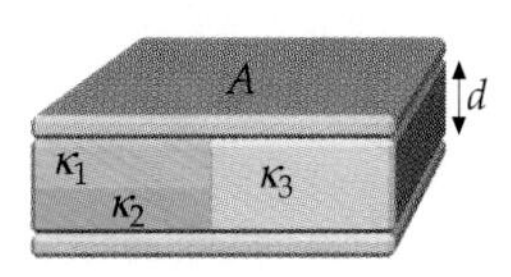

71 •• A parallel-plate capacitor has plates of area 600 cm^2 and a separation of 4 mm. The capacitor is charged to 100 V and is then disconnected from the battery. (*a*) Find the electric field E_0 and the electrostatic energy *U*. A dielectric of constant $\kappa = 4$ is then inserted, completely filling the space between the plates. Find (*b*) the new electric field *E*, (*c*) the potential difference *V*, and (*d*) the new electrostatic energy.

72 ••• A parallel-plate capacitor is constructed using a dielectric whose constant varies with position. The plates have area *A*. The bottom plate is at $y = 0$ and the top plate is at $y = y_0$. The dielectric constant is given as a function of *y* according to $\kappa = 1 + (3/y_0)\,y$. (*a*) What is the capacitance? (*b*) Find σ_b/σ_f on the surfaces of the dielectric. (*c*) Use Gauss's law to find the induced volume charge density $\rho(y)$ within this dielectric. (*d*) Integrate the expression for the volume charge density found in (*c*) over the dielectric, and show that the

total induced bound charge, including that on the surfaces, is zero.

General Problems

73 • True or false:

(*a*) The capacitance of a capacitor is defined as the total amount of charge it can hold.
(*b*) The capacitance of a parallel-plate capacitor depends on the voltage difference between the plates.
(*c*) The capacitance of a parallel-plate capacitor is proportional to the charge on its plates.

74 •• Two identical capacitors are connected in series to a 100-V battery. When only one capacitor is connected to this battery the energy stored is U_0. What is the total energy stored in the two capacitors when the series combination is connected to the battery?

(*a*) $4U_0$
(*b*) $2U_0$
(*c*) U_0
(*d*) $U_0/2$
(*e*) $U_0/4$

75 • Three capacitors have capacitances of 2.0, 4.0, and 8.0 μF. Find the equivalent capacitance if (*a*) the capacitors are connected in parallel and (*b*) they are connected in series.

76 • A 1.0-μF capacitor is connected in parallel with a 2.0-μF capacitor, and the combination is connected in series with a 6.0-μF capacitor. What is the equivalent capacitance of this combination?

77 • The voltage across a parallel plate capacitor with plate separation 0.5 mm is 1200 V. The capacitor is disconnected from the voltage source and the separation between the plates is increased until the energy stored in the capacitor has been doubled. Determine the final separation between the plates.

78 •• Determine the capacitance of each of the networks shown in Figure 25-30.

Figure 25-30 Problem 78

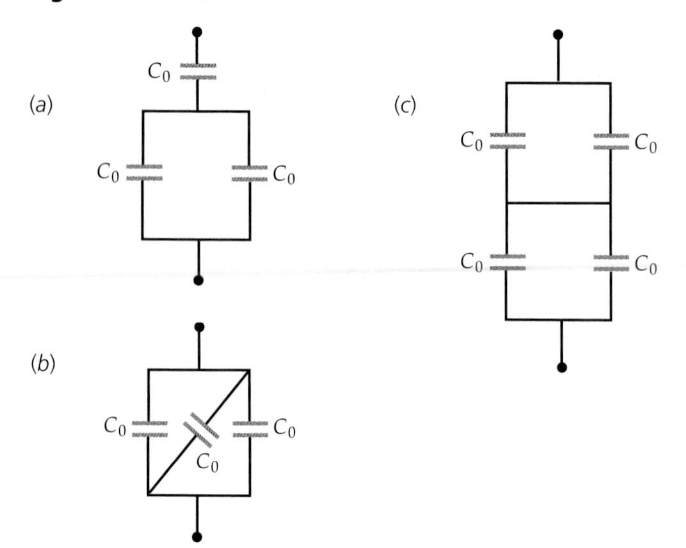

79 •• Figure 25-31 shows four capacitors connected in the arrangement known as a capacitance bridge. The capacitors are initially uncharged. What must be the relation between the four capacitances so that the potential between points *c* and *d* is zero when a voltage *V* is applied between points *a* and *b*?

Figure 25-31 Problem 79

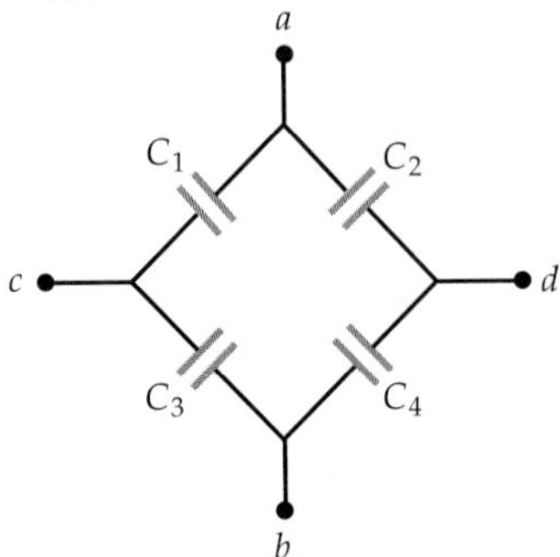

80 •• The plate areas and plate separations of the two parallel-plate capacitors shown in Figure 25-32 are identical. Half the region between the plates of capacitor C_1 is filled with a dielectric of dielectric constant κ. What fraction of the volume of capacitor C_2 should be filled with the same dielectric material so that the two capacitors have the same capacitance?

Figure 25-32 Problems 80 and 81

81 •• Repeat Problem 80 if the region filled with dielectric of capacitor (a) is two-thirds of the volume between the plates.

82 •• Two conducting spheres of radius *R* are separated by a distance large compared to their size. One initially has charge *Q* and the other is uncharged. A thin wire is then connected between them. What fraction of the initial energy is dissipated?

83 •• A parallel-plate capacitor of area *A* and separation *d* is charged to a potential difference *V* and then disconnected from the charging source. The plates are then pulled apart until the separation is 2*d*. Find expressions in terms of *A*, *d*, and *V* for (*a*) the new capacitance, (*b*) the new potential difference, and (*c*) the new stored energy. (*d*) How much work was required to change the plate separation from *d* to 2*d*?

84 •• A parallel-plate capacitor has capacitance C_0 with no dielectric. It is then filled with dielectric of constant κ. When a second capacitor of capacitance C' is connected in series with the first one, the capacitance of the series combination is C_0. Find C'.

85 •• Estimate the capacitance of a typical hot-air balloon.

86 •• A Leyden jar, the earliest type of capacitor, is a glass jar coated inside and out with metal foil. Suppose that a Leyden jar is a cylinder 40 cm high with 2.0-mm-thick walls and an inner diameter of 8 cm. Ignore any field fringing. (*a*) Find the capacitance of this Leyden jar if the dielectric constant κ of the glass is 5.0. (*b*) If the dielectric strength of the glass is 15 MV/m, what maximum charge can the Leyden jar carry without undergoing dielectric breakdown? (*Hint:* Treat the device as a parallel-plate capacitor.)

87 •• A parallel-plate capacitor is constructed from a layer of silicon dioxide of thickness 5×10^{-6} m between two conducting films. The dielectric constant of silicon dioxide is 3.8 and its dielectric strength is 8×10^{6} V/m. (*a*) What voltage can be applied across this capacitor without dielectric breakdown? (*b*) What should the surface area of the layer of silicon dioxide be for a 10-pF capacitor? (*c*) Estimate the number of these capacitors that can fit into a square 1 cm by 1 cm.

Figure 25-33 Problems 88 and 89

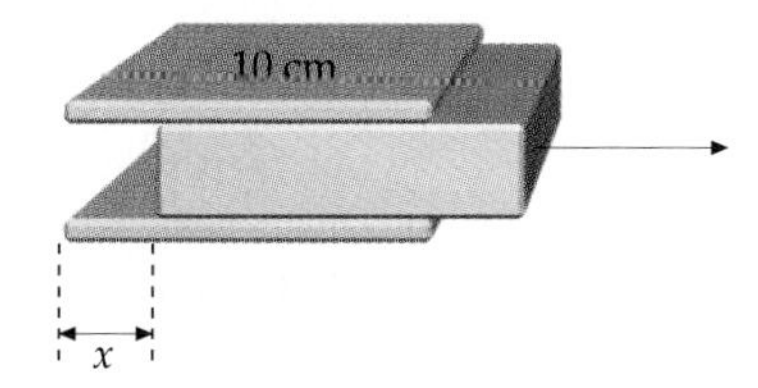

88 •• A parallel-plate capacitor has rectangular plates of length $L = 10$ cm and width $W = 4$ cm (Figure 25-33). The region between the plates is filled with a dielectric slab of dielectric constant $\kappa = 4$ which can slide along the length of the capacitor. Initially, the slab completely fills the rectangular region, and the capacitor holds a charge of 0.2 μC. How far should the dielectric slab be pulled so that the stored energy is double its initial value?

89 •• Suppose the capacitor of Problem 88 is connected to a constant voltage source of 20 V. How far should the dielectric slab be pulled so that the stored energy is reduced to half its initial value?

90 •• A parallel combination of two identical 2-μF parallel-plate capacitors is connected to a 100-V battery. The battery is then removed and the separation between the plates of one of the capacitors is doubled. Find the charge on each of the capacitors.

91 •• A parallel-plate capacitor has a capacitance C_0 and a plate separation d. Two dielectric slabs of constants κ_1 and κ_2, each of thickness $\frac{1}{2}d$ and having the same area as the plates, are inserted between the plates as shown in Figure 25-34. When the charge on the plates is Q, find (*a*) the electric field in each dielectric and (*b*) the potential difference between the plates. (*c*) Show that the new capacitance is given by $C = 2\kappa_1\kappa_2/(\kappa_1 + \kappa_2)\,C_0$. (*d*) Show that this system can be considered to be a series combination of two capacitors of thickness $\frac{1}{2}d$ filled with dielectrics of constant κ_1 and κ_2.

Figure 25-34 Problem 91

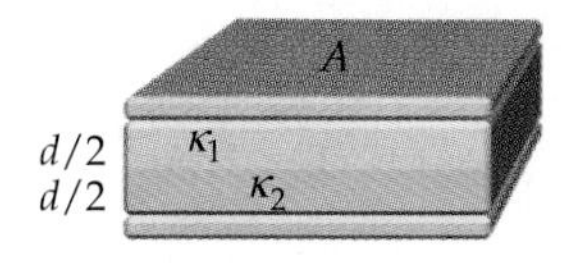

92 •• A parallel-plate capacitor has a plate area A and a separation d. A metal slab of thickness t and area A is inserted between the plates. (*a*) Show that the capacitance is given by $C = \epsilon_0 A/(d - t)$, regardless of where the metal slab is placed. (*b*) Show that this arrangement can be considered to be a capacitor of separation a in series with one of separation b, where $a + b + t = d$.

93 •• A parallel-plate capacitor is filled with two dielectrics of equal size as shown in Figure 25-35. (*a*) Show that this system can be considered to be two capacitors of area $\frac{1}{2}A$ connected in parallel. (*b*) Show that the capacitance is increased by the factor $(\kappa_1 + \kappa_2)/2$.

Figure 25-35 Problem 93

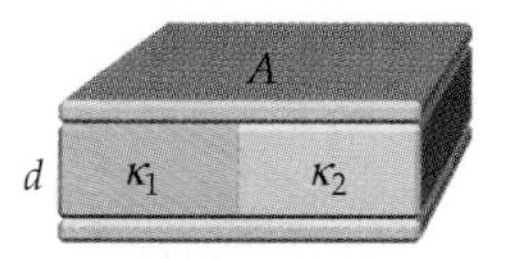

94 •• A parallel-plate capacitor of plate area A and separation x is given a charge Q and is then removed from the charging source. (*a*) Find the stored electrostatic energy as a function of x. (*b*) Find the increase in energy dU due to an increase in plate separation dx from $dU = (dU/dx)\,dx$. (*c*) If F is the force exerted by one plate on the other, the work needed to move one plate a distance dx is $F\,dx = dU$. Show that $F = Q^2/2\epsilon_0 A$. (*d*) Show that the force in part (*c*) equals $\frac{1}{2}EQ$, where Q is the charge on one plate and E is the electric field between the plates. Discuss the reason for the factor $\frac{1}{2}$ in this result.

95 •• A rectangular parallel-plate capacitor of length a and width b has a dielectric of width b partially inserted a distance x between the plates, as shown in Figure 25-36. (*a*) Find the capacitance as a function of x. Neglect edge effects. (*b*) Show that your answer gives the expected results for $x = 0$ and $x = a$.

Figure 25-36 Problems 95, 98, and 99

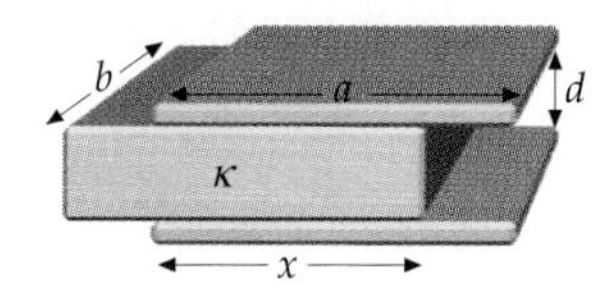

96 •• Two identical, 4-μF parallel-plate capacitors are connected in series across a 24-V battery. (*a*) What is the charge on each capacitor? (*b*) What is the total stored energy of the capacitors? A dielectric having a dielectric constant of 4.2 is inserted between the plates of one of the capacitors while the battery is still connected. (*c*) After the dielectric is inserted, what is the charge on each capacitor? (*d*) What is the potential difference across each capacitor? (*e*) What is the total stored energy of the capacitors?

97 •• A parallel-plate capacitor has a plate area of 1.0 m^2 and a plate separation distance of 0.5 cm. Completely filling the space between the conducting plates is a glass plate having a dielectric constant of 5.0. The capacitor is charged to a potential difference of 12.0 V and is then removed from its charging source. How much work is required to pull the glass plate out of the capacitor?

98 •• The capacitor shown in Figure 25-36 carries charges of $+Q$ and $-Q$ on its plates. (*a*) Find the stored energy as a function x. (*b*) Use the result of (*a*) to determine the force that acts on the dielectric slab.

99 •• The capacitor shown in Figure 25-36 is connected to a constant voltage source V. (*a*) Find the stored energy as a function of x. (*b*) Use the result of (*a*) to determine the force that acts on the dielectric slab.

100 •• A capacitor carries a charge of 15 μC when the potential between its plates is V. When the charge on the capacitor is increased to 18 μC, the potential between the plates increases by 6 V. Find the capacitance of the capacitor and the initial and final voltages.

101 ••• You are asked to construct a parallel-plate, air-gap capacitor that will store 100 kJ of energy. (*a*) What minimum volume is required between the plates of the capacitor? (*b*) Suppose you have developed a dielectric that can withstand 3×10^8 V/m and has a dielectric constant of 5. What volume of this dielectric between the plates of the capacitor is required for it to be able to store 100 kJ of energy?

102 ••• Consider two parallel-plate capacitors, C_1 and C_2, that are connected in parallel. The capacitors are identical except that C_2 has a dielectric inserted between its plates. A voltage source of 200 V is connected across the capacitors to charge them and is then disconnected. (*a*) What is the charge on each capacitor? (*b*) What is the total stored energy of the capacitors? (*c*) The dielectric is removed from C_2. What is the final stored energy of the capacitors? (*d*) What is the final voltage across the two capacitors?

103 ••• A capacitor is constructed of two concentric cylinders of radii a and b ($b > a$) having a length $L >> b$. A charge of $+Q$ is on the inner cylinder and a charge of $-Q$ is on the outer cylinder. The region between the two cylinders is filled with a dielectric having a dielectric constant κ. (*a*) Find the potential difference between the cylinders. (*b*) Find the density of the free charge σ_f on the inner cylinder and the outer cylinder. (*c*) Find the bound charge density σ_b on the inner cylindrical surface of the dielectric and the outer surface. (*d*) Find the total stored electrostatic energy. (*e*) If the dielectric will move without friction, how much mechanical work is required to remove the dielectric cylindrical shell?

104 ••• Two parallel-plate capacitors have the same separation and plate area. The capacitance of each is initially 10 μF. When a dielectric is inserted such that it completely fills the space between the plates of one of the capacitors, the capacitance of that capacitor increases to 35 μF. The 35- and 10-μF capacitors are connected in parallel and are charged to a potential difference of 100 V. The voltage source is then disconnected. (*a*) What is the stored energy of this system? (*b*) What are the charges on the two capacitors? (*c*) The dielectric is removed from the capacitor. What are the new charges on the plates of the capacitors? (*d*) What is the final stored energy of the system?

105 ••• A spherical weather balloon made of aluminized Mylar and filled with helium at atmospheric pressure can lift a payload of 0.2 kg. Determine the capacitance of the balloon. (Neglect the mass of the Mylar.)

106 ••• The two capacitors shown in Figure 25-37 have capacitances $C_1 = 0.4$ μF and $C_2 = 1.2$ μF. The voltages across the two capacitors are V_1 and V_2, respectively, and the total stored energy in the two capacitors is 1.14 mJ. If terminals b and c are connected together, the voltage $V_a - V_d = 80$ V; if terminal a is connected to b, and c is connected to d, the voltage $V_a - V_d = 20$ V. Find the initial voltages V_1 and V_2.

Figure 25-37 Problem 106

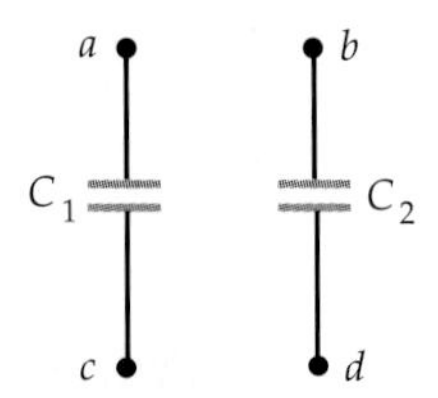

107 ••• Before switch S in Figure 25-38 is closed, the voltage across the terminals of the switch is 120 V and the voltage across the 0.2 μF capacitor is 40 V. The total energy stored in the two capacitors is 1440 μJ. After closing the switch, the voltage across each capacitor is 80 V, and the energy stored by the two capacitors has dropped to 960 μJ. Determine the capacitance of C_2 and the charge on that capacitor before the switch was closed.

Figure 25-38 Problem 107

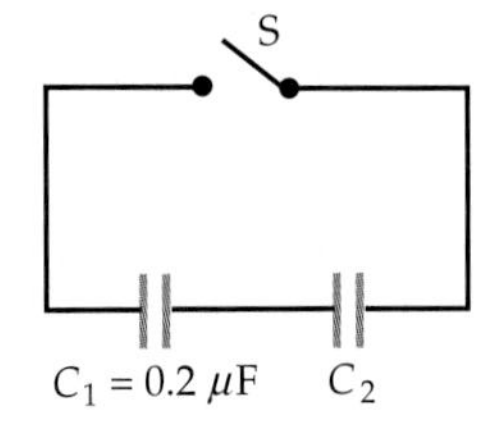

108 ••• A parallel-plate capacitor of area A and separation d is charged to a potential difference V and is then removed from the charging source. A dielectric slab of constant $\kappa = 2$, thickness d, and area $\frac{1}{2}A$ is inserted as shown in Figure 25-39. Let σ_1 be the free charge density at the conductor–dielectric surface and σ_2 be the free charge density at the conductor–air surface. (*a*) Why must the electric field have the same value inside the dielectric as in the free space between the plates? (*b*) Show that $\sigma_1 = 2\sigma_2$. (*c*) Show that the new capacitance is $3\epsilon_0 A/2d$ and that the new potential difference is $\frac{2}{3}V$.

Figure 25-39 Problem 108

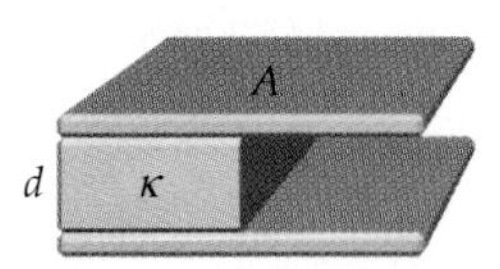

109 ••• Two identical, 10-μF parallel-plate capacitors are given equal charges of 100 μC each and are then removed from the charging source. The charged capacitors are connected by a wire between their positive plates and another wire between their negative plates. (*a*) What is the stored energy of the system? A dielectric having a dielectric constant of 3.2 is inserted between the plates of one of the capacitors such that it completely fills the region between the plates. (*b*) What is the final charge on each capacitor? (*c*) What is the final stored energy of the system?

110 ••• A capacitor has rectangular plates of length a and width b. The top plate is inclined at a small angle as shown in Figure 25-40. The plate separation varies from $d = y_0$ at the left to $d = 2y_0$ at the right, where y_0 is much less than a or b. Calculate the capacitance using strips of width dx and length b to approximate differential capacitors of area $b\,dx$ and separation $d = y_0 + (y_0/a)x$ that are connected in parallel.

Figure 25-40 Problem 110

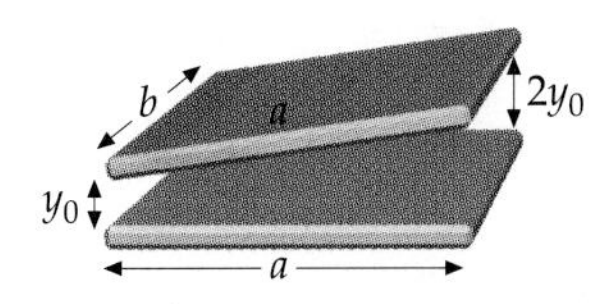

111 ••• Not all dielectrics that separate the plates of a capacitor are rigid. For example, the membrane of a nerve axon is a bilipid layer that has a finite compressibility. Consider a parallel-plate capacitor whose plate separation is maintained by a dielectric of dielectric constant $\kappa = 3.0$ and thickness $d = 0.2$ mm when the potential across the capacitor is zero. The dielectric, which has a dielectric strength of 40 kV/mm, is highly compressible, with a Young's modulus for compressive stress of 5×10^6 N/m^2. The capacitance of the capacitor in the limit $V \to 0$ is C_0. (*a*) Derive an expression for the capacitance as a function of voltage across the capacitor. (*b*) What is the maximum voltage that can be applied to the capacitor? (Assume that κ does not change under compression.) (*c*) What fraction of the total energy of the capacitor is electrostatic field energy and what fraction is mechanical stress energy stored in the compressed dielectric when the voltage across the capacitor is just below the breakdown voltage?

112 ••• A conducting sphere of radius R_1 is given a free charge Q. The sphere is surrounded by an uncharged concentric spherical dielectric shell having an inner radius R_1, an outer radius R_2, and a dielectric constant κ. The system is far removed from other objects. (*a*) Find the electric field everywhere in space. (*b*) What is the potential of the conducting sphere relative to $V = 0$ at infinity? (*c*) Find the total electrostatic potential energy of the system.

113 ••• A variable air capacitor like the one shown in the photograph on page 759 has a capacitance that changes between 0.02 and 0.12 μF as the shaft is rotated through an angle of 180°. A voltage of 100 V is maintained between the capacitor plates. Initially the capacitor is in its minimum capacitance position. (*a*) How much work must be done to rotate the shaft to the maximum capacitance position? (*b*) The shapes of the plates are designed so that the capacitance is a linear function of rotation angle. How much torque must be applied to rotate the capacitor to hold it in the position corresponding to $C = 0.07$ μF?

114 ••• Repeat Problem 113 with a voltage of 100 V applied and then disconnected when the capacitor is fully charged.

CHAPTER 26

Electric Current and Direct-Current Circuits

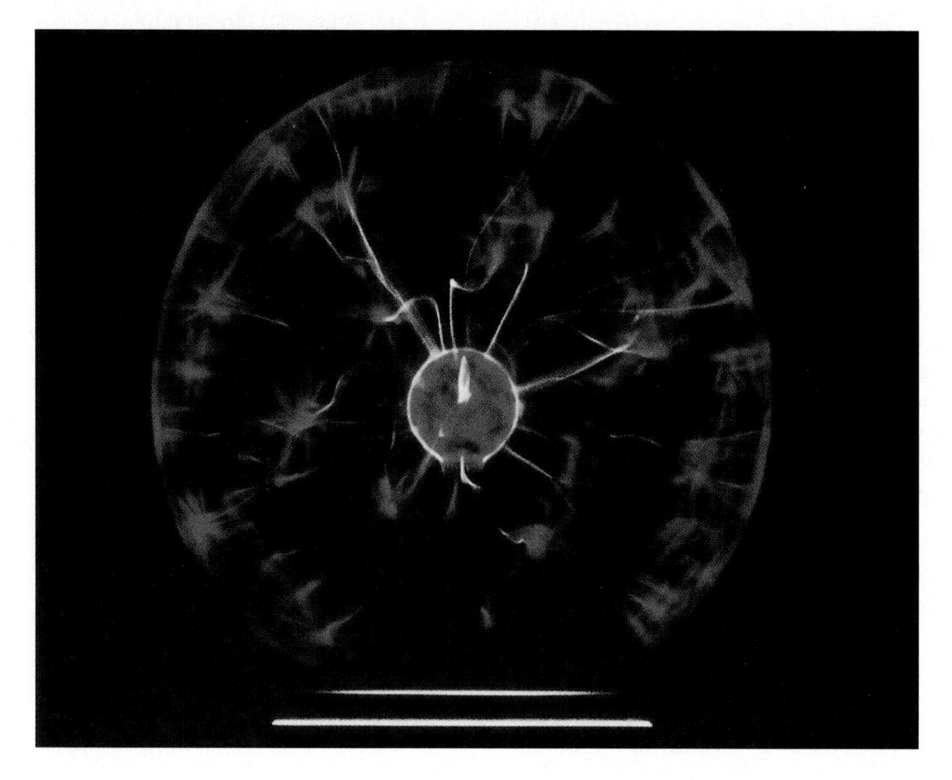

The plasma globe Eye of the Storm. In plasma globes, voltages ranging from 3 to 8 kV, at frequencies between 20 and 50 kHz, are applied between the sphere at the center of the globe and the outer shell, which is at ground. The region in between is filled with a mixture of inert gases. The applied voltage partially ionizes the gases, creating plasma filaments that conduct current. The colored displays are paths along which ionization is occurring.

When we turn on a light, we connect the wire filament in the light bulb across a potential difference that causes electric charge to flow through the wire, much the way a pressure difference in a garden hose causes water to flow through the hose. The flow of electric charge constitutes an electric current. Usually we think of currents as being in conducting wires, but the electron beam in a video monitor and a beam of charged ions from a particle accelerator also constitute electric currents.

When the direction of the current in a circuit element does not vary, the circuit is called a direct current (dc) circuit. Direct currents are usually produced by batteries connected to resistors and capacitors. In Chapter 31 we discuss alternating current (ac) circuits, in which the direction of the current alternates.

When a switch is thrown to turn on a circuit, an electric field propagates along the elements of the circuit at nearly the speed of light. Many complicated changes take place as the current builds up and charge accumulates at various points in the circuit, but an equilibrium or steady state is quickly es-

dent of the rate of flow of charge between them. The potential difference between the terminals of an ideal battery is equal in magnitude to the emf of the battery.

Figure 26-7 shows a simple circuit consisting of a resistance R connected to an ideal battery. In such diagrams, a battery is denoted by the symbol ⊣|⊢. The resistance is indicated by the symbol -\/\/\/-. The straight lines indicate connecting wires of negligible resistance. The source of emf maintains a constant potential difference equal to $\mathcal{E}$ between points a and b, with point a being at the higher potential. There is negligible potential difference between points a and c or between points d and b because the connecting wire is assumed to have negligible resistance. The potential difference from c and d is therefore equal in magnitude to the emf $\mathcal{E}$, and the current in the resistor is given by $I = \mathcal{E}/R$. The direction of the current in this circuit is clockwise, as shown in the figure.

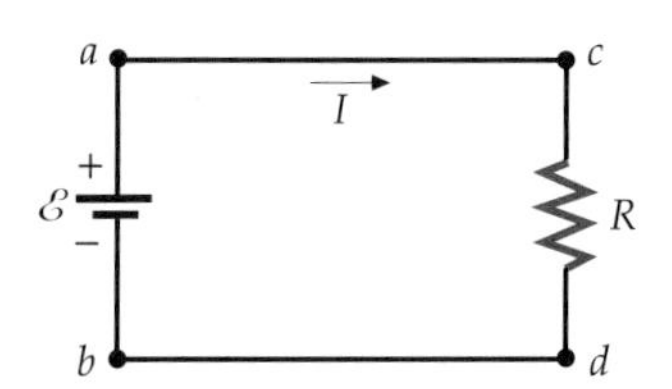

Figure 26-7 A simple circuit consisting of an ideal battery of emf $\mathcal{E}$, a resistance R, and connecting wires that are assumed to be without resistance.

Note that *inside* the source of emf, the charge flows from a region of low potential to a region of high potential, so it gains potential energy.* When charge ΔQ flows through the source of emf $\mathcal{E}$, its potential energy is increased by the amount $\Delta Q\, \mathcal{E}$. The charge then flows through the resistor, where this potential energy is converted into thermal energy. The rate at which energy is supplied by the source of emf is the power output:

$$P = \frac{\Delta Q\, \mathcal{E}}{\Delta t} = \mathcal{E}I \qquad \text{26-12}$$

In the simple circuit of Figure 26-7, the power put out by the source of emf equals that dissipated in the resistor.

A source of emf can be thought of as a charge pump that pumps the charge from a region of low electrical potential energy to a region of high electrical potential energy. Figure 26-8 shows a mechanical analog to the simple electric circuit just discussed.

In a **real battery**, the potential difference across the battery terminals, called the **terminal voltage**, is not simply equal to the emf of the battery. Consider the circuit consisting of a real battery and a resistor in Figure 26-9. If the current is varied by varying the resistance R and the terminal voltage is measured, the terminal voltage is found to decrease slightly as the current increases (Figure 26-10), just as if there were a small resistance within the battery.

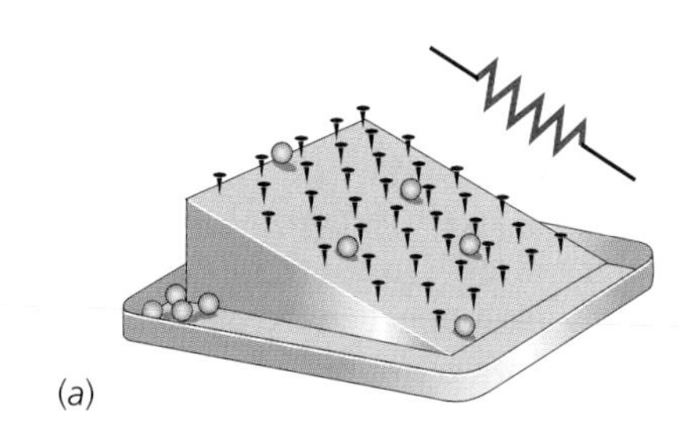

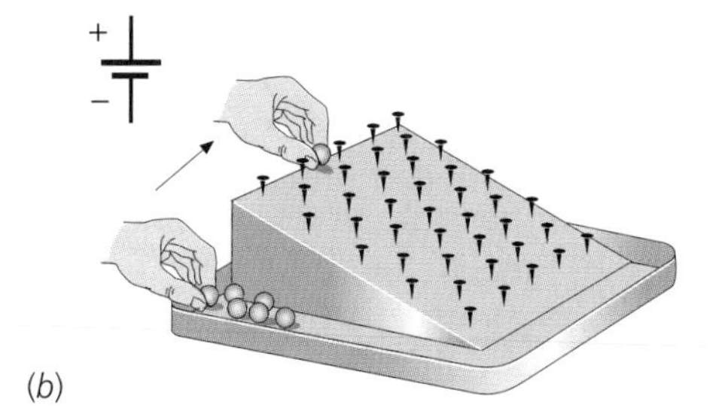

Figure 26-8 A mechanical analog of a simple circuit consisting of a resistance and source of emf. (*a*) The marbles start at some height h above the bottom and are accelerated between collisions with the nails by the gravitational field. The nails are analogous to the lattice ions in the resistor. During the collisions, the marbles transfer the kinetic energy they obtained between collisions to the nails. Because of the many collisions, the marbles have only a small, approximately constant, drift velocity toward the bottom. (*b*) When they reach the bottom, a child picks them up, lifts them to their original height h, and starts them again. The child, who does work mgh on each marble, is analogous to the source of emf. The energy source in this case is the internal chemical energy of the child.

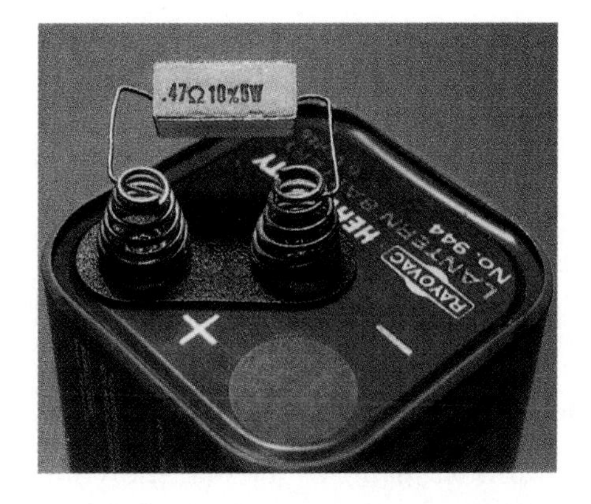

Figure 26-9 A simple circuit consisting of a real battery, a resistor, and connecting wires.

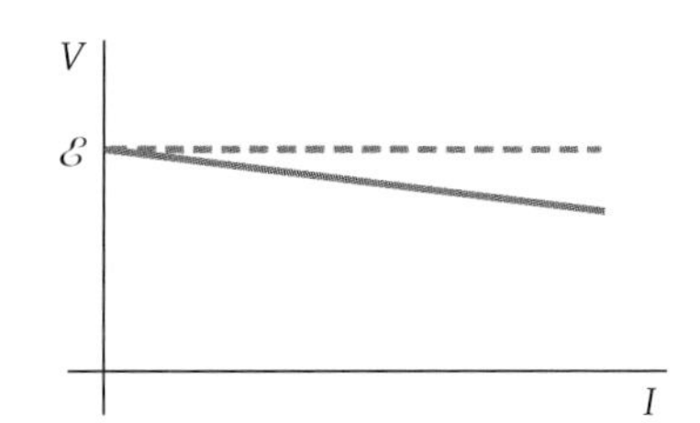

Figure 26-10 Terminal voltage V versus I for a real battery. The dashed line shows the terminal voltage of an ideal battery, which has the same magnitude as $\mathcal{E}$.

* When a battery is being charged by a generator or by another battery, the charge flows from a high-potential to a low-potential region within the battery being charged, thus losing electrostatic potential energy. The energy lost is converted to chemical energy and stored in the battery being charged.

Thus, we can consider a real battery to consist of an ideal battery of emf $\mathcal{E}$ plus a small resistance r, called the **internal resistance** of the battery.

The circuit diagram for a real battery and resistor is shown in Figure 26-11. If the current in the circuit is I, the potential at point a is related to that at point b by

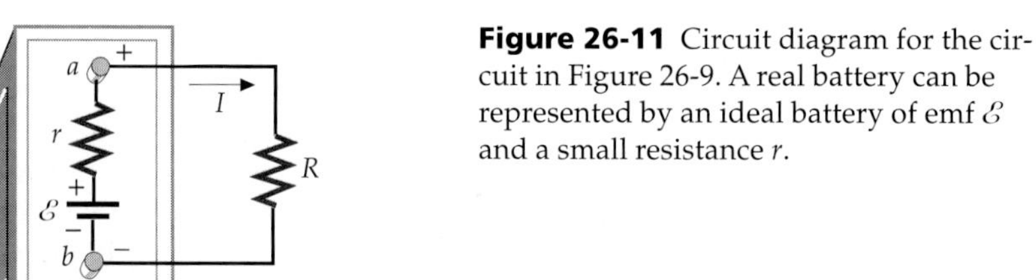

Figure 26-11 Circuit diagram for the circuit in Figure 26-9. A real battery can be represented by an ideal battery of emf $\mathcal{E}$ and a small resistance r.

$$V_a = V_b + \mathcal{E} - Ir$$

The terminal voltage is thus

$$V_a - V_b = \mathcal{E} - Ir \qquad \text{26-13}$$

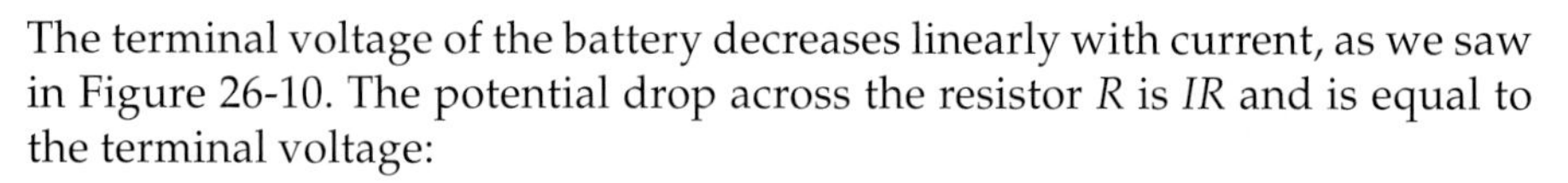

The terminal voltage of the battery decreases linearly with current, as we saw in Figure 26-10. The potential drop across the resistor R is IR and is equal to the terminal voltage:

$$IR = V_a - V_b = \mathcal{E} - Ir$$

Solving for the current I, we obtain

$$I = \frac{\mathcal{E}}{R + r} \qquad \text{26-14}$$

The terminal voltage given by Equation 26-13 is less than the emf of the battery because of the potential drop across the internal resistance of the battery. Real batteries such as a good car battery usually have an internal resistance of the order of a few hundredths of an ohm, so the terminal voltage is nearly equal to the emf unless the current is very large. One sign of a bad battery is an unusually high internal resistance. If you suspect that your car battery is bad, checking the terminal voltage with a voltmeter, which draws very little current, is not sufficient. You need to check the terminal voltage while current is being drawn from the battery, such as while you are trying to start your car. Then the terminal voltage may drop considerably, indicating a high internal resistance and a bad battery.

Batteries are often rated in ampere-hours (A·h), which is the total charge they can deliver:

$$1\ \text{A}\cdot\text{h} = 1\ \text{C/s}\ (3600\ \text{s}) = 3600\ \text{C}$$

The total energy stored in the battery is the total charge times the emf:

$$W = Q\mathcal{E} \qquad \text{26-15}$$

Example 26-7

An 11-Ω resistor is connected across a battery of emf 6 V and internal resistance 1 Ω. Find (*a*) the current, (*b*) the terminal voltage of the battery, (*c*) the power delivered by the emf source, (*d*) the power delivered to the external resistor, and (*e*) the power dissipated by the battery's internal resistance. (*f*) If the battery is rated at 150 A·h, how much energy does it store?

Picture the Problem The circuit diagram is the same as the one shown in Figure 26-11. We find the current from Equation 26-14 and then use it to find the terminal voltage and power delivered to the resistors.

(*a*) Equation 26-14 gives the current: $I = \dfrac{\mathcal{E}}{R + r} = \dfrac{6\text{ V}}{11\ \Omega + 1\ \Omega} = 0.5\text{ A}$

(*b*) Use the current to calculate the terminal voltage of the battery: $V_a - V_b = \mathcal{E} - Ir = 6\text{ V} - (0.5\text{ A})(1\ \Omega) = 5.5\text{ V}$

(*c*) The power delivered by the source of emf equals $\mathcal{E}I$: $P = \mathcal{E}I = (6\text{ V})(0.5\text{ A}) = 3\text{ W}$

(*d*) The power delivered to the external resistance equals I^2R: $I^2R = (0.5\text{ A})^2(11\ \Omega) = 2.75\text{ W}$

(*e*) The power dissipated in the internal resistance is I^2r: $I^2r = (0.5\text{ A})^2(1\ \Omega) = 0.25\text{ W}$

(*f*) The total energy stored is the emf times the total charge: $W = Q\mathcal{E} = 150\text{ A}\cdot\text{h}\,\dfrac{3600\text{ C}}{1\text{ A}\cdot\text{h}} \times 6\text{ V} = 3.24\text{ MJ}$

Remarks The value of the internal resistance is exaggerated in this example to simplify calculations. In other examples, we may simply ignore the internal resistance. Of the 3 W of power delivered by the battery, 2.75 W is dissipated as Joule heat in the resistor and 0.25 W is dissipated as Joule heat in the internal resistance of the battery.

Example 26-8 *try it yourself*

For a battery of given emf and internal resistance r, what value of external resistance R should be placed across the terminals to obtain the greatest Joule heating in R?

Picture the Problem The circuit diagram is the same as the one shown in Figure 26-11. The power input to R is I^2R, where $I = \mathcal{E}/(R + r)$. To find the maximum power, we compute dP/dR and set it equal to zero.

Cover the column to the right and try these on your own before looking at the answers.

Steps	***Answers***
1. Use Equation 26-14 to eliminate I from $P = I^2R$ so that P is written as a function of R and the constants $\mathcal{E}$ and r only.	$P = \dfrac{\mathcal{E}^2R}{(R + r)^2} = \mathcal{E}^2R(R + r)^{-2}$
2. Calculate the derivative dP/dR.	$\dfrac{dP}{dR} = \mathcal{E}^2(R + r)^{-2} - 2\mathcal{E}^2R(R + r)^{-3}$
3. Set $dP/dR = 0$ and solve for R in terms of r.	$R = r$

Remarks The maximum value of P occurs when $R = r$, that is, when the load resistance equals the internal resistance. A similar result holds for alternating-current circuits. Choosing $R = r$ to maximize the power delivered to the load is known as *impedance matching*. A graph of P versus R is shown in Figure 26-12.

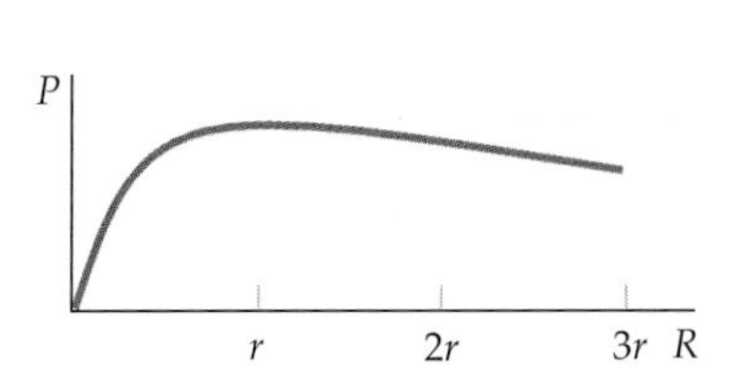

Figure 26-12

Exploring

Battery Technology

A battery is a set of chemical cells each of which consists of two metal electrodes immersed in a conducting solution called an electrolyte. Because of chemical reactions between the conductors and the electrolyte, one electrode, the cathode, becomes positively charged, and the other, the anode, becomes negatively charged. (*a*) In a dry cell, the electrolyte is a paste of ammonium chloride and other additives. The battery develops an emf of 1.5 V. (*b*) A 12-V storage battery consists of six cells that are rechargeable. Each cell has a lead dioxide cathode and a water solution of sulfuric acid for its electrolyte. (*c*) This giant battery, consisting of 200 cells, was built in 1870 in the basement of the Royal Institution, London, by Humphrey Davy.

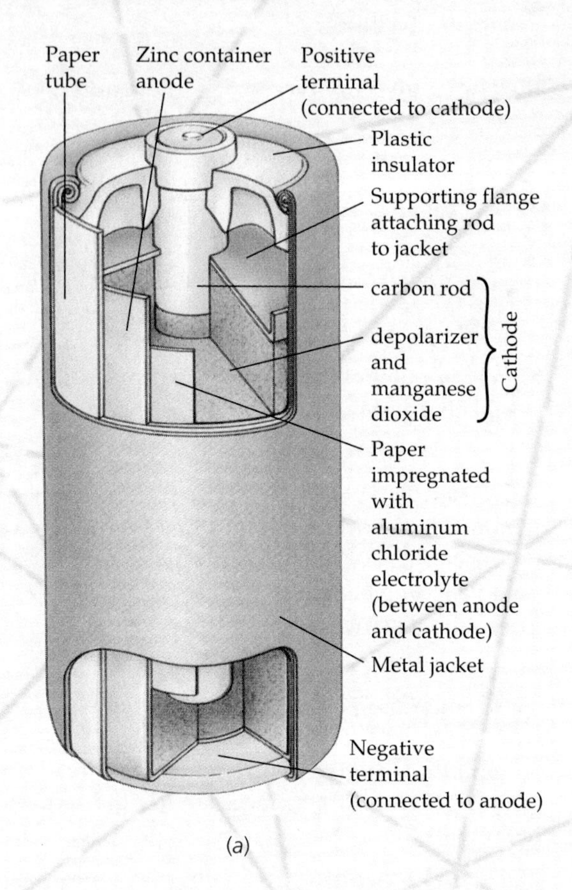

(*a*)

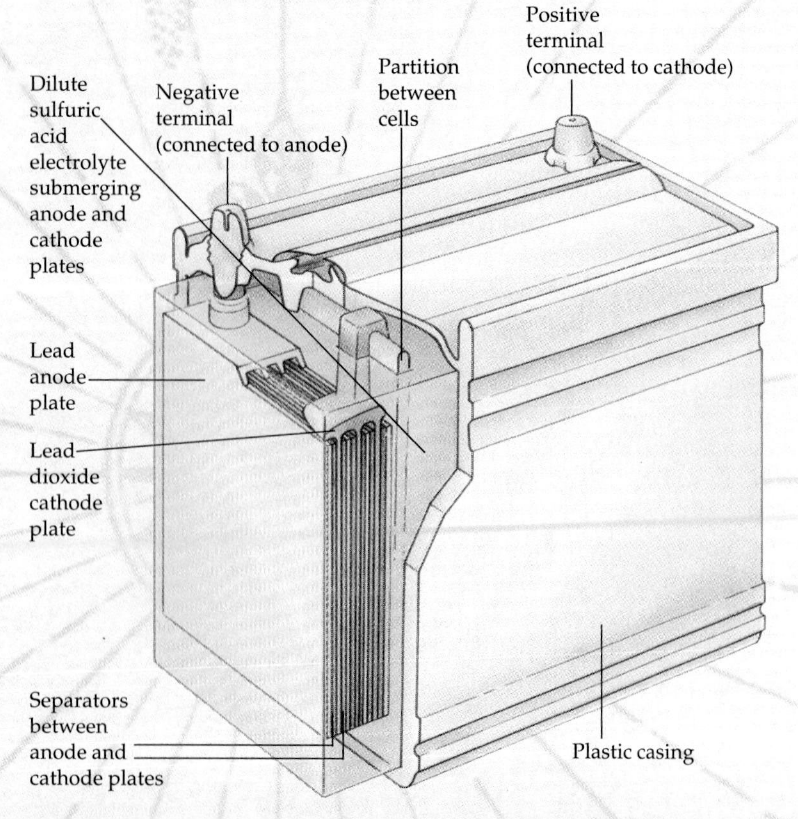

(*b*)

(*c*)

26-4 Combinations of Resistors

The analysis of a circuit can often be simplified by replacing two or more resistors by a single equivalent resistor that carries the same current with the same potential drop as the original resistors. The replacement of a set of resistors by an equivalent resistor is similar to the replacement of a set of capacitors by an equivalent capacitor, as discussed in Chapter 25.

Series Resistors

When two or more resistors are connected like R_1 and R_2 in Figure 26-13 so that they carry the same current I, the resistors are said to be connected in series. The potential drop across R_1 is IR_1 and that across R_2 is IR_2. The potential drop across the two resistors is the sum of the potential drops across the individual resistors:

$$V = IR_1 + IR_2 = I(R_1 + R_2) \qquad 26\text{-}16$$

The single equivalent resistance R_{eq} that gives the same total potential drop V when carrying the same current I is found by setting V equal to IR_{eq} (Figure 26-13*b*). Then R_{eq} is given by

$$R_{eq} = R_1 + R_2$$

When there are more than two resistors in series, the equivalent resistance is

$$R_{eq} = R_1 + R_2 + R_3 + \cdots \qquad 26\text{-}17$$

Equivalent resistance for resistors in series

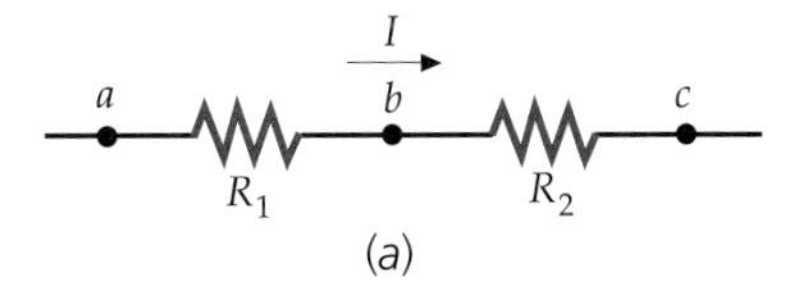

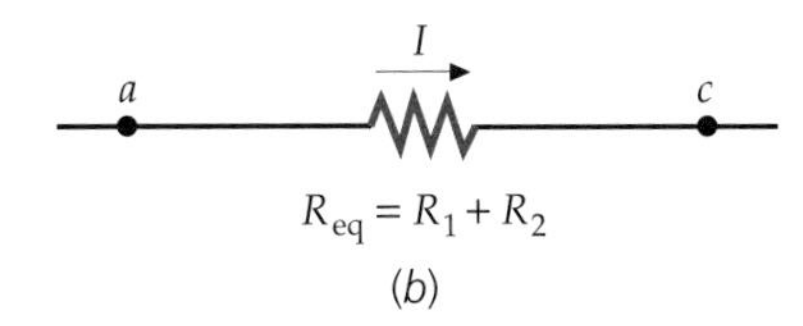

Figure 26-13 (*a*) Two resistors in series carry the same current. (*b*) The resistors in (*a*) can be replaced by a single equivalent resistance $R_{eq} = R_1 + R_2$ that gives the same total potential drop when carrying the same current as in (*a*).

Parallel Resistors

Two resistors that are connected as in Figure 26-14*a*, such that they have the same potential difference across them, are in parallel. Note that the resistors are connected at both ends by wires. Let I be the current from point a to point b. At point a the current splits into two parts, I_1 in resistor R_1 and I_2 in R_2. The total current is the sum of the individual currents:

$$I = I_1 + I_2 \qquad 26\text{-}18$$

The potential drop across either resistor, $V = V_a - V_b$, is related to the currents by

$$V = I_1R_1 = I_2R_2 \qquad 26\text{-}19$$

The equivalent resistance for parallel resistors is the resistance R_{eq} for which the same total current I produces the potential drop V (Figure 26-14*b*):

$$R_{eq} = \frac{V}{I}$$

Solving this equation for I and using $I = I_1 + I_2$, we have

$$I = \frac{V}{R_{eq}} = I_1 + I_2 = \frac{V}{R_1} + \frac{V}{R_2} \qquad 26\text{-}20$$

where we have used Equation 26-19 for I_1 and I_2. The equivalent resistance for two resistors in parallel is therefore given by

$$\frac{1}{R_{eq}} = \frac{1}{R_1} + \frac{1}{R_2}$$

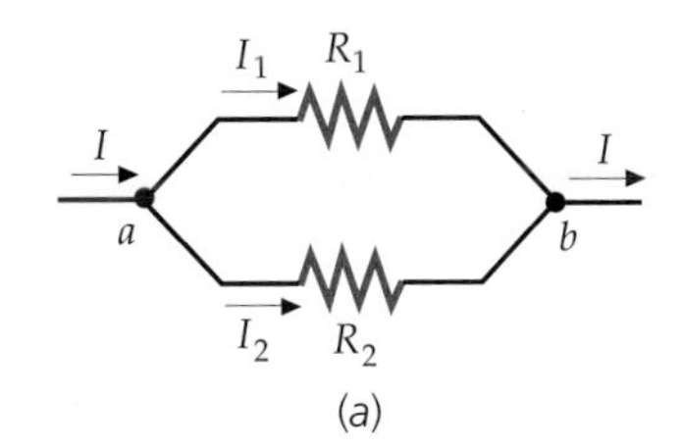

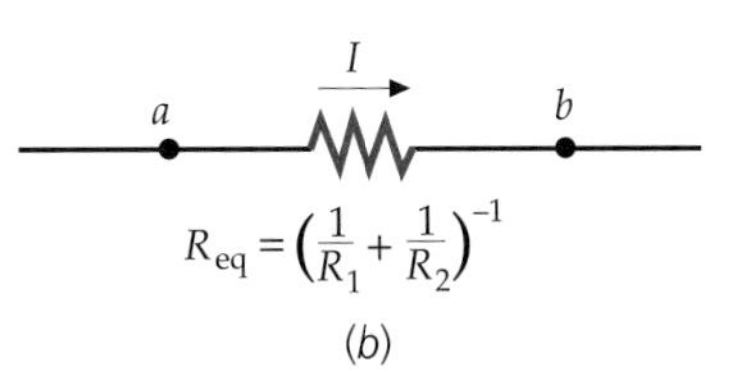

Figure 26-14 (*a*) Two resistors are in parallel when they are connected together at both ends so that the potential drop is the same across each. (*b*) The two resistors in (*a*) can be replaced by an equivalent resistance R_{eq} that is related to R_1 and R_2 by $1/R_{eq} = 1/R_1 + 1/R_2$.

This result can be generalized for combinations, such as that in Figure 26-15, in which three or more resistors are connected in parallel:

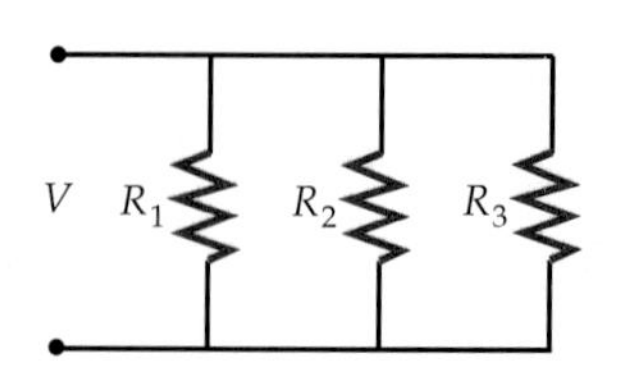

Figure 26-15 Three resistors in parallel.

$$\frac{1}{R_{eq}} = \frac{1}{R_1} + \frac{1}{R_2} + \frac{1}{R_3} + \cdots \qquad 26\text{-}21$$

Equivalent resistance for resistors in parallel

Exercise A 2-Ω resistor and a 4-Ω resistor are connected (*a*) in series and (*b*) in parallel. Find the equivalent resistances for both cases. (*Answers* (*a*) 6 Ω, (*b*) 1.33 Ω)

Example 26-9

A potential difference of 12 V is applied across the parallel combination of 4- and 6-Ω resistors shown in Figure 26-16. Find (*a*) the equivalent resistance, (*b*) the total current, (*c*) the current in each resistor, (*d*) the power dissipated in each resistor, and (*e*) the power delivered by the 12-V battery.

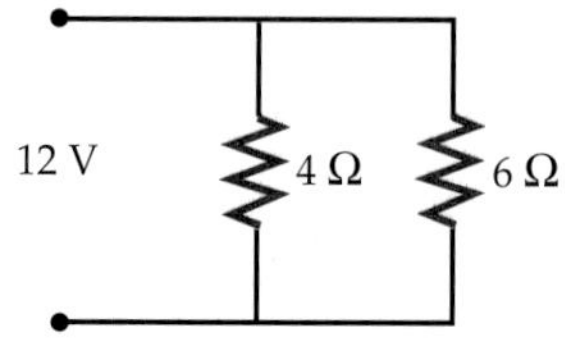

Figure 26-16

Picture the Problem Choose symbols and directions for the currents in Figure 26-17.

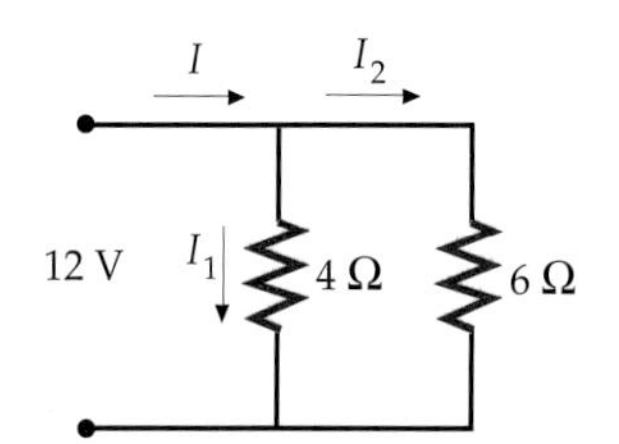

Figure 26-17

(*a*) Calculate the equivalent resistance:

$$\frac{1}{R_{eq}} = \frac{1}{4\ \Omega} + \frac{1}{6\ \Omega} = \frac{3}{12\ \Omega} + \frac{2}{12\ \Omega} = \frac{5}{12\ \Omega}$$

$$R_{eq} = \frac{12\ \Omega}{5} = 2.4\ \Omega$$

(*b*) The total current is the voltage drop divided by the equivalent resistance:

$$I = \frac{V}{R_{eq}} = \frac{12\ \text{V}}{2.4\ \Omega} = 5\ \text{A}$$

(*c*) We obtain the current in each resistor using Equation 26-19 and the fact that the potential drop is 12 V across each resistor:

$$V = IR$$

$$I_4 = \frac{12\ \text{V}}{4\ \Omega} = 3.0\ \text{A}$$

$$I_6 = \frac{12\ \text{V}}{6\ \Omega} = 2.0\ \text{A}$$

(*d*) Use these currents to find the power dissipated in each resistor:

$$P_4 = I_4^2R = (3.0\ \text{A})^2(4\ \Omega) = 36\ \text{W}$$

$$P_6 = I_6^2R = (2.0\ \text{A})^2(6\ \Omega) = 24\ \text{W}$$

(*e*) Use $P = VI$ to find the power delivered by the battery:

$$P = VI = (5\ \text{A})(12\ \text{V}) = 60\ \text{W}$$

Check the Result The power delivered by the battery equals the power dissipated in the two resistors $P = 60\ \text{W} = 36\ \text{W} + 24\ \text{W}$. In (*d*), we could have calculated the power dissipated in each resistor from $P_4 = VI_4 = (12\ \text{V})(3.0\ \text{A}) = 36\ \text{W}$ and $P_6 = VI_6 = (12\ \text{V})(2.0\ \text{A}) = 24\ \text{W}$.

Example 26-10 *try it yourself*

A 4-Ω resistor and a 6-Ω resistor are connected in series to a battery of emf 12 V and negligible internal resistance. Find (*a*) the equivalent resistance, (*b*) the current in the circuit, (*c*) the potential drop across each resistor, (*d*) the power dissipated in each resistor, and (*e*) the total power dissipated.

Cover the column to the right and try these on your own before looking at the answers.

Steps	*Answers*
(*a*)1. Draw a circuit diagram (Figure 26-18).	**Figure 26-18**
2. Calculate R_{eq} for the two series resistors.	$R_{eq} = 10\ \Omega$
(*b*) Use $V = IR_{eq}$ to find the current.	$I = 1.2$ A
(*c*) Use Ohm's law to find the potential drop across each resistor.	4.8 V, 7.2 V
(*d*) Find the power dissipated in each resistor using $P = I^2R$. Check your result using $P = VI$ for each resistor.	5.76 W, 8.64 W
(*e*) Add your results from (*d*) to find the total power. Check your result using $P = VI$ and $P = I^2R_{eq}$.	14.4 W

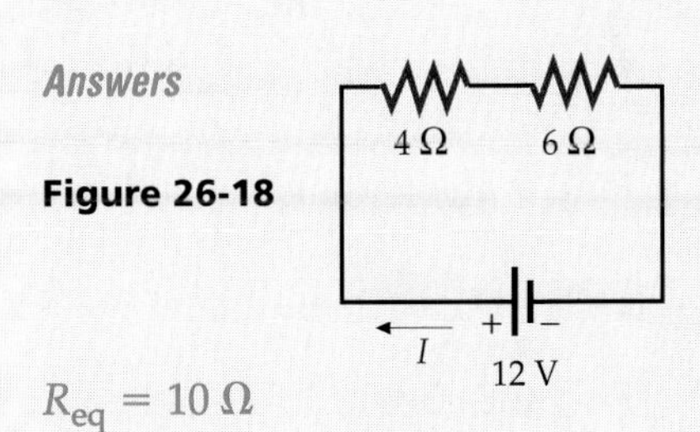

Remark Note that much less power is dissipated in the series circuit than in the corresponding parallel circuit of Example 26-9.

Note from Example 26-9 that the equivalent resistance of two parallel resistances is less than the resistance of either resistor alone. This is a general result. Suppose we have a single resistor R_1 carrying current I_1 with potential drop $V = I_1R_1$. If we add a second resistor in parallel, it will carry some additional current I_2 without affecting I_1. The equivalent resistance is $V/(I_1 + I_2)$, which is less than $R_1 = V/I_1$. Note also from Example 26-9 that the ratio of the currents in the two parallel resistors equals the inverse ratio of the resistances. This general result follows from Equation 26-19:

$$I_1R_1 = I_2R_2$$

$$\frac{I_1}{I_2} = \frac{R_2}{R_1} \quad \text{(parallel resistors)} \qquad 26\text{-}22$$

Example 26-11 *try it yourself*

For the circuit in Figure 26-19, find (*a*) the equivalent resistance of the circuit, (*b*) the total current in the source of emf, (*c*) the potential drop across each resistor, and (*d*) the current carried by each resistor.

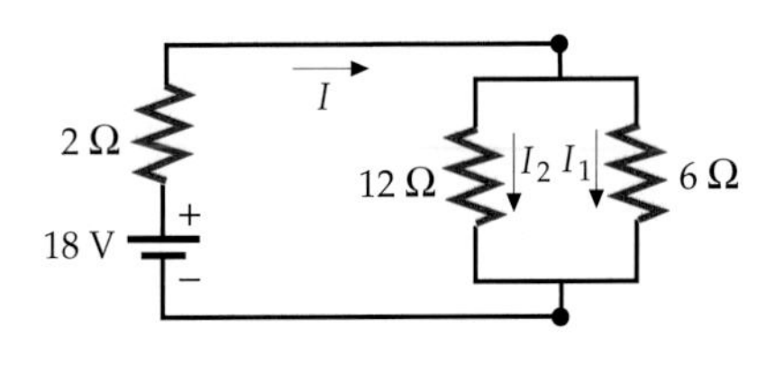

Figure 26-19

Picture the Problem To find the equivalent resistance of the circuit, first replace the two parallel resistors by their equivalent resistance. Ohm's law can then be used to find the current and potential drops.

Cover the column to the right and try these on your own before looking at the answers.

Steps	**Answers**
(*a*) 1. Find the equivalent resistance of the 6- and 12-Ω parallel combination.	$R_{eq} = 4\ \Omega$
2. Combine your result in step 1 with the 2-Ω resistor in series to find the total equivalent resistance of the circuit.	$R'_{eq} = 6\ \Omega$
(*b*) Find the total current using Ohm's law. This is the current in the 2-Ω resistor.	$I = 3$ A
(*c*) 1. Find the potential drop across the 2-Ω resistor from $V_2 = IR$.	$V_2 = 6$ V
2. Find the potential drop across the parallel combination using $V_p = IR_{eq}$.	$V_p = 12$ V
(*d*) Find the current in the 6- and 12-Ω resistors from $I = V_p/R$.	$I_6 = 2$ A, $I_{12} = 1$ A

Check the Result The current in the 6-Ω resistor is twice that in the 12-Ω resistor, as we should expect. Also, these two currents sum to give I, the total current in the circuit, as they must. Finally, note that the potential drops across the 2-Ω resistor and the parallel combination sum to the emf of the battery; $V_2 + V_p = 6\text{ V} + 12\text{ V} = 18\text{ V}$.

Exercise Repeat this example with the 6-Ω resistor replaced by a wire of negligible resistance. (*Answers* (*a*) $R'_{eq} = 2\ \Omega$, (*b*) $I = 9$ A, (*c*) $V_2 = 18$ V, $V_0 = 0$, $V_{12} = 0$, (*d*) $I_2 = 9$ A, $I_0 = 9$ A, $I_{12} = 0$)

Exercise Repeat this example for the case in which the 6-Ω resistor is increased to infinity. (*Answers* (*a*) $R'_{eq} = 14\ \Omega$, (*b*) $I = 1.29$ A, (*c*) $V_2 = 2.57$ V, $V_\infty = 15.4$ V, $V_{12} = 15.4$ V, (*d*) $I_2 = 1.29$ A, $I_\infty = 0$, $I_{12} = 1.29$ A)

Example 26-12 *try it yourself*

Find the equivalent resistance between points *a* and *b* for the combination of resistors shown in Figure 26-20.

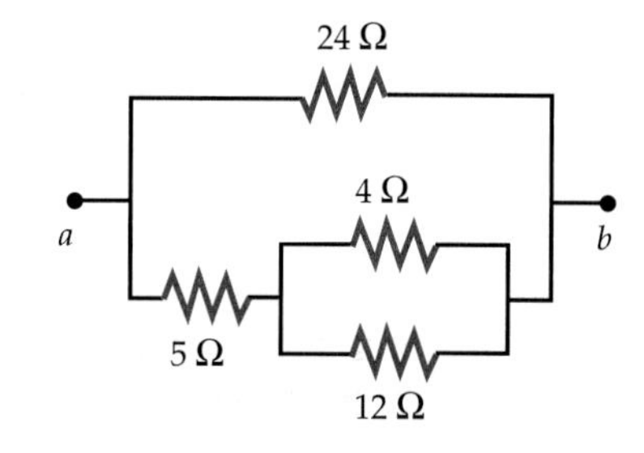

Figure 26-20

Picture the Problem You can analyze this complicated combination step by step. You first need to find the equivalent resistance R_{eq} of the 4- and 12-Ω parallel combination. Then combine your result with the 5-Ω resistance in series with the parallel combination. You are then left with two resistors in parallel.

Cover the column to the right and try these on your own before looking at the answers.

Steps	**Answers**
1. Find the equivalent resistance R_{eq} of the 4- and 12-Ω resistors in parallel.	$R_{eq} = 3\ \Omega$
2. Find the equivalent resistance R'_{eq} of R_{eq} in series with the 5-Ω resistor.	$R'_{eq} = 8\ \Omega$
3. Find the equivalent resistance of R'_{eq} in parallel with the 24-Ω resistor.	$R''_{eq} = 6\ \Omega$

26-5 Kirchhoff's Rules

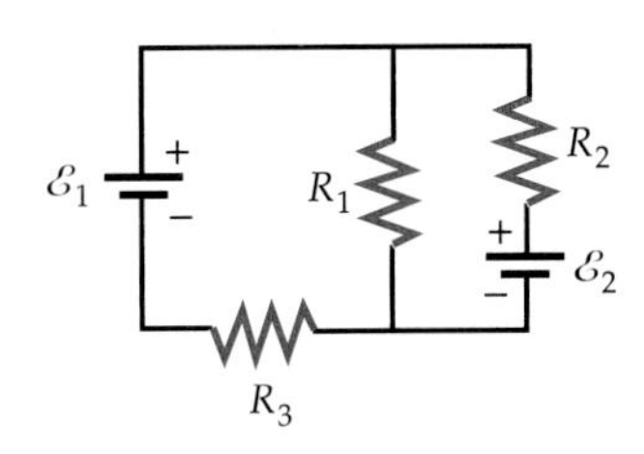

Figure 26-21 An example of a simple circuit that cannot be analyzed by replacing combinations of resistors in series or parallel with their equivalent resistances. The potential drops across R_1 and R_2 are not equal because of the emf source $\mathcal{E}_2$, so these resistors are not in parallel. (Parallel resistors would be connected together at both ends.) The resistors do not carry the same current, so they are not in series.

There are many simple circuits, such as that shown in Figure 26-21, that cannot be analyzed by merely replacing combinations of resistors by an equivalent resistance. The two resistors R_1 and R_2 in this circuit look as if they might be in parallel, but they are not. The potential drop is not the same across both resistors because of the presence of the emf source $\mathcal{E}_2$ in series with R_2. Nor are R_1 and R_2 in series, because they don't carry the same current.

Two rules, called **Kirchhoff's rules**, apply to this and any other circuit:

1. When any closed-circuit loop is traversed, the algebraic sum of the changes in potential must equal zero.
2. At any junction point in a circuit where the current can divide, the sum of the currents into the junction must equal the sum of the currents out of the junction.

Kirchhoff's rules

Kirchhoff's first rule, called the **loop rule**, follows directly from the conservation of energy. If we have a charge q at some point where the potential is V, the potential energy of the charge is qV. As the charge traverses a loop in a circuit, it loses or gains energy as it passes through resistors, batteries, or other devices, but when it arrives back at its starting point, its energy must again be qV. That is, *the net change in the potential must be zero.*

Kirchhoff's second rule, called the **junction rule**, follows from the conservation of charge. Figure 26-22 shows the junction of three wires carrying currents I_1, I_2, and I_3. Since charge does not originate or accumulate at this point, the conservation of charge implies the junction rule, which for this case gives

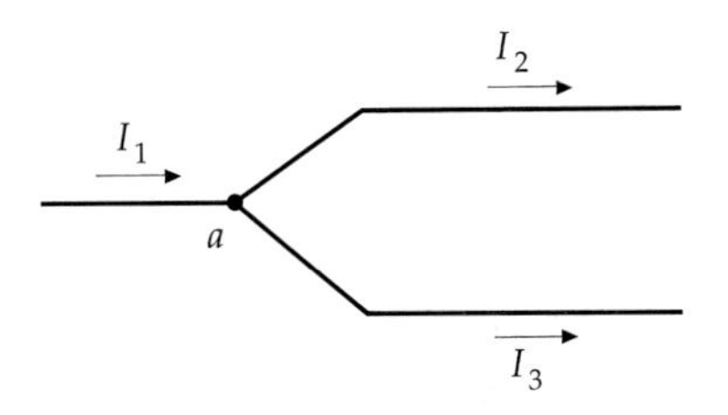

Figure 26-22 Illustration of Kirchhoff's junction rule. The current I_1 into point a equals the sum $I_2 + I_3$ of the currents out of point a.

$$I_1 = I_2 + I_3 \tag{26-23}$$

Single-Loop Circuits

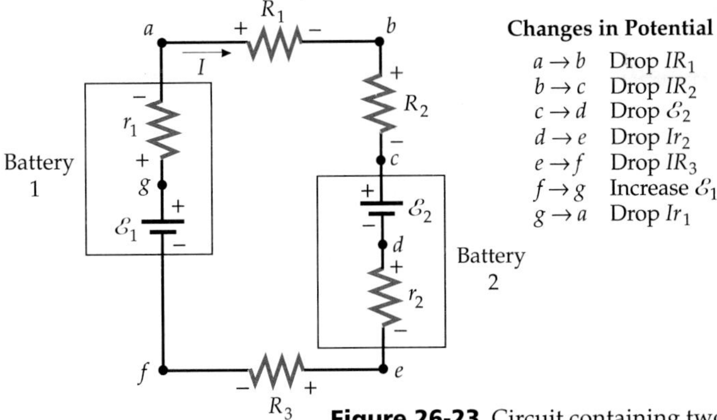

Figure 26-23 Circuit containing two batteries and three external resistors. The plus and minus signs on the resistors are there to help us remember which side of each resistor is at the higher potential for the current direction we have assumed.

As an example of using Kirchhoff's loop rule, consider the circuit shown in Figure 26-23 containing two batteries with internal resistances r_1 and r_2 and three external resistors. We wish to find the current in terms of the emfs.

Assume that I is clockwise, as indicated in Figure 26-23, and apply Kirchhoff's loop rule as we traverse the circuit in the assumed direction of the current, beginning at point a. The potential decreases and increases are given in the figure. Note that we encounter a potential drop as we traverse the source of emf between points c and d and a potential increase as we traverse the source of emf between f and g. Beginning at point a, we obtain from Kirchhoff's loop rule

$$-IR_1 - IR_2 - \mathcal{E}_2 - Ir_2 - IR_3 + \mathcal{E}_1 - Ir_1 = 0$$

Solving for the current I, we obtain

$$I = \frac{\mathcal{E}_1 - \mathcal{E}_2}{R_1 + R_2 + R_3 + r_1 + r_2} \tag{26-24}$$

If $\mathcal{E}_2$ is greater than $\mathcal{E}_1$, we get a negative value for the current I, indicating that we have assumed the wrong direction for I.

For this example, suppose that $\mathcal{E}_1$ is the greater emf. In battery 2, the charge flows from high potential to low potential. Therefore, a charge ΔQ moving through battery 2 from point c to point d loses energy $\mathcal{E}_2\,\Delta Q$. If battery 2 is a rechargeable battery, most of this electrical energy is converted into chemical energy and stored in the battery, which means that battery 2 is *charging*. (The rest of the energy is dissipated in the internal resistance.)

The analysis of a circuit is usually simplified if we choose one point to be at zero potential and then find the potentials of the other points relative to it. Since only potential differences are important, any point in a circuit can be chosen to have zero potential. In the following example, we choose point f in the figure to be at zero potential. This is indicated by the ground symbol ⏚ at point f.*

* As we saw in Section 18-2, the earth can be considered to be a very large conductor with a nearly unlimited supply of charge, which means that the potential of the earth remains essentially constant. In practice, electrical circuits are often grounded by connecting one point to the earth. The outside metal case of a washing machine, for example, is usually grounded by connecting it by a wire to a water pipe that is in contact with the earth. Since everything so grounded is at the same potential, it is convenient to designate this potential as zero.

Example 26-13

Suppose the elements in the circuit in Figure 26-23 have the values $\mathcal{E}_1 = 12$ V, $\mathcal{E}_2 = 4$ V, $r_1 = r_2 = 1\ \Omega$, $R_1 = R_2 = 5\ \Omega$, and $R_3 = 4\ \Omega$, as shown in Figure 26-24. (*a*) Find the potentials at points *a* through *g* in the figure, assuming that the potential at point *f* is zero. (*b*) Find the power input and output in the circuit.

Picture the Problem To find the potential differences, we first need to find the current I in the circuit. The voltage drop across each resistor is then IR. To discuss the energy balance, we calculate the power into or out of each element using Equations 26-10 and 26-11.

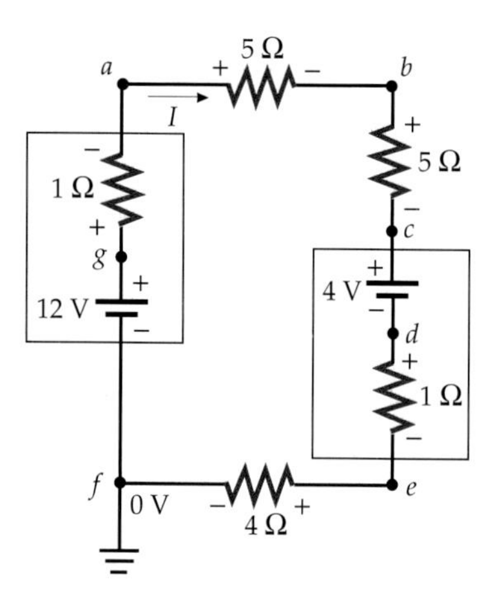

Figure 26-24

(*a*)1. The current I in the circuit is found using Equation 26-24:

$$I = \frac{12\text{ V} - 4\text{ V}}{5\ \Omega + 5\ \Omega + 4\ \Omega + 1\ \Omega + 1\ \Omega} = \frac{8\text{ V}}{16\ \Omega} = 0.5\text{ A}$$

2. We now find the potential at each labeled point in the circuit:

$$V_g = V_f + \mathcal{E}_1 = 0 + 12\text{ V} = 12\text{ V}$$
$$V_a = V_g - Ir_1 = 12\text{ V} - (0.5\text{ A})(1\ \Omega) = 11.5\text{ V}$$
$$V_b = V_a - IR_1 = 11.5\text{ V} - (0.5\text{ A})(5\ \Omega) = 9\text{ V}$$
$$V_c = V_b - IR_2 = 9\text{ V} - (0.5\text{ A})(5\ \Omega) = 6.5\text{ V}$$
$$V_d = V_c - \mathcal{E}_2 = 6.5\text{ V} - 4\text{ V} = 2.5\text{ V}$$
$$V_e = V_d - Ir_2 = 2.5\text{ V} - (0.5\text{ A})(1\ \Omega) = 2.0\text{ V}$$
$$V_f = V_e - IR_3 = 2.0\text{ V} - (0.5\text{ A})(4\ \Omega) = 0$$

(*b*)1. First, calculate the power delivered by the emf source $\mathcal{E}_1$:

$$P_{\mathcal{E}_1} = \mathcal{E}_1 I = (12\text{ V})(0.5\text{ A}) = 6\text{ W}$$

2. Part of this power is dissipated in the resistors, both internal and external:

$$P_R = I^2R_1 + I^2R_2 + I^2R_3 + I^2r_1 + I^2r_2$$
$$= (0.5\text{ A})^2(5\ \Omega + 5\ \Omega + 4\ \Omega + 1\ \Omega + 1\ \Omega) = 4.0\text{ W}$$

3. The remaining 2 W of power goes into charging battery 2:

$$P_{\mathcal{E}_2} = \mathcal{E}_2 I = (4\text{ V})(0.5\text{ A}) = 2\text{ W}$$

Remarks Figure 26-25 shows the potential at the labeled points of the circuit.

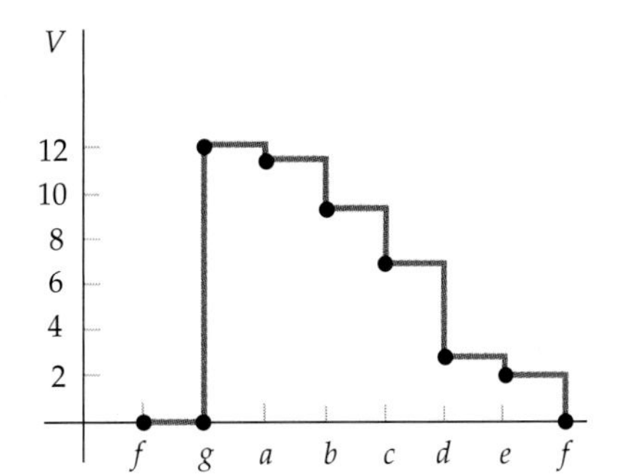

Figure 26-25

Note that the terminal voltage of the battery that is being charged in Example 26-13 is $V_c - V_e = 4.5$ V, which is greater than the emf of the battery. Because of its internal resistance, a battery is not completely reversible. If the same 4-V battery were to deliver 0.5 A to an external circuit, its terminal voltage would be 3.5 V (again assuming that its internal resistance is 1 Ω). If the internal resistance is very small, the terminal voltage of a battery is nearly equal to its emf, whether the battery is delivering current to an external circuit or is being charged. Some real batteries, such as those used in automobiles, are nearly reversible and can easily be recharged. Other types of batteries are not reversible. If you attempt to recharge one of these by driving current from its positive to its negative terminal, most, if not all, of the energy will go into heat rather than into the chemical energy of the battery, and the battery may explode.

Example 26-14

A good car battery is to be connected by jumper cables to a weak car battery to charge the weak one. (*a*) To which terminal of the weak battery should the positive terminal of the good battery be connected? (*b*) Assume that the good battery has an emf of $\mathcal{E}_1 = 12$ V and the weak battery has an emf of $\mathcal{E}_2 = 11$ V, that the internal resistances of the batteries are $r_1 = r_2 = 0.02\ \Omega$, and that the resistance of the jumper cables is $R = 0.01\ \Omega$. What will the charging current be? (*c*) What will the current be if the batteries are connected incorrectly?

(*a*) To charge the weak battery, we connect the terminals positive to positive and negative to negative to drive charge through the weak battery from the positive terminal to the negative terminal:

Figure 26-26

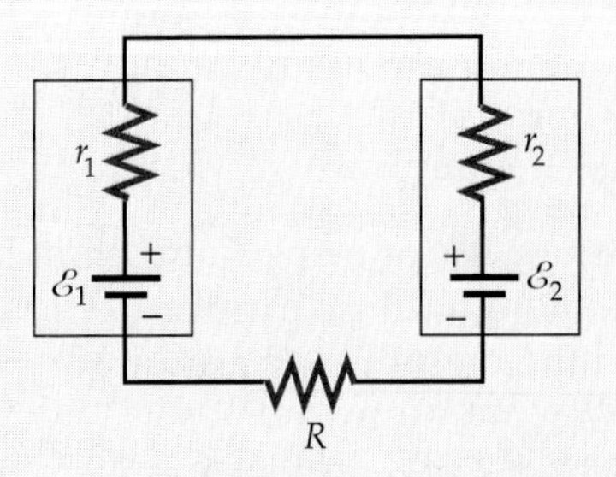

(*b*) Use Kirchhoff's loop rule to find the charging current:

$$I = \frac{\mathcal{E}_1 - \mathcal{E}_2}{R + r_1 + r_2} = \frac{12\text{ V} - 11\text{ V}}{0.05\ \Omega} = 20\text{ A}$$

(*c*) When the batteries are connected incorrectly, positive terminals to negative terminals, the emfs add:

$$I = \frac{\mathcal{E}_1 + \mathcal{E}_2}{R + r_1 + r_2} = \frac{12\text{ V} + 11\text{ V}}{0.05\ \Omega} = 460\text{ A}$$

Remarks If the batteries are connected incorrectly as shown in Figure 26-27, the total resistance of the circuit is of the order of hundredths of an ohm, the current is very large, and the batteries could explode in a shower of boiling battery acid.

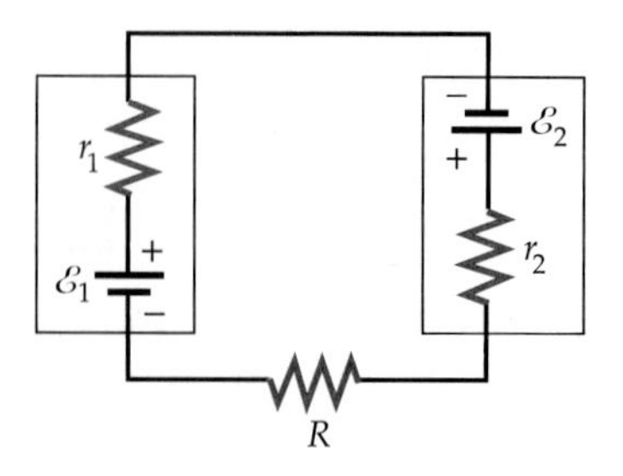

Figure 26-27 Two batteries connected incorrectly—dangerous!

Multiloop Circuits

To analyze circuits containing more than one loop, we need to use both of Kirchhoff's rules, with Kirchhoff's junction rule applied to points where the current splits into two or more parts.

Example 26-15

(*a*) Find the current in each part of the circuit shown in Figure 26-28. (*b*) Find the energy dissipated in the 4-Ω resistor in 3 s.

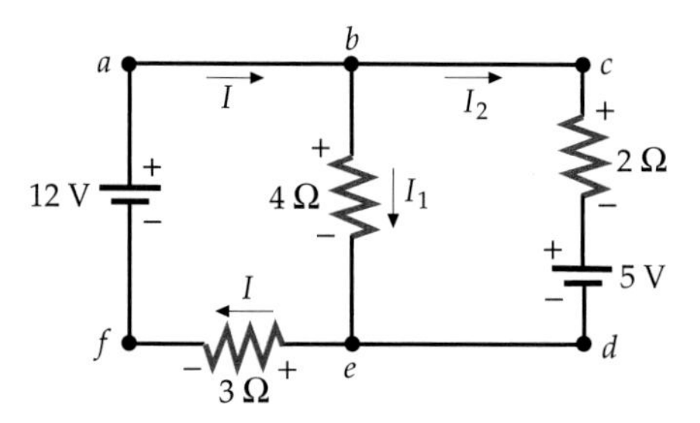

Figure 26-28

Picture the Problem There are three currents, I, I_1, and I_2, to be determined, so we need three conditions. One condition comes from applying the junction rule to point *b*. (We can also apply the junction rule to point *e*, the only other junction in the circuit, but it gives exactly the same information.) The other two conditions are obtained by applying the loop rule. There are three loops in the circuit: the two interior loops, *abefa* and *bcdeb*, and the exterior loop, *abcdefa*. We can use any two of these loops—the third will give redundant information.The direction of the current I_1 from *b* to *e* is not known before the circuit is analyzed. The plus and minus signs on the 4-Ω resistor are for the assumed direction of I_1 from *b* to *e*.

(*a*)1. Apply the junction rule to point *b*:

$$I = I_1 + I_2$$

2. Apply the loop rule to the outer loop, *abcdefa*:

$$12\text{ V} - (2\,\Omega)I_2 - 5\text{ V} - (3\,\Omega)(I_1 + I_2) = 0$$

3. Divide the above equation by 1 Ω, recalling that (1 V)/(1 Ω) = 1 A, then simplify:

$$7\text{ A} - 3I_1 - 5I_2 = 0$$

4. For the third condition, apply the loop rule to the loop on the left, *abefa*:

$$12\text{ V} - (4\,\Omega)I_1 - (3\,\Omega)(I_1 + I_2) = 0$$

$$12\text{ A} - 7I_1 - 3I_2 = 0$$

5. The results for steps 3 and 4 can be combined to solve for I_1 and I_2. To do so, first multiply the result for step 3 by 3, and then multiply the result for step 4 by 5:

$$21\text{ A} - 9I_1 - 15I_2 = 0$$

$$60\text{ A} - 35I_1 - 15I_2 = 0$$

6. Subtract the equations in step 5 to eliminate I_2, then solve for I_1:

$$39\text{ A} - 26I_1 = 0$$

$$I_1 = \frac{39\text{ A}}{26} = 1.5\text{ A}$$

7. Substitute I_1 in the results for step 3 or 4 to solve for I_2. Here, we choose step 3:

$$7\text{ A} - 3(1.5\text{ A}) - 5I_2 = 0$$

$$I_2 = \frac{2.5\text{ A}}{5} = 0.5\text{ A}$$

8. Finally, I_1 and I_2 determine I using the equation in step 1: $I = I_1 + I_2 = 1.5\text{ A} + 0.5\text{ A} = 2.0\text{ A}$

(*b*)1. The power dissipated in the 4-Ω resistor is found using $P = I_1^2R$: $P = I_1^2R = (1.5\text{ A})^2(4\ \Omega) = 9\text{ W}$

2. The total energy dissipated in a time t is $W = Pt$. In this case, $t = 3$ s: $W = Pt = (9\text{ W})(3\text{ s}) = 27\text{ J}$

Check the Result In Figure 26-29, we have chosen the potential to be zero at point f, and we have labeled the currents and the potentials at the other points. Note that $V_b - V_e = 6\text{ V}$ and $V_e - V_f = 6\text{ V}$.

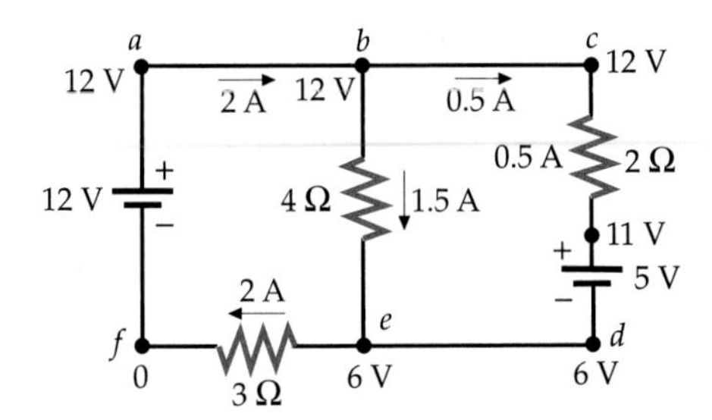

Figure 26-29

Remarks Applying the loop rule to the loop on the right, $bcde$, gives $-(2\ \Omega)I_2 - 5\text{ V} + (4\ \Omega)I_1 = 0$, or $-5\text{ A} + 4I_1 - 2I_2 = 0$. Note that this is just the result for step 3 minus the result for step 4 and hence contains no new information, as expected.

Exercise Find I_1 for the case in which the 3-Ω resistor approaches (*a*) zero resistance and (*b*) infinite resistance. (*Answers* (*a*) The potential drop across the 4-Ω resistor is 12 V; thus, $I_1 = 3$ A. (*b*) In this case, the loop on the left is an open circuit. Thus, $I_1 = (5\text{ V})/(2\ \Omega + 4\ \Omega) = 0.833$ A.)

Example 26-15 illustrates the general methods for the analysis of multiloop circuits:

1. Draw a sketch of the circuit.
2. Choose a direction for the current in each branch of the circuit, and label the currents in the circuit diagram. Add plus and minus signs to indicate the high- and low-potential sides of each resistor, capacitor, or source of emf.
3. Replace any combination of resistors in series or parallel with its equivalent resistance.
4. Apply the junction rule to each junction where the current divides.
5. Apply the loop rule to each loop until you obtain as many equations as unknowns.
6. Solve the equations to obtain the values of the unknowns.
7. Check your results by assigning a potential of zero to one point in the circuit and use the values of the currents found to determine the potentials at other points in the circuit.

General method for analyzing multiloop circuits

Example 26-16 *try it yourself*

(*a*) Find the current in each part of the circuit shown in Figure 26-30. Draw the circuit diagram with the correct magnitudes and directions for the current in each part. (*b*) Assign $V = 0$ to point *c* and then label the potential at each other point *a* through *f*.

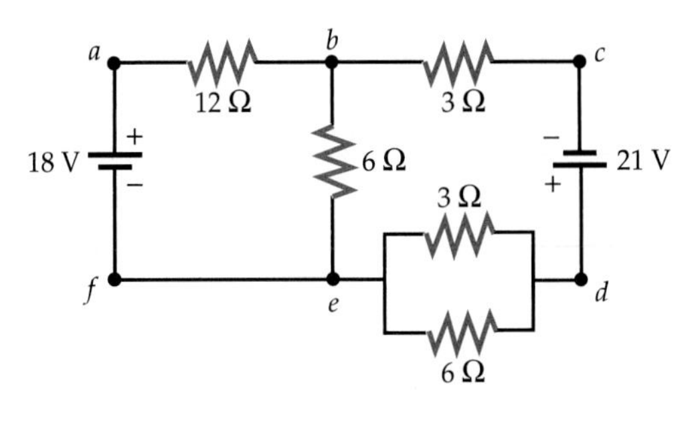

Figure 26-30

Picture the Problem First, replace the two parallel resistors by an equivalent resistance. Let I be the current through the 18-V battery, and I_1 be the current from b to e. The currents can then be found by applying the junction rule at points b and c and the loop rule to each loop.

Cover the column to the right and try these on your own before looking at the answers.

Steps	***Answers***
(*a*)1. Find the equivalent resistance of the 3- and 6-Ω parallel resistors.	$R_{eq} = 2\ \Omega$
2. Apply the junction rule at points *b* and *e* and redraw the circuit diagram with the currents indicated (Figure 26-31).	**Figure 26-31** 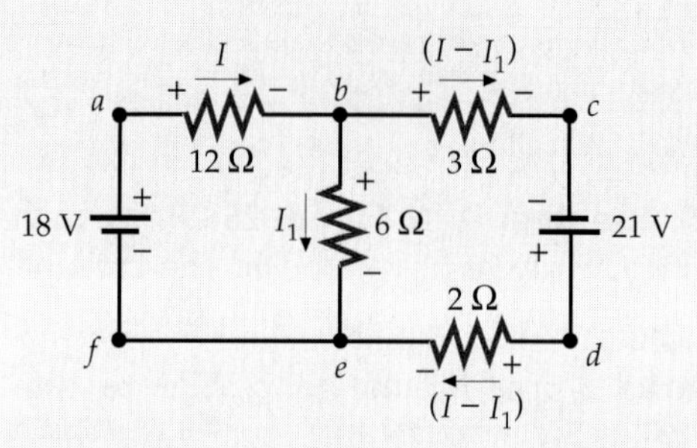
3. Apply Kirchhoff's loop rule to loop *abefa*.	$18\text{ V} - (12\ \Omega)I - (6\ \Omega)I_1 = 0$
4. Simplify your equation from step 3 to obtain an equation involving I and I_1.	$2I + I_1 = 3\text{ A}$
5. Apply Kirchhoff's loop rule to loop *bcdeb*.	$-(3\ \Omega)(I - I_1) + 21\text{ V} - (2\ \Omega)(I - I_1) + (6\ \Omega)I_1 = 0$
6. Simplify your equation in step 5.	$5I - 11I_1 = 21\text{ A}$
7. Solve your equations from steps 4 and 6 for I and I_1. One way to do this is to multiply the equation in step 4 by 11 and then add the equations to eliminate I_1.	$I = 2\text{ A},\quad I_1 = -1\text{ A}$
8. Find the current through the 21-V battery.	$I - I_1 = 3\text{ A}$
9. Use $V = (I - I_1)\,R_{eq}$ to find the potential drop across the parallel 3- and 6-Ω resistors.	$V = 6\text{ V}$
10. Use the result of step 9 to find the current in each of the parallel resistors.	$I_{3\Omega} = 2\text{ A},\ I_{6\Omega} = 1\text{ A}$
(*b*) Redraw Figure 26-31 showing the current through each part of the circuit (Figure 26-32). Begin with $V = 0$ at point *c* and calculate the potential at points *d*, *e*, *f*, *a*, and *b*.	**Figure 26-32** 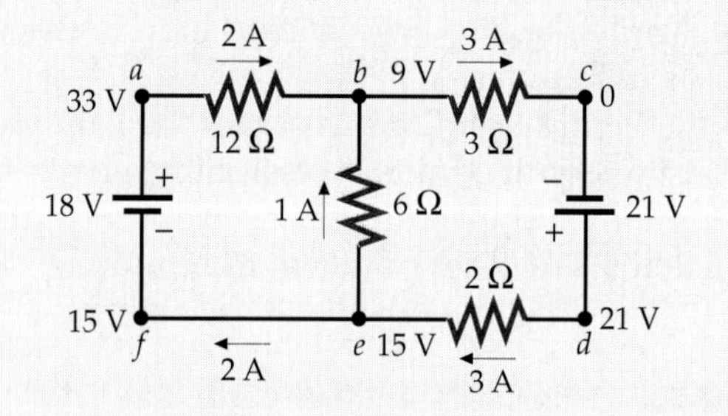

$$V_d = V_c + 21\text{ V} = 0 + 21\text{ V} = 21\text{ V}$$
$$V_e = V_d - (3\text{ A})(2\ \Omega) = 21\text{ V} - 6\text{ V} = 15\text{ V}$$
$$V_f = V_e = 15\text{ V}$$
$$V_a = V_f + 18\text{ V} = 15\text{ V} + 18\text{ V} = 33\text{ V}$$
$$V_b = V_a - (2\text{ A})(12\ \Omega) = 33\text{ V} - 24\text{ V} = 9\text{ V}$$

Check the Result From *b* to *c* the potential drops by $(3\text{ A})(3\ \Omega) = 9\text{ V}$, which gives $V_c = 0$, as assumed. From *e* to *b* the potential drops by $(1\text{ A})(6\ \Omega) = 6\text{ V}$ so $V_b = V_e - 6\text{ V} = 15\text{ V} - 6\text{ V} = 9\text{ V}$.

Ammeters, Voltmeters, and Ohmmeters

The devices that measure current, potential difference, and resistance are called **ammeters**, **voltmeters**, and **ohmmeters**, respectively. Often, all three of these meters are included in a single "multimeter" that can be switched from one use to another. You might use a voltmeter to measure the terminal voltage of your car battery and an ohmmeter to measure the resistance between two points in some electrical device at home (such as a toaster) when you suspect a short circuit or a broken wire.

To measure the current through a resistor in a simple circuit, we place an ammeter in series with the resistor, as shown in Figure 26-33, so that the ammeter and the resistor carry the same current. Since the ammeter has some resistance, the current in the circuit decreases slightly when the ammeter is inserted. Ideally, the ammeter should have a very small resistance so that the current to be measured is affected only slightly.

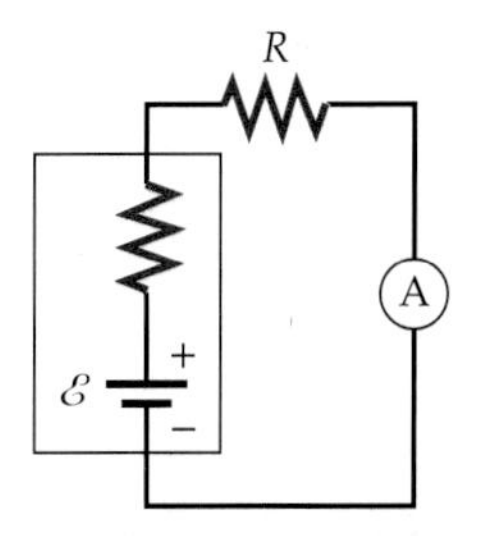

Figure 26-33 To measure the current in a resistor R, an ammeter Ⓐ is placed in series with the resistor so that it carries the same current as the resistor.

The potential difference across a resistor is measured by placing a voltmeter across the resistor in parallel with it, as shown in Figure 26-34, so that the potential drop across the voltmeter is the same as that across the resistor. The voltmeter reduces the resistance between points a and b, thus increasing the total current in the circuit and changing the potential drop across the resistor. A good voltmeter has a very large resistance so that its effect on the circuit is minimal.

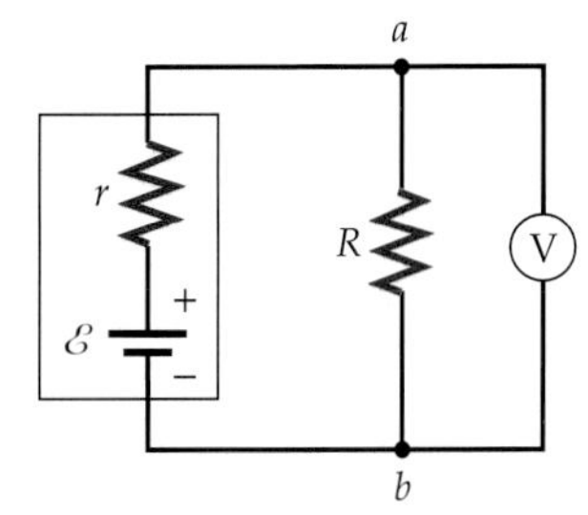

Figure 26-34 To measure the voltage drop across a resistor, a voltmeter Ⓥ is placed in parallel with the resistor so that the potential drops across the voltmeter and the resistor are the same.

The principal component of an ammeter and a voltmeter is a **galvanometer**, a device that detects small currents passing through it. The galvanometer is designed so that the scale reading is proportional to the current passing through. A typical galvanometer used in student laboratories consists of a coil of wire in the magnetic field of a permanent magnet. When the coil carries a current, the magnetic field exerts a torque on the coil causing it to rotate. A pointer attached to the coil indicates the reading on a scale. The coil itself contributes a small amount of resistance when the galvanometer is placed within a circuit. Many meters today have a digital readout rather than an indicator and a scale, but their basic operation is similar to that discussed here.

To construct an ammeter from a galvanometer, we place a small resistor called a **shunt resistor** in parallel with the galvanometer. The shunt resistance is usually much smaller than the resistance of the galvanometer so that most of the current is carried by the shunt resistor. The equivalent resistance of the ammeter is then approximately equal to the shunt resistance, and much smaller than the internal resistance of the galvanometer alone. To construct a voltmeter, we place a resistor with a large resistance in series with the galvanometer so that the equivalent resistance of the voltmeter is much larger than that of the galvanometer alone. Figure 26-35 illustrates the construction of an ammeter and voltmeter from a galvanometer. The resistance of the galvanometer R_g is shown separately in these schematic drawings, but it is actually part of the galvanometer.

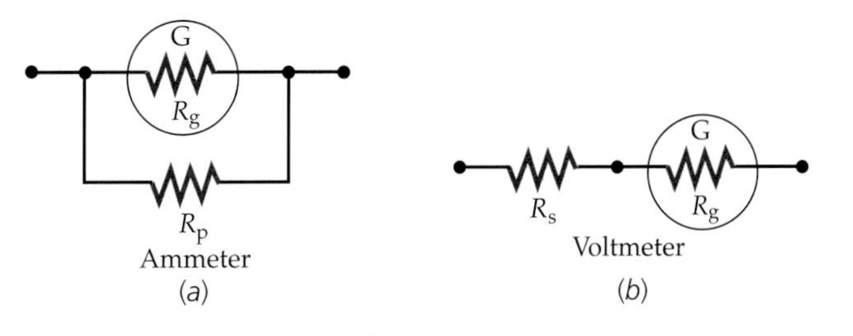

Figure 26-35 (*a*) An ammeter consists of a galvanometer Ⓖ whose resistance is R_g and a small parallel resistance R_p. (*b*) A voltmeter consists of a galvanometer Ⓖ and a large series resistance R_s.

A simple ohmmeter consists of a battery connected in series with a galvanometer and a resistor, as shown in Figure 26-36*a*. The resistance R_s is chosen such that when the terminals *a* and *b* are shorted (put in electrical contact, with negligible resistance between them), the current through the galvanometer gives a full-scale deflection. Thus, a full-scale deflection indicates no resistance between terminals *a* and *b*. A zero deflection indicates an infinite resistance between the terminals. When the terminals are connected across an unknown resistance *R*, the current through the galvanometer depends on *R*, so the scale can be calibrated to give a direct reading of *R*, as shown in Figure 26-36*b*. Because an ohmmeter sends a current through the resistance to be measured, some caution must be exercised when using this instrument. For example, you would not want to try to measure the resistance of a sensitive ammeter with an ohmmeter, because the current provided by the battery in the ohmmeter would probably damage the ammeter.

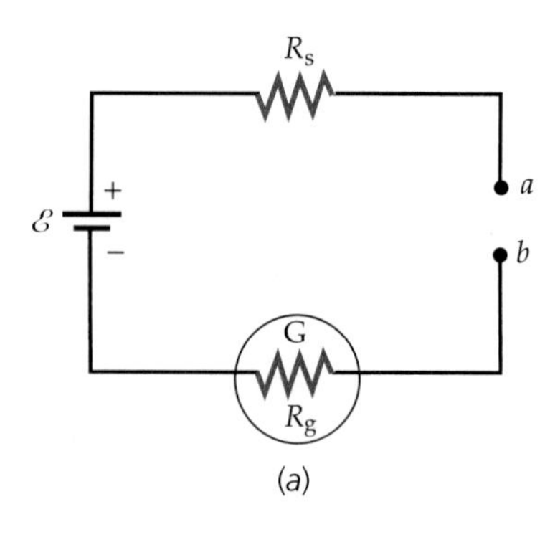

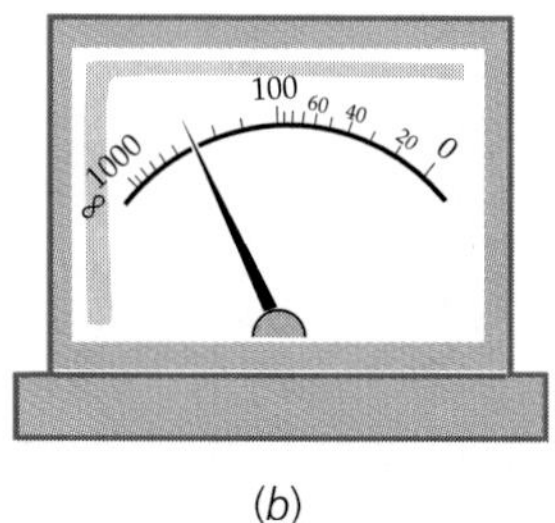

Figure 26-36 (*a*) An ohmmeter consists of a battery in series with a galvanometer and a resistor R_s, which is chosen such that the galvanometer gives full-scale deflection when points *a* and *b* are shorted. (*b*) When a resistor *R* is placed across *a* and *b*, the galvanometer needle deflects by an amount that depends on the value of *R*. The galvanometer scale is calibrated to give a readout in ohms.

26-6 *RC* Circuits

A circuit containing a resistor and capacitor is called an ***RC* circuit**. The current in an *RC* circuit flows in a single direction, as in all dc circuits, but the magnitude of the current varies with time. A practical example of an *RC* circuit is the circuit in the flash attachment of a camera. Before a flash photograph is taken, a battery in the flash attachment charges the capacitor through a resistor. When this is accomplished, the flash is ready. When the picture is taken, the capacitor discharges through the flash bulb. The capacitor is then recharged by the battery, and a short time later the flash is ready for another picture. Using Kirchhoff's rules, we can obtain equations for the charge *Q* and the current *I* as functions of time for both the charging and discharging of a capacitor through a resistor.

Discharging a Capacitor

Figure 26-37 shows a capacitor with initial charges of $+Q_0$ on the upper plate and $-Q_0$ on the lower plate. The capacitor is connected to a resistor *R* and a switch S, which is initially open. The potential difference across the capacitor is initially $V_0 = Q_0/C$, where *C* is the capacitance.

We close the switch at time $t = 0$. Since there is now a potential difference across the resistor, there must be a current in it. The initial current is

$$I_0 = \frac{V_0}{R} = \frac{Q_0}{RC} \qquad \text{26-25}$$

The current is due to the flow of charge from the positive plate of the capacitor to the negative plate through the resistor. After a time, the charge on the capacitor is reduced. If we choose the direction of *I* to be clockwise, then the current equals the rate of *decrease* of that charge. If *Q* is the charge on the capacitor at any time, the current at that time is

$$I = -\frac{dQ}{dt} \qquad \text{26-26}$$

Traversing the circuit in the direction of the current, we encounter a potential drop *IR* across the resistor and a potential increase *Q*/*C* across the capacitor. Thus, Kirchhoff's loop rule gives

$$\frac{Q}{C} - IR = 0 \qquad \text{26-27}$$

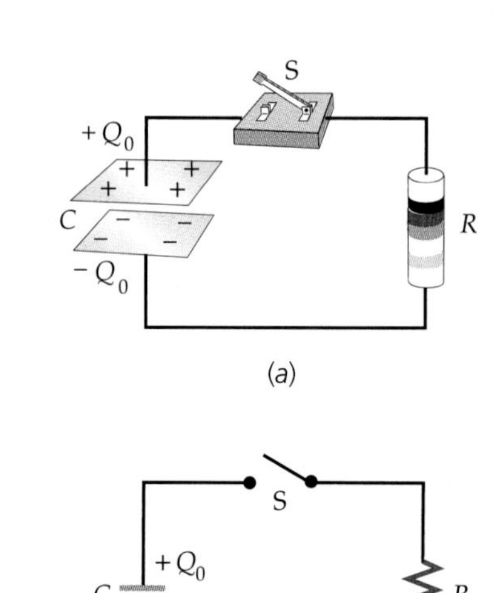

Figure 26-37 (*a*) A parallel-plate capacitor in series with a switch and a resistor *R*. (*b*) Circuit diagram for (*a*).

where both Q and I are functions of time and are related by Equation 26-26. Substituting $-dQ/dt$ for I in Equation 26-27, we have

$$\frac{Q}{C} + R\frac{dQ}{dt} = 0$$

or

$$\frac{dQ}{dt} = -\frac{1}{RC}Q \qquad 26\text{-}28$$

To solve this equation, we first separate the variables Q and t. Multiplying both sides of the equation by dt and dividing by Q, we obtain

$$\frac{dQ}{Q} = -\frac{dt}{RC} \qquad 26\text{-}29$$

Integrating from Q_0 at $t = 0$ to Q at time t gives

$$\ln\frac{Q}{Q_0} = -\frac{t}{RC}$$

Then

$$Q(t) = Q_0 e^{-t/RC} = Q_0 e^{-t/\tau} \qquad 26\text{-}30$$

where τ, called the **time constant**, is the time it takes for the charge to decrease to $1/e$ of its original value:

$$\tau = RC \qquad 26\text{-}31$$

Definition—Time constant

Figure 26-38 shows the charge on the capacitor in the circuit of Figure 26-37 as a function of time. After a time $t = \tau$, the charge is $Q = Q_0 e^{-1} = 0.37Q_0$, after a time $t = 2\tau$, the charge is $Q = Q_0 e^{-2} = 0.135Q_0$, and so forth. After a time equal to several time constants, the charge on the capacitor is negligible. This type of decrease, which is called an **exponential decrease**, is very common in nature. It occurs whenever the rate at which a quantity decreases is proportional to the quantity itself.*

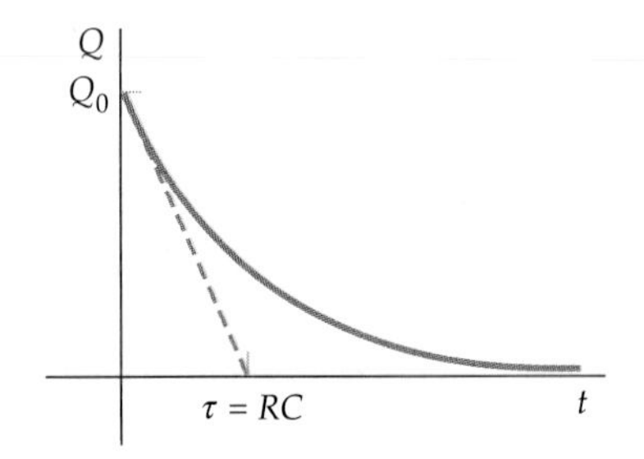

Figure 26-38 Plot of the charge on the capacitor versus time for the circuit in Figure 26-37 when the switch is closed at time $t = 0$. The time constant $\tau = RC$ is the time it takes for the charge to decrease to $e^{-1}Q_0$. The time constant is also the time it would take the capacitor to discharge fully if its discharge rate were constant, as indicated by the dashed line.

The decrease in the charge on a capacitor can be likened to the decrease in the amount of water in a bucket that has a small hole in the bottom. The rate at which the water flows out of the bucket is proportional to the pressure of the water, which is in turn proportional to the amount of water still in the bucket.

The current is obtained by differentiating Equation 26-30

$$I = -\frac{dQ}{dt} = \frac{Q_0}{RC}e^{-t/RC}$$

or

$$I = \frac{V_0}{R}e^{-t/RC} = I_0 e^{-t/\tau} \qquad 26\text{-}32$$

where $I_0 = Q_0/RC = V_0/R$ is the initial current. The current as a function of time is shown in Figure 26-39. The current also decreases exponentially with time and falls to $1/e$ of its initial value after a time $t = \tau = RC$.

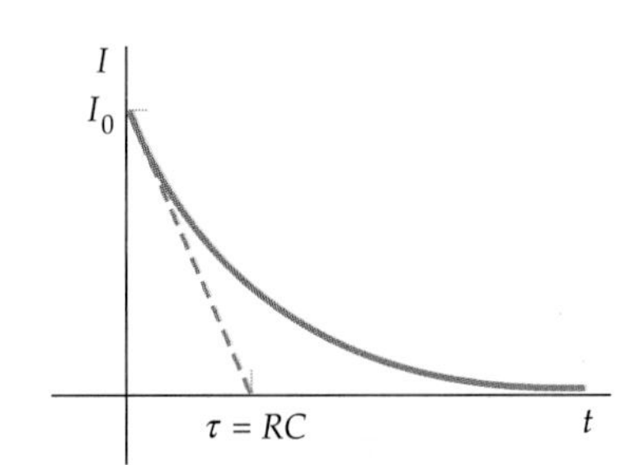

Figure 26-39 Plot of the current versus time for the circuit in Figure 26-37. The curve has the same shape as that in Figure 26-38. If the current decreased at a constant rate equal to its initial rate, it would reach zero after one time constant as indicated by the dashed line.

* We encountered exponential decreases in Chapter 14 when we studied the damped oscillator.

Example 26-17

A 4-μF capacitor is charged to 24 V and then connected across a 200-Ω resistor. Find (*a*) the initial charge on the capacitor, (*b*) the initial current through the 200-Ω resistor, (*c*) the time constant, and (*d*) the charge on the capacitor after 4 ms.

Picture the Problem The circuit diagram is the same as Figure 26-37.

(*a*) The initial charge is related to the capacitance and voltage:

$$Q_0 = CV = (4\ \mu\text{F})(24\ \text{V}) = 96\ \mu\text{C}$$

(*b*) The initial current is the initial voltage divided by the resistance:

$$I_0 = \frac{V_0}{R} = \frac{24\ \text{V}}{200\ \Omega} = 0.12\ \text{A}$$

(*c*) The time constant is *RC*:

$$\tau = RC = (200\ \Omega)(4\ \mu\text{F}) = 800\ \mu\text{s} = 0.8\ \text{ms}$$

(*d*) Substitute $t = 4$ ms into Equation 26-30 to find the charge on the capacitor at that time:

$$Q = Q_0 e^{-t/\tau} = (96\ \mu\text{C})e^{-(4\ \text{ms})/(0.8\ \text{ms})}$$
$$= (96\ \mu\text{C})e^{-5}$$
$$= 0.647\ \mu\text{C}$$

Exercise Find the current through the 200-Ω resistor at $t = 4$ ms. (*Answer* 0.809 mA)

Charging a Capacitor

Figure 26-40*a* shows a circuit for charging a capacitor. We will assume that the capacitor is initially uncharged. The switch, originally open, is closed at time $t = 0$. Charge immediately begins to flow through the resistor and onto the positive plate of the capacitor (Figure 26-40*b*). If the charge on the capacitor at some time is Q and the current in the circuit is I, Kirchhoff's loop rule gives

$$\mathcal{E} - V_R - V_C = 0$$

or

$$\mathcal{E} - IR - \frac{Q}{C} = 0 \qquad 26\text{-}33$$

In this circuit, we've chosen the direction of I so that the current equals the rate at which the charge on the capacitor is *increasing*:

$$I = +\frac{dQ}{dt}$$

Substituting $+dQ/dt$ for I in Equation 26-33 gives

$$\mathcal{E} = R\frac{dQ}{dt} + \frac{Q}{C} \qquad 26\text{-}34$$

At time $t = 0$, the charge on the capacitor is zero and the current is $I_0 = \mathcal{E}/R$. The charge then increases and the current decreases, as can be seen from Equation 26-33. The charge reaches a maximum value of $Q_f = C\mathcal{E}$ when the current I equals zero, as can also be seen from Equation 26-34.

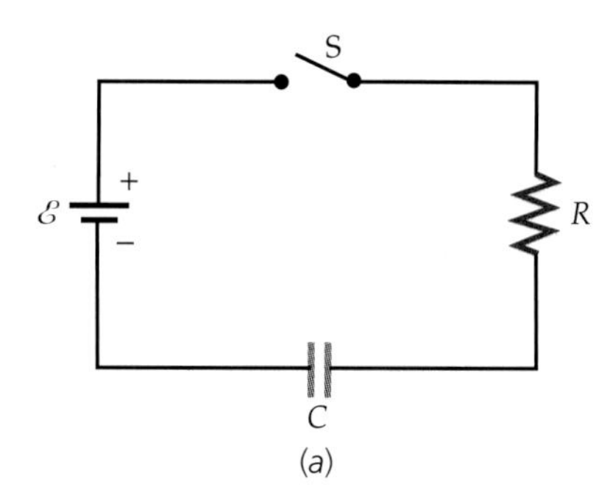

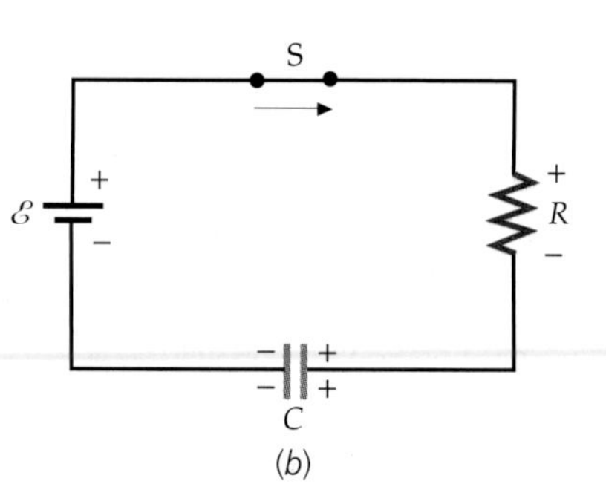

Figure 26-40 (*a*) Circuit for charging a capacitor to a potential difference $\mathcal{E}$. (*b*) After the switch is closed, there is a potential drop across the resistor and a charge on the capacitor.

Equation 26-34 can be solved in the same way as Equation 26-28. The details are left as a problem (see Problem 115). The result is

$$Q = C\mathcal{E}(1 - e^{-t/RC}) = Q_f(1 - e^{-t/\tau}) \qquad 26\text{-}35$$

where $Q_f = C\mathcal{E}$ is the final charge. The current is obtained from $I = dQ/dt$:

$$I = \frac{dQ}{dt} = -C\mathcal{E}e^{-t/RC}\left(\frac{-1}{RC}\right)$$

or

$$I = \frac{\mathcal{E}}{R}e^{-t/RC} = I_0 e^{-t/\tau} \qquad 26\text{-}36$$

where the initial current in this case is $I_0 = \mathcal{E}/R$.

Figures 26-41 and 26-42 show the charge and the current as functions of time.

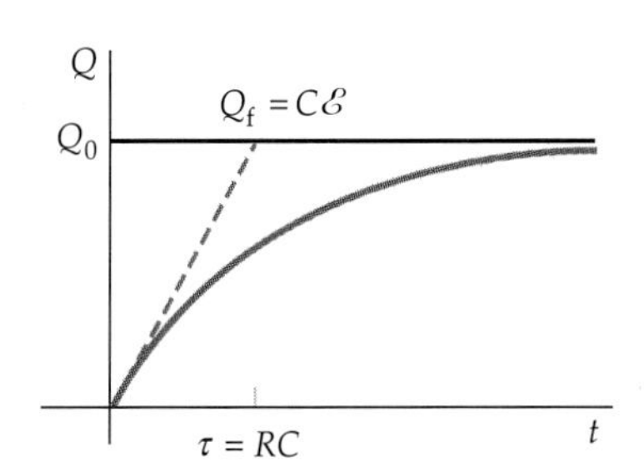

Figure 26-41 Plot of the charge on the capacitor versus time for the charging circuit of Figure 26-40 after the switch is closed at $t = 0$. After a time $t = \tau = RC$, the charge on the capacitor is $0.63C\mathcal{E}$, where $C\mathcal{E}$ is its final charge. If the charging rate were constant, the capacitor would be fully charged after a time $t = \tau$.

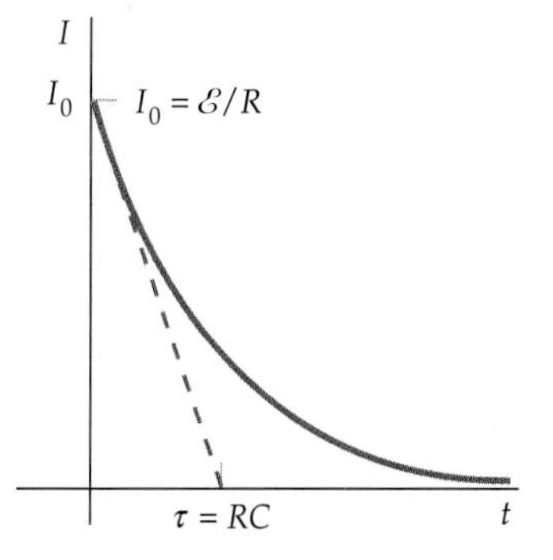

Figure 26-42 Plot of the current versus time for the charging circuit of Figure 26-40. The current is initially $\mathcal{E}/R$, and it decreases exponentially with time.

Exercise Show that Equation 26-35 does indeed satisfy Equation 26-34 by substituting $Q(t)$ and dQ/dt into Equation 26-34.

Exercise What fraction of the maximum charge is on the charging capacitor after a time $t = 2\tau$? (*Answer* 0.86)

Example 26-18 *try it yourself*

A 6-V battery of negligible internal resistance is used to charge a 2-μF capacitor through a 100-Ω resistor. Find (*a*) the initial current, (*b*) the final charge on the capacitor, and (*c*) the time required for the charge to reach 90% of its final value.

Cover the column to the right and try these on your own before looking at the answers.

Steps	***Answers***
(*a*) Find the initial current from $I_0 = \mathcal{E}/R$.	$I_0 = 0.06$ A
(*b*) Find the final charge from $Q = C\mathcal{E}$.	$Q_f = 12\ \mu$C
(*c*) 1. Set $Q = 0.9\,Q_f$ in Equation 26-35 and solve for $e^{+t/\tau}$.	$e^{t/\tau} = 10$
2. Take the natural logarithm of each side of your equation in the previous step.	$\ln(e^{t/\tau}) = t/\tau = \ln 10 = 2.3$
3. Solve for t.	$t = 2.3RC = 460\ \mu$s

Example 26-19

The capacitor in the circuit shown in Figure 26-43 is initially uncharged. Find the current through the battery (*a*) immediately after the switch is closed, and (*b*) a long time after the switch is closed.

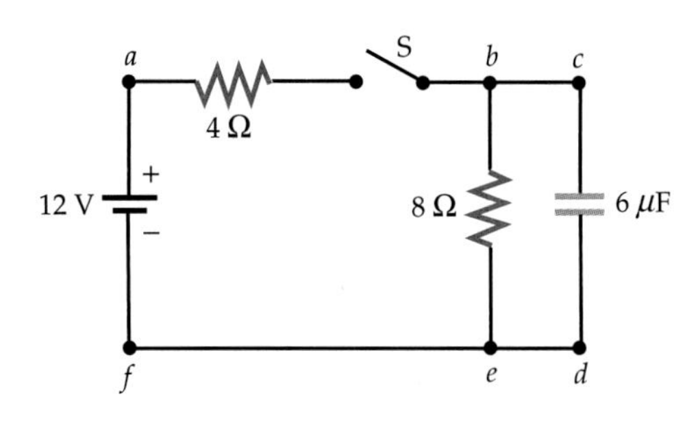

Figure 26-43

(*a*) Since the capacitor is initially uncharged, the potential is the same at points *d* and *c* just after the switch is closed. There is thus no initial current through the 8-Ω resistor between *b* and *e*. Apply the loop rule to the outer loop (*abcdefa*):

$12\ \text{V} - (4\ \Omega)I_0 = 0$

$I_0 = 3\ \text{A}$

(*b*) After a long time, the capacitor is fully charged, and no more charge flows onto or off of the plates. Apply the loop rule to the left loop (*abefa*):

$12\ \text{V} - (4\ \Omega)I_\text{f} - (8\ \Omega)I_\text{f} = 0$

$I_\text{f} = 1\ \text{A}$

Remarks The analysis of this circuit at the extreme times when the capacitor is either uncharged or fully charged is simple. When the capacitor is uncharged, it acts like a short circuit between points *c* and *d*, that is, the circuit is the same as the one shown in Figure 26-44*a*, where we have replaced the capacitor by a wire of zero resistance. When the capacitor is fully charged, it acts like an open circuit, as shown in Figure 26-44*b*.

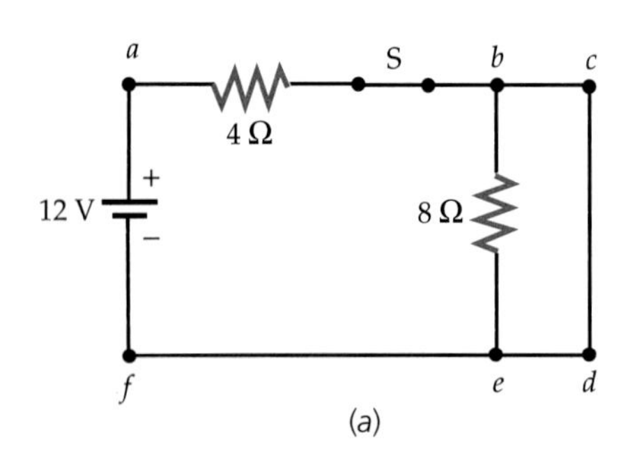

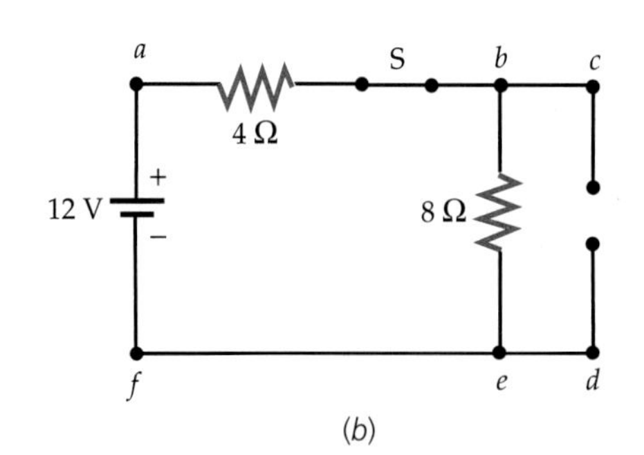

Figure 26-44

Energy Conservation in Charging a Capacitor

During the charging process, a total charge $Q_\text{f} = \mathcal{E}C$ flows through the battery. The battery therefore does work

$$W = Q_\text{f}\mathcal{E} = C\mathcal{E}^2$$

Half of this work is accounted for by the energy stored in the capacitor (see Equation 25-11):

$$U = \tfrac{1}{2}QV = \tfrac{1}{2}Q_f\mathcal{E} = \tfrac{1}{2}C\mathcal{E}^2$$

We now show that the other half of work done by the battery goes into Joule heat in the resistance of the circuit. The rate at which energy is put into the resistance R is

$$\frac{dW_R}{dt} = I^2R$$

Using Equation 26-36 for the current, we have

$$\frac{dW_R}{dt} = \left(\frac{\mathcal{E}}{R}e^{-t/RC}\right)^2 R = \frac{\mathcal{E}^2}{R}e^{-2t/RC}$$

We find the total Joule heat by integrating from $t = 0$ to $t = \infty$:

$$W_R = \int_0^\infty \frac{\mathcal{E}^2}{R}e^{-2t/RC}\,dt$$

The integration can be done by substituting $x = 2t/RC$. Then,

$$dt = \frac{RC}{2}\,dx$$

and

$$W_R = \frac{\mathcal{E}^2}{R}\frac{RC}{2}\int_0^\infty e^{-x}\,dx = \frac{\mathcal{E}^2C}{2}(-e^{-x})\Big|_0^\infty = \frac{\mathcal{E}^2C}{2}(-0+1)$$

The total Joule heat is thus

$$W_R = \frac{1}{2}\mathcal{E}^2C$$

This result is independent of the resistance R. Thus, when a capacitor is charged by a battery with a constant emf, half the energy provided by the battery is stored in the capacitor and half goes into thermal energy, independent of the resistance. The thermal energy includes the energy that goes into the internal resistance of the battery.

Summary

1. Ohm's law is an empirical law that holds only for certain materials.
2. Current, resistance, and emf are important *defined* quantities.
3. Kirchhoff's rules follow from the conservation of charge and the conservation of energy.

Topic	Remarks and Relevant Equations
1. Electric Current	Electric current is the rate of flow of charge through a cross-sectional area. $$I = \frac{\Delta Q}{\Delta t} = qnAv_d$$ **26-1, 26-3**
Drift velocity	In a conducting wire, electric current is the result of the slow drift of negatively charged electrons that are accelerated by an electric field in the wire and then collide with the lattice ions. Typical drift velocities of electrons in wires are of the order of a few millimeters per second.

2.	**Resistance**		
	Definition	$R = \frac{V}{I}$	**26-5**
	Resistivity ρ	$R = \rho\frac{L}{A}$	**26-8**
3.	**Ohm's Law**	For ohmic materials, the resistance does not depend on the current or voltage drop: $V = IR \quad (R \text{ constant})$	**26-7**
4.	**Power**		
	Supplied to a device or segment	$P = VI$	**26-10**
	Dissipated in a resistor	$P = VI = I^2R = \frac{V^2}{R}$	**26-11**
5.	**emf**	A device that supplies energy to a circuit is called a source of emf.	
	Power supplied by an emf	$P = \mathcal{E}I$	**26-12**
6.	**Battery**		
	Ideal	An ideal battery is a source of emf that maintains a constant potential difference between its two terminals, independent of the rate of flow of charge between them.	
	Real	A real battery can be considered to be an ideal battery in series with a small resistance called its internal resistance.	
	Terminal voltage	$V_a - V_b = \mathcal{E} - Ir$	**26-13**
	Total energy	$W = Q\mathcal{E}$	**26-15**
7.	**Equivalent Resistance**		
	Series	$R_{eq} = R_1 + R_2 + R_3 + \cdots$	**26-17**
	Parallel	$\frac{1}{R_{eq}} = \frac{1}{R_1} + \frac{1}{R_2} + \frac{1}{R_3} + \cdots$	**26-21**
8.	**Kirchhoff's rules**	1. When any closed circuit loop is traversed, the algebraic sum of the changes in potential must equal zero. 2. At any junction in a circuit where the current can divide, the sum of the currents into the junction must equal the sum of the currents out of the junction.	
9.	**Measuring Devices**		
	Ammeter	An ammeter is a very low resistance device that is placed in series with a circuit element to measure the current in the element.	
	Voltmeter	A voltmeter is a very high resistance device that is placed in parallel with a circuit element to measure the voltage drop across the element.	

Ohmmeter	An ohmmeter is a device containing a battery that is used to measure the resistance of a circuit element placed across its terminals.	
10. Discharging a Capacitor		
Charge on the capacitor	$Q(t) = Q_0 e^{-t/RC} = Q_0 e^{-t/\tau}$	**26-30**
Current in the circuit	$I = \frac{V_0}{R} e^{-t/RC} = I_0 e^{-t/\tau}$	**26-32**
Time constant	$\tau = RC$	**26-31**
11. Charging a Capacitor		
Charge on the capacitor	$Q = C\mathcal{E}(1 - e^{-t/RC}) = Q_f(1 - e^{-t/\tau})$	**26-35**
Current in the circuit	$I = \frac{\mathcal{E}}{R} e^{-t/RC} = I_0 e^{-t/\tau}$	**26-36**

Problem-Solving Guide

To solve circuit problems

1. Draw a sketch of the circuit.
2. Choose a direction for the current in each branch of the circuit, and label the currents in the circuit diagram. Add plus and minus signs to indicate the high- and low-potential sides of each resistor, capacitor, and source of emf.
3. Replace any combination of resistors in series or parallel with its equivalent resistance.
4. Apply the junction rule to each junction where the current divides.
5. Apply the loop rule to several loops to obtain as many equations as unknowns.
6. Solve the equations to obtain the values of the unknowns.
7. Check your results by assigning a potential of zero to one point in the circuit and then using the values of the currents found to determine the potentials at other points in the circuit.

Summary of Worked Examples

Type of Calculation	Procedure and Relevant Examples
1. Resistance	
Find the resistance of a wire.	Use $R = \rho L/A$. **Examples 26-3, 26-4**
2. Energy	
Calculate the energy dissipated in a circuit.	Use $P = I^2R = V^2/R = VI$ for each resistor and $P = \mathcal{E}I$ for each emf. **Examples 26-6, 26-7, 26-8, 26-9, 26-10, 26-13, 26-15**
3. Equivalent Resistance	
Find the equivalent resistance.	Use $R_{eq} = R_1 + R_2 + \cdots$ (series) and $1/R_{eq} = 1/R_1 + 1/R_2 + \cdots$ (parallel). **Examples 26-6, 26-10, 26-11, 26-12**

4. Circuits Without Capacitors		
Find the current in a single-loop circuit.	Apply Kirchhoff's loop rule.	**Examples 26-13, 26-14**
Find the currents in a multiloop circuit.	Apply Kirchhoff's loop and junction rules.	**Examples 26-15, 26-16**
5. *RC* Circuits		
Find the charge on a capacitor.	Use $Q(t) = Q_0 e^{-t/RC}$ for a capacitor that is discharging, and $Q = C\mathcal{E}(1 - e^{-t/RC})$ for a capacitor that is charging.	**Examples 26-17, 26-18, 26-19**
Find the time constant.	Use $\tau = RC$.	**Examples 26-17, 26-18**
Find the initial and final current.	An uncharged capacitor is momentarily equivalent to a resistor of zero resistance. For a simple series circuit containing an uncharged capacitor, $I_i = \mathcal{E}/R$. A fully charged capacitor is equivalent to an open circuit. For a simple series circuit containing a fully charged capacitor, $I_f = 0$.	**Examples 26-17, 26-19**

Problems

Conceptual Problems

Problems from Optional and Exploring sections

In a few problems, you are given more data than you actually need; in a few other problems, you are required to supply data from your general knowledge, outside sources, or informed estimates.

• Single-concept, single-step, relatively easy
•• Intermediate-level, may require synthesis of concepts
••• Challenging, for advanced students

Current and Motion of Charges

1 • In our study of electrostatics, we concluded that there is no electric field within a conductor in electrostatic equilibrium. How is it that we can now discuss electric fields inside a conductor?

2 • A physics professor has assembled his class at the baggage-claim carousel of the local airport to demonstrate an analog of electrical current. "Think of each suitcase on the conveyor belt as a package of electrons carrying one coulomb of charge," he says. Counting and timing the suitcases reveals that the conveyor belt represents a wire carrying a constant 2-A current (constant as long as annoyed travelers could be kept away from their baggage by some of the huskier students). (*a*) How many suitcases will go by a given point in 5.0 min? (*b*) How many electrons does that represent?

3 • A 10-gauge copper wire carries a current of 20 A. Assuming one free electron per copper atom, calculate the drift velocity of the electrons.

4 • In a fluorescent tube of diameter 3.0 cm, 2.0×10^{18} electrons and 0.5×10^{18} positive ions (with a charge of $+e$) flow through a cross-sectional area each second. What is the current in the tube?

5 • In a certain electron beam, there are 5.0×10^6 electrons per cubic centimeter. Suppose the kinetic energy of each electron is 10.0 keV, and the beam is cylindrical, with a diameter of 1.00 mm. (*a*) What is the velocity of an electron in the beam? (*b*) Find the beam current.

6 •• A charge $+q$ moves in a circle of radius r with speed v. (*a*) Express the frequency f with which the charge passes a particular point in terms of r and v. (*b*) Show that the average current is qf and express it in terms of v and r.

7 •• A ring of radius a with a linear charge density λ rotates about its axis with angular velocity ω. Find an expression for the current.

8 •• A 10-gauge copper wire and a 14-gauge copper wire are welded together end to end. The wires carry a current of 15 A. If there is one free electron per copper atom in each wire, find the drift velocity of the electrons in each wire.

9 •• In a certain particle accelerator, a proton beam with a diameter of 2.0 mm constitutes a current of 1.0 mA. The kinetic energy of each proton is 20 MeV. The beam strikes a metal target and is absorbed by it. (*a*) What is the number n of protons per unit volume in the beam? (*b*) How many protons strike the target in 1.0 min? (*c*) If the target is initially uncharged, express the charge of the target as a function of time.

10 •• The current in a wire varies with time according to the relation $I = 20 + 3t^2$, where I is in amperes and t is in seconds. (*a*) How many coulombs are transported by the wire between $t = 0$ and $t = 10$ s? (*b*) What constant current would transport the same charge in the same time interval?

11 •• In a proton supercollider, the protons in a 5-mA beam move with nearly the speed of light. (*a*) How many protons are there per meter of the beam? (*b*) If the cross-sectional area of the beam is 10^{-6} m^2, what is the number density of protons?

Resistance and Ohm's Law

12 • Figure 26-8 illustrates a mechanical analog of a simple electric circuit. Devise another mechanical analog in which the current is represented by a flow of water instead of marbles.

13 • Two wires of the same material with the same length have different diameters. Wire A has twice the diameter of wire B. If the resistance of wire B is R, then what is the resistance of wire A?
(*a*) R
(*b*) $2R$
(*c*) $R/2$
(*d*) $4R$
(*e*) $R/4$

14 •• Discuss the difference between an emf and a potential difference.

15 •• Name several common sources of emf. What sort of energy is converted into electrical energy in each?

16 •• A metal bar is to be used as a resistor. Its dimensions are 2 by 4 by 10 units. To get the smallest resistance from this bar, one should attach leads to the opposite sides that have the dimensions of

(*a*) 2 by 4 units.
(*b*) 2 by 10 units.
(*c*) 4 by 10 units.
(*d*) All connections will give the same resistance.
(*e*) None of the above is correct.

17 •• Two cylindrical copper wires have the same mass. Wire A is twice as long as wire B. Their resistances are related by

(*a*) $R_A = 8R_B$.
(*b*) $R_A = 4R_B$.
(*c*) $R_A = 2R_B$.
(*d*) $R_A = R_B$.

18 • A 10-m-long wire of resistance 0.2 Ω carries a current of 5 A. (*a*) What is the potential difference across the wire? (*b*) What is the magnitude of the electric field in the wire?

19 • A potential difference of 100 V produces a current of 3 A in a certain resistor. (*a*) What is the resistance of the resistor? (*b*) What is the current when the potential difference is 25 V?

20 • A block of carbon is 3.0 cm long and has a square cross-sectional area with sides of 0.5 cm. A potential difference of 8.4 V is maintained across its length. (*a*) What is the resistance of the block? (*b*) What is the current in this resistor?

21 • A carbon rod with a radius of 0.1 mm is used to make a resistor. The resistivity of this material is 3.5×10^{-5} Ω·m. What length of the carbon rod will make a 10-Ω resistor?

22 • The third (current-carrying) rail of a subway track is made of steel and has a cross-sectional area of about 55 cm^2. What is the resistance of 10 km of this track? (Use ρ for iron.)

23 • What is the potential difference across one wire of a 30-m extension cord made of 16-gauge copper wire carrying a current of 5.0 A?

24 • How long is a 14-gauge copper wire that has a resistance of 2 Ω?

25 •• A cylinder of glass 1 cm long has a resistivity of 10^{12} Ω·m. How long would a copper wire of the same cross-sectional area need to be to have the same resistance as the glass cylinder?

26 •• An 80.0-m copper wire 1.0 mm in diameter is joined end to end with a 49.0-m iron wire of the same diameter. The current in each is 2.0 A. (*a*) Find the electric field in each wire. (*b*) Find the potential drop across each wire.

27 •• A copper wire and an iron wire with the same length and diameter carry the same current I. (*a*) Find the ratio of the potential drops across these wires. (*b*) In which wire is the electric field greater?

28 •• A variable resistance R is connected across a potential difference V that remains constant. When $R = R_1$, the current is 6.0 A. When R is increased to $R_2 = R_1 + 10.0$ Ω, the current drops to 2.0 A. Find (*a*) R_1 and (*b*) V.

29 •• A rubber tube 1 m long with an inside diameter of 4 mm is filled with a salt solution that has a resistivity of 10^{-3} Ω·m. Metal plugs form electrodes at the ends of the tube. (*a*) What is the resistance of the filled tube? (*b*) What is the resistance of the filled tube if it is uniformly stretched to a length of 2 m?

30 •• A wire of length 1 m has a resistance of 0.3 Ω. It is uniformly stretched to a length of 2 m. What is its new resistance?

31 •• Currents up to 30 A can be carried by 10-gauge copper wire. (*a*) What is the resistance of 100 m of 10-gauge copper wire? (*b*) What is the electric field in the wire when the current is 30 A? (*c*) How long does it take for an electron to travel 100 m in the wire when the current is 30 A?

32 •• A cube of copper has sides of 2.0 cm. If it is drawn out to form a 14-gauge wire, what will its resistance be?

33 ••• A semiconducting diode is a nonlinear device whose current I is related to the voltage V across the diode by $I = I_0(e^{eV/kT} - 1)$, where k is Boltzmann's constant, e is the magnitude of the charge on an electron, and T is the absolute temperature. If $I_0 = 10^{-9}$ A and $T = 293$ K, (*a*) what is the resistance of the diode for $V = 0.5$ V? (*b*) What is the resistance for $V = 0.6$ V?

34 ••• Find the resistance between the ends of the half ring shown in Figure 26-45. The resistivity of the material of the ring is ρ.

Figure 26-45 Problem 34

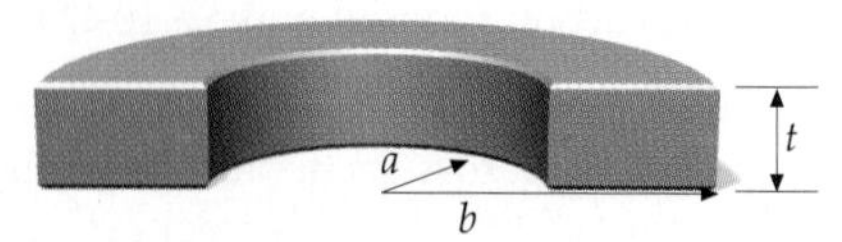

35 ••• The radius of a wire of length L increases linearly along its length according to $r = a + [(b - a)/L]\,x$, where x is the distance from the small end of radius a. What is the resistance of this wire in terms of its resistivity ρ, length L, radius a, and radius b?

36 ••• The space between two concentric spherical-shell conductors is filled with a material that has a resistivity of $10^9\ \Omega\cdot$m. If the inner shell has a radius of 1.5 cm and the outer shell has a radius of 5 cm, what is the resistance between the conductors? (*Hint:* Find the resistance of a spherical-shell element of the material of area $4\pi r^2$ and length dr, and integrate to find the total resistance of the set of shells in series.)

37 ••• The space between two metallic coaxial cylinders of length L and radii a and b is completely filled with a material having a resistivity ρ. (*a*) What is the resistance between the two cylinders? (See the hint in Problem 36.) (*b*) Find the current between the two cylinders if $\rho = 30\ \Omega\cdot$m, $a = 1.5$ cm, $b = 2.5$ cm, $L = 50$ cm, and a potential difference of 10 V is maintained between the two cylinders.

Temperature Dependence of Resistance

38 • A tungsten rod is 50 cm long and has a square cross-sectional area with sides of 1.0 mm. (*a*) What is its resistance at 20°C? (*b*) What is its resistance at 40°C?

39 • At what temperature will the resistance of a copper wire be 10% greater than it is at 20°C?

40 •• A toaster with a Nichrome heating element has a resistance of 80 Ω at 20°C and an initial current of 1.5 A. When the heating element reaches its final temperature, the current is 1.3 A. What is the final temperature of the heating element?

41 •• An electric space heater has a Nichrome heating element with a resistance of 8 Ω at 20°C. When 120 V are applied, the electric current heats the Nichrome wire to 1000°C. (*a*) What is the initial current drawn by the cold heating element? (*b*) What is the resistance of the heating element at 1000°C? (*c*) What is the operating wattage of this heater?

42 •• A 10-Ω Nichrome resistor is wired into an electronic circuit using copper leads (wires) of diameter 0.6 mm with a total length of 50 cm. (*a*) What additional resistance is due to the copper leads? (*b*) What percentage error in the total added resistance is produced by neglecting the resistance of the copper leads? (*c*) What change in temperature would produce a change in resistance of the Nichrome-wire equal to the resistance of the copper leads?

43 •• The filament of a certain lamp has a resistance that increases linearly with temperature. When a constant voltage is switched on, the initial current decreases until the filament reaches its steady-state temperature. The temperature coefficient of resistivity of the filament is $4 \times 10^{-3}\ \mathrm{K}^{-1}$. The final current through the filament is one-eighth the initial current. What is the change in temperature of the filament?

44 ••• A wire of cross-sectional area A, length L_1, resistivity ρ_1, and temperature coefficient α_1 is connected end to end to a second wire of the same cross-sectional area, length L_2, resistivity ρ_2, and temperature coefficient α_2, so that the wires carry the same current. (*a*) Show that if $\rho_1 L_1 \alpha_1 + \rho_2 L_2 \alpha_2 = 0$, the total resistance R is independent of temperature for small temperature changes. (*b*) If one wire is made of carbon and the other is copper, find the ratio of their lengths for which R is approximately independent of temperature.

Energy in Electric Circuits

45 • A resistor carries a current I. The power dissipated in the resistor is P. What is the power dissipated if the same resistor carries current $3I$? (Assume no change in resistance.)

(*a*) P (*b*) $3P$ (*c*) $P/3$
(*d*) $9P$ (*e*) $P/9$

46 • The power dissipated in a resistor is P when the voltage drop across it is V. If the voltage drop is increased to 2 V (with no change in resistance), what is the power dissipated?

(*a*) P (*b*) $2P$ (*c*) $4P$
(*d*) $P/2$ (*e*) $P/4$

47 • A heater consists of a variable resistance connected across a constant voltage supply. To increase the heat output, should you decrease the resistance or increase it?

48 •• Two resistors dissipate the same amount of power. The potential drop across resistor A is twice that across resistor B. If the resistance of resistor B is R, what is the resistance of A?

(*a*) R (*b*) $2R$ (*c*) $R/2$
(*d*) $4R$ (*e*) $R/4$

49 • Find the power dissipated in a resistor connected across a constant potential difference of 120 V if its resistance is (*a*) 5 Ω and (*b*) 10 Ω.

50 • A 10,000-Ω carbon resistor used in electronic circuits is rated at 0.25 W. (*a*) What maximum current can this resistor carry? (*b*) What maximum voltage can be placed across this resistor?

51 • A 1-kW heater is designed to operate at 240 V. (*a*) What is its resistance, and what current does it draw? (*b*) What is the power dissipated in this resistor if it operates at 120 V? Assume that its resistance is constant.

52 • A battery has an emf of 12.0 V. How much work does it do in 5 s if it delivers a current of 3 A?

53 • A battery with 12-V emf has a terminal voltage of 11.4 V when it delivers a current of 20 A to the starter of a car. What is the internal resistance r of the battery?

54 • (*a*) How much power is delivered by the emf of the battery in Problem 53 when it delivers a current of 20 A? (*b*) How much of this power is delivered to the starter? (*c*) By how much does the chemical energy of the battery

decrease when it delivers a current of 20 A to the starter for 3 min? (*d*) How much heat is developed in the battery when it delivers a current of 20 A for 3 min?

55 • A physics student runs a 1200-W electric heater constantly in her basement bedroom during the winter time. If electric energy costs 9 cents per kilowatt-hour, how much does this electric heating cost per 30-day month?

56 • A battery with an emf of 6 V and an internal resistance of 0.3 Ω is connected to a variable resistance R. Find the current and power delivered by the battery when R is (*a*) 0, (*b*) 5 Ω, (*c*) 10 Ω, and (*d*) infinite.

57 •• Staying up late to study, and having no stove to heat water, you use a 200-W heater from the lab to make coffee throughout the night. If 90% of the energy produced by the heater goes toward heating the water in your cup, (*a*) how long does it take to heat 0.25 kg of water from 15 to 100°C? (*b*) If you fall asleep while the water is heating, how long will it take to boil away after it reaches 100°C?

58 •• Suppose the bulb in a two-cell flashlight draws 4 W of power. The batteries go dead in 45 min and cost $7.99. (*a*) How many kilowatt-hours of energy can be supplied by the two batteries? (*b*) What is the cost per kilowatt-hour of energy if the batteries cannot be recharged? (*c*) If the batteries can be recharged at a cost of 9 cents per kilowatt-hour, what is the cost of recharging them?

59 •• A 12-V automobile battery with negligible internal resistance can deliver a total charge of 160 A·h. (*a*) What is the total stored energy in the battery? (*b*) How long could this battery provide 150 W to a pair of headlights?

60 •• A space heater in an old home draws a 12.5-A current. A pair of 12-gauge copper wires carries the current from the fuse box to the wall outlet, a distance of 30 m. The voltage at the fuse box is exactly 120 V. (*a*) What is the voltage delivered to the space heater? (*b*) If the fuse will blow at a current of 20 A, how many 60-W bulbs can be supplied by this line when the space heater is on? (Assume that the wires from the wall to the space heater and to the light fixtures have negligible resistance.)

61 •• A lightweight electric car is powered by ten 12-V batteries. At a speed of 80 km/h, the average frictional force is 1200 N. (*a*) What must be the power of the electric motor if the car is to travel at a speed of 80 km/h? (*b*) If each battery can deliver a total charge of 160 A·h before recharging, what is the total charge in coulombs that can be delivered by the 10 batteries before charging? (*c*) What is the total electrical energy delivered by the 10 batteries before recharging? (*d*) How far can the car travel at 80 km/h before the batteries must be recharged? (*e*) What is the cost per kilometer if the cost of recharging the batteries is 9 cents per kilowatt-hour?

62 ••• A 100-W heater is designed to operate with an emf of 120 V. (*a*) What is its resistance, and what current does it draw? (*b*) Show that if the potential difference across the heater changes by a small amount ΔV, the power changes by a small amount ΔP, where $\Delta P/P \approx 2\,\Delta V/V$. (*Hint:* Approximate the changes with differentials.) (*c*) Find the approximate power dissipated in the heater if the potential difference is decreased to 115 V.

Combinations of Resistors

63 • Two resistors are connected in parallel across a potential difference. The resistance of resistor A is twice that of resistor B. If the current carried by resistor A is I, then what is the current carried by B?

(*a*) I (*b*) $2I$ (*c*) $I/2$ (*d*) $4I$ (*e*) $I/4$

64 • Two resistors are connected in series across a potential difference. Resistor A has twice the resistance of resistor B. If the current carried by resistor A is I, then what is the current carried by B?

(*a*) I (*b*) $2I$ (*c*) $I/2$ (*d*) $4I$ (*e*) $I/4$

65 •• When two identical resistors are connected in series across the terminals of a battery, the power delivered by the battery is 20 W. If these resistors are connected in parallel across the terminals of the same battery, what is the power delivered by the battery?

(*a*) 5 W (*b*) 10 W (*c*) 20 W (*d*) 40 W (*e*) 80 W

66 • (*a*) Find the equivalent resistance between points a and b in Figure 26-46. (*b*) If the potential drop between a and b is 12 V, find the current in each resistor.

Figure 26-46 Problem 66

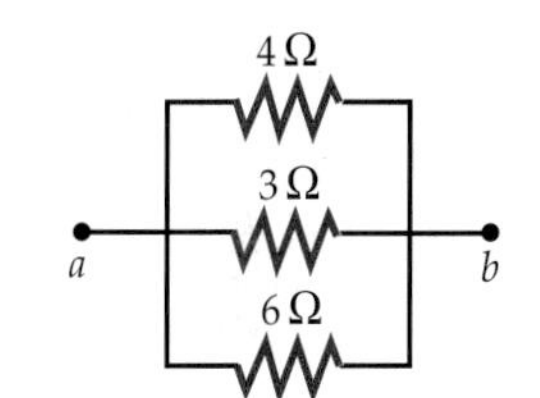

67 • Repeat Problem 66 for the resistor network shown in Figure 26-47.

Figure 26-47 Problem 67

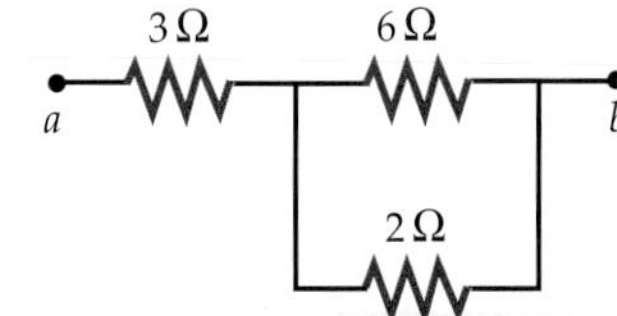

68 • Repeat Problem 66 for the resistor network shown in Figure 26-48.

Figure 26-48 Problems 68 and 69

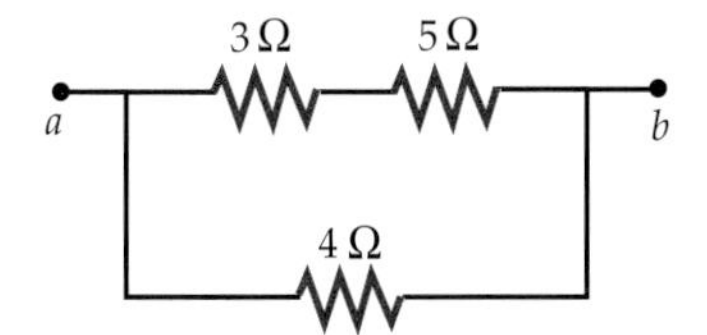

69 • In Figure 26-48, the current in the 4-Ω resistor is 4 A. (*a*) What is the potential drop between a and b? (*b*) What is the current in the 3-Ω resistor?

70 • (*a*) Show that the equivalent resistance between points a and b in Figure 26-49 is R. (*b*) What would be the effect of adding a resistance R between points c and d?

Figure 26-49 Problem 70

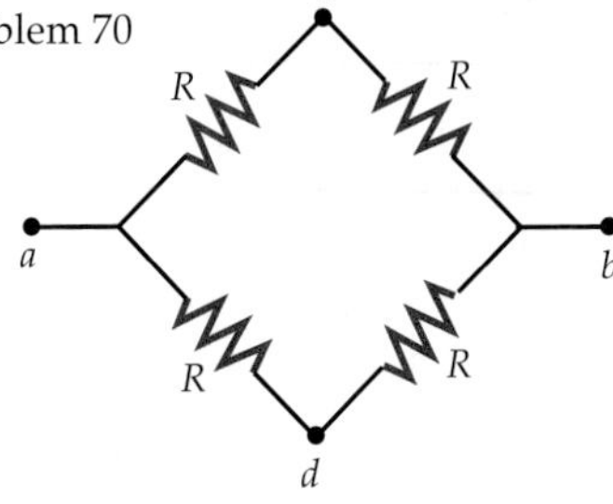

71 •• The battery in Figure 26-50 has negligible internal resistance. Find (*a*) the current in each resistor and (*b*) the power delivered by the battery.

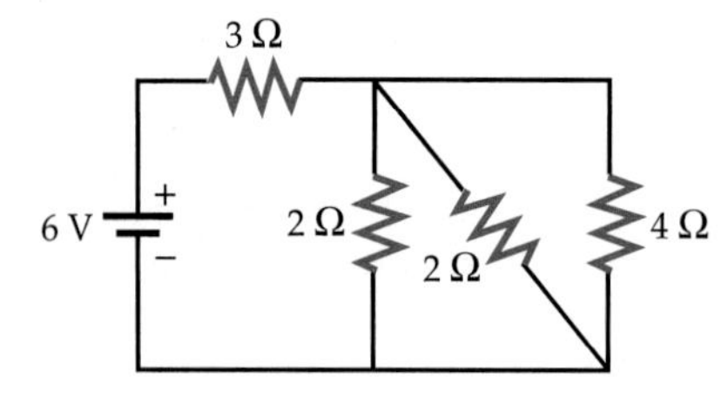

Figure 26-50
Problem 71

72 •• A battery has an emf $\mathcal{E}$ and an internal resistance r. When a 5.0-Ω resistor is connected across the terminals, the current is 0.5 A. When this resistor is replaced by an 11.0-Ω resistor, the current is 0.25 A. Find (*a*) the emf $\mathcal{E}$ and (*b*) the internal resistance r.

73 •• Consider the equivalent resistance of two resistors R_1 and R_2 connected in parallel as a function of the ratio $x = R_2/R_1$. (*a*) Show that $R_{eq} = R_1x/(1 + x)$. (*b*) Sketch a plot of R_{eq} as a function of x.

74 •• Repeat Problem 66 for the resistor network shown in Figure 26-51.

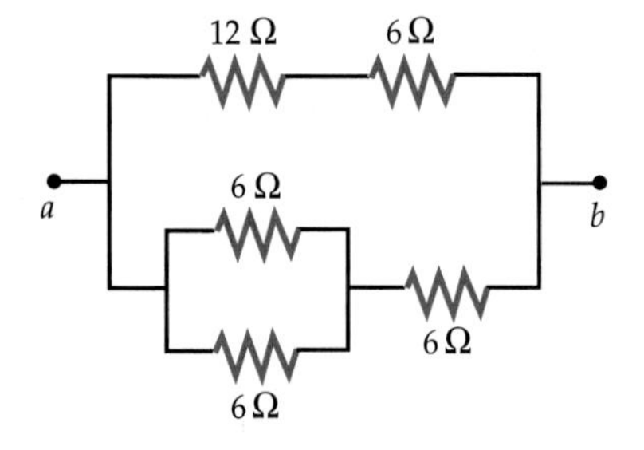

Figure 26-51
Problem 74

75 •• Repeat Problem 66 for the resistor network shown in Figure 26-52.

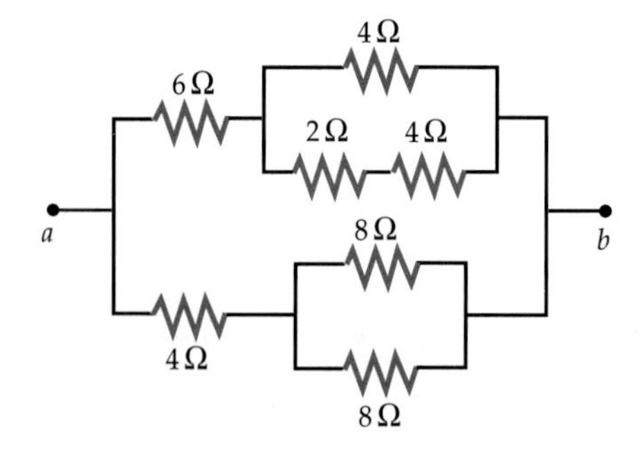

Figure 26-52
Problem 75

76 •• A length of wire has a resistance of 120 Ω. The wire is cut into N identical pieces which are then connected in parallel. The resistance of the parallel arrangement is 1.875 Ω. Find N.

77 •• A parallel combination of an 8-Ω resistor and an unknown resistor R is connected in series with a 16-Ω resistor and a battery. This circuit is then disassembled and the three resistors are then connected in series with each other and the same battery. In both arrangements, the current through the 8-Ω resistor is the same. What is the unknown resistance R?

78 •• For the resistance network shown in Figure 26-53, find (*a*) R_3 such that $R_{ab} = R_1$; (*b*) R_2 such that $R_{ab} = R_3$; and (*c*) R_1 such that $R_{ab} = R_1$.

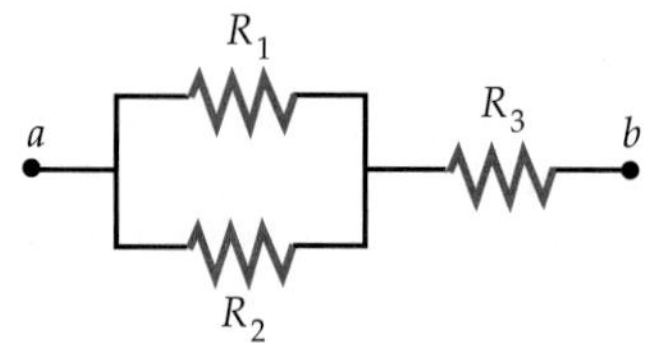

Figure 26-53
Problems 78 and 79

79 •• Check your results for Problem 78 using (*a*) $R_1 = 4\ \Omega$, $R_2 = 6\ \Omega$; (*b*) $R_1 = 4\ \Omega$, $R_3 = 3\ \Omega$; and (*c*) $R_2 = 6\ \Omega$, $R_3 = 3\ \Omega$.

80 ••• Nine 10-Ω resistors are connected as shown in Figure 26-54, and a potential difference of 20 V is applied between points a and b. (*a*) What is the equivalent resistance of this network? (*b*) Find the current in each of the nine resistors.

Figure 26-54
Problem 80

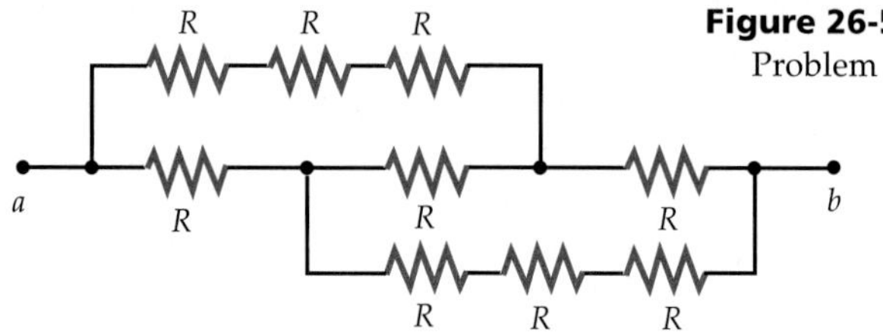

Kirchhoff's Rules

81 • Kirchoff's loop rule follows from

(*a*) conservation of charge.
(*b*) conservation of energy.
(*c*) Newton's laws.
(*d*) Coulomb's law.
(*e*) quantization of charge.

82 • In Figure 26-55, the emf is 6 V and $R = 0.5\ \Omega$. The rate of Joule heating in R is 8 W. (*a*) What is the current in the circuit? (*b*) What is the potential difference across R? (*c*) What is r?

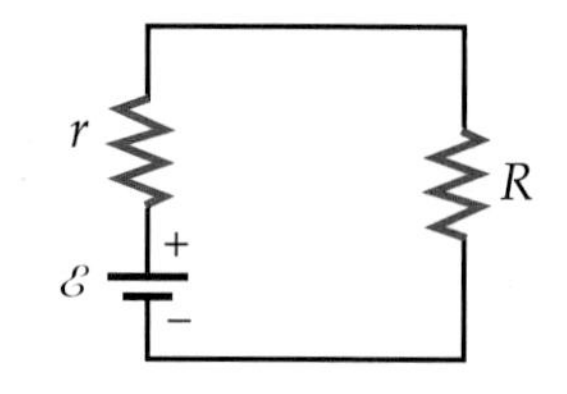

Figure 26-55 Problem 82

83 • For the circuit in Figure 26-56, find (*a*) the current, (*b*) the power delivered or absorbed by each emf, and (*c*) the rate of Joule heating in each resistor. (Assume that the batteries have negligible internal resistance.)

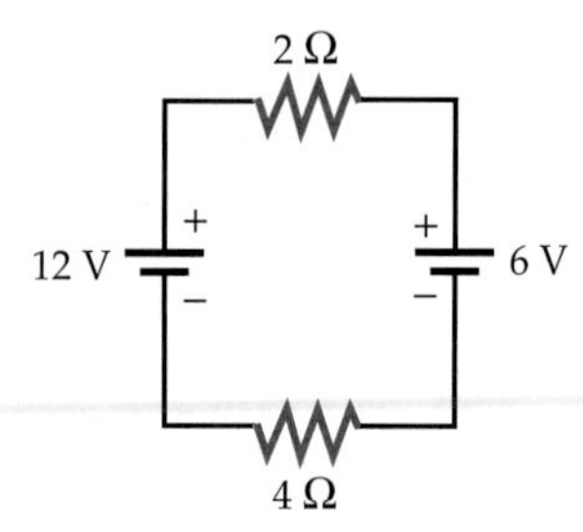

Figure 26-56 Problem 83

84 •• A sick car battery with an emf of 11.4 V and an internal resistance of 0.01 Ω is connected to a load of 2.0 Ω. To help the ailing battery, a second battery with an emf of 12.6 V and an internal resistance of 0.01 Ω is connected by jumper cables to the terminals of the first battery. (*a*) Draw a diagram of this circuit. (*b*) Find the current in each part of the circuit. (*c*) Find the power delivered by the second battery and discuss where this power goes, assuming that the emfs and internal resistances of both batteries remain constant.

85 •• In the circuit in Figure 26-57, the reading of the ammeter is the same with both switches open and both closed. Find the resistance R.

Figure 26-57 Problem 85

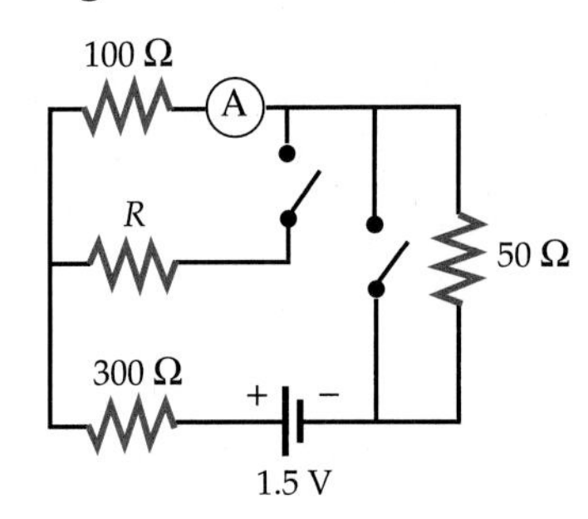

86 •• In the circuit in Figure 26-58, the batteries have negligible internal resistance, and the ammeter has negligible resistance. (*a*) Find the current through the ammeter. (*b*) Find the energy delivered by the 12-V battery in 3 s. (*c*) Find the total Joule heat produced in 3 s. (*d*) Account for the difference in your answers to parts (*b*) and (*c*).

Figure 26-58 Problem 86

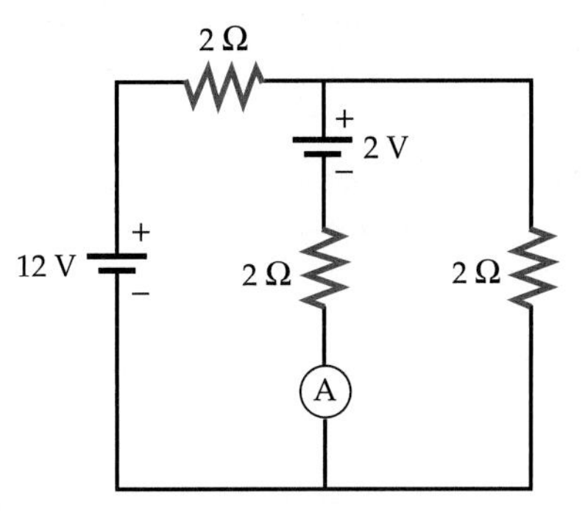

87 •• In the circuit in Figure 26-59, the batteries have negligible internal resistance. Find (*a*) the current in each resistor, (*b*) the potential difference between points *a* and *b*, and (*c*) the power supplied by each battery.

Figure 26-59 Problem 87

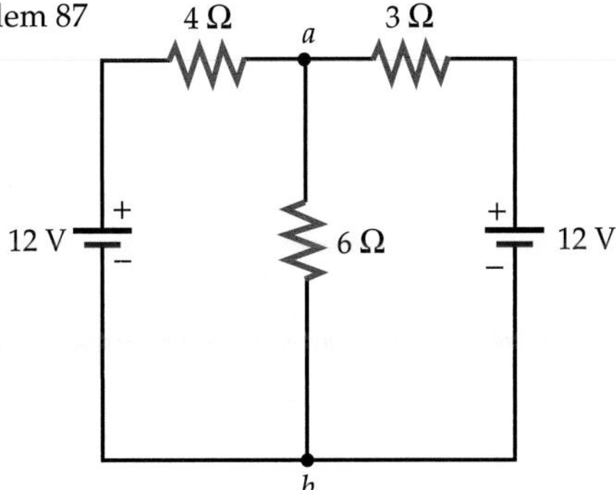

88 •• Repeat Problem 87 for the circuit in Figure 26-60.

Figure 26-60 Problem 88

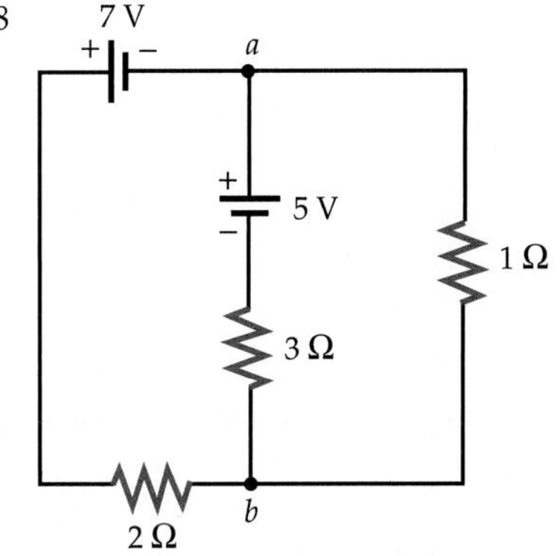

89 •• Two identical batteries, each with an emf $\mathcal{E}$ and an internal resistance r, can be connected across a resistance R either in series or in parallel. Is the power supplied to R greater when $R < r$ or when $R > r$?

90 •• For the circuit in Figure 26-61, find (*a*) the current in each resistor, (*b*) the power supplied by each emf, and (*c*) the power dissipated in each resistor.

Figure 26-61 Problem 90

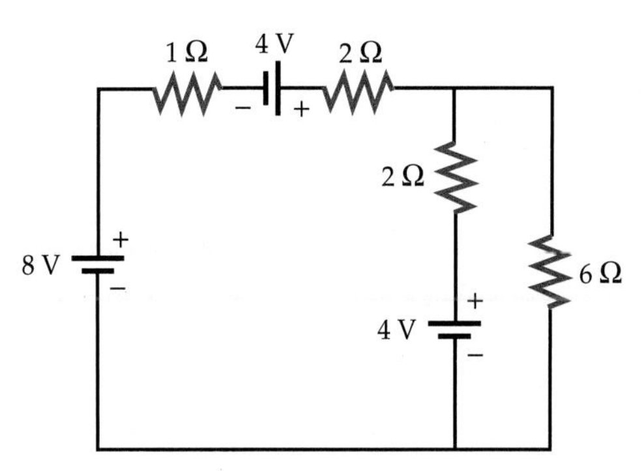

91 •• For the circuit in Figure 26-62, find the potential difference between points *a* and *b*.

Figure 26-62 Problem 91

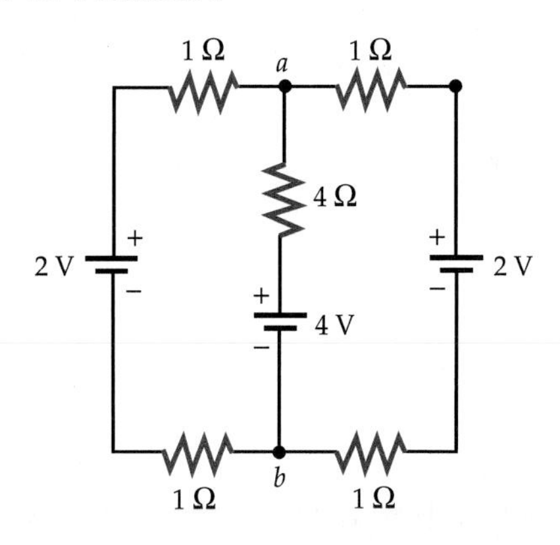

92 •• The battery in the circuit shown in Figure 26-63 has an internal resistance of 0.01 Ω. (*a*) An ammeter with a resistance of 0.01 Ω is inserted in series with the 0.74-Ω resistor at point *a*. What is the reading of the ammeter? (*b*) By what percentage is the current changed because of the ammeter? (*c*) The ammeter is removed and a voltmeter with a resistance of 1 kΩ is connected in parallel with the 0.74-Ω resistor from *a* to *b*. What is the reading of the voltmeter? (*d*) By what percentage is the voltage drop from *a* to *b* changed by the presence of the voltmeter?

Figure 26-63 Problem 92

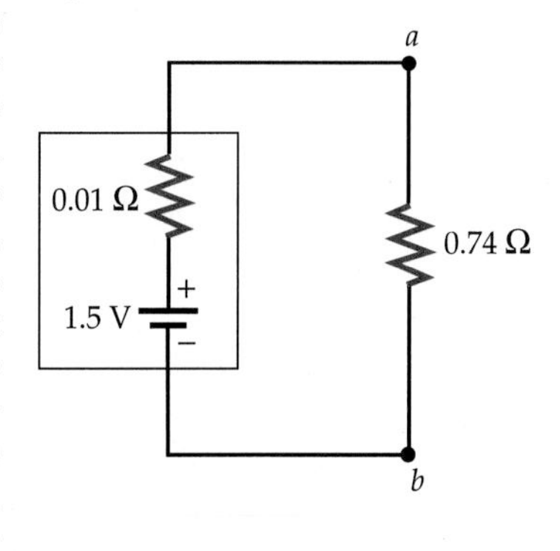

93 •• You have two batteries, one with $\mathcal{E} = 9.0$ V and $r = 0.8$ Ω and the other with $\mathcal{E} = 3.0$ V and $r = 0.4$ Ω. (*a*) Show how you would connect the batteries to give the largest current through a resistor R. Find the current for (*b*) $R = 0.2$ Ω, (*c*) $R = 0.6$ Ω, (*d*) $R = 1.0$ Ω, and (*e*) $R = 1.5$ Ω.

94 •• (*a*) Find the current in each part of the circuit shown in Figure 26-64. (*b*) Use your results from (*a*) to assign a potential at each indicated point assuming the potential at point *a* is zero.

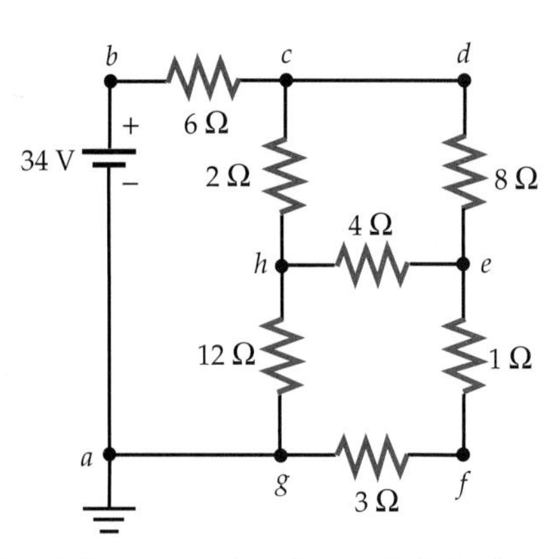

Figure 26-64 Problem 94

95 ••• In Problem 84, assume that the emf of the first battery increases at a constant rate of 0.2 V/h while the emf of the second battery and the internal resistances remain constant. (*a*) Find the current in each part of the circuit as a function of time. (*b*) Sketch a graph of the power delivered to the first battery as a function of time.

96 ••• (*a*) Find the current in each part of the circuit shown in Figure 26-65. (*b*) Use your results from (*a*) to assign a potential at each indicated point assuming the potential at point *a* is zero.

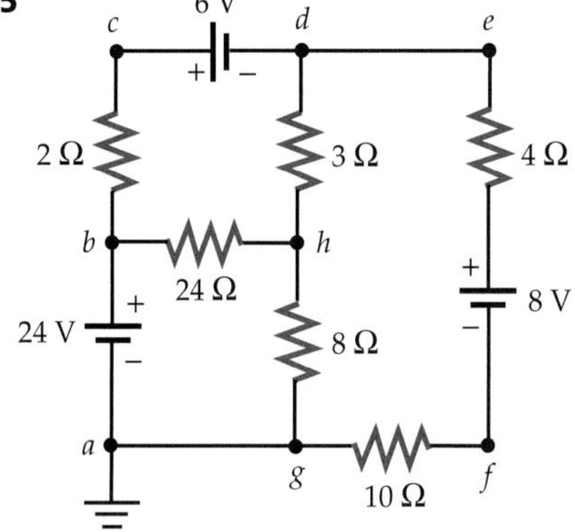

Figure 26-65 Problem 96

97 ••• Find the current in each resistor of the circuit shown in Figure 26-66.

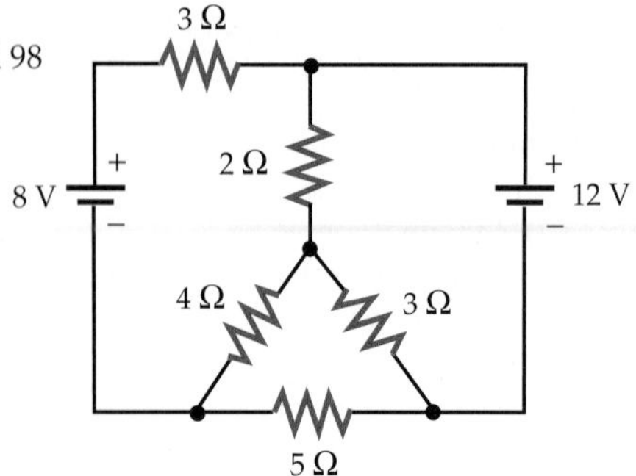

Figure 26-66 Problems 97 and 98

98 ••• Suppose that the emf of the left battery in Figure 26-66 is unknown but that the current delivered by the 12-V battery is known to be 0.6 A. Find the emf of the left battery and the current delivered by it.

RC Circuits

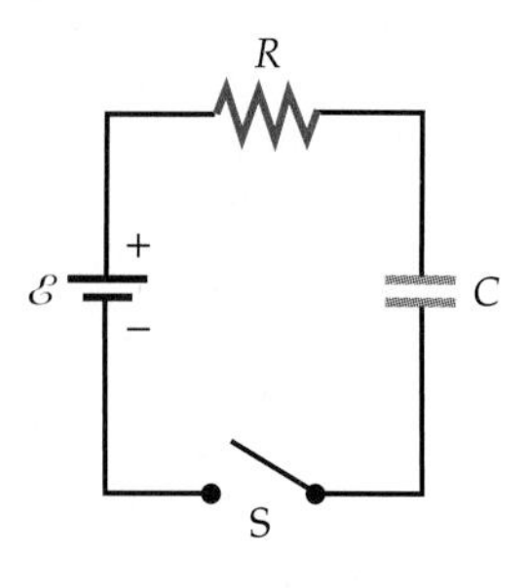

Figure 26-67 Problem 99

99 • The capacitor C in Figure 26-67 is initially uncharged. Just after the switch S is closed,

(*a*) the voltage across C equals $\mathcal{E}$.
(*b*) the voltage across R equals $\mathcal{E}$.
(*c*) the current in the circuit is zero.
(*d*) both (*a*) and (*c*) are correct.

100 •• During the time it takes to fully charge the capacitor of Figure 26-67,

(*a*) the energy supplied by the battery is $\frac{1}{2}C\mathcal{E}^2$.
(*b*) the energy dissipated in the resistor is $\frac{1}{2}C\mathcal{E}^2$.
(*c*) energy is dissipated in the resistor at a constant rate.
(*d*) the total charge flowing through the resistor is $\frac{1}{2}C\mathcal{E}$.

101 •• A battery is connected to a series combination of a switch, a resistor, and an initially uncharged capacitor. The switch is closed at $t = 0$. Which of the following statements is true?

(*a*) As the charge on the capacitor increases, the current increases.
(*b*) As the charge on the capacitor increases, the voltage drop across the resistor increases.
(*c*) As the charge on the capacitor increases, the current remains constant.
(*d*) As the charge on the capacitor increases, the voltage drop across the capacitor decreases.
(*e*) As the charge on the capacitor increases, the voltage drop across the resistor decreases.

102 •• A capacitor is discharging through a resistor. If it takes a time T for the charge on a capacitor to drop to half its initial value, how long does it take for the energy to drop to half its initial value?

103 •• A capacitor, resistor, and battery are connected in series. If R is doubled, how does this affect (*a*) the total energy stored, (*b*) the rate of energy storage, and (*c*) the time required to store $1/e$ of the final energy?

104 •• A capacitor, resistor, and battery are connected in series. If C is doubled, how does this affect (*a*) the total energy stored, (*b*) the rate of energy storage, and (*c*) the time required to store $1/e$ of the final energy?

105 • A 6-μF capacitor is charged to 100 V and is then connected across a 500-Ω resistor. (*a*) What is the initial charge on the capacitor? (*b*) What is the initial current just after the capacitor is connected to the resistor? (*c*) What is the time constant of this circuit? (*d*) How much charge is on the capacitor after 6 ms?

106 • (*a*) Find the initial energy stored in the capacitor of Problem 105. (*b*) Show that the energy stored in the capacitor is given by $U = U_0 e^{-2t/\tau}$, where U_0 is the initial energy and $\tau = RC$ is the time constant. (*c*) Sketch a plot of the energy U in the capacitor versus time t.

107 •• In the circuit of Figure 26-40, emf $\mathcal{E} = 50$ V and $C = 2.0$ μF; the capacitor is initially uncharged. At 4.0 s after switch S is closed, the voltage drop across the resistor is 20 V. Find the resistance of the resistor.

108 •• A 0.12-μF capacitor is given a charge Q_0. After 4 s, its charge is $\frac{1}{2}Q_0$. What is the effective resistance across this capacitor?

109 •• A 1.6-μF capacitor, initially uncharged, is connected in series with a 10-kΩ resistor and a 5.0-V battery of negligible internal resistance. (*a*) What is the charge on the capacitor after a very long time? (*b*) How long does it take the capacitor to reach 99% of its final charge?

110 •• Consider the circuit shown in Figure 26-68. From your knowledge of how capacitors behave in circuits, find (*a*) the initial current through the battery just after the switch is closed, (*b*) the steady-state current through the battery when the switch has been closed for a long time, and (*c*) the maximum voltage across the capacitor.

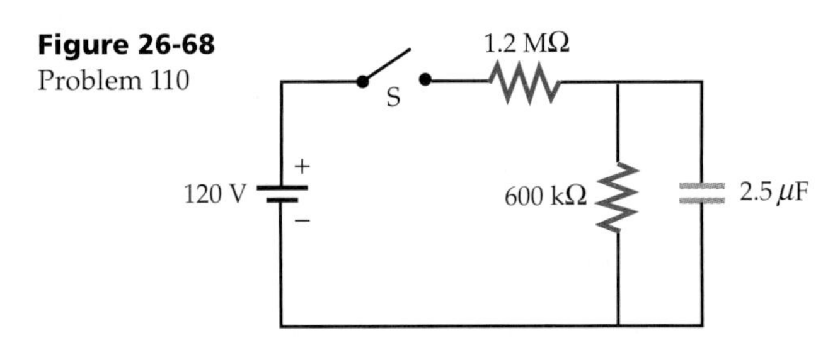

Figure 26-68 Problem 110

111 •• A 2-MΩ resistor is connected in series with a 1.5-μF capacitor and a 6.0-V battery of negligible internal resistance. The capacitor is initially uncharged. After a time $t = \tau = RC$, find (*a*) the charge on the capacitor, (*b*) the rate at which the charge is increasing, (*c*) the current, (*d*) the power supplied by the battery, (*e*) the power dissipated in the resistor, and (*f*) the rate at which the energy stored in the capacitor is increasing.

112 •• Repeat Problem 111 for the time $t = 2\tau$.

113 •• In the steady state, the charge on the 5-μF capacitor in the circuit in Figure 26-69 is 1000 μC. (*a*) Find the battery current. (*b*) Find the resistances R_1, R_2, and R_3.

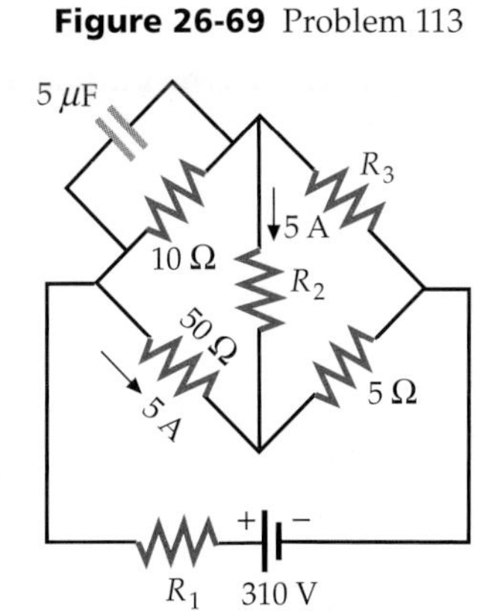

Figure 26-69 Problem 113

114 •• (*a*) What is the voltage across the capacitor in the circuit in Figure 26-70? (*b*) If the battery is disconnected, give the capacitor current as a function of time. (*c*) How long does it take the capacitor to discharge until the potential difference across it is 1 V?

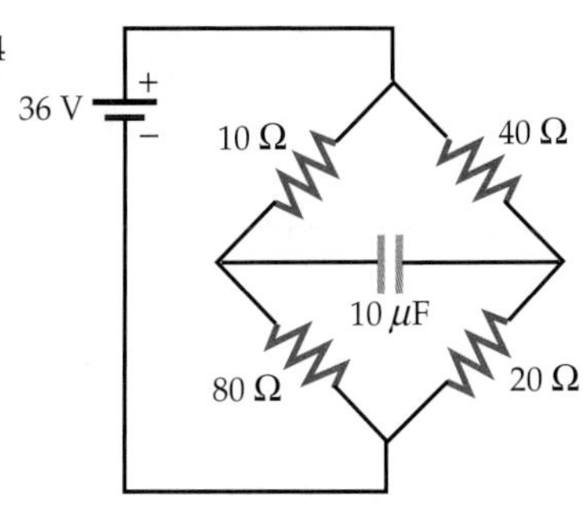

Figure 26-70 Problem 114

115 •• Show that Equation 26-34 can be written

$$\frac{dQ}{\mathcal{E}C - Q} = \frac{dt}{RC}$$

Integrate this equation to derive the solution given by Equation 26-35.

116 ••• A photojournalist's flash unit uses a 9.0-V battery pack to charge a 0.15-μF capacitor, which is then discharged through the flash lamp of 10.5-Ω resistance when a switch is closed. The minimum voltage necessary for the flash discharge is 7.0 V. The capacitor is charged through a 18-kΩ resistor. (*a*) How much time is required to charge the capacitor to the required 7.0 V? (*b*) How much energy is released when the lamp flashes? (*c*) How much energy is supplied by the battery during the charging cycle and what fraction of that energy is dissipated in the resistor?

117 ••• For the circuit in Figure 26-71, (*a*) what is the initial battery current immediately after switch S is closed? (*b*) What is the battery current a long time after switch S is closed? (*c*) What is the current in the 600-Ω resistor as a function of time?

Figure 26-71 Problem 117

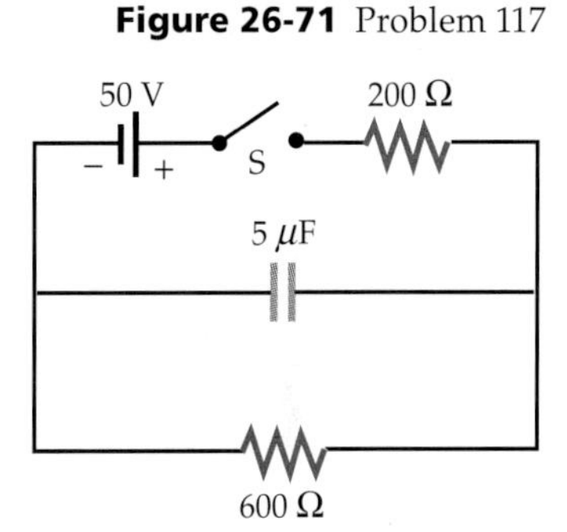

118 ••• For the circuit in Figure 26-72, (*a*) what is the initial battery current immediately after switch S is closed? (*b*) What is the battery current a long time after switch S is closed? (*c*) If the switch has been closed for a long time and is then opened, find the current through the 600-kΩ resistor as a function of time.

Figure 26-72 Problem 118

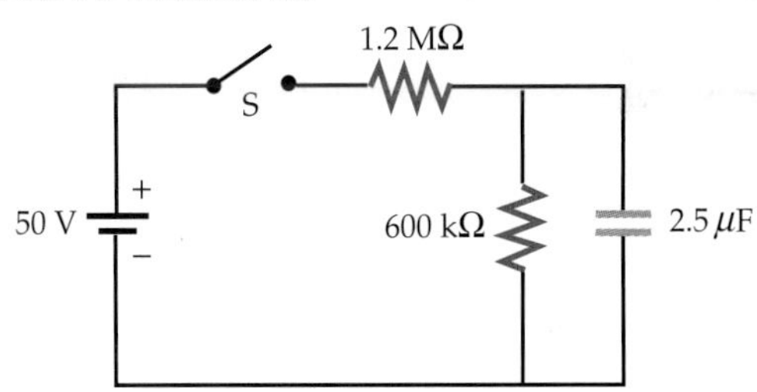

119 ••• In the circuit shown in Figure 26-73, the capacitor has a capacitance of 2.5 μF and the resistor a resistance of 0.5 MΩ. Before the switch is closed, the potential drop across the capacitor is 12 V, as shown. Switch S is closed at $t = 0$. (*a*) What is the current in R immediately after S is closed? (*b*) At what time t is the voltage across the capacitor 24 V?

Figure 26-73 Problems 119 and 120

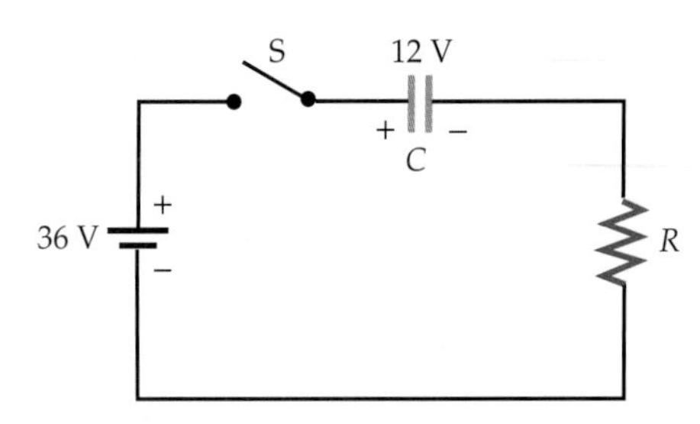

120 ••• Repeat Problem 119 if the capacitor is connected with reversed polarity.

General Problems

121 • A flash lamp is set off by the discharge of a capacitor that has been charged by a battery. Why not just connect the battery directly to the lamp?

122 • Which will produce more thermal energy when connected across an ideal battery, a small resistance or a large resistance?

123 • Do Kirchhoff's rules apply to circuits containing capacitors?

124 •• True or false:

(*a*) Ohm's law is $R = V/I$.
(*b*) Electrons drift in the direction of the current.
(*c*) A source of emf supplies power to an electrical circuit.
(*d*) When the potential drops by V in a segment of a circuit, the power supplied to that segment is IV.
(*e*) The equivalent resistance of two resistors in parallel is always less than the resistance of either resistor alone.
(*f*) The terminal voltage of a battery always equals its emf.
(*g*) The terminal voltage of a battery is always less than its emf.

125 •• In Figure 26-74, all three resistors are identical. The power dissipated is

(*a*) the same in R_1 as in the parallel combination of R_2 and R_3.
(*b*) the same in R_1 and R_2.
(*c*) greatest in R_1.
(*d*) smallest in R_1.

Figure 26-74 Problems 125 and 126

126 •• In Figure 26-74, $R_1 = 4\ \Omega$, $R_2 = 6\ \Omega$, and $R_3 = 12\ \Omega$. If we denote the currents through these resistors by I_1, I_2, and I_3, respectively, then

(*a*) $I_1 > I_2 > I_3$. (*b*) $I_2 = I_3$. (*c*) $I_3 > I_2$.
(*d*) none of the above is correct.

127 •• A 25-W light bulb is connected in series with a 100-W light bulb and a voltage V is placed across the combination. Which bulb is brighter? Explain.

128 • If the battery emf in Figure 26-74 is 24 V, then

(*a*) $I_2 = 4$ A. (*b*) $I_2 = 2$ A. (*c*) $I_2 = 1$ A.
(*d*) none of the above is correct.

129 • A 10.0-Ω resistor is rated as being capable of dissipating 5.0 W of power. (*a*) What maximum current can this resistor tolerate? (*b*) What voltage across this resistor will produce the maximum current?

130 • Margaret is economizing by turning off her space heater and warming herself with a toaster. She pushes the toaster plunger down and dozes off, but after 4 min it pops up again. Eventually the cold wakes her up, so she pushes the plunger down again and gets a little more sleep. This happens once every 15 min, with the toaster engaged for 4 min each time. It is a poor night's sleep, but she is determined to save money. Energy costs 9 cents per kilowatt-hour, and a 120-V source is used. (*a*) How much does it cost to operate an electric toaster for 4 min if its resistance is 11.0 Ω? (*b*) How much would it cost to operate a 5.0-Ω-heater connected across 120 V for 8 h?

131 • A 12-V car battery has an internal resistance of 0.4 Ω. (*a*) What is the current if the battery is shorted momentarily? (*b*) What is the terminal voltage when the battery delivers a current of 20 A to start the car?

132 •• The current drawn from a battery is 1.80 A when a 7.0-Ω resistor is connected across the battery terminals. If a second 12-Ω resistor is connected in parallel with the 7-Ω resistor, the battery delivers a current of 2.20 A. What are the emf and internal resistance of the battery?

133 •• A 16-gauge copper wire insulated with rubber can safely carry a maximum current of 6 A. (*a*) How great a potential difference can be applied across 40 m of this wire? (*b*) Find the electric field in the wire when it carries a current of 6 A. (*c*) Find the power dissipated in the wire when it carries a current of 6 A.

134 •• An automobile jumper cable 3 m long is constructed of multiple strands of copper wire that has an equivalent cross-sectional area of 10.0 mm^2. (*a*) What is the resistance of the jumper cable? (*b*) When the cable is used to start a car, it carries a current of 90 A. What is the potential drop that occurs across the jumper cable? (*c*) How much power is dissipated in the jumper cable?

135 •• A coil of Nichrome wire is to be used as the heating element in a water boiler that is required to generate 8.0 g of steam per second. The wire has a diameter of 1.80 mm and is connected to a 120-V power supply. Find the length of wire required.

136 •• A closed box has two metal terminals a and b. The inside of the box contains an unknown emf $\mathcal{E}$ in series with a resistance R. When a potential difference of 21 V is maintained between a and b, there is a current of 1 A between the terminals a and b. If this potential difference is reversed, a current of 2 A in the reverse direction is observed. Find $\mathcal{E}$ and R.

137 •• The capacitors in the circuit in Figure 26-75 are initially uncharged. (*a*) What is the initial value of the battery current when switch S is closed? (*b*) What is the battery current after a long time? (*c*) What are the final charges on the capacitors?

Figure 26-75 Problem 137

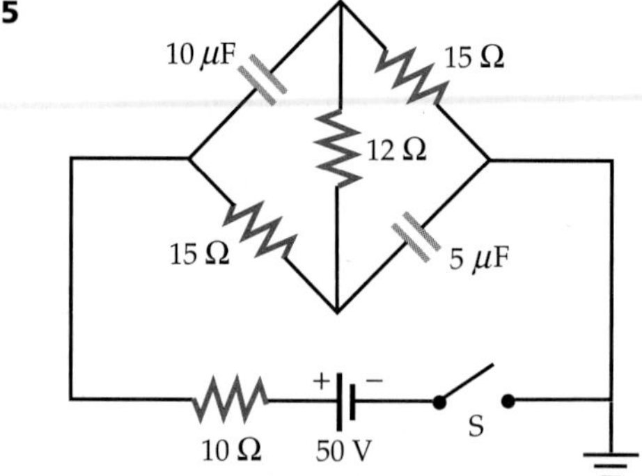

138 •• The circuit in Figure 26-76 is a slide-type *Wheatstone bridge.* It is used for determining an unknown resistance R_x in terms of the known resistances R_1, R_2, and R_0. The resistances R_1 and R_2 comprise a wire 1 m long. Point a is a sliding contact that is moved along the wire to vary these resistances. Resistance R_1 is proportional to the distance from the left end of the wire (labeled 0 cm) to point a, and R_2 is proportional to the distance from point a to the right end of the wire (labeled 100 cm). The sum of R_1 and R_2 remains constant. When points a and b are at the same potential, there is no current in the galvanometer and the bridge is said to be balanced. (Since the galvanometer is used to detect the absence of a current, it is called a *null detector.*) If the fixed resistance $R_0 = 200\ \Omega$, find the unknown resistance R_x if (*a*) the bridge balances at the 18-cm mark, (*b*) the bridge balances at the 60-cm mark, and (*c*) the bridge balances at the 95-cm mark.

Figure 26-76 Problems 138 and 139

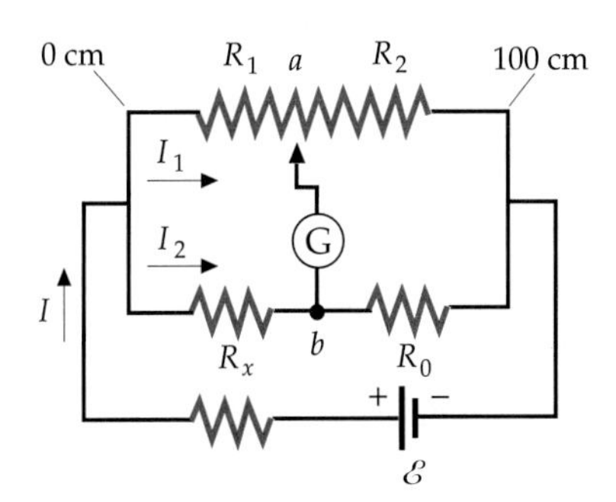

139 •• For the Wheatstone bridge of Problem 138, the bridge balances at the 98-cm mark when $R_0 = 200\ \Omega$. (*a*) What is the unknown resistance? (*b*) What effect would an error of 2 mm in the location of the balance point have on the measured value of the unknown resistance? (*c*) How should R_0 be changed so that the balance point for this unknown resistor will be nearer the 50-cm mark?

140 •• The wires in a house must be large enough in diameter so that they do not get hot enough to start a fire. Suppose a certain wire is to carry a current of 20 A, and it is determined that the Joule heating of the wire should not exceed 2 W/m. What diameter must a copper wire have to be safe for this current?

141 •• You are given n identical cells, each with emf $\mathcal{E}$ and internal resistance $r = 0.2\ \Omega$. When these cells are connected in parallel to form a battery, and a resistance R is connected to the battery terminal, the current through R is the same as when the cells are connected in series and R is attached to the terminals of that battery. Find the value of the resistor R.

142 •• A cyclotron produces a 3.50-μA proton beam of 60-MeV energy. The protons impinge and come to rest inside a 50-g copper target within the vacuum chamber. (*a*) Determine the number of protons that strike the target per second. (*b*) Find the energy deposited in the target per second. (*c*) How much time elapses before the target temperature rises 300°C? (Neglect cooling by radiation.)

143 •• Compact fluorescent light bulbs cost \$6 each and have an expected lifetime of 8000 h. These bulbs consume 20 W of power, but produce the illumination equivalent to 75-W incandescent bulbs. Incandescent bulbs cost about \$1.50 each and have an expected lifetime of 1200 h. If the average household has, on the average, six 75-W incandescent light bulbs on constantly, and if energy costs 11.5 cents per kilowatt-hour, how much money would a consumer save each year by installing the energy-efficient fluorescent light bulbs?

144 •• The space between the plates of a parallel-plate capacitor is filled with a dielectric of constant κ and resistivity ρ. (*a*) Show that the time constant for the decrease of charge on the plates is $\tau = \epsilon_0 \kappa \rho$. (*b*) If the dielectric is mica, for which $\kappa = 5.0$ and $\rho = 9 \times 10^{13}\ \Omega\cdot$m, find the time it takes for the charge to decrease to $1/e^2 \approx 14\%$ of its initial value.

145 •• The belt of a Van de Graaff generator carries a surface charge density of 5 mC/m^2. The belt is 0.5 m wide and moves at 20 m/s. (*a*) What current does it carry? (*b*) If this charge is raised to a potential of 100 kV, what is the minimum power of the motor needed to drive the belt?

146 •• Conventional large electromagnets use water cooling to prevent excessive heating of the magnet coils. A large laboratory electromagnet draws 100 A when a voltage of 240 V is applied to the terminals of the energizing coils. To cool the coils, water at an initial temperature of 15°C is circulated through the coils. How many liters per second must pass through the coils if their temperature should not exceed 50°C?

147 ••• We show in Figure 26-77 the basis of the sweep circuit used in an oscilloscope. S is an electronic switch that closes whenever the potential across its terminals reaches a value V_c and opens when the potential has dropped to 0.2 V. The emf $\mathcal{E}$, much greater than V_c, charges the capacitor C through a resistor R_1. The resistor R_2 represents the small but finite resistance of the electronic switch. In a typical circuit, $\mathcal{E} = 800$ V, $V_c = 4.2$ V, $R_2 = 0.001\ \Omega$, $R_1 = 0.5$ MΩ ($0.5 \times 10^6\ \Omega$), and $C = 0.02\ \mu$F. (*a*) What is the time constant for charging of the capacitor C? (*b*) Show that in the time required to bring the potential across S to the critical potential $V_c = 4.2$ V, the voltage across the capacitor increases almost linearly with time. (*Hint:* Use the expansion of the exponential for small values of exponent.) (*c*) What should be the value of R_1 so that C charges from 0.2 to 4.2 V in 0.1 s? (*d*) How much time elapses during the discharge of C through switch S? (*e*) At what rate is power dissipated in the resistor R_1 and in the switch resistance?

Figure 26-77 Problem 147

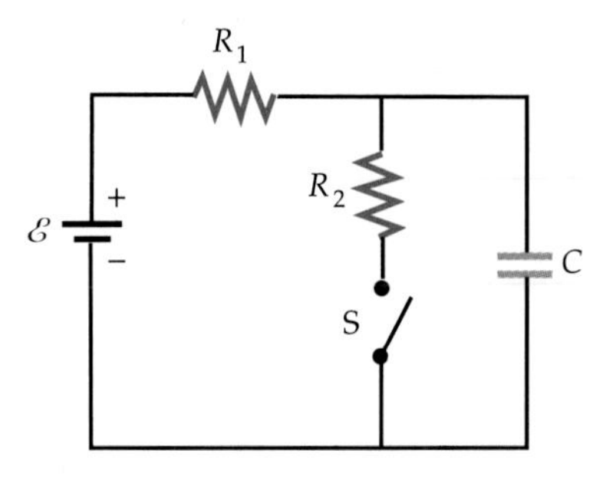

148 ••• In the circuit shown in Figure 26-78, $R_1 = 2.0\ \text{M}\Omega$, $R_2 = 5.0\ \text{M}\Omega$, and $C = 1.0\ \mu\text{F}$. At $t = 0$, switch S is closed, and at $t = 2.0$ s, switch S is opened. (*a*) Sketch the voltage across C and the current through R_2 between $t = 0$ and $t = 10$ s. (*b*) Find the voltage across the capacitor at $t = 2$ s and at $t = 8$ s.

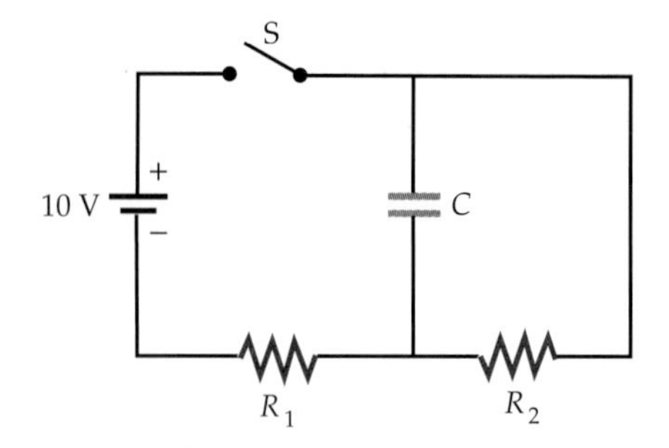

Figure 26-78 Problem 148

149 ••• If the capacitor in the circuit in Figure 26-70 is replaced by a 30-Ω resistor, what currents flow through the resistors?

150 ••• Two batteries with emfs $\mathcal{E}_1$ and $\mathcal{E}_2$ and internal resistances r_1 and r_2 are connected in parallel. Prove that if a resistor is connected in parallel with this combination, the optimal load resistance (the resistance at which maximum power is delivered) is $R = r_1r_2/(r_1 + r_2)$.

151 ••• Capacitors C_1 and C_2 are connected in parallel by a resistor and two switches as shown in Figure 26-79. Capacitor C_1 is initially charged to a voltage V_0, and capacitor C_2 is uncharged. The switches S are then closed. (*a*) What are the final charges on C_1 and C_2? (*b*) Compare the initial and final stored energies of the system. (*c*) What caused the decrease in the capacitor-stored energy?

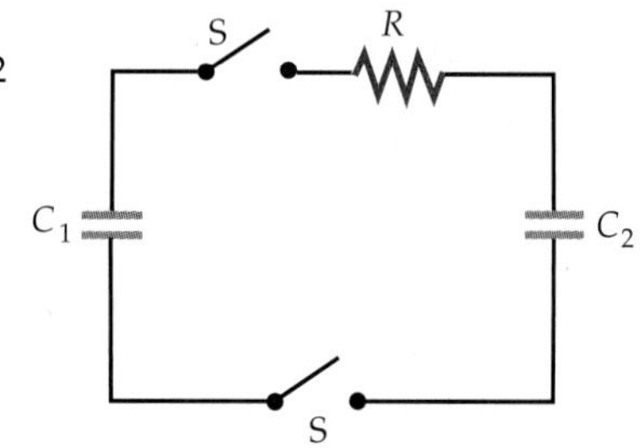

Figure 26-79 Problems 151 and 152

152 ••• (*a*) In Problem 151, find the current through R after the switches S are closed as a function of time. (*b*) Find the energy dissipated in the resistor as a function of time. (*c*) Find the total energy dissipated in the resistor and compare it with the loss of stored energy found in part (*b*) of Problem 151.

153 ••• In the circuit in Figure 26-80, the capacitors are initially uncharged. Switch S_2 is closed and then switch S_1 is closed. (*a*) What is the battery current immediately after S_1 is closed? (*b*) What is the battery current a long time after both switches are closed? (*c*) What is the final voltage across C_1? (*d*) What is the final voltage across C_2? (*e*) Switch S_2 is opened again after a long time. Give the current in the 150-Ω resistor as a function of time.

Figure 26-80 Problem 153

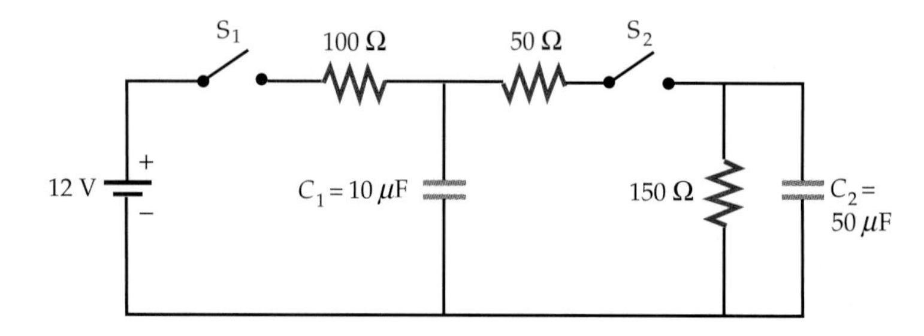

154 ••• In the *RC* circuit in Figure 26-40*a*, the capacitor is initially uncharged and the switch is closed at time $t = 0$. (*a*) What is the power supplied by the battery as a function of time? (*b*) What is the power dissipated in the resistor as a function of time? (*c*) What is the rate at which energy is stored in the capacitor as a function of time? Plot your answers to parts (*a*), (*b*), and (*c*) versus time on the same graph. (*d*) Find the maximum rate at which energy is stored in the capacitor as a function of the battery voltage *e* and the resistance *R*. At what time does this maximum occur?

155 ••• A linear accelerator produces a pulsed beam of electrons. The current is 1.6 A for the 0.1-μs duration of each pulse. (*a*) How many electrons are accelerated in each pulse? (*b*) What is the average current of the beam if there are 1000 pulses per second? (*c*) If each electron acquires an energy of 400 MeV, what is the average power output of the accelerator? (*d*) What is the peak power output? (*e*) What fraction of the time is the accelerator actually accelerating electrons? (This is called the *duty factor* of the accelerator.)

tribution. Because the behavior of this electron gas is so different from a gas of molecules, the electron gas is often called a **Fermi electron gas**. The main features of a Fermi electron gas can be understood by considering an electron in a metal to be a particle in a box, a problem whose one-dimensional version we studied extensively in Chapter 17. We discuss the main features of a Fermi electron gas semiquantitatively in this section and leave the details of the Fermi–Dirac distribution to Section 27-6.

Energy Quantization in a Box

In Chapter 17 we found that the wavelength associated with an electron of momentum p is given by the de Broglie relation:

$$\lambda = \frac{h}{p} \tag{27-10}$$

where h is Planck's constant. When a particle is confined to a finite region of space such as a box, only certain wavelengths λ_n given by standing-wave conditions are allowed. For a one-dimensional box of length L, the standing-wave condition is

$$n\frac{\lambda_n}{2} = L \tag{27-11}$$

This results in the quantization of energy:

$$E_n = \frac{p_n^2}{2m} = \frac{(h/\lambda_n)^2}{2m} = \frac{h^2}{2m}\frac{1}{\lambda_n^2} = \frac{h^2}{2m}\frac{1}{(2L/n)^2}$$

or

$$E_n = n^2\frac{h^2}{8mL^2} \tag{27-12}$$

The wave function for the nth state is given by

$$\psi_n(x) = \sqrt{\frac{2}{L}}\sin\frac{n\pi x}{L} \tag{27-13}$$

The quantum number n characterizes the wave function for a particular state and the energy of that state. In three-dimensional problems, three quantum numbers arise, one associated with each dimension.

The Pauli Exclusion Principle

The distribution of electrons among the possible energy states is dominated by the exclusion principle, which was first enunciated by Wolfgang Pauli in 1925 to explain the electronic structure of atoms:

> No two electrons in an atom can be in the same quantum state; that is, they cannot have the same set of values for their quantum numbers.

Pauli exclusion principle

The exclusion principles applies to all "spin one-half" particles, which include electrons, protons, and neutrons.* These particles have a *spin* quantum number m_s which has two possible values, $+\frac{1}{2}$ and $-\frac{1}{2}$. The quantum state of

* Intrinsic spin is discussed briefly in Chapter 10, Section 5.

optional

a particle is characterized by the spin quantum number m_s, plus the quantum numbers associated with the spatial part of the wave function. Because the spin quantum numbers have just two possible values, the exclusion principle can be stated in terms of the spatial states:

> There can be at most two electrons with the same set of values for their *spatial* quantum numbers.

Exclusion principle in terms of spatial states

When there are more than two electrons in a system such as an atom or metal, only two can be in the lowest energy state. The third and fourth must go into the second-lowest state, and so on.

Particles that obey the exclusion principle are called **fermions**. Other particles such as α particles, deuterons, photons, and mesons do not obey the exclusion principle. These particles are called **bosons** and have either zero intrinsic spin, or integral spin quantum numbers.

Example 27-1

Compare the total energy of the ground state of 5 identical bosons of mass m in a one-dimensional box with that of 5 identical fermions of mass m in the same box.

Picture the Problem The ground state is the lowest possible energy state. The energy levels in a one-dimensional box are given by $E_n = n^2E_1$, where $E_1 = (h^2/8mL^2)$. The lowest energy for 5 bosons occurs when all the bosons are in the state $n = 1$ as shown in Figure 27-2*a*. For fermions, the lowest state occurs with two in state $n = 1$, two in $n = 2$, and one in $n = 3$ as shown in Figure 27-2*b*.

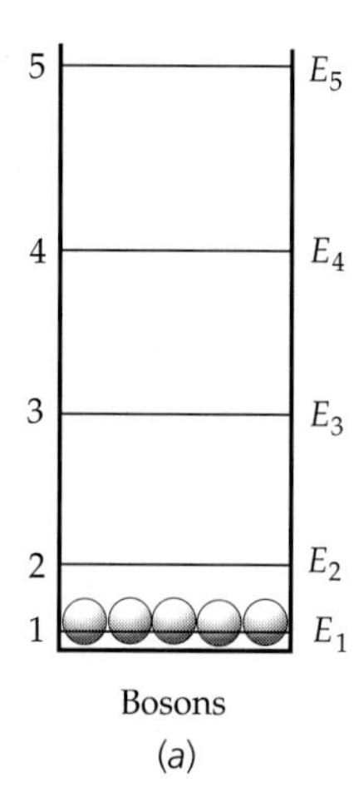

(*a*)

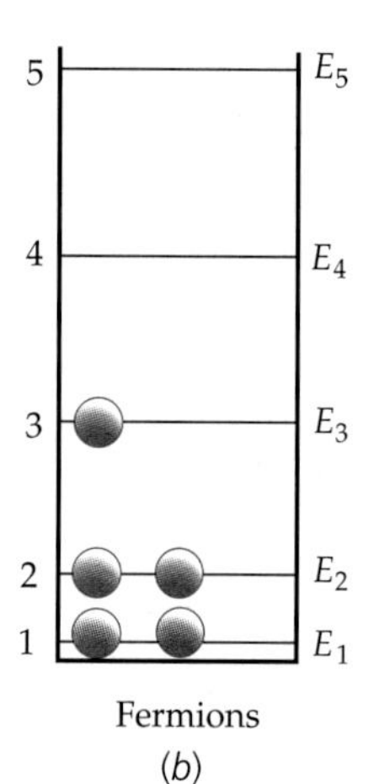

(*b*)

Figure 27-2

1. The energy of 5 bosons in the state $n = 1$ is: $E = 5E_1$

2. The energy of two fermions in the state $n = 1$, two in state $n = 2$, and one in state $n = 3$ is:

$$E = 2E_1 + 2E_2 + 1E_3 = 2E_1 + 2(2)^2E_1 + 1(3)^2E_1$$
$$= 2E_1 + 8E_1 + 9E_1 = 19E_1$$

Remark We see that the exclusion principle has a large effect on the energy of a multiple-particle system.

The Fermi Energy

When there are many electrons in a box, at $T = 0$ the electrons will occupy the lowest energy states consistent with the exclusion principle. If we have N electrons, we can put two electrons in the lowest energy level, two in the next lowest, and so on. The N electrons thus fill up the lowest $N/2$ energy levels

optional

(Figure 27-3). The energy of the last filled (or half-filled) level at $T = 0$ is called the **Fermi energy** E_F. If the electrons moved in a one-dimensional box, the Fermi energy would be given by Equation 27-12 with $n = N/2$:

$$E_F = \left(\frac{N}{2}\right)^2 \frac{h^2}{8m_eL^2} = \frac{h^2}{32m_e}\left(\frac{N}{L}\right)^2 \qquad 27\text{-}14$$

Fermi energy at $T = 0$ in one dimension

In a one-dimensional box, the Fermi energy depends on the number of free electrons per unit length of the box.

Exercise Suppose there is an ion, and therefore a free electron, every 0.1 nm in a one-dimensional box. Calculate the Fermi energy. (*Hint:* Write Equation 27-14 as

$$E_F = \frac{(hc)^2}{32m_ec^2}\left(\frac{N}{L}\right)^2 = \frac{(1240\text{ eV}\cdot\text{nm})^2}{32(0.511\text{ MeV})}\left(\frac{N}{L}\right)^2$$

(*Answer* $E_F = 9.4$ eV)

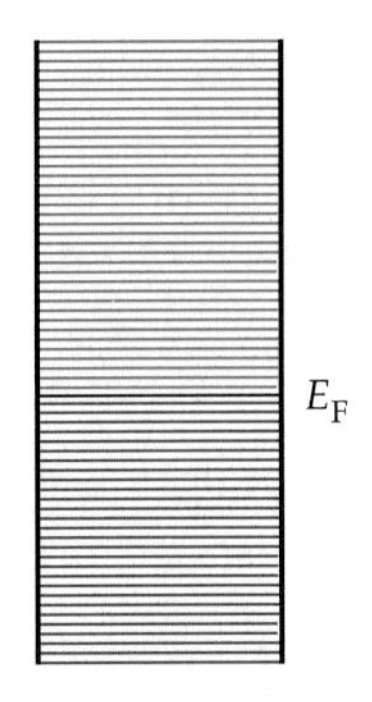

Figure 27-3 At $T = 0$ the electrons fill up the allowed energy states to the Fermi energy E_F. The levels are so closely spaced they can be assumed to be continuous.

In our model of conduction, the free electrons move in a *three-dimensional* box of volume V. The derivation of the Fermi energy in three dimensions is somewhat difficult, so we will just give the result. In three dimensions, the Fermi energy at $T = 0$ is given by

$$E_F = \frac{h^2}{8m_e}\left(\frac{3N}{\pi V}\right)^{2/3} \qquad 27\text{-}15a$$

Fermi energy at $T = 0$ in three dimensions

The Fermi energy depends on the number of electrons per unit volume (the number density) N/V. Substituting numerical values for the constants gives

$$E_F = (0.365\text{ eV}\cdot\text{nm}^2)\left(\frac{N}{V}\right)^{2/3} \qquad 27\text{-}15b$$

Fermi energy at $T = 0$ in three dimensions

Table 27-1 lists the free-electron number densities and Fermi energies at $T = 0$ for several metals.

Table 27-1

Free-Electron Number Densities and Fermi Energies at $T = 0$ for Selected Elements

	Element	N/V, Electrons/cm^3	E_F, eV
Al	Aluminum	18.1×10^{22}	11.7
Ag	Silver	5.86×10^{22}	5.50
Au	Gold	5.90×10^{22}	5.53
Cu	Copper	8.47×10^{22}	7.04
Fe	Iron	17.0×10^{22}	11.2
K	Potassium	1.4×10^{22}	2.11
Li	Lithium	4.70×10^{22}	4.75
Mg	Magnesium	8.60×10^{22}	7.11
Mn	Manganese	16.5×10^{22}	11.0
Na	Sodium	2.65×10^{22}	3.24
Sn	Tin	14.8×10^{22}	10.2
Zn	Zinc	13.2×10^{22}	9.46

optional

Example 27-2

The number density for electrons in copper was calculated in Example 26-1 and found to be $8.47 \times 10^{22}/\text{cm}^3$. Calculate the Fermi energy at $T = 0$ for copper.

1. The Fermi energy is given by Equation 27-15: $E_F = (0.365\ \text{eV}\cdot\text{nm}^2)\left(\frac{N}{V}\right)^{2/3}$

2. Substitute the given number density for copper: $E_F = (0.365\ \text{eV}\cdot\text{nm}^2)(84.7/\text{nm}^3)^{2/3} = 7.04\ \text{eV}$

Remark Note that the Fermi energy is much greater than kT at ordinary temperatures. For example, at $T = 300$ K, kT is only about 0.026 eV.

Exercise Use Equation 27-15*b* to calculate the Fermi energy at $T = 0$ for gold, which has a number density of 5.90×10^{22}. (*Answer* 5.53 eV)

The average energy of a free electron can be calculated from the complete energy distribution of the electrons, which is discussed in Section 27-6. At $T = 0$, the average energy turns out to be

$$E_{av} = \tfrac{3}{5}E_F \qquad \text{27-16}$$

Average energy of electrons in a Fermi gas at $T = 0$

For copper, E_{av} is about 4 eV. This average energy is huge compared with typical thermal energies of about $kT \approx 0.026$ eV at a normal temperature of $T = 300$ K. This result is very different from the classical, Maxwell–Boltzmann distribution result that at $T = 0$, $E = 0$, and that at some temperature T, E is of the order of kT.

The Fermi Factor at $T = 0$

The probability of an energy state being occupied is called the **Fermi factor**, $f(E)$. At $T = 0$ all the states below E_F are filled, whereas all those above this energy are empty, as shown in Figure 27-4. Thus, at $T = 0$ the Fermi factor is simply

$$f(E) = 1, \quad E < E_F$$
$$f(E) = 0, \quad E > E_F \qquad \text{27-17}$$

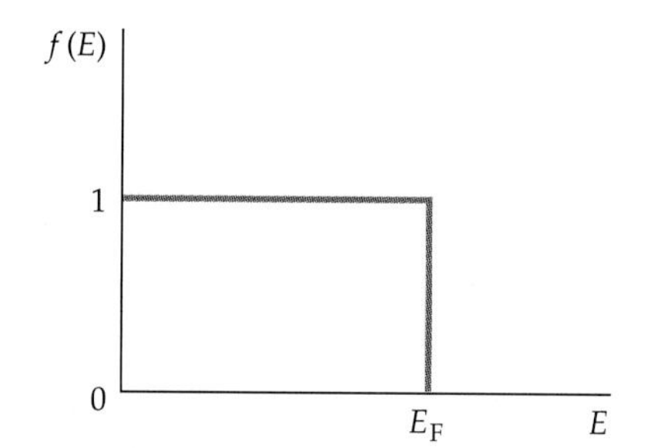

Figure 27-4 Fermi factor versus energy at $T = 0$.

The Fermi Factor for $T > 0$

At temperatures greater than $T = 0$, some electrons will occupy higher energy states because of thermal energy gained during collisions with the lattice. However, an electron cannot move to a higher or lower state unless it is unoccupied. Since the kinetic energy of the lattice ions is of the order of kT, electrons cannot gain much more energy than kT in collisions with the lattice ions. Therefore, only those electrons with energies within about kT of the Fermi energy can gain energy as the temperature is increased. At 300 K, kT is only 0.026 eV, so the exclusion principle prevents all but a very few electrons near the top of the energy distribution from gaining energy through random collisions with the lattice ions. Figure 27-5 shows the Fermi factor for some temperature T. Since for $T > 0$ there is no distinct energy that separates filled from unfilled levels, the definition of the Fermi energy must be slightly modified. At temperature T, the Fermi energy is defined to be that energy for which the probability of being occupied is $\frac{1}{2}$. For all but extremely high temperatures, the difference between the Fermi energy at temperature T and that at $T = 0$ is very small.

The **Fermi temperature** T_F is defined by

$$kT_F = E_F \qquad \text{27-18}$$

For temperatures much lower than the Fermi temperature, the average energy of the lattice ions will be much less than the Fermi energy, and the electron energy distribution will not differ greatly from that at $T = 0$.

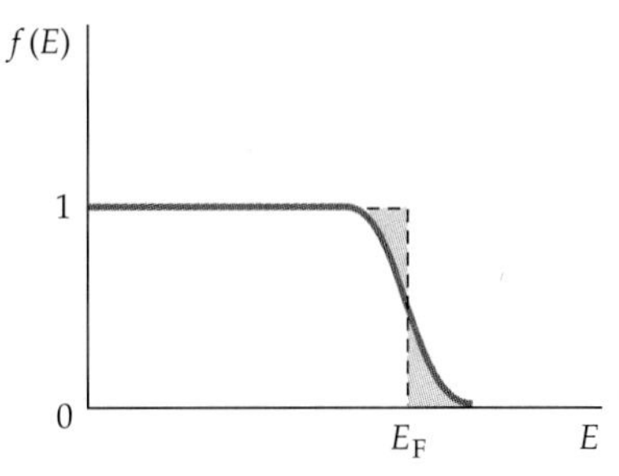

Figure 27-5 The Fermi factor for some temperature T. Some electrons with energies near the Fermi energy are excited, as indicated by the shaded regions. The Fermi energy is that value of E for which $f(E) = \frac{1}{2}$.

Example 27-3

Find the Fermi temperature for copper.

Use $E_F = 7.04$ eV and $k = 8.62 \times 10^{-5}$ eV/K in Equation 27-18:

$$T_F = \frac{E_F}{k} = \frac{7.04 \text{ eV}}{8.62 \times 10^{-5} \text{ eV/K}} = 81{,}700 \text{ K}$$

Remark We can see from this example that the Fermi temperature of copper is much greater than any temperature T for which copper remains a solid.

Because an electric field in a conductor accelerates all of the conduction electrons together, the exclusion principle does not prevent the free electrons in filled states from participating in conduction. Figure 27-6 shows the Fermi factor in one dimension versus *velocity* for an ordinary temperature. The factor is approximately 1 for speeds v_x in the range $-u_F < v_x < u_F$, where the Fermi speed u_F is related to the Fermi energy by $E_F = \frac{1}{2}mu_F^2$. Then

$$u_F = \sqrt{\frac{2E_F}{m_e}} \qquad \text{27-19}$$

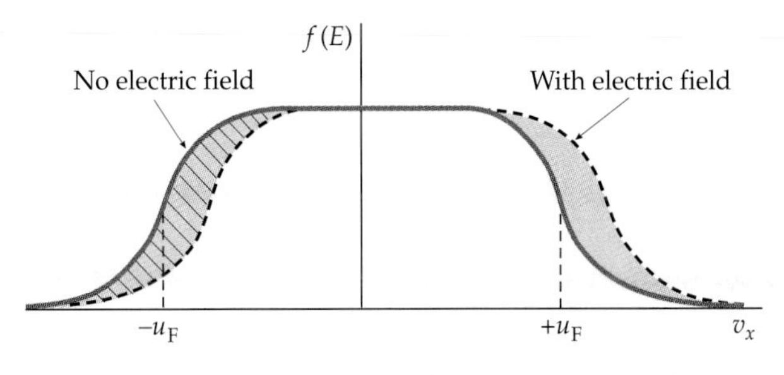

Figure 27-6 Fermi factor versus velocity in one dimension with no electric field (solid) and with an electric field in the $+x$ direction (dashed). The difference is greatly exaggerated.

Example 27-4

Calculate the Fermi speed for copper.

Use Equation 27-19 with $E_F = 7.04$ eV:

$$u_F = \sqrt{\frac{2(7.04 \text{ eV})}{9.11 \times 10^{-31} \text{ kg}}\left(\frac{1.6 \times 10^{-19} \text{ J}}{1 \text{ eV}}\right)} = 1.57 \times 10^6 \text{ m/s}$$

The dashed curve in Figure 27-6 shows the Fermi factor after the electric field has been acting for some time t. Although all of the electrons have been shifted to higher velocities, the net effect is equivalent to shifting only the electrons near the Fermi energy.

optional

Contact Potential

When two different metals are placed in contact, a potential difference $V_{contact}$ called the **contact potential** develops between them. The contact potential depends on the work functions of the two metals, ϕ_1 and ϕ_2 (we encountered work functions when the photoelectric effect was introduced in Chapter 17), and the Fermi energies of the two metals. When the metals are in contact, the total energy of the system is lowered if electrons near the boundary move from the metal with the higher Fermi energy into the metal with the lower Fermi energy until the Fermi energies of the two metals are the same, as shown in Figure 27-7. When equilibrium is established, the metal with the lower initial Fermi energy is negatively charged and the other is positively charged, so that between them there is a potential difference $V_{contact}$ given by

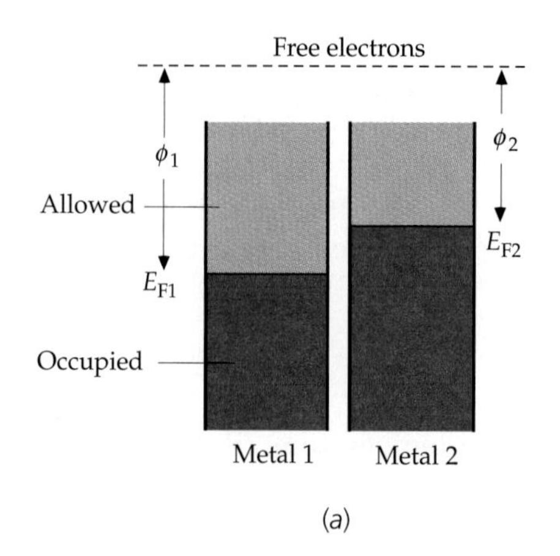

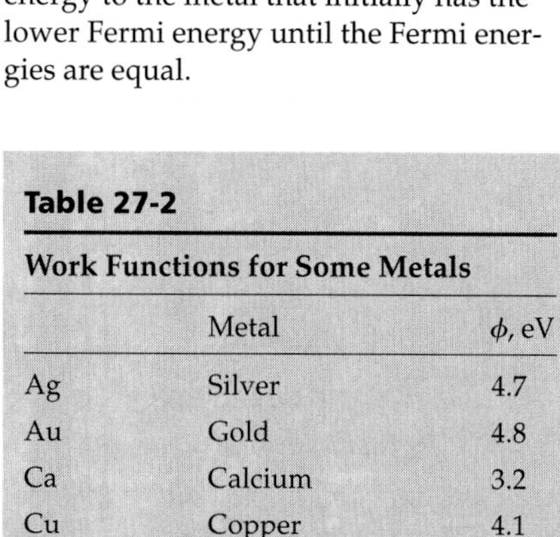

Figure 27-7 (*a*) Energy levels for two different metals with different Fermi energies and work functions. (*b*) When the metals are in contact, electrons flow from the metal that initially has the higher Fermi energy to the metal that initially has the lower Fermi energy until the Fermi energies are equal.

$$V_{contact} = \frac{\phi_1 - \phi_2}{e} \qquad 27\text{-}20$$

Table 27-2 lists the work functions for several metals.

Table 27-2

Work Functions for Some Metals

	Metal	ϕ, eV
Ag	Silver	4.7
Au	Gold	4.8
Ca	Calcium	3.2
Cu	Copper	4.1
K	Potassium	2.1
Mn	Manganese	3.8
Na	Sodium	2.3
Ni	Nickel	5.2

Example 27-5

The threshold wavelength for the photoelectric effect is 271 nm for tungsten and 262 nm for silver. What is the contact potential developed when silver and tungsten are placed in contact?

Picture the Problem The contact potential is proportional to the difference in the work functions for the two metals. The work function ϕ can be found from the given threshold wavelengths using $\phi = hc/\lambda_t$ (Equation 17-4).

1. The contact potential is given by Equation 27-20: $V_{contact} = \dfrac{\phi_1 - \phi_2}{e}$

2. The work function is related to the threshold wavelength: $\phi = \dfrac{hc}{\lambda_t}$

3. Substitute $\lambda_t = 271$ nm for tungsten: $\phi_W = \dfrac{hc}{\lambda_t} = \dfrac{1240\ \text{eV}\cdot\text{nm}}{271\ \text{nm}} = 4.58\ \text{eV}$

4. Substitute $\lambda_t = 262$ nm for silver: $\phi_{Ag} = \dfrac{1240\ \text{eV}\cdot\text{nm}}{262\ \text{nm}} = 4.73\ \text{eV}$

5. The contact potential is thus: $V_{contact} = \dfrac{\phi_{Ag} - \phi_W}{e} = 4.73\ \text{V} - 4.58\ \text{V}$

$= 0.15\ \text{V}$

Heat Capacity Due to Electrons in a Metal

The quantum-mechanical modification of the electron distribution in metals allows us to understand why the contribution of the electron gas to the heat capacity of a metal is much less that of the ions. According to the classical equipartition theorem, the energy of the lattice ions in n moles of a solid is $3nRT$, and thus the molar heat capacity is $C' = 3R$, where R is the universal gas constant (see Section 19-7). In a metal there is a free electron gas containing a number of electrons approximately equal to the number of lattice ions. If these electrons obey the classical equipartition theorem, they should have an energy of $\frac{3}{2}nRT$ and contribute an additional $\frac{3}{2}R$ to the molar heat capacity. But measured heat capacities of metals are just slightly greater than those of insulators. We can understand this because at some temperature T, only those electrons with energies near the Fermi energy can be excited by random collisions with the lattice ions. The number of these electrons is of the order of $(kT/E_F)N$, where N is the total number of electrons. The energy of these electrons is increased from that at $T = 0$ by an amount that is of the order of kT. So the total increase in thermal energy is of the order of $(kT/E_F)N \times kT$. We can thus express the energy of N electrons at temperature T as

$$E = NE_{av}(0) + \alpha N \frac{kT}{E_F} kT \qquad \text{27-21}$$

where α is some constant that we expect to be of the order of 1 if our reasoning is correct. The calculation of α is quite difficult. The result is $\alpha = \pi^2/4$. Using this result and writing E_F in terms of the Fermi temperature, $E_F = kT_F$, we obtain the following for the contribution of the electron gas to the heat capacity at constant volume:

$$C_v = \frac{dU}{dT} = 2\alpha Nk \frac{kT}{E_F} = \frac{\pi^2}{2} nR \frac{T}{T_F}$$

where we have written Nk in terms of the gas constant R ($Nk = nR$). The molar heat capacity at constant volume is then

$$C'_v = \frac{\pi^2}{2} R \frac{T}{T_F} \qquad \text{27-22}$$

We can see that because of the large value of T_F, the contribution of the electron gas is a small fraction of R at ordinary temperatures. Because $T_F =$ 81,700 K for copper, the molar heat capacity of the electron gas at $T = 300$ K is

$$C'_v = \frac{\pi^2}{2}\left(\frac{300\text{ K}}{81{,}700}\right) R \approx 0.02R$$

which is in good agreement with experiment.

27-3 Quantum Theory of Electrical Conduction

We can use Equation 27-7 for the resistivity if we use the Fermi speed u_F in place of v_{av}:

$$\rho = \frac{m_e u_F}{ne^2\lambda} \qquad \text{27-23}$$

We now have two problems. First, since the Fermi speed u_F is approximately independent of temperature, the resistivity given by Equation 27-23 is independent of temperature unless the mean free path depends on it. The second problem concerns magnitudes. As mentioned earlier, the classical expression for resistivity using v_{av} calculated from the Maxwell–Boltzmann distribution gives values that are about 6 times too large at $T = 300$ K. Since the Fermi speed u_F is about 16 times the Maxwell–Boltzmann value of v_{av}, the magnitude of ρ predicted by Equation 27-23 will be about 100 times greater than the experimentally determined value. The resolution of both of these problems lies in the calculation of the mean free path λ.

The Scattering of Electron Waves

In Equation 27-9 for the classical mean free path ($\lambda = 1/n_{ion}A$), the quantity $A = \pi r^2$ is the area of the lattice ion as seen by an electron. In the quantum calculation, the mean free path is related to the scattering of electron waves by the crystal lattice. Detailed calculations show that, for a *perfectly* ordered crystal, $\lambda = \infty$, that is, there is no scattering of the electron waves. The scattering of electron waves arises because of *imperfections* in the crystal lattice, which have nothing to do with the actual area of the lattice ions. According to the quantum theory of electron scattering, A depends merely on *deviations* of the lattice ions from a perfectly ordered array and not on the size of the ions. The most common causes of such deviations are thermal vibrations of the lattice ions or impurities.

We can use $\lambda = 1/n_{ion}A$ for the mean free path if we reinterpret the area A. Figure 27-8 compares the classical and quantum pictures of this area. In the quantum picture, the lattice ions are points that have no size but present an area $A = \pi r_0^2$, where r_0 is the amplitude of thermal vibrations. In Chapter 14 we saw that the energy of vibration in simple harmonic motion is proportional to the square of the amplitude, which is r_0^2. Thus, the effective area A is proportional to the energy of vibration of the lattice ions. From the equipartition theorem, we know that the average energy of vibration is proportional to kT.* Thus, A is proportional to T, and λ is proportional to $1/T$. Then the resistivity given by Equation 27-7 is proportional to T, in agreement with experiment.

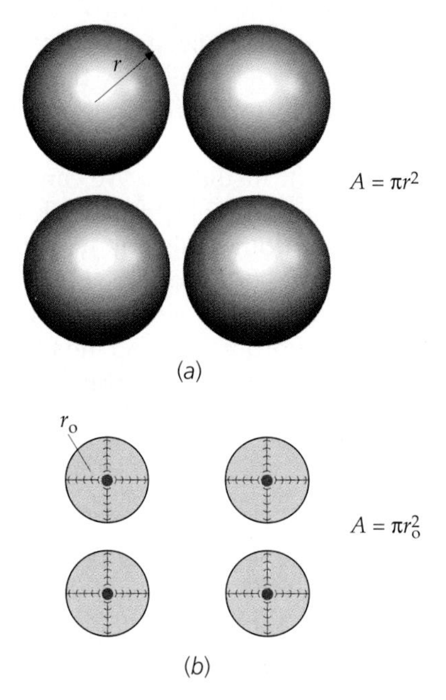

Figure 27-8 (*a*) Classical picture of the lattice ions as spherical balls of radius r that present an area πr^2 to the electrons. (*b*) Quantum-mechanical picture of the lattice ions as points that are vibrating in three dimensions. The area presented to the electrons is πr_0^2, where r_0 is the amplitude of oscillation of the ions.

The effective area A due to thermal vibrations can be calculated, and the results give values for the resistivity that are in agreement with experiment. At $T = 300$ K, for example, the effective area turns out to be about 100 times smaller than the actual area of a lattice ion. We see, therefore, that the free-electron model of metals gives a good account of electrical conduction if the classical mean speed v_{av} is replaced by the Fermi speed u_F and if the collisions between electrons and the lattice ions are interpreted in terms of the scattering of electron waves, for which only deviations from a perfectly ordered lattice are important.

The presence of impurities in a metal also causes deviations from perfect regularity in the crystal lattice. The effects of impurities on resistivity are approximately independent of temperature. The resistivity of a metal containing impurities can be written $\rho = \rho_t + \rho_i$, where ρ_t is the resistivity due to the thermal motion of the lattice ions and ρ_i is the resistivity due to impurities. Figure 27-9 shows typical resistance-versus-temperature curves for metals with impurities. As the temperature approaches zero, ρ_t approaches zero and the resistivity approaches the constant ρ_i due to impurities.

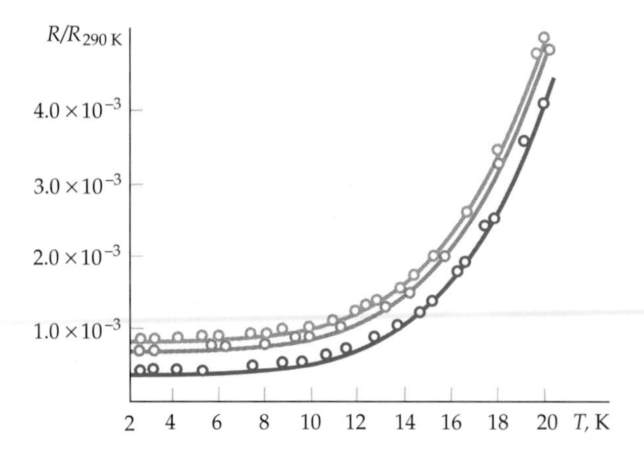

Figure 27-9 Relative resistance versus temperature for three samples of sodium. The three curves have the same temperature dependence but different magnitudes because of differing amounts of impurities in the samples.

* The equipartition theorem *does* hold for the lattice ions, which obey the Maxwell–Boltzmann energy distribution.

Note that the energy gap for a typical superconductor is much smaller than the energy gap for a typical semiconductor, which is of the order of 1 eV. As the temperature is increased from $T = 0$, some of the Cooper pairs are broken. Then there are fewer pairs available for each pair to interact with, and the energy gap is reduced until at $T = T_c$ the energy gap is zero (Figure 27-14).

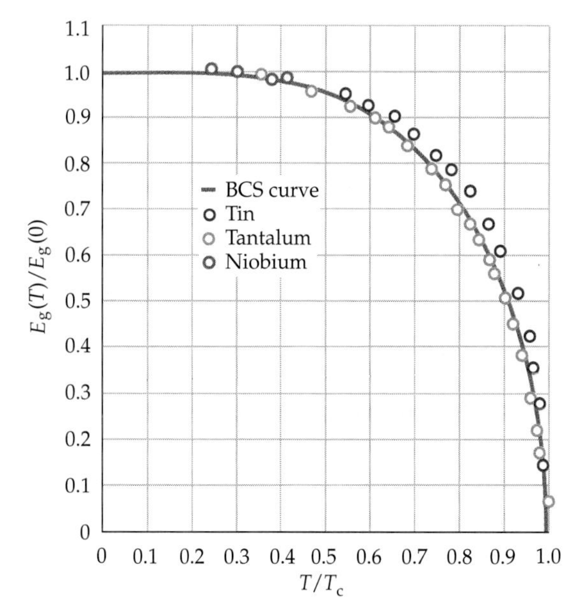

Figure 27-14 Ratio of the energy gap at temperature T to that at $T = 0$ as a function of the relative temperature T/T_c. The solid curve is that predicted by the BCS theory.

The Josephson Effect

When two superconductors are separated by a thin insulating barrier (for example, a layer of aluminum oxide a few nanometers thick), the junction is called a **Josephson junction**, based on the prediction in 1962 by Brian Josephson that Cooper pairs could tunnel across such a junction from one superconductor to the other with no resistance. The tunneling of Cooper pairs constitutes a current, which is observed even when there is no voltage applied across the junction. The current depends on the difference in phase of the wave functions that describe the Cooper pairs. Let ϕ_1 be the phase constant for the wave function of a Cooper pair in one superconductor. All the Cooper pairs in a superconductor act coherently, so they all have the same phase constant. If ϕ_2 is the phase constant for the Cooper pairs in the second superconductor, the current across the junction is given by

$$I = I_{max} \sin(\phi_2 - \phi_1) \qquad \text{27-25}$$

where I_{max} is the maximum current, which depends on the thickness of the barrier. This result has been observed experimentally and is known as the **dc Josephson effect**.

Josephson also predicted that if a dc voltage V were applied across a Josephson junction, there would be a current that alternates with frequency f given by

$$f = \frac{2eV}{h} \qquad \text{27-26}$$

This result, known as the **ac Josephson effect**, has been observed experimentally, and careful measurement of the frequency allows a precise determination of the ratio e/h. Because frequency can be measured very accurately, the ac Josephson effect is also used to establish precise voltage standards. The inverse effect, in which the application of an alternating voltage across a Josephson junction results in a dc current, has also been observed.

Using $e = 1.602 \times 10^{-19}$ C and $h = 6.626 \times 10^{-34}$ J·s, calculate the frequency of the Josephson current if the applied voltage is 1 μV.

Substitute the given values into Equation 27-25 to calculate f:

$$f = \frac{2eV}{h} = \frac{2(1.602 \times 10^{-19}\ \text{C})(10^{-6}\ \text{V})}{6.626 \times 10^{-34}\ \text{J·s}}$$

$$= 4.835 \times 10^{8}\ \text{Hz} = 483.5\ \text{MHz}$$

optional

27-6 The Fermi–Dirac Distribution*

*This material is somewhat complicated and may be skipped over on a first reading.

The classical Maxwell–Boltzmann distribution (Equation 18-39) gives the number of molecules with energy E in the range between E and $E + dE$. It is the product of $g(E)\,dE$ where $g(E)$ is the density of states (number of energy states in the range dE) and the Boltzmann factor $e^{-E/kT}$, which is the probability of a state being occupied. The distribution function for free electrons in a metal is called the **Fermi–Dirac distribution**. The Fermi–Dirac distribution can be written in the same form as the Maxwell–Boltzmann distribution with the density of states calculated from quantum theory and the Boltzmann factor replaced by the Fermi factor. Let $n(E)\,dE$ be the number of electrons with energies between E and $E + dE$. This number is written

$$n(E)\,dE = g(E)\,dE\,f(E) \tag{27-27}$$

Energy distribution function

where $g(E)\,dE$ is the number of states between E and $E + dE$, $g(E)$ is called the density of states, and $f(E)$ is the probability of a state being occupied, which is the Fermi factor. The density of states in three dimensions is somewhat difficult to calculate, so we just give the result. For electrons in a metal of volume V, the density of states is

$$g(E) = \frac{8\pi\sqrt{2}m_e^{3/2}V}{h^3}E^{1/2} \tag{27-28}$$

Density of states

As in the classical Maxwell–Boltzmann distribution, the density of states is proportional to $E^{1/2}$.

At $T = 0$, the Fermi factor is given by Equation 27-17:

$$f(E) = 1, \qquad E < E_F$$

$$f(E) = 0, \qquad E > E_F$$

The integral of $n(E)\,dE$ over all energies gives the total number of electrons N. We can derive Equation 27-15*a* for the Fermi energy at $T = 0$ by integrating $n(E)\,dE$ from $E = 0$ to $E = \infty$. We obtain

$$N = \int_0^\infty n(E)\,dE = \int_0^{E_F} g(E)\,dE = \frac{8\pi\sqrt{2}m_e^{3/2}V}{h^3}\int_0^{E_F} E^{1/2}\,dE = \frac{16\pi\sqrt{2}m_e^{3/2}V}{3h^3}E_F^{3/2}$$

Note that at $T = 0$, $n(E)$ is zero for $E > E_F$ so we had to integrate only from $E = 0$ to $E = E_F$. Solving for E_F gives the Fermi energy at $T = 0$:

$$E_F = \frac{h^2}{8m_e}\left(\frac{3N}{\pi V}\right)^{2/3} \tag{27-29}$$

which is Equation 27-15*a*. In terms of the Fermi energy, the density of states (Equation 27-28) is

$$g(E) = \frac{8\pi\sqrt{2}m_e^{3/2}V}{h^3}E^{1/2} = \frac{3N}{2}E_F^{-3/2}E^{1/2} \tag{27-30}$$

Density of states in terms of E_F

The average energy at $T = 0$ is calculated from

$$E_{av} = \frac{\int_0^{E_F} Eg(E)\,dE}{\int_0^{E_F} g(E)\,dE} = \frac{1}{N}\int_0^{E_F} Eg(E)\,dE$$

where $N = \int_0^{E_F} g(E)\,dE$ is the total number of electrons. Performing the integration, we obtain Equation 27-16

$$E_{av} = \tfrac{3}{5}E_F \tag{27-31}$$

Average energy at $T = 0$

At $T > 0$, the Fermi factor is more complicated. It can be shown to be given by

$$f(E) = \frac{1}{e^{(E - E_F)/kT} + 1} \tag{27-32}$$

Fermi factor

We can see from this equation that for E greater than E_F, $e^{(E - E_F)/kT}$ becomes very large as T approaches zero, so at $T = 0$, the Fermi factor is zero for $E > E_F$. On the other hand, for E less than E_F, $e^{(E - E_F)/kT}$ approaches 0 as T approaches zero, so at $T = 0$, $f(E) = 1$ for $E < E_F$. Thus the Fermi factor given by Equation 27-32 holds for all temperatures. Note also that for any nonzero value of T, $f(E) = \frac{1}{2}$ at $E = E_F$.

The complete Fermi–Dirac distribution function is thus

$$n(E)\,dE = \frac{8\pi\sqrt{2}m_e^{3/2}V}{h^3} E^{1/2} \frac{1}{e^{(E - E_F)/kT} + 1}\,dE \tag{27-33}$$

Fermi–Dirac distribution

We can see that for those few electrons with energies much greater than the Fermi energy, the Fermi factor approaches $1/e^{(E - E_F)/kT} = e^{(E_F - E)/kT} = e^{E_F}e^{-E/kT}$, which is proportional to $e^{-E/kT}$. Thus, the high-energy tail of the Fermi–Dirac energy distribution decreases as $e^{-E/kT}$, just like the classical Maxwell–Boltzmann energy distribution. The reason is that in this high-energy region, there are many unoccupied energy states and few electrons, so the Pauli exclusion principle is not important, and the distribution approaches the classical distribution. This result has practical importance because it applies to the conduction electrons in semiconductors.

Example 27-8

At what energy is the Fermi factor equal to 0.1 for copper at $T = 300$ K?

Picture the Problem We set $f(E) = 0.1$ in Equation 27-32 using $T = 300$ K and $E_F = 7.04$ eV from Table 27-1 and solve for E.

1. Solve Equation 27-32 for $e^{(E - E_F)/kT}$:

$$f(E) = \frac{1}{e^{(E - E_F)/kT} + 1} = 0.1$$

$$e^{(E - E_F)/kT} = \frac{1}{f(E)} - 1 = \frac{1}{0.1} - 1 = 9$$

2. Take the natural logarithm of both sides:

$$\frac{E - E_F}{kT} = \ln 9 = 2.20$$

3. Solve for $E - E_F$:

$$E - E_F = 2.20kT = 2.20(8.62 \times 10^{-5}\ \text{eV/K})(300\ \text{K})$$

$$= 2.20(0.0259\ \text{eV}) = 0.0570\ \text{eV}$$

4. Solve for E using $E_F = 7.04$ eV from Table 27-1:

$$E = 7.04\ \text{eV} + 0.0570\ \text{eV} = 7.10\ \text{eV}$$

Remark The Fermi factor drops from about 1 to 0.1 at just 0.06 eV above the Fermi energy of about 7 eV.

optional

Example 27-9

Find the probability that an energy state in copper 0.1 eV above the Fermi energy is occupied at $T = 300$ K.

Picture the Problem The probability is the Fermi factor given in Equation 27-32, with $E_F = 7.04$ eV, and $E = 7.14$ eV.

1. The probability of a state being occupied equals the Fermi factor:

$$P = f(E) = \frac{1}{e^{(E-E_F)/kT} + 1}$$

2. Calculate the dimensionless exponent in the Fermi factor:

$$\frac{E - E_F}{kT} = \frac{7.14 \text{ eV} - 7.04 \text{ eV}}{(8.62 \times 10^{-5} \text{ eV/K})(300 \text{ K})} = 3.87$$

3. Use this result to calculate the Fermi factor:

$$f = \frac{1}{e^{(E-E_F)/kT} + 1} = \frac{1}{e^{3.87} + 1}$$

$$= \frac{1}{48 + 1} = 0.0204 = 2.04\%$$

Remark The probability of an electron having an energy of 0.1 eV above the Fermi energy at 300 K is only about 2%.

Example 27-10 *try it yourself*

Find the probability that an energy state in copper 0.1 eV *below* the Fermi energy is occupied at $T = 300$ K.

Picture the Problem The probability is the Fermi factor given in Equation 27-32, with $E_F = 7.04$ eV, and $E = 6.94$ eV.

Cover the column to the right and try these on your own before looking at the answers.

Steps / ***Answers***

1. Write the Fermi factor.

$$f(E) = \frac{1}{e^{(E-E_F)kT} + 1}$$

2. Calculate the dimensionless exponent in the Fermi factor.

$$\frac{E - E_F}{kT} = \frac{6.94 \text{ eV} - 7.04 \text{ eV}}{(8.62 \times 10^{-5} \text{ eV/K})(300 \text{ K})} = -3.87$$

3. Use your result in step 2 to calculate the Fermi factor.

$$f = \frac{1}{e^{(E-E_F)/kT} + 1} = \frac{1}{e^{-3.87} + 1}$$

$$= \frac{1}{0.021 + 1} = 0.979 \approx 98\%$$

Remark The probability of an electron having an energy of 0.1 eV *below* the Fermi energy at 300 K is about 98%.

Exercise What is the probability of an energy state 0.1 eV below the Fermi energy being unoccupied at 300 K? (*Answer* $1 - 0.98 = 0.02$ or 2%. This is the probability of there being a hole at this energy.)

optional

Summary

Topic	Remarks and Relevant Equations	
1. Microscopic Picture of Conduction		
Resistivity	$\rho = \frac{m_e v_{av}}{n_e e^2 \lambda}$	27-7
	where v_{av} is the average speed of the electrons and λ is their mean free path between collisions with the lattice ions.	
Mean free path	$\lambda = \frac{1}{n_{ion}\pi r^2} = \frac{1}{n_{ion}A}$	27-9
	where n_{ion} is the number of lattice ions per unit volume, r is their effective radius, and A is their effective cross-sectional area.	
2. Classical Interpretation of v_{av} and λ	v_{av} is determined from the Maxwell–Boltzmann distribution, and r is the actual radius of a lattice ion.	
3. Quantum Interpretation of v_{av} and λ	v_{av} is determined from the Fermi–Dirac distribution and is approximately constant independent of temperature. The mean free path is determined from the scattering of electron waves, which occurs only because of deviations from a perfectly ordered array. The radius r is the amplitude of vibration of the lattice ion, which is proportional to $\sqrt{T}$, so A is proportional to T.	
4. Fermi Electron Gas		
Fermi energy E_F at $T = 0$	E_F is the energy of the highest filled (or half-filled) energy state.	
E_F at $T > 0$	E_F is the energy at which the probability of being occupied is $\frac{1}{2}$.	
Approximate magnitude of E_F	E_F is about 5 to 10 eV for most metals.	
Dependence of E_F on the number density of free electrons	$E_F = \frac{h^2}{8m_e}\left(\frac{3N}{\pi V}\right)^{2/3} = (0.365\text{ eV}\cdot\text{nm}^2)\left(\frac{N}{V}\right)^{2/3}$	27-15*a*,*b*
Average energy at $T = 0$	$E_{av} = \frac{3}{5}E_F$	27-16
Fermi factor at $T = 0$	The Fermi factor $f(E)$ is the probability of a state being occupied $f(E) = 1, \quad E < E_F$ $f(E) = 0, \quad E > E_F$	27-17
Fermi temperature	$T_F = \frac{E_F}{k}$	27-18
Fermi speed	$u_F = \sqrt{\frac{2E_F}{m_e}}$	27-19
Contact potential	When two different metals are placed in contact, electrons flow from the metal with the higher Fermi energy to the one with the lower Fermi energy until the Fermi energies of the two metals are equal. In equilibrium, there is a potential difference be-	

optional

tween the metals that is equal to the difference in the work function of the two metals divided by the electronic charge e:

$$V_{\text{contact}} = \frac{\phi_1 - \phi_2}{e} \quad \text{27-20}$$

Heat capacity due to electrons	$C'_v = \frac{\pi^2}{2} R \frac{T}{T_F}$	27-22

5. Band Theory of Solids

When many atoms are brought together to form a solid, the individual energy levels are split into bands of allowed energies. The splitting depends on the type of bonding and the lattice separation. The highest-energy band containing electrons is called the valence band. In a conductor, the valence band is only partially full, so there are many available states for excited electrons. In an insulator, the valence band is completely full and there is a large energy gap between it and the next allowed band, the conduction band. In a semiconductor, the energy gap between the filled valence band and the empty conduction band is small; so at ordinary temperatures, an appreciable number of electrons are thermally excited into the conduction band.

6. Superconductivity

In a superconductor, the resistance drops suddenly to zero below a critical temperature T_c. Superconductors with critical temperatures as high as 125 K have been discovered.

BCS theory

Superconductivity is described by a theory of quantum mechanics called the BCS theory in which the free electrons form Cooper pairs. The energy needed to break up a Cooper pair is called the energy gap E_g. When all the electrons are paired, individual electrons cannot be scattered by a lattice ion, so the resistance is zero.

Tunneling

When a normal conductor is separated from a superconductor by a thin layer of oxide, electrons can tunnel through the energy barrier if the applied voltage across the layer is $E_g/2e$, where E_g is the energy needed to break up a Cooper pair. The energy gap E_g can be determined by a measurement of the tunneling current versus the applied voltage.

Josephson junction

A system of two superconductors separated by a thin layer of insulating material is called a Josephson junction.

dc Josephson effect

A dc current is observed to tunnel through a Josephson junction even in the absence of voltage across the junction.

ac Josephson effect

When a dc voltage V is applied across a Josephson junction, an ac current is observed with a frequency

$$f = \frac{2eV}{h} \quad \text{27-26}$$

Measurement of the frequency of this current allows a precise determination of the ratio e/h.

7. Fermi–Dirac Distribution

The number of electrons with energies between E and $E + dE$ is given by

$$n(E)\,dE = g(E)\,dE\,f(E) \quad \text{27-27}$$

where $g(E)$ is the density of states and $f(E)$ is the Fermi factor.

Density of states	$g(E) = \frac{8\pi\sqrt{2}m_e^{3/2}V}{h^3} E^{1/2}$	27-28
Fermi factor at temperature T	$f(E) = \frac{1}{e^{(E-E_F)/kT} + 1}$	27-32

Problem-Solving Guide

Summary of Worked Examples

Type of Calculation	Procedure and Relevant Examples	
1. The Fermi Electron Gas		
Calculate the Fermi energy from the number density of free electrons.	Use $E_F = (0.365\ \text{eV}\cdot\text{nm}^2)(N/V)^{2/3}$.	**Example 27-2**
Calculate the Fermi temperature.	Use $kT_F = E_F$ and Table 27-1 for E_F.	**Example 27-3**
Calculate the Fermi speed.	Use $u_F = \sqrt{2E_F/m_e}$ and Table 27-1 for E_F.	**Example 27-4**
Calculate the contact potential developed when two metals are placed in contact.	Find the work functions from the photoelectric thresholds and use $V_{\text{contact}} = (\phi_1 - \phi_2)/e$.	**Example 27-5**
2. Superconductors		
Find the superconducting gap from the critical temperature.	Use $E_g = 3.5kT_c$.	**Example 27-6**
Find the Josephson frequency for a given applied voltage.	Use $f = 2eV/h$.	**Example 27-7**
3. The Fermi–Dirac Distribution		
Find the energy for a given value of $f(E)$ at a given temperature.	Use $f(E) = 1/(e^{(E-E_F)/kT} + 1)$ and solve for E.	**Example 27-8**
Find the probability that a state of a given energy will be occupied.	Calculate the Fermi factor for the given energy.	**Examples 27-9, 27-10**

Problems

Conceptual Problems

Problems from Optional and Exploring sections

In a few problems, you are given more data than you actually need; in a few other problems, you are required to supply data from your general knowledge, outside sources, or informed estimates.

- • Single-concept, single-step, relatively easy
- •• Intermediate-level, may require synthesis of concepts
- ••• Challenging, for advanced students

Use Table 27-1 for the Fermi energies and electron number densities when needed.

Microscopic Picture of Conduction

1 • In the classical model of conduction, the electron loses energy on average in a collision because it loses the drift velocity it had picked up since the last collision. Where does this energy appear?

2 • A measure of the density of the free-electron gas in a metal is the distance r_s, which is defined as the radius of the sphere whose volume equals the volume per conduction electron. (*a*) Show that $r_s = (3/4\pi n)^{1/3}$, where n is the free-electron number density. (*b*) Calculate r_s for copper in nanometers.

3 • (*a*) Given a mean free path $\lambda = 0.4$ nm and a mean speed $v_{av} = 1.17 \times 10^5$ m/s for the current flow in copper at a temperature of 300 K, calculate the classical value for the resistivity ρ of copper. (*b*) The classical model suggests that the mean free path is temperature independent and that v_{av} depends on temperature. From this model, what would ρ be at 100 K?

The Fermi Electron Gas

4 • Calculate the number density of free electrons in (*a*) Ag ($\rho = 10.5\ \text{g/cm}^3$) and (*b*) Au ($\rho = 19.3\ \text{g/cm}^3$), assuming one free electron per atom, and compare your results with the values listed in Table 27-1.

5 • The density of aluminum is 2.7 g/cm^3. How many free electrons are present per aluminum atom?

6 • The density of tin is 7.3 g/cm^3. How many free electrons are present per tin atom?

7 • Calculate the Fermi temperature for (*a*) Al, (*b*) K, and (*c*) Sn.

8 • What is the speed of a conduction electron whose energy is equal to the Fermi energy E_F for (*a*) Na, (*b*) Au, and (*c*) Sn?

9 • Calculate the Fermi energy for (*a*) Al, (*b*) K, and (*c*) Sn using the number densities given in Table 27-1.

10 • Find the average energy of the conduction electrons at $T = 0$ in (*a*) copper and (*b*) lithium.

11 • Calculate (*a*) the Fermi temperature and (*b*) the Fermi energy at $T = 0$ for iron.

12 •• The pressure of an ideal gas is related to the average energy of the gas particles by $PV = \frac{2}{3}NE_{av}$, where N is the number of particles and E_{av} is the average energy. Use this to calculate the pressure of the Fermi electron gas in copper in newtons per square meter, and compare your result with atmospheric pressure, which is about 10^5 N/m^2. (*Note:* The units are most easily handled by using the conversion factors $1\ N/m^2 = 1\ J/m^3$ and $1\ eV = 1.6 \times 10^{-19}$ J.)

13 •• The bulk modulus B of a material can be defined by

$$B = -V\frac{\partial P}{\partial V}$$

(*a*) Use the ideal-gas relation $PV = \frac{2}{3}NE_{av}$ and Equations 27-15 and 27-16 to show that

$$P = \frac{2NE_F}{5V} = CV^{-5/3}$$

where C is a constant independent of V. (*b*) Show that the bulk modulus of the Fermi electron gas is therefore

$$B = \frac{5}{3}P = \frac{2NE_F}{3V}$$

(*c*) Compute the bulk modulus in newtons per square meter for the Fermi electron gas in copper and compare your result with the measured value of 140×10^9 N/m^2.

Contact Potential

14 • Thomas refuses to believe that a potential difference can be created simply by bringing two different metals into contact with each other. John talks him into making a small wager, and is about to cash in. (*a*) Which two metals from Table 27-2 would demonstrate his point most effectively? (*b*) What is the value of that contact potential?

15 • (*a*) In Problem 14, which choices of different metals would make the least impressive demonstration? (*b*) What is the value of that contact potential?

16 • Calculate the contact potential between (*a*) Ag and Cu, (*b*) Ag and Ni, and (*c*) Ca and Cu.

Quantum Theory of Electrical Conduction

17 • When the temperature of pure copper is lowered from 300 K to 4 K, its resistivity drops by a much greater factor than that of brass when it is cooled the same way. Why?

18 • The resistivities of Na, Au, and Sn at $T = 273$ K are 4.2 $\mu\Omega\cdot$cm, 2.04 $\mu\Omega\cdot$cm, and 10.6 $\mu\Omega\cdot$cm, respectively. Use these values and the Fermi speeds calculated in Problem 8 to find the mean free paths λ for the conduction electrons in these elements.

19 •• The resistivity of pure copper is increased by about $1 \times 10^{-8}\ \Omega\cdot$m by the addition of 1% (by number of atoms) of an impurity throughout the metal. The mean free path depends on both the impurity and the oscillations of the lattice ions according to the equation

$$\frac{1}{\lambda} = \frac{1}{\lambda_t} + \frac{1}{\lambda_i}$$

(*a*) Estimate λ_i from data given in Table 27-1. (*b*) If r is the effective radius of an impurity lattice ion seen by an electron, the scattering cross section is πr^2. Estimate this area using the fact that r is related to λ_i by Equation 27-9.

Band Theory of Solids

20 • A metal is a good conductor because the valence energy band for electrons is

(*a*) completely full.
(*b*) full, but there is only a small gap to a higher empty band.
(*c*) partly full.
(*d*) empty.
(*e*) None of these is correct.

21 • Insulators are poor conductors of electricity because

(*a*) there is a small energy gap between the valence band and the next higher band where electrons can exist.
(*b*) there is a large energy gap between the full valence band and the next higher band where electrons can exist.
(*c*) the valence band has a few vacancies for electrons.
(*d*) the valence band is only partly full.
(*e*) None of these is correct.

22 • You are an electron sitting at the top of the valence band in a silicon atom, longing to jump across the 1.14-eV energy gap that separates you from the bottom of the conduction band and all of the adventures that it may contain. What you need, of course, is a photon. What is the maximum photon wavelength that will get you across the gap?

23 • Work Problem 22 for germanium, for which the energy gap is 0.74 eV.

24 • Work Problem 22 for diamond, for which the energy gap is 7.0 eV.

25 •• A photon of wavelength 3.35 μm has just enough energy to raise an electron from the valence band to the conduction band in a lead sulfide crystal. (*a*) Find the energy gap between these bands in lead sulfide. (*b*) Find the temperature T for which kT equals this energy gap.

BCS Theory of Superconductivity

26 • (*a*) Use Equation 27-24 to calculate the superconducting energy gap for tin and compare your result with the measured value of 6×10^{-4} eV. (*b*) Use the measured value to calculate the wavelength of a photon having sufficient energy to break up Cooper pairs in tin at $T = 0$.

27 • Repeat Problem 26 for lead, which has a measured energy gap of 2.73×10^{-3} eV.

The Fermi–Dirac Distribution

28 •• The number of electrons in the conduction band of an insulator or intrinsic semiconductor is governed chiefly by the Fermi factor. Since the valence band in these materials is nearly filled and the conduction band is nearly empty, the Fermi energy E_F is generally midway between the top of the valence band and bottom of the conduction band, i.e., at $E_g/2$, where E_g is the band gap between the two bands and the energy is measured from the top of the valence band. (*a*) In silicon, $E_g \approx 1.0$ eV. Show that in this case the Fermi factor for electrons at the bottom of the conduction band is given by $\exp(-E_g/2kT)$ and evaluate this factor. Discuss the significance of this result if there are 10^{22} valence electrons per cubic centimeter and the probability of finding an electron in the conduction band is given by the Fermi factor. (*b*) Repeat the calculation in (*a*) for an insulator with a band gap of 6.0 eV.

29 •• Show that at $E = E_F$, the Fermi factor is $F = 0.5$.

30 •• What is the difference between the energies at which the Fermi factor is 0.9 and 0.1 at 300 K in (*a*) copper, (*b*) potassium, and (*c*) aluminum.

31 •• What is the probability that a conduction electron in silver will have a kinetic energy of 4.9 eV at $T = 300$ K?

32 •• Show that $g(E) = (3N/2)E_F^{-3/2}\, E^{1/2}$ (Equation 27-30) follows from Equation 27-28 for $g(E)$, and Equation 27-15a for E_F.

33 •• Carry out the integration $E_{av} = (1/N)\int_0^{E_F} Eg(E)\, dE$ to show that the average energy at $T = 0$ is $\frac{3}{5}E_F$.

34 •• The density of the electron states in a metal can be written $g(E) = AE^{1/2}$, where A is a constant and E is measured from the bottom of the conduction band. (*a*) Show that the total number of states is $\frac{2}{3}AE_F^{3/2}$. (*b*) Approximately what fraction of the conduction electrons are within kT of the Fermi energy? (*c*) Evaluate this fraction for copper at $T = 300$ K.

35 •• What is the probability that a conduction electron in silver will have a kinetic energy of 5.49 eV at $T = 300$ K?

36 •• Use the density-of-states function, Equation 27-28, to estimate the fraction of the conduction electrons in copper that can absorb energy from collisions with the vibrating lattice ions at (*a*) 77 K and (*b*) 300 K.

37 •• In an intrinsic semiconductor, the Fermi energy is about midway between the top of the valence band and the bottom of the conduction band. In germanium, the forbidden energy band has a width of 0.7 eV. Show that at room temperature the distribution function of electrons in the conduction band is given by the Maxwell–Boltzmann distribution function.

38 ••• (*a*) Show that for $E \geq 0$, the Fermi factor may be written as

$$f(E) = \frac{1}{Ce^{E/kT} + 1}$$

(*b*) Show that if $C >> e^{-E/kT}$, $f(E) = Ae^{-E/kT} \ll 1$; in other words, show that the Fermi factor is a constant times the classical Boltzmann factor if $A \ll 1$. (*c*) Use $\int n(E)\, dE = N$ and Equation 27-28 to determine the constant A. (*d*) Using the result obtained in part (*c*), show that the classical approximation is applicable when the electron concentration is very small and/or the temperature is very high. (*e*) Most semiconductors have impurities added in a process called doping, which increases the free electron concentration so that it is about $10^{17}/\text{cm}^3$ at room temperature. Show that for these systems, the classical distribution function is applicable.

39 ••• Show that the condition for the applicability of the classical distribution function for an electron gas ($A \ll 1$ in Problem 38) is equivalent to the requirement that the average separation between electrons is much greater than their de Broglie wavelength.

40 ••• The root-mean-square (rms) value of a variable is obtained by calculating the average value of the square of that variable and then taking the square root of the result. Use this procedure to determine the rms energy of a Fermi distribution. Express your result in terms of E_F and compare it to the average energy. Why do E_{av} and E_{rms} differ?

41 ••• When a star with a mass of about twice that of the sun exhausts its nuclear fuel, it collapses to a neutron star, a dense sphere of neutrons of about 10 km diameter. Neutrons are spin-$\frac{1}{2}$ particles and, like electrons, are subject to the exclusion principle. (*a*) Determine the neutron density of such a neutron star. (*b*) Find the Fermi energy of the neutron distribution.

General Problems

42 • True or false:

(*a*) Solids that are good electrical conductors are usually good heat conductors.
(*b*) The classical free-electron theory adequately explains the heat capacity of metals.
(*c*) At $T = 0$, the Fermi factor is either 1 or 0.
(*d*) The Fermi energy is the average energy of an electron in a solid.
(*e*) The contact potential between two metals is proportional to the difference in the work functions of the two metals.
(*f*) At $T = 0$, an intrinsic semiconductor is an insulator.
(*g*) Semiconductors conduct current in one direction only.

43 • How does the change in the resistivity of copper compare with that of silicon when the temperature increases?

44 • The density of potassium is 0.851 g/cm³. How many free electrons are there per potassium atom?

optional

45 • Calculate the number density of free electrons for (*a*) Mg ($\rho = 1.74\ \mathrm{g/cm^3}$) and (*b*) Zn ($\rho = 7.1\ \mathrm{g/cm^3}$), assuming two free electrons per atom, and compare your results with the values listed in Table 27-1.

46 •• Estimate the fraction of free electrons in copper that are in excited states above the Fermi energy at (*a*) room temperature of 300 K and (*b*) 1000 K.

47 ••• A 2-$\mathrm{cm^2}$ wafer of pure silicon is irradiated with light having a wavelength of 775 nm. The intensity of the light beam is 4.0 $\mathrm{W/m^2}$ and every photon that strikes the sample is absorbed and creates an electron–hole pair. (*a*) How many electron–hole pairs are produced in one second? (*b*) If the number of electron–hole pairs in the sample is 6.25×10^{11} in the steady state, at what rate do the electron–hole pairs recombine? (*c*) If every recombination event results in the radiation of one photon, at what rate is energy radiated by the sample?

CHAPTER 28

The Magnetic Field

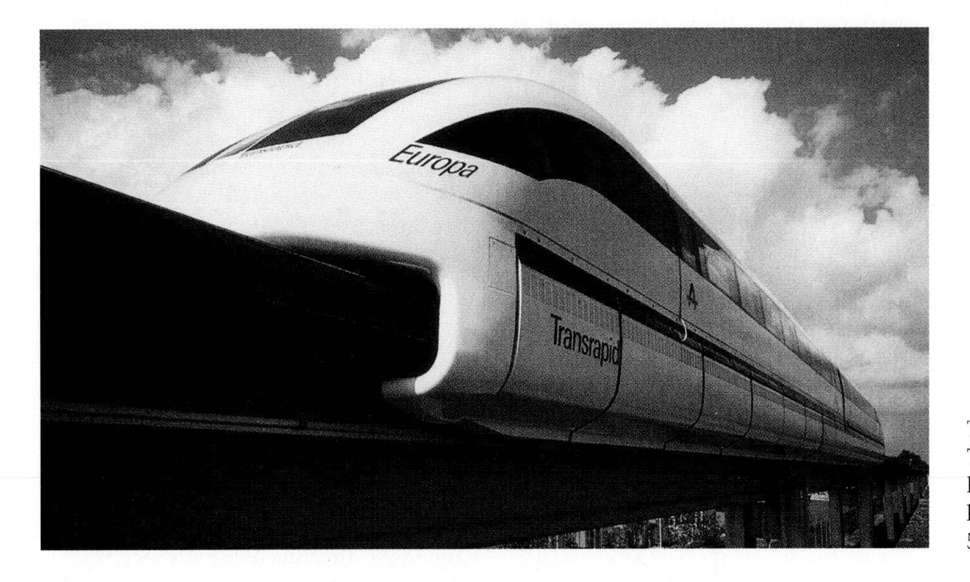

The superspeed Maglev System Transrapid train, using magnetic levitation, guidance, and propulsion, has achieved a peak speed greater than 500 km/h.

More than 2000 years ago, the Greeks were aware that a certain type of stone (now called magnetite) attracts pieces of iron, and there are written references to the use of magnets for navigation dating from the twelfth century.

In 1269, Pierre de Maricourt discovered that a needle laid at various positions on a spherical natural magnet orients itself along lines that pass through points at opposite ends of the sphere. He called these points the poles of the magnet. Subsequently, many experimenters noted that every magnet of whatever shape has two poles, designated the north and south poles, where the force exerted by the magnet is strongest. It was also noted that the like poles of two magnets repel each other and the unlike poles attract each other.

In 1600, William Gilbert discovered that the earth itself is a natural magnet with magnetic poles near the north and south geographic poles. Since the north pole of a compass needle points toward the south pole of a given magnet, what we call the north pole of the earth is actually a south magnetic pole, as illustrated in Figure 28-1.

Although electric charges and magnetic poles are similar in many respects, there is an important difference: Magnetic poles always occur in pairs. When a magnet is broken in half, equal and opposite poles appear at either side of the break point. The result is two magnets, each with a north and south pole. There has long been speculation as to the existence of an isolated magnetic pole, and in recent years considerable experimental effort has been made to find such an object. Thus far, there is no conclusive evidence that an isolated magnetic pole exists.

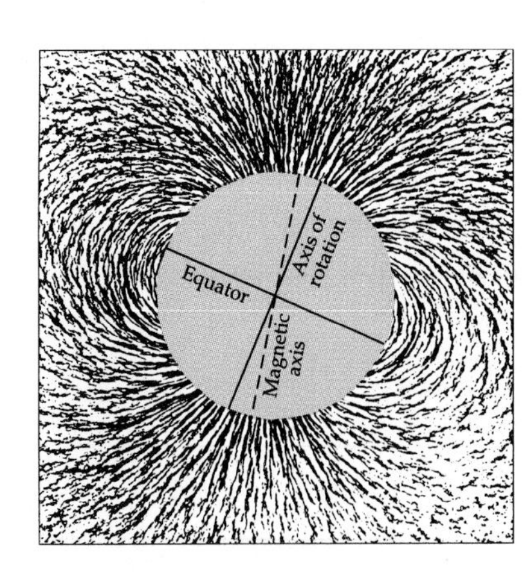

Figure 28-1 Magnetic field lines of the earth indicated by iron filings around a uniformly magnetized sphere. The field lines exit from the north magnetic pole, which is near the south geographic pole, and enter the south magnetic pole, which is near the north geographic pole.

In this chapter, we consider the effects of a given magnetic field on moving charges and on wires carrying currents. The sources of magnetic fields are discussed in the next chapter.

28-1 The Force Exerted by a Magnetic Field

The existence of a magnetic field $\vec{B}$ at some point in space can be demonstrated with a compass needle. If there is a magnetic field, the needle will align itself in the direction of the field.

Experimentally it is observed that, when a charge q has velocity $\vec{v}$ in a magnetic field, there is a force on it that is proportional to q and to v, and to the sine of the angle between $\vec{v}$ and $\vec{B}$. Surprisingly, the force is perpendicular to both the velocity and the field. These experimental results can be summarized as follows. When a charge q moves with velocity $\vec{v}$ in a magnetic field $\vec{B}$, the magnetic force $\vec{F}$ on the charge is

$$\vec{F} = q\vec{v} \times \vec{B} \qquad 28\text{-}1$$

Magnetic force on a moving charge

Since $\vec{F}$ is perpendicular to both $\vec{v}$ and $\vec{B}$, it is perpendicular to the plane defined by these two vectors. The direction of $\vec{F}$ is given by the right-hand rule as $\vec{v}$ is rotated into $\vec{B}$, as illustrated in Figure 28-2.

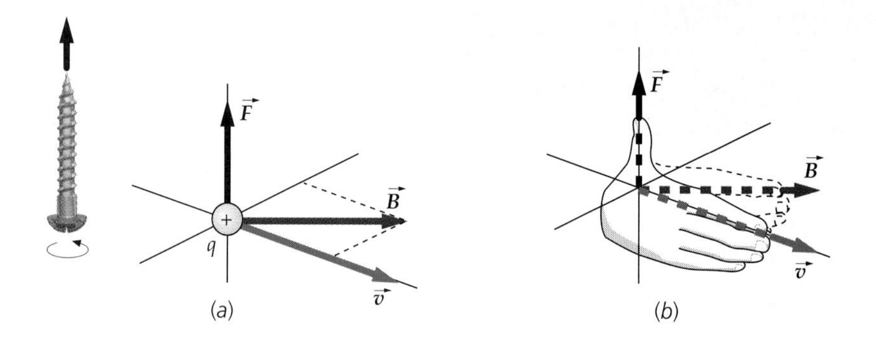

Figure 28-2 Right-hand rule for determining the direction of a force exerted on a charge moving in a magnetic field. (*a*) The force is perpendicular to both $\vec{v}$ and $\vec{B}$ and in the direction of the advance of a right-hand-threaded screw if turned in the same direction as to rotate $\vec{v}$ into $\vec{B}$. (*b*) If the fingers of the right hand are in the direction of $\vec{v}$ such that they can be curled into $\vec{B}$, the thumb points in the direction of $\vec{F}$.

Examples of the direction of the forces exerted on moving charges when the magnetic field vector $\vec{B}$ is in the vertical direction are given in Figure 28-3. Note that the direction of any particular magnetic field $\vec{B}$ can be found experimentally by measuring $\vec{F}$ and $\vec{v}$ for several velocities in different directions and then applying Equation 28-1.

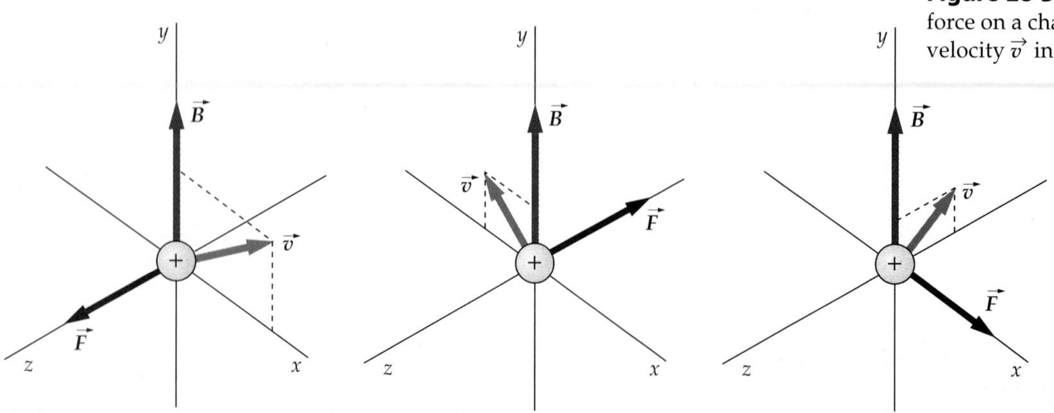

Figure 28-3 Direction of the magnetic force on a charged particle moving with velocity $\vec{v}$ in a magnetic field $\vec{B}$.

Equation 28-1 defines the **magnetic field** $\vec{B}$ in terms of the force exerted on a moving charge. The SI unit of magnetic field is the **tesla** (T). A charge of one coulomb moving with a velocity of one meter per second perpendicular to a magnetic field of one tesla experiences a force of one newton:

$$1\,\mathrm{T} = 1\frac{\mathrm{N/C}}{\mathrm{m/s}} = 1\,\mathrm{N/A\cdot m} \qquad \text{28-2}$$

This unit is rather large. The magnetic field of the earth has a magnitude of less than 10^{-4} T. The magnetic fields near powerful permanent magnets are about 0.1 to 0.5 T, and powerful laboratory and industrial electromagnets produce fields of 1 to 2 T. Fields greater than 10 T are quite difficult to produce because the resulting magnetic forces will tear the magnets apart or crush them. A commonly used unit, derived from the cgs system, is the **gauss** (G), which is related to the tesla as follows:

$$1\,\mathrm{G} = 10^{-4}\,\mathrm{T} \qquad \text{28-3}$$

Definition—Gauss

Since magnetic fields are often given in gauss, which is not an SI unit, remember to convert from gauss to teslas when making calculations.

Example 28-1

The magnetic field of the earth is measured at a point on the surface to have a magnitude of 0.6 G and is directed downward and northward, making an angle of about 70° with the horizontal, as shown in Figure 28-4. (The earth's magnetic field varies from place to place. These data are approximately correct for the central United States.) A proton of charge $q = 1.6 \times 10^{-19}$ C is moving horizontally in the northward direction with speed $v = 10$ Mm/s $= 10^7$ m/s. Calculate the magnetic force on the proton (*a*) using $F = qvB \sin\theta$ and (*b*) by expressing $\vec{v}$ and $\vec{B}$ in terms of the unit vectors $\hat{i}, \hat{j}, \hat{k}$, and computing $\vec{F} = q\vec{v} \times \vec{B}$.

Figure 28-4

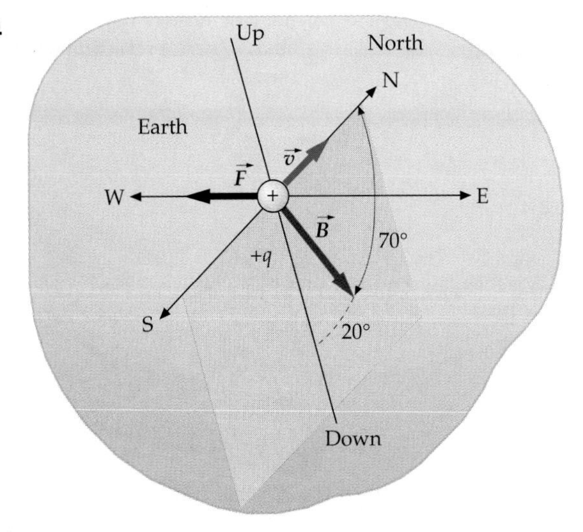

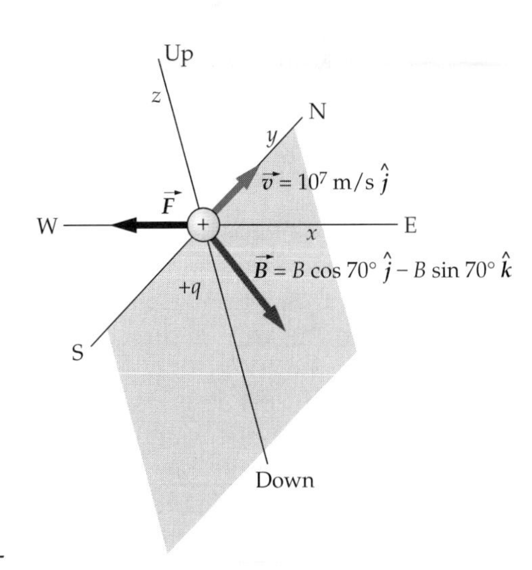

Figure 28-5

Picture the Problem Let the x and y directions be east and north, respectively, and let the z direction be upward (Figure 28-5). The velocity vector is then in the y direction.

(*a*) Calculate $F = qvB \sin\theta$ using $\theta = 70°$. From Figure 28-4 we see that the direction of the force is westward:

$$F = qvB \sin 70°$$
$$= (1.6 \times 10^{-19}\,\mathrm{C})(10^7\,\mathrm{m/s})(0.6 \times 10^{-4}\,\mathrm{T})(0.94)$$
$$= 9.02 \times 10^{-17}\,\mathrm{N}$$

(*b*)1. The magnetic force is the vector product of $q\vec{v}$ and $\vec{B}$: $\vec{F} = q\vec{v} \times \vec{B}$

2. Express $\vec{v}$ and $\vec{B}$ in terms of their components:
$$\vec{v} = v_y\hat{j}$$
$$\vec{B} = B_y\hat{j} + B_z\hat{k}$$

3. Write $\vec{F} = q\vec{v} \times \vec{B}$ in terms of these components:
$$\vec{F} = q\vec{v} \times \vec{B} = q(v_y\hat{j}) \times (B_y\hat{j} + B_z\hat{k})$$
$$= qv_yB_y(\hat{j} \times \hat{j}) + qv_yB_z(\hat{j} \times \hat{k}) = qv_yB_z\hat{i}$$

4. Evaluate $\vec{F}$:
$$\vec{F} = qv_y(-B\sin 70°)\hat{i}$$
$$= -(1.6 \times 10^{-19}\,\mathrm{C})(10^7\,\mathrm{m/s})(0.6 \times 10^{-4}\,\mathrm{T})(0.94)$$
$$= -9.02 \times 10^{-17}\,\mathrm{N}\,\hat{i}$$

Remark Note that the direction of $-\hat{i}$ is westward, so the force is directed westward as shown in Figure 28-5.

Exercise Find the force on a proton moving with velocity $\vec{v} = 4 \times 10^6\,\mathrm{m/s}\,\hat{i}$ in a magnetic field $\vec{B} = 2.0\,\mathrm{T}\,\hat{k}$. (*Answer* $-1.28 \times 10^{-12}\,\mathrm{N}\,\hat{j}$)

When a wire carries a current in a magnetic field, there is a force on the wire that is equal to the sum of the magnetic forces on the charged particles whose motion produces the current. Figure 28-6 shows a short segment of wire of cross-sectional area A and length ℓ carrying a current I. If the wire is in a magnetic field $\vec{B}$, the magnetic force on each charge is $q\vec{v}_\mathrm{d} \times \vec{B}$, where $\vec{v}_\mathrm{d}$ is the drift velocity of the charge carriers, which is the same as their average velocity. The number of charges in the wire segment is the number n per unit volume times the volume $A\ell$. Thus, the total force on the wire segment is

$$\vec{F} = (q\vec{v}_\mathrm{d} \times \vec{B})nA\ell$$

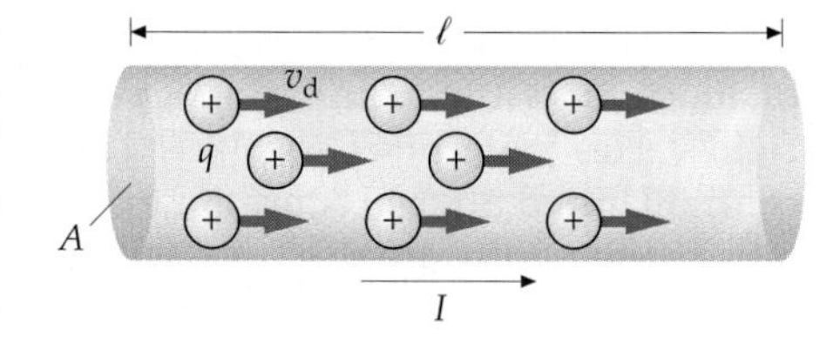

Figure 28-6 Wire segment of length ℓ carrying current I. If the wire is in a magnetic field there will be a force on each charge carrier resulting in a force on the wire.

From Equation 26-3, the current in the wire is

$$I = nqv_\mathrm{d}A$$

Hence the force can be written

$$\vec{F} = I\vec{\ell} \times \vec{B} \tag{28-4}$$

Magnetic force on a segment of current-carrying wire

where $\vec{\ell}$ is a vector whose magnitude is the length of the wire and whose direction is parallel to the current. For the current in the positive x direction and the magnetic field in the xy plane, the force on the wire is directed along the positive z axis, as shown in Figure 28-7.

In Equation 28-4 it is assumed that the wire segment is straight and that the magnetic field does not vary over its length. The equation can be generalized for an arbitrarily shaped wire in any magnetic field. If we choose a very small wire segment $d\vec{\ell}$ and write the force on this segment as $d\vec{F}$, we have

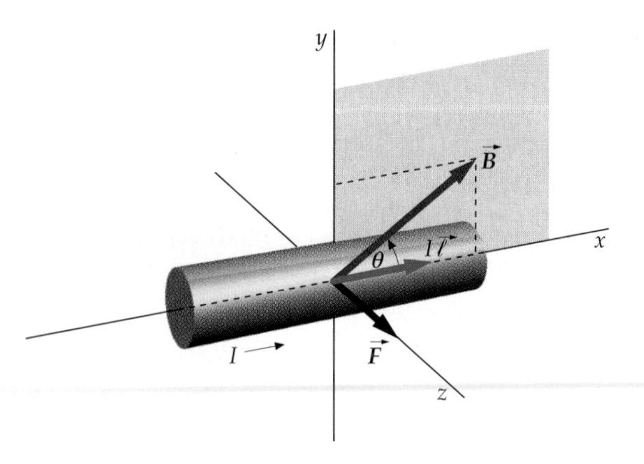

Figure 28-7 Magnetic force on a current-carrying segment of wire in a magnetic field. The current is in the x direction, and the magnetic field is in the xy plane and makes an angle θ with the x axis. The force $\vec{F}$ is in the z direction perpendicular to both $\vec{B}$ and $\vec{\ell}$ and has magnitude $I\ell B\sin\theta$.

$$d\vec{F} = I\,d\vec{\ell} \times \vec{B} \tag{28-5}$$

Magnetic force on a current element

where $\vec{B}$ is the magnetic field vector at the segment. The quantity $I\,d\vec{\ell}$ is called a **current element**. We find the total force on a current-carrying wire by summing (integrating) the forces due to all the current elements in the wire. Equation 28-5 is the same as Equation 28-1 with the current element $I\,d\vec{\ell}$ replacing $q\vec{v}$.

Just as the electric field $\vec{E}$ can be represented by electric field lines, the magnetic field $\vec{B}$ can be represented by **magnetic field lines.** In both cases, the direction of the field is indicated by the direction of the field lines and the magnitude of the field is indicated by their density. There are, however, two important differences between electric field lines and magnetic field lines:

1. Electric field lines are in the direction of the electric force on a positive charge, but the magnetic field lines are perpendicular to the magnetic force on a moving charge.
2. Electric field lines begin on positive charges and end on negative charges; magnetic field lines form closed loops. Since isolated magnetic poles apparently do not exist, there are no points in space where magnetic field lines begin or end.

Figure 28-8 shows the magnetic field lines both inside and outside a bar magnet.

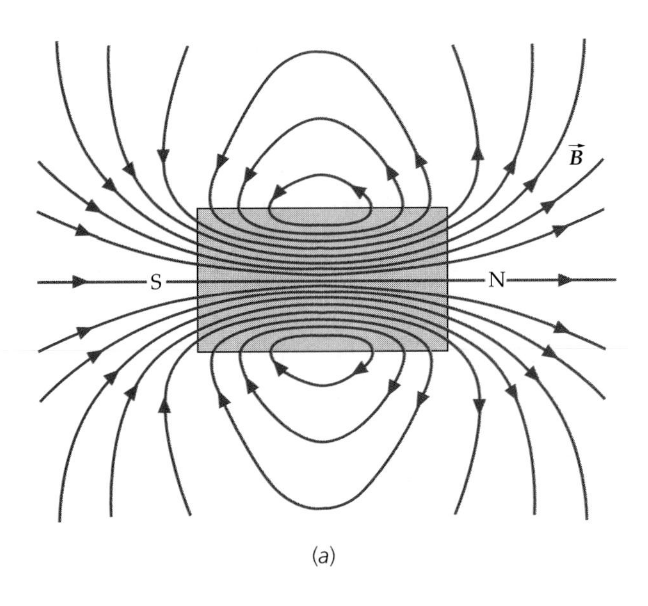

Figure 28-8 (*a*) Magnetic field lines inside and outside a bar magnet. The lines emerge from the north pole and enter the south pole, but they have no beginning or end. Instead, they form closed loops. (*b*) Magnetic field lines outside a bar magnet as indicated by iron filings.

Example 28-2

A wire segment 3 mm long carries a current of 3 A in the x direction. It lies in a magnetic field of magnitude 0.02 T that is in the xy plane and makes an angle of 30° with the x axis, as shown in Figure 28-9. What is the magnetic force exerted on the wire segment?

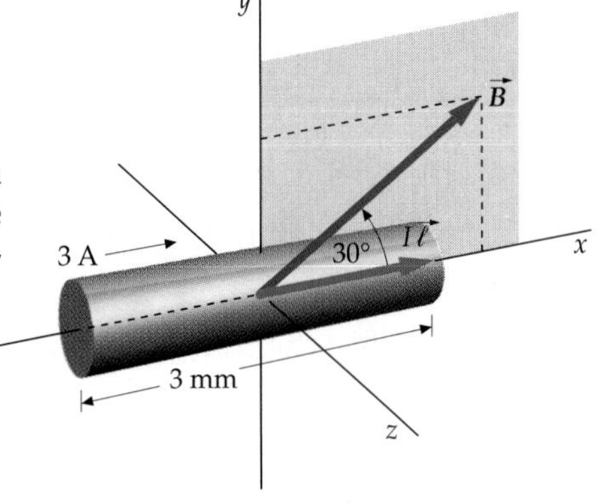

Figure 28-9

1. The magnetic force is in the direction of $\vec{\ell} \times \vec{B}$, which we see from Figure 28-9 is in the z direction. The magnetic force is given by Equation 28-4:

$$\vec{F} = I\vec{\ell} \times \vec{B} = I\ell B \sin 30^\circ\,\hat{k}$$
$$= (3.0\text{ A})(0.003\text{ m})(0.02\text{ T})(\sin 30^\circ)\hat{k}$$
$$= 9 \times 10^{-5}\text{ N}\,\hat{k}$$

Example 28-3

A wire bent into a semicircular loop of radius R lies in the xy plane. It carries a current I from point a to point b, as shown in Figure 28-10. There is a uniform magnetic field $\vec{B} = B\hat{k}$ perpendicular to the plane of the loop. Find the force acting on the semicircular loop part of the wire.

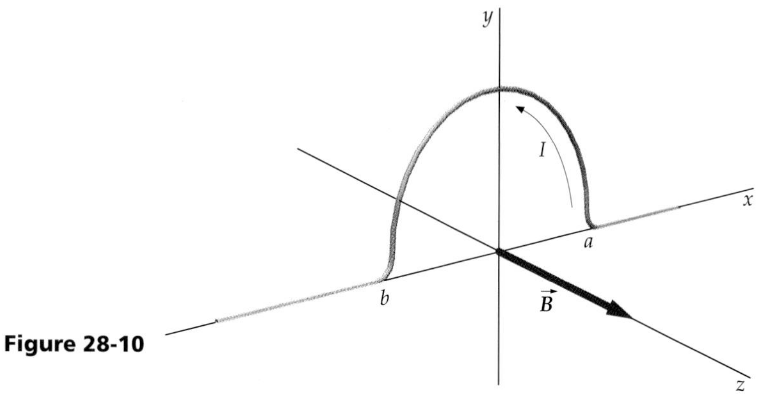

Figure 28-10

Picture the Problem The force $d\vec{F}$ exerted on a segment of the semicircular wire lies in the xy plane, as shown in Figure 28-11. We find the total force by expressing the x and y components of $d\vec{F}$ in terms of θ and integrating them separately from $\theta = 0$ to $\theta = \pi$.

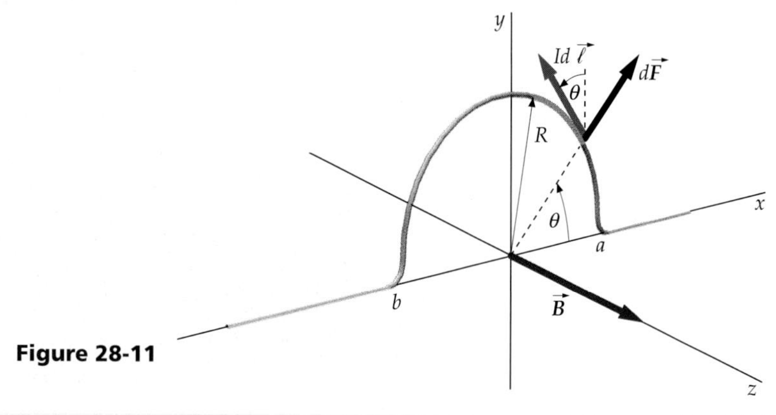

Figure 28-11

1. Write the force $d\vec{F}$ on a current element $d\vec{\ell}$:
$$d\vec{F} = I\,d\vec{\ell} \times \vec{B}$$

2. Express $d\vec{\ell}$ in terms of the unit vectors $\hat{i}$ and $\hat{j}$:
$$d\vec{\ell} = -d\ell \sin\theta\,\hat{i} + d\ell \cos\theta\,\hat{j}$$

3. Compute $I\,d\vec{\ell} \times \vec{B}$ using $d\ell = R\,d\theta$, and $\vec{B} = B\hat{k}$:
$$d\vec{F} = I\,d\vec{\ell} \times \vec{B}$$
$$= (-IR\sin\theta\,d\theta\,\hat{i} + IR\cos\theta\,d\theta\,\hat{j}) \times B\hat{k}$$
$$= IRB\sin\theta\,d\theta\,\hat{j} + IRB\cos\theta\,d\theta\,\hat{i}$$

4. Integrate each component of $d\vec{F}$ from $\theta = 0$ to $\theta = \pi$:
$$\vec{F} = IRB\hat{i}\int_0^\pi \cos\theta\,d\theta + IRB\hat{j}\int_0^\pi \sin\theta\,d\theta$$
$$\vec{F} = IRB\hat{i}(0) + IRB\hat{j}(2) = 2IRB\hat{j}$$

Check the Result The result that the x component of $\vec{F}$ is zero can be seen from symmetry. For the right half of the loop, $d\vec{F}$ points to the right; for the left half, $d\vec{F}$ points to the left.

Remark The net force on the semicircular wire is the same as if the semicircle were replaced by a straight-line segment of length $2R$ connecting points a and b. (This is a general result, as shown in Problem 16.)

28-2 Motion of a Point Charge in a Magnetic Field

The magnetic force on a charged particle moving through a magnetic field is always perpendicular to the velocity of the particle. The magnetic force thus changes the direction of the velocity but not its magnitude. *Therefore, magnetic fields do no work on particles and do not change their kinetic energy.*

In the special case where the velocity of a particle is perpendicular to a uniform magnetic field, as shown in Figure 28-12, the particle moves in a circular orbit. The magnetic force provides the centripetal force necessary for the centripetal acceleration v^2/r in circular motion. We can use Newton's second law to relate the radius of the circle to the magnetic field and the speed of the particle. If the velocity is $\vec{v}$, the magnitude of the net force is qvB, since $\vec{v}$ and $\vec{B}$ are perpendicular. Newton's second law gives

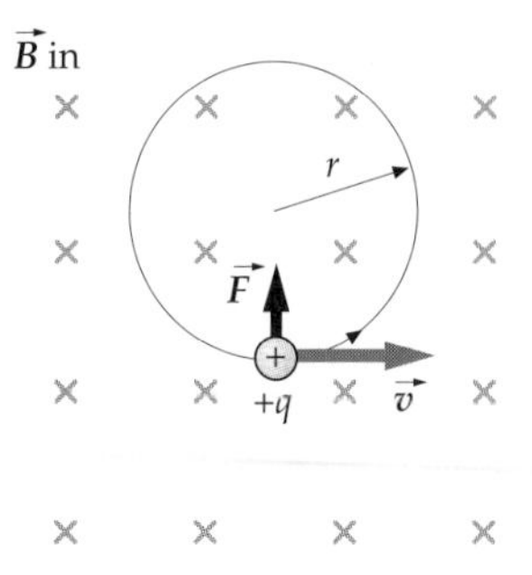

Figure 28-12 Charged particle moving in a plane perpendicular to a uniform magnetic field. The magnetic field is into the page as indicated by the crosses. (A field out of the plane of the page would be indicated by dots). The magnetic force is perpendicular to the velocity of the particle, causing it to move in a circular orbit.

$$F = ma = m\frac{v^2}{r}$$

$$qvB = \frac{mv^2}{r}$$

or

$$r = \frac{mv}{qB} \tag{28-6}$$

The period of the circular motion is the time it takes the particle to travel once around the circumference of the circle. The period is related to the speed by

$$T = \frac{2\pi r}{v}$$

Substituting in $r = mv/qB$ from Equation 28-6, we obtain the period of the particle's circular motion, called the **cyclotron period**:

$$T = \frac{2\pi(mv/qB)}{v} = \frac{2\pi m}{qB} \tag{28-7}$$

Cyclotron period

The frequency of the circular motion, called the **cyclotron frequency,** is the reciprocal of the period:

$$f = \frac{1}{T} = \frac{qB}{2\pi m} \tag{28-8}$$

Cyclotron frequency

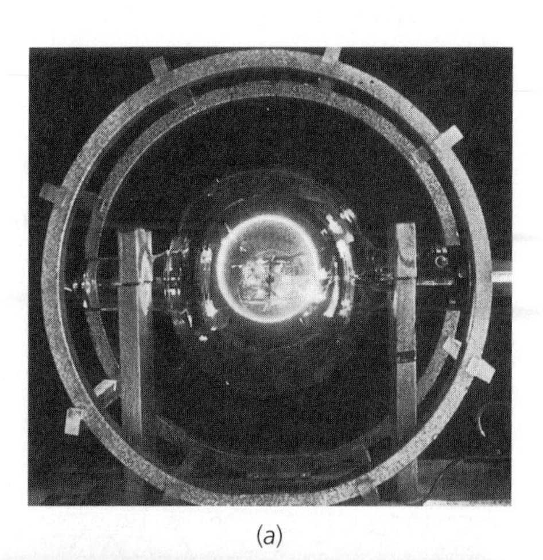

(a)

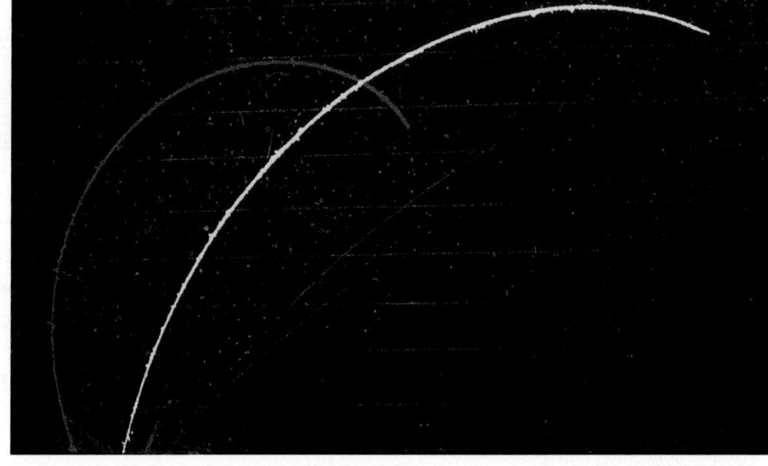

(b)

(*a*) Circular path of electrons moving in the magnetic field produced by two large coils. The electrons ionize the gas in the tube, causing it to give off a bluish glow that indicates the path of the beam. (*b*) False-color photograph showing tracks of a 1.6-MeV proton (red) and a 7-MeV α particle (yellow) in a cloud chamber. The radius of curvature is proportional to the momentum and inversely proportional to the charge of the particle. For these energies, the momentum of the α particle, which has twice the charge of the proton, is about four times that of the proton and so its radius of curvature is greater.

Note that the period and the frequency given by Equations 28-7 and 28-8 depend on the charge-to-mass ratio q/m but are independent of the radius r or velocity v. Two important applications of the circular motion of charged particles in a uniform magnetic field, the mass spectrometer and the cyclotron, are discussed later in this section.

Example 28-4

A proton of mass $m = 1.67 \times 10^{-27}$ kg and charge $q = e = 1.6 \times 10^{-19}$ C moves in a circle of radius 21 cm perpendicular to a magnetic field $B = 4000$ G. Find (*a*) the period of the motion and (*b*) the speed of the proton.

(*a*) Calculate the period T from Equation 28-7 with $B = 0.4$ T:

$$T = \frac{2\pi m}{qB} = \frac{2\pi(1.67 \times 10^{-27}\ \text{kg})}{(1.6 \times 10^{-19}\ \text{C})(0.4\ \text{T})} = 1.64 \times 10^{-7}\ \text{s}$$

(*b*) Calculate the speed v from Equation 28-6:

$$v = \frac{rqB}{m} = \frac{(0.21\ \text{m})(1.6 \times 10^{-19}\ \text{C})(0.4\ \text{T})}{1.67 \times 10^{-27}\ \text{kg}}$$

$$= 8.05 \times 10^{6}\ \text{m/s}$$

Remark The radius of the circular motion is proportional to the speed, but the period is independent of both the speed and radius.

Check the Result Note that the product of the speed v and the period T equals the circumference of the circle $2\pi r$ as expected: $vT = (8.05 \times 10^{6}\ \text{m/s})(1.64 \times 10^{-7}\ \text{s}) = 1.32\ \text{m}$; $2\pi r = 2\pi(0.21\ \text{m}) = 1.32\ \text{m}$.

Suppose that a charged particle enters a uniform magnetic field with a velocity that is not perpendicular to $\vec{B}$. We can resolve the velocity of the particle into components $v_{\parallel}$ parallel to $\vec{B}$ and $v_{\perp}$ perpendicular to $\vec{B}$. The motion due to the perpendicular component is the same as that just discussed. The component of the velocity parallel to $\vec{B}$ is not affected by the magnetic field and therefore remains constant. The path of the particle is thus a helix, as shown in Figure 28-13.

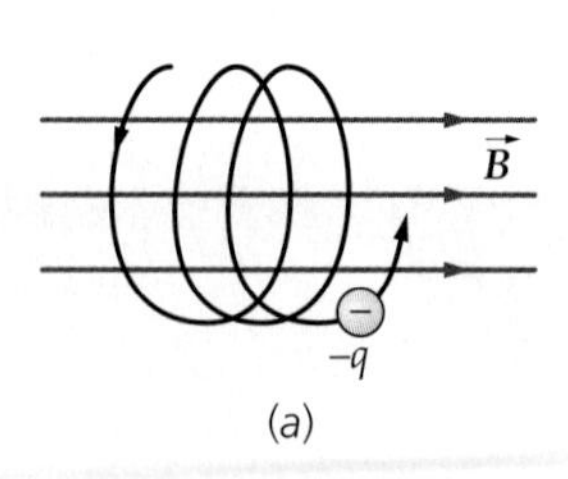

(*a*)

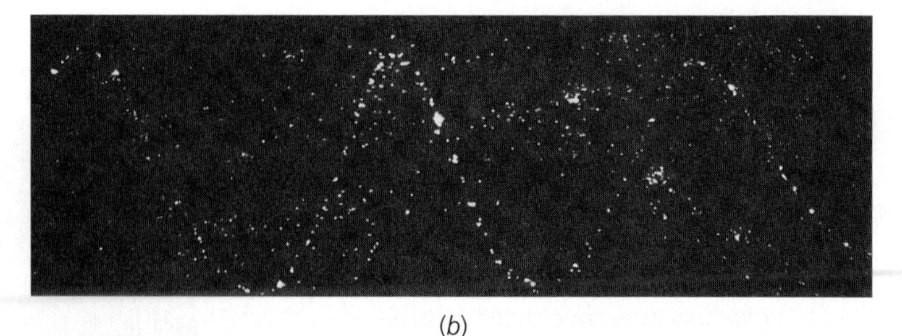

(*b*)

Figure 28-13 (*a*) When a particle has a velocity component parallel to a magnetic field as well as one perpendicular to the field, it moves in a helical path around the field lines. (*b*) Cloud-chamber photograph of the helical path of an electron moving in a magnetic field. The path of the electron is made visible by the condensation of water droplets in the cloud chamber.

The motion of charged particles in nonuniform magnetic fields can be quite complex. Figure 28-14 shows a **magnetic bottle,** an interesting magnetic field configuration in which the field is weak at the center and strong at both ends. A detailed analysis of the motion of a charged particle in such a field shows that the particle spirals around the field lines and becomes trapped, oscillating back and forth between points P_1 and P_2 in the figure. Such magnetic field configurations are used to confine dense beams of

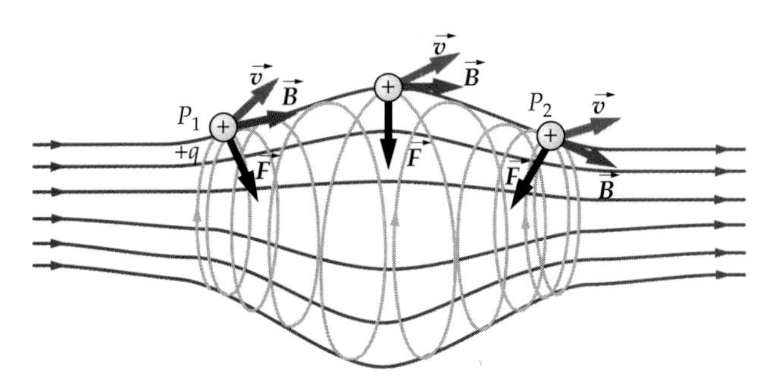

Figure 28-14 Magnetic bottle. When a charged particle moves in such a field, which is strong at both ends and weak in the middle, the particle becomes trapped and moves back and forth spiraling around the field lines.

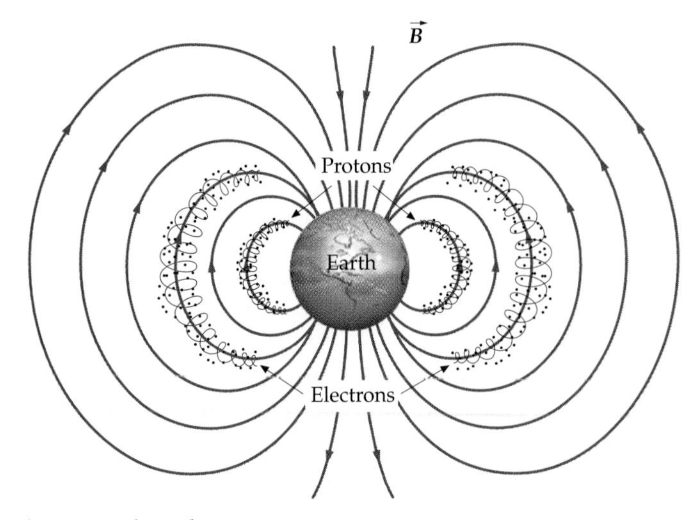

Figure 28-15 Van Allen belts. Protons (inner belts) and electrons (outer belts) are trapped in the earth's magnetic field and spiral around the field lines between the north and south poles.

charged particles, called *plasmas,* in nuclear fusion research. A similar phenomenon is the oscillation of ions back and forth between the earth's magnetic poles in the Van Allen belts (Figure 28-15).

The Velocity Selector

The magnetic force on a charged particle moving in a uniform magnetic field can be balanced by an electric force if the magnitudes and directions of the magnetic and electric fields are properly chosen. Since the electric force is in the direction of the electric field (for positive particles) and the magnetic force is perpendicular to the magnetic field, the electric and magnetic fields in the region through which the particle is moving must be perpendicular to each other if the forces are to balance. Such a region is said to have **crossed fields.**

Figure 28-16 shows a region of space between the plates of a capacitor where there is an electric field and a perpendicular magnetic field (produced by a magnet with poles above and below the paper). Consider a particle of charge q entering this space from the left. The net force on the particle is

$$\vec{F} = q\vec{E} + q\vec{v} \times \vec{B}$$

If q is positive, the electric force of magnitude qE is down and the magnetic force of magnitude qvB is up. If the charge is negative, each of these forces is reversed. The two forces balance if $qE = qvB$ or

$$v = \frac{E}{B} \qquad \text{28-9}$$

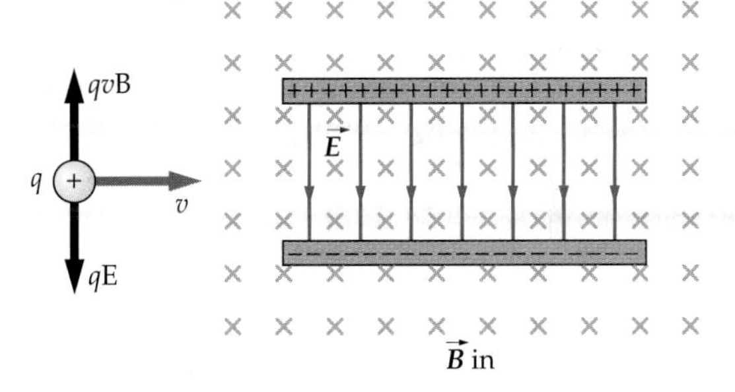

Figure 28-16 Crossed electric and magnetic fields. When a positive particle moves to the right, it experiences a downward electric force and an upward magnetic force. These forces balance if the speed of the particle is related to the field strengths by $vB = E$.

For given magnitudes of the electric and magnetic fields, the forces balance only for particles with the speed given by Equation 28-9. Any particle with this speed, regardless of its mass or charge, will traverse the space undeflected. A particle with a greater speed will be deflected in the direction of the magnetic force, and one with less speed will be deflected in the direction of the electric force. This arrangement of fields is often used as a **velocity selector,** a device that allows only particles with speed given by Equation 28-9 to pass.

Exercise A proton is moving in the x direction in a region of crossed fields where $\vec{E} = 2 \times 10^5$ N/C $\hat{k}$ and $\vec{B} = -3000$ G $\hat{j}$. (*a*) What is the speed of the proton if it is not deflected? (*b*) If the proton moves with twice this speed, in which direction will it be deflected? (*Answers* (*a*) 667 km/s (*b*) in the negative z direction)

optional

optional

Thomson's Measurement of q/m for Electrons

An example of the use of crossed electric and magnetic fields is the famous experiment performed by J. J. Thomson in 1897 in which he showed that the rays of a cathode-ray tube can be deflected by electric and magnetic fields, indicating that they must consist of charged particles. By measuring the deflections of these particles Thomson showed that all the particles have the same charge-to-mass ratio q/m. He also showed that particles with this charge-to-mass ratio can be obtained using any material for a source, which means that these particles, now called electrons, are a fundamental constituent of all matter.

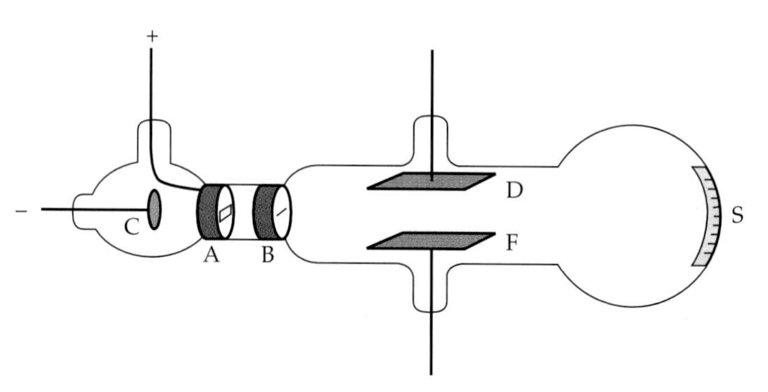

Figure 28-17 Thomson's tube for measuring q/m for the particles of cathode rays (electrons). Electrons from the cathode C pass through the slits at A and B and strike a phosphorescent screen S. The beam can be deflected by an electric field between plates D and F or by a magnetic field (not shown).

Figure 28-17 shows a schematic diagram of the cathode-ray tube Thomson used. Electrons are emitted from the cathode C, which is at a negative potential relative to the slits A and B. An electric field in the direction from A to C accelerates the electrons, and they pass through slits A and B into a field-free region. The electrons then enter the electric field between the capacitor plates D and F that is perpendicular to the velocity of the electrons. This field accelerates them vertically for the short time they are between the plates. The electrons are deflected and strike the phosphorescent screen S at the far right side of the tube at some deflection Δy from the point at which they strike when there is no field between the plates. The screen glows where the electrons strike it, indicating the location of the beam. The initial speed of the electrons v_0 is determined by introducing a magnetic field $\vec{B}$ between the plates in a direction that is perpendicular to both the electric field and the initial velocity of the electrons. The magnitude of $\vec{B}$ is adjusted until the beam is not deflected. The speed is then found from Equation 28-9.

With the magnetic field turned off, the beam is deflected by an amount Δy, which consists of two parts: the deflection Δy_1, which occurs while the electrons are between the plates, and the deflection Δy_2, which occurs after the electrons leave the region between the plates (Figure 28-18).

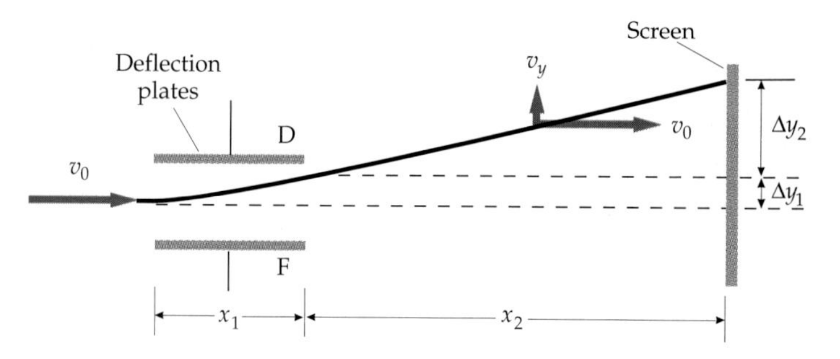

Figure 28-18 The total deflection of the beam in the J.J. Thomson experiments consists of the deflection y_1 while the electrons are between the plates plus the deflection y_2 that occurs in the field-free region between the plates and the screen.

Let x_1 be the horizontal distance across the deflection plates D and F. If the electron is moving horizontally with speed v_0 when it enters the plates, the time spent between the plates is $t_1 = x_1/v_0$, and the vertical velocity when it leaves the plates is

$$v_y = at_1 = \frac{qE}{m}t_1 = \frac{qE}{m}\frac{x_1}{v_0}$$

where E is the electric field between the plates. The deflection in this region is

$$\Delta y_1 = \frac{1}{2}at_1^2 = \frac{1}{2}\frac{qE}{m}\left(\frac{x_1}{v_0}\right)^2$$

The electron then travels an additional horizontal distance x_2 in the field-free region from the deflection plates to the screen. Since the velocity of the electron is constant in this region, the time to reach the screen is $t_2 = x_2/v_0$, and the additional vertical deflection is

$$\Delta y_2 = v_y t_2 = \frac{qE}{m}\frac{x_1}{v_0}\frac{x_2}{v_0}$$

The total deflection at the screen is therefore

$$\Delta y = \Delta y_1 + \Delta y_2 = \frac{1}{2}\frac{qE}{m}\left(\frac{x_1}{v_0}\right)^2 + \frac{qE}{m}\frac{x_1 x_2}{v_0^2} \qquad 28\text{-}10$$

The measured deflection Δy can be used to determine the charge-to-mass ratio, q/m, from Equation 28-10.

Example 28-5

Electrons pass undeflected through the plates of Thomson's apparatus when the electric field is 3000 V/m and there is a crossed magnetic field of 1.40 G. If the plates are 4 cm long and the ends of the plates are 30 cm from the screen, find the deflection on the screen when the magnetic field is turned off.

Picture the Problem The mass and charge of the electron are known: $m = 9.11 \times 10^{-31}$ kg and $q = e = 1.6 \times 10^{-19}$ C. The speed of the electron can be found from the ratio of the magnetic and electric fields.

1. The total deflection of the electron is given by Equation 28-10:

$$\Delta y = \Delta y_1 + \Delta y_2 = \frac{1}{2}\frac{qE}{m}\left(\frac{x_1}{v_0}\right)^2 + \frac{qE}{m}\frac{x_1 x_2}{v_0^2}$$

2. The speed v_0 equals E/B:

$$v_0 = \frac{E}{B} = \frac{3000\ \text{V/m}}{1.40 \times 10^{-4}\ \text{T}} = 2.14 \times 10^7\ \text{m/s}$$

3. Substitute this value for v_0, the given value of E, and the known values for m and q to find Δy:

$$\Delta y = \frac{1}{2}\frac{(1.6 \times 10^{-19}\ \text{C})(3000\ \text{V/m})}{9.11 \times 10^{-31}\ \text{kg}}\left(\frac{0.04\ \text{m}}{2.14 \times 10^7\ \text{m/s}}\right)^2$$

$$+ \frac{(1.6 \times 10^{-19}\ \text{C})(3000\ \text{V/m})}{9.11 \times 10^{-31}\ \text{kg}}\frac{(0.04\ \text{m})(0.30\ \text{m})}{(2.14 \times 10^7\ \text{m/s})^2}$$

$$= 9.20 \times 10^{-4}\ \text{m} + 1.38 \times 10^{-2}\ \text{m}$$

$$= 0.92\ \text{mm} + 13.8\ \text{mm} = 14.7\ \text{mm}$$

The Mass Spectrometer

The **mass spectrometer,** first designed by Francis William Aston in 1919, was developed as a means of measuring the masses of isotopes. Such measurements are an important way of determining both the presence of isotopes and their abundance in nature. For example, natural magnesium has been found to consist of 78.7% ^{24}Mg, 10.1% ^{25}Mg, and 11.2% ^{26}Mg. These isotopes have masses in the approximate ratio 24:25:26.

Figure 28-19 shows a simple schematic drawing of a mass spectrometer. Ions from an ion source are accelerated by an electric field and enter a uniform magnetic field. If the ions start from rest and move through a potential drop ΔV, their kinetic energy when they enter the magnetic field equals their loss in potential energy, $q\,\Delta V$:

$$\tfrac{1}{2}mv^2 = q\,\Delta V \qquad 28\text{-}11$$

The ions move in a semicircle of radius r given by Equation 28-6, $r = mv/qB$, and strike a photographic plate at point P_2, a distance $2r$ from the point P_1 where they entered the magnetic field.

The speed v can be eliminated from Equations 28-6 and 28-11 to find m/q in terms of the known quantities ΔV, B, and r. We first solve Equation 28-6 for v

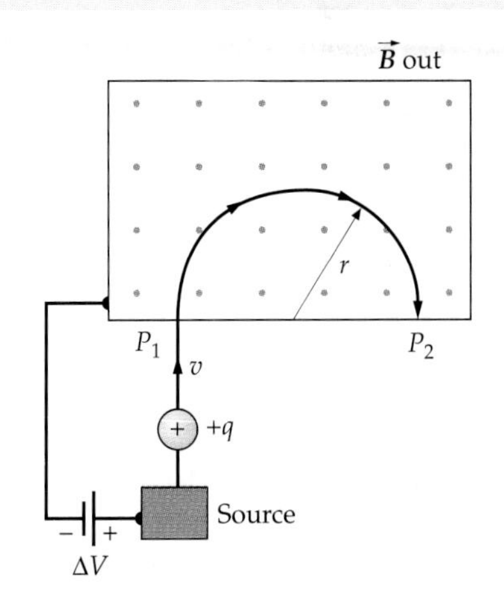

Figure 28-19 Schematic drawing of a mass spectrometer. Ions from an ion source are accelerated through a potential difference ΔV and enter a uniform magnetic field. The magnetic field is out of the plane of the page as indicated by the dots. The ions are bent into a circular arc and emerge at P_2. The radius of the circle varies with the mass of the ion.

optional

and square each term, which gives

$$v^2 = \frac{r^2q^2B^2}{m^2}$$

Substituting this expression for v^2 into Equation 28-11, we obtain

$$\frac{1}{2}m\left(\frac{r^2q^2B^2}{m^2}\right) = q\,\Delta V$$

Simplifying this equation and solving it for m/q, we obtain

$$\frac{m}{q} = \frac{B^2r^2}{2\,\Delta V} \qquad \text{28-12}$$

In Aston's original mass spectrometer, mass differences could be measured to a precision of about 1 part in 10,000. The precision has been improved by introducing a velocity selector between the ion source and the magnet, which increases the degree of accuracy with which the velocities of the incoming ions can be determined.

Example 28-6

A ^{58}Ni ion of charge $+e$ and mass 9.62×10^{-26} kg is accelerated through a potential difference of 3 kV and deflected in a magnetic field of 0.12 T. (*a*) Find the radius of curvature of the orbit of the ion. (*b*) Find the difference in the radii of curvature of ^{58}Ni ions and ^{60}Ni ions. (Assume that the mass ratio is 58/60.)

Picture the Problem The radius of curvature r can be found using Equation 28-12. Using the mass dependence of r, we can find the radius for ^{60}Ni ions from the radius for ^{58}Ni ions, and then take the difference.

(*a*) Solve Equation 28-12 for r:

$$r = \sqrt{\frac{2m\,\Delta V}{qB^2}} = \left[\frac{2(9.62 \times 10^{-26}\text{ kg})(3000\text{ V})}{(1.6 \times 10^{-19}\text{ C})(0.12\text{ T})^2}\right]^{1/2} = 0.501\text{ m}$$

(*b*)1. Let r_1 and r_2 be the radius of the orbit of the ^{58}Ni ion and the ^{60}Ni ion, respectively. Use the result in (*a*) to find the ratio of r_2 to r_1:

$$\frac{r_2}{r_1} = \sqrt{\frac{m_2}{m_1}} = \sqrt{\frac{60}{58}} = 1.017$$

2. Use the result of the previous step to calculate r_2 for ^{60}Ni:

$$r_2 = 1.017r_1 = (1.017)(0.501\text{ m}) = 0.510\text{ m}$$

3. The difference in orbital radii is $r_2 - r_1$:

$$r_2 - r_1 = 0.510\text{ m} - 0.501\text{ m} = 0.009\text{ m} = 9\text{ mm}$$

The Cyclotron

The cyclotron was invented by E. O. Lawrence and M. S. Livingston in 1934 to accelerate particles such as protons or deuterons to high kinetic energies.* The high-energy particles are used to bombard atomic nuclei, causing nuclear reactions that are then studied to obtain information about the nucleus. High-energy protons and deuterons are also used to produce radioactive materials and for medical purposes.

Figure 28-20 is a schematic drawing of a cyclotron. The particles move in two semicircular metal containers called *dees*, after their shape. The dees are housed in a vacuum chamber that is in a uniform magnetic field provided by

* A deuteron is the nucleus of heavy hydrogen, ^{2}H, which consists of a proton and neutron tightly bound together.

an electromagnet. The region in which the particles move must be evacuated so that the particles will not be scattered in collisions with air molecules and lose energy. A potential difference ΔV, which alternates in time with a period T, is maintained between the dees. The period is chosen to be the cyclotron period $T = 2\pi m/qB$ (Equation 28-7). The potential difference creates an electric field across the gap between the dees. At the same time, there is no electric field within each dee because the metal dees are shielded.

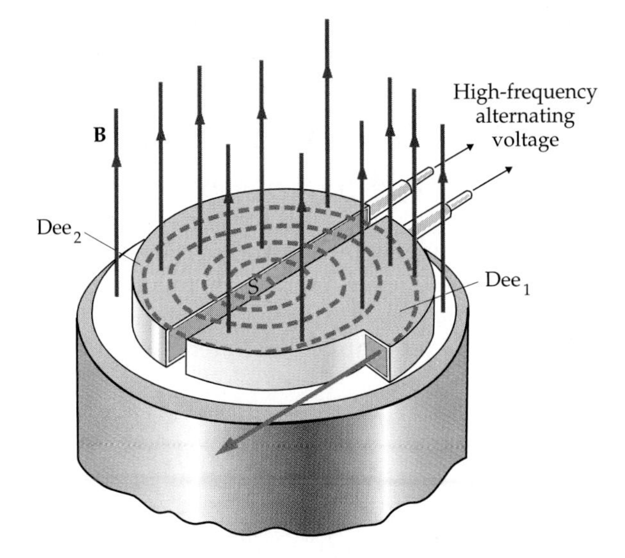

Figure 28-20 Schematic drawing of a cyclotron. The upper pole face of the magnet has been omitted. Charged particles such as protons are accelerated from a source at the center by the potential difference across the gap between the dees. When they arrive at the gap again, the potential difference has changed sign so they are again accelerated across the gap, and move in a larger circle. The potential difference across the gap alternates with the cyclotron frequency of the particle, which is independent of the radius of the circle.

Positively charged particles are initially injected into dee$_1$ with a small velocity from an ion source S near the center of the dees. They move in a semicircle in dee$_1$ and arrive at the gap between dee$_1$ and dee$_2$ after a time $\frac{1}{2}T$. The potential is adjusted so that dee$_1$ is at a higher potential than dee$_2$ when the particles arrive at the gap between them. Each particle is therefore accelerated across the gap by the electric field and gains kinetic energy equal to $q\,\Delta V$.

Because it now has more kinetic energy, the particle moves in a semicircle of larger radius in dee$_2$. It arrives at the gap again after a time $\frac{1}{2}T$, because the period is independent of the particle's speed. By this time, the potential between the dees has been reversed so that dee$_2$ is now at the higher potential. Once more the particle is accelerated across the gap and gains additional kinetic energy equal to $q\,\Delta V$. Each time the particle arrives at the gap, it is accelerated and gains kinetic energy equal to $q\,\Delta V$. Thus, it moves in larger and larger semicircular orbits until it eventually leaves the magnetic field. In the typical cyclotron, each particle may make 50 to 100 revolutions and exit with energies of up to several hundred mega-electron-volts.

The kinetic energy of a particle leaving a cyclotron can be calculated by setting r in Equation 28-6 equal to the maximum radius of the dees and solving the equation for v:

$$r = \frac{mv}{qB}, \qquad v = \frac{qBr}{m}$$

Then, $$K = \frac{1}{2}mv^2 = \frac{1}{2}\left(\frac{q^2B^2}{m}\right)r^2 \qquad \text{28-13}$$

Example 28-7

A cyclotron for accelerating protons has a magnetic field of 1.5 T and a maximum radius of 0.5 m. (*a*) What is the cyclotron frequency? (*b*) What is the kinetic energy of the protons when they emerge?

(*a*) The cyclotron frequency is given by Equation 28-8:

$$f = \frac{qB}{2\pi m} = \frac{(1.6 \times 10^{-19}\,\text{C})(1.5\,\text{T})}{2\pi(1.67 \times 10^{-27}\,\text{kg})} = 2.29 \times 10^{7}\,\text{Hz}$$

$$= 22.9\,\text{MHz}$$

(*b*)1. The kinetic energy of the emerging protons is given by Equation 28-13:

$$K = \frac{1}{2}\left[\frac{(1.6 \times 10^{-19}\,\text{C})^2\,(1.5\,\text{T})^2}{1.67 \times 10^{-27}\,\text{kg}}\right](0.5\,\text{m})^2 = 4.31 \times 10^{-12}\,\text{J}$$

2. The energies of protons and other elementary particles are usually expressed in electron volts. Use 1 eV $= 1.6 \times 10^{-19}$ J to convert to eV:

$$K = 4.31 \times 10^{-12}\,\text{J} \times \frac{1\,\text{eV}}{1.6 \times 10^{-19}\,\text{J}} = 26.9\,\text{MeV}$$

optional

28-3 Torques on Current Loops and Magnets

A current-carrying loop experiences no net force in a uniform magnetic field, but it does experience a torque that tends to twist it. The orientation of the loop can be described conveniently by a unit vector $\hat{n}$ that is perpendicular to the plane of the loop as illustrated in Figure 28-21.

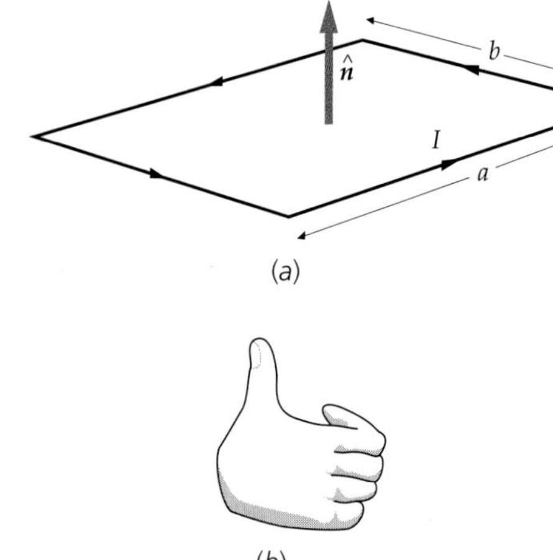

Figure 28-21 (*a*) The orientation of a current loop is described by the unit vector $\hat{n}$ perpendicular to the plane of the loop. (*b*) Right-hand rule for determining the sense of $\hat{n}$. When the fingers of the right hand curl around the loop in the direction of the current, the thumb points in the direction of $\hat{n}$.

Figure 28-22 shows the forces exerted by a uniform magnetic field on a rectangular loop whose normal unit vector $\hat{n}$ makes an angle θ with the magnetic field $\vec{B}$. The net force on the loop is zero. The forces F_1 and F_2 have the magnitude

$$F_1 = F_2 = IaB$$

These forces form a couple so the torque is the same about any point. Point P in Figure 28-22 is a convenient point about which to compute the torque. The magnitude of the torque is

$$\tau = F_2 b \sin\theta = IaBb \sin\theta = IAB \sin\theta$$

where $A = ab$ is the area of the loop. For a loop with N turns, the torque has the magnitude

$$\tau = NIAB \sin\theta$$

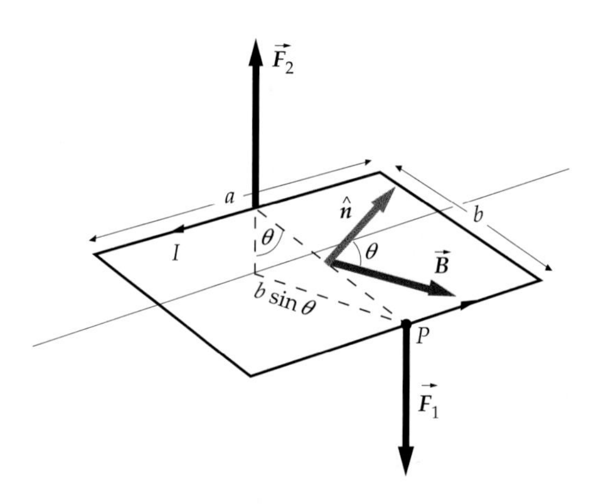

Figure 28-22 Rectangular current loop whose unit normal $\hat{n}$ makes an angle θ with a uniform magnetic field **B**. The torque on the loop has magnitude $IAB \sin\theta$ and is in the direction such that $\hat{n}$ tends to rotate into **B**.

This torque tends to twist the loop so that its plane is perpendicular to $\vec{B}$ (i.e., so that $\vec{n}$ is in the same direction as $\vec{B}$).

The torque can be written conveniently in terms of the **magnetic dipole moment** μ (also referred to simply as the **magnetic moment**) of the current loop, which is defined as

$$\vec{\mu} = NIA\hat{n} \tag{28-14}$$

Magnetic dipole moment of a current loop

The SI unit of magnetic moment is the ampere-meter^2 (A·m^2). In terms of the magnetic dipole moment, the torque on the current loop is given by

$$\vec{\tau} = \vec{\mu} \times \vec{B} \tag{28-15}$$

Torque on a current loop

Equation 28-15, which we have derived for a rectangular loop, holds in general for a flat loop of any shape. The torque on any loop is the cross product of the magnetic moment $\vec{\mu}$ of the loop and the magnetic field $\vec{B}$, where the magnetic moment is defined to be a vector that is perpendicular to the area of the loop (Figure 28-23) and has magnitude equal to NIA. Comparing Equation 28-15 with Equation 22-11 for the torque on an electric dipole, we see that a current loop in a magnetic field acts like an electric dipole in an electric field.

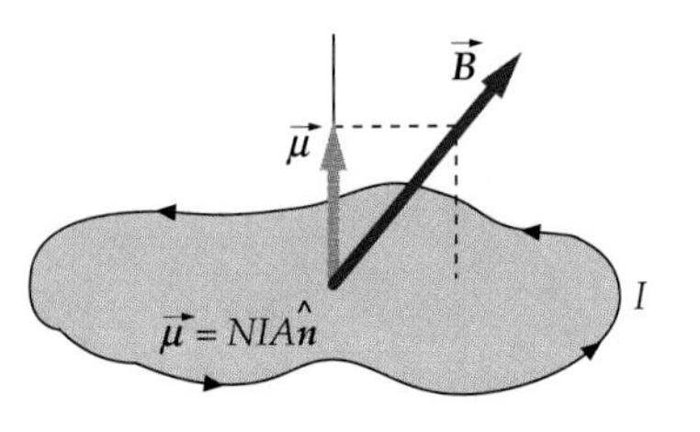

Figure 28-23 A current loop of arbitrary shape is described by its magnetic moment $\vec{\mu} = NIA\hat{n}$. In a magnetic field $\vec{B}$, it experiences a torque $\vec{\mu} \times \vec{B}$.

Example 28-8

A circular loop of radius 2 cm has 10 turns of wire and carries a current of 3 A. The axis of the loop makes an angle of 30° with a magnetic field of 8000 G. Find the magnitude of the torque on the loop.

1. The magnitude of the torque is given by Equation 28-15: $\tau = |\vec{\mu} \times \vec{B}| = \mu B \sin\theta = \mu B \sin 30°$

2. Calculate the magnitude of the magnetic moment of the loop: $\mu = NIA = (10)(3\text{ A})\pi(0.02\text{ m})^2 = 3.77 \times 10^{-2}\text{ A}\cdot\text{m}^2$

3. Substitute this value of μ and the given value of B to calculate τ: $\tau = \mu B \sin\theta = (3.77 \times 10^{-2}\text{ A}\cdot\text{m}^2)(0.8\text{ T})(\sin 30°)$
$= 1.51 \times 10^{-2}\text{ N}\cdot\text{m}$

Example 28-9 *try it yourself*

A circular wire loop of radius R, mass m, and current I lies on a rough surface (Figure 28-24). There is a horizontal magnetic field $\vec{B}$. How large can the current I be before one edge of the loop will lift off the surface?

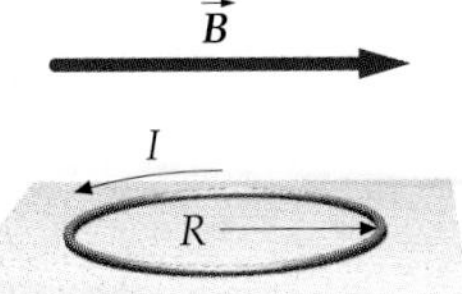

Figure 28-24

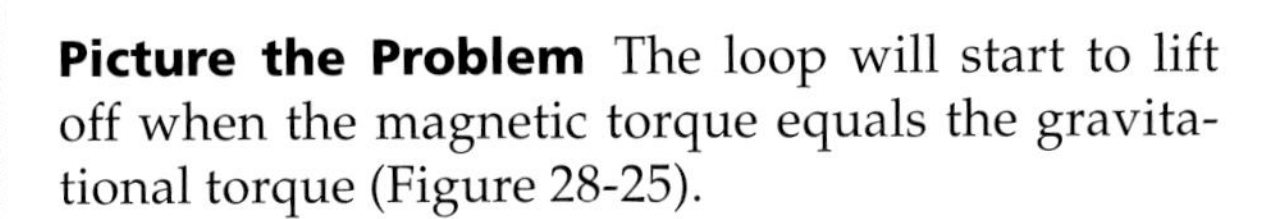

Picture the Problem The loop will start to lift off when the magnetic torque equals the gravitational torque (Figure 28-25).

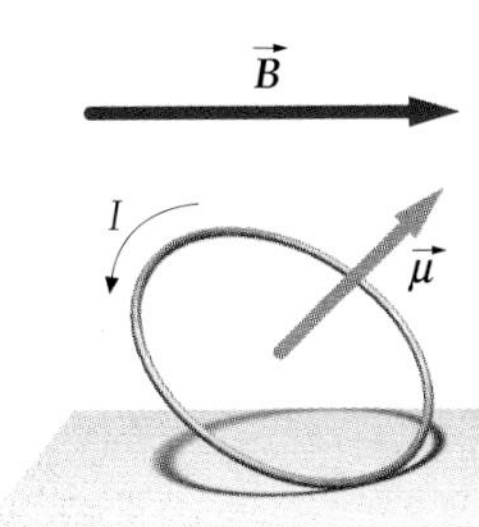

Figure 28-25

Cover the column to the right and try these on your own before looking at the answers.

Steps	*Answers*
1. Find the magnetic torque acting on the loop.	$\tau_m = \mu B = I\pi R^2 B$
2. Find the gravitational torque exerted on the loop.	$\tau_g = mgR$
3. Equate the torques and solve for I.	$I = \dfrac{mg}{\pi RB}$

Potential Energy of a Magnetic Dipole in a Magnetic Field

When a torque is exerted through an angle, work is done. When a dipole is rotated through an angle $d\theta$, the work done is

$$dW = -\tau\, d\theta = -\mu B \sin\theta\, d\theta$$

The minus sign arises because the torque tends to decrease θ. Setting this work equal to the decrease in potential energy, we have

$$dU = -dW = +\mu B \sin\theta\, d\theta$$

Integrating, we obtain

$$U = -\mu B \cos\theta + U_0$$

If we choose the potential energy to be zero when $\theta = 90°$, then $U_0 = 0$ and the potential energy of the dipole is

$$U = -\mu B \cos\theta = -\vec{\boldsymbol{\mu}}\cdot\vec{\boldsymbol{B}} \qquad \text{28-16}$$

Equation 28-16 gives the potential energy of a magnetic dipole at an angle θ to a magnetic field.

Example 28-10

A square 12-turn coil with sides of length 40 cm carries a current of 3 A. It lies in the xy plane as shown in a uniform magnetic field $\vec{B} = 0.3\,\mathrm{T}\hat{i} + 0.4\,\mathrm{T}\hat{k}$. Find (*a*) the magnetic moment of the coil and (*b*) the torque exerted on the coil. (c) Find the potential energy of the coil.

Picture the Problem From Figure 28-26 we see that the magnetic moment of the loop is in the positive z direction.

Figure 28-26

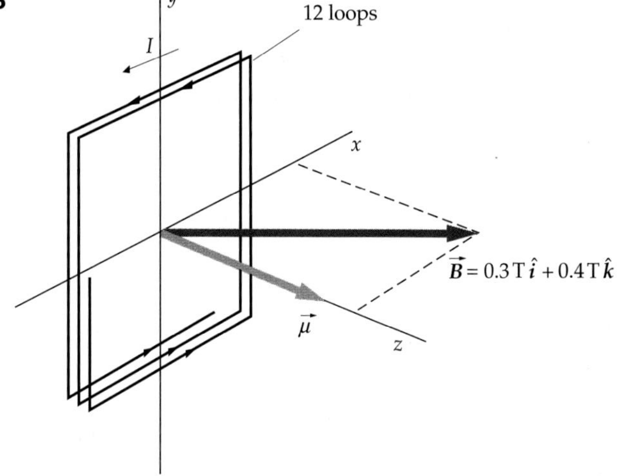

(*a*) Calculate the magnetic moment of the loop:

$$\vec{\boldsymbol{\mu}} = NIA\hat{k} = (12)(3\text{ A})(0.40\text{ m})^2\hat{k} = 5.76\text{ A}\cdot\text{m}^2\,\hat{k}$$

(*b*) The torque on the current loop is given by Equation 28-15:

$$\vec{\boldsymbol{\tau}} = \vec{\boldsymbol{\mu}}\times\vec{\boldsymbol{B}} = (5.76\text{ A}\cdot\text{m}^2\,\hat{k})\times(0.3\text{ T}\,\hat{i} + 0.4\text{ T}\,\hat{k}) = 1.73\text{ N}\cdot\text{m}\,\hat{j}$$

(*c*) The potential energy is the negative dot product of $\vec{\boldsymbol{\mu}}$ and $\vec{\boldsymbol{B}}$:

$$U = -\vec{\boldsymbol{\mu}}\cdot\vec{\boldsymbol{B}} = -(5.76\text{ A}\cdot\text{m}^2\,\hat{k})\cdot(0.3\text{ T}\,\hat{i} + 0.4\text{ T}\,\hat{k}) = -2.30\text{ J}$$

Remarks We have used $\hat{k} \times \hat{k} = 0$ and $\hat{k} \times \hat{i} = \hat{j}, \hat{k} \cdot \hat{i} = 0$ and $\hat{k} \cdot \hat{k} = 1$. The torque is in the y direction.

Exercise Calculate U if the coil rotates so that $\vec{\mu}$ is aligned with $\vec{B}$. (Answer $U = -\mu B = -(5.76\ \text{A}\cdot\text{m}^2)(0.5\ \text{T}) = -2.88\ \text{J}$. Note that this potential energy is lower than that found in the example. The torque tends to rotate the loop toward a position of lower potential energy.)

When a small permanent magnet such as a compass needle is placed in a magnetic field $\vec{B}$, the field exerts a torque on the magnet that tends to rotate the magnet so that it lines up with the field. This effect also occurs with previously unmagnetized iron filings, which become magnetized in the presence of a $\vec{B}$ field. The bar magnet is characterized by a magnetic moment $\vec{\mu}$ that points from the south pole to the north pole. A small bar magnet thus behaves like a current loop. This is not a coincidence. The origin of the magnetic moment of a bar magnet is, in fact, microscopic current loops that result from the motion of electrons in the atoms of the magnet.

Example 28-11

A nonconducting disk of mass M and radius R has a surface charge density σ and rotates with angular velocity ω about its axis. Find the magnetic moment of the rotating disk.

Picture the Problem We find the magnetic moment of a circular element of radius r and width dr and integrate (Figure 28-27). The charge on the element is $dq = \sigma\ dA = \sigma 2\pi r\ dr$. If the charge is positive, the magnetic moment is in the direction of $\vec{\omega}$, so we need only calculate the magnitude.

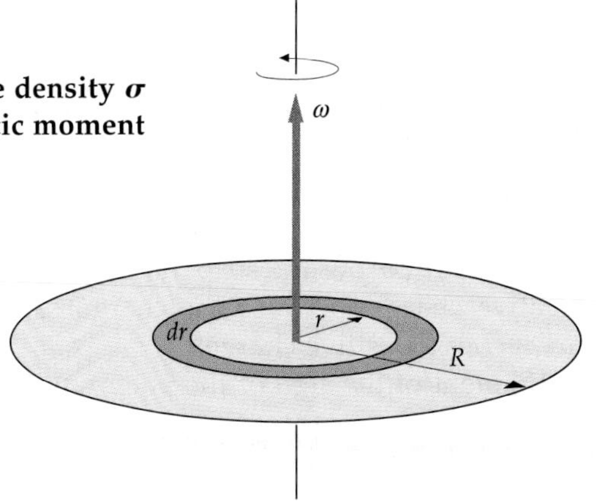

Figure 28-27

1. The magnetic moment of the strip shown is the current times the area of the loop: $d\mu = (dI)A = (dI)\pi r^2$

2. The current in the strip is the total charge on the strip times its frequency of rotation $f = \omega/2\pi$: $dI = (dq)f = (\sigma\, dA)\dfrac{\omega}{2\pi} = (\sigma\, 2\pi r\, dr)\dfrac{\omega}{2\pi} = \sigma\omega r\, dr$

3. Substitute to obtain the magnetic moment of the strip $d\mu$ in terms of r and dr: $d\mu = (dI)\pi r^2 = (\sigma\omega r\, dr)\pi r^2 = \pi\sigma\omega r^3\, dr$

4. Integrate from $r = 0$ to $r = R$: $\mu = \int d\mu = \int_0^R \pi\sigma\omega r^3\, dr = \dfrac{1}{4}\pi\sigma\omega R^4$

5. Use the fact that $\vec{\mu}$ is parallel to $\vec{\omega}$ if σ is positive to write the magnetic moment as a vector: $\vec{\mu} = \frac{1}{4}\pi\sigma R^4\vec{\omega}$

Remarks In terms of the total charge $Q = \sigma\pi R^2$, the magnetic moment is $\vec{\mu} = \frac{1}{4}QR^2\vec{\omega}$. The angular momentum of the disk is $\vec{L} = (\frac{1}{2}MR^2)\vec{\omega}$, so the magnetic moment can be written $\vec{\mu} = (Q/2M)\vec{L}$, which is a general result. (See Problem 53.)

28-4 The Hall Effect

As we have seen, charges moving in a magnetic field experience a force perpendicular to their motion. When these charges are traveling in a conducting wire, they will be pushed to one side of the wire. This results in a separation of charge in the wire called the **Hall effect**. This phenomenon allows us to determine the sign of the charge on the charge carriers and the number of charge carriers per unit volume n in a conductor. It also provides a convenient method for measuring magnetic fields.

Figure 28-28 shows two conducting strips, each of which carries a current I to the right because the left sides of the strips are connected to the positive terminal of a battery and the right sides are connected to the negative terminal. The strips are in a magnetic field that is directed into the paper. Let us assume for the moment that the current in the strip consists of positively charged particles moving to the right as shown in Figure 28-28*a*. The magnetic force on these particles is $q\vec{v}_d \times \vec{B}$ (where $\vec{v}_d$ is the drift velocity of the charge carriers). This force is directed upward. The positive particles therefore move up to the top of the strip, leaving the bottom of the strip with an excess negative charge. This separation of charge produces an electrostatic field in the strip that opposes the magnetic force on the charge carriers. When the electrostatic and magnetic forces balance, the charge carriers no longer move upward. In this equilibrium situation, the upper part of the strip is positively charged, so it is at a greater potential than the negatively charged lower part. If the current consists of negatively charged particles, as shown in Figure 28-28*b*, the charge carriers in the strip must move to the left (since the current is still to the right). The magnetic force $q\vec{v}_d \times \vec{B}$ is again up because the signs of both q and $\vec{v}_d$ have been changed. Again the carriers are forced to the upper part of the strip, but the upper part of the strip now carries a negative charge (because the charge carriers are negative) and the lower part carries a positive charge.

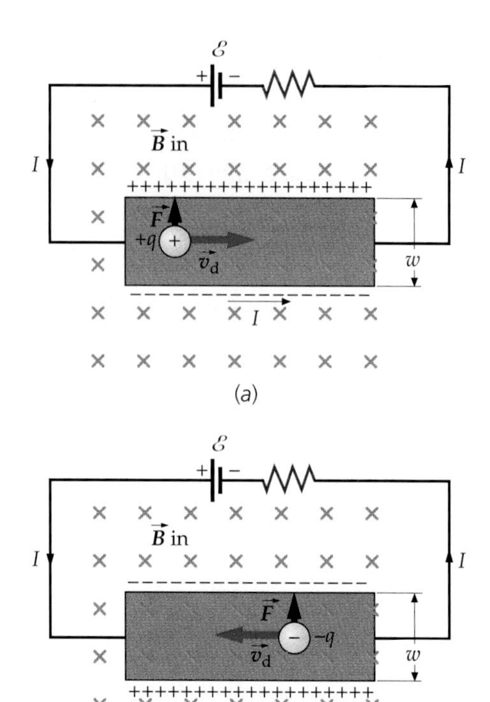

Figure 28-28 The Hall effect. The magnetic field is directed into the plane of the page as indicated by the crosses. The magnetic force on a charged particle is upward for a current to the right whether the current is due to (*a*) positive particles moving to the right or (*b*) negative particles moving to the left.

A measurement of the sign of the potential difference between the upper and lower parts of the strip tells us the sign of the charge carriers. In semiconductors, the charge carriers may be negative electrons or positive holes. A measurement of the sign of the potential difference tells us which are dominant for a particular semiconductor. For a normal metallic conductor, we find that the upper part of the strip in Figure 28-28 is at a lower potential than the lower part—which means that the upper part must carry a negative charge. Thus, Figure 28-28*b* is the correct illustration of the current in a normal conductor. It was this type of experiment that led to the discovery that the charge carriers in metallic conductors are negative.

The potential difference between the top and bottom of the strip is called the **Hall voltage**. We can calculate the magnitude of the Hall voltage in terms of the drift velocity. The magnitude of the magnetic force on the charge carriers in the strip is qv_dB. This magnetic force is balanced by the electrostatic force of magnitude qE, where E is the electric field due to the charge separation. Thus, we have $E = v_dB$. If the width of the strip is w, the potential difference is Ew. The Hall voltage is therefore

$$V_H = Ew = v_dBw \qquad 28\text{-}17$$

Exercise A conducting strip of width $w = 2.0$ cm is placed in a magnetic field of 8000 G. The Hall voltage is measured to be 0.64 μV. Calculate the drift velocity of the electrons. (*Answer* 4.0×10^{-5} m/s)

Since the drift velocity for ordinary currents is very small, we can see from Equation 28-17 that the Hall voltage is very small for ordinary-sized strips and magnetic fields. From measurements of the Hall voltage for a strip of a given size, we can determine the number of charge carriers per unit volume in the strip. The current is given by Equation 25-3:

$$I = nqv_dA$$

where A is the cross-sectional area of the strip. For a strip of width w and thickness t, the cross-sectional area is $A = wt$. Since the charge carriers are electrons, the quantity q is the charge on one electron e. The number density of charge carriers n is thus given by

$$n = \frac{I}{Aqv_d} = \frac{I}{wtev_d} \qquad \text{28-18}$$

Substituting $v_dw = V_H/B$ from Equation 28-17, we have

$$n = \frac{IB}{teV_H} \qquad \text{28-19}$$

Example 28-12

A silver slab of thickness 1 mm and width 1.5 cm carries a current of 2.5 A in a region in which there is a magnetic field of magnitude 1.25 T perpendicular to the slab. The Hall voltage is measured to be 0.334 μV. (*a*) Calculate the number density of the charge carriers. (*b*) Compare your answer in part (*a*) to the number density of atoms in silver, which has a mass density of $\rho = 10.5\ \text{g/cm}^3$ and a molar mass of $M = 107.9$ g/mol.

(*a*) Substitute numerical values into Equation 28-19 to find n:

$$n = \frac{IB}{teV_H} = \frac{(2.5\ \text{A})(1.25\ \text{T})}{(0.001\ \text{m})(1.6 \times 10^{-19}\ \text{C})(3.34 \times 10^{-7}\ \text{V})}$$

$$= 5.85 \times 10^{28}\ \text{electrons/m}^3$$

(*b*) The number of atoms per unit volume is $\rho N_A/M$:

$$n_a = \rho\frac{N_A}{M} = (10.5\ \text{g/cm}^3)\frac{6.02 \times 10^{23}\ \text{atoms/mol}}{107.9\ \text{g/mol}}$$

$$= 5.86 \times 10^{22}\ \text{atoms/cm}^3 = 5.86 \times 10^{28}\ \text{atoms/m}^3$$

Remark These results indicate that the number of charge carriers in silver is very nearly one per atom.

The Hall voltage provides a convenient method for measuring magnetic fields. If we rearrange Equation 28-19, we can write for the Hall voltage

$$V_H = \frac{I}{nte}B \qquad \text{28-20}$$

A given strip can be calibrated by measuring the Hall voltage for a given current in a known magnetic field. The strip can then be used to measure an unknown magnetic field B by measuring the Hall voltage for a given current.

The Quantum Hall Effect

According to Equation 28-20, the Hall voltage should increase linearly with magnetic field B for a given current in a given slab. In 1980, while studying the Hall effect in semiconductors at very low temperatures and very large magnetic fields, the German physicist Klaus von Klitzing discovered that a

optional

optional

plot of V_H versus B resulted in a series of plateaus, as shown in Figure 28-29, rather than a straight line. That is, the Hall voltage is quantized. For this discovery of the quantum Hall effect, von Klitzing won the Nobel Prize in physics in 1985.

In the theory of the quantum Hall effect, the Hall resistance, defined as $R_H = V_H/I$, can take on only the values

$$R_H = \frac{V_H}{I} = \frac{R_K}{n}, \qquad n = 1, 2, 3, \ldots \tag{28-21}$$

where n is an integer, and R_K, called the **von Klitzing constant**, is related to the fundamental electronic charge e and Planck's constant h by

$$R_K = \frac{h}{e^2} \tag{28-22}$$

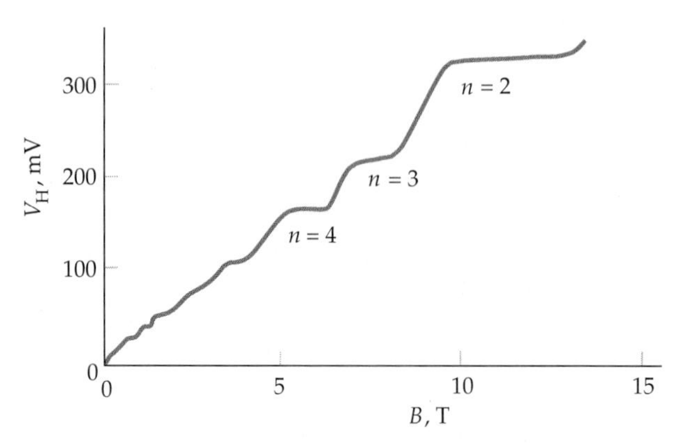

Figure 28-29 A plot of the Hall voltage versus applied magnetic field shows plateaus, indicating that the Hall voltage is quantized. These data were taken at a temperature of 1.39 K with the current I held fixed at 25.52 μA.

Because the von Klitzing constant can be measured to an accuracy of a few parts per billion, the quantum Hall effect is now used to define a standard of resistance. As of January 1990, the ohm is now defined so that R_K has the value 25,812.807 Ω exactly.

Recent experiments have shown that under certain special conditions, the Hall resistance is given by Equation 28-22 with the integer n replaced by a series of rational fractions. At present, the theory of this fractional quantum Hall effect is incomplete.

Summary

1. The magnetic field describes the condition in space in which moving charges experience a force perpendicular to their velocity.
2. The magnetic force is part of the electromagnetic force, one of the four fundamental forces of nature.
3. The magnitude and direction of a magnetic field $\vec{B}$ are defined by the force $\vec{F} = q\vec{v} \times \vec{B}$ exerted on moving charges.

Topic	Remarks and Relevant Equations	
1. Magnetic Force		
On a moving charge	$\vec{F} = q\vec{v} \times \vec{B}$	**28-1**
On a current element	$d\vec{F} = I\,d\vec{\ell} \times \vec{B}$	**28-5**
Unit of the magnetic field	The SI unit of magnetic fields is the tesla (T). A commonly used unit is the gauss (G), which is related to the tesla by $1\text{ T} = 10^4\text{ G}$	**28-3**
2. Motion of Point Charges	A particle of mass m and charge q moving with speed v in a plane perpendicular to a magnetic field moves in a circular orbit. The period and frequency of this circular motion are independent of the radius of the orbit and of the speed of the particle.	
Radius of circular orbit	$r = \frac{mv}{qB}$	**28-6**

Cyclotron period	$T = \frac{2\pi m}{qB}$	28-7
Cyclotron frequency	$f = \frac{1}{T} = \frac{qB}{2\pi m}$	28-8
Velocity selector (optional)	A velocity selector consists of crossed electric and magnetic fields such that the electric and magnetic forces balance for a particle whose speed is given by $v = \frac{E}{B}$	28-9
Measurement of q/m (optional)	The deflection of a charged particle in an electric field depends on the speed of the particle and is proportional to the charge-to-mass ratio q/m of the particle. J.J. Thomson used crossed electric and magnetic fields to measure the speed of cathode rays and then measured q/m for these particles by deflecting them in an electric field. He showed that all cathode rays consist of particles that all have the same charge-to-mass ratio. These particles are now called electrons.	
Mass spectrometer (optional)	The mass-to-charge ratio of an ion of known speed can be determined by measuring the radius of the circular path taken by the ion in a known magnetic field.	

3. Current Loops

Magnetic moment	$\vec{\mu} = NIA\hat{n}$	28-14
Torque	$\vec{\tau} = \vec{\mu} \times \vec{B}$	28-15
Potential energy	$U = -\mu B \cos\theta = -\vec{\mu} \cdot \vec{B}$	28-16
Net force	The net force on a current loop in a uniform magnetic field is zero.	

4. The Hall Effect

When a conducting strip carrying a current is placed in a magnetic field, the magnetic force on the charge carriers causes a separation of charge called the Hall effect. This results in a voltage V_H, called the Hall voltage. The sign of the charge carriers can be determined from a measurement of the sign of the Hall voltage, and the number of carriers per unit volume can be determined from the magnitude of V_H.

Hall voltage	$V_H = v_d B w = \frac{I}{nte} B$	28-17, 28-20
Quantum Hall effect (optional)	Measurements at very low temperatures in very large magnetic fields indicate that the Hall resistance $R_H = V_H/I$ is quantized and can take on only the values given by $R_H = \frac{V_H}{I} = \frac{R_K}{n}, \quad n = 1, 2, 3, \ldots$	28-21
von Klitzing constant (optional)	$R_K = \frac{h}{e^2} \approx 25{,}813\ \Omega$	28-22

Problem-Solving Guide

1. Begin by drawing a neat diagram that includes the important features of the problem.
2. The vector nature of the magnetic force is clarified with a sketch showing $\vec{v}$ and $\vec{B}$. Your sketch will help greatly in applying the right-hand rule.
3. For problems involving current-carrying wires, show the direction of I and again apply the right-hand rule.

Summary of Worked Examples

Type of Calculation	Procedure and Relevant Examples	
1. Magnetic Force		
Find the magnetic force on a charged particle.	Use $\vec{F} = q\vec{v} \times \vec{B}$.	**Example 28-1**
Find the magnetic force on a current-carrying wire.	Use $\vec{F} = I\vec{\ell} \times \vec{B}$.	**Example 28-2**
2. Motion of Particles		
Determine the radius and frequency of motion for a charged particle in a magnetic field.	Applying Newton's second law—equating the magnetic force and the centripetal force—yields the relations $r = mv/qB$ and $f = 1/T = qB/2\pi m$.	**Examples 28-3, 28-5, 28-6**
3. Magnetic Moments		
Find the magnetic moment of a loop.	The magnitude of the magnetic moment is $\mu = NIA$. The direction is found using the right-hand rule.	**Examples 28-8, 28-9, 28-10**
Find the torque acting on a current loop.	Use $\vec{\tau} = \vec{\mu} \times \vec{B}$.	**Examples 28-8, 28-9, 28-10**
4. The Hall Effect		
Calculate the density of charge carriers in a conductor.	Use $n = \dfrac{IB}{teV_H}$	**Example 28-12**

Problems

Conceptual Problems

Problems from Optional and Exploring sections

In a few problems, you are given more data than you actually need; in a few other problems, you are required to supply data from your general knowledge, outside sources, or informed estimates.

• Single-concept, single-step, relatively easy
•• Intermediate-level, may require synthesis of concepts
••• Challenging, for advanced students

Force Exerted by a Magnetic Field

1 • When a cathode-ray tube is placed horizontally in a magnetic field that is directed vertically upward, the electrons emitted from the cathode follow one of the dashed paths to the face of the tube in Figure 28-30. The correct path is ________.

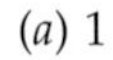

(*a*) 1 (*b*) 2 (*c*) 3
(*d*) 4 (*e*) 5

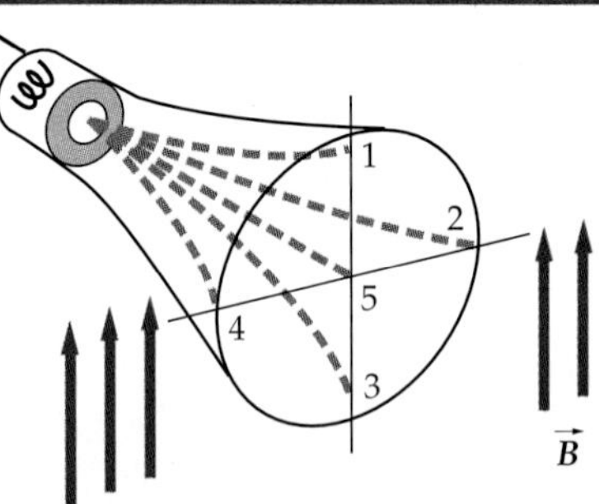

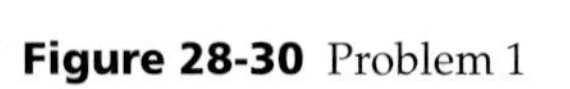

Figure 28-30 Problem 1

2 • Why not define $\vec{B}$ to be in the direction of $\vec{F}$, as we do for $\vec{E}$?

3 • Find the magnetic force on a proton moving with velocity 4.46 Mm/s in the positive x direction in a magnetic field of 1.75 T in the positive z direction.

4 • A charge $q = -3.64$ nC moves with a velocity of 2.75×10^6 m/s $\hat{i}$. Find the force on the charge if the magnetic field is (*a*) $\vec{B} = 0.38 \text{ T } \hat{j}$, (*b*) $\vec{B} = 0.75 \text{ T } \hat{i} + 0.75 \text{ T } \hat{j}$, (*c*) $\vec{B} = 0.65 \text{ T } \hat{i}$, (*d*) $\vec{B} = 0.75 \text{ T } \hat{i} + 0.75 \text{ T } \hat{k}$.

5 • A uniform magnetic field of magnitude 1.48 T is in the positive z direction. Find the force exerted by the field on a proton if the proton's velocity is (*a*) $\vec{v} = 2.7 \text{ Mm/s } \hat{i}$, (*b*) $\vec{v} = 3.7 \text{ Mm/s } \hat{j}$, (*c*) $\vec{v} = 6.8 \text{ Mm/s } \hat{k}$, and (*d*) $\vec{v} = 4.0 \text{ Mm/s } \hat{i} + 3.0 \text{ Mm/s } \hat{j}$.

6 • An electron moves with a velocity of 2.75 Mm/s in the xy plane at an angle of 60° to the x axis and 30° to the y axis. A magnetic field of 0.85 T is in the positive y direction. Find the force on the electron.

7 • A straight wire segment 2 m long makes an angle of 30° with a uniform magnetic field of 0.37 T. Find the magnitude of the force on the wire if it carries a current of 2.6 A.

8 • A straight wire segment $I\vec{\ell} = (2.7 \text{ A})(3 \text{ cm } \hat{i} + 4 \text{ cm } \hat{j})$ is in a uniform magnetic field $\vec{B} = 1.3 \text{ T } \hat{i}$. Find the force on the wire.

9 • What is the force (magnitude and direction) on an electron with velocity $\vec{v} = (2\hat{i} - 3\hat{j}) \times 10^6$ m/s in a magnetic field $\vec{B} = (0.8\hat{i} + 0.6\hat{j} - 0.4\hat{k})$ T?

10 •• The wire segment in Figure 28-31 carries a current of 1.8 A from a to b. There is a magnetic field $\vec{B} = 1.2 \text{ T } \hat{k}$. Find the total force on the wire and show that it is the same as if the wire were a straight segment from a to b.

Figure 28-31 Problem 10

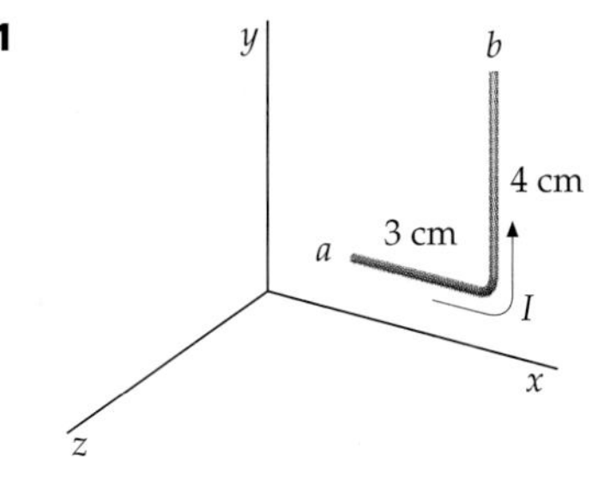

11 •• A straight, stiff, horizontal wire of length 25 cm and mass 50 g is connected to a source of emf by light, flexible leads. A magnetic field of 1.33 T is horizontal and perpendicular to the wire. Find the current necessary to float the wire, that is, the current such that the magnetic force balances the weight of the wire.

12 •• A simple gaussmeter for measuring horizontal magnetic fields consists of a stiff 50-cm wire that hangs from a conducting pivot so that its free end makes contact with a pool of mercury in a dish below. The mercury provides an electrical contact without constraining the movement of the wire. The wire has a mass of 5 g and conducts a current downward. (*a*) What is the equilibrium angular displacement of the wire from vertical if the horizontal magnetic field is 0.04 T and the current is 0.20 A? (*b*) If the current is 20 A and a displacement from vertical of 0.5 mm can be detected for the free end, what is the horizontal magnetic field sensitivity of this gaussmeter?

13 •• A current-carrying wire is bent into a semicircular loop of radius R that lies in the xy plane. There is a uniform magnetic field $\vec{B} = B\hat{k}$ perpendicular to the plane of the loop (Figure 28-32). Show that the force acting on the loop is $\vec{F} = 2IRB\hat{j}$.

Figure 28-32 Problem 13

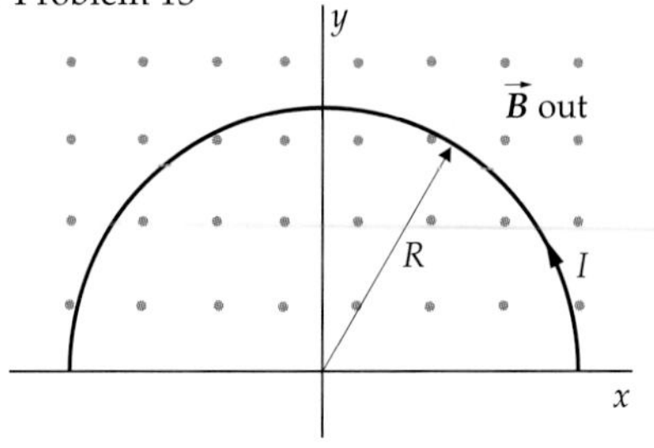

14 •• A 10-cm length of wire carries a current of 4.0 A in the positive z direction. The force on this wire due to a magnetic field $\vec{B}$ is $\vec{F} = (-0.2\hat{i} + 0.2\hat{j})$ N. If this wire is rotated so that the current flows in the positive x direction, the force on the wire is $\vec{F} = 0.2\hat{k}$ N. Find the magnetic field $\vec{B}$.

15 •• A 10-cm length of wire carries a current of 2.0 A in the positive x direction. The force on this wire due to the presence of a magnetic field $\vec{B}$ is $\vec{F} = (3.0\hat{j} + 2.0\hat{k})$ N. If this wire is now rotated so that the current flows in the positive y direction, the force on the wire is $\vec{F} = (-3.0\hat{i} - 2.0\hat{k})$ N. Determine the magnetic field $\vec{B}$.

16 ••• A wire bent in some arbitrary shape carries a current I in a uniform magnetic field $\vec{B}$. Show explicitly that the total force on the part of the wire from some point a to some point b is $\vec{F} = I\vec{\ell} \times \vec{B}$, where $\vec{\ell}$ is the vector from a to b.

Motion of a Point Charge in a Magnetic Field

17 • True or false: The magnetic force does not accelerate a particle because the force is perpendicular to the velocity of the particle.

18 • A moving charged particle enters a region in which it is suddenly deflected perpendicular to its motion. How can you tell if the deflection was caused by a magnetic field or an electric field?

19 • A proton moves in a circular orbit of radius 65 cm perpendicular to a uniform magnetic field of magnitude 0.75 T. (*a*) What is the period for this motion? (*b*) Find the speed of the proton. (*c*) Find the kinetic energy of the proton.

20 • An electron of kinetic energy 45 keV moves in a circular orbit perpendicular to a magnetic field of 0.325 T. (*a*) Find the radius of the orbit. (*b*) Find the frequency and period of the motion.

21 • An electron from the sun with a speed of 1×10^7 m/s enters the earth's magnetic field high above the equator where the magnetic field is 4×10^{-7} T. The electron moves nearly in a circle except for a small drift along the direction of the earth's magnetic field that will take it toward the north pole. (*a*) What is the radius of the circular motion? (*b*) What is the radius of the circular motion near the north pole where the magnetic field is 2×10^{-5} T?

22 •• Protons and deuterons (each with charge $+e$) and alpha particles (with charge $+2e$) of the same kinetic energy enter a uniform magnetic field $\vec{B}$ that is perpendicular to their velocities. Let r_p, r_d, and r_α be the radii of their circular orbits. Find the ratios r_d/r_p and r_α/r_p. Assume that $m_\alpha = 2m_d = 4m_p$.

23 •• A proton and an alpha particle move in a uniform magnetic field in circles of the same radii. Compare (*a*) their velocities, (*b*) their kinetic energies, and (*c*) their angular momenta. (See Problem 22.)

24 •• A particle of charge q and mass m has momentum $p = mv$ and kinetic energy $K = \frac{1}{2}mv^2 = p^2/2m$. If the particle moves in a circular orbit of radius r perpendicular to a uniform magnetic field B, show that (*a*) $p = Bqr$ and (*b*) $K = B^2q^2r^2/2m$.

25 •• A beam of particles with velocity $\vec{v}$ enters a region of uniform magnetic field $\vec{B}$ that makes a small angle θ with $\vec{v}$. Show that after a particle moves a distance $2\pi(m/qB)v\cos\theta$ measured along the direction of $\vec{B}$, the velocity of the particle is in the same direction as it was when it entered the field.

26 •• A proton with velocity $v = 10^7$ m/s enters a region of uniform magnetic field $B = 0.8$ T, which is into the page, as shown in Figure 28-33. The angle $\theta = 60°$. Find the angle ϕ and the distance d.

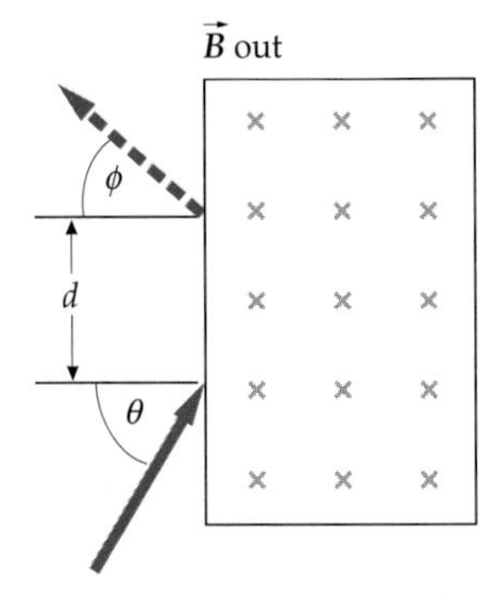

Figure 28-33
Problems 26 and 27

27 •• Suppose that in Figure 28-33 $B = 0.6$ T, the distance $d = 0.4$ m, and $\theta = 24°$. Find the speed v and the angle ϕ if the particles are (*a*) protons and (*b*) deuterons.

Velocity Selectors (optional)

28 • A beam of positively charged particles passes undeflected from left to right through a velocity selector in which the electric field is up. The beam is then reversed so that it travels from right to left. Will the beam now be deflected in the velocity selector? If so, in which direction?

29 • A velocity selector has a magnetic field of magnitude 0.28 T perpendicular to an electric field of magnitude 0.46 MV/m. (*a*) What must the speed of a particle be for it to pass through undeflected? What energy must (*b*) protons and (*c*) electrons have to pass through undeflected?

30 • A beam of protons moves along the x axis in the positive x direction with a speed of 12.4 km/s through a region of crossed fields balanced for zero deflection. (*a*) If there is a magnetic field of magnitude 0.85 T in the positive y direction, find the magnitude and direction of the electric field. (*b*) Would electrons of the same velocity be deflected by these fields? If so, in what direction?

Measuring q/m (optional)

31 •• The plates of a Thomson q/m apparatus are 6.0 cm long and are separated by 1.2 cm. The end of the plates is 30.0 cm from the tube screen. The kinetic energy of the electrons is 2.8 keV. (*a*) If a potential of 25.0 V is applied across the deflection plates, by how much will the beam deflect? (*b*) Find the magnitude of the crossed magnetic field that will allow the beam to pass through undeflected.

Mass Spectrometer (optional)

32 •• Chlorine has two stable isotopes, ^{35}Cl and ^{37}Cl, whose natural abundances are about 76% and 24%, respectively. Singly ionized chlorine gas is to be separated into its isotopic components using a mass spectrometer. The magnetic field in the spectrometer is 1.2 T. What is the minimum value of the potential through which these ions must be accelerated so that the separation between them is 1.4 cm?

33 •• A singly ionized ^{24}Mg ion (mass 3.983×10^{-26} kg) is accelerated through a 2.5-kV potential difference and deflected in a magnetic field of 557 G in a mass spectrometer. (*a*) Find the radius of curvature of the orbit for the ion. (*b*) What is the difference in radius for ^{26}Mg and ^{24}Mg ions? (Assume that their mass ratio is 26/24.)

34 •• A beam of ^{6}Li and ^{7}Li ions passes through a velocity selector and enters a magnetic spectrometer. If the diameter of the orbit of the ^{6}Li ions is 15 cm, what is the diameter of that for ^{7}Li ions?

35 •• In Example 28-6, determine the time required for a ^{58}Ni ion and a ^{60}Ni ion to complete the semicircular path.

36 •• Before entering a mass spectrometer, ions pass through a velocity selector consisting of parallel plates separated by 2.0 mm and having a potential difference of 160 V. The magnetic field between the plates is 0.42 T. The magnetic field in the mass spectrometer is 1.2 T. Find (*a*) the speed of the ions entering the mass spectrometer and (*b*) the difference in the diameters of the orbits of singly ionized ^{238}U and ^{235}U. (The mass of a ^{235}U ion is 3.903×10^{-25} kg.)

The Cyclotron (optional)

37 •• A cyclotron for accelerating protons has a magnetic field of 1.4 T and a radius of 0.7 m. (*a*) What is the cyclotron frequency? (*b*) Find the maximum energy of the protons when they emerge. (*c*) How will your answers change if deuterons, which have the same charge but twice the mass, are used instead of protons?

38 •• A certain cyclotron with magnetic field of 1.8 T is designed to accelerate protons to 25 MeV. (*a*) What is the cyclotron frequency? (*b*) What must the minimum radius of the magnet be to achieve a 25-MeV emergence energy? (*c*) If the alternating potential applied to the dees has a maximum value of 50 kV, how many revolutions must the protons make before emerging with an energy of 25 MeV?

39 •• Show that the cyclotron frequencies of deuterons and alpha particles are the same and are half that of a proton in the same magnetic field. (See Problem 22.)

40 •• Show that the radius of the orbit of a charged particle in a cyclotron is proportional to the square root of the number of orbits completed.

Torques on Current Loops and Magnets

41 • What orientation of a current loop gives maximum torque?

42 • A small circular coil of 20 turns of wire lies in a uniform magnetic field of 0.5 T such that the normal to the plane of the coil makes an angle of 60° with the direction of $\vec{B}$. The radius of the coil is 4 cm, and it carries a current of 3 A. (*a*) What is the magnitude of the magnetic moment of the coil? (*b*) What is the magnitude of the torque exerted on the coil?

43 • What is the maximum torque on a 400-turn circular coil of radius 0.75 cm that carries a current of 1.6 mA and resides in a uniform magnetic field of 0.25 T?

44 • A current-carrying wire is bent into the shape of a square of sides $L = 6$ cm and is placed in the xy plane. It carries a current $I = 2.5$ A. What is the torque on the wire if there is a uniform magnetic field of 0.3 T (*a*) in the z direction, and (*b*) in the x direction?

45 • Repeat Problem 44 if the wire is bent into an equilateral triangle of sides 8 cm.

46 •• A rigid, circular loop of radius R and mass M carries a current I and lies in the xy plane on a rough, flat table. There is a horizontal magnetic field of magnitude B. What is the minimum value of B such that one edge of the loop will lift off the table?

47 •• A rectangular, 50-turn coil has sides 6.0 and 8.0 cm long and carries a current of 1.75 A. It is oriented as shown in Figure 28-34 and pivoted about the z axis. (*a*) If the wire in the xy plane makes an angle $\theta = 37°$ with the y axis as shown, what angle does the unit normal $\hat{n}$ make with the x axis? (*b*) Write an expression for $\hat{n}$ in terms of the unit vectors $\hat{i}$ and $\hat{j}$. (*c*) What is the magnetic moment of the coil? (*d*) Find the torque on the coil when there is a uniform magnetic field $\vec{B} = 1.5\text{ T}\,\hat{j}$. (*e*) Find the potential energy of the coil in this field.

Figure 28-34
Problems 47 and 48

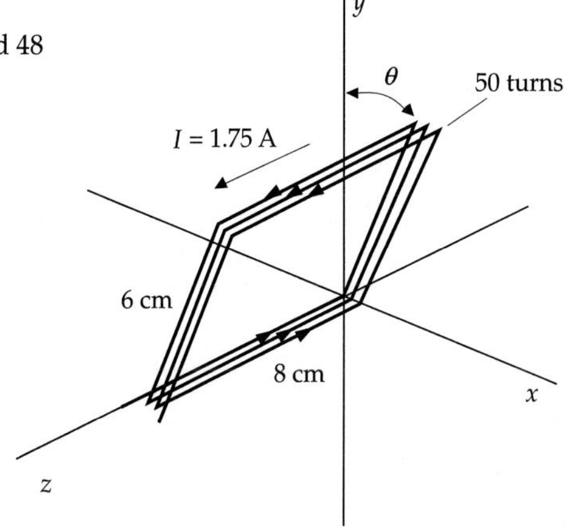

48 •• The coil in Problem 47 is pivoted about the z axis and held at various positions in a uniform magnetic field $\vec{B} = 2.0\text{ T}\,\hat{j}$. Sketch the position of the coil and find the torque exerted when the unit normal is (*a*) $\hat{n} = \hat{i}$, (*b*) $\hat{n} = \hat{j}$, (*c*) $\hat{n} = -\hat{j}$, and (*d*) $\hat{n} = (\hat{i} + \hat{j})/\sqrt{2}$.

Magnetic Moments

49 • The SI unit for the magnetic moment of a current loop is A·m^2. Use this to show that 1 T = 1 N/A·m.

50 •• A small magnet of length 6.8 cm is placed at an angle of 60° to the direction of a uniform magnetic field of magnitude 0.04 T. The observed torque has a magnitude of 0.10 N·m. Find the magnetic moment of the magnet.

51 •• A wire loop consists of two semicircles connected by straight segments (Figure 28-35). The inner and outer radii are 0.3 and 0.5 m, respectively. A current of 1.5 A flows in this loop with the current in the outer semicircle in the clockwise direction. What is the magnetic moment of this current loop?

Figure 28-35 Problem 51

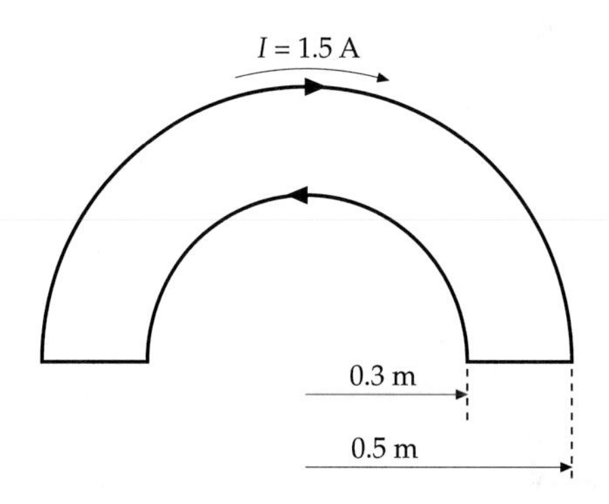

52 •• A wire of length L is wound into a circular coil of N loops. Show that when this coil carries a current I, its magnetic moment has the magnitude $IL^2/4\pi N$.

53 •• A particle of charge q and mass m moves in a circle of radius r and with angular velocity ω. (*a*) Show that the average current is $I = q\omega/2\pi$ and that the magnetic moment has the magnitude $\mu = \frac{1}{2}q\omega r^2$. (*b*) Show that the angular momentum of this particle has the magnitude $L = mr^2\omega$ and that the magnetic moment and angular momentum vectors are related by $\vec{\mu} = (q/2m)\vec{L}$.

54 ••• A single loop of wire is placed around the circumference of a rectangular piece of cardboard whose length and width are 70 and 20 cm, respectively. The cardboard is now folded along a line perpendicular to its length and midway between the two ends so that the two planes formed by the folded cardboard make an angle of 90°. If the wire loop carries a current of 0.2 A, what is the magnitude of the magnetic moment of this system?

55 ••• Repeat Problem 54 if the line along which the cardboard is folded is 40 cm from one end.

56 ••• A hollow cylinder has length L and inner and outer radii R_i and R_o, respectively (Figure 28-36). The cylinder carries a uniform charge density ρ. Derive an expression for the magnetic moment as a function of ω, the angular velocity of rotation of the cylinder about its axis.

Figure 28-36 Problem 56

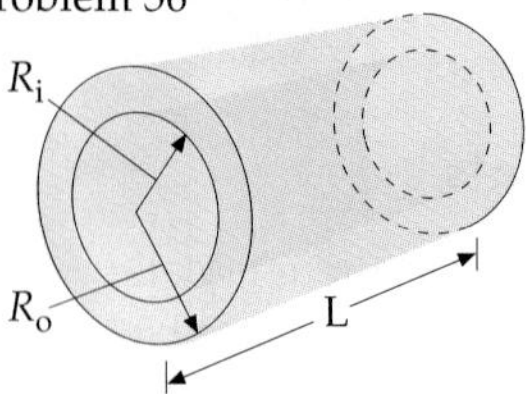

57 ••• A nonconducting rod of mass M and length ℓ has a uniform charge per unit length λ and rotates with angular velocity $\vec{\omega}$ about an axis through one end and perpendicular to the rod. (*a*) Consider a small segment of the rod of length dx and charge $dq = \lambda\, dx$ at a distance x from the pivot (Figure 28-37). Show that the magnetic moment of this segment is $\frac{1}{2}\lambda\omega x^2\, dx$. (*b*) Integrate your result to show that the total magnetic moment of the rod is $\mu = \frac{1}{6}\lambda\omega\,\ell^3$. (*c*) Show that the magnetic moment $\vec{\mu}$ and angular momentum $\vec{L}$ are related by $\vec{\mu} = (Q/2M)\vec{L}$, where Q is the total charge on the rod.

Figure 28-37 Problem 57

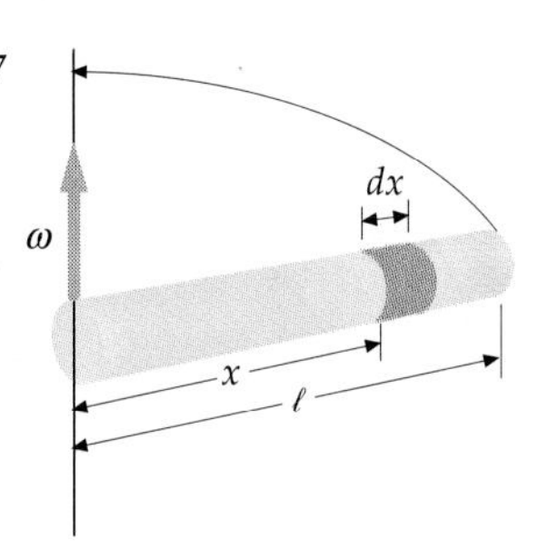

58 ••• A nonuniform, nonconducting disk of mass M, radius R, and total charge Q has a surface charge density $\sigma = \sigma_0 r/R$ and a mass per unit area $\sigma_m = (M/Q)\sigma$. The disk rotates with angular velocity ω about its axis. (*a*) Show that the magnetic moment of the disk has a magnitude $\mu = \frac{1}{5}\pi\omega\sigma_0 R^4 = \frac{3}{10}Q\omega R^2$. (*b*) Show that the magnetic moment $\vec{\mu}$ and angular momentum $\vec{L}$ are related by $\vec{\mu} = (Q/2M)\vec{L}$.

59 ••• A spherical shell of radius R carries a surface charge density σ. The sphere rotates about its diameter with angular velocity ω. Find the magnetic moment of the rotating sphere.

60 ••• A solid sphere of radius R carries a uniform volume charge density ρ. The sphere rotates about its diameter with angular velocity ω. Find the magnetic moment of this rotating sphere.

61 ••• A solid cylinder of radius R and length L carries a uniform charge density $+\rho$ between $r = 0$ and $r = R_s$ and an equal charge density of opposite sign, $-\rho$, between $r = R_s$ and $r = R$. What must be the radius R_s so that on rotation of the cylinder about its axis the magnetic moment is zero?

62 ••• A solid cylinder of radius R and length L carries a uniform charge density $\rho = -\rho_0$ between $r = 0$ and $r = \frac{1}{2}R$ and a positive charge density of equal magnitude, $+\rho_0$, between $r = \frac{1}{2}R$ and $r = R$ (Figure 28-38). The cylinder rotates about its axis with angular velocity $\vec{\omega}$. Derive an expression for the magnetic moment of the cylinder.

Figure 28-38 Problem 62

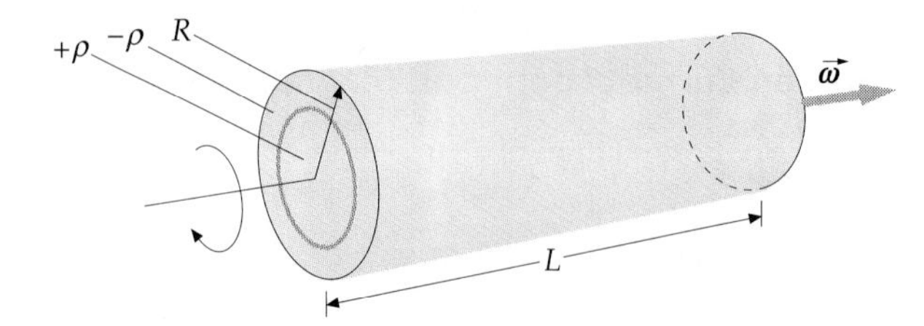

63 ••• A cylindrical shell of length L with inner radius R_i and outer radius R_o carries a uniform charge density, $+\rho_0$, between R_i and radius R_s and an equal charge density of opposite sign, $-\rho_0$, between R_s and R_o. The cylinder rotates about its axis with angular velocity $\vec{\omega}$. Derive an expression for the magnetic moment of this cylinder.

64 ••• A solid sphere of radius R carries a uniform charge density, $+\rho_0$, between $r = 0$ and $r = R_s$ and an equal charge density of opposite sign, $-\rho_0$, between $r = R_s$ and $r = R$. The sphere rotates about its diameter with angular velocity ω. Find R_s such that magnetic moment of the sphere is zero. What is the net charge carried by the sphere?

65 ••• A solid sphere of radius R carries a uniform charge density, $+\rho_0$, between $r = 0$ and $r = \frac{1}{2}R$ and an equal charge density of opposite sign, $-\rho_0$, between $r = \frac{1}{2}R$ and $r = R$. The sphere rotates about its diameter with angular velocity ω. Derive an expression for the magnetic moment of this rotating sphere.

The Hall Effect

66 • A metal strip 2.0 cm wide and 0.1 cm thick carries a current of 20 A in a uniform magnetic field of 2.0 T, as shown in Figure 28-39. The Hall voltage is measured to be 4.27 μV. (*a*) Calculate the drift velocity of the electrons in the strip. (*b*) Find the number density of the charge carriers in the strip. (*c*) Is point *a* or *b* at the higher potential?

Figure 28-39 Problems 66 and 67

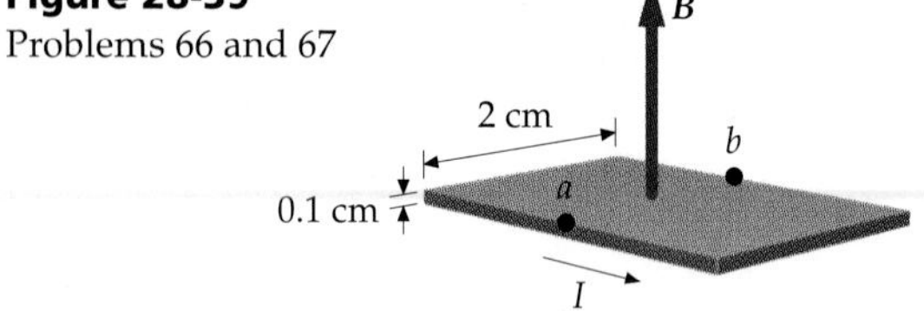

67 •• The number density of free electrons in copper is 8.47×10^{22} electrons per cubic centimeter. If the metal strip in Figure 28-39 is copper and the current is 10 A, find (*a*) the drift velocity v_d and (*b*) the Hall voltage. (Assume that the magnetic field is 2.0 T.)

68 •• A copper strip ($n = 8.47 \times 10^{22}$ electrons per cubic centimeter) 2 cm wide and 0.1 cm thick is used to measure the magnitudes of unknown magnetic fields that are perpendicular to the strip. Find the magnitude of B when $I = 20$ A and the Hall voltage is (*a*) 2.00 μV, (*b*) 5.25 μV, and (*c*) 8.00 μV.

69 •• Because blood contains charged ions, moving blood develops a Hall voltage across the diameter of an artery. A large artery with a diameter of 0.85 cm has a flow speed of 0.6 m/s. If a section of this artery is in a magnetic field of 0.2 T, what is the potential difference across the diameter of the artery?

70 •• The Hall coefficient R is defined as $R = E_y/J_xB_z$, where J_x is the current per unit area in the x direction in the slab, B_z is the magnetic field in the z direction, and E_y is the resulting Hall field in the y direction. Show that the Hall coefficient is $1/nq$, where q is the charge of the charge carriers, -1.6×10^{-19} C if they are electrons. (The Hall coefficients of monovalent metals, such as copper, silver, and sodium, are therefore negative.)

71 •• Aluminum has a density of 2.7×10^3 kg/m^3 and a molar mass of 27 g/mol. The Hall coefficient of aluminum is $R = -0.3 \times 10^{-10}$ m^3/C. (See Problem 70 for the definition of R.) Find the number of conduction electrons per aluminum atom.

72 •• Magnesium is a divalent metal. Its density is 1.74×10^3 kg/m^3 and its molar mass is 24.3 g/mol. Assuming that each magnesium atom contributes two conduction electrons, what should be the Hall coefficient of magnesium? How does your result compare to the measured value of -0.94×10^{-10} m^3/C?

General Problems

73 • True or false:

(*a*) The magnetic force on a moving charged particle is always perpendicular to the velocity of the particle.
(*b*) The torque on a magnet tends to align the magnetic moment in the direction of the magnetic field.
(*c*) A current loop in a uniform magnetic field behaves like a small magnet.
(*d*) The period of a particle moving in a circle in a magnetic field is proportional to the radius of the circle.
(*e*) The drift velocity of electrons in a wire can be determined from the Hall effect.

74 • Show that the force on a current element is the same in direction and magnitude regardless of whether positive charges, negative charges, or a mixture of positive and negative charges create the current.

75 • A proton with a charge $+e$ is moving with a speed v at 50° to the direction of a magnetic field $\vec{B}$. The component of the resulting force on the proton in the direction of $\vec{B}$ is

(*a*) $evB \sin 50° \cos 50°$.
(*b*) $evB \cos 50°$.
(*c*) zero.
(*d*) $evB \sin 50°$.
(*e*) none of these.

76 • If the magnetic field vector is directed toward the north and a positively charged particle is moving toward the east, what is the direction of the magnetic force on the particle?

77 • A positively charged particle is moving northward in a magnetic field. The magnetic force on the particle is toward the northeast. What is the direction of the magnetic field?

(*a*) Up
(*b*) West
(*c*) South
(*d*) Down
(*e*) This situation cannot exist.

78 • A ^{7}Li nucleus with a charge of $+3e$ and a mass of 7 u (1 u = 1.66×10^{-27} kg) and a proton with charge $+e$ and mass 1 u are both moving in a plane perpendicular to a magnetic field $\vec{B}$. The two particles have the same momentum. The ratio of the radius of curvature of the path of the proton R_p to that of the ^{7}Li nucleus, R_{Li} is

(*a*) $R_p/R_{Li} = 3$.
(*b*) $R_p/R_{Li} = 1/3$.
(*c*) $R_p/R_{Li} = 1/7$.
(*d*) $R_p/R_{Li} = 3/7$.
(*e*) none of these.

79 • An electron moving with velocity v to the right enters a region of uniform magnetic field that points out of the paper. After the electron enters this region, it will be

(*a*) deflected out of the plane of the paper.
(*b*) deflected into the plane of the paper.
(*c*) deflected upward.
(*d*) deflected downward.
(*e*) undeviated in its motion.

80 • How are magnetic field lines similar to electric field lines? How are they different?

81 • A long wire parallel to the x axis carries a current of 6.5 A in the positive x direction. There is a uniform magnetic field $\vec{B} = 1.35\ \mathrm{T}\ \hat{j}$. Find the force per unit length on the wire.

82 • An alpha particle (charge $+2e$) travels in a circular path of radius 0.5 m in a magnetic field of 1.0 T. Find (*a*) the period, (*b*) the speed, and (*c*) the kinetic energy (in electron volts) of the alpha particle. Take $m = 6.65 \times 10^{-27}$ kg for the mass of the alpha particle.

83 • If a current I in a given wire and a magnetic field $\vec{B}$ are known, the force $\vec{F}$ on the current is uniquely determined. Show that knowing $\vec{F}$ and I does not provide complete knowledge of $\vec{B}$.

84 •• The pole strength q_m of a bar magnet is defined by $q_m = |\vec{\mu}|/L$, where L is the length of the magnet. Show that the torque exerted on a bar magnet in a uniform magnetic field $\vec{B}$ is the same as if a force $+q_m\vec{B}$ is exerted on the north pole and a force $-q_m\vec{B}$ is exerted on the south pole.

85 •• A particle of mass m and charge q enters a region where there is a uniform magnetic field B along the x axis. The initial velocity of the particle is $\vec{v} = v_{0x}\hat{i} + v_{0y}\hat{j}$ so the particle moves in a helix. (*a*) Show that the radius of the helix is $r = mv_{0y}/qB$. (*b*) Show that the particle takes a time $t = 2\pi m/qB$ to make one orbit around the helix.

86 •• A metal crossbar of mass M rides on a pair of long, horizontal conducting rails separated by a distance ℓ and connected to a device that supplies constant current I to the circuit, as shown in Figure 28-40. A uniform magnetic field B is established as shown. (*a*) If there is no friction and the bar starts from rest at $t = 0$, show that at time t the bar has velocity $v = (BI\ell/M)t$. (*b*) In which direction will the bar move? (*c*) If the coefficient of static friction is μ_s, find the minimum field B necessary to start the bar moving.

Figure 28-40 Problems 86 and 87

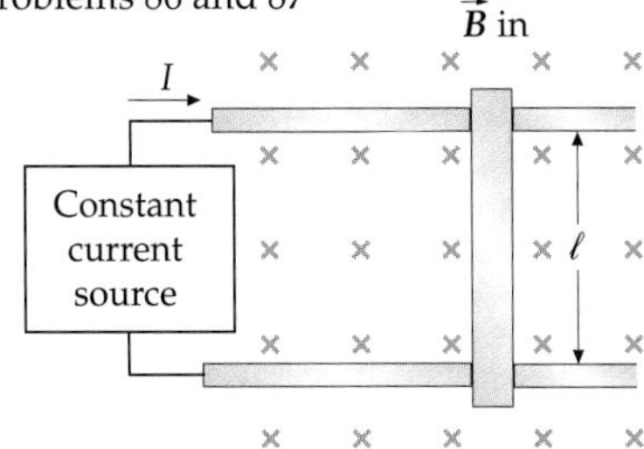

87 •• Assume that the rails in Figure 28-40 are frictionless but tilted upward so that they make an angle θ with the horizontal. (*a*) What vertical magnetic field B is needed to keep the bar from sliding down the rails? (*b*) What is the acceleration of the bar if B has twice the value found in part (*a*)?

88 •• A long, narrow bar magnet that has magnetic moment $\vec{\mu}$ parallel to its long axis is suspended at its center as a frictionless compass needle. When placed in a magnetic field $\vec{B}$, the needle lines up with the field. If it is displaced by a small angle θ, show that the needle will oscillate about its equilibrium position with frequency $f = \frac{1}{2\pi}\sqrt{\mu B/I}$, where I is the moment of inertia about the point of suspension.

89 •• A conducting wire is parallel to the y axis. It moves in the positive x direction with a speed of 20 m/s in a magnetic field $\vec{B} = 0.5\text{ T}\,\hat{k}$. (*a*) What are the magnitude and direction of the magnetic force on an electron in the conductor? (*b*) Because of this magnetic force, electrons move to one end of the wire leaving the other end positively charged, until the electric field due to this charge separation exerts a force on the electrons that balances the magnetic force. Find the magnitude and direction of this electric field in the steady state. (*c*) Suppose the moving wire is 2 m long. What is the potential difference between its two ends due to this electric field?

90 ••• The rectangular frame in Figure 28-41 is free to rotate about the axis A–A on the horizontal shaft. The frame is 10 cm long and 6 cm wide and the rods that make up the frame have a mass per unit length of 20 g/cm. A uniform magnetic field $B = 0.2$ T is directed as shown. A current may be sent around the frame by means of the wires attached at the top. (*a*) If no current passes through the frame, what is the period of this physical pendulum for small oscillations? (*b*) If a current of 8.0 A passes through the frame in the direction indicated by the arrow, what is then the period of this physical pendulum? (*c*) Suppose the direction of the current is opposite to that shown. The frame is displaced from the vertical by some angle θ. What must be the magnitude of the current so that this frame will be in equilibrium?

Figure 28-41 Problem 90

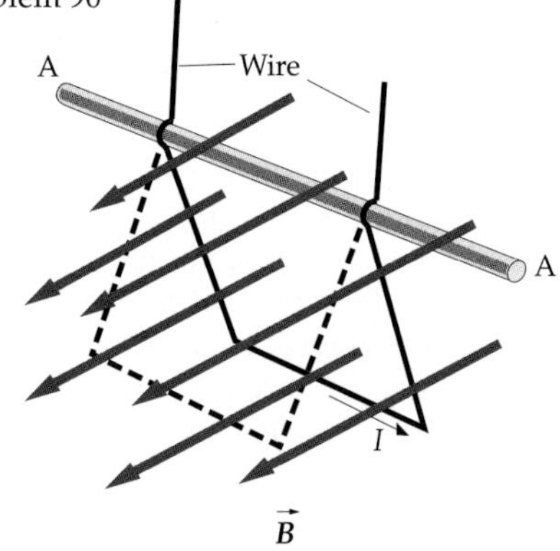

91 ••• A stiff, straight, horizontal wire of length 25 cm and mass 20 g is supported by electrical contacts at its ends, but is otherwise free to move vertically upward. The wire is in a uniform, horizontal magnetic field of magnitude 0.4 T perpendicular to the wire. A switch connecting the wire to a battery is closed and the wire is shot upward, rising to a maximum height h. The battery delivers a total charge of 2 C during the short time it makes contact with the wire. Find the height h.

92 ••• A solid sphere of radius R carries a charge density $-\rho_0$ in the region $r = 0$ to $r = R_s$ and an equal charge density of opposite sign, $+\rho_0$, between $r = R_s$ and $r = R$. The net charge carried by the sphere is zero. (*a*) What must be the ratio R/R_s? (*b*) If this sphere rotates with angular velocity ω about its diameter, what is its magnetic moment?

93 ••• A circular loop of wire with mass M carries a current I in a uniform magnetic field. It is initially in equilibrium with its magnetic moment vector aligned with the magnetic field. The loop is given a small twist about a diameter and then released. What is the period of the motion? (Assume that the only torque exerted on the loop is due to the magnetic field.)

94 ••• A small bar magnet has a magnetic moment $\vec{\mu}$ that makes an angle θ with the x axis and lies in a nonuniform magnetic field given by $\vec{B} = B_x(x)\hat{i} + B_y(y)\hat{j}$. Use $F_x = -dU/dx$ and $F_y = -dU/dy$ to show that there is a net force on the magnet that is given by

$$\vec{F} \approx \mu_x \frac{\partial B_x}{\partial x}\hat{i} + \mu_y \frac{\partial B_y}{\partial_y}\hat{j}$$

CHAPTER 29

Sources of the Magnetic Field

These coils at the Kettering Magnetics Laboratory at Oakland University are called Helmholtz coils. They are used to cancel the earth's magnetic field and to provide a uniform magnetic field in a small region of space for studying the magnetic properties of matter.

The earliest known sources of magnetism were permanent magnets. One month after Oersted announced his discovery that a compass needle is deflected by an electric current, Jean Baptiste Biot and Felix Savart announced the results of their measurements of the force on a magnet near a long, current-carrying wire and analyzed these results in terms of the magnetic field produced by each element of the current. André-Marie Ampère extended these experiments and showed that current elements also experience a force in the presence of a magnetic field and that two currents exert forces on each other.

We begin by considering the magnetic field produced by a single moving charge and by the moving charges in a current element. We then calculate the magnetic fields produced by some common current configurations, such as a straight wire segment, a long, straight wire, a current loop, and a solenoid.

Next we discuss Ampère's law, which relates the line integral of the magnetic field around a closed loop to the total current that passes through the loop. Finally, we consider the magnetic properties of matter.

29-1 The Magnetic Field of Moving Point Charges

When a point charge q moves with velocity $\vec{v}$, it produces a magnetic field $\vec{B}$ in space given by

$$\vec{B} = \frac{\mu_0}{4\pi}\frac{q\vec{v} \times \hat{r}}{r^2} \qquad 29\text{-}1$$

Magnetic field of a moving charge

where $\hat{r}$ is a unit vector that points from the charge q to the field point P (Figure 29-1) and μ_0 is a constant of proportionality called the **permeability of free space***, which has the value

$$\mu_0 = 4\pi \times 10^{-7}\,\mathrm{T{\cdot}m/A} = 4\pi \times 10^{-7}\,\mathrm{N/A^2} \qquad 29\text{-}2$$

The units of μ_0 are such that B is in teslas when q is in coulombs, v is in meters per second, and r is in meters. The unit $\mathrm{N/A^2}$ comes from the fact that $1\,\mathrm{T} = 1\,\mathrm{N/A{\cdot}m}$. The constant $1/4\pi$ is arbitrarily included in Equation 29-1 so that the factor 4π will not appear in Ampere's law (Equation 29-15), which we will study in Section 29-3.

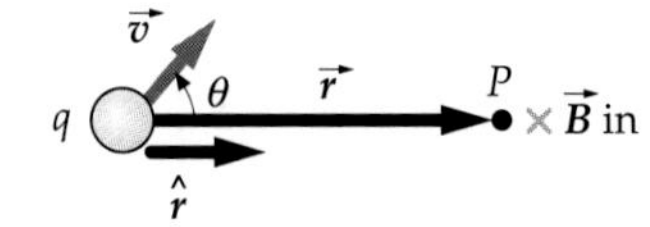

Figure 29-1 A point charge q moving with velocity $\vec{v}$ produces a magnetic field $\vec{B}$ at a field point P that is in the direction $\vec{v} \times \hat{r}$, where $\hat{r}$ is the unit vector pointing from the charge to the field point. The field varies inversely as the square of the distance from the charge to the field point and is proportional to the sine of the angle between $\vec{v}$ and $\hat{r}$. (The blue x at the field point indicates that the direction of the field is into the page.)

* Some care must be taken not to confuse the constant μ_0 with the magnetic moment μ.

Example 29-1

A point charge of magnitude $q = 4.5$ nC is moving with speed $v = 3.6 \times 10^7$ m/s parallel to the x axis along the line $y = 3$ m. Find the magnetic field at the origin produced by this charge when the charge is at the point $x = -4$ m, $y = 3$ m, as shown in Figure 29-2.

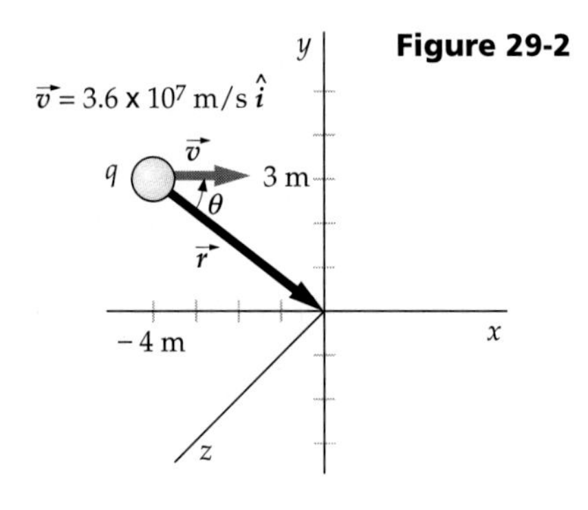

Figure 29-2

1. The magnetic field is given by Equation 29-1: $\vec{B} = \dfrac{\mu_0}{4\pi}\dfrac{q\vec{v} \times \hat{r}}{r^2}$, with $\vec{v} = v\hat{i}$

2. Find $\vec{r}$ and r from Figure 29-2, and write $\hat{r}$ in terms of $\hat{i}$ and $\hat{j}$:

$$\vec{r} = 4\,\mathrm{m}\,\hat{i} - 3\,\mathrm{m}\,\hat{j}$$

$$r = \sqrt{4^2 + 3^2}\,\mathrm{m} = 5\,\mathrm{m}$$

$$\hat{r} = \frac{\vec{r}}{r} = \frac{4\,\mathrm{m}\,\hat{i} - 3\,\mathrm{m}\,\hat{j}}{5\,\mathrm{m}} = 0.8i - 0.6\hat{j}$$

3. Evaluate the cross product:

$$\vec{v} \times \hat{r} = (v\hat{i}) \times (0.8\hat{i} - 0.6\hat{j}) = -0.6v\hat{k}$$

4. Substitute the above results in Equation 29-1 to obtain $\vec{B}$:

$$\vec{B} = \frac{\mu_0}{4\pi}\frac{q\vec{v} \times \hat{r}}{r^2} = \frac{\mu_0}{4\pi}\frac{q(-0.6v\hat{k})}{r^2}$$

$$= -(10^{-7}\,\text{T}\cdot\text{m/A})\frac{(4.5 \times 10^{-9}\,\text{C})(0.6)(3.6 \times 10^7\,\text{m/s})}{(5\,\text{m})^2}\hat{k}$$

$$= -3.89 \times 10^{-10}\,\text{T}\,\hat{k}$$

Remarks: It is also possible to obtain $\vec{B}$ without finding an explicit expression for the unit vector $\hat{r}$. From Figure 29-2 we note that $\vec{v} \times \hat{r}$ is in the negative z direction. In addition, the magnitude of $\vec{v} \times \hat{r}$ is $v \sin\theta$, where $\sin\theta = 3\text{ m}/5\text{ m} = 0.6$. Combining these results, we have $\vec{v} \times \hat{r} = v\sin\theta(-\hat{k}) = -v(0.6)\hat{k}$, in agreement with our result in step 2. Finally, this example shows that the magnetic field due to a moving charge is quite small. For comparison, the earth's magnetic field near its surface has a magnitude of about 10^{-4} T.

Exercise Find the magnetic field on the y axis at $y = 3$ m and at $y = 6$ m. (*Answers* $\vec{B} = 0$, $\vec{B} = 3.89 \times 10^{-10}\,\text{T}\,\hat{k}$)

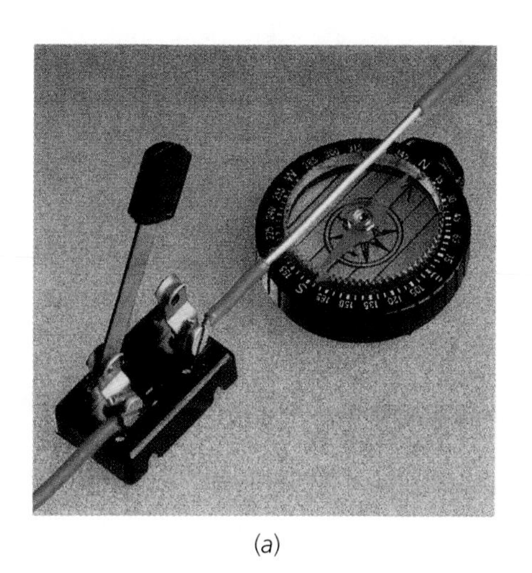
(a)

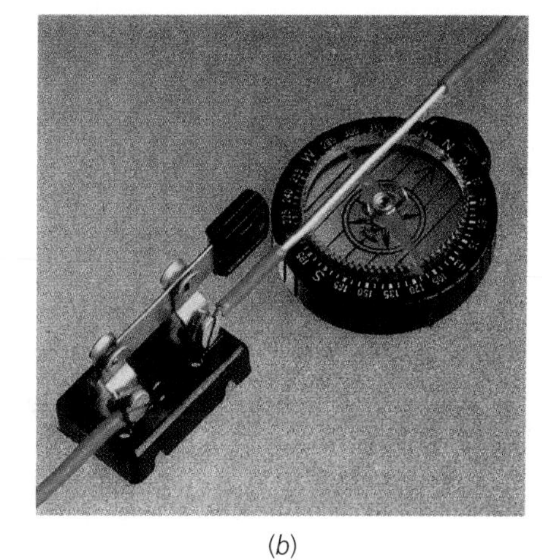
(b)

Oersted's experiment. (*a*) With no current in the wire, the compass needle points north. (*b*) When the wire carries a current, the needle is deflected in the direction of the resultant magnetic field. The current in the wire is directed upward from left to right. The insulation has been stripped from the wire to improve the contrast of the photograph.

29-2 The Magnetic Field of Currents: The Biot–Savart Law

In the previous chapter we extended our discussion of forces on point charges to forces on current elements by replacing $q\vec{v}$ with the current element $I\,d\vec{\ell}$. We do the same for the magnetic field produced by a current element. The magnetic field $d\vec{B}$ produced by a current element $I\,d\vec{\ell}$ is given by Equation 29-1 with $q\vec{v}$ replaced by $I\,d\vec{\ell}$:

$$d\vec{B} = \frac{\mu_0}{4\pi}\frac{I\,d\vec{\ell} \times \hat{r}}{r^2} \qquad 29\text{-}3$$

Biot–Savart law

Equation 29-3, known as the **Biot–Savart law,** was also deduced by Ampère. The Biot–Savart law and Equation 29-1 are analogous to Coulomb's law for the electric field of a point charge. The source of the magnetic field is a moving charge $q\vec{v}$ or a current element $I\,d\vec{\ell}$, just as the charge q is the source of the electrostatic field. The magnetic field decreases with the square of the distance from the moving charge or current element, just as the electric field decreases with the square of the distance from a point charge. However, the directional aspects of the electric and magnetic fields are quite different. Whereas the electric field points in the radial direction $\hat{r}$ from the point charge to the field point (for a positive charge), the magnetic field is perpendicular both to $\hat{r}$ and to the direction of motion $\vec{v}$ of the charges, which is along the direction of the current element. At a point along the line of a current element, such as point P_2 in Figure 29-3, the magnetic field due to that element is zero, because $I\,d\vec{\ell}$ is parallel to the vector $\hat{r}$.

The magnetic field due to the total current in a circuit can be calculated by using the Biot–Savart law to find the field due to each current element and then summing (integrating) over all the current elements in the circuit. This calculation is difficult for all but the simplest circuit geometries.

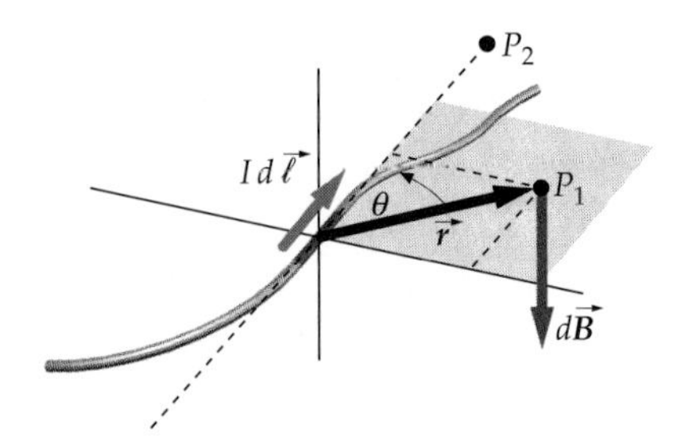

Figure 29-3 The current element $I\,d\vec{\ell}$ produces a magnetic field at point P_1 that is perpendicular to both $I\,d\vec{\ell}$ and $\hat{r}$. It produces no magnetic field at point P_2, which is along the line of $I\,d\vec{\ell}$.

$\vec{B}$ due to a Current Loop

Figure 29-4 shows a current element $I\,d\vec{\ell}$ of a current loop of radius R and the unit vector $\hat{r}$ that is directed from the element to the center of the loop. The magnetic field at the center of the loop due to this element is directed along the axis of the loop, and its magnitude is given by

$$dB = \frac{\mu_0}{4\pi}\frac{I\,d\ell\,\sin\theta}{R^2}$$

where θ is the angle between $I\,d\vec{\ell}$ and $\hat{r}$, which is 90° for each current element, so $\sin\theta = 1$. The magnetic field due to the entire current is found by integrating over all the current elements in the loop. Since R is the same for all elements, we obtain

$$B = \int dB = \frac{\mu_0}{4\pi}\frac{I}{R^2}\oint d\ell$$

The integral of $d\ell$ around the complete loop gives the total length $2\pi R$, the circumference of the loop. The magnetic field due to the entire loop is thus

$$B = \frac{\mu_0}{4\pi}\frac{I2\pi R}{R^2} = \frac{\mu_0 I}{2R} \qquad \text{29-4}$$

B at the center of a current loop

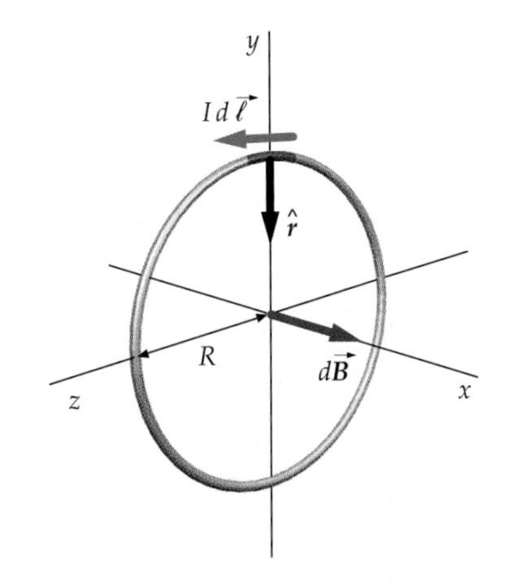

Figure 29-4 Current element for calculating the magnetic field at the center of a circular current loop. Each element produces a magnetic field that is directed along the axis of the loop.

Exercise Find the current in a circular loop of radius 8 cm that will give a magnetic field of 2 G at the center of the loop. (*Answer* 25.5 A)

Figure 29-5 shows the geometry for calculating the magnetic field at a point on the axis of a circular current loop a distance x from its center. We first consider the current element at the top of the loop. Here, as everywhere around the loop, $I\,d\vec{\ell}$ is tangent to the loop and perpendicular to the vector $\hat{r}$ from the current element to the field point P. The magnetic field $d\vec{B}$ due to this element is in the direction shown in the figure, perpendicular to $\vec{r}$ and also perpendicular to $I\,d\vec{\ell}$. The magnitude of $d\vec{B}$ is

$$|d\vec{B}| = \frac{\mu_0}{4\pi}\frac{I|d\vec{\ell}\times\hat{r}|}{r^2} = \frac{\mu_0}{4\pi}\frac{I\,d\ell}{x^2+R^2}$$

where we have used the facts that $r^2 = x^2 + R^2$ and that $d\vec{\ell}$ and $\hat{r}$ are perpendicular, so $|d\vec{\ell}\times\hat{r}| = d\ell$.

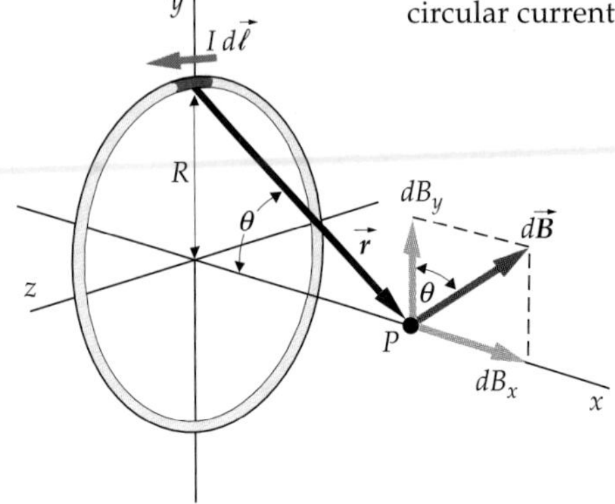

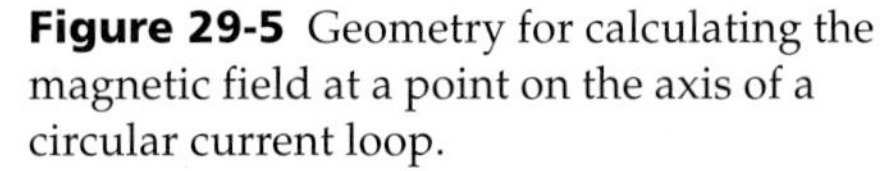

Figure 29-5 Geometry for calculating the magnetic field at a point on the axis of a circular current loop.

When we sum around all the current elements in the loop, the components of $d\vec{B}$ perpendicular to the axis of the loop, such as dB_y in Figure 29-5, sum to zero, leaving only the components dB_x that are parallel to the axis. We thus compute only the x component of the field. From Figure 29-5, we have

$$dB_x = dB\sin\theta = dB\left(\frac{R}{\sqrt{x^2+R^2}}\right) = \frac{\mu_0}{4\pi}\frac{I\,d\ell}{x^2+R^2}\frac{R}{\sqrt{x^2+R^2}}$$

To find the field due to the entire loop of current, we integrate dB_x around the loop:

$$B_x = \oint dB_x = \oint \frac{\mu_0}{4\pi}\frac{IR}{(x^2+R^2)^{3/2}}\,d\ell$$

Since neither x nor R varies as we sum over the elements in the loop, we can remove these quantities from the integral. Then,

$$B_x = \frac{\mu_0 IR}{4\pi(x^2+R^2)^{3/2}}\oint d\ell$$

The integral of $d\ell$ around the loop gives $2\pi R$. Thus,

$$B_x = \frac{\mu_0}{4\pi}\frac{IR(2\pi R)}{(x^2+R^2)^{3/2}} = \frac{\mu_0}{4\pi}\frac{2\pi R^2 I}{(x^2+R^2)^{3/2}} \qquad 29\text{-}5$$

B on the axis of a current loop

Exercise Show that Equation 29-5 reduces to $B_x = \frac{\mu_0 I}{2R}$ (Equation 29-4) at the center of the loop.

At great distances from the loop, x is much greater than R, so $(x^2 + R^2)^{3/2} \approx (x^2)^{3/2} = |x|^3$. Then,

$$B_x = \frac{\mu_0}{4\pi}\frac{2I\pi R^2}{|x^3|}$$

or

$$B_x = \frac{\mu_0}{4\pi}\frac{2\mu}{|x^3|} \qquad 29\text{-}6$$

Magnetic-dipole field on the axis of the dipole

where $\mu = I\pi R^2$ is the magnitude of the magnetic moment of the loop. Note the similarity of this expression and the electric field on the axis of an electric dipole of moment p (Equation 22-10):

$$E_x = \frac{1}{4\pi\epsilon_0}\frac{2p}{|x^3|}$$

Although it has not been demonstrated, our result that a current loop produces a magnetic dipole field far away holds in general for any point whether it is on or off of the axis of the loop. Thus, a current loop behaves as a magnetic dipole both in that it experiences a torque $\vec{\mu} \times \vec{B}$ when placed in an external magnetic field (as was shown in Chapter 28) and in that it produces a magnetic dipole field at a great distance from it. Figure 29-6 shows the magnetic field lines for a current loop.

Figure 29-6 The magnetic field lines of a circular current loop indicated by iron filings.

Example 29-2

A circular loop of radius 5.0 cm has 12 turns and lies in the yz plane. It carries a current of 4 A in the direction such that the magnetic moment of the loop is along the x axis. Find the magnetic field on the x axis at (*a*) $x = 0$, (*b*) $x = 15$ cm, and (*c*) $x = 3$ m.

Picture the Problem The magnetic field due to a loop with N turns is N times that due to a single turn. (*a*) At $x = 0$ (center of the loops) $B = \mu_0 NI/2R$ (from Equation 29-4). Equation 29-5 gives the magnetic field on axis due to the current in a single turn. Far from the loop, as in part (*c*), the field can be found using Equation 29-6. In this case, since we have N loops, the magnetic moment is $\mu = NI\pi R^2$.

(*a*) B_x at the center is N times that given by Equation 29-4 for a single loop:

$$B_x = \frac{\mu_0 NI}{2R} = \frac{\mu_0}{4\pi}\frac{2\pi NI}{R} = (10^{-7}\,\mathrm{T\cdot m/A})\frac{2\pi(12)(4\ \mathrm{A})}{0.05\ \mathrm{m}}$$

$$= 6.03 \times 10^{-4}\,\mathrm{T}$$

(*b*) B_x on the axis is N times that given by Equation 29-5:

$$B_x = \frac{\mu_0}{4\pi}\frac{2\pi R^2 NI}{(x^2 + R^2)^{3/2}}$$

$$= (10^{-7}\,\mathrm{T\cdot m/A})\frac{2\pi(0.05\ \mathrm{m})^2(12)(4\ \mathrm{A})}{[(0.15\ \mathrm{m})^2 + (0.05\ \mathrm{m})^2]^{3/2}}$$

$$= 1.91 \times 10^{-5}\,\mathrm{T}$$

(*c*)1. Since 3 m is much greater than the radius $R = 0.05$ m, we can use Equation 29-6 for the magnetic field far from the loop:

$$B_x = \frac{\mu_0}{4\pi}\frac{2\mu}{x^3}$$

2. The magnitude of the magnetic moment of the loop is NIA:

$$\mu = NI\pi R^2 = (12)(4\ \mathrm{A})\pi(0.05\ \mathrm{m})^2 = 0.377\ \mathrm{A\cdot m^2}$$

3. Substitute μ and $x = 3$ m into B_x in step 1:

$$B_x = \frac{\mu_0}{4\pi}\frac{2\mu}{x^3} = (10^{-7}\,\mathrm{T\cdot m/A})\frac{2(0.377\ \mathrm{A\cdot m^2})}{(3\ \mathrm{m})^3}$$

$$= 2.79 \times 10^{-9}\,\mathrm{T}$$

Remarks: Note in step 1 that the field produced by a current loop is typically much larger than the field due to a single moving charge (see Example 29-1). Since $x = 60R$ in (*c*), we used an approximation that is valid for $x \gg R$.

Exercise Find the magnetic field on the x axis at $x = -15$ cm. (*Answer* From Equation 29-5 we see that B_x is symmetric in x, thus $B_x = 1.91 \times 10^{-5}$ T.)

Example 29-3 *try it yourself*

A small bar magnet of magnetic moment $\mu = 0.03\ \mathrm{A\cdot m^2}$ is placed at the center of the loop of Example 29-2 so that its magnetic moment lies in the xy plane and makes an angle of 30° with the x axis. Neglecting any variation in B over the region of the magnet, find the torque on the magnet.

Picture the Problem The torque on a magnetic moment is given by $\vec{\tau} = \vec{\mu} \times \vec{B}$. Since $\vec{B}$ is in the x direction, you can see from Figure 29-7 that $\vec{\mu} \times \vec{B}$ is in the negative z direction.

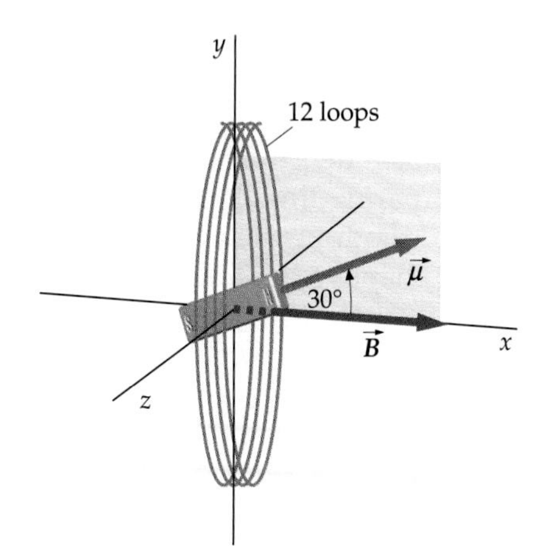

Figure 29-7

Cover the column to the right and try these on your own before looking at the answers.

Steps	*Answers*
1. Compute the magnitude of the torque from $\vec{\tau} = \vec{\mu} \times \vec{B}$.	$\tau = 9.04 \times 10^{-6}\ \mathrm{N \cdot m}$
2. Indicate the direction with a unit vector.	$\vec{\tau} = -(9.04 \times 10^{-6}\ \mathrm{N \cdot m})\hat{k}$

$\vec{B}$ due to a Current in a Solenoid

A **solenoid** is a wire tightly wound into a helix of closely spaced turns, as illustrated in Figure 29-8. It is used to produce a strong, uniform magnetic field in the region surrounded by its loops. Its role in magnetism is analogous to that of the parallel-plate capacitor, which produces a strong, uniform electric field between its plates. The magnetic field of a solenoid is essentially that of a set of N identical current loops placed side by side. Figure 29-9 shows the magnetic field lines for two such loops.

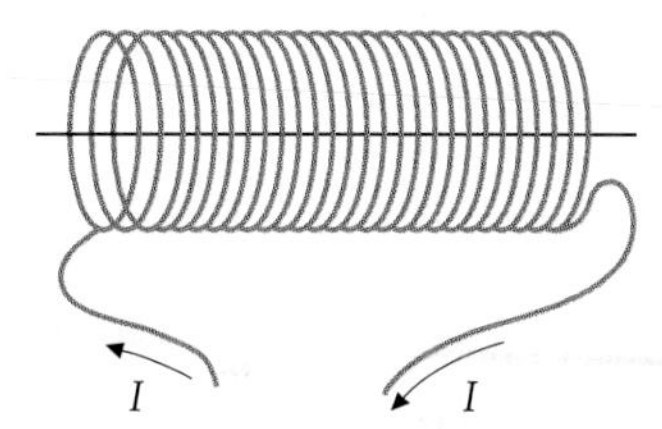

Figure 29-8 A tightly wound solenoid can be considered a set of circular current loops placed side by side that carry the same current. It produces a uniform magnetic field inside the loops.

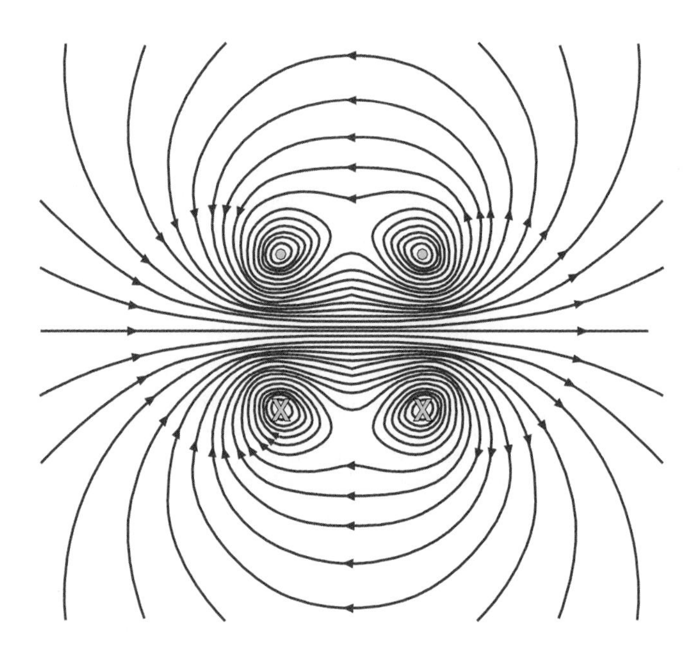

Figure 29-9 Magnetic field lines due to two loops carrying the same current in the same sense. The points where the loops intersect the plane of the page are marked by an X where the current enters and a dot where the current emerges. In the region between the loops, the magnetic fields of the individual loops add so the resultant field is strong. In the regions away from the loops, the resultant field is weak.

Figure 29-10 shows the magnetic field lines for a long, tightly wound solenoid. Inside the solenoid, the field lines are approximately parallel to the axis and are closely and uniformly spaced, indicating a strong, uniform magnetic field. Outside the solenoid, the lines are much less dense. They diverge from one end and converge at the other end. Comparing this figure with Figure 28-8, we see that the field lines of a solenoid, both inside and outside, are identical to those of a bar magnet of the same shape as the solenoid.

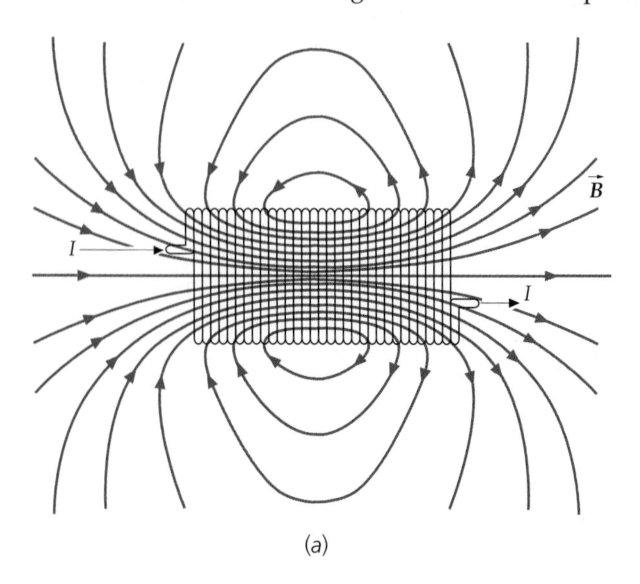

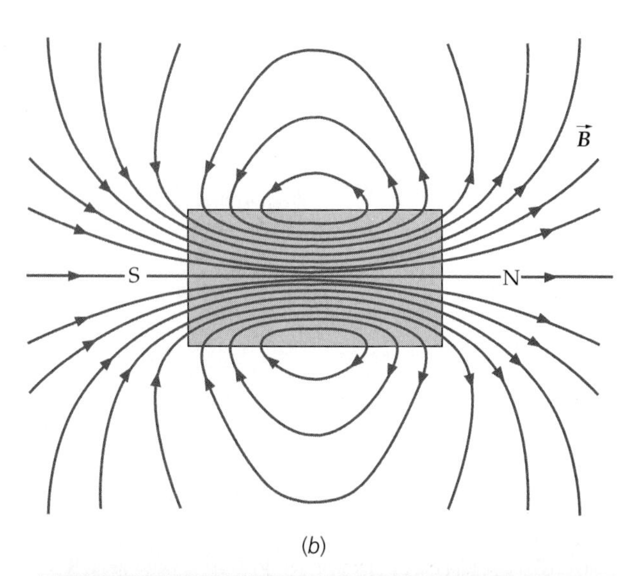

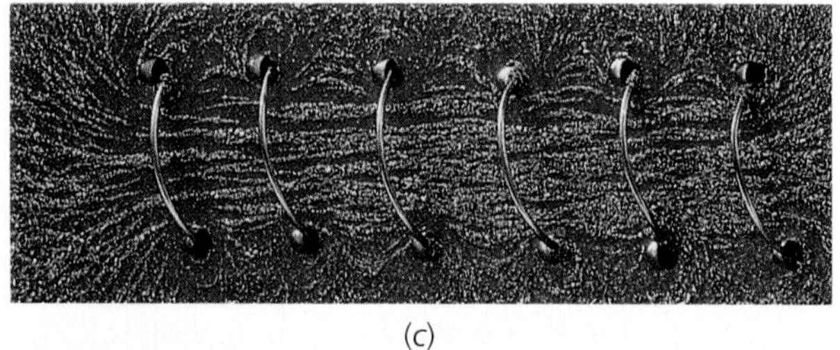

Figure 29-10 (*a*) Magnetic field lines of a solenoid. The lines are identical to those of a bar magnet of the same shape (*b*). (*c*) Magnetic field lines of a solenoid shown by iron filings.

Consider a solenoid of length L consisting of N turns of wire carrying a current I. We choose the axis of the solenoid to be along the x axis, with the left end at $x = -a$ and the right end at $x = +b$ as shown in Figure 29-11. We will calculate the magnetic field at the origin. The figure shows an element of the solenoid of length dx at a distance x from the origin. If $n = N/L$ is the number of turns per unit length, there are $n\,dx$ turns of wire in this element, with each turn carrying a current I. The element is thus equivalent to a single loop carrying a current $di = nI\,dx$. The magnetic field at a point on the x axis due to a loop at the origin carrying a current $nI\,dx$ is given by Equation 29-5 with I replaced by $nI\,dx$:

$$dB_x = \frac{\mu_0}{4\pi}\frac{2\pi R^2 nI\,dx}{(x^2 + R^2)^{3/2}}$$

This expression also gives the magnetic field at the origin due to a current loop at x. We find the magnetic field at the origin due to the entire solenoid by integrating this expression from $x = -a$ to $x = b$:

$$B_x = \frac{\mu_0}{4\pi} 2\pi R^2 nI \int_{-a}^{b} \frac{dx}{(x^2 + R^2)^{3/2}} \qquad \text{29-7}$$

The integral in Equation 29-7 can be found in standard tables of integrals. Its value is

$$\int_{-a}^{b} \frac{dx}{(x^2 + R^2)^{3/2}} = \left.\frac{x}{R^2\sqrt{x^2 + R^2}}\right|_{-a}^{b} = \frac{b}{R^2\sqrt{b^2 + R^2}} + \frac{a}{R^2\sqrt{a^2 + R^2}}$$

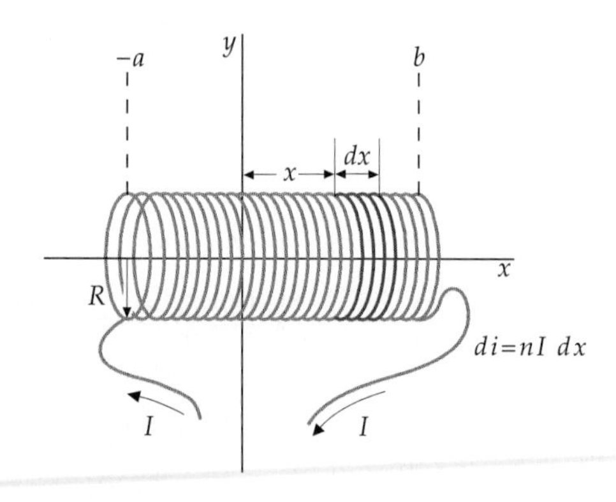

Figure 29-11 Geometry for calculating the magnetic field inside a solenoid on its axis. The number of turns in the element dx is $n\,dx$, where $n = N/\ell$ is the number of turns per unit length. The element dx is treated as a current loop carrying a current $di = nI\,dx$.

Substituting this into Equation 29-7, we obtain

$$B_x = \frac{1}{2}\mu_0 nI\left(\frac{b}{\sqrt{b^2 + R^2}} + \frac{a}{\sqrt{a^2 + R^2}}\right) \qquad \text{29-8}$$

For a long solenoid for which a and b are much larger than R, the two terms in the parentheses each tend toward 1. For this approximation, the magnetic field is

$$B_x = \mu_0 nI \qquad \text{29-9}$$

B inside a long solenoid

If the origin is at one end of the solenoid, either a or b is zero. Then, if the other end is far away compared with the radius, one of the terms in the parentheses of Equation 29-9 is zero and the other is 1, so $B \approx \frac{1}{2}\mu_0 nI$. Thus, the magnitude of $\vec{B}$ at a point near either end of a long solenoid is about half that at points within the solenoid away from the ends. Figure 29-12 gives a plot of the magnetic field on the axis of a solenoid versus position (with the origin at the center of the solenoid). The approximation that the field is constant independent of the position along the axis is quite good except for very near the ends.

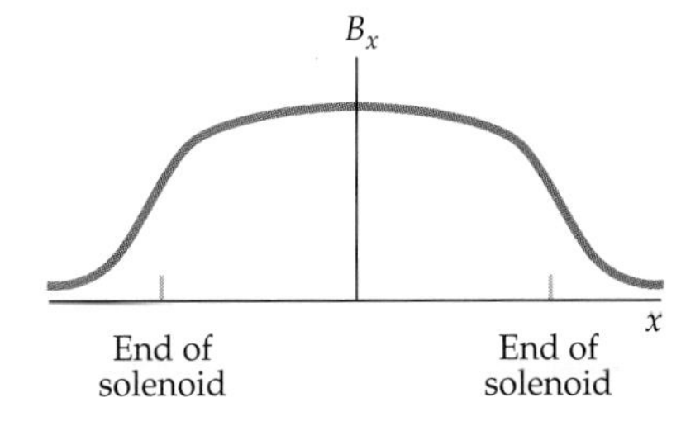

Figure 29-12 Graph of the magnetic field on the axis inside a solenoid versus the position x on the axis. The field inside the solenoid is nearly constant except near the ends.

Example 29-4

Find the magnetic field at the center of a solenoid of length 20 cm, radius 1.4 cm, and 600 turns that carries a current of 4 A.

1. We will calculate the field exactly using Equation 29-8.

$$B_x = \frac{1}{2}\mu_0 nI\left(\frac{b}{\sqrt{b^2 + R^2}} + \frac{a}{\sqrt{a^2 + R^2}}\right)$$

2. For a point at the center of the solenoid, $a = b = 10$ cm. Thus, each term in the parentheses in Equation 29-8 has the value:

$$\frac{a}{\sqrt{a^2 + R^2}} = \frac{b}{\sqrt{b^2 + R^2}} = \frac{10 \text{ cm}}{\sqrt{(10 \text{ cm})^2 + (1.4 \text{ cm})^2}} = 0.990$$

3. Substitute these results into B_x in step 1:

$$B_x = \tfrac{1}{2}\mu_0 nI(0.990 + 0.990)$$
$$= (0.5)(4\pi \times 10^{-7}\,\text{T}\cdot\text{m/A})(600 \text{ turns}/0.2 \text{ m})(4 \text{ A})(0.990 + 0.990)$$
$$= 1.50 \times 10^{-2}\,\text{T}$$

Remarks: Note that the approximation obtained using Equation 29-9 amounts to replacing 0.99 by 1.00, which differs by only 1%. Note also that the magnitude of the magnetic field inside this solenoid is fairly large—about 250 times the magnetic field of the earth.

Exercise: Calculate B_x using the long-solenoid approximation. (*Answer* 1.51×10^{-2} T)

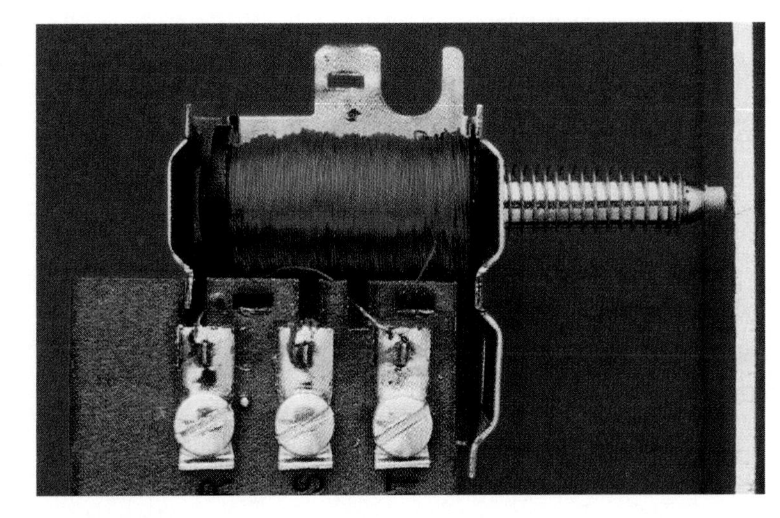

A cross section of a doorbell. When the outer solenoid is energized, its magnetic field causes the inner plunger to strike the bell.

$\vec{B}$ due to a Current in a Straight Wire

Figure 29-13 shows the geometry for calculating the magnetic field $\vec{B}$ at a point P due to the current in the straight wire segment shown. We choose the x axis to be along the wire and point P to be on the y axis. Because of the symmetry in this problem, any direction perpendicular to the wire could be chosen for the y axis.

A typical current element $I\,d\vec{\ell}$ at a distance x from the origin is shown. The vector $\vec{r}$ points from the element to the field point P. The direction of the magnetic field at P due to this element is the direction of $I\,d\vec{\ell} \times \vec{r}$, which is out of the paper. Note that the magnetic fields due to all the current elements of the wire are in this same direction. Thus, we need to compute only the magnitude of the field. The field due to the current element shown has the magnitude (Equation 29-3)

$$dB = \frac{\mu_0}{4\pi}\frac{I\,dx}{r^2}\sin\phi$$

It is more convenient to write this in terms of θ rather than ϕ:

$$dB = \frac{\mu_0}{4\pi}\frac{I\,dx}{r^2}\cos\theta \qquad \text{29-10}$$

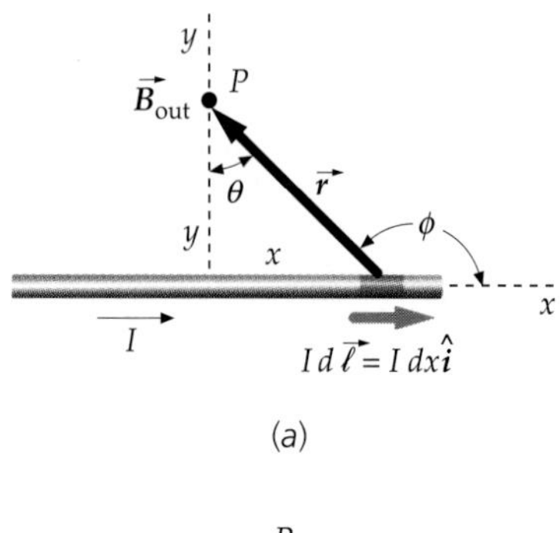

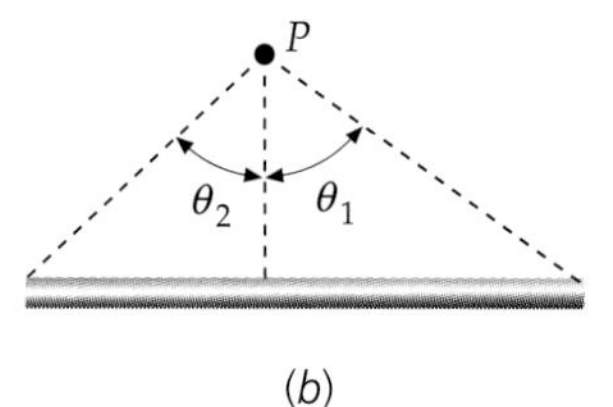

Figure 29-13 (*a*) Geometry for calculating the magnetic field at point P due to a straight current segment. Each element of the segment contributes to the total magnetic field at point P, which is directed out of the paper. (*b*) The result is expressed in terms of the angles θ_1 and θ_2.

To sum over all the current elements, we need to relate the variables θ, r, and x. It turns out to be easiest to express x and r in terms of θ. We have

$$x = y\tan\theta$$

Then,

$$dx = y\sec^2\theta\,d\theta = y\frac{r^2}{y^2}\,d\theta = \frac{r^2}{y}\,d\theta$$

where we have used $\sec\theta = r/y$. Substituting this expression for dx into Equation 29-10, we obtain

$$dB = \frac{\mu_0}{4\pi}\frac{I}{r^2}\frac{r^2\,d\theta}{y}\cos\theta = \frac{\mu_0}{4\pi}\frac{I}{y}\cos\theta\,d\theta$$

Let us first calculate the contribution from the current elements to the right of the point $x = 0$. We sum over these elements by integrating from $\theta = 0$ to $\theta = \theta_1$, where θ_1 is the angle between the line perpendicular to the wire and the line from P to the right end of the wire, as shown in Figure 29-13*b*. For this contribution, we have

$$B_1 = \int_0^{\theta_1}\frac{\mu_0}{4\pi}\frac{I}{y}\cos\theta\,d\theta$$

$$= \frac{\mu_0}{4\pi}\frac{I}{y}\int_0^{\theta_1}\cos\theta\,d\theta = \frac{\mu_0}{4\pi}\frac{I}{y}\sin\theta_1$$

Similarly, the contribution from elements to the left of $x = 0$ is

$$B_2 = \frac{\mu_0}{4\pi}\frac{I}{y}\sin\theta_2$$

The total magnetic field due to the wire segment is the sum of B_1 and B_2. Writing R instead of y for the perpendicular distance from the wire segment to the field point, we obtain

$$B = \frac{\mu_0}{4\pi}\frac{I}{R}(\sin\theta_1 + \sin\theta_2) \qquad \text{29-11}$$

B due to a straight wire segment

This result gives the magnetic field due to any wire segment in terms of the perpendicular distance R and the angles subtended at the field point by the ends of the wire. If the wire is very long, these angles are nearly 90°. The result for a very long wire is obtained from Equation 29-11 by setting $\theta_1 = \theta_2 = 90°$:

$$B = \frac{\mu_0}{4\pi}\frac{2I}{R} \qquad \text{29-12}$$

B due to a long, straight wire

At any point in space, the magnetic field lines of a long, straight, current-carrying wire are tangent to a circle of radius R about the wire, where R is the perpendicular distance from the wire to the field point. The direction of $\vec{B}$ can be determined by applying the right-hand rule as shown in Figure 29-14*a*. The magnetic field lines thus encircle the wire as shown in Figure 29-14*b*.

The result expressed by Equation 29-12 was found experimentally by Biot and Savart in 1820. From an analysis of it, they were able to discover the expression given in Equation 29-3 for the magnetic field due to a current element.

(*a*)

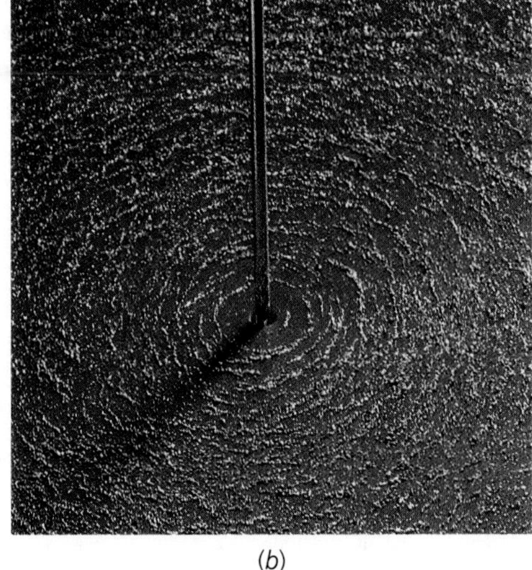

(*b*)

Figure 29-14 (*a*) Right-hand rule for determining the direction of the magnetic field due to a long, straight, current-carrying wire. The magnetic field lines encircle the wire in the direction of the fingers of the right hand when the thumb points in the direction of the current. (*b*) Magnetic field lines due to a long wire indicated by iron filings.

Example 29-5

Find the magnetic field at the center of a square current loop of side L = 50 cm carrying a current of 1.5 A.

Picture the Problem The magnetic field at the center of the loop is the sum of the contributions from each of the four sides of the loop. From Figure 29-15 we can see that each side of the loop produces a field of equal magnitude pointing out of the page. Thus, we use Equation 29-11 for a given side, then multiply by 4 for the total field.

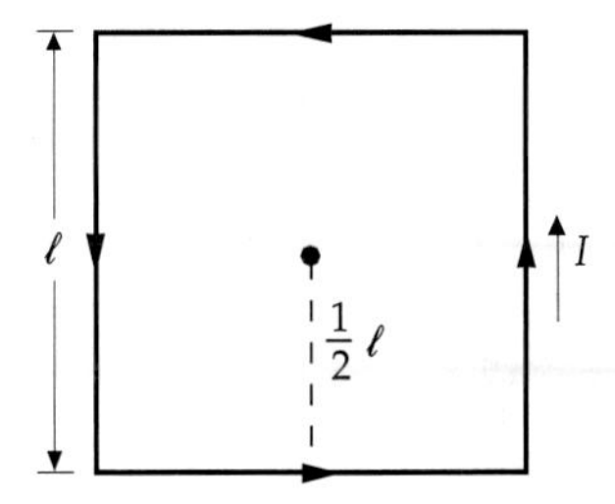

Figure 29-15

1. The total field is 4 times the field B_s due to a side: $\quad B = 4B_s$

2. Calculate the magnetic field B_s due to a given side of the loop. Note from the figure that $R = \frac{1}{2}L$ and $\theta_1 = \theta_2 = 45°$:

$$B_s = \frac{\mu_0}{4\pi}\frac{I}{\frac{1}{2}L}(\sin 45° + \sin 45°)$$
$$= (10^{-7}\,\text{T·m/A})\frac{1.5\text{ A}}{0.25\text{ m}}(2\sin 45°) = 8.49 \times 10^{-7}\,\text{T}$$

3. Multiply this value by 4 to find the total field: $\quad B = 4B_s = 4(8.49 \times 10^{-7}\,\text{T}) = 3.39 \times 10^{-6}\,\text{T}$

Exercise Compare the magnetic field at the center of a circular current loop of radius R with that at the center of a square current loop of side $L = 2R$. Which is larger? (*Answer* B is larger for the circle by about 10%)

Exercise Find the magnetic field at a distance of 20 cm from a long, straight wire carrying a current of 5 A. (Answer $B = 5.00 \times 10^{-6}$ T)

We note from the exercise above that the magnetic field near a wire carrying a current of ordinary size is small. At 20 cm from a long, straight wire carrying 5 A, it is only about 10% of the magnetic field due to the earth.

A current gun for measuring electric current. The jaws of the current gun clamp around a current-carrying wire without touching the wire. The magnetic field produced by the wire is measured with a Hall-effect device mounted in the current gun. The Hall-effect device puts out a voltage proportional to the magnetic field, which in turn is proportional to the current in the wire.

Example 29-6

A long, straight wire carrying a current of 1.7 A in the positive z direction lies along the line $x = -3$ cm, $y = 0$. A similar wire carrying a current of 1.7 A in the positive z direction lies along the line $x = +3$ cm, $y = 0$ as shown in Figure 29-16. Find the magnetic field at a point P on the y axis at $y = 6$ cm.

Figure 29-16

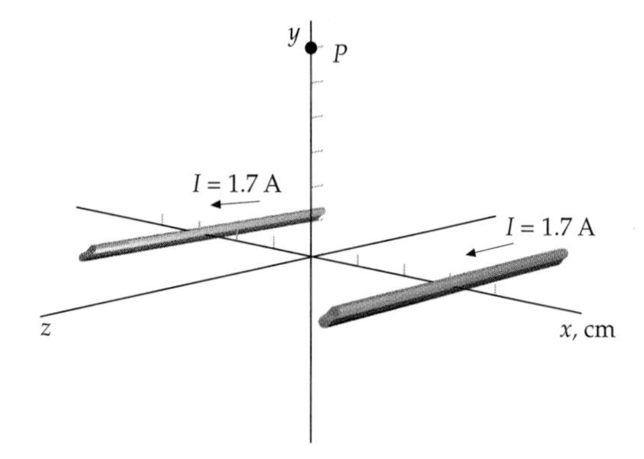

Picture the Problem The magnetic field at point P is the vector sum of the field $\vec{B}_L$ due to the wire on the left in Figure 29-17, and the field $\vec{B}_R$ due to the wire on the right. Since each wire carries the same current, and is the same distance from P, the magnitudes $\vec{B}_L$ and $\vec{B}_R$ are equal. $\vec{B}_L$ is perpendicular to the radius from the left wire to point P, and $\vec{B}_R$ is perpendicular to the radius from the right wire to the point P.

Figure 29-17

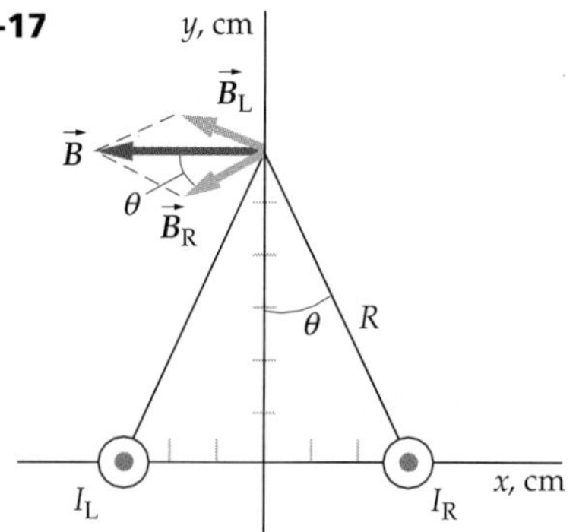

1. The field at P is the vector sum of the fields $\vec{B}_L$ and $\vec{B}_R$: $\vec{B} = \vec{B}_L + \vec{B}_R$

2. From Figure 29-17 we see that the resultant magnetic field is in the negative x direction and has the magnitude $2B_L \cos\theta$: $\vec{B} = -2B_L \cos\theta\, \hat{i}$

3. The magnitudes of $\vec{B}_L$ and $\vec{B}_R$ are given by Equation 29-12: $B_L = B_R = \dfrac{\mu_0}{4\pi}\dfrac{2I}{R}$

4. R is the distance from each wire to the point P. We find R from the figure and substitute into the expression for B_L and B_R:

$$R = \sqrt{(3\text{ cm})^2 + (6\text{ cm})^2} = 6.71\text{ cm}$$

$$B_L = B_R = (10^{-7}\,\text{T}\cdot\text{m/A})\frac{2(1.7\text{ A})}{0.0671\text{ m}} = 5.07 \times 10^{-6}\,\text{T}$$

5. We obtain $\cos\theta$ from the figure:

$$\cos\theta = \frac{6\text{ cm}}{R} = \frac{6\text{ cm}}{6.71\text{ cm}} = 0.894$$

6. Substitute the values of $\cos\theta$ and B_L into the equation in step 2 for $\vec{B}$:

$$\vec{B} = -2(5.07 \times 10^{-6}\,\text{T})(0.894)\hat{i} = -9.07 \times 10^{-6}\,\text{T}$$

Exercise Find $\vec{B}$ at the origin. (*Answer* 0)

Exercise Find $\vec{B}$ at the origin assuming that I_R goes *into* the page. (*Answer* $\vec{B} = 2.27 \times 10^{-5}\,\text{T}\,\hat{j}$)

Definition of the Ampere

We can use Equation 29-12 for the magnetic field due to a long, straight, current-carrying wire and $d\vec{F} = I\,d\vec{\ell} \times \vec{B}$ (Equation 28-5) for the force exerted by a magnetic field on a segment of a current-carrying wire to find the force exerted by one long, straight current on another. Figure 29-18 shows two long, parallel wires carrying currents in the same direction. We consider the force on a segment $d\vec{\ell}_2$ carrying current I_2 as shown. The magnetic field $\vec{B}_1$ at this segment due to current I_1 is perpendicular to the segment $I_2\,d\vec{\ell}_2$ as shown. This is true for all current elements along the wire. The magnetic force $d\vec{F}_2$ on current segment $I_2\,d\vec{\ell}_2$ is directed toward current I_1. Similarly, a current segment $I_1\,d\vec{\ell}_1$ will experience a magnetic force directed toward current I_2 due to a magnetic field arising from current I_2. Thus, two parallel currents attract each other. If one of the currents is reversed, the force will be reversed, so two antiparallel currents will repel each other. The attraction or repulsion of parallel or antiparallel currents was discovered experimentally by Ampère one week after he heard of Oersted's discovery of the effect of a current on a compass needle.

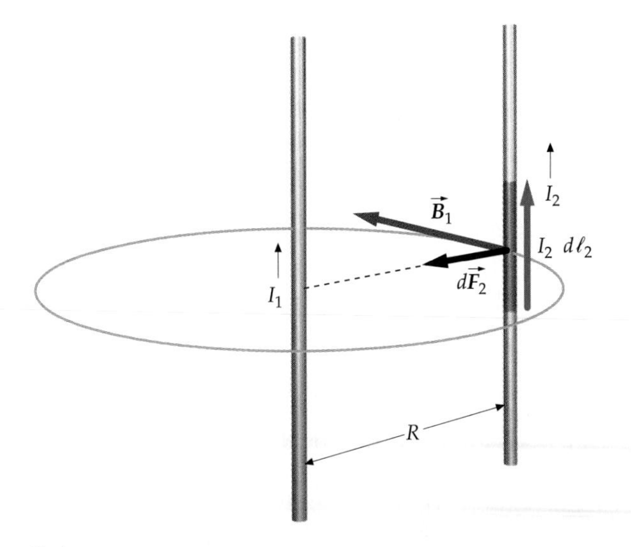

Figure 29-18 Two long, straight wires carrying parallel currents. The magnetic field $\vec{B}_1$ due to current I_1 is perpendicular to current I_2. The force on current I_2 is toward current I_1. There is an equal and opposite force exerted by current I_2 on I_1. The currents thus attract each other.

The magnitude of the magnetic force on the segment $I_2\,d\vec{\ell}_2$ is

$$dF_2 = |I\,d\vec{\ell}_2 \times \vec{B}_1|$$

Since the magnetic field at segment $I_2\,d\vec{\ell}_2$ is perpendicular to the current segment, we have

$$dF_2 = I_2\,d\ell_2 B_1$$

If the distance R between the wires is much less than their length, the field at $I_2\,d\vec{\ell}_2$ due to current I_1 will approximate the field due to an infinitely long, current-carrying wire, which is given by Equation 29-12. The magnitude of the force on the segment $I_2\,d\vec{\ell}_2$ is therefore

$$dF_2 = I_2\,d\ell_2\frac{\mu_0 I_1}{2\pi R}$$

The force per unit length is

$$\frac{dF_2}{d\ell_2} = I_2\frac{\mu_0 I_1}{2\pi R} = 2\frac{\mu_0}{4\pi}\frac{I_1 I_2}{R} \qquad 29\text{-}13$$

In Chapter 18, the coulomb was defined in terms of the ampere, but the definition of the ampere was deferred. The ampere is defined as follows:

> The ampere is that constant current which, if maintained in two straight, parallel conductors of infinite length, of negligible circular cross section, and placed one meter apart in a vacuum, would produce between these conductors a force equal to 2×10^{-7} newtons per meter of length.

Definition—Ampere

This definition of the ampere makes the permeability of free space μ_0 equal to exactly $4\pi \times 10^{-7}\ \mathrm{N/A^2}$. It also allows the unit of current (and therefore the unit of electric charge) to be determined by a mechanical measurement. In practice, currents much closer together than 1 m are used so that the force can be measured accurately with long but finite wires.

Figure 29-19 shows a **current balance,** a device that can be used to calibrate an ammeter from the definition of the ampere. The upper conductor, directly above the lower conductor, is free to rotate about knife edge contacts and is balanced so that the wires (or conducting rods) are a small distance apart. The conductors are connected in series to carry the same current but in opposite directions so that they will repel each other. Weights are placed on the upper conductor until it balances again at the original separation. The force of repulsion is thus determined by measuring the total weight needed to balance the upper conductor.

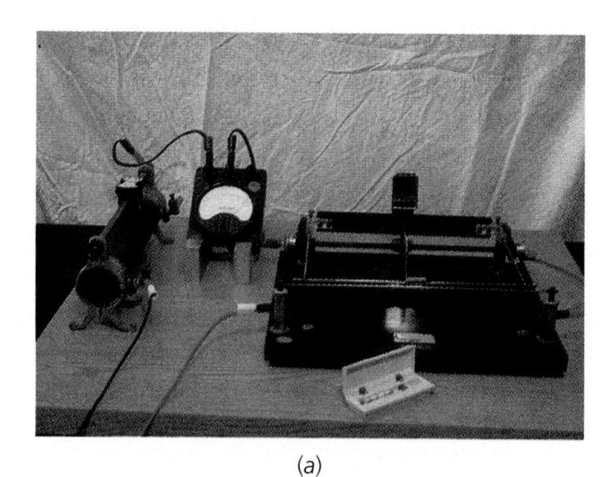
(*a*)

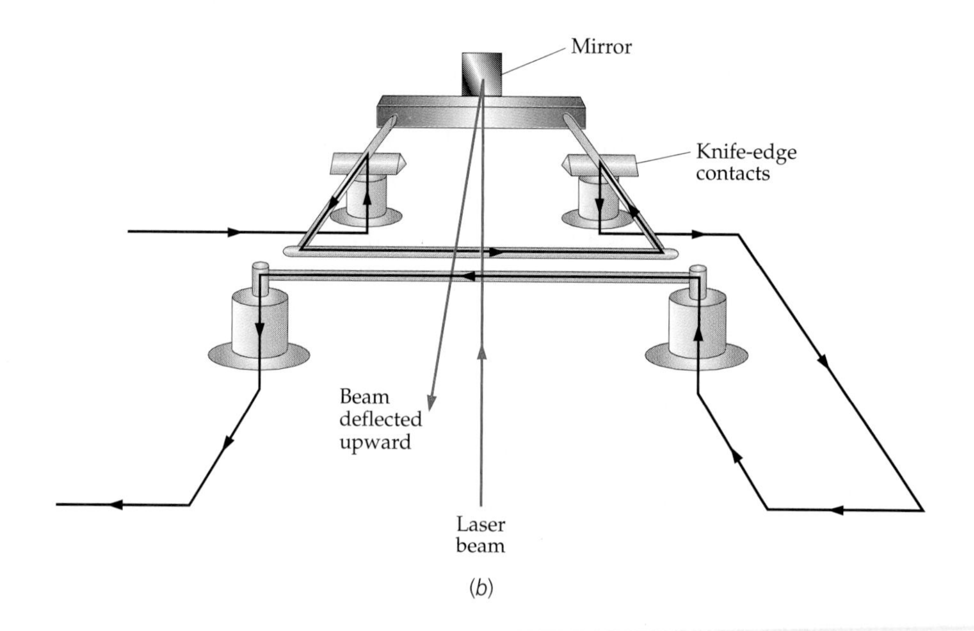

(*b*)

Figure 29-19 (*a*) Current balance used in an elementary physics laboratory to calibrate an ammeter. (*b*) A schematic diagram of the current balance in (*a*). The two parallel rods in front carry equal but oppositely directed currents and therefore repel each other. The force of repulsion is balanced by weights placed on the upper rod, which is part of a rectangle that is balanced on knife edges at the back. The mirror on top is used to reflect a beam of laser light for accurately determining the position of the upper rod.

Example 29-7 *try it yourself*

Two straight rods 50 cm long and 1.5 mm apart in a current balance carry currents of 15 A each in opposite directions. What mass must be placed on the upper rod to balance the magnetic force of repulsion?

Picture the Problem Equation 29-13 gives the magnitude of the magnetic force per unit length exerted by the lower rod on the upper rod. Find this force for a rod of length L and set it equal to the weight mg.

Cover the column to the right and try these on your own before looking at the answers.

Steps	*Answers*
1. Set the weight mg equal to the magnetic force of repulsion of the rods.	$mg = 2\left(\frac{\mu_0}{4\pi}\right)\frac{I_1 I_2}{R}L$
2. Solve for the mass m.	$m = 1.53 \times 10^{-3}\text{ kg} = 1.53\text{ g}$

Remark Since only 1.53 g are required to balance the system, we see that the magnetic force between two current-carrying wires is relatively small, even for currents as large as 15 A separated by only 1.5 mm.

29-3 Gauss's Law for Magnetism

The magnetic field lines as shown in Figures 29-6, 29-9, and 29-10 differ from electric field lines in that the lines of $\vec{B}$ form closed curves, whereas lines of $\vec{E}$ begin and end on electric charges. The magnetic equivalent of an electric charge is a magnetic pole, such as appears to be at the ends of a bar magnet. However, although the magnetic field lines appear to diverge from the north pole outside a bar magnet (Figure 29-10*b*), the lines inside the magnet point toward the pole. These lines enter the south pole of the magnet from the outside, but on the inside of the magnet the lines leave the south pole. If one end of a bar magnet is enclosed by a surface, the number of magnetic field lines that leave the surface is exactly equal to the number that enter the surface. That is, the net flux of the field through any closed surface is zero.*

$$\phi_{\text{m,net}} = \oint_S B_n\, dA = 0 \qquad \text{29-14}$$

Gauss's law for magnetism

where the definition of the magnetic flux ϕ_{m} is exactly analogous to the electric flux with $\vec{B}$ replacing $\vec{E}$. This result is called Gauss's law for magnetism. It is the mathematical statement that there are no points in space from which magnetic field lines diverge, or to which they converge. That is, isolated magnetic poles do not exist. The fundamental unit of magnetism is the magnetic dipole. Figure 29-20 compares the lines of $\vec{B}$ for a magnetic dipole with the lines of $\vec{E}$ for an electric dipole. Note that far from the dipoles the lines are identical. But inside the dipole, the lines of $\vec{E}$ are opposite the lines of $\vec{B}$. The lines of $\vec{E}$ diverge from the positive charge and converge to the negative charge, whereas the lines of $\vec{B}$ are continuous loops.

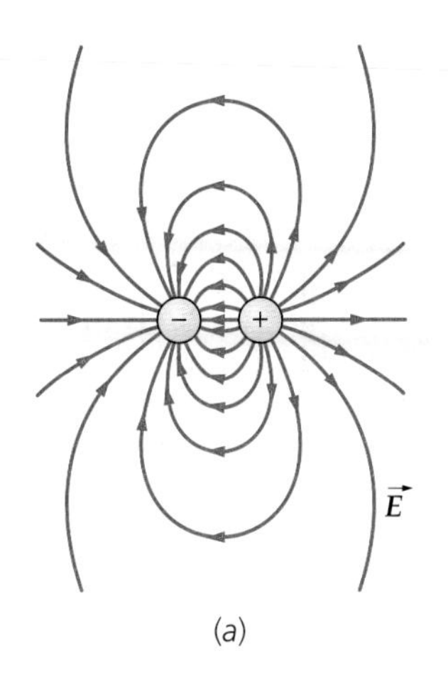

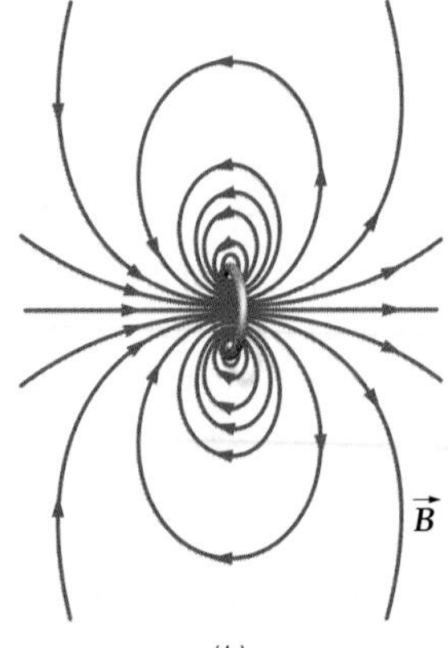

Figure 29-20 (*a*) Electric field lines of an electric dipole. (*b*) Magnetic field lines of a magnetic dipole. Far from the dipoles, the field lines are identical. In the region between the charges in (*a*), the electric field is opposite the dipole moment, whereas inside the loop in (*b*), the magnetic field is parallel to the dipole moment.

* Recall that the net flux of the electric field is a measure of the net number of lines that leave a closed surface and is equal to $Q_{\text{inside}}/\epsilon_0$.

29-4 Ampère's Law

In Chapter 23 we found that for highly symmetric charge distributions, we could calculate the electric field more easily using Gauss's law than Coulomb's law. A similar situation exists in magnetism. Ampère's law, which relates the tangential component of $\vec{B}$ summed around a closed curve C to the current I_C that passes through the curve can be used to obtain an expression for the magnetic field in situations that have a high degree of symmetry. In mathematical form, **Ampère's law** is

$$\oint_C \vec{B}\cdot d\vec{\ell} = \mu_0 I_C, \quad C \text{ is any closed curve} \qquad \text{29-15}$$

Ampère's law

where I_C is the net current that penetrates the area bounded by the curve C. Ampère's law holds for any curve C as long as the currents are continuous, that is, they do not begin or end at any finite point. It is useful in calculating the magnetic field $\vec{B}$ in situations that have a high degree of symmetry so that the line integral $\oint_C \vec{B}\cdot d\vec{\ell}$ can be written as the product of B and some distance. Ampere's law and Gauss's law are both of considerable theoretical importance, and both hold whether or not there is symmetry, but if there is no symmetry, neither is useful in calculating electric or magnetic fields.

The simplest application of Ampère's law is to find the magnetic field of an infinitely long, straight, current-carrying wire. Figure 29-21 shows a circular curve around a point on a long wire with its center at the wire. If we assume that we are far from the ends of the wire, we can use symmetry to rule out the possibility of any component of $\vec{B}$ parallel to the wire. We may then assume that the magnetic field is tangent to this circle and has the same magnitude B at any point on the circle. Ampère's law then gives

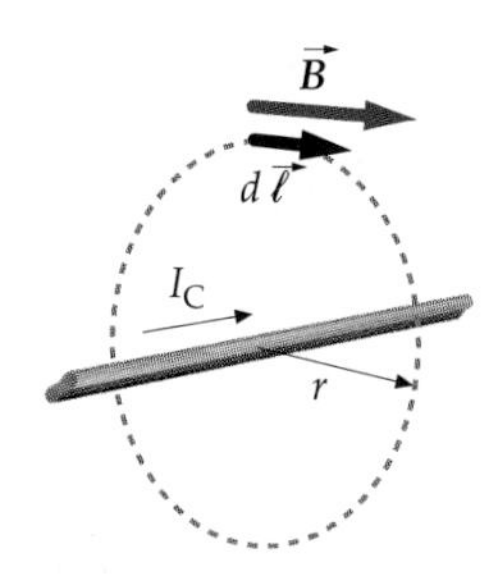

Figure 29-21 Geometry for calculating the magnetic field of a long, straight, current-carrying wire using Ampère's law. On a circle around the wire, the magnetic field is constant and tangent to the circle.

$$\oint_C \vec{B}\cdot d\vec{\ell} = B\oint_C d\ell = \mu_0 I_C$$

where we have taken B out of the integral because it has the same value everywhere on the circle. The integral of $d\ell$ around the circle equals $2\pi r$, the circumference of the circle. The current I_C is the current I in the wire. We thus obtain

$$B(2\pi r) = \mu_0 I$$

$$B = \frac{\mu_0 I}{2\pi R} = \frac{\mu_0}{4\pi}\frac{2I}{R}$$

which is Equation 29-12.

Example 29-8

A long, straight wire of radius R carries a current I that is uniformly distributed over the cross-sectional area of the wire. Find the magnetic field both outside and inside the wire.

Picture the Problem We can use Ampère's law to calculate $\vec{B}$ because of the high degree of symmetry. At a distance r (Figure 29-22), we know that $\vec{B}$ is tangent to the circle of radius r about the wire and constant in magnitude everywhere on the circle. The current through C depends on whether r is less than or greater than the radius of the wire a.

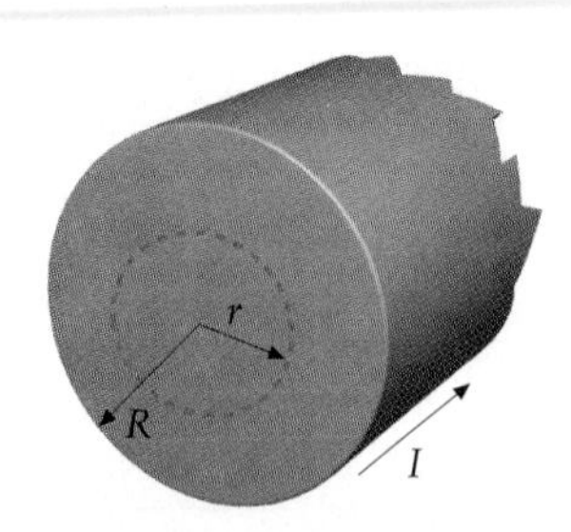

Figure 29-22

1. Apply Ampère's law to a circle of radius r:

$$\oint_C \vec{B}\cdot d\vec{\ell} = B\oint_C d\ell = B2\pi r = \mu_0 I_C$$

$$B = \frac{\mu_0}{2\pi}\frac{I_C}{r}$$

2. Ouside the wire, $r > R$, and the total current passes through the curve C:

$$I_C = I$$

$$B = \frac{\mu_0}{2\pi}\frac{I}{r}$$

3. Inside the wire, $r < R$ and the current passing through C is $(\pi r^2/\pi R^2)$ times the total current I:

$$I_C = \frac{\pi r^2}{\pi R^2}I = \frac{r^2}{R^2}I$$

$$B = \frac{\mu_0}{2\pi}\frac{I_C}{r} = \frac{\mu_0}{2\pi}\frac{(r^2/R^2)I}{r} = \frac{\mu_0}{2\pi R^2}Ir$$

Remark Inside the wire, the field increases with distance from the center of the wire. Figure 29-23 shows the graph of B versus r for this example.

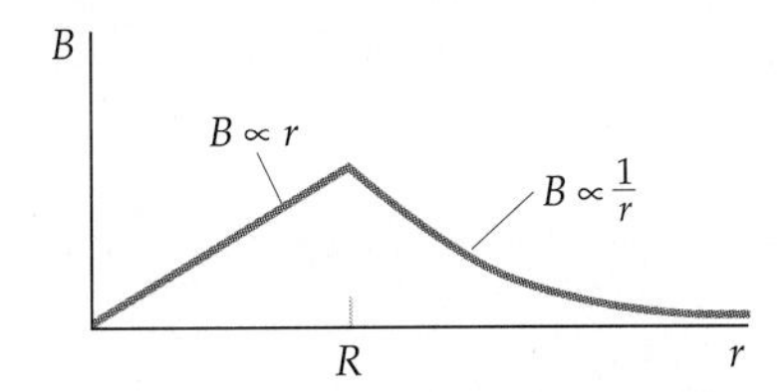

Figure 29-23

We see from Example 29-8 that the magnetic field due to a current uniformly distributed over a wire of radius R is given by

$$B = \frac{\mu_0}{2\pi R^2}Ir, \qquad r \le R$$

$$B = \frac{\mu_0}{2\pi}\frac{I}{r}, \qquad r \ge R \tag{29-16}$$

For our next application of Ampère's law, we calculate the magnetic field of a tightly wound **toroid,** which consists of loops of wire wound around a doughnut-shaped form as shown in Figure 29-24. There are N turns of wire, each carrying a current I. To calculate B, we evaluate the line integral $\oint_C \vec{B}\cdot d\vec{\ell}$ around a circle of radius r centered in the middle of the toroid. By symmetry, $\vec{B}$ is tangent to this circle and constant in magnitude at every point on the circle. Then,

$$\oint_C \vec{B}\cdot d\vec{\ell} = B2\pi r = \mu_0 I_C$$

Let a and b be the inner and outer radii of the toroid, respectively. The total current through the circle of radius r for $a < r < b$ is NI. Ampère's law then gives

$$\oint_C \vec{B}\cdot d\vec{\ell} = B2\pi r = \mu_0 I_C = \mu_0 NI$$

or

$$B = \frac{\mu_0 NI}{2\pi r}, \qquad a < r < b \tag{29-17}$$

B inside a tightly wound toroid

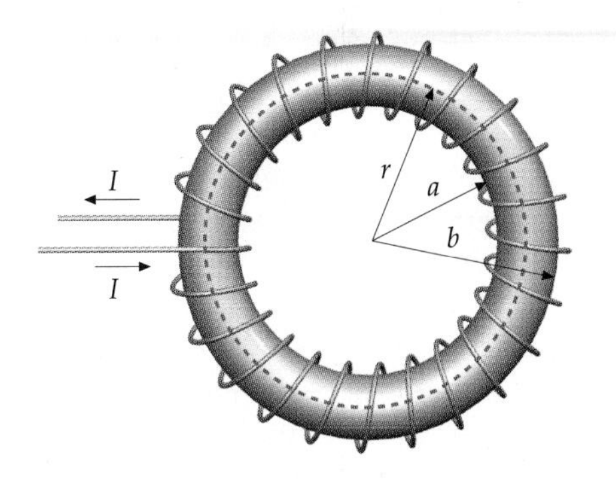

Figure 29-24 A toroid consists of loops of wire wound around a doughnut-shaped form. The magnetic field at any distance r can be found by applying Ampère's law to the circle of radius r.

If r is less than a, there is no current through the circle of radius r. If r is greater than b, the total current through r is zero because for each current I

into the page at the inner surface of the toroid in Figure 29-24, there is an equal current I out of the page at the outer surface. Thus, the magnetic field is zero for both $r < a$ and $r > b$:

$$B = 0, \qquad r < a \quad \text{or} \quad r > b$$

The magnetic field inside the toroid is not uniform but decreases with r. However, if the diameter of the loops of the toroid, $b - a$, is much less than the radius of the doughnut, the variation in r from $r = a$ to $r = b$ is small, and B is approximately uniform, as it is in a solenoid.

(a)

(b)

(a) The Tokamak fusion test reactor is a large toroid that produces a magnetic field for confining charged particles. Coils containing over 10 km of water-cooled copper wire carry a pulsed current, which has a peak value of 73,000 A and produces a magnetic field of 5.2 T for about 3 s. (b) Inspection of the assembly of the Tokamak from inside the toroid.

Limitations of Ampère's Law

Ampère's law is useful for calculating the magnetic field only when there is a high degree of symmetry. Consider the current loop shown in Figure 29-25. According to Ampère's law, the line integral $\oint_C \vec{B} \cdot d\vec{\ell}$ around a curve such as curve C in the figure equals μ_0 times the current I in the loop. Although Ampère's law is valid for this curve, the magnetic field $\vec{B}$ is not constant along any curve encircling the current, nor is it everywhere tangent to any such curve. Thus, there is not enough symmetry in this situation to allow us to calculate $\vec{B}$ using Ampère's law.

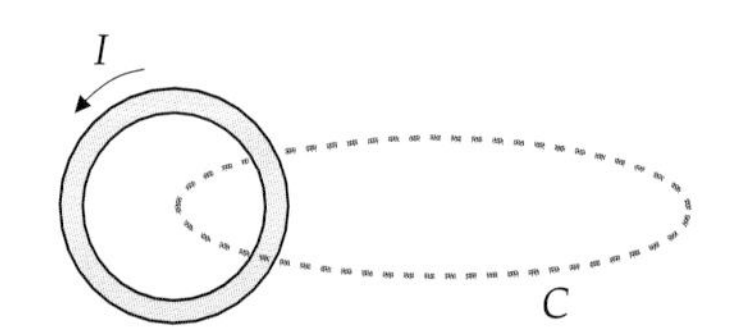

Figure 29-25 Ampère's law holds for the curve C encircling the current in the circular loop, but it is not useful for finding $\vec{B}$, because $\vec{B}$ is neither constant along the curve nor tangent to it.

Figure 29-26 shows a finite current segment of length ℓ. We wish to find the magnetic field at point P, which is equidistant from the ends of the segment and at a distance r from the center of the segment. A direct application of Ampère's law gives

$$B = \frac{\mu_0}{2\pi} \frac{I}{r}$$

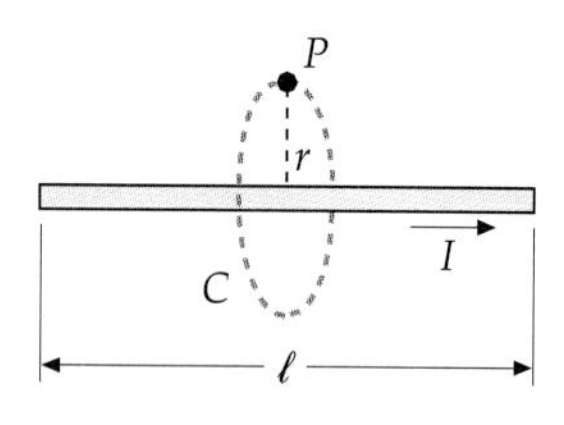

Figure 29-26 The application of Ampère's law to find the magnetic field on the bisector of a finite current segment gives an incorrect result.

This result is the same as for an infinitely long wire, since the same symmetry arguments apply. It does not agree with the result obtained from the Biot–Savart law, which depends on the length of the current segment and which agrees with experiment. If the current segment is just one part of a continuous circuit carrying a current, as shown in Figure 29-27, Ampère's law for curve C is valid, but it cannot be used to find the magnetic field at point P because there is no symmetry.

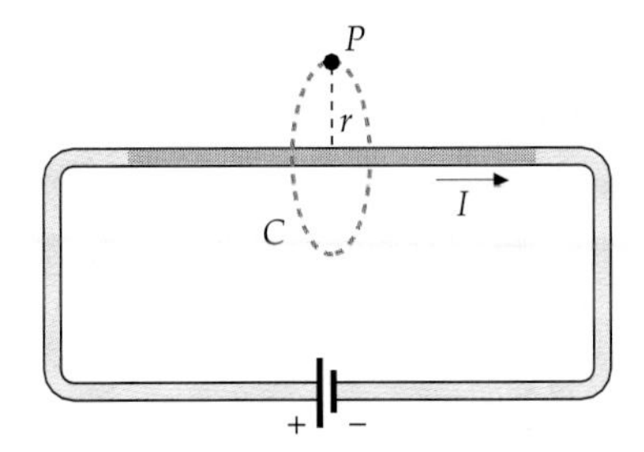

Figure 29-27 If the current segment in Figure 29-25 is part of a complete circuit, Ampère's law for the curve C is valid, but there is not enough symmetry to use it to find the magnetic field at point P.

In Figure 29-28, the current in the segment arises from a small spherical conductor with initial charge $+Q$ at the left of the segment and another one at the right with charge $-Q$. When they are connected, a current $I = -dQ/dt$ exists in the segment for a short time, until the spheres are uncharged. For this case, we *do* have the symmetry needed to assume that $\vec{B}$ is tangential to the curve and constant in magnitude along the curve. For a situation like this, in which the current is discontinuous in space, Ampère's law is not valid. In Chapter 32, we will see how Maxwell was able to modify Ampère's law so that it holds for all currents. When Maxwell's generalized form of Ampère's law is used to calculate the magnetic field for a current segment, such as that shown in Figure 29-28, the result agrees with that found from the Biot–Savart law.

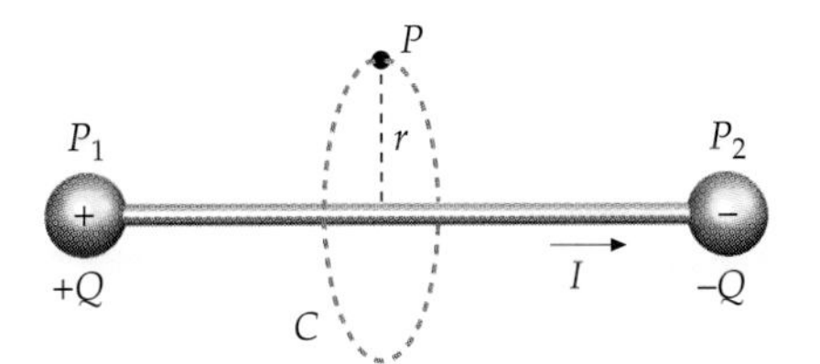

Figure 29-28 If the current segment in Figure 29-26 is due to a momentary flow of charge from a small conductor on the left to the one at the right, there is enough symmetry to use Ampère's law to compute the magnetic field at P, but Ampère's law is not valid because the current is not continuous in space.

29-5 Magnetism in Matter

Atoms have magnetic dipole moments due to the motion of their electrons and due to the intrinsic magnetic dipole moment associated with the spin of the electrons. Unlike the situation with electric dipoles, the alignment of magnetic dipoles parallel to an external magnetic field tends to *increase* the field. We can see this difference by comparing the electric field lines of an electric dipole with the magnetic field lines of a magnetic dipole, such as a small current loop, as in Figure 29-20. Far from the dipoles, the field lines are identical. However, between the charges of the electric dipole, the electric field lines are opposite the direction of the dipole moment, whereas inside the current loop, the magnetic field lines are parallel to the magnetic dipole moment. Thus, inside a magnetically polarized material, the magnetic dipoles create a magnetic field that is *parallel* to the magnetic-dipole-moment vectors.

The Magnetic Force and Conservation of Momentum

The magnetic force exerted by one moving charge on another is found by combining Equation 28-1 for the force on a moving charge in a magnetic field and Equation 29-1 for the magnetic field of a charge. The force $\vec{F}_{1,2}$ exerted by a charge q_1 moving with velocity v_1 on a charge q_2 moving with velocity v_2 is given by

$$\vec{F}_{1,2} = q_2\vec{v}_2 \times \vec{B}_1$$

$$= q_2\vec{v}_2 \times \left(\frac{\mu_0}{4\pi}\frac{q_1\vec{v}_1 \times \hat{r}_{1,2}}{r_{1,2}^2}\right) \qquad \text{1a}$$

where $\vec{B}_1$ is the magnetic field at the position of charge q_2 due to charge q_1, and $\hat{r}_{1,2}$ is the unit vector pointing from q_1 to q_2. Similarly, the force $\vec{F}_{2,1}$ exerted by a charge q_2 moving with velocity v_2 on a charge q_1 moving with velocity v_1 is given by

$$\vec{F}_{2,1} = q_1\vec{v}_1 \times \vec{B}_2$$

$$= q_1\vec{v}_1 \times \left(\frac{\mu_0}{4\pi}\frac{q_2\vec{v}_2 \times \hat{r}_{2,1}}{r_{2,1}^2}\right) \qquad \text{1b}$$

These relations are remarkable in that the force exerted by charge q_1 on charge q_2 is generally not equal and opposite to that exerted by charge q_2 on charge q_1. That is, these forces do not obey Newton's third law, as can be demonstrated by considering the special case illustrated in Figure 1. Here, the magnetic field $\vec{B}_1$ due to charge q_1 at charge q_2 is in the negative z direction, and the force on q_2 is to the left in the negative x direction. However, the magnetic field $\vec{B}_2$ due to q_2 at q_1 is zero because q_1 lies along the line of motion of q_2. Thus, there is no magnetic force exerted by q_2 on q_1. There is thus a net force $\vec{F}_{1,2}$ acting on the two-charge system. The system will accelerate in the direction of this force, and linear momentum will not be conserved.

This apparent violation of the law of conservation of linear momentum results from our treating the force exerted by one charge on another as an action-at-a-distance force and neglecting the momentum carried by the electric and magnetic fields of the moving charges. We saw in Chapter 22 that there is energy associated with an electric field, and we will see later that there is also energy associated with a magnetic field. Advanced treatments of the electric and magnetic fields of moving charges show that there is also momentum associated with these fields. When the charges move, as in Figure 1, the linear momentum produced when the system accelerates to the left is balanced by momentum in the opposite direction carried by the fields. Thus, when we include the momentum of the fields, the total momentum of the system is conserved.

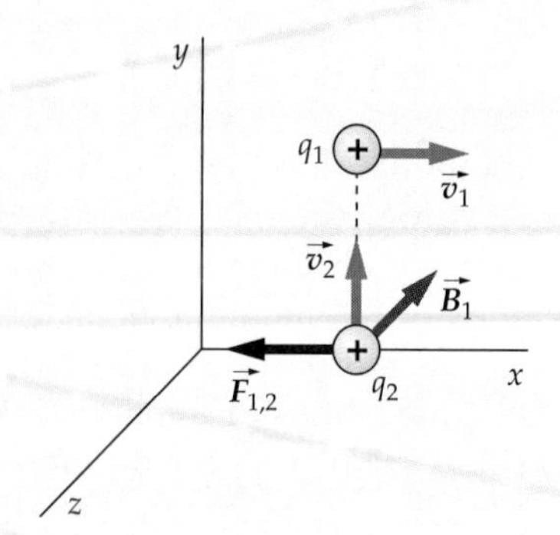

Figure 1 The forces exerted by moving charges on each other are not equal and opposite. The magnetic field $\vec{B}_1$ at charge q_2 due to charge q_1 is in the negative z direction, so it exerts a force $\vec{F}_{1,2}$ on q_2 to the left in the negative x direction. However, $\vec{B}_2$ at charge q_1 due to charge q_2 is zero, so there is no force on q_1.

Angular momentum is also carried by the electromagnetic fields produced by moving charges. Consider Figure 2, which shows a point charge q_1, $\vec{R} = x\hat{i} + y\hat{j}$, moving parallel to the x axis with velocity $\vec{v}_1 = v_1\hat{i}$, and a second point charge q_2 at the origin and moving along the x axis with velocity $\vec{v}_2 = v_2\hat{i}$. Let us calculate the magnetic force exerted by each charge on the other, assuming both charges to be positive.

We first find the force on charge q_1. We note that the vector $\vec{r}_{2,1}$ from charge q_2 to charge q_1 is just $\vec{R}$. Using $\hat{r}_{2,1} = \vec{R}/R$ we have

$$\frac{\vec{v}_2 \times \hat{r}_{2,1}}{r_{2,1}^2} = \frac{\vec{v}_2 \times \hat{R}}{R^3} = \frac{v_2\hat{i} \times (x\hat{i} + y\hat{j})}{R^3} = \frac{yv_2}{R^3}\hat{k}$$

so the magnetic field produced by charge q_2 at the position of charge q_1 is

$$\vec{B}_2 = \frac{\mu_0}{4\pi}\frac{q_2\vec{v}_2 \times \hat{r}_{2,1}}{r_{2,1}^2} = \frac{\mu_0}{4\pi}\frac{q_2 y v_2}{R^3}\hat{k}$$

The magnetic force exerted by charge q_2 on charge q_1 is then

$$\vec{F}_{2,1} = q_1\vec{v}_1 \times \vec{B}_2 = q_1 v_1\hat{i} \times \frac{\mu_0}{4\pi}\frac{q_2 y v_2}{R^3}\hat{k}$$

$$= -\frac{\mu_0}{4\pi}\frac{q_1 q_2 v_1 v_2 y}{R^3}\hat{j}$$

This force is downward and parallel to the y axis, as shown.

To find the magnetic force exerted by charge q_1 on charge q_2, we note that the vector $\vec{r}_{1,2}$ from charge q_1 to charge q_2 is $-\vec{R}$. Then the magnetic field produced by charge q_1 at the position of charge q_2 is

$$\vec{B}_1 = \frac{\mu_0}{4\pi}\frac{q_1\vec{v}_1 \times \hat{r}_{1,2}}{r_{1,2}^2}$$

$$= \frac{\mu_0}{4\pi}\frac{q_1\vec{v}_1 \times (-\vec{R})}{R^3} = -\frac{\mu_0}{4\pi}\frac{q_1 y v_1}{R^3}\hat{k}$$

The magnetic force exerted by charge q_1 on charge q_2 is then

$$\vec{F}_{1,2} = q_2\vec{v}_2 \times \vec{B}_1 = q_2 v_2\hat{i} \times \left(-\frac{\mu_0}{4\pi}\frac{q_1 y v_1}{R^3}\hat{k}\right)$$

$$= +\frac{\mu_0}{4\pi}\frac{q_1 q_2 v_1 v_2 y}{R^3}\hat{j}$$

In this case, the forces are equal and opposite, as shown in Figure 2*b*, but they are not along the line joining the two particles. The magnetic forces thus exert a torque on the two-particle system. Here, the apparent lack of conservation of angular momentum implied by the existence of this torque is resolved by the consideration of the angular momentum carried by the electromagnetic field.

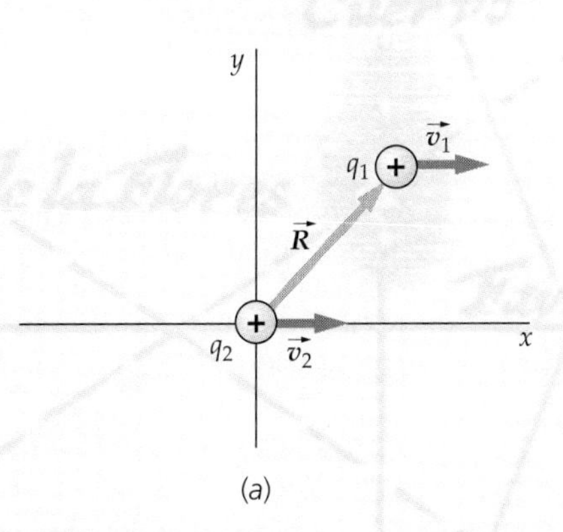

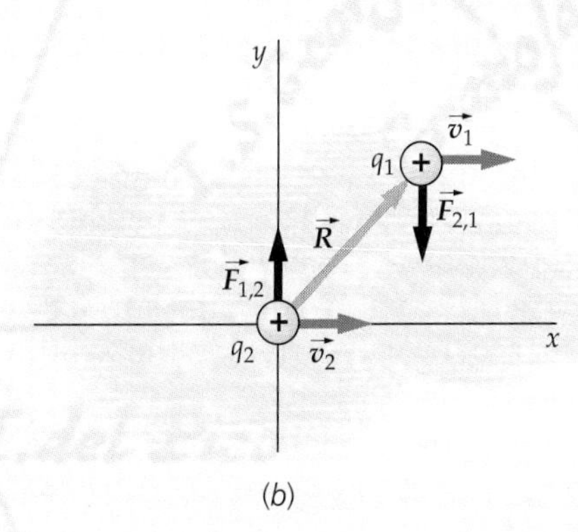

Figure 2 (*a*) Two charges moving in parallel directions. (*b*) The magnetic forces exerted by the charges on each other are equal and opposite, but they are not along the line joining the charges.

Materials fall into three categories—**paramagnetic**, **diamagnetic**, and **ferromagnetic**—according to the behavior of their magnetic moments in an external magnetic field. Paramagnetism arises from the partial alignment of the electron spins (in metals) or of atomic or molecular magnetic moments by an applied magnetic field in the direction of the field. In paramagnetic materials, the magnetic dipoles do not interact strongly with each other and are normally randomly oriented. In the presence of an external magnetic field, the dipoles are partially aligned in the direction of the field, thereby increasing the field. However, in external magnetic fields of ordinary strength at ordinary temperatures, only a very small fraction of the molecules are aligned because thermal motion tends to randomize their orientation. The increase in the total magnetic field is therefore very small. Ferromagnetism is much more complicated. Because of a strong interaction between neighboring magnetic dipoles, a high degree of alignment occurs even in weak external magnetic fields, causing a very large increase in the total field. Even when there is no external magnetic field, a ferromagnetic material may have its magnetic dipoles aligned, as in permanent magnets. Diamagnetism arises from the orbital magnetic dipole moments induced by an applied magnetic field. These magnetic moments are opposite the direction of the applied magnetic field so they decrease the total magnetic field B. This effect actually occurs in all materials, but because the induced magnetic moments are very small compared to the permanent magnetic moments, diamagnetism is masked by paramagnetic or ferromagnetic effects. Diamagnetism is thus observed only in materials that have no permanent magnetic moments.

Magnetization and Magnetic Susceptibility

When some material is placed in a strong magnetic field, such as that of a solenoid, the magnetic field of the solenoid tends to align the magnetic dipole moments (either permanent or induced) inside the material, and the material is said to be magnetized. We describe a magnetized material by its **magnetization $\vec{M}$,** which is defined as the net magnetic dipole moment per unit volume of the material:

$$\vec{M} = \frac{d\vec{\mu}}{dV} \qquad 29\text{-}18$$

Long before we had any understanding of atomic or molecular structure, Ampère proposed a model of magnetism in which the magnetization of materials is due to microscopic current loops inside the magnetized material. We now know that these current loops are the result of the intrinsic motion of atomic charges. Consider a cylinder of magnetized material. Figure 29-29 shows atomic current loops in the cylinder aligned with their magnetic moments along the axis of the cylinder. Because of cancellation of neighboring current loops, the net current at any point inside the material is zero, leaving a net current on the surface of the material (Figure 29-30). This surface current, called an **amperian current,** is similar to the real current in the windings of the solenoid.

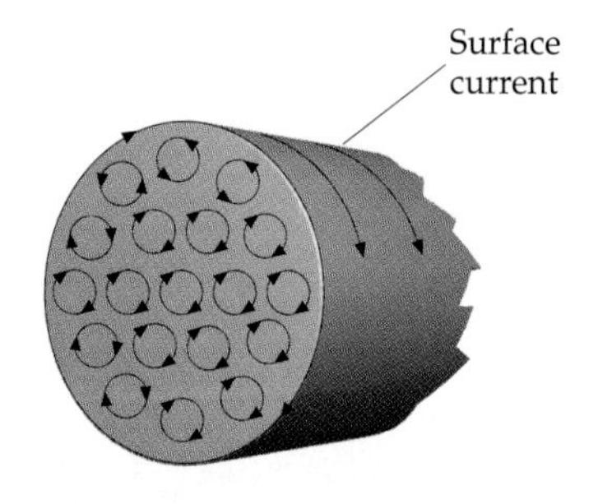

Figure 29-29 A model of atomic current loops in which all the atomic dipoles are parallel to the axis of the cylinder. The net current at any point inside the material is zero due to cancellation of neighboring atoms. The result is a surface current similar to that of a solenoid.

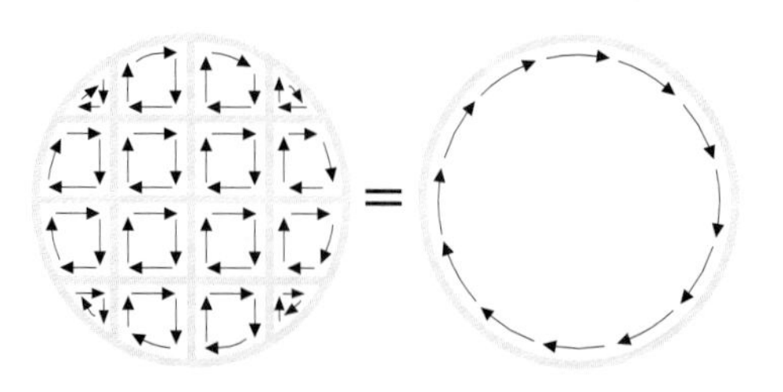

Figure 29-30 The currents in the adjacent current loops in the interior of a uniformly magnetized material cancel, leaving only a surface current. Cancellation occurs at every interior point independent of the shape of the loops.

Figure 29-31 shows a small disk of cross-sectional area A, length $d\ell$, and volume $dV = A\,d\ell$. Let di be the amperian current on the surface of the disk. The magnitude of the magnetic dipole moment of the disk is the same as that of a current loop of area A carrying a current di:

$$d\mu = A\,di$$

The magnetization of the disk is the magnetic moment per unit volume:

$$M = \frac{d\mu}{dV} = \frac{A\,di}{A\,d\ell} = \frac{di}{d\ell} \qquad 29\text{-}19$$

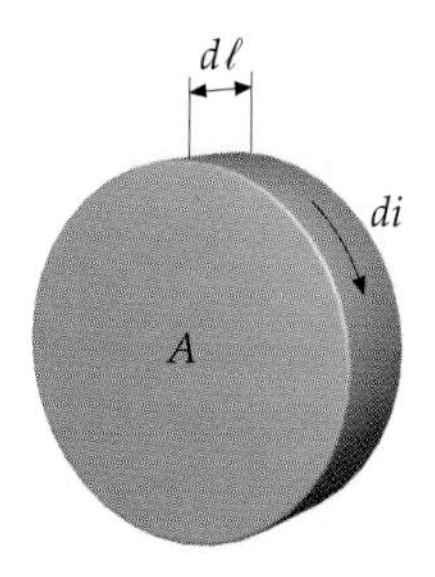

Figure 29-31 Disk element for relating the magnetization M to the surface current per unit length.

Thus, the magnitude of the magnetization vector is the amperian current per unit length along the surface of the magnetized material. We see from this result that the units of M are amperes per meter.

Consider a cylinder that has a uniform magnetization $\vec{M}$ parallel to its axis. The effect of the magnetization is the same as if the cylinder carried a surface current per unit length of magnitude M. This current is similar to the current carried by a tightly wound solenoid. For a solenoid, the current per unit length is nI, where n is the number of turns per unit length and I is the current in each turn. The magnitude of the magnetic field B_m inside the cylinder and far from its ends is thus given by Equation 29-9 for a solenoid with nI replaced by M:

$$B_m = \mu_0 M \qquad 29\text{-}20$$

Suppose we place a cylinder of magnetic material inside a long solenoid with n turns per unit length that carries a current I. The applied field of the solenoid $\vec{B}_{app}$ ($B_{app} = \mu_0 nI$) magnetizes the material so that it has a magnetization $\vec{M}$. The resultant magnetic field at a point inside the solenoid and far from its ends due to the current in the solenoid plus the magnetized material is

$$\vec{B} = \vec{B}_{app} + \mu_0\vec{M} \qquad 29\text{-}21$$

For paramagnetic and ferromagnetic materials, $\vec{M}$ is in the same direction as $\vec{B}_{app}$; for diamagnetic materials, $\vec{M}$ is opposite to $\vec{B}_{app}$. For paramagnetic and diamagnetic materials, the magnetization is found to be proportional to the applied magnetic field that produces the alignment of the magnetic dipoles in the material. We can thus write

$$\vec{M} = \chi_m \frac{\vec{B}_{app}}{\mu_0} \qquad 29\text{-}22$$

where χ_m is a dimensionless number called the **magnetic susceptibility.** Equation 29-21 is then

$$\vec{B} = \vec{B}_{app} + \mu_0\vec{M} = \vec{B}_{app}(1 + \chi_m) = K_m\vec{B}_{app} \qquad 29\text{-}23$$

where

$$K_m = 1 + \chi_m \qquad 29\text{-}24$$

is called the **relative permeability** of the material. For paramagnetic materials, χ_m is a small positive number that depends on temperature. For diamagnetic materials, it is a small negative number independent of temperature. Table 29-1 lists the magnetic susceptibility of various paramagnetic and diamagnetic materials. We see that the magnetic susceptibility for the solids listed is of the order of 10^{-5}, and $K_m \approx 1$.

The magnetization of ferromagnetic materials, which we discuss shortly, is much more complicated. The relative permeability K_m defined as the ratio B/B_{app} is not constant, and has maximum values ranging from 5000 to 100,000. In the case of permanent magnets, K_m is not even defined, since such materials exhibit magnetization even in the absence of an applied field.

Table 29-1

Magnetic Susceptibility of Various Materials at 20°C

Material	χ_m
Aluminum	2.3×10^{-5}
Bismuth	-1.66×10^{-5}
Copper	-0.98×10^{-5}
Diamond	-2.2×10^{-5}
Gold	-3.6×10^{-5}
Magnesium	1.2×10^{-5}
Mercury	-3.2×10^{-5}
Silver	-2.6×10^{-5}
Sodium	-0.24×10^{-5}
Titanium	7.06×10^{-5}
Tungsten	6.8×10^{-5}
Hydrogen (1 atm)	-9.9×10^{-9}
Carbon dioxide (1 atm)	-2.3×10^{-9}
Nitrogen (1 atm)	-5.0×10^{-9}
Oxygen (1 atm)	2090×10^{-9}

Atomic Magnetic Moments

The magnetization of a paramagnetic or ferromagnetic material can be related to the permanent magnetic moments of the individual atoms or electrons of the material. The orbital magnetic moment of an atomic electron can be derived semiclassically, even though it is quantum mechanical in origin. Consider a particle of mass m and charge q moving with speed v in a circle of radius r as shown in Figure 29-32. The magnitude of the angular momentum of the particle is

$$L = mvr \qquad 29\text{-}25$$

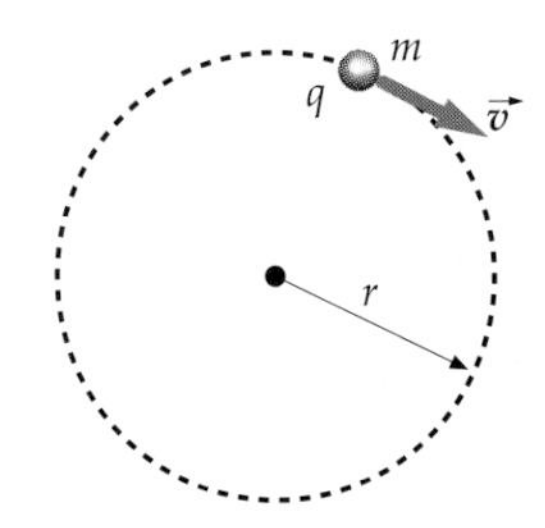

Figure 29-32 Particle of charge q and mass m moving in a circle of radius r. The angular momentum is into the paper and has a magnitude mvr, and the magnetic moment is into the paper (if q is positive) and has a magnitude $\frac{1}{2}qvr$.

The magnitude of the magnetic moment is the product of the current and the area of the circle:

$$\mu = IA = I\pi r^2$$

If T is the time for the charge to complete one revolution, the current (charge passing a point per unit time) is q/T. Since the period T is the distance $2\pi r$ divided by the velocity v, the current is

$$I = \frac{q}{T} = \frac{qv}{2\pi r}$$

The magnetic moment is then

$$\mu = IA = \frac{qv}{2\pi r}\,\pi r^2 = \frac{1}{2}\,qvr \qquad 29\text{-}26$$

Using $vr = L/m$ from Equation 29-25, we have for the magnetic moment

$$\mu = \frac{q}{2m}\,L$$

If the charge q is positive, the angular momentum and magnetic moment are in the same direction. We can therefore write

$$\vec{\mu} = \frac{q}{2m}\,\vec{L} \qquad 29\text{-}27$$

Classical relation between magnetic moment and angular momentum

Equation 29-27 is the general classical relation between magnetic moment and angular momentum. It also holds in the quantum theory of the atom for orbital angular momentum, but not for the intrinsic spin angular momentum of the electron. For electron spin, the magnetic moment is twice that predicted by this equation.* The extra factor of 2 is a result from quantum theory that has no analog in classical mechanics.

Since angular momentum is quantized, the magnetic moment of an atom is also quantized. The quantum of angular momentum is $\hbar = h/2\pi$, where h is Planck's constant, so we express the magnetic moment in terms of $\vec{L}/\hbar$

$$\vec{\mu} = \frac{q\hbar}{2m}\frac{\vec{L}}{\hbar}$$

For an electron, $m = m_e$ and $q = -e$, so the magnetic moment of the electron due to its orbital motion is

$$\vec{\mu}_\ell = -\frac{e\hbar}{2m_e}\frac{\vec{L}}{\hbar} = -\mu_B\frac{\vec{L}}{\hbar} \qquad 29\text{-}28$$

Magnetic moment due to orbital motion of an electron

* Precise measurements indicate that the magnetic moment of the electron due to its spin is 2.00232 times that predicted by Equation 29-25. This result, and the phenomenon of electron spin itself, was predicted in 1927 by P. Dirac, who combined special relativity and quantum mechanics into a relativistic wave equation called the Dirac equation. The fact that the intrinsic magnetic moment of the electron is approximately twice what we would expect makes it clear that the simple model of the electron as a spinning ball is not to be taken literally.

where

$$\mu_B = \frac{e\hbar}{2m_e} = 9.27 \times 10^{-24}\,\text{A}\cdot\text{m}^2 = 9.27 \times 10^{-24}\,\text{J/T} \qquad \text{29-29}$$
$$= 5.79 \times 10^{-5}\,\text{eV/T}$$

Bohr magneton

is the quantum unit of magnetic moment called a **Bohr magneton.** The magnetic moment of an electron due to its intrinsic spin angular momentum $\vec{S}$ is

$$\vec{\mu}_s = -2 \times \frac{e\hbar}{2m_e}\frac{\vec{S}}{\hbar} = -2\mu_B\frac{\vec{S}}{\hbar} \qquad \text{29-30}$$

Magnetic moment due to electron spin

Although the calculation of the magnetic moment of any atom is a complicated problem in quantum theory, the result for all electrons, according to both theory and experiment, is that the magnetic moment is of the order of a few Bohr magnetons. For atoms with zero net angular momentum, the net magnetic moment is zero.*

If all the atoms or molecules in some material have their magnetic moments aligned, the magnetic moment per unit volume of the material is the product of the number of molecules per unit volume n and the magnetic moment μ of each molecule. For this extreme case, the **saturation magnetization** M_s is

$$M_s = n\mu \qquad \text{29-31}$$

The number of molecules per unit volume can be found from the molecular mass $\mathcal{M}$, the density ρ of the material, and Avogadro's number N_A:

$$n = \frac{N_A\,(\text{atoms/mol})}{\mathcal{M}(\text{kg/mol})}\rho(\text{kg/m}^3) \qquad \text{29-32}$$

*The shell structure of atoms is discussed in Chapter 37.

Example 29-9

Find the saturation magnetization and the magnetic field it produces for iron, assuming that each iron atom has a magnetic moment of 1 Bohr magneton.

Picture the Problem We find the number of molecules per unit volume from the density of iron, $\rho = 7.9 \times 10^3$ kg/m^3, and its molecular mass $\mathcal{M} = 55.8 \times 10^{-3}$ kg/mol.

1. The saturation magnetic field is the product of the number of molecules per unit volume and the magnetic moment of each molecule:

$$M_s = n\mu$$

2. Calculate the number of molecules per unit volume from Avogadro's number, the molecular mass, and the density:

$$n = \frac{N_A}{\mathcal{M}}\rho = \frac{6.02 \times 10^{23}\,\text{atoms/mol}}{55.8 \times 10^{-3}\,\text{kg/mol}}(7.9 \times 10^3\,\text{kg/m}^3)$$
$$= 8.52 \times 10^{28}\,\text{atoms/m}^3$$

3. Substitute this result and $\mu = 1$ Bohr magneton to calculate the saturation magnetization:

$$M_s = n\mu$$
$$= (8.52 \times 10^{28}\,\text{atoms/m}^3)(9.27 \times 10^{-24}\,\text{A}\cdot\text{m}^2)$$
$$= 7.90 \times 10^5\,\text{A/m}$$

4. The magnetic field on the axis inside a long iron cylinder resulting from this maximum magnetization is given by $B = \mu_0 M_s$:

$$B = \mu_0 M_s$$
$$= (4\pi \times 10^{-7}\,\text{T}\cdot\text{m/A})(7.90 \times 10^5\,\text{A/m})$$
$$= 0.993\,\text{T} \approx 1\,\text{T}$$

Remarks The measured saturation magnetic field of annealed iron is about 2.16 T, indicating that the magnetic moment of an iron atom is slightly greater than 2 Bohr magnetons. This magnetic moment is due mainly to the spins of two unpaired electrons in the iron atom.

optional

Paramagnetism

Paramagnetism occurs in materials whose atoms have permanent magnetic moments that interact with each other only very weakly, resulting in a very small, positive magnetic susceptibility χ_m. When there is no external magnetic field, these magnetic moments are randomly oriented. In the presence of an external magnetic field, they tend to line up parallel to the field, but this is counteracted by the tendency for the magnetic moments to be randomly oriented due to thermal motion. The fraction of the moments that line up with the field depends on the strength of the field and on the temperature. This fraction is usually small because the energy of a magnetic moment in an external magnetic field is typically much smaller than the thermal energy of an atom of the material, which is of the order of kT, where k is Boltzmann's constant and T is the absolute temperature.

The potential energy of a magnetic dipole of moment $\vec{\mu}$ in an external magnetic field $\vec{B}$ is given by Equation 28-16:

$$U = -\mu B \cos\theta = -\vec{\mu}\cdot\vec{B}$$

The potential energy when the moment is parallel with the field ($\theta = 0$) is thus lower than when it is antiparallel ($\theta = 180°$) by the amount $2\mu B$. For a typical magnetic moment of 1 Bohr magneton and a typical strong magnetic field of 1 T, the difference in potential energy is

$$\Delta U = 2\mu_B B = 2(5.79 \times 10^{-5}\ \text{eV/T})(1\ \text{T}) = 1.16 \times 10^{-4}\ \text{eV}$$

At a normal temperature of $T = 300$ K, the typical thermal energy kT is

$$kT = (8.62 \times 10^{-5}\ \text{eV/K})(300\ \text{K}) = 2.59 \times 10^{-2}\ \text{eV}$$

which is about 200 times greater than $2\mu_B B$. Thus, even in a very strong magnetic field of 1 T, most of the magnetic moments will be randomly oriented because of thermal motions unless the temperature is very low.

Figure 29-33 shows a plot of the magnetization M versus an applied external magnetic field B_{app} at a given temperature. In very strong fields, nearly all the magnetic moments are aligned with the field and $M \approx M_s$. (For magnetic fields attainable in the laboratory, this can occur only for very low temperatures.) When $B_{app} = 0$, $M = 0$, indicating that the orientation of the moments is completely random. In weak fields, the magnetization is approximately proportional to the applied field, as indicated by the orange dashed line in the figure. In this region, the magnetization is given by

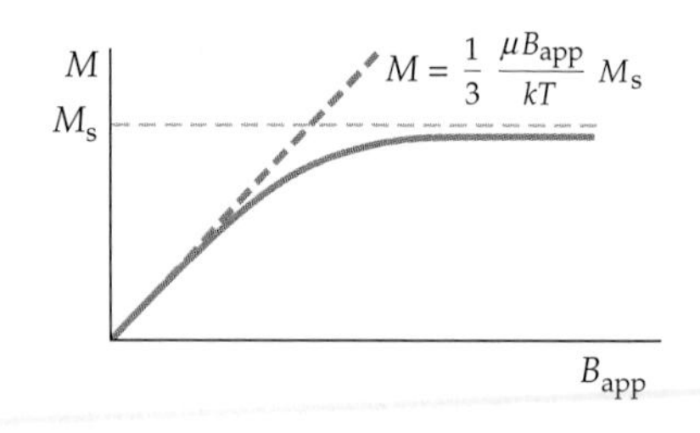

Figure 29-33 Plot of magnetization M versus applied field B_{app}. In very strong fields, the magnetization approaches the saturation value M_s. This can be achieved only at very low temperatures. In weak fields, the magnetization is approximately proportional to B_{app}, a result known as Curie's law.

$$M = \frac{1}{3}\frac{\mu B_{app}}{kT} M_s \qquad 29\text{-}33$$

Curie's law

Note that $(\mu B_{\text{app}}/kT)$ is the ratio of the maximum energy of a dipole in the magnetic field to the characteristic thermal energy. The result that the magnetization varies inversely with the absolute temperature was discovered experimentally by Pierre Curie and is known as **Curie's law.**

Liquid oxygen, which is paramagnetic, is attracted by the magnetic field of a permanent magnet. A net force is exerted on the magnetic dipoles because the magnetic field is not uniform.

Example 29-10

If $\mu = \mu_B$, at what temperature will the magnetization be 1% of the saturation magnetization in an applied magnetic field of 1 T?

1. Substitute $M = 0.01\,M_s$ into Curie's law:

$$M = \frac{1}{3}\frac{\mu B_{\text{app}}}{kT}M_s = 0.01 M_s$$

2. Solve for T:

$$T = \frac{\mu B_{\text{app}}}{0.03k} = \frac{(5.79 \times 10^{-5}\text{ eV/T})(1\text{ T})}{(0.03)(8.62 \times 10^{-5}\text{ eV/K})} = 22.4\text{ K}$$

Remark From this example, we see that even in a strong applied magnetic field of 1 T, the magnetization is less than 1% of saturation at temperatures above 22.4 K.

Exercise If $\mu = \mu_B$, what fraction of the saturation magnetization is M at 300 K for an external magnetic field of 15,000 G? (*Answer* $M/M_s = 1.12 \times 10^{-3}$)

Ferromagnetism

Ferromagnetism occurs in pure iron, cobalt, and nickel, and in alloys of these metals with each other. It also occurs in gadolinium, dysprosium, and a few compounds. Ferromagnetism arises from a strong interaction between the electrons in a partially full band in a metal or between the localized electrons that form magnetic moments on neighboring atoms or molecules. This interaction, called the **exchange interaction**, lowers the energy of a pair of electrons with parallel spins.

Ferromagnetic materials have very large, positive values of magnetic susceptibility χ_m (as measured under conditions described below). In these substances, a small external magnetic field can produce a very large degree of alignment of the atomic magnetic dipole moments. In some cases, the align-

A chunk of magnetite (lodestone) attracts the needle of a compass.

ment can persist even when the external magnetizing field is removed. This occurs because the magnetic dipole moments exert strong forces on their neighbors so that over a small region of space the moments are aligned with each other even when there is no external field. The region of space over which the magnetic dipole moments are aligned is called a **magnetic domain.** The size of a domain is usually microscopic. Within the domain, all the magnetic moments are aligned, but the direction of alignment varies from domain to domain so that the net magnetic moment of a macroscopic piece of ferromagnetic material is zero in the normal state. Figure 29-34 illustrates this situation. The dipole forces that produce this alignment are predicted by quantum theory but cannot be explained with classical physics. At temperatures above a critical temperature, called the **Curie temperature,** thermal agitation is great enough to break up this alignment, and ferromagnetic materials become paramagnetic.

(a)

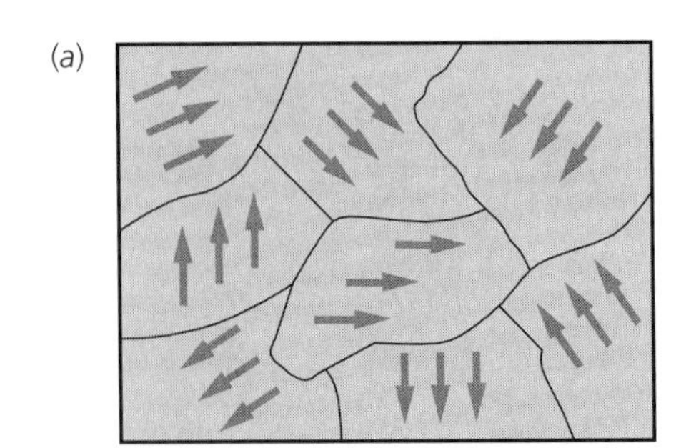

(b)

Figure 29-34 (*a*) Schematic illustration of ferromagnetic domains. Within a domain, the magnetic dipoles are aligned, but the direction of alignment varies from domain to domain so that the net magnetic moment is zero. A small external magnetic field may cause the enlargement of those domains that are aligned parallel to the field, or it may cause the alignment within a domain to rotate. In either case, the result is a net magnetic moment parallel to the field. (*b*) Magnetic domains on the surface of an Fe–3% Si crystal observed using a scanning electron microscope with polarization analysis. The four colors indicate four possible domain orientations.

When an external magnetic field is applied, the boundaries of the domains may shift or the direction of alignment within a domain may change so that there is a net macroscopic magnetic moment in the direction of the applied field. Since the degree of alignment is large for even a small external field, the magnetic field produced in the material by the dipoles is often much greater than the external field.

Let us consider what happens when we magnetize a long iron rod by placing it inside a solenoid and gradually increase the current in the solenoid windings. We assume that the rod and the solenoid are long enough to permit us to neglect end effects. Since the induced magnetic moments are in the same direction as the applied field, $\vec{B}_{app}$ and $\vec{M}$ are parallel. Then

$$B = B_{app} + \mu_0 M = \mu_0 nI + \mu_0 M \qquad 29\text{-}34$$

In ferromagnetic materials, the magnetic field $\mu_0 M$ due to the magnetic moments is often greater than the magnetizing field B_{app} by a factor of several thousand.

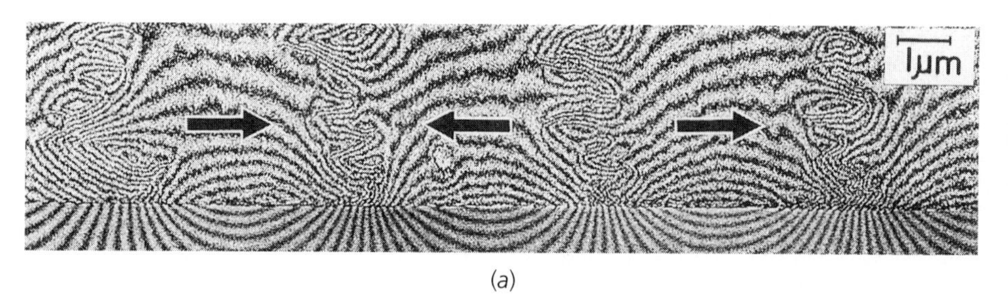

(*a*)

(*b*)

(*a*) Magnetic field lines on a cobalt magnetic recording tape. The solid arrows indicate the encoded magnetic bits. (*b*) Cross section of a magnetic tape recording head. Current from an audio amplifier is sent to wires around a magnetic core in the recording head where it produces a magnetic field. When the tape passes over a gap in the core of the recording head, the fringing magnetic field encodes information on the tape.

Figure 29-35 shows a plot of B versus the magnetizing field B_{app}. As the current is gradually increased from zero, B increases from zero along the part of the curve from the origin O to point P_1. The flattening of this curve near point P_1 indicates that the magnetization M is approaching its saturation value M_s, at which all the atomic magnetic moments are aligned. Above saturation, B increases only because the magnetizing field $B_{\text{app}} = \mu_0 nI$ increases. When B_{app} is gradually decreased from point P_1, there is not a corresponding decrease in the magnetization. The shift of the domains in a ferromagnetic material is not completely reversible, and some magnetization remains even when B_{app} is reduced to zero, as indicated in the figure. This effect is called **hysteresis,** from the Greek word *hysteros* meaning later or behind, and the curve in Figure 29-35 is called a **hysteresis curve.** The value of the magnetic field at point r when B_{app} is zero is called the **remnant field** B_r. At this point, the iron rod is a permanent magnet. If the current in the solenoid is now reversed so that B_{app} is in the opposite direction, the magnetic field B is gradually brought to zero at point c. The remaining part of the hysteresis curve is obtained by further increasing the current in the opposite direction until point P_2 is reached, which corresponds to saturation in the opposite direction, and then decreasing the current to zero at point P_3 and increasing it again in its original direction.

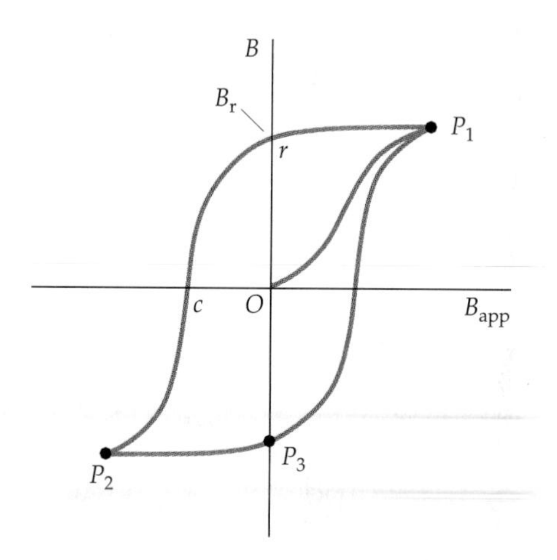

Figure 29-35 Plot of B versus the applied field B_{app}. The outer curve is called a hysteresis curve. The field B_r is called the remnant field. It remains when the applied field returns to zero.

Since the magnetization M depends on the previous history of the material, and since it can have a large value even when the applied field is zero, it is not simply related to the applied field B_{app}. However, if we confined ourselves to that part of the magnetization curve from the origin to point P_1 in Figure 29-35, $\vec{M}$ and $\vec{B}_{\text{app}}$ are parallel and M is zero when B_{app} is zero. We can then define the magnetic susceptibility as in Equation 29-22,

$$M = \chi_m \frac{B_{\text{app}}}{\mu_0}$$

and

$$B = B_{\text{app}} + \mu_0 M = B_{\text{app}}(1 + \chi_m) = K_m \mu_0 nI = \mu nI \qquad \text{29-35}$$

where

$$\mu = (1 + \chi_m)\mu_0 = K_m \mu_0 \qquad \text{29-36}$$

is called the **permeability** of the material. (For paramagnetic and diamagnetic materials, χ_m is much less than 1 so the permeability μ and the permeability of free space μ_0 are very nearly equal.)

Since B does not vary linearly with B_{app}, as can be seen from Figure 29-35, the relative permeability is not constant. The maximum value of K_m occurs at

optional

Table 29-2

Maximum Values of $\mu_0 M$ and K_m for Some Ferromagnetic Materials

Material	$\mu_0 M_s$, T	K_m
Iron (annealed)	2.16	5,500
Iron-silicon (96% Fe, 4% Si)	1.95	7,000
Permalloy (55% Fe, 45% Ni)	1.60	25,000
Mu-metal (77% Ni, 16% Fe, 5% Cu, 2% Cr)	0.65	100,000

a magnetization that is considerably less than the saturation magnetization. Table 29-2 lists the saturation magnetic field $\mu_0 M_s$ and the maximum values of K_m for some ferromagnetic materials. Note that the maximum values of K_m are much greater than 1.

The area enclosed by the hysteresis curve is proportional to the energy dissipated as heat in the irreversible process of magnetizing and demagnetizing. If the hysteresis effect is small, so that the area inside the curve is small, indicating a small energy loss, the material is called **magnetically soft.** Soft iron is an example. The hysteresis curve for a magnetically soft material is shown in Figure 29-36. Here the remnant field B_r is nearly zero, and the energy loss per cycle is small. Magnetically soft materials are used for transformer cores to allow the magnetic field B to change without incurring large energy losses as the field alternates. On the other hand, a large remnant field is desirable in a permanent magnet. **Magnetically hard** materials, such as carbon steel and the alloy Alnico 5, are used for permanent magnets.

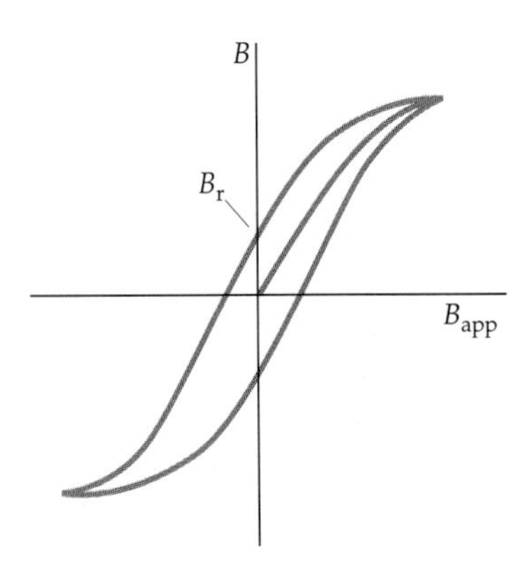

Figure 29-36 Hysteresis curve for a magnetically soft material. The remnant field is very small compared with that for a magnetically hard material such as that in Figure 29-35.

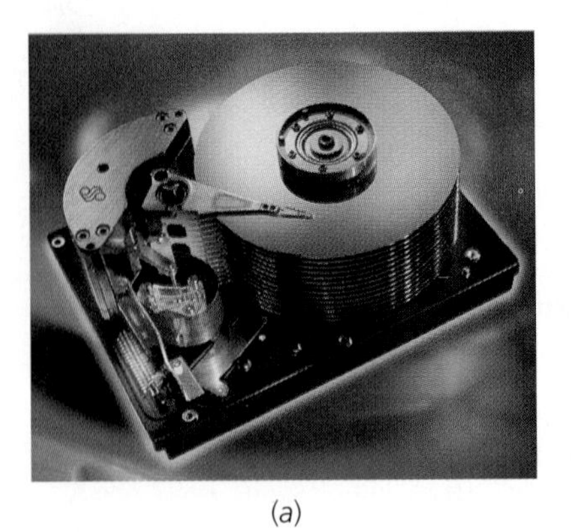

(a)

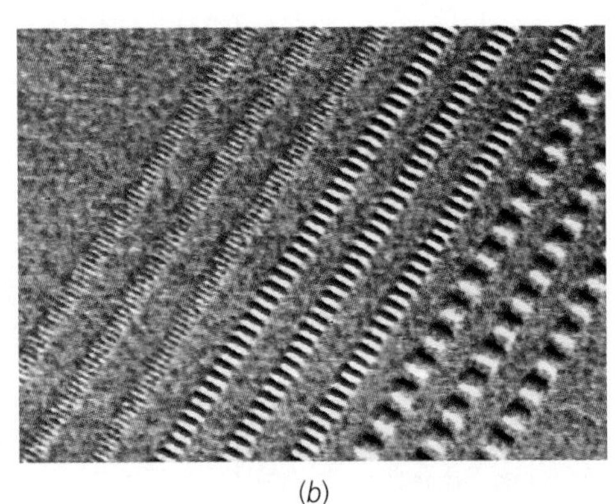

(b)

(*a*) An extremely high-capacity hard-disk drive for magnetic storage of information, capable of storing over 47 gigabytes of information. (*b*) A magnetic test pattern on a hard disk, magnified 2400 times. The light and dark regions correspond to oppositely directed magnetic fields. The smooth region just outside the pattern is a region of the disk that has been erased just prior to writing.

Example 29-11

A long solenoid with 12 turns per centimeter has a core of annealed iron. When the current is 0.50 A, the magnetic field inside the iron core is 1.36 T. Find (*a*) the applied field B_{app}, (*b*) the relative permeability K_m, and (*c*) the magnetization M.

Picture the Problem The applied field is just that of a long solenoid given by $B_{app} = \mu_0 nI$. Since the total magnetic field is given, we can find the relative permeability from its definition ($K_m = B/B_{app}$) and we can find M from $B = B_{app} + \mu_0 M$.

optional

(*a*) The applied field is given by Equation 29-10:

$$B_{app} = \mu_0 nI$$
$$= (4\pi \times 10^{-7}\ \mathrm{T\cdot m/A})(1200\ \mathrm{turns/m})(0.50\ \mathrm{A})$$
$$= 7.54 \times 10^{-4}\ \mathrm{T}$$

(*b*) The relative permeability is the ratio of B to B_{app}:

$$K_m = \frac{B}{B_{app}} = \frac{1.36\ \mathrm{T}}{7.54 \times 10^{-4}\ \mathrm{T}} = 1.80 \times 10^3 = 1800$$

(*c*) The magnetization M is found from Equation 29-34:

$$\mu_0 M = B - B_{app}$$
$$= 1.36\ \mathrm{T} - 7.54 \times 10^{-4}\ \mathrm{T} \approx B = 1.36\ \mathrm{T}$$
$$M = \frac{B}{\mu_0} = \frac{1.36\ \mathrm{T}}{4\pi \times 10^{-7}\ \mathrm{T\cdot m/A}} = 1.08 \times 10^6\ \mathrm{A/m}$$

Remarks The applied magnetic field of 7.54×10^{-4} T is a negligible fraction of the total field of 1.36 T. Note that the value for K_m of 1800 is considerably smaller than the maximum value of 5500 in Table 29-2. Note also that the susceptibility $\chi_m = K_m - 1 \approx K_m$ to the three-place accuracy with which we calculated K_m.

Diamagnetism

Diamagnetic materials are those having very small, negative values of magnetic susceptibility χ_m. Diamagnetism was discovered by Faraday in 1846 when he found that a piece of bismuth is repelled by either pole of a magnet, indicating that the external field of the magnet induces a magnetic moment in bismuth in the direction opposite the field.

We can understand this effect qualitatively from Figure 29-37, which shows two positive charges moving in circular orbits with the same speed but in opposite directions. Their magnetic moments are in opposite directions and therefore cancel.* In the presence of an external magnetic field $\vec{B}$ directed into the paper, the charges experience an extra force $q\vec{v} \times \vec{B}$, which is along the radial direction. For the charge on the left, this extra force is inward, increasing the centripetal force. If the charge is to remain in the same circular orbit, it must speed up so that mv^2/r equals the total centripetal force.† Its magnetic moment, which is outward, is thus increased. For the charge on the right, the additional force is outward, so the particle must slow down to maintain its circular orbit. Its magnetic moment, which is inward, is decreased. In each case, the *change* in the magnetic moment of the charges in the direction out of the page, opposite that of the external applied field. Since the permanent magnetic moments of the two charges are equal and oppositely directed, they add to zero, leaving only the induced magnetic moments, which are both opposite the direction of the applied magnetic field.

A material will be diamagnetic if its atoms have zero net angular momentum and therefore no permanent magnetic moment. (The net angular momentum of an atom depends on the electronic structure of the atom, a subject we study in Chapter 37.) The induced magnetic moments that cause diamag-

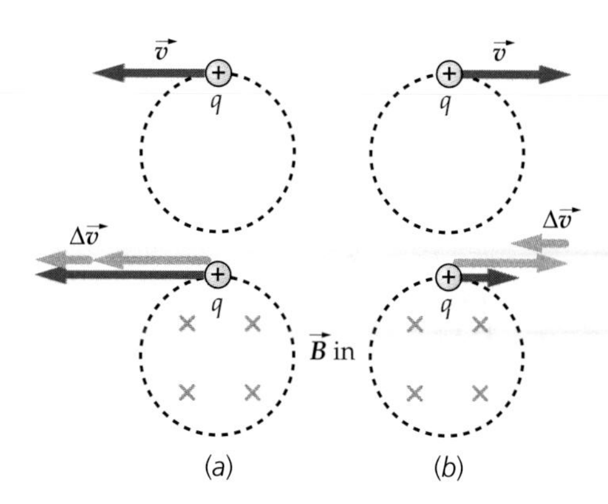

Figure 29-37 (*a*) A positive charge moving counterclockwise in a circle has its magnetic moment outward. When an external, inward magnetic field is turned on, the magnetic force increases the centripetal force so the speed of the particle must increase. The change in the magnetic moment is outward. (*b*) A positive charge moving clockwise in a circle has its magnetic moment inward. When an external, inward magnetic field is turned on, the magnetic force decreases the centripetal force so the speed of the particle must decrease. As in (*a*), the change in the magnetic moment is outward.

* It is simpler to consider positive charges even though it is the negatively charged electrons that provide the magnetic moments in matter.

† The electron speeds up because of an electric field induced by the changing magnetic field, an effect called induction, which we discuss in Chapter 30.

optional

netism have magnitudes of the order of 10^{-5} Bohr magnetons. Since this is much smaller than the permanent magnetic moments of the atoms of paramagnetic or ferromagnetic materials, the diamagnetic effect in these atoms is masked by the alignment of their permanent magnetic moments. However, since this alignment decreases with temperature, all materials are theoretically diamagnetic at sufficiently high temperatures.

When a superconductor is placed in an external magnetic field, electric currents are induced on its surface so that the net magnetic field in the superconductor is zero. Consider a superconducting rod inside a solenoid of n turns per unit length. When the solenoid is connected to a source of emf so that it carries a current I, the magnetic field due to the solenoid is $\mu_0 nI$. A surface current of $-nI$ per unit length is induced on the superconducting rod that cancels out the field due to the solenoid so that the net field inside the superconductor is zero. From Equation 29-23,

$$\vec{B} = \vec{B}_{\text{app}}(1 + \chi_{\text{m}}) = 0$$

so

$$\chi_{\text{m}} = -1$$

A superconductor is thus a perfect diamagnet with a magnetic susceptibility of -1.

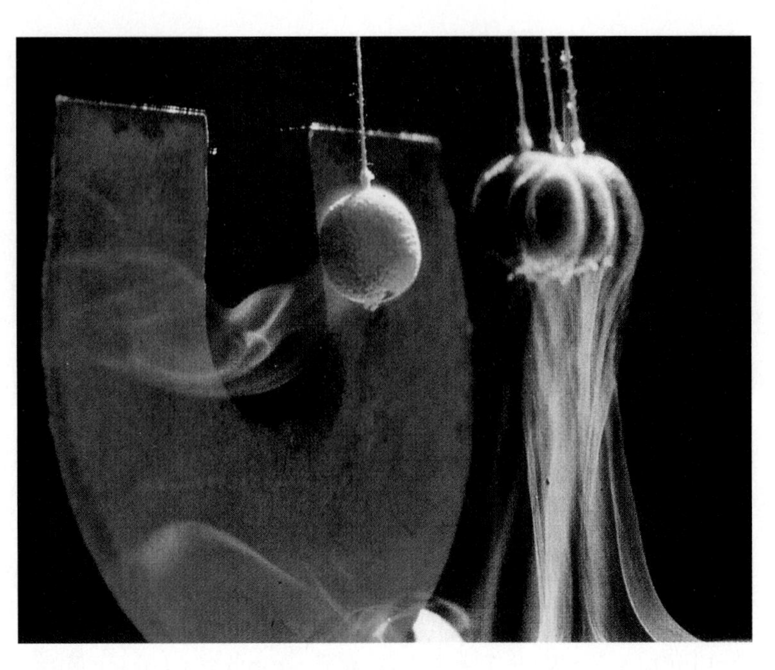

A superconductor is a perfect diamagnet. Here, the superconducting pendulum bob is repelled by the permanent magnet.

Summary

1. Magnetic fields arise from moving charges, and therefore from currents.
2. The Biot–Savart law describes the magnetic field produced by a current element.
3. Ampere's law relates the line integral of the magnetic field along some curve to the current that passes through any area bounded by the curve.
4. The magnetization vector $\vec{M}$ describes the magnetic moment per unit volume of matter.
5. The classical relation $\vec{\mu} = (q/2m)\vec{L}$ is derived from the definitions of angular momentum and magnetic moment.
6. The Bohr magneton is a convenient unit for atomic and nuclear magnetic moments.

Topic	Remarks and Relevant Equations	
1. Magnetic Field $\vec{B}$		
Due to a moving point charge	$\vec{B} = \frac{\mu_0}{4\pi}\frac{q\vec{v} \times \hat{r}}{r^2}$	**29-1**
	where $\hat{r}$ is a unit vector that points from the charge to the field point, and μ_0 is the permeability of free space:	
	$\mu_0 = 4\pi \times 10^{-7}\,\text{T·m/A} = 4\pi \times 10^{-7}\,\text{N/A}^2$	**29-2**

	Due to a current element (Biot–Savart law)	$d\vec{B} = \frac{\mu_0}{4\pi}\frac{I\,d\vec{\ell} \times \hat{r}}{r^2}$	29-3
	On the axis of a current loop	$B_x = \frac{\mu_0}{4\pi}\frac{2\pi R^2 I}{(x^2 + R^2)^{3/2}}$	29-5
	On the axis of a current loop far from the loop	$B_x = \frac{\mu_0}{4\pi}\frac{2\mu}{\lvert x^3 \rvert}$ where μ is the magnetic moment.	29-6
	Inside a solenoid, far from the ends	$B = \mu_0 n I$ where n is the number of turns per unit length.	29-9
	Due to a straight, current-carrying wire	$B = \frac{\mu_0}{4\pi}\frac{I}{R}(\sin\theta_1 + \sin\theta_2)$ where R is the perpendicular distance to the wire and θ_1 and θ_2 are the angles subtended at the field point by the ends of the wire.	29-11
	Due to an infinitely long, straight wire	$B = \frac{\mu_0}{4\pi}\frac{2I}{R}$ The direction of $\vec{B}$ is such that the lines of $\vec{B}$ encircle the wire in the direction of the fingers of the right hand if the thumb points in the direction of the current.	29-12
	Inside a toroid	$B = \frac{\mu_0 NI}{2\pi r}, \quad a < r < b$	29-17
2.	**Magnetic Field Lines**	The magnetic field is indicated by lines parallel to $\vec{B}$ at any point whose density is proportional to the magnitude of $\vec{B}$. Magnetic lines do not begin or end at any point in space. Instead, they form continuous loops.	
3.	**Gauss's Law for Magnetism**	$\phi_{m,net} = \oint_S B_n\,dA = 0$	29-14
4.	**Magnetic Poles**	Magnetic poles always occur in pairs. Isolated magnetic poles have not been found.	
5.	**Ampère's Law**	$\oint_C \vec{B}\cdot d\vec{\ell} = \mu_0 I_C$ where C is any closed curve.	29-15
	Validity of Ampère's law	Ampère's law is valid only if the currents are continuous. It can be used to derive expressions for the magnetic field for situations with a high degree of symmetry, such as a long, straight, current-carrying wire, or a long, tightly wound solenoid.	
6.	**Magnetic Matter**	Matter can be classified as either paramagnetic, ferromagnetic, or diamagnetic.	
	Magnetization	A magnetized material is described by its magnetization vector $\vec{M}$, which is defined to be the net magnetic dipole moment per unit volume of the material: $\vec{M} = \frac{d\vec{\mu}}{dV}$	29-18
		The magnetic field due to a uniformly magnetized cylinder is the same as if the cylinder carried a current per unit length of magnitude M on its surface. This current,	

which is due to the intrinsic motion of the atomic charges in the cylinder, is called an amperian current.

7. $\vec{B}$ in Magnetic Materials

$$\vec{B} = \vec{B}_{app} + \mu_0\vec{M} \quad (29\text{-}21)$$

Magnetic susceptibility χ_m

$$\vec{M} = \chi_m \frac{\vec{B}_{app}}{\mu_0} \quad (29\text{-}22)$$

For paramagnetic materials, χ_m is a small positive number that depends on temperature. For diamagnetic materials (other than superconductors), it is a small negative constant independent of temperature. For superconductors, $\chi_m = -1$. For ferromagnetic materials, the magnetization depends not only on the magnetizing current but also on the past history of the material.

Relative permeability

$$\vec{B} = K_m\vec{B}_{app} \quad (29\text{-}23)$$

where

$$K_m = 1 + \chi_m \quad (29\text{-}24)$$

8. Atomic Magnetic Moments

$$\vec{\mu} = \frac{q}{2m}\vec{L} \quad \text{(classical)} \quad (29\text{-}27)$$

where $\vec{L}$ is the angular momentum of the particle.

Orbital

$$\vec{\mu}_\ell = -\frac{e\hbar}{2m_e}\frac{\vec{L}}{\hbar} = -\mu_B\frac{\vec{L}}{\hbar} \quad (29\text{-}28)$$

Spin

$$\vec{\mu}_s = -2 \times \frac{e\hbar}{2m_e}\frac{\vec{S}}{\hbar} = -2\mu_B\frac{\vec{S}}{\hbar} \quad (29\text{-}30)$$

Bohr magneton

$$\mu_B = \frac{e\hbar}{2m_e} = 9.27 \times 10^{-24}\ \text{A}\cdot\text{m}^2$$
$$= 9.27 \times 10^{-24}\ \text{J/T} = 5.79 \times 10^{-5}\ \text{eV/T} \quad (29\text{-}29)$$

where

$$\hbar = \frac{h}{2\pi} = 1.05 \times 10^{-34}\ \text{J}\cdot\text{s}$$

and $h = 6.67 \times 10^{-34}$ J·s is Planck's constant.

9. Paramagnetism (optional)

Paramagnetic materials have permanent atomic magnetic moments that have random directions in the absence of an external magnetic field. In an external field, some of these dipoles are aligned, producing a small contribution to the total field that adds to the external field. The degree of alignment is small except in very strong fields and at very low temperatures. At ordinary temperatures, thermal motion tends to maintain the random directions of the magnetic moments.

Curie's law

In weak fields, the magnetization is approximately proportional to the applied field and inversely proportional to the absolute temperature.

$$M = \frac{1}{3}\frac{\mu B_{app}}{kT}M_s \quad (29\text{-}33)$$

10. Ferromagnetism (optional)

Ferromagnetic materials have small regions of space called magnetic domains in which the permanent atomic magnetic moments are aligned. When the material is unmagnetized, the direction of alignment in one domain is independent of that in another so that no net magnetic field is produced. When the material is magnetized, the domains of a ferromagnetic material are aligned, producing a very strong contribution to the magnetic field. This alignment can persist even when the external field is removed, thus leading to permanent magnetism.

11. Diamagnetism (optional)

Diamagnetic materials are those in which the magnetic moments of all electrons in each atom cancel, leaving each atom with zero magnetic moment in the absence of an external field. In an external field, a very small magnetic moment is induced that tends to weaken the field. This effect is independent of temperature. Superconductors are diamagnetic with susceptibility equal to -1.

Problem-Solving Guide

1. Begin by drawing a neat diagram that includes the important features of the problem. The vector nature of the magnetic field can be clarified with a diagram showing $\vec{v}$ and $\hat{r}$ (for moving point charges) or showing $d\vec{\ell}$ and $\hat{r}$ (for a current element). Indicate the direction of $\vec{B}$.
2. Ampère's law can be used to calculate B in problems with cylindrical symmetry such as a very long, current-carrying wire.

Summary of Worked Examples

Type of Calculation	Procedure and Relevant Examples
1. $\vec{B}$ due to Moving Charges	
Find the magnetic field produced by a moving point charge.	Use $\vec{B} = \frac{\mu_0}{4\pi}\frac{q\vec{v} \times \hat{r}}{r^2}$ where $\hat{r}$ points from the charge to the field point. **Example 29-1**
2. $\vec{B}$ due to Currents	
Find the magnetic field on the axis of a current loop.	At a distance x from the loop, B_x is given by $B_x = \frac{\mu_0}{4\pi}\frac{2\pi R^2 I}{(x^2 + R^2)^{3/2}}$ At the center of the loop ($x = 0$) $B_x = \mu_0 I/2R$. At great distances ($x >> R$), $B_x = \frac{\mu_0}{4\pi}\frac{2\mu}{x^3}$ where $\mu = NI\pi R^2$. **Example 29-2**
Find the magnetic field inside a solenoid.	The general result is $B_x = \frac{1}{2}\mu_0 nI\left(\frac{b}{\sqrt{b^2 + R^2}} + \frac{a}{\sqrt{a^2 + R^2}}\right)$ where a and b are the distances from the field point to either end of the solenoid. For a long solenoid, $B = \mu_0 nI$. **Example 29-4**
Find the magnetic field due to a current in a straight wire.	Use $B = \frac{\mu_0}{4\pi}\frac{I}{R}(\sin\theta_1 + \sin\theta_2)$ where θ_1 and θ_2 are the angles from the field point to either end of the wire. For a long wire, $B = \frac{\mu_0 I}{2\pi R} = \frac{\mu_0}{4\pi}\frac{2I}{R}$ **Examples 29-5, 29-6, 29-7**

Find the force exerted by one wire on another.	Calculate $\vec{B}$ at one wire due to the current in the other wire. Then use $$d\vec{F} = I\,d\vec{\ell} \times \vec{B} \quad \text{or} \quad \frac{F}{d\ell} = 2\frac{\mu_0}{4\pi}\frac{I_1 I_2}{R}$$ for the force between two parallel current-carrying wires. **Example 29-7**
3. Torques on Magnets	The torque exerted on a magnet in a field $\vec{B}$ is given by $$\vec{\tau} = \vec{\mu} \times \vec{B}$$ **Example 29-3**
4. Ampère's Law	
Use Ampère's law to find the magnetic field.	Find a curve C such that $\vec{B}$ is tangential to the curve and constant along it. Then use $\oint_C \vec{B}\cdot d\vec{\ell} = \mu_0 I_C$ and remove B from the integral. **Example 29-8**
5. $\vec{B}$ in Matter	
Find the saturation magnetic field in some material.	Multiply the number of molecules per unit volume by the magnetic moment of each molecule to obtain the magnetization M. Then use $B_m = \mu_0 M$. **Example 29-9**
Find the relative permeability and magnetization of a ferrromagnetic material given the total field (optional).	Calculate the applied field B_{app} and use $K_m = B/B_{app}$. Then use $\mu_0 M = B - B_{app}$. **Example 29-11**

Problems

Conceptual Problems

Problems from Optional and Exploring sections

In a few problems, you are given more data than you actually need; in a few other problems, you are required to supply data from your general knowledge, outside sources, or informed estimates.

• Single-concept, single-step, relatively easy
•• Intermediate-level, may require synthesis of concepts
••• Challenging, for advanced students

Moving Point Charges

1 • Compare the directions of the electric and magnetic forces between two positive charges, which move along parallel paths (*a*) in the same direction, and (*b*) in opposite directions.

2 • At time $t = 0$, a particle with charge $q = 12\ \mu\text{C}$ is located at $x = 0$, $y = 2$ m; its velocity at that time is $\vec{v} = 30\ \text{m/s}\ \hat{i}$. Find the magnetic field at (*a*) the origin; (*b*) $x = 0, y = 1$ m; (*c*) $x = 0, y = 3$ m; and (*d*) $x = 0, y = 4$ m.

3 • For the particle in Problem 2, find the magnetic field at (*a*) $x = 1$ m, $y = 3$ m; (*b*) $x = 2$ m, $y = 2$ m; and (*c*) $x = 2$ m, $y = 3$ m.

4 • A proton (charge $+e$) traveling with a velocity of $\vec{v} = 1 \times 10^4\ \text{m/s}\ \hat{i} + 2 \times 10^4\ \text{m/s}\ \hat{j}$ is located at $x = 3$ m, $y = 4$ m at some time t. Find the magnetic field at the following positions: (*a*) $x = 2$ m, $y = 2$ m; (*b*) $x = 6$ m, $y = 4$ m; and (*c*) $x = 3$ m, $y = 6$ m.

5 • An electron orbits a proton at a radius of 5.29×10^{-11} m. What is the magnetic field at the proton due to the orbital motion of the electron?

6 •• Two equal charges q located at $(0, 0, 0)$ and $(0, b, 0)$ at time zero are moving with speed v in the positive x direction ($v \ll c$). Find the ratio of the magnitudes of the magnetic and electrostatic force on each.

The Biot–Savart Law

7 • The Biot–Savart law is similar to Coulomb's law in that both

(*a*) are inverse square laws.
(*b*) deal with forces on charged particles.
(*c*) deal with excess charges.
(*d*) include the permeability of free space.
(*e*) are not electrical in nature.

8 • A small current element $I\,d\vec{\ell}$, with $d\vec{\ell} = 2\ \text{mm}\ \hat{k}$ and $I = 2$ A, is centered at the origin. Find the magnetic field $d\vec{B}$ at the following points: (*a*) on the x axis at $x = 3$ m, (*b*) on the x axis at $x = -6$ m, (*c*) on the z axis at $z = 3$ m, and (*d*) on the y axis at $y = 3$ m.

9 • For the current element in Problem 8, find the magnitude and direction of $d\vec{B}$ at $x = 0$, $y = 3$ m, $z = 4$ m.

10 • For the current element in Problem 8, find the magnitude of $d\vec{B}$ and indicate its direction on a diagram at (*a*) $x = 2$ m, $y = 4$ m, $z = 0$ and (*b*) $x = 2$ m, $y = 0$, $z = 4$ m.

Current Loops

11 • Is $\vec{B}$ uniform everywhere within a current loop? Explain.

12 • A single loop of wire of radius 3 cm carries a current of 2.6 A. What is the magnitude of B on the axis of the loop at (*a*) the center of the loop, (*b*) 1 cm from the center, (*c*) 2 cm from the center, and (*d*) 35 cm from the center?

13 • A single-turn, circular loop of radius 10.0 cm is to produce a field at its center that will just cancel the earth's magnetic field at the equator, which is 0.7 G directed north. Find the current in the loop and make a sketch showing the orientation of the loop and the current.

14 •• For the loop of wire in Problem 13, at what point along the axis of the loop is the magnetic field (*a*) 10% of the field at the center, (*b*) 1% of the field at the center, and (*c*) 0.1% of the field at the center?

15 •• A single-turn circular loop of radius 8.5 cm is to produce a field at its center that will just cancel the earth's field of magnitude 0.7 G directed at 70° below the horizontal north direction. Find the current in the loop and make a sketch showing the orientation of the loop and the current.

16 •• A circular current loop of radius R carrying a current I is centered at the origin with its axis along the x axis. Its current is such that it produces a magnetic field in the positive x direction. (*a*) Sketch a graph of B_x versus x for points on the x axis. Include both positive and negative values of x. Compare this graph with that for E_x due to a charged ring of the same size. (*b*) A second, identical current loop, carrying an equal current in the same sense, is in a plane parallel to the yz plane with its center at $x = d$. Sketch graphs of the magnetic field on the x axis due to each loop separately and the resultant field due to the two loops. Show from your sketch that dB_x/dx is zero midway between the two loops.

17 •• Two coils that are separated by a distance equal to their radius and that carry equal currents such that their axial fields add are called Helmholtz coils. A feature of Helmholtz coils is that the resultant magnetic field between the coils is very uniform. Let $R = 10$ cm, $I = 20$ A, and $N = 300$ turns for each coil. Place one coil in the yz plane with its center at the origin and the other in a parallel plane at $x = 10$ cm. (*a*) Calculate the resultant field B_x at $x = 5$ cm, $x = 7$ cm, $x = 9$ cm, and $x = 11$ cm. (*b*) Use your results and the fact that B_x is symmetric about the midpoint of the coils to sketch B_x versus x. (See also Problem 18.)

18 ••• Two Helmholtz coils with radii R have their axes along the x axis (see Problem 17). One coil is in the yz plane and the other is in a parallel plane at $x = R$. Show that at the midpoint of the coils ($x = \frac{1}{2}R$), $dB_x/dx = 0$, $d^2B_x/dx^2 = 0$, and $d^3B_x/dx^3 = 0$. This shows that the magnetic field at points near the midpoint is approximately equal to that at the midpoint.

Straight-Line Current Segments

19 • Two wires lie in the plane of the paper and carry equal currents in opposite directions, as shown in Figure 29-38. At a point midway between the wires, the magnetic field is

(*a*) zero.
(*b*) into the page.
(*c*) out of the page.
(*d*) toward the top or bottom of the page.
(*e*) toward one of the two wires.

Figure 29-38 Problem 19

20 • Two parallel wires carry currents I_1 and $I_2 = 2I_1$ in the same direction. The forces F_1 and F_2 on the wires are related by

(*a*) $F_1 = F_2$. (*b*) $F_1 = 2F_2$. (*c*) $2F_1 = F_2$.
(*d*) $F_1 = 4F_2$. (*e*) $4F_1 = F_2$.

21 • A wire carries an electrical current straight up. What is the direction of the magnetic field due to the wire a distance of 2 m north of the wire?

(*a*) North (*b*) East (*c*) West
(*d*) South (*e*) Upward

22 • Two current-carrying wires are perpendicular to each other. The current in one flows vertically upward and the current in the other flows horizontally toward the east. The horizontal wire is one meter south of the vertical wire. What is the direction of the net magnetic force on the horizontal wire?

(*a*) North (*b*) East (*c*) West (*d*) South
(*e*) There is no net magnetic force on the horizontal wire.

23 • A long, straight wire carries a current of 10 A. Find the magnitude of B at (*a*) 10 cm, (*b*) 50 cm, and (*c*) 2 m from the center of the wire.

Problems 24 to 29 refer to Figure 29-39, which shows two long, straight wires in the xy plane and parallel to the x axis. One wire is at $y = -6$ cm and the other is at $y = +6$ cm. The current in each wire is 20 A.

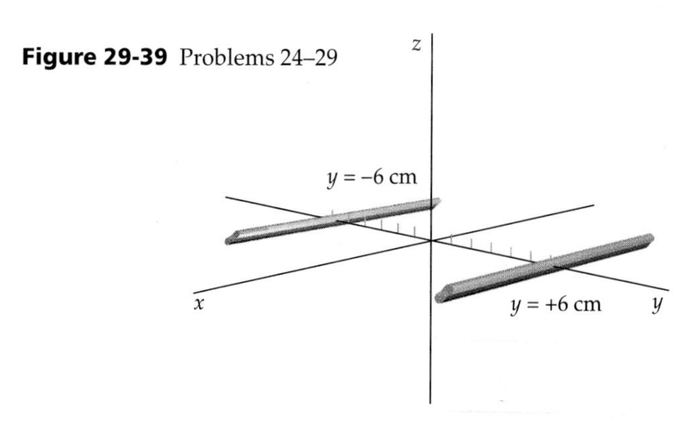

Figure 29-39 Problems 24–29

24 • If the currents in Figure 29-39 are in the negative x direction, find $\vec{B}$ at the points on the y axis at (*a*) $y = -3$ cm, (*b*) $y = 0$, (*c*) $y = +3$ cm, and (*d*) $y = +9$ cm.

25 • Sketch B_z versus y for points on the y axis when both currents are in the negative x direction.

26 • Find $\vec{B}$ at points on the y axis as in Problem 24 when the current in the wire at $y = -6$ cm is in the negative x direction and the current in the wire at $y = +6$ cm is in the positive x direction.

27 • Sketch B_z versus y for points on the y axis when the directions of the currents are opposite to those in Problem 26.

28 • Find $\vec{B}$ on the z axis at $z = +8$ cm if (*a*) the currents are parallel, as in Problem 24 and (*b*) the currents are antiparallel, as in Problem 26.

29 • Find the magnitude of the force per unit length exerted by one wire on the other.

30 • Two long, straight, parallel wires 8.6 cm apart carry currents of equal magnitude I. They repel each other with a force per unit length of 3.6 nN/m. (*a*) Are the currents parallel or antiparallel? (*b*) Find I.

31 •• The current in the wire of Figure 29-40 is 8.0 A. Find B at point P due to each wire segment and sum to find the resultant B.

Figure 29-40 Problem 31

32 •• A wire of length 16 cm is suspended by flexible leads above a long, straight wire. Equal but opposite currents are established in the wires such that the 16-cm wire floats 1.5 mm above the long wire with no tension in its suspension leads. If the mass of the 16-cm wire is 14 g, what is the current?

33 •• Three long, parallel, straight wires pass through the corners of an equilateral triangle of sides 10 cm as shown in Figure 29-41, where a dot means that the current is out of the paper and a cross means that it is into the paper. If each current is 15.0 A, find (*a*) the force per unit length on the upper wire, and (*b*) the magnetic field B at the upper wire due to the two lower wires.

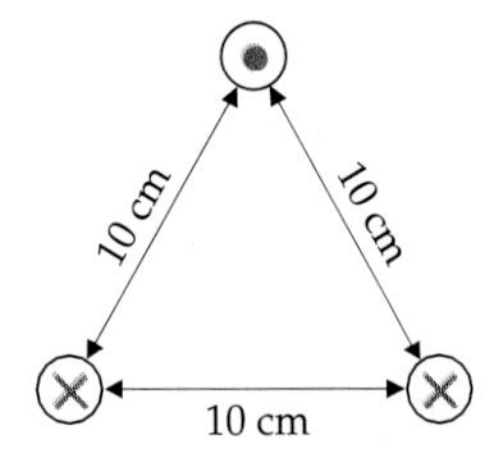

Figure 29-41 Problems 33 and 34

34 •• Work Problem 33 with the current in the lower right corner of Figure 29-41 reversed.

35 •• An infinitely long, insulated wire lies along the x axis and carries current I in the positive x direction. A second infinitely long, insulated wire lies along the y axis and carries current I in the positive y direction. Where in the xy plane is the resultant magnetic field zero?

36 •• An infinitely long wire lies along the z axis and carries a current of 20 A in the positive z direction. A second infinitely long wire is parallel to the z axis at $x = 10$ cm. (*a*) Find the current in the second wire if the magnetic field at $x = 2$ cm is zero. (*b*) What is the magnetic field at $x = 5$ cm?

37 •• Three very long, parallel wires are at the corners of a square, as shown in Figure 29-42. They each carry a current of magnitude I. Find the magnetic field B at the unoccupied corner of the square when (*a*) all the currents are into the paper, (*b*) I_1 and I_3 are in and I_2 is out, and (*c*) I_1 and I_2 are in and I_3 is out.

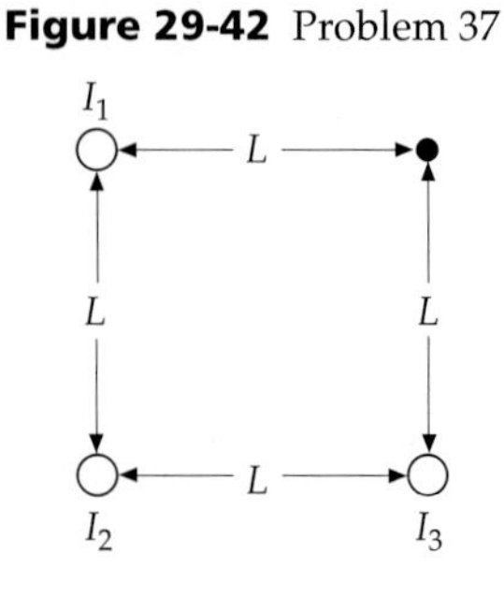

Figure 29-42 Problem 37

38 •• Four long, straight, parallel wires each carry current I. In a plane perpendicular to the wires, the wires are at the corners of a square of side a. Find the force per unit length on one of the wires if (*a*) all the currents are in the same direction, and (*b*) the currents in the wires at adjacent corners are oppositely directed.

39 •• An infinitely long, nonconducting cylinder of radius R lies along the z axis. Five long, conducting wires are parallel to the cylinder and spaced equally on the upper half of its surface. Each wire carries a current I in the positive z direction. Find the magnetic field on the z axis.

Solenoids

40 • A solenoid with length 30 cm, radius 1.2 cm, and 300 turns carries a current of 2.6 A. Find B on the axis of the solenoid (*a*) at the center, (*b*) inside the solenoid at a point 10 cm from one end, and (*c*) at one end.

41 • A solenoid 2.7 m long has a radius of 0.85 cm and 600 turns. It carries a current I of 2.5 A. What is the approximate magnetic field B on the axis of the solenoid?

42 ••• A solenoid has n turns per unit length and radius R and carries a current I. Its axis is along the x axis with one end at $x = -\frac{1}{2}\ell$ and the other end at $x = +\frac{1}{2}\ell$, where ℓ is the total length of the solenoid. Show that the magnetic field B at a point on the axis outside the solenoid is given by

$$B = \frac{1}{2}\mu_0 nI(\cos\theta_1 - \cos\theta_2) \qquad \text{29-35}$$

where

$$\cos\theta_1 = \frac{x + \frac{1}{2}\ell}{[R^2 + (x + \frac{1}{2}\ell)^2]^{1/2}}$$

and

$$\cos\theta_2 = \frac{x - \frac{1}{2}\ell}{[R^2 + (x - \frac{1}{2}\ell)^2]^{1/2}}$$

43 ••• In Problem 42, a formula for the magnetic field along the axis of a solenoid is given. For $x >> \ell$ and $\ell > R$, the angles θ_1 and θ_2 in Equation 29-35 are very small, so the small-angle approximation $\cos\theta \approx 1 - \theta^2/2$ is valid. (*a*) Draw a diagram and show that

$$\theta_1 \approx \frac{R}{x + \frac{1}{2}\ell} \quad \text{and} \quad \theta_2 \approx \frac{R}{x - \frac{1}{2}\ell}$$

(*b*) Show that the magnetic field at a point far from either end of the solenoid can be written

$$B = \frac{\mu_0}{4\pi}\left(\frac{q_m}{r_1^2} - \frac{q_m}{r_2^2}\right) \qquad 29\text{-}37$$

where $r_1 = x - \frac{1}{2}\ell$ is the distance to the near end of the solenoid, $r_2 = x + \frac{1}{2}\ell$ is the distance to the far end, and $q_m = nI\pi R^2 = \mu/\ell$, where $\mu = NI\pi R^2$ is the magnetic moment of the solenoid.

44 ••• In this problem, you will derive Equation 29-37 by another method. Consider a long, tightly wound solenoid of length ℓ and radius $R \ll \ell$ lying along the axis with its center at the origin. It has N turns and carries a current I. Consider an element of the solenoid of length dx. (*a*) What is the magnetic moment of this element? (*b*) Show that the magnetic field dB due to this element at a point on the x axis x_0 far from the element is given by

$$dB = \frac{\mu_0}{2\pi} nIA \frac{dx}{x'^3}$$

where $A = \pi R^2$ and $x' = x_0 - x$ is the distance from the element to the field point. (*c*) Integrate this expression from $x = -\frac{1}{2}\ell$ to $x = +\frac{1}{2}\ell$ to obtain Equation 29-37.

Ampère's Law

45 • Ampère's law is valid

(*a*) when there is a high degree of symmetry.
(*b*) when there is no symmetry.
(*c*) when the current is constant.
(*d*) when the magnetic field is constant.
(*e*) in all of these situations if the current is continuous.

46 • A long, straight, thin-walled, cylindrical shell of radius R carries a current I. Find B inside and outside the cylinder.

47 • In Figure 29-43, one current is 8 A into the paper, the other current is 8 A out of the paper, and each curve is a circular path. (*a*) Find $\oint_C \vec{B}\cdot d\vec{\ell}$ for each path indicated. (*b*) Which path, if any, can be used to find B at some point due to these currents?

Figure 29-43 Problem 47

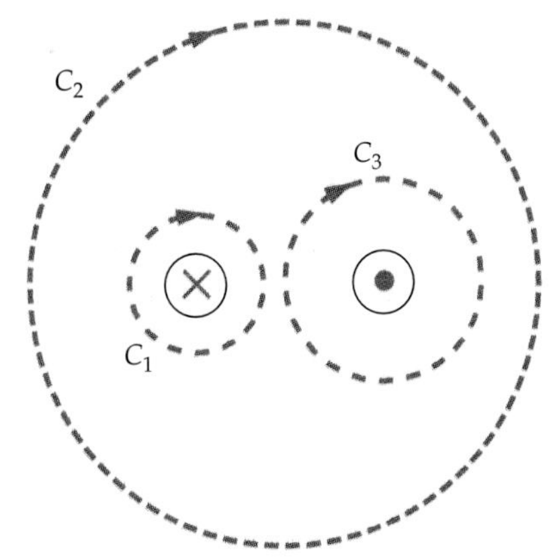

48 • A very long, coaxial cable consists of an inner wire and a concentric outer cylindrical conducting shell of radius R. At one end, the wire is connected to the shell. At the other end, the wire and shell are connected to opposite terminals of a battery, so there is a current down the wire and back up the shell. Assume that the cable is straight. Find B (*a*) at points between the wire and the shell far from the ends, and (*b*) outside the cable.

49 •• A wire of radius 0.5 cm carries a current of 100 A that is uniformly distributed over its cross-sectional area. Find B (*a*) 0.1 cm from the center of the wire, (*b*) at the surface of the wire, and (*c*) at a point outside the wire 0.2 cm from the surface of the wire. (*d*) Sketch a graph of B versus the distance from the center of the wire.

50 •• Show that a uniform magnetic field with no fringing field, such as that shown in Figure 29-44, is impossible because it violates Ampère's law. Do this by applying Ampère's law to the rectangular curve shown by the dashed lines.

Figure 29-44 Problem 50

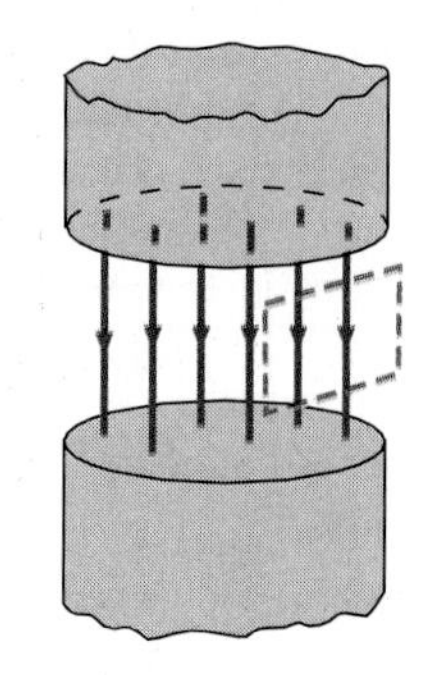

51 •• A coaxial cable consists of a solid inner cylindrical conductor of radius 1.00 mm and an outer cylindrical shell conductor of inner radius 2.00 mm and outer radius 3.00 mm. There is a current of 18 A down the inner wire and an equal return current in the outer conductor. The currents are uniform over the cross section of each conductor. Find the numerical value of $\oint_C \vec{B}\cdot d\vec{\ell}$ for a closed circular path (centered on the axis of the cable and in a plane perpendicular to the axis) which has a radius r for (*a*) $r = 1.50$ mm, (*b*) $r = 2.50$ mm, and (*c*) $r = 3.50$ mm.

52 •• An infinitely long, thick, cylindrical shell of inner radius a and outer radius b carries a current I uniformly distributed across a cross section of the shell. Find the magnetic field for (*a*) $r < a$, (*b*) $a < r < b$, and (*c*) $r > b$.

53 •• Figure 29-45 shows a solenoid carrying a current I with n turns per unit length. Apply Ampère's law to the rectangular curve shown to derive an expression for B assuming that it is uniform inside the solenoid and zero outside it.

Figure 29-45 Problem 53

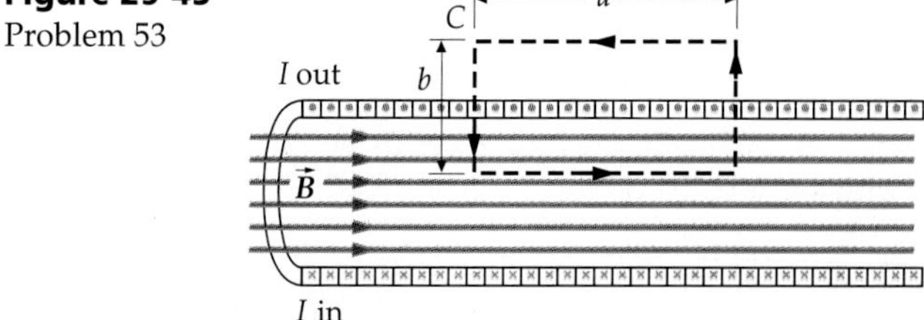

54 •• A tightly wound toroid of inner radius 1 cm and outer radius 2 cm has 1000 turns of wire and carries a current of 1.5 A. (*a*) What is the magnetic field at a distance of 1.1 cm from the center? (*b*) What is the field 1.5 cm from the center?

55 •• The xz plane contains an infinite sheet of current in the positive z direction. The current per unit length (along

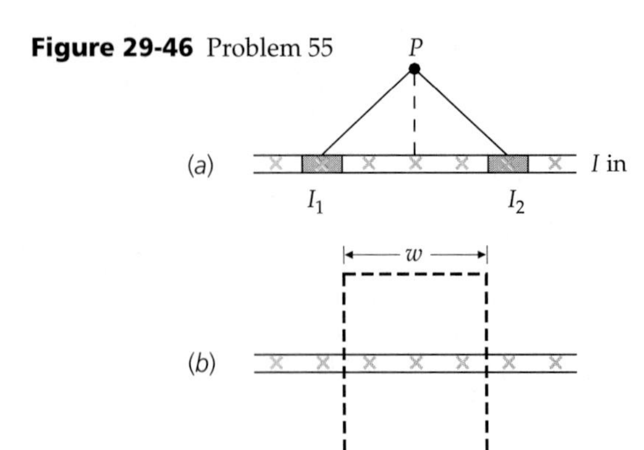

Figure 29-46 Problem 55

the x direction) is λ. Figure 29-46*a* shows a point P above the sheet ($y > 0$) and two portions of the current sheet labeled I_1 and I_2. (*a*) What is the direction of the magnetic field $\vec{B}$ at P due to the two portions of the current shown? (*b*) What is the direction of the magnetic field $\vec{B}$ at P due to the entire sheet? (*c*) What is the direction of $\vec{B}$ at a point below the sheet ($y < 0$)? (*d*) Apply Ampère's law to the rectangular curve shown in Figure 29-46*b* to show that the magnetic field at any point above the sheet is given by $\vec{B} = -\frac{1}{2}\mu_0\lambda\hat{i}$.

Magnetization and Magnetic Susceptibility

56 • True or false:

(*a*) Diamagnetism is the result of induced magnetic dipole moments.
(*b*) Paramagnetism is the result of the partial alignment of permanent magnetic dipole moments.

57 • If the magnetic susceptibility is positive,

(*a*) paramagnetic effects or ferromagnetic effects must be greater than diamagnetic effects.
(*b*) diamagnetic effects must be greater than paramagnetic effects.
(*c*) diamagnetic effects must be greater than ferromagnetic effects.
(*d*) ferromagnetic effects must be greater than paramagnetic effects.
(*e*) paramagnetic effects must be greater than ferromagnetic effects.

58 • A tightly wound solenoid 20 cm long has 400 turns and carries a current of 4 A such that its axial field is in the z direction. Neglecting end effects, find B and B_{app} at the center when (*a*) there is no core in the solenoid, and (*b*) there is an iron core with a magnetization $M = 1.2 \times 10^6$ A/m.

59 • Which of the four gases listed in Table 29-1 are diamagnetic and which are paramagnetic?

60 • If the solenoid of Problem 58 has an aluminum core, find B_{app}, M, and B at the center, neglecting end effects.

61 • Repeat Problem 60 for a tungsten core.

62 • A long solenoid is wound around a tungsten core and carries a current. (*a*) If the core is removed while the current is held constant, does the magnetic field inside the solenoid decrease or increase? (*b*) By what percentage?

63 • When a sample of liquid is inserted into a solenoid carrying a constant current, the magnetic field inside the solenoid decreases by 0.004%. What is the magnetic susceptibility of the liquid?

64 • A long solenoid carrying a current of 10 A has 50 turns/cm. What is the magnetic field in the interior of the solenoid when the interior is (*a*) a vacuum, (*b*) filled with aluminum, and (*c*) filled with silver?

65 •• An engineer intends to fill a solenoid with a mixture of oxygen and nitrogen at room temperature and 1 atmosphere pressure such that K_m is exactly 1. Assume that the magnetic dipole moments of the gas molecules are all aligned and that the susceptibility of a gas is proportional to the number density of its molecules. What should the ratio of the number densities of oxygen to nitrogen molecules be so that $K_m = 1$?

66 •• A cylinder of magnetic material is placed in a long solenoid of n turns per unit length and current I. Table 29-3 gives the magnetic field B versus nI. Use these values to plot B versus B_{app} and K_m versus nI.

Table 29-3

nI, A/m	0	50	100	150	200	500	1000	10,000
B, T	0	0.04	0.67	1.00	1.2	1.4	1.6	1.7

67 •• A small magnetic sample is in the form of a disk having a radius of 1.4 cm, a thickness of 0.3 cm, and a uniform magnetization along its axis throughout its volume. The magnetic moment of the sample is 1.5×10^{-2} A·m². (*a*) What is the magnetization $\vec{M}$ of the sample? (*b*) If this magnetization is due to the alignment of N electrons each with a magnetic moment of 1 μ_B, what is N? (*c*) If the magnetization is along the axis of the disk, what is the magnitude of the amperian surface current?

68 •• The magnetic moment of the earth is about 9×10^{22} A·m². (*a*) If the magnetization of the earth's core were 1.5×10^9 A/m, what is the core volume? (*b*) What is the radius of such a core if it were spherical and centered with the earth?

Atomic Magnetic Moments

69 •• Nickel has a density of 8.7 g/cm³ and molecular mass of 58.7 g/mol. Its saturation magnetization is given by $\mu_0 M_s = 0.61$ T. Calculate the magnetic moment of a nickel atom in Bohr magnetons.

70 •• Repeat Problem 69 for cobalt, which has a density of 8.9 g/cm³, a molecular mass of 58.9 g/mol, and a saturation magnetization given by $\mu_0 M_s = 1.79$ T.

Paramagnetism (optional)

71 • Show that Curie's law predicts that the magnetic susceptibility of a paramagnetic substance is $\chi_m = \mu\mu_0 M_s/3kT$.

72 •• In a simple model of paramagnetism, we can consider that some fraction f of the molecules have their magnetic moments aligned with the external magnetic field and that the rest of the molecules are randomly oriented and so do not contribute to the magnetic field. (*a*) Use this model and Curie's law to show that at temperature T and external magnetic field B the fraction of aligned molecules is $f = \mu B/3kT$. (*b*) Calculate this fraction for $T = 300$ K, $B = 1$ T, assuming μ to be 1 Bohr magneton.

73 •• Assume that the magnetic moment of an aluminum atom is 1 Bohr magneton. The density of aluminum is 2.7 g/cm^3, and its molecular mass is 27 g/mol. (*a*) Calculate M_s and $\mu_0 M_s$ for aluminum. (*b*) Use the results of Problem 71 to calculate χ_m at $T = 300$ K. (*c*) Explain why the result for part (*b*) is larger than the value listed in Table 29-1.

74 •• A toroid with N turns carrying a current I has mean radius R and cross-sectional radius r, where $r \ll R$ (Figure 29-47). When the toroid is filled with material, it is called a *Rowland ring*. Find B_{app} and B in such a ring, assuming a magnetization $\vec{M}$ everywhere parallel to $\vec{B}_{app}$.

Figure 29-47 Problem 74

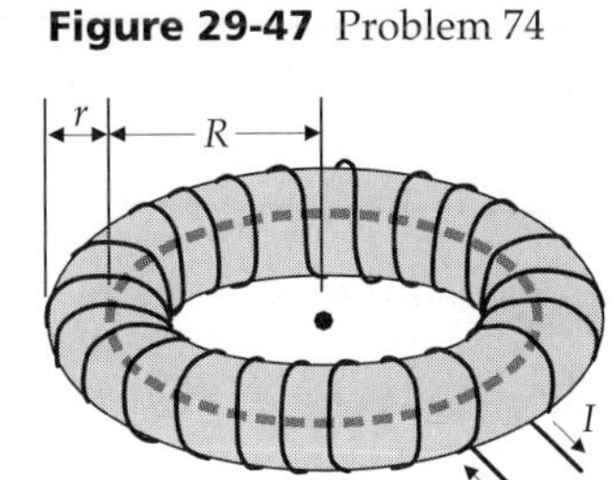

75 •• A toroid is filled with liquid oxygen that has a susceptibility of 4×10^{-3}. The toroid has 2000 turns and carries a current of 15 A. Its mean radius is 20 cm, and the radius of its cross section is 0.8 cm. (*a*) What is the magnetization M? (*b*) What is the magnetic field B? (*c*) What is the percentage increase in B produced by the liquid oxygen?

76 •• A toroid has an average radius of 14 cm and a cross-sectional area of 3 cm^2. It is wound with fine wire, 60 turns/cm measured along its mean circumference, and the wire carries a current of 4 A. The core is filled with a paramagnetic material of magnetic susceptibility 2.9×10^{-4}. (*a*) What is the magnitude of the magnetic field within the substance? (*b*) What is the magnitude of the magnetization? (*c*) What would the magnitude of the magnetic field be if there were no paramagnetic core present?

Ferromagnetism (optional)

77 • For annealed iron, the relative permeability K_m has its maximum value of about 5500 at $B_{app} = 1.57 \times 10^{-4}$ T. Find M and B when K_m is maximum.

78 •• The saturation magnetization for annealed iron occurs when $B_{app} = 0.201$ T. Find the permeability μ and the relative permeability K_m of annealed iron at saturation. (See Table 29-2.)

79 •• The coercive force is defined to be the applied magnetic field needed to bring B back to zero along the hysteresis curve (point c in Figure 29-35). For a certain permanent bar magnet, the coercive force $B_{app} = 5.53 \times 10^{-2}$ T. The bar magnet is to be demagnetized by placing it inside a 15-cm-long solenoid with 600 turns. What minimum current is needed in the solenoid to demagnetize the magnet?

80 •• A long solenoid with 50 turns/cm carries a current of 2 A. The solenoid is filled with iron, and B is measured to be 1.72 T. (*a*) Neglecting end effects, what is B_{app}? (*b*) What is M? (*c*) What is the relative permeability K_m?

81 •• When the current in Problem 80 is 0.2 A, the magnetic field is measured to be 1.58 T. (*a*) Neglecting end effects, what is B_{app}? (*b*) What is M? (*c*) What is the relative permeability K_m?

82 •• A long, iron-core solenoid with 2000 turns/m carries a current of 20 mA. At this current, the relative permeability of the iron core is 1200. (*a*) What is the magnetic field within the solenoid? (*b*) With the iron core removed, what current will produce the same field within the solenoid?

83 •• Two long, straight wires 4.0 cm apart are embedded in a uniform insulator having relative permeability of $K_m = 120$. The wires carry 40 A in opposite directions. (*a*) What is the magnetic field at the midpoint of the plane of the wires? (*b*) What is the force per unit length on the wires?

84 •• The toroid of Problem 75 has its core filled with iron. When the current is 10 A, the magnetic field in the toroid is 1.8 T. (*a*) What is the magnetization M? (*b*) Find the values for K_m, μ, and χ_m for the iron sample.

85 •• Find the magnetic field in the toroid of Problem 76 if the current in the wire is 0.2 A and soft iron, having a relative permeability of 500, is substituted for the paramagnetic core?

86 •• A long, straight wire with a radius of 1.0 mm is coated with an insulating ferromagnetic material that has a thickness of 3.0 mm and a relative magnetic permeability of $K_m = 400$. The coated wire is in air and the wire itself is nonmagnetic. The wire carries a current of 40 A. (*a*) Find the magnetic field inside the wire as a function of radius r. (*b*) Find the magnetic field inside the ferromagnetic material as a function of radius r. (*c*) Find the magnetic field outside the ferromagnetic material as a function of r. (*d*) What must the magnitudes and directions of the amperian currents be on the surfaces of the ferromagnetic material to account for the magnetic fields observed?

General Problems

87 • True or false:

(*a*) The magnetic field due to a current element is parallel to the current element.
(*b*) The magnetic field due to a current element varies inversely with the square of the distance from the element.
(*c*) The magnetic field due to a long wire varies inversely with the square of the distance from the wire.
(*d*) Ampère's law is valid only if there is a high degree of symmetry.
(*e*) Ampère's law is valid only for continuous currents.

88 • Can a particle have angular momentum and not have a magnetic moment?

89 • Can a particle have a magnetic moment and not have angular momentum?

90 • A circular loop of wire carries a current I. Is there angular momentum associated with the magnetic moment of the loop? If so, why is it not noticed?

91 • A hollow tube carries a current. Inside the tube, $\vec{B} = 0$. Why is this the case, since $\vec{B}$ is strong inside a solenoid?

92 • When a current is passed through the wire in Figure 29-48, will it tend to bunch up or form a circle?

Figure 29-48 Problem 92

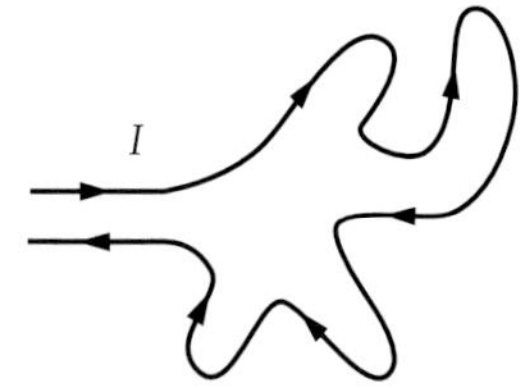

93 • Find the magnetic field at point P in Figure 29-49.

Figure 29-49 Problem 93

94 • In Figure 29-50, find the magnetic field at point P, which is at the common center of the two semicircular arcs.

Figure 29-50 Problem 94

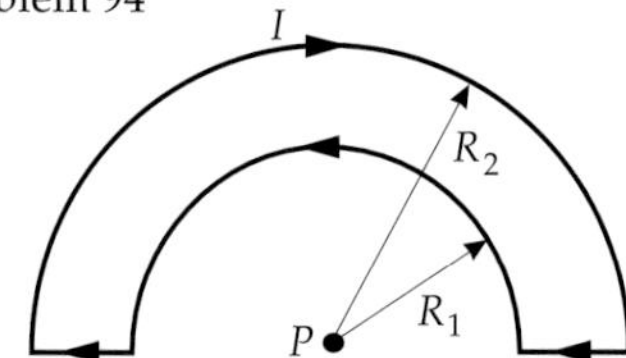

95 •• A wire of length ℓ is wound into a circular coil of N loops and carries a current I. Show that the magnetic field at the center of the coil is given by $B = \mu_0 \pi N^2 I/\ell$.

96 •• A very long wire carrying a current I is bent into the shape shown in Figure 29-51. Find the magnetic field at point P.

Figure 29-51 Problem 96

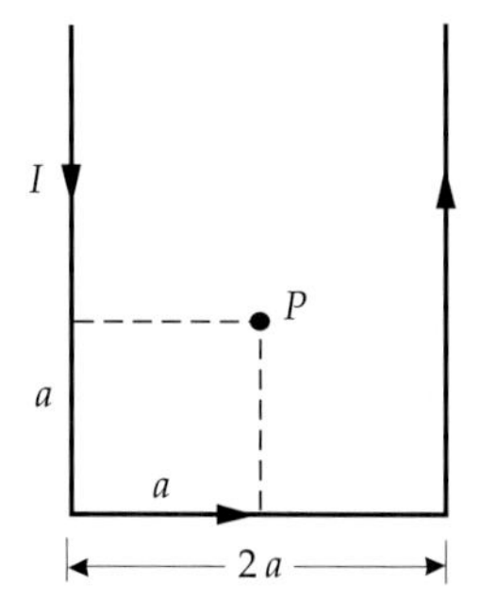

97 •• A loop of wire of length ℓ carries a current I. Compare the magnetic fields at the center of the loop when it is (*a*) a circle, (*b*) a square, and (*c*) an equilateral triangle. Which field is largest?

98 •• A power cable carrying 50.0 A is 2.0 m below the earth's surface, but its direction and precise position are unknown. Show how you could locate the cable using a compass. Assume that you are at the equator, where the earth's magnetic field is 0.7 G north.

99 •• A long, straight wire carries a current of 20 A as shown in Figure 29-52. A rectangular coil with two sides parallel to the straight wire has sides 5 cm and 10 cm with the near side a distance 2 cm from the wire. The coil carries a current of 5 A. (*a*) Find the force on each segment of the rectangular coil. (*b*) What is the net force on the coil?

Figure 29-52 Problem 99

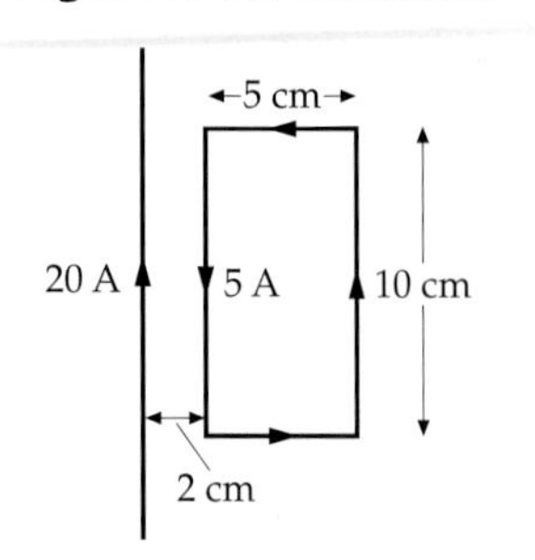

100 •• The closed loop shown in Figure 29-53 carries a current of 8.0 A in the counterclockwise direction. The radius of the outer arc is 60 cm, that of the inner arc is 40 cm. Find the magnetic field at point P.

Figure 29-53 Problem 100

101 •• A closed circuit consists of two semicircles of radii 40 and 20 cm that are connected by straight segments as shown in Figure 29-54. A current of 3.0 A flows around this circuit in the clockwise direction. Find the magnetic field at point P.

Figure 29-54 Problem 101

102 •• A very long, straight wire carries a current of 20.0 A. An electron 1.0 cm from the center of the wire is moving with a speed of 5.0×10^6 m/s. Find the force on the electron when it moves (*a*) directly away from the wire, (*b*) parallel to the wire in the direction of the current, and (*c*) perpendicular to the wire and tangent to a circle around the wire.

103 •• A current I is uniformly distributed over the cross section of a long, straight wire of radius 1.40 mm. At the surface of the wire, the magnitude of the magnetic field is $B = 2.46$ mT. Find the magnitude of the magnetic field at (*a*) 2.10 mm from the axis and (*b*) 0.60 mm from the axis. (*c*) Find the current I.

104 •• A large, 50-turn circular coil of radius 10.0 cm carries a current of 4.0 A. At the center of the large coil is a small 20-turn coil of radius 0.5 cm carrying a current of 1.0 A. The planes of the two coils are perpendicular. Find the torque exerted by the large coil on the small coil. (Neglect any variation in B due to the large coil over the region occupied by the small coil.)

105 •• Figure 29-55 shows a bar magnet suspended by a thin wire that provides a restoring torque $-\kappa\theta$. The magnet is 16 cm long, has a mass of 0.8 kg, a dipole moment of $\mu = 0.12$ A·m², and it is located in a region where a uniform magnetic field B can be established. When the external magnetic field is 0.2 T and the magnet is given a small angular displacement $\Delta\theta$, the bar magnet oscillates about its equilibrium position with a period of 0.500 s. Determine the constant κ and the period of this torsional pendulum when $B = 0$.

Figure 29-55 Problem 105

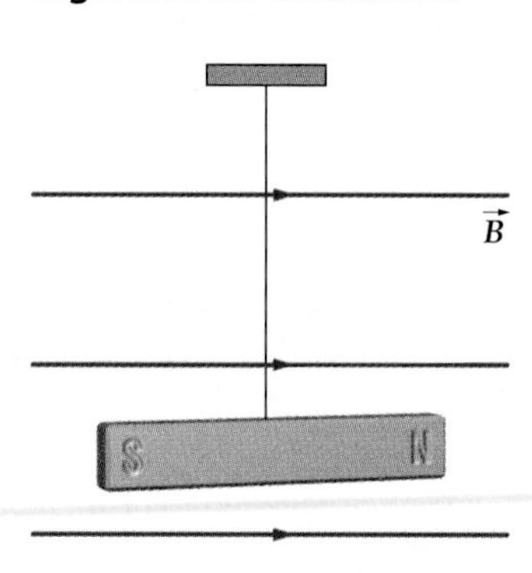

106 •• A long, narrow bar magnet that has magnetic moment $\vec{\mu}$ parallel to its long axis is suspended at its center as a frictionless compass needle. When placed in a magnetic field $\vec{B}$, the needle lines up with the field. If it is displaced by a small angle θ, show that the needle will oscillate about its equilibrium position with frequency $f = (1/2\pi)\sqrt{\mu B/I}$, where I is the moment of inertia about the point of suspension.

107 •• A small bar magnet of mass 0.1 kg, length 1 cm, and magnetic moment $\mu = 0.04\ \mathrm{A{\cdot}m^2}$ is located at the center of a 100-turn loop of 0.2 m diameter. The loop carries a current of 5.0 A. At equilibrium, the bar magnet is aligned with the field due to the current loop. The bar magnet is given a displacement along the axis of the loop and released. Show that if the displacement is small, the bar magnet executes simple harmonic motion, and find the period of this motion.

108 •• Suppose the needle in Problem 106 is a uniformly magnetized iron rod that is 8 cm long and has a cross-sectional area of 3 mm^2. Assume that the magnetic dipole moment for each iron atom is $2.2\mu_B$ and that all the iron atoms have their dipole moments aligned. Calculate the frequency of small oscillations about the equilibrium position when the magnetic field is 0.5 G.

109 •• The needle of a magnetic compass has a length of 3 cm, a radius of 0.85 mm, and a density of $7.96 \times 10^3\ \mathrm{kg/m^3}$. It is free to rotate in a horizontal plane, where the horizontal component of the earth's magnetic field is 0.6 G. When disturbed slightly, the compass executes simple harmonic motion about its midpoint with a frequency of 1.4 Hz. (*a*) What is the magnetic dipole moment of the needle? (*b*) What is the magnetization M? (*c*) What is the amperian current on the surface of the needle? (See Problem 106.)

110 •• An iron bar of length 1.4 m has a diameter of 2 cm and a uniform magnetization of 1.72×10^6 A/m directed along the bar's length. The bar is stationary in space and is suddenly demagnetized so that its magnetization disappears. What is the rotational angular velocity of the bar if its angular momentum is conserved? (Assume that Equation 29-27 holds where m is the mass of an electron and $q = -e$.)

111 •• The magnetic dipole moment of an iron atom is $2.219\mu_B$. (*a*) If all the atoms in an iron bar of length 20 cm and cross-sectional area 2 cm^2 have their dipole moments aligned, what is the dipole moment of the bar? (*b*) What torque must be supplied to hold the iron bar perpendicular to a magnetic field of 0.25 T?

112 •• A relatively inexpensive ammeter called a *tangent galvanometer* can be made using the earth's field. A plane circular coil of N turns and radius R is oriented such that the field B_c it produces in the center of the coil is either east or west. A compass is placed at the center of the coil. When there is no current in the coil, the compass needle points north. When there is a current I, the compass needle points in the direction of the resultant magnetic field $\vec{B}$ at an angle θ to the north. Show that the current I is related to θ and the horizontal component of the earth's field B_e by

$$I = \frac{2RB_e}{\mu_0 N} \tan\theta$$

113 •• An infinitely long, straight wire is bent as shown in Figure 29-56. The circular portion has a radius of 10 cm with its center a distance r from the straight part. Find r such that the magnetic field at the center of the circular portion is zero.

Figure 29-56 Problem 113

114 •• (*a*) Find the magnetic field at point P for the wire carrying current I shown in Figure 29-57. (*b*) Use your result from (*a*) to find the field at the center of a polygon of N sides. Show that when N is very large, your result approaches that for the magnetic field at the center of a circle.

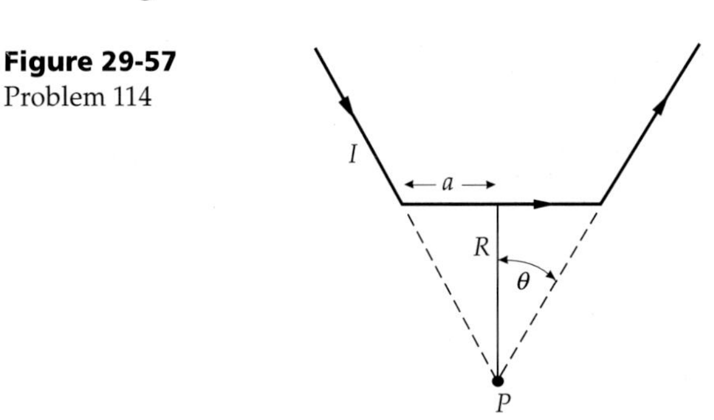

Figure 29-57 Problem 114

115 •• The current in a long, cylindrical conductor of radius $R = 10$ cm varies with distance from the axis of the cylinder according to the relation $I(r) = (50\ \mathrm{A/m})r$. Find the magnetic field at (a) $r = 5$ cm (b) at $r = 10$ cm, and (c) $r = 20$ cm.

116 •• Figure 29-58 shows a square loop, 20 cm per side, in the xy plane with its center at the origin. The loop carries a current of 5 A. Above it at $y = 0$, $z = 10$ cm is an infinitely long wire parallel to the x axis carrying a current of 10 A. (*a*) Find the torque on the loop. (*b*) Find the net force on the loop.

Figure 29-58 Problem 116

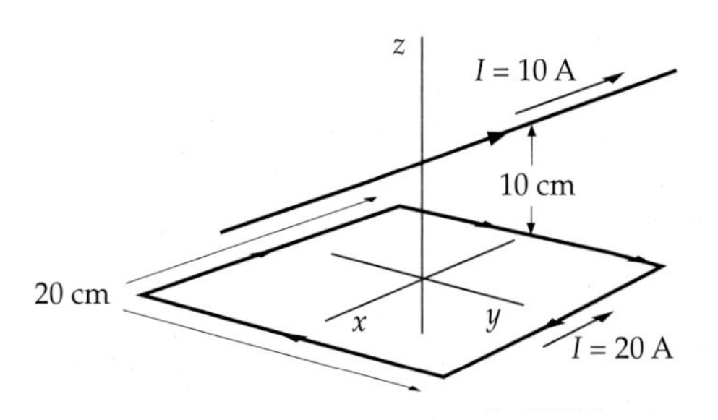

117 •• In the Bohr model of the hydrogen atom, an electron in the ground state orbits a proton at a radius of 5.29×10^{-11} m. In a reference frame in which the orbiting electron is at rest, the proton circulates about the electron at a radius of 5.29×10^{-11} m with the same angular velocity as that of the electron in the reference frame in which the proton is at rest.

Consequently, in the rest frame of the electron, the magnetic field due to the motion of the proton has the same magnitude as that calculated in Problem 5. The electron has an intrinsic magnetic moment of magnitude μ_B. Find the energy difference between the two possible orientations of the electron's intrinsic magnetic moment, either parallel or antiparallel to the magnetic field due to the apparent motion of the proton. (This energy difference is readily observed spectroscopically and is known as the *fine structure splitting*.)

118 •• The proton also has an intrinsic magnetic moment whose magnitude is $1.52 \times 10^{-3}\mu_B$. The orientation of the proton's magnetic moment is quantized; it can only be parallel or antiparallel to the magnetic field at the location of the proton. Using the result of Problem 117, determine the energy difference of the hydrogen atom in the Bohr model for the two possible orientations of the proton's magnetic moment; neglect the magnetic field at the proton due to the electron's intrinsic magnetic moment. (This energy difference is referred to as *hyperfine structure splitting*.)

119 •• In the calculation of the hyperfine structure splitting in Problem 117, you neglected the magnetic field at the proton's position due to the intrinsic magnetic moment of the electron. Calculate the magnetic field due to the intrinsic magnetic moment of the electron at a distance of 5.29×10^{-11} m and compare its magnitude at the location of the proton with that due to the orbital motion of the electron. (*Hint*: Assume the electron spin is perpendicular to the plane of its orbit, and show that the magnitude of the magnetic field at a great distance from a magnetic dipole on a line perpendicular to the dipole is given by $B = (\mu_0/4\pi)\mu/r^3$.)

120 ••• A disk of radius R carries a fixed charge density σ and rotates with angular velocity ω. (*a*) Consider a circular strip of radius r and width dr with charge dq. Show that the current produced by this strip $dI = (\omega/2\pi)\,dq = \omega\sigma r\,dr$. (*b*) Use your result from part (*a*) to show that the magnetic field at the center of the disk is $B = \frac{1}{2}\mu_0\sigma\omega R$. (*c*) Use your result from part (*a*) to find the magnetic field at a point on the axis of the disk a distance x from the center.

121 ••• A very long, straight conductor with a circular cross section of radius R carries a current I. Inside the conductor, there is a cylindrical hole of radius a whose axis is parallel to the axis of the conductor a distance b from it (Figure 29-59). Let the z axis be the axis of the conductor, and let the axis of the hole be at $x = b$. Find the magnetic field $\vec{B}$ at the point (*a*) on the x axis at $x = 2R$, and (*b*) on the y axis at $y = 2R$. (*Hint*: Consider a uniform current distribution throughout the cylinder of radius R plus a current in the opposite direction in the hole.)

Figure 29-59 Problem 121

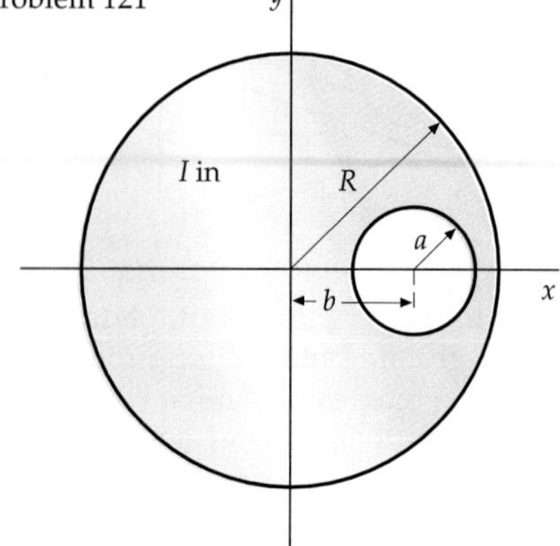

122 ••• For the cylinder with the hole in Problem 120, show that the magnetic field inside the hole is uniform, and find its magnitude and direction.

123 ••• A square loop of side ℓ lies in the yz plane with its center at the origin. It carries a current I. Find the magnetic field B at any point on the x axis and show from your expression that for x much larger than ℓ,

$$B \approx \frac{\mu_0}{4\pi}\frac{2\mu}{x^3}$$

where $\mu = I\ell^2$ is the magnetic moment of the loop.

124 ••• A circular loop carrying current I lies in the yz plane with its axis along the x axis. (*a*) Evaluate the line integral $\oint_C \vec{B}\cdot d\vec{\ell}$ along the axis of the loop from $x = -\ell_1$ to $x = +\ell_1$. (*b*) Show that when $\ell_1 \to \infty$, the line integral approaches $\mu_0 I$. This result can be related to Ampère's law by closing the curve of integration with a semicircle of radius ℓ on which $B \approx 0$ for ℓ very large.

125 ••• The current in a long cylindrical conductor of radius R is given by $I(r) = I_0(1 - e^{r/a})$. Derive expressions for the magnetic field for $r < R$ and for $r > R$.

126 ••• In Example 29-8 we calculated the magnetic field inside and outside a wire of radius R carrying a uniform current I. Consider a filament of current at a distance r from the center of the wire. Show that this filament experiences a force directed toward the center of the wire and that, therefore, the current distribution cannot be truly uniform. This so-called *pinch effect* depends on the magnitude of the current, but is generally negligibly small. Consider a copper wire of 2.0 cm diameter that carries a nominally uniform current of 400 A. Calculate the force per meter exerted on a 0.1-mm-thick annular region of current at the periphery of the wire. Assume that the current is carried by the conduction electrons. How far toward the center of the wire would this annular current region have to move so that the electrostatic force between the electrons and the fixed positive ions at the periphery just balances the magnetic force?

CHAPTER 30

Magnetic Induction

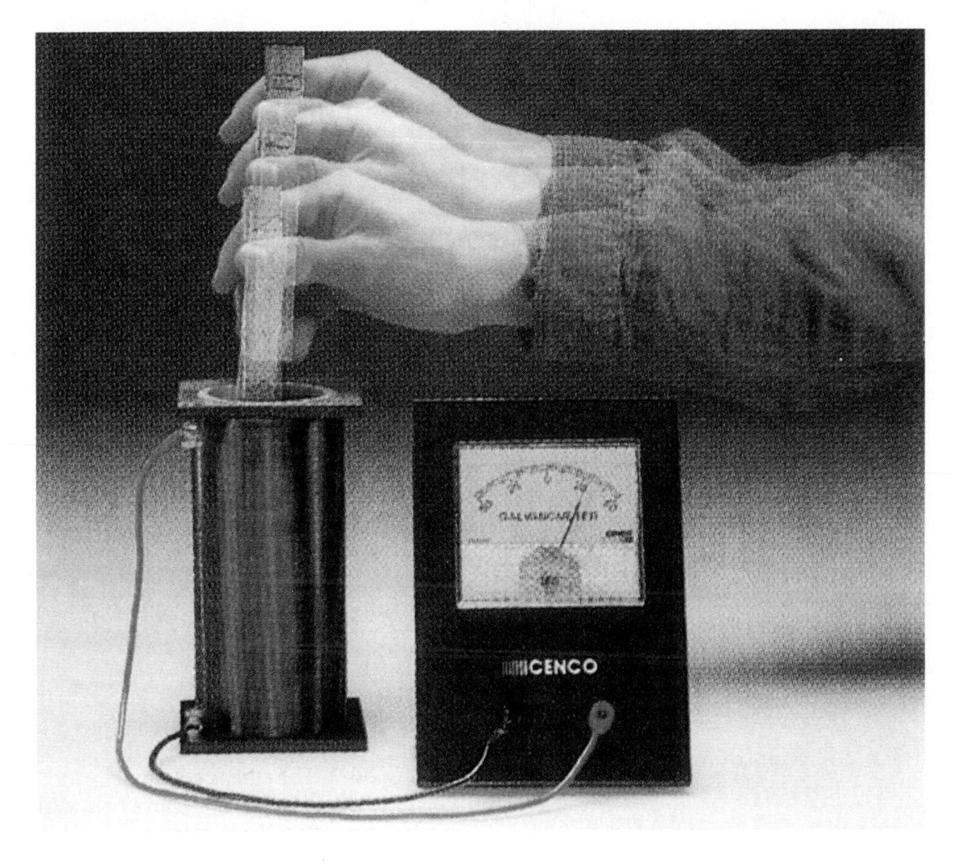

Demonstration of induced emf. When the magnet is moving toward or away from the coil, an emf is induced in the coil, as shown by the galvanometer's deflection. No deflection is observed when the magnet is stationary.

In the early 1830s, Michael Faraday in England and Joseph Henry in America independently discovered that a *changing* magnetic field induces a current in a wire. The emfs and currents caused by changing magnetic fields are called **induced emfs** and **induced currents.** The process itself is referred to as **magnetic induction.**

When you pull the plug of an electric cord from its socket, you sometimes observe a small spark. Before the cord is disconnected, it carries a current, which produces a magnetic field encircling the current. When the cord is disconnected, the current abruptly ceases and the magnetic field around it collapses. The changing magnetic field produces an emf that tries to maintain the original current, resulting in a spark across the plug. Once the magnetic field reaches zero it is no longer changing, and the emf is zero.

Changing magnetic fields can result from changing currents or from moving magnets. The chapter-opening photo illustrates a simple classroom demonstration of induced currents. The ends of a coil are attached to a galvanometer and a strong magnet is moved toward or away from the coil. The momentary deflection shown by the galvanometer *during* the motion indicates that there is an induced electric current in the coil–galvanometer circuit. A current is also induced if the coil is moved toward or away from the magnet, or if the coil is rotated in a fixed magnetic field. A coil rotating in a magnetic field is the basic element of a generator, which converts mechanical energy into electrical energy.

All of the various methods of magnetic induction can be summarized by a single relation known as Faraday's law, which relates the induced emf in a circuit to the change in magnetic flux through the circuit.

30-1 Magnetic Flux

The flux of a magnetic field through a surface is defined similarly to the flux of an electric field (Section 22-2). Let dA be an element of area on the surface and $\hat{n}$ be the unit vector perpendicular to the element (Figure 30-1). The magnetic flux ϕ_m is then defined to be

$$\phi_m = \int_S \vec{B}\cdot\hat{n}\, dA = \int_S B_n\, dA \qquad 30\text{-}1$$

Definition—Magnetic flux

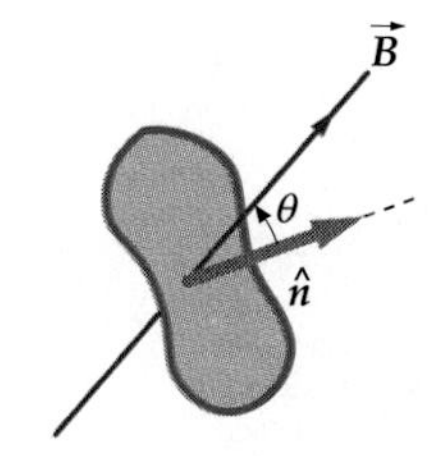

Figure 30-1 When $\vec{B}$ makes an angle θ with the normal to the area of a loop, the flux through the loop is $B \cos \theta\, A$.

The unit of magnetic flux is that of magnetic field times area, tesla-meter squared, which is called a **weber** (Wb):

$$1\ \text{Wb} = 1\ \text{T}\cdot\text{m}^2 \qquad 30\text{-}2$$

Since B is proportional to the number of field lines per unit area, the magnetic flux is proportional to the number of lines through the area.

Exercise Show that a weber per second is a volt.

If the surface is a plane with area A, and $\vec{B}$ is constant in magnitude and direction over the surface and makes an angle θ with the unit normal vector, the flux is

$$\phi_m = BA \cos \theta$$

We are often interested in the flux through a coil containing several turns of wire. If the coil contains N turns, the flux through the coil is N times the flux through each turn (Figure 30-2):

$$\phi_m = NBA \cos \theta \qquad 30\text{-}3$$

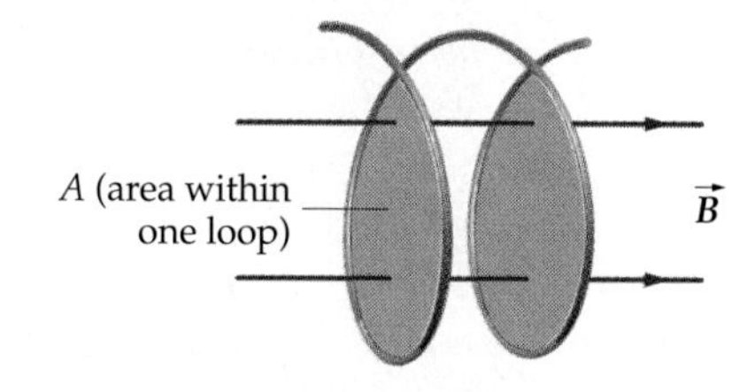

Figure 30-2 The area A bounded by a coil of two turns is twice the area bounded by each turn. In general, the area bounded by a coil of N turns is N times that of each turn.

Example 30-1

Find the magnetic flux through a solenoid that is 40 cm long, has a radius of 2.5 cm, has 600 turns, and carries a current of 7.5 A.

Picture the Problem The magnetic field $\vec{B}$ inside the solenoid is uniform and along the axis of the solenoid. It is therefore perpendicular to the plane of the coils. We thus need to find B inside the solenoid and then multiply B by NA.

1. The magnetic flux is the product of the number of turns, the magnetic field, and the area of the coils: $\phi_m = NBA$

2. The magnetic field inside the solenoid is given by $B = \mu_0 nI$, where $n = N/\ell$ is the number of turns per unit length: $\phi_m = N\mu_0(N/\ell)IA = \mu_0 N^2 IA/\ell$

3. Express the area of the coils in terms of its radius: $A = \pi r^2$

4. Substitute the given values to calcuate the flux:
$$\phi_m = \mu_0 N^2 IA/\ell$$
$$= (4\pi \times 10^{-7}\ \text{T}\cdot\text{m/A})(600\ \text{turns})^2(7.5\ \text{A})\pi(0.025\ \text{m})^2/(0.40\ \text{m})$$
$$= 1.66 \times 10^{-2}\ \text{Wb}$$

Remark Note that since $\phi_m = NBA$ and B is proportional to the number of turns N, the flux is proportional to N^2.

30-2 Induced emf and Faraday's Law

Experiments by Faraday, Henry, and others showed that if the magnetic flux through an area bounded by a circuit is changed by any means, an emf equal in magnitude to the rate of change of the flux is induced in the circuit. We usually detect the emf by observing a current in the circuit, but it is present even when the circuit is incomplete (not closed) and there is no current. Previously we considered emfs that were localized in a specific part of the circuit, such as between the terminals of the battery. However, induced emfs can be considered to be distributed throughout the circuit.

The magnetic flux through a circuit can be changed in many different ways. The current producing the magnetic field may be increased or decreased, permanent magnets may be moved toward the circuit or away from it, the circuit itself may be moved toward or away from the source of the flux, the orientation of the circuit may be changed, or the area of the circuit in a fixed magnetic field may be increased or decreased. In every case, an emf is induced in the circuit that is equal in magnitude to the rate of change of the magnetic flux.

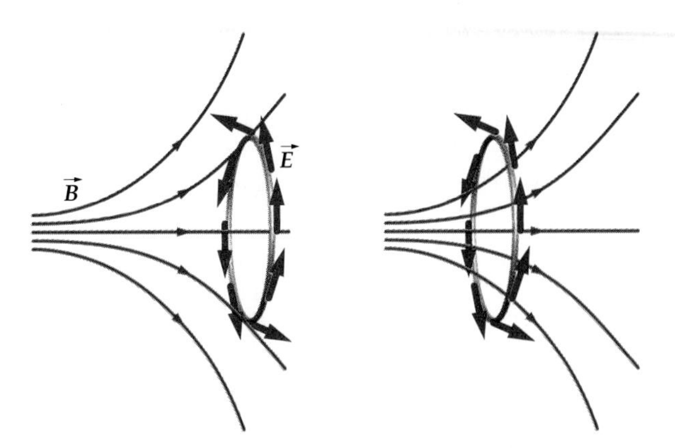

Figure 30-3 When the magnetic flux through the wire loop is changing, an emf is induced in the loop. The emf is distributed throughout the loop and is equivalent to a nonconservative electric field $\vec{E}$ parallel to the wire. In this figure, the direction of $\vec{E}$ corresponds to the case in which the flux through the loop is increasing.

Figure 30-3 shows a single loop of wire in a magnetic field. If the flux through the loop is changing, an emf is induced in the loop. Since emf is the work done per unit charge, there must be a force exerted on the charge associated with the emf. The force per unit charge is the electric field $\vec{E}$, which in this case is induced by the changing flux. The line integral of the electric field around a complete circuit equals the work done per unit charge, which, by definition, is the emf in the circuit:

$$\mathcal{E} = \oint_C \vec{E}\cdot d\vec{\ell} \qquad \text{30-4}$$

Definition—emf

The electric fields that we studied previously resulted from static electric charges. These fields are conservative, meaning that the work done by the electrostatic field around a closed curve is zero. The electric field resulting from changing magnetic flux is not conservative. Its line integral around a closed curve equals the induced emf, which equals the rate of change of the magnetic flux:

$$\mathcal{E} = \oint_C \vec{E} \cdot d\vec{\ell} = -\frac{d\phi_m}{dt} \qquad 30\text{-}5$$

Faraday's law

This result is known as **Faraday's law.** The negative sign in Faraday's law has to do with the direction of the induced emf, which we will discuss shortly.

Example 30-2

A uniform magnetic field makes an angle of 30° with the axis of a circular coil of 300 turns and a radius of 4 cm. The field changes at a rate of 85 T/s. Find the magnitude of the induced emf in the coil.

Picture the Problem The induced emf equals N times the rate of change of the flux through each turn. Since B is uniform, the flux through each turn is simply $\phi_m = BA\cos\theta$, where $A = \pi r^2$ is the area of the coil.

1. The magnitude of the induced emf is given by Faraday's law: $|\mathcal{E}| = \frac{d\phi_m}{dt}$

2. For a uniform field, the flux is: $\phi_m = NBA\cos\theta$

3. Substitute this expression for ϕ_m and calculate $|\mathcal{E}|$:

$$|\mathcal{E}| = \frac{d\phi_m}{dt} = \frac{d}{dt}(NBA\cos\theta) = NA\cos\theta\frac{dB}{dt}$$

$$= (300)(3.14)(0.04\text{ m})^2\cos 30°\ (85\text{ T/s}) = 111\text{ V}$$

Exercise If the resistance of the coil is 200 Ω, what is the induced current? (*Answer* 0.555 A)

Example 30-3 *try it yourself*

An 80-turn coil has a radius of 5.0 cm and a resistance of 30 Ω. At what rate must a perpendicular magnetic field change to produce a current of 4.0 A in the coil?

Picture the Problem The rate of change of the magnetic field is related to the rate of change of the flux, which is related to the induced emf by Faraday's law. The emf in the coil equals IR.

Cover the column to the right and try these on your own before looking at the answers.

Steps	***Answers***
1. Write the magnetic flux in terms of B, N, and the radius r, and solve for B.	$\phi_m = N\pi r^2 B$ $B = \dfrac{\phi_m}{N\pi r^2}$
2. Take the time derivative of B.	$\dfrac{dB}{dt} = \dfrac{1}{N\pi r^2}\dfrac{d\phi_m}{dt}$
3. Use Faraday's law to relate the rate of change of the flux to the emf.	$\dfrac{d\phi_m}{dt} = \mathcal{E}$
4. Calculate the emf in the coil from the current and resistance of the coil.	$\mathcal{E} = IR = 120\text{ V}$
5. Substitute numerical values of E, N, and r to calculate dB/dt.	$\dfrac{dB}{dt} = \dfrac{1}{N\pi r^2}\dfrac{d\phi_m}{dt} = 191\text{ T/s}$

Example 30-4

A magnetic field $\vec{B}$ is perpendicular to the plane of the page and uniform in a circular region of radius R as shown in Figure 30-4. Outside of the circular region, $\vec{B}$ decreases to 0. The rate of change of the magnitude of $\vec{B}$ is dB/dt. What is the magnitude of the induced electric field in the plane of the page (*a*) at a distance $r < R$ from the center of the circular region, and (*b*) at a distance $r > R$, where $B = 0$.

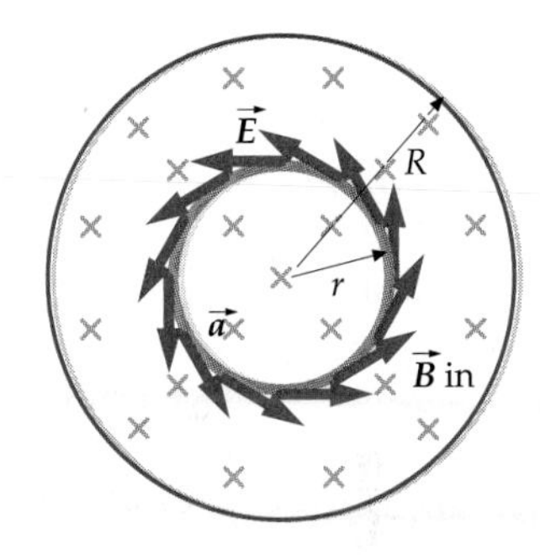

Figure 30-4

Picture the Problem The magnetic field $\vec{B}$ is into the page and uniform over a circular region of radius R. When B changes, the magnetic flux changes and an emf $\mathcal{E} = \oint_C \vec{E}\cdot d\vec{\ell}$ is induced around any curve enclosing the flux. The induced electric field is found by applying Faraday's law. Since we are interested only in magnitudes, we neglect the minus sign and use $\oint_C \vec{E}\cdot d\vec{\ell} = d\phi_m/dt$. To take advantage of the system's symmetry we choose a circular curve of radius r to compute the line integral. By symmetry, $\vec{E}$ is tangent to this curve and has the same magnitude at any point on it. We then calculate the magnetic flux ϕ_m and take its time derivative. Setting the integral and the time derivative equal yields an expression for E.

(*a*)1. E is found from the line integral for a circle of radius $r < R$. $\vec{E}$ is tangent to the circle and has a constant magnitude:	$\oint_C \vec{E}\cdot d\vec{\ell} = E(2\pi r)$
2. The line integral is calculated from Faraday's law:	$\oint_C \vec{E}\cdot d\vec{\ell} = \dfrac{d\phi_m}{dt}$
3. For $r < R$, B is constant over the circle. Since $\vec{B}$ is perpendicular to the plane of the circle, the flux is simply BA:	$\phi_m = BA = B\pi r^2$
4. Calculate the time derivative of ϕ_m:	$\dfrac{d\phi_m}{dt} = \pi r^2 \dfrac{dB}{dt}$

5. Substitute these results for $\oint_C \vec{E}\cdot d\vec{\ell}$ and solve for E:

$$2\pi rE = \pi r^2 \frac{dB}{dt}$$

$$E = \frac{r}{2}\frac{dB}{dt}, \quad r < R$$

(*b*)1. For a circle of radius $r > R$, where the magnetic field is zero, the line integral is the same as before:

$$\oint_C \vec{E}\cdot d\vec{\ell} = E(2\pi r)$$

2. Since $B = 0$ for $r > R$, the magnetic flux is $\pi R^2 B$:

$$\phi_m = \pi R^2 B$$

3. Apply Faraday's law to find E:

$$2\pi rE = \pi R^2 \frac{dB}{dt}$$

$$E = \frac{R^2}{2r}\frac{dB}{dt}, \quad r > R$$

Remarks Note that the electric field in this example is produced by a changing magnetic field rather than by electric charges. If charges had caused the field, $\vec{E}$ would have to start on positive charges and end on negative charges. Since charges are not present, however, $\vec{E}$ forms circles that have no beginning and no end. Note also that the emf exists in any closed curve bounding the area through which the magnetic flux is changing whether or not there is a wire or circuit along the curve.

Example 30-5

A small coil of N turns has its plane perpendicular to a uniform magnetic field $\vec{B}$ as shown in Figure 30-5. The coil is connected to a current integrator Ⓒ, a device designed to measure the total charge passing through it. Find the charge passing through the coil if the coil is rotated through 180° about its diameter.

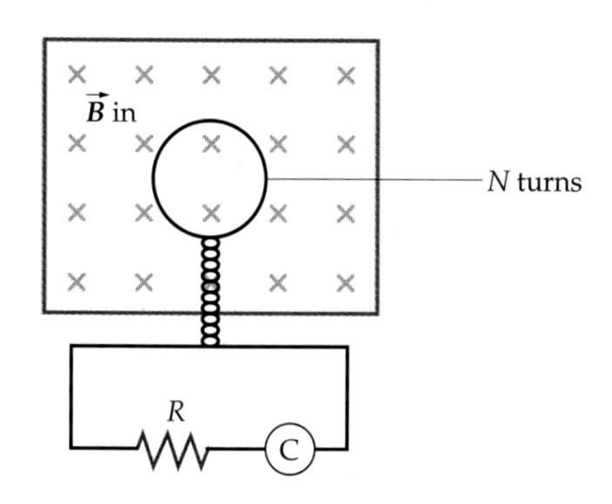

Figure 30-5

Picture the Problem When the coil in Figure 30-5 is rotated, the magnetic flux through it changes, causing an induced emf, $\mathcal{E}$. The emf in turn causes a current $I = \mathcal{E}/R$, where R is the total resistance of the circuit. Since $I = dQ/dt$, we can find the charge passing through the coil by integrating I; that is, $Q = \int dQ = \int I\,dt$.

1. The total charge is the integral of the current:

$$Q = \int dQ = \int I\,dt$$

2. The current is related to the emf by Ohm's law:

$$I = \frac{\mathcal{E}}{R}$$

3. The magnitude of the emf $\mathcal{E}$ is given by Faraday's law:

$$|\mathcal{E}| = \frac{d\phi_m}{dt}$$

4. Substitute these results to find the charge:

$$Q = \int I\,dt = \int \frac{\mathcal{E}}{R}\,dt = \frac{1}{R}\int \frac{d\phi_m}{dt}\,dt = \frac{1}{R}\int d\phi_m$$

$$= \frac{\Delta\phi_m}{R}$$

inductance of a circuit is constant, the change in flux is related to the change in current by

$$\frac{d\phi_m}{dt} = \frac{d(LI)}{dt} = L\frac{dI}{dt}$$

According to Faraday's law, we have

$$\mathcal{E} = -\frac{d\phi_m}{dt} = -L\frac{dI}{dt} \qquad \text{30-10}$$

Thus, the self-induced emf is proportional to the rate of change of the current. A coil or solenoid with many turns has a large self-inductance and is called an **inductor**. In circuits it is denoted by the symbol ‿‿‿. We can often neglect the self-inductance of the rest of the circuit compared with that of an inductor.

Exercise At what rate must the current in the solenoid of Example 30-9 change to induce an emf of 20 V? (*Answer* 3.18×10^5 A/s)

Mutual Inductance

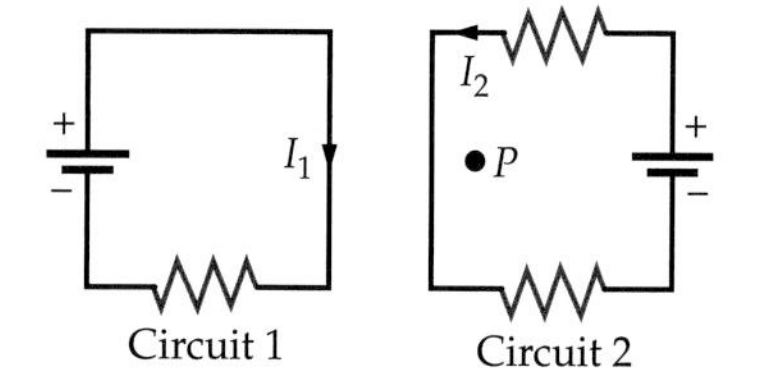

Figure 30-18 Two adjacent circuits. The magnetic field at point P is partly due to current I_1 and partly due to I_2. The flux through either circuit is the sum of two terms, one proportional to I_1 and the other to I_2.

When two or more circuits are close to each other, as in Figure 30-18, the magnetic flux through one circuit depends not only on the current in that circuit but also on the current in the nearby circuits. Let I_1 be the current in circuit 1 on the left in Figure 30-18, and let I_2 be the current in circuit 2 on the right. The magnetic field at some point P is due partly to I_1 and partly to I_2. The contribution to the overall field from each circuit is proportional to the current in each circuit. We can therefore write the flux through circuit 2, ϕ_{m2}, as the sum of two parts, one proportional to the current I_1 and the other proportional to the current I_2:

$$\phi_{m2} = L_2 I_2 + M_{2,1} I_1 \qquad \text{30-11}a$$

Definition—Mutual inductance

where L_2 is the self-inductance of circuit 2 and $M_{2,1}$ is called the **mutual inductance** of the two circuits. The mutual inductance depends on the geometrical arrangement of the two circuits. For instance, if the circuits are far apart, the flux through circuit 2 due to the current I_1 will be small and the mutual inductance will be small. An equation similar to Equation 30-11*a* can be written for the flux through circuit 1:

$$\phi_{m1} = L_1 I_1 + M_{1,2} I_2 \qquad \text{30-11}b$$

where L_1 is the self-inductance of circuit 1.

We can calculate the mutual inductance for two tightly wound concentric solenoids like the ones shown in Figure 30-19. Let ℓ be the length of both solenoids, and let the inner solenoid have N_1 turns and radius r_1 and the outer solenoid have N_2 turns and radius r_2. We will first calculate the mutual inductance $M_{2,1}$ by assuming that the inner solenoid carries a current I_1 and finding the magnetic flux ϕ_{m2} due to this current through the outer solenoid.

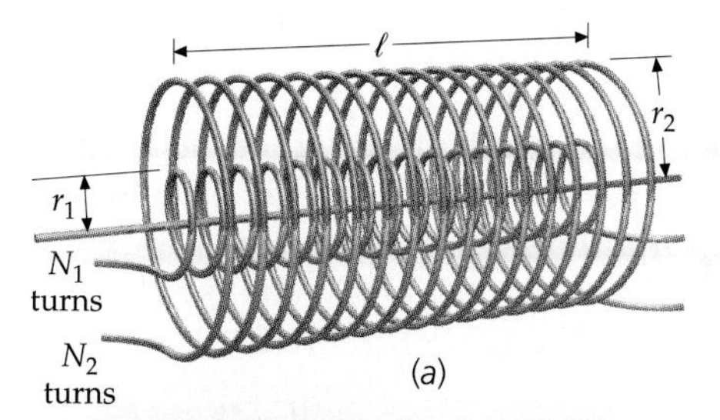

Figure 30-19 (*a*) A long, narrow solenoid inside a second solenoid of the same length. A current in either solenoid produces magnetic flux in the other. (*b*) Tesla coil illustrating the geometry of the wires in part (*a*). Such a device functions as a transformer (Chapter 31). Here, low-voltage alternating current in the outer winding is transformed into a higher-voltage alternating current in the inner winding. Induced alternating voltage from the changing fields is great enough to light the bulb above the coil.

The magnetic field due to the current in the inner solenoid is constant in the space within the solenoid and has magnitude

$$B_1 = \mu_0(N_1/\ell)I_1 = \mu_0 n_1 I_1 \qquad 30\text{-}12$$

The flux through the outer solenoid due to this magnetic field is therefore

$$\phi_{m2} = N_2 B_1(\pi r_1^2) = n_2 \ell B_1(\pi r_1^2) = \mu_0 n_2 n_1\, \ell(\pi r_1^2) I_1$$

Note that the area used to compute the flux through the outer solenoid is not the area of that solenoid, πr_2^2, but rather is the area of the inner solenoid, πr_1^2, because the magnetic field due to the inner solenoid is zero outside the inner solenoid. The mutual inductance $M_{1,2}$ is thus

$$M_{2,1} = \frac{\phi_{m2}}{I_1} = \mu_0 n_2 n_1 \ell \pi r_1^2 \qquad 30\text{-}13$$

Exercise Calculate the mutual inductance $M_{1,2}$ of the concentric solenoids of Figure 30-19 by finding the flux through the inner solenoid due to a current I_2 in the outer solenoid. (*Answer* $M_{1,2} = M_{2,1} = \mu_0 n_2 n_1 \ell \pi r_1^2$)

Note from the exercise above that $M_{1,2} = M_{2,1}$. It can be shown that this is a general result. We will therefore drop the subscripts for mutual inductance and simply write M.

30-7 Magnetic Energy

An inductor stores magnetic energy, just as a capacitor stores electrical energy. Consider the circuit consisting of an inductance L and a resistance R in series with a battery of emf $\mathcal{E}_0$ and a switch S shown in Figure 30-20. We assume that R and L are the resistance and inductance of the entire circuit. The switch is initially open, so there is no current in the circuit. A short time after the switch is closed, there is a current I in the circuit, a potential drop IR across the resistor, and a back emf of magnitude $L\,dI/dt$ in the inductor. In the circuit diagram, we put plus and minus signs on the inductor to indicate the direction of the emf when the current is increasing, that is, when dI/dt is positive. Applying Kirchhoff's loop rule to this circuit gives

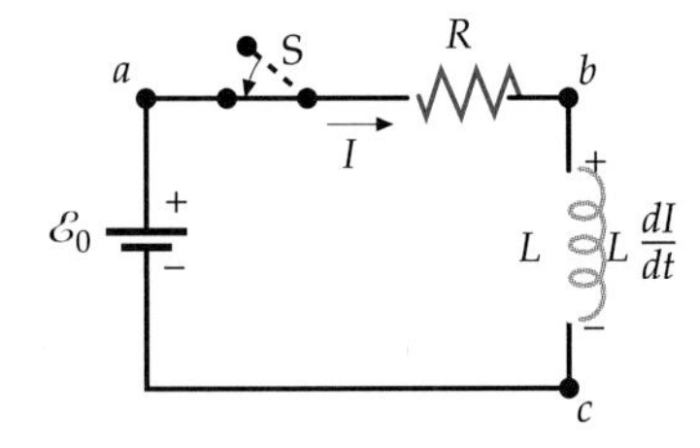

Figure 30-20 Just after the switch S is closed in this circuit, the current begins to increase and a back emf of magnitude $L\,dI/dt$ is generated in the inductor. The potential drop across the resistor IR plus the potential drop across the inductor equals the emf of the battery.

$$\mathcal{E}_0 - IR - L\frac{dI}{dt} = 0 \qquad 30\text{-}14$$

If we multiply each term by the current I and rearrange, we obtain

$$\mathcal{E}_0 I = I^2 R + LI\frac{dI}{dt} \qquad 30\text{-}15$$

The term $\mathcal{E}_0 I$ is the power output of the battery. The term I^2R is the power dissipated as heat in the resistance of the circuit. The term $LI\,dI/dt$ is the rate at which energy is put into the inductor. If U_m is the energy in the inductor, then

$$\frac{dU_m}{dt} = LI\frac{dI}{dt}$$

or

$$dU_m = LI\,dI$$

Integrating this equation from time $t = 0$, when the current is zero, to $t = \infty$, when the current has reached its final value I_f, we obtain

$$U_m = \int dU_m = \int_0^{I_f} LI\,dI = \frac{1}{2}LI_f^2$$

The energy stored in an inductor carrying a current I is thus given by

$$U_m = \tfrac{1}{2}LI^2 \tag{30-16}$$

Energy stored in an inductor

When a current is produced in an inductor, a magnetic field is created in the space within the inductor coil. We can think of the energy stored in an inductor as energy stored in the magnetic field. For the special case of a solenoid, the magnetic field is related to the current I and the number of turns per unit length n by

$$B = \mu_0 nI$$

and the self-inductance is given by Equation 30-9:

$$L = \mu_0 n^2 A\ell$$

where A is the cross-sectional area and ℓ is the length. Substituting $B/\mu_0 n$ for I and $\mu_0 n^2 A\ell$ for L in Equation 30-16, we obtain

$$U_m = \frac{1}{2}LI^2 = \frac{1}{2}\mu_0 n^2 A\ell\left(\frac{B}{\mu_0 n}\right)^2 = \frac{B^2}{2\mu_0}A\ell$$

The quantity $A\ell$ is the volume of the space within the solenoid containing the magnetic field. The energy per unit volume is the **magnetic energy density** u_m:

$$u_m = \frac{B^2}{2\mu_0} = \tfrac{1}{2}\left(\frac{1}{\mu_0}\right)B^2 \tag{30-17}$$

Magnetic energy density

Although we derived this by considering the special case of the magnetic field in a solenoid, it is a general result. Whenever there is a magnetic field in space, the magnetic energy per unit volume is given by Equation 30-17. Note the similarity to the energy density in an electric field (Equation 25-13):

$$u_e = \tfrac{1}{2}\epsilon_0 E^2 \tag{30-18}$$

Example 30-10

A certain region of space contains a magnetic field of 200 G and an electric field of 2.5×10^6 N/C. Find (*a*) the total energy density and (*b*) the energy in a cubical box of side $\ell = 12$ cm.

Picture the Problem The total energy density u is the sum of the electrical and magnetic energy densities, $u = u_e + u_m$. The energy in a volume $\mathcal{V}$ is given by $U = u\mathcal{V}$.

(*a*)1. Calculate the electrical energy density:

$$u_e = \tfrac{1}{2}\epsilon_0 E^2 = \tfrac{1}{2}(8.85 \times 10^{-12}\,\mathrm{C^2/N\cdot m^2})(2.5 \times 10^6\,\mathrm{N/C})^2 = 27.7\,\mathrm{J/m^3}$$

2. Calculate the magnetic energy density:

$$u_m = \frac{1}{2}\frac{B^2}{\mu_0} = \frac{1}{2}\frac{(0.02\,\mathrm{T})^2}{(4\pi \times 10^{-7}\,\mathrm{N/A^2})} = 159\,\mathrm{J/m^3}$$

3. The total energy density is the sum of the above two contributions:

$$u = u_e + u_m = 27.7\,\mathrm{J/m^3} + 159\,\mathrm{J/m^3} = 187\,\mathrm{J/m^3}$$

(*b*) The total energy in the box is $U = u\mathcal{V}$, where $\mathcal{V} = \ell^3$ is the volume of the box:

$$U = u\mathcal{V} = u\ell^3 = (187\,\mathrm{J/m^3})(0.12\,\mathrm{m})^3 = 0.323\,\mathrm{J}$$

30-8 *RL* Circuits

A circuit containing a resistor and an inductor such as that shown in Figure 30-20 is called an ***RL* circuit.** Since all circuits have resistance and self-inductance, the analysis of an *RL* circuit can be applied to some extent to all circuits.* For the circuit in Figure 30-20, application of Kirchhoff's loop rule gave us (Equation 30-14):

$$\mathcal{E}_0 - IR - L\frac{dI}{dt} = 0$$

Let's look at some general features of the current before we solve this equation. Just after we close the switch in the circuit, the current is zero so IR is zero, and the back emf $L\,dI/dt$ equals the emf of the battery, $\mathcal{E}_0$. Setting $I = 0$ in Equation 30-14, we get

$$\left(\frac{dI}{dt}\right)_0 = \frac{\mathcal{E}_0}{L} \qquad \text{30-19}$$

As the current increases, IR increases, and dI/dt decreases. Note that the current cannot jump suddenly from zero to some finite value as it would if there were no inductance. When there is some inductance ($L \neq 0$), dI/dt is finite and therefore the current must be continuous in time. After a short time, the current has reached a positive value I, and the rate of change of the current is

$$\frac{dI}{dt} = \frac{\mathcal{E}_0}{L} - \frac{IR}{L}$$

At this time the current is still increasing, but its rate of increase is less than at $t = 0$. The final value of the current can be obtained by setting dI/dt equal to zero:

$$I_f = \frac{\mathcal{E}_0}{R} \qquad \text{30-20}$$

Figure 30-21 shows the current in this circuit as a function of time. This figure is similar to that for the charge on a capacitor when the capacitor is charged in an *RC* circuit (Figure 26-41).

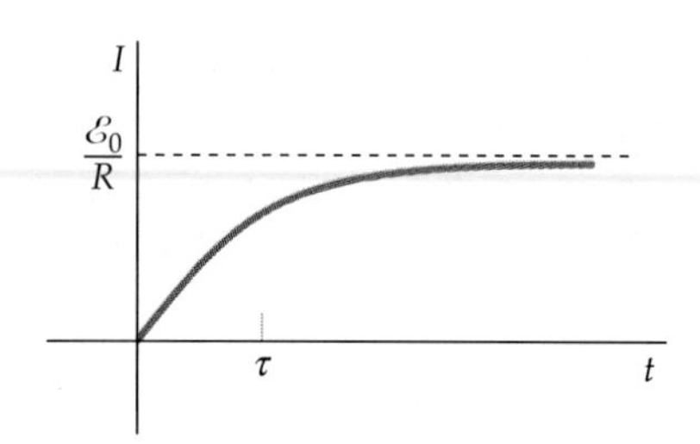

Figure 30-21 Current versus time in an *LR* circuit. At a time $t = \tau = L/R$, the current is at 63% of its maximum value $\mathcal{E}_0/R$.

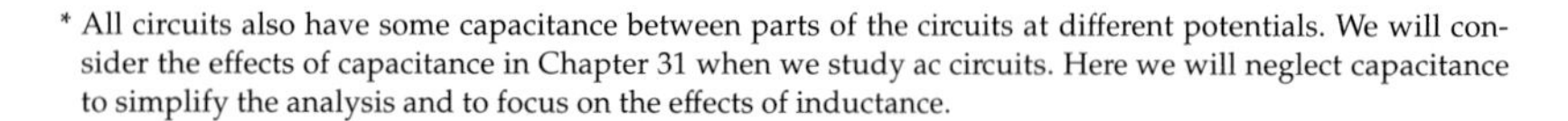
* All circuits also have some capacitance between parts of the circuits at different potentials. We will consider the effects of capacitance in Chapter 31 when we study ac circuits. Here we will neglect capacitance to simplify the analysis and to focus on the effects of inductance.

Equation 30-14 is of the same form as Equation 26-34 for the charging of a capacitor and can be solved in the same way. The result is

$$I = \frac{\mathcal{E}_0}{R}(1 - e^{-Rt/L}) = \frac{\mathcal{E}_0}{R}(1 - e^{-t/\tau}) = I_f(1 - e^{-t/\tau}) \qquad \text{30-21}$$

where

$$\tau = \frac{L}{R} \qquad \text{30-22}$$

is the **time constant** of the circuit. The larger the self-inductance L or the smaller the resistance R, the longer it takes for the current to build up.

Example 30-11

A coil of self-inductance 5.0 mH and a resistance of 15.0 Ω is placed across the terminals of a 12-V battery of negligible internal resistance. (*a*) What is the final current? (*b*) What is the current after 100 μs?

Picture the Problem The final current is $I_f = \mathcal{E}_0/R$ as given in Equation 30-20. The current as a function of time is given by Equation 30-21, $I = I_f(1 - e^{-t/\tau})$, where $\tau = L/R$.

(*a*) Use Equation 30-20 to find the final current, I_f:

$$I_f = \frac{\mathcal{E}_0}{R} = \frac{12\text{ V}}{15\ \Omega} = 0.800\text{ V}$$

(*b*)1. Use Equation 30-21 to write the current I at any time t:

$$I = I_f(1 - e^{-t/\tau})$$

2. Calculate the time constant τ:

$$\tau = \frac{L}{R} = \frac{5 \times 10^{-3}\text{ H}}{15\ \Omega} = 333\ \mu\text{s}$$

3. Use this result for τ and calculate I for $t = 300\ \mu$s:

$$I = I_f(1 - e^{-t/\tau}) = (0.800\text{ A})(1 - e^{-100/333})$$
$$= (0.800\text{ A})(1 - 0.741) = 0.207\text{ A}$$

Exercise How much energy is stored in this inductor when the final current has been attained? (*Answer* $U_m = \frac{1}{2}LI_f^2 = 1.6 \times 10^{-3}$ J)

In Figure 30-22, the circuit has an additional switch that allows us to remove the battery, and an additional resistor R_1 to protect the battery so that it is not shorted when both switches are momentarily closed. When S_2 is open and S_1 is closed, the current builds up in the circuit just as discussed, except that the total resistance is now $R_1 + R$ and the final current is $\mathcal{E}_0/(R + R_1)$. Suppose that S_1 has been closed for a long time, so that the current is approximately steady at its final value, which we will call I_0. At time $t = 0$ we close switch S_2 and open switch S_1 (to remove the battery from consideration completely). We now have a circuit with just a resistor and an inductor (loop *abcda*) carrying an initial current I_0. Applying Kirchhoff's loop rule to this circuit gives

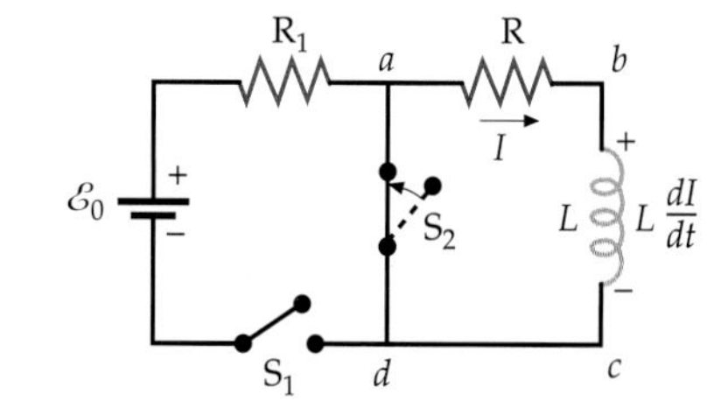

Figure 30-22 An *RL* circuit with two switches so that the battery can be removed from the circuit. After the current in the inductor reaches its maximum value with S_1 closed, S_2 is closed and S_1 is opened.

$$-IR - L\frac{dI}{dt} = 0$$

or

$$\frac{dI}{dt} = -\frac{R}{L}I \qquad \text{30-23}$$

optional

Equation 30-23 is of the same form as Equation 26-28 for the discharge of a capacitor. It can be solved by direct integration. We will omit the details and merely state the solution. The current I is given by

$$I = I_0 e^{-Rt/L} = I_0 e^{-t/\tau} \qquad \text{30-24}$$

where $\tau = L/R$ is the time constant. Figure 30-23 shows the current as a function of time.

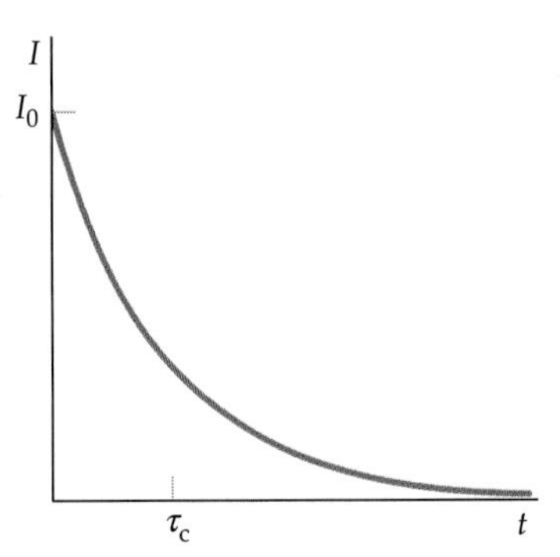

Figure 30-23 Current versus time for the circuit in Figure 30-22. The current decreases exponentially with time.

Exercise What is the time constant of a circuit of resistance 85 Ω and inductance 6 mH? (*Answer* 70.6 μs)

Example 30-12

Find the total heat produced in the resistor R in Figure 30-22 when the current in the inductor decreases from its initial value of I_0 to 0.

Picture the Problem The rate of heat production I^2R varies with time so we must integrate.

1. The rate of heat production is I^2R: $\quad P = \dfrac{dW}{dt} = I^2R$

2. The total energy dissipated as heat in the resistor is the integral of dW from $t = 0$ to $t = \infty$: $\quad W = \displaystyle\int_0^\infty I^2R\,dt$

3. The current I is given by Equation 30-24: $\quad I = I_0 e^{-Rt/L}$

4. Substitute this current into the integral: $\quad W = \displaystyle\int_0^\infty I^2R\,dt = \int_0^\infty I_0^2 e^{-2Rt/L} R\,dt$

5. The integration can be done by substituting $x = 2Rt/L$:

$$x = 2Rt/L$$

$$dt = \frac{L}{2R}dx$$

$$W = I_0^2 R \frac{L}{2R}\int_0^\infty e^{-x}\,dx$$

6. The integral in step 5 is 1. Then: $\quad W = \frac{1}{2}LI_0^2$

Remark The total heat produced equals the energy $\frac{1}{2}LI_0^2$ originally stored in the inductor.

Example 30-13

For the circuit shown in Figure 30-24, find the currents I_1, I_2, and I_3 (*a*) immediately after switch S is closed and (*b*) a long time after switch S has been closed. After the switch has been closed for a long time, it is opened. Find the three currents (*c*) immediately after switch S is opened and (*d*) a long time after switch S was opened.

Figure 30-24

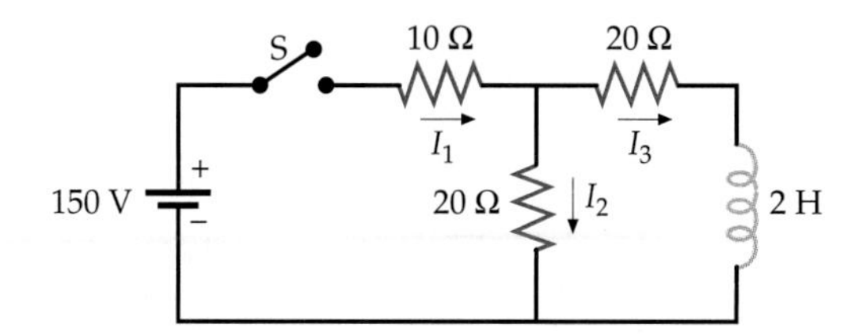

Picture the Problem (*a*) We simplify our calculations by using the fact that the current in an inductor cannot change abruptly. Thus, the current in the inductor must be zero just after the switch is closed, because it is zero before. (*b*) When the current reaches its final value, dI/dt equals zero, so there is no potential drop across the inductor. The inductor thus acts like a short circuit, that is, like a wire with zero resistance. (*c*) Immediately after the switch is opened, the current in the inductor is the same as it was before. (*d*) A long time after the switch is opened, all the currents must be zero.

(*a*) The current through the inductor is zero, just as it was before the switch was closed. The current in the left loop equals the emf divided by the equivalent resistance of the two resistors in series:

$$I_1 = I_2 = \frac{150\text{ V}}{10\ \Omega + 20\ \Omega} = 5\text{ A}$$

$$I_3 = 0$$

(*b*)1. After a long time, the current is steady and the inductor acts like a short circuit, so we have two 20-Ω resistors in parallel, with the combination in series with the 10-Ω resistor. Redraw the circuit (Figure 30-25) and find the equivalent resistance of the parallel resistors:

$$\frac{1}{R_{eq}} = \frac{1}{20} + \frac{1}{20} = \frac{2}{20} = \frac{1}{10}$$

$$R_{eq} = 10\ \Omega$$

Figure 30-25

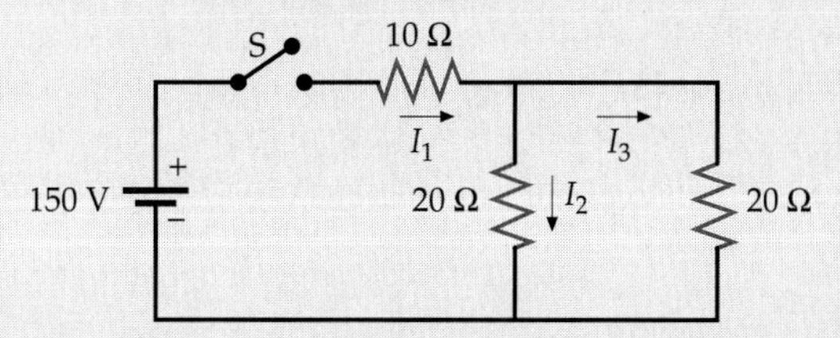

2. Find the current I_1:

$$I_1 = \frac{150\text{ V}}{10\ \Omega + 10\ \Omega} = 7.5\text{ A}$$

3. Find I_2 and I_3 using the fact that the current in the parallel 20-Ω resistors must be equal:

$$I_2 = I_3 = 3.75\text{ A}$$

(*c*) When the switch is reopened, I_1 must be zero, and the current in the inductor I_3 momentarily remains the same at 3.75 A. Then I_2 must equal $-I_3$:

$$I_1 = 0$$

$$I_3 = 3.75\text{ A}$$

$$I_2 = -I_3 = -3.75\text{ A}$$

(*d*) A long time after the switch is opened, all the currents must be zero.

$$I_1 = I_2 = I_3 = 0$$

30-9 Magnetic Properties of Superconductors

Superconductors have resistivities of zero below a critical temperature I_c, which varies from material to material. In the presence of a magnetic field B, the critical temperature is lower than it is when there is no field. As the magnetic field increases, the critical temperature decreases. If the magnetic field is greater than some critical field B_c, superconductivity does not exist at any temperature.

Meissner Effect

When a superconductor is cooled below the critical temperature in an external magnetic field, the magnetic field lines are expelled from the superconductor so the magnetic field inside the superconductor is zero (Figure 30-26). This effect was discovered by Meissner and Ochsenfeld in 1933 and is now known as the **Meissner effect.** The mechanism by which the magnetic field lines are expelled is an induced superconducting current on the surface of the superconductor. The magnetic levitation shown on page 656 results from the repulsion between the permanent magnet producing the external field and the magnetic field produced by the currents induced in the superconductor. Only certain superconductors called **type I superconductors** exhibit the complete Meissner effect. Figure 30-27*a* shows a plot of the magnetization M times μ_0 versus the applied magnetic field B_{app} for a type I superconductor. For a magnetic field less than the critical field B_c, the magnetic field $\mu_0 M$ induced in the superconductor is equal and opposite to the external magnetic field. The values of B_c for type I superconductors are always too small for such materials to be useful in the coils of a superconducting magnet.

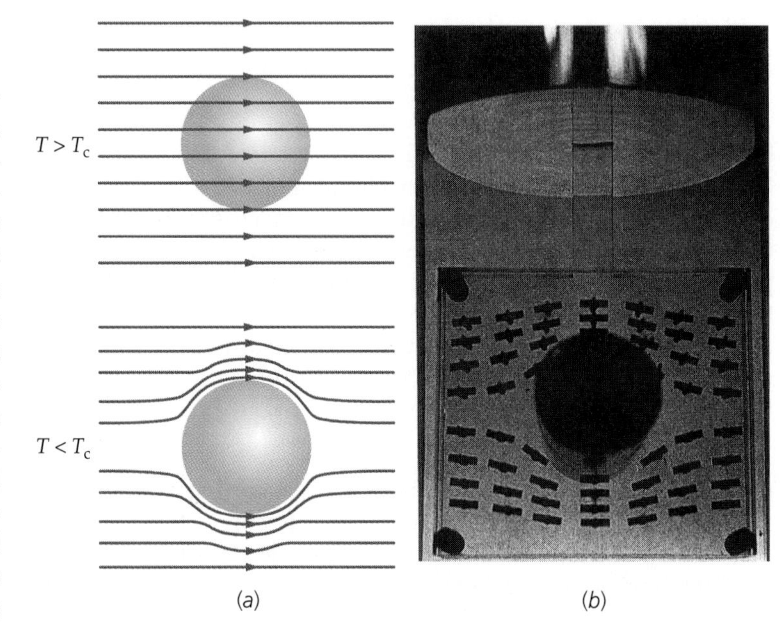

Figure 30-26 (*a*) The Meissner effect in a superconducting sphere cooled in a constant applied magnetic field. As the temperature drops below the critical temperature T_c, the magnetic field lines are expelled from the sphere. (*b*) Demonstration of the Meissner effect. A superconducting tin cylinder is situated with its axis perpendicular to a horizontal magnetic field. The directions of the field lines are indicated by weakly magnetized compass needles mounted in a Lucite sandwich so that they are free to turn.

Other materials, known as **type II superconductors,** have a magnetization curve similar to that in Figure 30-27*b*. Such materials are usually alloys or metals that have large resistivities in the normal state. Type II superconductors exhibit the electrical properties of superconductors except for the Meissner effect up to the critical field B_{c2}, which may be several hundred times the

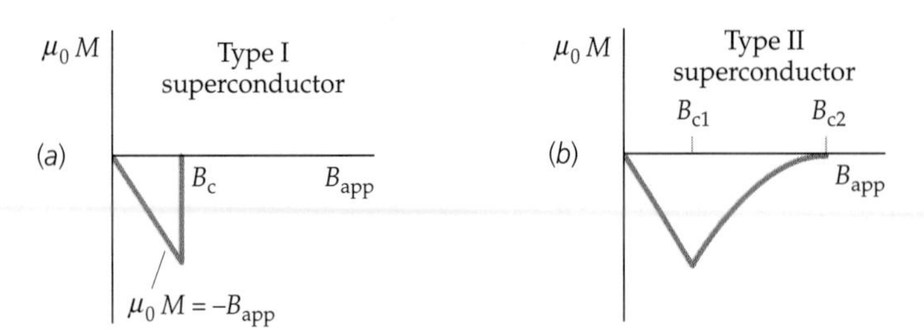

Figure 30-27 Plots of μ_0 times the magnetization M versus applied magnetic field for type I and type II superconductors. (*a*) In a type I superconductor, the resultant magnetic field is zero below a critical applied field B_c because the field due to induced currents on the surface of the superconductor exactly cancels the applied field. Above the critical field, the material is a normal conductor and the magnetization is too small to be seen on this scale. (*b*) In a type II superconductor, the magnetic field starts to penetrate the superconductor at a field B_{c1}, but the material remains superconducting up to a field B_{c2}, after which it becomes a normal conductor.

typical values of critical fields for type I superconductors. For example, the alloy Nb_3Ge has a critical field $B_{c2} = 34$ T. Such materials can be used for high-field superconducting magnets. Below the critical field B_{c1}, the behavior of a type II superconductor is the same as that of a type I superconductor. In the region between fields B_{c1} and B_{c2}, the superconductor is said to be in a vortex state.

optional

Flux Quantization

Consider a superconducting ring of area A carrying a current. There can be a magnetic flux $\phi_m = B_n A$ through the ring due to the current in the ring and due also perhaps to other currents external to the ring. According to Faraday's law, if the flux changes, an emf will be induced in the ring that is proportional to the rate of change of the flux. But there can be no emf in a superconducting ring because it is resistanceless. The flux through the ring is thus frozen and cannot change.

Another effect, which results from the quantum-mechanical treatment of superconductivity, is that the total flux through the loop is quantized and is given by

$$\phi_m = n\frac{h}{2e}, \qquad n = 1, 2, 3, \ldots \qquad 30\text{-}25$$

The smallest unit of flux, called a **fluxon,** is

$$\phi_0 = \frac{h}{2e} = 2.0678 \times 10^{-15}\ \text{T}\cdot\text{m}^2 \qquad 30\text{-}26$$

Summary

1. Faraday's law and Lenz' law are fundamental laws of physics.
2. Inductance is a property of a circuit element that relates the flux through the element to the current.

Topic	Remarks and Relevant Equations	
1. Magnetic Flux ϕ_m		
General definition	$\phi_m = \int_S \vec{B}\cdot\hat{n}\, dA$	30-1
Constant field, coil of n turns	$\phi_m = NBA\cos\theta$	30-3
Units	$1\ \text{Wb} = 1\ \text{T}\cdot\text{m}^2$	30-2
Due to current in a circuit	$\phi_m = LI$	30-7
Due to current in two circuits	$\phi_{m1} = L_1I_1 + MI_2 \qquad \phi_{m2} = L_2I_2 + MI_1$	30-11
Quantization (optional)	$\phi_m = n\frac{h}{2e}, \quad n = 1, 2, 3, \ldots$	30-25
Fluxon (optional)	$\phi_0 = \frac{h}{2e} = 2.0678 \times 10^{-15}\ \text{T}\cdot\text{m}^2$	30-26

2.	**emf**		
	Definition	$\mathcal{E} = \oint_C \vec{E} \cdot d\vec{\ell}$	**30-4**
	Motional (optional)	$\lvert\mathcal{E}\rvert = vB\ell$	**30-6**
	Self-induced (Back emf)	$\mathcal{E} = -L\frac{dI}{dt}$	**30-10**
3.	**Faraday's Law**	$\mathcal{E} = \oint_C \vec{E} \cdot d\vec{\ell} = -\frac{d\phi_m}{dt}$	**30-5**
4.	**Lenz's Law**	The induced emf and induced current are in such a direction as to oppose the change that produces them.	
5.	**Inductance**		
	Self-inductance	$\phi_m = LI$	**30-7**
	Self-inductance of a solenoid	$L = \mu_0 n^2 A\ell$	**30-9**
	Mutual inductance	$\phi_{m1} = MI_2 \qquad \phi_{m2} = MI_1$	
	Units	$1\ \mathrm{H} = 1\frac{\mathrm{Wb}}{\mathrm{A}} = 1\frac{\mathrm{T \cdot m^2}}{\mathrm{A}}$ $\mu_0 = 4\pi \times 10^{-7}\ \mathrm{H/m}$	
6.	**Magnetic Energy**		
	Energy in an inductor	$U_m = \frac{1}{2}LI^2$	**30-16**
	Energy density in a magnetic field	$u_m = \frac{B^2}{2\mu_0}$	**30-17**
7.	***RL* Circuits** (optional)	In an *RL* circuit, which consists of a resistance R, an inductance L, and a battery of emf $\mathcal{E}_0$ in series, the current does not reach its maximum value I instantaneously but rather takes some time to build up. If the current is initially zero, its value at some later time t is given by	
	Current	$I = \frac{\mathcal{E}_0}{R}(1 - e^{-Rt/L}) = \frac{\mathcal{E}_0}{R}(1 - e^{-t/\tau}) = I_f(1 - e^{-t/\tau})$	**30-21**
	Time constant τ	$\tau = \frac{L}{R}$	**30-22**

Problem-Solving Guide

Begin by drawing a neat diagram that includes the important features of the problem. Show the direction of $\vec{B}$ in order to correctly calculate the magnetic flux. When B changes with time, it is helpful to include in your drawing the direction of the induced electric field $\vec{E}$. In circuit problems, be sure to show the direction of the current I and to show which side of the inductor is at higher potential.

Summary of Worked Examples

Type of Calculation	Procedure and Relevant Examples
1. Flux	
Find the magnetic flux through a coil of N turns.	Use $\phi_m = NBA \cos \theta$ if B is uniform, or $\phi_m = \int_S NB_n\, dA$ if B varies. **Examples 30-1, 30-2**
2. Induced emf	
Find the induced electric field $\vec{E}$, or the induced emf $\mathcal{E}$ due to a changing magnetic flux.	Use Faraday's law $\mathcal{E} = \oint_C \vec{E} \cdot d\vec{\ell} = -\frac{d\phi_m}{dt}$ **Examples 30-2, 30-3, 30-4, 30-5, 30-6**
Find the emf produced by motion through a magnetic field.	The magnitude of the motional emf is $\mathcal{E} = vB\ell$. The direction is given by Lenz's law. **Examples 30-7, 30-8**
3. Inductance	
Calculate the inductance of a solenoid.	Use $L = \mu_0 n^2 A\ell$. **Example 30-9**
4. Magnetic Energy	
Calculate the magnetic energy and the magnetic energy density.	The magnetic energy is $U_m = \frac{1}{2}LI^2$ and the magnetic energy density is $u_m = \frac{B^2}{2\mu_0}$. **Example 30-10**
5. *RL* Circuits	
Find the current in a simple *RL* circuit.	Use $I = I_f(1 - e^{-t/\tau})$ for a switch closed at $t = 0$ and $I = I_0 e^{-t/\tau}$ for a switch opened at $t = 0$. In each case, $\tau = L/R$. **Example 30-11**
Determine the current in a circuit containing inductors immediately after a switch is opened or closed, or long after a switch is closed.	Immediately after a switch is opened or closed the current in an inductor remains the same. Long after a switch is closed, an inductor acts like a wire with no resistance. **Example 30-11**

Problems

Conceptual Problems

Problems from Optional and Exploring sections

In a few problems, you are given more data than you actually need; in a few other problems, you are required to supply data from your general knowledge, outside sources, or informed estimates.

• Single-concept, single-step, relatively easy

•• Intermediate-level, may require synthesis of concepts

••• Challenging, for advanced students

Magnetic Flux

1 • A uniform magnetic field of magnitude 2000 G is parallel to the x axis. A square coil of side 5 cm has a single turn and makes an angle θ with the z axis as shown in Figure 30-28. Find the magnetic flux through the coil when (*a*) $\theta = 0°$, (*b*) $\theta = 30°$, (*c*) $\theta = 60°$, and (*d*) $\theta = 90°$.

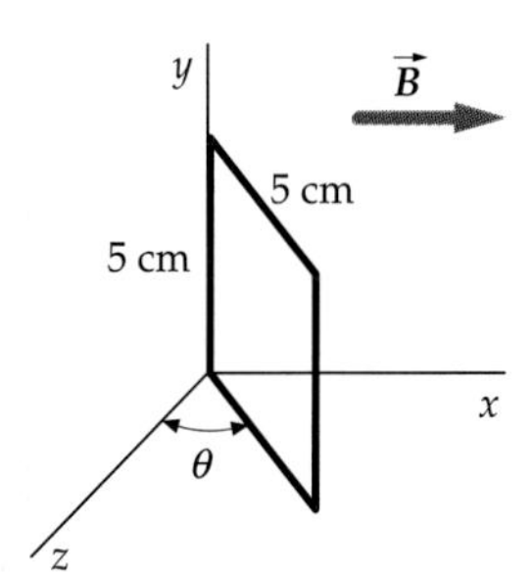

Figure 30-28 Problem 1

2 • A circular coil has 25 turns and a radius of 5 cm. It is at the equator, where the earth's magnetic field is 0.7 G north. Find the magnetic flux through the coil when its plane

is (*a*) horizontal, (*b*) vertical with its axis pointing north, (*c*) vertical with its axis pointing east, and (*d*) vertical with its axis making an angle of 30° with north.

3 • A magnetic field of 1.2 T is perpendicular to a square coil of 14 turns. The length of each side of the coil is 5 cm. (*a*) Find the magnetic flux through the coil. (*b*) Find the magnetic flux through the coil if the magnetic field makes an angle of 60° with the normal to the plane of the coil.

4 • A circular coil of radius 3.0 cm has its plane perpendicular to a magnetic field of 400 G. (*a*) What is the magnetic flux through the coil if the coil has 75 turns? (*b*) How many turns must the coil have for the flux to be 0.015 Wb?

5 • A uniform magnetic field $\vec{B}$ is perpendicular to the base of a hemisphere of radius R. Calculate the magnetic flux through the spherical surface of the hemisphere.

6 •• Find the magnetic flux through a solenoid of length 25 cm, radius 1 cm, and 400 turns that carries a current of 3 A.

7 •• Work Problem 6 for an 800-turn solenoid of length 30 cm, and radius 2 cm, carrying a current of 2 A.

8 •• A circular coil of 15 turns of radius 4 cm is in a uniform magnetic field of 4000 G in the positive x direction. Find the flux through the coil when the unit vector perpendicular to the plane of the coil is (*a*) $\hat{n} = \hat{i}$, (*b*) $\hat{n} = \hat{j}$, (*c*) $\hat{n} = (\hat{i} + \hat{j})/\sqrt{2}$, (*d*) $\hat{n} = \hat{k}$, and (*e*) $\hat{n} = 0.6\hat{i} + 0.8\hat{j}$.

9 •• A solenoid has n turns per unit length, radius R_1, and carries a current I. (*a*) A large circular loop of radius $R_2 > R_1$ and N turns encircles the solenoid at a point far away from the ends of the solenoid. Find the magnetic flux through the loop. (*b*) A small circular loop of N turns and radius $R_3 < R_1$ is completely inside the solenoid, far from its ends, with its axis parallel to that of the solenoid. Find the magnetic flux through this small loop.

10 •• A long, straight wire carries a current I. A rectangular loop with two sides parallel to the straight wire has sides a and b with its near side a distance d from the straight wire, as shown in Figure 30-29. (*a*) Compute the magnetic flux through the rectangular loop. (*Hint:* Calculate the flux through a strip of area $dA = b\,dx$ and integrate from $x = d$ to $x = d + a$.) (*b*) Evaluate your answer for a = 5 cm, b = 10 cm, d = 2 cm, and I = 20 A.

Figure 30-29
Problems 10 and 46

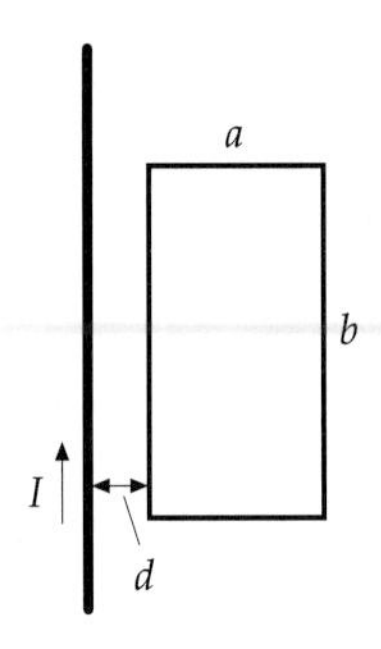

11 ••• A long, cylindrical conductor of radius R carries a current I that is uniformly distributed over its cross-sectional area. Find the magnetic flux per unit length through the area indicated in Figure 30-30.

Figure 30-30 Problem 11

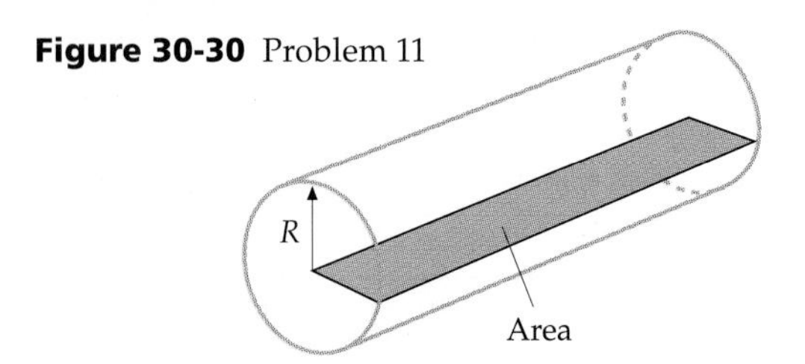

12 ••• A rectangular coil in the plane of the page has dimensions a and b. A long wire that carries a current I is placed directly above the coil (Figure 30-31). (*a*) Obtain an expression for the magnetic flux through the coil as a function of x for $0 \le x \le 2b$. (*b*) For what value of x is flux through the coil a maximum? For what value of x is the flux a minimum?

Figure 30-31 Problem 12

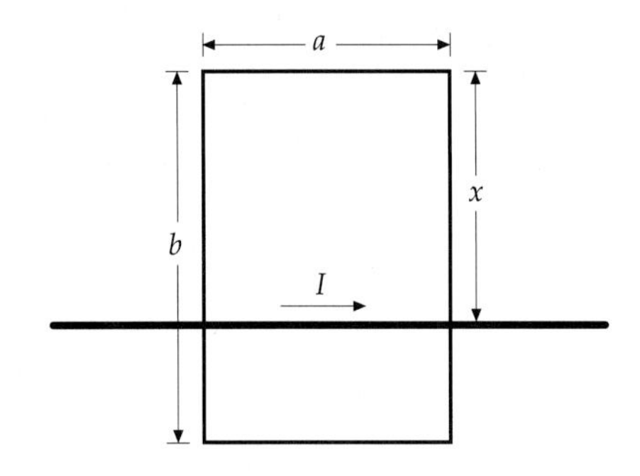

Induced emf and Faraday's Law

13 • A conducting loop lies in the plane of this page and carries a clockwise induced current. Which of the following statements could be true?

(*a*) A constant magnetic field is directed into the page.
(*b*) A constant magnetic field is directed out of the page.
(*c*) An increasing magnetic field is directed into the page.
(*d*) A decreasing magnetic field is directed into the page.
(*e*) A decreasing magnetic field is directed out of the page.

14 • A uniform magnetic field $\vec{B}$ is established perpendicular to the plane of a loop of radius 5.0 cm, resistance 0.4 Ω, and negligible self-inductance. The magnitude of $\vec{B}$ is increasing at a rate of 40 mT/s. Find (*a*) the induced emf in the loop, (*b*) the induced current in the loop, and (*c*) the rate of Joule heating in the loop.

15 • The flux through a loop is given by $\phi_m = (t^2 - 4t) \times 10^{-1}$ Wb, where t is in seconds. (*a*) Find the induced emf $\mathcal{E}$ as a function of time. (*b*) Find both ϕ_m and $\mathcal{E}$ at $t = 0$, $t = 2$ s, $t = 4$ s, and $t = 6$ s.

16 • (*a*) For the flux given in Problem 15, sketch graphs of ϕ_m and $\mathcal{E}$ versus t. (*b*) At what time is the flux minimum? What is the emf at this time? (*c*) At what times is the flux zero? What is the emf at these times?

17 • The magnetic field in Problem 4 is steadily reduced to zero in 0.8 s. What is the magnitude of the emf induced in the coil of part (*b*)?

18 • A solenoid of length 25 cm and radius 0.8 cm with 400 turns is in an external magnetic field of 600 G that makes an angle of 50° with the axis of the solenoid. (*a*) Find the magnetic flux through the solenoid. (*b*) Find the magnitude of the emf induced in the solenoid if the external magnetic field is reduced to zero in 1.4 s.

19 •• A 100-turn circular coil has a diameter of 2.0 cm and resistance of 50 Ω. The plane of the coil is perpendicular to a uniform magnetic field of magnitude 1.0 T. The direction of the field is suddenly reversed. (*a*) Find the total charge that passes through the coil. If the reversal takes 0.1 s, find (*b*) the average current in the coil and (*c*) the average emf in the coil.

20 •• At the equator, a 1000-turn coil with a cross-sectional area of 300 cm^2 and a resistance of 15.0 Ω is aligned with its plane perpendicular to the earth's magnetic field of 0.7 G. If the coil is flipped over, how much charge flows through it?

21 •• A circular coil of 300 turns and radius 5.0 cm is connected to a current integrator. The total resistance of the circuit is 20 Ω. The plane of the coil is originally aligned perpendicular to the earth's magnetic field at some point. When the coil is rotated through 90°, the charge that passes through the current integrator is measured to be 9.4 μC. Calculate the magnitude of the earth's magnetic field at that point.

22 •• An elastic circular conducting loop is expanding at a constant rate so that its radius is given by $R = R_0 + vt$. The loop is in a region of constant magnetic field perpendicular to the loop. What is the emf generated in the expanding loop? Neglect possible effects of self-inductance.

23 •• The wire in Problem 12 is placed at $x = b/4$. (*a*) Obtain an expression for the emf induced in the coil if the current varies with time according to $I = 2t$. (*b*) If $a = 1.5$ m and $b = 2.5$ m, what should be the resistance of the coil so that the induced current is 0.1 A? What is the direction of this current?

24 •• Repeat Problem 23 if the wire is placed at $x = b/3$.

Lenz's Law

25 • Give the direction of the induced current in the circuit on the right in Figure 30-32 when the resistance in the circuit on the left is suddenly (*a*) increased and (*b*) decreased.

Figure 30-32 Problem 25

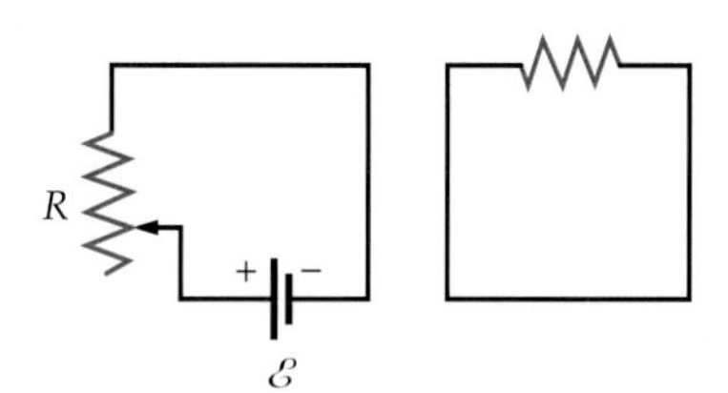

26 •• The two circular loops in Figure 30-33 have their planes parallel to each other. As viewed from A toward B, there is a counterclockwise current in loop A. Give the direction of the current in loop B and state whether the loops attract or repel each other if the current in loop A is (*a*) increasing and (*b*) decreasing.

Figure 30-33 Problem 26

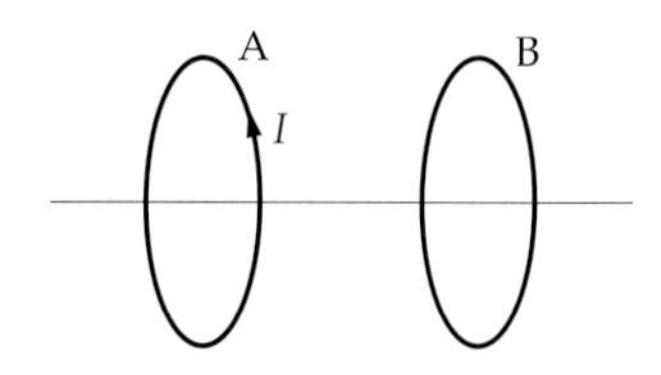

27 •• A bar magnet moves with constant velocity along the axis of a loop as shown in Figure 30-34. (*a*) Make a qualitative graph of the flux ϕ_m through the loop as a function of time. Indicate the time t_1 when the magnet is halfway through the loop. (*b*) Sketch a graph of the current *I* in the loop versus time, choosing *I* to be positive when it is counterclockwise as viewed from the left.

Figure 30-34 Problem 27

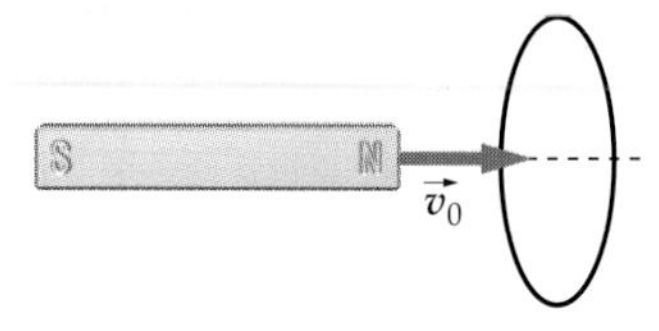

28 •• A bar magnet is mounted on the end of a coiled spring in such a way that it moves with simple harmonic motion along the axis of a loop as shown in Figure 30-35. (*a*) Make a qualitative graph of the flux ϕ_m through the loop as a function of time. Indicate the time t_1 when the magnet is halfway through the loop. (*b*) Sketch the current *I* in the loop versus time, choosing *I* to be positive when it is counterclockwise as viewed from above.

Figure 30-35 Problem 28

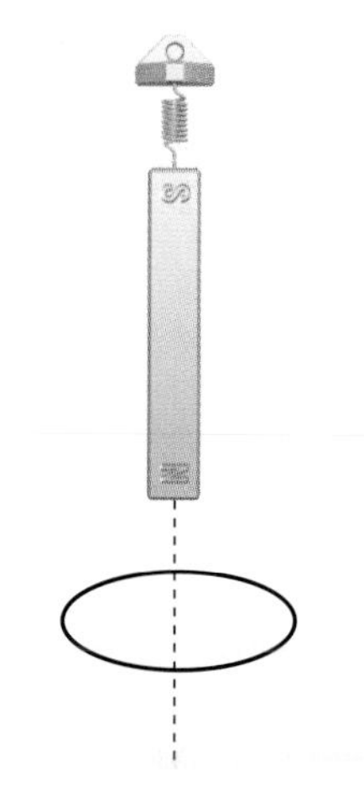

Motional emf (optional)

29 • A rod 30 cm long moves at 8 m/s in a plane perpendicular to a magnetic field of 500 G. The velocity of the rod is perpendicular to its length. Find (*a*) the magnetic force on an electron in the rod, (*b*) the electrostatic field $\vec{E}$ in the rod, and (*c*) the potential difference *V* between the ends of the rod.

30 • Find the speed of the rod in Problem 29 if the potential difference between the ends is 6 V.

31 • In Figure 30-14, let *B* be 0.8 T, $v = 10.0$ m/s, $\ell = 20$ cm, and $R = 2$ Ω. Find (*a*) the induced emf in the circuit, (*b*) the current in the circuit, and (*c*) the force needed to move the rod with constant velocity assuming negligible friction. Find (*d*) the power input by the force found in part (*c*), and (*e*) the rate of Joule heat production I^2R.

32 • Work Problem 31 for $B = 1.5$ T, $v = 6$ m/s, $\ell = 40$ cm, and $R = 1.2$ Ω.

33 •• A 10-cm by 5-cm rectangular loop with resistance 2.5 Ω is pulled through a region of uniform magnetic field $B = 1.7$ T (Figure 30-36) with constant speed $v = 2.4$ cm/s. The front of the loop enters the region of the magnetic field at time $t = 0$. (*a*) Find and graph the flux through the loop as a function of time. (*b*) Find and graph the induced emf and the current in the loop as functions of time. Neglect any self-inductance of the loop and extend your graphs from $t = 0$ to $t = 16$ s.

Figure 30-36 Problem 33

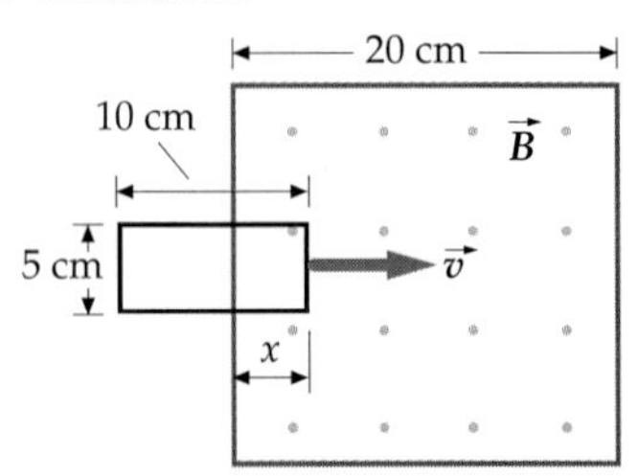

34 •• A uniform magnetic field of magnitude 1.2 T is in the *z* direction. A conducting rod of length 15 cm lies parallel to the *y* axis and oscillates in the *x* direction with displacement given by $x = (2 \text{ cm}) \cos 120\pi t$. What is the emf induced in the rod?

35 •• In Figure 30-37, the rod has a resistance *R* and the rails are horizontal and have negligible resistance. A battery of emf $\mathcal{E}$ and negligible internal resistance is connected between points *a* and *b* such that the current in the rod is downward. The rod is placed at rest at $t = 0$. (*a*) Find the force on the rod as a function of the speed *v* and write Newton's second law for the rod when it has speed *v*. (*b*) Show that the rod will approach a terminal speed and find an expression for it. (*c*) What is the current when the rod moves at its terminal speed?

Figure 30-37
Problems 35 and 38

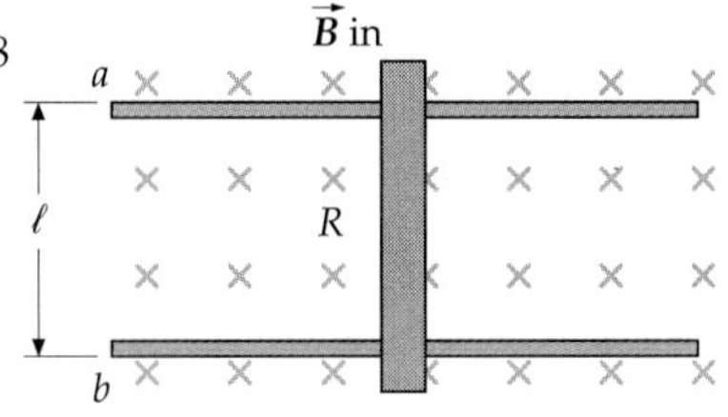

36 •• In Example 30-8, find the total energy dissipated in the resistance and show that it is equal to $\frac{1}{2}mv_0^2$.

37 •• Find the total distance traveled by the rod in Example 30-8.

38 •• In Figure 30-37, the rod has a resistance *R* and the rails have negligible resistance. A capacitor with charge Q_0 and capacitance *C* is connected between points *a* and *b* such that the current in the rod is downward. The rod is placed at rest at $t = 0$. (*a*) Write the equation of motion for the rod on the rails. (*b*) Show that the terminal speed of the rod down the rails is related to the final charge on the capacitor.

39 •• In Figure 30-38, a conducting rod of mass *m* and negligible resistance is free to slide without friction along two parallel rails of negligible resistance separated by a distance ℓ and connected by a resistance *R*. The rails are attached to a long, inclined plane that makes an angle θ with the horizontal. There is a magnetic field *B* directed upward. (*a*) Show that there is a retarding force directed up the incline given by $F = (B^2\ell^2 v \cos^2 \theta)/R$. (*b*) Show that the terminal speed of the rod is $v_t = (mgR \sin \theta)/(B^2\ell^2 \cos^2 \theta)$.

Figure 30-38 Problems 39, 42, and 43

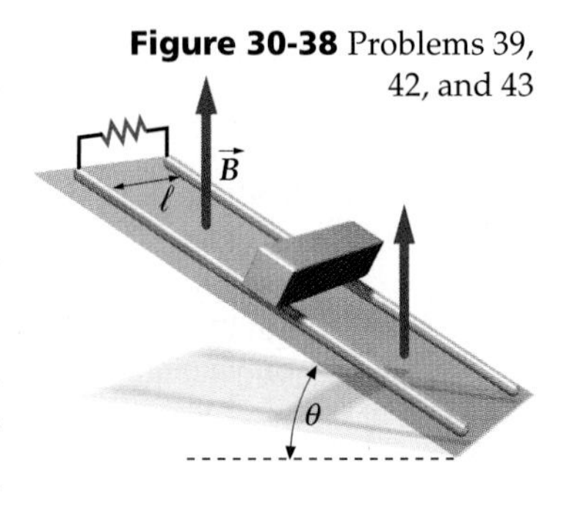

40 •• A simple pendulum has a wire of length ℓ supporting a metal ball of mass *m*. The wire has negligible mass and moves in a uniform horizontal magnetic field *B*. This pendulum executes simple harmonic motion having angular amplitude θ_0. What is the emf generated along the wire?

41 •• A wire lies along the *z* axis and carries current $I = 20$ A in the positive *z* direction. A small conducting sphere of radius $R = 2$ cm is initially at rest on the *y* axis at a distance $h = 45$ m above the wire. The sphere is dropped at time $t = 0$. (*a*) What is the electric field at the center of the sphere at $t = 3$ s? Assume that the only magnetic field is that produced by the wire. (*b*) What is the voltage across the sphere at $t = 3$ s?

42 •• In Figure 30-38, let $\theta = 30°$, $m = 0.4$ kg, $\ell = 15$ m, and $R = 2.0$ Ω. The rod starts from rest at the top of the inclined plane at $t = 0$. The rails have negligible resistance. There is a constant, vertically directed magnetic field of magnitude $B = 1.2$ T. (*a*) Find the emf induced in the rod as a function of its velocity down the rails. (*b*) Write Newton's law of motion for the rod; show that the rod will approach a terminal speed and determine its value.

43 •• When the rod of Problem 42 moves at its terminal speed, what is the power dissipated in the resistor? What is the rate of change of the potential energy of the rod?

44 ••• A solid conducting cylinder of radius 0.1 m and mass of 4 kg rests on horizontal conducting rails (Figure 30-39). The rails, separated by a distance $a = 0.4$ m, have a rough surface so the cylinder rolls rather than slides. A 12-V battery is connected to the rails as shown. The only significant resistance in the circuit is the contact resistance of 6 Ω between the cylinder and rails. The system is in a uniform vertical magnetic field. The cylinder is initially at rest next to the battery. (*a*) What must be the magnitude and direction of $\vec{B}$ so that the cylinder has an initial acceleration of 0.1 m/s^2 to the right? (*b*) Find the force on the cylinder as a function of its speed *v*. (*c*) Find the terminal velocity of the cylinder. (*d*) What is the kinetic energy of the cylinder when it has reached its terminal velocity? (Neglect the magnetic field due to the current in the battery–rails–cylinder loop and assume that the current density in the cylinder is uniform.)

Figure 30-39 Problem 44

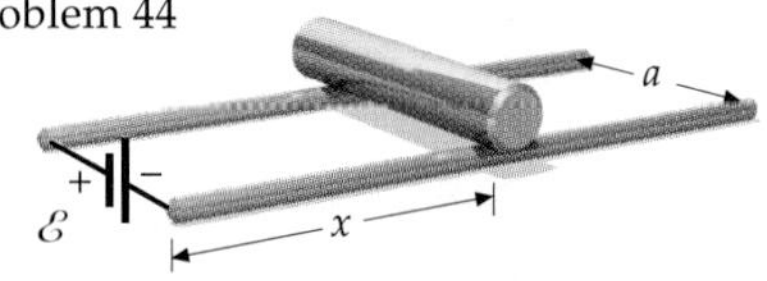

45 ••• A rod of length ℓ is perpendicular to a long wire carrying current I, as shown in Figure 30-40. The near end of the rod is a distance d away from the wire. The rod moves with a speed v in the direction of the current I. (a) Show that the potential difference between the ends of the rod is given by

$$V = \frac{\mu_0 I}{2\pi} v \ln\frac{d + \ell}{d}$$

(b) Use Faraday's law to obtain this result by considering the flux through a rectangular area $A = \ell vt$ swept out by the rod.

Figure 30-40 Problem 45

46 ••• The loop in Problem 10 moves away from the wire with a constant speed v. At time $t = 0$, the left side of the loop is a distance d from the long straight wire. (a) Compute the emf in the loop by computing the motional emf in each segment of the loop that is parallel to the long wire. Explain why you can neglect the emf in the segments that are perpendicular to the wire. (b) Compute the emf in the loop by first computing the flux through the loop as a function of time and then using $\mathcal{E} = -d\phi_m/dt$ and compare your answer with that obtained in part (a).

47 ••• A conducting rod of length ℓ rotates at constant angular velocity about one end, in a plane perpendicular to a uniform magnetic field B (Figure 30-41). (a) Show that the magnetic force on a charge q at a distance r from the pivot is $Bqr\omega$. (b) Show that the potential difference between the ends of the rod is $V = \frac{1}{2}B\omega\ell^2$. ($c$) Draw any radial line in the plane from which to measure $\theta = \omega t$. Show that the area of the pie-shaped region between the reference line and the rod is $A = \frac{1}{2}\ell^2\theta$. Compute the flux through this area, and show that $\mathcal{E} = \frac{1}{2}B\omega\ell^2$ follows when Faraday's law is applied to this area.

Figure 30-41 Problem 47

Inductance

48 • How would the self-inductance of a solenoid be changed if (a) the same length of wire were wound onto a cylinder of the same diameter but twice the length; (b) twice as much wire were wound onto the same cylinder; and (c) the same length of wire were wound onto a cylinder of the same length but twice the diameter?

49 • A coil with a self-inductance of 8.0 H carries a current of 3 A that is changing at a rate of 200 A/s. Find (a) the magnetic flux through the coil and (b) the induced emf in the coil.

50 • A coil with self-inductance L carries a current I, given by $I = I_0 \sin 2\pi ft$. Find and graph the flux ϕ_m and the self-induced emf as functions of time.

51 •• A solenoid has a length of 25 cm, a radius of 1 cm, and 400 turns, and carries a 3-A current. Find (a) B on the axis at the center of the solenoid; (b) the flux through the solenoid, assuming B to be uniform; (c) the self-inductance of the solenoid; and (d) the induced emf in the solenoid when the current changes at 150 A/s.

52 •• Two solenoids of radii 2 cm and 5 cm are coaxial. They are each 25 cm long and have 300 and 1000 turns, respectively. Find their mutual inductance.

53 •• A long, insulated wire with a resistance of 18 Ω/m is to be used to construct a resistor. First, the wire is bent in half, and then the doubled wire is wound in a cylindrical form as shown in Figure 30-42. The diameter of the cylindrical form is 2 cm, its length is 25 cm, and the total length of wire is 9 m. Find the resistance and inductance of this wire-wound resistor.

Figure 30-42 Problem 53

54 •• Figure 30-43 shows two long solenoids each with 2000 turns of wire. The outer solenoid is 20 cm long and has a diameter of 2 cm. The inner solenoid is 10 cm long and has a diameter of 1 cm. Find the effective inductance of this arrangement.

Figure 30-43 Problem 54

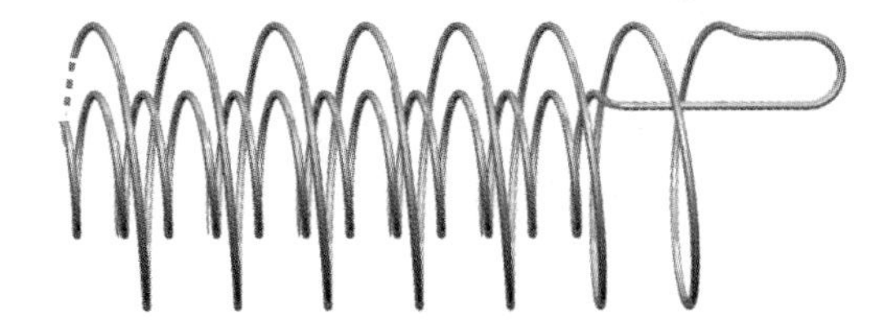

55 ••• In Figure 30-44, circuit 2 has a total resistance of 300 Ω. A total charge of 2×10^{-4} C flows through the galvanometer in circuit 2 when switch S in circuit 1 is closed. After a long time, the current in circuit 1 is 5 A. What is the mutual inductance between the two coils?

Figure 30-44 Problem 55

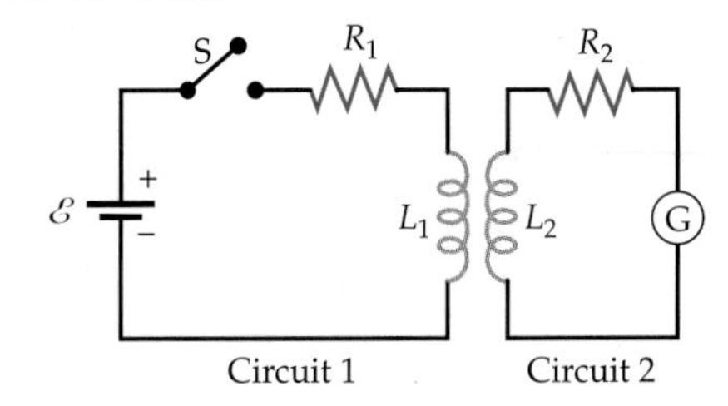

56 ••• Show that the inductance of a toroid of rectangular cross section as shown in Figure 30-45 is given by

$$L = \frac{\mu_0 N^2 H \ln(b/a)}{2\pi}$$

where N is the total number of turns, a is the inside radius, b is the outside radius, and H is the height of the toroid.

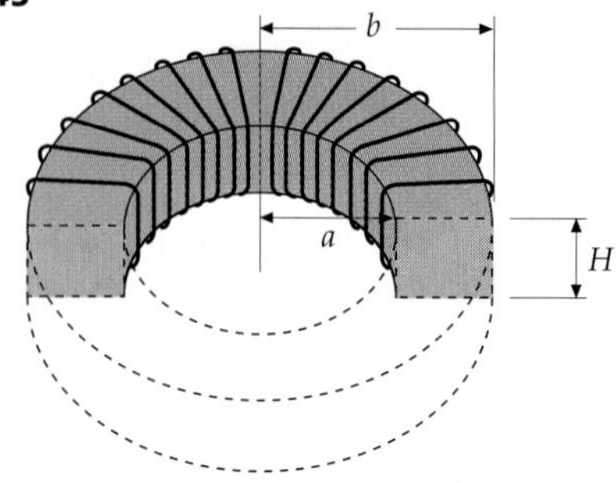

Figure 30-45
Problem 56

Magnetic Energy

57 • If the current through an inductor were doubled, the energy stored in the inductor would be

(*a*) the same. (*b*) doubled.
(*c*) quadrupled. (*d*) halved.
(*e*) quartered.

58 • A coil with a self-inductance of 2.0 H and a resistance of 12.0 Ω is connected across a 24-V battery of negligible internal resistance. (*a*) What is the final current? (*b*) How much energy is stored in the inductor when the final current is attained?

59 • Find (*a*) the magnetic energy, (*b*) the electric energy, and (*c*) the total energy in a volume of 1.0 m^3 in which there is an electric field of 10^4 V/m and a magnetic field of 5000 G.

60 •• In a plane electromagnetic wave such as a light wave, the magnitudes of the electric and magnetic fields are related by $E = cB$, where $c = 1/\sqrt{\epsilon_0 \mu_0}$ is the speed of light. Show that in this case the electric and the magnetic energy densities are equal.

61 •• A solenoid of 2000 turns, area 4 cm^2, and length 30 cm carries a current of 4.0 A. (*a*) Calculate the magnetic energy stored in the solenoid from $\frac{1}{2}LI^2$. (*b*) Divide your answer in part (*a*) by the volume of the solenoid to find the magnetic energy per unit volume in the solenoid. (*c*) Find B in the solenoid. (*d*) Compute the magnetic energy density from $u_m = B^2/2\mu_0$, and compare your answer with your result for part (*b*).

62 •• A long, cylindrical wire of radius a = 2 cm carries current I = 80 A uniformly distributed over its cross-sectional area. Find the magnetic energy per unit length within the wire.

63 •• A toroid of mean radius 25 cm and circular cross section of radius 2 cm is wound with a superconducting wire of length 1000 m that carries a current of 400 A. (*a*) What is the number of turns on the coil? (*b*) What is the magnetic field at the mean radius? (*c*) Assuming that B is constant over the area of the coil, calculate the magnetic energy density and the total energy stored in the toroid.

RL Circuits

64 • A coil of resistance 8.0 Ω and self-inductance 4.0 H is suddenly connected across a constant potential difference of 100 V. Let $t = 0$ be the time of connection, at which the current is zero. Find the current I and its rate of change dI/dt at times (*a*) $t = 0$, (*b*) $t = 0.1$ s, (*c*) $t = 0.5$ s, and (*d*) $t = 1.0$ s.

65 • The current in a coil with a self-inductance of 1 mH is 2.0 A at $t = 0$, when the coil is shorted through a resistor. The total resistance of the coil plus the resistor is 10.0 Ω. Find the current after (*a*) 0.5 ms and (*b*) 10 ms.

66 •• In the circuit of Figure 30-20, let $\mathcal{E}_0 = 12.0$ V, R = 3.0 Ω, and L = 0.6 H. The switch is closed at time $t = 0$. At time $t = 0.5$ s, find (*a*) the rate at which the battery supplies power, (*b*) the rate of Joule heating, and (*c*) the rate at which energy is being stored in the inductor.

67 •• Do Problem 66 for the times $t = 1$ s and $t = 100$ s.

68 •• The current in an *RL* circuit is zero at time $t = 0$ and increases to half its final value in 4.0 s. (*a*) What is the time constant of this circuit? (*b*) If the total resistance is 5 Ω, what is the self-inductance?

69 •• How many time constants must elapse before the current in an *RL* circuit that is initially zero reaches (*a*) 90%, (*b*) 99%, and (*c*) 99.9% of its final value?

70 •• A coil with inductance 4 mH and resistance 150 Ω is connected across a battery of emf 12 V and negligible internal resistance. (*a*) What is the initial rate of increase of the current? (*b*) What is the rate of increase when the current is half its final value? (*c*) What is the final current? (*d*) How long does it take for the current to reach 99% of its final value?

71 •• A large electromagnet has an inductance of 50 H and a resistance of 8.0 Ω. It is connected to a dc power source of 250 V. Find the time for the current to reach (*a*) 10 A and (*b*) 30 A.

72 •• Given the circuit shown in Figure 30-46, assume that switch S has been closed for a long time so that steady currents exist in the circuit and that the inductor L is made of superconducting wire so that its resistance may be considered to be zero. (*a*) Find the battery current, the current in the 100-Ω resistor, and the current through the inductor. (*b*) Find the initial voltage across the inductor when switch S is opened. (*c*) Give the current in the inductor as a function of time measured from the instant of opening switch S.

Figure 30-46
Problem 72

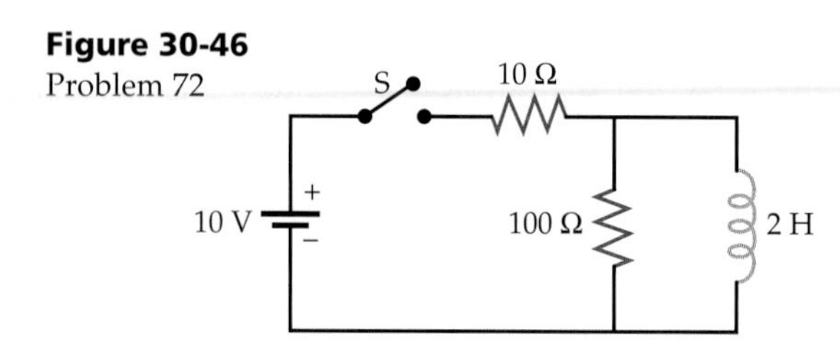

73 •• Compute the initial slope dI/dt at $t = 0$ from Equation 30-24, and show that if the current decreased steadily at this rate, it would be zero after one time constant.

74 •• An inductance L and resistance R are connected in series with a battery as in Figure 30-22. A long time after switch S_1 is closed, the current is 2.5 A. When the battery is switched out of the circuit by opening switch S_1 and closing S_2, the current drops to 1.5 A in 45 ms. (*a*) What is the time constant for this circuit? (*b*) If $R = 0.4\ \Omega$, what is L?

75 •• When the current in a certain coil is 5.0 A and is increasing at the rate of 10.0 A/s, the potential difference across the coil is 140 V. When the current is 5.0 A and is decreasing at the rate of 10 A/s, the potential difference is 60 V. Find the resistance and self-inductance of the coil.

76 •• For the circuit of Figure 30-47, (*a*) find the rate of change of the current in each inductor and in the resistor just after the switch is closed. (*b*) What is the final current? (Use the result from Problem 92.)

Figure 30-47 Problem 76

S
24 V
8 mH
4 mH
15 Ω

77 •• For the circuit of Example 30-11, find the time at which the power dissipation in the resistor equals the rate at which magnetic energy is stored in the inductor.

78 ••• In the circuit of Figure 30-21, let $\mathcal{E}_0 = 12.0$ V, $R = 3.0\ \Omega$, and $L = 0.6$ H. The switch is closed at time $t = 0$. From time $t = 0$ to $t = \tau$, find (*a*) the total energy that has been supplied by the battery, (*b*) the total energy that has been dissipated in the resistor, and (*c*) the energy that has been stored in the inductor. (*Hint:* Find the rates as functions of time and integrate from $t = 0$ to $t = \tau = L/R$.)

General Problems

79 • Two identical bar magnets are dropped from equal heights. Magnet A is dropped from above bare earth, whereas magnet B is dropped from above a metal plate. Which magnet strikes first?

(*a*) Magnet A
(*b*) Magnet B
(*c*) Both strike at the same time.
(*d*) Whichever has the N pole toward the ground.
(*e*) Whichever has the S pole toward the ground.

80 • True or false:

(*a*) The induced emf in a circuit is proportional to the magnetic flux through the circuit.
(*b*) There can be an induced emf at an instant when the flux through the circuit is zero.
(*c*) Lenz's law is related to the conservation of energy.
(*d*) The inductance of a solenoid is proportional to the rate of change of the current in it.
(*e*) The magnetic energy density at some point in space is proportional to the square of the magnetic field at that point.

81 • A bar magnet is dropped inside a long vertical tube. If the tube is made of metal, the magnet quickly approaches a terminal speed, but if the tube is made of cardboard, it does not. Explain.

82 • A circular coil of radius 3.0 cm has 6 turns. A magnetic field $B = 5000$ G is perpendicular to the coil. (*a*) Find the magnetic flux through the coil. (*b*) Find the magnetic flux through the coil if the coil makes an angle of 20° with the magnetic field.

83 • The magnetic field in Problem 82 is steadily reduced to zero in 1.2 s. Find the emf induced in the coil when (*a*) the magnetic field is perpendicular to the coil and (*b*) the magnetic field makes an angle of 20° with the normal to the coil.

84 • A 100-turn coil has a radius of 4.0 cm and a resistance of 25 Ω. At what rate must a perpendicular magnetic field change to produce a current of 4.0 A in the coil?

85 •• Figure 30-48 shows an ac generator. It consists of a rectangular loop of dimensions a and b with N turns connected to slip rings. The loop rotates with an angular velocity ω in a uniform magnetic field $\vec{B}$. (*a*) Show that the potential difference between the two slip rings is $E = NBab\omega \sin \omega t$. (*b*) If $a = 1.0$ cm, $b = 2.0$ cm, $N = 1000$, and $B = 2$ T, at what angular frequency ω must the coil rotate to generate an emf whose maximum value is 110 V?

Figure 30-48 Problems 85 and 86

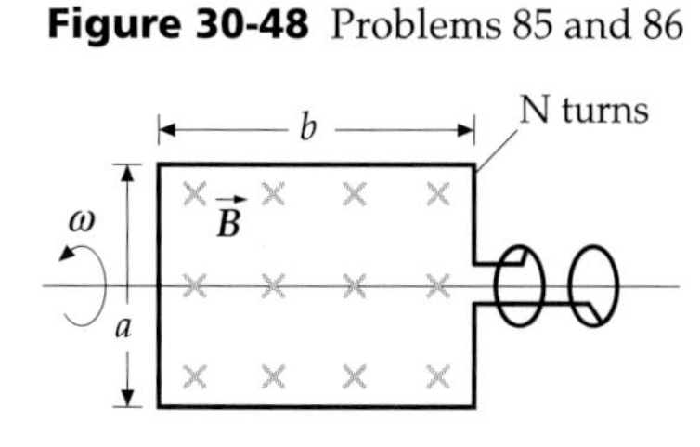

86 •• Prior to about 1960, magnetic field strength was measured by means of a rotating coil gaussmeter. This device used a small loop of many turns rotating on an axis perpendicular to the magnetic field at fairly high speed and connected to an ac voltmeter by means of slip rings like those shown in Figure 30-48. The sensing coil for a rotating coil gaussmeter has 400 turns and an area of 1.4 cm^2. The coil rotates at 180 rpm. If the magnetic field strength is 0.45 T, find the maximum induced emf in the coil and the orientation of the coil relative to the field for which this maximum induced emf occurs.

87 •• Show that the effective inductance for two inductors L_1 and L_2 connected in series such that none of the flux from either passes through the other is given by $L_{\text{eff}} = L_1 + L_2$.

88 •• The rectangular coil in Figure 30-49 has 80 turns, is 25 cm wide and 30 cm long, and is located in a magnetic field $B = 1.4$ T directed out of the page as shown, with only half of the coil in the region of the magnetic field. The resistance of the coil is 24 Ω. Find the magnitude and direction of the induced current if the coil is moved with a speed of 2 m/s (*a*) to the right, (*b*) up, (*c*) to the left, and (*d*) down.

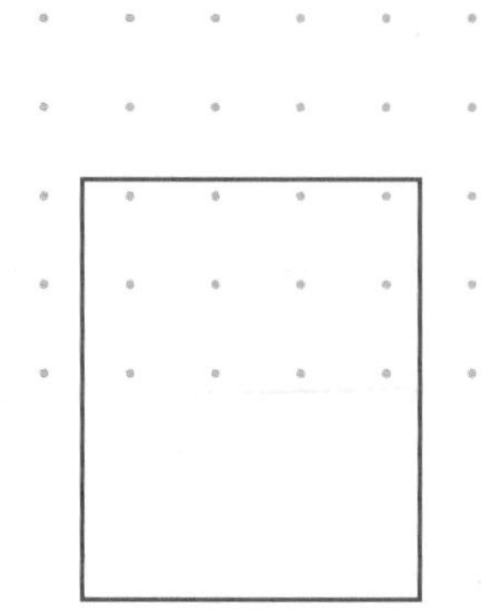

Figure 30-49 Problems 88–90

89 •• Suppose the coil of Problem 88 is rotated about its vertical centerline at constant angular velocity of 2 rad/s. Find the induced current as a function of time.

90 •• Suppose the coil of Problem 88 is rotated about its horizontal centerline at constant angular velocity of 2 rad/s. Find the induced current as a function of time.

91 •• Show that if the flux through each turn of an N-turn coil of resistance R changes from ϕ_{m1} to ϕ_{m2} the total charge passing through the coil is given by $Q = N(\phi_{m1} - \phi_{m2})/R$.

92 •• Show that the effective inductance for two inductors L_1 and L_2 connected in parallel such that none of the flux from either passes through the other is given by

$$\frac{1}{L_{\text{eff}}} = \frac{1}{L_1} + \frac{1}{L_2}$$

93 •• A long solenoid has n turns per unit length and carries a current given by $I = I_0 \sin \omega t$. The solenoid has a circular cross section of radius R. Find the induced electric field at a radius r from the axis of the solenoid for (*a*) $r < R$ and (*b*) $r > R$.

94 ••• A thin-walled hollow wire of radius a lies with its axis along the z axis and carries current I in the positive z direction. A second identical wire is parallel to the first with its axis along the line $x = d$. The second wire carries current I in the negative z direction. (*a*) Find the magnetic flux per unit length through the space in the xz plane between the wires. (*b*) If the far ends of the wires are connected together so that the parallel wires form two sides of a loop, find the self-inductance per unit length of the loop.

95 ••• A coaxial cable consists of two very thin-walled conducting cylinders of radii r_1 and r_2 (Figure 30-50). Current I goes in one direction down the inner cylinder and in the opposite direction in the outer cylinder. (*a*) Use Ampère's law to find B. Show that $B = 0$ except in the region between the conductors. (*b*) Show that the magnetic energy density in the region between the cylinders is

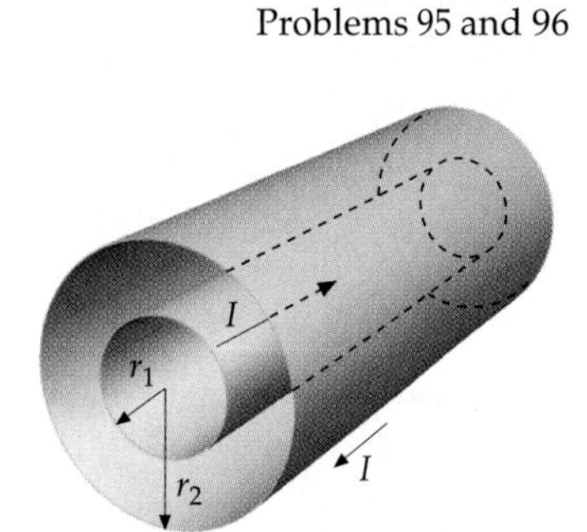

Figure 30-50
Problems 95 and 96

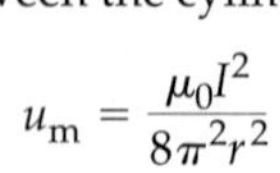

$$u_m = \frac{\mu_0 I^2}{8\pi^2 r^2}$$

(*c*) Find the magnetic energy in a cylindrical shell volume element of length ℓ and volume $dV = \ell 2\pi r\, dr$, and integrate your result to show that the total magnetic energy in the volume of length ℓ is

$$U_m = \frac{\mu_0}{4\pi} I^2 \ell \ln \frac{r_2}{r_1}$$

(*d*) Use the result in part (*c*) and $U_m = \frac{1}{2}LI^2$ to show that the self-inductance per unit length is

$$\frac{L}{\ell} = \frac{\mu_0}{2\pi} \ln \frac{r_2}{r_1}$$

96 ••• In Figure 30-50, compute the flux through a rectangular area of sides ℓ and $r_2 - r_1$ between the conductors. Show that the self-inductance per unit length can be found from $\phi_m = LI$ (see part (*d*) of Problem 95).

97 ••• Figure 30-51 shows a rectangular loop of wire, 0.30 m wide and 1.50 m long, in the vertical plane and perpendicular to a uniform magnetic field $B = 0.40$ T, directed inward as shown. The portion of the loop not in the magnetic field is 0.10 m long. The resistance of the loop is 0.20 Ω and its mass is 0.50 kg. The loop is released from rest at $t = 0$. (*a*) What is the magnitude and direction of the induced current when the loop has a downward velocity v? (*b*) What is the force that acts on the loop as a result of this current? (*c*) What is the net force acting on the loop? (*d*) Write the equation of motion of the loop. (*e*) Obtain an expression for the velocity of the loop as a function of time. (*f*) Integrate the expression obtained in part (*e*) to find the displacement y as a function of time. (*g*) From the result obtained in part (*f*) find t for $y = 1.40$ m, i.e., the time when the loop leaves the region of magnetic field. (*h*) Find the velocity of the loop at that instant. (*i*) What would be the velocity of the loop after it has dropped 1.40 m if $B = 0$?

Figure 30-51
Problems 97 and 98

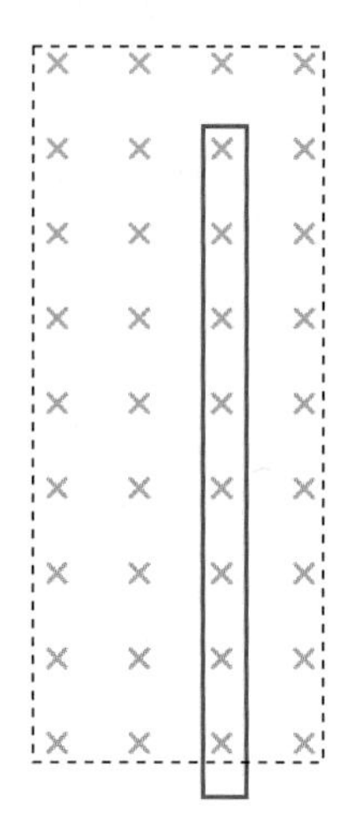

98 ••• The loop of Problem 97 is attached to a plastic spring of spring constant κ (Figure 30-52). (*a*) When $B = 0$, the period of small-amplitude vertical oscillations of the mass–spring system is 0.8 s. Find the spring constant κ. (*b*) When $B \neq 0$, a current is induced in the loop as a result of its up and down motion. Obtain an expression for the induced current as a function of time when $B = 0.40$ T. (*c*) Show that the induced current acts as a damping mechanism. (*d*) Determine the value of the magnetic field for which the Q of the mass–spring system is 100.

Figure 30-52
Problems 98 and 99

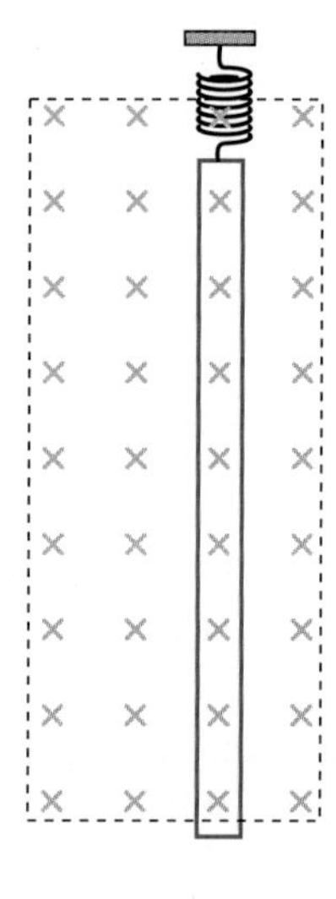

99 ••• Show that the effective inductance of two inductors, L_1 and L_2, connected in series and in close proximity, is $L_{\text{eff}} = L_1 + L_2 \pm 2M$. When should the plus sign be used in this expression, and when should the minus sign be used?

CHAPTER 31

Alternating-Current Circuits

Most high-voltage long-distance power transmission uses alternating current because converting ac to dc at the sending end (such as a hydroelectric power plant) and from dc back to ac at the consuming end has been expensive. However, recent advances in technology have revived interest in high-voltage direct current (HVDC). Shown here is a dc-to-ac conversion station near Boston, linked by HVDC lines to a 2000-MW hydroelectric generator unit in James Bay, Quebec. This particular station is used to convert direct current to alternating current.

More than 99% of the electrical energy used today is produced by electrical generators in the form of alternating current, which has a great advantage over direct current in that electrical energy can be transported over long distances at very high voltages and low currents to reduce energy losses due to Joule heat. It can then be transformed, with almost no energy loss, to lower and safer voltages and correspondingly higher currents for everyday use. The transformer that accomplishes this change in voltage and current works on the basis of magnetic induction. In North America, power is delivered by a sinusoidal current of frequency 60 Hz. Devices such as radios, television sets, and microwave ovens detect or generate alternating currents of much greater frequencies.

Alternating current is easily produced by magnetic induction in an ac generator, which is designed to put out a sinusoidal emf. We will see that when the generator output is sinusoidal, the current in an inductor, a capacitor, or a resistor is also sinusoidal, though it is generally not in phase with the generator's emf. When the emf and current are both sinusoidal, their maximum values can be easily related. The study of sinusoidal currents is important because even currents that are not sinusoidal can be analyzed in terms of sinusoidal components using Fourier analysis.

31-1 ac Generators

Figure 31-1 shows a simple **generator** consisting of a coil of area A and N turns rotating in a uniform magnetic field. The ends of the coil are connected to rings called slip rings that rotate with the coil. They make electrical contact through stationary conducting brushes in contact with the rings.

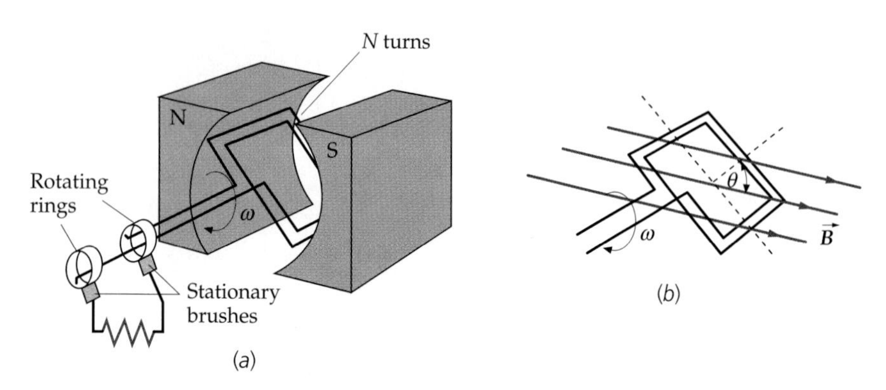

Figure 31-1 (*a*) An ac generator. A coil rotating with constant angular frequency ω in a magnetic field $\vec{B}$ generates a sinusoidal emf. Energy from a waterfall or a steam turbine is used to rotate the coil to produce electrical energy. The emf is supplied to an external circuit by the brushes in contact with the rings. (*b*) At this instant, the normal to the plane of the coil makes an angle θ with the magnetic field and the flux is $BA \cos\theta$.

When the line perpendicular to the plane of the coil makes an angle θ with a uniform magnetic field $\vec{B}$, as shown in the figure, the magnetic flux through the coil is

$$\phi_m = NBA \cos\theta \qquad \text{31-1}$$

When the coil is mechanically rotated, the flux through it will change, and an emf will be induced. If ω is the angular velocity of rotation and the initial angle is δ, the angle at some later time t is given by

$$\theta = \omega t + \delta$$

Then

$$\phi_m = NBA \cos(\omega t + \delta) = NBA \cos(2\pi f t + \delta)$$

The emf in the coil will then be

$$\mathcal{E} = -\frac{d\phi_m}{dt} = -NBA\frac{d}{dt}\cos(\omega t + \delta) = +NBA\omega \sin(\omega t + \delta) \qquad \text{31-2}$$

or

$$\mathcal{E} = \mathcal{E}_{max} \sin(\omega t + \delta) \qquad \text{31-3}$$

where

$$\mathcal{E}_{max} = NBA\omega \qquad \text{31-4}$$

We can thus produce a sinusoidal emf in a coil by rotating it with constant angular velocity in a magnetic field. Although practical generators are considerably more complicated, they work on the same principle that an alternating emf is produced in a coil rotating in a magnetic field, and they are designed so that the emf produced is sinusoidal. In circuit diagrams, an ac generator is represented by the symbol ⊖.

The same coil in a magnetic field that can be used to generate an alternating emf can also be used as an ac **motor.** Instead of mechanically rotating the coil to generate an emf, we apply an alternating current to the coil from another ac generator. The torque due to the magnetic force on the wire rotates the coil. As the coil rotates in the magnetic field, a back emf is generated that tends to counter the emf that supplies the current. When the motor is first turned on, there is no back emf and the current is very large, being limited only by the resistance in the circuit. As the motor begins to rotate, the back emf increases and the current decreases.

Multiplying both sides by dt and dividing by L, we obtain

$$dI = \frac{\mathcal{E}_{max}}{L} \cos \omega t \, dt \qquad 31\text{-}19$$

We solve for the current I by integrating both sides of the equation:

$$I = \frac{\mathcal{E}_{max}}{L} \int \cos \omega t \, dt = \frac{\mathcal{E}_{max}}{\omega L} \sin \omega t + C \qquad 31\text{-}20$$

where the constant of integration C is the dc component of the current. Setting the dc component of the current to be zero, we have

$$I = \frac{\mathcal{E}_{max}}{\omega L} \sin \omega t = I_{max} \sin \omega t \qquad 31\text{-}21$$

where

$$I_{max} = \frac{\mathcal{E}_{max}}{\omega L} \qquad 31\text{-}22$$

The current $I = I_{max} \sin \omega t$ is 90° out of phase with the voltage across the inductor $V_L = \mathcal{E}_{max} \cos \omega t$. From Figure 31-6, which shows I and V_L as functions of time, we can see that the maximum value of the voltage occurs 90° or one-fourth period before the corresponding maximum value of the current. The voltage drop across an inductor is said to *lead the current by 90°*. We can understand this physically. When I is zero but increasing, dI/dt is maximum, so the back emf induced in the inductor is at its maximum. One-quarter cycle later, I is maximum. At this time, dI/dt is zero, so V_L is zero. Using the trigonometric identity $\sin \omega t = \cos (\omega t - \pi/2)$, Equation 31-21 for the current can be written

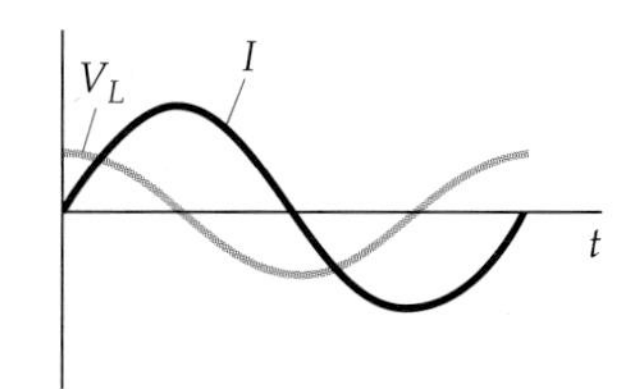

Figure 31-6 Current and voltage across the inductor in Figure 31-5 as functions of time. The maximum voltage occurs one-fourth period before the maximum current. Thus, the voltage is said to lead the current by one-fourth period or 90°.

$$I = I_{max} \cos (\omega t - \pi/2) \qquad 31\text{-}23$$

The relation between the maximum current and the maximum voltage (or between the rms current and rms voltage) for an inductor can be written in a form similar to Equation 31-15 for a resistor. From Equation 31-22, we have

$$I_{max} = \frac{\mathcal{E}_{max}}{\omega L} = \frac{\mathcal{E}_{max}}{X_L} \qquad 31\text{-}24$$

where

$$X_L = \omega L \qquad 31\text{-}25$$

Definition—Inductive reactance

is called the **inductive reactance.** Since $I_{rms} = I_{max}/\sqrt{2}$ and $\mathcal{E}_{rms} = \mathcal{E}_{max}/\sqrt{2}$ the rms current is given by

$$I_{rms} = \frac{\mathcal{E}_{rms}}{X_L} \qquad 31\text{-}26$$

Like resistance, inductive reactance has units of ohms. As we can see from Equation 31-26, the larger the reactance for a given emf, the smaller the current. Unlike resistance, the inductive reactance depends on the frequency of the current—the greater the frequency, the greater the reactance.

The instantaneous power input to the inductor from the generator is

$$P = \mathcal{E}I = (\mathcal{E}_{max} \cos \omega t)(I_{max} \sin \omega t) = \mathcal{E}_{max} I_{max} \cos \omega t \sin \omega t$$

The average power into the inductor is zero. We can see this by using

$$\cos \omega t \sin \omega t = \tfrac{1}{2} \sin 2 \omega t$$

The value of this term oscillates twice during each cycle and is negative as often as it is positive. Thus, on the average, no energy is dissipated in an inductor. (This is true only if the resistance of the inductor can be neglected.)

Example 31-2

A 40-mH inductor is placed across an ac generator that has a maximum emf of 120 V. Find the inductive reactance and the maximum current when the frequency is (*a*) 60 Hz and (*b*) 2000 Hz.

Picture the Problem We calculate the inductive reactance at each frequency and use Equation 31-24 to find the maximum current.

(*a*)1. The maximum current equals the maximum emf divided by the inductive reactance: $I_{\max} = \dfrac{\mathcal{E}_{\max}}{X_L}$

2. Compute the inductive reactance at 60 Hz: $X_{L1} = \omega_1 L = 2\pi f_1 L = (2\pi)(60\text{ Hz})(40 \times 10^{-3}\text{ H})$
$= 15.1\ \Omega$

3. Use this value of X_L to compute the maximum current at 60 Hz: $I_{1,\max} = \dfrac{120\text{ V}}{15.1\ \Omega} = 7.95\text{ A}$

(*b*)1. Compute the inductive reactance at 2000 Hz: $X_{L2} = \omega_2 L = 2\pi f_2 L$
$= (2\pi)(2000\text{ Hz})(40 \times 10^{-3}\text{ H}) = 503\ \Omega$

2. Use this value of X_L to compute the maximum current at 2000 Hz: $I_{2,\max} = \dfrac{120\text{ V}}{503\ \Omega} = 0.239\text{ A}$

Capacitors in ac Circuits

When a capacitor is connected across the terminals of a generator (Figure 31-7), the voltage drop across the capacitor is

$$V_C = V_+ - V_- = \frac{Q}{C} \tag{31-27}$$

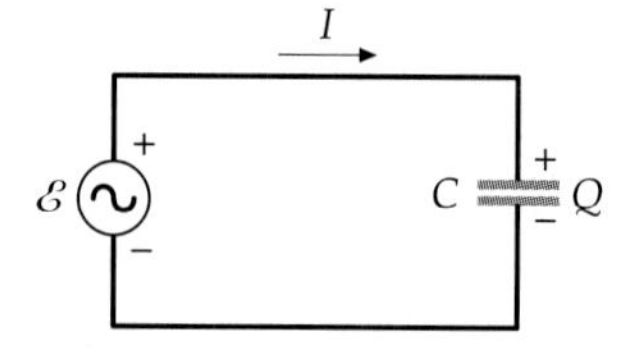

Figure 31-7 An ac generator in series with a capacitor *C*. Again plus and minus signs on the capacitor plates indicate a positive charge on the plate where the current enters and a negative charge on the plate where the current leaves. The current is related to the charge by $I = dQ/dt$.

From Kirchhoff's loop rule, we have

$$\mathcal{E} - V_C = 0$$

or

$$V_C = \mathcal{E} = \mathcal{E}_{\max} \cos \omega t = \frac{Q}{C}$$

Thus

$$Q = \mathcal{E}_{\max} C \cos \omega t$$

The current is

$$I = \frac{dQ}{dt} = -\omega \mathcal{E}_{\max} C \sin \omega t = -I_{\max} \sin \omega t$$

where

$$I_{\max} = \omega \mathcal{E}_{\max} C \tag{31-28}$$

Using the trigonometric identity $\sin \omega t = -\cos(\omega t + \pi/2)$, we obtain

$$I = -\omega C \mathcal{E}_{\max} \sin \omega t = I_{\max} \cos(\omega t + \pi/2) \tag{31-29}$$

As with the inductor, the current $I = I_{max} \cos(\omega t + \pi/2)$ is not in phase with the voltage drop across the capacitor, $V_C = \mathcal{E}_{max} \cos \omega t$. From Figure 31-8, we see that the maximum value of the voltage occurs 90° or one-fourth period *after* the maximum value of the current. Thus, *the voltage drop across a capacitor lags the current by 90°*. Again, we can understand this physically. The maximum rate of charge buildup $dQ/dt = I$ occurs when the charge Q is zero and therefore when V_C is zero ($\omega t = 3\pi/2$ in Figure 31-7). As the charge on the capacitor plate increases, the current decreases until the charge is a maximum (so that V_C is a maximum) and the current is zero ($\omega t = 2\pi$ or $\omega t = 0$ in Figure 31-7). The current then becomes negative as the charge flows back in the opposite direction, off the capacitor.

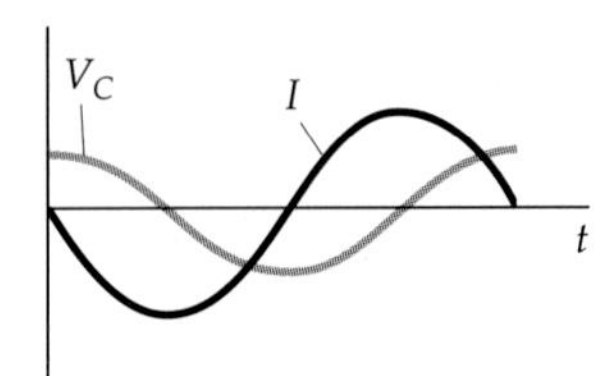

Figure 31-8 Current and voltage across the capacitor in Figure 31-7 versus time. The maximum voltage occurs one-fourth period after the maximum current. Thus, the voltage is said to lag the current by 90°.

Again, we can relate the current to the emf in a form similar to Equation 31-8 for a resistor. From Equation 31-28, we have

$$I_{max} = \omega C \mathcal{E}_{max} = \frac{\mathcal{E}_{max}}{1/\omega C} = \frac{\mathcal{E}_{max}}{X_C}$$

and, similarly,

$$I_{rms} = \frac{\mathcal{E}_{rms}}{X_C} \qquad 31\text{-}30$$

where

$$X_C = \frac{1}{\omega C} \qquad 31\text{-}31$$

Definition—Capacitive reactance

is called the **capacitive reactance** of the circuit. Like resistance and inductive reactance, capacitive reactance has units of ohms, and like inductive reactance, capacitive reactance depends on the frequency of the current. In this case, the greater the frequency, the smaller the reactance. The average power input to a capacitor from an ac generator is zero, as it is for an inductor. This is because the emf is proportional to cos wt and the current is proportional to sin wt and $(\cos wt \sin wt)_{av} = 0$. Thus, like inductors with no resistance, capacitors dissipate no energy.

Since charge cannot pass across the space between the plates of a capacitor, it may seem strange that there is a continuing alternating current in the circuit of Figure 31-7. Consider an initially uncharged capacitor across a source of emf, with the upper plate attached to the positive terminal. Initially, positive charge flows to the upper plate and away from the lower plate. The effect is the same as if the charge actually flows *across* the space between the plates. If the source of emf is an ac generator, the charge on one plate and the potential difference changes sign every half-period. If we double the frequency, we double the amount of charge that flows onto and off of the plate in a given time, thus we double the current. Hence, the greater the frequency, the less the capacitor impedes the flow of charge.

Example 31-3

A 20-μF capacitor is placed across a generator that has a maximum emf of 100 V. Find the capacitive reactance and the maximum current when the frequency is 60 Hz and when it is 5000 Hz.

Picture the Problem The capacitive reactance is $X_C = 1/\omega C$ and the maximum current is $I_{max} = \mathcal{E}_{max}/X_C$.

1. Calculate the capacitive reactance at 60 Hz and at 5000 Hz:

$$X_{C1} = \frac{1}{\omega_1 C} = \frac{1}{2\pi f_1 C} = \frac{1}{2\pi(60\text{ Hz})(20 \times 10^{-6}\text{ F})} = 133\ \Omega$$

$$X_{C2} = \frac{1}{\omega_2 C} = \frac{1}{2\pi f_2 C} = \frac{1}{2\pi(5000\text{ Hz})(20 \times 10^{-6}\text{ F})} = 1.59\ \Omega$$

2. Use these values of X_C to find the maximum currents:

$$I_{1,\text{max}} = \frac{\mathcal{E}_{\text{max}}}{X_{C1}} = \frac{100\text{ V}}{133\ \Omega} = 0.752\text{ A}$$

$$I_{2,\text{max}} = \frac{\mathcal{E}_{\text{max}}}{X_{C2}} = \frac{100\text{ V}}{1.59\ \Omega} = 62.9\text{ A}$$

Remark Note that the current increases with frequency, as expected.

In the circuits of Figures 31-5 and 31-7, which contain only a generator and an inductor or capacitor, the voltage drop across the inductor or capacitor equals the voltage of the generator. In more complicated circuits containing three or more elements, the voltage drop across each element is usually not equal to the generator voltage. It is useful, therefore, to write Equations 31-26 and 31-30 in terms of the voltage drops across the inductor and capacitor, respectively. If $V_{L,\text{rms}}$ is the rms voltage drop across an inductor, the rms current in the inductor is given by

$$I_{\text{rms}} = \frac{V_{L,\text{rms}}}{\omega L} = \frac{V_{L,\text{rms}}}{X_L} \qquad \text{31-32}$$

The voltage drop across the inductor leads the current by 90°. Similarly, if $V_{C,\text{rms}}$ is the rms voltage across a capacitor, the rms current in the capacitor is given by

$$I_{\text{rms}} = \frac{V_{C,\text{rms}}}{1/\omega C} = \frac{V_{C,\text{rms}}}{X_C} \qquad \text{31-33}$$

The voltage drop across the capacitor lags the current by 90°. Equations 31-32 and 31-33 can also be written in terms of the maximum voltages and maximum currents.

optional

31-4 Phasors

The phase relations between the current and the voltage drop in a capacitor or inductor can be represented by two-dimensional vectors called **phasors.** In Figure 31-9, the voltage across a resistor V_R is represented by a vector $\vec{V}_R$ that has magnitude $I_{\text{max}}R$ and makes an angle θ with the x axis. This voltage is in phase with the current. In general, a steady-state current in an ac circuit varies with time as

$$I = I_{\text{max}} \cos \theta = I_{\text{max}} \cos (\omega t - \delta) \qquad \text{31-34}$$

where ω is the angular frequency and δ is some phase constant. The voltage drop across a resistor is then given by

$$V_R = IR = I_{\text{max}} R \cos (\omega t - \delta) \qquad \text{31-35}$$

The instantaneous value of the voltage drop across a resistor is thus equal to the x component of the phasor vector $\vec{V}_R$, which rotates counterclockwise

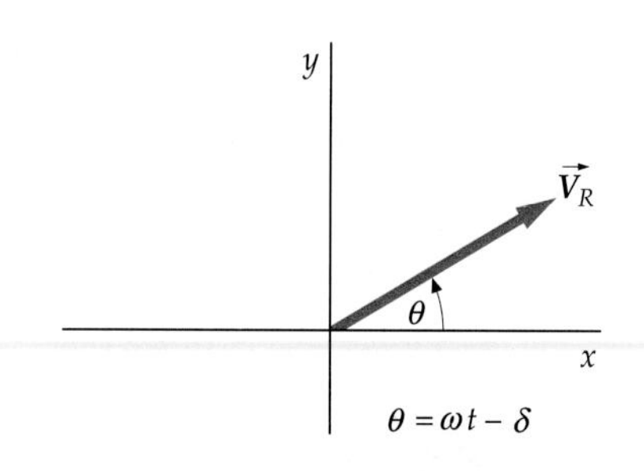

Figure 31-9 The voltage across a resistor can be represented by a vector $\vec{V}_R$, called a phasor, that has magnitude $I_{\text{max}}R$ and makes an angle $\theta = \omega t - \delta$ with the x axis. The phasor rotates with an angular frequency ω. The voltage $V_R = IR$ is the x component of $\vec{V}_R$.

with an angular frequency ω. The current I may be written as the x component of a phasor $\vec{I}$ having the same orientation as $\vec{V}_R$.

When several components are connected together in a series circuit, their voltages add. When they are connected in parallel, their currents add. Adding sines or cosines of different amplitudes and phases algebraically is awkward. It is much easier to do this by vector addition.

Phasors are used as follows. Any ac voltage or current is written in the form $A \cos(\omega t - \delta)$, which in turn is treated as the x component A_x of a phasor $\vec{A}$ that makes an angle $(\omega t - \delta)$ with the x axis. Instead of adding two voltages or currents algebraically as $A \cos(\omega t - \delta_1) + B \cos(\omega t - \delta_2)$, we represent these quantities as phasors $\vec{A}$ and $\vec{B}$ and find the phasor sum $\vec{C} - \vec{A} + \vec{B}$ gcometrically. The resultant voltage or current is then the x component of the resultant phasor, $C_x = A_x + B_x$. The geometric representation conveniently shows the relative amplitudes and phases of the phasors.

Consider a circuit containing an inductor L, a capacitor C, and a resistor R all connected in series. They all carry the same current, which is represented as the x component of the current phasor $\vec{I}$. The voltage across the inductor V_L is represented by a phasor $\vec{V}_L$ that has magnitude $I_{max}X_L$ and leads the current phasor $\vec{I}$ by 90°. Similarly, the voltage across the capacitor V_C is represented by a phasor $\vec{V}_C$ that has magnitude $I_{max}X_C$ and lags the current by 90°. Figure 31-10 shows the three phasors $\vec{V}_R$, $\vec{V}_L$, and $\vec{V}_C$. As time goes on, the three phasors rotate counterclockwise with an angular frequency ω, so the relative positions of the vectors do not change. At any time, the instantaneous value of the voltage drop across any of these elements equals the x component of the corresponding phasor.

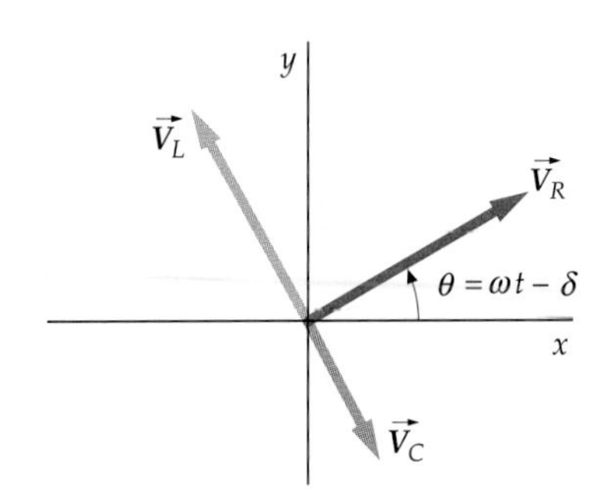

Figure 31-10 Phasor representations of the voltages V_R, V_L, and V_C. Each vector rotates in the counterclockwise direction with an angular frequency ω. At any instant, the voltage across an element equals the x component of the corresponding phasor, and the sum of the voltages equals the x component of the vector sum $\vec{V}_R + \vec{V}_L + \vec{V}_C$.

31-5 *LC* and *RLC* Circuits Without a Generator

Figure 31-11 shows a simple circuit with inductance and capacitance but no resistance. Such a circuit is called an ***LC* circuit**. We assume that the capacitor carries an initial charge Q_0 and that the switch is initially open. After the switch is closed at $t = 0$, the charge begins to flow through the inductor. In the figure, the signs of Q on the capacitor and the direction of the current I have been chosen such that

$$I = \frac{dQ}{dt}$$

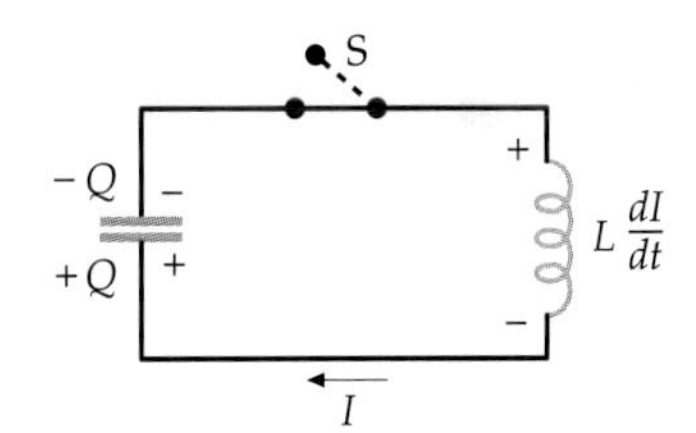

Figure 31-11 An *LC* circuit. When the switch is closed, the initially charged capacitor discharges through the inductor, producing a back emf.

Applying Kirchhoff's loop rule to the circuit for the assumed signs of Q and I, we have

$$L\frac{dI}{dt} + \frac{Q}{C} = 0 \qquad \text{31-36}$$

Substituting dQ/dt for I gives

$$L\frac{d^2Q}{dt^2} + \frac{Q}{C} = 0 \qquad \text{31-37}$$

This is of the same form as the equation for the acceleration of a mass on a spring:

$$m\frac{d^2x}{dt^2} + kx = 0$$

optional

The behavior of an LC circuit is thus analogous to that of a mass on a spring, with L analogous to the mass m, Q analogous to the position x, and $1/C$ analogous to the spring constant k. Also, the current I is analogous to the velocity v, since $v = dx/dt$ and $I = dQ/dt$. In mechanics, the mass of an object describes the inertia of the object. The greater the mass, the more difficult it is to change the velocity of the object. Similarly, the inductance L can be thought of as the inertia of an ac circuit. The greater the inductance, the more difficult it is to change the current I.

If we divide each term in Equation 31-37 by L and rearrange, we obtain

$$\frac{d^2Q}{dt^2} = -\frac{1}{LC}Q \qquad \text{31-38}$$

which is analogous to

$$\frac{d^2x}{dt^2} = -\frac{k}{m}x = -\omega^2 x \qquad \text{31-39}$$

where $\omega^2 = k/m$. In Chapter 14, we found that we could write the solution of Equation 31-39 for simple harmonic motion in the form

$$x = A\cos(\omega t - \delta)$$

where $\omega = \sqrt{k/m}$ is the angular frequency, A is the amplitude, and δ is the phase constant, which depends on the initial conditions. We can put Equation 31-38 in this same form by writing ω^2 for $1/LC$. Then

$$\frac{d^2Q}{dt^2} = -\omega^2 Q \qquad \text{31-40}$$

$$\omega = \frac{1}{\sqrt{LC}} \qquad \text{31-41}$$

The solution of Equation 31-40 is

$$Q = A\cos(\omega t - \delta)$$

The current is found by differentiating:

$$I = \frac{dQ}{dt} = -\omega A\sin(\omega t - \delta)$$

If we choose our initial conditions to be $Q = Q_0$ and $I = 0$ at $t = 0$, the phase constant δ is zero and $A = Q_0$. Our solutions are then

$$Q = Q_0\cos\omega t \qquad \text{31-42}$$

and

$$I = -\omega Q_0\sin\omega t = -I_{\max}\sin\omega t \qquad \text{31-43}$$

where $I_{\max} = \omega Q_0$.

Figure 31-12 shows graphs of Q and I versus time. The charge oscillates between the values $+Q_0$ and $-Q_0$ with angular frequency $\omega = \sqrt{1/LC}$. The current oscillates between $+\omega Q_0$ and $-\omega Q_0$ with the same frequency and is 90° out of phase with the charge. The current is maximum when the charge is zero and zero when the charge is maximum.

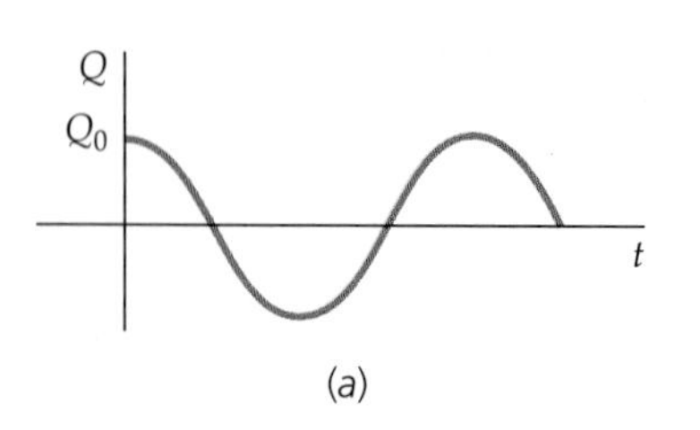

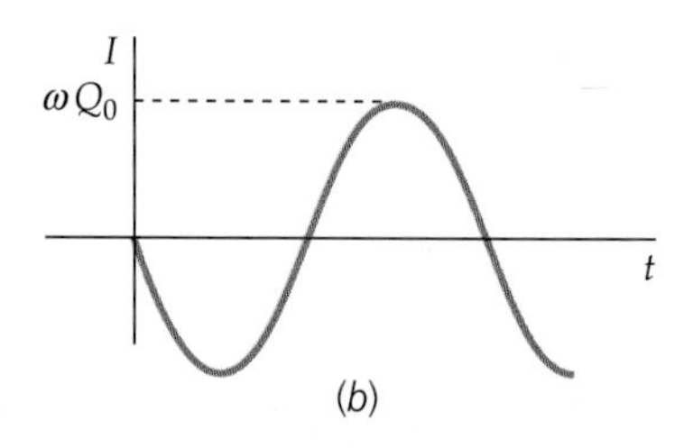

Figure 31-12 Graphs of (*a*) Q versus t and (*b*) I versus t for the LC circuit of Figure 31-11.

In our study of the oscillations of a mass on a spring, we found that the total energy is constant, and that it oscillates between potential and kinetic energy. In our LC circuit, we also have two kinds of energy, electric energy and magnetic energy. The electric energy stored in the capacitor is

$$U_e = \frac{1}{2}QV_C = \frac{1}{2}\frac{Q^2}{C}$$

Substituting $Q_0 \cos \omega t$ for Q, we have for the electric energy

$$U_e = \frac{1}{2}\frac{Q_0^2}{C}\cos^2 \omega t \qquad \text{31-44}$$

The electric energy oscillates between its maximum value $Q_0^2/2C$ and zero. The magnetic energy stored in the inductor is

$$U_m = \frac{1}{2}LI^2 \qquad \text{31-45}$$

Substituting $I = -\omega Q_0 \sin \omega t$ (Equation 31-43), we get

$$U_m = \frac{1}{2}L\omega^2 Q_0^2 \sin^2 \omega t = \frac{1}{2}\frac{Q_0^2}{C}\sin^2 \omega t \qquad \text{31-46}$$

where we have used $\omega^2 = 1/LC$. The magnetic energy also oscillates between its maximum value of $Q_0^2/2C$ and zero. The sum of the electrostatic and magnetic energies is the total energy, which is constant in time:

$$U_{\text{total}} = U_e + U_m = \frac{1}{2}\frac{Q_0^2}{C}\cos^2 \omega t + \frac{1}{2}\frac{Q_0^2}{C}\sin^2 \omega t = \frac{1}{2}\frac{Q_0^2}{C}$$

This equals the energy initially stored on the capacitor.

Example 31-4

A 2-μF capacitor is charged to 20 V and is then connected across a 6-μH inductor. (*a*) What is the frequency of oscillation? (*b*) What is the maximum value of the current?

(*a*) The frequency of oscillation depends only on the values of the capacitance and inductance:

$$f = \frac{\omega}{2\pi} = \frac{1}{2\pi\sqrt{LC}} = \frac{1}{2\pi\sqrt{(6 \times 10^{-6}\text{ H})(2 \times 10^{-6}\text{ F})}}$$

$$= 4.59 \times 10^4 \text{ Hz}$$

(*b*)1. The maximum value of the current is related to the maximum value of the charge:

$$I_{\text{max}} = \omega Q_0 = \frac{Q_0}{\sqrt{LC}}$$

2. Find the initial charge on the capacitor from the initial voltage:

$$Q_0 = CV_0 = (2\ \mu\text{F})(20\text{ V}) = 40\ \mu\text{C}$$

3. Use the value of Q_0 to calculate I_{max}:

$$I_{\text{max}} = \frac{40\ \mu\text{C}}{\sqrt{(6\ \mu\text{H})(2\ \mu\text{F})}} = 11.5\text{ A}$$

Exercise A 5-μF capacitor is charged and is then discharged through an inductor. What should the value of the inductance be so that the current oscillates with frequency 8 kHz? (*Answer* 79.2 μH)

If we include a resistor in series with the capacitor and inductor as in Figure 31-13, we have an ***RLC* circuit**. Kirchhoff's loop rule gives

$$L\frac{dI}{dt} + \frac{Q}{C} + IR = 0 \qquad \text{31-47}a$$

or

$$L\frac{d^2Q}{dt^2} + \frac{Q}{C} + R\frac{dQ}{dt} = 0 \qquad \text{31-47}b$$

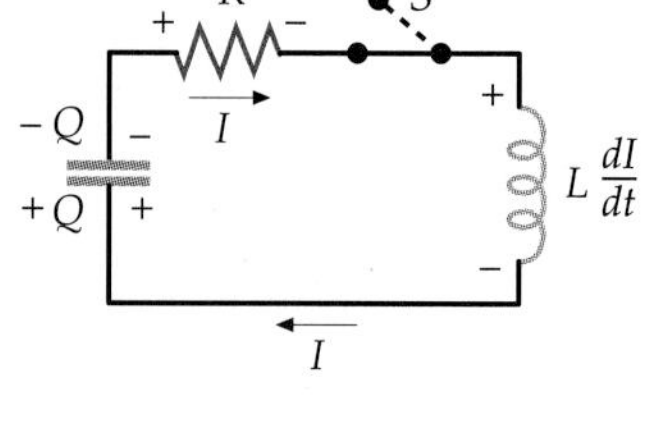

Figure 31-13 An *RLC* circuit.

where we have used $I = dQ/dt$ as before. Equations 31-47*a* and *b* are analogous to the equation for a damped harmonic oscillator (see Equation 14-32):

$$m\frac{d^2x}{dt^2} + kx + b\frac{dx}{dt} = 0$$

The first term, $L\,dI/dt = L\,d^2Q/dt^2$, is analogous to the mass times the acceleration, $m\,dv/dt = m\,d^2x/dt^2$; the second term, Q/C, is analogous to the restoring force kx; and the third term, $IR = R\,dQ/dt$, is analogous to the damping term, $bv = b\,dx/dt$. In the oscillation of a mass on a spring, the damping constant b leads to a dissipation of mechanical energy as heat. In an *RLC* circuit, the resistance R is analogous to the damping constant b and leads to a dissipation of electrical energy as Joule heat.

If the resistance is small, the charge and current oscillate with (angular) frequency* that is very nearly equal to $\omega_0 = 1/\sqrt{LC}$, which is called the natural frequency of the circuit, but the oscillations are damped. We can understand this qualitatively from energy considerations. If we multiply each term in Equation 31-47*a* by the current I, we obtain

$$IL\frac{dI}{dt} + I\frac{Q}{C} + I^2R = 0 \qquad \text{31-48}$$

The first term in this equation is the current times the voltage across the inductor. This is the rate at which energy is put into the inductor or taken out of it; that is, it is the rate of change of the magnetic energy, $d(\frac{1}{2}LI^2)/dt$, which is positive or negative depending on whether I and dI/dt have the same sign or different signs. Similarly, the second term is the current times the voltage across the capacitor. This is the rate of change of the energy of the capacitor, which may be positive or negative. The last term, I^2R, is the rate at which energy is dissipated in the resistor as Joule heat and is always positive. The sum of the electric and magnetic energies is not constant for this circuit because energy is continually dissipated in the resistor. Figure 31-14 shows graphs of Q versus t and I versus t for a small resistance R. If we increase R, the oscillations become more heavily damped until a critical value of R is reached for which there is not even one oscillation. Figure 31-15 shows Q versus t when the value of R is greater than the critical damping value.

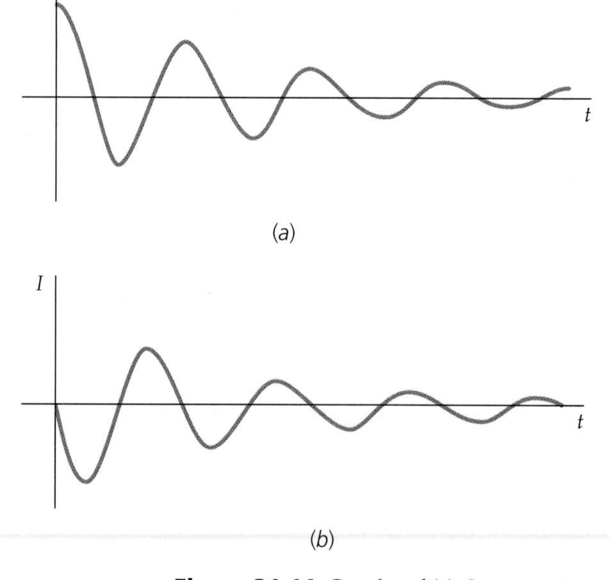

Figure 31-14 Graphs of (*a*) Q versus t and (*b*) I versus t for the *RLC* circuit of Figure 31-13 when R is small enough so that the oscillations are underdamped.

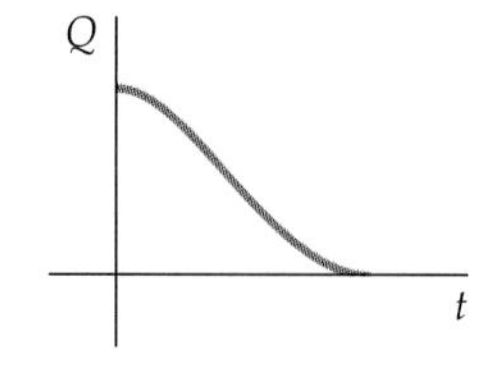

Figure 31-15 Graph of Q versus t for the *RLC* circuit of Figure 31-14 when R is so large that the oscillations are overdamped.

* As we did in Chapter 14 when we discussed mechanical oscillations, we usually omit the word *angular* when the omission will not cause confusion.

31-6 *RLC* Circuits With a Generator

Series *RLC* Circuit

Figure 31-16 shows a series *RLC* circuit with a generator. If the generator emf varies with time as $\mathcal{E} = \mathcal{E}_{\max} \cos \omega t$, Kirchhoff's loop rule gives

$$\mathcal{E}_{\max} \cos \omega t - L\frac{dI}{dt} - \frac{Q}{C} - IR = 0$$

Using $I = dQ/dt$ and rearranging, we obtain

$$L\frac{d^2Q}{dt^2} + R\frac{dQ}{dt} + \frac{Q}{C} = \mathcal{E}_{\max} \cos \omega t \qquad \text{31-49}$$

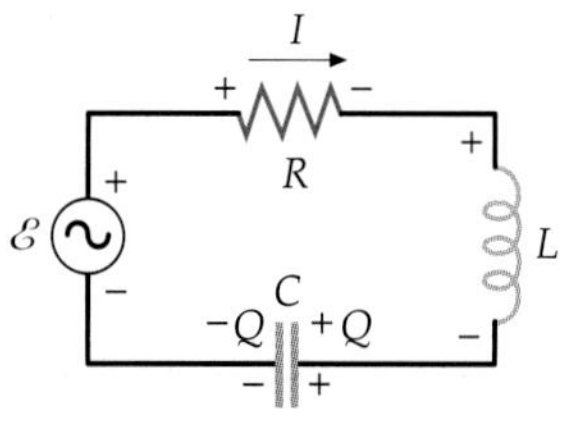

Figure 31-16 A series *RLC* circuit with an ac generator.

This equation is analogous to Equation 14-47 for the forced oscillation of a mass on a spring*:

$$m\frac{d^2x}{dt^2} + b\frac{dx}{dt} + m\omega_0^2 x = F_0 \cos \omega t$$

We will discuss the solution of Equation 31-49 qualitatively as we did with Equation 14-47 for the forced oscillator. The current in the circuit consists of a transient current that depends on the initial conditions (such as the initial phase of the generator and the initial charge on the capacitor) and a steady-state current that does not. We will ignore the transient current, which decreases exponentially with time and is eventually negligible, and concentrate on the steady-state current. The steady-state current obtained by solving Equation 31-49 is

$$I = I_{\max} \cos(\omega t - \delta) \qquad \text{31-50}$$

where the phase angle δ is given by

$$\tan \delta = \frac{X_L - X_C}{R} \qquad \text{31-51}$$

The maximum current is

$$I_{\max} = \frac{\mathcal{E}_{\max}}{\sqrt{R^2 + (X_L - X_C)^2}} = \frac{\mathcal{E}_{\max}}{Z} \qquad \text{31-52}$$

Current in a series RLC circuit

where

$$Z = \sqrt{R^2 + (X_L - X_C)^2} \qquad \text{31-53}$$

Impedance of a series RLC circuit

The quantity $X_L - X_C$ is called the **total reactance,** and Z is called the **impedance.** Combining these results, we have

$$I = \frac{\mathcal{E}_{\max}}{Z} \cos(\omega t - \delta) \qquad \text{31-54}$$

Equation 31-54 can also be obtained from a simple diagram using the phasor representations. Figure 31-17 shows the phasors representing the voltage drops across the resistance, the inductance, and the capacitance. The x component of each of these vectors equals the instantaneous voltage drop across the corresponding element. Since the sum of the x components equals the x component of the sum of the vectors, the sum of the x components equals the sum of the voltage drops across these elements, which by Kirchhoff's loop rule equals the instantaneous emf.

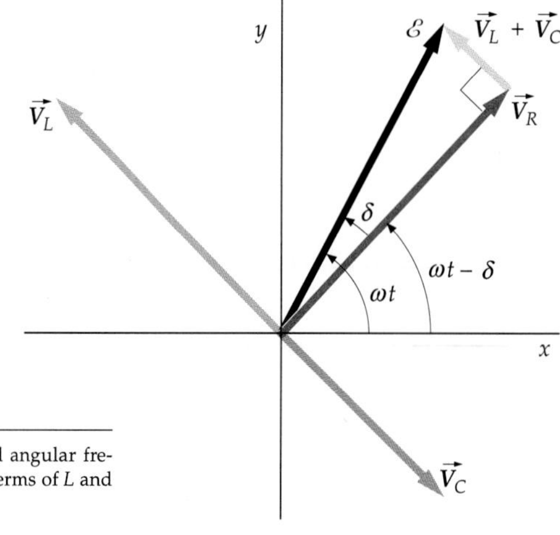

Figure 31-17 Phase relations among voltages in a series *RLC* circuit. The voltage across the resistor is in phase with the current. The voltage across the inductor V_L leads the current by 90°. The voltage across the capacitor lags the current by 90°. The sum of the vectors representing these voltages gives a vector at an angle δ with the current representing the applied emf. For the case shown here, V_L is greater than V_C and the current lags the emf by δ.

* In Equation 14-53 the force constant k was written in terms of the mass m and the natural angular frequency ω_0 using $k = m\omega_0^2$. The capacitance in Equation 31-49 could be similarly written in terms of L and the natural angular frequency using $1/C = L\omega_0^2$.

optional

If we represent the applied emf, $\mathcal{E} = \mathcal{E}_{max} \cos \omega t$, as a phasor $\vec{\mathcal{E}}$ that has the magnitude $\mathcal{E}_{max}$, we have

$$\vec{\mathcal{E}} = \vec{V}_R + \vec{V}_L + \vec{V}_C \tag{31-55}$$

In terms of the magnitudes,

$$\mathcal{E} = |\vec{V}_R + \vec{V}_L + \vec{V}_C| = \sqrt{V_{R,max}^2 + (V_{L,max} - V_{C,max})^2}$$

But $V_R = I_{max}R$, $V_L = I_{max}X_L$, and $V_C = I_{max}X_C$. Thus,

$$\mathcal{E}_{max} = I_{max}\sqrt{R^2 + (X_L - X_C)^2} = I_{max}Z$$

The phasor $\vec{\mathcal{E}}$ makes an angle δ with $\vec{V}_R$ as shown in Figure 31-17. From the figure, we can see that

$$\tan \delta = \frac{|\vec{V}_L + \vec{V}_C|}{|\vec{V}_R|} = \frac{I_{max}X_L - I_{max}X_C}{I_{max}R} = \frac{X_L - X_C}{R}$$

in agreement with Equation 31-51. Since $\vec{\mathcal{E}}$ makes an angle ωt with the x axis, $\vec{V}_R$ makes an angle $\omega t - \delta$ with the x axis. This voltage is in phase with the current, which is therefore given by

$$I = I_{max} \cos (\omega t - \delta) = \frac{\mathcal{E}_{max}}{Z} \cos (\omega t - \delta)$$

This is Equation 31-54. The relation between the impedance Z and the resistance R and the total reactance $X_L - X_C$ can be remembered using the right triangle shown in Figure 31-18.

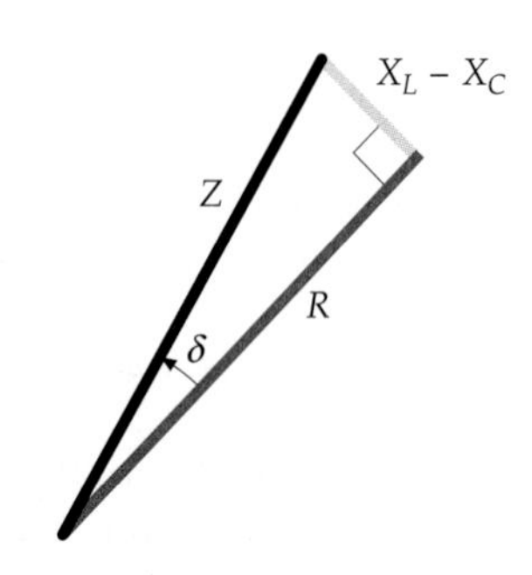

Figure 31-18 Triangle relating capacitive and inductive reactance, resistance, impedance, and the phase angle in an *RLC* circuit.

Resonance

When X_L and X_C are equal, the total reactance is zero, and the impedance Z has its smallest value R. Then I_{max} has its greatest value and the phase angle δ is zero, which means that the current is in phase with the applied emf. The value of ω for which X_L and X_C are equal is obtained from

$$X_L = X_C$$

$$\omega L = \frac{1}{\omega C}$$

or

$$\omega = \frac{1}{\sqrt{LC}} = \omega_0$$

When the generator frequency ω equals the natural frequency ω_0, the impedance is smallest, I_{max} is greatest, and the circuit is said to be at **resonance**. The natural frequency ω_0 is therefore also called the **resonance frequency**. This resonance condition in a driven *RLC* circuit is similar to that in a driven simple harmonic oscillator.

Since neither an inductor nor a capacitor dissipates energy, the average power delivered to a series *RLC* circuit is the average power supplied to the resistor. The instantaneous power supplied to the resistor is

$$P = I^2R = [I_{max} \cos (\omega t - \delta)]^2R$$

Averaging over one or more cycles and using $(\cos^2 \theta)_{av} = \frac{1}{2}$, we obtain for the average power

$$P_{av} = \tfrac{1}{2}I_{max}^2R = I_{rms}^2R \tag{31-56}$$

Using $R/Z = \cos\delta$ from Figure 31-18 and $I = \mathcal{E}/Z$ this can be written

$$P_{av} = \tfrac{1}{2}\mathcal{E}_{max}I_{max}\cos\delta = \mathcal{E}_{rms}I_{rms}\cos\delta \qquad \text{31-57}$$

The quantity $\cos\delta$ is called the **power factor** of the *RLC* circuit. At resonance, δ is zero, and the power factor is 1.

The power can also be expressed as a function of the angular frequency ω. Using $I_{rms} = \mathcal{E}_{rms}/Z$, Equation 31-56 becomes

$$P_{av} = I_{rms}^2 R = \mathcal{E}_{rms}^2 \frac{R}{Z^2}$$

From the definition of impedance Z, we have

$$Z^2 = (X_L - X_C)^2 + R^2 = \left(\omega L - \frac{1}{\omega C}\right)^2 + R^2$$

$$= \frac{L^2}{\omega^2}\left(\omega^2 - \frac{1}{LC}\right)^2 + R^2$$

$$= \frac{L^2}{\omega^2}(\omega^2 - \omega_0^2)^2 + R^2$$

where we have used $\omega_0^2 = 1/LC$. Using this expression for Z^2, we obtain the average power as a function of ω:

$$P_{av} = \frac{\mathcal{E}_{rms}^2 R\omega^2}{L^2(\omega^2 - \omega_0^2)^2 + \omega^2R^2} \qquad \text{31-58}$$

Figure 31-19 shows the average power supplied by the generator to the circuit as a function of generator frequency for two different values of the resistance R. These curves, called **resonance curves,** are the same as the power-versus-frequency curves for a driven damped oscillator (see Section 14-5). The average power is maximum when the generator frequency equals the resonance frequency. When the resistance is small, the resonance curve is narrow; when it is large, the curve is broad. A resonance curve can be characterized by the **resonance width** $\Delta\omega$. As shown in Figure 31-19, the resonance width is the frequency difference between the two points on the curve where the power is half its maximum value. When the width is small compared with the resonance frequency, the resonance is sharp, that is, the resonance curve is narrow.

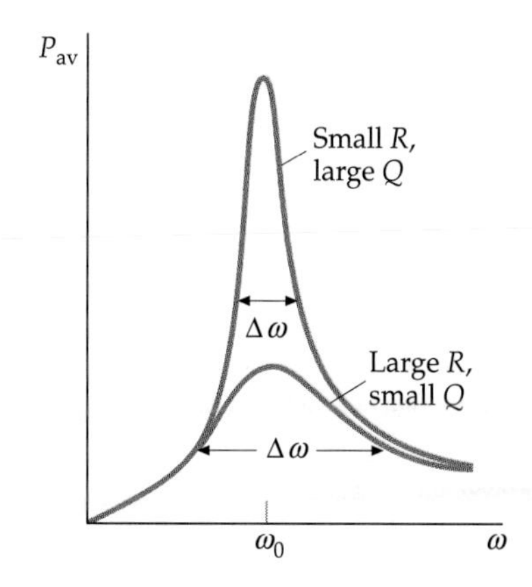

Figure 31-19 Plot of average power versus frequency for a series *RLC* circuit. The power is maximum when the frequency of the generator ω equals the natural frequency of the circuit $\omega_0 = 1/\sqrt{LC}$. If the resistance is small, the Q factor is large and the resonance is sharp. The resonance width $\Delta\omega$ of the curves is measured between points where the power is half its maximum value.

In Chapter 14, the Q factor for a mechanical oscillator was defined as $Q = \omega_0 m/b$ (Equation 14-39) where m is the mass and b is the damping constant. We then saw that $Q = 2\pi E/|\Delta E|$ (Equation 14-41), where E is the total energy of the system and ΔE is the energy lost in one cycle. The **Q factor** for an *RLC* circuit can be defined in a similar way. Since L is analogous to the mass m and R is analogous to the damping constant b, the Q factor for an *RLC* circuit is given by

$$Q = \frac{2\pi E}{|\Delta E|} = \frac{\omega_0 L}{R} \qquad \text{31-59}$$

When the resonance curve is reasonably narrow (that is, when Q is greater than about 2 or 3), the Q factor can be approximated by

$$Q \approx \frac{\omega_0}{\Delta\omega} = \frac{f_0}{\Delta f} \qquad \text{31-60}$$

Q factor for an RLC circuit

Resonance circuits are used in radio receivers, where the resonance frequency of the circuit is varied by varying the capacitance. Resonance occurs when the natural frequency of the circuit equals one of the frequencies of the radio waves picked up at the antenna. At resonance, there is a relatively large

optional

current in the antenna circuit. If the Q factor of the circuit is sufficiently high, currents due to other station frequencies off resonance will be negligible compared with those due to the station frequency to which the circuit is tuned.

Example 31-5

A series RLC circuit with $L = 2$ H, $C = 2\ \mu$F, and $R = 20\ \Omega$ is driven by a generator with a maximum emf of 100 V and a variable frequency. Find (*a*) the resonance frequency f_0, (*b*) the Q value, (*c*) the width of the resonance Δf, and (*d*) the maximum current at resonance.

Picture the Problem The resonance frequency is found from $\omega_0 = 1/\sqrt{LC}$ and the Q value is found from $Q = \omega_0 L/R$.

(*a*) The resonance frequency is $f_0 = \omega_0/2\pi$:

$$f_0 = \frac{\omega_0}{2\pi} = \frac{1}{2\pi\sqrt{LC}}$$

$$= \frac{1}{2\pi\sqrt{(2\text{ H})(2 \times 10^{-6}\text{ F})}} = 79.6\text{ Hz}$$

(*b*) Use this result to calculate Q:

$$Q = \frac{\omega_0 L}{R} = \frac{2\pi(79.6\text{ Hz})(2\text{ H})}{20\ \Omega} = 50$$

(*c*) Use the value of Q to find the width of the resonance Δf:

$$\Delta f = \frac{f_0}{Q} = \frac{79.6\text{ Hz}}{50} = 1.59\text{ Hz}$$

(*d*) At resonance the impedance is just R and I_{max} is $\mathcal{E}/R$:

$$I_{max} = \frac{\mathcal{E}_{max}}{R} = \frac{100\text{ V}}{20\ \Omega} = 5\text{ A}$$

Remark The width is only 1.59 Hz, which is small compared with the resonance frequency of 79.6 Hz, so the resonance peak is quite sharp.

Example 31-6 *try it yourself*

If the generator in Example 31-5 has a frequency of 60 Hz, find (*a*) the maximum current I_{max}, (*b*) the phase angle δ, (*c*) the power factor, and (*d*) the average power delivered.

Picture the Problem The maximum current is the maximum emf divided by the total impedance of the circuit. The phase angle δ is found from $\tan \delta = (X_L - X_C)/R$. You can use either Equation 31-56 or 31-57 to find the average power delivered.

Cover the column to the right and try these on your own before looking at the answers.

Steps	*Answers*
(*a*) 1. Write the maximum current in terms of the maximum emf and the impedance.	$I_{max} = \mathcal{E}_{max}/Z$
2. Calculate the capacitive and inductive reactances and the total reactance.	$X_C = 1326\ \Omega,\ \ X_L = 754\ \Omega$ $X_L - X_C = -572\ \Omega$

3. Calculate the total impedance Z. — $Z \approx 572\ \Omega$

4. Use the results of steps 2 and 3 to calculate I_{max}. — $I_{max} = 0.175$ A

(*b*) Use the results of steps 2 and 3 above to calculate tan δ. Then find δ. — $\tan\delta = \dfrac{X_L - X_C}{R} = -28.6, \quad \delta = -88.0°$

(*c*) Use your value of δ to compute the power factor. — $\cos\delta = 0.0349$

(*d*) Calculate the average power delivered from Equation 31-56, and check your result using the power factor found in (*c*). — $P_{av} = \frac{1}{2}I_{max}^2 R = 0.306$ W; $P_{av} = \frac{1}{2}\mathcal{E}_{max}I_{max}\cos\delta = 0.305$ W

Remarks The slight discrepancy in the two calculations of power in (*d*) is due to rounding error in the current.

The generator frequency of 60 Hz is well below the resonance frequency of 79.6 Hz. (Recall that the width as calculated in Example 31-5 is only 1.59 Hz.) As a result, the total reactance is much greater in magnitude than the resistance. This is always the case far from resonance. Similarly, the maximum current of 0.175 A is much less than I_{max} at resonance, which was found to be 5 A. Finally, we see from Figure 31-17 that a negative phase angle means that the current leads the generator voltage.

Example 31-7 *try it yourself*

Find the maximum voltage across the resistor, the inductor, and the capacitor at resonance for the circuit in Example 31-5.

Picture the Problem The maximum voltage across the resistor is I_{max} times R. Similarly, the maximum voltage across the inductor or capacitor is I_{max} times the appropriate reactance. We found $I_{max} = 5$ A and $f_0 = 79.6$ Hz in Example 31-5.

Cover the column to the right and try these on your own before looking at the answers.

Steps	**Answers**
1. Calculate $V_{R,max} = I_{max}R$.	$V_{R,max} = I_{max}R = 100$ V
2. Express $V_{L,max}$ in terms of I_{max} and X_L.	$V_{L,max} = I_{max}X_L$
3. Calculate X_L at resonance.	$X_L = \omega_0 L = 2\pi f_0 L = 1000\ \Omega$
4. Use your result in step 3 to calculate $V_{L,max}$.	$V_{L,max} = I_{max}X_L = 5000$ V
5. Express $V_{C,max}$ in terms of I_{max} and X_C.	$V_{C,max} = I_{max}X_C$
6. Calculate X_C at resonance.	$X_C = \dfrac{1}{\omega_0 C} = 1000\ \Omega$
7. Use your result in step 6 to calculate $V_{C,max}$.	$V_{C,max} = I_{max}X_C = 5000$ V

optional

Remarks The inductive and capacitive reactances are equal, as we would expect since we found the resonance frequency by setting them equal. The phasor diagram for the voltages across the resistor, capacitor, and inductor is shown in Figure 31-20. The maximum voltage across the resistor is a relatively safe 100 V, equal to the maximum emf of the generator. However, the maximum voltages across the inductor and the capacitor are a dangerously high 5000 V. These voltages are 180° out of phase. At resonance, the voltage across the inductor at any instant is the negative of that across the capacitor, so they always sum to zero, leaving the voltage across the resistor equal to the emf in the circuit.

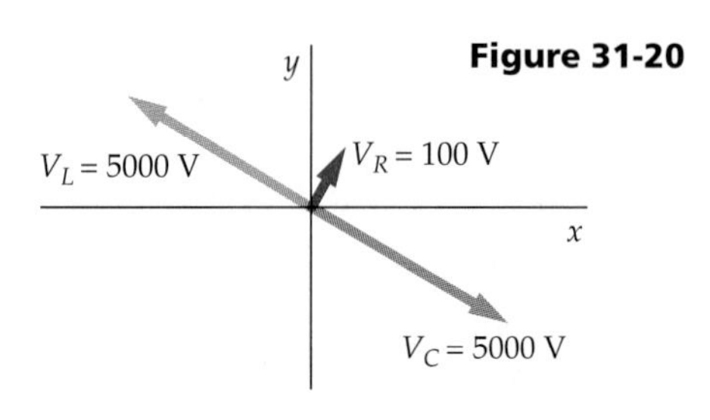

Figure 31-20

Example 31-8

A resistor R and capacitor C are in series with a generator as shown in Figure 31-21. The generator voltage is given by $V_{in} = V_0 \cos \omega t$. Find the rms voltage across the capacitor $V_{out,rms}$ as a function of frequency ω.

Figure 31-21

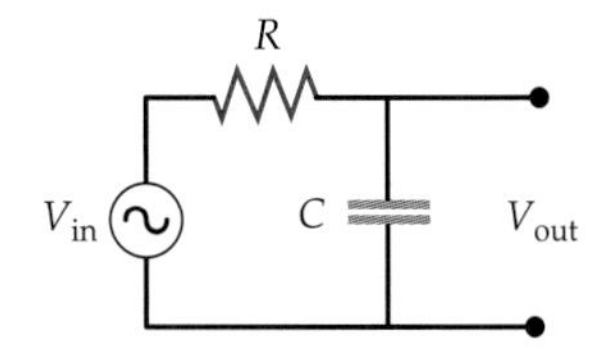

Picture the Problem The rms voltage across the capacitor is related to the rms current and the capacitive reactance. The rms current is found from the input voltage and the impedance.

1. The voltage across the capacitor is I_{rms} times X_C: $V_{out\ rms} = I_{rms}X_C$

2. The rms current depends on the input voltage and the impedance: $I_{rms} = \dfrac{V_{in\ rms}}{Z}$

3. In this circuit, only R and X_C contribute to the total impedance: $Z = \sqrt{R^2 + X_C^2}$

4. Substitute these values and $X_C = 1/\omega C$ to find the output voltage:

$$V_{out\ rms} = I_{rms}X_C = \frac{V_{in\ rms}X_C}{\sqrt{R^2 + X_C^2}}$$

$$= \frac{V_{in\ rms}(1/\omega C)}{\sqrt{R^2 + (1/\omega C)^2}} = \frac{V_{in\ rms}}{\sqrt{\omega^2 C^2 R^2 + 1}}$$

Remarks This circuit is called an *RC low-pass filter*, since it transmits low frequencies with greater amplitude than high frequencies. In fact, the output voltage equals the input voltage in the limit that $\omega \to 0$, but approaches zero for $\omega \to \infty$, as shown in the graph of the ratio of output voltage to input voltage in Figure 31-22.

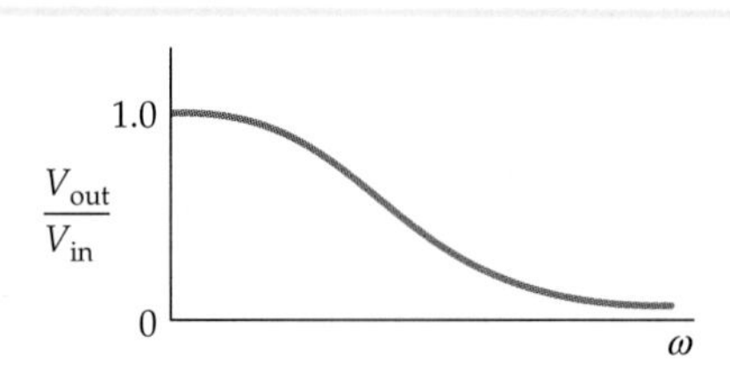

Figure 31-22

Exercise Find the output voltage for this circuit if the capacitor is replaced by an inductor L. (*Answer* $V_{out\ rms} = V_{in\ rms}(\omega L)/\sqrt{R^2 + (\omega L)^2}$. This circuit is a *high-pass filter*.)

A shipboard radio, circa 1920. Exposed at the operator's left are the inductance coils and capacitor plates of the tuning circuit.

Parallel *RLC* Circuit

Figure 31-23 shows a resistor R, capacitor C, and an inductor L connected in parallel across an ac generator. The total current I from the generator divides into three currents, the current I_R in the resistor, the current I_C in the capacitor, and the current I_L in the inductor. The instantaneous voltage V is the same across each element. The current in the resistor is in phase with the voltage and has magnitude V/R. Since the voltage drop across an inductor *leads* the current in the inductor by 90°, I_L *lags* the voltage by 90° and has magnitude V/X_L. Similarly, the I_C leads the voltage by 90° and has magnitude V/X_C. These currents are represented by phasors in Figure 31-24. The total current I is the x component of the vector sum of the individual currents as shown in the figure. The magnitude of the total current is

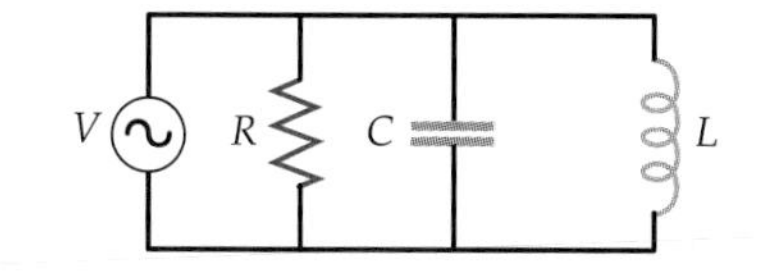

Figure 31-23 A parallel *RLC* circuit.

$$I = \sqrt{I_R^2 + (I_L - I_C)^2} = \sqrt{\left(\frac{V}{R}\right)^2 + \left(\frac{V}{X_L} - \frac{V}{X_C}\right)^2} = \frac{V}{Z} \qquad \text{31-61}$$

where the impedance Z is related to the resistance and the capacitive and inductive reactances by

$$\frac{1}{Z} = \sqrt{\left(\frac{1}{R}\right)^2 + \left(\frac{1}{X_L} - \frac{1}{X_C}\right)^2} \qquad \text{31-62}$$

At resonance, the generator frequency ω equals the natural frequency $\omega_0 = 1/\sqrt{LC}$, and the inductive and capacitive reactances are equal. Then from Equation 31-62 we see that $1/Z$ has its minimum value $1/R$, so the impedance Z is maximum and the total current is minimum. We can understand this if we note that at resonance $X_C = X_L$, and the currents in the inductor and capacitor are equal but 180° out of phase, so the total current is just the current in the resistor.

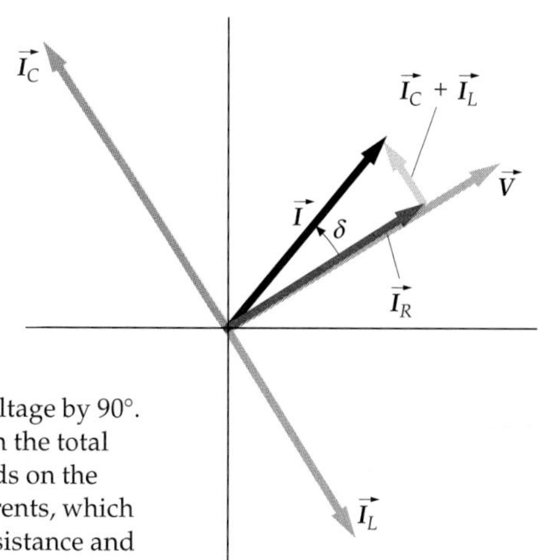

Figure 31-24 Phasor diagram for the currents in the parallel *RLC* circuit of Figure 31-23. The voltage is the same across each element. The current in the resistor is in phase with the voltage. The current in the capacitor leads the voltage by 90° and that in the inductor lags the voltage by 90°. The phase difference δ between the total current and the voltage depends on the relative magnitudes of the currents, which depend on the values of the resistance and of the capacitive and inductive reactances.

Solving *RLC* Circuits Using Complex Numbers

The methods discussed in the essay "Exploring . . . Using Complex Numbers to Solve the Oscillator Equations" in Chapter 14 for solving the driven oscillator equations apply equally well to the solution of *RLC* circuits. Equation 31-49 is

$$L\frac{d^2Q}{dt^2} + R\frac{dQ}{dt} + \frac{Q}{C} = \mathcal{E}_{\max}\cos\omega t$$

Since we are interested in the current, we write $I = dQ/dt$ and $Q = \int I\,dt$. Then

$$L\frac{dI}{dt} + RI + \frac{1}{C}\int I\,dt = \mathcal{E}_{\max}\cos\omega t \qquad 1$$

We note that the right side of this equation is the real part of $\mathcal{E}_{\max}e^{i\omega t}$. To solve Equation 1 we then find the complex function z that satisfies

$$L\frac{dz}{dt} + Rz + \frac{1}{C}\int z\,dt = \mathcal{E}_{\max}e^{i\omega t} \qquad 2$$

The real part of our solution will be the current I. As we did in Chapter 14, we try

$$z = I_0 e^{i\omega t} \qquad 3$$

Then $dz/dt = i\omega z$, and $\int z\,dt = (1/i\omega)z$. Substituting these results into Equation 2 gives

$$i\omega Lz + Rz + \frac{1}{i\omega C}z = \mathcal{E}_{\max}e^{i\omega t} = \frac{\mathcal{E}_{\max}}{I_0}z \qquad 4$$

Divide each term by z and write $\omega L = X_L$ and $1/\omega C = X_C$. Then, using $X_C/i = -iX_C$, we obtain

$$i(X_L - X_C) + R = \frac{\mathcal{E}_{\max}}{I_0} \qquad 5$$

Solve for I_0

$$I_0 = \frac{\mathcal{E}_{\max}}{i(X_L - X_C) + R} \qquad 6$$

Again, we put the denominator of Equation 6 in polar form (Figure 1):

$$i(X_L - X_C) + R = \sqrt{(X_L - X_C)^2 + R^2}\,e^{i\delta}$$

$$= Ze^{i\delta} \qquad 7$$

where $Z = \sqrt{(X_L - X_C)^2 + R^2}$ is the impedance and

$$\tan\delta = \frac{X_L - X_C}{R} \qquad 8$$

Then

$$I_0 = \frac{\mathcal{E}_{\max}}{Ze^{i\delta}} = \frac{\mathcal{E}_{\max}}{Z}e^{-i\delta}$$

and

$$z = I_0 e^{i\omega t} = \frac{\mathcal{E}_{\max}}{Z}e^{-i\delta}e^{i\omega t} = \frac{\mathcal{E}_{\max}}{Z}e^{i(\omega t - \delta)}$$

The current is thus given by

$$I = \mathrm{Re}(z) = \frac{\mathcal{E}_{\max}}{Z}\cos(\omega t - \delta)$$

which is the same as Equation 31-54.

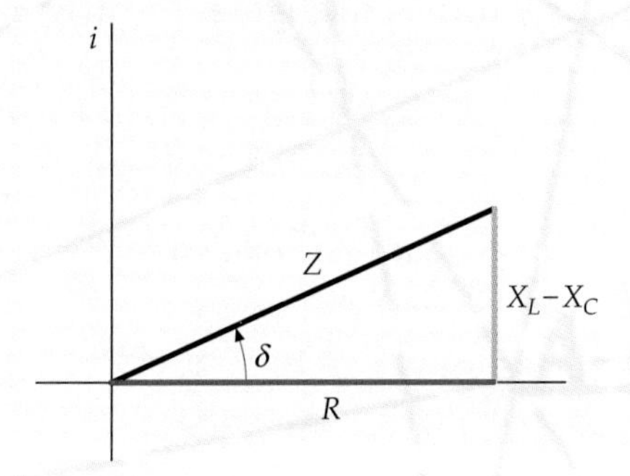

Figure 1 Polar representation of $i(X_L - X_C) + R$. The magnitude is $Z = \sqrt{(X_L - X_C)^2 + R^2}$ and the phase angle is δ.

Complex Impedances

The complex number technique described above can be extended into a generalized form of Ohm's

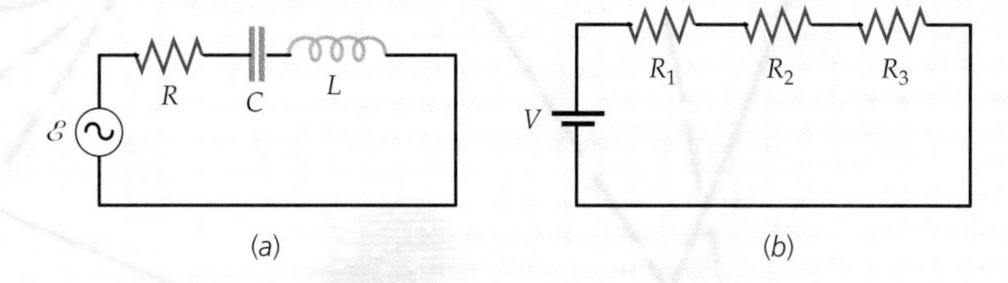

Figure 2 (*a*) *RLC* series circuit. (*b*) Analogous dc circuit.

law that makes ac problems look like dc problems. Consider the simple dc circuit shown next to our series *RLC* circuit in Figure 2. In the dc circuit, the applied voltage equals the sum of the voltage drops:

$$V = I(R_1 + R_2 + R_3)$$

If we multiply each term in Equation 4 by I_0/z, we obtain

$$\mathcal{E}_{max} = I_0\left(R + \frac{1}{i\omega C} + i\omega L\right) \qquad 9$$

The quantities R, $1/i\omega C$, and $i\omega L$ are the complex impedances for these circuit elements. Just as the resistance is the ratio of the voltage to current for a resistor, the complex impedance is the ratio of the complex voltage to the complex current for a circuit element. We illustrate the use of complex impedance by considering a circuit with an inductor and capacitor in parallel, and with the combination in series with a resistor, as shown in Figure 3*a*. The analogous dc circuit is shown in Figure 3*b*.

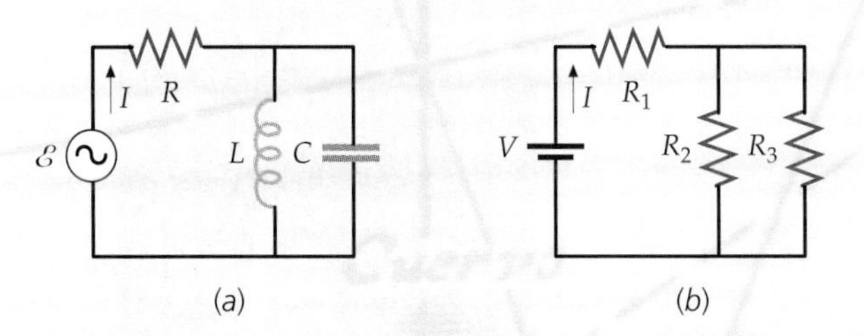

Figure 3 (*a*) An ac circuit with inductance and capacitance in parallel, and with the combination in series with a resistance. (*b*) An analogous dc circuit.

To find the current in the dc circuit, we first find the total resistance by adding R_1 to the resistance of the parallel combination of R_2 and R_3. The result is

$$R_{eq} = R_1 + \frac{R_2R_3}{R_2 + R_3}$$

The total impedance of the ac circuit is found in the same way.

$$Z_{eq} = Z_1 + \frac{Z_2Z_3}{Z_2 + Z_3} = R + \frac{(i\omega L)(1/i\omega C)}{i\omega L + 1/i\omega C}$$

$$= R + i\frac{L/C}{(1/\omega C) - \omega L} = R + i\frac{L/C}{X_C - X_L} \qquad 10$$

In polar form, the complex equivalent impedance is

$$Z_{eq} = \sqrt{R^2 + \left(\frac{L/C}{X_C - X_L}\right)^2}\, e^{i\delta} \qquad 11$$

where

$$\tan\delta = \frac{L/C}{R(X_C - X_L)} \qquad 12$$

If the emf is given by $\mathcal{E} = \mathcal{E}_{max}\cos\omega t = \text{Re}\,(\mathcal{E}_{max}e^{i\omega t})$, the current is given by

$$I = \text{Re}\left(\frac{\mathcal{E}_{max}e^{i\omega t}}{Z_{eq}}\right) = I_{max}\cos(\omega t - \delta) \qquad 13$$

where

$$I_{max} = \frac{\mathcal{E}_{max}}{|Z_{eq}|} = \frac{\mathcal{E}_{max}}{\sqrt{R^2 + \left(\frac{L/C}{X_C - X_L}\right)^2}} \qquad 14$$

and the phase angle δ is given by Equation 12. Note that the current goes to zero when $X_C = X_L$, that is, at the frequency $\omega = 1/\sqrt{LC}$ for which $\omega L = 1/\omega C$. At this frequency, the parallel combination is equivalent to an *open* circuit rather than a *short* circuit in the series combination.

optional

31-7 The Transformer

A transformer is a device used to raise or lower the voltage in a circuit without an appreciable loss of power. Figure 31-25 shows a simple transformer consisting of two wire coils around a common iron core. The coil carrying the input power is called the primary, and the other coil is called the secondary. Either coil of a transformer can be used for the primary or secondary. The transformer operates on the principle that an alternating current in one circuit induces an alternating emf in a nearby circuit due to the mutual inductance of the two circuits. The iron core increases the magnetic field for a given current and guides it so that nearly all the magnetic flux through one coil goes through the other coil. If no power were lost, the product of the voltage and the current in the secondary circuit would equal the product of the voltage and the current in the primary circuit. Thus, if the voltage is raised, the current is lowered, and vice versa. Power losses arise because of the Joule heating in the small resistances in both coils, or in current loops within the core* and from hysteresis in the iron cores. We will neglect these losses and consider an ideal transformer of 100% efficiency, for which all of the power supplied to the primary coil appears in the secondary coil. Actual transformers are often 90 to 95% efficient.

V_1 N_1 N_2

Figure 31-25 Transformer with N_1 turns in the primary and N_2 turns in the secondary.

Consider a transformer with an emf V_1 across the primary of N_1 turns; the secondary coil of N_2 turns is an open circuit. Because of the iron core, there is a large flux through each coil even when the magnetizing current I_{m} in the primary circuit is very small. We can ignore the resistances of the coils, which are negligible in comparison with their inductive reactances. The primary is then a simple circuit consisting of an ac generator and a pure inductance like that discussed in Section 31-2. The (magnetizing) current and the voltage in the primary are out of phase by 90°, and the average power dissipated in the primary coil is zero. If ϕ_{turn} is the magnetic flux in one turn of the primary coil, the voltage drop across the primary coil is $V_{L1} = N_1\, d\phi_{\text{turn}}/dt$. Applying Kirchhoff's loop rule to the primary circuit then gives

$$V_1 - N_1\frac{d\phi_{\text{turn}}}{dt} = 0$$

or

$$V_1 = N_1\frac{d\phi_{\text{turn}}}{dt} \qquad \text{31-63}$$

If there is no flux leakage out of the iron core, the flux through each turn is the same for both coils. Thus, the total flux through the secondary coil is $N_2\,\phi_{\text{turn}}$, and the voltage across the secondary coil is

$$V_2 = N_2\frac{d\phi_{\text{turn}}}{dt} \qquad \text{31-64}$$

Comparing equations 31-63 and 31-64, we can see that

$$V_2 = \frac{N_2}{N_1}V_1 \qquad \text{31-65}$$

If N_2 is greater than N_1, the voltage in the secondary coil is greater than that in the primary coil, and the transformer is called a step-up transformer. If N_2 is less than N_1, the voltage in the secondary coil is less than that in the primary coil, and the transformer is called a step-down transformer.

(a)

(b)

(*a*) A suburban power substation where transformers step down voltage from high-voltage transmission lines. (*b*) A power box with transformer for stepping down voltage for distribution to homes.

* The induced currents, called eddy currents, can be greatly reduced by using a core of laminated metal to break up current paths.

When we put a resistance R, called a load resistance, across the secondary coil, there will then be a current I_2 in the secondary circuit that is in phase with the voltage V_2 across the resistance. This current sets up an additional flux ϕ'_{turn} through each turn that is proportional to N_2I_2. This flux opposes the original flux set up by the original magnetizing current I_m in the primary. However, the voltage across the primary coil is determined by the generator emf, which is unaffected by the secondary circuit. According to Equation 31-64, the flux in the iron core must change at the original rate; that is, the total flux in the iron core must be the same as when there is no load across the secondary. The primary coil thus draws an additional current I_1 to maintain the original flux ϕ_{turn}. The flux through each turn produced by this additional current is proportional to N_1I_1. Since this flux equals $-\phi'_{turn}$, the additional current I_1 in the primary is related to the current I_2 in the secondary by

$$N_1I_1 = -N_2I_2 \qquad \text{31-66}$$

These currents are 180° out of phase and produce counteracting fluxes. Since I_2 is in phase with V_2, the additional current I_1 is in phase with the applied emf. The power input from the generator is $V_{1,rms}\,I_{1,rms}$, and the power output is $V_{2,rms}I_{2,rms}$. (The magnetizing current does not contribute to the power input because it is 90° out of phase with the generator voltage.) If there are no losses,

$$V_{1,rms}I_{1,rms} = V_{2,rms}I_{2,rms} \qquad \text{31-67}$$

In most cases the additional current in the primary I_1 is much greater than the original magnetizing current I_m that is drawn from the generator when there is no load. This can be demonstrated by putting a light bulb in series with the primary coil. The bulb is much brighter when there is a load across the secondary than when the secondary circuit is open. If I_m can be neglected, Equation 31-67 relates the total currents in the primary and secondary circuits.

Example 31-9

A doorbell requires 0.4 A at 6 V. It is connected to a transformer whose primary, containing 2000 turns, is connected to a 120-V ac line. (*a*) How many turns should there be in the secondary? (*b*) What is the current in the primary?

Picture the Problem We can find the number of turns from the turns ratio, which equals the voltage ratio. The primary current can be found by equating the power out to the power in.

(*a*)1. The turns ratio can be obtained from Equation 31-65:

$$\frac{N_2}{N_1} = \frac{V_2}{V_1} = \frac{6\text{ V}}{120\text{ V}} = \frac{1}{20}$$

2. Solve for the number of turns in the secondary, N_2:

$$N_2 = \frac{1}{20}(2000\text{ turns}) = 100\text{ turns}$$

(*b*)1. Since we are assuming 100% efficiency in power transmission, the input and output currents are related by Equation 31-66:

$$V_2I_2 = V_1I_1$$

2. Solve for the current in the primary, I_1:

$$I_1 = \frac{V_2}{V_1}I_2 = \frac{6\text{ V}}{120\text{ V}}(0.4\text{ A}) = 0.02\text{ A}$$

An important use of transformers is in the transport of electrical power. To minimize the I^2R heat loss in transmission lines, it is economical to use a high voltage and a low current. On the other hand, safety and other considerations require that power be delivered to consumers at lower voltages and therefore with higher currents. Suppose, for example, that each person in a city with a population of 50,000 uses 1.2 kW of electric power. (The per capita consumption of power in the United States is actually somewhat higher than this). At 120 V, the current required for each person would be

$$I = \frac{1200\text{ W}}{120\text{ V}} = 10\text{ A}$$

The total current for 50,000 people would then be 500,000 A. The transport of such a current from a power-plant generator to a city many kilometers away would require conductors of enormous size, and the I^2R power loss would be substantial. Rather than transmit the power at 120 V, step-up transformers are used at the power plant to step up the voltage to some very large value, such as 600,000 V. For this voltage, the current needed is only

$$I = \frac{120\text{ V}}{600{,}000\text{ V}}(500{,}000\text{ A}) = 100\text{ A}$$

To reduce the voltage to a safer level for transport within a city, power stations are located just outside the city to step down the voltage to a safer value, such as 10,000 V. Transformers in boxes attached to the power poles outside each house again step down the voltage to 120 V (or 240 V) for distribution to the house. It is because of the ease of stepping the voltage up or down with transformers that alternating current rather than direct current is in common use.

Example 31-10

A transmission line has a resistance of 0.02 Ω/km. Calculate the I^2R power loss if 200 kW of power is transmitted from a power generator to a city 10 km away at (*a*) 240 V and (*b*) 4.4 kV.

Picture the Problem First, note that the total resistance of 10 km of wire is $R = (0.02\ \Omega/\text{km})(10\text{ km}) = 0.2\ \Omega$. In each case, begin by finding the current needed to transmit 200 kW using $P = IV$, then find the power loss using I^2R.

(*a*)1. Find the current needed to transmit 200 kW of power at 240 V:

$$I = \frac{P}{V} = \frac{200\text{ kW}}{240\text{ V}} = 833\text{ A}$$

2. Calculate the power loss:

$$I^2R = (833\text{ A})^2(0.2\ \Omega) = 139{,}000\text{ W}$$

(*b*)1. Now find the current needed to transmit 200 kW of power at 4.4 kV:

$$I = \frac{P}{V} = \frac{200\text{ kW}}{4.4\text{ kV}} = 45.5\text{ A}$$

2. Calculate the power loss:

$$I^2R = (45.5\text{ A})^2(0.2\ \Omega) = 414\text{ W}$$

Remark Note that with a transmission voltage of 240 V almost 70% of the power is wasted through heat loss, but with transmission at 4.4 kV only about 0.2% is lost. This illustrates the advantage of high-voltage power transmission.

Electric Motors

John Dentler
United States Naval Academy

The wide variety of applications for electric motors requires many different designs. Electric-clock motors must operate at a precise speed. Automobile starters must deliver a tremendous torque from a standstill. A hand-held hair dryer must be lightweight and operate at several different speeds. Engineers design motors for various applications using models derived from the physical principles discussed in this text. These models are equations that predict a motor's performance for a set of specific applications or loads.

The rotating motor (Figure 1) has current-carrying conductors that react with an external field. The field, called the *stator field,* is created and controlled by the coil of wire visible at the bottom of Figure 1. Flux from the wire coil passes through the core, creating a north pole on the left and a south pole on the right of the rotating element. The rotating assembly, called the *armature,* consists of an iron cylinder with eight slots that contain conductors. If current can be driven through these conductors in the direction shown (front to back near the south pole and back to front near the north pole), then a net clockwise torque (down on the south-pole side and up on the north-pole side) will be developed to turn the armature.

Constructing a device that maintains the proper current direction in each conductor as the armature turns is a complicated task. Such devices are called *commutator brush assemblies.* The commutator brush assembly shown in Figure 2 consists of four segments that protrude along the motor shaft and two brushes that conduct current from a source to the segments. Each segment is connected to two conductors, which run through the slots of the rotating assembly. The conductors are interconnected through wires in the rear of the rotating assembly and by the commutator segments in the front of the assembly. This method of connection results in two parallel paths between the brushes; thus all of the conductors are used all of the time.

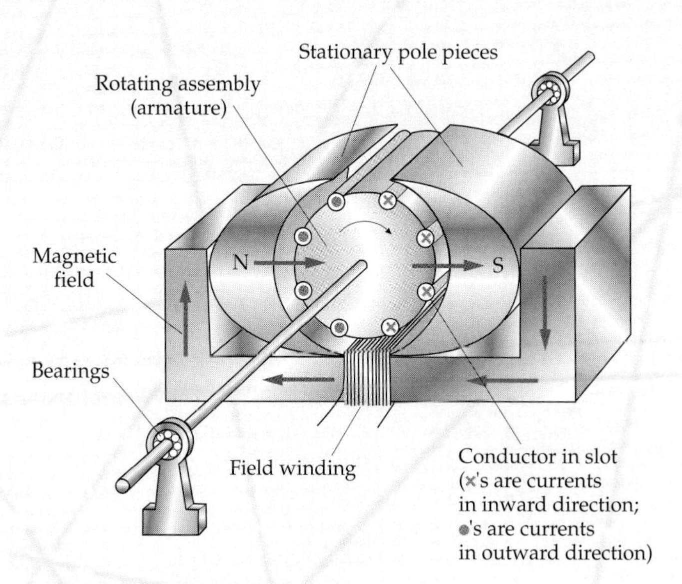

Figure 1 A simple rotating electric motor.

In the commutator shown in Figure 2, current delivered from the brush on the right follows one of the two parallel paths through the armature. The conductors in slots 2 and 5 both carry the current from the front to the back of the armature.

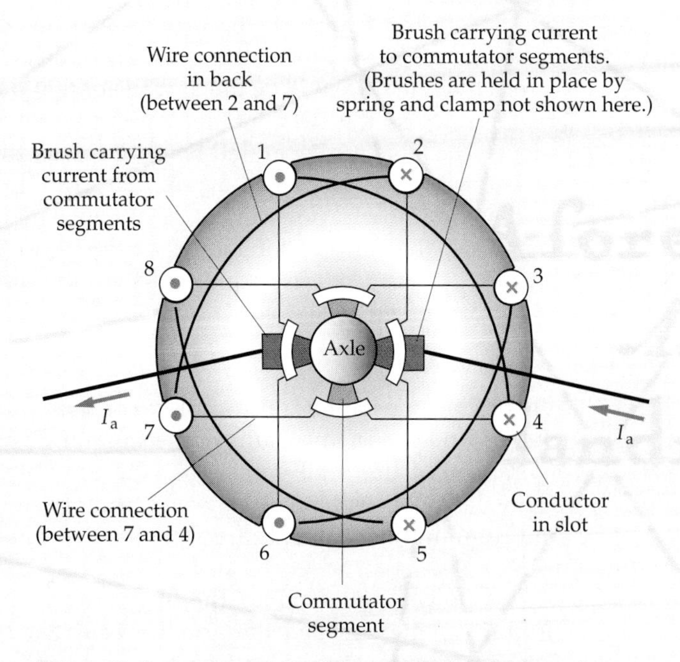

Figure 2 A commutator brush assembly for the motor in Figure 1. A commutator, in its most general sense, is a switching device. The device shown switches current direction through the armature to maintain the clockwise rotation.

Conductors 2 and 5 are connected to conductors 7 and 8 via wires in the rear of the armature. The current returns to the front through slots 7 and 8, which are connected to slots 3 and 4 via the common connection on the commutator segments. The current is carried to the back along slots 3 and 4 and is then returned to the front through slots 1 and 6, where it is picked off by the brush on the left. The commutator assembly rotates with the armature and moves under the brushes. The brushes are stationary and will contact different commutator segments when the armature has moved 90°. Since the armature is symmetrically wound, the slots on the right will always carry current from front to back and the slots on the left will always carry current from back to front, thus maintaining the clockwise torque.

The total torque turning the motor is the sum of the torques exerted by the conductors in each slot. In any position there are four armature conductors acting on the right and four on the left; therefore, the torque is approximately constant. Similarly, the total emf developed between the brushes is the sum of the emfs on each conductor. At any position there are two parallel paths, each consisting of four conductors. From Section 30-4, the emf developed across a single length of wire can be shown to be

$$\mathcal{E} = B\ell r\omega \sin(\omega t + \delta) \qquad 1$$

where ℓ is the length of the armature (front to back) and r is the radius of the rotor. The total emf across the armature will be the average of that developed across the two parallel paths described above. The slots are separated by only 45°, so the variation in the total emf across the armature as the armature turns will be relatively small. Therefore, the time-varying term of Equation 1, $\sin(\omega t + \delta)$, can be discarded and the total emf can be expressed as

$$\mathcal{E}_{\text{total}} = BK\omega \qquad 2$$

where the constant K, the motor constant, includes ℓ and r and the results of the summing and averaging of the total emf across the armature. The validity of Equation 2 improves as more slots and commutator segments are added to the armature.

The power delivered to the armature is the product of the emf and the armature current I_a. For a rotating motor, the load is a torque τ applied to the shaft opposing the direction of rotation. The mechanical power delivered to the load is the product of the torque and the angular velocity. At equilibrium, the driving torque from the motor is equal and opposite to the load torque. Thus,

$$P = \mathcal{E}I_a = \tau\omega \qquad 3$$

Substituting $BK\omega$ for the emf from Equation 7, we obtain

$$P = BKI_a\omega = \tau\omega \qquad 4$$

We can represent the armature by a simple voltage source with an external resistance R_a. The field winding connections of the coil shown at the bottom of Figure 1 can be connected either in series or in parallel (shunt) with the armature. These two methods of connection yield motors with extremely different characteristics.

Parallel (Shunt) Connection

Figure 3 shows the circuit for the shunt or parallel field connection. A variable resistance controls the field and thereby the speed of the motor. Applying Kirchhoff's loop rule to this circuit yields

$$V - I_aR_a - BK\omega = 0 \qquad 5$$

which can be rearranged to express the rotational speed ω in terms of the armature current I_a:

$$\omega = -\frac{R_a}{BK}I_a + \frac{V}{BK} \qquad 6$$

If we substitute τ/BK for the current from Equation 4, the rotational speed is

$$\omega = -\frac{R_a}{(BK)^2}\tau + \frac{V}{BK} \qquad 7$$

Equation 7 is a linear equation relating the rotational speed to the load. The speed can be controlled either by varying the voltage V or, more commonly, by varying the current into the coil by varying the resistance.

At high armature currents, the armature core saturates, the voltage drop due to the armature inductance becomes significant, and the relationship between the torque and speed becomes nonlinear. However, for normal loads, Equation 7

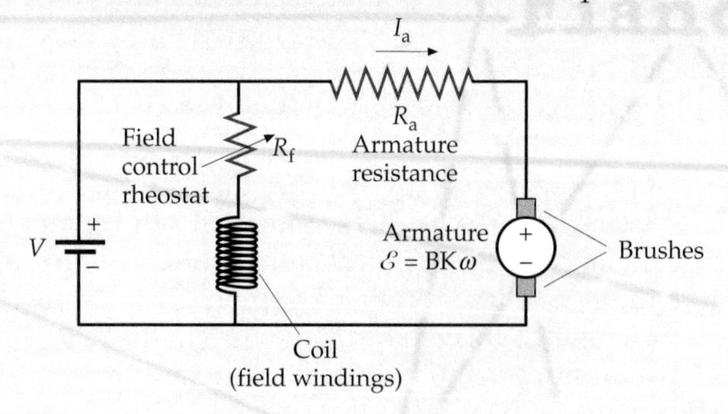

Figure 3 Circuit for a typical dc shunt motor.

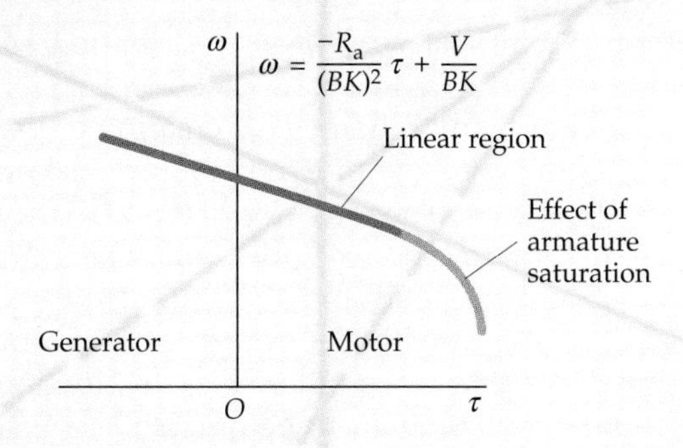

Figure 4 Graph of torque versus rotational speed showing the effect of armature saturation on the performance characteristic of a typical dc shunt motor.

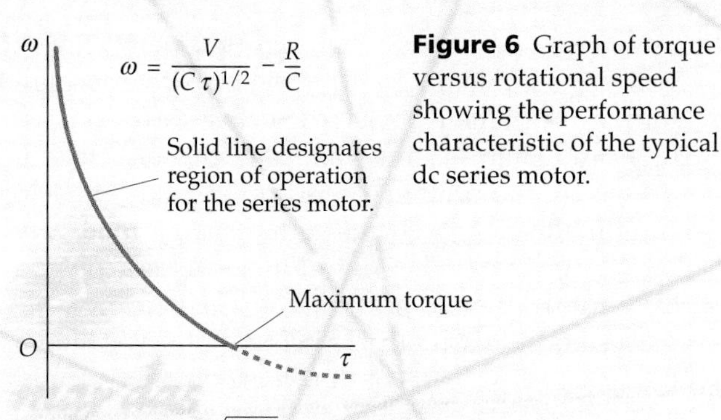

Figure 6 Graph of torque versus rotational speed showing the performance characteristic of the typical dc series motor.

accurately describes the motor's operation. Figure 4 shows how the *performance characteristic* of the motor, or its speed versus torque, is affected by armature saturation.

Series Connection

In the motor circuit of Figure 5, the coil is connected in series with the armature, so the field strength is a function of the armature current.

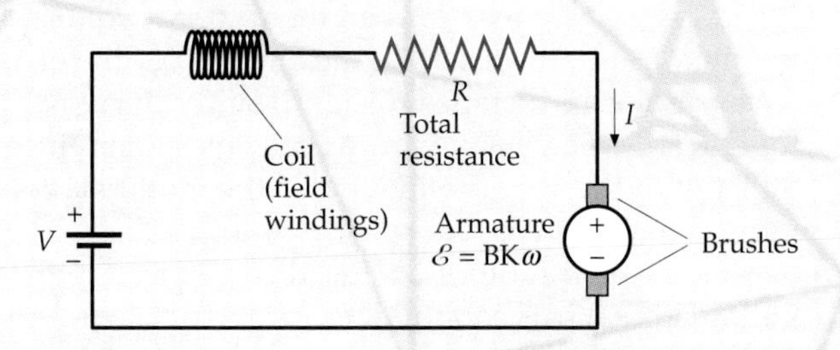

Figure 5 Circuit for a typical dc series motor.

If the armature current is small and the field does not saturate, the product of the field strength and the motor constant K can be expressed as a linear function of the armature current:

$$BK = CI$$

where C is some constant. Substituting this value of BK in the expressions for the armature emf, power, and torque yields

$$\mathcal{E} = CI\omega P = CI^2\omega \quad \text{and} \quad \tau = CI^2$$

The Kirchhoff loop rule then gives

$$V - IR - CI\omega = 0$$

where R represents the total resistance of the coil and armature, and I represents the only current in the circuit. This gives the following speed-versus-current equation:

$$\omega = \frac{V}{CI} - \frac{R}{C}$$

Substituting $\sqrt{\tau/C}$ for I, we obtain the equation for the speed versus torque, which is the performance characteristic for the series motor:

$$\omega = \frac{V}{(C\tau)^{1/2}} - \frac{R}{C}$$

Figure 6 shows the performance characteristic for the series motor. Comparing this performance characteristic with that for the shunt motor reveals striking differences. At low torques, the series motor runs very fast, almost without limit (the only load is the friction of its bearings and the air around the rotor) whereas the shunt motor is regulated to run close to the speed V/BK. At high torques, the speed of the shunt motor tapers off and the motor stalls, but the series motor delivers its greatest torque when the motor is stopped. A series motor is therefore the best choice to start a car engine, which requires a high torque at $\omega = 0$. On the other hand, a shunt motor is the better choice to drive a speed-sensitive load, like a tape recorder.

With only minor modifications, the principles of dc motor construction and operation also apply to ac motors. The torque of a series motor is proportional to I^2 and is thus independent of current direction. This is because the same current is in both the stationary field and rotating armature. With this cursory examination, it might be concluded that any series dc motor would run on alternating current. However, an assumption made to simplify the analysis of the dc motor was that inductance could be ignored. Inductance cannot be ignored when driving a motor from an ac source. Inductance has two effects: (1) It acts as a throttle limiting the amount of ac current for a given input voltage, and (2) it changes the phase relationship of the current and voltage.

A dc shunt motor typically has field windings with high resistance and an armature with high inductance. Applying alternating current to such a motor would create a phase difference between the field and the armature currents resulting in unsatisfactory performance.

A series motor has a very tight magnetic circuit with close tolerances to develop a very high torque in a small package. Such a device has a high inductance, thus limiting the ac current drawn by the motor. A series motor designed to run on alternating current must have a relatively low inductance. The low inductance is achieved by limiting the amount of iron used in the pole pieces and the armature. Such a motor is called a *universal motor.* By its nature, it is both lightweight and limited to driving devices with relatively light loads such as vacuum cleaners, food blenders, hair dryers, and sewing machines. Its performance characteristic is similar to that for the dc series motor shown in Figure 6.

The most common ac motor is the *induction motor.* This motor has a rotating assembly like the one shown in Figure 1, but unlike the dc motor, the commutator and interconnecting wires are replaced with shorting plates, connecting all the slotted conductors, mounted on the front and back. The challenge becomes how to make the shorted rotor rotate. The solution is to make the field from the stator appear to rotate. If the field rotates, there will be a relative velocity between the rotor and the stator field. An emf develops across the shorted rotor, driving current through the conductors in the slots. The rotating stator field produces a torque on the induced current in the rotor. The rotor moves such as to minimize the relative motion between it and the field. Thus, the rotor turns almost as fast as the rotating stator field.

There are many schemes for creating apparent field rotation. The one shown in Figure 7 is known as the *shaded pole.* The motor is identical to the one in Figure 1 with the exception that the rotor is shorted on either end and the stator pole pieces have been sliced, with a conducting band around the small pieces of each pole. This construction allows the magnetic field to be quickly established through the faces of the large pole pieces and to be delayed through the small faces by the inductance of the conducting band. The phase delay between the field through the large pole faces and the field through the small pole faces creates the appearance of a rotating field.

The performance characteristic of a typical induction motor is shown in Figure 8. Normal operation is close to the speed of field rotation. If the motor in Figure 7 were connected to a 60-Hz source, the rotational speed would be somewhat less than 60 rev/s. The maximum torque shown on the performance characteristic curve occurs where the difference between the rotor speed and the field rotation speed is large enough for the effects of rotor inductance to significantly delay rotor currents. The delayed rotor currents cannot interact with the stator field, and so the motor stalls if the load is increased.

Shaded-pole motors are used in devices with light loads such as cooling fans in electrical equipment. More complex schemes for creating field rotation are used in the induction motors for refrigerators and air conditioners. Large industrial induction motors use three-phase electricity to rotate the field.

Figure 7 A shaded-pole induction motor.

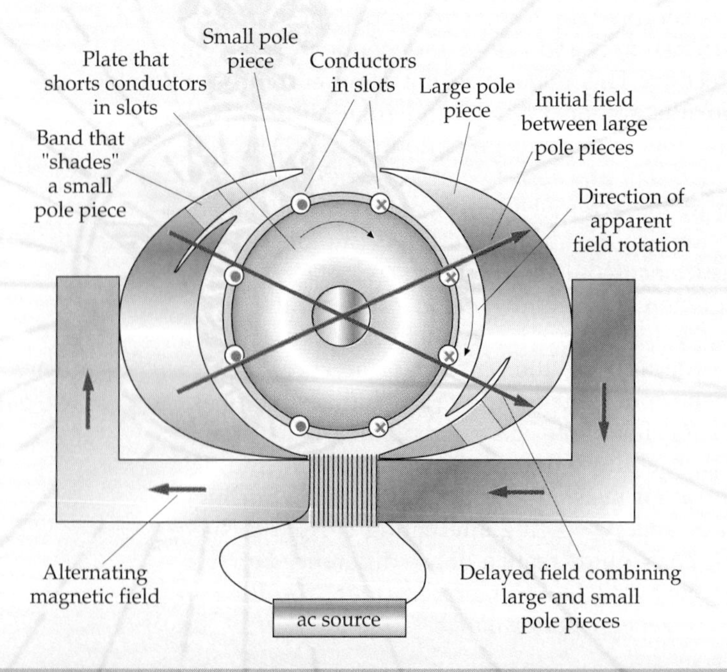

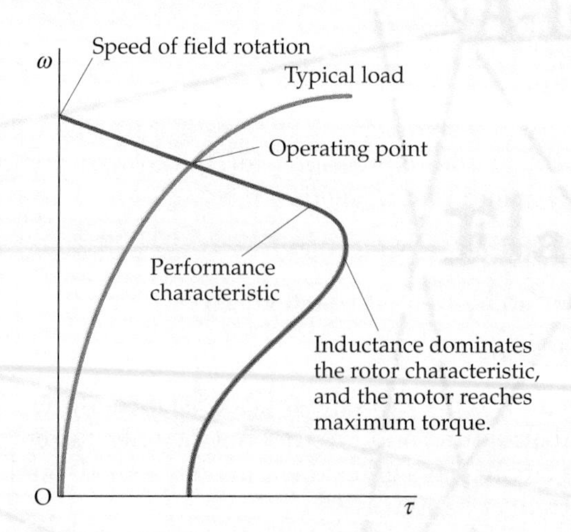

Figure 8 Graph of torque versus speed showing the performance characteristic of a typical induction motor. The load line indicated is typical for a centrifugal pump.

Phasors (optional)

21 • Draw the resultant phasor diagram for a series *RLC* circuit when $V_L < V_C$. Show on your diagram that the emf will lag the current by the phase angle δ given by

$$\tan \delta = \frac{V_C - V_L}{V_R}$$

22 •• Two ac voltage sources are connected in series with a resistor $R = 25\ \Omega$. One source is given by $V_1 = (5.0\text{ V}) \cos(\omega t - \alpha)$, and the other source is $V_2 = (5.0\text{ V}) \cos(\omega t + \alpha)$, with $\alpha = \pi/6$. (*a*) Find the current in *R* using a trigonometric identity for the sum of two cosines. (*b*) Use phasor diagrams to find the current in *R*. (*c*) Find the current in *R* if $\alpha = \pi/4$ and the amplitude of V_2 is increased from 5.0 to 7.0 V.

LC and *RLC* Circuits Without a Generator (optional)

23 • The SI units of inductance times capacitance are

(*a*) seconds squared. (*b*) hertz.
(*c*) volts. (*d*) amperes.
(*e*) ohms.

24 •• Making *LC* circuits with oscillation frequencies of thousands of hertz or more is easy, but making *LC* circuits that have small frequencies is difficult. Why?

25 • Show from the definitions of the henry and the farad that $1/\sqrt{LC}$ has the unit s^{-1}.

26 • (*a*) What is the period of oscillation of an *LC* circuit consisting of a 2-mH coil and a 20-μF capacitor? (*b*) What inductance is needed with an 80-μF capacitor to construct an *LC* circuit that oscillates with a frequency of 60 Hz?

27 •• An *LC* circuit has capacitance C_1 and inductance L_1. A second circuit has $C_2 = \frac{1}{2}C_1$ and $L_2 = 2L_1$, and a third circuit has $C_3 = 2C_1$ and $L_3 = \frac{1}{2}L_1$. (*a*) Show that each circuit oscillates with the same frequency. (*b*) In which circuit would the maximum current be greatest if the capacitor in each were charged to the same potential *V*?

28 •• A 5-μF capacitor is charged to 30 V and is then connected across a 10-mH inductor. (*a*) How much energy is stored in the system? (*b*) What is the frequency of oscillation of the circuit? (*c*) What is the maximum current in the circuit?

RL Circuits With a Generator (optional)

29 • A coil can be considered to be a resistance and an inductance in series. Assume that $R = 100\ \Omega$ and $L = 0.4$ H. The coil is connected across a 120-V-rms, 60-Hz line. Find (*a*) the power factor, (*b*) the rms current, and (*c*) the average power supplied.

30 •• A resistance *R* and a 1.4-H inductance are in series across a 60-Hz ac voltage. The voltage across the resistor is 30 V and the voltage across the inductor is 40 V. (*a*) What is the resistance *R*? (*b*) What is the ac input voltage?

31 •• A coil has a dc resistance of 80 Ω and an impedance of 200 Ω at a frequency of 1 kHz. One may neglect the wiring capacitance of the coil at this frequency. What is the inductance of the coil?

32 •• A single transmission line carries two voltage signals given by $V_1 = (10\text{ V}) \cos 100t$ and $V_2 = (10\text{ V}) \cos 10{,}000t$, where *t* is in seconds. A series inductor of 1 H and a shunting resistor of 1 kΩ is inserted into the transmission line as indicated in Figure 31-29. (*a*) What is the voltage signal observed at the output side of the transmission line? (*b*) What is the ratio of the low-frequency amplitude to the high-frequency amplitude?

Figure 31-29 Problem 32

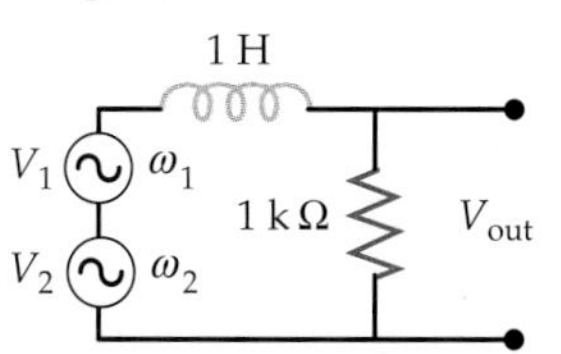

33 •• A coil with resistance and inductance is connected to a 120-V-rms, 60-Hz line. The average power supplied to the coil is 60 W, and the rms current is 1.5 A. Find (*a*) the power factor, (*b*) the resistance of the coil, and (*c*) the inductance of the coil. (*d*) Does the current lag or lead the voltage? What is the phase angle δ?

34 •• A 36-mH inductor with a resistance of 40 Ω is connected to a source whose voltage is $\mathcal{E} = (345\text{ V}) \cos(150\pi t)$, where *t* is in seconds. Determine the maximum current in the circuit, the maximum and rms voltages across the inductor, the average power dissipation, and the maximum and average energy stored in the magnetic field of the inductor.

35 •• A coil of resistance *R*, inductance *L*, and negligible capacitance has a power factor of 0.866 at a frequency of 60 Hz. What is the power factor for a frequency of 240 Hz?

36 •• A resistor and an inductor are connected in parallel across an emf $\mathcal{E} = \mathcal{E}_{max} \cos \omega t$ as shown in Figure 31-30. Show that (*a*) the current in the resistor is $I_R = (\mathcal{E}_{max}/R) \cos \omega t$, (*b*) the current in the inductor is $I_L = (\mathcal{E}_{max}/X_L) \cos(\omega t - 90°)$, and (*c*) $I = I_R + I_L = I_{max} \cos(\omega t - \delta)$, where $\tan \delta = R/X_L$ and $I_{max} = \mathcal{E}_{max}/Z$ with $Z^{-2} = R^{-2} + X_L^{-2}$.

Figure 31-30 Problem 36

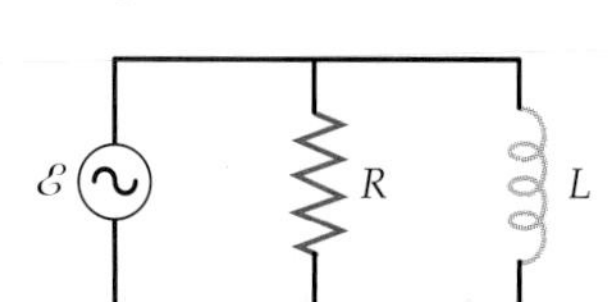

37 •• Figure 31-31 shows a load resistor $R_L = 20\ \Omega$ connected to a high-pass filter consisting of an inductor $L = 3.2$ mH and a resistor $R = 4\ \Omega$. The input voltage is $\mathcal{E} = (100\text{ V}) \cos(2\pi ft)$. Find the rms currents in *R*, *L*, and R_L if (*a*) $f = 500$ Hz and (*b*) $f = 2000$ Hz. (*c*) What fraction of the total power delivered by the voltage source is dissipated in the load resistor if the frequency is 500 Hz and if the frequency is 2000 Hz?

Figure 31-31 Problems 37 and 96

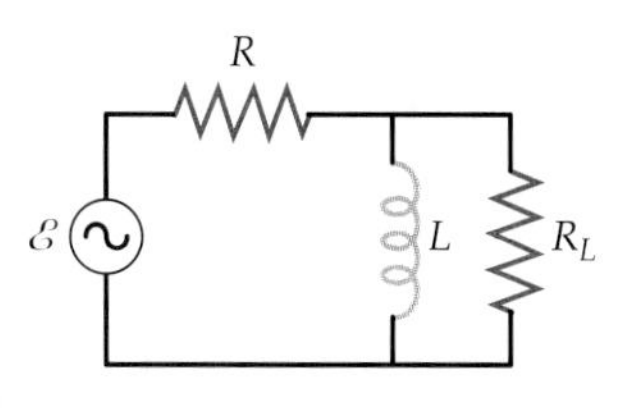

38 •• An ac source $\mathcal{E}_1 = (20\text{ V}) \cos(2\pi ft)$ in series with a battery $\mathcal{E}_2 = 16$ V is connected to a circuit consisting of resistors $R_1 = 10\ \Omega$ and $R_2 = 8\ \Omega$ and an inductor $L = 6$ mH (Figure 31-32). Find the power dissipated in R_1 and R_2 if (*a*) $f = 100$ Hz, (*b*) $f = 200$ Hz, and (*c*) $f = 800$ Hz.

Figure 31-32 Problem 38

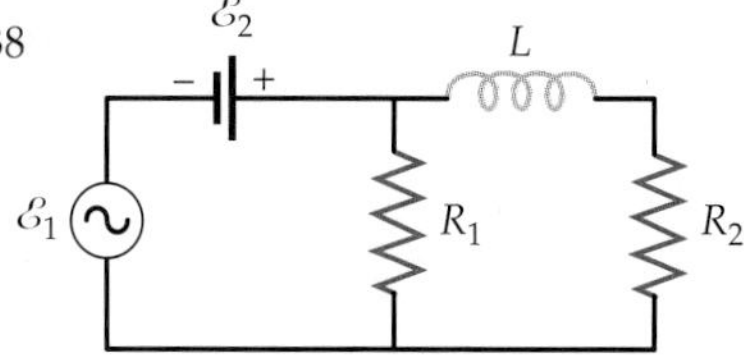

RC and *RL* Circuits With a Generator (optional)

39 •• A 100-V-rms voltage is applied to a series *RC* circuit. The rms voltage across the capacitor is 80 V. What is the voltage across the resistor?

40 •• The circuit shown in Figure 31-33 is called an *RC* high-pass filter because high input frequencies are transmitted with greater amplitude than low input frequencies. (*a*) If the input voltage is $V_{in} = V_0 \cos \omega t$, show that the output voltage is

$$V_{out} = \frac{V_0}{\sqrt{(1/\omega RC)^2 + 1}}$$

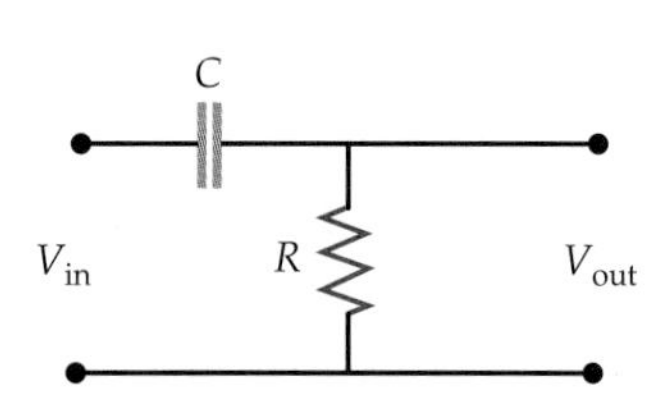

Figure 31-33 Problem 40

(*b*) At what angular frequency is the output voltage half the input voltage? (*c*) Sketch a graph of V_{out}/V_0 as a function of ω.

41 •• A coil draws 15 A when connected to a 220-V 60-Hz ac line. When it is in series with a 4-Ω resistor and the combination is connected to a 100-V battery, the battery current after a long time is observed to be 10 A. (*a*) What is the resistance in the coil? (*b*) What is the inductance of the coil?

42 •• Figure 31-34 shows a load resistor $R_L = 20\ \Omega$ connected to a low-pass filter consisting of a capacitor $C = 8\ \mu F$ and resistor $R = 4\ \Omega$. The input voltage is $\mathcal{E} = (100\ V) \cos (2\pi ft)$. Find the rms currents in R, C, and R_L if (*a*) $f = 500$ Hz and (*b*) $f = 2000$ Hz. (*c*) What fraction of the total power delivered by the voltage source is dissipated in the load resistor if the frequency is 500 Hz and if the frequency is 2000 Hz?

Figure 31-34 Problems 42 and 97

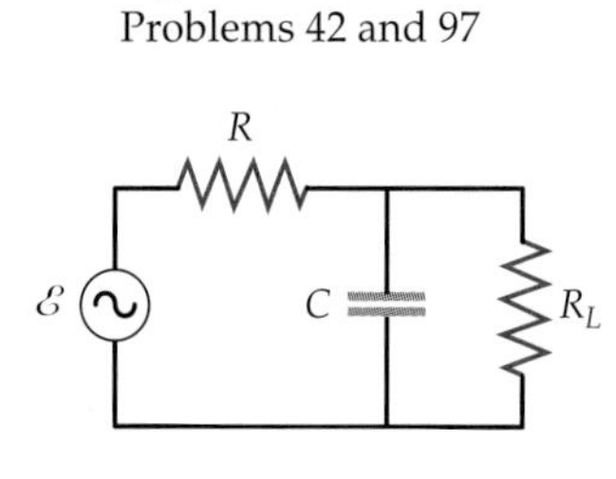

LC Circuits With a Generator (optional)

43 •• The generator voltage in Figure 31-35 is given by $\mathcal{E} = (100\ V) \cos (2\pi ft)$. (*a*) For each branch, what is the amplitude of the current and what is its phase relative to the applied voltage? (*b*) What is the angular frequency ω such that the current in the generator vanishes? (*c*) At this resonance, what is the current in the inductor? What is the current in the capacitor? (*d*) Draw a phasor diagram showing the general relationships between the applied voltage, the generator current, the capacitor current, and the inductor current for the case where the inductive reactance is larger than the capacitive reactance.

Figure 31-35 Problem 43

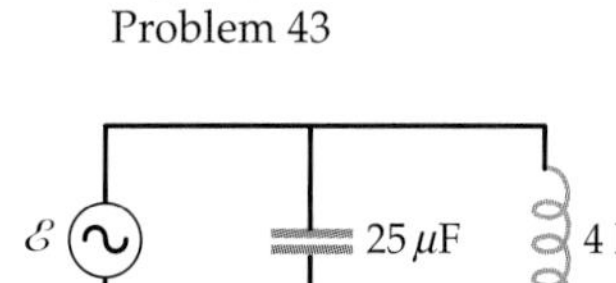

44 •• The charge on the capacitor of a series *LC* circuit is given by $Q = (15\ \mu C) \cos (1250t + \pi/4)$ where t is in seconds. (*a*) Find the current as a function of time. (*b*) Find C if $L = 28$ mH. (*c*) Write expressions for the electrical energy U_e, the magnetic energy U_m, and the total energy U.

45 ••• One method for measuring the compressibility of a dielectric material uses an *LC* circuit with a parallel-plate capacitor. The dielectric is inserted between the plates and the change in resonance frequency is determined as the capacitor plates are subjected to a compressive stress. In such an arrangement, the resonance frequency is 120 MHz when a dielectric of thickness 0.1 cm and dielectric constant $\kappa = 6.8$ is placed between the capacitor plates. Under a compressive stress of 800 atm, the resonance frequency decreases to 116 MHz. Find Young's modulus of the dielectric material.

46 ••• Figure 31-36 shows an inductance L and a parallel plate capacitor of width $w = 20$ cm and thickness 0.2 cm. A dielectric with dielectric constant $\kappa = 4.8$ that can completely fill the space between the capacitor plates can be slid between the plates. The inductor has an inductance $L = 2$ mH. When half the dielectric is between the capacitor plates, i.e, when $x = \frac{1}{2}w$, the resonant frequency of this *LC* combination is 90 MHz. (*a*) What is the capacitance of the capacitor without the dielectric? (*b*) Find the resonance frequency as a function of x.

Figure 31-36 Problem 46

RLC Circuits With a Generator (optional)

47 • True or false:

(*a*) An *RLC* circuit with a high *Q* factor has a narrow resonance curve.
(*b*) At resonance, the impedance of an *RLC* circuit equals the resistance *R*.
(*c*) At resonance, the current and generator voltage are in phase.

48 • Does the power factor depend on the frequency?

49 • Are there any disadvantages to having a radio tuning circuit with an extremely large *Q* factor?

50 • What is the power factor for a circuit that has inductance and capacitance but no resistance?

51 • A series *RLC* circuit in a radio receiver is tuned by a variable capacitor so that it can resonate at frequencies from 500 to 1600 kHz. If $L = 1.0\ \mu H$, find the range of capacitances necessary to cover this range of frequencies.

52 • (*a*) Find the power factor for the circuit in Example 31-5 when $\omega = 400$ rad/s. (*b*) At what angular frequency is the power factor 0.5?

53 • An ac generator with a maximum emf of 20 V is connected in series with a 20-μF capacitor and an 80-Ω resistor. There is no inductance in the circuit. Find (*a*) the power factor, (*b*) the rms current, and (*c*) the average power if the angular frequency of the generator is 400 rad/s.

54 •• Show that the formula $P_{av} = R\mathcal{E}_{rms}^2/Z^2$ gives the correct result for a circuit containing only a generator and (*a*) a resistor, (*b*) a capacitor, and (*c*) an inductor.

55 •• A series RLC circuit with $L = 10$ mH, $C = 2\ \mu$F, and $R = 5\ \Omega$ is driven by a generator with a maximum emf of 100 V and a variable angular frequency ω. Find (*a*) the resonant frequency ω_0 and (*b*) I_{rms} at resonance. When $\omega = 8000$ rad/s, find (*c*) X_C and X_L, (*d*) Z and I_{rms}, and (*e*) the phase angle δ.

56 •• For the circuit in Problem 55, let the generator frequency be $f = \omega/2\pi = 1$ kHz. Find (*a*) the resonance frequency $f_0 = \omega_0/2\pi$, (*b*) X_C and X_L, (*c*) the total impedance Z and I_{rms}, and (*d*) the phase angle δ.

57 •• Find the power factor and the phase angle δ for the circuit in Problem 55 when the generator frequency is (*a*) 900 Hz, (*b*) 1.1 kHz, and (*c*) 1.3 kHz.

58 •• Find (*a*) the Q factor and (*b*) the resonance width for the circuit in Problem 55. (*c*) What is the power factor when $\omega = 8000$ rad/s?

59 •• FM radio stations have carrier frequencies that are separated by 0.20 MHz. When the radio is tuned to a station, such as 100.1 MHz, the resonance width of the receiver circuit should be much smaller than 0.2 MHz so that adjacent stations are not received. If $f_0 = 100.1$ MHz and $\Delta f = 0.05$ MHz, what is the Q factor for the circuit?

60 •• A coil is connected to a 60-Hz, 100-V ac generator. At this frequency the coil has an impedance of 10 Ω and a reactance of 8 Ω. (*a*) What is the current in the coil? (*b*) What is the phase angle between the current and the applied voltage? (*c*) What series capacitance is required so that the current and voltage are in phase? (*d*) What then is the voltage measured across the capacitor?

61 •• A 0.25-H inductor and a capacitor C are connected in series with a 60-Hz ac generator. An ac voltmeter is used to measure the rms voltages across the inductor and capacitor separately. The rms voltage across the capacitor is 75 V and that across the inductor is 50 V. (*a*) Find the capacitance C and the rms current in the circuit. (*b*) What would be the measured rms voltage across both the capacitor and inductor together?

62 •• (*a*) Show that Equation 31-51 can be written as

$$\tan\delta = \frac{L(\omega^2 - \omega_0^2)}{\omega R}$$

Find δ approximately at (*b*) very low frequencies and (*c*) very high frequencies.

63 •• (*a*) Show that in a series RC circuit with no inductance, the power factor is given by

$$\cos\delta = \frac{RC\omega}{\sqrt{1 + (RC\omega)^2}}$$

(*b*) Sketch a graph of the power factor versus ω.

64 •• In the circuit in Figure 31-37, the ac generator produces an rms voltage of 115 V when operated at 60 Hz. What is the rms voltage across points (*a*) AB, (*b*) BC, (*c*) CD, (*d*) AC, and (*e*) BD?

Figure 31-37 Problem 64

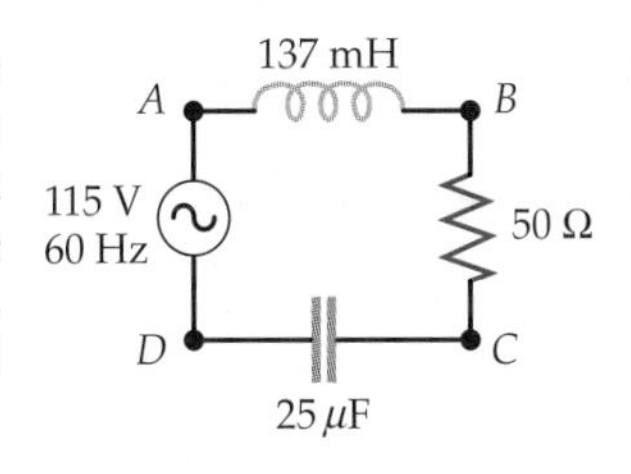

65 •• A variable-frequency ac generator is connected to a series RLC circuit for which $R = 1$ kΩ, $L = 50$ mH, and $C = 2.5\ \mu$F. (*a*) What is the resonance frequency of the circuit? (*b*) What is the Q value? (*c*) At what frequencies is the value of the average power delivered by the generator half of its maximum value?

66 •• An experimental physicist wishes to design a series RLC circuit with a Q value of 10 and a resonance frequency of 33 kHz. She has a 45-mH inductor with negligible resistance. What values for the resistance R and capacitance C should she use?

67 •• When an RLC series circuit is connected to a 120-V rms, 60-Hz line, the current is $I_{rms} = 11.0$ A and the current leads the voltage by 45°. (*a*) Find the power supplied to the circuit. (*b*) What is the resistance? (*c*) If the inductance $L = 0.05$ H, find the capacitance C. (*d*) What capacitance or inductance should you add to make the power factor 1?

68 •• A series RLC circuit is driven at a frequency of 500 Hz. The phase angle between the applied voltage and current is determined from an oscilloscope measurement to be $\delta = 75°$. If the total resistance is known to be 35 Ω and the inductance is 0.15 H, what is the capacitance of the circuit?

69 •• A series RLC circuit with $R = 400\ \Omega$, $L = 0.35$ H, and $C = 5\ \mu$F is driven by a generator of variable frequency f. (*a*) What is the resonance frequency f_0? Find f and f/f_0 when the phase angle δ is (*b*) 60°, and (*c*) −60°.

70 •• Sketch the impedance Z versus ω for (*a*) a series LR circuit, (*b*) a series RC circuit, and (*c*) a series RLC circuit.

71 •• Given the circuit shown in Figure 31-38, (*a*) find the power loss in the inductor. (*b*) Find the resistance r of the inductor. (*c*) Find the inductance L.

Figure 31-38 Problem 71

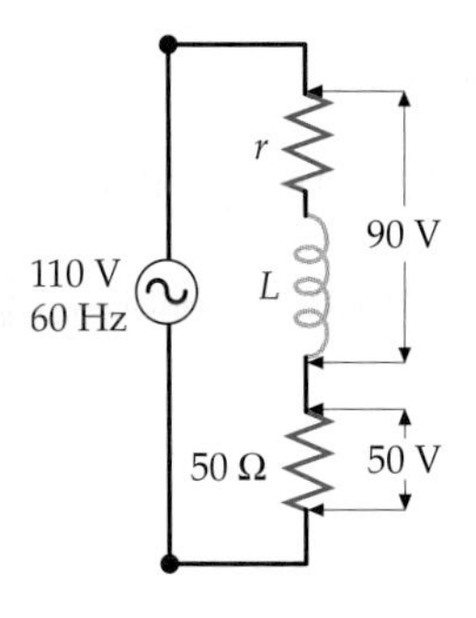

72 •• Show that Equation 31-52 can be written as

$$I_{max} = \frac{\omega\mathcal{E}_{max}}{\sqrt{L^2(\omega^2 - \omega_0^2)^2 + \omega^2R^2}}$$

73 •• In a series RLC circuit, $X_C = 16\ \Omega$ and $X_L = 4\ \Omega$ at some frequency. The resonance frequency is $\omega_0 = 10^4$ rad/s. (*a*) Find L and C. If $R = 5\ \Omega$ and $\mathcal{E}_{max} = 26$ V, find (*b*) the Q factor and (*c*) the maximum current.

74 •• In a series RLC circuit connected to an ac generator whose maximum emf is 200 V, the resistance is 60 Ω and the capacitance is 8.0 μF. The inductance can be varied from 8.0 mH to 40.0 mH by the insertion of an iron core in the solenoid. The angular frequency of the generator is 2500 rad/s. If the capacitor voltage is not to exceed 150 V, find (*a*) the maximum current and (*b*) the range of inductance that is safe to use.

75 •• A certain electrical device draws 10 A rms and has an average power of 720 W when connected to a 120-V-rms, 60-Hz power line. (*a*) What is the impedance of the device? (*b*) What series combination of resistance and reactance is this device equivalent to? (*c*) If the current leads the emf, is the reactance inductive or capacitive?

76 •• A method for measuring inductance is to connect the inductor in series with a known capacitance, a known resistance, an ac ammeter, and a variable-frequency signal generator. The frequency of the signal generator is varied and the emf is kept constant until the current is maximum. (*a*) If $C = 10\ \mu F$, $\mathcal{E}_{max} = 10$ V, $R = 100\ \Omega$, and I is maximum at $\omega = 5000$ rad/s, what is L? (*b*) What is I_{max}?

77 •• A resistor and a capacitor are connected in parallel across a sinusoidal emf $\mathcal{E} = \mathcal{E}_{max} \cos \omega t$ as shown in Figure 31-39. (*a*) Show that the current in the resistor is $I_R = (\mathcal{E}_{max}/R) \cos \omega t$. (*b*) Show that the current in the capacitor branch is $I_C = (\mathcal{E}_{max}/X_C) \cos (\omega t + 90°)$. (*c*) Show that the total current is given by $I = I_R + I_C = I_{max} \cos (\omega t + \delta)$, where $\tan \delta = R/X_C$ and $I_{max} = \mathcal{E}_{max}/Z$ with $Z^{-2} = R^{-2} + X_C^{-2}$.

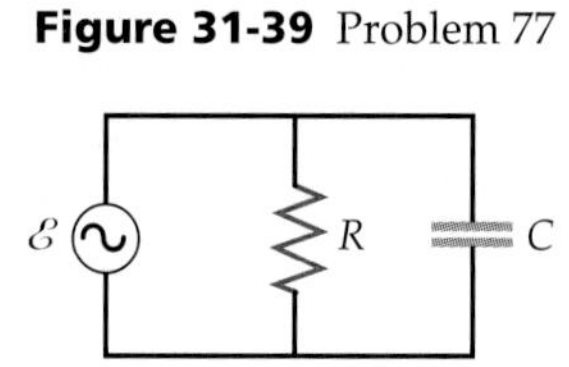

Figure 31-39 Problem 77

78 •• The impedances of motors, transformers, and electromagnets have inductive reactance. Suppose that the phase angle of the total impedance of a large industrial plant is 25° when the plant is under full operation and using 2.3 MW of power. The power is supplied to the plant from a substation 4.5 km from the plant; the 60 Hz rms line voltage at the plant is 40,000 V. The resistance of the transmission line from the substation to the plant is 5.2 Ω. The cost per kilowatt-hour is 0.07 dollars. The plant pays only for the actual energy used. (*a*) What are the resistance and inductive reactance of the plant's total load? (*b*) What is the current in the power lines and what must be the rms voltage at the substation to maintain the voltage at the plant at 40,000 V? (*c*) How much power is lost in transmission? (*d*) Suppose that the phase angle of the plant's impedance were reduced to 18° by adding a bank of capacitors in series with the load. How much money would be saved by the electric utility during one month of operation, assuming the plant operates at full capacity for 16 h each day? (*e*) What must be the capacitance of this bank of capacitors?

79 •• In the circuit shown in Figure 31-40, $R = 10\ \Omega$, $R_L = 30\ \Omega$, $L = 150$ mH, and $C = 8\ \mu F$; the frequency of the ac source is 10 Hz and its amplitude is 100 V. (*a*) Using phasor diagrams, determine the impedance of the circuit when switch S is closed. (*b*) Determine the impedance of the circuit when switch S is open. (*c*) What are the voltages across the load resistor R_L when switch S is closed and when it is open? (*d*) Repeat parts (*a*), (*b*), and (*c*) with the frequency of the source changed to 1000 Hz. (*e*) Which arrangement is a better low-pass filter, S open or S closed?

Figure 31-40
Problem 79

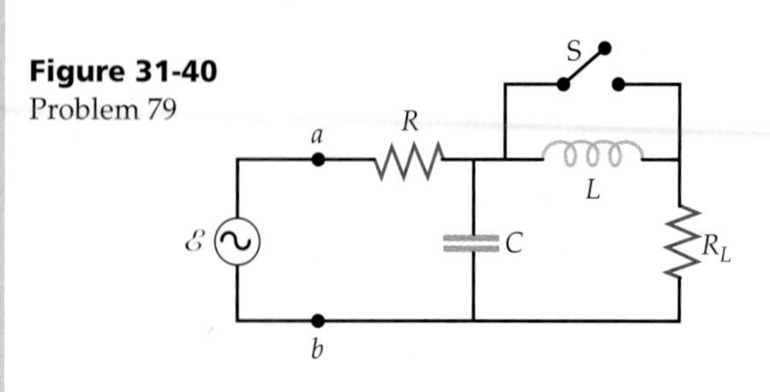

80 •• In the circuit shown in Figure 31-41, $R_1 = 2\ \Omega$, $R_2 = 4\ \Omega$, $L = 12$ mH, $C = 30\ \mu F$, and $\mathcal{E} = (40\text{ V}) \cos (\omega t)$. (*a*) Find the resonance frequency. (*b*) At the resonance frequency, what are the rms currents in each resistor and the rms current supplied by the source emf?

Figure 31-41
Problems 80, 98, and 99

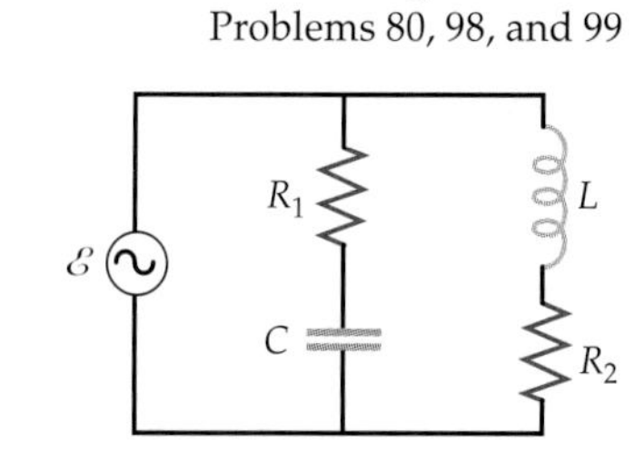

81 •• For the circuit in Figure 31-23, derive an expression for the Q of the circuit assuming the resonance is sharp.

82 •• For the circuit in Figure 31-23, $L = 4$ mH. (*a*) What capacitance C will result in a resonance frequency is 4 kHz? (*b*) When C has the value found in (*a*), what should be the resistance R so that the Q of the circuit is 8?

83 •• If the capacitance of C in Problem 82 is reduced to half the value found in Problem 82, what then are the resonance frequency and the Q of the circuit? What should be the resistance R to give $Q = 8$?

84 •• A series circuit consists of a 4.0-nF capacitor, a 36-mH inductor, and a 100-Ω resistor. The circuit is connected to a 20-V ac source whose frequency can be varied over a wide range. (*a*) Find the resonance frequency f_0 of the circuit. (*b*) At resonance, what is the rms current in the circuit and what are the rms voltages across the inductor and capacitor? (*c*) What is the rms current and what are the rms voltages across the inductor and capacitor at $f = f_0 + \frac{1}{2}\Delta f$, where Δf is the width of the resonance?

85 •• Repeat Problem 84 with the 100-Ω resistor replaced by a 40-Ω resistor.

86 ••• In the parallel circuit shown in Figure 31-42, $V_{max} = 110$ V. (*a*) What is the impedance of each branch? (*b*) For each branch, what is the current amplitude and its phase relative to the applied voltage? (*c*) Give the current phasor diagram, and use it to find the total current and its phase relative to the applied voltage.

Figure 31-42
Problem 86

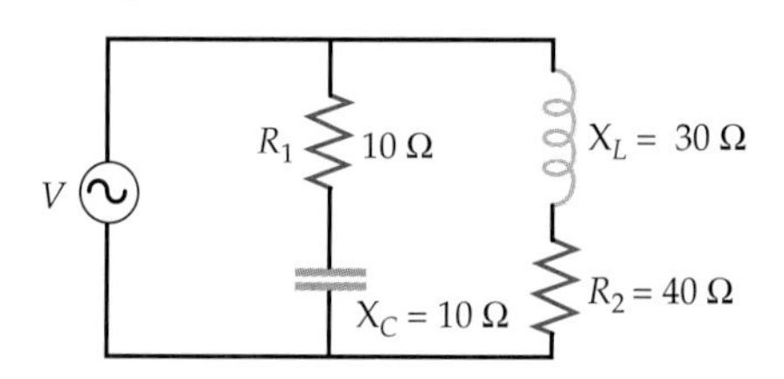

87 ••• (*a*) Show that Equation 31-51 can be written as

$$\tan \delta = \frac{Q(\omega^2 - \omega^2 0)}{\omega \omega_0}$$

(*b*) Show that near resonance

$$\tan \delta \approx \frac{2Q(\omega - \omega_0)}{\omega}$$

(*c*) Sketch a plot of δ versus x, where $x = \omega/\omega_0$, for a circuit with high Q and for one with low Q.

88 ••• Show by direct substitution that the current given by Equation 31-50 with δ and I_{max} given by Equations 31-51 and 31-52, respectively, satisfies Equation 31-49. (*Hint:* Use trigonometric identities for the sine and cosine of the sum of two angles, and write the equation in the form

$$A \sin \omega t + B \cos \omega t = 0$$

Since this equation must hold for all times, $A = 0$ and $B = 0$.)

89 ••• An ac generator is in series with a capacitor and an inductor in a circuit with negligible resistance. (*a*) Show that the charge on the capacitor obeys the equation

$$L\frac{d^2Q}{dt^2} + \frac{Q}{C} = \mathscr{E}_{\text{max}} \cos \omega t$$

(*b*) Show by direct substitution that this equation is satisfied by $Q = Q_{\text{max}} \cos \omega t$ if

$$Q_{\text{max}} = -\frac{\mathscr{E}_{\text{max}}}{L(\omega^2 - \omega_0^2)}$$

(*c*) Show that the current can be written as $I = I_{\text{max}} \cos(\omega t - \delta)$, where

$$I_{\text{max}} = \frac{\omega\mathscr{E}_{\text{max}}}{L|\omega^2 - \omega_0^2|} = \frac{\mathscr{E}_{\text{max}}}{|X_L - X_C|}$$

and $\delta = -90°$ for $\omega < \omega_0$ and $\delta = 90°$ for $\omega > \omega_0$.

90 ••• Figure 31-19 shows a plot of average power P_{av} versus generator frequency ω for an *RLC* circuit with a generator. The average power P_{av} is given by Equation 31-58. The "full width at half-maximum" $\Delta\omega$ is the width of the resonance curve between the two points where P_{av} is one-half its maximum value. Show that, for a sharply peaked resonance, $\Delta\omega \approx R/L$ and, hence, that $Q \approx \omega_0/\Delta\omega$ in this case (Equation 31-60). [*Hint:* At resonance, the denominator of the expression on the right of Equation 31-58 is ω^2R^2. The half-power points will occur when the denominator is twice the value near resonance, that is, when $L^2(\omega^2 - \omega_0^2)^2 = \omega^2R^2 \approx \omega_0^2R^2$. Let ω_1 and ω_2 be the solutions of this equation. For a sharply peaked resonance, $\omega_1 \approx \omega_0$ and $\omega_2 \approx \omega_0$. Then, using the fact that $\omega + \omega_0 \approx 2\omega_0$, one finds that $\Delta\omega = \omega_2 - \omega_1 \approx R/L$.]

91 ••• Show by direct substitution that Equation 31-47*b* is satisfied by $Q = Q_0e^{-Rt/2L} \cos \omega' t$ where $\omega' = \sqrt{(1/LC) - (R/2L)^2}$ and Q_0 is the charge on the capacitor at $t = 0$.

92 ••• (*a*) Compute the current $I = dQ/dt$ from the solution of Equation 31-47*b* given in Problem 91, and show that

$$I = -I_0\left(\sin \omega' t + \frac{R}{2L\omega'} \cos \omega' t\right)e^{-Rt/2L}$$

where $I_0 = \omega' Q_0$. (*b*) Show that this can be written

$$I = -\frac{I_0}{\cos \delta}(\cos \delta \sin \omega' t + \sin \delta \cos \omega' t)e^{-Rt/2L}$$

$$= -\frac{I_0}{\cos \delta} \sin(\omega' t + \delta)e^{-Rt/2L}$$

where $\tan \delta = R/2L\omega'$. When $R/2L\omega'$ is small, $\cos \delta \approx 1$, and $I \approx I_0 \sin(\omega' t + \delta)e^{-Rt/2L}$.

93 ••• One method for measuring the magnetic susceptibility of a sample uses an *LC* circuit consisting of an air-core solenoid and a capacitor. The resonant frequency of the circuit without the sample is determined and then measured again with the sample inserted in the solenoid. Suppose the solenoid is 4.0 cm long, 0.3 cm in diameter, and has 400 turns of fine wire. Assume that the sample that is inserted in the solenoid is also 4.0 cm long and fills the air space. Neglect end effects. (In practice, a test sample of known susceptibility of the same shape as the unknown is used to calibrate the instrument.) (*a*) What is the inductance of the empty solenoid? (*b*) What should be the capacitance of the capacitor so that the resonance frequency of the circuit without a sample is 6.0000 MHz? (*c*) When a sample is inserted in the solenoid, the resonance frequency drops to 5.9989 MHz. Determine the sample's susceptibility.

94 ••• A concentric cable of cylindrical cross section has an inner conductor of 0.4 cm diameter and an outer conductor of 2.0 cm diameter. Air fills the space between the conductors. (*a*) Find the resonance frequency of a one-meter length of this conductor. (*b*) What length of conductor will result in a resonance frequency of 18 GHz?

95 ••• Repeat Problem 94 if the inner and outer conductors of the cable are separated by a dielectric of dielectric constant $\kappa = 5.8$.

96 ••• At what frequency will the voltage across the load resistor of Problem 37 be half the source voltage?

97 ••• At what frequency will the voltage across the load resistor of Problem 42 be half the source voltage?

98 ••• (*a*) Find the angular frequency ω for the circuit in Problem 80 such that the magnitude of the reactance of the two parallel branches are equal. (*b*) At that frequency, what is the power dissipation in each of the two resistors?

99 ••• (*a*) For the circuit of Problem 80, find the angular frequency ω for which the power dissipation in the two resistors is the same. (*b*) At that angular frequency, what is the reactance of each of the two parallel branches? (*c*) Draw a phasor diagram showing the current through each of the two parallel branches. (*d*) What is the impedance of the circuit?

The Transformer (optional)

100 • A transformer is used to change

(*a*) capacitance. (*b*) frequency. (*c*) voltage.
(*d*) power. (*e*) none of these.

101 • True or false: If a transformer increases the current, it must decrease the voltage.

102 •• An ideal transformer has N_1 turns on its primary and N_2 turns on its secondary. The power dissipated in a load resistance R connected across the secondary is P_2 when the primary voltage is V_1. The current in the primary windings is then

(*a*) P_2/V_1. (*b*) $(N_1/N_2)(P_2/V_1)$.
(*c*) $(N_2/N_1)(P_2/V_1)$. (*d*) $(N_2/N_1)^2(P_2/V_1)$.

103 • An ac voltage of 24 V is required for a device whose impedance is 12 Ω. (*a*) What should the turn ratio of a transformer be so the device can be operated from a 120-V line? (*b*) Suppose the transformer is accidentally connected reversed, i.e., with the secondary winding across the 120-V line and the 12-Ω load across the primary. How much current will then flow in the primary winding?

104 • A transformer has 400 turns in the primary and 8 turns in the secondary. (*a*) Is this a step-up or step-down transformer? (*b*) If the primary is connected across 120 V rms, what is the open-circuit voltage across the secondary? (*c*) If the primary current is 0.1 A, what is the secondary current, assuming negligible magnetization current and no power loss?

105 • The primary of a step-down transformer has 250 turns and is connected to a 120-V-rms line. The secondary is to supply 20 A at 9 V. Find (*a*) the current in the primary and (*b*) the number of turns in the secondary, assuming 100% efficiency.

106 • A transformer has 500 turns in its primary, which is connected to 120 V rms. Its secondary coil is tapped at three places to give outputs of 2.5, 7.5, and 9 V. How many turns are needed for each part of the secondary coil?

107 • The distribution circuit of a residential power line is operated at 2000 V rms. This voltage must be reduced to 240 V rms for use within the residences. If the secondary side of the transformer has 400 turns, how many turns are in the primary?

108 •• An audio oscillator (ac source) with an internal resistance of 2000 Ω and an open-circuit rms output voltage of 12 V is to be used to drive a loudspeaker with a resistance of 8 Ω. What should be the ratio of primary to secondary turns of a transformer so that maximum power is transferred to the speaker? Suppose a second identical speaker is connected in parallel with the first speaker. How much power is then supplied to the two speakers combined?

109 •• One use of a transformer is for *impedance matching*. For example, the output impedance of a stereo amplifier is matched to the impedance of a speaker by a transformer. In Equation 31-67, the currents I_1 and I_2 can be related to the impedance Z in the secondary since $I_2 = V_2/Z$. Using Equations 31-65 and 31-66, show that

$$I_1 = \frac{\mathcal{E}}{(N_1/N_2)^2 Z}$$

and, therefore, $Z_{\text{eff}} = (N_1/N_2)^2 Z$.

General Problems

110 • True or false:

(*a*) Alternating current in a resistance dissipates no power because the current is negative as often as it is positive.
(*b*) At very high frequencies, a capacitor acts like a short circuit.

111 • A 5.0-kW electric clothes dryer runs on 240 V rms. Find (*a*) I_{rms} and (*b*) I_{max}. (*c*) Find the same quantities for a dryer of the same power that operates at 120 V rms.

112 • Find the reactance of a 10.0-μF capacitor at (*a*) 60 Hz, (*b*) 6 kHz, and (*c*) 6 MHz.

113 • Sketch a graph of X_L versus f for $L = 3$ mH.

114 • Sketch a graph of X_C versus f for $C = 100$ μF.

115 •• A resistance R carries a current $I = (5.0\text{ A})\sin 120\pi t + (7.0\text{ A})\sin 240\pi t$. (*a*) What is the rms current? (*b*) If the resistance R is 12 Ω, what is the power dissipated in the resistor? (*c*) What is the rms voltage across the resistor?

116 •• Figure 31-43 shows the voltage V versus time t for a "square-wave" voltage. If $V_0 = 12$ V, (*a*) what is the rms voltage of this waveform? (*b*) If this alternating waveform is rectified by eliminating the negative voltages so that only the positive voltages remain, what now is the rms voltage of the rectified waveform?

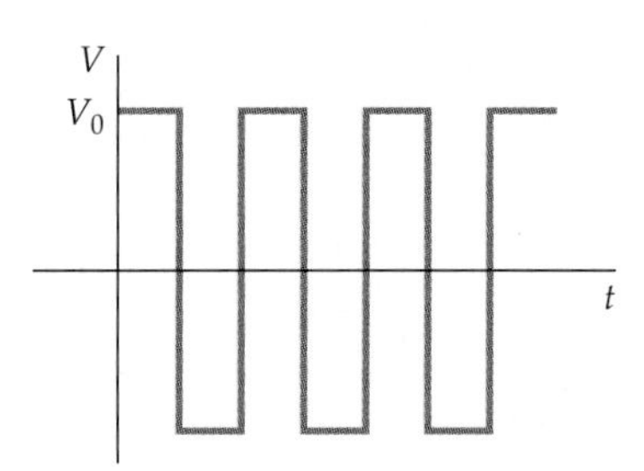

Figure 31-43 Problem 116

117 •• A pulsed current has a constant value of 15 A for the first 0.1 s of each second and is then 0 for the next 0.9 s of each second. (*a*) What is the rms value for this current waveform? (*b*) Each current pulse is generated by a voltage pulse of maximum value 100 V. What is the average power delivered by the pulse generator?

118 •• A circuit consists of two capacitors, a 24-V battery, and an ac voltage connected as shown in Figure 31-44. The ac voltage is given by $\mathcal{E} = (20\text{ V})\cos(120\pi t)$ where t is in seconds. (*a*) Find the charge on each capacitor as a function of time. Assume that transient effects have had sufficient time to decay. (*b*) What is the steady-state current? (*c*) What is the maximum energy stored in the capacitors? (*d*) What is the minimum energy stored in the capacitors?

Figure 31-44 Problem 118

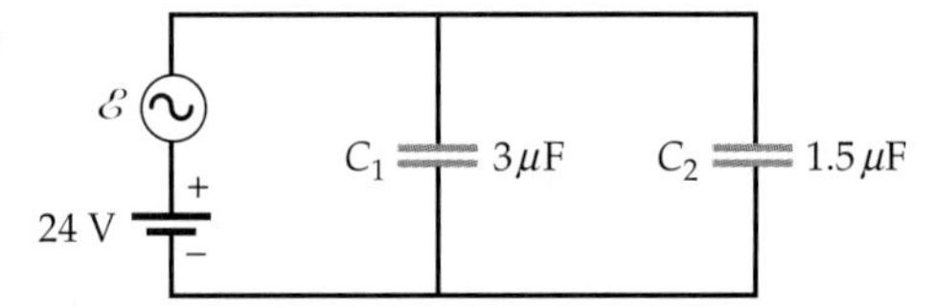

119 •• What are the average and rms values of current for the two current waveforms shown in Figure 31-45?

Figure 31-45 Problem 119

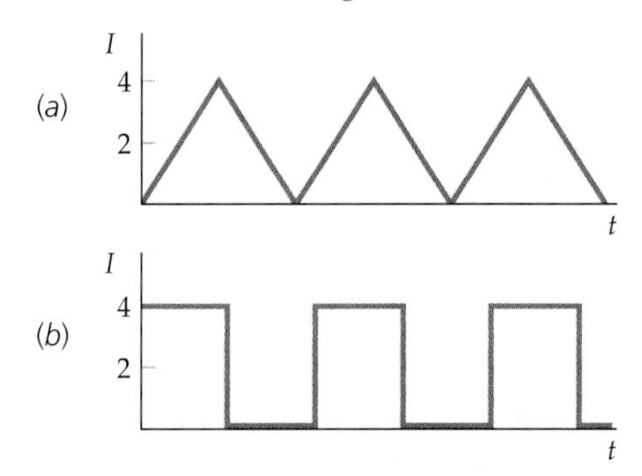

120 •• In the circuit shown in Figure 31-46, $\mathcal{E}_1 = (20\text{ V})\cos(2\pi ft)$, $f = 180$ Hz, $\mathcal{E}_2 = 18$ V, and $R = 36$ Ω. Find the maximum, minimum, average, and rms values of the current through the resistor.

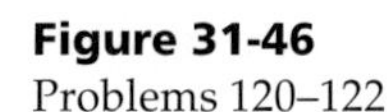

Figure 31-46 Problems 120–122

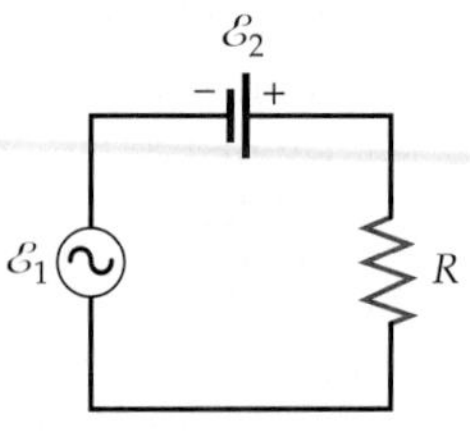

121 •• Repeat Problem 120 if the resistor R is replaced by a 2-μF capacitor.

122 •• Repeat Problem 120 if the resistor R is replaced by a 12-mH inductor.

CHAPTER 32

Maxwell's Equations and Electromagnetic Waves

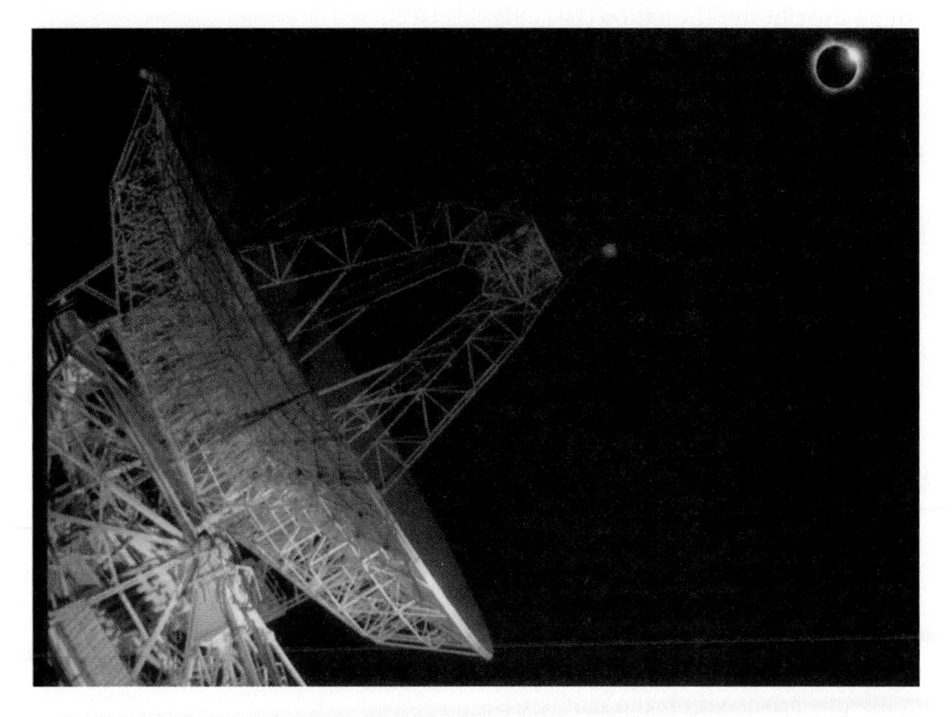

A multiple-exposure view showing the 26-m tracking antenna at Wallops Station, Virginia, and a total solar eclipse. Electromagnetic radiation at radio wavelengths, like that at optical wavelengths, is not readily absorbed by the earth's atmosphere, making it a viable means of communication between distant points on the ground or between ground and space.

Maxwell's equations, first proposed by the great Scottish physicist James Clerk Maxwell, relate the electric and magnetic field vectors $\vec{E}$ and $\vec{B}$ to their sources, which are electric charges, currents, and changing fields. These equations summarize the experimental laws of electricity and magnetism—the laws of Coulomb, Gauss, Biot–Savart, Ampère, and Faraday. These experimental laws hold in general except for Ampère's law, which does not apply to discontinuous currents such as those that occur when charging or discharging a capacitor. Maxwell was able to generalize Ampère's law with the invention of the displacement current (Section 32-1). He was then able to show that the generalized laws of electricity and magnetism imply the existence of electromagnetic waves.

Maxwell's equations play a role in classical electromagnetism analogous to that of Newton's laws in classical mechanics. In principle, all problems in classical electricity and magnetism can be solved using Maxwell's equations, just as all problems in classical mechanics can be solved using Newton's laws. Maxwell's equations are considerably more complicated than Newton's laws, however, and their application to most problems involves mathematics beyond the scope of this book. Nevertheless, Maxwell's equations are of great theoretical importance. For example, Maxwell showed that these equations can be combined to yield a wave equation for the electric and magnetic field vectors $\vec{E}$ and $\vec{B}$. Such **electromagnetic waves** are caused by accelerating charges, for example, the charges in an alternating current in an antenna. They were first produced in the laboratory by Heinrich Hertz in 1887. Maxwell showed that the speed of electromagnetic waves in free space should be

$$c = \frac{1}{\sqrt{\mu_0 \epsilon_0}} \qquad 32\text{-}1$$

where ϵ_0, the permittivity of free space, is the constant appearing in Coulomb's and Gauss's laws and μ_0, the permeability of free space, is the constant appearing in the Biot–Savart law and Ampère's law. When the measured value of ϵ_0 and the defined value of μ_0 are put into Equation 32-1, the speed of electromagnetic waves is found to be about 3×10^8 m/s, the same as the measured speed of light. Maxwell noted this "coincidence" with great excitement and correctly surmised that light itself is an electromagnetic wave.

32-1 Maxwell's Displacement Current

Ampère's law (Equation 29-15) relates the line integral of the magnetic field around some closed curve C to the current that passes through any area bounded by that curve:

$$\oint_C \vec{B} \cdot d\vec{\ell} = \mu_0 I, \quad \text{for any closed curve } C \qquad 32\text{-}2$$

Maxwell recognized a flaw in Ampère's law. Figure 32-1 shows two different surfaces bounded by the same curve C, which encircles a wire carrying current into a capacitor plate. The current through surface S_1 is I, but there is no current through surface S_2 because the charge stops on the capacitor plate. There is thus ambiguity in the phrase "the current through any surface bounded by the curve." Such a problem always arises when the current is discontinuous.

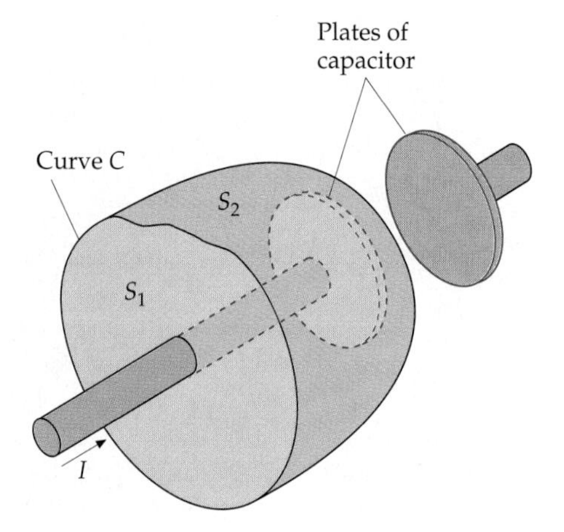

Figure 32-1 Two surfaces S_1 and S_2 bounded by the same curve C. The current I passes through surface S_1 but not S_2. Ampère's law, which relates the line integral of the magnetic field $\vec{B}$ around the curve C to the total current passing through any surface bounded by C, is not valid when the current is not continuous, as when it stops at the capacitor plate here.

Maxwell showed that the law can be generalized to include all situations if the current I in the equation is replaced by the sum of the conduction current I and another term I_d, called **Maxwell's displacement current,** defined as

$$I_d = \epsilon_0 \frac{d\phi_e}{dt} \qquad 32\text{-}3$$

Definition—Displacement current

where ϕ_e is the flux of the electric field through the same surface bounded by the curve C. The generalized form of Ampère's law is then

$$\oint_C \vec{B} \cdot d\vec{\ell} = \mu_0(I + I_d) = \mu_0 I + \mu_0 \epsilon_0 \frac{d\phi_e}{dt} \qquad 32\text{-}4$$

Generalized form of Ampère's law

We can understand this generalization by considering Figure 32-1 again. Let us call the sum $I + I_d$ the generalized current. According to the argument just stated, the same generalized current must cross any area bounded by the curve C. Thus, there can be no net generalized current into or out of the closed volume. If there is a net true current I into the volume, there must be an equal net displacement current I_d out of the volume. In the volume in the figure, there is a net conduction current I into the volume that increases the charge within the volume:

$$I = \frac{dQ}{dt}$$

The flux of the electric field out of the volume is related to the charge by Gauss's law:

$$\phi_{net} = \oint_S E_n\, dA = \frac{1}{\epsilon_0} Q_{inside}$$

The rate of increase of the charge is thus proportional to the rate of increase of the net flux out of the volume:

$$\frac{dQ}{dt} = \epsilon_0 \frac{d\phi_{e,net}}{dt} = I_d$$

Thus, the net conduction current into the volume equals the net displacement current out of the volume. The generalized current is always continuous.

It is interesting to compare Equation 32-4 to Faraday's law (Equation 30-5):

$$\mathcal{E} = \oint_C \vec{E} \cdot d\vec{\ell} = -\frac{d\phi_m}{dt} \qquad 32\text{-}5$$

According to Faraday's law, a changing magnetic flux produces an electric field whose line integral around a closed curve is proportional to the rate of change of magnetic flux through the curve. Maxwell's modification of Ampère's law shows that a changing electric flux produces a magnetic field whose line integral around a curve is proportional to the rate of change of the electric flux. We thus have the interesting reciprocal result that a changing magnetic field produces an electric field (Faraday's law) and a changing electric field produces a magnetic field (generalized form of Ampère's law). Note that there is no magnetic analog of a conduction current I. This is because the magnetic monopole, the magnetic analog of an electric charge, does not exist.

Example 32-1

A parallel-plate capacitor has closely spaced circular plates of radius R. Charge is flowing onto the positive plate and off of the negative plate at the rate $I = dQ/dt = 2.5$ A. Compute the displacement current between the plates.

Picture the Problem The displacement current is $I_d = \epsilon_0\, d\phi_{e/dt}$, where ϕ_e is the electric flux between the plates. Since the parallel plates are closely spaced, we can consider the electric field to be uniform and perpendicular to the plates within the capacitor, and zero outside the capacitor. Thus, the electric flux is simply $\phi_e = EA$, where E is the electric field between the plates and A is the area of the plates.

1. The displacement current is found by taking the time derivative of the electric flux: $\quad I_d = \epsilon_0 \dfrac{d\phi_e}{dt}$

2. The flux equals the electric field times the area: $\quad \phi_e = EA$

3. The electric field is proportional to the charge density on the plates, which we treat as infinite sheets:
$$E = \frac{\sigma}{\epsilon_0} = \frac{Q/A}{\epsilon_0}$$

4. Substitute these results to calculate I_d:
$$I_d = \epsilon_0 \frac{d(EA)}{dt} = \epsilon_0 A \frac{d}{dt}\left(\frac{Q}{A\epsilon_0}\right)$$
$$= \frac{dQ}{dt} = 2.5 \text{ A}$$

Remark Note that the displacement current across the gap of the capacitor is equal to the conduction current in the wires leading in and out of the capacitor.

Example 32-2

The circular plates in Example 32-1 have a radius of R = 3.0 cm. Find the magnetic field at a point between the plates at a distance r = 2.0 cm from the axis of the plates when the current into the positive plate is 2.5 A.

Picture the Problem We find B from the generalized form of Ampère's law (Equation 32-4). We chose a circular path of radius r = 2.0 cm about the center line joining the plates as shown in Figure 32-2 to compute $\oint \vec{B}\cdot d\vec{\ell}$. By symmetry, $\vec{B}$ is tangent to this circle and has the same magnitude everywhere on it.

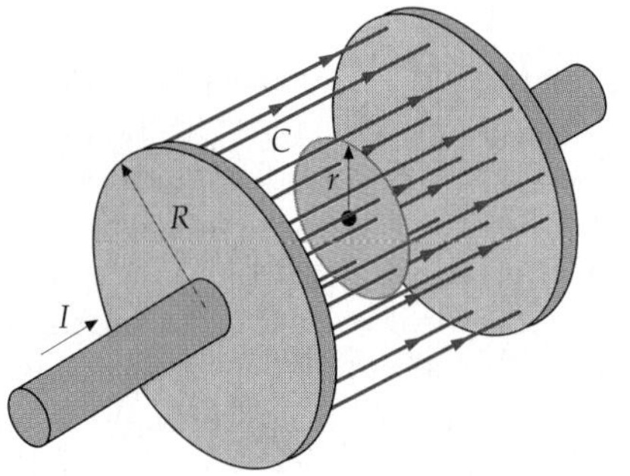

Figure 32-2

1. We find B from the generalized form of Ampère's law:
$$\oint \vec{B}\cdot d\vec{\ell} = \mu_0 I + \mu_0\epsilon_0 \frac{d\phi_e}{dt}$$

2. By symmetry the line integral is merely B times the circumference of the circle:
$$\oint \vec{B}\cdot d\vec{\ell} = B(2\pi r)$$

3. Since there is no conduction current between the plates of the capacitor, the generalized current is just the displacement current:
$$\oint \vec{B}\cdot d\vec{\ell} = B(2\pi r) = \mu_0\epsilon_0 \frac{d\phi_e}{dt}$$

4. The electric flux equals the product of the constant field E and the area bounded by the curve:
$$\phi_e = \pi r^2 E = \pi r^2 \frac{\sigma}{\epsilon_0} = \pi r^2 \frac{Q}{\epsilon_0 \pi R^2} = \frac{r^2 Q}{\epsilon_0 R^2}$$

5. Substitute these results into step 1 and solve for B:
$$B(2\pi r) = \mu_0\epsilon_0 \frac{d\phi_e}{dt} = \mu_0\epsilon_0 \frac{d}{dt}\left(\frac{r^2 Q}{\epsilon_0 R^2}\right) = \mu_0 \frac{r^2}{R^2}\frac{dQ}{dt}$$
$$B = \frac{\mu_0}{2\pi}\frac{r}{R^2}\frac{dQ}{dt} = \frac{\mu_0}{2\pi}\frac{r}{R^2} I$$
$$= (2 \times 10^{-7}\ \text{T}\cdot\text{m/A})\left(\frac{0.02 \text{ m}}{(0.03 \text{ m})^2}\right)(2.5 \text{ A}) = 1.11 \times 10^{-5} \text{ T}$$

32-2 Maxwell's Equations

Maxwell's equations are

$$\oint_S E_n \, dA = \frac{1}{\epsilon_0} Q_{\text{inside}} \qquad \text{32-6}a$$

$$\oint_S B_n \, dA = 0 \qquad \text{32-6}b$$

$$\oint_C \vec{E} \cdot d\vec{\ell} = -\frac{d}{dt} \int_S B_n \, dA \qquad \text{32-6}c$$

$$\oint_C \vec{B} \cdot d\vec{\ell} = \mu_0 I + \mu_0 \epsilon_0 \frac{d}{dt} \int_S E_n \, dA \qquad \text{32-6}d$$

Maxwell's equations

Equation 32-6*a* is Gauss's law; it states that the flux of the electric field through any closed surface equals $1/\epsilon_0$ times the net charge inside the surface. As discussed in Chapter 23, Gauss's law implies that the electric field due to a point charge varies inversely as the square of the distance from the charge. This law describes how electric field lines diverge from a positive charge and converge on a negative charge. Its experimental basis is Coulomb's law.

Equation 32-6*b*, sometimes called Gauss's law for magnetism, states that the flux of the magnetic field vector $\vec{B}$ is zero through any closed surface. This equation describes the experimental observation that magnetic field lines do not diverge from any point in space or converge on any point; that is, it implies that isolated magnetic poles do not exist.

Equation 32-6*c* is Faraday's law; it states that the integral of the electric field around any closed curve *C*, which is the emf, equals the (negative) rate of change of the magnetic flux through any surface *S* bounded by the curve. (*S* is not a closed surface, so the magnetic flux through *S* is not necessarily zero.) Faraday's law describes how electric field lines encircle any area through which the magnetic flux is changing, and it relates the electric field vector $\vec{E}$ to the rate of change of the magnetic field vector $\vec{B}$.

Equation 32-6*d*, Ampère's law modified to include Maxwell's displacement current, states that the line integral of the magnetic field $\vec{B}$ around any closed curve *C* equals μ_0 times the current through any surface bounded by the curve plus $\mu_0\epsilon_0$ times the rate of change of the electric flux through the surface. This law describes how the magnetic field lines encircle an area through which a current is passing or through which the electric flux is changing.

In Section 32-4 we show how wave equations for both the electric field $\vec{E}$ and the magnetic field $\vec{B}$ can be derived from Maxwell's equations.

32-3 Electromagnetic Waves

Figure 32-3 shows the electric and magnetic field vectors of an electromagnetic wave. The electric and magnetic fields are perpendicular to each other and perpendicular to the direction of propagation of the wave. Electromagnetic waves are thus transverse waves. The magnitudes of $\vec{E}$ and $\vec{B}$ are in phase and are related by

$$E = cB \qquad \text{32-7}$$

where $c = 1/\sqrt{\mu_0\epsilon_0}$ is the speed of

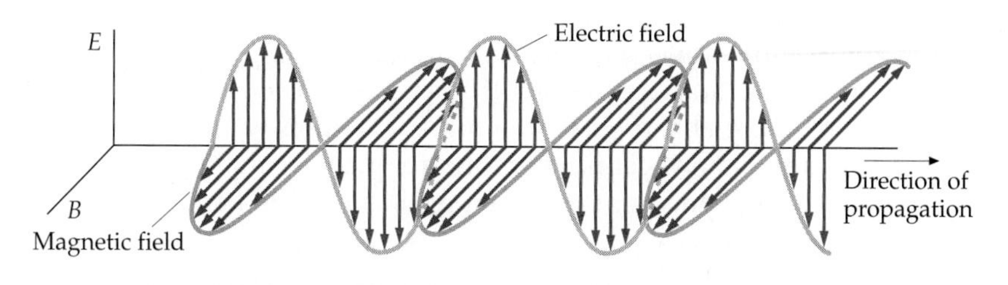

Figure 32-3 The electric and magnetic field vectors in an electromagnetic wave. The fields are in phase, perpendicular to each other, and perpendicular to the direction of propagation of the wave.

the wave. In general, the direction of propagation of an electromagnetic wave is the direction of the cross product $\vec{E} \times \vec{B}$.

The Electromagnetic Spectrum

The various types of electromagnetic waves—light, radio waves, X rays, gamma rays, microwaves, and others—differ only in wavelength and frequency, which are related to the speed c in the usual way, $f = c/\lambda$. Table 32-1 gives the **electromagnetic spectrum** and the names usually associated with the various frequency and wavelength ranges. These ranges are often not well defined and sometimes overlap. For example, electromagnetic waves with wavelengths of about 0.1 nm are usually called X rays, but if they originate from nuclear radioactivity, they are called gamma rays.

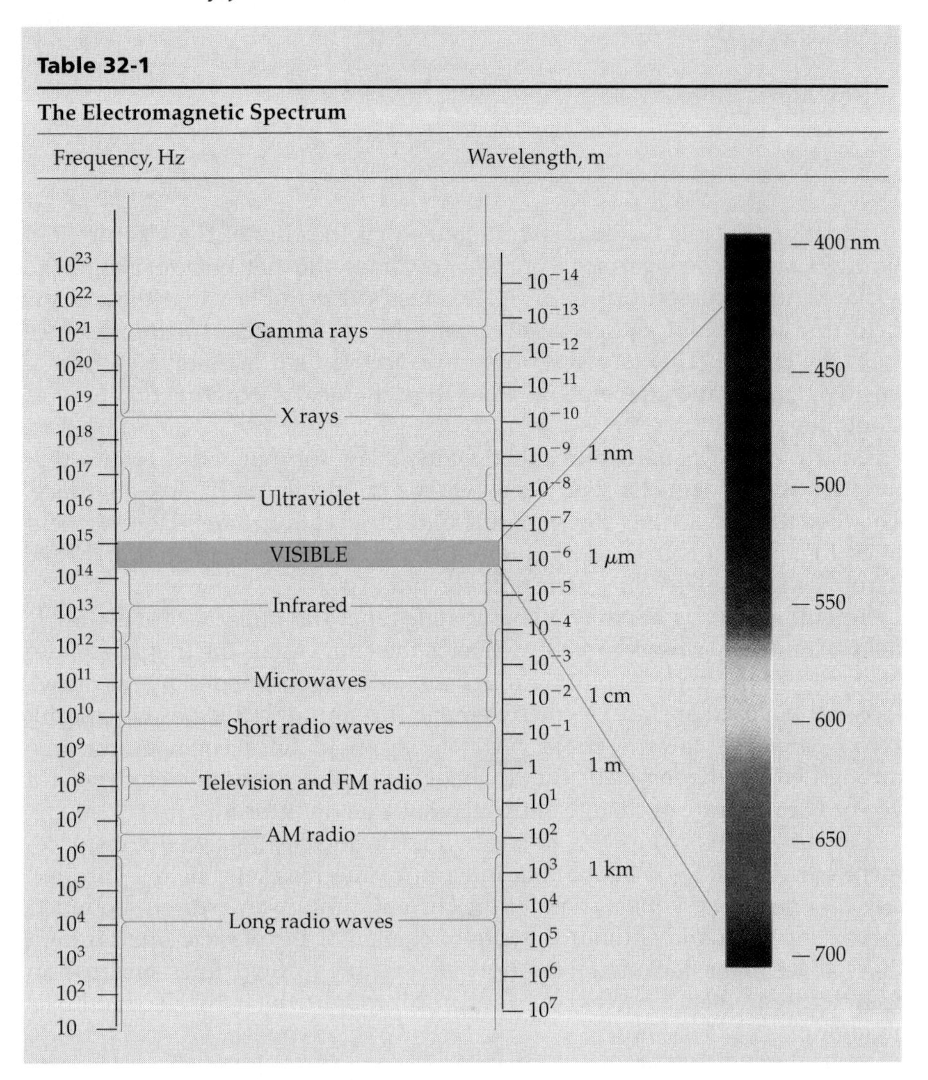

The human eye is sensitive to electromagnetic radiation with wavelengths from about 400 to 700 nm, the range called **visible light**. The shortest wavelengths in the visible spectrum correspond to violet light and the longest to red light, with all the colors of the rainbow falling between these extremes. Electromagnetic waves with wavelengths just beyond the visible spectrum on the short-wavelength side are called **ultraviolet rays**, and those with wavelengths just beyond the visible spectrum on the long-wavelength side are called **infrared waves**. Heat radiation given off by bodies at ordinary temperatures is in the infrared region of the electromagnetic spectrum. There are no limits on the wavelengths of electromagnetic radiation; that is, all wavelengths (or frequencies) are theoretically possible.

The differences in wavelengths of the various kinds of electromagnetic waves have important physical consequences. As we know, the behavior of waves depends strongly on the relative sizes of the wavelengths and the physical objects or apertures the waves encounter. Since the wavelengths of light are in the rather narrow range from about 400 to 700 nm, they are much smaller than most obstacles, so the ray approximation (introduced in Section 15-4) is often valid. The wavelength and frequency are also important in determining the kinds of interactions between electromagnetic waves and matter. X rays, for example, have very short wavelengths and high frequencies. They easily penetrate many materials that are opaque to lower-frequency light waves, which are absorbed by the materials. Microwaves have wavelengths of the order of a few centimeters and frequencies that are close to the natural resonance frequencies of water molecules in solids and liquids. Mi-

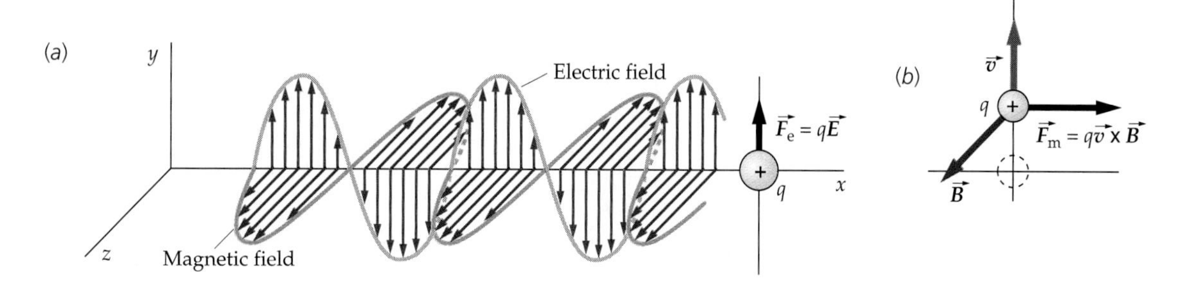

Figure 32-9 An electromagnetic wave incident on a point charge that is initially at rest on the x axis. (a) The electric force $q\vec{E}$ accelerates the charge in the upward direction. (b) When the velocity of the charge is $\vec{v}$ upward, the magnetic force $q\vec{v} \times \vec{B}$ accelerates the charge in the direction of the wave.

the time dependence of the fields. The particle experiences a force $q\vec{E}$ in the y direction and is thus accelerated by the electric field. At any time t, the velocity in the y direction is

$$v_y = at = \frac{qE}{m}t$$

After a short time t_1, the charge has acquired kinetic energy equal to

$$K = \frac{1}{2}mv_y^2 = \frac{1}{2}\frac{mq^2E^2t_1^2}{m^2} = \frac{1}{2}\frac{q^2E^2}{m}t_1^2 \qquad \text{32-11}$$

When the charge is moving in the y direction, it experiences a magnetic force

$$\vec{F}_m = q\vec{v} \times \vec{B} = qv_y\hat{j} \times B\hat{k} = qv_yB\hat{i} = \frac{q^2EB}{m}t\hat{i}$$

Note that this force is in the direction of propagation of the wave. Using $F_x = dp_x/dt$, we find for the momentum p_x transferred by the wave to the particle in time t_1:

$$p_x = \int_0^{t_1} F_x\,dt = \int_0^{t_1} \frac{q^2EB}{m}t\,dt = \frac{1}{2}\frac{q^2EB}{m}t_1^2$$

If we use $B = E/c$, this becomes

$$p_x = \frac{1}{c}\left(\frac{1}{2}\frac{q^2E^2}{m}t_1^2\right) \qquad \text{32-12}$$

Comparing Equations 32-11 and 32-12, we see that the momentum acquired by the charge in the direction of the wave is $1/c$ times the energy. Although our simple calculation was not rigorous, the results are correct. The magnitude of the momentum carried by an electromagnetic wave is $1/c$ times the energy carried by the wave:

$$p = \frac{U}{c} \qquad \text{32-13}$$

Momentum and energy in an electromagnetic wave

Since the intensity is the energy per unit time per unit area, the intensity divided by c is the momentum carried by the wave per unit time per unit area. The momentum carried per unit time is a force. The intensity divided by c is thus a force per unit area, which is a pressure. This pressure is the radiation pressure P_r:

$$P_r = \frac{I}{c} \qquad \text{32-14}$$

Radiation pressure and intensity

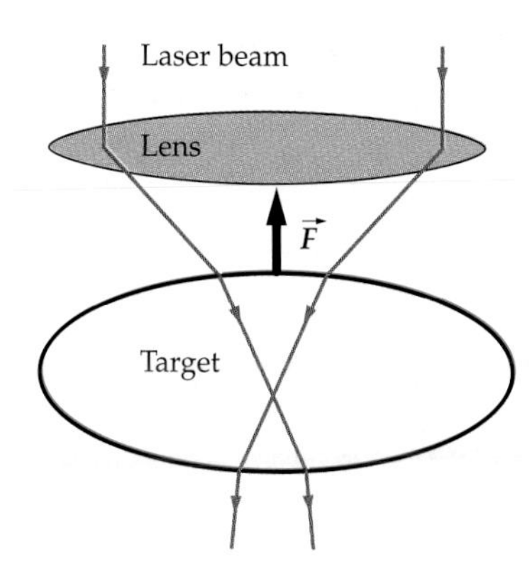

"Laser tweezers" make use of the momentum carried by electromagnetic waves to manipulate targets on a molecular scale. The two rays shown are refracted as they pass through a transparent target, such as a biological cell, or on an even smaller scale, a tiny transparent bead attached to a large molecule within a cell. At each refraction, the rays are bent downward, which increases the downward component of momentum of the rays. The target thus exerts a downward force on the laser beams, and the laser beams exert an upward force on the target, which pulls the target toward the laser source. The force is typically of the order of piconewtons. Laser tweezers have been used to accomplish such astonishing feats as stretching out coiled DNA.

We can relate the radiation pressure to the electric or magnetic fields by using Equation 32-9 to relate I to E and B, and Equation 32-7 to eliminate either E or B:

$$P_r = \frac{I}{c} = \frac{E_0 B_0}{2\mu_0 c} = \frac{E_{rms} B_{rms}}{\mu_0 c} = \frac{E_0^2}{2\mu_0 c^2} = \frac{B_0^2}{2\mu_0} \qquad \text{32-15}$$

Radiation pressure in terms of E and B

Consider an electromagnetic wave incident normally on some surface. If the surface absorbs energy U from the electromagnetic wave, it also absorbs momentum p given by Equation 32-14, and the pressure exerted on the surface equals the radiation pressure. If the wave is reflected, the momentum transferred is $2p$ because the wave now carries momentum in the opposite direction. The pressure exerted on the surface by the wave is then twice the radiation pressure.

Example 32-4

A light bulb emits spherical electromagnetic waves uniformly in all directions. Find (*a*) the intensity, (*b*) the radiation pressure, and (*c*) the electric and magnetic fields at a distance of 3 m from the bulb, assuming that 50 W of electromagnetic radiation is emitted.

Picture the Problem At a distance r from the bulb, the energy is spread uniformly over an area $4\pi r^2$. The intensity is the power divided by the area. The radiation pressure can then be found from $P_r = I/c$.

(*a*)1. Divide the power output by the area to find the intensity:	$I = \frac{50\ \text{W}}{4\pi r^2}$
2. Substitute $r = 3$ m:	$I = \frac{50\ \text{W}}{4\pi(3\ \text{m})^2} = 0.442\ \text{W/m}^2$
(*b*) The radiation pressure is the intensity divided by the speed of light:	$P_r = \frac{I}{c} = \frac{0.442\ \text{W/m}^2}{3 \times 10^8\ \text{m/s}} = 1.47 \times 10^{-9}\ \text{Pa}$
(*c*)1. B_0 is related to P_r by Equation 32-15:	$B_0 = (2\mu_0 P_r)^{1/2} = [2(4\pi \times 10^{-7}\ \text{T·m/A})(1.47 \times 10^{-9}\ \text{Pa})]^{1/2}$ $= 6.08 \times 10^{-8}\ \text{T}$
2. The maximum value of the electric field E_0 is c times B_0:	$E_0 = cB_0 = (3 \times 10^8\ \text{m/s})(6.08 \times 10^{-8}\ \text{T})$ $= 18.2\ \text{V/m}$
3. The electric and magnetic fields at that point are of the form:	$E = E_0 \sin \omega t$ and $B = B_0 \sin \omega t$ with $E_0 = 18.2\ \text{V/m}$ and $B_0 = 6.08 \times 10^{-8}\ \text{T}$.

Remark Note that the pressure calculated in (*b*) is very small compared with atmospheric pressure, which is of the order of 10^5 Pa.

Example 32-5

You are stranded in space a distance of 20 m from your spaceship. You carry a 1-kW laser. If your total mass, including your space suit and laser, is 95 kg, how long will it take you to reach the ship if you point the laser directly away from it?

Picture the Problem The laser emits light, which carries with it momentum. By momentum conservation, you are given an equal and opposite momentum toward the ship. The momentum carried by light is $p = U/c$, where U is the energy of the light. If the power of the laser is $P = dU/dt$, then the rate of change of momentum produced by the laser is $dp/dt = (dU/dt)/c = P/c$. This is the force exerted on you, which is constant.

1. The time taken is related to the distance and acceleration:

$$x = \tfrac{1}{2}at^2; \qquad t = \sqrt{\frac{2x}{a}}$$

2. Your acceleration is the force divided by your mass:

$$a = \frac{F}{m} = \frac{P/c}{m} = \frac{1000\ \text{W}}{(95\ \text{kg})(3 \times 10^8\ \text{m/s})} = 3.51 \times 10^{-8}\ \text{m/s}^2$$

3. Use this acceleration to calculate the time t:

$$t = \sqrt{\frac{2x}{a}} = \sqrt{\frac{2(20\ \text{m})}{3.51 \times 10^{-8}\ \text{m/s}^2}} = 3.38 \times 10^4\ \text{s} = 9.38\ \text{h}$$

Remarks Note that the acceleration found here is extremely small—only about one-billionth the acceleration of gravity. Your speed when you reach the ship would be $v = at = 1.19$ mm/s, which is practically imperceptible.

32-4 The Wave Equation for Electromagnetic Waves

optional

In Section 15-1, we saw that waves on a string obey a partial differential equation called the **wave equation**:

$$\frac{\partial^2 y(x,t)}{\partial x^2} = \frac{1}{v^2}\frac{\partial^2 y(x,t)}{\partial t^2} \qquad \text{32-16}$$

where $y(x, t)$ is the wave function, which for string waves is the displacement of the string.* The velocity of the wave is given by $v = \sqrt{F/\mu}$, where F is the tension and μ is the linear mass density. The solutions to this equation are harmonic wave functions of the form

$$y(x, t) = y_0 \sin(kx - \omega t)$$

where $k = 2\pi/\lambda$ is the wave number and $\omega = 2\pi f$ is the angular frequency.

Maxwell's equations imply that both $\vec{E}$ and $\vec{B}$ obey wave equations similar to Equation 32-16. We consider only free space, in which there are no charges or currents, and we assume that the electric and magnetic fields $\vec{E}$ and $\vec{B}$ are functions of time and one space coordinate only, which we will take to be the x coordinate. Such a wave is called a **plane wave,** because field

* The derivatives are partial derivatives because the wave function depends on both x and t.

optional

quantities are constant across any plane perpendicular to the x axis. For a plane electromagnetic wave traveling parallel to the x axis, the x components of the fields are zero, so the vectors $\vec{E}$ and $\vec{B}$ are perpendicular to the x axis and each obeys the wave equation:

$$\frac{\partial^2 \vec{E}}{\partial x^2} = \frac{1}{c^2}\frac{\partial^2 \vec{E}}{\partial t^2} \qquad \text{32-17}a$$

Wave equation for $\vec{E}$

$$\frac{\partial^2 \vec{B}}{\partial x^2} = \frac{1}{c^2}\frac{\partial^2 \vec{B}}{\partial t^2} \qquad \text{32-17}b$$

Wave equation for $\vec{B}$

where $c = 1/\sqrt{\mu_0 \epsilon_0}$ is the speed of the waves.

Derivation of the Wave Equation

We can relate the space derivative of one of the field vectors to the time derivative of the other by applying Equations 32-6*c* and 32-6*d* to appropriately chosen curves in space. We first relate the space derivative of E_y to the time derivative of B_z by applying Equation 32-6*c* (which is Faraday's law) to the rectangular curve of sides Δx and Δy lying in the xy plane (Figure 32-10). If Δx and Δy are very small, the line integral of $\vec{E}$ around this curve is approximately

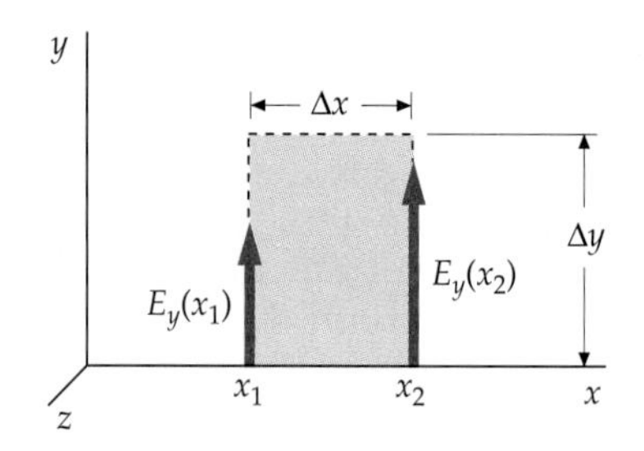

Figure 32-10 A rectangular curve in the xy plane for the derivation of Equation 32-18.

$$\oint \vec{E} \cdot d\vec{\ell} = E_y(x_2)\Delta y - E_y(x_1)\Delta y$$

where $E_y(x_1)$ is the value of E_y at the point x_1 and $E_y(x_2)$ is the value of E_y at the point x_2. The contributions of the type $E_x\,\Delta x$ from the top and bottom of this curve are zero because $E_x = 0$. Since Δx is very small, we can approximate the difference in E_y at the points x_1 and x_2 by

$$E_y(x_2) - E_y(x_1) = \Delta E \approx \frac{\partial E_y}{\partial x}\Delta x$$

Then

$$\oint \vec{E} \cdot d\vec{\ell} \approx \frac{\partial E_y}{\partial x}\Delta x\,\Delta y$$

The flux of the magnetic field through this curve is approximately

$$\int_S B_n\,dA = B_z\,\Delta x\,\Delta y$$

Faraday's law then gives

$$\frac{\partial E_y}{\partial x}\Delta x\,\Delta y = -\frac{\partial B_z}{\partial t}\Delta x\,\Delta y$$

or

$$\frac{\partial E_y}{\partial x} = -\frac{\partial B_z}{\partial t} \qquad \text{32-18}$$

Equation 32-18 implies that if there is a component of the electric field E_y that depends on x, there must be a component of the magnetic field B_z that depends on time or, conversely, that if there is a component of the magnetic field B_z that depends on time, there must be a component of the electric field E_y that depends on x. We can get a similar equation relating the space derivative of the magnetic field B_z to the time derivative of the electric field E_y by applying Equation 32-6*d* to the curve of sides Δx and Δz in the xz plane shown in Figure 32-11.

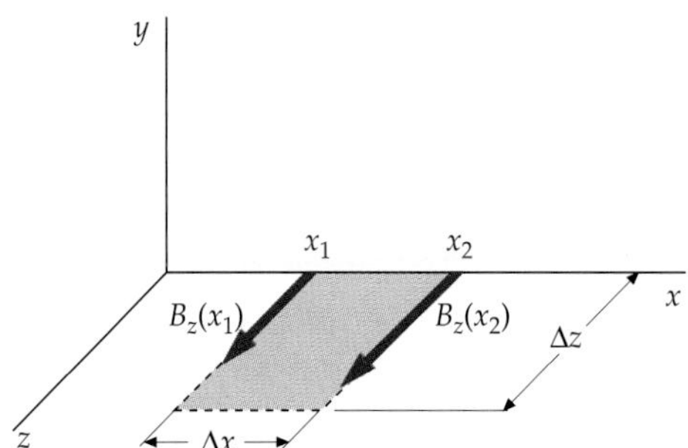

Figure 32-11 A rectangular curve in the xz plane for the derivation of Equation 32-19.

For the case of no conduction currents, Equation 32-6*d* is

$$\oint \vec{B} \cdot d\vec{\ell} = \mu_0 \epsilon_0 \frac{d}{dt} \int_S E_n \, dA$$

The details of this calculation are similar to those for Equation 32-18. The result is

$$\frac{\partial B_z}{\partial x} = -\mu_0 \epsilon_0 \frac{\partial E_y}{\partial t} \qquad \text{32-19}$$

We can eliminate either B_z or E_y from Equations 32-18 and 32-19 by differentiating either equation with respect to x or t. If we differentiate both sides of Equation 32-18 with respect to x, we obtain

$$\frac{\partial}{\partial x}\left(\frac{\partial E_y}{\partial x}\right) = -\frac{\partial}{\partial x}\left(\frac{\partial B_z}{\partial t}\right)$$

or

$$\frac{\partial^2 E_y}{\partial x^2} = -\frac{\partial}{\partial t}\left(\frac{\partial B_z}{\partial x}\right)$$

where the order of the time and space derivatives on the right side have been interchanged. We now use Equation 32-19 for $\partial B_z/\partial x$:

$$\frac{\partial^2 E_y}{\partial x^2} = -\frac{\partial}{\partial t}\left(-\mu_0 \epsilon_0 \frac{\partial E_y}{\partial t}\right)$$

which yields the wave equation

$$\frac{\partial^2 E_y}{\partial x^2} = \mu_0 \epsilon_0 \frac{\partial^2 E_y}{\partial t^2} \qquad \text{32-20}$$

Comparing this equation with Equation 32-16, we see that E_y obeys a wave equation for waves with speed $c = 1/\sqrt{\mu_0 \epsilon_0}$, which is Equation 32-1.

If we had instead chosen to eliminate E_y from Equations 32-18 and 32-19 (by differentiating Equation 32-18 with respect to t, for example), we would have obtained an equation identical to Equation 32-20 except with B_z replacing E_y. We can thus see that both the electric field E_y and the magnetic field B_z obey a wave equation for waves traveling with the velocity $1/\sqrt{\mu_0 \epsilon_0}$, which is the velocity of light.

optional

By following the same line of reasoning as used above, and applying Equation 32-6*c* (Faraday's law) to the curve in the *xz* plane (Figure 32-11), we would obtain

$$\frac{\partial E_z}{\partial x} = \frac{\partial B_y}{\partial t} \qquad \text{32-21}$$

Similarly, the application of Equation 32-6*d* to the curve in the *xy* plane (Figure 32-10) gives

$$\frac{\partial B_y}{\partial x} = \mu_0 \epsilon_0 \frac{\partial E_z}{\partial t} \qquad \text{32-22}$$

We can use these results to show that, for a wave propagating in the x direction, the components E_z and B_y also obey the wave equation.

Consider the harmonic wave function of the form

$$E_y = E_{y0} \sin (kx - \omega t) \qquad \text{32-23}$$

If we substitute this solution into Equation 32-18, we can see that the magnetic field B_z is in phase with the electric field E_y. We have

$$\frac{\partial B_z}{\partial t} = -\frac{\partial E_y}{\partial x} = -kE_{y0} \cos (kx - \omega t)$$

Solving for B_z gives

$$B_z = \frac{k}{\omega} E_{y0} \sin (kx - \omega t) = B_{z0} \sin (kx - \omega t) \qquad \text{32-24}$$

where

$$B_{z0} = \frac{k}{\omega} E_{y0} = \frac{E_{y0}}{c}$$

and $c = \omega/k$ is the velocity of the wave. (We have omitted the arbitrary constant of integration because it plays no part in the wave that we are interested in.) Since the electric and magnetic fields oscillate in phase with the same frequency, we have the general result that the magnitude of the electric field is c times the magnitude of the magnetic field for an electromagnetic wave:

$$E = cB$$

which is Equation 32-7.

We see that Maxwell's equations imply wave equations 32-17*a* and 32-17*b* for the electric and magnetic fields; and that if E_y varies harmonically, as in Equation 32-23, the magnetic field B_z is in phase with E_y and has an amplitude related to the amplitude of E_y by $B_z = E_y/c$. The electric and magnetic fields are perpendicular to each other and to the direction of the wave propagation, as shown in Figure 32-3.

Example 32-6

The electric field vector of an electromagnetic wave is given by $\vec{E}(x, t) = E_0 \sin (kx - \omega t)\hat{j} + E_0 \cos (kx - \omega t)\hat{k}$. (*a*) Find the corresponding magnetic field. (*b*) Compute $\vec{E} \cdot \vec{B}$ and $\vec{E} \times \vec{B}$.

Picture the Problem We find B_y using either Equation 32-21 or 32-22, and B_z using Equation 32-18 or 32-19. The products $\vec{E}\cdot\vec{B}$ and $\vec{E}\times\vec{B}$ are found using standard vector operations.

(*a*)1. Use Equation 32-21 to obtain a relation for B_y:

$$\frac{\partial B_y}{\partial t} = \frac{\partial E_z}{\partial x}$$

$$= \frac{\partial}{\partial x}[E_0 \cos(kx - \omega t)] = -kE_0 \sin(kx - \omega t)$$

2. Integrate the above result (neglecting the arbitrary constant of integration) to find B_y. Let $B_0 = kE_0/\omega = E_0/c$:

$$B_y = [kE_0 \cos(kx - \omega t)]\left(\frac{-1}{\omega}\right) = -B_0 \cos(kx - \omega t)$$

3. Equation 32-18 gives us a relation for B_z:

$$\frac{\partial B_z}{\partial t} = -\frac{\partial E_y}{\partial x}$$

$$= -\frac{\partial}{\partial x}[E_0 \sin(kx - \omega t)] = -kE_0 \cos(kx - \omega t)$$

4. Integrating and using $B_0 = kE_0/\omega = E_0/c$, we find B_z:

$$B_z = [-kE_0 \sin(kx - \omega t)]\left(\frac{-1}{\omega}\right) = B_0 \sin(kx - \omega t)$$

(*b*)1. Let $\theta = kx - \omega t$ to simplify the notation and calculate $\vec{E}\cdot\vec{B}$:

$$\vec{E}\cdot\vec{B} = [E_0 \sin\theta\hat{j} + E_0\cos\theta\,\hat{k}]\cdot[-B_0\cos\theta\,\hat{j} + B_0\sin\theta\,\hat{k}]$$

$$= -E_0B_0\sin\theta\cos\theta\,\hat{j}\cdot\hat{j} + E_0B_0\sin^2\theta\,\hat{j}\cdot\hat{k}$$

$$-E_0B_0\cos^2\theta\,\hat{k}\cdot\hat{j} + E_0B_0\cos\theta\sin\theta\,\hat{k}\cdot\hat{k}$$

$$= -E_0B_0\sin\theta\cos\theta + 0 - 0 + E_0B_0\cos\theta\sin\theta = 0$$

2. Calculate $\vec{E}\times\vec{B}$:

$$\vec{E}\times\vec{B} = [E_0\sin\theta\,\hat{j} + E_0\cos\theta\,\hat{k}]\times[-B_0\cos\theta\,\hat{j} + B_0\sin\theta\,\hat{k}]$$

$$= E_0B_0\sin^2\theta\,\hat{j}\times\hat{k} + (-E_0B_0\cos^2\theta\,\hat{k}\times\hat{j})$$

$$= E_0B_0\sin^2\theta\,\hat{i} + E_0B_0\cos^2\theta\,\hat{i} = E_0B_0\hat{i}$$

Remarks We see that $\vec{E}$ and $\vec{B}$ are perpendicular to one another, and that $\vec{E}\times\vec{B}$ is in the direction of propagation of the wave. This type of electromagnetic wave is said to be *circularly polarized*. At a fixed value of x, both $\vec{E}$ and $\vec{B}$ rotate in a circle in a plane perpendicular to x with angular frequency ω. The fields $\vec{E}$ and $\vec{B}$ are constant in magnitude, as can be seen by noting that $\vec{E}\cdot\vec{E}$ and $\vec{B}\cdot\vec{B}$ are constant.

Exercise Calculate $\vec{E}\cdot\vec{E}$ and $\vec{B}\cdot\vec{B}$. [*Answers* $\vec{E}\cdot\vec{E} = E_y^2 + E_z^2 = E_0^2\sin^2(kx - \omega t) + E_0^2\cos^2(kx - \omega t) = E_0^2$ and $\vec{B}\cdot\vec{B} = B_y^2 + B_z^2 = B_0^2\cos^2(kx - \omega t) + B_0^2\sin^2(kx - \omega t) = B_0^2$]

James Clerk Maxwell (1831–1879)

C. W. F. Everitt
Stanford University

One day in 1877 a Scottish undergraduate named Donald MacAlister wrote home from Cambridge University that he had just had dinner with a professor who was "one of the best of our men, and a thorough old Scotch laird in ways and speech." James Clerk Maxwell was wealthy, an expert swimmer and horseman, proprietor of an estate of 2000 acres in Scotland—and a scientist whose writings remain astonishingly up to date. The greatest mathematical physicist since Newton, he created the electromagnetic theory of light, predicted the existence of radio waves, wrote the first significant paper on control theory, and was joint inventor with Ludwig Boltzmann of statistical mechanics. He also performed with his wife's aid a brilliant series of experiments on color vision and took the first color photograph. In the 2 years before his death due to cancer in 1879, at the age of 48, he would lay the foundations of another new subject that was to reach fruition in the twentieth century, rarefied gas dynamics.

Maxwell's electromagnetic theory of light was rooted in the work of two men, Michael Faraday and William Thomson. Faraday's invention of the electric motor and his research on electromagnetic induction, electrochemistry, dielectric and diamagnetic action, and magneto-optical rotation made him in Maxwell's words "the nucleus of everything electric since 1830." His contributions to theory lay in his progressively advancing ideas about lines of electric and magnetic force, in particular the geometrical relations governing electromagnetic phenomena and the idea that magnetic forces might be accounted for not by direct attractions and repulsions between elements of current but by attributing to lines of force the property of shortening themselves and repelling each other sideways (Figure 1). Thomson's role was to relate lines of force to existing theories in electrostatics and magnetostatics, to invent a number of highly ingenious analytical techniques for solving electrical problems, and to emphasize the cardinal importance of energy principles in electromagnetism. Maxwell then introduced a series of new concepts: the *electrotonic function* (vector potential), the energy density of the field, and the displacement current; he organized the subject into a coherent structure and in 1861 made the momentous discovery of the equivalence between light and electromagnetic waves.

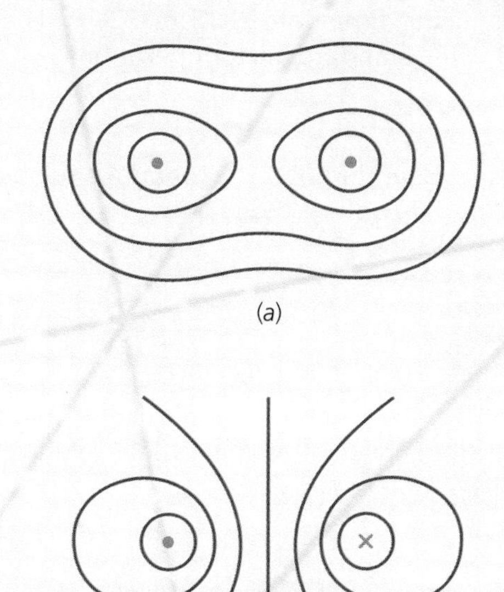

Figure 1 Faraday's explanation of forces between current-carrying wires. The two diagrams show the lines of force observed when currents are flowing in parallel wires. Faraday assumed that the lines of force tend to shorten and repel each other sideways. (*a*) For wires with currents flowing in the same direction, the lines of force pull the two wires together. (*b*) For wires with currents flowing in opposite directions, the lines of force push the wires apart.

The first part of Maxwell's paper "On Faraday's Lines of Force" (1855–1856) developed an analogy, due in essence to Thomson, between lines of electric and magnetic force and streamlines in a moving incompressible fluid. Maxwell applied this to interpret many of Faraday's observations, prefacing his paper with a luminous discussion of the significance of analogies in physics.

Next, still building on Faraday and Thomson, Maxwell extended the discussion to electromagnetism. He formulated a group of equations summarizing the relations of the electric and magnetic fields to the charges and currents producing them—the beginnings of what we now call Maxwell's equations. They described the phenomena with elegant precision from a point of view completely different from the then-popular action-at-a-distance theories of André-Marie Ampère and Wilhelm Weber.

After such a brilliant start one might have expected a rush of papers following up the new ideas. But other physicists ignored them, and Maxwell had the habit of investigating different subjects in turn, often with long intervals between successive papers in the same field. Six years elapsed before the appearance of his next paper, "On Physical Lines of Force," published in four parts in 1861–1862. During the interval Maxwell made brilliant contributions to three distinct subjects before returning to electromagnetism: color vision, the theory of Saturn's rings, and the kinetic theory of gases.

"On Physical Lines of Force" contained Maxwell's extraordinary molecular-vortex model of the electromagnetic field. To account for the pattern of stresses associated with lines of force by Faraday, Maxwell investigated the properties of a medium occupying all space in which tiny molecular vortices rotate with their axes parallel to the lines of force. The closer together the lines are, the faster the rotation of the vortices. In a medium of this kind the lines of force do tend to shorten themselves and repel each other sideways, yielding the right forces between currents and magnets. The question is, what makes the vortices rotate? Here Maxwell put forward an idea as ingenious as it was weird. He postulated that an electric current consists in the motion of tiny particles that mesh like gear wheels with the vortices, and that the medium is filled with similar particles between the vortices. Figure 2 gives the picture. Maxwell remarks:

> I do not bring [this hypothesis] forward as a mode of connexion existing in nature . . . [but] I venture to say that anyone who understands [its] provisional and temporary character . . . will find himself helped rather than hindered by it in his search for the true interpretation of [electromagnetic] phenomena.

The question then was how to fit electrostatic phenomena into the model. Maxwell made the medium an elastic one. Thus, magnetic forces were accounted for by rotations in the medium, and electric forces by its elastic distortion. Any elastic medium will transmit waves. In Maxwell's medium the velocity of the waves turned out to be related to the ratio of electric to magnetic forces. Putting in numbers from an experiment of 1856 by G. Kohlrausch and W. Weber, Maxwell found to his astonishment that the propagation velocity was equal to the velocity of light. With excitement he wrote, "we can scarcely avoid the inference that *light consists in the transverse undulation of the same medium which is the cause of electric and magnetic phenomena.*"

Having made the great discovery, Maxwell promptly jettisoned his model. Instead of attempting a more refined mechanical explanation of the phenomena, he formulated a system of electromagnetic equations from which he deduced that waves of electric and magnetic force would propagate through space with the velocity of light. That is why his is called an *electromagnetic* theory of light, in contrast to the theories of the mechanical ether that preceded it. The theory appeared in two papers of 1865 and 1868, and in its most general form in the great *Treatise on Electricity and Magnetism*, published in 1873. This was a

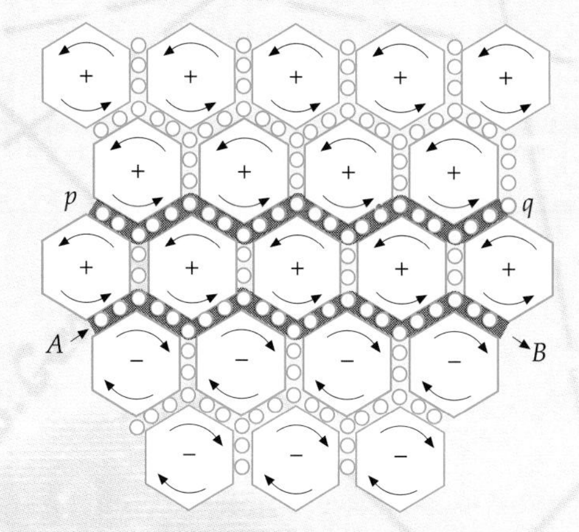

Figure 2 Maxwell's vortex model of the magnetic field. The rotating vortices represent lines of magnetic force. They mesh with small particles that act like gear wheels. In free space the particles are restrained from moving, except for a small elastic reaction (the displacement current), but in a conducting wire they are free to move. Their motion constitutes an electric current, which in turn sets the vortices in rotation, creating the magnetic field around the wire. *A* and *B* represent current through a wire, and *p* and *q* represent an induced current in an adjacent wire. (Redrawn from *The Scientific Papers of James Clerk Maxwell*, Vol. I, Figure 2 after p. 488.)

work of such scope that Robert Andrews Millikan, author of the famous oil-drop experiment to measure the charge on the electron, ranked it with Newton's *Principia* in considering them the two most influential books in the history of physics, "the one creating our modern mechanical world and the other our modern electrical world."

Equally profound were Maxwell's contributions to statistical and molecular physics. They began with a paper in 1859 on the kinetic theory of gases, in which Maxwell introduced the velocity distribution function and enunciated the *equipartition theorem* (Section 18-5), which in its original form stated that the average translational and rotational energies of large numbers of colliding molecules, whether of the same or different species, are equal. One result was Maxwell's estimate of the mean free path of a gas molecule, which Loschmidt in 1865 applied to make the first serious estimates of the diameters of molecules. Later, Maxwell developed the general theory of transport phenomena, from which the Boltzmann equation is derived; invented the concept of ensemble averaging; created rarefied-gas dynamics; and conceived that "very small BUT lively being" the Maxwell demon.

The demon, so named by Kelvin, is one of the earliest examples in physics of a "thought experiment." Maxwell imagined two chambers of gas, A and B, separated by a wall in which there was a trap door guarded by a tiny being with eyesight so acute that it could discern the motion of individual molecules. By opening the door when a fast molecule approached from chamber A or as a slow molecule approached from chamber B, the demon could redistribute the velocities to make B hotter than A without doing any work and thereby defeat the second law of thermodynamics. Maxwell's point was to demonstrate that the second law of thermodynamics is inherently a statistical law and not a dynamical one.

The work by Maxwell and Boltzmann on statistical mechanics had profound implications for modern physics. Brilliant as its successes were, the failures were—as Maxwell saw—in some ways even more striking. The equipartition theorem gave an answer for the ratio of the specific heats of gases that disagreed with experiment, while some of Boltzmann's theorems "proved too much" because they would apply to the properties of solids and liquids as well as gases. These questions remained shrouded in mystery until the emergence in 1900 of Planck's quantum hypothesis. Writing about them in 1877, Maxwell confessed his bewilderment and stated that nothing remained but to adopt the attitude of "thoroughly conscious ignorance that is the prelude to every real advance in knowledge."

Maxwell was an unusually sensitive man, with strong religious feeling and a fascinating and astonishing sense of humor. Many of his letters reveal a delightfully sly irony. He also had some talent for writing poetry, usually light, but occasionally touching a deeper note. The last stanza of one poem to his wife, written in 1867, was

All powers of mind, all force of will
May lie in dust when we are dead,
But love is ours, and shall be still
When earth and seas are fled.

James Clerk Maxwell (1831–1879) with his wife, Katherine Mary, and their dog.

Summary

1. Maxwell's equations summarize the fundamental laws of physics that govern electricity and magnetism.
2. Electromagnetic waves include light, radio and television waves, X rays, gamma rays, microwaves, and others.

Topic	Remarks and Relevant Equations	
1. Maxwell's Displacement Current	Ampère's law can be generalized to apply to discontinuous currents if the conduction current I is replaced by $I + I_d$, where I_d is Maxwell's displacement current: $I_d = \epsilon_0 \frac{d\phi_e}{dt}$	32-3
Generalized form of Ampère's law	$\oint_C \vec{B} \cdot d\vec{\ell} = \mu_0(I + I_d) = \mu_0 I + \mu_0\epsilon_0 \frac{d\phi_e}{dt}$	32-4
2. Maxwell's Equations	The laws of electricity and magnetism are summarized by Maxwell's equations.	
Gauss's law	$\oint_S E_n\, dA = \frac{1}{\epsilon_0} Q_{inside}$	32-6*a*
Gauss's law for magnetism (isolated magnetic poles do not exist)	$\oint_S B_n\, dA = 0$	32-6*b*
Faraday's law	$\oint_C \vec{E} \cdot d\vec{\ell} = -\frac{d}{dt}\int_S B_n\, dA$	32-6*c*
Ampère's law modified	$\oint_C \vec{B} \cdot d\vec{\ell} = \mu_0 I + \mu_0\epsilon_0 \frac{d}{dt}\int_S E_n\, dA$	32-6*d*
3. Electromagnetic Waves	In an electromagnetic wave, the electric and magnetic field vectors are perpendicular to each other and to the direction of propagation. Their magnitudes are related by $E = cB$	32-7
Wave speed	$c = \frac{1}{\sqrt{\mu_0\epsilon_0}} \approx 3 \times 10^8 \text{ m/s}$	32-1
Electromagnetic spectrum	The various types of electromagnetic waves—light, radio waves, X rays, gamma rays, microwaves, and others—differ only in wavelength and frequency. The human eye is sensitive to the range from about 400 to 700 nm.	
Electric-dipole radiation	Electromagnetic waves are produced when electric charges accelerate. Oscillating charges in an electric-dipole antenna radiate electromagnetic waves with an intensity that is maximum in directions perpendicular to the antenna and zero along the axis of the antenna. Perpendicular to the antenna and far away from it, the electric field of the electromagnetic wave is parallel to the antenna.	
Energy density in an electromagnetic wave	$u = u_e + u_m = \epsilon_0 E^2 = \frac{B^2}{\mu_0} = \frac{EB}{\mu_0 c}$	32-8
Intensity of an electromagnetic wave	$I = u_{av}c = \frac{E_{rms}B_{rms}}{\mu_0} = \frac{1}{2}\frac{E_0B_0}{\mu_0} = \lvert\vec{S}\rvert_{av}$	32-9

Poynting vector	$\vec{S} = \dfrac{\vec{E} \times \vec{B}}{\mu_0}$	**32-10**
Momentum in an electromagnetic wave	$p = \dfrac{U}{c}$	**32-13**
Radiation pressure	$P_r = \dfrac{I}{c}$	**32-14**

4. Wave Equation (optional)

Maxwell's equations imply that the electric and magnetic field vectors in free space obey a wave equation

$$\frac{\partial^2 \vec{E}}{\partial x^2} = \frac{1}{c^2}\frac{\partial^2 \vec{E}}{\partial t^2} \quad \textbf{32-17}a$$

$$\frac{\partial^2 \vec{B}}{\partial x^2} = \frac{1}{c^2}\frac{\partial^2 \vec{B}}{\partial t^2} \quad \textbf{32-17}b$$

Problem-Solving Guide

Begin by drawing a neat diagram that includes the important features of the problem. In problems involving Maxwell's displacement current, it is useful to draw the surface through which the electric flux is to be calculated and the bounding curve C.

Summary of Worked Examples

Type of Calculation	Procedure and Relevant Examples
1. Displacement Current	
Find the displacement current due to a time-dependent electric field.	The displacement current I_d is given by Equation 32-3, $I_d = \epsilon_0\, d\phi_e/dt$, where ϕ_e is the electric flux. **Example 32-1**
Find the magnetic field due to a displacement current.	The magnetic field can be calculated from the generalized form of Ampère's law, $\oint_C \vec{B} \cdot d\vec{\ell} = \mu_0(I + I_d)$, where I is the conduction current and I_d is the displacement current. **Example 32-2**
2. Electromagnetic Waves	
Calculate the rms-induced emf in a loop antenna due to electromagnetic waves of frequency *f*.	Use $\mathcal{E}_{rms} = \pi r^2\,(dB/dt)_{rms}$ and $B = E/c$. **Example 32-3**
Find the intensity, momentum, and radiation pressure associated with an electromagnetic wave.	Use $I = E_0 B_0/2\mu_0$, $p = U/c$, and $P_r = I/c$. **Examples 32-4, 32-5**

Problems

Conceptual Problems

Problems from Optional and Exploring sections

In a few problems, you are given more data than you actually need; in a few other problems, you are required to supply data from your general knowledge, outside sources, or informed estimates.

• Single-concept, single-step, relatively easy
•• Intermediate-level, may require synthesis of concepts
••• Challenging, for advanced students

Maxwell's Displacement Current

1 • A parallel-plate capacitor in air has circular plates of radius 2.3 cm separated by 1.1 mm. Charge is flowing onto the upper plate and off the lower plate at a rate of 5 A. (*a*) Find the time rate of change of the electric field between the plates. (*b*) Compute the displacement current between the plates and show that it equals 5 A.

2 • In a region of space, the electric field varies according to $E = (0.05\ \text{N/C})\sin 2000t$, where t is in seconds. Find the maximum displacement current through a 1-m^2 area perpendicular to $\vec{E}$.

3 •• For Problem 1, show that at a distance r from the axis of the plates the magnetic field between the plates is given by $B = (1.89 \times 10^{-3}\ \text{T/m})r$ if r is less than the radius of the plates.

4 •• (*a*) Show that for a parallel-plate capacitor the displacement current is given by $I_d = C\, dV/dt$, where C is the capacitance and V the voltage across the capacitor. (*b*) A parallel plate capacitor $C = 5$ nF is connected to an emf $\mathcal{E} = \mathcal{E}_0 \cos \omega t$, where $\mathcal{E}_0 = 3$ V and $\omega = 500\pi$. Find the displacement current between the plates as a function of time. Neglect any resistance in the circuit.

5 •• Current of 10 A flows into a capacitor having plates with areas of 0.5 m^2. (*a*) What is the displacement current between the plates? (*b*) What is dE/dt between the plates for this current? (*c*) What is the line integral of $\vec{B}\cdot d\vec{\ell}$ around a circle of radius 10 cm that lies within and parallel to the plates?

6 •• A parallel-plate capacitor with circular plates is given a charge Q_0. Between the plates is a leaky dielectric having a dielectric constant of κ and a resistivity ρ. (*a*) Find the conduction current between the plates as a function of time. (*b*) Find the displacement current between the plates as a function of time. What is the total (conduction plus displacement) current? (*c*) Find the magnetic field produced between the plates by the leakage discharge current as a function of time. (*d*) Find the magnetic field between the plates produced by the displacement current as a function of time. (*e*) What is the total magnetic field between the plates during discharge of the capacitor?

7 •• The leaky capacitor of Problem 6 is charged such that the voltage across the capacitor is given by $V(t) = (0.01\ \text{V/s})t$. (*a*) Find the conduction current as a function of time. (*b*) Find the displacement current. (*c*) Find the time for which the displacement current is equal to the conduction current.

8 •• The space between the plates of a capacitor is filled with a material of resistivity $\rho = 10^4\ \Omega \cdot \text{m}$ and dielectric constant $\kappa = 2.5$. The parallel plates are circular with a radius of 20 cm and are separated by 1 mm. The voltage across the plates is given by $V_0 \cos \omega t$, with $V_0 = 40$ V and $\omega = 120\pi$ rad/s. (*a*) What is the displacement current density? (*b*) What is the conduction current between the plates? (*c*) At what angular frequency is the total current 45° out of phase with the applied voltage?

9 ••• In this problem, you are to show that the generalized form of Ampère's law (Equation 32-4) and the Biot–Savart law give the same result in a situation in which they both can be used. Figure 32-12 shows two charges $+Q$ and $-Q$ on the x axis at $x = -a$ and $x = +a$, with a current $I = -dQ/dt$ along the line between them. Point P is on the y axis at $y = R$.

Figure 32-12 Problem 9

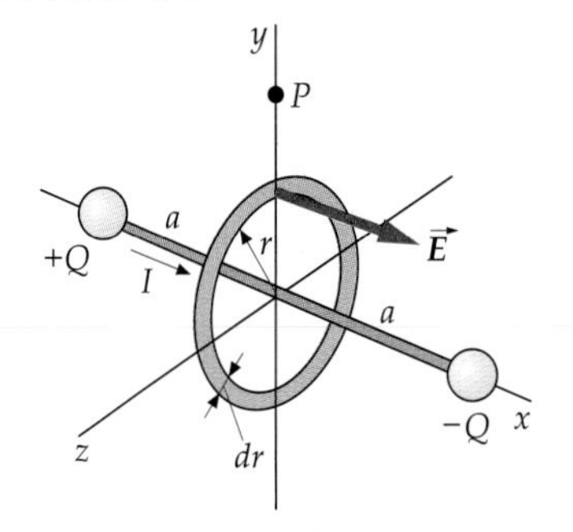

(*a*) Use the Biot–Savart law to show that the magnitude of B at point P is

$$B = \frac{\mu_0 I a}{2\pi R} \frac{1}{\sqrt{R^2 + a^2}}$$

(*b*) Consider a circular strip of radius r and width dr in the yz plane with its center at the origin. Show that the flux of the electric field through this strip is

$$E_x\, dA = \frac{Q}{\epsilon_0} a(r^2 + a^2)^{-3/2}\, r\, dr$$

(*c*) Use your result for part (*b*) to find the total flux ϕ_e through a circular area of radius R. Show that

$$\epsilon_0 \phi_e = Q\left(1 - \frac{a}{\sqrt{a^2 + R^2}}\right)$$

(*d*) Find the displacement current I_d, and show that

$$I + I_d = I \frac{a}{\sqrt{a^2 + R^2}}$$

(*e*) Then show that Equation 32-4 gives the same result for B as that found in part (*a*).

Maxwell's Equations

10 •• Theorists have speculated on the possible existence of magnetic monopoles, and there have been several, as yet unsuccessful, experimental searches for such monopoles. Suppose magnetic monopoles were found and that the magnetic field at a distance r from a monopole of strength q_m is given by $B = (\mu_0/4\pi)q_m/r^2$. How would Maxwell's equations have to be modified to be consistent with such a discovery?

11 •• Show that the normal component of the magnetic field $\vec{B}$ is continuous across a surface. Do this by applying Gauss's law for $\vec{B}$ ($\int B_n\, dA = 0$) to a pillbox Gaussian surface that has a face on each side of the surface.

The Electromagnetic Spectrum

12 • Which waves have greater frequencies, light waves or X rays?

13 • Are the frequencies of ultraviolet radiation greater or less than those of infrared radiation?

14 • What kind of waves have wavelengths of the order of a few meters?

15 • Find the wavelength for (*a*) a typical AM radio wave with a frequency of 1000 kHz and (*b*) a typical FM radio wave of 100 MHz.

16 • What is the frequency of a 3-cm microwave?

17 • What is the frequency of an X ray with a wavelength of 0.1 nm?

Electric-Dipole Radiation

18 • The detection of radio waves can be accomplished with either a dipole antenna or a loop antenna. The dipole antenna detects the (pick one) [*electric*] [*magnetic*] field of the wave, and the loop antenna detects the [*electric*] [*magnetic*] field of the wave.

19 • A transmitter uses a loop antenna with the loop in the horizontal plane. What should be the orientation of a dipole antenna at the receiver for optimum signal reception?

20 •• The intensity of radiation from an electric dipole is proportional to $(\sin^2\theta)/r^2$, where θ is the angle between the electric dipole moment and the position vector $\vec{r}$. A radiating electric dipole lies along the z axis (its dipole moment is in the z direction). Let I_1 be the intensity of the radiation at a distance $r = 10$ m and at angle $\theta = 90°$. Find the intensity (in terms of I_1) at (*a*) $r = 30$ m, $\theta = 90°$; (*b*) $r = 10$ m, $\theta = 45°$; and (*c*) $r = 20$ m, $\theta = 30°$.

21 •• (*a*) For the situation described in Problem 20, at what angle is the intensity at $r = 5$ m equal to I_1? (*b*) At what distance is the intensity equal to I_1 at $\theta = 45°$?

22 •• The transmitting antenna of a radio station is a dipole located atop a mountain 2000 m above sea level. The intensity of the signal on a nearby mountain 4 km distant and also 2000 m above sea level is 4×10^{-12} W/m². What is the intensity of the signal at sea level and 1.5 km from the transmitter? (See Problem 20.)

23 ••• A radio station that uses a vertical dipole antenna broadcasts at a frequency of 1.20 MHz with total power output of 500 kW. The radiation pattern is as shown in Figure 32-8, i.e., the intensity of the signal varies as $\sin^2\theta$, where θ is the angle between the direction of propagation and the vertical, and is independent of azimuthal angle. Calculate the intensity of the signal at a horizontal distance of 120 km from the station. What is the intensity at that point as measured in photons per square centimeter per second?

24 ••• At a distance of 30 km from a radio station broadcasting at a frequency of 0.8 MHz, the intensity of the electromagnetic wave is 2×10^{-13} W/m². The transmitting antenna is a vertical dipole. What is the total power radiated by the station?

25 ••• A small private plane approaching an airport is flying at an altitude of 2500 m above ground. The airport's flight control system transmits 100 W at 24 MHz, using a vertical dipole antenna. What is the intensity of the signal at the plane's receiving antenna when the plane's position on a map is 4 km from the airport?

Energy and Momentum in an Electromagnetic Wave

26 • An electromagnetic wave has an intensity of 100 W/m². Find (*a*) the radiation pressure P_r, (*b*) E_{rms}, and (*c*) B_{rms}.

27 • The amplitude of an electromagnetic wave is $E_0 = 400$ V/m. Find (*a*) E_{rms}, (*b*) B_{rms}, (*c*) the intensity I, and (*d*) the radiation pressure P_r.

28 • The rms value of the electric field in an electromagnetic wave is $E_{rms} = 400$ V/m. (*a*) Find B_{rms}, (*b*) the average energy density, and (*c*) the intensity.

29 • Show that the units of $E = cB$ are consistent; that is, show that when B is in teslas and c is in meters per second, the units of cB are volts per meter or newtons per coulomb.

30 • The root-mean-square value of the magnitude of the magnetic field in an electromagnetic wave is $B_{rms} = 0.245\ \mu$T. Find (*a*) E_{rms}, (*b*) the average energy density, and (*c*) the intensity.

31 •• (*a*) An electromagnetic wave of intensity 200 W/m² is incident normally on a rectangular black card with sides of 20 and 30 cm that absorbs all the radiation. Find the force exerted on the card by the radiation. (*b*) Find the force exerted by the same wave if the card reflects all the radiation incident on it.

32 •• Find the force exerted by the electromagnetic wave on the reflecting card in part (*b*) of Problem 31 if the radiation is incident at an angle of 30° to the normal.

33 •• An AM radio station radiates an isotropic sinusoidal wave with an average power of 50 kW. What are the amplitudes of E_{max} and B_{max} at a distance of (*a*) 500 m, (*b*) 5 km, and (*c*) 50 km?

34 •• The intensity of sunlight striking the earth's upper atmosphere (called the solar constant) is 1.35 kW/m^2. (*a*) Find E_{rms} and B_{rms} due to the sun at the upper atmosphere of the earth. (*b*) Find the average power output of the sun. (*c*) Find the intensity and the radiation pressure at the surface of the sun.

35 •• A demonstration laser has an average output power of 0.9 mW and a beam diameter of 1.2 mm. What is the force exerted by the laser beam on (*a*) a 100% absorbing black surface? (*b*) a 100% reflecting surface?

36 •• A laser beam has a diameter of 1.0 mm and average power of 1.5 mW. Find (*a*) the intensity of the beam, (*b*) E_{rms}, (*c*) B_{rms}, and (*d*) the radiation pressure.

37 •• Instead of sending power by a 750-kV, 1000-A transmission line, one desires to beam this energy via an electromagnetic wave. The beam has a uniform intensity within a cross-sectional area of 50 m^2. What are the rms values of the electric and the magnetic fields?

38 •• A laser pulse has an energy of 20 J and a beam radius of 2 mm. The pulse duration is 10 ns and the energy density is constant within the pulse. (*a*) What is the spatial length of the pulse? (*b*) What is the energy density within the pulse? (*c*) Find the electric and magnetic amplitudes of the laser pulse.

39 •• The electric field of an electromagnetic wave oscillates in the y direction and the Poynting vector is given by

$$\vec{S}(x,t) = (100\ \mathrm{W/m^2})\cos^2[10x - (3 \times 10^9)t]\hat{i}$$

where x is in meters and t is in seconds. (*a*) What is the direction of propagation of the wave? (*b*) Find the wavelength and the frequency. (*c*) Find the electric and magnetic fields.

40 •• A pulsed laser fires a 1000-MW pulse of 200-ns duration at a small object of mass 10 mg suspended by a fine fiber 4 cm long. If the radiation is completely absorbed without other effects, what is the maximum angle of deflection of this pendulum?

41 •• A 10- by 15-cm card has a mass of 2 g and is perfectly reflecting. The card hangs in a vertical plane and is free to rotate about a horizontal axis through the top edge. The card is illuminated uniformly by an intense light that causes the card to make an angle of 1° with the vertical. Find the intensity of the light.

42 •• A valuable 0.08-kg gem and a 105-kg spaceperson are separated by 95 m. Both objects are initially at rest. The spaceperson has a 1.5-kW laser that can be used as a photon rocket motor to propel the person toward the diamond. How long would it take the spaceperson to move 95 m using the laser rocket propulsion?

43 •• It has been suggested that spacecraft could be propelled by the radiation pressure from the sun. What must be the surface mass density (kg/m^2) of a perfectly reflecting sheet so that at a distance of one astronomical unit the force due to radiation pressure is twice that due to the gravitational attraction between the reflecting sheet and the sun? (*Note:* One astronomical unit is the average radius of the earth's orbit.) How will the ratio of radiation force to gravitational force change as the reflecting sheet accelerates away from the sun?

44 •• Suppose a mass of 50 kg is attached to a perfectly reflecting sheet whose surface mass density is that obtained in Problem 43. What must be the surface area of the sheet so that at a distance of one astronomical unit the acceleration of the system away from the sun is 0.4 mm/s^2? How does the acceleration vary with distance from the sun?

Blackbody Radiation

A blackbody is an object that is a perfect absorber; that is, it absorbs all radiation incident on it. It is also a perfect radiator. The power radiated by a blackbody of area A at temperature T is given by the Stefan–Boltzmann law (Equation 21-17 with e = 1),

$$P_r = \sigma A T^4$$

where $\sigma = 5.6703 \times 10^{-8}\ W/m^2 \cdot K^4$.

45 •• A very long wire of radius 4 mm is heated to 1000 K. The surface of the wire is an ideal blackbody radiator. (*a*) What is the total power radiated per unit length? Find (*b*) the magnitude of the Poynting vector S, (*c*) E_{rms}, and (*d*) B_{rms} at a distance of 25 cm from the wire.

46 •• A blackbody sphere of radius R is a distance 2×10^{11} m from the sun. The effective area of the body for absorption of energy from the sun is πR^2, but the area for radiation by the object is $4\pi R^2$. The power output of the sun is 3.83×10^{26} W. What is the temperature of the sphere?

47 •• (*a*) If the earth were an ideal blackbody with infinite thermal conductivity and no atmosphere, what would be the temperature of the earth? (*b*) If 40% of the incident sun's energy were reflected, what then would be the temperature of the earth? (See Problem 46.)

The Wave Equation for Electromagnetic Waves (optional)

48 • Show by direct substitution that Equation 32-17*a* is satisfied by the wave function

$$E_y = E_0 \sin(kx - \omega t) = E_0 \sin k(x - ct)$$

where $c = \omega/k$.

49 • Use the known values of μ_0 and ϵ_0 in SI units to compute $c = 1/\sqrt{\epsilon_0\mu_0}$ and show that it is approximately 3×10^8 m/s.

50 ••• (*a*) Using arguments similar to those given in the text, show that for a plane wave, in which E and B are independent of y and z,

$$\frac{\partial E_z}{\partial x} = \frac{\partial B_y}{\partial t}$$

and

$$\frac{\partial B_y}{\partial x} = \mu_0\epsilon_0\frac{\partial E_z}{\partial t}$$

(*b*) Show that E_z and B_y also satisfy the wave equation.

General Problems

51 • True or false:

(*a*) Maxwell's equations apply only to fields that are constant over time.
(*b*) The wave equation can be derived from Maxwell's equations.
(*c*) Electromagnetic waves are transverse waves.
(*d*) In an electromagnetic wave in free space, the electric and magnetic fields are in phase.
(*e*) In an electromagnetic wave in free space, the electric and magnetic field vectors $\vec{E}$ and $\vec{B}$ are equal in magnitude.
(*f*) In an electromagnetic wave in free space, the electric and magnetic energy densities are equal.

52 • (*a*) Show that if E is in volts per meter and B is in teslas, the units of the Poynting vector $\vec{S} = \vec{E} \times \vec{B}/\mu_0$ are watts per square meter. (*b*) Show that if the intensity I is in watts per square meter, the units of radiation pressure $P_r = I/c$ are newtons per square meter.

53 •• A loop antenna that may be rotated about a vertical axis is used to locate an unlicensed amateur radio transmitter. If the output of the receiver is proportional to the intensity of the received signal, how does the output of the receiver vary with the orientation of the loop antenna?

54 •• An electromagnetic wave has a frequency of 100 MHz and is traveling in a vacuum. The magnetic field is given by $\vec{B}(z,t) = (10^{-8}\text{ T})\cos(kz - \omega t)\hat{i}$. (*a*) Find the wavelength, and the direction of propagation of this wave. (*b*) Find the electric vector $\vec{E}(z, t)$. (*c*) Give Poynting's vector, and find the intensity of this wave.

55 •• A circular loop of wire can be used to detect electromagnetic waves. Suppose a 100-MHz FM station radiates 50 kW uniformly in all directions. What is the maximum rms voltage induced in a loop of radius 30 cm at a distance of 10^5 m from the station?

56 •• The electric field from a radio station some distance from the transmitter is given by $E = (10^{-4}\text{ N/C}) \times \cos 10^6 t$, where t is in seconds. (*a*) What voltage is picked up on a 50-cm wire oriented along the electric field direction? (*b*) What voltage can be induced in a loop of radius 20 cm?

57 •• A circular capacitor of radius a has a thin wire of resistance R connecting the centers of the two plates. A voltage $V_0 \sin \omega t$ is applied between the plates. (*a*) What is the current drawn by this capacitor? (*b*) What is the magnetic field as a function of radial distance r from the centerline within the plates of this capacitor? (*c*) What is the phase angle between current and applied voltage?

58 •• A 20-kW beam of radiation is incident normally on a surface that reflects half of the radiation. What is the force on this surface?

59 •• Show that the relation between the momentum carried by an electromagnetic wave and the energy, Equation 32-13, can also be derived using the Einstein–Planck relation, $E = hf$; the de Broglie equation, $p = h/\lambda$; and $c = f\lambda$.

60 •• The electric fields of two harmonic waves of angular frequency ω_1 and ω_2 are given by $\vec{E}_1 = E_{1,0}\cos(k_1x - \omega_1 t)\hat{j}$ and $\vec{E}_2 = E_{2,0}\cos(k_2x - \omega_2 t + \delta)\hat{j}$. Find (*a*) the instantaneous Poynting vector for the resultant wave motion and (*b*) the time-average Poynting vector. If $\vec{E}_2 = E_{2,0}\cos(k_2x + \omega_2 t + \delta)\hat{j}$, find (*c*) the instantaneous Poynting vector for the resultant wave motion and (*d*) the time-average Poynting vector.

61 •• At the surface of the earth, there is an approximate average solar flux of 0.75 kW/m^2. A family wishes to construct a solar energy conversion system to power their home. If the conversion system is 30% efficient and the family needs a maximum of 25 kW, what effective surface area is needed for perfectly absorbing collectors?

62 •• Suppose one has an excellent radio capable of detecting a signal as weak as 10^{-14} W/m^2. This radio has a 2000-turn coil antenna having a radius of 1 cm wound on an iron core that increases the magnetic field by a factor of 200. The radio frequency is 140 kHz. (*a*) What is the amplitude of the magnetic field in this wave? (*b*) What is the emf induced in the antenna? (*c*) What would be the emf induced in a 2-m wire oriented in the direction of the electric field?

63 •• A 654-nm laser whose beam diameter is 0.4 mm points upward. A small, perfectly reflecting spherical particle having a diameter of 10 μm and a density of 0.2 g/cm^3 is supported against gravity by the radiation pressure from the laser beam. Determine the power output of this laser.

The following two problems do not concern waves, but illustrate the use of the Poynting vector to describe the flow of electromagnetic energy.

64 ••• A long, cylindrical conductor of length L, radius a, and resistivity ρ carries a steady current I that is uniformly distributed over its cross-sectional area. (*a*) Use Ohm's law to relate the electric field E in the conductor to I, ρ, and a. (*b*) Find the magnetic field B just outside the conductor. (*c*) Use the results for parts (*a*) and (*b*) to compute the Poynting vector $\vec{S} = \vec{E} \times \vec{B}/\mu_0$ at $r = a$ (the edge of the conductor). In what direction is $\vec{S}$? (*d*) Find the flux $\oint S_n\, dA$ through the surface of the conductor into the conductor, and show that the rate of energy flow into the conductor equals I^2R, where R is the resistance. (Here S_n is the *inward* component of $\vec{S}$ perpendicular to the surface of the conductor.)

65 ••• A long solenoid of n turns per unit length has a current that slowly increases with time. The solenoid has radius r, and the current in the windings has the form $I(t) = at$. (*a*) Find the induced electric field at a distance $r < R$ from the solenoid axis. (*b*) Find the magnitude and direction of the Poynting vector $\vec{S}$ at the cylindrical surface $r = R$ just inside the solenoid windings. (*c*) Calculate the flux $\oint S_n\, dA$ into the solenoid, and show that it equals the rate of increase of the magnetic energy inside the solenoid. (Here S_n is the *inward* component of $\vec{S}$ perpendicular to the surface of the solenoid.)

66 ••• Small particles might be blown out of solar systems by the radiation pressure of sunlight. Assume that the particles are spherical with a radius r and a density of 1 g/cm^3 and that they absorb all the radiation in a cross-sectional area of πr^2. They are a distance R from the sun, which has a power output of 3.83×10^{26} W. What is the radius r for which the radiation force of repulsion just balances the gravitational force of attraction to the sun?

67 ••• Some science fiction writers have used solar sails to propel interstellar spaceships. Imagine a giant sail erected on a spacecraft subjected to the solar radiation pressure. (*a*) Show that the spacecraft's acceleration is given by

$$a = \frac{P_S A}{4\pi r^2\, cm}$$

where P_s is the power output of the sun and is equal to 3.8×10^{26} W, A is the surface area of the sail, m is the total mass of the spacecraft, r is the distance from the sun, and c is the speed of light. (*b*) Show that the velocity of the spacecraft at a distance r from the sun is found from

$$v^2 = v_0^2 + \left(\frac{P_S A}{2\pi mc}\right)\left(\frac{1}{r_0} - \frac{1}{r}\right)$$

where v_0 is the initial velocity at r_0. (*c*) Compare the relative accelerations due to the radiation pressure and the gravitational force. Use reasonable values for A and m. Will such a system work?

68 ••• Novelty stores sell a device called a radiometer (Figure 32-13), in which a balanced vane spins rapidly. A card is mounted on each arm of the vane. One side of each card is white and the other is black. Assume that the mass of each card is 2 g, that the light-collecting area for each card is 1 cm^2, and that each arm of the vane has a length of 2 cm. (*a*) If a 100-W light bulb produces 50 W of electromagnetic energy and the bulb is 50 cm from the radiometer, find the maximum angular acceleration of the vane. (Estimate the moment of inertia of the vane by assuming that all the mass of each card is at the end of the arms.) (*b*) How long will it take for the vane to accelerate to 10 rev/min if it starts from rest and is subject to the maximum angular acceleration at all times? (*c*) Can the radiation pressure account for the rapid motion of the radiometer?*

69 ••• When an electromagnetic wave is reflected at normal incidence on a perfectly conducting surface, the electric field vector of the reflected wave at the reflecting surface is the negative of that of the incident wave. (*a*) Explain why this should be. (*b*) Show that the superposition of incident and reflected waves result in a standing wave. (*c*) What is the relationship between the magnetic field vector of the incident and reflected waves at the reflecting surface?

70 ••• An intense point source of light radiates 1 MW isotropically. The source is located 1.0 m above an infinite perfectly reflecting plane. Determine the force that acts on the plane.

Figure 32-13 Problem 68

* The radiometer actually spins in the opposite direction from what would be expected if the force were due to radiation pressure. The reason is that the air near the black side is warmer than that near the white side, so the air molecules hitting the black side have greater energy than those hitting the white side.

PART V

light

Stress patterns around a crack in a sheet of transparent plastic are revealed by polarized light. The stress is perpendicular to the crack. Two smaller cracks have propagated from the lower end of the large one, creating additional patterns of stress. Smaller circular flaws surround the upper end of the large crack.

CHAPTER 33

Properties of Light

A bright primary rainbow and the fainter secondary rainbow in a sheet of rain over Lake Michigan. The primary bow is formed by light rays that enter spherical drops of water and are reflected once internally before leaving the drops. The secondary bow results from rays that experience two internal reflections before leaving the drops.

The human eye is sensitive to electromagnetic radiation with wavelengths from about 400 to 700 nm. The shortest wavelengths in the visible spectrum correspond to violet light and the longest to red light. The perceived colors of light are the result of the physiological and psychological response of the eye–brain sensing system to the different frequencies of visible light. Although the correspondence between perceived color and frequency is quite good, there are many interesting deviations. For example, a mixture of red light and green light is perceived by the eye–brain sensing system as yellow even in the absence of light in the yellow region of the spectrum. In this chapter, we study how light is produced, how its speed is measured, and how it is scattered, reflected, refracted, and polarized.

33-1 Wave–Particle Duality

The wave nature of light was first demonstrated by Thomas Young, who observed the interference pattern of two coherent light sources produced by illuminating a pair of narrow, parallel slits with a single source.* The wave

* The wave–particle duality of light and electrons is discussed in detail in Chapter 17. General wave properties such as propagation, reflection, refraction, interference, and coherence are discussed in Chapters 15 and 16.

theory of light culminated in 1860 with Maxwell's prediction of electromagnetic waves. The particle nature of light was first proposed by Albert Einstein in 1905 in his explanation of the photoelectric effect. A particle of light called a **photon** has energy E that is related to the frequency f and wavelength λ of the light wave by the Einstein equation

$$E = hf = \frac{hc}{\lambda} \tag{33-1}$$

Einstein equation for photon energy

where c is the speed of light and h is Planck's constant:

$$h = 6.626 \times 10^{-34}\ \mathrm{J{\cdot}s} = 4.136 \times 10^{-15}\ \mathrm{eV{\cdot}s}$$

Since energies are often given in electron volts and wavelengths are given in nanometers, it is convenient to express the combination hc in eV·nm. We have

$$hc = (4.14 \times 10^{-15}\ \mathrm{eV{\cdot}s})(3 \times 10^{8}\ \mathrm{m/s}) = 1.24 \times 10^{-6}\ \mathrm{eV{\cdot}m}$$

or

$$hc = 1240\ \mathrm{eV{\cdot}nm} \tag{33-2}$$

The propagation of light is governed by its wave properties, whereas the exchange of energy between light with matter is governed by its particle properties. This wave–particle duality is a general property of nature. For example, electrons (and other so-called "particles") also propagate as waves and exchange energy as particles.

33-2 Light Spectra

Newton was the first to recognize that white light is a mixture of light of all colors of approximately equal intensity. He demonstrated this by letting sunlight fall on a glass prism and observing the spectrum of refracted light (Figure 33-1). Because the angle of refraction of a glass prism depends slightly on wavelength, the refracted beam is spread out in space into its component col-

Figure 33-1 Newton demonstrating the spectrum of sunlight with a glass prism.

ors or wavelengths, like a rainbow. Figure 33-2 shows a spectroscope, which is a device for analyzing the spectra of a light source. Light from the source passes through a narrow slit, traverses a lens to make the beam parallel, and falls on a glass prism. The refracted beam is viewed with a telescope, which is mounted on a rotating platform so that the angle of the refracted beam, which depends on the wavelength, can be measured. The spectrum of the light source can thus be analyzed in terms of its component wavelengths. The spectrum of sunlight contains a continuous range of wavelengths and is therefore called a **continuous spectrum**. The light emitted by the atoms in low-pressure gases contains only a discrete set of wavelengths. Each wavelength emitted by the source produces a separate image of the collimating slit in the spectroscope. Such a spectrum is called a **line spectrum**. The continuous visible spectrum and the line spectra from several elements are shown in the photograph below.

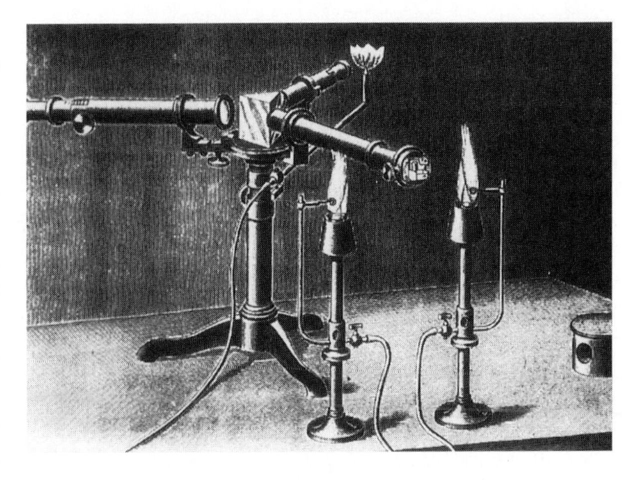

Figure 33-2 A late nineteenth-century spectroscope belonging to Gustave Kirchhoff. Modern student spectroscopes usually share the same general design.

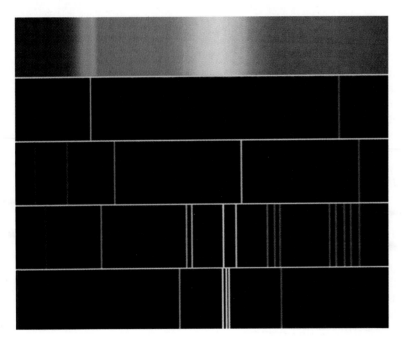

The continuous visible spectrum (*top*) and the line spectra of *(from top to bottom)* hydrogen, helium, barium, and mercury.

33-3 Sources of Light

Line Spectra The most common sources of visible light are transitions of the outer electrons in atoms. Normally an atom is in its ground state with its electrons at their lowest allowed energy levels, consistent with the exclusion principle. (The exclusion principle, which was first enunciated by Wolfgang Pauli in 1925 to explain the electronic structure of atoms, states that no two electrons in an atom can be in the same quantum state.) The lowest energy electrons are closest to the nucleus and are tightly bound, forming a stable inner core. The one or two electrons in the highest energy states are much farther from the nucleus and are relatively easily excited to vacant higher energy states. These outer electrons are responsible for the energy changes in the atom that result in the emission or absorption of visible light.

When an atom collides with another atom or with a free electron, or when it absorbs electromagnetic energy, the outer electrons can be excited to higher energy states. After a time of about 10^{-8} s, these outer electrons spontaneously make transitions to lower energy states with the emission of a photon. This process, called **spontaneous emission**, is random; the photons emitted from two different atoms are not correlated. The emitted light is thus incoherent. By conservation of energy, the energy of an emitted photon is the energy difference ΔE between the initial and final state. The frequency of the light wave is related to the energy by the Einstein equation, $\Delta E = hf$. The wavelength of the emitted light is then

$$\lambda = \frac{c}{f} = \frac{hc}{hf} = \frac{hc}{\Delta E} \qquad \text{33-3}$$

The photon energies corresponding to shortest (400 nm) and longest (700 nm) wavelengths in the visible spectrum are

$$E_{400\text{nm}} = \frac{hc}{\lambda} = \frac{1240\ \text{eV}\cdot\text{nm}}{400\ \text{nm}} = 3.1\ \text{eV} \qquad \text{33-4}a$$

and

$$E_{700\text{nm}} = \frac{hc}{\lambda} = \frac{1240\ \text{eV}\cdot\text{nm}}{700\ \text{nm}} = 1.77\ \text{eV} \qquad \text{33-4}b$$

Because the energy levels in atoms form a discrete set, the emission spectrum of light from single atoms or atoms in low-pressure gases consists of a set of sharp discrete lines that are characteristic of the element. These narrow lines are broadened somewhat by Doppler shifts due to the motion of the atom relative to the observer and by collisions with other atoms, but, generally, if the gas density is low enough, the lines are narrow and well separated from one another. The study of the line spectra of hydrogen and other atoms led to the first understanding of the energy levels of atoms.

Continuous Spectra When atoms are close together and interact strongly, as in liquids and solids, the energy levels of the individual atoms are spread out into energy bands, resulting in essentially continuous bands of energy levels. When the bands overlap, as they often do, the result is a continuous spectrum of possible energies and a continuous emission spectrum. In an incandescent material such as a hot metal filament, electrons are randomly accelerated by frequent collisions, resulting in a broad spectrum of thermal radiation. The rate at which an object radiates thermal energy is proportional to the fourth power of its absolute temperature.* The radiation emitted by an object at temperatures below about 600°C is concentrated in the infrared and is not visible. As an object is heated, the energy radiated extends to shorter and shorter wavelengths. Between about 600 and 700°C, enough of the radiated energy is in the visible spectrum for the object to glow a dull red. At higher and higher temperatures, the object becomes bright red and then white. The wavelength at which the power is a maximum varies inversely with the temperature, a result known as Wien's displacement law. The surface of the sun at $T = 6000$ K emits a continuous spectrum of approximately constant intensity over the visible range of wavelengths.

---

* This is known as the Stefan–Boltzmann law. This and other properties of thermal radiation such as Wien's displacement law were discussed more fully in Section 21-4.

Absorption, Scattering, and Stimulated Emission

Radiation is emitted when an atom makes a transition from an excited state to a state of lower energy; radiation is absorbed when an atom makes a transition from a lower state to a higher state. When atoms are irradiated with a continuous spectrum of radiation, the transmitted spectrum shows dark lines corresponding to absorption of light at discrete wavelengths. The absorption spectra of atoms were the first line spectra observed. Since atoms and molecules at normal temperatures are in either their ground states or low-lying excited states, absorption spectra are usually simpler than emission spectra.

Figure 33-3 illustrates several interesting phenomena that can occur when a photon is incident on an atom. In Figure 33-3*a*, the energy of the incoming photon is too small to excite the atom to an excited state, so the atom remains in its ground state and the photon is said to be scattered. Since the incoming and outgoing or scattered photons have the same energy, the scattering is said to be elastic. If the wavelength of the incident light is large compared with the size of the atom, the scattering can be described in terms of classical electromagnetic theory and is called **Rayleigh scattering** after Lord Rayleigh, who worked out the theory in 1871. The probability of Rayleigh scattering varies as $1/\lambda^4$. This means that blue light is scattered much more readily than red light, which accounts for the bluish color of the sky. The removal of blue light by Rayleigh scattering also accounts for the reddish color of the transmitted light seen in sunsets.

Figure 33-3*b* shows **inelastic scattering,** which occurs when the incident photon has enough energy to cause the atom to make a transition to an excited state. The energy of the scattered photon hf' is less than that of the incident photon hf by ΔE, the difference between the energy of the ground state and the energy of the excited state. Inelastic scattering of light from molecules was first observed by the Indian physicist C. V. Raman and is often referred to as **Raman scattering.**

In Figure 33-3*c*, the energy of the incident photon is just equal to the difference in energy between the ground state and the first excited state of the atom. The atom makes a transition to its first excited state and then after a short delay decays by spontaneous emission back to the ground state with the emission of a photon whose energy is equal to that of the incident photon. The phase of the emitted photon is not correlated with the phase of the incident photon. This multistep process is called **resonance absorption.**

In Figure 33-3*d*, the energy of the incident photon is great enough to excite the atom to one of its higher excited states. The atom then loses its energy by spontaneous emission as it makes one or more transitions to lower energy states. A common example occurs when the atom is excited by ultraviolet light and emits visible light as it returns to its ground state. This process is called **fluorescence.** Since the lifetime of a typical excited atomic energy state is of the order of 10^{-8} s, this process appears to occur instantaneously. However, some excited states have much longer lifetimes—of the order of milliseconds or occasionally seconds or even minutes. Such a state is called a **metastable state. Phosphorescent materials** have very long-lived metastable states, and so emit light long after the original excitation.

Figure 33-3*e* illustrates the photoelectric effect, in which the absorption of the photon ionizes the atom by causing the emis-

Figure 33-3 Photon–atom interactions.

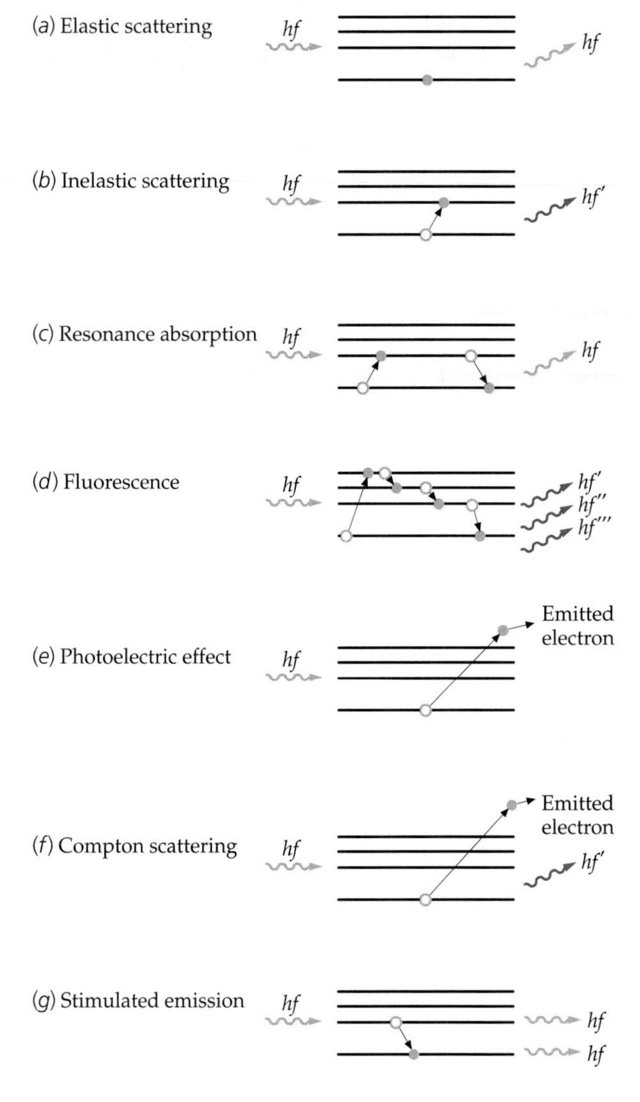

sion of an electron. Figure 33-3*f* illustrates Compton scattering, which occurs if the energy of the incident photon is much greater than the ionization energy. Note that in Compton scattering, a photon is emitted, whereas in the photoelectric effect, the photon is absorbed with none emitted.

Figure 33-3*g* illustrates **stimulated emission.** This process occurs if the atom or molecule is initially in an excited state of energy E_2, and the energy of the incident photon is equal to $E_2 - E_1$, where E_1 is the energy of a lower state or the ground state. In this case, the oscillating electromagnetic field associated with the incident photon stimulates the excited atom or molecule, which then emits a photon in the same direction as the incident photon and in phase with it. In stimulated emission, the phase of the light emitted from one atom is related to that emitted by every other atom, so the resulting light is coherent. As a result, interference of the light from different atoms can be observed.

(*a*)

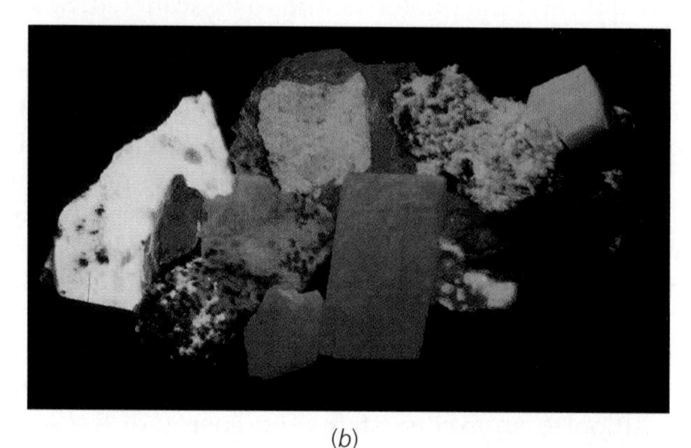

(*b*)

A collection of minerals in (*a*) daylight and (*b*) ultraviolet light (sometimes called "black light"). Identified by number in the schematic (*c*), they are 1, powellite; 2, willemite; 3, scheelite; 4, calcite; 5, calcite and willemite composite; 6, optical calcite; 7, willemite; and 8, opal. The change in color is due to the minerals fluorescing under the ultraviolet light. In optical calcite, both fluorescence and phosphorescence occur.

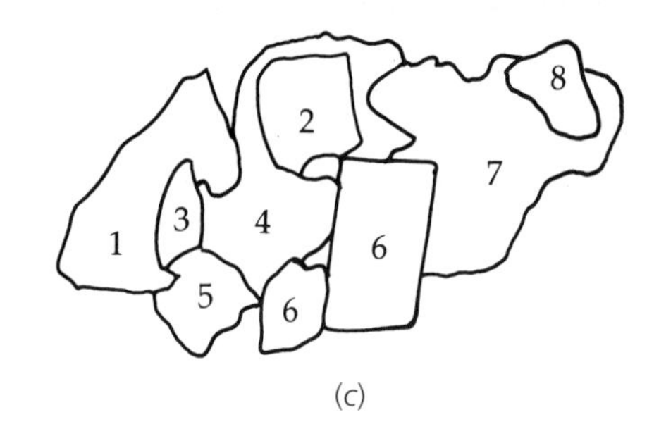

(*c*)

Example 33-1

The first excited state of potassium is $E_2 = 1.62$ eV above the ground state E_1, which we take to be zero. Potassium also has energy levels at $E_3 = 2.61$ eV and $E_4 = 3.07$ eV. (*a*) What is the maximum wavelength of radiation that can be absorbed by potassium in its ground state? Calculate the wavelength when the atom makes a transition from (*b*) E_4 to the ground state, and (*c*) from E_4 to E_3.

Picture the Problem The ground state E_1 and the first three excited energy levels are shown in Figure 33-4. (*a*) Since the wavelength is related to the energy of a photon by $\lambda = hc/\Delta E$, longer wavelengths correspond to smaller energy differences. The smallest energy difference for a transition originating at the ground state is from the ground state to the first excited state. (*b*) The wavelengths of the photons given off when the atom de-excites are related to the energy differences by $\lambda = hc/\Delta E$.

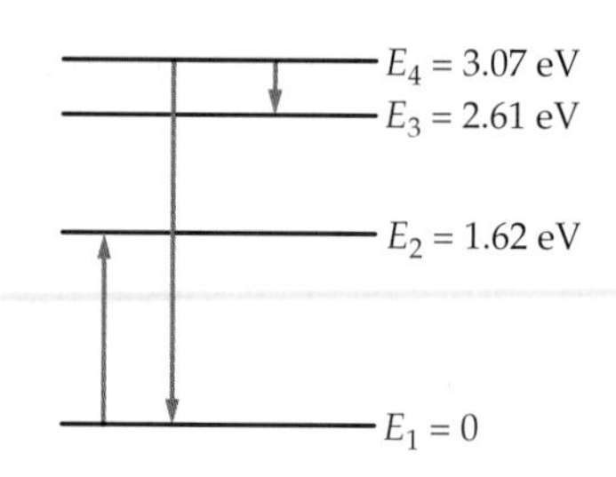

Figure 33-4

(*a*) Calculate the wavelength of radiation absorbed in a transition from the ground state to the first excited state:

$$\lambda = \frac{hc}{\Delta E} = \frac{1240 \text{ eV}\cdot\text{nm}}{1.62 \text{ eV} - 0} = 765 \text{ nm}$$

(*b*) For the transition from E_4 to the ground state, the photon energy is $E_4 - E_1 = E_4$. Calculate the wavelength of radiation emitted in this transition:

$$\lambda = \frac{hc}{\Delta E} = \frac{hc}{E_4 - E_1} = \frac{1240 \text{ eV}\cdot\text{nm}}{3.07 \text{ eV} - 0} = 404 \text{ nm}$$

(*c*) For the transition from E_4 to E_3, the photon energy is $E_4 - E_3$. Calculate the wavelength of radiation emitted in this transition:

$$\lambda = \frac{hc}{\Delta E} = \frac{hc}{E_4 - E_3} = \frac{1240 \text{ eV}\cdot\text{nm}}{3.07 \text{ eV} - 2.61 \text{ eV}} = 2700 \text{ nm}$$

Remarks The wavelength of radiation emitted in the transition from E_2 to the ground state E_1 is 765 nm, the same as that for radiation absorbed in the transition from the ground state to E_2. This transition and the one from E_4 to the ground state both result in photons in the visible spectrum.

Lasers

The laser (*l*ight *a*mplification by *s*timulated *e*mission of *r*adiation) is a device that produces a strong beam of coherent photons by stimulated emission. Consider a system consisting of atoms that have a ground state of energy E_1 and an excited metastable state of energy E_2. If these atoms are irradiated by photons of energy $E_2 - E_1$, those atoms in the ground state can absorb a photon and make the transition to state E_2, whereas those atoms already in the excited state may be stimulated to decay back to the ground state. The relative probabilities of absorption and stimulated emission were first worked out by Einstein, who showed them to be equal. Ordinarily, nearly all the atoms of the system at normal temperature will initially be in the ground state, so absorption will be the main effect. To produce more stimulated-emission transitions than absorption transitions, we must arrange to have more atoms in the excited state than in the ground state. This condition, called population inversion, can be achieved by a method called optical pumping in which atoms are "pumped" up to levels of energy greater than E_2 by the absorption of an intense auxiliary radiation. The atoms then decay down to state E_2 either by spontaneous emission or by nonradiative transitions such as those due to collisions.

Figure 33-5 shows a schematic diagram of the first laser, a ruby laser built by Theodore Maiman in 1960. It consists of a ruby rod a few centimeters long surrounded by a helical gaseous flashtube that emits a broad spectrum of light. The ends of the ruby rod are flat and perpendicular to the axis of the rod. Ruby is a transparent crystal of Al_2O_3 with a small amount (about 0.05%) of chromium. It appears red because the chromium ions (Cr^{3+}) have strong absorption bands in the blue and green regions of the visible spectrum, as shown in Figure 33-6. The energy levels of chromium that are important for the operation of a ruby laser are shown in Figure 33-7. When the

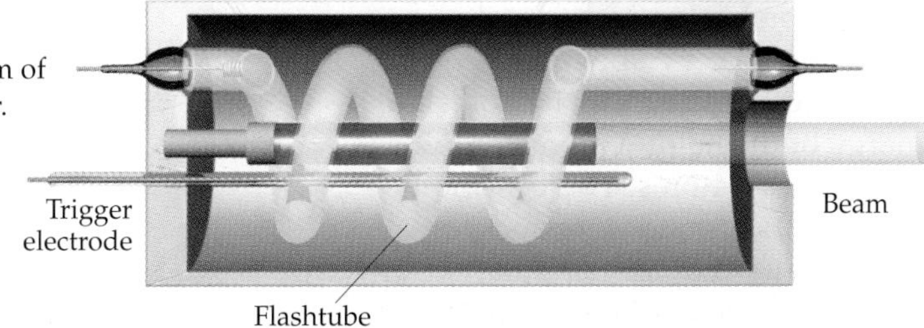

Figure 33-5 Schematic diagram of the first ruby laser.

Absorption
300 400 500 600 700 λ, nm

Figure 33-6 Absorption versus wavelength for Cr^{3+} in ruby. Ruby appears red because of the strong absorption of green and blue light by the chromium ions.

flashtube is fired, there is an intense burst of light lasting a few milliseconds. Absorption excites many of the chromium ions to the bands of energy levels indicated by the shading in Figure 33-7. The chromium ions then relax, giving up their energy to the crystal in nonradiative transitions as they drop down to a pair of metastable states labeled E_2 in the figure. These metastable states are about 1.79 eV above the ground state. If the flash is intense enough, more atoms will make the transition to the states E_2 than remain in the ground state. As a result, the populations of the ground state and the metastable states become inverted. When some of the atoms in the states E_2 decay to the ground state by spontaneous emission, they emit photons of energy 1.79 eV and wavelength 694.3 nm. Some of these photons stimulate other excited atoms to emit photons of the same energy and wavelength.

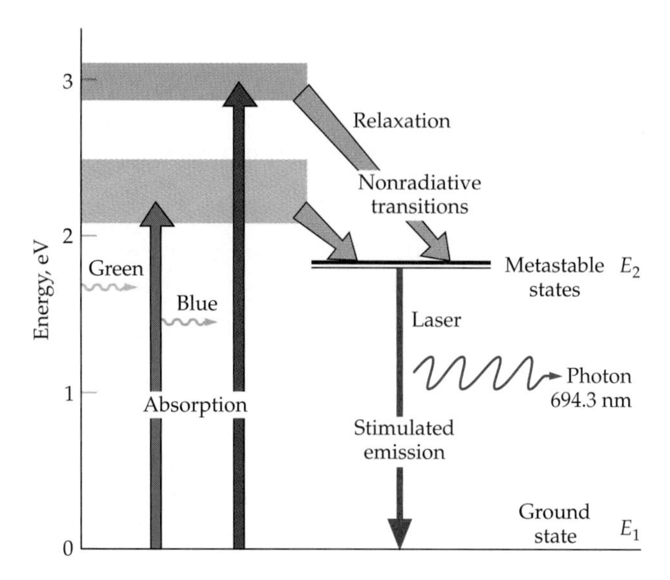

Figure 33-7 Energy levels in a ruby laser. To make the population of the metastable states greater than that of the ground state, the ruby crystal is subjected to intense radiation that contains energy in the green and blue wavelengths. This excites atoms from the ground state to the bands of energy levels indicated by the shading, from which they decay to the metastable states by nonradiative transitions.

In the ruby laser, both ends of the crystal are silvered such that one end is almost totally reflecting and the other end is only partially reflecting (about 99%). When photons traveling parallel to the axis of the crystal strike the silvered ends, all are reflected from the back face and most are reflected from the front face with a few photons escaping through the partially silvered front face. During each pass through the crystal, the photons stimulate more and more atoms so that the photon beam builds up and an intense beam is emitted (Figure 33-8). Modern ruby lasers generate intense light beams with energies ranging from 50 to 100 J in pulses lasting a few milliseconds. The beam can have a diameter as small as 1 mm and an angular divergence as small as 0.25 milliradian to about 7 milliradians.

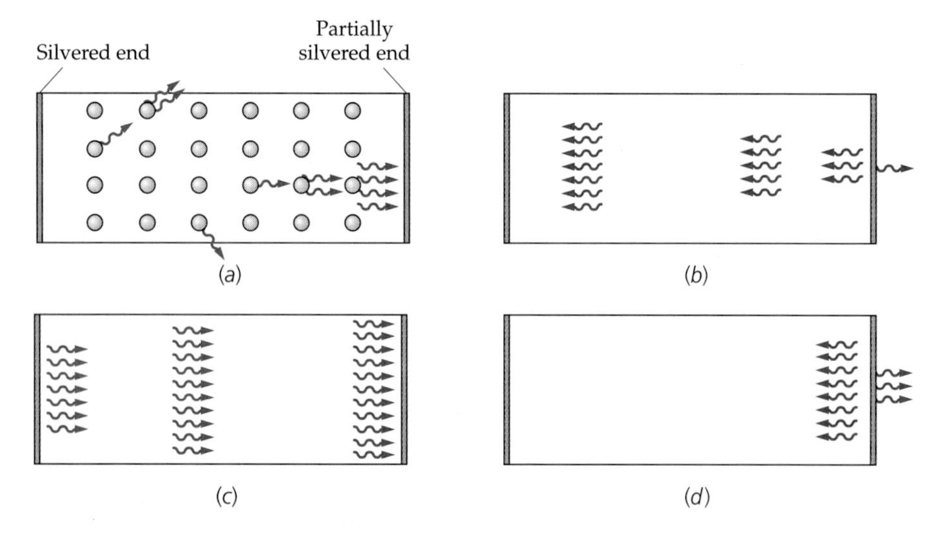

Figure 33-8 Buildup of photon beam in a laser. (*a*) When irradiated, some atoms spontaneously emit photons, some of which travel to the right and stimulate other atoms to emit photons parallel to the axis of the crystal. (*b*) Of the four photons that strike the right face, one is transmitted and three are reflected. As the reflected photons traverse the laser crystal, they stimulate other atoms to emit photons and the beam builds up. By the time the beam reaches the right face again (*c*), it comprises many photons. (*d*) Some of these photons are transmitted, the rest are reflected.

Population inversion is achieved somewhat differently in the continuous helium–neon laser. The energy levels of helium and neon that are important for operation of the laser are shown in Figure 33-9. Helium has an excited energy state $E_{2,\mathrm{He}}$ that is 20.61 eV above its ground state. Helium atoms are excited to state $E_{2,\mathrm{He}}$ by an electric discharge. Neon has an excited state $E_{3,\mathrm{Ne}}$ that is 20.66 eV above its ground state. This is just 0.05 eV above the first excited state of helium. The neon atoms are excited to state $E_{3,\mathrm{Ne}}$ by collisions with excited helium atoms. The kinetic energy of the helium atoms provides the extra 0.05 eV of energy needed to excite the neon atoms. There is another excited state of neon $E_{2,\mathrm{Ne}}$ that is 18.70 eV above its ground state and 1.96 eV below state $E_{3,\mathrm{Ne}}$. Since state $E_{2,\mathrm{Ne}}$ is normally unoccupied, population inversion between states $E_{3,\mathrm{Ne}}$ and $E_{2,\mathrm{Ne}}$ is obtained immediately. The stimu-

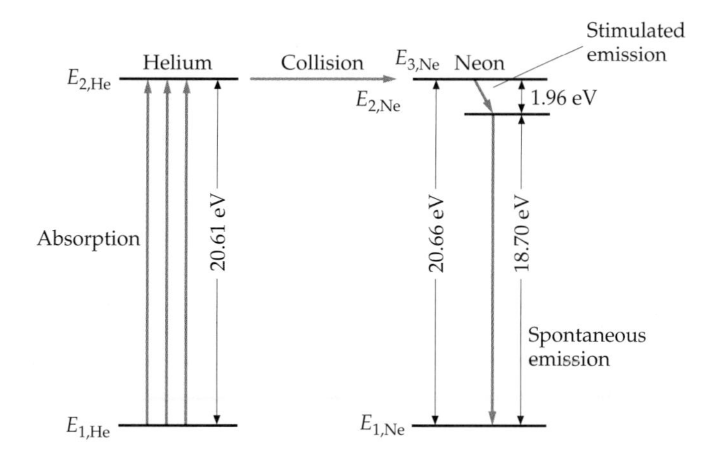

Figure 33-9 Energy levels of helium and neon that are important for the helium–neon laser. The helium atoms are excited by electrical discharge to an energy state 20.61 eV above the ground state. They collide with neon atoms, exciting some neon atoms to an energy state 20.66 eV above the ground state. Population inversion is thus achieved between this level and one 1.96 eV below it. The spontaneous emission of photons of energy 1.96 eV stimulates other atoms in the upper state to emit photons of energy 1.96 eV.

lated emission that occurs between these states results in photons of energy 1.96 eV and wavelength 632.8 nm, which produces a bright red light. After stimulated emission, the atoms in state $E_{2,Ne}$ decay to the ground state by spontaneous emission.

Note that there are four energy levels involved in the helium–neon laser, whereas the ruby laser involved only three levels. In a three-level laser, population inversion is difficult to achieve because more than half the atoms in the ground state must be excited. In a four-level laser, population inversion is easily achieved because the state after stimulated emission is not the ground state but an excited state that is normally unpopulated.

Figure 33-10 shows a schematic diagram of a helium–neon laser commonly used for physics demonstrations. It consists of a gas tube containing 15% helium gas and 85% neon gas. A totally reflecting flat mirror is mounted at one end of the gas tube and a partially reflecting concave mirror is placed at the other end. The concave mirror focuses parallel light at the flat mirror and also acts as a lens that transmits part of the light so that it emerges as a parallel beam.

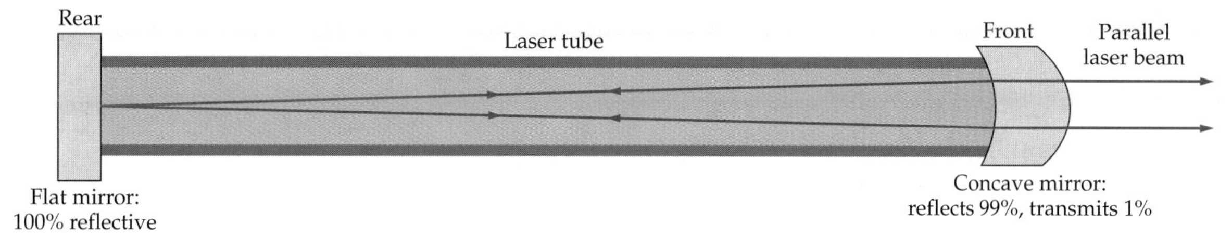

Figure 33-10 Schematic drawing of a helium–neon laser. The use of a concave mirror rather than a second plane mirror makes the alignment of the mirrors less critical than it is for the ruby laser. The concave mirror on the right also serves as a lens that focuses the emitted light into a parallel beam.

A laser beam is coherent, very narrow, and intense. Its coherence makes the laser beam useful in the production of holograms, which we discuss in Chapter 35. The precise direction and small angular spread of the beam make it useful as a surgical tool for destroying cancer cells or reattaching a detached retina. Lasers are also used by surveyors for precise alignment over large distances. Distances can be accurately measured by reflecting a laser pulse from a mirror and measuring the time the pulse takes to travel to the mirror and back. The distance to the moon has been measured to within a few centimeters using a mirror placed on the moon for that purpose. Laser beams are also used in fusion research. An intense laser pulse is focused on tiny pellets of deuterium–tritium in a combustion chamber. The beam heats the pellets to temperatures of the order of 10^8 K in a very short time, causing the deuterium and tritium to fuse and release energy.

Laser technology is advancing so fast that it is possible to mention only a few of the recent developments. In addition to the ruby laser, there are many other solid-state lasers with output wavelengths ranging from about 170 nm to about 3900 nm. Lasers that generate more than 1 kW of continuous power have been constructed. Pulsed lasers can now deliver nanosecond pulses of power exceeding 10^{14} W. Various gas lasers can now produce beams of wavelengths ranging from the far infrared to the ultraviolet. Semiconductor lasers (also known as diode lasers or junction lasers) have shrunk in just 10 years from the size of a pinhead to mere billionths of a meter. Liquid lasers using chemical dyes can be tuned over a range of wavelengths (about 70 nm for continuous lasers and more than 170 nm for pulsed lasers). A relatively new laser, the free-electron laser, extracts light energy from a beam of free electrons moving through a spatially varying magnetic field. The free-electron laser has the potential for very high power and high efficiency and can be tuned over a large range of wavelengths. There appears to be no limit to the variety and uses of modern lasers.

33-4 The Speed of Light

The first effort to measure the speed of light was made by Galileo. He and a partner stood on hilltops about three kilometers apart, each with a lantern and a shutter to cover it. Galileo proposed to measure the time it took for light to travel back and forth between the experimenters. First, one would uncover his lantern, and when the other saw the light, he would uncover his. The time between the first partner's uncovering his lantern and his seeing the light from the other lantern would be the time it took for light to travel back and forth between the experimenters. Though this method is sound in principle, the speed of light is so great that the time interval to be measured is much smaller than fluctuations in human response time, so Galileo was unable to obtain any value for the speed of light.

The first indication of the true magnitude of the speed of light came from astronomical observations of the period of Io, one of the moons of Jupiter. This period is determined by measuring the time between eclipses (when the moon Io disappears behind Jupiter). The eclipse period is about 42.5 h, but measurements made when the earth is moving away from Jupiter along path ABC in Figure 33-11 give a greater time for this period than do measurements made when the earth is moving toward Jupiter along path CDA in the figure. Since these measurements differ from the average value by only about 15 s, the discrepancies were difficult to measure accurately. In 1675, the astronomer Ole Römer attributed these discrepancies to the fact that the speed of light is not infinite. During the 42.5 h between eclipses of Jupiter's moon, the distance between the earth and Jupiter changes, making the path for the light longer or shorter. Römer devised the following method for measuring the cumulative effect of these discrepancies. Because Jupiter moves much more slowly than earth, we can neglect its motion. When the earth is at point A, nearest to Jupiter, the distance between the earth and Jupiter is changing negligibly. The period of Io's eclipse is measured, providing the time between the beginnings of successive eclipses. Based on this measurement, the number of eclipses in 6 months is computed, and the time when an eclipse should begin a half-year later when the earth is at point C is predicted. When the earth is actually at C, the observed beginning of the eclipse is about 16.6 min later than predicted. This is the time it takes light to travel a distance equal to the diameter of the earth's orbit.

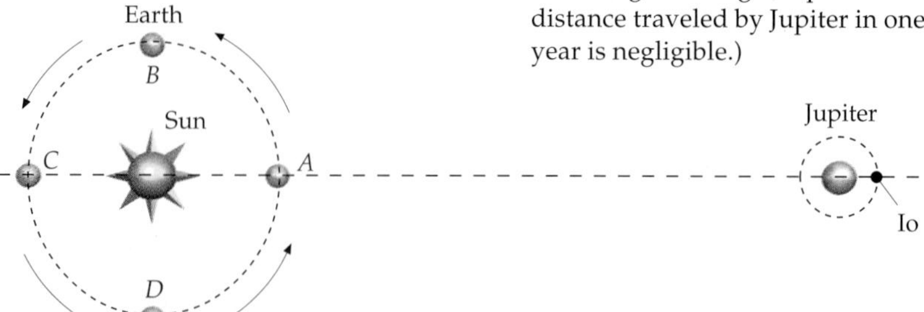

Figure 33-11 Römer's method of measuring the speed of light. The time between eclipses of Jupiter's moon Io appears to be greater when the earth is moving along path ABC than when it is moving along path CDA. The difference is due to the time it takes light to travel the distance traveled by the earth along the line of sight during one period of Io. (The distance traveled by Jupiter in one earth year is negligible.)

(a) Beams from a krypton and an argon laser, split into their component wavelengths. In these gas lasers, krypton and argon atoms have been stripped of multiple electrons, forming positive ions. The light-emitting energy transitions occur when excited electrons in the ions decay from one upper energy level to another. Here, several energy transitions are occurring at once, each corresponding to emitted light of a different wavelength.
(b) A femtosecond pulsed laser. By a technique known as "modelocking," different excited modes within a laser's cavity can be made to interfere with one another and create a series of ultrashort pulses, picoseconds long, that correspond to the time it takes light to bounce back and forth once within the cavity. Ultrashort pulses have been used as probes to study the behavior of molecules during chemical reactions.
(c) A carbon dioxide laser takes just two minutes to cut out a steel saw blade.
(d) A groove etched in the zona pellucida (protective outer covering) of a mouse egg by a "laser scissor" facilitates implantation. This technique has already been applied in human fertility therapies. Several effects contribute to the ability of the finely focused laser to cut on such a delicate scale—photon absorption may heat the target, break molecular bonds, or drive chemical reactions.
(e) The so-called nanolasers shown are semiconductor disks mere microns in diameter and fractions of a micron in width. These tiny lasers work like their larger counterparts, generating and trapping photons until stimulated emission creates enough intensity for the photons to break out from the perimeters of the disks. Exploiting quantum effects that prevail on this microscopic scale, nanolasers promise great efficiency and are being explored as ultrafast, low-energy switching devices.

Example 33-2

The diameter of the earth's orbit is 3.00×10^{11} m. If light takes 16.6 min to travel this distance, what is the speed of light in meters per second?

Picture the Problem We divide the distance traveled by the time and convert the time from minutes to seconds.

1. The speed is the distance divided by the time: $c = \frac{\Delta x}{\Delta t} = \frac{3.00 \times 10^{11}\text{ m}}{16.6\text{ min}} \times \frac{1\text{ min}}{60\text{ s}} = 3.01 \times 10^{8}\text{ m/s}$

Remark Römer obtained a considerably smaller value for c because he used 22 min for Δt.

The first nonastronomical measurement of the speed of light was made by the French physicist Fizeau in 1849. On a hill in Paris, Fizeau placed a light source and a system of lenses arranged such that the light reflected from a semitransparent mirror was focused on a gap in a toothed wheel as shown in Figure 33-12. On a distant hill (about 8.63 km away), he placed a mirror to reflect the light back to be viewed by an observer as shown. The toothed wheel was rotated, and the speed of rotation was varied. At low speeds of rotation, no light was visible because the light that passed through a gap in the rotating wheel and was reflected back by the mirror was obstructed by the next tooth of the wheel. The speed of rotation was then increased. The light suddenly became visible when the rotation speed was such that the reflected light passed through the next gap in the wheel. The time for the wheel to rotate through the angle between successive gaps equals the time for the light to make the round trip to the distant mirror and back.

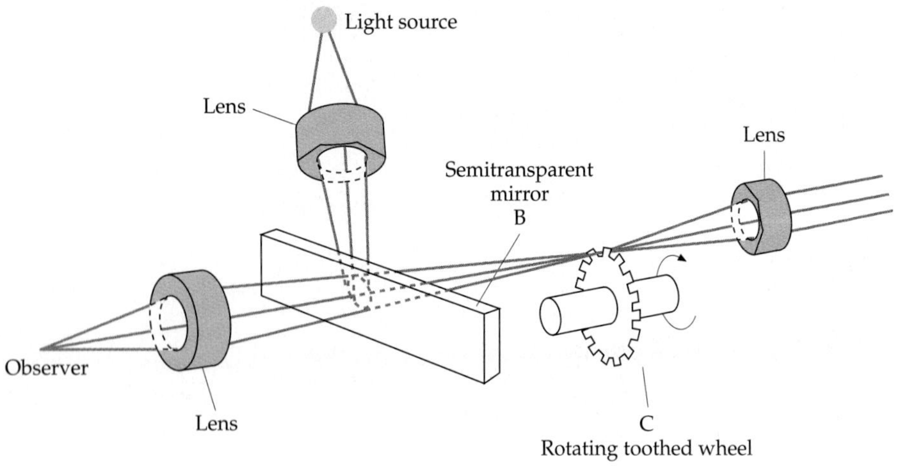

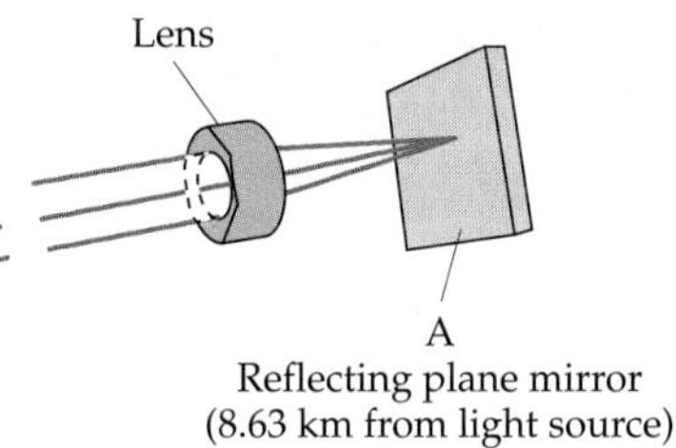

Figure 33-12 Fizeau's method of measuring the speed of light. Light from the source is reflected by mirror B and is transmitted through a gap in the toothed wheel to mirror A. The speed of light is determined by measuring the angular speed of the wheel that will permit the reflected light to pass through the next gap in the toothed wheel so that an image of the source is observed.

Fizeau's method was improved upon by Foucault, who replaced the toothed wheel with an eight-sided rotating mirror as shown in Figure 33-13. Light strikes one face of the mirror, is reflected to a distant fixed mirror, and is then reflected back to another face of the rotating mirror and then to an observing telescope. When the mirror rotates through one-eighth of a turn during the time it takes for the light to travel to the fixed mirror and back (or $n/8$ turns, where n is an integer) another face of the mirror is in the right position for the reflected light to enter the telescope. In about 1850, Foucault measured the speed of light in air and in water and showed that it is less in water. Using essentially the same method, the American physicist A. A. Michelson made precise measurements of the speed of light from 1880 to 1930.

Another method of determining the speed of light involves the measurement of the electrical constants ϵ_0 and μ_0 to determine c from $c = 1/\sqrt{\epsilon_0\mu_0}$.

The various methods we have discussed for measuring the speed of light are all in general agreement. Today, the speed of light is defined to be exactly

$$c = 299{,}792{,}457 \text{ m/s} \qquad \text{33-5}$$

and the standard unit of length, the meter, is defined in terms of this speed. A measurement of the speed of light is therefore now a measurement of the size of the meter, which is the distance light travels in 1/299,792,457 s. The value 3×10^8 m/s for the speed of light is accurate enough for nearly all calculations. The speed of radio waves and all other electromagnetic waves (in a vacuum) is the same as the speed of light.

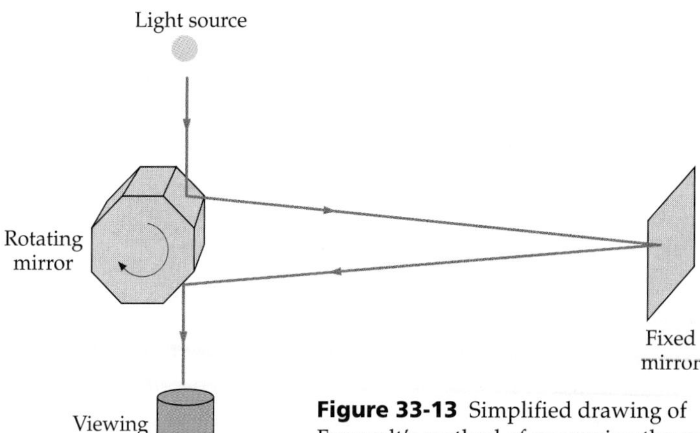

Figure 33-13 Simplified drawing of Foucault's method of measuring the speed of light. Essentially, Fizeau's rotating toothed wheel is replaced by a rotating octagonal mirror. When the mirror makes an eighth of a revolution during the time it takes for the light to travel to the fixed mirror and back, the next face of the mirror is in the proper position to reflect the light into the telescope.

Example 33-3

In Fizeau's experiment, the wheel had 720 teeth, and light was observed when the wheel rotated at 25.2 revolutions per second. If the distance from the wheel to the distant mirror was 8.63 km, what was Fizeau's value for the speed of light?

Picture the Problem The time taken for the light to travel from the wheel to the mirror and back is the time for the wheel to rotate from one tooth to the next, which is the time for the wheel to make $\frac{1}{720}$ revolution.

1. The speed is the distance divided by the time: $c = \dfrac{\Delta x}{\Delta t}$

2. The distance is twice the distance from the wheel to the mirror: $\Delta x = 2 \times 8.63 \text{ km} = 17.3 \text{ km}$

3. Calculate the time to make $\frac{1}{720}$ revolution: $\Delta t = \dfrac{1 \text{ s}}{25.2 \text{ rev}}\left(\dfrac{1}{720}\text{ rev}\right) = 5.51 \times 10^{-5} \text{ s}$

4. Substitute these values to calculate c: $c = \dfrac{\Delta x}{\Delta t} = \dfrac{17.3 \times 10^3 \text{ m}}{5.51 \times 10^{-5} \text{ s}} = 3.14 \times 10^8 \text{ m/s}$

Remark This result is about 5% too high.

Exercise Space travelers on the moon use electromagnetic waves to communicate with the space control center on earth. Use $c = 3 \times 10^8$ m/s to calculate the time delay for their signal to reach the earth, which is 3.84×10^8 m away. (*Answer* 1.28 s each way)

Large distances are often given in terms of the distance traveled by light in a given time. For example, the distance to the sun is 8.33 light-minutes, written 8.33 c-min. A light-year is the distance light travels in one year. We can easily find a conversion factor between light-years and meters. The number of seconds in one year is

$$1\text{ y} = 1\text{ y} \times \frac{365.24\text{ d}}{1\text{ y}} \times \frac{24\text{ h}}{1\text{ d}} \times \frac{3600\text{ s}}{1\text{ h}} = 3.156 \times 10^7\text{ s}$$

The number of meters in one light-year is thus

$$1\ c\text{-year} = (2.998 \times 10^8\text{ m/s})(3.156 \times 10^7\text{ s}) = 9.46 \times 10^{15}\text{ m} \qquad 33\text{-}6$$

33-5 The Propagation of Light

The propagation of light is governed by the wave equation discussed in Chapter 32. But long before Maxwell's theory of electromagnetic waves, the propagation of light and other waves was described empirically by two interesting and very different principles attributed to the Dutch physicist Christian Huygens (1629–1695) and the French mathematician Pierre de Fermat (1601–1665).

Huygens' Principle

Figure 33-14 shows a portion of a spherical wavefront emanating from a point source. The wavefront is the locus of points of constant phase. If the radius of the wavefront is r at time t, its radius at time $t + \Delta t$ is $r + c\,\Delta t$, where c is the speed of the wave. However, if a part of the wave is blocked by some obstacle or if the wave passes through a different medium, as in Figure 33-15, the determination of the new wavefront at time $t + \Delta t$ is much more difficult.

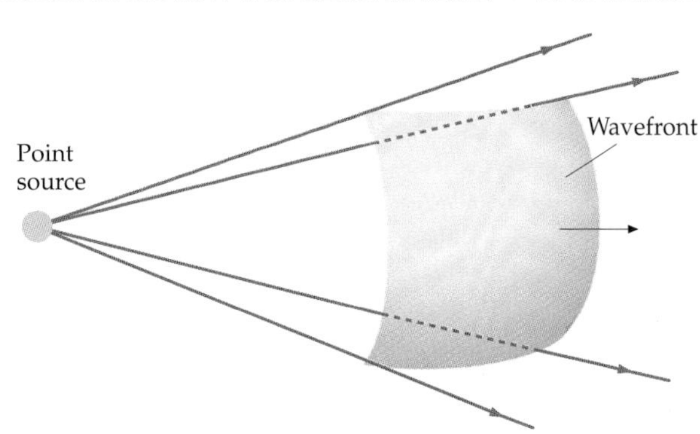

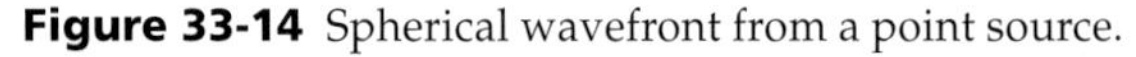

Figure 33-14 Spherical wavefront from a point source.

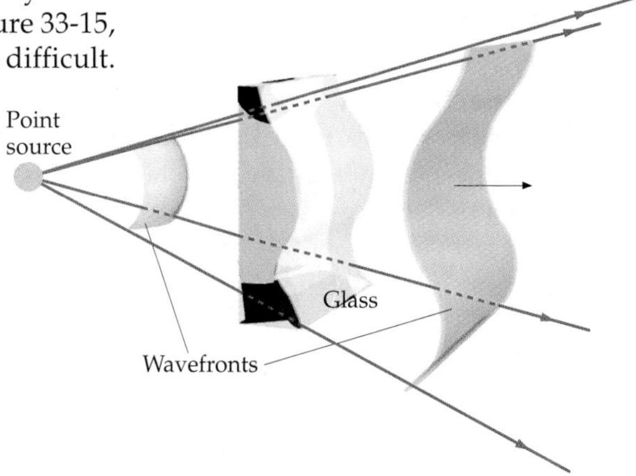

Figure 33-15 Wavefront from a point source before and after passing through an irregularly shaped piece of glass.

The propagation of any wave through space can be described using a geometric method discovered by Huygens about 1678, which is now known as **Huygens' principle** or **Huygens' construction:**

> Each point on a primary wavefront serves as the source of spherical secondary wavelets that advance with a speed and frequency equal to those of the primary wave. The primary wavefront at some later time is the envelope of these wavelets.

Huygens' principle

Figure 33-16 shows the application of Huygens' principle to the propagation of a plane wave and of a spherical wave. Of course, if each point on a wavefront were really a point source, there would be waves in the backward direction as well. Huygens ignored these back waves.

Huygens' principle was later modified by Fresnel so that the new wavefront was calculated from the old wavefront by superposition of the wavelets considering their relative amplitudes and phases. Kirchhoff later showed that the Huygens–Fresnel principle was a consequence of the wave equation, thus putting it on a firm mathematical basis. Kirchhoff showed that the intensity of the wavelets depends on the angle and is zero in the backward direction.

We will use Huygens' principle to derive the laws of reflection and refraction in Section 33-8. In Chapter 35, we apply Huygens' principle with Fresnel's modification to calculate the diffraction pattern of a single slit. Because the wavelength of light is so small, we can often use the ray approximation to describe its propagation.

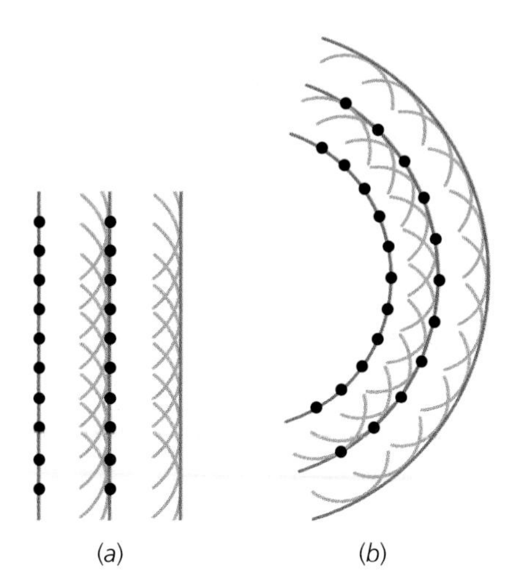

Figure 33-16 Huygens' construction for the propagation to the right of (*a*) a plane wave and (*b*) an outgoing spherical, or circular, wave.

Fermat's Principle

The propagation of light can also be described by Fermat's principle:

> The path taken by light in traveling from one point to another is such that the time of travel is a minimum.*

Fermat's principle

In Section 33-8 we will use Fermat's principle to derive the laws of reflection and refraction.

33-6 Reflection and Refraction

The speed of light in a transparent medium such as air, water, or glass is less than the speed $c = 3 \times 10^8$ m/s in vacuum. A transparent medium is characterized by the **index of refraction** n, which is defined as the ratio of the speed of light in a vacuum, c, to the speed in the medium, v:

$$n = \frac{c}{v} \qquad \text{33-7}$$

Definition—Index of refraction

For water, $n = 1.33$, whereas for glass n ranges from about 1.5 to 1.66 depending on the type of glass. Diamond has a very high index of refraction of about 2.4. The index of refraction of air is about 1.0003 so for most purposes we can assume the speed of light in air is the same as in vacuum.

When a beam of light strikes a boundary surface separating two different media, such as an air–glass interface, part of the light energy is reflected and part enters the second medium. If the incident light is not perpendicular to the surface, then the transmitted light is not parallel to the incident light. The change in direction of the transmitted ray is called **refraction.** Figure 33-17 shows a light ray striking a smooth air–glass interface. The angle θ_1 between the incident ray and the normal (the line perpendicular to the surface) is called the **angle of incidence,** and the plane defined by these two lines is called the **plane of incidence.** The reflected ray lies in the plane of incidence and makes an angle θ_1' with the normal that is equal to the angle of incidence as shown in the figure:

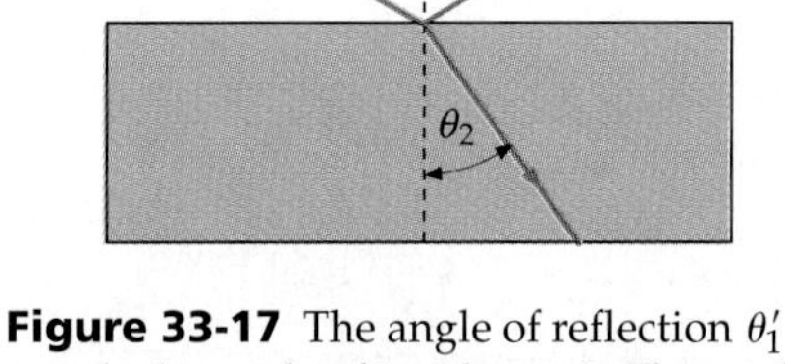

Figure 33-17 The angle of reflection θ_1' equals the angle of incidence θ_1. The angle of refraction θ_2 is less than the angle of incidence if the light speed in the second medium is less than that in the incident medium.

$$\theta_1' = \theta_1 \qquad \text{33-8}$$

Law of reflection

* A more complete and general statement is that the time of travel is stationary with respect to variations in path; that is, if t is expressed in terms of some parameter x, the path taken will be such that $dt/dx = 0$. The important characteristic of a stationary path is that the time taken along nearby paths will be approximately the same as that along the true path.

This result is known as the **law of reflection.** The law of reflection holds for any type of wave. Figure 33-18 illustrates the law of reflection for rays of light and for wavefronts of ultrasonic waves.

The ray that enters the glass in Figure 33-17 is called the refracted ray, and the angle θ_2 is called the angle of refraction. When a wave crosses a boundary at which the wave speed is reduced, as in the case of light entering glass from air, the angle of refraction is less than the angle of incidence θ_1, as shown in Figure 33-17; that is, the refracted ray is bent toward the normal. If, on the other hand, the light beam originates in the glass and is refracted into the air, then the refracted ray is bent away from the normal.

The angle of refraction θ_2 depends on the angle of incidence and on the relative speed of light waves in the two mediums. If v_1 is the wave speed in the incident medium and v_2 is the wave speed in the transmission medium, the angles of incidence and refraction are related by

$$\frac{1}{v_1}\sin\theta_1 = \frac{1}{v_2}\sin\theta_2 \qquad \text{33-9}a$$

Equation 33-9*a* holds for the refraction of any kind of wave incident on a boundary interface separating two media.

In terms of the indexes of refraction of the two media n_1 and n_2, Equation 33-9*a* is

$$n_1\sin\theta_1 = n_2\sin\theta_2 \qquad \text{33-9}b$$

Snell's law of refraction

This result was discovered experimentally in 1621 by the Dutch scientist Willebrod Snell and is known as **Snell's law** or the **law of refraction.** It was independently discovered a few years later by René Descartes.

(*a*)

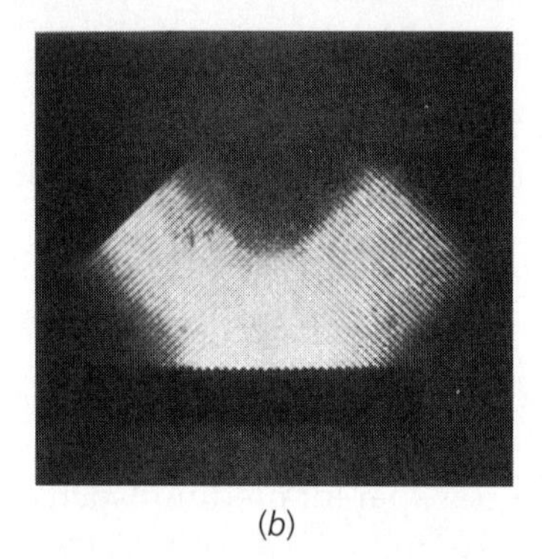

(*b*)

Figure 33-18 (*a*) Light rays reflecting from an air–glass interface showing equal angles of incidence and reflection. (*b*) Ultrasonic plane waves in water reflecting from a steel plate.

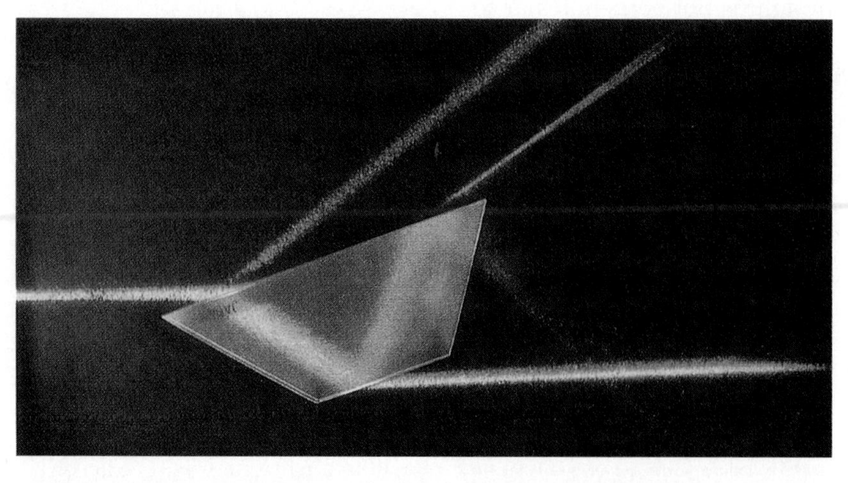

Reflection and refraction of a beam of light incident on a glass slab.

Mirages

When the index of refraction of a medium changes gradually, the refraction is continuous, leading to a gradual bending of the light. An interesting example of this is the formation of a mirage. On a hot day, there is often a layer of air near the ground that is warmer and therefore less dense than the air just above it. The speed of light is slightly greater in this less dense layer, so a light beam passing from the cooler layer into the warmer layer is bent. Figure 33-25*a* shows the light from a tree when the surrounding air is all at the same temperature. The wavefronts are spherical, and the rays are straight lines. In Figure 33-25*b*, the air near the ground is warmer, resulting in a greater speed of light there. The portions of the wavefronts near the ground travel faster and get ahead of the higher portions, creating a nonspherical wavefront and causing a curving of the rays. Thus, the ray shown initially heading for the ground is bent upward. As a result, the viewer sees an image of the tree looking as if it were reflected off a water surface on the ground. When driving on a very hot day, you may have noticed apparent wet spots on the highway that disappear when you get to them. This is due to the refraction of light from a hot air layer near the pavement.

Light

(*a*)

Light

Air warmer near ground

(*b*)

(*c*)

Figure 33-25 A mirage. (*a*) When the air is at a uniform temperature, the wavefronts of the light from the tree are spherical. (*b*) When the air near the ground is warmer, the wavefronts are not spherical and the light from the tree is continuously refracted into a curved path. (*c*) Apparent reflections of motorcycles on a hot road.

Dispersion

The index of refraction of a material has a slight dependence on wavelength. For many materials, n decreases slightly as the wavelength increases, as shown in Figure 33-26. The dependence of the index of refraction on wavelength (and therefore on frequency) is called **dispersion.** When a beam of white light is incident at some angle on the surface of a glass prism, the angle of refraction (which is measured relative to the normal) for the

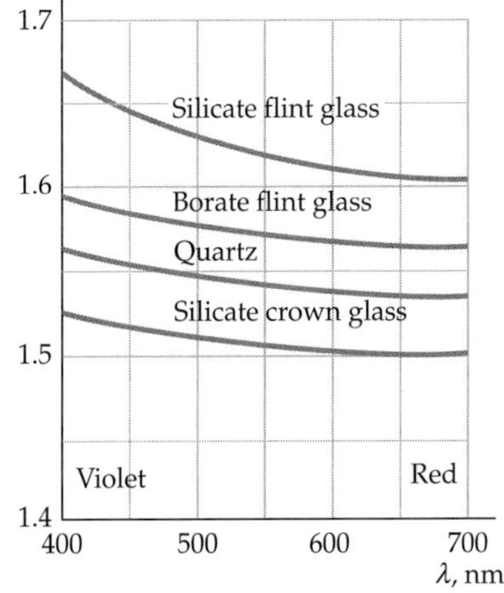

Figure 33-26 Index of refraction versus wavelength for various materials.

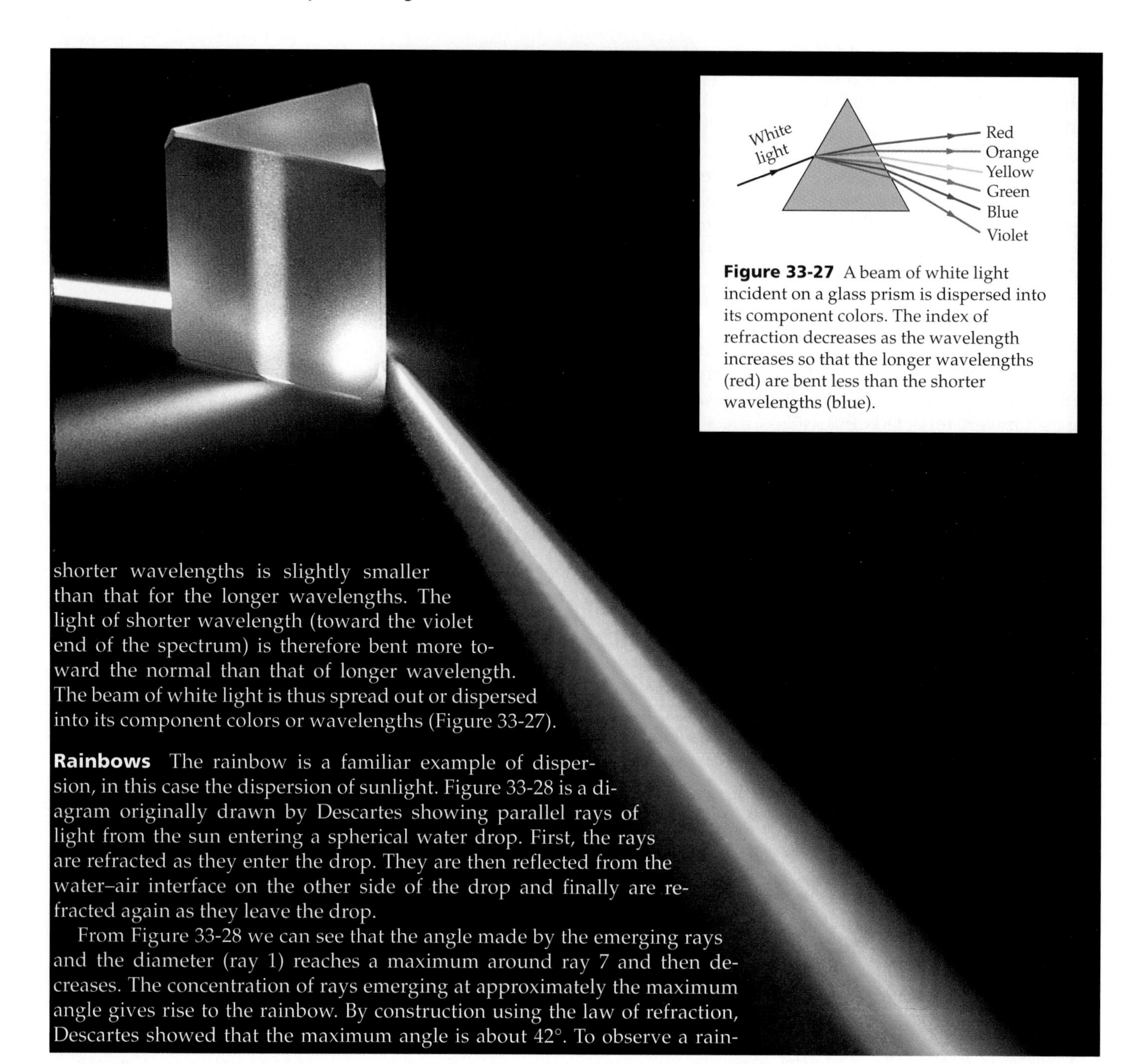

Figure 33-27 A beam of white light incident on a glass prism is dispersed into its component colors. The index of refraction decreases as the wavelength increases so that the longer wavelengths (red) are bent less than the shorter wavelengths (blue).

shorter wavelengths is slightly smaller than that for the longer wavelengths. The light of shorter wavelength (toward the violet end of the spectrum) is therefore bent more toward the normal than that of longer wavelength. The beam of white light is thus spread out or dispersed into its component colors or wavelengths (Figure 33-27).

Rainbows The rainbow is a familiar example of dispersion, in this case the dispersion of sunlight. Figure 33-28 is a diagram originally drawn by Descartes showing parallel rays of light from the sun entering a spherical water drop. First, the rays are refracted as they enter the drop. They are then reflected from the water–air interface on the other side of the drop and finally are refracted again as they leave the drop.

From Figure 33-28 we can see that the angle made by the emerging rays and the diameter (ray 1) reaches a maximum around ray 7 and then decreases. The concentration of rays emerging at approximately the maximum angle gives rise to the rainbow. By construction using the law of refraction, Descartes showed that the maximum angle is about 42°. To observe a rain-

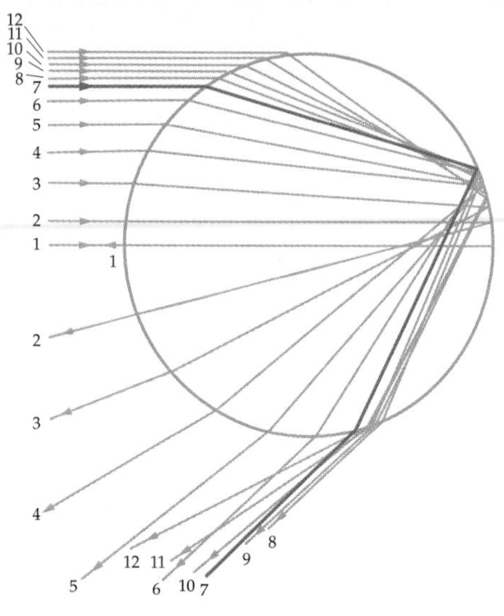

Figure 33-28 Descartes' construction of parallel rays of light entering a spherical water drop. Ray 1 enters the drop along a diameter and is reflected back along its incident path. Ray 2 enters slightly above the diameter and emerges below the diameter at a small angle with it. The rays entering farther and farther away from the diameter emerge at greater and greater angles up to ray 7, shown as the heavy line. Rays entering above ray 7 emerge at smaller and smaller angles with the diameter.

bow, we must therefore look at the water drops at an angle of 42° relative to the line back to the sun, as shown in Figure 33-29. The angular radius of the rainbow is therefore 42°.

The separation of the colors in the rainbow results from the fact that the index of refraction of water depends slightly on the wavelength of light. The angular radius of the bow will therefore depend slightly on the wavelength of the light. The observed rainbow is made up of light rays from many different droplets of water (Figure 33-30). The color seen at a particular angular radius corresponds to the wavelength of light that allows the light to reach the eye from the droplets at that angular radius. Because n_{water} is smaller for red light than for blue light, the red part of the rainbow is at a slightly greater angular radius than the blue part of the rainbow, so red is at the outer side of the rainbow.

When a light ray strikes a surface separating water and air, part of the light is reflected and part is refracted. A secondary rainbow results from the light rays that are reflected twice within a droplet (Figure 33-31). The secondary bow has an angular radius of 51°, and its color sequence is the reverse of that of the primary bow; that is, the violet is on the outside in the secondary bow. Because of the small fraction of light reflected from a water–air interface, the secondary bow is considerably fainter than the primary bow.

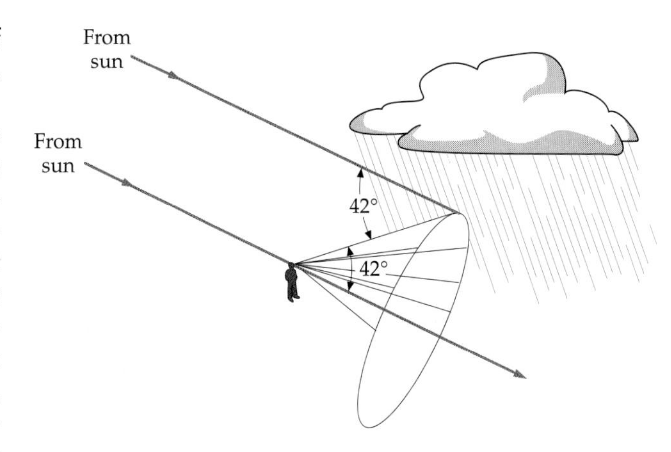

Figure 33-29 A rainbow is viewed at an angle of 42° from the line to the sun, as predicted by Descartes' construction in Figure 33-28.

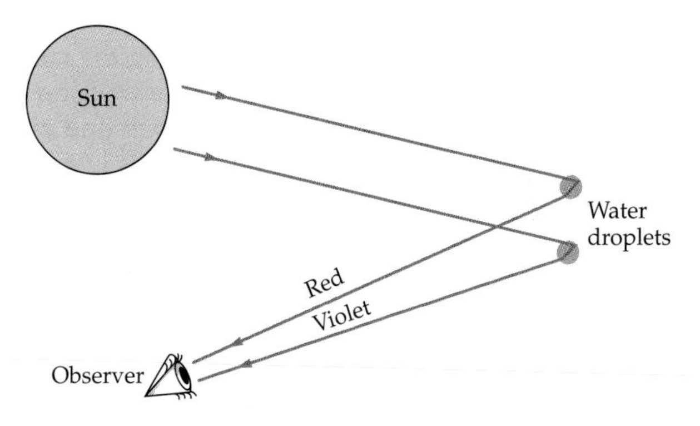

Figure 33-30 The rainbow results from light from many different water droplets.

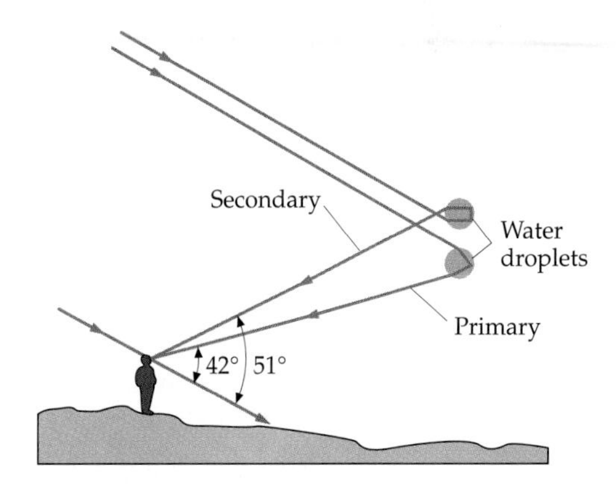

Figure 33-31 The secondary rainbow results from light rays that are reflected twice within a water droplet.

(*a*)

(*b*)

(*a*) This 22° halo around the sun results from reflection and refraction from hexagonal ice crystals that are randomly oriented in the atmosphere. (*b*) When the ice crystals are not randomly oriented but are falling with their flat bases horizontal, only parts of the halo on each side of the sun, called "sun dogs," are seen.

Beyond the (Visible) Rainbow*

Robert Greenler
University of Wisconsin, Milwaukee

One day while I was sitting at my desk woolgathering rather than addressing the task at hand, the question occurred to me: Is there an infrared rainbow in the sky? How does one explore such a question? Here is the process I went through. For there to be an infrared rainbow, a number of conditions must be met. First, the source of light must emit infrared radiation (the sun emits light over the entire electromagnetic spectrum from X rays to radio waves). Second, the infrared radiation must pass through the earth's atmosphere (water vapor and carbon dioxide in the atmosphere absorb some infrared wavelengths, but others pass through unimpeded). The rainbow is caused by light rays that enter a droplet of water and are reflected internally before emerging from the drop. For there to be an infrared rainbow, the third requirement is that the infrared rays would have to pass through a water droplet. This is a serious consideration. Just because a droplet of water appears transparent in visible light, we cannot assume that it is transparent to infrared "light"; indeed, liquid water does absorb over a broad range of infrared wavelengths. However, the measured spectral transmittance of water shows that water drops should be quite transparent from the visible region out to an infrared wavelength of about 1300 nm. Finally, after emerging from the raindrop, the infrared rays that have survived all these losses must again pass through air to the (unseeing) eye of the would-be observer.

* This essay is adapted from an article that first appeared in *Optic News*, published by the Optical Society of America, in November 1988. Robert Greenler has been professor of physics at the University of Wisconsin–Milwaukee since 1962, where he has been instrumental in the development of the Laboratory for Surface Studies at Milwaukee. His research interests include the study of the structure of molecules adsorbed on solid surfaces, optical effects of the sky, and the understanding of the iridescent colors seen in many biological organisms.

The Search

This line of reasoning produced a tentative answer to the question that prompted the speculation: Yes, there should be an infrared rainbow in the sky and it should lie in a band just outside of the red of the visible rainbow.

I decided to try to photograph this invisible bow using film that is sensitive to a portion of the infrared spectrum. Figure 1 shows the curve of the sensitivity of the film. The figure also shows a curve of the sensitivity of the human eye, as a way of defining the limits of the visible spectral region (extending from about 400 nm at the violet end of the spectrum to 700 nm at the red end). Note that the infrared film has a sensitivity extending out to about 930 nm. Because the film is sensitive not only to the infrared but to the entire visible region (it is *very* sensitive to blue light), we used a filter that is opaque to visible light and transmits only wavelengths longer than about 800 nm. As can be seen from Figure 1, this combination of film and filter will permit the recording of only those wavelengths in a band between 800 and 930 nm, well removed from the visible spectral region.

The Capture

Anyone who has tried to photograph rainbows knows that they usually occur when a camera is not at hand and fade just before one is located. I decided to first try an easier subject—the rainbow in a water spray. Figure 2 shows one of the first photographic results. A garden hose with many holes was wrapped back and forth across a board resting on top of the ladder. And in the spray of the hose—the infrared rainbow! You can also see the fainter, secondary rainbow outside the

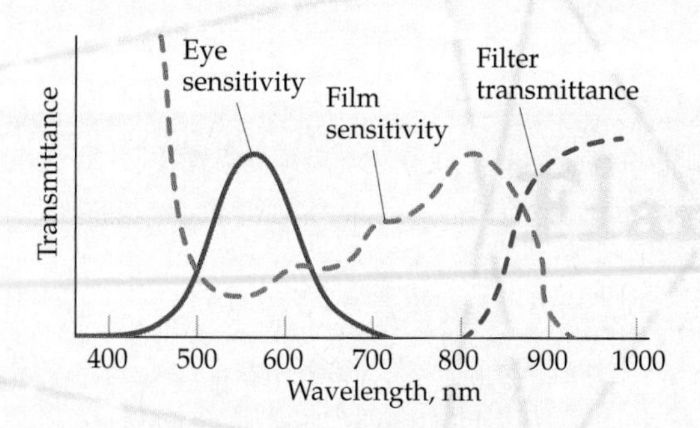

Figure 1 The infrared film has a sensitivity extending through the visible and into the near infrared range. The filter is opaque to visible light but transmits in the infrared for wavelengths longer than 800 nm. The combination of film and filter records an image with wavelengths between 800 and 930 nm, well outside the visible spectrum.

Figure 2 An infrared rainbow photographed in the spray of a garden hose. The fainter secondary bow is shown outside (to the left of) the primary bow. The fringes seen inside the primary bow are caused by interference effects.

brighter, primary bow. This corresponds to the secondary bow seen rather commonly in visible light and it results from rays that enter a water droplet and experience two internal reflections before leaving the drop (see Figure 33-31).

There is another interesting feature in this infrared photograph: immediately *inside* (to the right side of) the bright primary bow there is another bright band—or perhaps two bands. Such fringes, sometimes seen inside a visible bow, are called supernumerary bows and result from the interference of light waves (see Chapter 35).

Close inspection of the negative of Figure 2 reveals yet another feature, which is difficult to reproduce in a printed picture. There is a faint fringe just *outside* the secondary bow. A process similar to the one that produces supernumerary bows inside the primary should, in theory, produce a similar set of fringes outside the secondary. I have never seen any of those fringes associated with any rainbow or with any rainbow photograph, but they are visible in the original of this photograph. They are even visible when I project the slide onto a screen. For the first attempt, that was quite an exciting collection of effects.

Other Effects in the Infrared Photos

Some other features of these infrared photographs are worth considering. If the only radiation that produced these photographic images is invisible, infrared radiation, is it surprising that we can see the ladder, trees, and grass? One should not be too surprised. These objects absorb some wavelengths and reflect or scatter others. Objects that absorb the infrared appear dark in the photos and those that scatter it strongly appear bright. To make it clear just what these photos show, we need to understand the difference between reflected (or scattered) radiation and emitted radiation.

Normally when you look at objects in your landscape, you see them only by the light they scatter. However, if the temperature of an object is high enough, it emits light. If it is very hot—you might call it "white hot"—it emits a broad spectrum of wavelengths with the peak of the emission curve in the visible spectrum. If the object cools down a bit, the peak in the emission curve moves to longer wavelengths. The result is that there is more red light than blue light being emitted, and the appropriate description for its temperature is "red hot." At a lower temperature, you might see a dull red glow. At this point, the peak of the emission curve is in the infrared with just a small amount of emission in the red end of the visible region. At a slightly lower temperature, the object appears dark; the emission peak has moved further into the infrared, so no visible radiation can be seen. If the object cools to where it is warm to the touch, its emission peak is far out in the infrared—perhaps at 10,000 nm—and it is emitting almost nothing in the visible or in the near infrared region to which photographic films are sensitive.

If, however, you could produce a picture with 10,000-nm radiation, objects slightly warmer than their surroundings would appear to be bright—they would be glowing with emitted infrared radiation. There *are* ways to produce such pictures; they are used to show sources of heat loss in homes or to record relatively warm spots on a human body (thermograms) that may indicate the site of some physical disorder. These pictures are usually described as infrared pictures, but they are quite different from the photographs taken with infrared-sensitive film. This film is sensitive only to the near infrared, but the "heat pictures" result from emitted radiation in the far infrared. So the infrared photographs shown here show only the infrared radiation from the sun that is scattered by the leaves or ladder or transformed by raindrop spheres into an invisible rainbow.

Another interesting feature of the photographs is the darkness of the clear sky background. We see light in the clear, clean sky, away from the sun, due to scattering from the molecules of gases in the air. These small scattering particles (much smaller than the wavelength of the light) scatter the shorter waves more effectively than the longer waves. Thus, more blue light is scattered than red light, giving the sky its blue color. This same effect, which makes the sky darker in red light than in blue, makes it even darker in the infrared light sampled by these photographs.

optional

Calculating the Angular Radius of the Rainbow We can calculate the angular radius of the rainbow from the laws of reflection and refraction. Figure 33-32 shows a ray of light incident on a spherical water droplet at point A. The angle of refraction θ_2 is related to the angle of incidence θ_1 by Snell's law:

$$n_{\text{air}} \sin \theta_1 = n_{\text{water}} \sin \theta_2 \qquad \text{33-13}$$

Point P in Figure 33-32 is the intersection of the line of the incident ray and the line of the emerging ray. The angle ϕ_d is called the angle of deviation of the ray. The angle 2β is the angular radius of the rainbow. 2β is related to ϕ_d by:

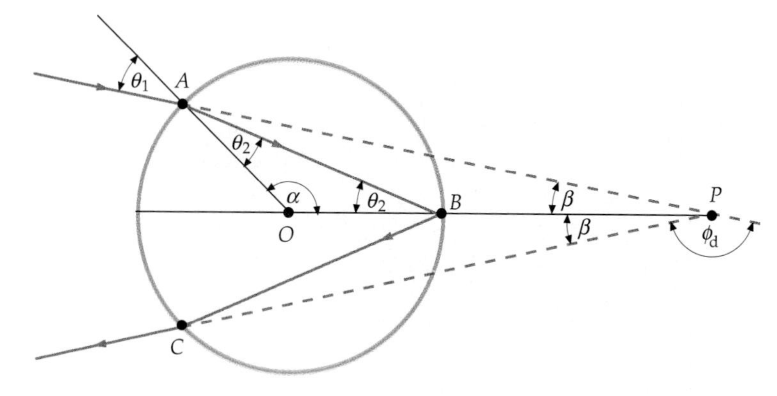

Figure 33-32 Light ray incident on a spherical water drop. The refracted ray strikes the back of the water droplet at point B. It makes an angle θ_2 with the radial line OB and is reflected at an equal angle. The ray is refracted again at point C, where it leaves the droplet.

$$\phi_d + 2\beta = \pi \qquad \text{33-14}$$

We wish to relate the angle of deviation ϕ_d to the angle of incidence θ_1. From the triangle AOB, we have

$$2\theta_2 + \alpha = \pi \qquad \text{33-15}$$

Similarly, from the triangle AOP, we have

$$\theta_1 + \beta + \alpha = \pi \qquad \text{33-16}$$

Eliminating α from Equations 33-15 and 33-16 and solving for β gives

$$\beta = \pi - \theta_1 - \alpha = \pi - \theta_1 - (\pi - 2\theta_2) = 2\theta_2 - \theta_1$$

Substituting this value for β into Equation 33-14 gives the angle of deviation:

$$\phi_d = \pi - 2\beta = \pi - 4\theta_2 + 2\theta_1 \qquad \text{33-17}$$

Equation 33-17 can be combined with Snell's law to eliminate θ_2 and give the angle of deviation ϕ_d in terms of the angle of incidence θ_1:

$$\phi_d = \pi + 2\theta_1 - 4 \arcsin\left(\frac{n_{\text{air}} \sin \theta_1}{n_{\text{water}}}\right) \qquad \text{33-18}$$

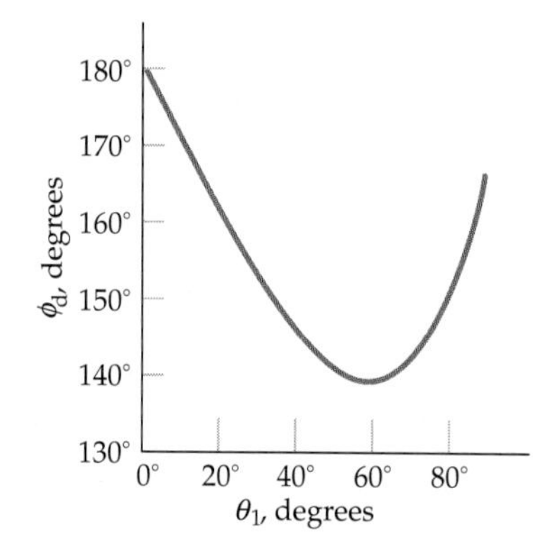

Figure 33-33 Plot of the angle of deviation ϕ_d as a function of incident angle θ_1. The angle of deviation has its minimum value of 138° when the angle of incidence is 60°. Since $d\phi_d/d\theta_1 = 0$ at minimum deviation, the deviation of rays with incident angles slightly less or slightly greater than 60° will be approximately the same.

Figure 33-33 shows a plot of ϕ_d versus θ_1. The angle of deviation ϕ_d has its minimum value when $\theta_1 = 60°$. At this angle of incidence, the angle of deviation is $\phi_d = 138°$. This angle is called the **angle of minimum deviation.** At incident angles that are slightly greater or slightly smaller than 60°, the angle of deviation is approximately the same. Therefore, the light reflected by the water droplet will be concentrated near the angle of minimum deviation. The angular radius of the rainbow is thus

$$2\beta = \pi - \phi_d = 180° - 138° = 42° \qquad \text{33-19}$$

33-7 Polarization

In a transverse wave, the vibration is perpendicular to the direction of propagation of the wave. For example, in a light wave traveling in the z direction, the electric field is perpendicular to the z direction. (The magnetic field of a light wave is also perpendicular to the z direction.) If the vibration remains parallel to a fixed line in space, the wave is said to be **linearly polarized.** We can visualize polarization most easily by considering mechanical waves on a string. If one end of a string held horizontal is moved up and down, the resulting waves on the string are linearly polarized with each element of the string vibrating in the vertical direction. Similarly, if one end is moved along a horizontal line, the displacements of the string are linearly polarized in the horizontal direction. If one end of the string is moved with constant speed in a circle, the resulting wave is said to be **circularly polarized.** In this case, each element of the string moves in a circle. Unpolarized waves can be produced by moving the end of the string vertically and horizontally in a random way. Then, if the string itself is in the z direction, the vibrations will have both x and y components that vary randomly.

Most waves produced by a single source are polarized. For example, electromagnetic waves produced by a single dipole antenna are linearly polarized with the electric field vector parallel to the antenna. Waves produced by many sources are usually unpolarized. A typical light source, for example, contains millions of atoms acting independently. The electric field for such a wave can be resolved into x and y components that vary randomly because there is no correlation between the individual atoms producing the light.

The polarization of electromagnetic waves can be demonstrated with microwaves, which have wavelengths on the order of centimeters. In a typical microwave generator, polarized waves are radiated by a dipole antenna. In Figure 33-34, the dipole antenna is vertical, so the electric field vector $\vec{E}$ of the radiated waves is vertical. An absorber can be made of a screen of parallel straight wires. When the wires are vertical, as in Figure 33-34*a*, the electric field parallel to the wires sets up currents in the wires and energy is absorbed. When the wires are horizontal and therefore perpendicular to $\vec{E}$, as in Figure 33-34*b*, no currents are set up and the waves are transmitted.

There are four phenomena that produce polarized electromagnetic waves from unpolarized waves: (1) absorption, (2) reflection, (3) scattering, and (4) birefringence (also called double refraction), each of which is examined in the upcoming sections.

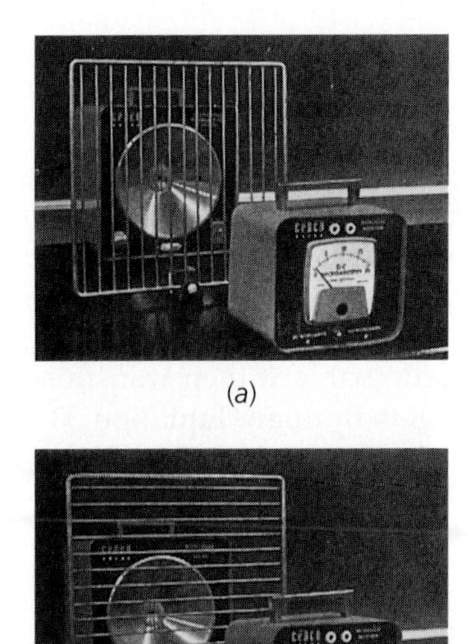

(*a*)

(*b*)

Figure 33-34 Demonstration showing the polarization of microwaves. The electric field of the microwaves is vertical, parallel to the vertical dipole antenna. (*a*) When the metal wires of the absorber are vertical, electric currents are set up in the wires and energy is absorbed, as indicated by the low reading on the microwave detector. (*b*) When the wires are horizontal, no currents are set up, and the microwaves are transmitted, as indicated by the high reading on the detector.

Polarization by Absorption

Several naturally occurring crystals, when cut into appropriate shapes, absorb and transmit light differently depending on the polarization of the light. These crystals can be used to produce linearly polarized light. In 1938, E. H. Land invented a simple commercial polarizing film called Polaroid. This material contains long-chain hydrocarbon molecules that are aligned when the sheet is stretched in one direction during the manufacturing process. These chains become conducting at optical frequencies when the sheet is dipped in a solution containing iodine. When light is incident with its electric field vector parallel to the chains, electric currents are set up along the chains, and the light energy is absorbed, just as the microwaves are absorbed by the wires in Figure 33-34. If the electric field is perpendicular to the chains, the light is transmitted. The direction perpendicular to the chains is called the **transmission axis.** We will make the simplifying assumption that all the light is transmitted when the electric field is parallel to the transmission axis and all is absorbed when it is perpendicular to the transmission axis.

Consider an unpolarized light beam traveling in the z direction incident on a polarizing film with its transmission axis in the x direction. On the average, half of the incident light has its electric field in the y direction and half has it in the x direction. Thus, half the intensity is transmitted, and the transmitted light is linearly polarized with its electric field in the x direction.

Suppose we have a second polarizing film whose transmission axis makes an angle θ with that of the first, as shown in Figure 33-35. If $\vec{E}$ is the electric field between the films, its component along the direction of the transmission axis of the sec-

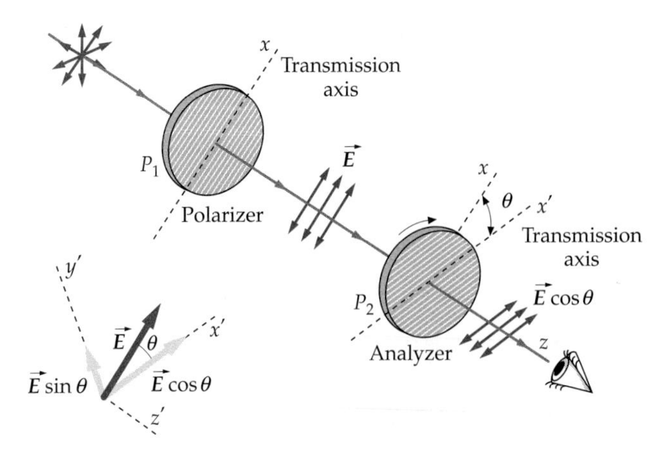

Figure 33-35 Two polarizing films with their transmission axes making an angle θ with each other. Only the component $E\cos\theta$ is transmitted through the second film. If the intensity between the films is I_0, that transmitted by both films is $I_0\cos^2\theta$.

ond film is $E \cos \theta$. Since the intensity of light is proportional to E^2, the intensity of light transmitted by both films will be given by

$$I = I_0 \cos^2 \theta \qquad \text{33-20}$$

where I_0 is the intensity incident on the second film and is, of course, half the intensity incident on the first film. When two polarizing elements are placed in succession in a beam of light as described here, the first is called the **polarizer** and the second is called the **analyzer.** If the polarizer and analyzer are crossed, that is, if their transmission axes are perpendicular to each other, no light gets through. Equation 33-20 is known as **Malus's law** after its discoverer, E. L. Malus (1775–1812). It applies to any two polarizing elements whose transmission axes make an angle θ with each other.

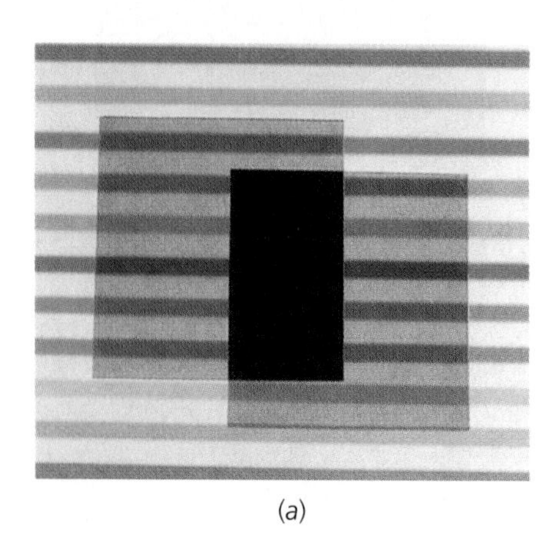

(a)

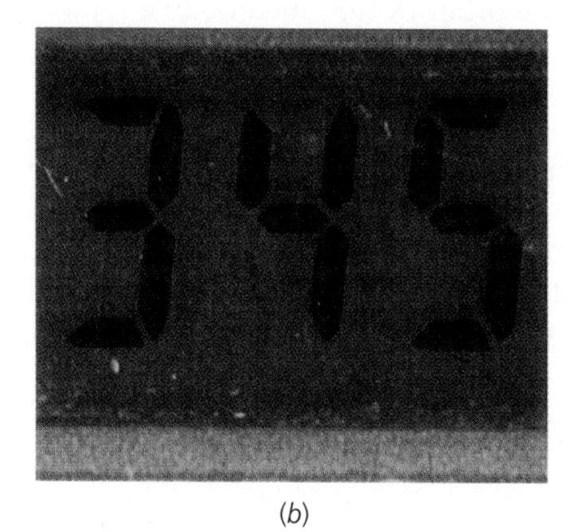

(b)

(*a*) Cross polarizers block out all of the light. (*b*) In a liquid crystal display, the crystal is between crossed polarizers. Light incident on the crystal is transmitted because the crystal rotates the direction of polarization of the light 90°. The light is reflected back out through the crystal by a mirror behind the crystal, and a uniform background is seen. When a voltage is applied across a small segment of the crystal, the polarization is not rotated, so no light is transmitted and the segment appears black.

Example 33-6

Unpolarized light of intensity 3.0 W/m² is incident on two polarizing films whose transmission axes make an angle of 60° (Figure 33-36). What is the intensity of light transmitted by the second film?

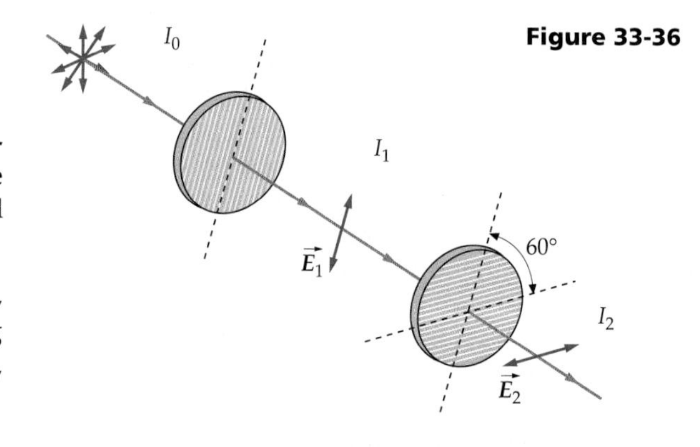

Figure 33-36

Picture the Problem Since the incident light is unpolarized, half the incident intensity is transmitted by the first polarizing film. The second film reduces the intensity by a factor of $\cos^2 \theta$, with $\theta = 60°$.

1. The intensity transmitted by the second film is related to the incident intensity I_1 by Equation 33-18:

$$I_2 = I_1 \cos^2 \theta$$

2. The intensity incident on the second film I_1 is half the intensity of unpolarized light incident on the first film:

$$I_1 = \tfrac{1}{2} I_0$$

3. Combine these results and substitute the given data:

$$I_2 = \tfrac{1}{2} I_0 \cos^2 60° = \tfrac{1}{2}(3.0\ \text{W/m}^2)(0.500)^2 = 0.375\ \text{W/m}^2$$

Remarks Half the intensity passes through the first film no matter what its orientation is. Note that the second film rotates the direction of polarization.

Polarization by Reflection

When unpolarized light is reflected from a plane surface boundary between two transparent media, such as air and glass or air and water, the reflected light is partially polarized. The degree of polarization depends on the angle between the incident ray and the normal to the surface (called the angle of incidence) and on the ratio of the wave speeds in the two media. For a certain angle of incidence called the polarizing angle θ_p, the reflected light is completely polarized. At the polarizing angle the reflected and refracted rays are perpendicular to each other. This result was discovered experimentally by Sir David Brewster in 1812.

Figure 33-37 shows light incident at the polarizing angle θ_p for which the reflected light is completely polarized. The electric field of the incident light can be resolved into components parallel and perpendicular to the plane of incidence. The reflected light is completely polarized with its electric field perpendicular to the plane of incidence. We can relate the polarizing angle to the indexes of refraction of the media using Snell's law. If n_1 is the index of refraction of the first medium and n_2 is that of the second medium, Snell's law gives

$$n_1 \sin \theta_p = n_2 \sin \theta_2$$

where θ_2 is the angle of refraction. From Figure 33-37, we can see that the sum of the angle of reflection and the angle of refraction is 90°. Since the angle of reflection equals the angle of incidence, we have

$$\theta_2 = 90° - \theta_p$$

Then

$$n_1 \sin \theta_p = n_2 \sin(90° - \theta_p) = n_2 \cos \theta_p$$

or

$$\tan \theta_p = \frac{n_2}{n_1} \qquad 33\text{-}21$$

Brewster's law

Equation 33-21 is known as **Brewster's law.**

Although the reflected light is completely polarized for this angle of incidence, the transmitted light is only partially polarized because only a small fraction of the incident light is reflected. If the incident light itself is polarized with $\vec{E}$ in the plane of incidence, there is no reflected light when the angle of incidence is θ_p. We can understand this qualitatively from Figure 33-38. If we consider the molecules in the second medium to be oscillating parallel to the electric field of the refracted ray, there can be no reflected ray because no energy is radiated along the line of oscillation.

Because of the polarization of reflected light, sunglasses made of polarizing material can be very effective in cutting out glare. If light is reflected from a horizontal surface such as a lake or snow on the ground, the plane of incidence will be vertical and the electric field of the reflected light will be predominantly horizontal. Polarized sunglasses with a vertical transmission axis will then reduce glare by absorbing much of the reflected light. If you have polarized sunglasses, you can observe this effect by looking through the glasses at reflected light and then rotating the glasses 90°; much more of the light will be transmitted.

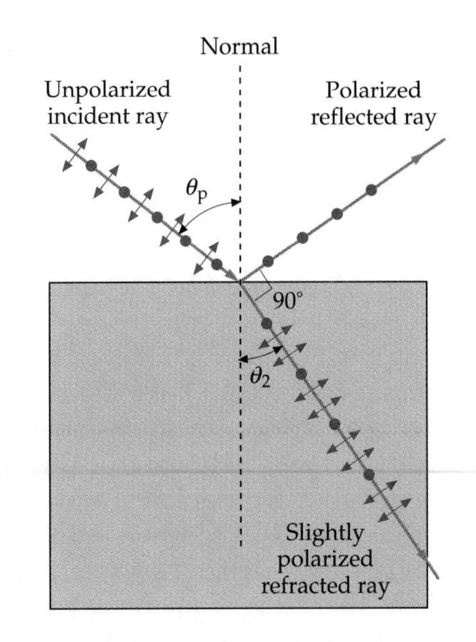

Figure 33-37 Polarization by reflection. The incident wave is unpolarized and has components of the electric field parallel to the plane of incidence (arrows) and components perpendicular to this plane (dots). For incidence at the polarizing angle, the reflected wave is completely polarized, with its electric field perpendicular to the plane of incidence.

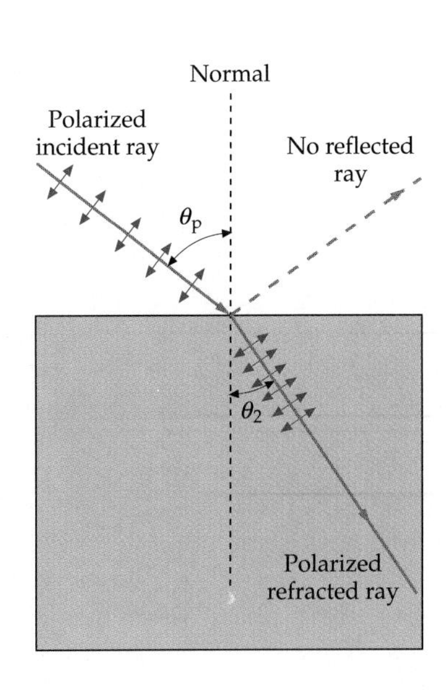

Figure 33-38 Polarized light incident at the polarizing angle. When the light is polarized with $\vec{E}$ in the plane of incidence, there is no reflected ray.

optional

Polarization by Scattering

The phenomenon of absorption and reradiation is called **scattering.** Scattering can be demonstrated by passing a light beam through a container of water to which a small amount of powdered milk has been added. The milk particles absorb light and reradiate it, making the light beam visible. Similarly, laser beams can be made visible by introducing chalk or smoke particles into the air to scatter the light. A familiar example of light scattering is that from air molecules, which tend to scatter short wavelengths more than long wavelengths, thereby giving the sky its blue color.

We can understand polarization by scattering if we think of an absorbing molecule as an electric-dipole antenna that radiates waves with a maximum intensity in the direction perpendicular to the antenna with the electric field vector parallel to the antenna and zero intensity in the direction along the antenna. Figure 33-39 shows a beam of unpolarized light that initially travels along the z axis, striking a scattering center at the origin. The electric field in the light beam has components in both the x and y directions perpendicular to the direction of motion of the light beam. These fields set up oscillations of the scattering center in both the x and y directions, but there is no oscillation in the z direction. The oscillation of the scattering center in the x direction produces light along the y axis but not along the x axis, which is along the line of oscillation. The light radiated along the y axis is thus polarized in the x direction. Similarly, the light radiated along the x axis is polarized in the y direction. This can be seen easily by examining the scattered light with a piece of polarizing film.

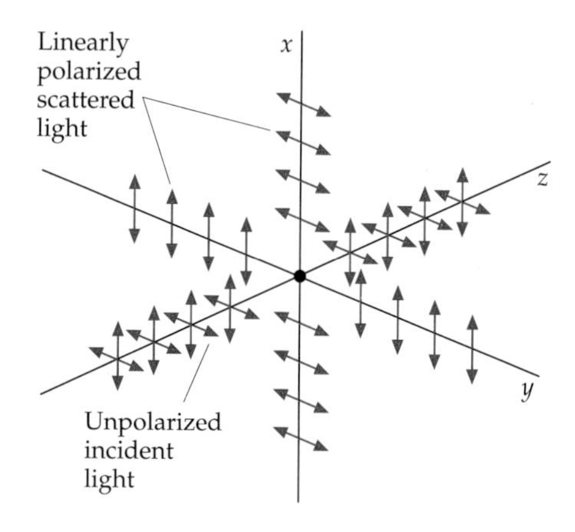

Figure 33-39 Polarization by scattering. Unpolarized light propagating in the z direction is incident on a scattering center at the origin. The light scattered in the x direction is polarized in the y direction and that scattered in the y direction is polarized in the x direction.

Polarization by Birefringence

Birefringence is a complicated phenomenon that occurs in calcite and other noncubic crystals and in some stressed plastics such as cellophane. Most materials are **isotropic**, that is, the speed of light passing through the material is the same in all directions. Because of their atomic structure, birefringent materials are **anisotropic.** The speed of light depends on its direction of propagation through the material. When a light ray is incident on such materials, it may be separated into two rays called the *ordinary ray* and the *extraordinary ray.* These rays are polarized in mutually perpendicular directions, and they travel with different speeds. Depending on the relative orientation of the material and the incident light, the rays may also travel in different directions.

There is one particular direction in a birefringent material in which both rays propagate with the same speed. This direction is called the **optic axis** of the material. (The optic axis is actually a *direction* rather than a line in the material.) Nothing unusual happens when light travels along the optic axis. However, when light is incident at an angle to the optic axis, as shown in Figure 33-40, the rays travel in different directions and emerge separated in space. If the material is rotated, the extraordinary ray (the e ray in the figure) rotates in space.

If light is incident on a birefringent plate perpendicular to its crystal face and perpendicular to the optic axis, the two rays travel in the same direction but at different speeds. The number of wavelengths in the two rays in the plate is different because the wavelengths ($\lambda = v/f$) of the rays differ. The rays emerge with a phase difference that depends on the thickness of the plate and on the wavelength of the incident light. In a **quarter-wave plate,** the thickness is such that there is a 90° phase difference between the waves of a particular wavelength when they emerge. In a **half-wave plate,** the rays emerge with a phase difference of 180°.

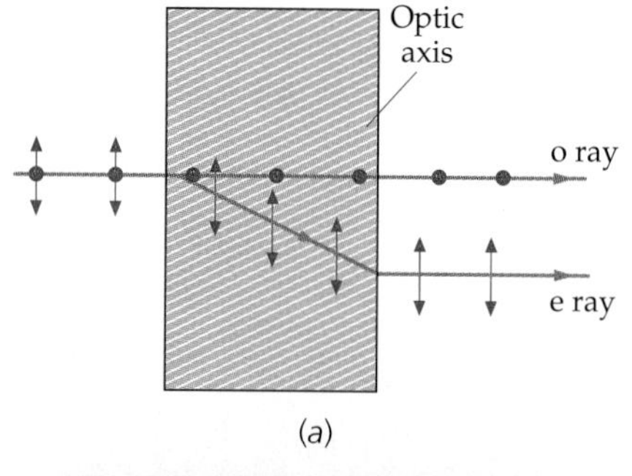

(*a*)

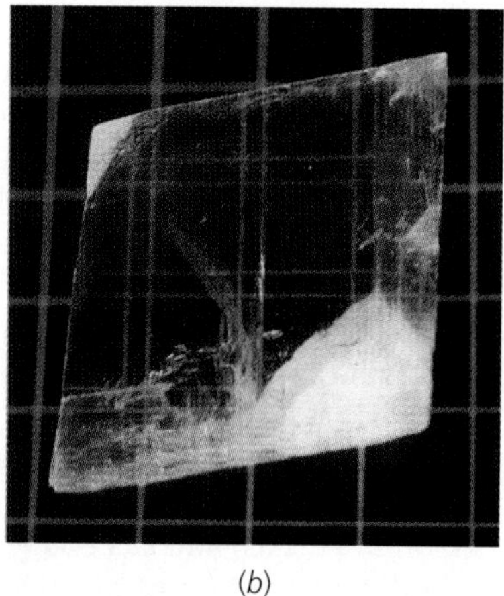

(*b*)

Figure 33-40 (*a*) A narrow beam of light incident on a birefringent crystal such as calcite is split into two beams, called the ordinary ray (o ray) and the extraordinary ray (e ray), that have mutually perpendicular polarizations. If the crystal is rotated, the extraordinary ray rotates in space. (*b*) A double image of the cross hatching is produced by this birefringent crystal of calcium carbonate.

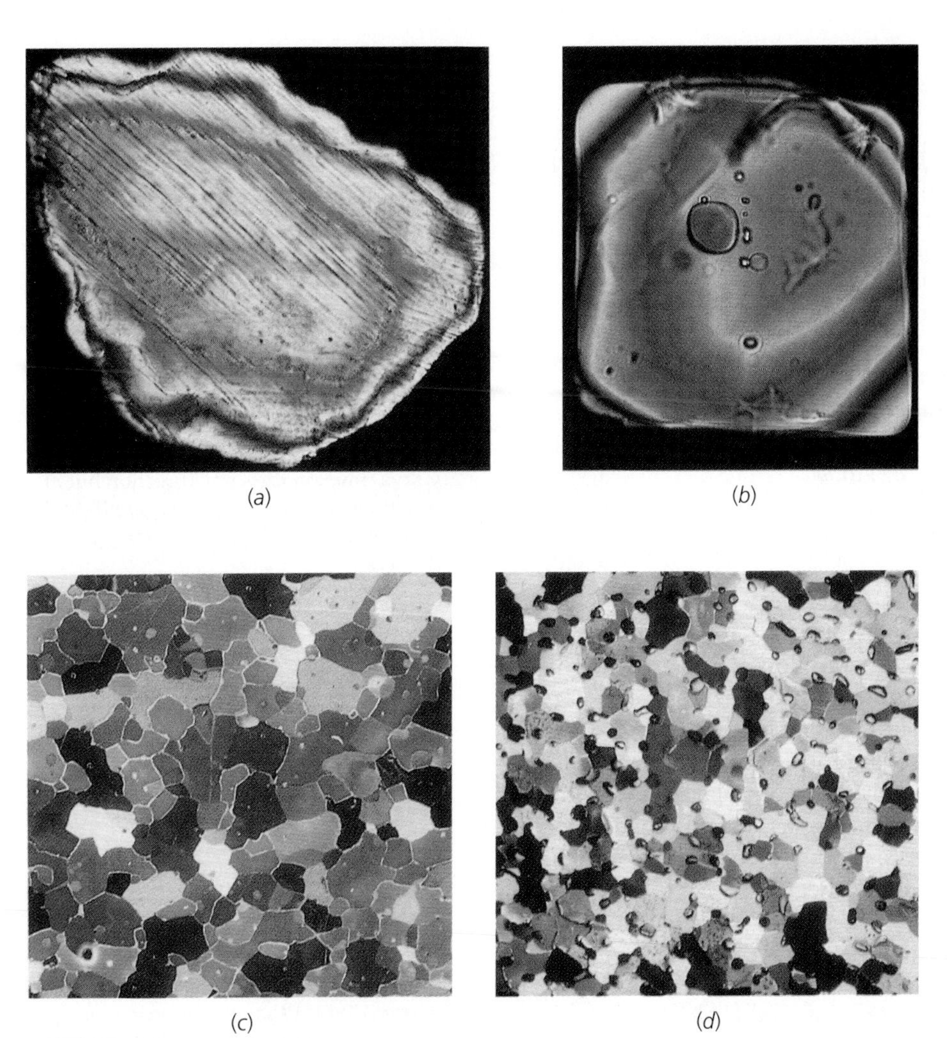

(a) (b) (c) (d)

(e)

When the transmission axes of two polarizing films are perpendicular, the polarizers are said to be crossed and no light is transmitted. However, many materials are birefringent or become so under stress. Such materials rotate the direction of polarization of the light so that light of a particular wavelength is transmitted through both polarizers. When a birefringent material is viewed between crossed polarizers, information about its internal structure is revealed. (*a*) A shocked quartz grain from the site of a meteorite crater. The layered structure, evidenced by the parallel lines, arises from the shock of the impact of the meteor. (*b*) A grain of quartz typically found in silicic volcanic rocks. No shock lines are seen. (*c*) Thin sections of ice core from the antarctic ice sheet reveal bubbles of trapped CO_2, which appear amber-colored. This sample was taken from a depth of 194 m, corresponding to air trapped 1600 years ago, whereas that in (*d*) is from a depth of 56 m, corresponding to air trapped 450 years ago. Ice core measurements have replaced the less reliable technique of analyzing carbon in tree rings to compare current atmospheric CO_2 levels with those of the recent past. (*e*) Robert Mark of the Princeton School of Architecture examines the stress patterns in a plastic model of the nave structure of Chartres Cathedral.

Suppose that the incident light is linearly polarized such that the electric field vector is at 45° to the optic axis, as illustrated in Figure 33-41. The ordinary and extraordinary rays start out in phase and have equal amplitudes. With a quarter-wave plate, the waves emerge with a phase difference of 90°, so the resultant electric field has components $E_x = E_0 \sin \omega t$ and $E_y = E_0 \sin(\omega t + 90°) = E_0 \cos \omega t$. The electric field vector thus rotates in a circle and the wave is circularly polarized.

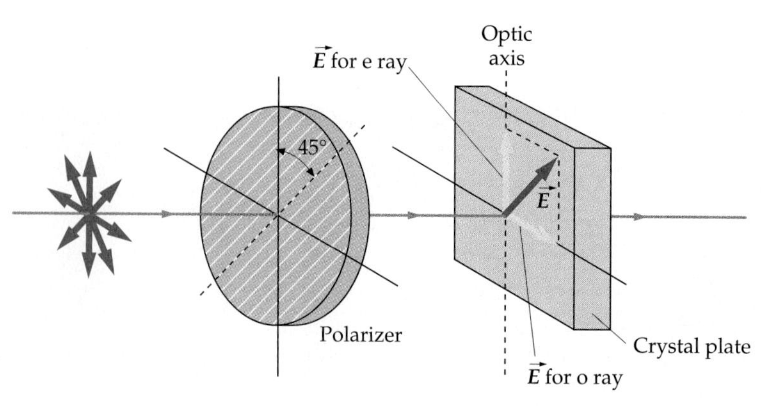

Figure 33-41 Polarized light emerging from the polarizer is incident on a birefringent crystal such that the electric field vector makes a 45° angle with the optic axis, which is perpendicular to the light beam. The ordinary and extraordinary rays travel in the same direction but at different speeds. The polarization of the emerging light depends on the thickness of the crystal and the wavelength of the light.

With a half-wave plate, the waves emerge with a phase difference of 180°, so the resultant electric field is linearly polarized with components $E_x = E_0 \sin \omega t$ and $E_y = E_0 \sin(\omega t + 180°) = -E_0 \sin \omega t$. The net effect is that the direction of polarization of the wave is rotated by 90° relative to that of the incident light, as shown in Figure 33-42.

Interesting and beautiful patterns, like those on page 1026, can be observed by placing birefringent materials such as cellophane or stressed plastic between two polarizing sheets with their transmission axes perpendicular to each other. Ordinarily, no light is transmitted through crossed polarizing sheets. However, if we place a birefringent material between the crossed Polaroids, the material acts as a half-wave plate for light of a certain color depending on the material's thickness. The direction of polarization is rotated and some light gets through both films. Various glasses and plastics become birefringent when under stress. The stress patterns can be observed when the material is placed between crossed polarizing sheets.

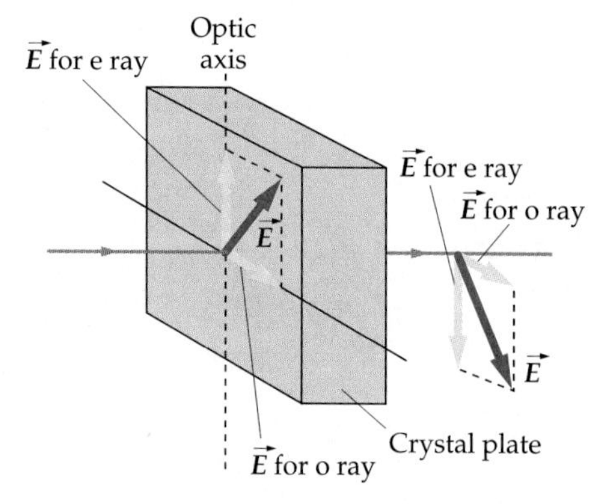

Figure 33-42 When the birefringent crystal in Figure 33-41 is a half-wave plate, the direction of polarization of the emerging light is rotated by 90°.

33-8 Derivation of the Laws of Reflection and Refraction

The laws of reflection and refraction can be derived from either Huygens' principle or Fermat's principle.

Huygens' Principle

Reflection Figure 33-43 shows a plane wavefront AA' striking a mirror at point A. As can be seen from the figure, the angle ϕ_1 between the wavefront and the mirror is the same as the angle of incidence θ_1, which is the angle between the perpendicular to the mirror and the rays that are perpendicular to the wavefronts. According to Huygens' principle, each point on a given wavefront can be considered to be a point source of secondary wavelets. The position of the wavefront after a time t is found by constructing wavelets of radius ct with their centers on the wavefront AA'. Wavelets that do not strike the mirror form the portion of the new wavefront BB'. Wavelets that do strike the mirror are reflected and form the portion of the new wavefront BB''. By a similar construction, the wavefront $C''C$ is obtained from the Huygens' wavelets originating on the wavefront $B''B$. Figure 33-44 is an enlargement of a portion of Figure 33-43 showing AP, which is part of the original wavefront. During the time t, the wavelet from point P reaches the mirror at point B, and the wavelet from point A reaches point B''. The reflected wave BB'' makes an angle ϕ_1' with the mirror that

Figure 33-43 Plane wave reflected at a plane mirror. The angle θ_1 between the incident ray and the normal to the mirror is the angle of incidence. It is equal to the angle ϕ_1 between the incident wavefront and the mirror.

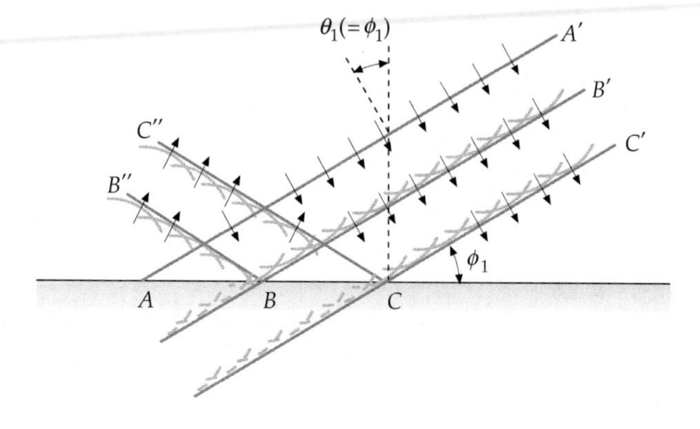

is equal to the angle of reflection θ_1' between the reflected ray and the normal to the mirror. The triangles ABP and BAB'' are both right triangles with a common side AB and equal sides $AB'' = BP = ct$. Hence, these triangles are congruent, and the angles ϕ_1 and ϕ_1' are equal, implying that the angle of reflection θ_1' equals the angle of incidence θ_1.

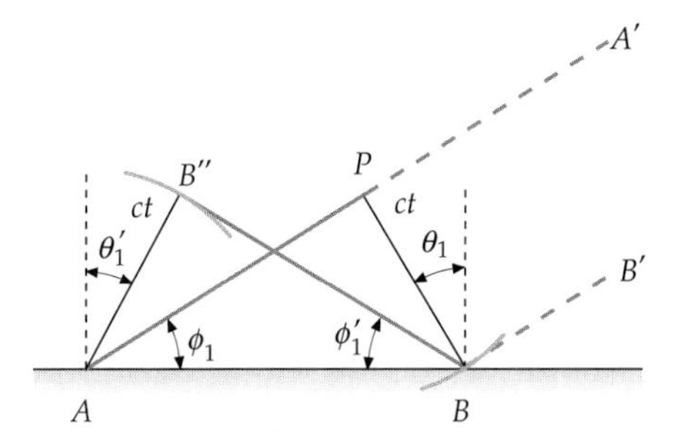

Figure 33-44 Geometry of Huygen's construction for the calculation of the law of reflection. The wavefront AP initially strikes the mirror at point A. After a time t, the Huygens' wavelet from P strikes the mirror at point B, and the one from A reaches point B'.

Refraction Figure 33-45 shows a plane wave incident on an air–glass interface. We apply Huygens' construction to find the wavefront of the transmitted wave. Line AP indicates a portion of the wavefront in medium 1 that strikes the glass surface at an angle ϕ_1. In time t the wavelet from P travels the distance v_1t and reaches the point B on the line AB separating the two media, while the wavelet from point A travels a shorter distance v_2t into the second medium. The new wavefront BB' is not parallel to the original wavefront AP because the speeds v_1 and v_2 are different. From the triangle APB,

$$\sin\phi_1 = \frac{v_1t}{AB}$$

or

$$AB = \frac{v_1t}{\sin\phi_1} = \frac{v_1t}{\sin\theta_1}$$

using the fact that the angle ϕ_1 equals the angle of incidence θ_1. Similarly, from triangle $AB'B$,

$$\sin\phi_2 = \frac{v_2t}{AB}$$

or

$$AB = \frac{v_2t}{\sin\phi_2} = \frac{v_2t}{\sin\theta_2}$$

where $\theta_2 = \phi_2$ is the angle of refraction. Equating the two values for AB, we obtain

$$\frac{\sin\theta_1}{v_1} = \frac{\sin\theta_2}{v_2} \tag{33-22}$$

Substituting $v_1 = c/n_1$ and $v_2 = c/n_2$ in this equation and multiplying by c, we obtain $n_1 \sin\theta_1 = n_2 \sin\theta_2$, which is Snell's law.

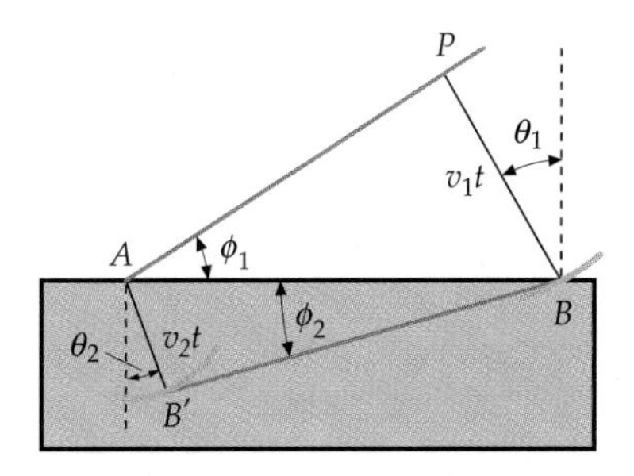

Figure 33-45 Application of Huygens' principle to the refraction of plane waves at the surface separating a medium in which the wave speed is v_1 from a medium in which the wave speed v_2 is less than v_1. The angle of refraction in this case is less than the angle of incidence.

Fermat's Principle

Reflection Figure 33-46 shows two paths in which light leaves point A, strikes the plane surface, which we can consider to be a mirror, and travels to point B. The problem for the application of Fermat's principle to reflection can be stated as follows: At what point P in the figure must the light strike the mirror so that it will travel from point A to point B in the least time? Since the light is traveling in the same medium for this problem, the time will be minimum when the distance is minimum. In Figure 33-46 the distance APB is the same as the distance $A'PB$, where A' is the image of the source A. Point A' lies along the perpendicular from A to the mirror and is equidistant behind the mirror. As we vary point P, the distance $A'PB$ is least when the points A', P, and B lie on a straight line. We can see from the figure that this occurs when the angle of incidence equals the angle of reflection.

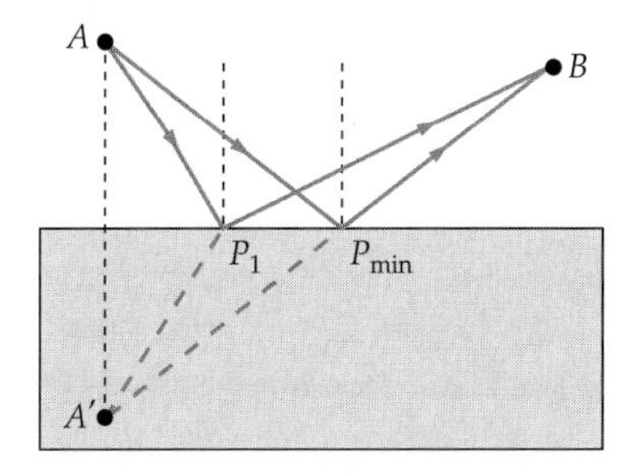

Figure 33-46 Geometry for deriving the law of reflection from Fermat's principle. The time it takes for the light to travel from point A to point B is a minimum when the light strikes the surface at point P.

optional

Refraction The derivation of Snell's law of refraction from Fermat's principle is slightly more complicated. Figure 33-47 shows the possible paths for light traveling from point A in air to point B in glass. Point P_1 is on the straight line between A and B, but this path is not the one for the shortest travel time because light travels with a smaller speed in the glass. If we move slightly to the right of P_1, the total path length is greater, but the distance traveled in the slower medium is less than for the path through P_1. It is not apparent from the figure which path is that of least time, but it is not surprising that a path slightly to the right of the straight-line path takes less time because the time gained by traveling a shorter distance in the glass more than compensates for the time lost traveling a longer distance in the air. As we move the point of intersection of the possible path to the right of point P_1, the total time of travel from A to B decreases until we reach a minimum at point P_{min}. Beyond this point, the time saved by traveling a shorter distance in the glass does not compensate for the greater time required for the greater distance traveled in the air.

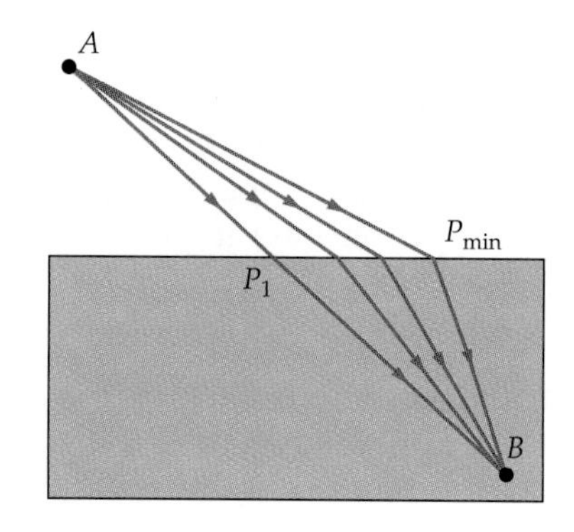

Figure 33-47 Geometry for deriving Snell's law from Fermat's principle. The point P_{min} is the point at which light must strike the glass in order that the travel time from A to B is a minimum.

Figure 33-48 shows the geometry for finding the path of least time. If L_1 is the distance traveled in medium 1 with index of refraction n_1, and L_2 is the distance traveled in medium 2 with index of refraction n_2, the time for light to travel the total path AB is

$$t = \frac{L_1}{v_1} + \frac{L_2}{v_2} = \frac{L_1}{c/n_1} + \frac{L_2}{c/n_2} = \frac{n_1 L_1}{c} + \frac{n_2 L_2}{c} \tag{33-23}$$

We wish to find the point P_{min} for which this time is a minimum. We do this by expressing the time in terms of a single parameter x, as shown in the figure, indicating the position of point P_{min}. In terms of the distance x,

$$L_1^2 = a^2 + x^2 \quad \text{and} \quad L_2^2 = b^2 + (d - x)^2 \tag{33-24}$$

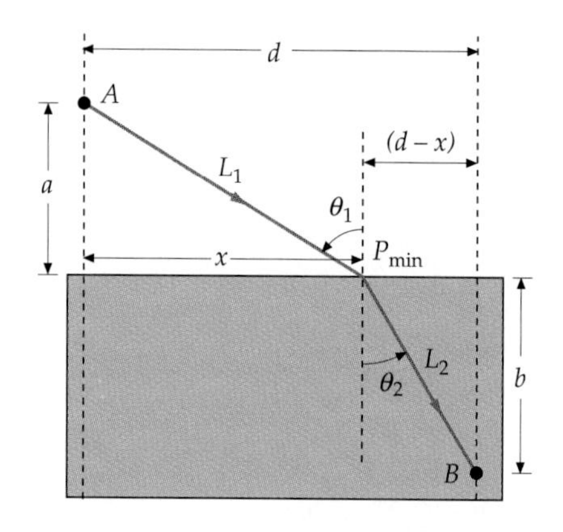

Figure 33-48 Geometry for calculating the minimum time in the derivation of Snell's law from Fermat's principle.

Figure 33-49 shows the time t as a function of x. At the value of x for which the time is a minimum, the slope of the graph of t versus x is zero:

$$\frac{dt}{dx} = 0$$

Differentiating each term in Equation 33-23 with respect to x and setting the result equal to zero, we obtain

$$\frac{dt}{dx} = \frac{1}{c}\left(n_1 \frac{dL_1}{dx} + n_2 \frac{dL_2}{dx}\right) = 0 \tag{33-25}$$

We can compute these derivatives from Equations 33-24. We have

$$2L_1 \frac{dL_1}{dx} = 2x \quad \text{or} \quad \frac{dL_1}{dx} = \frac{x}{L_1}$$

But x/L_1 is just $\sin\theta_1$, where θ_1 is the angle of incidence. Thus,

$$\frac{dL_1}{dx} = \sin\theta_1 \tag{33-26}$$

Similarly,

$$2L_2 \frac{dL_2}{dx} = 2(d - x)(-1)$$

or

$$\frac{dL_2}{dx} = -\frac{d - x}{L_2} = -\sin\theta_2 \tag{33-27}$$

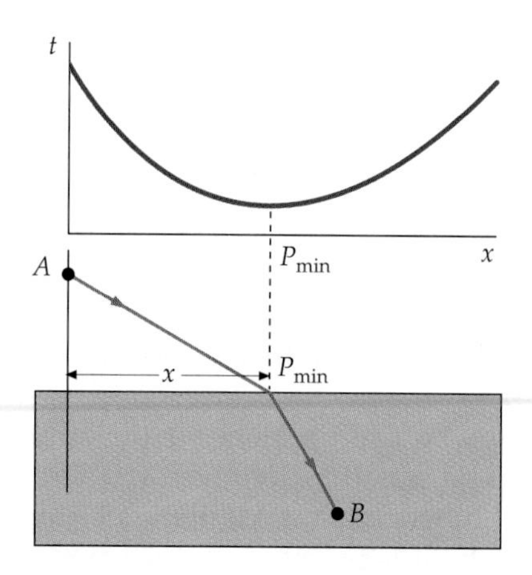

Figure 33-49 Graph of the time it takes for light to travel from A to B versus x, measured along the refracting surface. The time is a minimum at the point at which the angles of incidence and refraction obey Snell's law.

where θ_2 is the angle of refraction. From Equation 33-25,

$$n_1 \frac{dL_1}{dx} + n_2 \frac{dL_2}{dx} = 0 \qquad \text{33-28}$$

Substituting the results of Equations 33-26 and 33-27 for dL_1/dx and dL_2/dx gives

$$n_1 \sin \theta_1 + n_2(-\sin \theta_2) = 0$$

or

$$n_1 \sin \theta_1 = n_2 \sin \theta_2$$

which is Snell's law.

Summary

Topic	Remarks and Relevant Equations	
1. Visible Light	The human eye is sensitive to electromagnetic radiation with wavelengths from about 400 nm (violet) to about 700 nm (red). The photon energies range from about 1.8 eV to 3.1 eV. A uniform mixture of wavelengths, such as that emitted by the sun, appears white to our eyes.	
2. Wave–Particle Duality	Light propagates like a wave, but interacts with matter like a particle.	
Photon energy	$E = hf = \frac{hc}{\lambda}$	**33-1**
Planck's constant	$h = 6.626 \times 10^{-34}\ \text{J}\cdot\text{s} = 4.136 \times 10^{-15}\ \text{eV}\cdot\text{s}$	
hc	$hc = 1240\ \text{eV}\cdot\text{nm}$	**33-2**
3. Emission of Light	Light is emitted when an outer atomic electron makes a transition from an excited state to a state of lower energy.	
Line spectra	Atoms in dilute gases emit a discrete set of wavelengths called a line spectra. The photon energy $E = hf = hc/\lambda$ equals the difference in energy of the initial and final states of the atom.	
Continuous spectra	Atoms in high-density gases, liquids, or solids have continuous bands of energy levels so they emit a continuous spectrum of light. Thermal radiation is visible if the temperature of the emitting object is above about 600°C.	
Spontaneous emission	An atom in an excited state will spontaneously make a transition to a lower state with the emission of a photon. This process is random with a characteristic lifetime of about 10^{-8} s. The photons from two or more atoms are not correlated, so the light is incoherent.	
Stimulated emission	Stimulated emission occurs if an atom is initially in an excited state and a photon of energy equal to the energy difference between that state and a lower state is incident on the atom. The oscillating electromagnetic field of the incident photon stimulates the excited atom to emit another photon in the same direction and in phase with the incident photon. The emitted light is coherent.	

4. Lasers	A laser produces an intense, coherent, and narrow beam of photons as the result of stimulated emission. The operation of a laser depends on population inversion, in which there are more atoms in an excited state than in the ground state or a lower state.
5. Speed of Light	The SI unit of length, the meter, is defined so that the speed of light in vacuum is exactly $c = 299{,}792{,}457 \text{ m/s}$ **33-5**
c in a medium	$v = \frac{c}{n}$ **33-7** where n is the index of refraction
6. Huygens' Principle	Each point on a primary wavefront serves as the source of spherical secondary wavelets that advance with a speed and frequency equal to that of the primary wave. The primary wavefront at some later time is the envelope of these wavelets.
7. Reflection and Refraction	When light is incident on a surface separating two media in which the speed of light differs, part of the light energy is transmitted and part is reflected.
Law of reflection	The reflected ray lies in the plane of incidence and makes an angle θ_1' with the normal that is equal to the angle of incidence $\theta_1' = \theta_1$ **33-8**
Reflected intensity, normal incidence	$I = \left(\frac{n_1 - n_2}{n_1 + n_2}\right)^2 I_0$ **33-11**
Index of refraction	$n = \frac{c}{v}$ **33-7**
Law of refraction (Snell's law)	$n_1 \sin \theta_1 = n_2 \sin \theta_2$ **33-9*b***
Total internal reflection	When light is traveling in a medium with an index of refraction n_1 and is incident on the boundary of a second medium with a lower index of refraction $n_2 < n_1$, the light is totally reflected if the angle of incidence is greater than the critical angle θ_c given by
Critical angle	$\sin \theta_c = \frac{n_2}{n_1}$ **33-12**
Dispersion	The speed of light in a medium and therefore the index of refraction of that medium depends on the wavelength of light. Because of dispersion, a beam of white light incident on a refracting prism is dispersed into its component colors. Similarly, the reflection and refraction of sunlight by raindrops produces a rainbow.
8. Polarization	Transverse waves can be polarized. The four phenomena that produce polarized electromagnetic waves from unpolarized waves are (1) absorption, (2) scattering, (3) reflection, and (4) birefringence.
Malus's Law	When two polarizers have their transmission axes at an angle θ, the intensity transmitted by the second polarizer is reduced by the factor $\cos^2 \theta$: $I = I_0 \cos^2 \theta$ **33-20**

Problem-Solving Guide

Ray diagrams are particularly useful in problems involving reflection and refraction. Be sure to label the appropriate angles and indicate the index of refraction.

Summary of Worked Examples

Type of Calculation	Procedure and Relevant Examples
1. Absorption and Emission of Light	
Find the wavelength of light absorbed or emitted.	Use $\lambda = hc/\Delta E$, where ΔE is the difference in energy between the initial and final states of the atom. **Example 33-1**
2. Speed of Light	
Find c given the distance and time.	Use $c = \Delta x/\Delta t$. **Example 33-2**
Find c from Fizeau's or Foucault's experiment.	Use $c = \Delta x/\Delta t$ and calculate the time Δt for the wheel or mirror to rotate from one gap or from one mirror to the next. **Example 33-3**
3. Reflection	
Find the angle of reflection.	The angle of reflection is always equal to the angle of incidence, $\theta_1' = \theta_1$.
4. Refraction	
Find the angle of refraction.	Use Snell's law, $n_1 \sin \theta_1 = n_2 \sin \theta_2$. **Example 33-4**
Determine the critical angle for total internal reflection.	The critical angle is given by $\sin \theta_c = n_2/n_1$. **Example 33-5**
5. Polarization	
Find the intensity of light transmitted through a polarizer.	If the incident light is unpolarized, half the intensity will be transmitted through a polarizing film. If the incident light is polarized, the transmitted intensity will be $I_0 \cos^2 \theta$. **Example 33-6**

Problems

Conceptual Problems

Problems from Optional and Exploring sections

In a few problems, you are given more data than you actually need; in a few other problems, you are required to supply data from your general knowledge, outside sources, or informed estimates.

- • Single-concept, single-step, relatively easy
- •• Intermediate-level, may require synthesis of concepts
- ••• Challenging, for advanced students

Use $n = 1.33$ for water and $n = 1.5$ for glass unless otherwise specified.

Light Sources

1 •• Why is helium needed in a helium–neon laser? Why not just use neon?

2 •• When a beam of visible white light passes through a gas of atomic hydrogen and is viewed with a spectroscope, dark lines are observed at the wavelengths of the emission series. The atoms that participate in the resonance absorption then emit this same wavelength light as they return to the ground state. Explain why the observed spectrum nevertheless exhibits pronounced dark lines.

3 • A pulse from a ruby laser has an average power of 10 MW and lasts 1.5 ns. (*a*) What is the total energy of the pulse? (*b*) How many photons are emitted in this pulse?

4 • A helium–neon laser emits light of wavelength 632.8 nm and has a power output of 4 mW. How many photons are emitted per second by this laser?

5 • The first excited state of an atom of a gas is 2.85 eV above the ground state. (*a*) What is the wavelength of radiation for resonance absorption? (*b*) If the gas is irradiated with monochromatic light of 320 nm wavelength, what is the wavelength of the Raman scattered light?

6 •• A gas is irradiated with monochromatic ultraviolet light of 368 nm wavelength. Scattered light of the same wavelength and of 658 nm wavelength is observed. Assuming that the gas atoms were in their ground state prior to irradiation, find the energy difference between the ground state and the atomic state excited by the irradiation.

7 •• Sodium has excited states 2.11 eV, 3.2 eV, and 4.35 eV above the ground state. (*a*) What is the maximum wavelength of radiation that will result in resonance fluorescence? What is the wavelength of the fluorescent radiation? (*b*) What wavelength will result in excitation of the state 4.35 eV above the ground state? If that state is excited, what are the possible wavelengths of resonance fluorescence that might be observed?

8 •• Singly ionized helium is a hydrogen-like atom with a nuclear charge of $2e$. Its energy levels are given by $E_n = -4E_0/n^2$, where $E_0 = 13.6$ eV. If a beam of visible white light is sent through a gas of singly ionized helium, at what wavelengths will dark lines be found in the spectrum of the transmitted radiation?

The Speed of Light

9 • Estimate the time required for light to make the round trip in Galileo's experiment to determine the speed of light.

10 • Mission Control sends a brief wake-up call to astronauts in a far away spaceship. Five seconds after the call is sent, Mission Control can hear the groans of the astronauts. How far away (at most) from the earth is the spaceship?

(*a*) 7.5×10^8 m
(*b*) 15×10^8 m
(*c*) 30×10^8 m
(*d*) 45×10^8 m
(*e*) The spaceship is on the moon.

11 • The spiral galaxy in the Andromeda constellation is about 2×10^{19} km away from us. How many light-years is this?

12 • On a spacecraft sent to Mars to take pictures, the camera is triggered by radio waves, which like all electromagnetic waves travel with the speed of light. What is the time delay between sending the signal from the earth and receiving it on Mars? (Take the distance to Mars to be 9.7×10^{10} m.)

13 • The distance from a point on the surface of the earth to one on the surface of the moon is measured by aiming a laser light beam at a reflector on the surface of the moon and measuring the time required for the light to make a round trip. The uncertainty in the measured distance Δx is related to the uncertainty in the time Δt by $\Delta x = c\,\Delta t$. If the time intervals can be measured to ± 1.0 ns, find the uncertainty of the distance in meters.

14 •• In Galileo's attempt to determine the speed of light, he and his assistant were located on hilltops about 3 km apart. Galileo flashed a light and received a return flash from his assistant. (*a*) If his assistant had an instant reaction, what time difference would Galileo need to be able to measure for this method to be successful? (*b*) How does this time compare with human reaction time, which is about 0.2 s?

Reflection and Refraction

15 • How does a thin layer of water on the road affect the light you see reflected off the road from your own headlights? How does it affect the light you see reflected from the headlights of an oncoming car?

16 • A ray of light passes from air into water, striking the surface of the water with an angle of incidence of 45°. Which of the following four quantities change as the light enters the water: (1) wavelength, (2) frequency, (3) speed of propagation, (4) direction of propagation?

(*a*) 1 and 2 only
(*b*) 2, 3, and 4 only
(*c*) 1, 3, and 4 only
(*d*) 3 and 4 only
(*e*) 1, 2, 3, and 4

17 •• The density of the atmosphere decreases with height, as does the index of refraction. Explain how one can see the sun after it has set. Why does the setting sun appear flattened?

18 • Calculate the fraction of light energy reflected from an air–water interface at normal incidence.

19 • Find the angle of refraction of a beam of light in air that hits a water surface at an angle of incidence of (*a*) 20°, (*b*) 30°, (*c*) 45°, and (*d*) 60°. Show these rays on a diagram.

20 • Repeat Problem 18 for a beam of light initially in water that is incident on a water–air interface.

21 • Find the speed of light in water and in glass.

22 • The index of refraction for silicate flint glass is 1.66 for light with a wavelength of 400 nm and 1.61 for light with a wavelength of 700 nm. Find the angles of refraction for light of these wavelengths that is incident on this glass at an angle of 45°.

23 •• A slab of glass with an index of refraction of 1.5 is submerged in water with an index of refraction of 1.33. Light in the water is incident on the glass. Find the angle of refraction if the angle of incidence is (*a*) 60°, (*b*) 45°, and (*c*) 30°.

24 •• Repeat Problem 23 for a beam of light initially in the glass that is incident on the glass–water interface at the same angles.

25 •• Light is incident normally on a slab of glass with an index of refraction $n = 1.5$. Reflection occurs at both surfaces of the slab. About what percentage of the incident light energy is transmitted by the slab?

26 •• This problem is a refraction analogy. A band is marching down a football field with a constant speed v_1. About midfield, the band comes to a section of muddy ground that has a sharp boundary making an angle of 30° with the 50-yd line as shown in Figure 33-50. In the mud, the marchers move with speed $v_2 = \frac{1}{2} v_1$. Diagram how each line of marchers is bent as it encounters the muddy section of the field so that the band is eventually marching in a different direction. Indicate the original direction by a ray and the final direction by a second ray, and find the angles between the rays and the line perpendicular to the boundary. Is their direction of motion bent toward the perpendicular to the boundary or away from it?

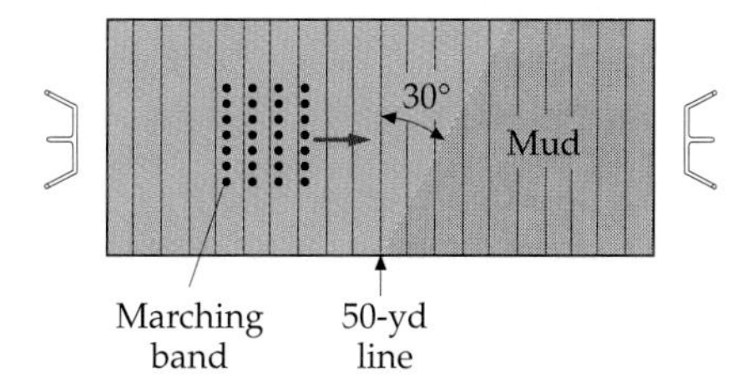

Figure 33-50 Problem 26

27 •• A point source of light is 5 cm above a plane reflecting surface (such as a mirror). Draw a ray from the source that strikes the surface at an angle of incidence of 45° and two more rays that strike the surface at angles slightly less than 45°, and draw the reflected ray for each. The reflected rays appear to diverge from a point called the image of the light source. Draw dotted lines extending the reflected rays back until they meet at a point behind the surface to locate the image point.

28 •• In Figure 33-51, light is initially in a medium (such as air) of index of refraction n_1. It is incident at angle θ_1 on the surface of a liquid (such as water) of index of refraction n_2. The light passes through the layer of water and enters glass of index of refraction n_3. If θ_3 is the angle of refraction in the glass, show that $n_1 \sin \theta_1 = n_3 \sin \theta_3$. That is, show that the second medium can be neglected when finding the angle of refraction in the third medium.

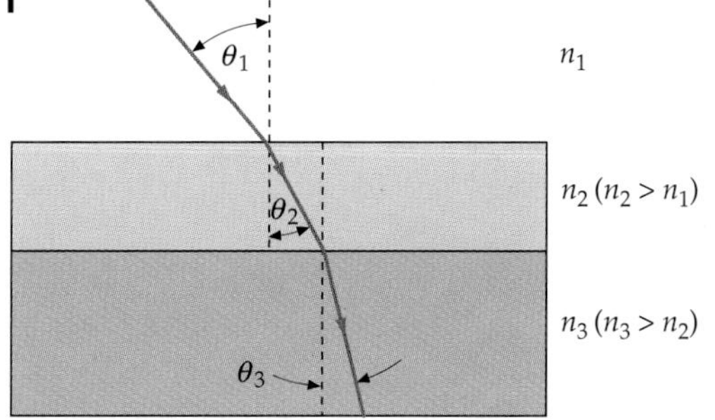

Figure 33-51 Problem 28

29 ••• Figure 33-52 shows a beam of light incident on a glass plate of thickness d and index of refraction n. (*a*) Find the angle of incidence such that the perpendicular separation between the ray reflected from the top surface and that reflected from the bottom surface and exiting the top surface is a maximum. (*b*) What is this angle of incidence if the index of refraction of the glass is 1.60? What is the separation of the two beams if the thickness of the glass plate is 4.0 cm?

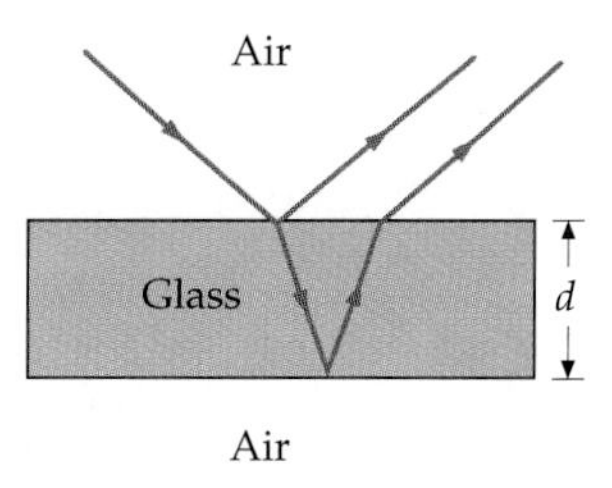

Figure 33-52 Problem 29

30 ••• Consider the situation shown in Figure 33-53. The index of refraction of the glass plate is n. Find the angle of incidence such that the perpendicular separation between the two beams emerging from the top surface is the same as the perpendicular displacement of the beam emerging from the bottom surface from the incident beam.

Figure 33-53 Problem 30
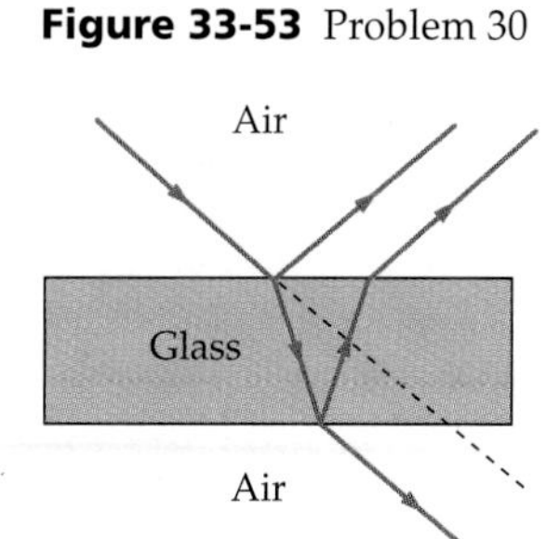

Deriving the Laws of Reflection and Refraction (optional)

31 • A physics student playing pocket billiards wants to strike her cue ball so that it hits a cushion and then hits the eight ball squarely. She chooses several points on the cushion and for each point measures the distance from it to the cue ball and to the eight ball. She aims at the point for which the sum of these distances is least. (*a*) Will her cue ball hit the eight ball? (*b*) How is her method related to Fermat's principle?

32 • A swimmer at S in Figure 33-54 develops a leg cramp while swimming near the shore of a calm lake and calls for help. A lifeguard at L hears the call. The lifeguard can run 9 m/s and swim 3 m/s. He knows physics and chooses a path that will take the least time to reach the swimmer. Which of the paths shown in Figure 33-54 does he take?

Figure 33-54 Problem 32
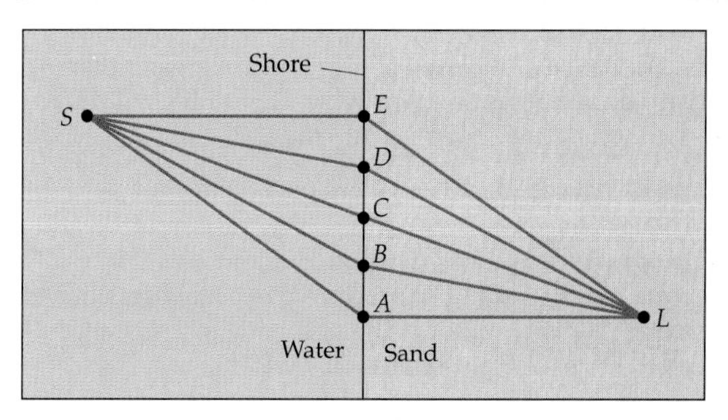

Total Internal Reflection

33 • What is the critical angle for total internal reflection for light traveling initially in water that is incident on a water–air interface?

34 •• A glass surface (n = 1.50) has a layer of water (n = 1.33) on it. Light in the glass is incident on the glass–water interface. Find the critical angle for total internal reflection.

35 •• A point source of light is located 5 m below the surface of a large pool of water. Find the area of the largest circle

on the pool's surface through which light coming directly from the source can emerge.

36 •• Light is incident normally on the largest face of an isosceles-right-triangle prism. What is the speed of light in this prism if the prism is just barely able to produce total internal reflection?

37 •• A point source of light is located at the bottom of a steel tank, and an opaque circular card of radius 6.0 cm is placed over it. A transparent fluid is gently added to the tank so that the card floats on the surface with its center directly above the light source. No light is seen by an observer above the surface until the fluid is 5 cm deep. What is the index of refraction of the fluid?

38 •• A grain of sand is embedded at the center of the base of a cube of transparent material. The grain of sand is visible when viewed through the top surface but cannot be seen when looking into any of the sides of the cube. What is the minimum index of refraction of the material of which the cube is made?

39 ••• Light is incident normally upon one face of a prism of glass with an index of refraction n (Figure 33-55). The light is totally reflected at the right side. (*a*) What is the minimum value n can have? (*b*) When the prism is immersed in a liquid whose index of refraction is 1.15, there is still total reflection, but when it is immersed in water, whose index of refraction is 1.33, there is no longer total reflection. Use this information to establish limits for possible values of n.

Figure 33-55 Problem 39

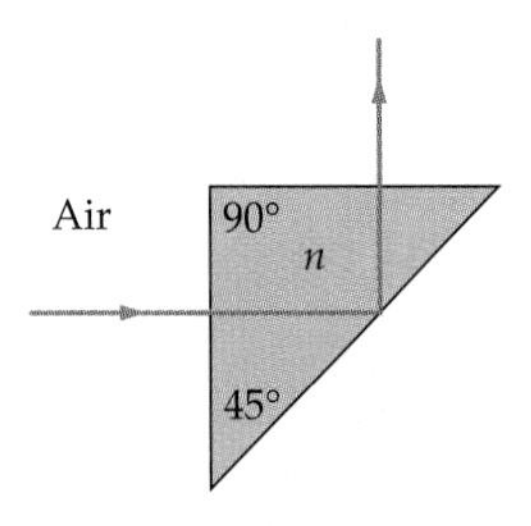

40 ••• Investigate how a thin film of water on a glass surface affects the critical angle for total reflection. Take $n = 1.5$ for glass and $n = 1.33$ for water. (*a*) What is the critical angle for total internal reflection at the glass–water interface? (*b*) Is there any range of incident angles that are greater than θ_c for glass-to-air refraction and for which light rays will leave the glass and the water and pass into the air?

41 ••• A laser beam is incident on a plate of glass of thickness 3 cm. The glass has an index of refraction of 1.5 and the angle of incidence is 40°. The top and bottom surfaces of the glass are parallel and both produce reflected beams of nearly the same intensity. What is the perpendicular distance d between the two adjacent reflected beams?

42 ••• Figure 33-56 shows a glass prism of index of refraction $n = 1.52$ in the shape of an isosceles triangle with base angles of 45°. (*a*) Find the maximum angle of incidence of the beam incident on the side face so that it suffers total internal reflection at the base. (*b*) What is the maximum value of the index of refraction of the prism so that the light beam will suffer total internal reflection at the base whatever the angle of incidence?

Figure 33-56 Problem 42

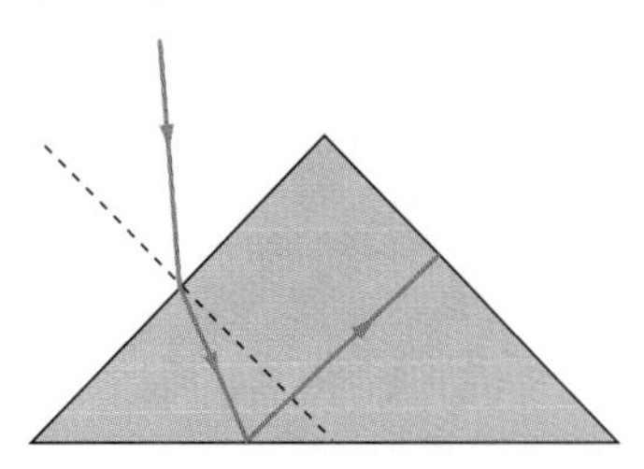

Dispersion

43 •• A beam of light strikes the plane surface of silicate flint glass at an angle of incidence of 45°. The index of refraction of the glass varies with wavelength as shown in the graph in Figure 33-26. How much smaller is the angle of refraction for violet light of wavelength 400 nm than that for red light of wavelength 700 nm?

44 •• Repeat Problem 43 for quartz.

Polarization

45 • Two polarizers have their transmission axes at an angle θ. Unpolarized light of intensity I is incident upon the first polarizer. What is the intensity of the light transmitted by the second polarizer?

(*a*) $I \cos^2 \theta$
(*b*) $(I \cos^2 \theta)/2$
(*c*) $(I \cos^2 \theta)/4$
(*d*) $I \cos \theta$
(*e*) $(I \cos \theta)/4$
(*f*) None of the above.

46 • Which of the following is *not* a phenomenon whereby polarized light can be produced from unpolarized light?

(*a*) absorption
(*b*) reflection
(*c*) birefringence
(*d*) diffraction
(*e*) scattering

47 • What is the polarizing angle for (*a*) water with $n = 1.33$ and (*b*) glass with $n = 1.5$?

48 • Light known to be polarized in the horizontal direction is incident on a polarizing sheet. It is observed that only 15.0% of the intensity of the incident light is transmitted through the sheet. What angle does the transmission axis of the sheet make with the horizontal?

(*a*) 8.6°
(*b*) 21°
(*c*) 23°
(*d*) 67°
(*e*) 81°

49 • Two polarizing sheets have their transmission axes crossed so that no light gets through. A third sheet is inserted between the first two so that its transmission axis makes an angle θ with that of the first sheet. Unpolarized light of intensity I_0 is incident on the first sheet. Find the intensity of the light transmitted through all three sheets if (*a*) $\theta = 45°$ and (*b*) $\theta = 30°$.

50 •• The polarizing angle for a certain substance is 60°. (*a*) What is the angle of refraction of light incident at this angle? (*b*) What is the index of refraction of this substance?

51 •• Two polarizing sheets have their transmission axes crossed and a third sheet is inserted so that its transmission axis makes an angle θ with that of the first sheet as in Problem 49. Find the intensity of the transmitted light as a function of θ. Show that the intensity transmitted through all three sheets is maximum when $\theta = 45°$.

52 •• If the middle polarizing sheet in Problem 51 is rotating at an angular velocity ω about an axis parallel to the light beam, find the intensity transmitted through all three sheets as a function of time. Assume that $\theta = 0$ at time $t = 0$.

53 •• A stack of $N + 1$ ideal polarizing sheets is arranged with each sheet rotated by an angle of $\pi/2N$ rad with respect to the preceding sheet. A plane linearly polarized light wave of intensity I_0 is incident normally on the stack. The incident light is polarized along the transmission axis of the first sheet and therefore perpendicular to the transmission axis of the last sheet in the stack. (*a*) What is the transmitted intensity through the stack? (*b*) For 3 sheets ($N = 2$), what is the transmitted intensity? (*c*) For 101 sheets, what is the transmitted intensity? (*d*) What is the direction of polarization of the transmitted beam in each case?

54 •• Show that a linearly polarized wave can be thought of as a superposition of a right and a left circularly polarized wave.

55 •• Suppose that in Problem 49 the middle sheet is replaced by two polarizing sheets. If the angles between the directions of polarization of adjacent sheets is 30°, what is the intensity of the transmitted light? How does this compare with the intensity obtained in Problem 49*a*?

56 •• In a circularly polarized wave, the magnitude of the electric field is constant. If the wave propagates along the z axis, the angle between $\vec{E}$ and the x axis changes by 2π radians over one wavelength. Write expressions for the electric and magnetic fields of a circularly polarized wave of angular frequency ω propagating in vacuum in the positive z direction.

57 •• Show that the electric field of a circularly polarized wave propagating in the x direction can be expressed by

$$\vec{E} = E_0 \sin(kx - \omega t)\,\hat{j} + E_0 \cos(kx - \omega t)\,\hat{k}$$

58 •• For the wave whose electric field is given by the expression in Problem 57, what is the corresponding expression for the magnetic field $\vec{B}$?

59 •• Find expressions for the electric field $\vec{E}$ and magnetic field $\vec{B}$ for a circularly polarized wave propagating in the negative z direction. (See Problems 57 and 58.)

60 •• A circularly polarized wave is said to be *right circularly polarized* if the electric and magnetic fields rotate clockwise when viewed along the direction of propagation and *left circularly polarized* if the field rotate counterclockwise. What is the sense of the circular polarization for the wave described by the expression in Problem 57? What would be the corresponding expression for a circularly polarized wave of the opposite sense?

61 •• Vertically polarized light of intensity I_0 is incident on a stack of N ideal polarizing sheets whose angles with respect to the vertical are $\theta_n = n\pi/2N$. Determine the direction of polarization of the transmitted light and its intensity. Show that as $N \to \infty$ the direction of polarization is rotated without loss of intensity.

General Problems

62 • True or false:

(*a*) Light and radio waves travel with the same speed through a vacuum.
(*b*) Most of the light incident normally on an air–glass interface is reflected.
(*c*) The angle of refraction of light is always less than the angle of incidence.
(*d*) The index of refraction of water is the same for all wavelengths in the visible spectrum.
(*e*) Longitudinal waves cannot be polarized.

63 •• Of the following statements about the speeds of the various colors of light in glass, which are true?

(*a*) All colors of light have the same speed in glass.
(*b*) Violet has the highest speed, red the lowest.
(*c*) Red has the highest speed, violet the lowest.
(*d*) Green has the highest speed, red and violet the lowest.
(*e*) Red and violet have the highest speed, green the lowest.

64 •• It is a common experience that on a calm, sunny day one can hear voices of persons in a boat over great distances. Explain this phenomenon, keeping in mind that sound is reflected from the surface of the water and that the temperature of the air just above the water's surface is usually less than that at a height of 10 or 20 m above the water.

65 • A beam of monochromatic red light with a wavelength of 700 nm in air travels in water. (*a*) What is the wavelength in water? (*b*) Does a swimmer underwater observe the same color or a different color for this light?

66 • As the speed of computer operations increases, computer architecture acquires greater importance; the time required to transfer a signal between the central processing unit (CPU) and memory can be a limiting factor in determining the time required for computation. What is the maximum separation between a memory chip and the CPU to allow transfer information between these units in less than 0.5 ns?

67 •• The critical angle for total internal reflection for a substance is 45°. What is the polarizing angle for this substance?

68 •• Figure 33-57 shows two plane mirrors that make an angle θ with each other. Show that the angle between the incident and reflected rays is 2θ.

Figure 33-57 Problem 68

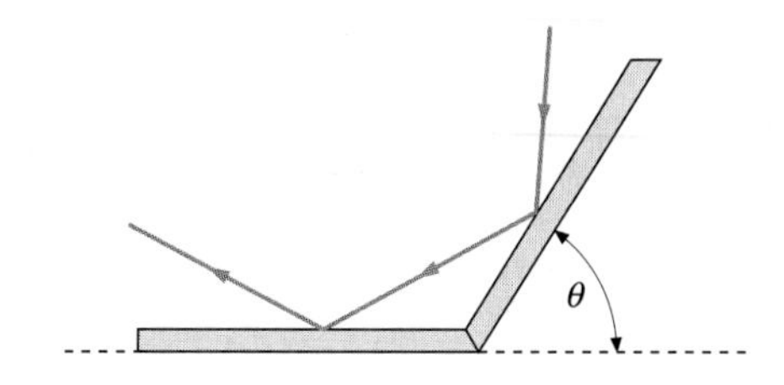

69 •• A silver coin sits on the bottom of a swimming pool that is 4 m deep. A beam of light reflected from the coin emerges from the pool making an angle of 20° with respect to the water's surface and enters the eye of an observer. Draw a ray from the coin to the eye of the observer. Extend this ray, which goes from the water–air interface to the eye, straight back until it intersects with the vertical line drawn through the coin. What is the apparent depth of the swimming pool to this observer?

70 •• Two affluent students decide to improve on Galileo's experiment to measure the speed of light. One student goes to London and calls the other in New York on the telephone. The telephone signals are transmitted by reflecting electromagnetic waves from a satellite that is 37.9 Mm above the earth's surface. If the distance between London and New York is neglected, the distance traveled is twice this distance. One student claps his hands, and when the other student hears the sound over the phone, she claps her hands. The first student measures the time between his clap and his hearing the second one. Calculate this time lapse, neglecting the students' response times. Do you think this experiment would be successful? What improvements for measuring this time interval would you suggest? (Time delays in the electronic circuits that are greater than those due to the light traveling to the satellite and back make this experiment not feasible.)

71 •• Fishermen always insist on silence because noise on shore will scare fish away. Suppose a fisherman cast a baited hook 20 m from the shore of a calm lake to a point where the depth is 15 m. Show that noise on shore cannot possibly be sensed by fish at that point. *Note:* The speed of sound in air is 330 m/s; the speed of sound in water is 1450 m/s.

72 •• A swimmer at the bottom of a pool 3 m deep looks up and sees a circle of light. If the index of refraction of the water in the pool is 1.33, find the radius of the circle.

73 •• Show that when a mirror is rotated through an angle θ, the reflected beam of light is rotated through 2θ.

74 •• Use Figure 33-26 to calculate the critical angles for total internal reflection for light initially in silicate flint glass that is incident on a glass–air interface if the light is (*a*) violet light of wavelength 400 nm, and (*b*) red light of wavelength 700 nm.

75 •• Show that for normally incident light, the intensity transmitted through a glass slab with an index of refraction of n is approximately given by

$$I_T = I_0\left[\frac{4n}{(n+1)^2}\right]^2$$

76 •• A ray of light begins at the point $x = -2$ m, $y = 2$ m, strikes a mirror in the xz plane at some point x, and reflects through the point $x = 2$ m, $y = 6$ m. (*a*) Find the value of x that makes the total distance traveled by the ray a minimum. (*b*) What is the angle of incidence on the reflecting plane? What is the angle of reflection?

77 •• Light passes symmetrically through a prism having an apex angle of α as shown in Figure 33-58. (*a*) Show that the angle of deviation δ is given by

$$\sin\frac{\alpha+\delta}{2} = n\sin\frac{\alpha}{2}$$

(*b*) If the refractive index for red light is 1.48 and that for violet light is 1.52, what is the angular separation of visible light for a prism with an apex angle of 60°?

Figure 33-58 Problems 77, 88, and 89

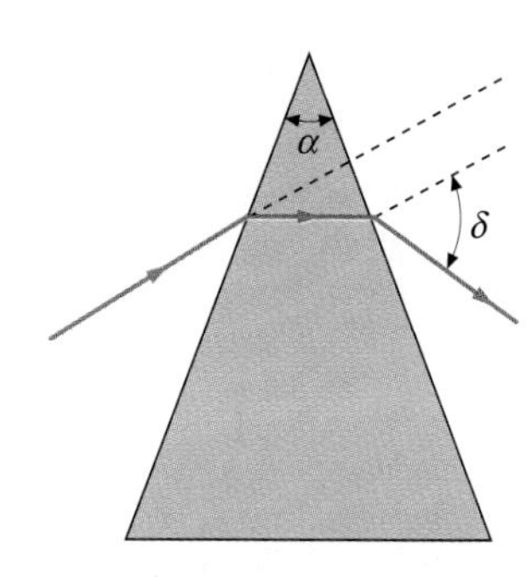

78 •• (*a*) For a light ray inside a transparent medium having a planar interface with a vacuum, show that the polarizing angle and the critical angle for internal reflection satisfy $\tan\theta_p = \sin\theta_c$. (*b*) Which angle is larger?

79 •• Light is incident from air on a transparent substance at an angle of 58.0° with the normal. The reflected and refracted rays are observed to be mutually perpendicular. (*a*) What is the index of refraction of the transparent substance? (*b*) What is the critical angle for total internal reflection in this substance?

80 •• A light ray in dense flint glass with an index of refraction of 1.655 is incident on the glass surface. An unknown liquid condenses on the surface of the glass. Total internal reflection on the glass–liquid interface occurs for an angle of incidence on the glass–liquid interface of 53.7°. (*a*) What is the refractive index of the unknown liquid? (*b*) If the liquid is removed, what is the angle of incidence for total internal reflection? (*c*) For the angle of incidence found in part (*b*), what is the angle of refraction of the ray into the liquid film? Does a ray emerge from the liquid film into the air above? Assume that the glass and liquid have perfect planar surfaces.

81 •• Given that the index of refraction for red light in water is 1.3318 and that the index of refraction for blue light in water is 1.3435, find the angular separation of these colors in the primary rainbow. (Use the equation given in Problem 86.)

82 •• A ray of light falls on a rectangular glass block ($n = 1.5$) that is almost completely submerged in water ($n = 1.33$) as shown in Figure 33-59. (*a*) Find the angle θ for which total internal reflection just occurs at point P. (*b*) Would total internal reflection occur at point P for the value of θ found in part (*a*) if the water were removed? Explain.

Figure 33-59 Problem 82

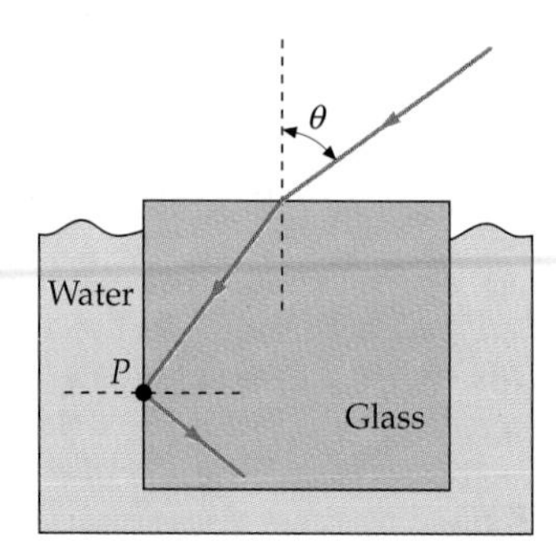

83 •• (*a*) Use the result for Problem 75 to find the ratio of the transmitted intensity to the incident intensity through N

parallel slabs of glass for light of normal incidence. (*b*) Find this ratio for three slabs of glass with $n = 1.5$. (*c*) How many slabs of glass with $n = 1.5$ will reduce the intensity to 10% of the incident intensity?

84 •• Light is incident on a slab of transparent material at an angle θ_1 as shown in Figure 33-60. The slab has a thickness t and an index of refraction n. Show that

$$n = \frac{\sin \theta_1}{\sin[\arctan(d/t)]}$$

where d is the distance shown in the figure and $\arctan(d/t)$ is the angle whose tangent is d/t.

Figure 33-60 Problem 84

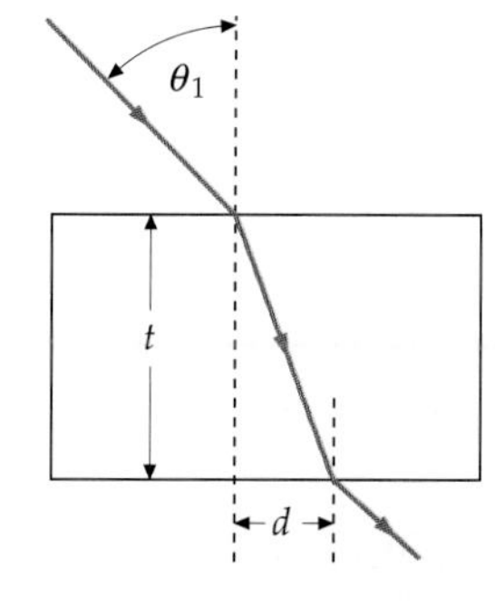

85 •• Suppose rain falls vertically from a stationary cloud 10,000 m above a confused marathoner running in a circle with constant speed of 4 m/s. The rain has a terminal speed of 9 m/s. (*a*) What is the angle that the rain appears to make with the vertical to the marathoner? (*b*) What is the apparent motion of the cloud as observed by the marathoner? (*c*) A star on the axis of the earth's orbit appears to have a circular orbit of angular diameter of 41.2 seconds of arc. How is this angle related to the earth's speed in its orbit and the velocity of photons received from this distant star? (*d*) What is the speed of light as determined from the data in part (*c*)?

86 ••• Equation 33-18 gives the relation between the angle of deviation ϕ_d of a light ray incident on a spherical drop of water in terms of the incident angle θ_1 and the index of refraction of water. (*a*) Assume that $n_{air} = 1$, and differentiate ϕ_d with respect to θ_1. [*Hint:* If $y = \arcsin x$, then $dy/dx = (1 - x^2)^{-1/2}$.] (*b*) Set $d\phi_d/d\theta_1 = 0$ and show that the angle of incidence θ_{1m} for minimum deviation is given by

$$\cos \theta_{1m} = \sqrt{\frac{n^2 - 1}{3}}$$

and find θ_{1m} for water, where the index of refraction for water is 1.33.

87 ••• (*a*) Show that a light ray transmitted through a glass slab emerges parallel to the incident ray but displaced from it. (*b*) For an incident angle of 60°, glass of index of refraction $n = 1.5$, and a slab of thickness 10 cm, find the displacement measured perpendicularly from the incident ray.

88 ••• Show that if the apex angle α of the prism of Problem 77 is small, the angle of deviation δ is given by $\delta = (n - 1)\alpha$, independent of the angle of incidence.

89 ••• Show that the angle of deviation δ is a minimum if the angle of incidence is such that the ray passes through the prism symmetrically as shown in Figure 33-58.

CHAPTER 34

Optical Images

The focusing of rays by reflection and refraction is illustrated by these laser beams incident on a glass lens.

Because the wavelength of light is very small compared with most obstacles and openings, diffraction—the bending of waves around corners—is often negligible, and the ray approximation, in which waves are considered to propagate in straight lines, is valid. In this chapter we apply the laws of reflection and refraction to the formation of images by mirrors and lenses.

34-1 Mirrors

Plane Mirrors

Figure 34-1 shows a bundle of light rays emanating from a point source P and reflected from a plane mirror. After reflection, the rays diverge exactly as if they came from a point P' behind the plane of the mirror. The point P' is called the **image** of the **object** P. When these reflected rays enter the eye, they cannot be distinguished from rays diverging from a source at P' with no mirror present. This image is called a **virtual image** because the light does not actually emanate from it. The image point P' lies on the line through the ob-

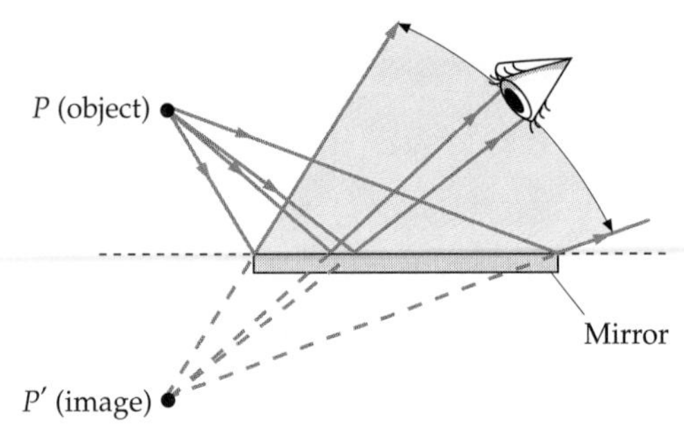

Figure 34-1 Image formed by a plane mirror. The rays from point P that strike the mirror and enter the eye appear to come from the image point P' behind the mirror. The image can be seen by the eye placed anywhere in the shaded region.

ject P perpendicular to the plane of the mirror at a distance behind the plane equal to that from the plane to the object, as shown in Figure 34-1. The image can be seen by an eye anywhere in the shaded region indicated, in which a straight line from the image to the eye passes through the mirror. The object need not be directly in front of the mirror. As long as the object is not behind the plane of the mirror, there is some position at which the eye can be placed to view the image.

If you hold up your right hand and look in the mirror, the image you see is the same size, but it looks like a left hand (Figure 34-2). This right-to-left reversal is a result of **depth inversion**—the hand is transformed from a right hand to a left hand because the front and back of the hand are reversed by the mirror. Depth inversion is also illustrated in Figure 34-3. Figure 34-4 shows the image of a simple rectangular coordinate system. The mirror transforms a right-handed coordinate system for which $\hat{i} \times \hat{j} = \hat{k}$, into a left-handed coordinate system for which $\hat{i} \times \hat{j} = -\hat{k}$.

Figure 34-2 The image of a right hand in a plane mirror is a left hand. This right-to-left reversal is a result of depth inversion.

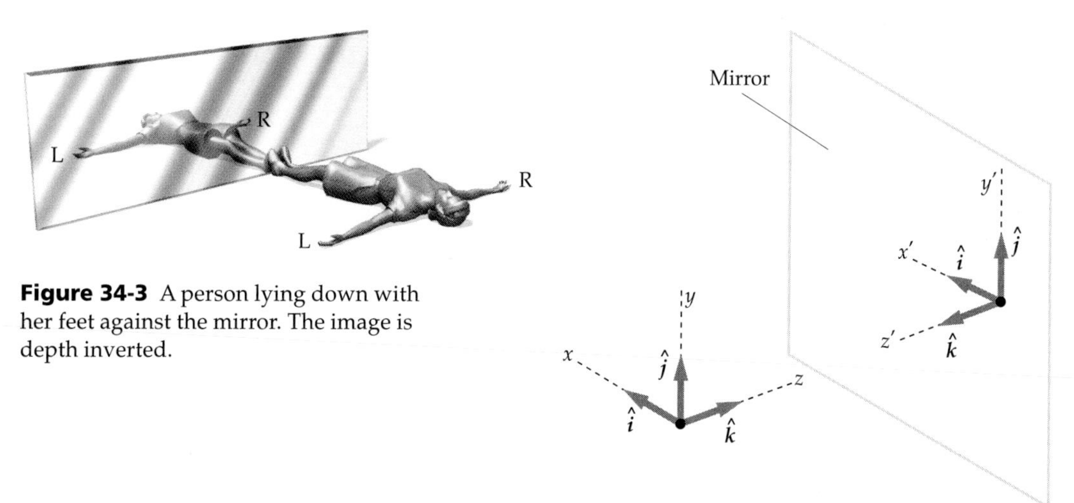

Figure 34-3 A person lying down with her feet against the mirror. The image is depth inverted.

Figure 34-4 Image of a rectangular coordinate system in a plane mirror. The arrow along the z axis is reversed in the image. The image of the original right-handed coordinate system, for which $\hat{i} \times \hat{j} = \hat{k}$, is a left-handed coordinate system, for which $\hat{i} \times \hat{j} = -\hat{k}$.

Figure 34-5 shows an arrow of height y standing parallel to a plane mirror a distance s from it. We can locate the image of the arrowhead (and of any other point on the arrow) by drawing two rays. One ray, drawn perpendicular to the mirror, hits the mirror at point A and is reflected back onto itself. The other ray, making an angle θ with the normal to the mirror, is reflected, making an equal angle θ with the x axis. The extension of these two rays back behind the mirror locates the image of the arrowhead, as shown by the dashed lines in the figure. We can see from this figure that the image is the same distance behind the mirror as the object is in front of the mirror, and that the image is erect (points in the same direction as the object) and is the same size as the object.

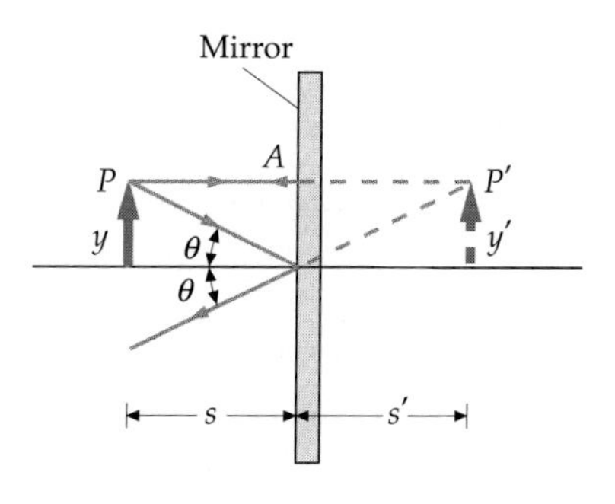

Figure 34-5 Ray diagram for locating the image of an arrow in a plane mirror.

The formation of multiple images by two plane mirrors making an angle with each other is illustrated in Figure 34-6. We frequently see this phenomenon in clothing stores that provide adjacent mirrors. Light reflected from mirror 1 strikes mirror 2 just as if it came from the image point P_1. The image P_1' is the object for mirror 2. Its image is at point $P_{1,2}''$. This image will be formed whenever the image point P_1 is in front of the plane of mirror 2. The image at point P_2' is due to rays from the object that reflect directly from mirror 2. Since P_2' is behind the plane of mirror 1, it cannot serve as an object point for a further image in mirror 1. The number of multiple images formed by two mirrors depends on the angle between the mirrors and the position of the object.

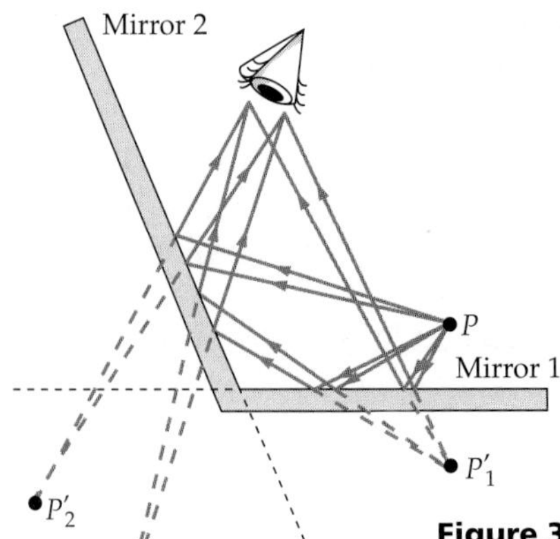

Figure 34-6 Images formed by two plane mirrors. P_1' is the image of the object P in mirror 1, and P_2' is the image of the object in mirror 2. Point $P_{1,2}''$ is the image of P_1' in mirror 2 seen when light rays from the object reflect first from mirror 1 and then from mirror 2. The image P_2' does not have an image in mirror 1 because it is behind that mirror.

Figure 34-7 shows two mirrors at right angles to each other. Rays from the object to the eye that strike mirror 1 and then mirror 2 are shown in Figure 34-7*a*. In this case, the image point $P_{1,2}''$ is the same as that for rays that strike mirror 2 first and then mirror 1, as can be seen from Figure 34-7*b*. If you stand in front of two vertical mirrors that are perpendicular to each other, such as in the corner of a room, the image you see is the same as that seen by others who are facing you because depth inversion occurs twice, once in each mirror.

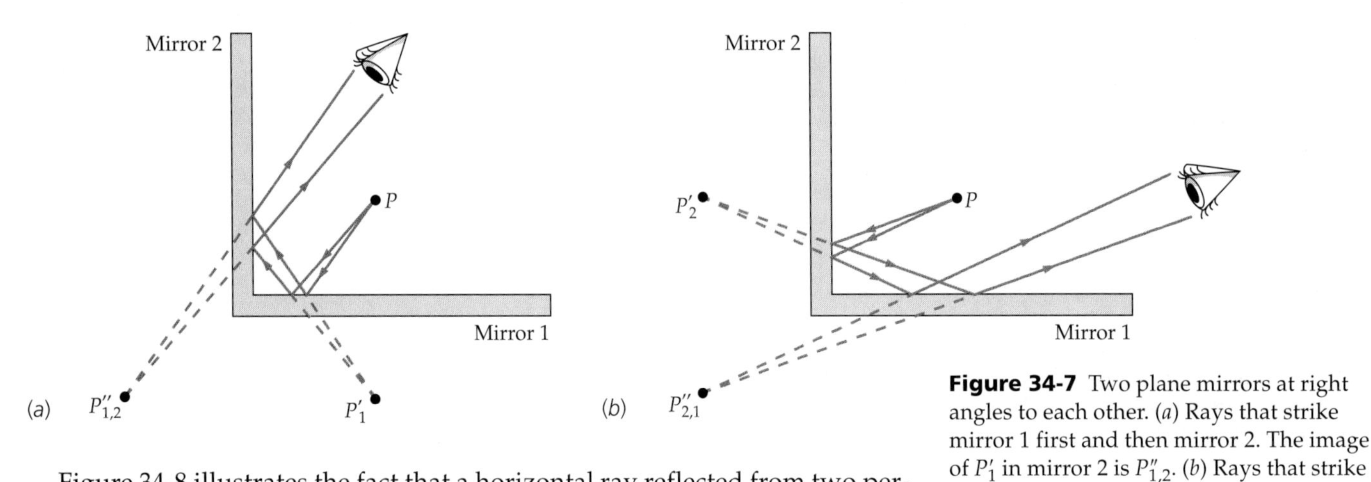

Figure 34-7 Two plane mirrors at right angles to each other. (*a*) Rays that strike mirror 1 first and then mirror 2. The image of P_1' in mirror 2 is $P_{1,2}''$. (*b*) Rays that strike mirror 2 first and then mirror 1. The image of P_2' in mirror 1 is $P_{2,1}''$, which coincides with $P_{1,2}''$ for perpendicular mirrors.

Figure 34-8 illustrates the fact that a horizontal ray reflected from two perpendicular vertical mirrors is exactly reversed in direction no matter what angle the ray makes with the mirrors. If three mirrors are placed perpendicular to each other like the sides of an inside corner of a box, any ray incident on any of the mirrors from any direction is exactly reversed. A set of mirrors of this type was placed on the moon facing the earth. A laser beam from earth directed at the mirrors is reflected back to the same place on the earth. Such a beam has been used to measure the distance to the mirrors to within a few centimeters by measuring the time it takes for the light to reach the mirrors and return.

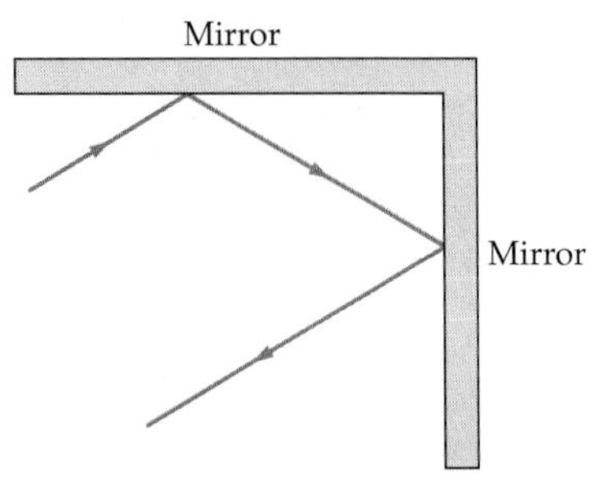

Figure 34-8 A ray striking one of two perpendicular plane mirrors is reflected from the second mirror in the direction opposite the original direction for any angle of incidence.

Spherical Mirrors

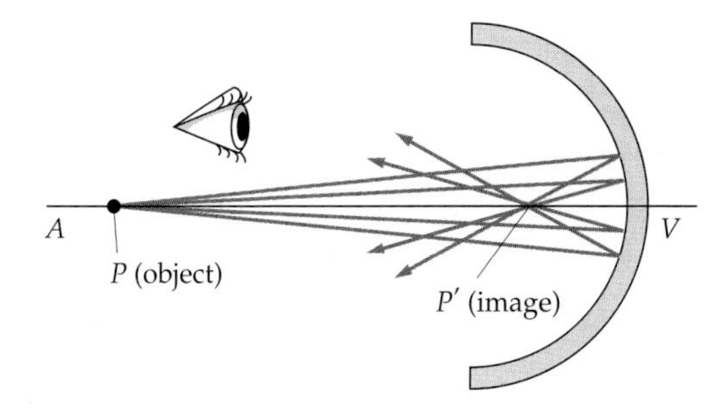

Figure 34-9 Rays from a point object P on the axis AV of a concave spherical mirror form an image at P'. The image is sharp if the rays strike the mirror near the axis.

Figure 34-9 shows a bundle of rays from a point source P on the axis of a concave spherical mirror reflecting from the mirror and converging at point P'. The rays then diverge from this point just as if there were an object at that point. This image is called a **real image,** because light actually does emanate from the image point. The image can be seen by an eye at the left of the image looking into the mirror. It could also be observed on a ground-glass viewing screen or photographic film placed at the image point. A virtual image, such as that formed by a plane mirror as discussed in the previous section, cannot be observed on a screen at the image point because there is no light there. Despite this distinction between real and virtual images, the eye makes no distinction between them. The light rays diverging from a real image and those appearing to diverge from a virtual image are the same to the eye.

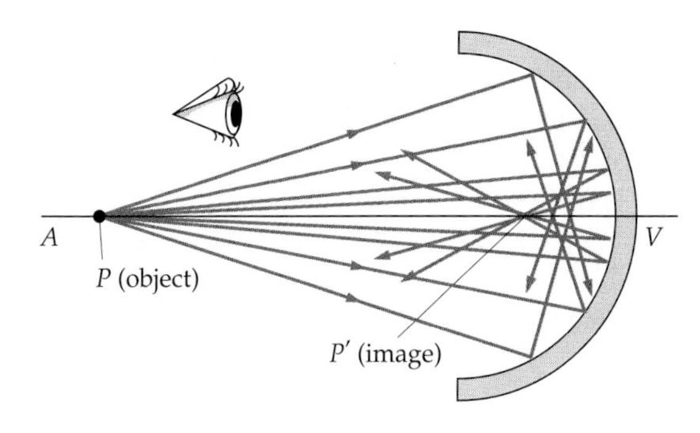

Figure 34-10 Spherical aberration of a mirror. Nonparaxial rays that strike the mirror at points far from the axis AV are not reflected through the image point P'. These rays blur the image.

From Figure 34-10, we can see that only rays that strike the spherical mirror at points near the axis AV are reflected through the image point. Such rays are called **paraxial rays.** Rays that strike the mirror at points far from the axis, called *nonparaxial rays,* converge to different points near the image point. Such rays cause the image to appear blurred, an effect called **spherical aberration.** The image can be sharpened by blocking off all but the central part of the mirror so that nonparaxial rays do not strike it. The image is then sharper, but its brightness is reduced because less light is reflected to the image point.

Figure 34-11 shows a ray from an object point P reflecting off the mirror and passing through the image point P'. Point C is the center of curvature of the mirror. The incident and reflected rays make equal angles with the radial line CA, which is perpendicular to the surface of the mirror.

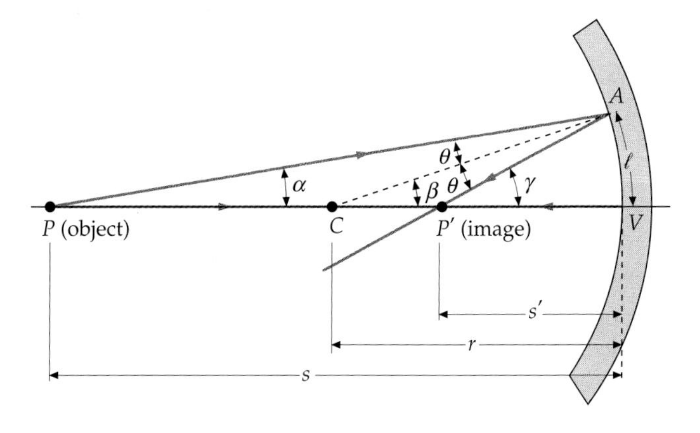

Figure 34-11 Geometry for calculating the image distance s' from the object distance s and the radius of curvature r. The angle β is an exterior angle to the triangle PAC, therefore $\beta = \alpha + \theta$. Similarly, from the triangle PAP', $\gamma = \alpha + 2\theta$. Eliminating θ from these equations gives $2\beta = \alpha + \gamma$. Equation 34-1 follows directly if we assume the following small-angle approximations: $\alpha \approx \ell/s$, $\beta \approx \ell/r$, and $\gamma \approx \ell/s'$.

The image distance s' from the vertex of the mirror V to P' can be related to the object distance s from the vertex V to point P and the radius of curvature r of the mirror by elementary geometry. The result is

$$\frac{1}{s} + \frac{1}{s'} = \frac{2}{r} \qquad \text{34-1}$$

The derivation of this equation assumes that the rays are paraxial.

When the object distance is large compared with the radius of curvature of the mirror, the term $1/s$ in Equation 34-1 is much smaller than $2/r$ and can be neglected. For $s = \infty$, the image distance is $s' = \frac{1}{2}r$. This distance is called the **focal length** f of the mirror, and the point at which parallel rays incident on the mirror are focused is called the **focal point** F as illustrated in Figure 34-12*a*. (Again, only paraxial rays are focused at a single point.)

$$f = \frac{1}{2}r \tag{34-2}$$

Focal length for a mirror

The focal length of a spherical mirror is half the radius of curvature. In terms of the focal length f, Equation 34-1 is

$$\frac{1}{s} + \frac{1}{s'} = \frac{1}{f} \tag{34-3}$$

Mirror equation

Equation 34-3 is called the **mirror equation**.

When an object is very far from the mirror, the rays are parallel, and the wavefronts are approximately planes (Figure 34-12*b*). In Figure 34-12*b*, note how the edges of the wavefront hit the concave mirror surface before the central portion near the axis, resulting in a spherical wavefront upon reflection. Figure 34-13 shows the wavefronts and rays for plane waves striking a convex mirror. In this case, the central part of the wavefront strikes the mirror first, and the reflected waves appear to come from the focal point behind the mirror.

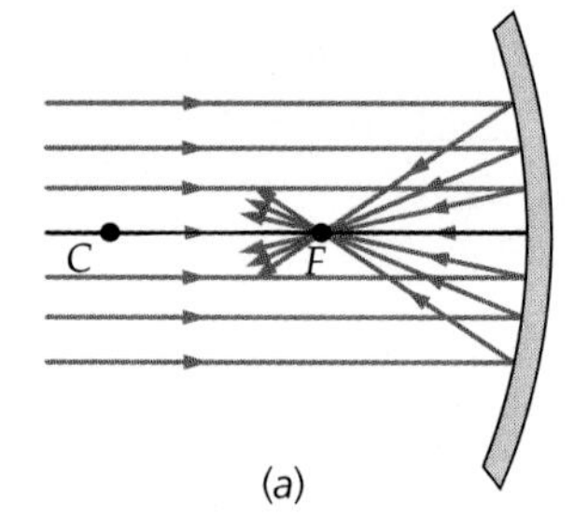

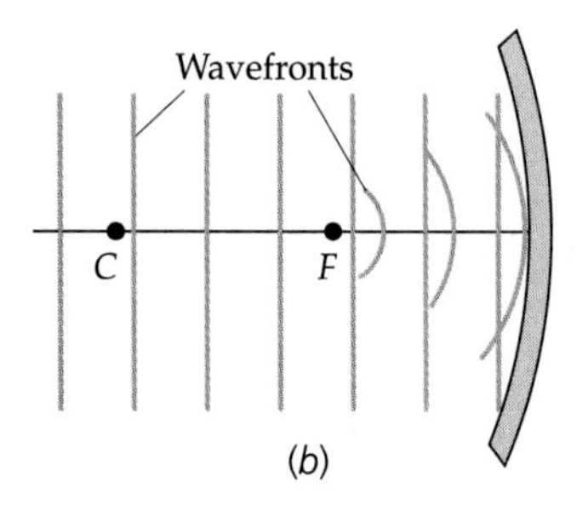

Figure 34-12 (*a*) Parallel rays strike a concave mirror and are reflected through the focal point F at a distance $r/2$. (*b*) The incoming wavefronts are plane waves; upon reflection, they become spherical waves that converge at the focal point.

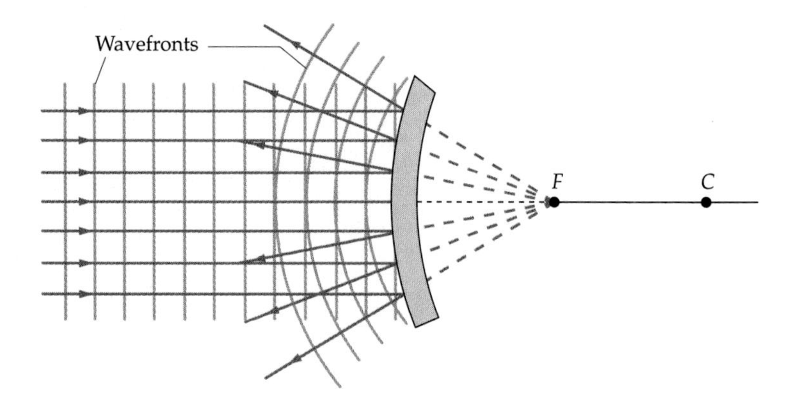

Figure 34-13 Reflection of plane waves from a convex mirror. The outgoing wavefronts are spherical as if emanating from the focal point F behind the mirror. The rays are perpendicular to the wavefronts, and appear to diverge from F.

Figure 34-14 illustrates a property of waves called **reversibility.** If we reverse the direction of a reflected ray, the law of reflection assures us that the reflected ray will be along the original incoming ray but in the opposite direction. (Reversibility holds also for refracted rays, which are discussed in later sections.) Thus, if we have a real image of an object formed by a reflecting (or refracting) surface, we can place an object at the image point and a new image will be formed at the position of the original object.

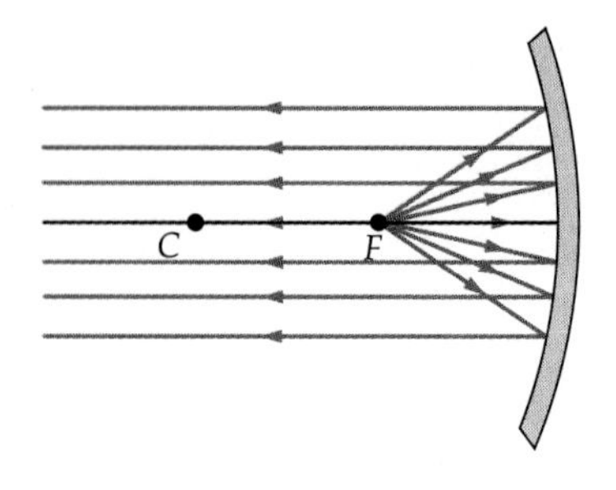

Figure 34-14 Reversibility. Rays diverging from a point source at the focal point of a concave mirror are reflected from the mirror as parallel rays. The rays are the same as in Figure 34-12*a* but in the reverse direction.

Example 34-1

An object is 12 cm from a concave mirror with a radius of curvature of 6 cm. Find (*a*) the focal length of the mirror and (*b*) the image distance.

Picture the Problem The focal length of a spherical mirror is half the radius of curvature. Once the focal length is known, the image distance can be found using the mirror equation (Equation 34-3).

(*a*) The focal length is half the radius of curvature:	$f = \frac{1}{2}r = \frac{1}{2}(6\text{ cm}) = 3\text{ cm}$
(*b*)1. Use the mirror equation to find a relation for the image distance s':	$\frac{1}{s} + \frac{1}{s'} = \frac{1}{f}$ or $\frac{1}{12\text{ cm}} + \frac{1}{s'} = \frac{1}{3\text{ cm}}$
2. Solve for s':	$\frac{1}{s'} = \frac{4}{12\text{ cm}} - \frac{1}{12\text{ cm}} = \frac{3}{12\text{ cm}}$ $s' = 4\text{ cm}$

Exercise A concave mirror has a focal length of 4 cm. (*a*) What is its radius of curvature? (*b*) Find the image distance for an object 2 cm from the mirror. (*Answers* (*a*) 8 cm; (*b*) $s' = -4$ cm)

Ray Diagrams for Mirrors

A useful method to locate images is by geometric construction of a **ray diagram**, as illustrated in Figure 34-15, where the object is a human figure perpendicular to the axis a distance s from the mirror. By the judicious choice of rays from the head of the figure, we can quickly locate the image. There are three **principal rays** that are convenient to use:

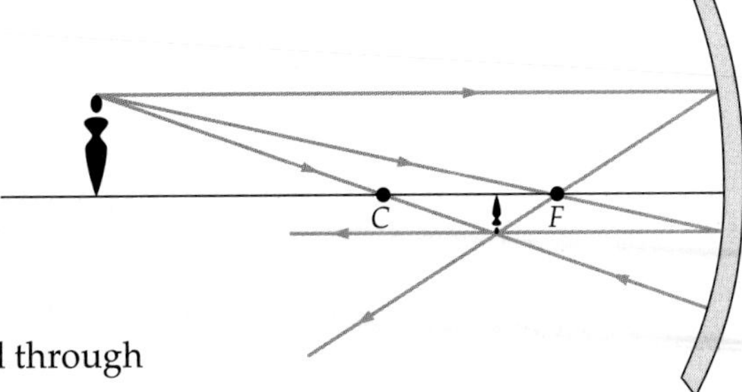

Figure 34-15 Ray diagram for the location of the image by geometric construction.

1. The **parallel ray,** drawn parallel to the axis. This ray is reflected through the focal point.
2. The **focal ray,** drawn through the focal point. This ray is reflected parallel to the axis.
3. The **radial ray,** drawn through the center of curvature. This ray strikes the mirror perpendicular to its surface and is thus reflected back on itself.

Principal rays for a mirror

These rays are shown in Figure 34-15. The intersection of any two rays locates the image point of the head. The third ray can be used to provide a check. Ray diagrams are easier to draw if the mirror is replaced by a straight line that extends as far as necessary to intercept the rays, as shown in Figure 34-16. Note that the image in this case is inverted and smaller than the object.

When the object is between the mirror and its focal point, the rays reflected from the mirror do not converge but appear to diverge from a point

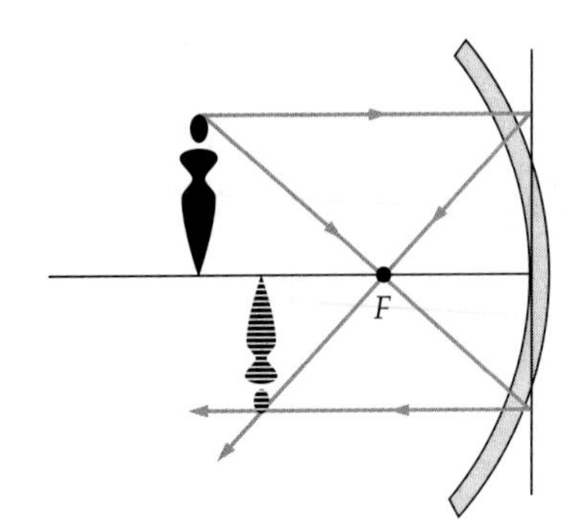

Figure 34-16 Ray diagrams are easier to construct if the curved surface is replaced by a plane.

behind the mirror, as illustrated in Figure 34-17. In this case, the image is virtual and erect (*erect* meaning not inverted relative to the object). For an object between the mirror and the focal point, s is less than $\frac{1}{2}r$, so the image distance s' calculated from Equation 34-1 turns out to be negative. We can apply Equations 34-1, 34-2, and 34-3 to this case and to convex mirrors if we adopt a convenient sign convention. Whether the mirror is convex or concave, real images can be formed only in front of the mirror, that is, on the same side of the mirror as the object. Virtual images are formed behind the mirror where there are no actual light rays. Our sign convention is as follows:

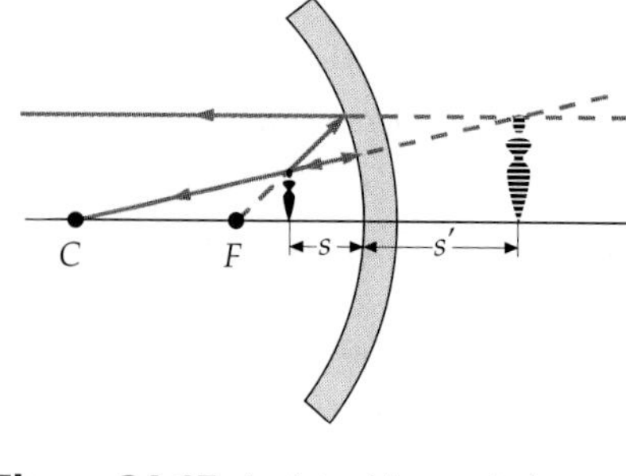

Figure 34-17 A virtual image is formed by a concave mirror when the object is inside the focal point. Here the image is located by the radial ray, which is reflected back on itself, and the focal ray, which is reflected parallel to the axis. These rays appear to diverge from a point behind the mirror found by extending them.

s	+ if the object is in front of the mirror (real object) − if the object is behind the mirror (virtual object)*
s'	+ if the image is in front of the mirror (real image) − if the image is behind the mirror (virtual image)
r, f	+ if the center of curvature is in front of the mirror (concave mirror) − if the center of curvature is behind the mirror (convex mirror)

Sign conventions for reflection

With these sign conventions, Equations 34-1, 34-2, and 34-3 can be used for all situations with any type of mirror.

The ratio of the image size to the object size is defined as the **lateral magnification** of the image. From Figure 34-18 we see that the lateral magnification is

$$m = \frac{y'}{y} = -\frac{s'}{s} \qquad 34\text{-}4$$

Lateral magnification

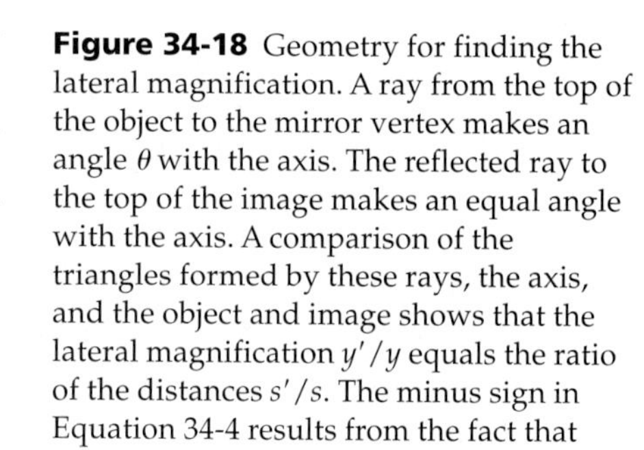

Figure 34-18 Geometry for finding the lateral magnification. A ray from the top of the object to the mirror vertex makes an angle θ with the axis. The reflected ray to the top of the image makes an equal angle with the axis. A comparison of the triangles formed by these rays, the axis, and the object and image shows that the lateral magnification y'/y equals the ratio of the distances s'/s. The minus sign in Equation 34-4 results from the fact that y'/y is negative when s and s' are both positive.

A negative magnification, which occurs when both s and s' are positive, indicates that the image is inverted.

For plane mirrors, the radius of curvature is infinite. The focal length given by Equation 34-2 is then also infinite. Equation 34-3 then gives $s' = -s$, indicating that the image is behind the mirror at a distance equal to the object distance. The magnification given by Equation 34-4 is then +1, indicating that the image is erect and the same size as the object.

Although the preceding equations coupled with our sign conventions are relatively easy to use, we often need to know only the approximate location and magnification of the image and whether it is real or virtual, and erect or inverted. This knowledge is usually easiest to obtain by constructing a ray diagram. It is always a good idea to use both the graphical method and the algebraic method to locate an image so that one method serves as a check on the results of the other.

Convex Mirrors Figure 34-19 shows a ray diagram for an object in front of a convex mirror. The central ray heading toward the center of curvature C is perpendicular to the mirror and is reflected back on itself. The parallel ray is reflected as if it came from the focal point F behind the mirror. The focal ray (not shown) would be drawn toward the focal point and would be reflected parallel to the axis. We can see from the figure that the image is behind the mirror and is therefore virtual. It is also erect and smaller than the object.

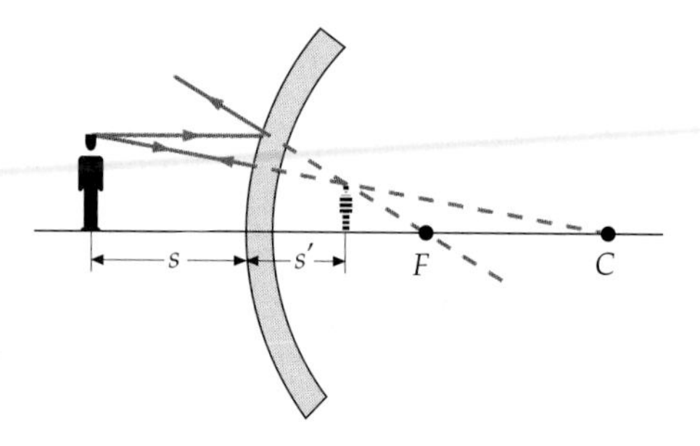

Figure 34-19 Ray diagram for a convex mirror.

* You may wonder how an object can be behind a mirror. This occurs when there is a lens in front of the mirror and the rays to the image of the lens are intercepted by the mirror. The image of the lens is then never formed, but the distance to the unformed image behind the mirror is taken as the object distance for the mirror, and the object is called a virtual object. We will discuss examples of this in Section 34-2 when we discuss lenses.

Example 34-2

An object 2 cm high is 10 cm from a convex mirror with a radius of curvature of 10 cm. (*a*) Locate the image and (*b*) find its height.

Picture the Problem The ray diagram for this problem is the same as Figure 34-19. From this we see that the image is erect, virtual, and smaller than the object. To find the exact location and size, we use the mirror equation with $s = 10$ cm and $r = -10$ cm.

(*a*)1. The image distance s' is related to the object distance s and the focal length f by the mirror equation:

$$\frac{1}{s} + \frac{1}{s'} = \frac{1}{f}$$

2. Calculate the focal length of the mirror:

$$f = \frac{1}{2}r = \frac{1}{2}(-10\text{ cm}) = -5\text{ cm}$$

3. Substitute $s = 10$ cm and $f = -5$ cm into the mirror equation to find the image distance:

$$\frac{1}{10\text{ cm}} + \frac{1}{s'} = \frac{1}{f} = \frac{1}{-5\text{ cm}}$$

4. Solve for s':

$$\frac{1}{s'} = -\frac{2}{10\text{ cm}} - \frac{1}{10\text{ cm}} = -\frac{3}{10\text{ cm}}$$

$$s' = -3.33\text{ cm}$$

(*b*)1. The height of the image is m times the height of the object:

$$y' = my$$

2. Calculate the magnification m:

$$m = -\frac{s'}{s} = -\frac{-3.33\text{ cm}}{10\text{ cm}} = +0.333$$

3. Use m to find the height of the image:

$$y' = my = (0.333)(2\text{ cm}) = 0.666\text{ cm}$$

Remarks The image distance is negative, indicating a virtual image behind the mirror. The magnification is positive, indicating that the image is erect.

Exercise Find the image distance and magnification for an object 5 cm away from the mirror in Example 34-2 and draw a ray diagram. (*Answers* $s' = -2.5$ cm, $m = +0.5$, the image is erect, virtual, and reduced in size.)

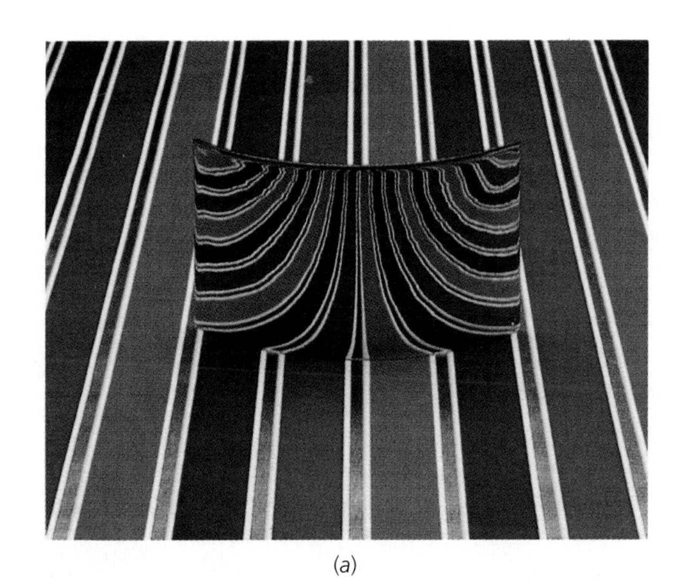

(*a*)

(*b*)

(*a*) A convex mirror resting on paper with equally spaced parallel stripes. Note the large number of lines imaged in a small space and the reduction in size and distortion in shape of the image. (*b*) A convex mirror is used for security in a store.

34-2 Lenses

Images Formed by Refraction

Figure 34-20 illustrates the formation of an image by refraction at a spherical surface separating two media with indexes of refraction n_1 and n_2. In this figure, n_2 is greater than n_1, so the waves travel more slowly in the second medium.

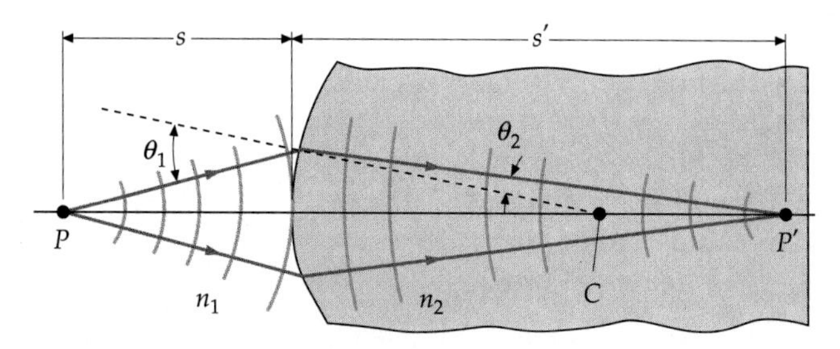

Figure 34-20 Image formed by refraction at a spherical surface between two media where the waves move slower in the second medium.

Again, only paraxial rays converge to one point. An equation relating the image distance to the object distance, the radius of curvature, and the indexes of refraction can be derived by applying Snell's law of refraction to these rays and using small-angle approximations. The geometry is shown in Figure 34-21. The result is

$$\frac{n_1}{s} + \frac{n_2}{s'} = \frac{n_2 - n_1}{r} \quad 34\text{-}5$$

Refraction at a single surface

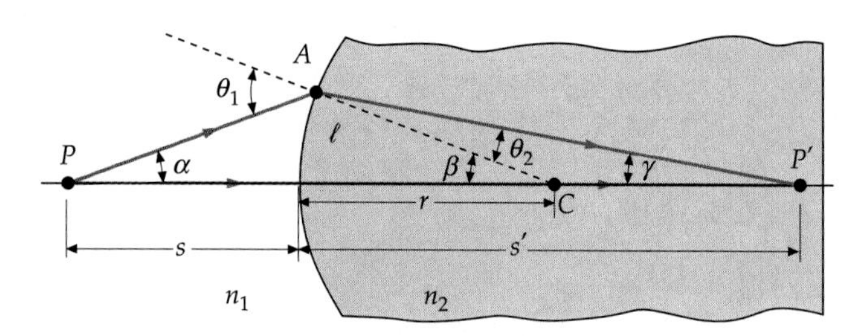

Figure 34-21 Geometry for relating the image position to the object position for refraction at a single spherical surface. The angles θ_1 and θ_2 are related by Snell's law: $n_1 \sin\theta_1 = n_2 \sin\theta_2$. The small-angle approximation $\sin\theta = \theta$ gives $n_1\theta_1 = n_2\theta_2$. From triangle ACP', we have $\beta = \theta_2 + \gamma = (n_1/n_2)\theta_1 + \gamma$. We can obtain another relation for θ_1 from triangle PAC: $\theta_1 = \alpha + \beta$. Eliminate θ_1 from these two equations: $n_1\alpha + n_1\beta + n_2\gamma = n_2\beta$. Simplify: $n_1\alpha + n_2\gamma = (n_2 - n_1)\beta$. Using the small-angle approximations $\alpha \approx \ell/s$, $\beta \approx \ell/r$, and $\gamma \approx \ell/s'$ gives Equation 34-5.

In refraction, real images are formed in back of the surface, which we will call the transmission side, whereas virtual images occur on the incident side in front of the surface. The sign conventions we use for refraction are similar to those for reflection:

s	+ (real object) for objects in front of the surface (incident side) − (virtual object) for objects in back of the surface (transmission side)
s'	+ (real image) for images in back of the surface (transmission side) − (virtual image) for images in front of the surface (incident side)
r, f	+ if the center of curvature is on the transmission side − if the center of curvature is on the incident side

Sign conventions for refraction

If we compare these sign conventions with those for reflection, we see that s' is positive and the image is real when the image is on the side of the surface traversed by the reflected or refracted light. For reflection, this side is in front of the mirror, whereas for refraction, it is behind the refracting surface. Similarly, r and f are positive when the center of curvature is on the side traversed by the reflected or refracted light.

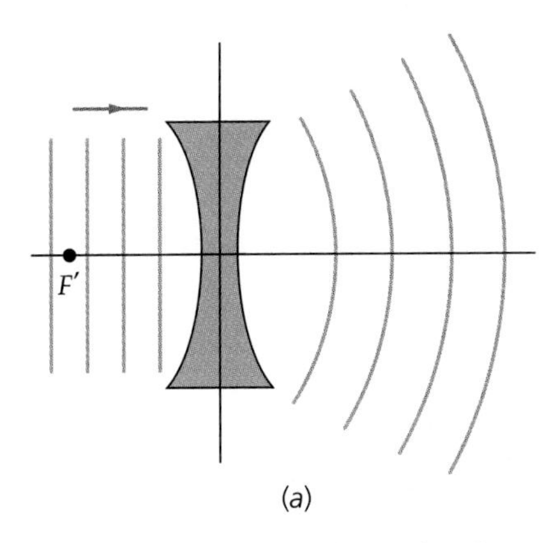

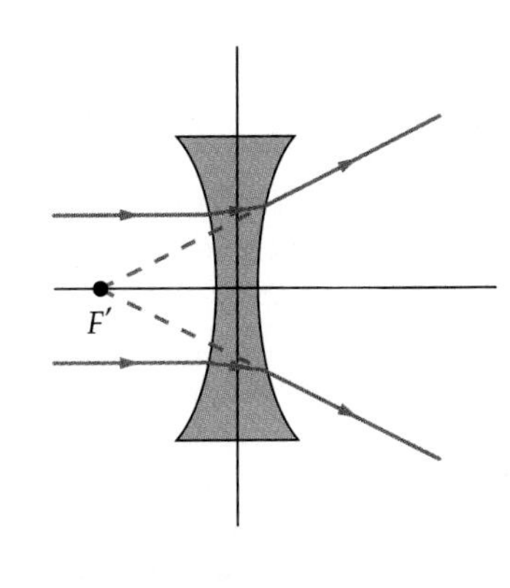

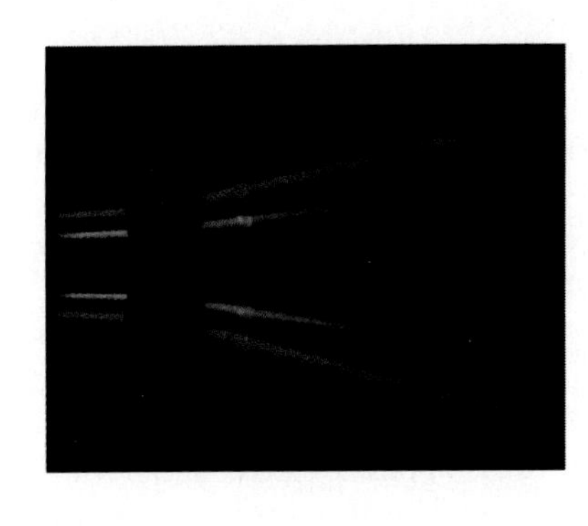

Figure 34-27 (*a*) Wavefronts for plane waves striking a diverging lens. Here, the outer parts of the wavefronts are retarded more than the central part, resulting in a spherical wave that diverges as it moves out as if it came from the focal point F' in front of the lens. (*b*) *Left:* Rays for plane waves striking the same diverging lens. The rays are bent outward and diverge as if they came from the focal point F'. *Right:* Photograph of rays passing through a diverging lens.

wavefronts and rays for plane waves incident on a double concave lens. In this case, the outer part of the wavefronts lag behind the central parts, resulting in outgoing spherical waves that diverge from a focal point on the incident side of the lens. The focal length of this lens is negative. Any lens (with index of refraction greater than that of the surrounding medium) that is thinner in the middle than at the edges is a **diverging,** or **negative, lens.**

Example 34-6

A double convex thin glass lens with index of refraction $n = 1.5$ has radii of curvature of magnitude 10 cm and 15 cm as shown in Figure 34-28. Find its focal length.

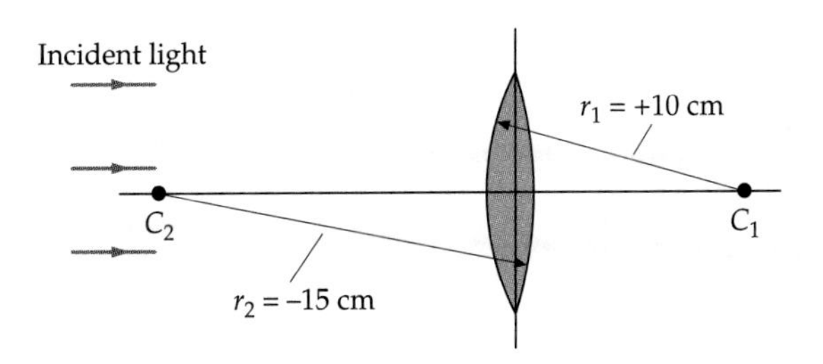

Figure 34-28

Picture the Problem We can find the focal length using the lens-maker's equation (Equation 34-11). Here, light is incident on the surface with the smaller radius of curvature. The center of curvature of this surface, C_1, is on the transmission side of the lens, thus $r_1 = +10$ cm. For the second surface, the center of curvature, C_2, is on the incident side, hence $r_2 = -15$ cm.

Numerical substitution in Equation 34-11 yields the focal length f:

$$\frac{1}{f} = (1.5 - 1)\left(\frac{1}{10\text{ cm}} - \frac{1}{-15\text{ cm}}\right) = 0.5\left(\frac{1}{6\text{ cm}}\right)$$

$$f = 12\text{ cm}$$

Remark Note that both surfaces tend to converge the light rays, hence they both make a positive contribution to the focal length of the lens.

Exercise A double convex thin lens has an index of refraction $n = 1.6$ and radii of curvature of equal magnitude. If its focal length is 15 cm, what is the magnitude of the radius of curvature of each surface? (*Answer* 18 cm)

Exercise Show that if you reverse the direction of the incoming light so that it is incident on the surface with the greater radius of curvature, you get the same result for the focal length.

If parallel light strikes the lens of Example 34-6 from the left, it is focused at a point 12 cm to the right of the lens, whereas if parallel light strikes the lens from the right, it is focused at 12 cm to the left of the lens. Both of these points are focal points of the lens. Using the reversibility property of light rays, we can see that light diverging from a focal point and striking a lens will leave the lens as a parallel beam, as shown in Figure 34-29. In a particular lens problem in which the direction of the incident light is specified, the object point for which light emerges as a parallel beam is called the **first focal point** F and the point at which parallel light is focused is called the **second focal point** F'. For a positive lens, the first focal point is on the incident side and the second focal point is on the transmission side. If parallel light is incident on the lens at a small angle with the axis, as in Figure 34-30, it is focused at a point in the **focal plane** a distance f from the lens.

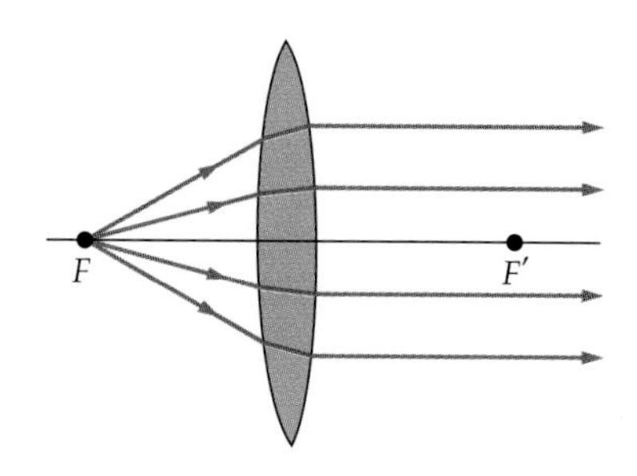

Figure 34-29 Light rays diverging from the focal point of a positive lens emerge parallel to the axis.

The reciprocal of the focal length is called the **power of a lens.** When the focal length is expressed in meters, the power is given in reciprocal meters called **diopters** (D):

$$P = \frac{1}{f}\text{ diopters} \qquad 34\text{-}13$$

The power of a lens measures its ability to focus parallel light at a short distance from the lens. The shorter the focal length, the greater the power. For example, a lens with a focal length of 25 cm = 0.25 m has a power of 4.0 D. A lens with a focal length of 10 cm = 0.10 m has a power of 10 D. Since the focal length of a diverging lens is negative, its power is also negative.

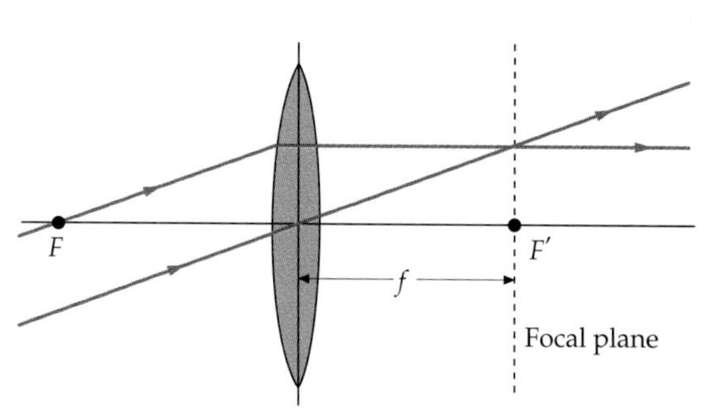

Figure 34-30 Parallel rays incident on the lens at an angle to its axis are focused at a point in the focal plane of the lens.

Example 34-7

A double concave lens has an index of refraction of 1.5 and radii of curvature of magnitude 10 cm and 15 cm. Find (*a*) its focal length (*b*) its power.

Picture the Problem For the orientation of the lens relative to the incident light shown in Figure 34-31, the radius of curvature of the first surface is $r_1 = -15$ cm and that of the second surface is $r_2 = +10$ cm.

Figure 34-31

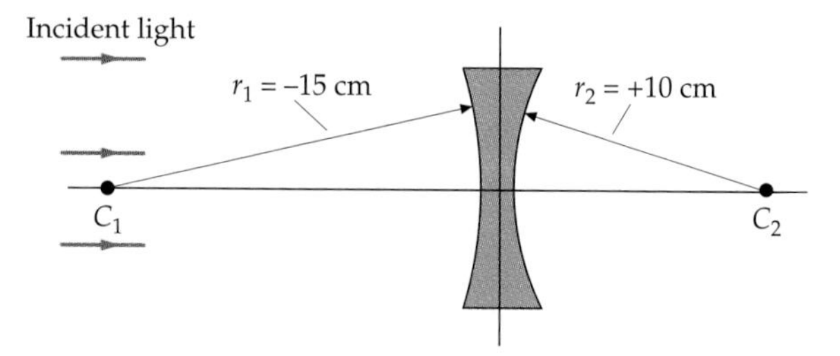

(*a*) Calculate f from the lens-maker's equation using the given value of n and the values of r_1 and r_2 for the orientation shown:

$$\frac{1}{f} = (1.5 - 1.0)\left(\frac{1}{-15\text{ cm}} - \frac{1}{+10\text{ cm}}\right)$$

$$f = -12\text{ cm}$$

(*b*) The power is the reciprocal of the focal length expressed in meters:

$$P = \frac{1}{f} = \frac{1}{-0.12\text{ m}} = -8.33\text{ D}$$

Remark We obtain the same result no matter which surface the light strikes first.

In laboratory experiments involving lenses, it is usually much easier to measure the focal length than to calculate it from the radii of curvature of the surfaces.

The weight and bulk of a large-diameter lens can be reduced by constructing the lens from annular segments at different angles such that light from a point is refracted by the segments into a parallel beam. Such an arrangement is called a Fresnel lens. Several Fresnel lenses are used in this lighthouse to produce intense parallel beams of light from a source at the focal point of the lenses.

Ray Diagrams for Lenses

As with images formed by mirrors, it is convenient to locate the images of lenses by graphical methods. Figure 34-32 illustrates the graphical method for a converging lens. For the sake of simplicity, we consider the rays to bend at the plane through the center of the lens. For a positive lens, the three principal rays are

1. The **parallel ray,** drawn parallel to the axis. This ray is bent through the second focal point of the lens.
2. The **central ray,** drawn through the center (the vertex) of the lens. This ray is undeflected. (The faces of the lens are parallel at this point, so the ray emerges in the same direction but displaced slightly. Since the lens is thin, the displacement is negligible.)
3. The **focal ray,** drawn through the first focal point. This ray emerges parallel to the axis.

Principal rays for a positive lens

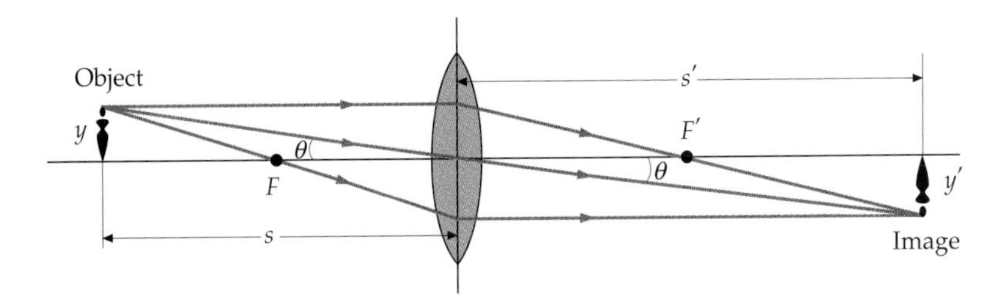

Figure 34-32 Ray diagram for a thin converging lens. For the sake of simplicity, we assume that all the bending of light takes place at the central plane. The ray through the center is undeflected because the lens surfaces there are parallel and close together.

These three rays converge to the image point, as shown in Figure 34-32. In this case, the image is real and inverted. From the figure, we have $\tan\theta = y/s = -y'/s'$. The lateral magnification is then

$$m = \frac{y'}{y} = -\frac{s'}{s} \qquad 34\text{-}14$$

This expression is the same as that for mirrors. Again, a negative magnification indicates that the image is inverted.

The principal rays for a negative, or diverging, lens are

1. The **parallel ray,** drawn parallel to the axis. This ray diverges from the lens as if it came from the second focal point.
2. The **central ray,** drawn through the center (the vertex) of the lens. This ray is undeflected.
3. The **focal ray,** drawn toward the first focal point. This ray emerges parallel to the axis.

Principal rays for a negative lens

The ray diagram for a diverging lens is shown in Figure 34-33.

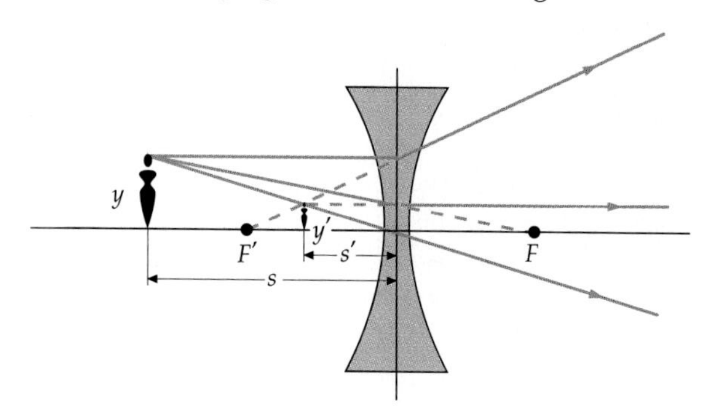

Figure 34-33 Ray diagram for a diverging lens. The parallel ray is bent away from the axis as if it came from the second focal point F'. The ray toward the first focal point F emerges parallel to the axis.

Example 34-8

An object 1.2 cm high is placed 4 cm from the double convex lens of Example 34-6. Locate the image both graphically and algebraically, state whether it is real or virtual, and find its height.

1. Draw the parallel ray. This ray leaves the object parallel to the axis, then is bent by the lens to pass through the second focal point, F' (Figure 34-34):

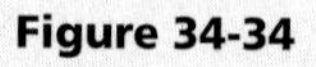

Figure 34-34

1.2 cm

F

4 cm

F′

2. Draw the central ray, which passes undeflected through the center of the lens. Since the two rays are diverging on the transmission side, we extend them back to the incident side to find the image (Figure 34-35):

Figure 34-35

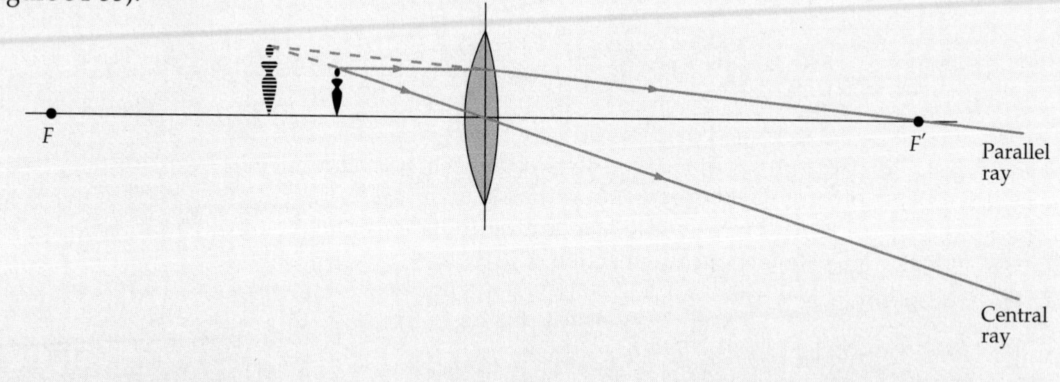

3. As a check, we also draw the focal ray. This ray leaves the object on a line passing through the first focal point, then emerges parallel to the axis. Note that the image is virtual, erect, and enlarged (Figure 34-36):

Figure 34-36

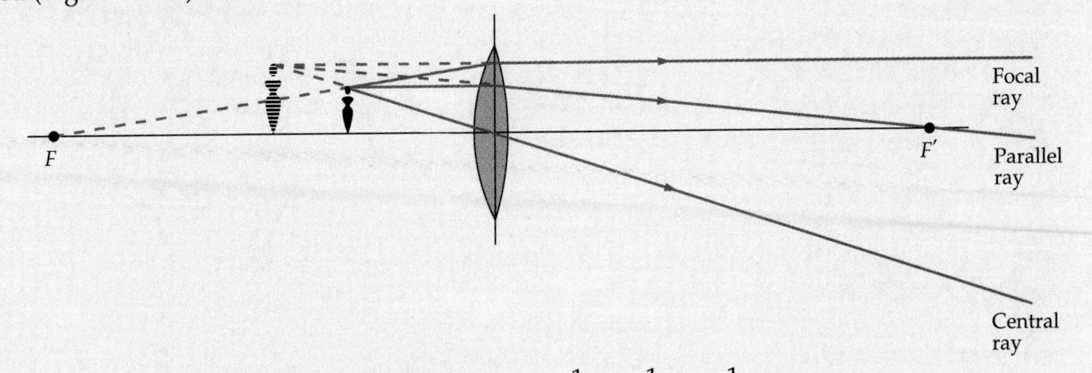

4. We now verify the results of the ray diagram algebraically. First, find the image distance using Equation 34-12:

$$\frac{1}{4\text{ cm}} + \frac{1}{s'} = \frac{1}{12\text{ cm}}$$

$$\frac{1}{s'} = \frac{1}{12\text{ cm}} - \frac{1}{4\text{ cm}} = -\frac{1}{6\text{ cm}}$$

$$s' = -6\text{ cm}$$

5. The height of the image is found from the height of the object and the magnification:

$$h' = mh$$

6. The magnification m is given by Equation 34-14:

$$m = -\frac{s'}{s} = -\frac{-6\text{ cm}}{4\text{ cm}} = +1.5$$

7. Using this result we find the height of the image, h':

$$h' = mh = (1.5)(1.2\text{ cm}) = 1.8\text{ cm}$$

Remarks Note the agreement between the algebraic and ray-diagram results. Algebraically, we find that the image is 6 cm from the lens on the incident side (since $s' < 0$); that is, the image is 2 cm to the left of the object. Since $m > 0$, it follows that the image is erect, and because $m > 1$, the image is enlarged. It is good practice to solve lens problems both ways and compare the results.

Exercise An object is placed 15 cm from a double convex lens of focal length 10 cm. Find the image distance and the magnification. Draw a ray diagram. Is the image real or virtual? Erect or inverted? (Answers $s' = 30$ cm, $m = -2$, real, inverted)

Exercise Work the previous exercise for an object placed 5 cm from a lens with a focal length of 10 cm. (Answers $s' = -10$ cm, $m = 2$, virtual, erect)

Combinations of Lenses

If we have two or more thin lenses, we can find the final image produced by the system by finding the image distance for the first lens and using it along with the distance between lenses to find the object distance for the second lens. That is, we consider each image, whether it is real or virtual and whether it is formed or not, as the object for the next lens.

Example 34-9

A second lens of focal length +6 cm is placed 12 cm to the right of the lens in Example 34-8. Locate the final image.

Picture the Problem The rays used to locate the image of the first lens will not necessarily be the principal rays for the second lens. If they are not, we merely draw additional rays from the first image that are principal rays for the second lens; for example, we draw a ray from the image parallel to the axis and one from the image through the first focal point of the second lens or one through the vertex of the second lens (Figure 34-37). In this example, two of the principal rays for the first lens are also principal rays for the second lens. The parallel ray for the first lens turns out to be the central ray for the second lens. Also, the focal ray for the first lens emerges parallel to the axis and is therefore refracted through the focal point of the second lens. (In the figure, we have extended the central ray for the first lens so that it passes through the image found from the other two rays.)

Algebraically, we use $s_2 = 18$ cm because the first image is 6 cm to the left of the first lens and therefore 18 cm to the left of the second lens.

Figure 34-37

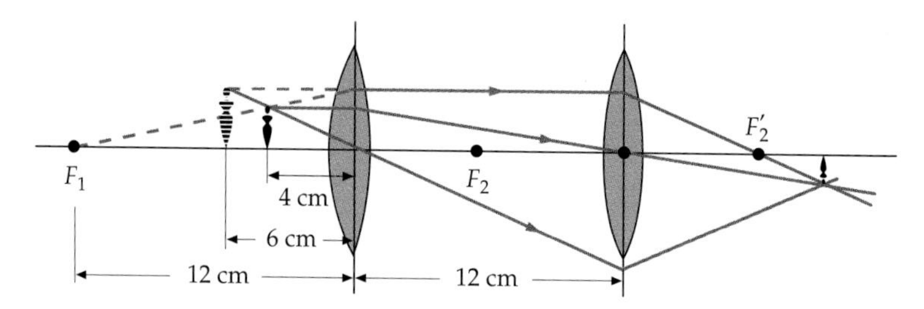

Use $s_2 = 18$ cm and $f = 6$ cm to calculate s_2':

$$\frac{1}{s_2} + \frac{1}{s_2'} = \frac{1}{f_2}$$

$$\frac{1}{18\text{ cm}} + \frac{1}{s_2'} = \frac{1}{6\text{ cm}}$$

$$s_2' = 9\text{ cm}$$

Example 34-10 *try it yourself*

Two lenses, each of focal length 10 cm, are 15 cm apart. Find the final image of an object 15 cm from one of the lenses.

Picture the Problem Use a ray diagram to find the approximate location of the image formed by lens 1. When these rays strike lens 2 they are further refracted, leading to the final image. Precise results are obtained algebraically using the thin lens equation for both lens 1 and lens 2.

Cover the column to the right and try these on your own before looking at the answers.

Steps	*Answers*
1. Draw the parallel (a), central (b), and focal (c) rays for lens 1 (Figure 34-38). If lens 2 did not alter these rays, they would form an image at I_1.	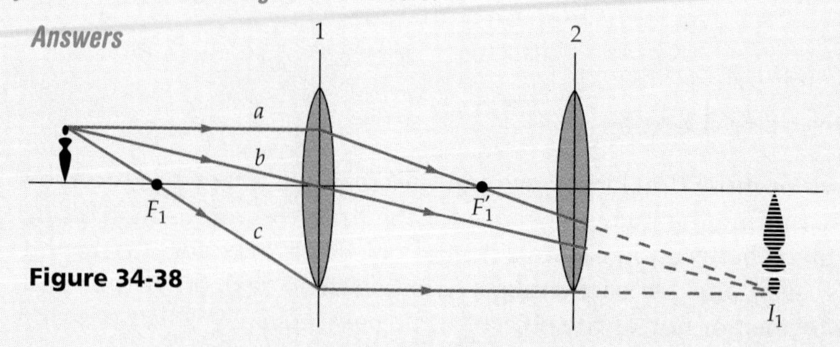

Figure 34-38

2. Note that the focal ray (c) strikes lens 2 parallel to the axis, hence it is refracted through the second focal point of lens 2. To find the final image, include one additional ray (d) that passes through the first focal point of lens 2. It would meet the other rays at I_1, but lens 2 refracts it parallel to the axis. The intersection of (c) and (d) gives the image location (Figure 34-39).

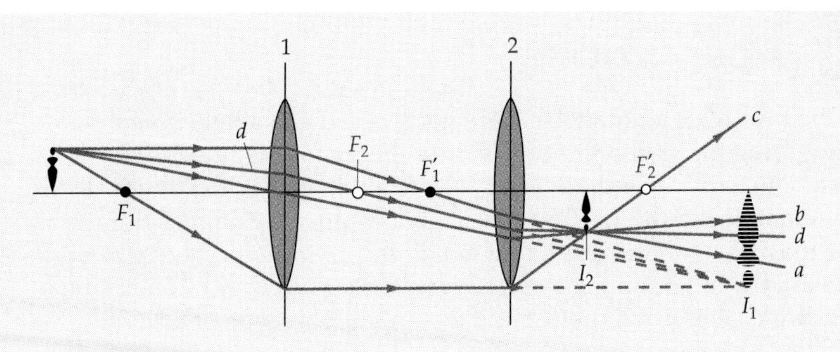

Figure 34-39

3. To proceed algebraically, use the thin-lens equation to find the image distance s_1' produced by lens 1. — $s_1' = 30$ cm

4. For lens 2, the image, I_1 is 15 cm from the lens on the transmission side, hence $s_2 = -15$ cm. Use this to find the final image distance s_2'. — $s_2' = 6$ cm

Remark From the ray diagram we see that the final image is real, inverted, and slightly reduced.

When two thin lenses of focal lengths f_1 and f_2 are placed together, the equivalent focal length of the combination f is given by

$$\frac{1}{f} = \frac{1}{f_1} + \frac{1}{f_2} \qquad \text{34-15}$$

The power of two lenses in contact is given by

$$P = P_1 + P_2 \qquad \text{34-16}$$

Example 34-11 *try it yourself*

For two lenses close together, derive the relation

$$\frac{1}{f} = \frac{1}{f_1} + \frac{1}{f_2}$$

Picture the Problem Apply the thin-lens equation to each lens using the fact that the distance between the lenses is zero so the object distance for the second lens is the negative of the image distance for the first lens.

Cover the column to the right and try these on your own before looking at the answers.

Steps	***Answers***
1. Write the thin-lens equation for lens 1.	$\frac{1}{s} + \frac{1}{s_1'} = \frac{1}{f_1}$
2. Using $s_2 = -s_1'$, write the thin-lens equation for lens 2.	$\frac{1}{-s_1'} + \frac{1}{s'} = \frac{1}{f_2}$
3. Add your two resulting equations to eliminate s_1'.	$\frac{1}{s} + \frac{1}{s'} = \frac{1}{f_1} + \frac{1}{f_2} = \frac{1}{f}$

optional

34-3 Aberrations

When all the rays from a point object are not focused at a single image point, the resulting blurring of the image is called **aberration.** Figure 34-40 shows rays from a point source on the axis traversing a thin lens with spherical surfaces. Rays that strike the lens far from the axis are bent much more than those near the axis, with the result that not all the rays are focused at a single point. Instead, the image appears as a circular disk. The **circle of least confusion** is at point C, where the diameter is minimum. This type of aberration in a lens is called **spherical aberration;** it is the same as the spherical aberration of mirrors discussed in Section 34-2. Similar but more complicated aberrations called *coma* (for the comet-shaped image) and *astigmatism* occur when objects are off axis. The aberration in the shape of the image of an extended object due to the fact that the magnification depends on the distance of the object point from the axis is called **distortion.** We will not discuss these aberrations further except to point out that they do not arise from any defect in the lens or mirror but instead result from the application of the laws of refraction and reflection to spherical surfaces. They are not evident in our simple equations because we used small-angle approximations in the derivation of these equations.

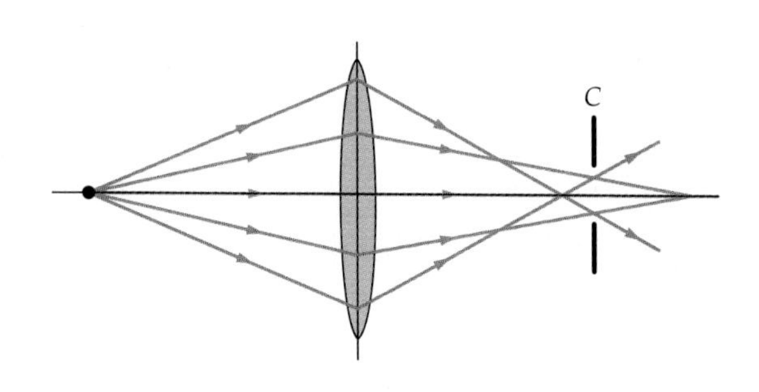

Figure 34-40 Spherical aberration in a lens. Rays from a point object on the axis are not focused at a point. Spherical aberration can be reduced by blocking off the outer parts of the lens, but this also reduces the amount of light reaching the image.

Some aberrations can be eliminated or partially corrected by using nonspherical surfaces for mirrors or lenses, but nonspherical surfaces are usually much more difficult and costly to produce than spherical surfaces. One example of a nonspherical reflecting surface is the parabolic mirror illustrated in Figure 34-41. Rays that are parallel to the axis of a parabolic surface are reflected and focused at a common point no matter how far they are from the axis. Parabolic reflecting surfaces are sometimes used in large astronomical telescopes, which need a large reflecting surface to gather as much light as possible to make the image as intense as possible (reflecting telescopes are described in the upcoming optional Section 34-4). Satellite dishes use parabolic surfaces to focus microwaves from communications satellites. A parabolic surface can also be used in a searchlight to produce a parallel beam of light from a small source placed at the focal point of the surface.

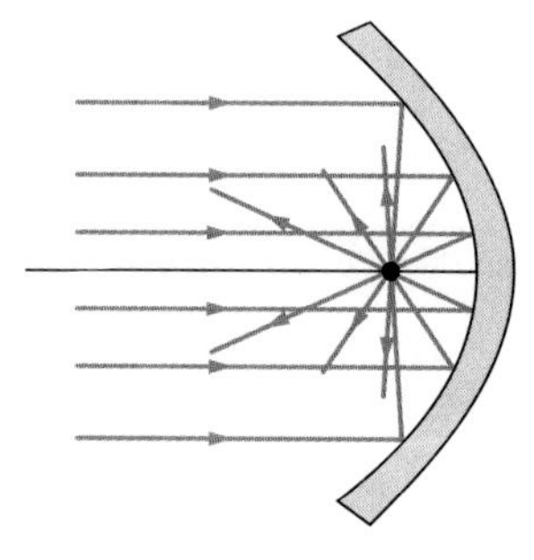

Figure 34-41 A parabolic mirror focuses all rays parallel to the axis to a single point with no spherical aberration.

An important aberration found with lenses but not with mirrors is **chromatic aberration,** which is due to variations in the index of refraction with wavelength. From Equation 34-11, we can see that the focal length of a lens depends on its index of refraction and is therefore different for different wavelengths. Since n is slightly greater for blue light than for red light, the focal length for blue light will be shorter than that for red light. Because chromatic aberration does not occur for mirrors, many large telescopes use a large mirror instead of the large, light-gathering (objective) lens.

Chromatic and other aberrations can be partially corrected by using combinations of lenses instead of a single lens. For example, a positive lens and a negative lens of greater focal length can be used together to produce a converging lens system that has much less chromatic aberration than a single lens of the same focal length. The lens of a good camera typically contains six elements to correct for the various aberrations that are present.

34-4 Optical Instruments

The Eye

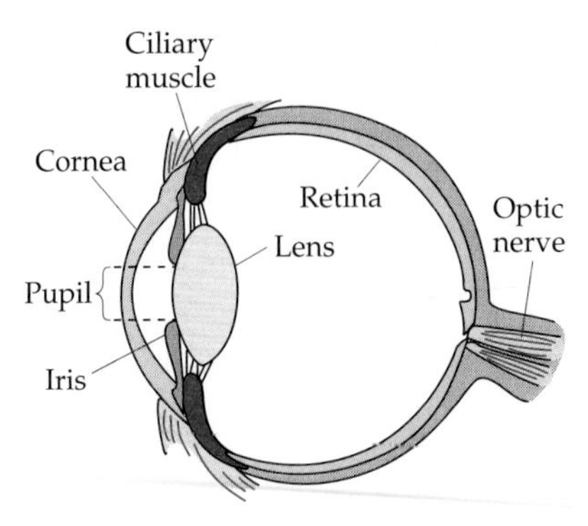

Figure 34-42 The human eye. The amount of light entering the eye is controlled by the iris, which regulates the size of the pupil. The lens thickness is controlled by the ciliary muscle. The cornea and lens together focus the image on the retina, which contains about 125 million receptors called rods and cones and about 1 million optic-nerve fibers.

The optical system of prime importance is the eye, shown in Figure 34-42. Light enters the eye through a variable aperture, the pupil, and is focused by the cornea–lens system on the retina, a film of nerve fibers covering the back surface. The retina contains tiny sensing structures called *rods* and *cones*, which receive the image and transmit the information along the optic nerve to the brain. The shape of the crystalline lens can be altered slightly by the action of the ciliary muscle. When the eye is focused on an object far away, the muscle is relaxed and the cornea–lens system has its maximum focal length, about 2.5 cm, the distance from the cornea to the retina. When the object is brought closer to the eye, the ciliary muscle increases the curvature of the lens slightly, thereby decreasing its focal length, so that the image is again focused on the retina. This process is called *accommodation*. If the object is too close to the eye, the lens cannot focus the light on the retina and the image is blurred. The closest point for which the lens can focus the image on the retina is called the **near point**. The distance from the eye to the near point varies greatly from one person to another and changes with age. At 10 years, the near point may be as close as 7 cm, whereas at 60 years it may recede to 200 cm because of the loss of flexibility of the lens. The standard value taken for the near point is 25 cm.

If the eye underconverges, resulting in the images being focused behind the retina, the person is said to be farsighted. A farsighted person can see distant objects where little convergence is required, but has trouble seeing close objects. Farsightedness is corrected with a converging (positive) lens (Figure 34-43).

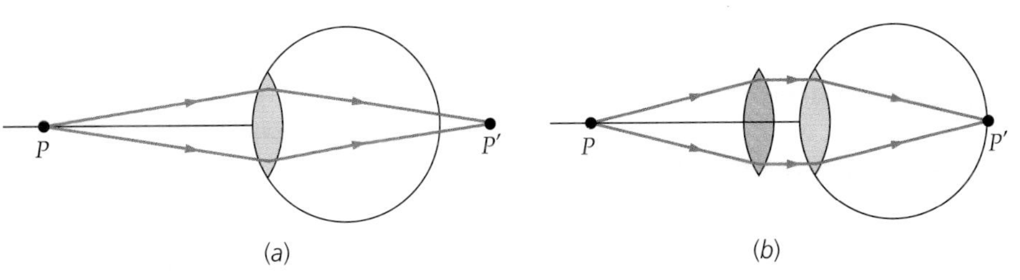

Figure 34-43 (*a*) A farsighted eye focuses rays from a nearby object to a point behind the retina. (*b*) A converging lens corrects this defect by bringing the image onto the retina. These diagrams and those following are drawn as if all the focusing of the eye is done at the lens, whereas in fact the lens and cornea system act more like a spherical refracting surface than a thin lens.

On the other hand, the eye of a nearsighted person overconverges and focuses light from distant objects in front of the retina. A nearsighted person can see nearby objects for which the widely diverging incident rays can be focused on the retina, but has trouble seeing distant objects. Nearsightedness is corrected with a diverging (negative) lens (Figure 34-44).

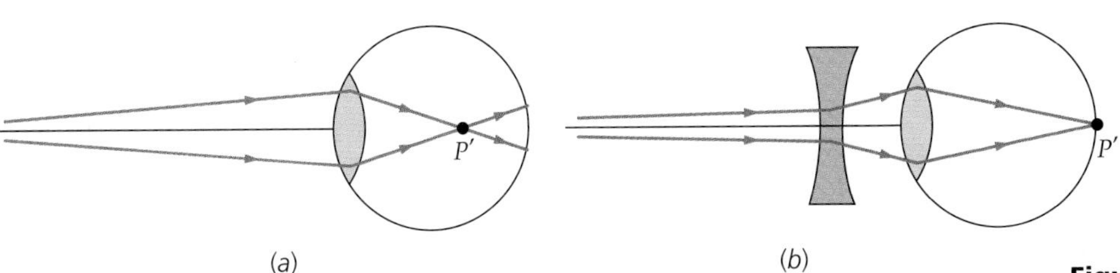

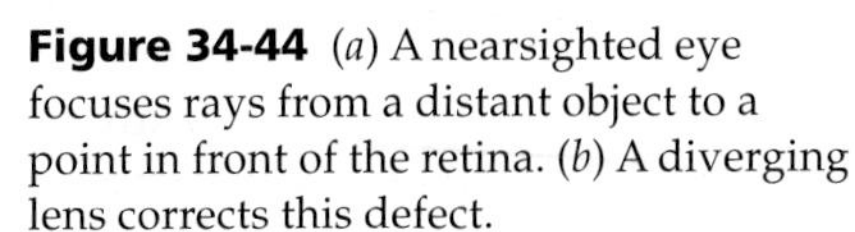

Figure 34-44 (*a*) A nearsighted eye focuses rays from a distant object to a point in front of the retina. (*b*) A diverging lens corrects this defect.

Another common defect of vision is astigmatism, which is caused by the cornea being not quite spherical but having a different curvature in one plane than in another. This results in a blurring of the image of a point object into a short line. Astigmatism is corrected by glasses using lenses of cylindrical rather than spherical shape.

optional

Example 34-12

By how much must the focal length of the cornea–lens system of the eye change when the object is moved from infinity to the near point at 25 cm? Assume that the distance from the cornea to the retina is 2.5 cm.

Picture the Problem At infinity the focal length is 2.5 cm. We use the thin-lens equation to calculate the focal length when $s = 25$ cm and $s' = 2.5$ cm.

1. Use the thin-lens equation to calculate f:

$$\frac{1}{f} = \frac{1}{25\text{ cm}} + \frac{1}{2.5\text{ cm}} = \frac{1}{25\text{ cm}} + \frac{10}{25\text{ cm}} = \frac{11}{25\text{ cm}}$$

$$f = 2.27\text{ cm}$$

2. Subtract the original focal length of 2.5 cm to find the change:

$$\Delta f = 2.27\text{ cm} - 2.5\text{ cm} = -0.23\text{ cm}$$

Remarks In terms of the power of the cornea–lens system, when the focal length is 2.5 cm $= 0.025$ m for distant objects, the power is $P = 1/f = 40$ D. When the focal length is 2.27 cm, the power is 44 D.

Exercise Find the change in the focal length of the eye when an object originally at 4 m is brought to 40 cm from the eye. (Assume that the distance from the cornea to the retina is 2.5 cm.) (*Answer* -0.13 cm)

The apparent size of an object is determined by the size of the image on the retina. The larger the image on the retina, the greater the number of rods and cones activated. From Figure 34-45 we see that the size of the image on the retina is greater when the object is close than it is when the object is far away. The apparent size of an object is thus greater when it is closer to the eye. The image size is proportional to the angle θ subtended by the object at the eye. For Figure 34-45,

$$\theta = \frac{y'}{2.5\text{ cm}} \tag{34-17}$$

The angle θ is related to the object size y and object distances. For small angles,

$$\theta \approx \tan\theta = \frac{y}{s} \tag{34-18}$$

Combining Equations 34-17 and 34-18 gives

$$y' = (2.5\text{ cm})\,\theta \approx (2.5\text{ cm})\,\frac{y}{s} \tag{34-19}$$

The size of the image on the retina is proportional to the size of the object and inversely proportional to the distance between the object and the eye. Since the near point is the closest point to the eye for which a sharp image can be formed on the retina, the distance to the near point is called the *distance of most distinct vision.*

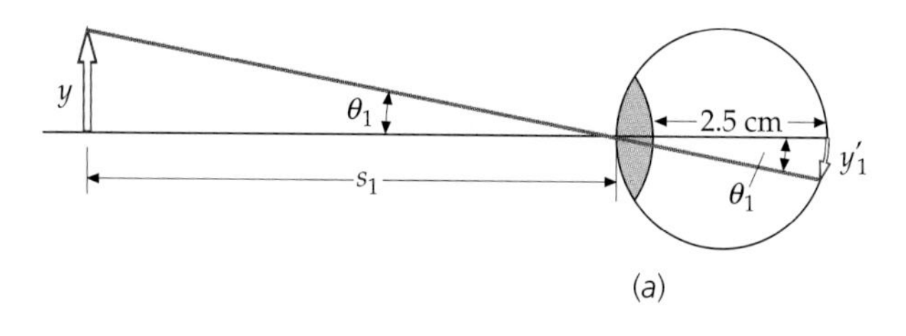

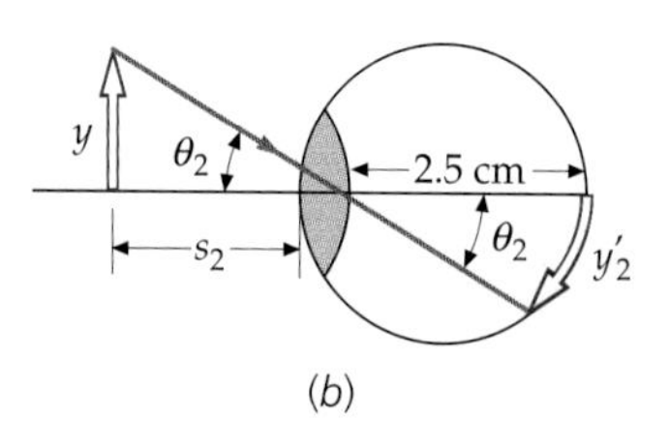

Figure 34-45 (*a*) A distant object of height y looks small because the image on the retina is small. (*b*) When the same object is closer, it looks larger because the image on the retina is larger. The angle subtended is $\theta = y'/(2.5\text{ cm})$.

optional

Example 34-13

The near point of a person's eye is 75 cm. What power reading glasses should be used to bring the near point to 25 cm? Assume that the lens of the glasses is in contact with the lens of the eye.

Picture the Problem Figure 34-46 shows a diagram of an object 25 cm from a converging lens that produces a virtual, erect image at $s' = -75$ cm.

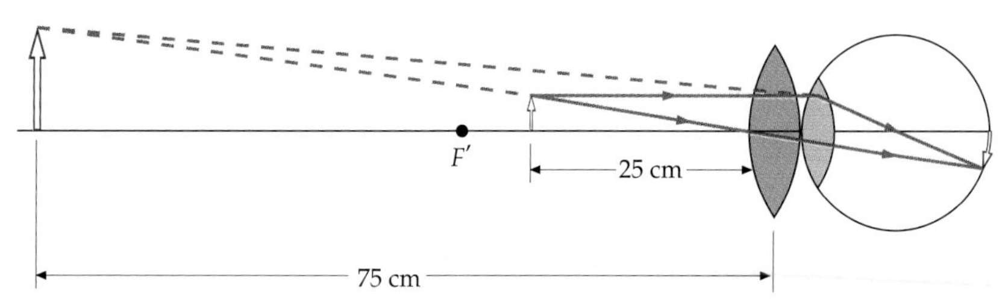

Figure 34-46

Use the thin-lens equation with $s = 25$ cm and $s' = -75$ cm to calculate the power, $1/f$:

$$\frac{1}{25\text{ cm}} + \frac{1}{-75\text{ cm}} = \frac{1}{f}$$

$$\frac{1}{f} = \frac{2}{75\text{ cm}} = \frac{1}{0.375\text{ m}} = 2.67\text{ diopter}$$

Remarks If your near point is 75 cm, you are farsighted. To read a book you must hold it about 75 cm from your eye to be able to focus on the print. The image of the print on your retina is then very small. A converging lens of the type found in reading glasses allows you to bring the book closer to the eye, which makes the retinal image of the print larger. In this example we assumed that the lens was in contact with the eye. For reading glasses, which are perched a short distance in front of the eye, the results are slightly different.

Exercise Calculate the power of the lens of the eye for which the near point is 75 cm, and calculate the combined power of the two lenses in contact. Compare this with the power of a lens for which $s' = 2.5$ cm when $s = 25$ cm. (*Answers* $P_{\text{eye}} = 41.33$ D; $P_c = 41.33\text{ D} + 2.67\text{ D} = 44$ D; $P = 44$ D)

The Simple Magnifier

We saw in Example 34-13 that the apparent size of an object can be increased by using a converging lens to allow the object to be brought closer to the eye, thus increasing the size of the image on the retina. Such a converging lens is called a simple magnifier. In Figure 34-47*a*, a small object of height y is at the near point of the eye at a distance x_{np}. The angle subtended, θ_o, is given approximately by

$$\theta_o = \frac{y}{x_{\text{np}}}$$

In Figure 34-47*b*, a converging lens of focal length f, smaller than x_{np}, is placed in front of the eye, and the object is placed at the focal point of

Figure 34-47 (*a*) An object at the near point subtends an angle θ_o at the eye. (*b*) When the object is at the focal point of the converging lens, the rays emerge from the lens parallel and enter the eye as if they came from an object a very large distance away. The image can thus be viewed at infinity by the relaxed eye. When f is less than the near point, the converging lens allows the object to be brought closer to the eye, increasing the angle subtended by the object to θ, thereby increasing the size of the image on the retina.

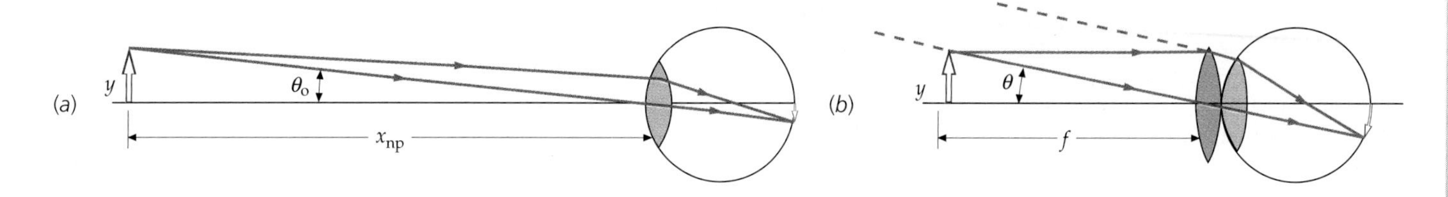

the lens. The rays emerge from the lens parallel, indicating that the image is at an infinite distance in front of the lens. The parallel rays are focused by the relaxed eye on the retina. Assuming that the lens is in contact with the eye, the angle subtended by the object is now approximately

$$\theta = \frac{y}{f}$$

The ratio θ/θ_o is called the *angular magnification* or *magnifying power M* of the lens:

$$M = \frac{\theta}{\theta_o} = \frac{x_{np}}{f} \qquad \text{34-20}$$

Simple magnifiers are used as eyepieces (called oculars) in microscopes and telescopes to view the image formed by another lens or lens system. To correct aberrations, combinations of lenses that result in a short positive focal length may be used in place of a single lens, but the principle of the simple magnifier is the same.

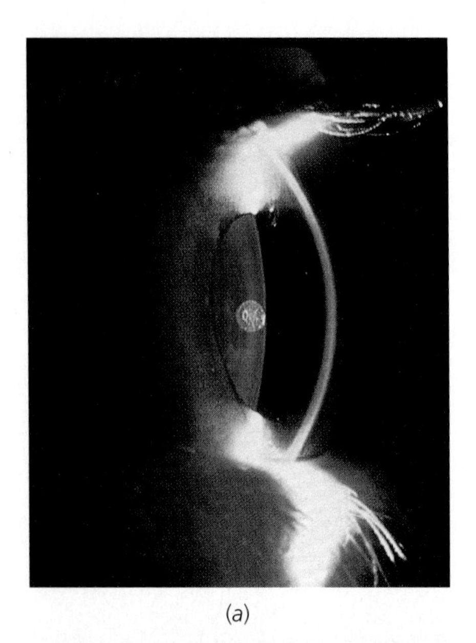

(*a*)

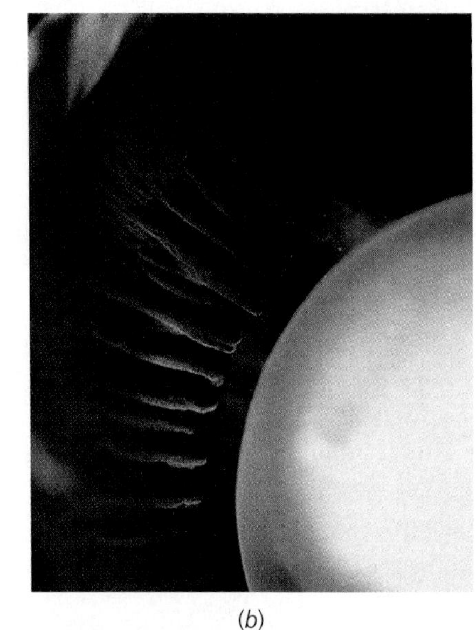

(*b*)

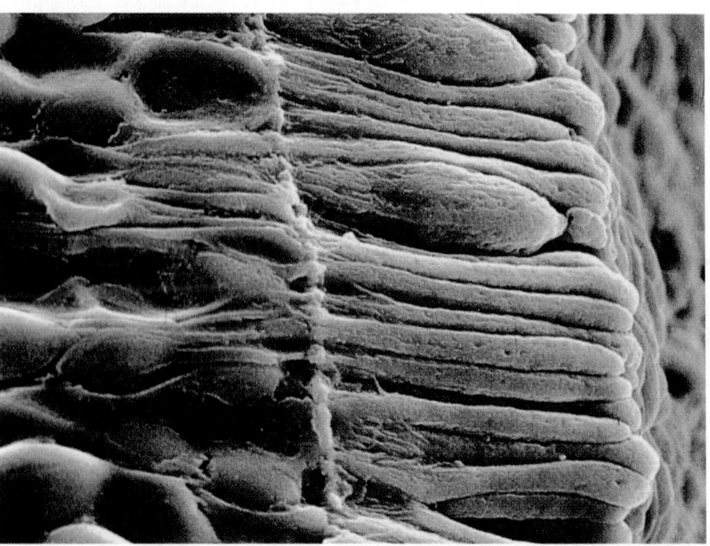

(*c*)

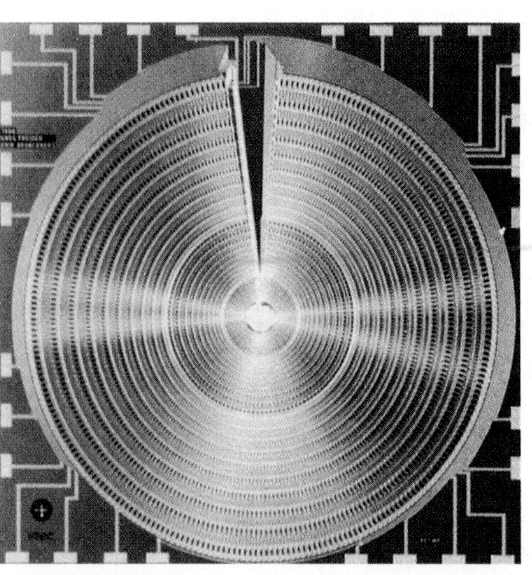

(*d*)

(*a*) The human eye in profile. (*b*) The lens of the eye is kept in place by the ciliary muscle (shown here in the upper left), which rings the lens. When the ciliary muscle contracts, the lens tends to bulge. The greater lens curvature enables the eye to focus on nearby objects. (*c*) Some of the 120 million rods and 7 million cones in the eye, magnified approximately 5000 times. The rods (the more slender of the two) are more sensitive in dim light, whereas the cones are more sensitive to color. The rods and cones form the bottom layer of the retina and are covered by nerve cells, blood vessels, and supporting cells. Most of the light entering the eye is reflected or absorbed before reaching the rods and cones. The light that does reach them triggers electrical impulses along nerve fibers that ultimately reach the brain. (*d*) A neural net used in the vision system of certain robots. Loosely modeled on the human eye, it contains 1920 sensors.

optional

Example 34-14 *try it yourself*

A person with a near point of 25 cm uses a 40-D lens as a simple magnifier. What angular magnification is obtained?

Picture the Problem The angular magnification is found from the focal length f (Equation 34-20), which is the reciprocal of the power.

Cover the column to the right and try these on your own before looking at the answers.

Steps	*Answers*
1. Calculate the focal length of the lens.	$f = 2.5$ cm
2. Use your result in Equation 34-20 to calculate the angular magnification.	$M = 10$

Remark The object looks 10 times larger because it can be placed at 2.5 cm rather than at 25 cm from the eye, thus increasing the image on the retina tenfold.

Exercise What is the magnification in this example if the near point of the person is 30 cm rather than 25 cm? (*Answer* $M = 12$)

The Compound Microscope

The compound microscope (Figure 34-48) is used to look at very small objects at short distances. In its simplest form, it consists of two converging lenses. The lens nearest the object, called the **objective,** forms a real image of the object. This image is enlarged and inverted. The lens nearest the eye, called the **eyepiece** or **ocular,** is used as a simple magnifier to view the image formed by the objective. The eyepiece is placed such that the image formed by the objective falls at the first focal point of the eyepiece. The light thus emerges from the eyepiece as a parallel beam as if it were coming from a point a great distance in front of the lens. (This is commonly called "viewing the image at infinity.")

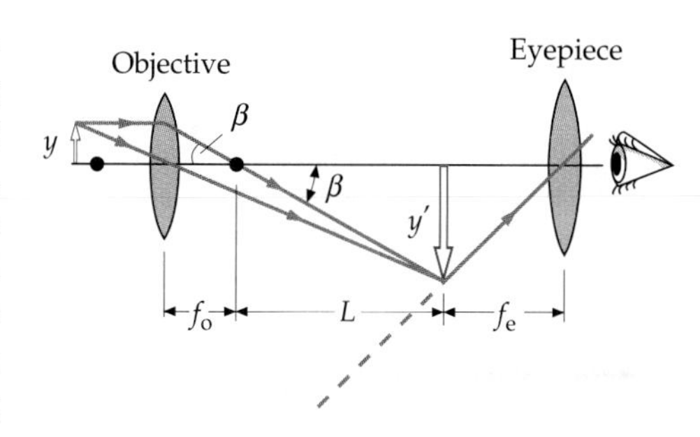

Figure 34-48 Schematic diagram of a compound microscope consisting of two positive lenses, the objective of focal length f_o and the ocular, or eyepiece, of focal length f_e. The real image of the object formed by the objective is viewed by the eyepiece, which acts as a simple magnifier. The final image is at infinity.

The distance between the second focal point of the objective and the first focal point of the eyepiece is called the **tube length** L. It is typically fixed at about 16 cm. The object is placed just outside the first focal point of the objective so that an enlarged image is formed at the first focal point of the eyepiece a distance $L + f_o$ from the objective, where f_o is the focal length of the objective. From Figure 34-48, $\tan \beta = y/f_o = -y'/L$. The lateral magnification of the objective is therefore

$$m_o = \frac{y'}{y} = -\frac{L}{f_o} \qquad \text{34-21}$$

The angular magnification of the eyepiece is

$$M_e = \frac{x_{np}}{f_e}$$

where x_{np} is the near point of the viewer (the nearest point at which the viewer can focus), and f_e is the focal length of the eyepiece. The magnifying power of the compound microscope is the product of the lateral magnification of the objective and the angular magnification of the eyepiece:

$$M = m_o M_e = -\frac{L}{f_o}\frac{x_{np}}{f_e} \qquad \text{34-22}$$

Magnifying power of a microscope

optional

Example 34-15

A microscope has an objective lens of focal length 1.2 cm and an eyepiece of focal length 2.0 cm separated by 20 cm. (*a*) Find the magnifying power if the near point of the viewer is 25 cm. (*b*) Where should the object be placed if the final image is to be viewed at infinity?

(*a*)1. The magnifying power is given by Equation 34-22:
$$M = -\frac{L}{f_o}\frac{x_{np}}{f_e}$$

2. The tube length L is the distance between the lenses minus the focal distances:
$$L = 20 \text{ cm} - 2 \text{ cm} - 1.2 \text{ cm} = 16.8 \text{ cm}$$

3. Substitute this value for L and the given values of x_{np}, f_o, and f_e to calculate M:
$$M = -\frac{L}{f_o}\frac{x_{np}}{f_e} = -\frac{16.8 \text{ cm}}{1.2 \text{ cm}}\frac{25 \text{ cm}}{2 \text{ cm}} = -175$$

(*b*)1. Calculate the object distance s in terms of the image distance for the objective s' and the focal length f_o:
$$\frac{1}{s} + \frac{1}{s'} = \frac{1}{f_o}$$

2. From Figure 34-48, the image distance for the image of the objective is $f_o + L$:
$$s' = f_o + L = 1.2 \text{ cm} + 16.8 \text{ cm} = 18 \text{ cm}$$

3. Substitute to calculate s:
$$\frac{1}{s} + \frac{1}{18 \text{ cm}} = \frac{1}{1.2 \text{ cm}}$$
$$s = 1.29 \text{ cm}$$

Remark The object should thus be placed at 1.29 cm from the objective or 0.09 cm outside its first focal point.

The Telescope

A telescope is used to view objects that are far away and often large. It works by creating an image of the object that is much closer than the object. The astronomical telescope, illustrated schematically in Figure 34-49, consists of two positive lenses—an objective lens that forms a real, inverted image and an eyepiece that is used as a simple magnifier to view that image. Because the object is very far away, the image of the objective lies at the focal point of the objective, and the image distance equals the focal length f_o. The image formed by the objective is much smaller than the object because the object distance is much larger than the focal length of the objective. For example, if we are looking at the moon, the image of the moon formed by the objective is much smaller than the moon itself. The purpose of the objective is not to magnify the object, but to produce an image that is close so it can be viewed by the eyepiece. The eyepiece is placed a distance f_e from the image, where f_e is the focal length of the eyepiece, so the final image can be viewed at infinity. Since this image is at the second focal point of the objective and at the first focal point of the ocular, the objective and ocular must be separated by the sum of the focal lengths of the objective and eyepiece, $f_o + f_e$.

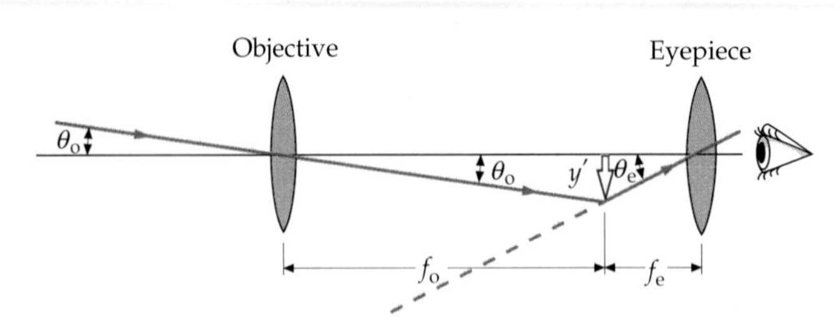

Figure 34-49 Schematic diagram of an astronomical telescope. The objective forms a real image of a distant object near its second focal point, which coincides with the first focal point of the eyepiece. The eyepiece serves as a simple magnifier to view the image.

The magnifying power of the telescope is the angular magnification θ_e/θ_o, where θ_e is the angle subtended by the final image as viewed through the eyepiece and θ_o is the angle subtended by the object when it is viewed directly by the unaided eye. The angle θ_o is the same as that subtended by the object at the objective shown in Figure 34-49. (The distance from a distant object, such as the moon, to the objective is essentially the same as the distance to the eye.) From this figure, we can see that

$$\tan\theta_o = -\frac{y'}{f_o} \approx \theta_o$$

where we have used the small-angle approximation $\tan\theta \approx \theta$ and have introduced a negative sign to make θ_o positive when y' is negative. The angle θ_e in the figure is that subtended by the final image:

$$\tan\theta_e = -\frac{y'}{f_e} \approx \theta_e$$

Since y' is negative, θ_e is negative, indicating that the image is inverted. The magnifying power of the telescope is then

$$M = \frac{\theta_e}{\theta_o} = -\frac{f_o}{f_e} \qquad \text{34-23}$$

Magnifying power of a telescope

From Equation 34-23, we can see that a large magnifying power is obtained with an objective of large focal length and an eyepiece of short focal length.

Exercise The world's largest refracting telescope is at the Yerkes Observatory of the University of Chicago at Williams Bay, Wisconsin. The objective has a diameter of 102 cm and a focal length of 19.5 m. The focal length of the eyepiece is 10 cm. What is its magnifying power? (*Answer* −195)

The main consideration with an astronomical telescope is not its magnifying power but its light-gathering power, which depends on the size of the objective. The larger the objective, the brighter the image. Very large lenses without aberrations are difficult to produce. In addition, there are mechanical problems in supporting very large, heavy lenses by their edges. A reflecting telescope (Figures 34-50 and 34-51) uses a concave mirror instead of a lens for its objective. This offers several advantages. For one, a mirror does not produce chromatic aberration. In addition, mechanical support is much simpler, since the mirror weighs far less than a lens of equivalent optical quality and can be supported over its entire back surface. In modern telescopes, the objective mirror consists of several dozen adaptive mirror segments that can be adjusted individually to correct for minute variations in gravitational stress when the telescope is tilted, and to compensate for thermal expansions and contractions and other changes caused by climate conditions.

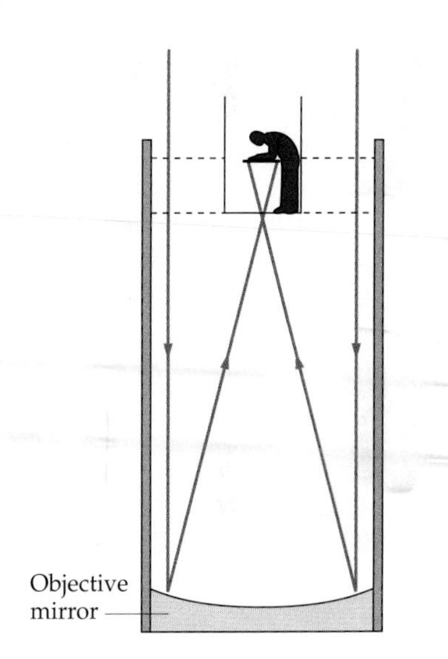

Figure 34-50 A reflecting telescope uses a mirror for its objective. Because the viewer compartment blocks off some of the incoming light, the arrangement shown here is used only in telescopes with very large objective mirrors.

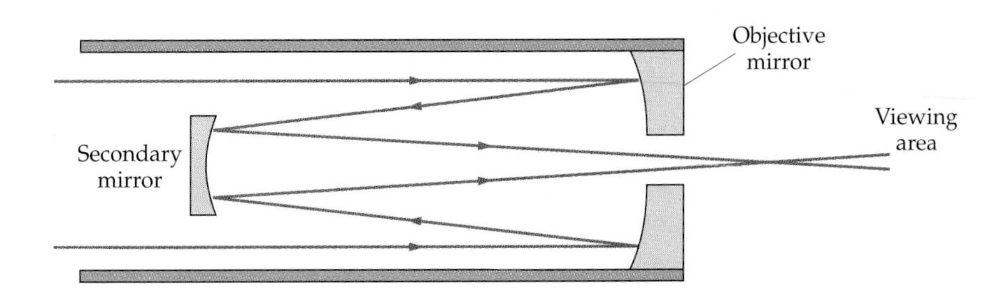

Figure 34-51 This reflecting telescope has a secondary mirror to redirect the light through a small hole in the objective mirror, thus providing more room for auxiliary instruments in the viewing area.

optional

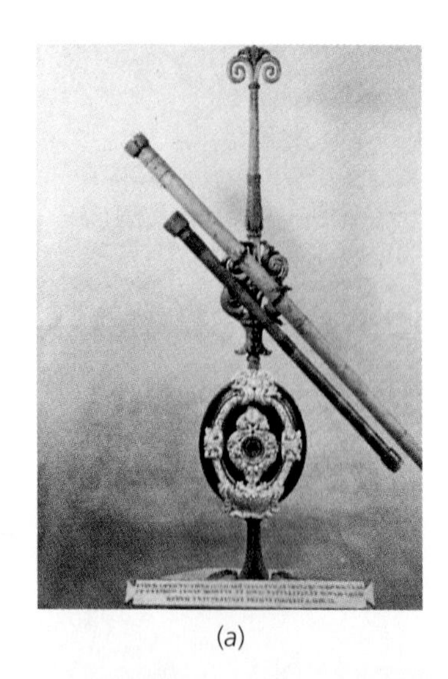

(*a*)

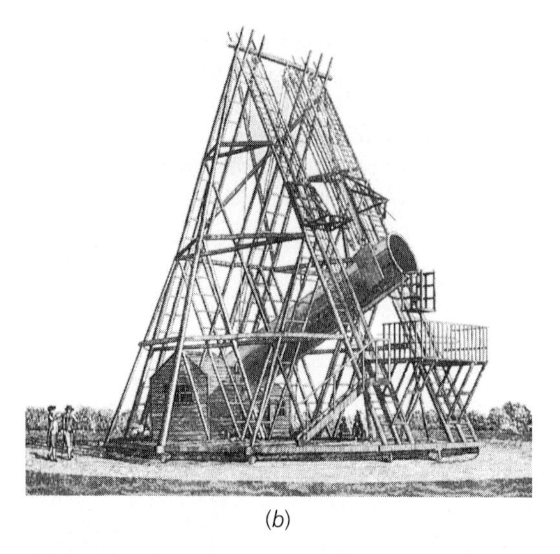

(*b*)

(*c*)

(*d*)

(*e*)

Astronomy at optical wavelengths began with Galileo approximately 400 years ago. In this century astronomers began to explore the electromagnetic spectrum at other wavelengths, beginning with radio astronomy in the 1940s, satellite-based x-ray astronomy in the early 1960s, and more recently, ultraviolet, infrared, and gamma-ray astronomy. (*a*) Galileo's seventeenth-century telescope, with which he discovered mountains on the moon, sunspots, Saturn's rings, and the bands and moons of Jupiter. (*b*) An engraving of the reflector telescope built in the 1780s and used by the great astronomer Hershel, who was the first to observe galaxies outside our own. (*c*) Because it is difficult to make large, flaw-free lenses, refractor telescopes like this 91.4-cm telescope at Lick Observatory have been superseded in light-gathering power by reflector telescopes. (*d*) The great astronomer Hubble, who discovered the apparent expansion of the universe, is shown seated in the observer's cage of the 5.08-m Hale reflecting telescope, which is large enough for the observer to sit at the prime focus itself. (*e*) This 10-m optical reflector at the Whipple Observatory in southern Arizona is the largest instrument designed exclusively for use in gamma-ray astronomy. High-energy gamma rays of unknown origin strike the upper atmosphere and create cascades of particles. Among these particles are high-energy electrons that emit Cerenkov radiation observable from the ground. According to one hypothesis, high-energy gamma rays are emitted as matter is accelerated toward ultradense rotating stars called pulsars.

(*a*)

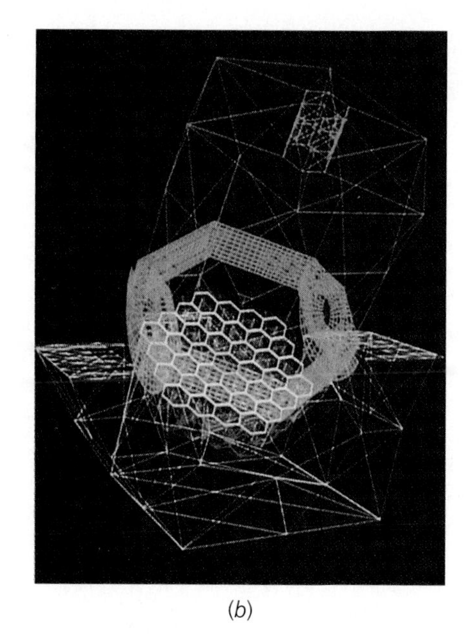

(*b*)

(*c*)

(*a*) The Keck Observatory, atop the inactive volcano of Mauna Kea, Hawaii, houses the world's largest optical telescope. The clear, dry air and lack of light pollution make the remote heights of Mauna Kea an ideal site for astronomical observations. (*b*) The Keck telescope is composed of 36 hexagonal mirror segments performing together as if they were a single mirror 10 m wide—roughly twice as large as the largest single-mirror telescope presently in operation.

(*c*) Beneath each Keck mirror is a system of computer-controlled sensors and motor-driven actuators that can continuously vary the mirror's shape. These variations, which are sensitive to within 100 nm, enable the system to compensate for variations in the alignments of the segments due to minute variations in gravitational stress when the telescope is tilted, thermal expansions and contractions, and fluctuations caused by gusts of wind on the mountain top.

The Hubble Space Telescope, high above the atmospheric turbulence that limits the ability of ground-based telescopes to resolve images at optical wavelengths.

Summary

Topic	Remarks and Relevant Equations	
1. Virtual and Real Images and Objects	An image is real if light actually converges at the image point and diverges (such as in front of a concave mirror, or behind a thin lens). It is virtual if light only appears to diverge from the image point (such as behind a plane mirror).	
Virtual object	Images due to reflection or refraction at one surface are often the objects for the next surface. If such an image is not formed because the light is interrupted by the surface, it is a virtual object.	
2. Mirrors		
Focal length	The focal length is the image distance when the object is at infinity so the incident light is parallel to the axis: $f = \frac{1}{2}r$	**34-2**
Equation for locating an image	$\frac{1}{s} + \frac{1}{s'} = \frac{1}{f}$	**34-3**
Lateral magnification	$m = \frac{y'}{y} = -\frac{s'}{s}$	**34-4**
Ray diagrams	Images can be located by a ray diagram using any two of three principle rays: 1. The parallel ray, drawn parallel to the axis, is reflected through the focal point. 2. The focal ray, drawn through the focal point, is reflected parallel to the axis. 3. The radial ray, drawn through the center of curvature, strikes the mirror perpendicular to its surface and is thus reflected back on itself.	
Sign conventions	s + if the object is in front of the mirror (real object) − if the object is behind the mirror (virtual object) s' + if the image is in front of the mirror (real image) − if the image is behind the mirror (virtual image) r, f + if the center of curvature is in front of the mirror (concave mirror) − if the center of curvature is behind the mirror (convex mirror)	
3. Images Formed by Refraction at a Spherical Surface		
Location	$\frac{n_1}{s} + \frac{n_2}{s'} = \frac{n_2 - n_1}{r}$	**34-5**
Magnification	$m = \frac{y'}{y} = -\frac{n_1 s'}{n_2 s}$	**34-6**
Sign convention	s + (real object) for objects in front of the surface (incident side) − (virtual object) for objects in back of the surface (transmission side) s' + (real image) for images in back of the surface (transmission side) − (virtual image) for images in front of the surface (incident side) r, f + if the center of curvature is on the transmission side − if the center of curvature is on the incident side	

4. Thin Lenses		
Focal length (lens-maker's equation)	$$\frac{1}{f} = (n - 1)\left(\frac{1}{r_1} - \frac{1}{r_2}\right)$$ A positive lens ($f > 0$) is a converging lens (like a double convex lens). A negative lens ($f < 0$) is a diverging lens (like a double concave lens).	34-11
Power	$$P = \frac{1}{f} \text{ diopters}$$	34-13
Equation for locating image	$$\frac{1}{s} + \frac{1}{s'} = \frac{1}{f}$$	34-12
Magnification	$$m = \frac{y'}{y} = -\frac{s'}{s}$$	34-14
Ray diagrams	Images can be located by a ray diagram using any two of the three principal rays:	
Positive lens	1. The parallel ray, drawn parallel to the axis, is bent through the second focal point of the lens. 2. The central ray, drawn through the center (the vertex) of the lens, is undeflected. 3. The focal ray, drawn through the first focal point, emerges parallel to the axis.	
Negative lens	1. The parallel ray, drawn parallel to the axis, diverges from the lens as if it came from the second focal point. 2. The central ray, drawn through the center (the vertex) of the lens, is undeflected. 3. The focal ray, drawn toward the first focal point, emerges parallel to the axis.	
Sign convention	The sign conventions are the same as for refraction at a spherical surface.	
5. Aberrations (optional)	Blurring of the image of a single object point is called aberration. Spherical aberration results from the fact that a spherical surface focuses only paraxial rays (those that travel close to the axis) at a single point. Nonparaxial rays are focused at nearby points depending on the angle made with the axis. Spherical aberration can be reduced by reducing the size of the spherical surface, which also reduces the amount of light reaching the image. Chromatic aberration, which occurs with lenses but not mirrors, results from the variation in the index of refraction with wavelength. Lens aberrations are most commonly reduced by using a series of lens elements.	
6. The Eye (optional)	The cornea–lens system of the eye focuses light on the retina, where it is sensed by the rods and cones that send information along the optic nerve to the brain. When the eye is relaxed, the focal length of the cornea–lens system is about 2.5 cm, the distance to the retina. When objects are brought near the eye, the lens changes shape to decrease the overall focal length so that the image remains focused on the retina. The closest distance for which the image can be focused on the retina is called the near point, typically about 25 cm. The apparent size of an object depends on the size of the image on the retina. The closer the object, the larger the image on the retina and therefore the larger the apparent size of the object.	
7. Simple Magnifier (optional)	A simple magnifier consists of a lens with a positive focal length that is smaller than the near point.	
Angular magnification (or magnifying power)	$$M = \frac{\theta}{\theta_o} = \frac{x_{np}}{f}$$	34-20

8. Compound Microscope (optional)	The compound microscope is used to look at very small objects that are nearby. Its simplest form consists of two lenses, an objective and an ocular or eyepiece. The object to be viewed is placed just outside the focal point of the objective, which forms an enlarged image of the object at the focal point of the eyepiece. The eyepiece acts as a simple magnifier to view the final image.
Magnifying power	$M = m_o M_e = -\frac{L}{f_o}\frac{x_{np}}{f_e}$ **34-22** where L is the tube length, the distance between the second focal point of the objective and the first focal point of the eyepiece.
9. Telescope (optional)	The telescope is used to view objects far away. The objective of the telescope forms a real image that is much smaller than the object but much closer. The eyepiece is then used as a simple magnifier to view the image. A reflecting telescope uses a mirror for its objective.
Magnifying power	$M = \frac{\theta_e}{\theta_o} = -\frac{f_o}{f_e}$ **34-23**

Problem-Solving Guide

The approximate location, size, and orientation of an image are most easily determined by a ray diagram.

Summary of Worked Examples

Type of Calculation	Procedure and Relevant Examples
1. Mirrors	
Find the image distance and size.	Use $1/s + 1/s' = 1/f$, where $f = \frac{1}{2}r$ is the focal length of the mirror. Draw a ray diagram to check your result. To find the size, use $m = y'/y = -s'/s$. **Examples 34-1, 34-2**
2. Refracting Surfaces	
Find the image distance and size.	Use $n_1/s + n_2/s' = (n_2 - n_1)/r$ to find the distance. To find the size, use $m = y'/y = -n_1 s'/n_2 s$. **Example 34-4**
Find the apparent depth of an object.	The apparent depth equals the real depth divided by the index of refraction of the medium. **Example 34-5**
3. Thin Lenses	
Find the focal length.	Use $1/f = (n - 1)(1/r_1 - 1/r_2)$. Be sure to use the correct sign for each radius. **Examples 34-6, 34-7**
Find the power.	The power is the reciprocal of the focal length expressed in meters. **Example 34-7**
Find the image distance and size.	Use $1/s + 1/s' = 1/f$ to locate the image and $m = y'/y = -s'/s$ to find the size. The image is real if s' is positive, virtual if s' is negative. Check your result with a ray diagram. **Example 34-8**
4. Lens Combinations	To find the final image, first find the image distance for the first lens, then use it to find the object distance for the second lens. **Examples 34-9, 34-10**

5. Optical Instruments (optional)

Find the change in focal length of the eye lens.	The image distance is fixed at $s' = 2.5$ cm, the distance from the lens to the retina. Use the thin-lens equation to find f for a given s.	**Example 34-12**
Find the power of reading glasses needed.	Use the same procedure as for lens combinations. If the glasses are right next to the eye, the object distance for the eye lens will be the negative of the image distance for the glasses lens.	**Example 34-13**
Find the magnifying power of a simple magnifier.	Use $M = x_{np}/f$, where x_{np} is the near point of the eye.	**Example 34-14**
Find the magnifying power of a compound microscope.	Use $M = -f_o/f_e$.	**Example 34-15**

Problems

Conceptual Problems

Problems from Optional and Exploring sections

- • Single-concept, single-step, relatively easy
- •• Intermediate-level, may require synthesis of concepts
- ••• Challenging, for advanced students

In a few problems, you are given more data than you actually need; in a few other problems, you are required to supply data from your general knowledge, outside sources, or informed estimates.

Use $n = 1.33$ for the index of refraction of water unless otherwise specified.

Plane Mirrors

1 • Can a virtual image be photographed?

2 • Suppose each axis of a coordinate system like the one in Figure 34-4 is painted a different color. One photograph is taken of the coordinate system and another is taken of its image in a plane mirror. Is it possible to tell that one of the photographs is of a mirror image rather than both being photographs of the real coordinate system from different angles?

3 • The image of the point object P in Figure 34-52 is viewed by an eye as shown. Draw a bundle of rays from the object that reflect from the mirror and enter the eye. For this object position and mirror, indicate the region of space in which the eye can see the image.

Figure 34-52 Problem 3

4 • A person 1.62 m tall wants to be able to see her full image in a plane mirror. (*a*) What must be the minimum height of the mirror? (*b*) How far above the floor should it be placed, assuming that the top of the person's head is 15 cm above her eye level? Draw a ray diagram.

5 • Two plane mirrors make an angle of 90°. Show by considering various object positions that there are three images for any position of an object. Draw appropriate bundles of rays from the object to the eye for viewing each image.

6 • (*a*) Two plane mirrors make an angle of 60° with each other. Show on a sketch the location of all the images formed of a point object on the bisector of the angle between the mirrors. (*b*) Repeat for an angle of 120°.

7 •• When two plane mirrors are parallel, such as on opposite walls in a barber shop, multiple images arise because each image in one mirror serves as an object for the other mirror. A point object is placed between parallel mirrors separated by 30 cm. The object is 10 cm in front of the left mirror and 20 cm in front of the right mirror. (*a*) Find the distance from the left mirror to the first four images in that mirror. (*b*) Find the distance from the right mirror to the first four images in that mirror.

Spherical Mirrors

8 •• True or False

(*a*) The virtual image formed by a concave mirror is always smaller than the object.
(*b*) A concave mirror always forms a virtual image.
(*c*) A convex mirror never forms a real image of a real object.
(*d*) A concave mirror never forms an enlarged real image of an object.

9 •• Under what condition will a concave mirror produce an erect image? A virtual image? An image smaller than the object? An image larger than the object?

10 •• Answer Problem 9 for a convex mirror.

11 •• Convex mirrors are often used for rear-view mirrors on cars and trucks to give a wide-angle view. Below the mirror is written, "Warning, objects are closer than they appear." Yet according to a ray diagram such as Figure 34-19, the image distance for distant objects is much smaller than the object distance. Why then do they appear more distant?

12 •• As an object is moved from a great distance toward the focal point of a concave mirror, the image moves from

(*a*) a great distance toward the focal point and is always real.
(*b*) the focal point to a great distance from the mirror and is always real.
(*c*) the focal point toward the center of curvature of the mirror and is always real.
(*d*) the focal point to a great distance from the mirror and changes from a real to a virtual image.

13 • A concave spherical mirror has a radius of curvature of 40 cm. Draw ray diagrams to locate the image (if one is formed) for an object at a distance of (*a*) 100 cm, (*b*) 40 cm, (*c*) 20 cm, and (*d*) 10 cm from the mirror. For each case, state whether the image is real or virtual; erect or inverted; and enlarged, reduced, or the same size as the object.

14 • Use the mirror equation to locate and describe the images for the object distances and mirror of Problem 13.

15 • Repeat Problem 13 for a convex mirror with the same radius of curvature.

16 • Use the mirror equation to locate and describe the images for the object distances and convex mirror of Problem 15.

17 • Show that a convex mirror cannot form a real image of a real object, no matter where the object is placed, by showing that s' is always negative for a positive s.

18 • A dentist wants a small mirror that will produce an upright image with a magnification of 5.5 when the mirror is located 2.1 cm from a tooth. (*a*) What should the radius of curvature of the mirror be? (*b*) Should it be concave or convex?

19 •• Convex mirrors are used in stores to provide a wide angle of surveillance for a reasonable mirror size. The mirror shown in Figure 34-53 allows a clerk 5 m away from the mirror to survey the entire store. It has a radius of curvature of 1.2 m. (*a*) If a customer is 10 m from the mirror, how far from the mirror surface is his image? (*b*) Is the image in front of or behind the mirror? (*c*) If the customer is 2 m tall, how high is his image?

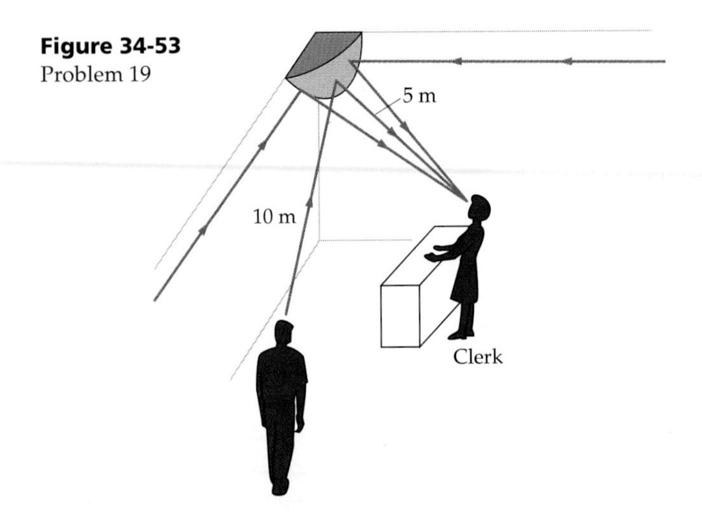

Figure 34-53 Problem 19

20 •• A certain telescope uses a concave spherical mirror with a radius of curvature of 8 m. Find the location and diameter of the image of the moon formed by this mirror. The moon has a diameter of 3.5×10^6 m and is 3.8×10^8 m from the earth.

21 •• A concave spherical mirror has a radius of curvature of 6.0 cm. A point object is on the axis 9 cm from the mirror. Construct a precise ray diagram showing rays from the object that make angles of 5°, 10°, 30°, and 60° with the axis, strike the mirror, and are reflected back across the axis. (Use a compass to draw the mirror, and use a protractor to measure the angles needed to find the reflected rays.) What is the spread δx of the points where these rays cross the axis?

22 •• A concave mirror has a radius of curvature 6.0 cm. Draw rays parallel to the axis at 0.5, 1.0, 2.0, and 4.0 cm above the axis and find the points at which the reflected rays cross the axis. (Use a compass to draw the mirror and a protractor to find the angle of reflection for each ray.) (*a*) What is the spread δx of the points where these rays cross the axis? (*b*) By what percentage could this spread be reduced if the edge of the mirror were blocked off so that parallel rays more than 2.0 cm from the axis could not strike the mirror?

23 •• An object placed 8 cm from a concave spherical mirror produces a virtual image 10 cm behind the mirror. (*a*) If the object is moved back to 25 cm from the mirror, where is the image located? (*b*) Is it real or virtual?

24 •• An object located 100 cm from a concave mirror forms a real image 75 cm from the mirror. The mirror is then turned around so that its convex side faces the object. The mirror is moved so that the image is now 35 cm behind the mirror. How far was the mirror moved? Was it moved toward or away from the object?

25 •• Parallel light from a distant object strikes the large mirror in Figure 34-54 ($r = 5$ m) and is reflected by the small mirror that is 2 m from the large mirror. The small mirror is actually spherical, not planar as shown. The light is focused at the vertex of the large mirror. (*a*) What is the radius of curvature of the small mirror? (*b*) Is it convex or concave?

Figure 34-54 Problem 25

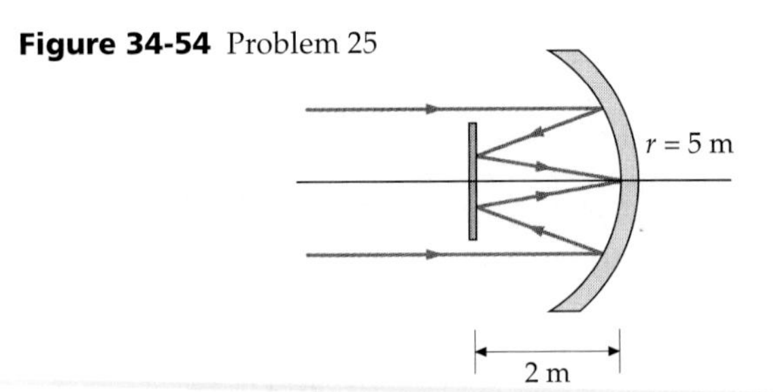

26 •• A woman uses a concave makeup mirror with a radius of curvature of 1.5 m. How far from the mirror should her face be for the image to be 80 cm from her face?

Images Formed by Refraction

27 • A bird above the water is viewed by a scuba diver submerged beneath the water's surface directly below the bird. Does the bird appear to the diver to be closer to or farther from the surface than it actually is?

28 • A sheet of paper with writing on it is protected by a thick glass plate having an index of refraction of 1.5. If the plate is 2 cm thick, at what distance beneath the top of the plate does the writing appear when it is viewed from directly overhead?

29 • A fish is 10 cm from the front surface of a fish bowl of radius 20 cm. (*a*) Where does the fish appear to be to someone in air viewing it from in front of the bowl? (*b*) Where does the fish appear to be when it is 30 cm from the front surface of the bowl?

30 •• A very long glass rod of 2-cm diameter has one end ground to a convex spherical surface of radius 5 cm. Its index of refraction is 1.5. (*a*) A point object in air is on the axis of the rod 20 cm from the surface. Find the image and state whether it is real or virtual. Repeat for (*b*) an object 5 cm from the surface and (*c*) an object very far from the surface. Draw a ray diagram for each case.

31 •• At what distance from the rod of Problem 30 should the object be placed so that the light rays in the rod are parallel? Draw a ray diagram for this situation.

32 •• Repeat Problem 30 for a glass rod with a concave hemispherical surface of radius -5 cm.

33 •• Repeat Problem 30 when the glass rod and objects are immersed in water.

34 •• Repeat Problem 30 for a glass rod with a concave hemispherical surface of radius -5 cm when it and the objects are immersed in water.

35 •• A glass rod 96 cm long with an index of refraction of 1.6 has its ends ground to convex spherical surfaces of radii 8 and 16 cm. A point object is in air on the axis of the rod 20 cm from the end with the 8-cm radius. (*a*) Find the image distance due to refraction at the first surface. (*b*) Find the final image due to refraction at both surfaces. (*c*) Is the final image real or virtual?

36 •• Repeat Problem 35 for a point object in air on the axis of the rod 20 cm from the end with the 16-cm radius.

Thin Lenses

37 • Under what conditions will the focal length of a thin lens be positive? Negative?

38 • The focal length of a simple lens is different for different colors of light. Why?

39 •• An object is placed 40 cm from a lens of focal length -10 cm. The image is

(*a*) real, inverted, and diminished.
(*b*) real, inverted, and enlarged.
(*c*) virtual, inverted, and diminished.
(*d*) virtual, upright, and diminished.
(*e*) virtual, upright, and enlarged.

40 •• If a real object is placed just inside the focal point of a converging lens, the image is

(*a*) real, inverted, and enlarged.
(*b*) virtual, erect, and diminished.
(*c*) virtual, erect, and enlarged.
(*d*) real, inverted, and diminished.

41 • The following thin lenses are made of glass with an index of refraction of 1.5. Make a sketch of each lens, and find its focal length in air: (*a*) double convex, $r_1 = 10$ cm and $r_2 = -21$ cm; (*b*) plano-convex, $r_1 = \infty$ and $r_2 = -10$ cm; (*c*) double concave, $r_1 = -10$ cm and $r_2 = +10$ cm; (*d*) plano-concave, $r_1 = \infty$ and $r_2 = +20$ cm.

42 • Glass with an index of refraction of 1.6 is used to make a thin lens that has radii of equal magnitude. Find the radii of curvature and make a sketch of the lens if the focal length in air is (*a*) $+5$ cm and (*b*) -5 cm.

43 • Find the focal length of a glass lens of index of refraction 1.62 that has a concave surface with radius of magnitude 100 cm and a convex surface with a radius of magnitude 40 cm.

44 • A double-concave lens of index of refraction 1.45 has radii of magnitudes 30 and 25 cm. An object is located 80 cm to the left of the lens. Find (*a*) the focal length of the lens, (*b*) the location of the image, and (*c*) the magnification of the image. (*d*) Is the image real or virtual? Upright or inverted?

45 • The following thin lenses are made of glass of index of refraction 1.6. Make a sketch of each lens, and find its focal length in air: (*a*) $r_1 = 20$ cm, $r_2 = 10$ cm; (*b*) $r_1 = 10$ cm, $r_2 = 20$ cm; (*c*) $r_1 = -10$ cm, $r_2 = -20$ cm.

46 • For the following object distances and focal lengths of thin lenses in air, find the image distance and the magnification and state whether the image is real or virtual and erect or inverted: (*a*) $s = 40$ cm, $f = 20$ cm; (*b*) $s = 10$ cm, $f = 20$ cm; (*c*) $s = 40$ cm, $f = -30$ cm; (*d*) $s = 10$ cm, $f = -30$ cm.

47 • An object 3.0 cm high is placed 20 cm in front of a thin lens of power 20 D. Draw a precise ray diagram to find the position and size of the image and check your results using the thin-lens equation.

48 • Repeat Problem 47 for an object 1.0 cm high placed 10 cm in front of a thin lens of power 20 D.

49 • Repeat Problem 47 for an object 1.0 cm high placed 10 cm in front of a thin lens whose power is -20 D.

50 •• (*a*) What is meant by a negative object distance? How can it occur? Find the image distance and magnification and state whether the image is virtual or real and erect or inverted for a thin lens in air when (*b*) $s = -20$ cm, $f = +20$ cm and (*c*) $s = -10$ cm, $f = -30$ cm. Draw a ray diagram for each of these cases.

51 •• Two converging lenses, each of focal length 10 cm, are separated by 35 cm. An object is 20 cm to the left of the first lens. (*a*) Find the position of the final image using both a ray diagram and the thin-lens equation. (*b*) Is the image real or virtual? Erect or inverted? (*c*) What is the overall lateral magnification of the image?

52 •• Work Problem 51 for a second lens that is a diverging lens of focal length -15 cm.

53 •• A thin lens of index of refraction 1.5 has one convex side with a radius of magnitude 20 cm. When an object 1 cm in height is placed 50 cm from this lens, an upright image 2.15 cm in height is formed. (*a*) Calculate the radius of the second side of the lens. Is it concave or convex? (*b*) Draw a sketch of the lens.

54 •• (*a*) Show that to obtain a magnification of magnitude m with a converging thin lens of focal length f, the object distance must be given by $s = (m - 1)f/m$. (*b*) A camera lens with 50-mm focal length is used to take a picture of a person 1.75 m tall. How far from the camera should the person stand so that the image size is 24 mm?

55 •• An object is 15 cm in front of a positive lens of focal length 15 cm. A second positive lens of focal length 15 cm is 20 cm from the first lens. Find the final image and draw a ray diagram.

56 •• Work Problem 55 for a second lens with a focal length of -15 cm.

57 ••• In a convenient form of the thin-lens equation used by Newton, the object and image distances are measured from the focal points. Show that if $x = s - f$ and $x' = s' - f$, the thin-lens equation can be written as $xx' = f^2$, and the lateral magnification is given by $m = -x'/f = -f/x$. Indicate x and x' on a sketch of a lens.

58 ••• An object is placed 2.4 m from a screen, and a lens of focal length f is placed between the object and the screen so that a real image of the object is formed on the screen. When the lens is moved 1.2 m toward the screen, another real image of the object is formed on the screen. (*a*) Where was the lens located before it was moved? (*b*) What is the focal length of the lens?

59 ••• An object is 17.5 cm to the left of a lens of focal length 8.5 cm. A second lens of focal length -30 cm is 5 cm to the right of the first lens. (*a*) Find the distance between the object and the final image formed by the second lens. (*b*) What is the overall magnification? (*c*) Is the final image real or virtual? Upright or inverted?

Aberrations (optional)

60 • Chromatic aberration is a common defect of

(*a*) concave and convex lenses.
(*b*) concave lenses only.
(*c*) concave and convex mirrors.
(*d*) all lenses and mirrors.

61 • True or false:

(*a*) Aberrations occur only for real images.
(*b*) Chromatic aberration does not occur with mirrors.

62 • A double-convex lens of radii $r_1 = +10$ cm and $r_2 = -10$ cm is made from glass with indexes of refraction of 1.53 for blue light and 1.47 for red light. Find the focal length of this lens for (*a*) red light and (*b*) blue light.

The Eye (optional)

In the following problems, take the distance from the cornea–lens system of the eye to the retina to be 2.5 cm.

63 • If an object is placed 25 cm from the eye of a farsighted person who does not wear corrective lenses, a sharp image is formed

(*a*) behind the retina, and the corrective lens should be convex.
(*b*) behind the retina, and the corrective lens should be concave.
(*c*) in front of the retina, and the corrective lens should be convex.
(*d*) in front of the retina, and the corrective lens should be concave.

64 •• Myopic (nearsighted) persons sometimes claim to see better under water without corrective lenses. Why?

(*a*) The accommodation of the eye's lens is better under water.
(*b*) Refraction at the water–cornea interface is less than at the air–cornea interface.
(*c*) Refraction at the water–cornea interface is greater than at the air–cornea interface.
(*d*) No reason; the effect is only an illusion and not really true.

65 •• A nearsighted person who wears corrective lenses would like to examine an object at close distance. Identify the correct statement.

(*a*) The corrective lenses give an enlarged image and should be worn while examining the object.
(*b*) The corrective lenses give a reduced image of the object and should be removed.
(*c*) The corrective lenses result in a magnification of unity; it does not matter whether they are worn or removed.

66 • Suppose the eye were designed like a camera with a lens of fixed focal length $f = 2.5$ cm that could move toward or away from the retina. Approximately how far would the lens have to move to focus the image of an object 25 cm from the eye onto the retina? (*Hint:* Find the distance from the retina to the image behind it for an object at 25 cm.)

67 • Find the change in the focal length of the eye when an object originally at 3 m is brought to 30 cm from the eye.

68 • Find (*a*) the focal length and (*b*) the power of a lens that will produce an image at 80 cm from the eye of a book that is 30 cm from the eye.

69 • A farsighted person requires lenses with a power of 1.75 D to read comfortably from a book that is 25 cm from the eye. What is that person's near point without the lenses?

70. • If two point objects close together are to be seen as two distinct objects, the images must fall on the retina on two different cones that are not adjacent. That is, there must be an unactivated cone between them. The separation of the cones is about 1 μm. (*a*) What is the smallest angle the two points can subtend? (See Figure 34-55.) (*b*) How close can two points be if they are 20 m from the eye?

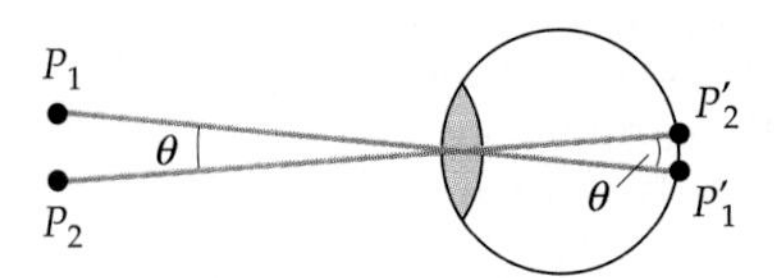

Figure 34-55 Problem 70. The two points will look like two separate points only if their images fall on two different, nonadjacent cones of the retina.

71 •• A person with a near point of 80 cm needs to read from a computer screen that is 45 cm from her eye. (*a*) Find the focal length of the lenses in reading glasses that will produce an image of the screen at 80 cm from her eye. (*b*) What is the power of the lenses?

72 •• A nearsighted person cannot focus clearly on objects more distant than 225 cm from her eye. What power lenses are required for her to see distant objects clearly?

73 •• Since the index of refraction of the lens of the eye is not very different from that of the surrounding material, most of the refraction takes place at the cornea, where n changes abruptly from 1.0 in air to about 1.4. Assuming the cornea to be a homogeneous sphere with an index of refraction of 1.4, calculate its radius if it focuses parallel light on the retina a distance 2.5 cm away. Do you expect your result to be larger or smaller than the actual radius of the cornea?

74 •• The near point of a certain person is 80 cm. Reading glasses are prescribed so that he can read a book at 25 cm from his eye. The glasses are 2 cm from the eye. What diopter lens should be used in the glasses?

75 ••• At age 45, a person is fitted for reading glasses of power 2.1 D in order to read at 25 cm. By the time she reaches 55, she discovers herself holding her newspaper at a distance of 40 cm in order to see it clearly with her glasses on. (*a*) Where was her near point at age 45? (*b*) Where is her near point at age 55? (*c*) What power is now required for the lenses of her reading glasses so that she can again read at 25 cm? (Assume that the glasses are 2.2 cm from her eyes.)

76 ••• An aging physics professor discovers that he can see objects clearly only between 0.75 m and 2.5 m so he decides he needs bifocals. The upper part of the lens allows him to see objects clearly at infinity, and the lower part allows him to see objects clearly at 25 cm. Assume that the lens is 2 cm from his eye. (*a*) Calculate the power of the lens required for the upper part of his bifocals. (*b*) Calculate the power of the lens required for the lower part of his bifocals. (*c*) Is there a range of distance over which he cannot see objects clearly no matter which part of the bifocals he looks through? If so, what is that range?

The Simple Magnifier (optional)

77 • A person with a near-point distance of 30 cm uses a simple magnifier of power 20 D. What is the magnification obtained if the final image is at infinity?

78 • A person with a near-point distance of 25 cm wishes to obtain a magnifying power of 5 with a simple magnifier. What should be the focal length of the lens used?

79 • What is the magnifying power of a lens of focal length 7 cm when the image is viewed at infinity by a person whose near point is at 35 cm?

80 •• A lens of focal length 6 cm is used as a simple magnifier with the image at infinity by one person whose near point is 25 cm and by another whose near point is 40 cm. What is the effective magnifying power of the lens for each person? Compare the size of the image on the retina when each looks at the same object with the magnifier.

81 •• A botanist examines a leaf using a convex lens of power 12 D as a simple magnifier. What is the expected angular magnification if (*a*) the final image is at infinity, and (*b*) the final image is at 25 cm?

82 •• (*a*) Show that if the final image of a simple magnifier is to be at the near point of the eye rather than at infinity, the angular magnification is given by

$$M = \frac{x_{\text{np}}}{f} + 1$$

(*b*) Find the magnification of a 20-D lens for a person with a near point of 30 cm if the final image is at the near point. Draw a ray diagram for this situation.

83 •• Show that when the image of a simple magnifier is viewed at the near point, the lateral and angular magnification of the magnifier are equal.

Microscopes (optional)

84 •• A microscope objective has a focal length of 0.5 cm. It forms an image at 16 cm from its second focal point. What is the magnifying power for a person whose near point is at 25 cm if the focal length of the eyepiece is 3 cm?

85 •• A microscope has an objective of focal length 16 mm and an eyepiece that gives an angular magnification of 5 for a person whose near point is 25 cm. The tube length is 18 cm. (*a*) What is the lateral magnification of the objective? (*b*) What is the magnifying power of the microscope?

86 •• A crude symmetric hand-held microscope consists of two converging 20-D lenses fastened in the ends of a tube 30 cm long. (*a*) What is the "tube length" of this microscope? (*b*) What is the lateral magnification of the objective? (*c*) What is the magnifying power of the microscope? (*d*) How far from the objective should the object be placed?

87 •• Repeat Problem 86 for the same two lenses separated by 40 cm.

88 •• A compound microscope has an object with a power of 45 D and an eyepiece with a power of 80 D. The lenses are separated by 28 cm. Assuming that the final image is formed 25 cm from the eye, what is the magnifying power?

89 ••• A microscope has a magnifying power of 600 and an eyepiece of angular magnification of 15. The objective lens is 22 cm from the eyepiece. Without making any approximations, calculate (*a*) the focal length of the eyepiece, (*b*) the location of the object such that it is in focus for a normal relaxed eye, and (*c*) the focal length of the objective lens.

Telescopes (optional)

90 • A simple telescope has an objective with a focal length of 100 cm and an eyepiece of focal length 5 cm. It is used to look at the moon, which subtends an angle of about 0.009 rad. (*a*) What is the diameter of the image formed by the objective? (*b*) What angle is subtended by the final image at infinity? (*c*) What is the magnifying power of the telescope?

91 • The objective lens of the refracting telescope at the Yerkes Observatory has a focal length of 19.5 m. When it is used to look at the moon, which subtends an angle of about 0.009 rad, what is the diameter of the image of the moon formed by the objective?

92 •• The 200-in (5.1-m) mirror of the reflecting telescope at Mt. Palomar has a focal length of 1.68 m. (*a*) By what factor is the light-gathering power increased over the 40-in (1.016-m) diameter refracting lens of the Yerkes Observatory telescope? (*b*) If the focal length of the eyepiece is 1.25 cm, what is the magnifying power of this telescope?

93 •• An astronomical telescope has a magnifying power of 7. The two lenses are 32 cm apart. Find the focal length of each lens.

94 •• A disadvantage of the astronomical telescope for terrestrial use (for example, at a football game) is that the image is inverted. A Galilean telescope uses a converging lens as its objective, but a diverging lens as its eyepiece. The image formed by the objective is behind the eyepiece at its focal point so that the final image is virtual, erect, and at infinity. (*a*) Show that the magnifying power is $M = -f_o/f_e$, where f_o is the focal length of the objective and f_e is that of the eyepiece (which is negative). (*b*) Draw a ray diagram to show that the final image is indeed virtual, erect, and at infinity.

95 •• A Galilean telescope (see Problem 94) is designed so that the final image is at the near point, which is 25 cm (rather than at infinity). The focal length of the objective is 100 cm and that of the eyepiece is −5 cm. (*a*) If the object distance is 30 m, where is the image of the objective? (*b*) What is the object distance for the eyepiece so that the final image is at the near point? (*c*) How far apart are the lenses? (*d*) If the object height is 1.5 m, what is the height of the final image? What is the angular magnification?

96 ••• A hunter lost in the mountains tries to make a telescope from two lenses of power 2.0 and 6.5 D, and a cardboard tube. (*a*) What is the maximum possible magnifying power? (*b*) How long must the tube be? (*c*) Which lens should be used as the eyepiece? Why?

97 ••• If you look into the wrong end of a telescope, that is, into the objective, you will see distant objects reduced in size. For a refracting telescope with an objective of focal length 2.25 m and an eyepiece of focal length 1.5 cm, by what factor is the angular size of the object reduced?

General Problems

98 • The image of a real object formed by a convex mirror

(*a*) is always real and inverted.
(*b*) is always virtual and enlarged.
(*c*) may be real.
(*d*) is always virtual and diminished.

99 • The glass of a converging lens has an index of refraction of 1.6. When the lens is in air, its focal length is 30 cm. If immersed in water, its focal length will be

(*a*) greater than 30 cm. (*b*) less than 30 cm.
(*c*) the same as before, 30 cm. (*d*) negative.

100 •• True or false:

(*a*) A virtual image cannot be displayed on a screen.
(*b*) A negative image distance implies that the image is virtual.
(*c*) All rays parallel to the axis of a spherical mirror are reflected through a single point.
(*d*) A diverging lens cannot form a real image from a real object.
(*e*) The image distance for a positive lens is always positive.

101 • Show that a diverging lens can never form a real image from a real object. (*Hint:* Show that s' is always negative.)

102 • A camera uses a positive lens to focus light from an object onto a film. Unlike the eye, the camera lens has a fixed focal length, but the lens itself can be moved slightly to vary the image distance to the image on the film. A telephoto lens has a focal length of 200 mm. By how much must it move to change from focusing on an object at infinity to one at a distance of 30 m?

103 • A wide-angle lens of a camera has a focal length of 28 mm. By how much must it move to change from focusing on an object at infinity to one at a distance of 5 m? (See Problem 102.)

104 • A converging lens made of polystyrene (index of refraction, 1.59) has a focal length of 50 cm. One surface is convex with radius of magnitude 50 cm. Find the radius of the second surface. Is it convex or concave?

105 • A thin converging lens of focal length 10 cm is used to obtain an image that is twice as large as a small object. Find the object and image distances if (*a*) the image is to be erect and (*b*) the image is to be inverted. Draw a ray diagram for each case.

106 •• A scuba diver wears a diving mask with a face plate that bulges outward with a radius of curvature of 0.5 m. There is thus a convex spherical surface between the water and the air in the mask. A fish is 2.5 m in front of the diving mask. (*a*) Where does the fish appear to be? (*b*) What is the magnification of the image of the fish?

107 •• You wish to see an image of your face for applying makeup or shaving. If you want the image to be upright, virtual, and magnified 1.5 times when your face is 30 cm from the mirror, what kind of mirror should you use, convex or concave, and what should its focal length be?

108 •• A small object is 20 cm from a thin positive lens of focal length 10 cm. To the right of the lens is a plane mirror that crosses the axis at the second focal point of the lens and is tilted so that the reflected rays do not go back through the lens (Figure 34-56). (*a*) Find the position of the final image. (*b*) Is this image real or virtual? (*c*) Sketch a ray diagram showing the final image.

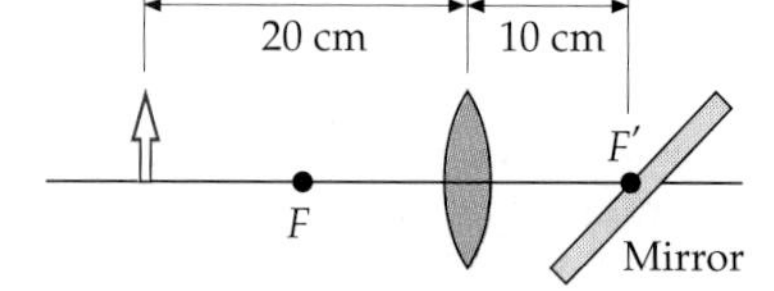

Figure 34-56
Problem 108

109 •• A 35-mm camera has a picture size of 24 mm by 36 mm. It is used to take a picture of a person 175 cm tall so that the image just fills the height (24 mm) of the film. How far should the person stand from the camera if the focal length of the lens is 50 mm?

110 •• A 35-mm camera with interchangeable lenses is used to take a picture of a hawk that has a wing span of 2 m. The hawk is 30 m away. What would be the ideal focal length of the lens used so that the image of the wings just fills the width of the film, which is 36 mm?

111 •• An object is placed 12 cm to the left of a lens of focal length 10 cm. A second lens of focal length 12.5 cm is placed 20 cm to the right of the first lens. (*a*) Find the position of the final image. (*b*) What is the magnification of the image? (*c*) Sketch a ray diagram showing the final image.

112 •• (*a*) Show that if f is the focal length of a thin lens in air, its focal length in water is

$$\frac{n_w(n-1)}{n-n_w}f$$

where n_w is the index of refraction of water and n is that of the lens. (*b*) Calculate the focal length in air and in water of a double-concave lens of index of refraction $n = 1.5$ that has radii of magnitudes 30 and 35 cm.

113 •• (*a*) Find the focal length of a *thick* double-convex lens with an index of refraction of 1.5, a thickness of 4 cm, and radii of +20 and −20 cm. (*b*) Find the focal length of this lens in water.

114 •• A 2-cm-thick layer of water ($n = 1.33$) floats on top of a 4-cm-thick layer of carbon tetrachloride ($n = 1.46$) in a tank. How far below the top surface of the water does the bottom of the tank appear to be to an observer looking from above at normal incidence?

115 •• While sitting in your car, you see a jogger in your side mirror, which is convex with a radius of curvature of magnitude 2 m. The jogger is 5 m from the mirror and is approaching at 3.5 m/s. How fast does the jogger appear to be running when viewed in the mirror?

116 •• In the seventeenth century, Antonie van Leeuwenhoek, the first great microscopist, used simple spherical lenses made first of water droplets and then of glass for his first instruments. He made staggering discoveries with these simple lenses. Consider a glass sphere of radius 2.0 mm with an index of refraction of 1.50. Find the focal length of this lens. *Hint:* Use the equation for refraction at a single spherical surface to find the image distance for an infinite object distance for the first surface. Then use this image point as the object point for the second surface.

117 ••• An object is 15 cm to the left of a thin convex lens of focal length 10 cm. A concave mirror of radius 10 cm is 25 cm to the right of the lens. (*a*) Find the position of the final image formed by the mirror and lens. (*b*) Is the image real or virtual? Erect or inverted? (*c*) Show on a diagram where your eye must be to see this image.

118 ••• Find the final image for the situation in Problem 108 when the mirror is not tilted. Assume that the image is viewed by an eye to the left of the object looking through the lens into the mirror.

119 ••• When a bright light source is placed 30 cm in front of a lens, there is an erect image 7.5 cm from the lens. There is also a faint inverted image 6 cm in front of the lens due to reflection from the front surface of the lens. When the lens is turned around, this weaker, inverted image is 10 cm in front of the lens. Find the index of refraction of the lens.

120 ••• A horizontal concave mirror with radius of curvature of 50 cm holds a layer of water with an index of refraction of 1.33 and a maximum depth of 1 cm. At what height above the mirror must an object be placed so that its image is at the same position as the object?

121 ••• A lens with one concave side with a radius of magnitude 17 cm and one convex side with a radius of magnitude 8 cm has a focal length in air of 27.5 cm. When placed in a liquid with an unknown index of refraction, the focal length increases to 109 cm. What is the index of refraction of the liquid?

122 ••• A glass ball of radius 10 cm has an index of refraction of 1.5. The back half of the ball is silvered so that it acts as a concave mirror (Figure 34-57). Find the position of the final image seen by an eye to the left of the object and ball for an object at (*a*) 30 cm and (*b*) 20 cm to the left of the front surface of the ball.

Figure 34-57 Problem 122

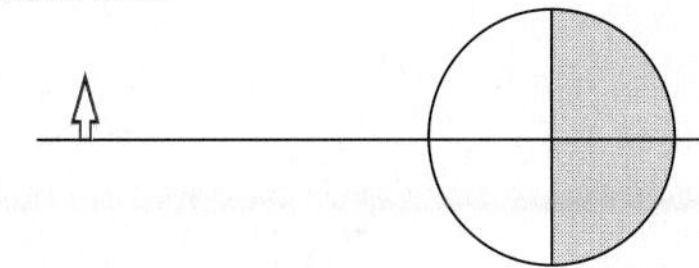

123 ••• (*a*) Show that a small change dn in the index of refraction of a lens material produces a small change in the focal length df given approximately by $df/f = -dn/(n-1)$. (*b*) Use this result to find the focal length of a thin lens for blue light, for which $n = 1.53$, if the focal length for red light, for which $n = 1.47$, is 20 cm.

124 ••• The lateral magnification of a spherical mirror or a thin lens is given by $m = -s'/s$. Show that for objects of small horizontal extent, the longitudinal magnification is approximately $-m^2$. (*Hint:* Show that $ds'/ds = s'^2/s^2$.)

125 ••• A thin double-convex lens has radii r_1 and r_2 and an index of refraction n_L. The surface of radius r_1 is in contact with a liquid of index of refraction n_1, and the surface of radius r_2 is in contact with a liquid of index of refraction n_2. Show that the thin-lens equation for this situation can be expressed as $n_1/s + n_2/s' = n_2/f$ where the focal length is given by

$$\frac{1}{f} = \frac{n_L - n_1}{n_2 r_1} - \frac{n_L - n_2}{n_2 r_2}$$

CHAPTER 35

Interference and Diffraction

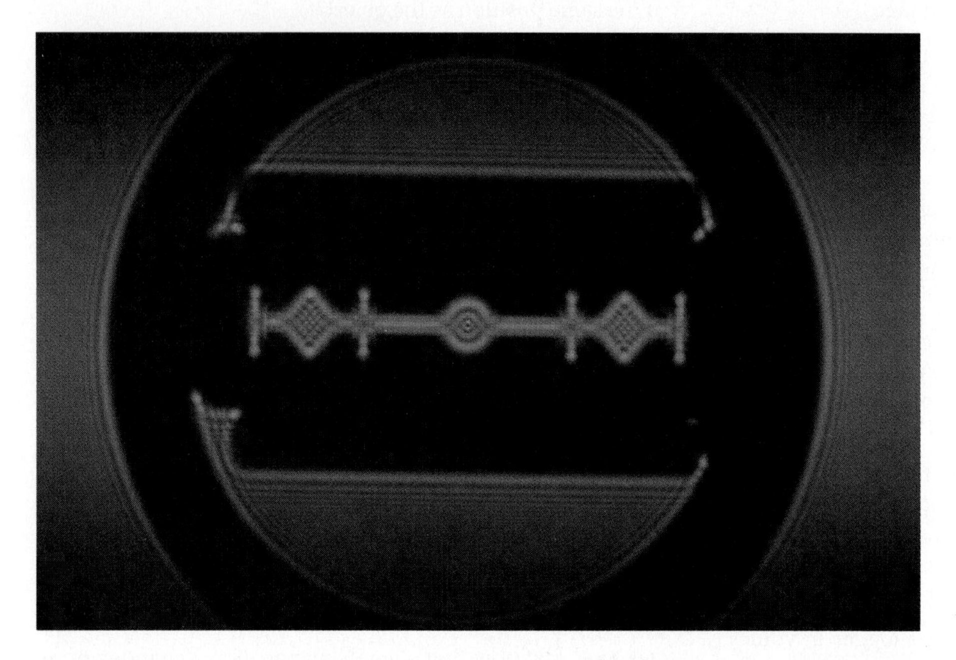

The diffraction of light incident on a razor blade.

Interference and diffraction are the important phenomena that distinguish waves from particles.* Interference is the combining by superposition of two or more waves that meet at one point in space. Diffraction is the bending of waves around corners that occurs when a portion of a wavefront is cut off by a barrier or obstacle. The pattern of the resulting wave can be calculated by treating each point on the original wavefront as a point source according to Huygens' principle and calculating the interference pattern resulting from these sources.

35-1 Phase Difference and Coherence

When two harmonic waves of the same frequency and wavelength but differing in phase combine, the resultant wave is a harmonic wave whose amplitude depends on the phase difference. If the phase difference is zero or an integer times 360°, the waves are in phase and interfere constructively. The resultant amplitude equals the sum of the individual amplitudes, and the intensity (which is proportional to the square of the amplitude) is maximum. If the phase difference is 180° or any odd integer times 180°, the waves are out of phase and interfere destructively. The resultant amplitude is then the dif-

*Before you study this chapter, you should review Chapters 15, 16, and 17, where the general topics of interference and diffraction of waves are discussed.

ference between the individual amplitudes, and the intensity is a minimum. If the amplitudes are equal, the maximum intensity is four times that of either source and the minimum intensity is zero.

A phase difference between two waves is often the result of a difference in path length. A path difference of one wavelength produces a phase difference of 360°, which is equivalent to no phase difference at all. A path difference of one-half wavelength produces a 180° phase difference. In general, a path difference of Δr contributes a phase difference δ given by

$$\delta = \frac{\Delta r}{\lambda} 2\pi = \frac{\Delta r}{\lambda} 360^\circ \qquad \text{35-1}$$

Example 35-1

(*a*) What is the minimum path difference that will produce a phase difference of 180° for light of wavelength 800 nm? (*b*) What phase difference will that path difference produce in light of wavelength 700 nm?

Picture the Problem For both questions we use the relationship between phase difference and path difference given in Equation 35-1.

(*a*) Set $\delta = 180^\circ$ in Equation 35-1 and solve for Δr:

$$\delta = \frac{\Delta r}{\lambda} 360^\circ = 180^\circ$$

$$\Delta r = \frac{\lambda\delta}{360^\circ} = \frac{1}{2}\lambda = \frac{1}{2}(800 \text{ nm}) = 400 \text{ nm}$$

(*b*) Set $\lambda = 700$ nm, $\Delta r = 400$ nm, and solve for δ:

$$\delta = \frac{\Delta r}{\lambda} 360^\circ = \frac{400 \text{ nm}}{700 \text{ nm}} 360^\circ = 206^\circ = 3.59 \text{ rad}$$

Another cause of phase difference is the 180° phase change a wave sometimes undergoes upon reflection from a boundary surface. This phase change is analogous to the inversion of a pulse on a string when it reflects from a point where the density suddenly increases, such as when a light string is attached to a heavier string or rope. The inversion of the reflected pulse is equivalent to a phase change of 180° for a sinusoidal wave, which can be thought of as a series of pulses. When light traveling in air strikes the surface of a medium in which light travels more slowly, such as glass or water, there is a 180° phase change in the reflected light. When light is originally traveling in glass or water, there is no phase change in the light reflected from the glass–air or water–air interface. This is analogous to the reflection without inversion of a pulse on a heavy string at a point where the heavy string is attached to a lighter string.

As we saw in Chapter 16, interference of waves from two sources is not observed unless the sources are coherent. Because a light beam is usually the result of millions of atoms radiating independently, the phase difference between the waves from such sources fluctuates randomly many times per second, so two light sources are usually not coherent. Coherence in optics is often achieved by splitting the light beam from a single source into two or more beams that can then be combined to produce an interference pattern. The light beam can be split by reflecting the light from the two closely spaced surfaces of a thin film (Section 35-2), by diffracting the beam through two small openings or slits in an opaque barrier (Section 35-3), or by using a single point source and its image in a plane mirror for the two sources (Section 35-3). Today, lasers are the most important sources of coherent light in the laboratory.

35-2 Interference in Thin Films

You have probably noticed the colored bands in a soap bubble or in the film on the surface of oily water. These bands are due to the interference of light reflected from the top and bottom surfaces of the film. The different colors arise because of variations in the thickness of the film, causing interference for different wavelengths at different points.

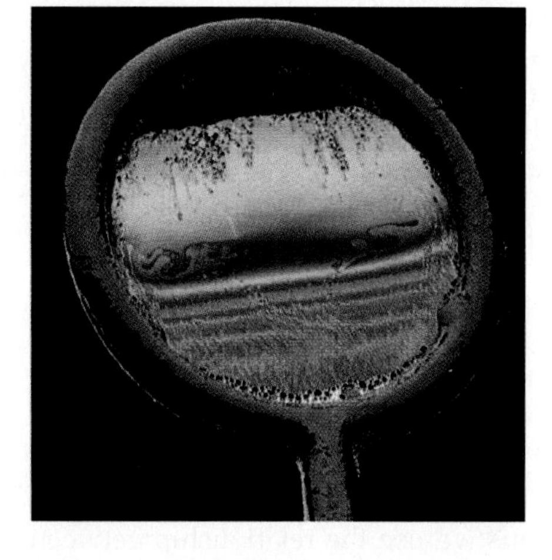

Interference of light from the front and back surface of a thin soap film. At the top where the film is very thin, the rays from the front surface of the film (which undergo a 180° phase change) and the rays from the back surface of the film (which do not change phase) interfere destructively and the film appears dark. At other parts of the film, the interference is destructive or constructive depending on the wavelength and on the thickness of the film.

Consider a thin film of water (such as a small section of a soap bubble) of uniform thickness viewed at small angles with the normal as shown in Figure 35-1. Part of the light is reflected from the upper, air–water interface where it undergoes a 180° phase change. Some of the light enters the film and is partially reflected by the bottom water–air interface. There is no phase change in this reflected light. If the light is nearly perpendicular to the surfaces, both the ray reflected from the top surface and the one reflected from the bottom surface can enter the eye at point P in the figure. The path difference between these two rays is $2t$, where t is the thickness of the film. This path difference produces a phase difference of $(2t/\lambda')360°$, where $\lambda' = \lambda/n$ is the wavelength of the light in the film, and n is the index of refraction of the film. The total phase difference between these two rays is thus 180° plus that due to the path difference. Destructive interference occurs when the path difference $2t$ is zero or a whole number of wavelengths λ' (in the film). Constructive interference occurs when the path difference is an odd number of half-wavelengths.

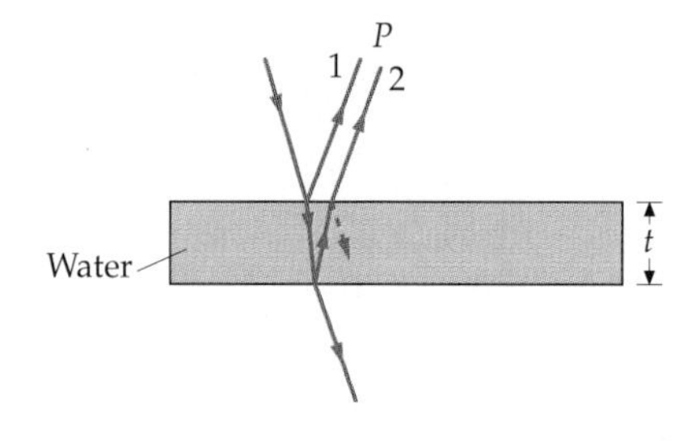

Figure 35-1 Light rays reflected from the top and bottom surfaces of a thin film are coherent because both rays come from the same source. If the light is incident nearly normally, the two reflected rays will be very close to each other and will produce interference.

When a thin water film lies on a glass surface as in Figure 35-2, the ray that reflects from the lower, water–glass interface also undergoes a 180° phase change because the index of refraction of glass (about 1.5) is greater than that of water (about 1.33). Thus, both the rays shown in the figure have undergone a 180° phase change upon reflection. The phase difference between these rays is due solely to the path difference and is given by $\delta = (2t/\lambda')360°$.

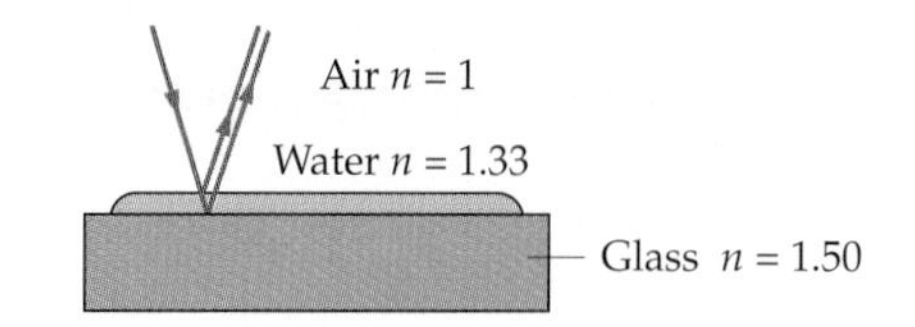

Figure 35-2 Interference of light reflected from a thin film of water resting on a glass surface. In this case, both rays undergo a change in phase of 180° upon reflection.

When a thin film of varying thickness is viewed with monochromatic light, such as the yellow light from a sodium lamp, alternating bright and dark bands or lines called **fringes** are observed. The distance between a bright fringe and a dark fringe is that distance over which the film's thickness changes such that the path difference $2t$ is $\lambda'/2$. Figure 35-3a shows the interference pattern observed when light is reflected from an air film between a spherical glass surface and a plane glass surface in contact. These circular interference fringes are known as **Newton's rings.** Typical rays re-

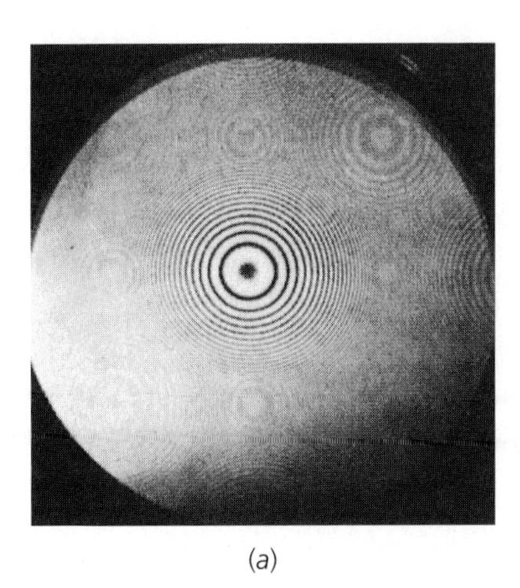

(a)

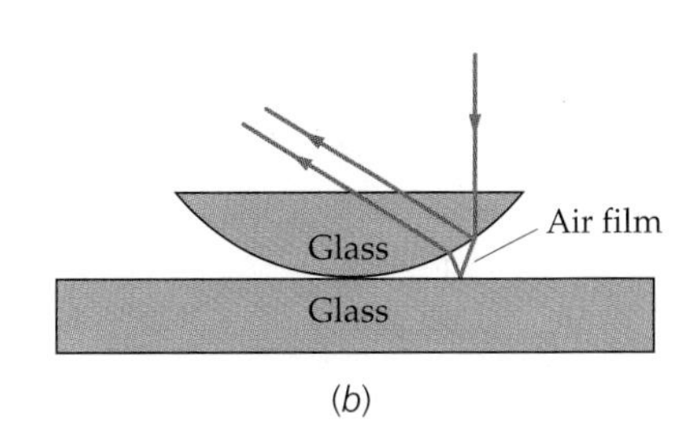

(b)

Figure 35-3 (*a*) Newton's rings observed with light reflected from a thin film of air between a plane glass surface and a spherical glass surface. At the center, the thickness of the air film is negligible and the interference is destructive because of the phase change of one of the rays. (*b*) Glass surfaces for the observation of Newton's rings shown in part (*a*). The thin film in this case is the film of air between the glass surfaces.

flected at the top and bottom of the air film are shown in Figure 35-3*b*. Near the point of contact of the surfaces, where the path difference between the ray reflected from the upper glass–air interface and the ray reflected from the lower air–glass interface is essentially zero or is at least small compared with the wavelength of light, the interference is perfectly destructive because of the 180° phase shift of the ray reflected from the lower air–glass interface. This central region in Figure 35-3*a* is therefore dark. The first bright fringe occurs at the radius at which the path difference is $\lambda/2$, which contributes a phase difference of 180°. This adds to the phase shift due to reflection to produce a total phase difference of 360°, which is equivalent to a zero phase difference. The second dark region occurs at the radius at which the path difference is λ, and so on.

Example 35-2

A wedge-shaped film of air is made by placing a small slip of paper between the edges of two flat pieces of glass as shown in Figure 35-4. Light of wavelength 500 nm is incident normally on the glass, and interference fringes are observed by reflection. If the angle θ made by the plates is 3×10^{-4} rad, how many interference fringes per centimeter are observed?

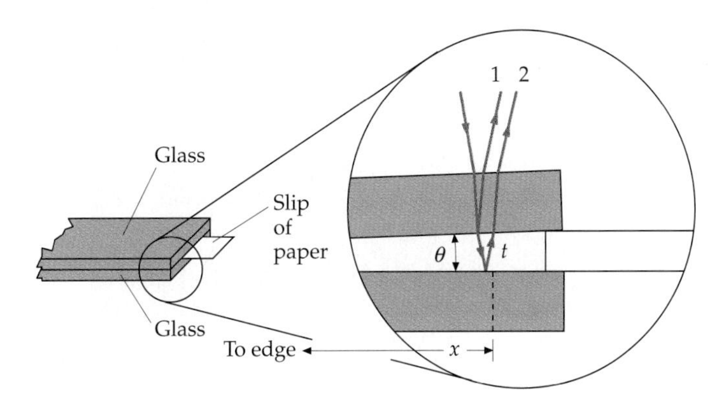

Figure 35-4 The angle θ, which is less than 0.02°, is exaggerated. The incoming and outgoing rays are essentially perpendicular to all air–glass interfaces.

Picture the Problem We find the number of fringes per centimeter by finding the horizontal distance x to the mth fringe and solving for m/x. Because the ray reflected from the bottom plate undergoes a 180° phase shift, the point of contact (where the path difference is zero) will be dark. The first dark fringe after this point occus when $2t = \lambda'$, where $\lambda' = \lambda$ is the wavelength in the air film, and t is the plate separation at x as shown in Figure 35-4. Since the angle q is small, we can use the small-angle approximation $\theta \approx t/x$.

1. The mth fringe occurs when the path difference $2t$ equals m wavelengths:
$$2t = m\lambda' = m\lambda$$
$$m = \frac{2t}{\lambda}$$

2. The thickness t is related to the angle θ:
$$\theta = \frac{t}{x}$$

3. Substitute $t = x\theta$ into the equation for m:
$$m = \frac{2x\theta}{\lambda}$$

4. Calculate m/x:
$$\frac{m}{x} = \frac{2\theta}{\lambda} = \frac{2(3 \times 10^{-4})}{5 \times 10^{-7}\ \text{m}} = 1200\ \text{m}^{-1} = 12\ \text{cm}^{-1}$$

Remarks We therefore observe 12 dark fringes per centimeter. In practice, the number of fringes per centimeter, which is easy to count, can be used to determine the angle. Note that if the angle of the wedge is increased, the fringes become more closely spaced.

Exercise How many fringes per centimeter are observed if light of wavelength 650 nm is used? (*Answer* 9.2 cm^{-1})

Figure 35-5*a* shows interference fringes produced by a wedge-shaped air film between two flat glass plates as in Example 35-2. Plates that produce straight fringes such as those in Figure 35-5*a* are said to be **optically flat.** A similar wedge-shaped air film formed by two ordinary glass plates yields the irregular fringe pattern in Figure 35-5*b*, which indicates that these plates are not optically flat.

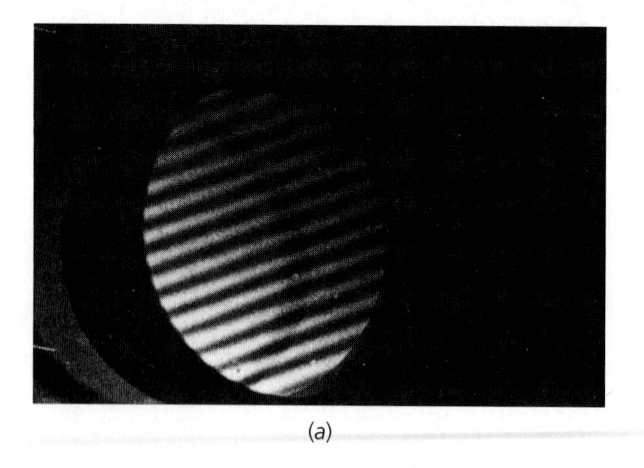

(*a*)

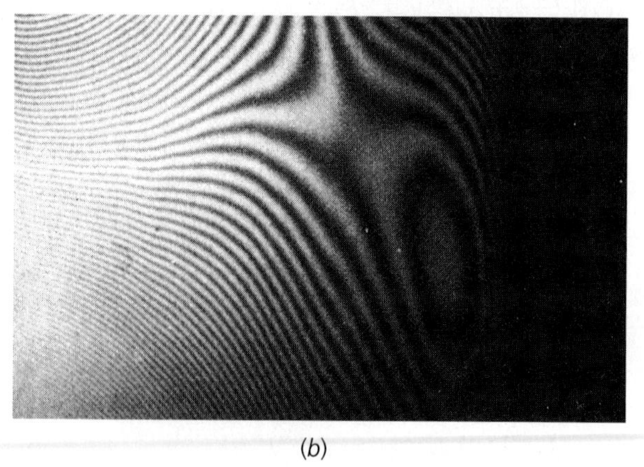

(*b*)

Figure 35-5 (*a*) Straight-line fringes from a wedge-shaped film of air like that in Figure 35-4. The straightness of the fringes indicates that the glass plates are optically flat. (*b*) Fringes from a wedge-shaped film of air between glass plates that are not optically flat.

One application of interference effects in thin films is in nonreflecting lenses, which are made by covering a lens with a thin film of a material that has an index of refraction of about 1.22, which is between that of glass and air. Then the intensities of the light reflected from the top and bottom surfaces of the film are approximately equal, and since both rays undergo a 180° phase change, there is no phase difference between the rays due to reflection. The thickness of the film is chosen to be $\lambda'/4 = \lambda/4n$ where λ is in the middle of the visible spectrum, so that there is a phase change of 180° due to the path difference of $\lambda'/2$. Reflection from the coated surface is thus minimized.

35-3 The Two-Slit Interference Pattern

Interference patterns of light from two or more sources can be observed only if the sources are coherent. The interference in thin films discussed previously can be observed because the two beams come from the same light source but are separated by reflection. In Young's famous experiment, in which he demonstrated the wave nature of light, two coherent light sources are produced by illuminating two very narrow parallel slits with a single light source. We saw in Chapter 15 that when a wave encounters a barrier with a very small opening, the opening acts as a point source of waves (Figure 35-6). In Young's experiment, diffraction causes each slit to act as a line source (which is equivalent to a point source in two dimensions). The interference pattern is observed on a screen far from the slits (Figure 35-7*a*). At very large distances from the slits, the lines from the two slits to some point P on the screen are approximately parallel, and the path difference is approximately $d \sin \theta$, where d is the separation of the slits as shown in Figure 35-7*b*. We thus have interference maxima at an angle given by

$$d \sin \theta = m\lambda, \qquad m = 0, 1, 2, \ldots \tag{35-2}$$

Two-slit interference maxima

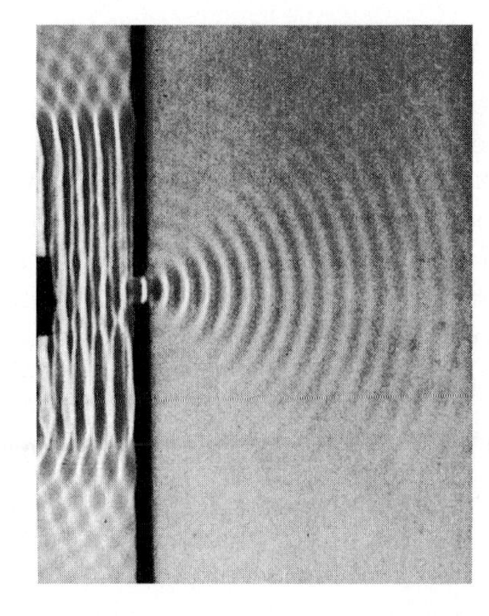

Figure 35-6 Plane water waves in a ripple tank encountering a barrier with a small opening. The waves to the right of the barrier are circular waves that are concentric about the opening just as if there were a point source at the opening.

where m is called the **order number**. The interference minima occur at

$$d \sin \theta = (m + \tfrac{1}{2})\lambda, \qquad m = 0, 1, 2, \ldots \tag{35-3}$$

Two-slit interference minima

The phase difference δ at a point P is

$$\delta = \frac{2\pi}{\lambda} d \sin \theta \tag{35-4}$$

We can relate the distance y_m measured along the screen from the central point to the mth bright fringe (see Figure 35-7*b*) to the distance L from the slits to the screen:

$$\tan \theta = \frac{y_m}{L}$$

For small angles (which is nearly always the case), $\tan \theta \approx \sin \theta$. Substituting y_m/L for $\sin \theta$ in Equation 35-2 and solving for y_m gives

$$y_m = m\frac{\lambda L}{d} \tag{35-5}$$

Distance on screen to the mth bright fringe

From this result we see that the fringes are equally spaced on the screen.

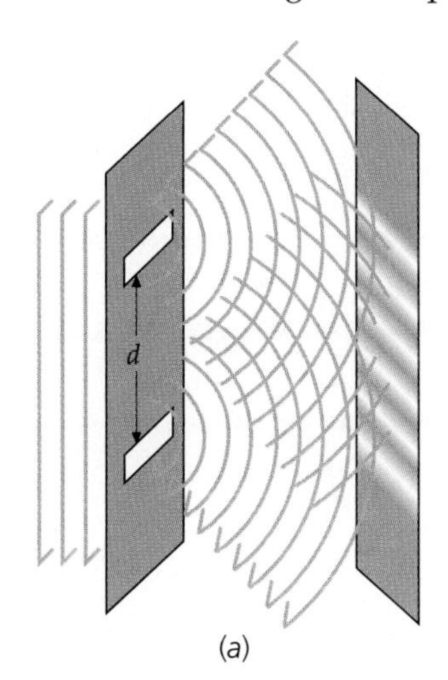

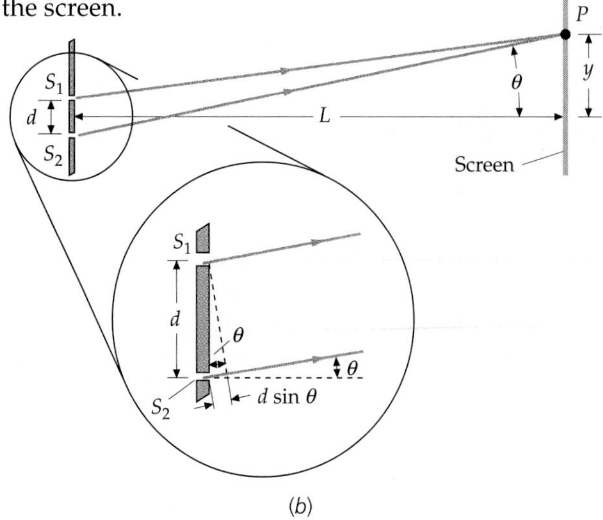

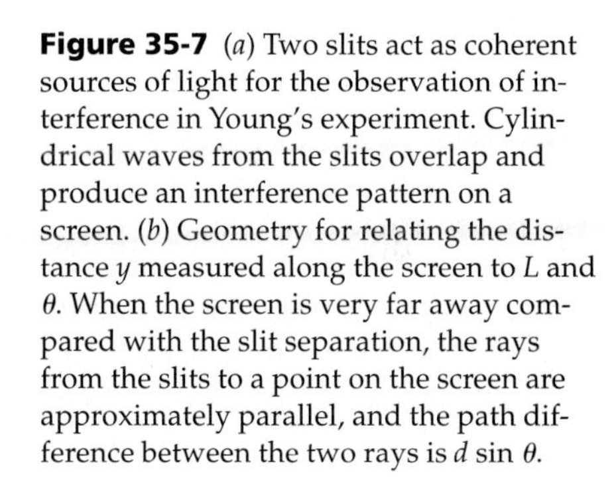

Figure 35-7 (*a*) Two slits act as coherent sources of light for the observation of interference in Young's experiment. Cylindrical waves from the slits overlap and produce an interference pattern on a screen. (*b*) Geometry for relating the distance y measured along the screen to L and θ. When the screen is very far away compared with the slit separation, the rays from the slits to a point on the screen are approximately parallel, and the path difference between the two rays is $d \sin \theta$.

Calculation of Intensity

To calculate the intensity of the light on the screen at a general point P, we need to add two harmonic wave functions that differ in phase.* The wave functions for electromagnetic waves are the electric field vectors. Let E_1 be the electric field at some point P on the screen due to the waves from slit 1, and let E_2 be the electric field at that point due to waves from slit 2. Since the angles of interest are small, we can assume that these fields are parallel and we therefore consider only their magnitudes. Both electric fields oscillate with the same frequency (they result from a single source that illuminates both slits) and they have the same amplitude. (The path difference is only of the order of a few wavelengths of light at most.) They have a phase difference δ given by Equation 35-4. If we represent these wave functions by

$$E_1 = A_0 \sin \omega t$$

and

$$E_2 = A_0 \sin (\omega t + \delta)$$

the resultant wave function is

$$E = E_1 + E_2 = A_0 \sin \omega t + A_0 \sin (\omega t + \delta)$$

$$= 2A_0 \cos \tfrac{1}{2}\delta \sin (\omega t + \tfrac{1}{2}\delta) \qquad \text{35-6}$$

where we used

$$\sin \alpha + \sin \beta = 2 \cos \tfrac{1}{2}(\alpha - \beta) \sin \tfrac{1}{2}(\alpha + \beta) \qquad \text{35-7}$$

The amplitude of the resultant wave is thus $2A_0 \cos \frac{1}{2}\delta$. It has its maximum value of $2A_0$ when the waves are in phase and is zero when they are 180° out of phase. Since the intensity is proportional to the square of the amplitude, the intensity at any point P is

$$I = 4I_0 \cos^2 \tfrac{1}{2}\delta \qquad \text{35-8}$$

Intensity in terms of phase difference

where I_0 is the intensity of the light on the screen from either slit separately. The phase angle δ is related to the position on the screen by Equation 35-4.

Figure 35-8*a* shows the intensity pattern as seen on a screen. A graph of the intensity as a function of $\sin \theta$ is shown in Figure 35-8*b*. For small θ, this is equivalent to a plot of intensity versus y since $y \approx L \sin \theta$. The intensity I_0 is that from each slit separately. The dashed line in Figure 35-8*b* shows the average intensity $2I_0$, which is the result of averaging over many interference maxima and minima. This is the intensity that would arise from the two sources if they acted independently without interference, that is, if they were not coherent. Then total phase difference between them would fluctuate randomly so that only the average intensity would be observed.

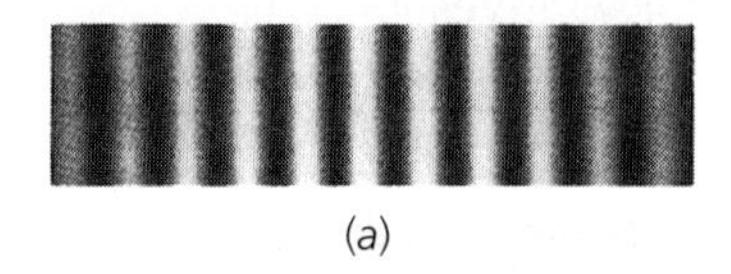

(a)

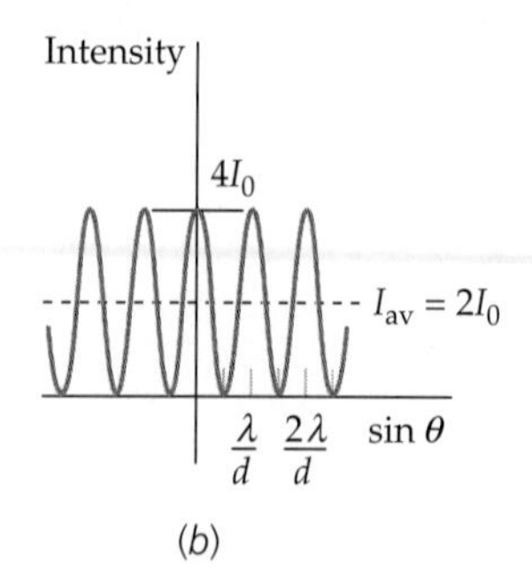

(b)

Figure 35-8 (*a*) The interference pattern observed on a screen far away from the two slits of Figure 35-7. (*b*) Plot of intensity versus $\sin \theta$. The maximum intensity is $4I_0$, where I_0 is the intensity due to each slit separately. The average intensity (dashed line) is $2I_0$.

* We did this in Chapter 16 where we discussed the general superposition of two waves.

Figure 35-9 shows another method of producing the two-slit interference pattern, an arrangement known as **Lloyd's mirror.** A single slit is placed at a distance $\frac{1}{2}d$ above the plane of a mirror. Light striking the screen directly from the source interferes with that reflected from the mirror. The reflected light can be considered to come from the virtual image of the slit formed by the mirror. Because of the 180° change in phase upon reflection at the mirror, the interference pattern is that of two coherent line sources that differ in phase by 180°. The pattern is the same as that shown in Figure 35-8 for two slits except that the maxima and minima are interchanged. The central fringe just above the mirror at a point equidistant from the sources is dark. Constructive interference occurs at points for which the path difference is a half-wavelength or any odd number of half-wavelengths. At these points, the 180° phase difference due to the path difference combines with the 180° phase difference of the sources to produce constructive interference.

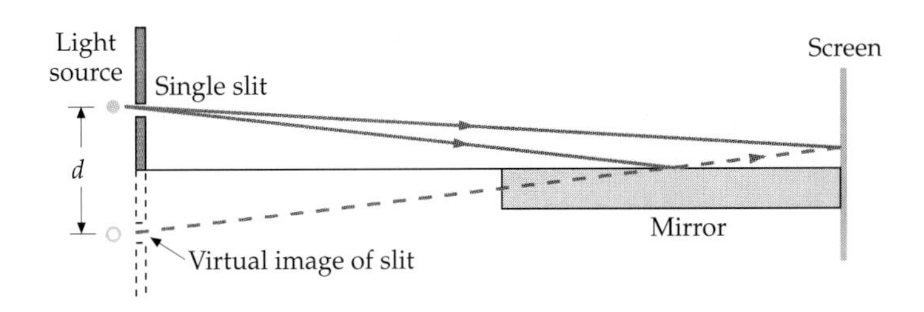

Figure 35-9 Lloyd's mirror for producing a two-slit interference pattern. The two sources (the slit and its image) are coherent and are 180° out of phase. The central interference band at the point equidistant from the sources is dark.

Example 35-3 *try it yourself*

Two narrow slits separated by 1.5 mm are illuminated by yellow light of wavelength 589 nm from a sodium lamp. Find the spacing of the fringes observed on a screen 3 m away.

Picture the Problem The distance y_m measured along the screen to the mth bright fringe is given by Equation 35-5, with $L = 3$ m, $d = 1.5$ mm, and $\lambda = 589$ nm. The spacing of the fringes is y_m/m, the distance divided by the number of fringes.

Cover the column to the right and try these on your own before looking at the answers.

Steps	***Answers***
1. Solve Equation 35-5 for y_m/m.	$\frac{y_m}{m} = \lambda\frac{L}{d}$
2. Substitute the given values.	$\frac{y_m}{m} = 1.18$ mm

Exercise A point source of light ($\lambda = 589$ nm) is placed 0.4 mm above a mirror. Interference fringes are observed on a screen 6 m away. Find the spacing of the fringes. (*Answer* 4.42 mm)

35-4 Diffraction Pattern of a Single Slit

In our discussion of the interference patterns produced by two or more slits, we assumed that the slits were very narrow so that we could consider them to be line sources of cylindrical waves, which in our two-dimensional diagrams are point sources of circular waves. We could therefore assume that the intensity due to one slit acting alone was the same (I_0) at any point P on the screen independent of the angle θ made between the ray to point P and the normal line between the slit and the screen. When the slit is not narrow, the intensity on a screen far away is not independent of angle but decreases as the angle increases. Consider a slit of width a. Figure 35-10 shows the intensity pattern on a screen far away from the slit of width a as a function of $\sin\theta$. We can see that the intensity is maximum in the forward direction ($\sin\theta = 0$) and decreases to zero at an angle that depends on the slit width a and the wavelength λ.

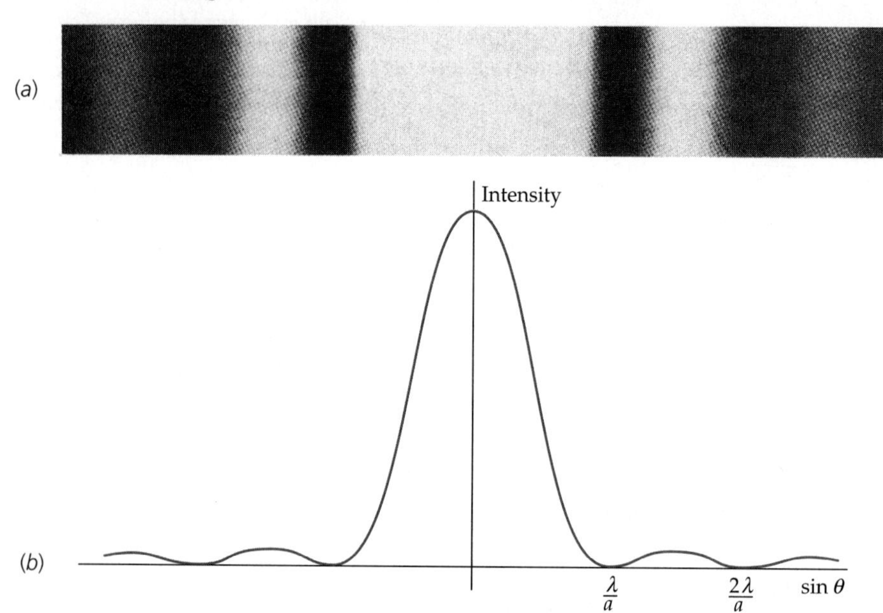

Figure 35-10 (*a*) Diffraction pattern of a single slit as observed on a screen far away. (*b*) Plot of intensity versus $\sin\theta$ for the pattern in (*a*).

Most of the light intensity is concentrated in the broad **central diffraction maximum,** though there are minor secondary maxima bands on either side of the central maximum. The first zeroes in the intensity occur at angles given by

$$\sin\theta = \frac{\lambda}{a} \qquad 35\text{-}9$$

Note that for a given wavelength λ, the width of the central maximum varies inversely with the width of the slit. If we *increase* the slit width a, the angle θ at which the intensity first becomes zero *decreases,* giving a more narrow central diffraction maximum. Conversely, if we *decrease* the slit width, the angle of the first zero *increases,* giving a wider central diffraction maximum. When a is very small, there are no points of zero intensity in the pattern, and the slit acts as a line source (a point source in two dimensions), radiating light energy essentially equally in all directions.

Multiplying both sides of Equation 35-9 by a gives

$$a\sin\theta = \lambda \qquad 35\text{-}10$$

The quantity $a\sin\theta$ is the path difference between a light ray leaving the top of the slit and one leaving the bottom of the slit. We see that the first diffraction *minimum* occurs when these two rays are in phase, that is, when their path difference equals 1 wavelength. We can understand this result by considering each point on a wavefront to be a point source of light in accordance with Huygens' principle. In Figure 35-11 we have placed a line of dots on the wavefront at the slit to represent these point sources schematically. Suppose, for example, that we have 100 such dots and that we look at an angle θ for which $a \sin \theta = \lambda$, that is, the angle for which the waves from the top and bottom of the slit are in phase. Let us consider the slit to be divided into two regions, with the first 50 sources in the first, upper region and sources 51 through 100 in the second, lower region. When the path difference between the top and bottom of the slit equals one wavelength, the path difference between source 1 (the first source in the upper region) and source 51 (the first source in the lower region) is $\frac{1}{2}\lambda$. The waves from these two sources will be out of phase by 180° and will thus cancel. Similarly, waves from the second source in each region (source 2 and source 52) will cancel. Continuing this argument, we can see that the waves from each pair of sources separated by $a/2$ will cancel. Thus, there will be no light energy at this angle. We can extend this argument to the second and third minima in the diffraction pattern of Figure 35-10. At an angle such that $a \sin \theta = 2\lambda$, we can divide the slit into four regions, two for the top half and two for the bottom half. Using this same argument, the light intensity from the top half is zero because of the cancellation of pairs of sources, and, similarly, the light intensity from the bottom half is zero. The general expression for the points of zero intensity in the diffraction pattern of a single slit is thus

$$a \sin \theta = m\lambda, \qquad m = 1, 2, 3, \ldots \tag{35-11}$$

Points of zero intensity for a single-slit diffraction pattern

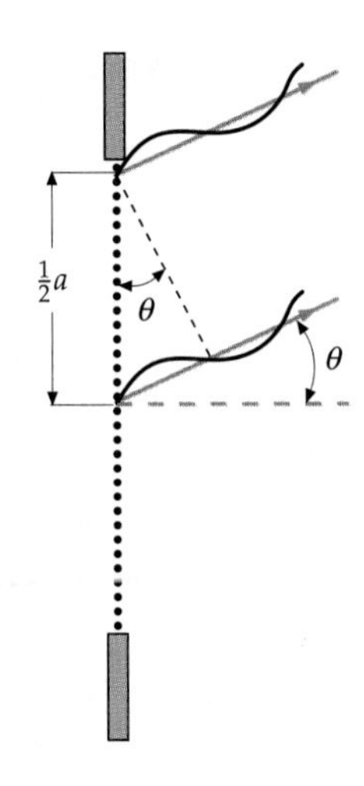

Figure 35-11 A single slit is represented by a large number of point sources of equal amplitude. At the first diffraction minimum of a single slit, the waves from the source near the top and those from the source just below the middle of the slit are 180° out of phase and cancel, as do all other pairs of sources.

Usually, we are just interested in the first occurrence of a minimum in the light intensity because nearly all of the light energy is contained in the central diffraction maximum.

In Figure 35-12, the distance y from the central maximum to the first diffraction minimum is related to the angle θ and the distance L from the slit to the screen by

$$\tan \theta = \frac{y}{L}$$

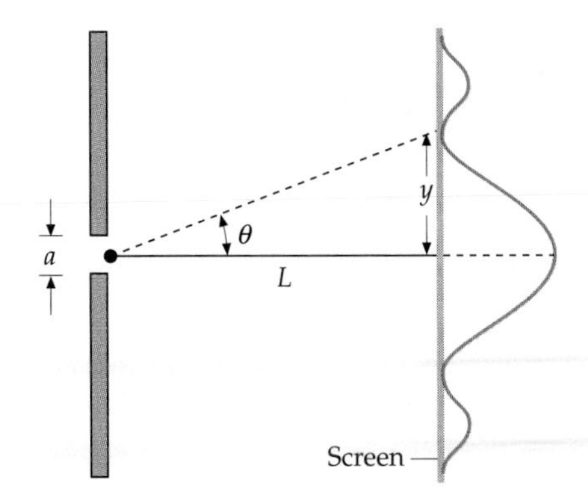

Figure 35-12 The distance y measured along the screen from the central maximum to the first diffraction minimum is related to the angle θ by $\tan \theta = y/L$, where L is the distance to the screen. Since the angle is very small, $\tan \theta \approx \sin \theta$. Then $y = L \tan \theta \approx L \sin \theta = L\lambda/a$.

Since this angle is very small, $\tan \theta \approx \sin \theta$. Then, according to Equation 35-11, we have $\sin \theta = \lambda/a \approx y/L$, or

$$y = \frac{L\lambda}{a} \tag{35-12}$$

Example 35-4

In a lecture demonstration of single-slit diffraction, a laser beam of wavelength 700 nm passes through a vertical slit 0.2 mm wide and hits a screen 6 m away. Find the width of the central diffraction maximum on the screen, that is, the distance between the first minimum on the left and the first minimum on the right of the central maximum.

Picture the Problem Referring to Figure 35-12, the width of the central diffraction maximum is $2y$.

Substitute the given data to calculate $2y$ from Equation 35-12:

$$2y = \frac{2L\lambda}{a} = \frac{2(6\text{ m})(700 \times 10^{-9}\text{ m})}{0.0002\text{ m}} = 4.2 \times 10^{-2}\text{ m} = 4.2\text{ cm}$$

Interference–Diffraction Pattern of Two Slits

When there are two or more slits, the intensity pattern on a screen far away is a combination of the single-slit diffraction pattern and the multiple-slit interference pattern we have studied. Figure 35-13 shows the intensity pattern on a screen far from two slits whose separation d is 10 times the width a of each slit. The pattern is the same as the two-slit pattern with very narrow slits (Figure 35-10) except that it is modulated by the single-slit diffraction pattern; that is, the intensity due to each slit separately is now not constant but decreases with angle as shown in Figure 35-13*b*.

Note that in Figure 35-13 the central diffraction maximum contains 19 interference maxima—the central interference maximum and 9 maxima on either side. The tenth interference maximum on either side of the central one is at the angle θ given by $\sin\theta = 10\lambda/d = \lambda/a$ since $d = 10a$. This coincides with the position of the first diffraction minimum, so this interference maximum is not seen. At these points, the light from the two slits would be in phase and would interfere constructively, but there is no light from either slit because the points are diffraction minima. We can see in general that if $m = d/a$, the mth interference maximum will fall at the first diffraction minimum. Since the mth fringe is not seen, there will be $m - 1$ fringes on each side of the central fringe for a total of N fringes in the central maximum, where N is given by

$$N = 2(m-1) + 1 = 2m - 1 \qquad \text{35-13}$$

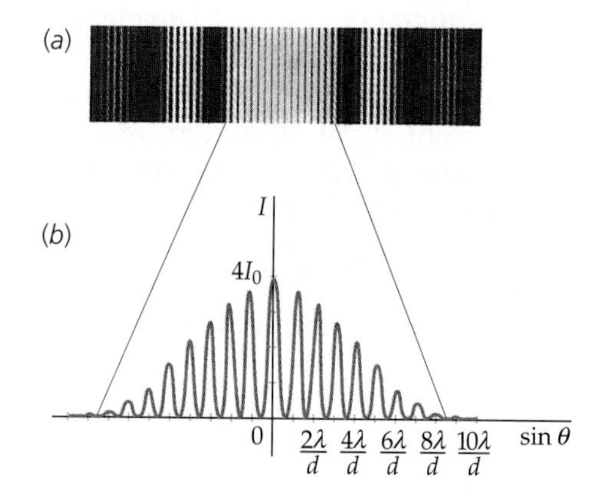

Figure 35-13 (*a*) Interference–diffraction pattern for two slits whose separation d is equal to 10 times their width a. The tenth interference maximum on either side of the central interference maximum is missing because it falls at the first diffraction minimum. (*b*) Plot of intensity versus $\sin\theta$ for the central band of the pattern in (*a*).

Example 35-5

Two slits of width $a = 0.015$ mm are separated by a distance $d = 0.06$ mm and illuminated by light of wavelength $\lambda = 650$ nm. How many bright fringes are seen in the central diffraction maximum?

Picture the Problem We need to find the value of m for which the mth interference maximum coincides with the first diffraction minimum. Then there will be $N = 2m - 1$ fringes in the central maximum.

1. Find the angle θ_1 of the first diffraction minimum: $\sin\theta_1 = \frac{\lambda}{a}$ (first diffraction minimum)

2. Find the angle θ_m of the mth interference maxima: $\sin\theta_m = \frac{m\lambda}{d}$ (mth interference maxima)

3. Set these angles equal and solve for m: $\frac{m\lambda}{d} = \frac{\lambda}{a}$

$$m = \frac{d}{a} = \frac{0.06\text{ mm}}{0.015\text{ mm}} = 4$$

4. Use $N = 2m - 1$ to find the number of bright fringes in the central maximum: $N = 2m - 1 = 2(4) - 1 = 7$ fringes

Remarks The first diffraction minimum coincides with the fourth bright fringe. Therefore, there are 3 bright fringes visible on either side of the central diffraction maximum. These six maxima, plus the central interference maximum, combine for a total of 7 bright fringes in the central diffraction maximum.

35-5 Using Phasors to Add Harmonic Waves

To calculate the interference pattern produced by three, four, or more coherent light sources and to calculate the diffraction pattern of a single slit, we need to combine several harmonic waves of the same frequency that differ in phase. A simple geometric interpretation of harmonic wave functions leads to a method of adding harmonic waves of the same frequency by geometric construction.

Let the wave functions for two waves at some point be $E_1 = A_1 \sin \alpha$ and $E_2 = A_2 \sin (\alpha + \delta)$, where $\alpha = \omega t$. Our problem is then to find the sum

$$E_1 + E_2 = A_1 \sin \alpha + A_2 \sin (\alpha + \delta)$$

We can represent each wave function by a two-dimensional vector as shown in Figure 35-14. The geometric method of addition is based on the fact that the y (or x) component of the resultant of two vectors equals the sum of the y (or x) components of the vectors as illustrated in the figure. The wave function y_1 is represented by the vector $\vec{A}_1$. As the time varies, this vector rotates in the xy plane with angular frequency ω. Such a vector is called a **phasor.** (We encountered phasors in our study of ac circuits in Section 31-4.) The wave function E_2 is the y component of a phasor of magnitude A_2 that makes an angle $\alpha + \delta$ with the x axis. By the laws of vector addition, the sum of these components equals the y component of the resultant phasor $\vec{A}$, as shown in Figure 35-14. The y component of the resultant phasor, $A \sin (\alpha + \delta')$, is a harmonic wave function that is the sum of the two original wave functions:

$$A_1 \sin \alpha + A_2 \sin (\alpha + \delta) = A \sin (\alpha + \delta') \qquad 35\text{-}14$$

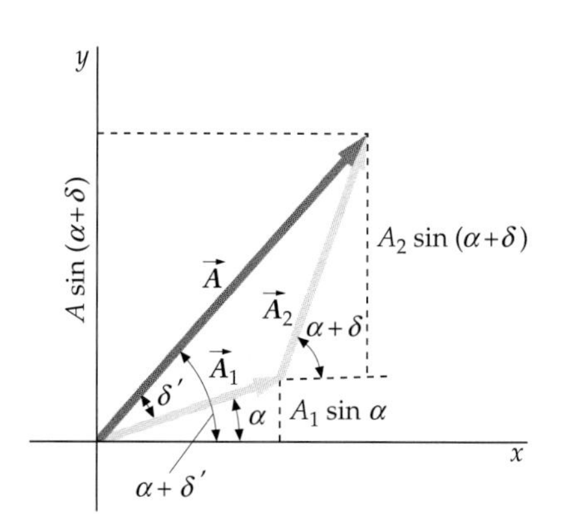

Figure 35-14 Phasor representation of wave functions.

where A (the amplitude of the resultant wave) and δ' (the phase of the resultant wave relative to the first wave) are found by adding the phasors representing the waves. As time varies, α varies. The phasors representing the two wave functions and the resultant phasor representing the resultant wave function rotate in space, but their relative positions do not change because they all rotate with the same angular velocity ω.

Example 35-6 try it yourself

Use the phasor method of addition to derive Equation 35-14 for the superposition of two waves of the same amplitude.

Picture the Problem Represent the waves $y_1 = A_0 \sin \alpha$ and $y_2 = A_0 \sin (\alpha + \delta)$ by phasors of length A_0 making an angle δ with one another. The resultant wave $y_r = A \sin (\alpha + \delta')$ is represented by the sum of these vectors, which form an isosceles triangle as shown in Figure 35-15.

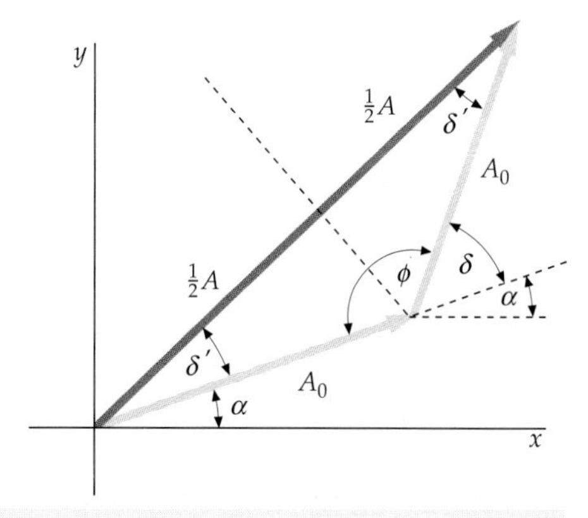

Figure 35-15

Cover the column to the right and try these on your own before looking at the answers.

Steps	*Answers*
1. Find the phase angle δ' in terms of ϕ from the fact that the three angles in the triangle must sum to 180°:	$\delta' + \delta' + \phi = 180°$
2. Relate ϕ to δ:	$\delta + \phi = 180°$

optional

3. Eliminate ϕ and solve for δ'. $\delta' = \frac{1}{2}\delta$

4. Write $\cos \delta'$ in terms of A and A_0. $\cos \delta' = \dfrac{\frac{1}{2}A}{A_0}$

5. Solve for A in terms of δ. $A = 2A_0 \cos \delta' = 2A_0 \cos \frac{1}{2}\delta$

6. Use your results for A and δ' to write the resultant wave function. $y_r = A \sin (\alpha + \delta') = (2A_0 \cos \frac{1}{2}\delta) \sin (\alpha + \frac{1}{2}\delta)$

Exercise Find the resultant of the two waves $E_1 = 4 \sin (\omega t)$ and $E_2 = 3 \sin (\omega t + 90°)$ (*Answer* $E_1 + E_2 = 5 \sin (\omega t + 37°)$)

The Interference Pattern of Three or More Equally Spaced Sources

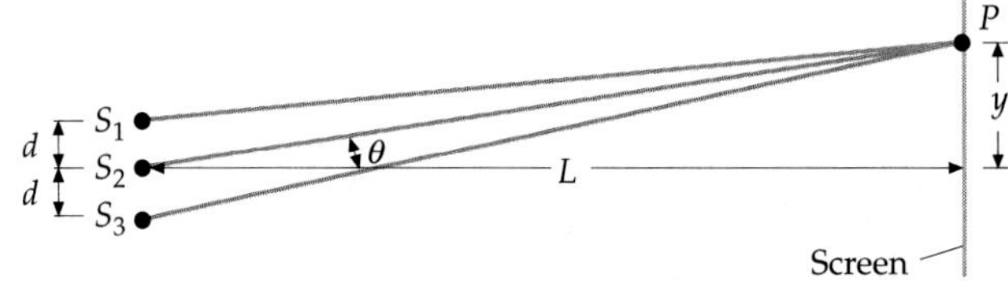

Figure 35-16 Geometry for calculating the intensity pattern far from three equally spaced sources that are in phase.

We can apply the phasor method of addition to calculate the interference pattern of three or more equally spaced, coherent sources in phase. We are most interested in the interference maxima and minima. Figure 35-16 illustrates the case of three sources. The geometry is the same as for two sources. At a great distance from the sources, the rays from the sources to a point P on the screen are approximately parallel. The path difference between the first and second source is then $d \sin \theta$, as before, and that between the first and third source is $2d \sin \theta$. The wave at point P is the sum of three waves. Let $\alpha = \omega t$ be the phase of the first wave at point P. We thus have the problem of adding three waves of the form

$$E_1 = A_0 \sin \alpha$$
$$E_2 = A_0 \sin (\alpha + \delta) \qquad \text{35-15}$$
$$E_3 = A_0 \sin (\alpha + 2\delta)$$

where

$$\delta = \frac{2\pi}{\lambda} d \sin \theta \approx \frac{2\pi}{\lambda} \frac{yd}{L} \qquad \text{35-16}$$

as in the two-slit problem.

At $\theta = 0$, $\delta = 0$, so all the waves are in phase. The amplitude of the resultant wave is 3 times that of each individual wave and the intensity is 9 times that due to each source acting separately. As the angle δ increases from $\theta = 0$, the phase angle δ increases and the intensity decreases. The position $\theta = 0$ is thus a position of maximum intensity.

Figure 35-17 shows the phasor addition of three waves for a phase angle δ of about 30°. (This corresponds to a point P on the screen for which θ is given by $\sin \theta = \lambda\delta/2\pi d = \lambda/12d$.) The resultant amplitude A is considerably less than 3 times that of each source. As the phase angle δ increases, the resultant amplitude decreases until the amplitude is zero at $\delta = 120°$. For this phase difference, the three phasors form an equilateral triangle (Figure 35-18). This first interference minimum for three sources occurs at a smaller phase angle (and therefore at a smaller space angle θ) than it does for only two sources (for which the first minimum occurs at $\delta = 180°$). As δ increases from 120°, the resultant amplitude increases, reaching a secondary maximum near $\delta =$

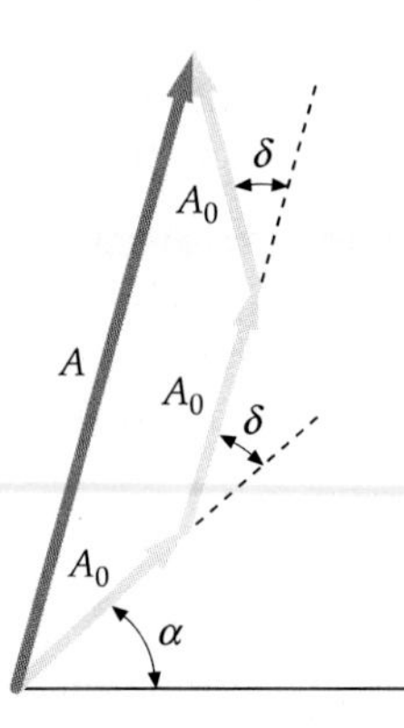

Figure 35-17 Phasor diagram for determining the resultant amplitude A due to three waves, each of amplitude A_0, that have phase differences of δ and 2δ due to path differences of $d \sin \theta$ and $2d \sin \theta$. The angle $\alpha = \omega t$ varies with time, but this does not affect the calculation of A.

180°. At the phase angle $\delta = 180°$, the amplitude is the same as that from a single source since the waves from the first two sources cancel each other, leaving only the third. The intensity of the secondary maximum is one-ninth that of the maximum at $\theta = 0$. As δ increases beyond 180°, the amplitude again decreases and is zero at $\delta = 180° + 60° = 240°$. For δ greater than 240°, the amplitude increases and is again 3 times that of each source when $\delta = 360°$. This phase angle corresponds to a path difference of 1 wavelength for the waves from the first two sources and 2 wavelengths for the waves from the first and third sources. Hence, the three waves are in phase at this point. The largest maxima, called the principal maxima, are at the same positions as for just two sources, which are those points corresponding to the angles θ given by

$$d \sin \theta = m\lambda, \qquad m = 0, 1, 2, \ldots \tag{35-17}$$

These maxima are stronger and narrower than those for two sources. They occur at points for which the path difference between adjacent sources is zero or an integral number of wavelengths.

These results can be generalized to more than three sources. For four equally spaced sources that are in phase, the principal interference maxima are again given by Equation 35-17, but these maxima are even more intense, they are narrower, and there are two small secondary maxima between each pair of principal maxima. At $\theta = 0$, the intensity is 16 times that due to a single source. The first interference minimum occurs when δ is 90°, as can be seen from the phasor diagram of Figure 35-19. The first secondary maximum is near $\delta = 120°$, where the waves from three of the sources cancel, leaving only the wave from the fourth source. The intensity of the secondary maximum is approximately one-sixteenth that of the central maximum. There is another minimum at $\delta = 180°$, another secondary maximum near $\delta = 240°$, and another minimum at $\delta = 270°$ before the next principal maximum at $\delta = 360°$.

Figure 35-20 shows the intensity patterns for two, three, and four equally spaced sources. Figure 35-21 shows a graph of I/I_0, where I_0 is the intensity due to each source acting separately. For three sources, there is a very small secondary maximum between each pair of principal maxima, and the principal maxima are sharper and more intense than those due to just two sources. For four sources, there are two small secondary maxima between each pair of principal maxima, and the principal maxima are even more narrow and intense.

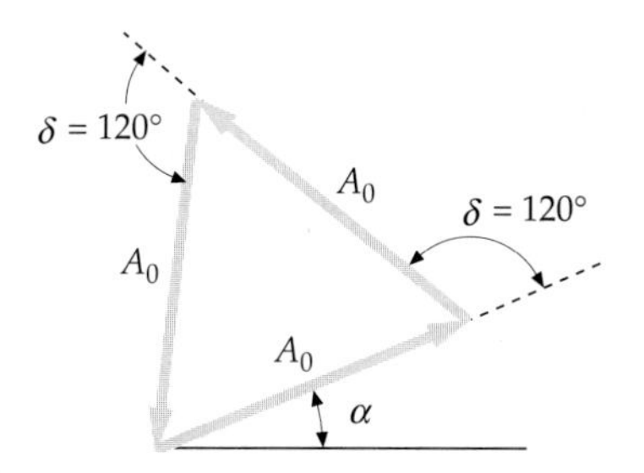

Figure 35-18 The resultant amplitude for the waves from three sources is zero when δ is 120°. This interference minimum occurs at a smaller angle θ than does the first minimum for two sources, which occurs when δ is 180°.

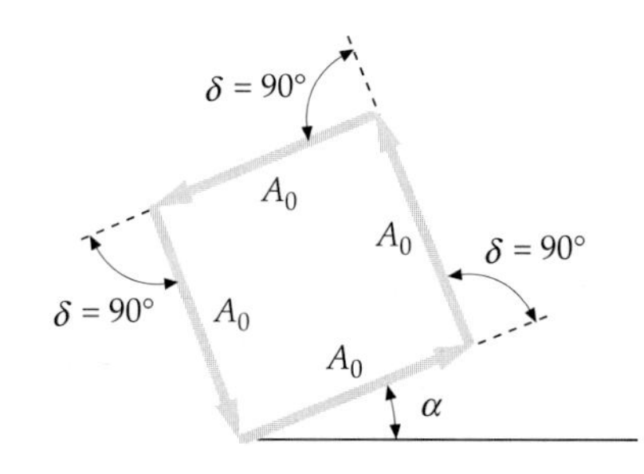

Figure 35-19 Phasor diagram for the first minimum for four equally spaced in-phase sources. The amplitude is zero when the phase difference of the waves from adjacent sources is 90°.

Two sources

Three sources

Four sources

Figure 35-20 Intensity patterns for two, three, and four equally spaced coherent sources. There is a secondary maximum between each pair of principal maxima for three sources, and two secondary maxima for four sources.

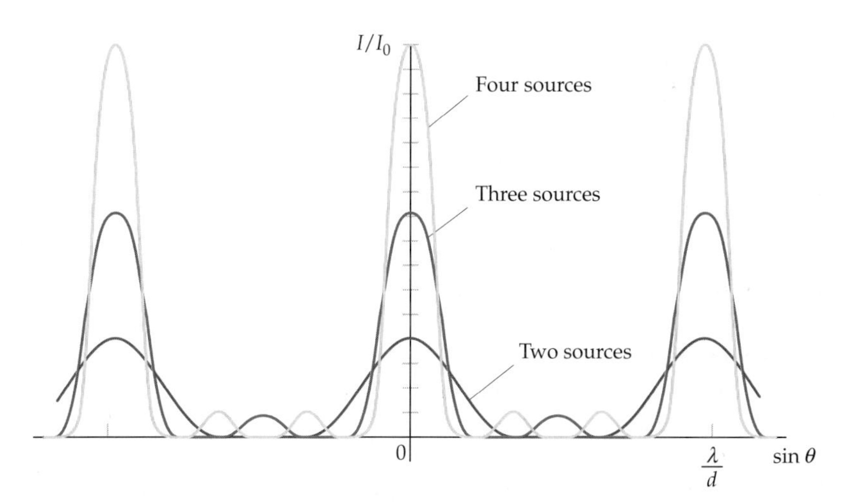

Figure 35-21 Plot of relative intensity versus sin θ for two, three, and four equally spaced coherent sources.

From this discussion, we can see that as we increase the number of sources, the intensity becomes more and more concentrated in the principal maxima given by Equation 35-17, and these maxima become narrower. For N sources, the intensity of the principal maxima is N^2 times that due to a single source. The first minimum occurs at a phase angle of $\delta = 360°/N$, for which the N phasors form a closed polygon of N sides. There are $N - 2$ secondary maxima between each pair of principal maxima. These secondary maxima are very weak compared with the principal maxima. As the number of sources is increased, the principal maxima become sharper and more intense, and the intensities of the secondary maxima become negligible compared to those of the principal maxima.

Calculating the Single-Slit Diffraction Pattern

We now use the phasor method of addition of harmonic waves to calculate the intensity pattern shown in Figure 35-10. We assume that the slit of width a is divided into N equal intervals and that there is a point source of waves at the midpoint of each interval (Figure 35-22). If d is the distance between two adjacent sources and a is the width of the opening, we have $d = a/N$. Since the screen on which we are calculating the intensity is far from the sources, the rays from the sources to a point P on the screen are approximately parallel. The path difference between any two adjacent sources is $d \sin\theta$, and the phase difference is

$$\delta = \frac{2\pi}{\lambda} d \sin\theta$$

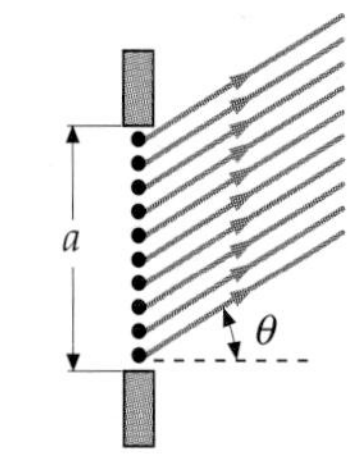

Figure 35-22 Diagram for calculating the diffraction pattern far away from a narrow slit. The slit width a is assumed to contain a large number of in-phase point sources separated by a distance d. The rays from these sources to a point far away are approximately parallel. The path difference for the waves from adjacent sources is $d \sin\theta$.

If A_0 is the amplitude due to a single source, the amplitude at the central maximum, where $\theta = 0$ and all the waves are in phase, is $A_{max} = NA_0$ (Figure 35-23).

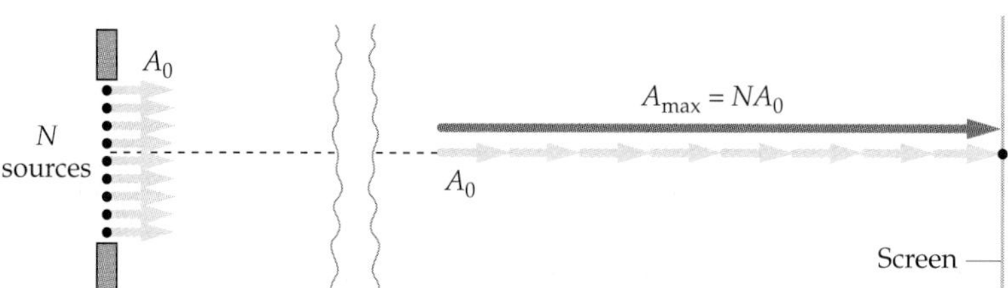

Figure 35-23 A single slit is represented by N sources, each of amplitude A_0. At the central maximum point at $\theta = 0$, the waves from the sources add in phase, giving a resultant amplitude $A_{max} = NA_0$.

We can find the amplitude at some other point at an angle θ by using the phasor method for the addition of harmonic waves. As in the addition of two, three, or four waves, the intensity is zero at any point where the phasors representing the waves form a closed polygon. In this case the polygon has N sides (Figure 35-24). At the first minimum, the wave from the first source just below the top of the opening and that from the source just below the middle of the opening are 180° out of phase. In this case, the waves from the source near the top of the opening differ from those from the bottom by nearly 360°. (The phase difference is, in fact, $360° - 360°/N$.) Thus, if the number of sources is very large, we get complete cancellation when the waves from the first and last sources are out of phase by 360°, corresponding to a path difference of 1 wavelength, in agreement with Equation 35-11.

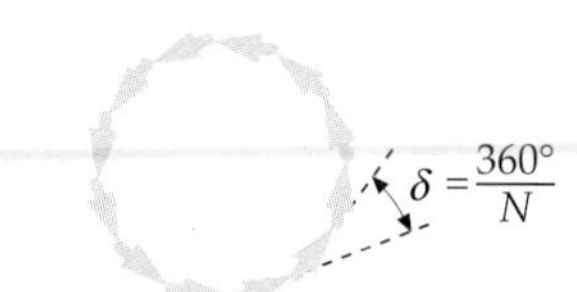

Figure 35-24 Phasor diagram for calculating the first minimum in the single-slit diffraction pattern. When the waves from the N sources completely cancel, the N phasors form a closed polygon. The phase difference between the waves from adjacent sources is then $\delta = 360°/N$. When N is very large, the waves from the first and last sources are approximately in phase.

We will now calculate the amplitude at a general point at which the waves from two adjacent sources differ in phase by δ. Figure 35-25 shows the phasor diagram for the addition of N waves where the subsequent waves differ in phase from the first wave by $\delta, 2\delta, \ldots, (N-1)\delta$. When N is very large and δ is very small, the phasor diagram approximates the arc of a circle. The resultant amplitude A is the length of the chord of this arc. We will calculate this resultant amplitude in terms of the phase difference ϕ between the first wave and the last wave. From Figure 35-25, we have

$$\sin \tfrac{1}{2}\phi = \frac{A/2}{r}$$

or

$$A = 2r \sin \tfrac{1}{2}\phi \qquad \text{35-18}$$

where r is the radius of the arc. Since the length of the arc is $A_{max} = NA_0$ and the angle subtended is ϕ, we have

$$\phi = \frac{A_{max}}{r} \qquad \text{35-19}$$

or

$$r = \frac{A_{max}}{\phi}$$

Substituting this into Equation 35-18 gives

$$A = \frac{2A_{max}}{\phi} \sin \tfrac{1}{2}\phi = A_{max} \frac{\sin \tfrac{1}{2}\phi}{\tfrac{1}{2}\phi}$$

Since the amplitude at the center of the central maximum ($\theta = 0$) is A_{max}, the ratio of the intensity at any other point to that at the center of the central maximum is

$$\frac{I}{I_0} = \frac{A^2}{A_{max}^2} = \left(\frac{\sin \tfrac{1}{2}\phi}{\tfrac{1}{2}\phi}\right)^2$$

or

$$I = I_0\left(\frac{\sin \tfrac{1}{2}\phi}{\tfrac{1}{2}\phi}\right)^2 \qquad \text{35-20}$$

Intensity for a single-slit diffraction pattern

The phase difference ϕ between the first and last waves is $2\pi/\lambda$ times the path difference $a \sin \theta$ between the top and bottom of the opening:

$$\phi = \frac{2\pi}{\lambda} a \sin \theta \qquad \text{35-21}$$

Equations 35-20 and 35-21 describe the intensity pattern shown in Figure 35-10. The first minimum occurs at $a \sin \theta = \lambda$, the point where the waves from the top and bottom of the opening have a path difference of λ and are in phase. The second minimum occurs at $a \sin \theta = 2\lambda$, where the waves from the top and bottom of the opening have a path difference of 2λ.

There is a secondary maximum approximately midway between the first and second minima at $a \sin \theta \approx \tfrac{3}{2}\lambda$. Figure 35-26 shows the phasor diagram for determining the approximate intensity of this secondary maximum. The phase difference between the first and last waves is approximately 360° + 180°. The phasors thus complete $1\tfrac{1}{2}$ circles. The resultant amplitude is the diameter of a circle with a circumference that is two-thirds the total length A_{max}. If $C = \tfrac{2}{3}A_{max}$ is the circumference, the diameter A is

$$A = \frac{C}{\pi} = \frac{\tfrac{2}{3}A_{max}}{\pi} = \frac{2}{3\pi}A_{max}$$

and

$$A^2 = \frac{4}{9\pi^2}A_{max}^2$$

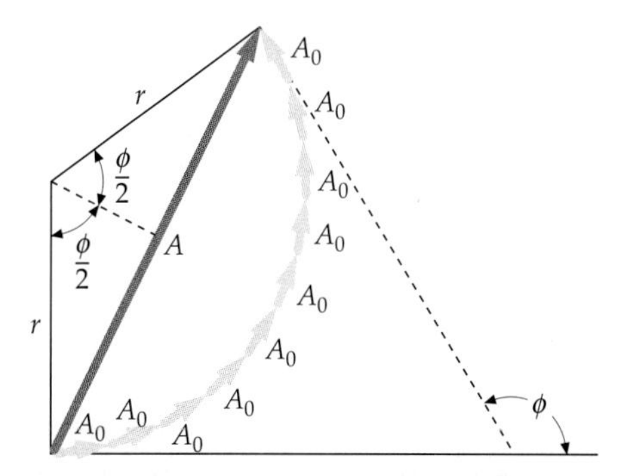

Figure 35-25 Phasor diagram for calculating the resultant amplitude due to the waves from N sources in terms of the phase difference ϕ between the wave from the first source just below the top of the slit and that from the last source just above the bottom of the slit. When N is very large, the resultant amplitude A is the chord of a circular arc of length $NA_0 = A_{max}$.

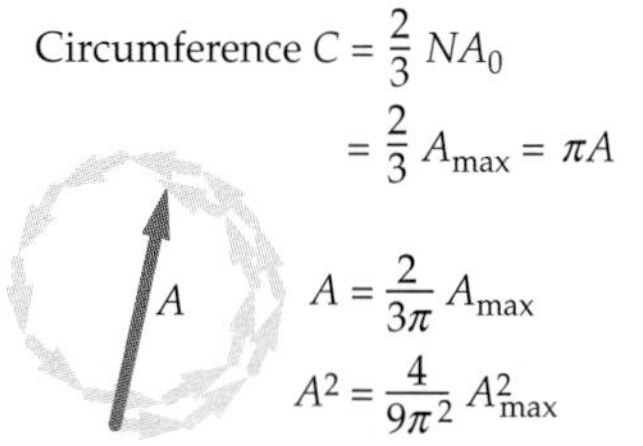

Figure 35-26 Phasor diagram for calculating the approximate amplitude of the first secondary maximum of the single-slit diffraction pattern. This secondary maximum occurs near the midpoint between the first and second minima when the N phasors complete $1\tfrac{1}{2}$ circles.

The intensity at this point is

$$I = \frac{4}{9\pi^2} I_0 = \frac{1}{22.2} I_0 \qquad \text{35-22}$$

Calculating the Interference–Diffraction Pattern of Two Slits

The intensity of the two-slit interference–diffraction pattern can be calculated from the two-slit pattern (Equation 35-8) with the intensity of each slit (I_0 in that equation) replaced by the diffraction-pattern intensity due to each slit, I, given by Equation 35-20. The intensity for the two-slit interference–diffraction pattern is thus

$$I = 4I_0\left(\frac{\sin \frac{1}{2}\phi}{\frac{1}{2}\phi}\right)^2 \cos^2 \tfrac{1}{2}\delta \qquad \text{35-23}$$

Interference–diffraction intensity for two slits

where ϕ is the difference in phase between rays from the top and bottom of each slit, which is related to the width of each slit by

$$\phi = \frac{2\pi}{\lambda} a \sin \theta$$

and δ is the difference in phase between rays from the centers of two adjacent slits, which is related to the slit separation by

$$\delta = \frac{2\pi}{\lambda} d \sin \theta$$

In Equation 35-23, the intensity I_0 is the intensity at $\theta = 0$ due to one slit alone.

35-6 Fraunhofer and Fresnel Diffraction

Diffraction patterns like the single-slit pattern in Figure 35-10 that are observed at points for which the rays from an aperture or obstacle are nearly parallel are called **Fraunhofer diffraction patterns.** Fraunhofer patterns can be observed at great distances from the obstacle or aperture so that the rays reaching any point are approximately parallel, or they can be observed using a lens to focus parallel rays on a viewing screen placed in the focal plane of the lens.

The diffraction pattern observed near an aperture or obstacle is called a **Fresnel diffraction pattern.** Because the rays from an aperture or obstacle close to a screen cannot be considered parallel, Fresnel diffraction is much more difficult to analyze. Figure 35-27 illustrates the difference between the Fresnel and Fraunhofer patterns for a single slit.*

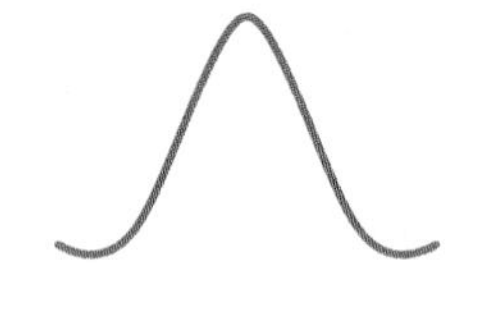
As the screen is moved closer,
the Fraunhofer pattern observed far from the slit...

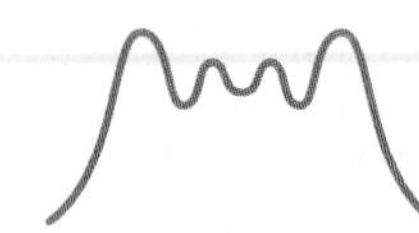
gradually changes into...
the Fresnel pattern observed near the slit.

Figure 35-27 Diffraction patterns for a single slit at various screen distances.

* See Richard E. Haskel, "A Simple Experiment on Fresnel Diffraction," *American Journal of Physics*, vol. 38, 1970, p. 1039.

Figure 35-28*a* shows the Fresnel diffraction pattern of an opaque disk. Note the bright spot at the center of the pattern caused by the constructive interference of the light waves diffracted from the edge of the disk. This pattern is of some historical interest. In an attempt to discredit Fresnel's wave theory of light, Poisson pointed out that it predicted a bright spot at the center of the shadow, which he assumed was a ridiculous contradiction of fact. However, Fresnel immediately demonstrated experimentally that such a spot does, in fact, exist. This demonstration convinced many doubters of the validity of the wave theory of light. The Fresnel diffraction pattern of a circular aperture is shown in Figure 35-28*b*. Comparing this with the pattern of the opaque disk in Figure 35-28*a*, we can see that the two patterns are complements of each other.

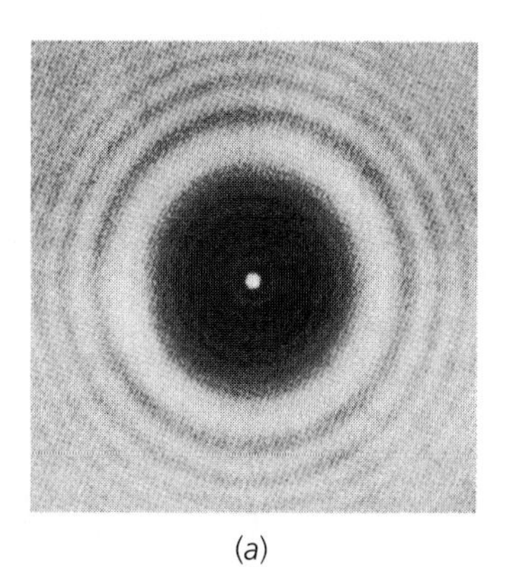
(*a*)

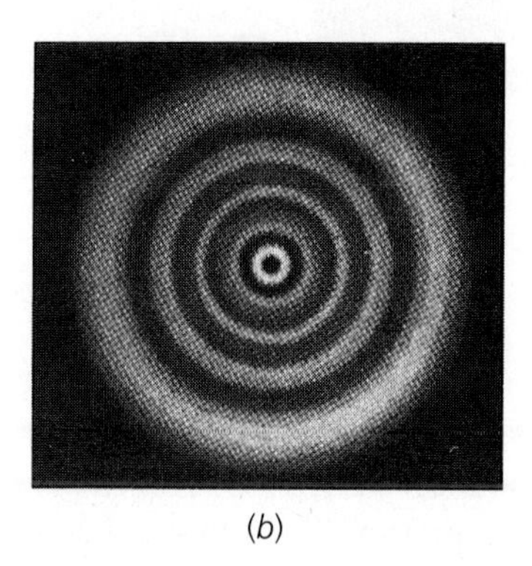
(*b*)

Figure 35-28 (*a*) Fresnel diffraction pattern of an opaque disk. At the center of the shadow, the light waves diffracted from the edge of the disk are in phase and produce a bright spot called the *Poisson spot.* (*b*) Fresnel diffraction pattern of a circular aperture. Compare this with part (*a*).

(*a*)

(*b*)

Figure 35-29 (*a*) Fresnel diffraction of a straightedge. (*b*) Intensity versus distance along a line perpendicular to the edge.

Figure 35-29*a* shows the Fresnel diffraction pattern of a straight edge illuminated by light from a point source. A graph of the intensity versus distance (measured along a line perpendicular to the edge) is shown in Figure 35-29*b*. The light intensity does not fall abruptly to zero in the geometric shadow, but it decreases rapidly and is negligible within a few wavelengths of the edge. The Fresnel diffraction pattern of a rectangular aperture is shown in Figure 35-30. These patterns cannot be seen with broad light sources like an ordinary light bulb because the dark fringes of the pattern produced by light from one point on the source overlap the bright fringes of the pattern produced by light from another point.

Figure 35-30 Fresnel diffraction of a rectangular aperture.

35-7 Diffraction and Resolution

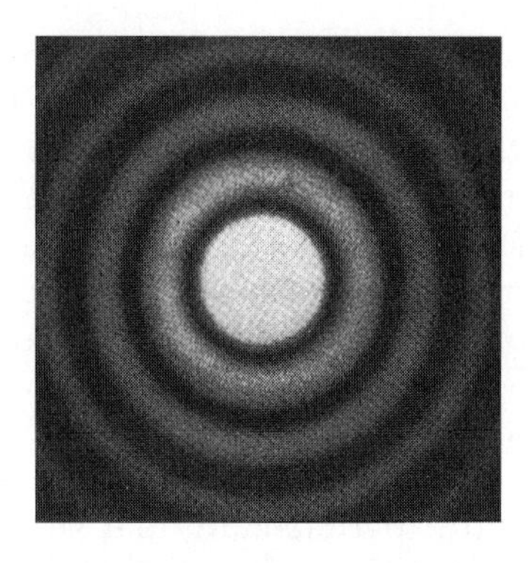

Figure 35-31 Fraunhofer diffraction pattern of a circular aperture.

Diffraction due to a circular aperture has important implications for the resolution of many optical instruments. Figure 35-31 shows the Fraunhofer diffraction pattern of a circular aperture. The angle θ subtended by the first diffraction minimum is related to the wavelength and the diameter of the opening D by

$$\sin\theta = 1.22\frac{\lambda}{D} \qquad 35\text{-}24$$

Equation 35-24 is similar to Equation 35-9 except for the factor 1.22, which arises from the mathematical analysis, which is similar to that for a single slit but more complicated because of the circular geometry. In many applications, the angle θ is small, so $\sin\theta$ can be replaced by θ. The first diffraction minimum is then at an angle θ given by

$$\theta \approx 1.22\frac{\lambda}{D} \qquad 35\text{-}25$$

Figure 35-32 shows two point sources that subtend an angle α at a circular aperture far from the sources. The intensities of the Fraunhofer diffraction pattern are also indicated in this figure. If α is much greater than $1.22\lambda/D$, the sources will be seen as two sources. However, as α is decreased, the overlap of the diffraction patterns increases, and it becomes difficult to distinguish the two sources from one source. At the critical angular separation, α_c, given by

$$\alpha_c = 1.22\frac{\lambda}{D} \qquad 35\text{-}26$$

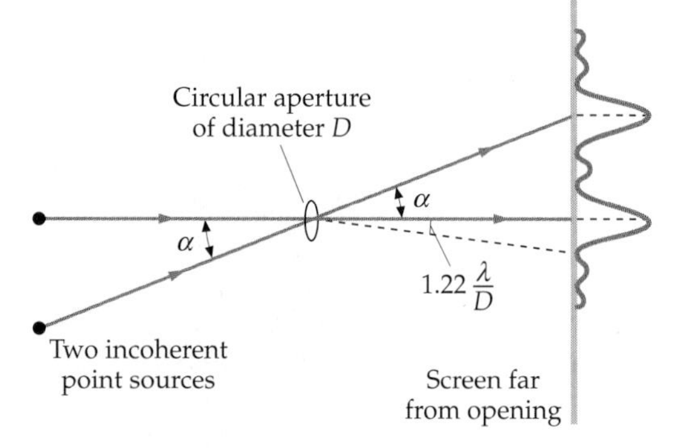

Figure 35-32 Two distant sources that subtend an angle α. If α is much greater than $1.22\lambda/D$, where λ is the wavelength of light and D is the diameter of the aperture, the diffraction patterns have little overlap and the sources are easily seen as two sources. If α is not much greater than $1.22\lambda/D$, the overlap of the diffraction patterns makes it difficult to distinguish two sources from one.

the first minimum of the diffraction pattern of one source falls on the central maximum of the other source. These objects are said to be just resolved by **Rayleigh's criterion for resolution.** Figure 35-33 shows the diffraction patterns for two sources when α is greater than the critical angle for resolution and when α is just equal to the critical angle for resolution.

Equation 35-26 has many applications. The *resolving power* of an optical instrument such as a microscope or telescope is the ability of the instrument to resolve two objects that are close together. The images of the objects tend to overlap because of diffraction at the entrance aperture of the instrument. We can see from Equation 35-26 that the resolving power can be increased either by increasing the diameter D of the lens (or mirror) or by decreasing the wavelength λ. Astronomical telescopes use large objective lenses or mirrors to increase their resolution as well as to increase their light-gathering power. In a microscope, a film of transparent oil with index of refraction of about 1.55 is sometimes used under the objective to decrease the wavelength of the light ($\lambda' = \lambda/n$). The wavelength can be reduced further by using ultraviolet light and photographic film; however, ordinary glass is opaque to ultraviolet light, so the lenses in an ultraviolet microscope must be made from quartz or fluorite. To obtain very high resolutions, electron microscopes are used—microscopes that

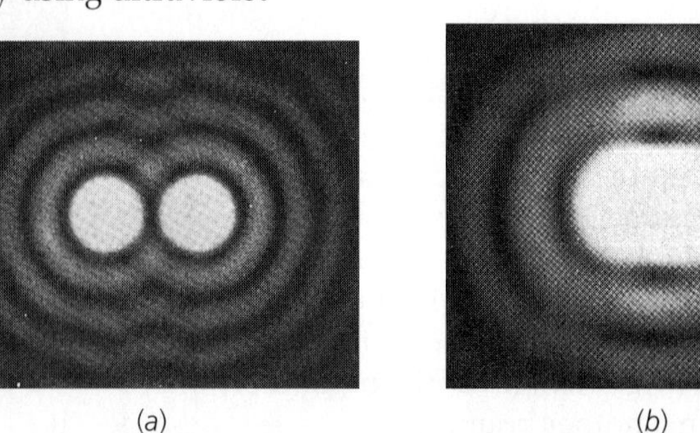

Figure 35-33 Diffraction patterns for a circular aperture and two incoherent point sources when (*a*) α is much greater than $1.22\lambda/D$ and (*b*) when α is at the limit of resolution, $\alpha_c = 1.22\lambda/D$.

use electrons rather than light. The wavelengths of electrons vary inversely with the square root of their kinetic energy and can be made as small as desired.*

*The wave properties of electrons are discussed in Chapter 17.

Example 35-7

(*a*) What minimum angular separation must two point objects have if they are to be just resolved by the eye? (*b*) How far apart must they be if they are 100 m away? Assume that the diameter of the pupil of the eye is 5 mm and that the wavelength of the light is 600 nm.

Picture the Problem (*a*) The critical angle is calculated from $\alpha_c = 1.22\ \lambda/D$. (*b*) If the objects are separated by a distance y and are 100 m away, they will be just resolved if $\tan \alpha_c = y/(100\text{ m})$.

(*a*) Use $\alpha_c = 1.22\lambda/D$ with $D = 5$ mm and $\lambda = 600$ nm:

$$\alpha_c = 1.22 \frac{6 \times 10^{-7}\text{ m}}{5 \times 10^{-3}\text{ m}} = 1.46 \times 10^{-4}\text{ rad}$$

(*b*) Set $\tan \alpha_c = y/100$ m and solve for y:

$$\tan \alpha_c = \frac{y}{100\text{ m}}$$

$$y = (100\text{ m}) \tan \alpha_c \approx (100\text{ m})\alpha_c$$

$$= (100\text{ m})(1.46 \times 10^{-4}) = 1.46 \times 10^{-2}\text{ m} = 1.46\text{ cm}$$

Exercise Two objects are 4 cm apart. How far away from them can you be and still resolve them with your eye if $\lambda = 600$ nm and the diameter of the pupil of your eye is 5 mm? (*Answer* 274 m)

It is instructive to compare the limitation on resolution of the eye due to diffraction as seen in Example 35-7 with that due to the separation of the receptors (cones) on the retina. To be seen as two distinct objects, the images of the objects must fall on the retina on two nonadjacent cones. (See Problem 70 in Chapter 34.) Because the retina is about 2.5 cm from the eye lens, the distance y on the retina corresponding to an angular separation of 1.5×10^{-4} rad is found from

$$\alpha_c = 1.5 \times 10^{-4}\text{ rad} = \frac{y}{2.5\text{ cm}}$$

or

$$y = 3.75 \times 10^{-4}\text{ cm} = 3.75 \times 10^{-6}\text{ m} = 3.75\ \mu\text{m}$$

The actual separation of the cones in the fovea centralis, where the cones are the most tightly packed, is about 1 μm. Outside this region, they are about 3 to 5 μm apart.

35-8 Diffraction Gratings

A useful tool for measuring the wavelength of light is the **diffraction grating,** which consists of a large number of equally spaced lines or slits on a flat surface. Such a grating can be made by cutting parallel, equally spaced grooves on a glass or metal plate with a precision ruling machine. With a reflection grating, light is reflected from the ridges between the lines. Phonograph

optional

records and compact disks exhibit some of the properties of reflection gratings. In a transmission grating, the light passes through the clear gaps between the rulings. Inexpensive plastic gratings with 10,000 or more slits per centimeter are common items. The spacing of the slits in a grating with 10,000 slits per centimeter is $d = (1 \text{ cm})/10{,}000 = 10^{-4}$ cm.

Compact disks act as reflection gratings.

Consider a plane light wave incident normally on a transmission grating (Figure 35-34). Assume that the width of each slit is very small so that it produces a widely diffracted beam. The interference pattern produced on a screen a large distance from the grating is that due to a large number of equally spaced light sources. Suppose we have N slits with separation d between adjacent slits. At $\theta = 0$, the light from each slit is in phase with that from all the other slits so the amplitude of the wave is NA_0, where A_0 is the amplitude from each slit, and the intensity is N^2I_0, where I_0 is the intensity due to each slit. At an angle θ such that $d \sin \theta = \lambda$, the path difference between any two successive slits is λ, so again the light from each slit is in phase with that from all the other slits and the intensity is N^2I_0. The interference maxima are thus at angles θ given by

$$d \sin \theta = m\lambda, \qquad m = 0, 1, 2, \ldots \qquad \text{35-27}$$

The position of an interference maximum does not depend on the number of sources, but the more sources there are, the sharper and more intense the maximum will be.

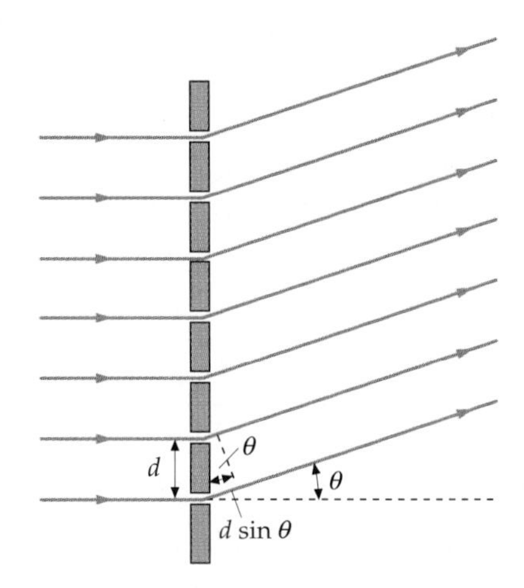

Figure 35-34 Light incident normally on a diffraction grating. At an angle θ, the path difference between rays from adjacent slits is $d \sin \theta$.

To see that the interference maxima will be sharper when there are many slits, consider the case of N slits. The distance from the first slit to the Nth slit is $(N - 1)d \approx Nd$. When the path difference for the light from the first slit and that from the Nth slit is λ, the resulting intensity will be zero. (We saw this in our discussion of single-slit diffraction.) Since the first and Nth slits are separated by approximately Nd, the intensity will be zero at angle θ_{min} given by

$$Nd \sin \theta_{min} = \lambda$$

$$\sin \theta_{min} \approx \theta_{min} = \frac{\lambda}{Nd}$$

The width of the interference maximum $2\theta_{min}$ is thus proportional to $1/N$. Thus, the greater the number of slits N, the sharper the maximum. Since the intensity in the maximum is proportional to N^2I_0, the amount of light in the maximum is proportional to NI_0.

Figure 35-35*a* shows a student spectroscope that uses a diffraction grating to analyze light. In student laboratories, the light source is typically a glass tube containing atoms of a gas, for example, helium or sodium vapor, that are excited by a bombardment of electrons accelerated by high voltage across the tube. The light emitted by such a source contains only certain wavelengths that are characteristic of the atoms in the source. Light from the source passes through a narrow collimating slit and is made parallel by a lens. Parallel light from the lens is incident on the grating. Instead of falling on a screen a large distance away, the parallel light from the grating is focused by a telescope and viewed by the eye. The telescope is mounted on a rotating platform that has been calibrated so that the angle θ can be measured. In the forward direction ($\theta = 0$), the central maximum for all wavelengths is seen. If light of a particular wavelength λ is emitted by the source, the first interference maximum is seen at the angle θ given by Equation 35-27

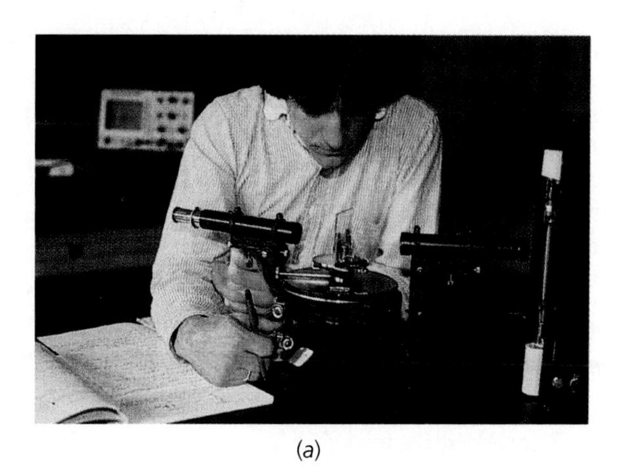

(a)

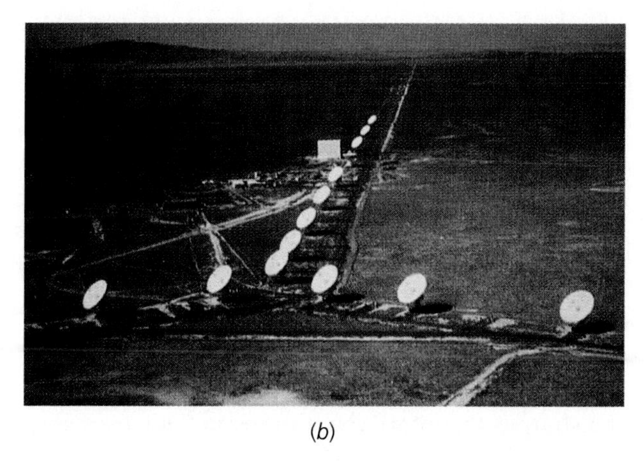

(b)

Figure 35-35 (*a*) Typical student spectroscope. Light from a collimating slit near the source is made parallel by a lens and falls on a grating. The diffracted light is viewed with a telescope at an angle that can be accurately measured. (*b*) Aerial view of the very large array (VLA) radio telescope in New Mexico. Radio signals from distant galaxies add constructively when Equation 35-27 is satisfied, where d is the distance between two adjacent telescopes.

with $m = 1$. Each wavelength emitted by the source produces a separate image of the collimating slit in the spectroscope called a **spectral line.** The set of lines corresponding to $m = 1$ is called the **first-order spectrum.** The **second-order spectrum** corresponds to $m = 2$ for each wavelength. Higher orders may be seen if the angle θ given by Equation 35-27 is less than 90°. Depending on the wavelengths, the orders may be mixed; that is, the third-order line for one wavelength may occur before the second-order line for another wavelength. If the spacing of the slits in the grating is known, the wavelengths emitted by the source can be determined by measuring the angle θ.

Example 35-8

Sodium light is incident on a diffraction grating with 12,000 lines per centimeter. At what angles will the two yellow lines of wavelengths 589.00 nm and 589.59 nm be seen in the first order?

Picture the Problem Apply $d \sin \theta = m\lambda$ to each wavelength, with $m = 1$ and $d = 1 \text{ cm}/12{,}000$.

1. The angle θ is given by $d \sin \theta = m\lambda$, with $m = 1$:

$$\sin \theta = \frac{m\lambda}{d} = \frac{\lambda}{d}$$

2. Calculate θ for $\lambda = 589$ nm:

$$\sin \theta = \frac{\lambda}{d} = \frac{589 \times 10^{-9} \text{ m}}{(1 \text{ cm}/12{,}000)} = 0.7068$$

$$\theta = \sin^{-1} 0.7068 = 44.98°$$

3. Repeat the calculation for $\lambda = 589.59$ nm:

$$\sin \theta = \frac{\lambda}{d} = \frac{589.59 \times 10^{-9} \text{ m}}{1 \text{ cm}/12{,}000} = 0.7075$$

$$\theta = \sin^{-1} 0.7075 = 45.03°$$

Remark Note that light of greater wavelength is diffracted through greater angles.

Exercise Find the angles for the two yellow lines if the grating has 15,000 lines per centimeter. (*Answers* 62.07° and 62.18°)

optional

An important feature of a spectroscope is its ability to resolve spectral lines of two nearly equal wavelengths λ_1 and λ_2. For example, the two prominent yellow lines in the spectrum of sodium have wavelengths 589.00 and 589.59 nm. These can be seen as two separate wavelengths if their interference maxima do not overlap. According to Rayleigh's criterion for resolution, these wavelengths are resolved if the angular separation of their interference maxima is greater than the angular separation between one interference maximum and the first interference minimum on either side of it. The **resolving power** of a diffraction grating is defined to be $\lambda/|\Delta\lambda|$, where $|\Delta\lambda|$ is the smallest difference between two nearby wavelengths, each approximately equal to λ, that may be resolved. The resolving power is proportional to the number of slits illuminated because the more slits illuminated, the sharper the interference maxima. The resolving power R can be shown to be

$$R = \frac{\lambda}{|\Delta\lambda|} = mN \qquad \text{35-28}$$

where N is the number of slits and m is the order number (see Problem 73). We can see from Equation 35-28 that to resolve the two yellow lines in the sodium spectrum the resolving power must be

$$R = \frac{589.00 \text{ nm}}{589.59 - 589.00 \text{ nm}} = 998$$

Thus, to resolve the two yellow sodium lines in the first order ($m = 1$), we need a grating containing about 1000 slits in the area illuminated by the light.

Holograms

An interesting application of diffraction gratings is the production of a three-dimensional photograph called a **hologram** (Figure 35-36). In an ordinary photograph, the intensity of reflected light from an object is recorded on a film. When the film is viewed by transmitted light, a two-dimensional image is produced. In a hologram, a beam from a laser is split into two beams, a reference beam and an object beam. The object beam reflects from the object to be photographed and the interference pattern between it and the reference beam is recorded on a photographic film. This can be done because the laser

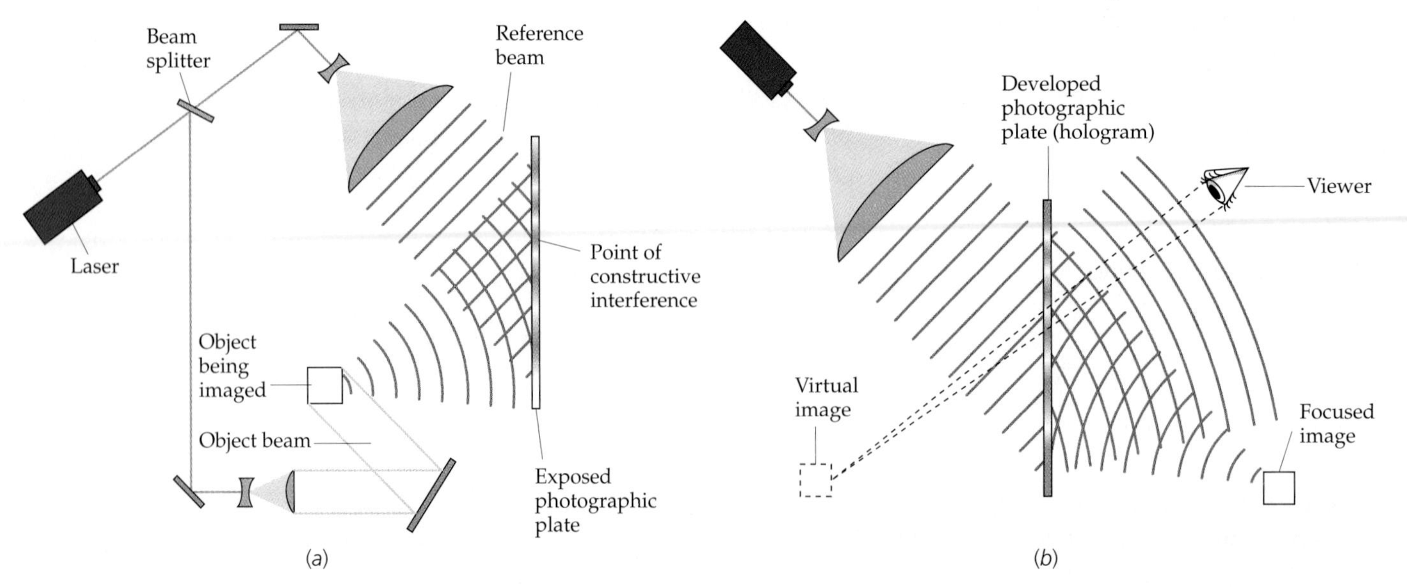

Figure 35-36 (*a*) The production of a hologram. The interference pattern produced by the reference beam and object beam is recorded on a photographic film. (*b*) When the film is developed and illuminated by coherent laser light, a three-dimensional image is seen.

beam is coherent so that the relative phase difference between the reference beam and object beam can be kept constant during the exposure. The interference fringes on the film act as a diffraction grating. When the film is illuminated with a laser, a three-dimensional replica of the object is produced.

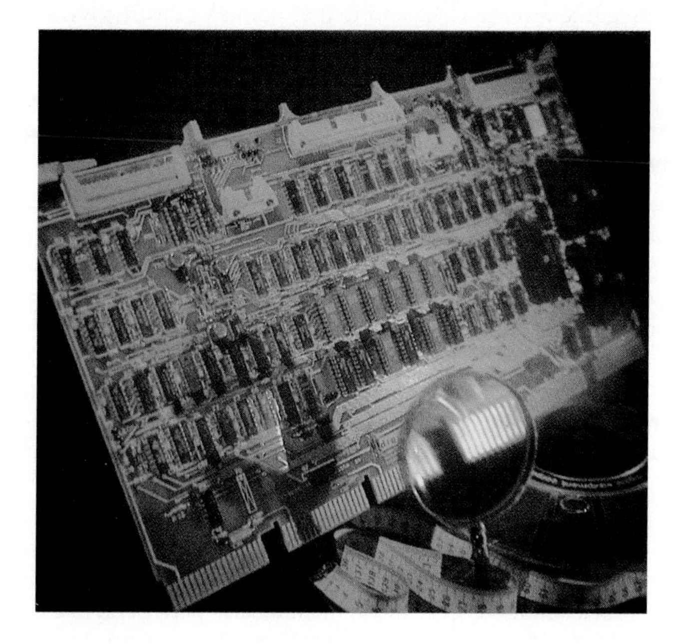

A hologram viewed from two different angles. Note that different parts of the circuit board appear behind the front magnifying lens.

Holograms that you see on credit cards or postage stamps, called rainbow holograms, are more complicated. A horizontal strip of the original hologram is used to make a second hologram. The three-dimensional image can be seen as the viewer moves from side to side, but if viewed with laser light, the image disappears when the viewer's eyes move above or below the slit image. When viewed with white light, the image is seen in different colors as the viewer moves in the vertical direction.

Summary

Topic	Remarks and Relevant Equations	
1. Interference	Two light rays interfere constructively if their phase difference is zero or an integer times 360°. They interfere destructively if their phase difference is 180° or an odd integer times 180°.	
Phase difference due to path difference	$\delta = \frac{\Delta r}{\lambda} 2\pi = \frac{\Delta r}{\lambda} 360°$	35-1
Phase difference due to reflection	A phase difference of 180° is introduced when a light wave is reflected from a boundary between two media for which the wave speed is greater in the original medium, such as the one between air and glass.	
Thin films	The interference of light rays reflected from the top and bottom surfaces of a thin film produces colored bands or fringes, commonly observed in soap films or oil films. The difference in phase between the two rays results from the path difference of twice	

the thickness of the film plus any phase change due to reflection of one or both of the rays.

Two slits	The path difference at an angle θ on a screen far away from two narrow slits separated by a distance d is $d \sin \theta$. If the intensity due to each slit separately is I_0, the intensity at points of constructive interference is $4I_0$ and that at points of destructive interference is 0.	
Interference maxima	$d \sin \theta = m\lambda, \quad m = 0, 1, 2, \ldots$	35-2
Interference minima	$d \sin \theta = (m + \frac{1}{2})\lambda, \quad m = 0, 1, 2, \ldots$	35-3
2. Diffraction	Diffraction occurs whenever a portion of a wavefront is limited by an obstacle or aperture. The intensity of light at any point in space can be computed using Huygens' principle by taking each point on the wavefront to be a point source and computing the resulting interference pattern.	
Fraunhofer patterns	Fraunhofer patterns are observed at great distances from the obstacle or aperture so that the rays reaching any point are approximately parallel, or they can be observed using a lens to focus parallel rays on a viewing screen placed in the focal plane of the lens.	
Fresnel patterns	Fresnel patterns are observed at points close to the source.	
Single slit	When light is incident on a single slit of width a, the intensity pattern on a screen far away shows a broad central diffraction maximum that decreases to zero at an angle θ given by $$\sin \theta = \frac{\lambda}{a} \qquad (35\text{-}9)$$ The width of the central maximum is inversely proportional to the width of the slit. Other zeros in the single-slit diffraction pattern occur at angles given by $$a \sin \theta = m\lambda, \quad m = 1, 2, 3, \ldots \qquad (35\text{-}11)$$ On each side of the central maximum are secondary maxima of much smaller intensity.	
Two slits	The interference–diffraction pattern of two slits of finite size is the two-slit interference pattern modulated by the single-slit diffraction pattern. If $d/a = m$, the mth interference maximum coincides with the first diffraction minimum and will not be seen. Then there will be $2m - 1$ interference maxima within the central diffraction maximum.	
Resolution of two sources	When light from two point sources that are close together passes through an aperture, the diffraction patterns of the sources may overlap. If the overlap is too great, the two sources cannot be resolved as two separate sources. When the central diffraction maximum of one source falls at the diffraction minimum of the other source, the two sources are said to be just resolved by Rayleigh's criterion for resolution. For a circular aperture of diameter D, the critical angular separation of two sources for resolution by Rayleigh's criterion is	
Rayleigh's criterion	$\alpha_c = 1.22 \frac{\lambda}{D}$	35-26
Gratings (optional)	A diffraction grating consisting of a large number of closely spaced lines or slits is used to measure the wavelength of light emitted by a source. The positions of the mth	

order interference maxima from a grating are at angles given by

$$d \sin \theta = m\lambda, \qquad m = 0, 1, 2, \ldots$$

The resolving power of a grating is

$$R = \frac{\lambda}{|\Delta\lambda|} = mN \qquad \text{35-27}$$

where N is the number of slits of the grating that are illuminated and m is the order number.

3. Phasors (optional)

Two or more harmonic waves can be added by representing each wave as a vector called a phasor. The phase difference between the waves is represented as the angle between the phasors.

Problem-Solving Guide

Summary of Worked Examples

Type of Calculation	Procedure and Relevant Examples	
1. Interference		
Relate phase difference to path difference.	Use $\delta = \frac{\Delta r}{\lambda} 2\pi = \frac{\Delta r}{\lambda} 360°$.	**Example 35-1**
Calculate the number of interference fringes in a thin film.	Find the distance x to the mth fringe. Then compute m/x.	**Example 35-2**
Locate fringes in a two-slit interference pattern.	The mth maximimum is at $y_m = m\lambda L/d$.	**Example 35-3**
2. Diffraction		
Locate the zero-intensity points for single-slit diffraction.	Use $a \sin \theta = m\lambda$, where $m = 1, 2, 3, \ldots$.	**Example 35-4**
Find the number of interference maxima in the central diffraction maximum for the two-slit pattern.	Use $N = 2m - 1$, where $m = d/a$.	**Example 35-5**
Find the critical angle for resolution.	Use $\alpha_c = 1.22\,\lambda/D$.	**Example 35-7**
Locate interference maxima produced by a diffraction grating.	Use $d \sin \theta = m\lambda$, where $m = 0, 1, 2, \ldots$.	**Example 35-8**
3. Phasors (optional)		
Find the resultant of the sum of two or more harmonic waves.	Use phasors to represent the waves and add vectorily.	**Example 35-6**

Problems

Conceptual Problems

Problems from Optional and Exploring sections

In a few problems, you are given more data than you actually need; in a few other problems, you are required to supply data from your general knowledge, outside sources, or informed estimates.

• Single-concept, single-step, relatively easy
•• Intermediate-level, may require synthesis of concepts
••• Challenging, for advanced students

Phase Difference and Coherence

1 • When destructive interference occurs, what happens to the energy in the light waves?

2 • Which of the following pairs of light sources are coherent: (*a*) two candles; (*b*) one point source and its image in a plane mirror; (*c*) two pinholes uniformly illuminated by the same point source; (*d*) two headlights of a car; (*e*) two images of a point source due to reflection from the front and back surfaces of a soap film.

3 • (*a*) What minimum path difference is needed to introduce a phase shift of 180° in light of wavelength 600 nm? (*b*) What phase shift will that path difference introduce in light of wavelength 800 nm?

4 • Light of wavelength 500 nm is incident normally on a film of water 10^{-4} cm thick. The index of refraction of water is 1.33. (*a*) What is the wavelength of the light in the water? (*b*) How many wavelengths are contained in the distance $2t$, where t is the thickness of the film? (*c*) What is the phase difference between the wave reflected from the top of the air–water interface and the one reflected from the bottom of the water–air interface after it has traveled this distance?

5 •• Two coherent microwave sources that produce waves of wavelength 1.5 cm are in the xy plane, one on the y axis at $y = 15$ cm and the other at $x = 3$ cm, $y = 14$ cm. If the sources are in phase, find the difference in phase between the two waves from these sources at the origin.

Interference in Thin Films

6 • The spacing between Newton's rings decreases rapidly as the diameter of the rings increases. Explain qualitatively why this occurs.

7 •• If the angle of a wedge-shaped air film such as that in Example 35-2 is too large, fringes are not observed. Why?

8 •• Why must a film used to observe interference colors be thin?

9 • A loop of wire is dipped in soapy water and held so that the soap film is vertical. (*a*) Viewed by reflection with white light, the top of the film appears black. Explain why. (*b*) Below the black region are colored bands. Is the first band red or violet? (*c*) Describe the appearance of the film when it is viewed by *transmitted* light.

10 • A wedge-shaped film of air is made by placing a small slip of paper between the edges of two flat plates of glass. Light of wavelength 700 nm is incident normally on the glass plates, and interference bands are observed by reflection. (*a*) Is the first band near the point of contact of the plates dark or bright? Why? (*b*) If there are five dark bands per centimeter, what is the angle of the wedge?

11 •• The diameters of fine wires can be accurately measured using interference patterns. Two optically flat pieces of glass of length L are arranged with the wire between them as shown in Figure 35-37. The setup is illuminated by monochromatic light, and the resulting interference fringes are detected. Suppose $L = 20$ cm and yellow sodium light ($\lambda \approx$ 590 nm) is used for illumination. If 19 bright fringes are seen along this 20-cm distance, what are the limits on the diameter of the wire? *Hint:* The nineteenth fringe might not be right at the end, but you do not see a twentieth fringe at all.

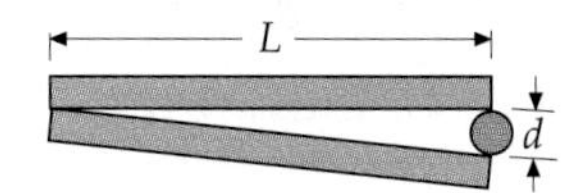

Figure 35-37 Problem 11

12 •• Light of wavelength 600 nm is used to illuminate normally two glass plates 22 cm in length that touch at one end and are separated at the other end by a wire of radius 0.025 mm. How many bright fringes appear along the total length of the plates?

13 •• A thin film having an index of refraction of 1.5 is surrounded by air. It is illuminated normally by white light and is viewed by reflection. Analysis of the resulting reflected light shows that the wavelengths 360, 450, and 602 nm are the only missing wavelengths in or near the visible portion of the spectrum. That is, for these wavelengths, there is destructive interference. (*a*) What is the thickness of the film? (*b*) What visible wavelengths are brightest in the reflected interference pattern? (*c*) If this film were resting on glass with an index of refraction of 1.6, what wavelengths in the visible spectrum would be missing from the reflected light?

14 •• A drop of oil ($n = 1.22$) floats on water ($n = 1.33$). When reflected light is observed from above as shown in Figure 35-38, what is the thickness of the drop at the point where the second red fringe, counting from the edge of the drop, is observed? Assume red light has a wavelength of 650 nm.

Figure 35-38 Problem 14

15 •• A film of oil of index of refraction $n = 1.45$ rests on an optically flat piece of glass of index of refraction $n = 1.6$. When illuminated with white light at normal incidence, light of wavelengths 690 and 460 nm is predominant in the reflected light. Determine the thickness of the oil film.

16 •• A film of oil of index of refraction $n = 1.45$ floats on water ($n = 1.33$). When illuminated with white light at normal incidence, light of wavelengths 700 nm and 500 nm is predominant in the reflected light. Determine the thickness of the oil film.

Newton's Rings

17 •• A Newton's-ring apparatus consists of a glass lens with radius of curvature R that rests on a flat glass plate as shown in Figure 35-39. The thin film is air of variable thickness. The pattern is viewed by reflected light. (*a*) Show that for a thickness t the condition for a bright (constructive) interference ring is

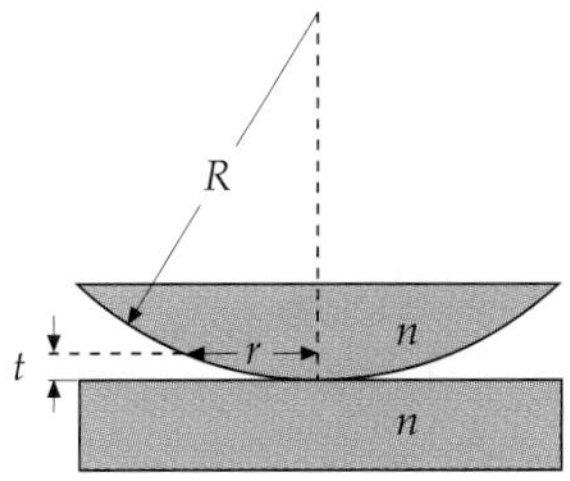

$$t = \left(m + \frac{1}{2}\right)\frac{\lambda}{2}, \quad m = 0, 1, 2, \ldots$$

Figure 35-39 Problem 17

(*b*) Apply the Pythagorean theorem to the triangle of sides r, $R - t$, and hypotenuse R to show that for $t \ll R$, the radius of a fringe is related to t by

$$r = \sqrt{2tR} \qquad \text{35-29}$$

(*c*) How would the transmitted pattern look in comparison with the reflected one? (*d*) Use $R = 10$ m and a diameter of 4 cm for the lens. How many bright fringes would you see if the apparatus were illuminated by yellow sodium light ($\lambda \approx$ 590 nm) and were viewed by reflection? (*e*) What would be the diameter of the sixth bright fringe? (*f*) If the glass used in the apparatus has an index of refraction $n = 1.5$ and water ($n_w = 1.33$) is placed between the two pieces of glass, what change will take place in the bright fringes?

18 •• A plano-convex glass lens of radius of curvature 2.0 m rests on an optically flat glass plate. The arrangement is illuminated from above with monochromatic light of 520-nm wavelength. The indexes of refraction of the lens and plate are 1.6. Determine the radii of the first and second bright fringe in the reflected light. (Use Equation 35-29 from Problem 17 to relate r to t.)

19 •• Suppose that before the lens of Problem 18 is placed on the plate a film of oil of refractive index 1.82 is deposited on the plate. What will then be the radii of the first and second bright fringes? (Use Equation 35-29 from Problem 17 to relate r to t.)

Two-Slit Interference Pattern

20 • A double-slit interference experiment is set up in a chamber that can be evacuated. Using monochromatic light, an interference pattern is observed when the chamber is open to air. As the chamber is evacuated one will note that

(*a*) the interference fringes remain fixed.
(*b*) the interference fringes move closer together.
(*c*) the interference fringes move farther apart.
(*d*) the interference fringes disappear completely.

21 • Two narrow slits separated by 1 mm are illuminated by light of wavelength 600 nm, and the interference pattern is viewed on a screen 2 m away. Calculate the number of bright fringes per centimeter on the screen.

22 • Using a conventional two-slit apparatus with light of wavelength 589 nm, 28 bright fringes per centimeter are observed on a screen 3 m away. What is the slit separation?

23 • Light of wavelength 633 nm from a helium–neon laser is shone normally on a plane containing two slits. The first interference maximum is 82 cm from the central maximum on a screen 12 m away. (*a*) Find the separation of the slits. (*b*) How many interference maxima can be observed?

24 •• Two narrow slits are separated by a distance d. Their interference pattern is to be observed on a screen a large distance L away. (*a*) Calculate the spacing y of the maxima on the screen for light of wavelength 500 nm when $L =$ 1 m and $d = 1$ cm. (*b*) Would you expect to observe the interference of light on the screen for this situation? (*c*) How close together should the slits be placed for the maxima to be separated by 1 mm for this wavelength and screen distance?

25 •• Light is incident at an angle ϕ with the normal to a vertical plane containing two slits of separation d (Figure 35-40). Show that the interference maxima are located at angles θ given by $\sin\theta + \sin\phi = m\lambda/d$.

Figure 35-40 Problems 25 and 26

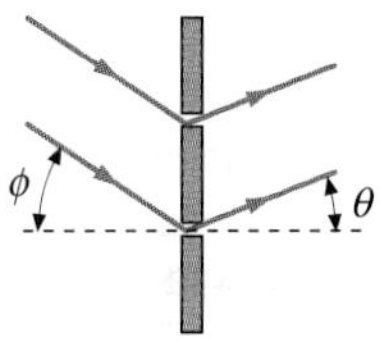

26 •• White light falls at an angle of 30° to the normal of a plane containing a pair of slits separated by 2.5 μm. What visible wavelengths give a bright interference maximum in the transmitted light in the direction normal to the plane? (See Problem 25.)

27 •• Laser light falls normally on three evenly spaced, very narrow slits. When one of the side slits is covered, the first-order maximum is at 0.60° from the normal. If the center slit is covered and the other two are open, find (*a*) the angle of the first-order maximum and (*b*) the order number of the maximum that now occurs at the same angle as the fourth-order maximum did before.

Diffraction Pattern of a Single Slit

28 • As the width of a slit producing a single-slit diffraction pattern is slowly and steadily reduced, how will the diffraction pattern change?

29 • Equation 35-2, $d \sin\theta = m\lambda$, and Equation 35-11, $a \sin\theta = m\lambda$, are sometimes confused. For each equation, define the symbols and explain the equation's application.

30 • Light of wavelength 600 nm is incident on a long, narrow slit. Find the angle of the first diffraction minimum if the width of the slit is (*a*) 1 mm, (*b*) 0.1 mm, and (*c*) 0.01 mm.

31 • The single-slit diffraction pattern of light is observed on a screen a large distance L from the slit. Note from Equation 35-12 that the width $2y$ of the central maximum varies inversely with the width a of the slit. Calculate the width $2y$ for $L = 2$ m, $\lambda = 500$ nm, and (*a*) $a = 0.1$ mm, (*b*) $a = 0.01$ mm, and (*c*) $a = 0.001$ mm.

32 • Plane microwaves are incident on a long, narrow metal slit of width 5 cm. The first diffraction minimum is observed at $\theta = 37°$. What is the wavelength of the microwaves?

33 •• For a ruby laser of wavelength 694 nm, the end of the ruby crystal is the aperture that determines the diameter of the light beam emitted. If the diameter is 2 cm and the laser is aimed at the moon, 380,000 km away, find the approximate diameter of the light beam when it reaches the moon, assuming the spread is due solely to diffraction.

Interference–Diffraction Pattern of Two Slits

34 • How many interference maxima will be contained in the central diffraction maximum in the diffraction–interference pattern of two slits if the separation d of the slits is 5 times their width a? How many will there be if $d = Na$ for any value of N?

35 •• A two-slit Fraunhofer interference–diffraction pattern is observed with light of wavelength 500 nm. The slits have a separation of 0.1 mm and a width of a. (*a*) Find the width a if the fifth interference maximum is at the same angle as the first diffraction minimum. (*b*) For this case, how many bright interference fringes will be seen in the central diffraction maximum?

36 •• A two-slit Fraunhofer interference–diffraction pattern is observed with light of wavelength 700 nm. The slits have widths of 0.01 mm and are separated by 0.2 mm. How many bright fringes will be seen in the central diffraction maximum?

37 •• Suppose that the *central* diffraction maximum for two slits contains 17 interference fringes for some wavelength of light. How many interference fringes would you expect in the first *secondary* diffraction maximum?

38 •• Light of wavelength 550 nm illuminates two slits of width 0.03 mm and separation 0.15 mm. (*a*) How many interference maxima fall within the full width of the central diffraction maximum? (*b*) What is the ratio of the intensity of the third interference maximum to the side of the centerline (not counting the center interference maximum) to the intensity of the center interference maximum?

Using Phasors to Add Harmonic Waves (optional)

39 • Find the resultant of the two waves $E_1 = 2 \sin \omega t$ and $E_2 = 3 \sin (\omega t + 270°)$.

40 • Find the resultant of the two waves $E_1 = 4 \sin \omega t$ and $E_2 = 3 \sin (\omega t + 60°)$.

41 •• At the second secondary maximum of the diffraction pattern of a single slit, the phase difference between the waves from the top and bottom of the slit is approximately 5π. The phasors used to calculate the amplitude at this point complete 2.5 circles. If I_0 is the intensity at the central maximum, find the intensity I at this second secondary maximum.

42 •• (*a*) Show that the positions of the interference minima on a screen a large distance L away from three equally spaced sources (spacing d, with $d >> \lambda$) are given approximately by

$$y = \frac{n\lambda L}{3d}, \quad \text{where } n = 1, 2, 4, 5, 7, 8, 10, \ldots$$

that is, n is not a multiple of 3. (*b*) For $L = 1$ m, $\lambda = 5 \times 10^{-7}$ m, and $d = 0.1$ mm, calculate the width of the principal interference maxima (the distance between successive minima) for three sources.

43 •• (*a*) Show that the positions of the interference minima on a screen a large distance L away from four equally spaced sources (spacing d, with $d >> \lambda$) are given approximately by

$$y = \frac{n\lambda L}{4d}, \quad \text{where } n = 1, 2, 3, 5, 6, 7, 9, 10, \ldots$$

that is, n is not a multiple of 4. (*b*) For $L = 2$ m, $\lambda = 6 \times 10^{-7}$ m, and $d = 0.1$ mm, calculate the width of the principal interference maxima (the distance between successive minima) for four sources. Compare this width with that for two sources with the same spacing.

44 •• Light of wavelength 480 nm falls normally on four slits. Each slit is 2 μm wide and is separated from the next by 6 μm. (*a*) Find the angle from the center to the first point of zero intensity of the single-slit diffraction pattern on a distant screen. (*b*) Find the angles of any bright interference maxima that lie inside the central diffraction maximum. (*c*) Find the angular spread between the central interference maximum and the first interference minimum on either side of it. (*d*) Sketch the intensity as a function of angle.

45 ••• Three slits, each separated from its neighbor by 0.06 mm, are illuminated by a coherent light source of wavelength 550 nm. The slits are extremely narrow. A screen is located 2.5 m from the slits. The intensity on the centerline is 0.05 W/m^2. Consider a location 1.72 cm from the centerline. (*a*) Draw the phasors, according to the phasor model for the addition of harmonic waves, appropriate for this location. (*b*) From the phasor diagram, calculate the intensity of light at this location.

46 ••• Four coherent sources are located on the y axis at $+3\lambda/4$, $+\lambda/4$, $-\lambda/4$, and $-3\lambda/4$. They emit waves of wavelength λ and intensity I_0. (*a*) Calculate the net intensity I as a function of the angle θ measured from the $+x$ axis. (*b*) Make a polar plot of $I(\theta)$.

47 ••• For single-slit diffraction, calculate the first three values of ϕ (the total phase difference between rays from each edge of the slit) that produce subsidiary maxima by (*a*) using the phasor model and (*b*) setting $dI/d\phi = 0$, where I is given by Equation 35-20.

Diffraction and Resolution

48 • Light of wavelength 700 nm is incident on a pinhole of diameter 0.1 mm. (*a*) What is the angle between the central maximum and the first diffraction minimum for a Fraunhofer diffraction pattern? (*b*) What is the distance between the central maximum and the first diffraction minimum on a screen 8 m away?

49 • Two sources of light of wavelength 700 nm are 10 m away from the pinhole of Problem 48. How far apart must the sources be for their diffraction patterns to be resolved by Rayleigh's criterion?

50 • Two sources of light of wavelength 700 nm are separated by a horizontal distance x. They are 5 m from a vertical slit of width 0.5 mm. What is the least value of x for which the diffraction pattern of the sources can be resolved by Rayleigh's criterion?

51 • The headlights on a small car are separated by 112 cm. At what maximum distance could you resolve them if the diameter of your pupil is 5 mm and the effective wavelength of the light is 550 nm?

52 • You are told not to shoot until you see the whites of their eyes. If their eyes are separated by 6.5 cm and the diameter of your pupil is 5 mm, at what distance can you resolve the two eyes using light of wavelength 550 nm?

53 •• (*a*) How far apart must two objects be on the moon to be resolved by the eye? Take the diameter of the pupil of the eye to be 5 mm, the wavelength of the light to be 600 nm, and the distance to the moon to be 380,000 km. (*b*) How far apart must the objects on the moon be to be resolved by a telescope that has a mirror of diameter 5 m?

54 •• The ceiling of your lecture hall is probably covered with acoustic tile, which has small holes separated by about 6 mm. (*a*) Using light with a wavelength of 500 nm, how far could you be from this tile and still resolve these holes? The diameter of the pupil of your eye is about 5 mm. (*b*) Could you resolve these holes better with red light or with violet light?

55 •• The telescope on Mount Palomar has a diameter of 200 inches. Suppose a double star were 4 lightyears away. Under ideal conditions, what must be the minimum separation of the two stars for their images to be resolved using light of wavelength 550 nm?

56 •• The star Mizar in Ursa Major is a binary system of stars of nearly equal magnitudes. The angular separation between the two stars is 14 seconds of arc. What is the minimum diameter of the pupil that allows resolution of the two stars using light of wavelength 550 nm?

Diffraction Gratings (optional)

57 • When a diffraction grating is illuminated with white light, the first-order maximum of green light

(*a*) is closer to the central maximum than that of red light.
(*b*) is closer to the central maximum than that of blue light.
(*c*) overlaps the second order maximum of red light.
(*d*) overlaps the second order maximum of blue light.

58 • A diffraction grating with 2000 slits per centimeter is used to measure the wavelengths emitted by hydrogen gas. At what angles θ in the first-order spectrum would you expect to find the two violet lines of wavelengths 434 and 410 nm?

59 • With the grating used in Problem 58, two other lines in the first-order hydrogen spectrum are found at angles $\theta_1 = 9.72 \times 10^{-2}$ rad and $\theta_2 = 1.32 \times 10^{-1}$ rad. Find the wavelengths of these lines.

60 • Repeat Problem 58 for a diffraction grating with 15,000 slits per centimeter.

61 • What is the longest wavelength that can be observed in the fifth-order spectrum using a diffraction grating with 4000 slits per centimeter?

62 •• A diffraction grating of 2000 slits per centimeter is used to analyze the spectrum of mercury. (*a*) Find the angular separation in the first-order spectrum of the two lines of wavelengths 579.0 and 577.0 nm. (*b*) How wide must the beam on the grating be for these lines to be resolved?

63 •• A diffraction grating with 4800 lines per centimeter is illuminated at normal incidence with white light (wavelength range 400 to 700 nm). For how many orders can one observe the complete spectrum in the transmitted light? Do any of these orders overlap? If so, describe the overlapping regions.

64 •• A square diffraction grating with an area of 25 cm^2 has a resolution of 22,000 in the fourth order. At what angle should you look to see a wavelength of 510 nm in the fourth order?

65 •• Sodium light of wavelength 589 nm falls normally on a 2-cm-square diffraction grating ruled with 4000 lines per centimeter. The Fraunhofer diffraction pattern is projected onto a screen at 1.5 m by a lens of focal length 1.5 m placed immediately in front of the grating. Find (*a*) the positions of the first two intensity maxima on one side of the central maximum, (*b*) the width of the central maximum, and (*c*) the resolution in the first order.

66 •• The spectrum of neon is exceptionally rich in the visible region. Among the many lines are two at wavelengths of 519.313 and 519.322 nm. If light from a neon discharge tube is normally incident on a transmission grating with 8400 lines per centimeter and the spectrum is observed in second order, what must be the width of the grating that is illuminated so that these two lines can be resolved?

67 •• Mercury has several stable isotopes, among them ^{198}Hg and ^{202}Hg. The strong spectral line of mercury at about 546.07 nm is a composite of spectral lines from the various mercury isotopes. The wavelengths of this line for ^{198}Hg and ^{202}Hg are 546.07532 and 546.07355 nm, respectively. What must be the resolving power of a grating capable of resolving these two isotopic lines in the third-order spectrum? If the grating is illuminated over a 2-cm-wide region, what must be the number of lines per centimeter of the grating?

68 •• A transmission grating is used to study the spectral region extending from 480 to 500 nm. The angular spread of

this region is 12° in third order. (*a*) Find the number of lines per centimeter. (*b*) How many orders are visible?

69 •• White light is incident normally on a transmission grating and the spectrum is observed on a screen 8.0 m from the grating. In the second-order spectrum, the separation between light of 520- and 590-nm wavelengths is 8.4 cm. (*a*) Determine the number of lines per centimeter of the grating. (*b*) What is the separation between these two wavelengths in the first-order and third-order spectra?

70 ••• A diffraction grating has n lines per meter. Show that the angular separation of two lines of wavelengths λ and $\lambda + \Delta\lambda$ meters is approximately

$$\Delta\theta = \frac{\Delta\lambda}{\sqrt{(1/nm)^2 - \lambda^2}}$$

71 ••• When assessing a diffraction grating, we are interested not only in its resolving power R, which is the ability of the grating to separate two close wavelengths, but also in the dispersion D of the grating. This is defined by $D = \Delta\theta_m/\Delta\lambda$ in the mth order. (*a*) Show that D can be written

$$D = \frac{m}{\sqrt{d^2 - m^2\lambda^2}}$$

where d is the slit spacing. (*b*) If a diffraction grating with 2000 slits per centimeter is to resolve the two yellow sodium lines in the second order (wavelengths 589.0 and 589.6 nm), how many slits must be illuminated by the beam? (*c*) What would the separation be between these resolved yellow lines if the pattern were viewed on a screen 4 m from the grating?

72 ••• For a diffraction grating in which all the surfaces are normal to the incident radiation, most of the energy goes into the zeroth order, which is useless from a spectroscopic point of view since in zeroth order all the wavelengths are at 0°. Therefore, modern gratings have shaped, or *blazed,* grooves as shown in Figure 35-41. This shifts the specular reflection, which contains most of the energy, from the zeroth order to some higher order. (*a*) Calculate the blaze angle ϕ in terms of a (the groove separation), λ (the wavelength), and m (the order in which specular reflection is to occur). (*b*) Calculate the proper blaze angle for the specular reflection to occur in the second order for light of wavelength 450 nm incident on a grating with 10,000 lines per centimeter.

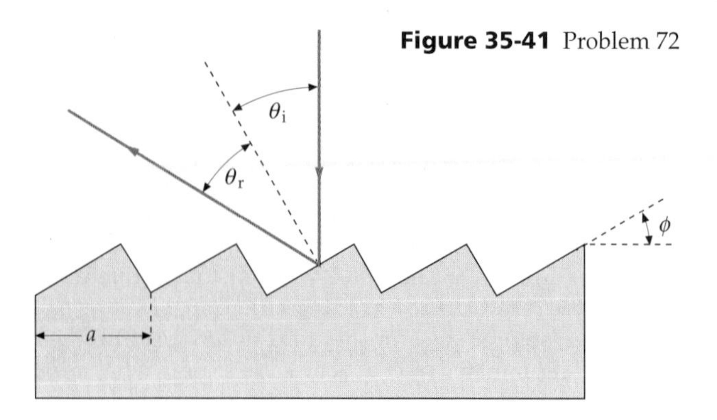

Figure 35-41 Problem 72

73 ••• In this problem you will derive Equation 35-28 for the resolving power of a diffraction grating containing N slits separated by a distance d. To do this you will calculate the angular separation between the maximum and minimum for some wavelength λ and set it equal to the angular separation of the mth-order maximum for two nearby wavelengths. (*a*) Show that the phase difference ϕ between the light from two adjacent slits is given by

$$\phi = \frac{2\pi d}{\lambda}\sin\theta$$

(*b*) Differentiate this expression to show that a small change in angle $d\theta$ results in a change in phase of $d\phi$ given by

$$d\phi = \frac{2\pi d}{\lambda}\cos\theta\, d\theta$$

(*c*) For N slits, the angular separation between an interference maximum and interference minimum corresponds to a phase change of $d\phi = 2\pi/N$. Use this to show that the angular separation $d\theta$ between the maximum and minimum for some wavelength λ is given by

$$d\theta = \frac{\lambda}{Nd\cos\theta} \qquad 35\text{-}30$$

(*d*) The angle of the mth-order interference maximum for wavelength λ is given by Equation 35-27. Compute the differential of each side of this equation to show that angular separation of the mth-order maximum for two nearly equal wavelengths differing by $d\lambda$ is given by

$$d\theta \approx \frac{m\, d\lambda}{d\cos\theta} \qquad 35\text{-}31$$

(*e*) According to Rayleigh's criterion, two wavelengths will be resolved in the mth order if the angular separation of the wavelengths given by Equation 35-31 equals the angular separation of the interference maximum and interference minimum given by Equation 35-30. Use this to derive Equation 35-28 for the resolving power of a grating.

General Problems

74 • True or false:

(*a*) When waves interfere destructively, the energy is converted into heat energy.
(*b*) Interference is observed only for waves from coherent sources.
(*c*) In the Fraunhofer diffraction pattern for a single slit, the narrower the slit, the wider the central maximum of the diffraction pattern.
(*d*) A circular aperture can produce both a Fraunhofer and a Fresnel diffraction pattern.
(*e*) The ability to resolve two point sources depends on the wavelength of the light.

75 • In a lecture demonstration, laser light is used to illuminate two slits separated by 0.5 mm, and the interference pattern is observed on a screen 5 m away. The distance on the screen from the centerline to the thirty-seventh bright fringe is 25.7 cm. What is the wavelength of the light?

76 • A long, narrow, horizontal slit lies 1 mm above a plane mirror, which is in the horizontal plane. The interference pattern produced by the slit and its image is viewed on a screen 1 m from the slit. The wavelength of the light is 600 nm. (*a*) Find the distance from the mirror to the first maximum. (*b*) How many dark bands per centimeter are seen on the screen?

77 • In a lecture demonstration, a laser beam of wavelength 700 nm passes through a vertical slit 0.5 mm wide and hits a screen 6 m away. Find the horizontal length of the principal diffraction maximum on the screen; that is, find the distance between the first minimum on the left and the first minimum on the right of the central maximum.

78 • What minimum aperture, in millimeters, is required for opera glasses (binoculars) if an observer is to be able to distinguish the soprano's individual eyelashes (separated by 0.5 mm) at an observation distance of 25 m? Assume the effective wavelength of the light to be 550 nm.

79 • The diameter of the aperture of the radio telescope at Arecibo, Puerto Rico, is 300 m. What is the resolving power of the telescope when tuned to detect microwaves of 3.2 cm wavelength?

80 •• A thin layer of a transparent material with an index of refraction of 1.30 is used as a nonreflective coating on the surface of glass with an index of refraction of 1.50. What should the thickness of the material be for it to be nonreflecting for light of wavelength 600 nm?

81 •• A *Fabry–Perot interferometer* consists of two parallel, half-silvered mirrors separated by a small distance a. Show that when light is incident on the interferometer with an angle of incidence θ, the transmitted light will have maximum intensity when $a = m\lambda/2\cos\theta$.

82 •• A mica sheet 1.2 μm thick is suspended in air. In reflected light, there are gaps in the visible spectrum at 421, 474, 542, and 633 nm. Find the index of refraction of the mica sheet.

83 •• A camera lens is made of glass with an index of refraction of 1.6. This lens is coated with a magnesium fluoride film ($n = 1.38$) to enhance its light transmission. This film is to produce zero reflection for light of wavelength 540 nm. Treat the lens surface as a flat plane and the film as a uniformly thick flat film. (*a*) How thick must the film be to accomplish its objective in the first order? (*b*) Would there be destructive interference for any other visible wavelengths? (*c*) By what factor would the reflection for light of wavelengths 400 and 700 nm be reduced by this film? Neglect the variation in the reflected light amplitudes from the two surfaces.

84 •• In a pinhole camera, the image is fuzzy because of geometry (rays arrive at the film through different parts of the pinhole) and because of diffraction. As the pinhole is made smaller, the fuzziness due to geometry is reduced, but the fuzziness due to diffraction is increased. The optimum size of the pinhole for the sharpest possible image occurs when the spread due to diffraction equals that due to the geometric effects of the pinhole. Estimate the optimum size of the pinhole if the distance from it to the film is 10 cm and the wavelength of the light is 550 nm.

85 •• The Impressionist painter Georges Seurat used a technique called "pointillism," in which his paintings are composed of small, closely spaced dots of pure color, each about 2 mm in diameter. The illusion of the colors blending together smoothly is produced in the eye of the viewer by diffraction effects. Calculate the minimum viewing distance for this effect to work properly. Use the wavelength of visible light that requires the *greatest* distance, so that you're sure the effect will work for *all* visible wavelengths. Assume that the pupil of the eye has a diameter of 3 mm.

86 ••• A *Jamin refractometer* is a device for measuring or comparing the indexes of refraction of fluids. A beam of monochromatic light is split into two parts, each of which is directed along the axis of a separate cylindrical tube before being recombined into a single beam that is viewed through a telescope. Suppose that each tube is 0.4 m long and that sodium light of wavelength 589 nm is used. Both tubes are initially evacuated, and constructive interference is observed in the center of the field of view. As air is slowly allowed to enter one of the tubes, the central field of view changes to dark and back to bright a total of 198 times. (*a*) What is the index of refraction of air? (*b*) If the fringes can be counted to ±0.25 fringe, where one fringe is equivalent to one complete cycle of intensity variation at the center of the field of view, to what accuracy can the index of refraction of air be determined by this experiment?

87 ••• Light of wavelength λ is diffracted through a single slit of width a, and the resulting pattern is viewed on a screen a long distance L away from the slit. (*a*) Show that the width of the central maximum on the screen is approximately $2L\lambda/a$. (*b*) If a slit of width $2L\lambda/a$ is cut in the screen and is illuminated, show that the width of its central diffraction maximum at the same distance L is a to the same approximation.

88 ••• Television viewers in rural areas often find that the picture flickers (fades in and out) as an airplane flies across the sky in the vicinity. The flickering arises from the interference between the signal directly from the transmitter and that reflected to the antenna from the airplane. Suppose the receiver is 36 km from the transmitter broadcasting at a frequency of 86.0 MHz and an airplane is flying at a height of about 600 m above the receiver toward the transmitter. The rate of oscillation of the picture's intensity is 4 Hz. (*a*) Determine the speed of the plane. (*b*) If the picture's intensity is a maximum when the plane is directly overhead, what is the exact height of the plane above the receiving antenna?

89 ••• For the situation described in Problem 88, show that the rate of oscillation of the picture's intensity is a minimum when the airplane is directly above the midpoint between the transmitter and receiving antenna.

90 ••• A double-slit experiment uses a helium–neon laser with a wavelength of 633 nm and a slit separation of 0.12 mm. When a thin sheet of plastic is placed in front of one of the slits, the interference pattern shifts by 5.5 fringes. When the experiment is repeated under water, the shift is 3.5 fringes. Calculate (*a*) the thickness of the plastic sheet and (*b*) the index of refraction of the plastic sheet.

91 ••• Two coherent sources are located on the y axis at $+\lambda/4$ and $-\lambda/4$. They emit waves of wavelength λ and intensity I_0. (*a*) Calculate the net intensity I as a function of the angle θ measured from the $+x$ axis. (*b*) Make a polar plot of $I(\theta)$.

APPENDIX A

SI Units and Conversion Factors

Basic Units	
Length	The *meter* (m) is the distance traveled by light in a vacuum in 1/299,792,458 s.
Time	The *second* (s) is the duration of 9,192,631,770 periods of the radiation corresponding to the transition between the two hyperfine levels of the ground state of the ^{133}Cs atom.
Mass	The *kilogram* (kg) is the mass of the international standard body preserved at Sèvres, France.
Current	The *ampere* (A) is that current in two very long parallel wires 1 m apart that gives rise to a magnetic force per unit length of 2×10^{-7} N/m.
Temperature	The *kelvin* (K) is 1/273.16 of the thermodynamic temperature of the triple point of water.
Luminous intensity	The *candela* (cd) is the luminous intensity, in the perpendicular direction, of a surface of area 1/600,000 m^2 of a blackbody at the temperature of freezing platinum at a pressure of 1 atm.

Derived Units		
Force	newton (N)	$1\ N = 1\ kg{\cdot}m/s^2$
Work, energy	joule (J)	$1\ J = 1\ N{\cdot}m$
Power	watt (W)	$1\ W = 1\ J/s$
Frequency	hertz (Hz)	$1\ Hz = s^{-1}$
Charge	coulomb (C)	$1\ C = 1\ A{\cdot}s$
Potential	volt (V)	$1\ V = 1\ J/C$
Resistance	ohm (Ω)	$1\ \Omega = 1\ V/A$
Capacitance	farad (F)	$1\ F = 1\ C/V$
Magnetic field	tesla (T)	$1\ T = 1\ N/A{\cdot}m$
Magnetic flux	weber (Wb)	$1\ Wb = 1\ T{\cdot}m^2$
Inductance	henry (H)	$1\ H = 1\ J/A^2$

Conversion Factors

Conversion factors are written as equations for simplicity; relations marked with an asterisk are exact.

Length

1 km = 0.6215 mi
1 mi = 1.609 km
1 m = 1.0936 yd = 3.281 ft = 39.37 in
*1 in = 2.54 cm
*1 ft = 12 in = 30.48 cm
*1 yd = 3 ft = 91.44 cm
1 lightyear = 1 $c \cdot y$ = 9.461×10^{15} m
*1 Å = 0.1 nm

Area

*1 $m^2 = 10^4$ cm^2
1 km^2 = 0.3861 mi^2 = 247.1 acres
*1 in^2 = 6.4516 cm^2
1 $ft^2 = 9.29 \times 10^{-2}$ m^2
1 m^2 = 10.76 ft^2
*1 acre = 43,560 ft^2
1 mi^2 = 640 acres = 2.590 km^2

Volume

*1 $m^3 = 10^6$ cm^3
*1 L = 1000 cm^3 = 10^{-3} m^3
1 gal = 3.786 L
1 gal = 4 qt = 8 pt = 128 oz = 231 in^3
1 in^3 = 16.39 cm^3
1 ft^3 = 1728 in^3 = 28.32 L = 2.832×10^4 cm^3

Time

*1 h = 60 min = 3.6 ks
*1 d = 24 h = 1440 min = 86.4 ks
1 y = 365.24 d = 31.56 Ms

Speed

1 km/h = 0.2778 m/s = 0.6215 mi/h
1 mi/h = 0.4470 m/s = 1.609 km/h
1 mi/h = 1.467 ft/s

Angle and Angular Speed

*π rad = 180°
1 rad = 57.30°
1° = 1.745×10^{-2} rad
1 rev/min = 0.1047 rad/s
1 rad/s = 9.549 rev/min

Mass

*1 kg = 1000 g
*1 tonne = 1000 kg = 1 Mg
1 u = 1.6606×10^{-27} kg
1 kg = 6.022×10^{23} u
1 slug = 14.59 kg
1 kg = 6.852×10^{-2} slug
1 u = 931.50 MeV/c^2

Density

*1 g/cm^3 = 1000 kg/m^3 = 1 kg/L
(1 g/cm^3)g = 62.4 lb/ft^3

Force

1 N = 0.2248 lb = 10^5 dyn
1 lb = 4.4482 N
(1 kg)g = 2.2046 lb

Pressure

*1 Pa = 1 N/m^2
*1 atm = 101.325 kPa = 1.01325 bars
1 atm = 14.7 lb/in^2 = 760 mmHg
= 29.9 inHg = 33.8 ftH_2O
1 lb/in^2 = 6.895 kPa
1 torr = 1 mmHg = 133.32 Pa
1 bar = 100 kPa

Energy

*1 kW·h = 3.6 MJ
*1 cal = 4.1840 J
1 ft·lb = 1.356 J = 1.286×10^{-3} Btu
*1 L·atm = 101.325 J
1 L·atm = 24.217 cal
1 Btu = 778 ft·lb = 252 cal = 1054.35 J
1 eV = 1.602×10^{-19} J
1 u·c^2 = 931.50 MeV
*1 erg = 10^{-7} J

Power

1 horsepower = 550 ft·lb/s = 745.7 W
1 Btu/min = 17.58 W
1 W = 1.341×10^{-3} horsepower
= 0.7376 ft·lb/s

Magnetic Field

*1 G = 10^{-4} T
*1 T = 10^4 G

Thermal Conductivity

1 W/m·K = 6.938 Btu·in/h·ft^2·F°
1 Btu·in/h·ft^2·F° = 0.1441 W/m·K

APPENDIX 

Numerical Data

Terrestrial Data	
Acceleration of gravity g	9.80665 m/s^2
Standard value	32.1740 ft/s^2
At sea level, at equator†	9.7804 m/s^2
At sea level, at poles†	9.8322 m/s^2
Mass of earth M_E	5.98×10^{24} kg
Radius of earth R_E, mean	6.37×10^6 m; 3960 mi
Escape speed $\sqrt{2R_E g}$	1.12×10^4 m/s; 6.95 mi/s
Solar constant‡	1.35 kW/m^2
Standard temperature and pressure (STP):	
Temperature	273.15 K
Pressure	101.325 kPa; 1.00 atm
Molar mass of air	28.97 g/mol
Density of air (STP), ρ_{air}	1.293 kg/m^3
Speed of sound (STP)	331 m/s
Heat of fusion of H_2O (0°C, 1 atm)	333.5 kJ/kg
Heat of vaporization of H_2O (100°C, 1 atm)	2.257 MJ/kg

†Measured relative to the earth's surface.
‡Average power incident normally on 1 m^2 outside the earth's atmosphere at the mean distance from the earth to the sun.

Astronomical Data	
Earth	
Distance to moon†	3.844×10^8 m; 2.389×10^5 mi
Distance to sun, mean†	1.496×10^{11} m; 9.30×10^7 mi; 1.00 AU
Orbital speed, mean	2.98×10^4 m/s
Moon	
Mass	7.35×10^{22} kg
Radius	1.738×10^6 m
Period	27.32 d
Acceleration of gravity at surface	1.62 m/s^2
Sun	
Mass	1.99×10^{30} kg
Radius	6.96×10^8 m

† Center to center.

Physical Constants

Gravitational constant	G	$6.672\,6 \times 10^{-11}$ N·m²/kg²
Speed of light	c	$2.997\,924\,58 \times 10^{8}$ m/s
Fundamental charge	e	$1.602\,177\,33 \times 10^{-19}$ C
Avogadro's number	N_A	$6.022\,136\,7 \times 10^{23}$ particles/mol
Gas constant	R	8.314 51 J/mol·K
		1.987 22 cal/mol·K
		$8.205\,78 \times 10^{-2}$ L·atm/mol·K
Boltzmann's constant	$k = R/N_A$	$1.380\,658 \times 10^{-23}$ J/K
		$8.617\,385 \times 10^{-5}$ eV/K
Unified mass unit	$u = (1/N_A)$ g	$1.660\,540 \times 10^{-24}$ g
Coulomb constant	$k = 1/4\pi\epsilon_0$	$8.987\,551\,788 \times 10^{9}$ N·m²/C²
Permittivity of free space	ϵ_0	$8.854\,187\,817 \times 10^{-12}$ C²/N·m²
Permeability of free space	μ_0	$4\pi \times 10^{-7}$ N/A²
		$1.256\,637 \times 10^{-6}$ N/A²
Planck's constant	h	$6.626\,075\,5 \times 10^{-34}$ J·s
		$4.135\,669\,2 \times 10^{-15}$ eV·s
	$\hbar = h/2\pi$	$1.054\,572\,66 \times 10^{-34}$ J·s
		$6.582\,122\,0 \times 10^{-16}$ eV·s
Mass of electron	m_e	$9.109\,389\,7 \times 10^{-31}$ kg
		510.999 1 keV/c^2
Mass of proton	m_p	$1.672\,623\,1 \times 10^{-27}$ kg
		938.272 3 MeV/c^2
Mass of neutron	m_n	$1.674\,929 \times 10^{-27}$ kg
		939.565 6 MeV/c^2
Bohr magneton	$m_B = e\hbar/2m_e$	$9.274\,015\,4 \times 10^{-24}$ J/T
		$5.788\,382\,63 \times 10^{-5}$ eV/T
Nuclear magneton	$m_n = e\hbar/2m_p$	$5.050\,786\,6 \times 10^{-27}$ J/T
		$3.152\,451\,66 \times 10^{-8}$ eV/T
Magnetic flux quantum	$\phi_0 = h/2e$	$2.067\,834\,6 \times 10^{-15}$ T·m²
Quantized Hall resistance	$R_K = h/e^2$	$2.581\,280\,7 \times 10^{4}$ Ω
Rydberg constant	R_H	$1.097\,373\,153\,4 \times 10^{7}$ m⁻¹
Josephson frequency–voltage quotient	$2e/h$	$4.835\,979 \times 10^{14}$ Hz/V
Compton wavelength	$\lambda_C = h/m_e c$	$2.426\,310\,58 \times 10^{-12}$ m

For additional data, see the last four pages in the book and the following tables in the text.

APPENDIX C

Periodic Table of Elements

1	2	3	4	5	6	7	8	9	10	11	12	13	14	15	16	17	18
1 **H** 1.00797																	2 **He** 4.003
3 **Li** 6.941	4 **Be** 9.012											5 **B** 10.81	6 **C** 12.011	7 **N** 14.007	8 **O** 15.9994	9 **F** 19.00	10 **Ne** 20.179
11 **Na** 22.990	12 **Mg** 24.31											13 **Al** 26.98	14 **Si** 28.09	15 **P** 30.974	16 **S** 32.064	17 **Cl** 35.453	18 **Ar** 39.948
19 **K** 39.102	20 **Ca** 40.08	21 **Sc** 44.96	22 **Ti** 47.88	23 **V** 50.94	24 **Cr** 52.00	25 **Mn** 54.94	26 **Fe** 55.85	27 **Co** 58.93	28 **Ni** 58.69	29 **Cu** 63.55	30 **Zn** 65.38	31 **Ga** 69.72	32 **Ge** 72.59	33 **As** 74.92	34 **Se** 78.96	35 **Br** 79.90	36 **Kr** 83.80
37 **Rb** 85.47	38 **Sr** 87.62	39 **Y** 88.906	40 **Zr** 91.22	41 **Nb** 92.91	42 **Mo** 95.94	43 **Tc** (98)	44 **Ru** 101.1	45 **Rh** 102.905	46 **Pd** 106.4	47 **Ag** 107.870	48 **Cd** 112.41	49 **In** 114.82	50 **Sn** 118.69	51 **Sb** 121.75	52 **Te** 127.60	53 **I** 126.90	54 **Xe** 131.29
55 **Cs** 132.905	56 **Ba** 137.33	57–71 **Rare Earths**	72 **Hf** 178.49	73 **Ta** 180.95	74 **W** 183.85	75 **Re** 186.2	76 **Os** 190.2	77 **Ir** 192.2	78 **Pt** 195.09	79 **Au** 196.97	80 **Hg** 200.59	81 **Tl** 204.37	82 **Pb** 207.19	83 **Bi** 208.98	84 **Po** (210)	85 **At** (210)	86 **Rn** (222)
87 **Fr** (223)	88 **Ra** (226)	89–103 **Actinides**	104 **Rf** (261)	105 **Ha** (260)	106 (263)	107 (262)	108 (265)	109 (266)									

Rare Earths (Lanthanides)	57 **La** 138.91	58 **Ce** 140.12	59 **Pr** 140.91	60 **Nd** 144.24	61 **Pm** (147)	62 **Sm** 150.36	63 **Eu** 152.0	64 **Gd** 157.25	65 **Tb** 158.92	66 **Dy** 162.50	67 **Ho** 164.93	68 **Er** 167.26	69 **Tm** 168.93	70 **Yb** 173.04	71 **Lu** 174.97
Actinides	89 **Ac** 227.03	90 **Th** 232.04	91 **Pa** 231.04	92 **U** 238.03	93 **Np** 237.05	94 **Pu** (244)	95 **Am** (243)	96 **Cm** (247)	97 **Bk** (247)	98 **Cf** (251)	99 **Es** (252)	100 **Fm** (257)	101 **Md** (258)	102 **No** (259)	103 **Lr** (260)

The 1–18 group designation has been recommended by the International Union of Pure and Applied Chemistry (IUPAC).

Atomic Numbers and Atomic Masses

Name	Symbol	Atomic Number	Mass	Name	Symbol	Atomic Number	Mass
Actinium	Ac	89	227.03	Mercury	Hg	80	200.59
Aluminum	Al	13	26.98	Molybdenum	Mo	42	95.94
Americium	Am	95	(243)	Neodymium	Nd	60	144.24
Antimony	Sb	51	121.75	Neon	Ne	10	20.179
Argon	Ar	18	39.948	Neptunium	Np	93	237.05
Arsenic	As	33	74.92	Nickel	Ni	28	58.69
Astatine	At	85	(210)	Niobium	Nb	41	92.91
Barium	Ba	56	137.3	Nitrogen	N	7	14.007
Berkelium	Bk	97	(247)	Nobelium	No	102	(259)
Beryllium	Be	4	9.012	Osmium	Os	76	190.2
Bismuth	Bi	83	208.98	Oxygen	O	8	15.9994
Boron	B	5	10.81	Palladium	Pd	46	106.4
Bromine	Br	35	79.90	Phosphorus	P	15	30.974
Cadmium	Cd	48	112.41	Platinum	Pt	78	195.09
Calcium	Ca	20	40.08	Plutonium	Pu	94	(244)
Californium	Cf	98	(251)	Polonium	Po	84	(210)
Carbon	C	6	12.011	Potassium	K	19	39.102
Cerium	Ce	58	140.12	Praseodymium	Pr	59	140.91
Cesium	Cs	55	132.905	Promethium	Pm	61	(147)
Chlorine	Cl	17	35.453	Protactinium	Pa	91	231.04
Chromium	Cr	24	52.00	Radium	Ra	88	(226)
Cobalt	Co	27	58.93	Radon	Rn	86	(222)
Copper	Cu	29	63.55	Rhenium	Re	75	186.2
Curium	Cm	96	(247)	Rhodium	Rh	45	102.905
Dysprosium	Dy	66	162.50	Rubidium	Rb	37	85.47
Einsteinium	Es	99	(252)	Ruthenium	Ru	44	101.1
Erbium	Er	68	167.26	Rutherfordium	Rf	104	(261)
Europium	Eu	63	152.0	Samarium	Sm	62	150.36
Fermium	Fm	100	(257)	Scandium	Sc	21	44.96
Fluorine	F	9	19.00	Selenium	Se	34	78.96
Francium	Fr	87	(223)	Silicon	Si	14	28.09
Gadolinium	Gd	64	157.25	Silver	Ag	47	107.870
Gallium	Ga	31	69.72	Sodium	Na	11	22.990
Germanium	Ge	32	72.59	Strontium	Sr	38	87.62
Gold	Au	79	196.97	Sulfur	S	16	32.064
Hafnium	Hf	72	178.49	Tantalum	Ta	73	180.95
Hahnium	Ha	105	(260)	Technetium	Tc	43	(98)
Helium	He	2	4.003	Tellurium	Te	52	127.60
Holmium	Ho	67	164.93	Terbium	Tb	65	158.92
Hydrogen	H	1	1.00797	Thallium	Tl	81	204.37
Indium	In	49	114.82	Thorium	Th	90	232.04
Iodine	I	53	126.90	Thulium	Tm	69	168.93
Iridium	Ir	77	192.2	Tin	Sn	50	118.69
Iron	Fe	26	55.85	Titanium	Ti	22	47.88
Krypton	Kr	36	83.80	Tungsten	W	74	183.85
Lanthanum	La	57	138.91	Uranium	U	92	238.03
Lawrencium	Lr	103	(260)	Vanadium	V	23	50.94
Lead	Pb	82	207.19	Xenon	Xe	54	131.29
Lithium	Li	3	6.941	Ytterbium	Yb	70	173.04
Lutetium	Lu	71	174.97	Yttrium	Y	39	88.906
Magnesium	Mg	12	24.31	Zinc	Zn	30	65.38
Manganese	Mn	25	54.94	Zirconium	Zr	40	91.22
Mendelevium	Md	101	(258)				

ILLUSTRATION CREDITS

Part Openers

Part IV p. 656 © 1988 Richard Megna/Fundamental Photographs
Part V p. 1026 Courtesy AT&T Archives

Chapter 22

Opener p. 657 Lawrence Berkeley Laboratory/Science Photo Library/Photo Researchers; **p. 658** *PSSC* Physics, 2nd ed., 1965. D.C. Heath & Co. and Education Development Center, Inc., Newton, MA; **p. 659** Bruce Terris/IBM Almaden Research Center; **p. 662 (left)** © Grant Heilman; **(right)** Ann Ronan Picture Library; **p. 663** Burndy Library, Norwalk, Connecticut; **p. 672 (Figure 22-17)** Harold M. Waage; **(Figure 22-18)** Harold M. Waage; **p. 673 (Figure 22-19)** Harold M. Waage; **p. 675** Courtesy of Hulon Forrester/Video Display Corporation, Tucker, Georgia.

Chapter 23

Opener p. 688 © Dagmar Hailer-Hamann/Peter Arnold, Inc.; **p. 691** Ben Damsky Electric Power Research Institute; **p. 702 (Figure 23-21)** Runk/Schoenberger from Grant Heilman; **p. 708** Harold M. Waage.

Chapter 24

Opener p. 720 © 1990 Richard Megna/Fundamental Photographs; **p. 739 (Figure 24-18) (b)** © Karen R. Preuss; **(c)** © Hank Morgan/Rainbow; **p. 741 (Figure 24-20)** Harold M. Waage.

Chapter 25

Opener p. 752 © Picturepoint, Ltd., London; **p. 756 (Figure 25-2)** Harold M. Waage; **p. 758 (left)** © Mark Antman/The Image Works; **(right)** © Bruce Iverson; **p. 759 (left)** © Bruce Iverson; **(right)** © Paul Brierly; (bottom) Courtesy Tusonix, Tucson, Arizona; **p. 764** © Lawrence Livermore National Laboratory; **p. 767** © Manfred Kage/Peter Arnold, Inc.

Chapter 26

Opener p. 786 © Rabbit System, Santa Monica, California. Photo caption by kind cooperation of Larry Albright; **p. 793** © Chris Rogers/The Stock Market; **p. 795 (Figure 26-9)** © Paul Silverman/Fundamental Photographs; **p. 798** Ann Ronan Picture Library.

Chapter 28

Opener p. 855 © Stadler GmbH, Courtesy Transrapid International, Munich; **(Figure 28-1)** Fred Weiss; **p. 859 (Figure 28-8)** © 1990 Richard Megna/Fundamental Photographics; **p. 861 (top)** Larry Langrill; **(bottom)** © Lawrence Berkeley Laboratory/Science Photo Library; **p. 862 (Figure 28-13)** Carl E. Nielsen.

Chapter 29

Opener p. 883 Bob Williamson, Oakland University, Rochester, Michigan; **p. 885 (left)** © 1990 Richard Megna/Fundamental Photographs; **(right)** © 1990 Richard Megna/Fundamental Photographs; **p. 887 (Figure 29-6)** © 1990 Richard Megna/ Fundamental Photographs; **p. 890 (Figure 29-10)** © 1990 Richard Megna/Fundamental Photographs; **p. 891** © Bruce Iverson; **p. 893 (Figure 29-14)** © 1990 Richard Megna/Fundamental Photographs; **p. 894** Courtesy F.W. Bell Co.; **p. 896 (Figure 29-19)** Clarence Bennett, Oakland University, Rochester, Michigan; **p. 900 (left)** Courtesy Princeton University Plasma Physics Laboratory; **(right)** Courtesy Princeton University Plasma Physics Laboratory; **p. 909 (top)** J. F. Allen, St. Andrews University, Scotland; **(bottom)** © Paul Silverman/Fundamental Photographs; **p. 910 (Figure 29-34)** Robert J. Celotta, National Institute of Standards and Technology; **p. 911 (left)** Akira Tonomura, Hitachi Advanced Research Laboratory, Hatomaya, Japan; **(right)** © Bruce Iverson; **p. 912 (left)** Courtesy Seagate Technology; **(right)** Tom Chang/IBM Storage Systems Division, San Jose, CA; **p. 914** © Bill Pierce/Time Magazines, Inc.

Chapter 30

Opener p. 927 © 1990 Richard Megna/Fundamental Photographs; **p. 941 (Figure 30-19)** © Michael Holford, Collection of the Science Museum, London; **p. 948 (Figure 30-26)** A. Leitner/Renselaer Polytechnic Institute.

Chapter 31

Opener p. 959 © 1990 C. J. Allen/New England Hydro Transmission Corporation; **p. 961 (left)** Courtesy U.S. Dept. of the Interior, Dept. of Reclamation; **(right)** © Lee Langum/Photo Researchers; **p. 979** © George H. Clark Radioana Collection-Archive Center, National Museum of American History; **p. 982 (top)** © Yoav/Phototake; **(right)** © Daniel S. Brody/Stock Boston.

Chapter 32

Opener p. 999 N.A.S.A.; **p. 1004** Wabash Instrument Corporation; **p. 1018** Courtesy Cavendish Laboratory, University of Cambridge; **p. 1025 (Figure 32-13)** Courtesy Central Scientific Company.

Chapter 33

Opener p. 1027 © Robert Greenler; **p. 1028 (Figure 33-1)** Corbis-Bettmann; **p. 1029 (Figure 33-2)** Corbis-Bettmann; **(bottom)** Adapted from Eastman Kodak and Wabash Instrument Corporation; **p. 1032 (left)** © 1991 Paul Silverman/Fundamental Photographs; **(right)** © 1991 Paul Silverman/Fundamental Photographs; **p. 1037 (a)** © Chuck O'Rear/West Light; **(b)** Courtesy of Ahmed H. Zewail, California Institute of Technology; **(c)** © Chuck O'Rear/West Light; **(d)** © Michael W. Berns/Scientific American; **(e)** © David Scharf, 1988. All rights reserved; **p. 1042 (Figure 33-18) (a)** © 1987 Ken Kay/ Fundamental Photographs; **(b)** Courtesy Battelle-Northwest

Laboratories; **(bottom)** © 1990 Richard Megna/ Fundamental Photographs; **p. 1043 (Figure 33-19)** Macduff Everton/Corbis; **(Figure 33-20)** © 1987 Pete Saloutos/The Stock Market; **p. 1044 (Figure 33-22)** © 1987 Ken Kay/Fundamental Photographs; **p. 1046 (Figure 33-24)** © Ted Horowitz/The Stock Market; **(bottom left)** © Dan Boyd/Courtesy Naval Research Laboratory; **(bottom right)** Courtesy AT&T Archives; **p. 1047 (Figure 33-25)** © Robert Greenler; **p. 1048 (Figure 33-27)** David Parker/Science Photo Library/Photo Researchers; **p. 1049 (top)** © Robert Greenler; **(bottom)** Giovanni DeAmici, NSF, Lawrence Berkeley Laboratory; **p. 1051** © Robert Greenler; **p. 1053 (Figure 33-34) (a)** Larry Langrill; **(b)** Larry Langrill; **p. 1054 (left)** © 1970 Fundamental Photographs; **(right)** © 1990 PAR/NYC, Inc./Photo by Elizabeth Algieri; **p. 1056 (Figure 33-40)** © 1987 Paul Silverman Photographs; **p. 1057 (top left)** Glen A. Izett, U.S. Geological Survey, Denver Colorado; **(top right)** Glen A. Izett, U.S. Geological Survey, Denver Colorado; **(center left)** Dr. Anthony J. Gow/Cold Regions Research and Engineering Laboratory, Hanover, New Hampshire; **(center right)** Dr. Anthony J. Gow/Cold Regions Research and Engineering Laboratory, Hanover, New Hampshire; **(bottom)** © Sepp Seitz/Woodfin Camp & Associates.

Chapter 34

Opener p. 1070 © Dagmar Heiler-Hamann/Peter Arnold, Inc.; **p. 1071 (Figure 34-2)** Demetrios Zangos; **p. 1077 (left)** © 1990 Richard Megna/Fundamental Photographs; **(right)** © 1990 Richard Megna/Fundamental Photographs; **p. 1080** © 1990 Richard Megna/Fundamental Photographs; **p. 1082 (Figure 34-26) (a)** Nils Abramson; **(b)** © 1974 Fundamental Photographs; **p. 1083 (Figure 34-27)** © Fundamental Photographs; **p. 1085** © Bohdan Hrynewych/Stock Boston; **p. 1094 (top left)** © Lennart Nilsson; **(top right)** © Lennart Nilsson; **(bottom left)** © Lennart Nilsson; **(bottom right)** Courtesy IMEC and University of Pennsylvania Department of Electrical Engineering; **p. 1098 (top left)** Scala/Art Resource; **(top center)** © Royal Astronomical Society Library; **(top right)** © Lick Observatory, University of California Regents; **(bottom left)** California Institute of Technology; **(bottom right)** © 1980 Gary Ladd; **p. 1099 (top left)** © California Association for Research in Astronomy; **(top right)** © California Association for Research in Astronomy; **(center)** © California Association for Research in Astronomy; **(bottom)** NASA/Corbis.

Chapter 35

Opener p. 1110 © Ken Kay/Fundamental Photographs; **p. 1112** © 1990 Richard Megna/Fundamental Photographs; **p. 1113 (Figure 35-3)** Courtesy Bausch & Lomb; **p. 1114 (Figure 35-5) (a)** Courtesy T. A. Wiggins; **(b)** Courtesy T. A. Wiggins; **p. 1115 (Figure 35-6)** *PSSC Physics*, 2nd ed., 1965. D.C. Heath & Co. and Education Development Center, Newton, MA; **p. 1116 (Figure 35-8)** Courtesy Michel Cagnet; **p. 1118 (Figure 35-10)** Courtesy Michel Cagnet; **p. 1120 (Figure 35-13)** Courtesy Michel Cagnet; **p. 1123 (Figure 35-20) (a)** Courtesy Michel Cagnet; **(b)** Courtesy Michel Cagnet; **(c)** Courtesy Michel Cagnet; **p. 1127 (Figure 35-28) (a)** M. Cagnet, M. Françon, J. C. Thrierr, *Atlas of Optical Phenomena*; **(b)** M. Cagnet, M. Françon, J. C. Thrierr, *Atlas of Optical Phenomena*; **(Figure 35-29)** Courtesy Battelle-Northwest Laboratories; **(Figure 35-30)** Courtesy Michel Cagnet; **p. 1128 (Figure 35-31)** Courtesy Michel Cagnet; **(Figure 35-33) (a)** Courtesy Michel Cagnet; **(b)** Courtesy Michel Cagnet; **p. 1130** Kevin R. Morris/Corbis; **p. 1131 (Figure 35-35) (a)** Clarence Bennett/Oakland University, Rochester, Michigan; **(b)** NRAO/AUI/Science Photo Library/Photo Researchers; **p. 1133 (left)** © Ronald R. Erickson, 1981. Hologram by Nicklaus Phillips, 1978, for Digital Equipment Corporation; **(right)** © Ronald R. Erickson, 1981. Hologram by Nicklaus Phillips, 1978, for Digital Equipment Corporation.

ANSWERS

Problem answers are calculated used $g = 9.81\text{ m/s}^2$ unless otherwise specified in the Problem. Differences in the last figure can easily result from differences in rounding the input data and are not important.

To help you master the techniques in Examples and to solve the intermediate-level problems at the end of each chapter, the problem maps preceding the answers for the chapters indicate which Examples and odd-numbered intermediate-level Problems deal with similar material.

Chapter 22

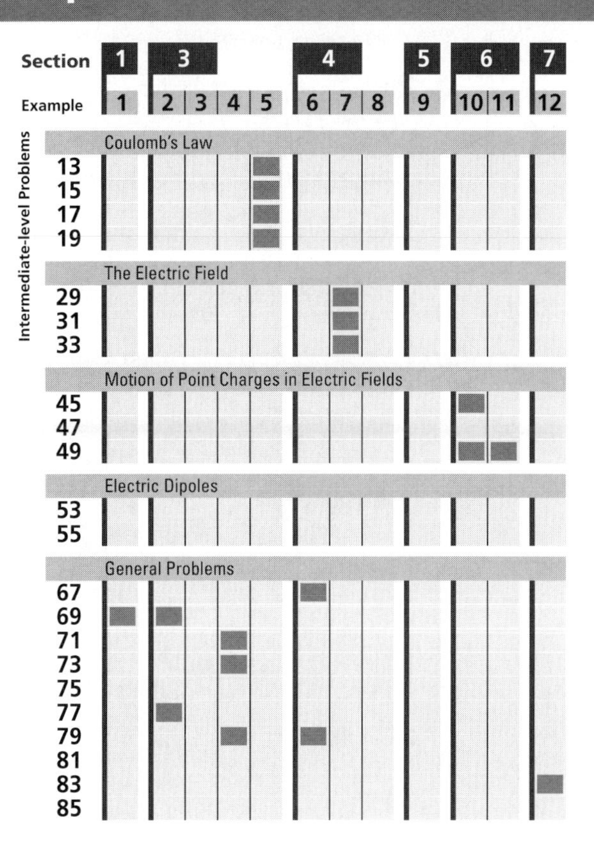

Section	1	3	3	3	3	4	4	4	5	6	6	7
Example	1	2	3	4	5	6	7	8	9	10	11	12
Coulomb's Law												
13					X							
15					X							
17					X							
19					X							
The Electric Field												
29							X					
31							X					
33							X					
Motion of Point Charges in Electric Fields												
45										X		
47												
49										X	X	
Electric Dipoles												
53												
55												
General Problems												
67						X						
69	X	X										
71				X								
73				X								
75												
77		X										
79				X		X						
81												
83												X
85												

(Intermediate-level Problems)

1. Yes

3. 5×10^{12} electrons

5. 4.82×10^7 C

7. (*c*)

9. (*a*)

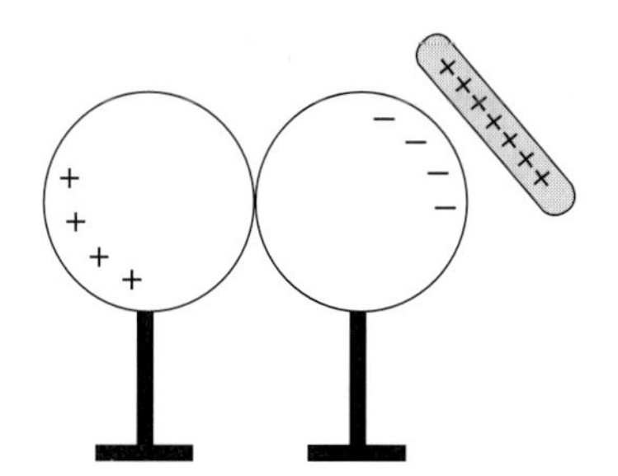

(*b*)

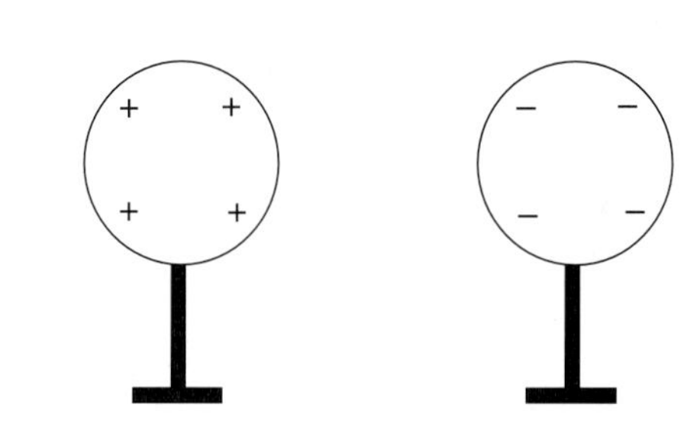

11. (*a*) $0.024\text{ N }\hat{i}$; (*b*) $-0.024\text{ N }\hat{i}$; (*c*) The direction of both forces would be reversed.

13. $1.27 \times 10^{-3}\text{ N }\hat{i} - 3.24 \times 10^{-4}\text{ N }\hat{j}$

15. $-8.66\text{ N }\hat{j}$

17. $0.899\text{ N }\hat{i} + 1.80\text{ N }\hat{j}$, $-1.29\text{ N }\hat{i} - 1.16\text{ N }\hat{j}$, $0.391\text{ N }\hat{i} - 0.643\text{ N }\hat{j}$

19. $(\sqrt{2} + 1)\dfrac{kqQ}{R^2}\,\hat{i}$

21. (*d*)

23. (*d*)

25. (*a*) $-9.35 \times 10^3\text{ N/C }\hat{i}$; (*b*) $7.99 \times 10^3\text{ N/C }\hat{i}$; (*c*) $-7.99 \times 10^3\text{ N/C }\hat{i}$; (*d*) $9.35 \times 10^3\text{ N/C }\hat{i}$; (*e*) 4 m

(*f*)

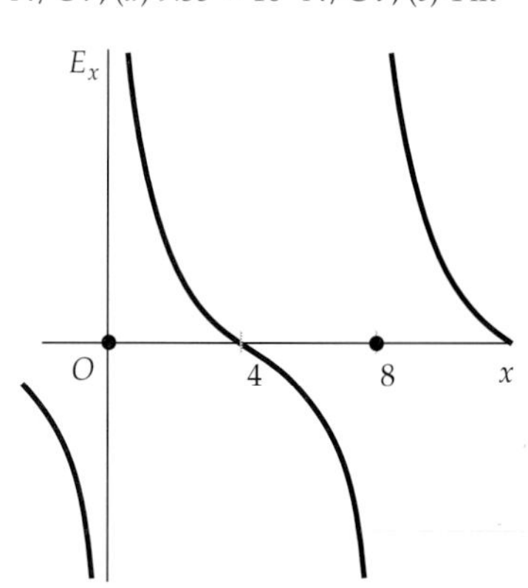

27. 8.18×10^5 N/C, upward

29. (*a*) $3.45 \times 10^4\text{ N/C }\hat{i}$; (*b*) $6.90 \times 10^{-5}\text{ N/C }\hat{i}$

31. (*a*) 1.30×10^4 N/C, 231° counterclockwise from the x axis; (*b*) 2.07×10^{-15} N, 51.3° counterclockwise from the x axis

33. (*a*) 1.91×10^3 N/C, 234° counterclockwise from the x axis; (*b*) 3.06×10^{-16} N, 234° counterclockwise from the x axis

35. (*a*) Unstable along the x axis, stable along the y axis; (*b*) Stable along the x axis, unstable along the y axis; (*c*) $-q/4$; (*d*) The system will be unstable.

37. (*d*)

39.

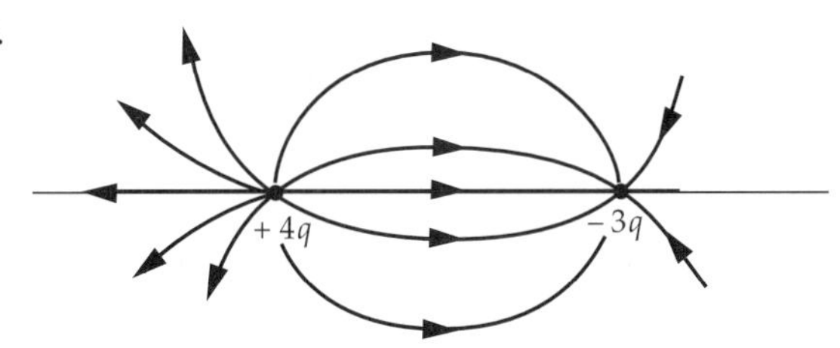

41.

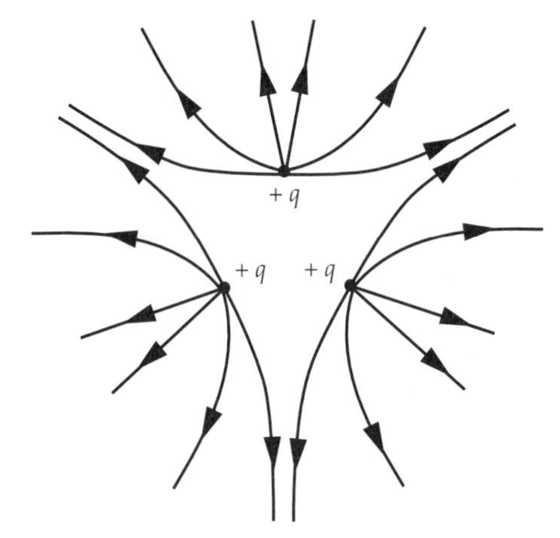

43. (*a*) 9.58×10^7 C/kg, 9.58×10^9 m/s^2; (*b*) 3.13×10^{-4} s

45. 3.75×10^7 m/s

47. 8×10^{-4} C

49. Bottom plate, 4.07 cm from the starting point

51. (*a*) 8×10^{-18} C·m
(*b*)

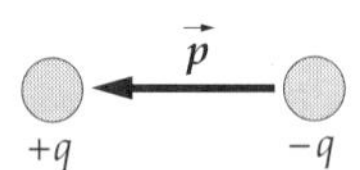

53. Answer given in the problem.

55. (*a*) The force on the positive charge is $qC(x_1 + a)\hat{i}$ and the force on the negative charge is $-qC(x_1 - a)\hat{i}$. Therefore, the net force on the dipole is $2aqC\hat{i} = pC\hat{i}$. (*b*) Answer given in the problem.

57. (*a*) $-3\,kqa^2x^{-4}\hat{i}$; (*b*) $6\,kqa^2y^{-4}\hat{j}$

59. (*a*)

61. The dipole will oscillate about its equilibrium orientation $\theta = 0$. In the nonuniform field, the dipole will accelerate in the x direction as it oscillates about $\theta = 0$.

63. The nonconducting ball is polarized with the side closest to the positive charge becoming negative. This produces a net attractive force. With a negative charge, the ball is polarized with the side closest to the charge becoming positive.

65. Yes. A positively charged ball will induce a dipole on the metal ball, and if the two are in close proximity, the net force can be attractive.

67. 1.14×10^8 N/C, 1.74×10^6 N/C, 0.46 cm

69. 3.3×10^{-7}%, 32.4 N

71. $3kq^2/d^2$, $9kq^2/d^2$

73. 3.0 μC

75. qE/g

77. (*a*) 28 μC, 172 μC; (*b*) 250 N

79. (*a*) -97 μC; (*b*) Between 0.16 m and 0.17 m

81. (*a*) 10.2°; (*b*) 9.9°

83. Answer given in the problem.

85. $\sqrt{ke^2/2mr}$

87. (*a*), (*b*) Answer given in the problem.

89. (*a*) 8.5°; (*b*) 7.8° and 8.6°

91. 1.1×10^4 N/C, upward

93. (*a*) $kQ[(L/2 + x)^{-2} - (L/2 - x)^{-2}]\hat{i}$; (*b*) Answer given in the problem; (*c*) Answer given in the problem; (*d*) $2\pi\sqrt{mL^3/32kqQ}$

Chapter 23

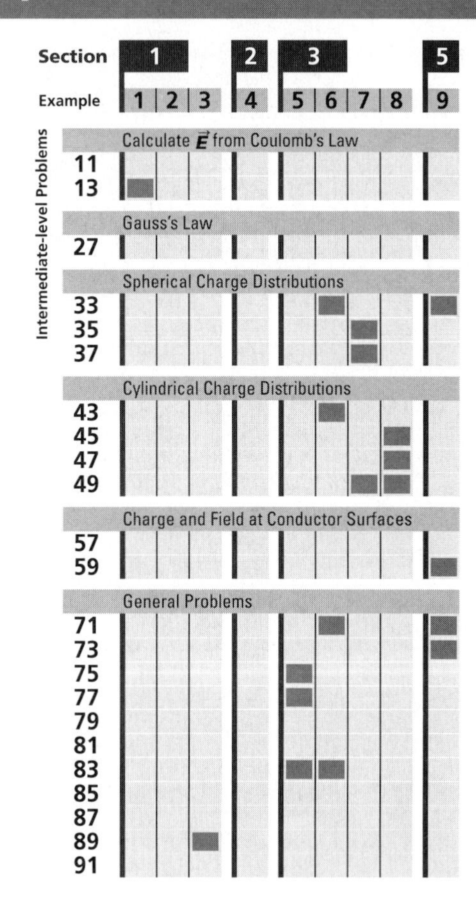

1. (*a*) 17.5 nC; (*b*) 26.2 N/C; (*c*) 4.37 N/C; (*d*) 2.57×10^{-3} N/C; (*e*) 2.52×10^{-3} N/C

3. (*a*) 4.69×10^5 N/C; (*b*) 1.13×10^6 N/C; (*c*) 1.54×10^3 N/C; (*d*) 1.55×10^3 N/C

5. (*a*) 2.00×10^5 N/C; (*b*) 2.54 N/C

7. $a/\sqrt{3}$

9. (*a*) $0.804(2\pi k\sigma)$; (*b*) $0.553(2\pi k\sigma)$; (*c*) $0.427(2\pi k\sigma)$; (*d*) $0.293(2\pi k\sigma)$; (*e*) $0.106(2\pi k\sigma)$

(*f*)

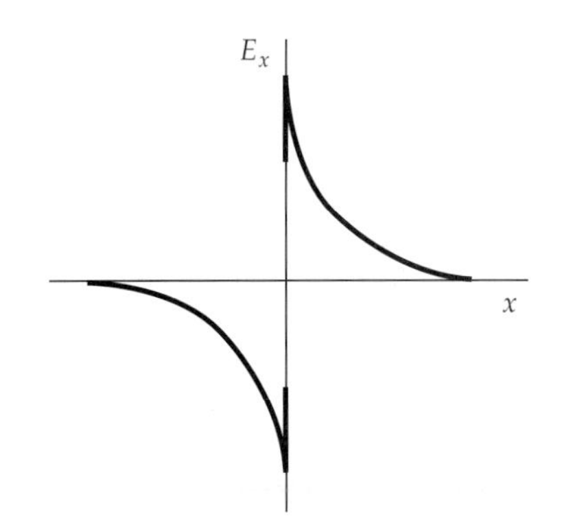

11.

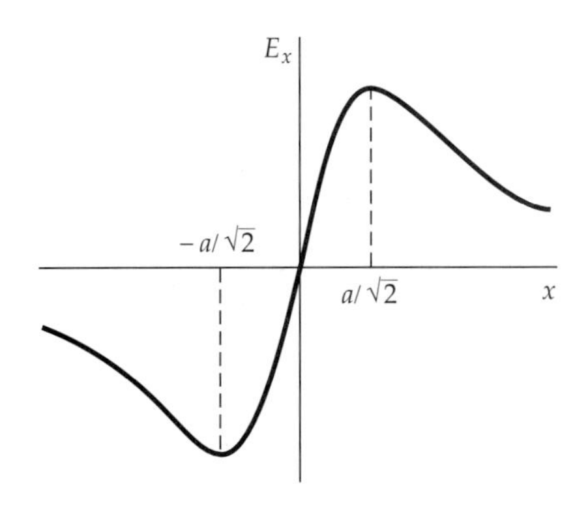

13. (*a*), (*b*) Answer given in the problem.

15. $(\sigma/2\epsilon_0)(1 - 1/\sqrt{2})$

17. (*a*) False; (*b*) False

19. *E* in Gauss's law is the total electric field due to all charges.

21. (*a*) Three lines enter the sphere.

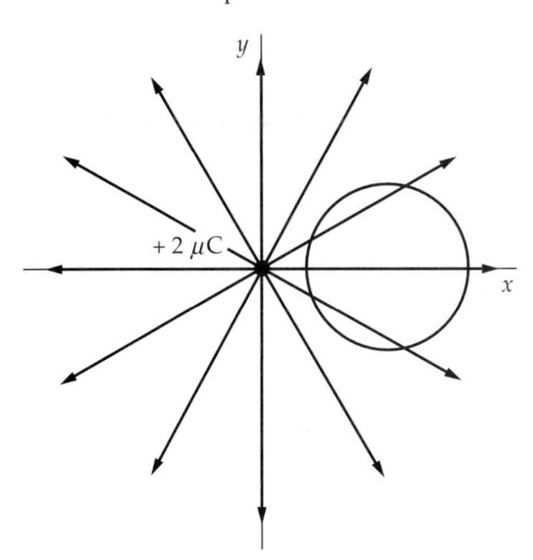

(*b*) 0; (*c*) 0

23. (*a*) *N*; (*b*) $N/6$; (*c*) q/ϵ_0; (*d*) $q/6\epsilon_0$; (*e*) (*b*) and (*d*)

25. (*a*) 3.14 m^2; (*b*) 7.19×10^4 N/C; (*c*) 2.26×10^5 N·m^2/C; (*d*) No; (*e*) 2.26×10^5 N·m^2/C

27. 1.13×10^5 N·m^2/C

29. The field at a distance *r* is proportional to Q'/r^2, where Q' is the charge within a sphere of radius *r*. The charge Q' is proportional to the volume of the sphere and therefore is proportional to r^3. Thus the field is proportional to $r^3/r^2 = r$.

31. (*a*) 0.407 nC; (*b*) 0; (*c*) 0; (*d*) 984 N/C; (*e*) 366 N/C

33. (*a*) $-5Q$, $-2Q$; (*b*) $-5Q$, $-5Q$, 0, goes to 0; (*c*) $-2Q$, $-2Q$, 0

35. (*a*) πAR^4; (*b*) $Ar^2/4\epsilon_0$, $AR^4/4\epsilon_0 r^2$

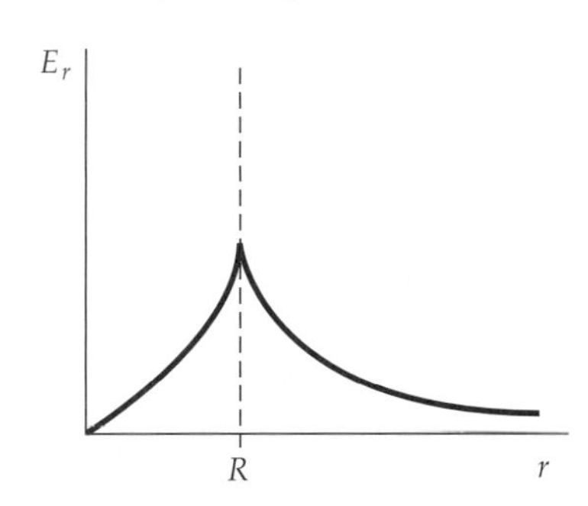

37. (*a*) $4\pi CR$; (*b*) $C/\epsilon_0 r$, $CR/\epsilon_0 r^2$

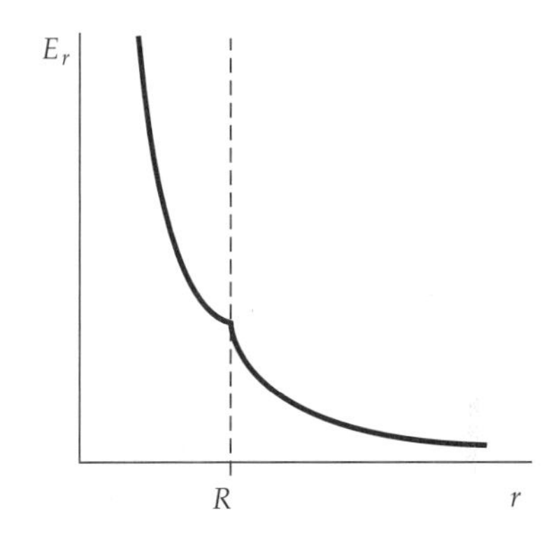

39. (*a*) $(4\pi\rho/3)(b^3 - a^3)$; (*b*) 0, $(\rho/3\epsilon_0 r^2)(r^3 - a^3)$, $(\rho/3\epsilon_0 r)(r^3 - a^3)$

41. Answer given in the problem.

43. (*a*) 6.79×10^{-7} C; (*b*) 0; (*c*) 0; (*d*) 1.00×10^3 N/C; (*e*) 6.11×10^2 N/C

45. (*a*) 6.79×10^{-7} C; (*b*) 3.39×10^2 N/C; (*c*) 1.00×10^3 N/C; (*d*) 1.00×10^3 N/C; (*e*) 6.11×10^2 N/C; The fields for the shell charge and solid cylinder of charge with the same total charge are identical outside the charges. Inside the shell, the field is zero.

47. (*a*) zero for $0 \le r \le 1.5$ cm and 4.5 cm $\le r < 6.5$ cm, $(108/r)$ N/C for 1.5 cm $< r <$ 4.5 cm and 6.5 cm $\le r$; (*b*) 2.10×10^{-8} C/m^2, 1.47×10^{-8} C/m^2

49. Answer given in the problem.

51. (*a*) 1.885×10^{-8} C/m; (*b*) 2.26×10^4 N/C for $0 \le r \le 1.5$ cm, $(339.3/r)$ N/C for 1.5 cm $< r <$ 4.5 cm, zero for 4.5 cm $< r <$ 6.5 cm, $(339.3/r)$ N/C for 6.5 cm $< r$

53. 9.41×10^3 N/C

55. (*a*) kq/r^2, 0, kq/r^2

(*b*)

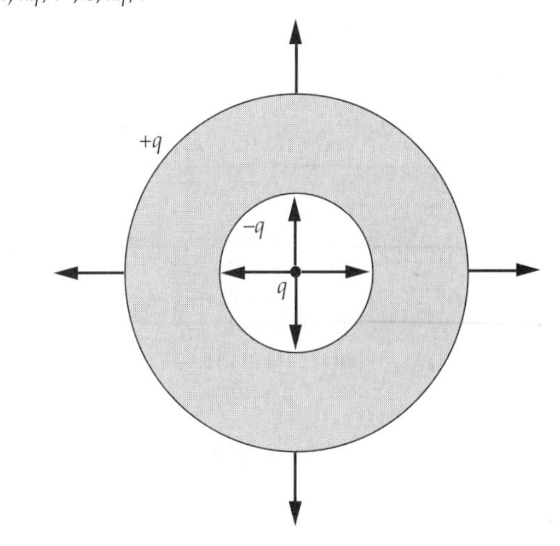

(c) $-q/4\pi a^2, q/4\pi b^2$

57. (a) $-0.553\ \mu\text{C/m}^2$, $0.246\ \mu\text{C/m}^2$, $2.5\ \mu\text{C}$ for outer surface and $-2.5\ \mu\text{C}$ for inner surface; (b) kq/r^2 for $0 \le r \le 60$ cm, zero for 60 cm $< r <$ 90 cm, kq/r^2 for $r \ge 90$ cm; (c) $-0.553\ \mu\text{C/m}^2$, $0.589\ \mu\text{C/m}^2$, $6\ \mu\text{C}$ for outer surface and $-2.5\ \mu\text{C}$ for inner surface; kq/r^2, 0, $(6\ \mu\text{C})k/r^2$

59. (a) 1.60×10^{-6} C/m^2, 1.81×10^5 N/C; (b) 6.78×10^4 N/C, 2.94×10^5 N/C, $0.6\ \mu\text{C/m}^2$, $2.6\ \mu\text{C/m}^2$

61. (a) False; (b) True; (c) True; (d) False

63. (a)

65. (b)

67. (c)

69. Gauss's law cannot be used to find the field of a finite line charge because there is not enough symmetry to find a Gaussian surface from which to calculate the flux.

71. (a) toward; (b) $+Q_0$; (c) $-Q_0$; (d) $+Q_0$; (e) 0

(f)

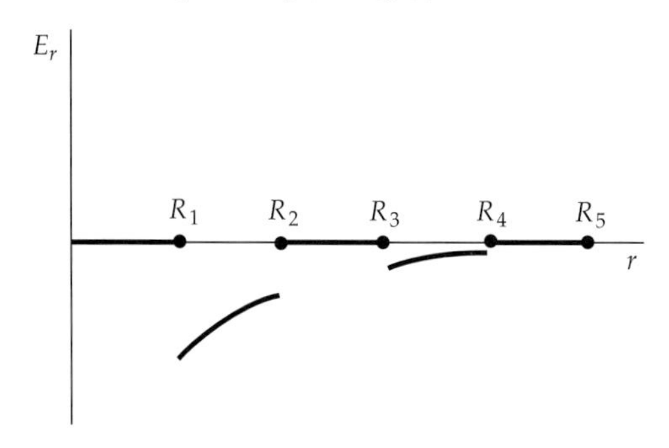

73. 1.15×10^5 N/C

75. (a) -1.41×10^5 N/C $\hat{i}$; (b) -5.37×10^5 N/C $\hat{i}$; (c) 1.41×10^5 N/C $\hat{i}$

77. (a) 1.27×10^3 N/C $\hat{i}$ + 1.47×10^3 N/C $\hat{j}$; (b) -1.27×10^3 N/C $\hat{i}$ + 5.87×10^3 N/C $\hat{j}$

79. (a) $3\sqrt{3}\, gR^2/k|Q|$; (b) $1.896R$

81. $(2kQ\lambda/L)(\ln(d+L)/d$

83. (a) $(1.13\hat{i} + 1.69\hat{j}) \times 10^5$ N/C; (b) $(2.43\hat{i} + 2.17\hat{j}) \times 10^5$ N/C

85. $2\pi(mR^2/2k|q\lambda|)^{1/2}$

87. 7.42 rad/s

89. (a) 1.19×10^5 N/C; (b) 1.03×10^5 N/C; (c) 1.12×10^3 N/C

91. 3.27×10^{-8} C/m

93. (a) $k\lambda^2 \ln[(d+L)^2/d(2L+d)]$
(b) Answer given in the problem.

95. (a) $12Q/L^3$; (b) $dF = [2ka\lambda y^2/(y+b)]\, dy$; (c) $F = 2ka\lambda(L/2+d)\{(L/2+d)\ln[(d+L)/d] - L\}$

Chapter 24

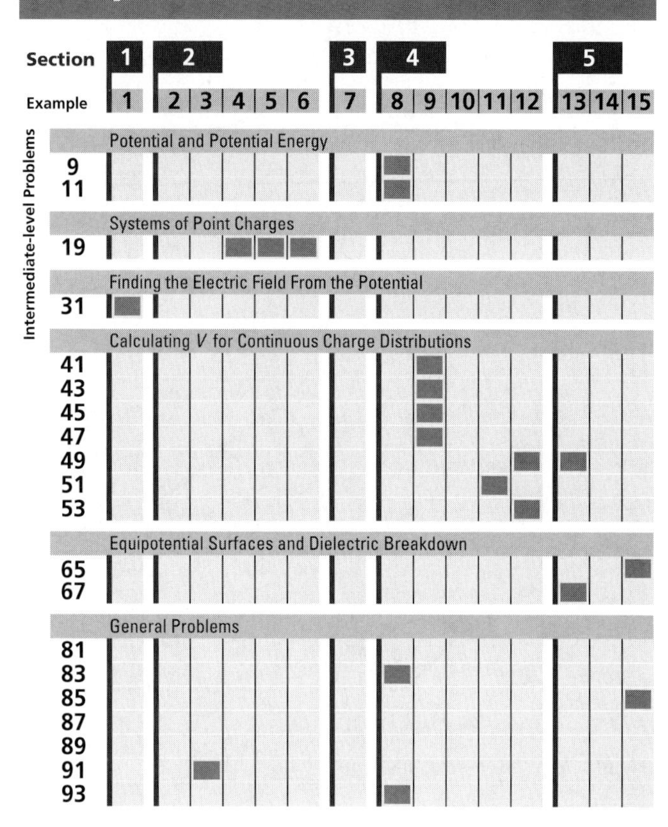

1. (a) -8000 V; (b) -2.4×10^{-2} J; (c) 2.4×10^{-2} J; (d) $(-2\text{ kV/m})x$; (e) 4000 V $-$ (2 kV/m)x; (f) 2000 V $-$ (2 kV/m)x

3. (a) 5 kV/m, positive; (b) 500 eV, 8×10^{-17} J; (c) -500 eV, 500 eV

5. smaller

7. (a) 4.5×10^3 V; (b) 1.35×10^{-2} J; (c) 1.35×10^{-2} J

9. Answer given in the problem.

11. (a) 30,000 eV; (b) 4.8×10^{-15} J; (c) 1.03×10^8 m/s

13. (a) 25.4 kV; (b) 12.7 kV; (c) 0

15. (a) 12,000 V; (b) 0.0599 J; (c) 0, 0

17. $-1/2$

19. (a)

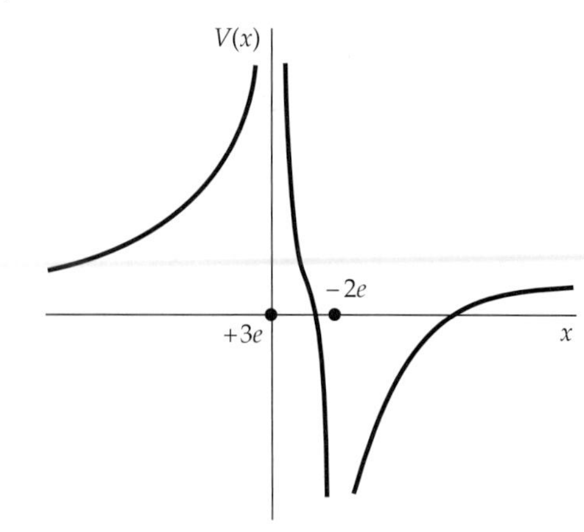

(b) $3a/5$; (c) $2ke^2/a$

21. No

23. (*a*) positive; (b) 25,000 V/m

25. (*a*) 8.99×10^3 V, 8.96×10^3 V; (*b*) decreases, 2.97×10^3 V/m; (*c*) 3.00×10^3 V/m; (*d*) 8.99×10^3 V. The displacement Δy is on an equipotential surface.

27. (*a*) positive; (*b*) 25,000 V/m

29. $-2C_2x$, $-\hat{i}$ for $x > 0$ and $\hat{i}$ for $x < 0$

31. −7500 V

33. $(-8\hat{i} - 2\hat{j} - \hat{k})$ V/m

35. Yes, both V and E_x would be different if Q were not uniformly distributed.

37. (*a*) 6.24×10^3 V/m, 0; (*b*) 749 V, 749 V; (*c*) 749 V, 0

39. (*a*) 6.02×10^3 V; (*b*) -1.27×10^4 V; (*c*) -4.23×10^4 V

41. (*a*) $\dfrac{kQ}{L}\ln\left(\dfrac{\sqrt{x^2 + L^2/4} + L/2}{\sqrt{x^2 + L^2/4} - L/2}\right)$
(*b*) Answer given in the problem.

43. (*a*) $\pi\sigma_0 R^2/2$; (*b*) $\sigma_0 k$

45. (*a*) $2\pi\sigma_0 k[2\sqrt{x^2 + R^2/2} - x - \sqrt{x^2 + R^2}]$; (*b*) $\pi\sigma_0 kR^4/8x^3$

47. (*a*) $(kQ/L)\ln[(x + L/2)/(x - L/2)]$; (*b*) Answer given in the problem.

49. $-2kq/L \ln(b/a)$

51. (*a*) zero for $0 < x < a$, $-(\sigma/\epsilon_0)(x - a)$ for $x > a$, $\sigma x/\epsilon_0$ for $x < 0$; (*b*) $-\sigma x/\epsilon_0$ for $0 < x < a$, $-\sigma a/\epsilon_0$ for $x > a$, zero for $x < 0$

53. (*a*) Qr^3/R^3, kQr^2/R^3; (*b*) $3kQr'\,dr'/R^3$; (*c*) $(3kQ/2R^3)(R^2 - r^2)$; (*d*) $(kQ)/2R)(3 - r^2/R^2)$

55. (*c*)

57.

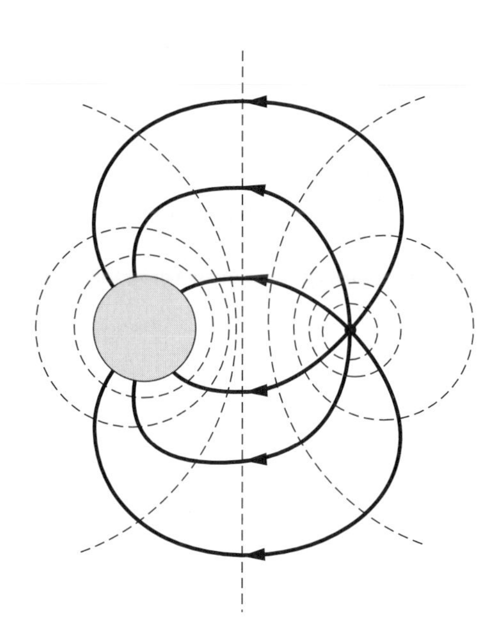

59.

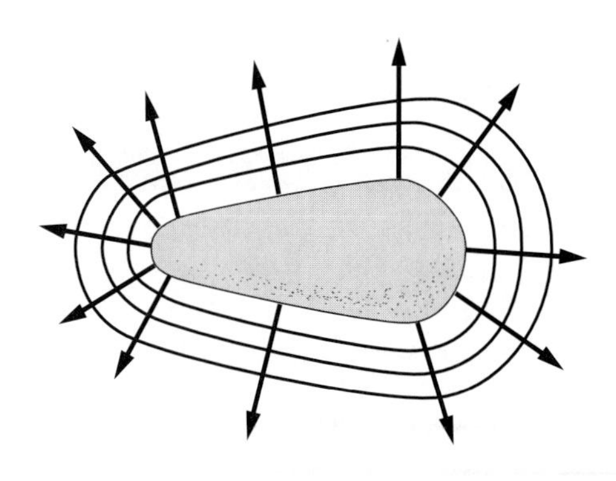

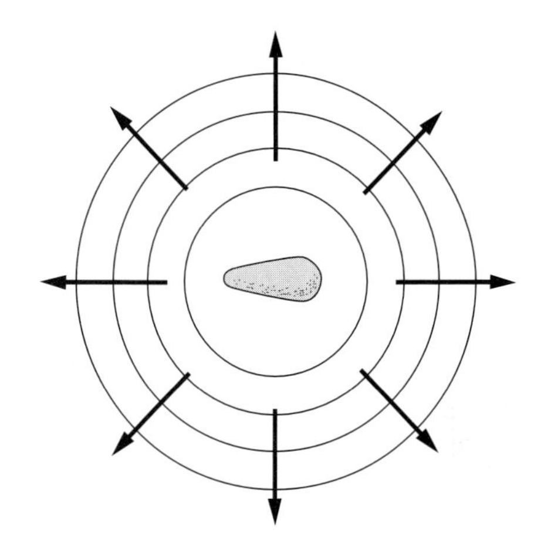

61. 5.06×10^{-4} m

63. (*a*) ± 8.54 N/C; (*b*) $\pm 4.80 \times 10^5$ V

65. The smaller sphere has the larger field by a factor of 2.

67. $kq(1/a - 1/b)$

69. (*b*)

71. (*c*)

73. (*a*) No; (*b*) Yes, yes

75. (*a*) kg/Cs^2; (*b*) $q_0ax^2/2$; (*c*) $-ax^2/2$

77. 3.33×10^{-3} m

79. 20 μC

81. 250 W

83. 1.45×10^{-7} J, 9.03×10^{11} eV

85. $2R_1/3$

87. 50 cm

89. 6000 V

91. 4.46×10^{-14} m

93. 55.1 J

95. (*a*) $kq_1/x^2 + kq_2/(x - a)^2$ for $x > a$, $kq_1/x^2 - kq_2/(a - x)^2$ for $0 < x < a$, $-kq_1/x^2 - kq_2/(x - a)^2$ for $x < 0$;
(*b*) $kq_1/|y| + kq_2/(y^2 + a^2)^{1/2}$; (*c*) $\pm kq_1/y^2 + kq_2y/(y^2 + a^2)^{3/2}$ (positive for $y > 0$ and negative for $y < 0$)

97. (*a*) $V(a) = kQ(1/b - 1/c)$, $V(b) = kQ(1/b - 1/c)$, $V(c) = 0$;
(*b*) $V(a) = 0$, $V(b) = kQ(b - a)(c - b)/b^2(c - a)$, $V(c) = 0$,
$Q_a = -Q\dfrac{a}{b}\left(\dfrac{c - b}{c - a}\right)$, $Q_b = Q$, $Q_c = -Q\dfrac{c}{b}\left(\dfrac{b - a}{c - a}\right)$

Chapter 25

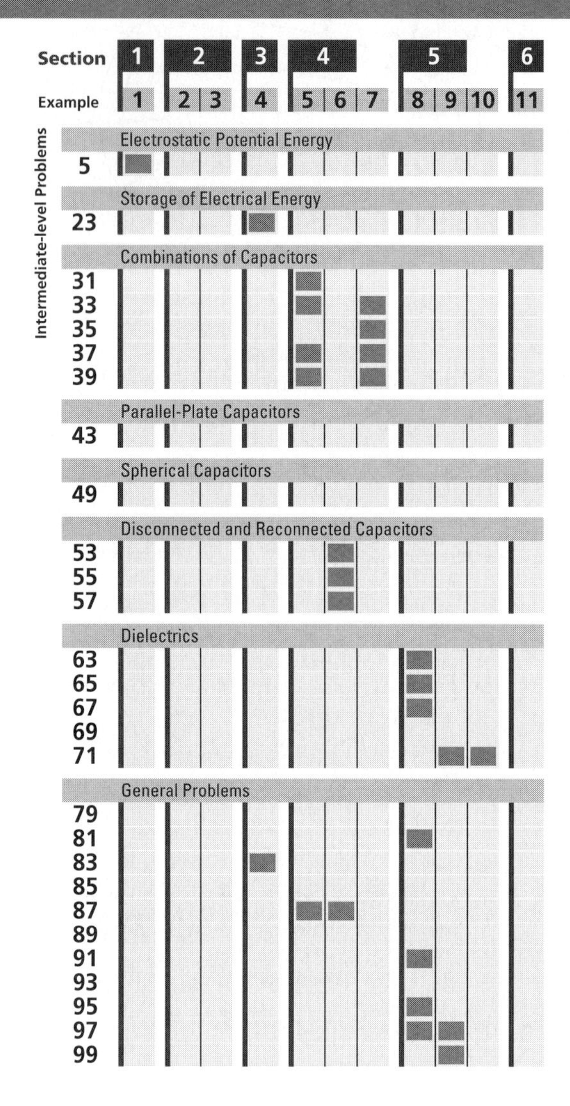

1. (*a*) 3.00×10^{-2} J; (*b*) -5.99×10^{-3} J; (*c*) -1.80×10^{-2} J

3. 2.23×10^{-5} J

5. $(6\sqrt{2}\,k/ma)^{1/2}\,q$

7. (*c*)

9. (*a*) 2.22×10^{-8} C; (*b*) 1.11×10^{-11} F; (*c*) no change

11. (*a*) 1.69×10^{7} m^2; (*b*) 4117 m

13. 2

15. (*d*)

17. (*a*) 1.5×10^{-2} J; (*b*) 4.5×10^{-2} J

19. (*a*) 0.625 J; (*b*) 1.875 J

21. (*a*) 10^5 V/m; (*b*) 4.43×10^{-2} J/m^3; (*c*) 8.85×10^{-5} J; (*d*) 1.77×10^{-8} F; (*e*) same

23. (*a*) 1.11×10^{-8} C; (*b*) 5.53×10^{-7} J

25. (*a*) True; (*b*) True

27. (*a*) 100; (*b*) 10 V; (*c*) 10^{-5} C

29. $C_2 + C_1C_3/(C_1 + C_3)$

31. (*a*) 4.00×10^{-5} C; (*b*) $V_{10} = 4$ V, $V_{20} = 2$ V

33. (*a*) 15.2 μF; (*b*) $Q_{12} = 2400\ \mu$C, $Q_4 = Q_{15} = 632\ \mu$C; (*c*) 0.303 J

35. (*a*) 0.242 μF; (*b*) $Q_{0.30} = 2.42\ \mu$C, $Q_{1.0} = 1.94\ \mu$C, $Q_{0.25} = 0.484\ \mu$C; (*c*) 1.21×10^{-5} J

37. (*a*) 5.0 μF; (*b*) 133 V

39. 3 μF, 5 μF, 6 μF, 7 μF (parallel); (4/7) μF, (4/5) μF, (4/3) μF, (2/3) μF (series); (6/7) μF, (10/7) μF, (12/7) μF (1 in series, 2 in parallel); (14/5) μF, (7/3) μF, (14/3) μF (2 in series, 1 in parallel)

41. (*a*) 40 V; (*b*) 4.24 m

43. (*a*) 0.333 mm; (*b*) 3.77 m^2

45. (*a*) 1.55×10^{-12} F; (*b*) 1.55×10^{-8} C/m

47. 1.79×10^{-10} F/m

49. (*a*) kQ/r^2 between spheres, zero elsewhere $\eta = \epsilon_0 k^2Q^2/2r^4$; (*b*) $(kQ^2/2r^2)\,dr$; (*c*) $(\frac{1}{2})\,kQ^2\,(R_2 - R_1)/R_1R_2$

51. $5R/3$

53. (*a*) 2.0 kV; (*b*) 0

55. (*a*) 2.4 μF; (*b*) 3.60×10^{-4} J

57. (*a*) 6 V; (*b*) 1.15×10^{-3} J, 2.88×10^{-4} J

59. (*a*) 200 V, 200 V, 200 V; (*b*) $Q_1 = -533\ \mu$C, $Q_2 = -133\ \mu$C, $Q_3 = 267\ \mu$C; (*c*) $V_1 = -267$ V, $V_2 = -133$ V, $V_3 = 400$ V

61. 2.71 nF

63. (*a*) $kQ/\kappa r^2$ between spheres, zero elsewhere $\eta = \epsilon_0 k^2Q^2/2\kappa^2 r^4$; (*b*) $(kQ^2/2\kappa r^2)dr$; (*c*) $kQ^2(R_2 - R_1)/2\kappa R_1 R_2$

65. $4\kappa_1\kappa_2C_0/(3\kappa_1 + \kappa_2)$

67. (*a*) 1.67×10^{-8} F; (*b*) 1.17×10^{-9} C; (*c*) 7×10^6 V/m

69. (*a*) 2.08; (*b*) 45.2 cm^2; (*c*) 5.2 nC

71. (*a*) 2.5×10^4 V/m 6.64×10^{-7} J; (*b*) 6.25×10^3 V/m; (*c*) 25 V; (*d*) 1.66×10^{-7} J

73. (*a*) False; (*b*) False; (*c*) False

75. (*a*) 14 μF; (*b*) (8/7) μF

77. 1.0 mm

79. $C_1C_4 = C_2C_3$

81. $2/(\kappa + 2)$

83. (*a*) $\epsilon_0A/2d$; (*b*) 2 *V*; (*c*) ϵ_0AV^2/d; (*d*) $\epsilon_0AV^2/2d$

85. For a balloon with a radius of 3 m, $C = 0.3$ nF.

87. (*a*) 40 V; (*b*) 1.49×10^{-5} m^2; (*c*) 6

89. 6.67 cm

91. (*a*) $Q/\kappa_1A\epsilon_0$, $Q/\kappa_2A\epsilon_0$; (*b*) $(Q/2C_0)\,(1/\kappa_1 + 1/\kappa_2)$; (*c*), (*d*) Answer given in the problem.

93. (*a*) Since the capacitor plates are conductors, the potentials are the same across the entire upper and lower plates. So the system is equivalent to two capacitors in parallel, each of area $A/2$.
(*b*) Answer given in the problem.

95. (*a*) $(\epsilon_0b/d)[(\kappa - 1)\,x + a]$; (*b*) Answer given in the problem.

97. 2.55×10^{-6} J

99. (*a*) $(\epsilon_0bV^2/2d)[(\kappa - 1)x + a]$; (*b*) $(\epsilon_0bV^2/2d)(\kappa - 1)$

101. (*a*) 2.51×10^3 m^3; (*b*) 5.02×10^{-2} m^3

103. (*a*) $2kQ/\kappa L\ln(b/a)$; (*b*) $Q/2\pi aL$, $-Q/2\pi bL$; (*c*) $-Q(\kappa - 1)/2\pi aL\kappa$, $Q(\kappa - 1)/2\pi bL\kappa$; (*d*) $kQ^2/\kappa L\ln(b/a)$; (*e*) $(kQ^2/L\kappa)\,(\kappa - 1)\ln(b/a)$

105. 3.89×10^{-11} F

107. 0.1 μF, 32 μC

109. (*a*) 0.001 J; (*b*) 47.6 μC, 152 μC; (*c*) 4.76×10^{-4} J

111. (*a*) $C(V) = C_0[1 + (\kappa\epsilon_0\,V^2/2Yd^2)]$ if $\Delta x \ll d$; $C(V) = 0.133(1 + 6.64 \times 10^{-11}\,V^2)\mu$ F/m^2; (*b*) 7.97 kV; (*c*) 99.8%, 0.2%

113. (*a*) 5×10^{-4} J; (*b*) 1.59×10^{-4} N·m

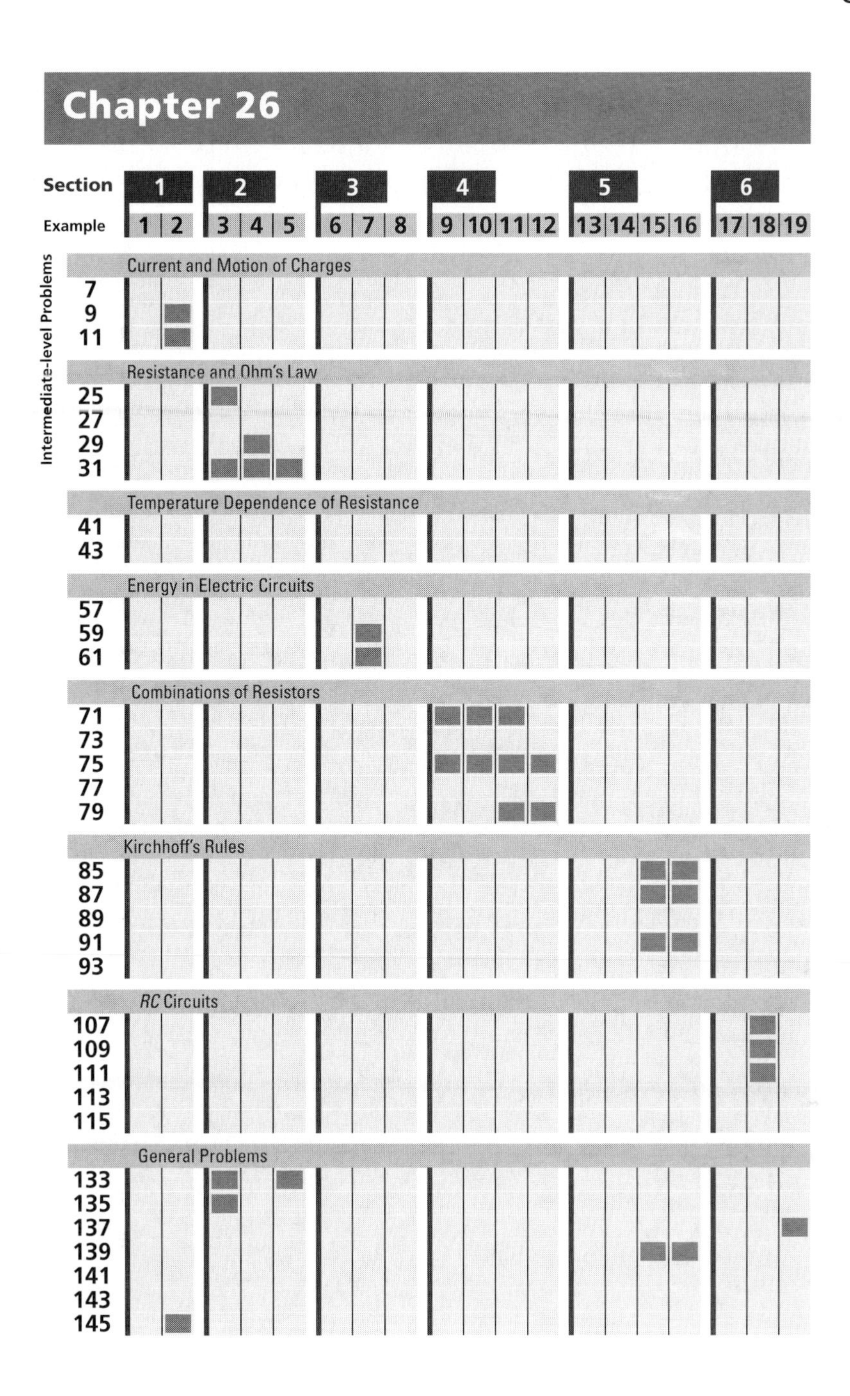

1. Electric fields inside a conductor cause free electrons in the conductor to move; the conductor is no longer in static equilibrium.

3. 2.81×10^{-4} m/s

5. (*a*) 5.93×10^{7} m/s; (*b*) 3.73×10^{-5} A

7. $\lambda\omega a$

9. (*a*) 3.21×10^{13} protons/m^3; (*b*) 3.75×10^{17} protons; (*c*) $(10^{-3}$ C/s$)t$

11. (*a*) 1.04×10^{8} protons/m; (*b*) 1.04×10^{14} protons/m^3

13. (*e*)

15. battery (chemical), generator (mechanical)

17. (*b*)

19. (*a*) 33.3 Ω; (*b*) (3/4) A

21. 8.98 mm

23. 1.95 V

25. 5.88×10^{17} m

27. (*a*) 5.88; (*b*) iron

29. (*a*) 79.6 Ω; (*b*) 318 Ω

31. (*a*) 0.323 Ω; (*b*) 9.69×10^{-2} V/m; (*c*) 2.38×10^{5} s

33. (*a*) 1.28 Ω; (*b*) 0.029 Ω

35. $\rho L/\pi ab$

37. (*a*) $(\rho/2\pi L)\ln(b/a)$; (*b*) 2.05 A

39. 45.6°C

41. (*a*) 15 A; (*b*) 11.1 Ω; (*c*) 1.29 kW

43. 1750 K

45. (*d*)

47. decrease

49. (*a*) 2.88 kW; (*b*) 1.44 kW

51. (*a*) 57.6 Ω, 4.17 A; (*b*) 250 W

53. 0.03 Ω

55. $77.76

57. (*a*) 8 min 13 s; (*b*) 52 min 13 s

59. (*a*) 6.91×10^6 J; (*b*) 12 h 48 min

61. (*a*) 26.7 kW; (*b*) 5.76×10^6 C; (*c*) 6.91×10^7 J; (*d*) 57.6 km; (*e*) 3 cents/km

63. (*b*)

65. (*e*)

67. (*a*) 4.5 Ω; (*b*) $I_3 = 2.67$ A, $I_6 = 0.67$ A, $I_2 = 2.0$ A

69. (*a*) 16 V; (*b*) 2 A

71. (*a*) I_2 (series) = 2.14 A, $I_4 = 0.43$ A, I_2 (parallel) = 0.86 A; (*b*) 12.8 W

73. (*a*) $R_{eq} = xR_1/(1 + x)$

(*b*)

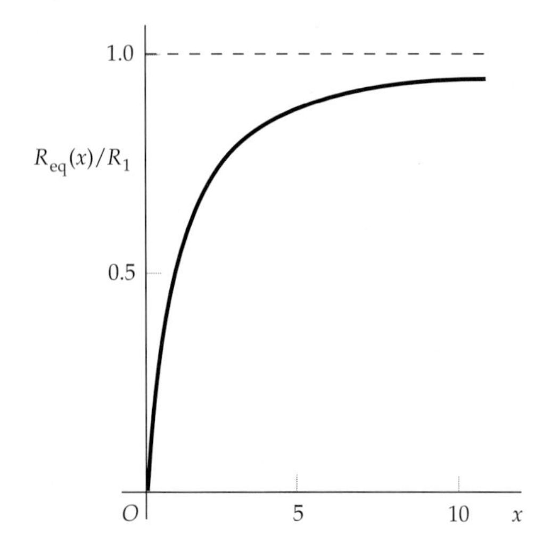

75. (*a*) 4.10 Ω; (*b*) $I_6 = 1.43$ A, I_4 (upper) = 0.86 A, $I_{2,4} = 0.57$ A, I_4 (lower) = 1.50 A, $I_8 = 0.75$ A

77. 11.3 Ω

79. Answer given in the problem.

81. (*b*)

83. (*a*) 1 A; (*b*) 12 W delivered by 12-V battery, 6 W absorbed by 6-V battery; (*c*) 2 W in the 2 Ω, 4 W in the 4 Ω

85. 600 Ω

87. (*a*) $I_4 = \frac{2}{3}$ A, $I_3 = \frac{8}{9}$ A, $I_6 = \frac{14}{9}$ A; (*b*) $\frac{28}{3}$ V; (*c*) 8 W (left battery), $\frac{32}{3}$ W (right battery)

89. Greater when $R < r$ (parallel) and $R > r$ (series)

91. 2.4 V

93. (*a*) For $R > 0.4$ Ω connect in series and for $R < 0.4$ Ω connect in parallel; (*b*) 10.7 A; (*c*) 6.67 A; (*d*) 5.45 A; (*e*) 4.44 A

95. (*a*) $I_1 = -57.0\text{ A} + (10\text{ A/h})t$, $I_2 = 63.0\text{ A} - (10\text{ A/h})t$

(*b*)

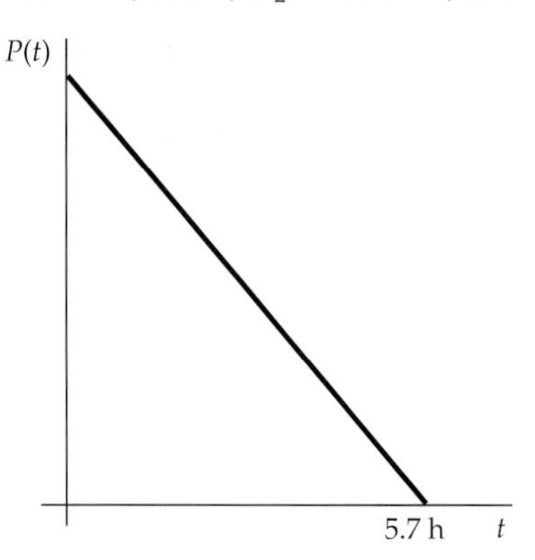

97. I_3 (parallel) = 2.13 A, $I_4 = 0.678$ A, $I_5 = 0.735$ A, $I_2 = 2.807$ A, I_3 (series) = 0.057 A

99. (*b*)

101. (*e*)

103. (*a*) not affected; (*b*) decreases; (*c*) doubles

105. (*a*) 6×10^{-4} C; (*b*) 0.2 A; (*c*) 3×10^{-3} s; (*d*) 8.12×10^{-5} C

107. 2.18×10^6 Ω

109. (*a*) 8 μC; (*b*) 0.0737 s

111. (*a*) 5.69 μC; (*b*) 1.10 μA; (*c*) 1.10 μA; (*d*) 6.62×10^{-6} W; (*e*) 2.44×10^{-6} W; (*f*) 4.19×10^{-6} W

113. (*a*) 25 A; (*b*) 0.4 Ω, 10 Ω, 6.67 Ω

115. Answer given in the problem.

117. (*a*) 0.25 A; (*b*) 0.0625 A; (*c*) $I_{600} = 0.0625(1 - e^{-t/0.00075})$

119. (*a*) 0; (*b*) never

121. To avoid a sizable fraction of the energy being dumped into the internal resistance of the battery

123. Yes

125. (*c*)

127. The 25-W bulb because it has the higher operating resistance

129. (*a*) 0.707 A; (*b*) 7.07 V

131. (*a*) 30 A; (*b*) 4 V

133. (*a*) 3.11 V; (*b*) 0.0779 V/m; (*c*) 18.7 W

135. 2.03 m

137. (*a*) 3.42 A; (*b*) 0.962 A; (*c*) 260 μC (10 μF), 130 μC (5 μF)

139. (*a*) 9800 Ω; (*b*) For $x = 98.2$ cm, $R_x = 10,911$ Ω. For $x = 97.8$ cm, $R_x = 8891$ Ω. The errors are 11.3% and 9.28% respectively. (*c*) Increase to 9800 Ω

141. 0.2 Ω

143. $359

145. (*a*) 0.05 A; (*b*) 5000 W

147. (*a*) 0.01 s; (*b*) 0.2 V + 79,980 (V/s)t; (*c*) 10^9 Ω; (*d*) 6.09×10^{-11} s; (*e*) 1.28 W, 1.76×10^4 W

149. $I_{10} = 0.740$ A, $I_{40} = 0.472$ A, $I_{30} = 0.383$ A, $I_{80} = 0.357$ A, $I_{20} = 0.855$ A

151. (*a*) $QC_1/(C_1 + C_2)$, $QC_2/(C_1 + C_2)$; (*b*) $\frac{1}{2}Q^2/C_1$, $\frac{1}{2}Q^2/(C_1 + C_2)$; (*c*) Heat generated in resistor R

153. (*a*) 0.12 A; (*b*) 0.04 A; (*c*) 8 V; (*d*) 6 V; (*e*) $0.04e^{-133.3t}$ A

155. (*a*) 10^{12} electrons; (*b*) 0.16 mA; (*c*) 6.4×10^4 W; (*d*) 6.4×10^8 W; (*e*) 10^{-4}

Chapter 27

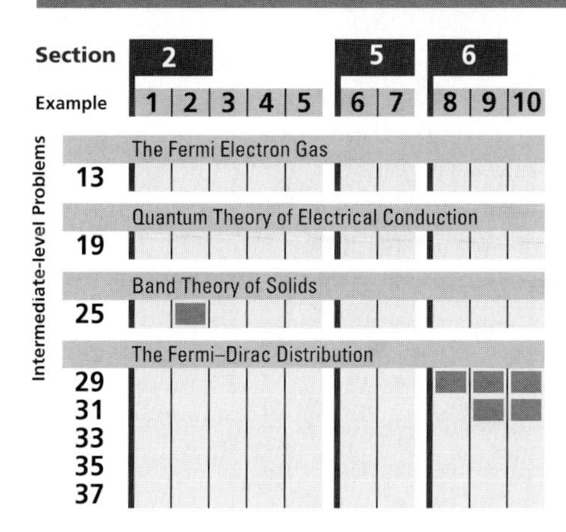

1. as heat energy

3. (*a*) $1.23 \times 10^{-7}\ \Omega\cdot\text{m}$; (*b*) $7.10 \times 10^{-8}\ \Omega\cdot\text{m}$

5. 3 electrons/A1 atoms

7. (*a*) 1.36×10^{5} K; (*b*) 2.46×10^{4} K; (*c*) 1.18×10^{5} K

9. (*a*) 11.7 eV; (*b*) 2.12 eV; (*c*) 10.2 eV

11. (*a*) 1.30×10^{5} K; (*b*) 11.2 eV

13. (*a*), (*b*) Answer given in the problem.
(*c*) $6.36 \times 10^{10}\ \text{N/m}^2$, less than half the measured value. In this calculation, the electron gas is treated as an ideal gas and thus the effect of electron-electron interactions are not taken into account.

15. (*a*) Au, Ag; (*b*) 0.1 eV

17. The impurity effect in brass is higher.

19. (*a*) 66.1 nm; (*b*) $1.77 \times 10^{-20}\ \text{m}^2$

21. (*b*)

23. 1.68×10^{-6} m

25. (*a*) 0.37 eV; (*b*) 4300 K

27. (*a*) 2.17×10^{-3} eV; (*b*) 0.455 mm

29. Answer given in the problem.

31. Approximately 1.0

33. Answer given in the problem.

35. 60%

37. Answer given in the problem.

39. Answer given in the problem.

41. (*a*) $4.57 \times 10^{45}\ \text{neutrons/m}^3$; (*b*) 5.48×10^{8} eV

43. The resistivity of copper increases and the resistivity of silicon decreases.

45. (*a*) $8.62 \times 10^{22}\ \text{electrons/cm}^3$; (*b*) $13.1 \times 10^{22}\ \text{electrons/cm}^3$

47. (*a*) 3.12×10^{15} pair/s; (*b*) 3.12×10^{15} pair/s; (*c*) 8×10^{-4} W

Chapter 28

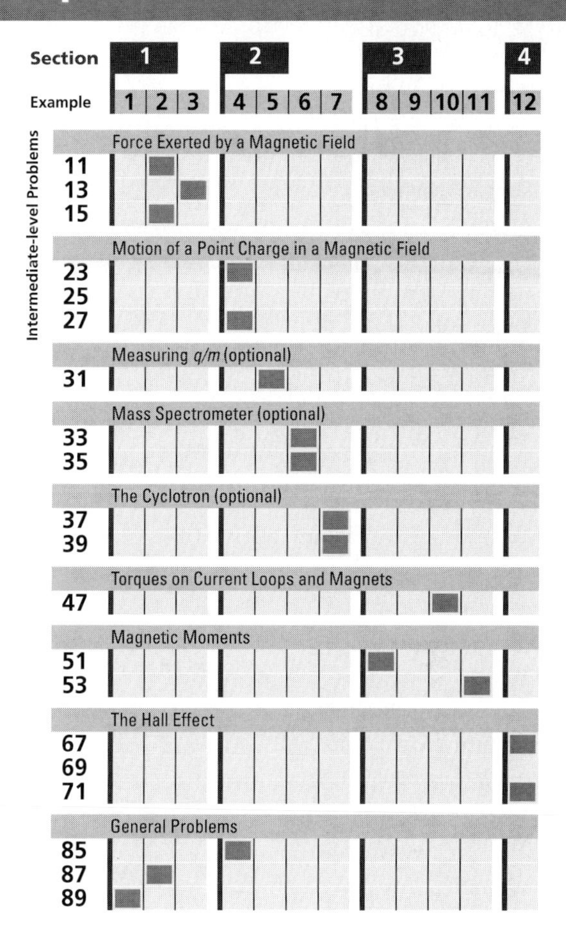

1. (*b*)

3. $-1.25 \times 10^{-12}\ \text{N}\ \hat{j}$

5. (*a*) $-6.39 \times 10^{-13}\ \text{N}\ \hat{j}$; (*b*) $8.76 \times 10^{-13}\ \text{N}\ \hat{i}$; (*c*) 0;
(*d*) $(7.10\hat{i} - 9.47\hat{j}) \times 10^{-13}$ N

7. 0.962 N

9. $(-1.92\hat{i} - 1.28\hat{j} - 5.76\hat{k}) \times 10^{-13}$ N

11. 1.48 A

13. Answer given in the problem.

15. $10\ \text{T}\ \hat{i} + 10\ \text{T}\ \hat{j} - 15\ \text{T}\ \hat{k}$

17. False

19. (*a*) 8.74×10^{-8} s; (*b*) 4.67×10^{7} m/s; (*c*) 1.82×10^{-12} J

21. (*a*) 142 m; (*b*) 2.85 m

23. (*a*) $v_p/v_\alpha = 2$; (*b*) $K_p/K_\alpha = 1$; (*c*) $L_p/L_\alpha = 1/2$

25. Answer given in the problem.

27. (*a*) 12.6×10^{6} m/s, 24°; (*b*) 6.28×10^{6} m/s, 24°

29. (*a*) 1.64×10^{6} m/s; (*b*) 14.1 keV; (*c*) 7.68 eV

31. (*a*) 7.35 mm; (*b*) 6.64×10^{-5} T

33. (*a*) 63.5 cm; (*b*) 2.59 cm

35. 1.57×10^{-5} s, 1.63×10^{-5} s

37. (*a*) 2.13×10^{7} Hz; (*b*) 46.0 MeV; (*c*) f and K are reduced by a factor of 2.

39. Answer given in the problem.

41. If the plane of current loop is parallel to the field.

43. 2.83×10^{-5} N·m

45. (*a*) 0; (*b*) 2.08×10^{-3} N·m

47. (*a*) 37° below the *x*-axis; (*b*) $0.799\hat{i} - 0.602\hat{j}$; (*c*) $[0.335\hat{i} - 0.253\hat{j}]$ A·m^2; (*d*) 0.503 N·m $\hat{k}$; (*e*) 0.380 J

49. Answer given in the problem.

51. 0.377 A·m^2 into the page

53. Answers given in the problem.

55. 0.02 A·m^2

57. Answers given in the problem.

59. $\frac{4}{3}\pi\sigma\omega R^4$

61. $2^{-1/4}R$

63. $(\pi\omega L\rho_0/4)(2R_s^4 - R_i^4 - R_o^4)$

65. $(\pi/4)\omega\rho_0 R^5$, opposite the direction of $\vec{\omega}$.

67. (*a*) 3.69×10^{-5} m/s; (*b*) 1.48 μV

69. 1.02×10^{-3} V

71. 3.5 electrons/A1 atom

73. (*a*) True; (*b*) True; (*c*) True; (*d*) False; (*e*) True

75. (*c*)

77. (*e*)

79. (*c*)

81. 8.78 N/m $\hat{k}$

83. Answer given in the problem.

85. Answers given in the problem.

87. (*a*) $(mg/I\ell)\tan\theta$; (*b*) $g\sin\theta$

89. (*a*) 1.6×10^{-18} N $\hat{j}$; (*b*) 10 V/m $\hat{j}$; (*c*) 20 V

91. 5.10 m

93. $2\pi(M/\pi IB)^{1/2}$

Chapter 29

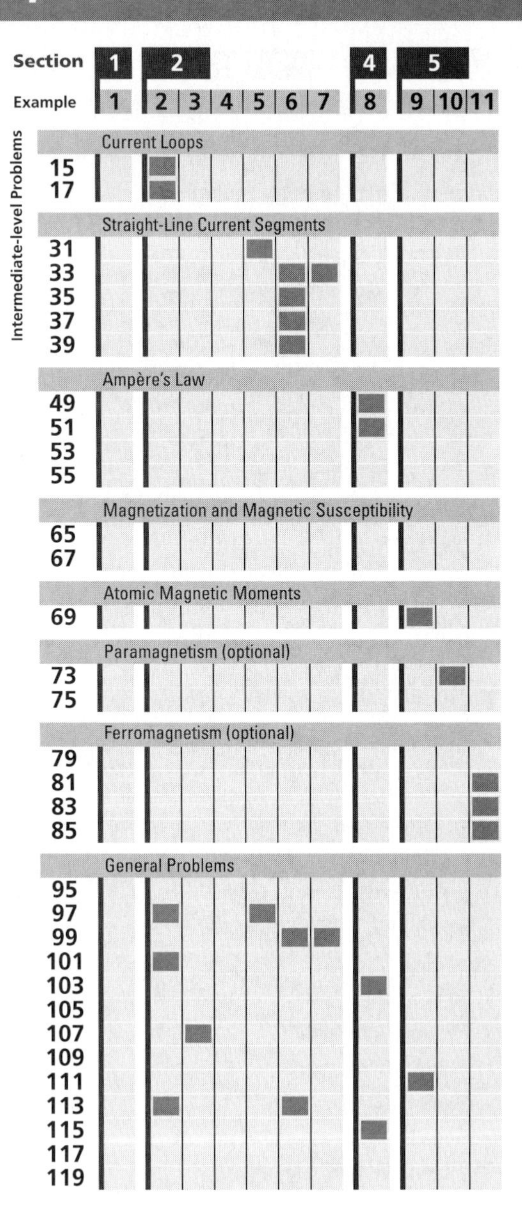

Section
1
2
4
5
Example
1
2
3
4
5
6
7
8
9
10
11
Intermediate-level Problems
Current Loops
15
17
Straight-Line Current Segments
31
33
35
37
39
Ampère's Law
49
51
53
55
Magnetization and Magnetic Susceptibility
65
67
Atomic Magnetic Moments
69
Paramagnetism (optional)
73
75
Ferromagnetism (optional)
79
81
83
85
General Problems
95
97
99
101
103
105
107
109
111
113
115
117
119

1. (*a*) opposite; (*b*) same

3. (*a*) 1.27×10^{-11} T $\hat{k}$; (*b*) 0; (*c*) 3.22×10^{-12} T $\hat{k}$

5. 12.5 T

7. (*a*), (*b*)

9. -9.6×10^{-12} T $\hat{i}$

11. No

13. 11.1 A

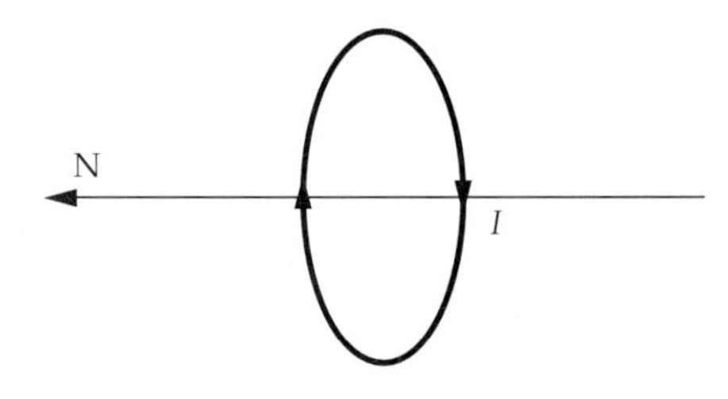

15. 9.47 A

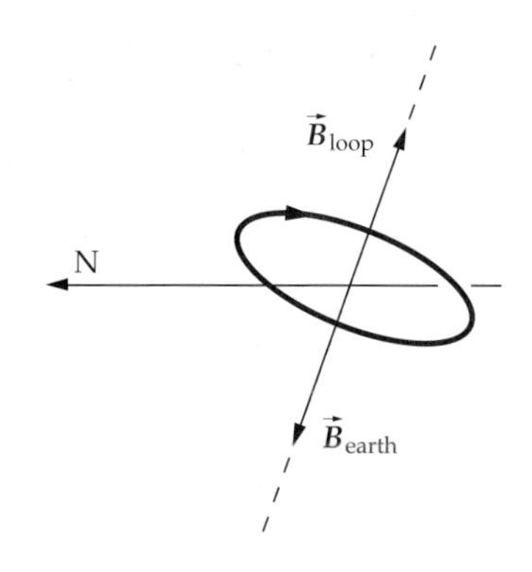

17. (*a*) 0.0540 T, 0.0539 T, 0.0526 T, 0.0486 T

(*b*)

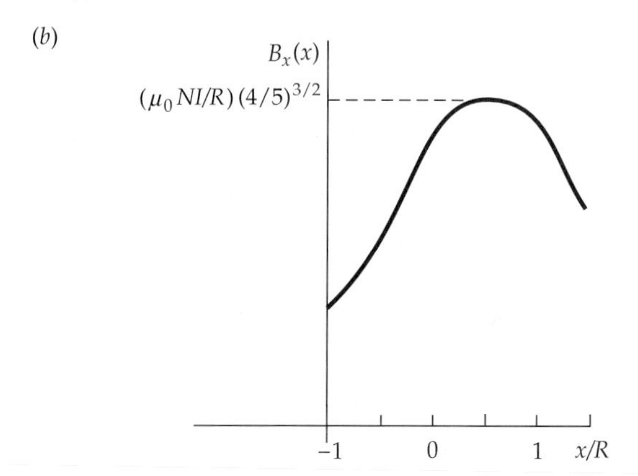

19. (*c*)

21. (*c*)

23. (*a*) 2×10^{-5} T; (*b*) 4×10^{-6} T; (*c*) 10^{-6} T

25.

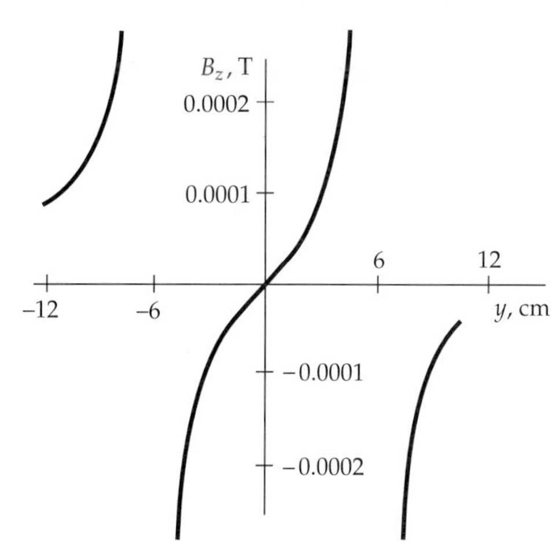

27.

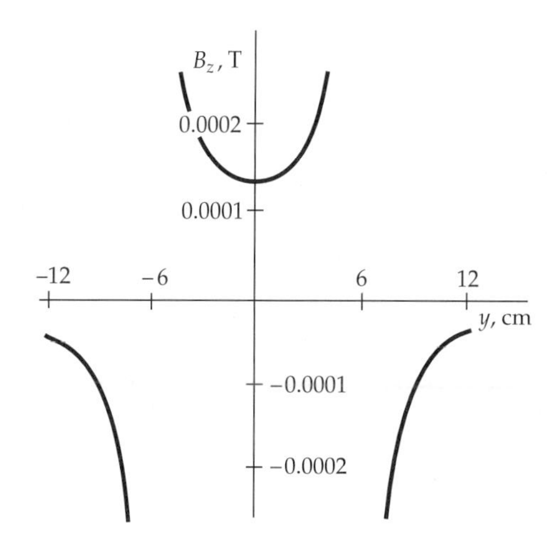

29. 6.67×10^{-4} N/m

31. $B_{1\text{ cm}} = 56.6\ \mu$T, into the page; $B_{2\text{ cm}} = 113\ \mu$T, into the page; $B_{total} = 226\ \mu$T

33. (*a*) 7.79×10^{-4} N/m, upward; (*b*) 5.20×10^{-5} T, to the right

35. Along line $y = x$ except at $x = y = 0$ the field is not defined.

37. (*a*) $(3\mu_0 I/4\pi L)(\hat{\imath} - \hat{\jmath})$; (*b*) $(\mu_0 I/4\pi L)(\hat{\imath} - \hat{\jmath})$; (*c*) $(\mu_0 I/4\pi L)(-\hat{\imath} - 3\hat{\jmath})$

39. $[(1 + \sqrt{2})\mu_0 I/2\pi R]\hat{\imath}$

41. 6.98×10^{-4} T

43. Answers given in the problem.

45. (*e*)

47. (*a*) C_1: 1.01×10^{-5} T·m, C_2: 0, C_3: -1.01×10^{-5} T·m; (*b*) none

49. (*a*) 8×10^{-4} T; (*b*) 4×10^{-3} T; (*c*) 2.86×10^{-3} T

(*d*)

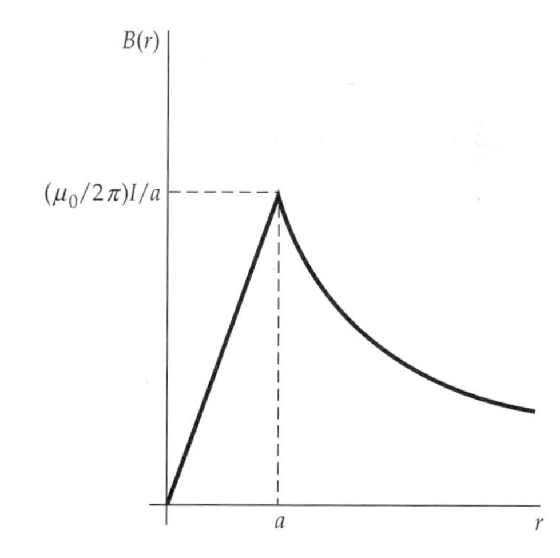

51. (*a*) 2.26×10^{-5} T·m; (*b*) 1.24×10^{-5} T·m; (*c*) 0

53. Answer given in the problem.

55. (*a*) $-\hat{\imath}$; (*b*) $-\hat{\imath}$; (*c*) $\hat{\imath}$; (*d*) Answer given in the problem.

57. (*a*)

59. Diamagnetic: hydrogen, carbon dioxide, nitrogen. Paramagnetic: oxygen.

61. 1.01×10^{-2} T, 0.547 A/m, 1.01×10^{-2} T

63. -4.00×10^{-5}

65. 2.38×10^{-3}

67. (*a*) 8.12×10^3 A/m; (*b*) 1.62×10^{21} electrons; (*c*) 24.4 A

69. $0.587\mu_B$

71. Answer given in the problem.

73. (*a*) 5.58×10^5 A/m, 0.701 T; (*b*) 5.23×10^{-4}; (*c*) μ is less then μ_B. Diamagnetic effects have been neglected, and these effects tend to reduce susceptibility.

75. (*a*) 95.5 A/m; (*b*) 0.0301 T; (*c*) 0.4%

77. 0.864 T, 6.87×10^5 A/m

79. 11.0 A

81. (*a*) 1.26×10^{-3} T; (*b*) 1.26×10^6 A/m; (*c*) 1250

83. (*a*) 0.096 T; (*b*) 0.96 N/m

85. 0.755 T

87. (*a*) False; (*b*) True; (*c*) False; (*d*) False; (*e*) True

89. No

91. If the current in the tube is uniform and parallel to the axis, the magnetic field is zero inside. In a solenoid the current encircles the axis.

93. 2.36×10^{-5} T

95. Answer given in the problem.

97. (*a*) $\pi\mu_0 I/\ell$; (*b*) $(8\sqrt{2}/\pi)\mu_0 I/\ell$; (*c*) $(27/2\pi)\mu_0 I/\ell$; The greatest field is produced by the triangle.

99. (*a*) The force on each of the horizontal segments is 0.251×10^{-4} N, down on the upper segment and up on the lower; the force on the left vertical segment is 1×10^{-4} N to the right, and the force on the right vertical segment is 0.286×10^{-4} N to the left.
(*b*) 0.714×10^{-4} N, to the right

101. 7.07×10^{-6} T, into the page

103. (*a*) 1.64 mT; (*b*) 1.05 mT; (*c*) 17.2 A

105. 1.05 N·m/rad, 0.524 s

107. 0.512 s

109. (*a*) 0.0524 A·m^2; (*b*) 7.70×10^5 A/m; (*c*) 2.31×10^4 A

111. (*a*) 70.5 A·m^2; (*b*) 17.6 N·m

113. 3.18 cm

115. (*a*) 10^{-5} T; (*b*) 10^{-5} T; (*c*) 5×10^{-6} T

117. 2.32×10^{-22} J $= 1.45 \times 10^{-3}$ eV

119. 6.26 T

121. (*a*) $\vec{B} = \dfrac{\mu_0 I}{2\pi(R^2 - a^2)}\left[\dfrac{R}{2} - \dfrac{a^2}{2R - b}\right]\hat{j}$;

(*b*) $\vec{B} = \dfrac{\mu_0 I}{2\pi(R^2 - a^2)}\left(\dfrac{R}{2} - \dfrac{2a^2R}{4R^2 + b^2}\right)\hat{i} - \dfrac{\mu_0 I}{2\pi(R^2 - a^2)}\left(\dfrac{a^2 b}{4R^2 + b^2}\right)\hat{j}$

123. $\vec{B} = 2\dfrac{\mu_0}{4\pi x^3}\left(\dfrac{I\ell^2}{[1 + (\ell^2/4x^2)][1 + (\ell^2/2x^2)]^{1/2}}\right)\hat{i}$.

In the limit $x >> 1$, this reduces to

$$\vec{B} = 2\frac{\mu_0 I\ell^2}{4\pi x^3}\hat{i} = \frac{\mu_0 2\vec{\mu}}{4\pi x^3} \text{ where } \vec{\mu} = I\ell^2\hat{i}.$$

125. $r < R$: $B = \dfrac{\mu_0 I_0}{2\pi r}(1 - e^{-r/a})$; $r > R$: $B = \dfrac{\mu_0 I_0}{2\pi r}(1 - e^{-R/a})$

Chapter 30

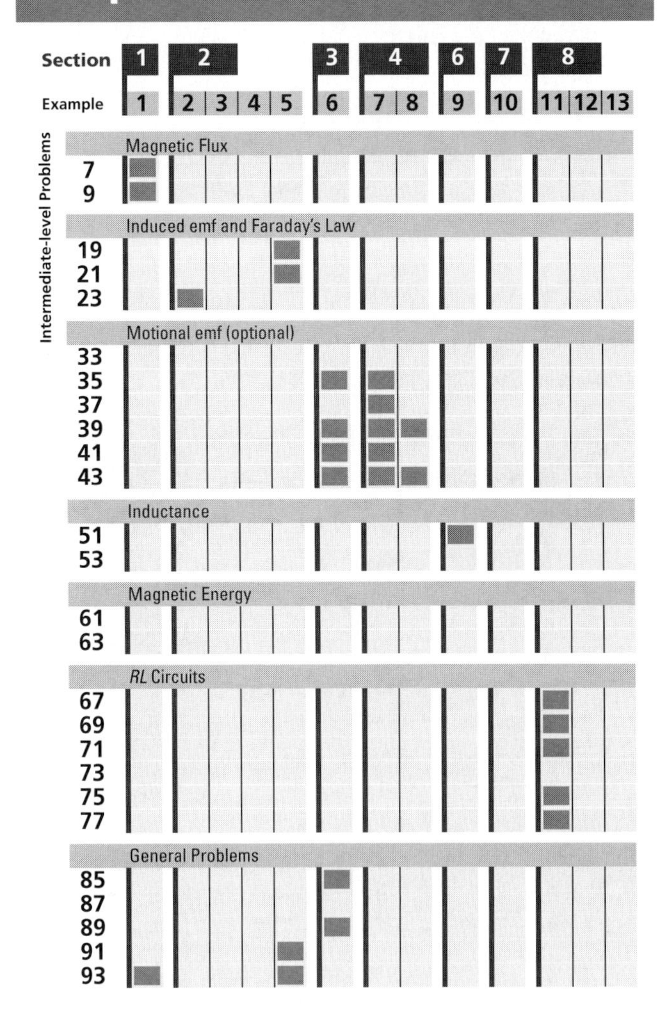

1. (*a*) 5×10^{-4} Wb; (*b*) 4.33×10^{-4} Wb; (*c*) 2.5×10^{-4} Wb; (*d*) 0

3. (*a*) 4.2×10^{-2} Wb; (*b*) 2.1×10^{-2} Wb

5. $\pi R^2 B$

7. 6.74×10^{-3} Wb

9. (*a*) $\mu_0 hNI\pi R_1^2$; (*b*) $\mu_0 hNI\pi R_3^2$

11. $\mu_0 I/4\pi$

13. (*d*)

15. (*a*) $(-0.2t + 0.4)$ V; (*b*) $t = 0$ s: 0 Wb and 0.4 V, $t = 2$ s: -0.4 Wb and 0 V, $t = 4$ s: 0 Wb and -0.4 V, $t = 6$ s: 1.2 Wb and -0.8 V

17. 1.88×10^{-2} V

19. (*a*) 1.26×10^{-3} C; (*b*) 12.6 mA; (*c*) 0.628 V

21. 0.798 G

23. (*a*) $\left(\dfrac{\mu_0}{4\pi}\right)4a \ln 3$; (*b*) 6.59 $\mu\Omega$, counterclockwise

25. (*a*) counterclockwise; (*b*) clockwise

11. 2.1×10^6 light years

13. ± 0.3 m

15. You will see less light from your headlights reflected off the road, but you will see more reflected light from an oncoming car.

17. Because of refraction due to the density variation of air, a ray from the bottom of the sun curves in the direction of the curvature of the earth so the apparent position of the bottom of the sun is raised and can be seen even after it is below the horizon. The curvature of the rays from the top of the sun is less so its position is raised less than that of the bottom, which results in the observed flattening.

19. (*a*) 14.9°; (*b*) 22.1°; (*c*) 32.1°; (*d*) 40.6°

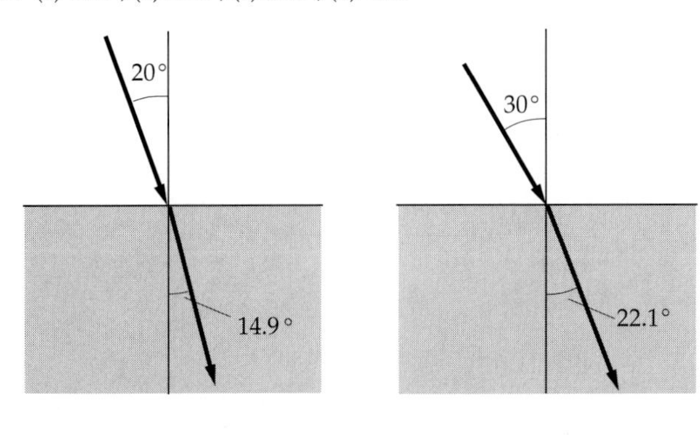

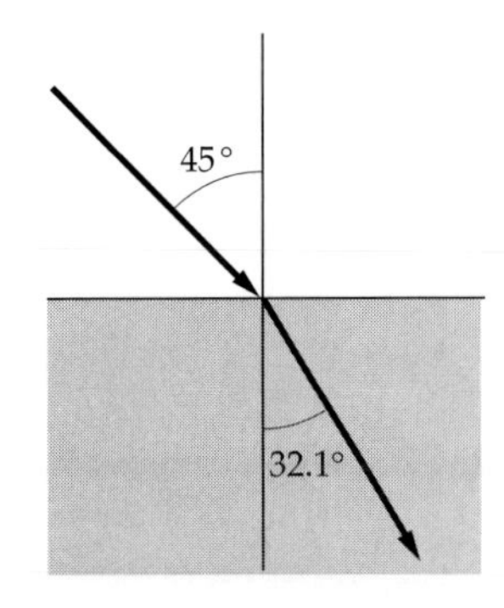

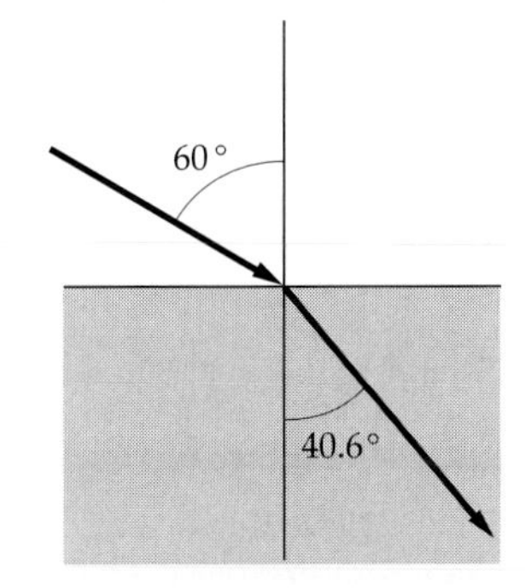

21. 2.26×10^8 m/s, 2×10^8 m/s

23. (*a*) 50.2°; (*b*) 38.8°; (*c*) 26.3°

25. $\approx 92\%$

27.

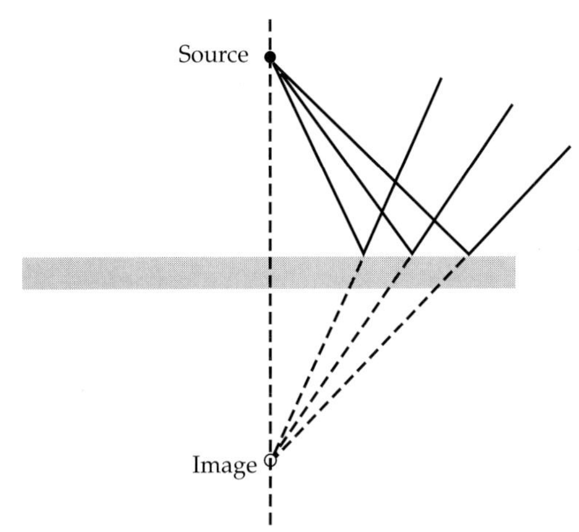

29. (*a*) $\sin^{-1}(n^2 - n\sqrt{n^2 - 1})^{1/2}$; (*b*) 48.5°, 2.81 cm

31. (*a*) Yes; (*b*) If the ball travels with constant speed, the path of least distance will be the path of least time, which is the path that would be taken by a light beam according to Fermat's principle.

33. 48.6°

35. 102 m^2

37. 1.30

39. (*a*) 1.41; (*b*) between 1.63 and 1.88

41. 2.18 cm

43. 1.1°

45. (*b*)

47. (*a*) 53.1°; (*b*) 56.3°

49. (*a*) $I_0/8$; (*b*) $3I_0/32$

51. $I = (1/8)I_0 \sin^2 2\theta$

53. (*a*) $I_0 [\cos(\pi/2N)]^{2N}$; (*b*) $I_0/4$; (*c*) $0.976I_0$; (*d*) Perpendicular to the original polarization.

55. $0.211I_0$, which is greater than $I = 0.125I_0$ for a single sheet in the middle at 45°.

57. Answer given in the problem.

59. $\vec{E} = E_0 \cos(kz + \omega t)\hat{i} + E_0 \sin(kz + \omega t)\hat{j}$, $\vec{B} = (E_0/c) \sin(kz + \omega t)\hat{i} - (E_0/c) \cos(kz + \omega t)\hat{j}$

61. Horizontally polarized, $I_0 [\cos(\pi/2N)]^{2N}$

63. (*c*)

65. (*a*) 526 nm; (*b*) the same color

67. 35.3°

69. 1.45 m

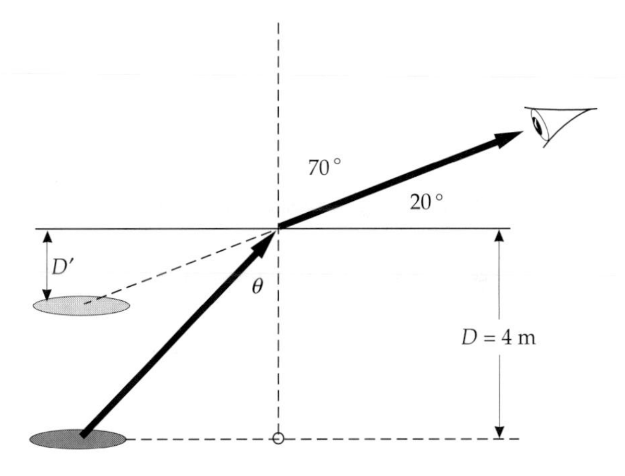

71. Answer given in the problem.

73. Answer given in the problem.

75. Answer given in the problem.

77. (*a*) Answer given in the problem. (*b*) 3.47°

79. (*a*) 1.60; (*b*) 38.7°

81. 1.68°

83. (*a*) $[4n/(n + 1)^2]^{2N}$; (*b*) 0.783; (*c*) 28

85. (*a*) 24.0°; (*b*) The cloud appears to move in a circle of angular radius 24.0°. (*c*) The angular radius θ is related to the earth's speed v by $\tan\theta = v/c$. (*d*) 2.99×10^8 m/s

87. (*a*) Answer given in the problem. (*b*) 5.12 cm

89. Answer given in the problem.

Chapter 34

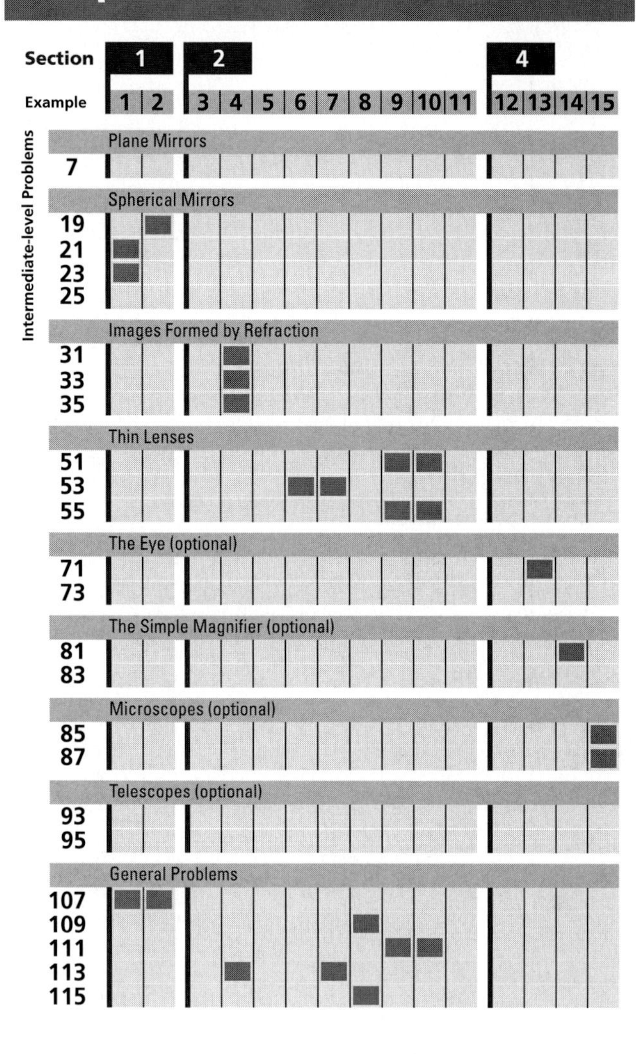

Intermediate-level Problems	Section 1: Ex. 1	2	Section 2: Ex. 3	4	5	6	7	8	9	10	11	Section 4: Ex. 12	13	14	15
Plane Mirrors															
7															
Spherical Mirrors															
19		■													
21	■														
23	■														
25															
Images Formed by Refraction															
31				■											
33				■											
35				■											
Thin Lenses															
51									■	■					
53						■	■								
55									■	■					
The Eye (optional)															
71													■		
73															
The Simple Magnifier (optional)															
81														■	
83															
Microscopes (optional)															
85															■
87															■
Telescopes (optional)															
93															
95															
General Problems															
107	■	■													
109								■							
111									■	■					
113				■			■								
115								■							

1. Yes

3.

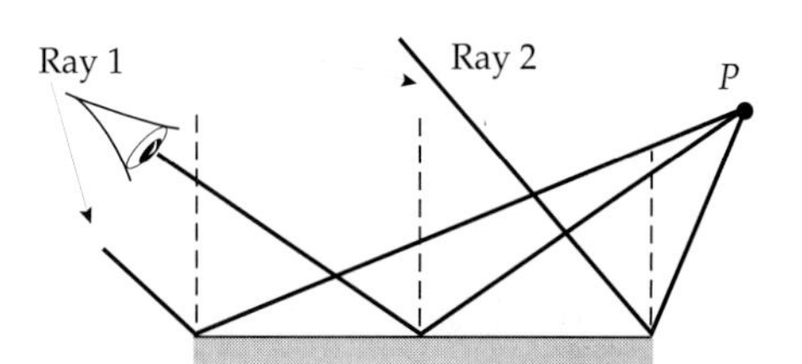

5.

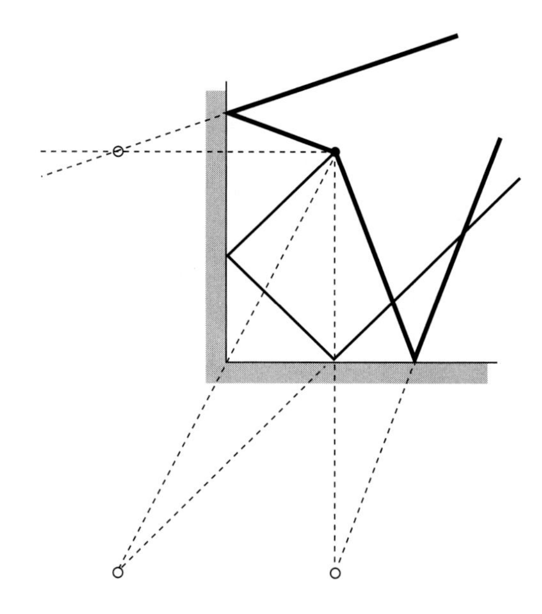

7. (*a*) 10 cm, 50 cm, 70 cm, 110 cm; (*b*) 20 cm, 40 cm, 80 cm, 100 cm

9. The image will be virtual and erect if $s < F$; the image will be smaller if $s > 2F$ and larger if $s < 2F$

11. While the image distance is smaller than the object distance, the size of the image is also smaller than the object. The second effect outweighs the first so that the angular size presented to the driver is smaller for the image than for the actual object. Thus the image appears more distant.

13. (*a*) real, inverted, reduced

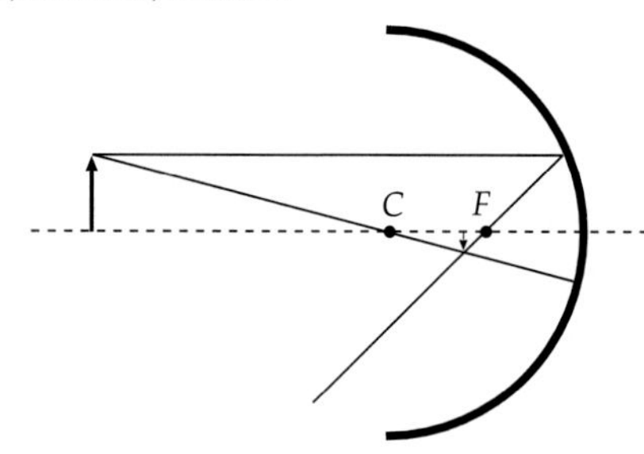

(*b*) real, inverted, same size

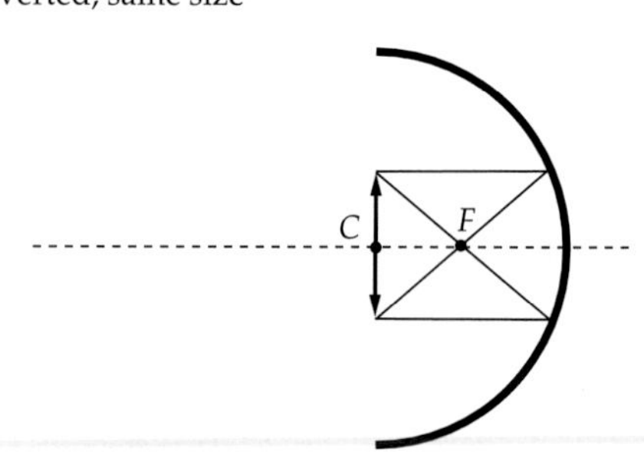

(*c*) no image is formed

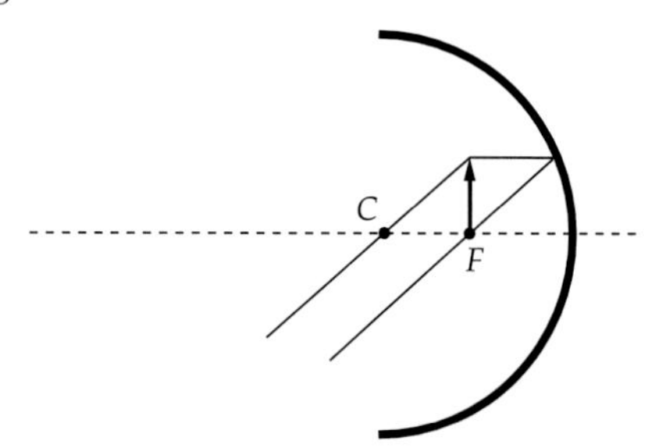

(*d*) virtual, erect, enlarged

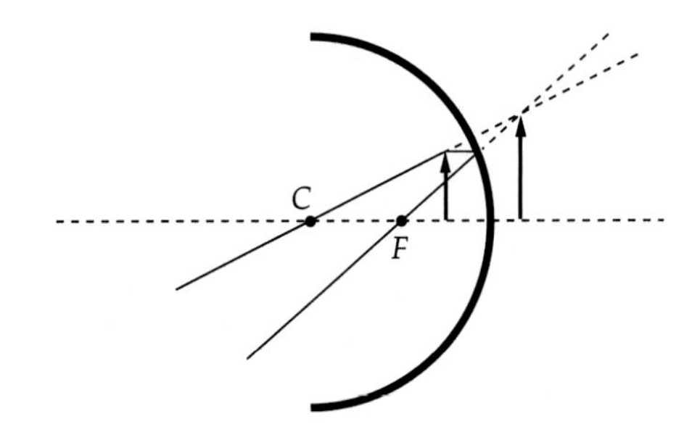

15. (*a*) virtual, erect, reduced

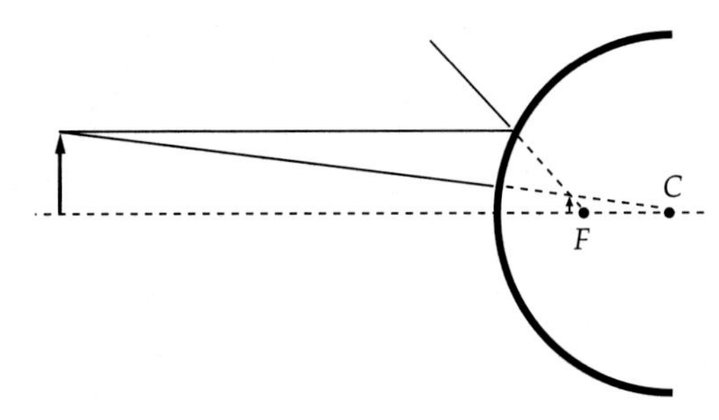

(*b*) virtual, erect, reduced

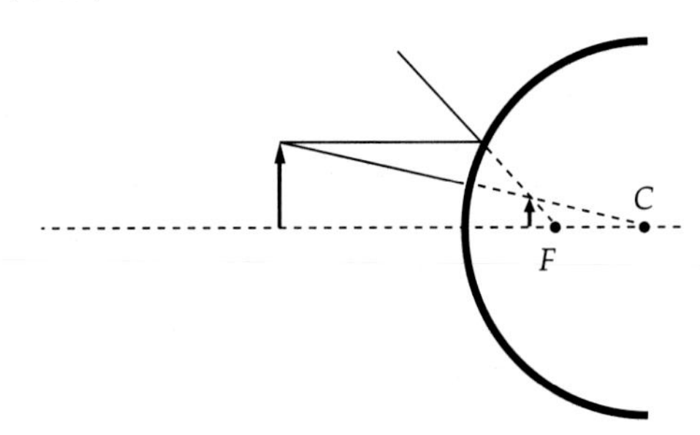

(*c*) virtual, erect, reduced

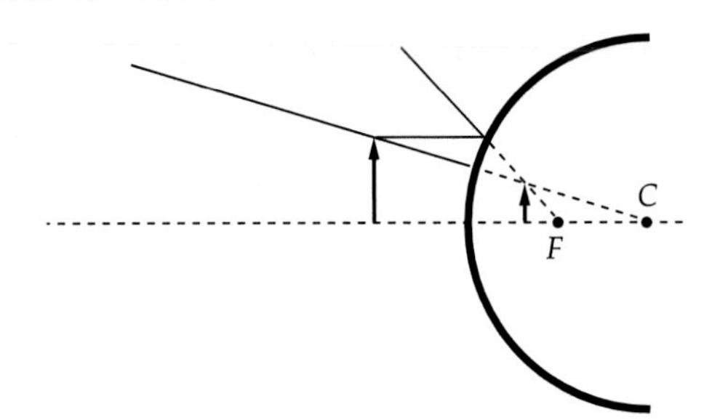

(*d*) virtual, erect, reduced

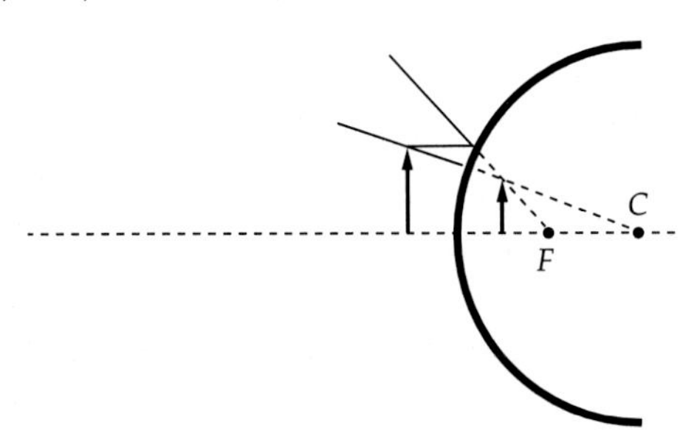

17. Answer given in the problem.

19. (*a*) 0.566 m; (*b*) behind; (*c*) 0.113 m

21. 1.5 cm

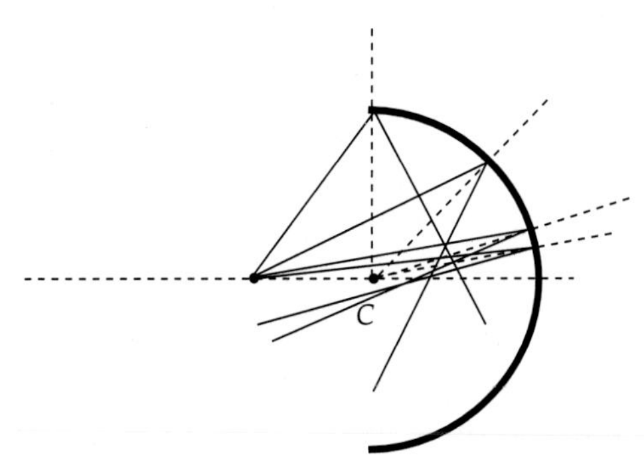

23. (*a*) −66.7 cm; (*b*) virtual

25. (*a*) −1.33 m; (*b*) convex

27. farther

29. (*a*) −8.58 cm; (*b*) −35.9 cm

31. 10 cm

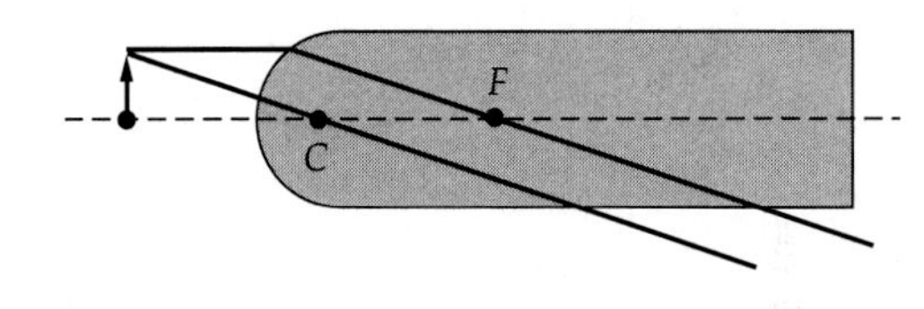

33. (*a*) −46.2 cm, virtual

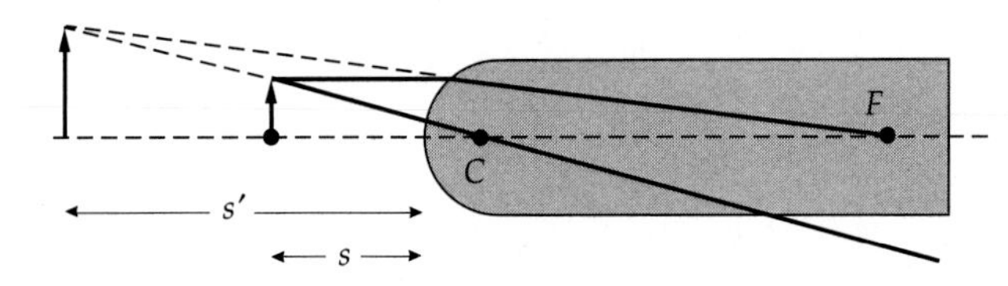

(*b*) −6.47 cm, virtual

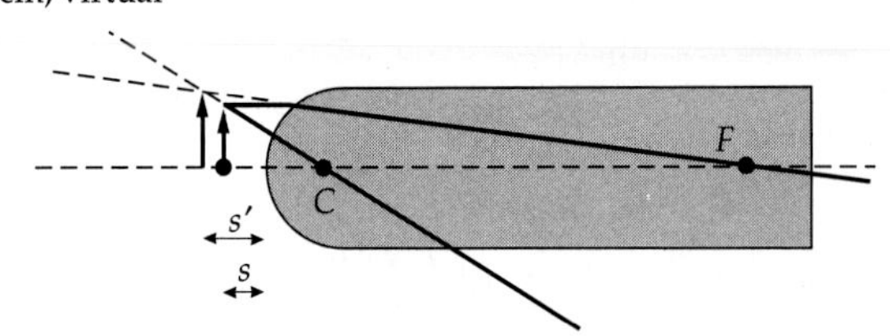

(*c*) 44.1 cm, real

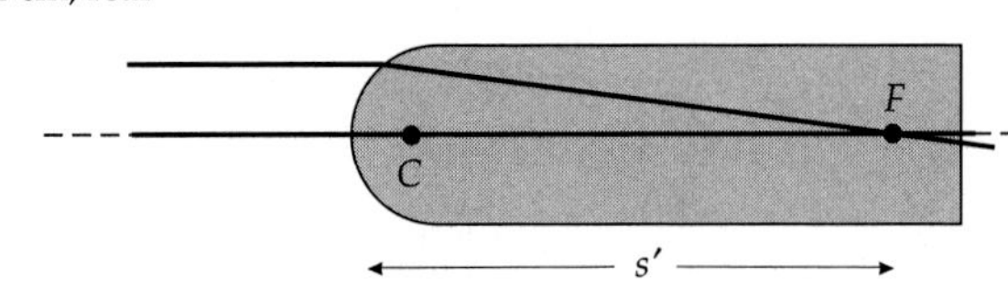

35. (*a*) 64 cm; (*b*) −80 cm; (*c*) virtual

37. Assuming that n is greater than 1, a thin lens will have a positive focal length if it is double convex, plano convex, or if the radius of curvature of the convex side is smaller in magnitude than the radius of curvature of the concave side. The focal length will be negative if the lens is double concave, plano concave, or if the radius of curvature of the concave side is smaller in magnitude than the radius of curvature of the convex side.

39. (*d*)

41. (*a*) 13.5 cm

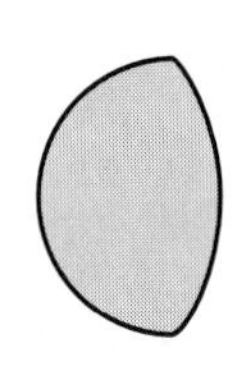

(*b*) 20 cm

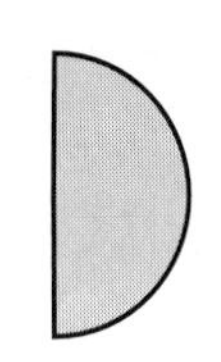

(*c*) −10 cm

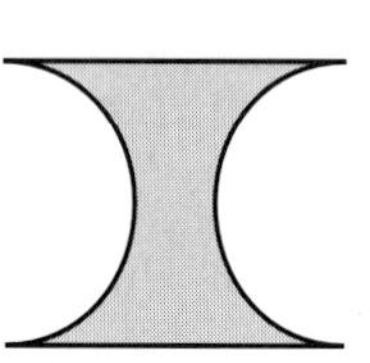

(*d*) −40 cm

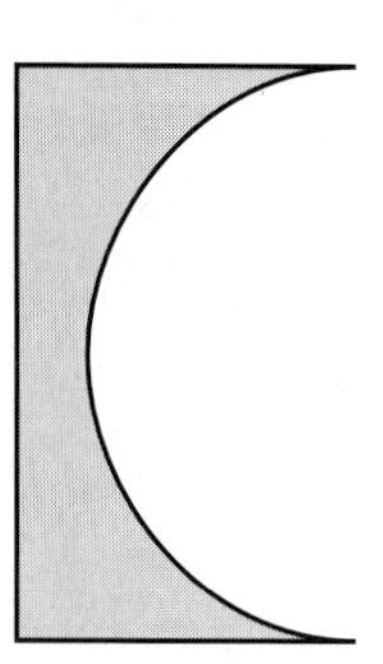

43. 108 cm

45. (*a*) −33.3 cm

(*b*) 33.3 cm

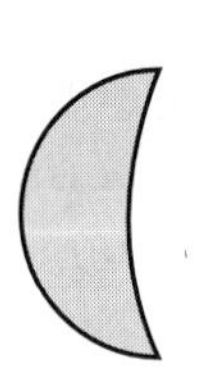

(*c*) −33.3 cm

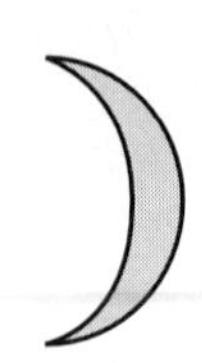

47. $S' = 6.67$ cm, $h = 1$ cm

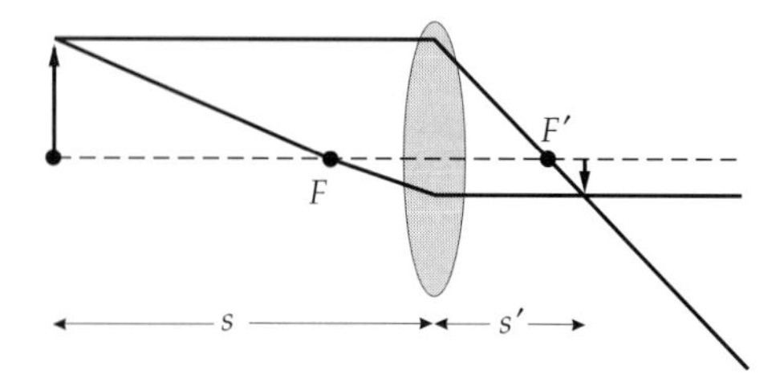

49. $S' = -3.33$ cm, $h = 0.333$ cm

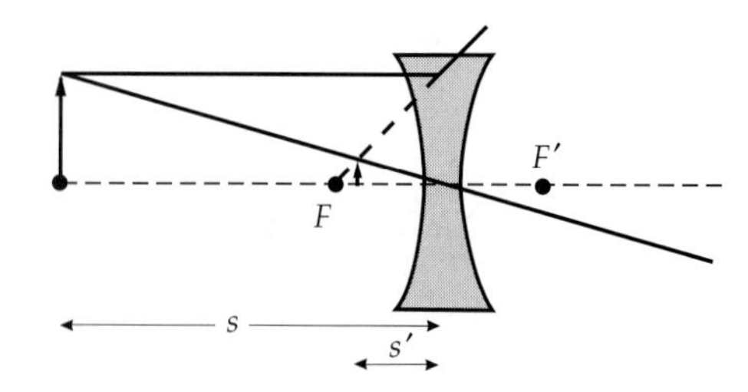

51. (*a*) 30 cm to the far side of the second lens. (*b*) real, erect; (*c*) 2

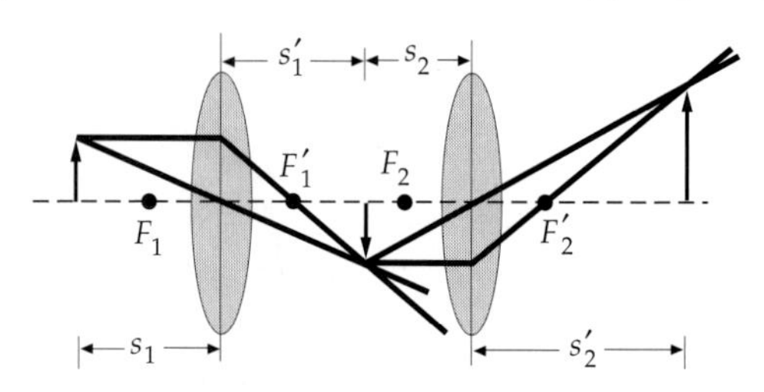

53. (*a*) 35.0 cm, concave

(*b*)

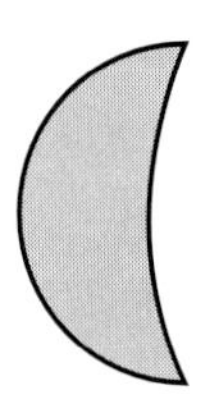

55. 15 cm to right of second lens, real, inverted, magnification of −1

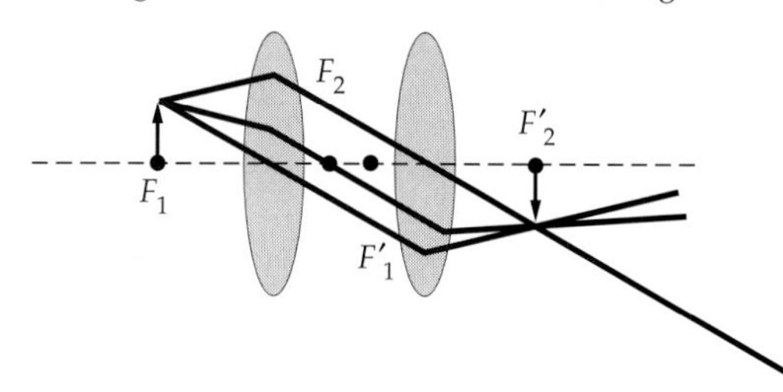

57. Answer given in the problem.

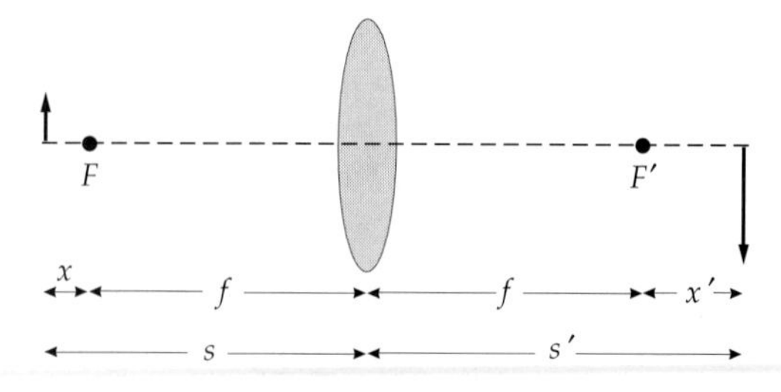

59. (*a*) 41.2 cm; (*b*) −1.53; (*c*) real, inverted

61. (*a*) False; (*b*) True

63. (*a*)

65. (*b*)

67. 0.172 cm

69. 44.4 cm

71. (*a*) 103 cm; (*b*) 0.972 diopters

73. 0.714, smaller

75. (*a*) 46.0 cm; (*b*) 186 cm; (*c*) 3.84 diopters

77. 6

79. 5

81. (*a*) 3; (*b*) 4

83. Answer given in the problem.

85. (*a*) −11.3; (*b*) −56.3

87. (*a*) 30 cm; (*b*) −6; (*c*) −30; (*d*) 5.83 cm

89. (*a*) 1.67 cm; (*b*) 0.508 cm; (*c*) 0.496 cm

91. 17.6 cm

93. 4 cm (eyepiece), 28 cm

95. (*a*) 103.4 cm; (*b*) −6.25 cm; (*c*) 97.2 cm; (*d*) 20.7 cm, 25.6

97. 0.00667

99. (*a*)

101. Answer given in the problem.

103. 0.158 mm

105. (*a*) 5 cm, −10 cm

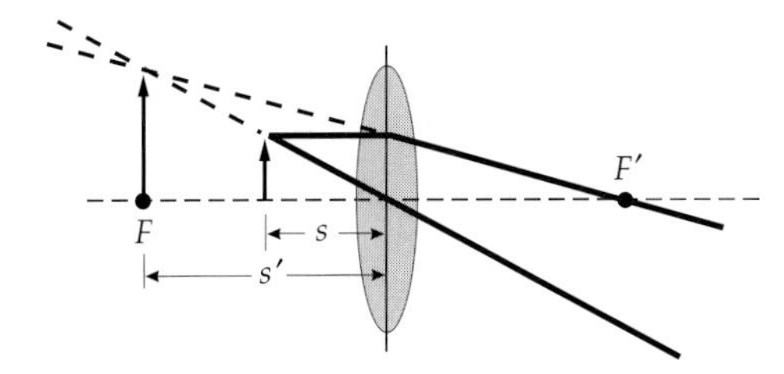

(*b*) 15 cm, 30 cm

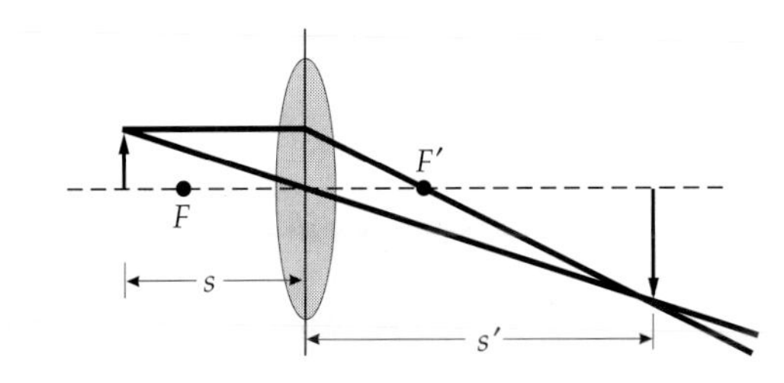

107. concave, 90 cm

109. 3.7 m

111. (*a*) 9.52 cm; (*b*) −1.19

(*c*)

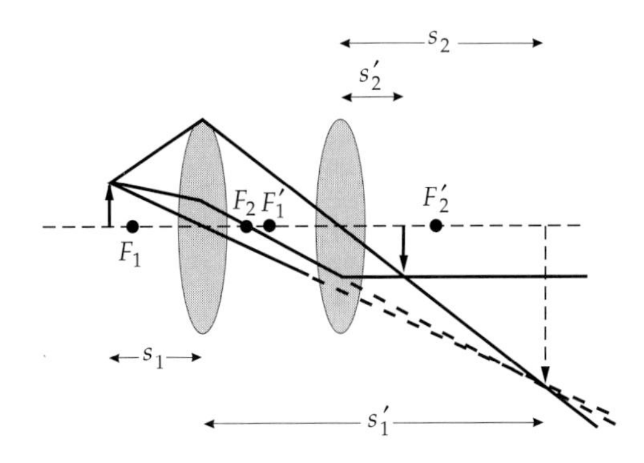

113. (*a*) 19.3 cm; (*b*) 77.3 cm

115. 9.7 cm/s

117. (*a*) 18 cm to left of lens; (*b*) real, erect

(*c*)

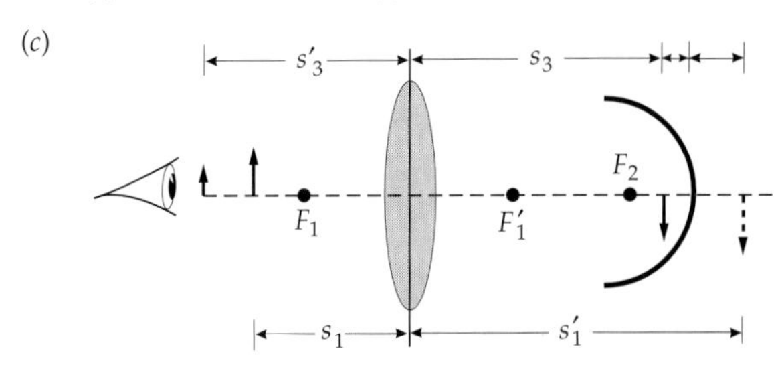

119. 1.6

121. 1.36

123. (*a*) Answer given in the problem. (*b*) 17.4 cm

125. Answer given in the problem.

Chapter 35

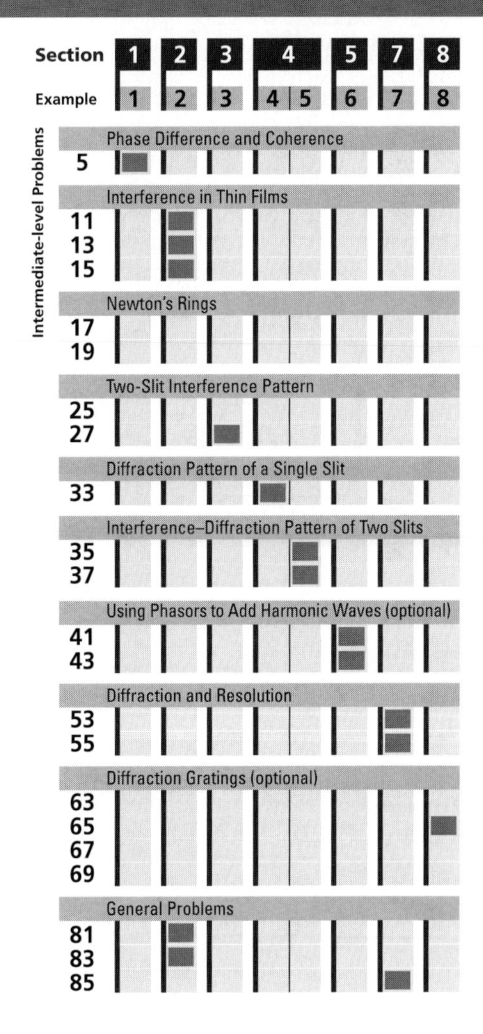

1. It goes to where constructive interference occurs.

3. (*a*) 300 nm; (*b*) 135°

5. 164°

7. The fringes are too close together to be resolved by the eye.

9. (*a*) The top of the film approaches zero thickness, so the phase difference approaches 180°. (*b*) violet; (*c*) The top of the film is white; the color of the first band is red.

11. 5.46 μm $< d <$ 5.75 μm

13. (*a*) 602 nm; (*b*) 401 nm, 516 nm, 722 nm; (*c*) 401 nm, 516 nm, 722 nm

15. 476 nm

17. (*a*) Answer given in the problem. (*b*) Answer given in the problem. (*c*) The transmitted pattern is reversed. (*d*) 68; (*e*) 1.14 cm; (*f*) The fringes will get closer.

19. 534 μm, 926 μm

21. 8.33 fringes/cm

23. (*a*) 9.29 μm; (*b*) 29

25. Answer given in the problem.

27. (*a*) 0.30°; (*b*) 8

29. The equation $d \sin \theta = m\lambda$ describes the angles at which the *interference* between *two slits separated by a distance d* produces maximum intensity. The equation $a \sin \theta = m\lambda$ describes the angles at which the *diffraction of light from a single slit of width a* results in zero intensity.

31. (*a*) 2 cm; (*b*) 20 cm; (*c*) 231 cm

33. 32.2 km

35. (*a*) 20 μm; (*b*) 9

37. 8

39. $3.61 \sin(\omega t - 56.3°)$

41. $0.0162 I_0$

43. (*a*) Answer given in the problem. (*b*) 4 sources: 6.00 mm, 2 sources: 12.0 mm

45. (*a*)

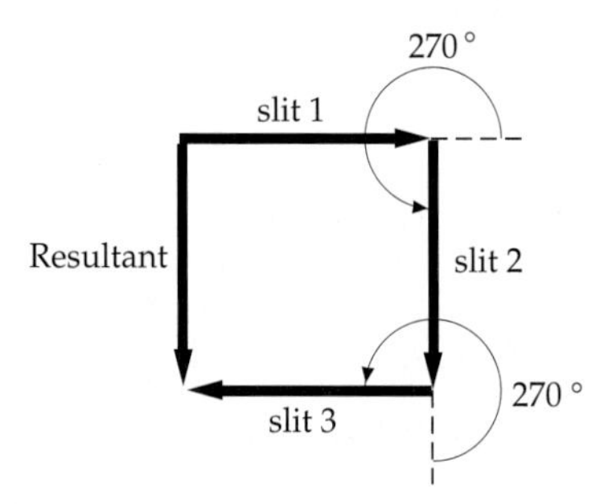

(*b*) 5.56×10^{-3} W/m^2

47. (*a*) $3\pi, 5\pi, 7\pi$; (*b*) $2.86\pi, 4.92\pi, 6.94\pi$

49. 8.54 cm

51. 8.35 km

53. (*a*) 55.6 km; (*b*) 55.6 m

55. 5×10^6 km

57. (*a*)

59. 485 nm, 658 nm

61. 500 nm

63. 2, The second-order long wavelength overlaps the third-order short wavelength.

65. (*a*) 0.353 m, 0.707 m; (*b*) 88.4 μm; (*c*) 8000

67. 3.09×10^5, 5.14×10^4 lines/cm

69. (*a*) 750 lines/cm; (*b*) 4.21 cm, 12.6 cm

71. (*a*) Answer given in the problem. (*b*) 491; (*c*) 0.988 mm

73. Answer given in the problem.

75. 695 nm

77. 1.68 cm

79. 1.30×10^{-4} rad

81. Answer given in the problem.

83. (*a*) 97.8 nm; (*b*) No; (*c*) 0.273, 0.124

85. 12.3 m

87. Answers given in the problem.

89. Answer given in the problem.

91. (*a*) $I_{max} \cos^2[(\pi/2)\sin\theta]$

(*b*)

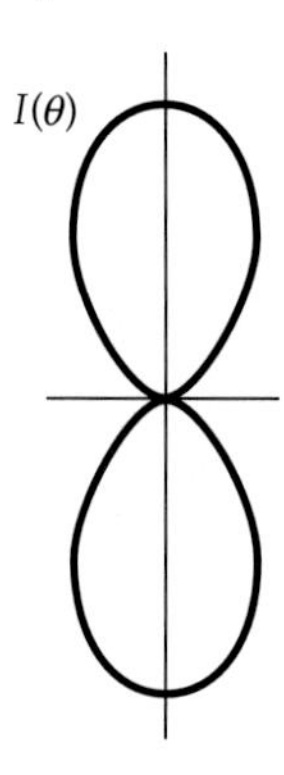

INDEX

Numbers in **bold** indicate additional display material, such as diagrams; *n* indicates a footnote; AP indicates material in the Appendixes.

Pages 1–655 are found in Volume 1; pages 656–1141 are found in Volume 2; pages 1142–1335 are found in Volume 3.

Prefixes for Powers of 10

Multiple	Prefix	Abbreviation
10^{18}	exa	E
10^{15}	peta	P
10^{12}	tera	T
10^{9}	giga	G
10^{6}	mega	M
10^{3}	kilo	k
10^{2}	hecto	h
10^{1}	deka	da
10^{-1}	deci	d
10^{-2}	centi	c
10^{-3}	milli	m
10^{-6}	micro	μ
10^{-9}	nano	n
10^{-12}	pico	p
10^{-15}	femto	f
10^{-18}	atto	a

Some Physical Data

Quantity	Symbol	Value
Acceleration of gravity at earth's surface	g	$9.81\ \mathrm{m/s^2} = 32.2\ \mathrm{ft/s^2}$
Radius of earth	R_E	6370 km = 3960 mi
Mass of earth	M_E	5.98×10^{24} kg
Mass of sun		1.99×10^{30} kg
Mass of moon		7.36×10^{22} kg
Escape speed at earth's surface		11.2 km/s = 6.95 mi/s
Standard temperature and pressure (STP)		0°C = 273.15 K 1 atm = 101.3 kPa
Earth–moon distance		3.84×10^{8} m = 2.39×10^{5} mi
Earth–sun distance (mean)		1.50×10^{11} m = 9.30×10^{7} mi
Speed of sound in dry air (at STP)		331 m/s
Density of air		$1.29\ \mathrm{kg/m^3}$
Density of water		$1000\ \mathrm{kg/m^3}$
Heat of fusion of water	L_f	333.5 kJ/kg
Heat of vaporization of water	L_v	2.257 MJ/kg

The Greek Alphabet

Alpha	A	α	Iota	I	ι	Rho	P	ρ
Beta	B	β	Kappa	K	κ	Sigma	Σ	σ
Gamma	Γ	γ	Lambda	Λ	λ	Tau	T	τ
Delta	Δ	δ	Mu	M	μ	Upsilon	Υ	υ
Epsilon	E	ϵ	Nu	N	ν	Phi	Φ	ϕ
Zeta	Z	ζ	Xi	Ξ	ξ	Chi	X	χ
Eta	H	η	Omicron	O	o	Psi	Ψ	ψ
Theta	Θ	θ	Pi	Π	π	Omega	Ω	ω

Abbreviations for Units

A	ampere
Å	angstrom (10^{-10} m)
atm	atmosphere
Btu	British thermal unit
Bq	becquerel
C	coulomb
°C	degree Celsius
cal	calorie
Ci	curie
cm	centimeter
dyn	dyne
eV	electron volt
°F	degree Fahrenheit
fm	femtometer, fermi (10^{-15} m)
ft	foot
Gm	gigameter (10^{9} m)
G	gauss
Gy	gray
g	gram
H	henry
h	hour
Hz	hertz
in	inch
J	joule
K	kelvin
kg	kilogram
km	kilometer
keV	kilo-electron volt
lb	pound
L	liter
m	meter
MeV	mega-electron volt
Mm	megameter (10^{6} m)
mi	mile
min	minute
mm	millimeter
ms	millisecond
N	newton
nm	nanometer (10^{-9} m)
pt	pint
qt	quart
rev	revolution
R	roentgen
Sv	seivert
s	second
T	tesla
u	unified mass unit
V	volt
W	watt
Wb	weber
y	year
yd	yard
μm	micrometer (10^{-6} m)
μs	microsecond
μC	microcoulomb
Ω	ohm

Some Conversion Factors

$1\ \text{m} = 39.37\ \text{in} = 3.281\ \text{ft} = 1.094\ \text{yd}$

$1\ \text{m} = 10^{15}\ \text{fm} = 10^{10}\ \text{Å} = 10^{9}\ \text{nm}$

$1\ \text{km} = 0.6215\ \text{mi}$

$1\ \text{mi} = 5280\ \text{ft} = 1.609\ \text{km}$

$1\ \text{lightyear} = 1\ c{\cdot}\text{y} = 9.461 \times 10^{15}\ \text{m}$

$1\ \text{in} = 2.540\ \text{cm}$

$1\ \text{L} = 10^{3}\ \text{cm}^{3} = 10^{-3}\ \text{m}^{3} = 1.057\ \text{qt}$

$1\ \text{h} = 3.6\ \text{ks}$

$1\ \text{y} = 365.24\ \text{d} = 3.156 \times 10^{7}\ \text{s}$

$1\ \text{km/h} = 0.278\ \text{m/s} = 0.6215\ \text{mi/h}$

$1\ \text{ft/s} = 0.3048\ \text{m/s} = 0.6818\ \text{mi/h}$

$1\ \text{rev} = 2\pi\ \text{rad} = 360^{\circ}$

$1\ \text{rad} = 57.30^{\circ}$

$1\ \text{rev/min} = 0.1047\ \text{rad/s}$

$1\ \text{slug} = 14.59\ \text{kg}$

$1\ \text{tonne} = 10^{3}\ \text{kg} = 1\ \text{Mg}$

$1\ \text{atm} = 101.3\ \text{kPa} = 1.013\ \text{bar} = 76.00\ \text{cmHg} = 14.70\ \text{lb/in}^{2}$

$1\ \text{N} = 10^{5}\ \text{dyn} = 0.2248\ \text{lb}$

$1\ \text{lb} = 4.448\ \text{N}$

$1\ \text{Pa}{\cdot}\text{s} = 10\ \text{poise}$

$1\ \text{J} = 10^{7}\ \text{erg} = 0.7373\ \text{ft}{\cdot}\text{lb} = 9.869 \times 10^{-3}\ \text{L}{\cdot}\text{atm}$

$1\ \text{kW}{\cdot}\text{h} = 3.6\ \text{MJ}$

$1\ \text{cal} = 4.184\ \text{J} = 4.129 \times 10^{-2}\ \text{L}{\cdot}\text{atm}$

$1\ \text{L}{\cdot}\text{atm} = 101.3\ \text{J} = 24.22\ \text{cal}$

$1\ \text{eV} = 1.602 \times 10^{-19}\ \text{J}$

$1\ \text{Btu} = 778\ \text{ft}{\cdot}\text{lb} = 252\ \text{cal} = 1054\ \text{J}$

$1\ \text{horsepower} = 550\ \text{ft}{\cdot}\text{lb/s} = 746\ \text{W}$

$1\ \text{W/m}{\cdot}\text{K} = 6.938\ \text{Btu}{\cdot}\text{in/h}{\cdot}\text{ft}^{2}{\cdot}{}^{\circ}\text{F}$

$1\ \text{T} = 10^{4}\ \text{G}$

1 kg weighs about 2.205 lb

Some Physical Constants

Avogadro's number	N_A	$6.022\,136\,7 \times 10^{23}$ particles/mol
Boltzmann's constant	k	$1.380\,658 \times 10^{-23}$ J/K
Bohr magneton	$m_B = e\hbar/2m_e$	$9.274\,015\,4 \times 10^{-24}$ J/T
Coulomb constant	$k = 1/4\pi\epsilon_0$	$8.987\,551\,788 \times 10^{9}$ N·m²/C²
Compton wavelength	$\lambda_C = h/2e$	$2.426\,310\,58 \times 10^{-12}$ m
Fundamental charge	e	$1.602\,177\,33 \times 10^{-19}$ C
Gas constant	$R = N_A k$	8.314 51 J/mol·K = 1.987 22 cal/mol·K $= 8.205\,78 \times 10^{-2}$ L·atm/mol·K
Gravitational constant	G	$6.672\,6 \times 10^{-11}$ N·m²/kg²
Mass, of electron	m_e	$9.109\,389\,7 \times 10^{-31}$ kg = 510.999 1 keV/c^2
of proton	m_p	$1.672\,623\,1 \times 10^{-27}$ kg = 938.272 3 MeV/c^2
of neutron	m_n	$1.674\,929 \times 10^{-27}$ kg = 939.565 6 MeV/c^2
Permeability of free space	μ_0	$4\pi \times 10^{-7}$ N/A²
Planck's constant	h	$6.626\,075\,5 \times 10^{-34}$ J·s $= 4.135\,669\,2 \times 10^{-15}$ eV·s
	$\hbar$	$1.054\,572\,66 \times 10^{-34}$ J·s $6.582\,122\,0 \times 10^{-16}$ eV·s
Speed of light	c	$2.997\,924\,58 \times 10^{8}$ m/s
Unified mass unit	u	$1.660\,540 \times 10^{-27}$ kg = 931.494 32 MeV/c^2

Mathematical Symbols

$=$	is equal to	Δx	change in x
$\neq$	is not equal to	$\lvert x \rvert$	absolute value of x
$\approx$	is approximately equal to	$n!$	$n(n-1)(n-2)\cdots 1$
$\sim$	is of the order of	Σ	sum
$\propto$	is proportional to	lim	limit
$>$	is greater than	$\Delta t \to 0$	Δt approaches zero
$\geq$	is greater than or equal to	$\frac{dx}{dt}$	derivative of x with respect to t
$\gg$	is much greater than		
$<$	is less than	$\frac{\partial x}{\partial t}$	partial derivative of x with respect to t
$\leq$	is less than or equal to		
$\ll$	is much less than	$\int$	integral

Chapter 32

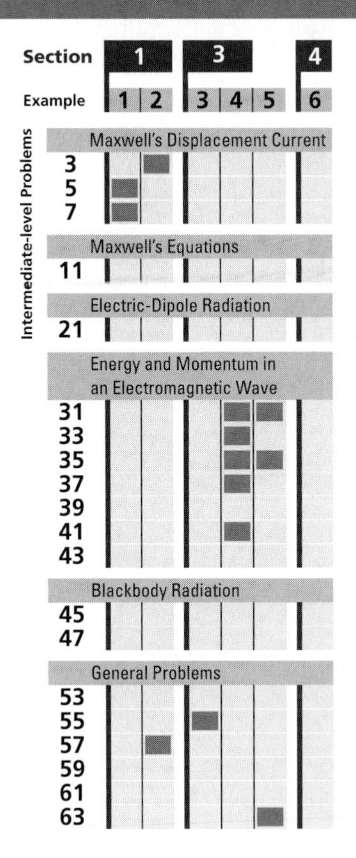

1. (*a*) 3.4×10^{14} V/m·s; (*b*) 5 A

3. Answer given in the problem.

5. (*a*) 10 A; (*b*) 2.26×10^{12} V/m·s; (*c*) 7.89×10^{-7} T·m

7. (*a*) $(A/100\rho d)t$; (*b*) $\kappa\epsilon_0 A/100d$; (*c*) $\kappa\epsilon_0\rho$

9. Answers given in the problem.

11. Answer given in the problem.

13. greater

15. (*a*) 300 m; (*b*) 3 m

17. 3×10^{18} Hz

19. Horizontal and normal to the direction from the antenna to the receiver.

21. (*a*) 30°; (*b*) 7.07 m

23. 4.14×10^{-6} W/m^2, 5.21×10^{17} photons/s·cm^2

25. 1.51×10^{-7} W/m^2

27. (*a*) 283 V/m; (*b*) 943 nT; (*c*) 212 W/m^2; (*d*) 707 nPa

29. Answer given in the problem.

31. (*a*) 4×10^{-8} N; (*b*) 8×10^{-8} N

33. (*a*) 12 V/m, 4×10^{-8} T; (*b*) 1.2 V/m, 4×10^{-9} T; (*c*) 0.12 V/m, 4×10^{-10} T

35. (*a*) 3×10^{-12} N; (*b*) 6×10^{-12} N

37. 7.52×10^4 V/m, 2.51×10^{-4} T

39. (*a*) positive *x* direction; (*b*) 0.628 m, 4.77×10^8 Hz; (*c*) $\vec{E} = (194\ \text{V/m}) \cos[10x - (3 \times 10^9)t]\,\hat{j}$, $\vec{B} = (0.647 \times 10^{-6}\ \text{T}) \times \cos[10x - (3 \times 10^9)t]\hat{k}$

41. 3.42×10^6 W/m^2

43. 7.59×10^{-4} kg/m^2, the ratio does not change

45. (*a*) 1425 W/m^2; (*b*) 907 W/m^2; (*c*) 585 V/m; (*d*) 1.95×10^{-6} T

47. (*a*) 278 K; (*b*) 245 K

49. Answer given in the problem.

51. (*a*) False; (*b*) True; (*c*) True; (*d*) True; (*e*) False; (*f*) True

53. The output of the receiver varies as cos θ, where θ is the angle between the normal to the plane of the loop and a line directed toward the source.

55. 7.25×10^{-3} V

57. (*a*) $V_0[(1/R) \sin \omega t + (\epsilon_0 \pi a^2 \omega/d) \cos \omega t]$; (*b*) $(\mu_0/2\pi)[(V_0/aR) \sin \omega t + (\epsilon_0\omega\pi V_0 a/d) \cos \omega t]$; (*c*) $\tan \delta = \epsilon_0 \pi a^2 \omega R/d$

59. Answer given in the problem.

61. 111 m^2

63. 247 mW

65. (*a*) $-\mu_0 nar/2$, tangent to the circle of radius *r* and opposite the sense of the current; (*b*) $\mu_0 n^2 a^2 Rt/2$, points inward toward the axis of the solenoid; (*c*) $\mu_0 \pi R^2 n^2 a^2 Lt$

67. Answers given in the problem.

69. Answers given in the problem.

Chapter 33

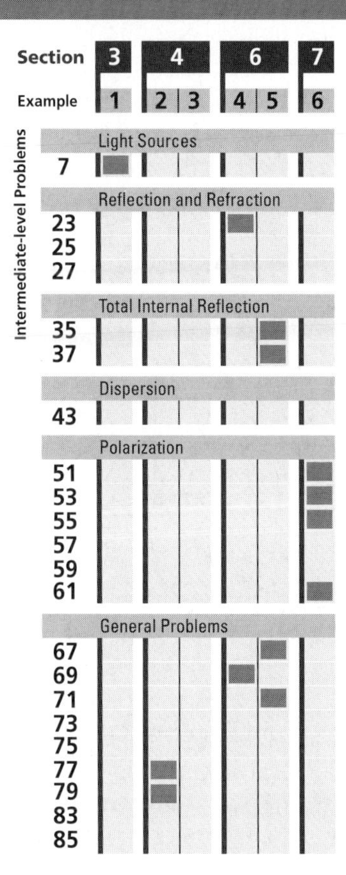

1. Helium atoms are needed to obtain the inverted energy population in neon atoms to produce a laser beam.

3. (*a*) 0.015 J; (*b*) 5.24×10^{16} photons

5. (*a*) 435 nm; (*b*) 1210 nm

7. (*a*) 1140 nm; (*b*) 1140 nm, 285 nm, 554 nm, 588 nm

9. 20 μs

63. (*a*) Answer given in the problem.
(*b*)

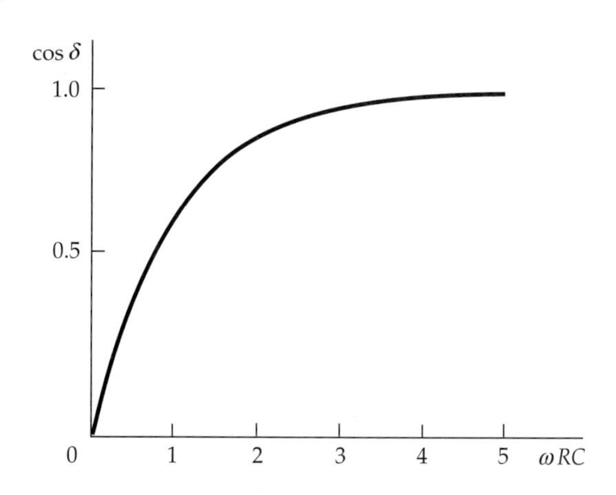

65. (*a*) 450 Hz; (*b*) 0.141; (*c*) 3.25 kHz, 62.4 Hz

67. (*a*) 933 W; (*b*) 7.71 Ω; (*c*) 99.9 μF; (*d*) 40.9 μF, 20.5 mH

69. (*a*) 120 Hz; (*b*) 356 Hz, 2.96; (*c*) 40.7 Hz, 0.338

71. (*a*) 15 W; (*b*) 15 Ω; (*c*) 0.235 H

73. (*a*) 0.8 mH, 12.5 μF; (*b*) 1.6; (*c*) 2 A

75. (*a*) 12 Ω; (*b*) 7.2 Ω of *R*, 9.6 Ω of *X*; (*c*) capacitive

77. Answers given in the problem.

79. (*a*) 40 Ω; (*b*) 41.1 Ω; (*c*) closed: 0.75ℰ, open: 0.73ℰ; (*d*) 23.7 Ω, 22.7 Ω, closed: 0.701ℰ, open: 0.028ℰ (*e*) open S

81. Assuming the resonance to be sharp, $Q = 1/\omega RC = \omega L/R$

83. 5.66 kHz, 5.66, 1.14 kΩ

85. (*a*) 13.3 MHz; (*b*) 0.5 A, $V_L = 1.5$ kV, $V_C = 1.5$ kV; (*c*) 0.354 A, $V_L = 1.07$ kV, $V_C = 1.05$ kV

87. (*a*), (*b*) Answer given in the problem.
(*c*)

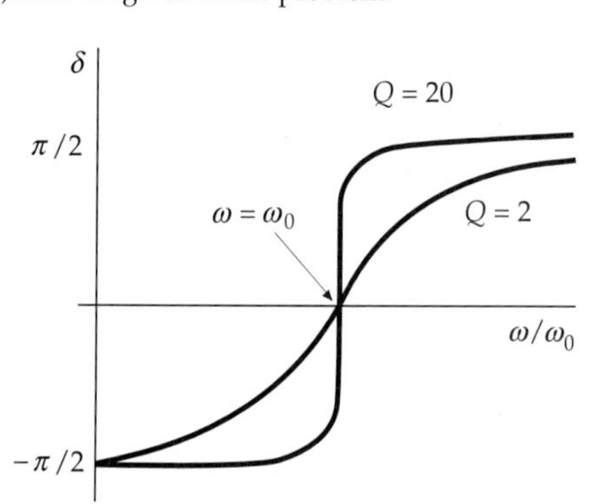

89. Answers given in the problem.

91. Answer given in the problem.

93. (*a*) 3.55×10^{-5} H; (*b*) 1.98×10^{-11} F; (*c*) 3.67×10^{-4}

95. (*a*) 19.8 MHz; (*b*) 1.10 mm

97. 7.96 kHz

99. (*a*) 1975 rad/s; (*b*) 16.9 Ω, 23.7 Ω;
(*c*)

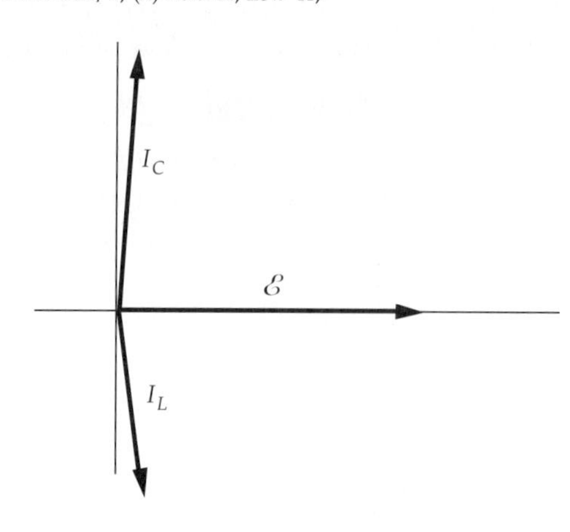

(*d*) 45.3 Ω

101. True

103. (*a*) 5; (*b*) 50 A

105. (*a*) 1.5 A; (*b*) 19

107. 3,333 turns

109. Answer given in the problem.

111. (*a*) 20.8 A; (*b*) 29.5 A; (*c*) 41.7 A, 58.9 A

113.

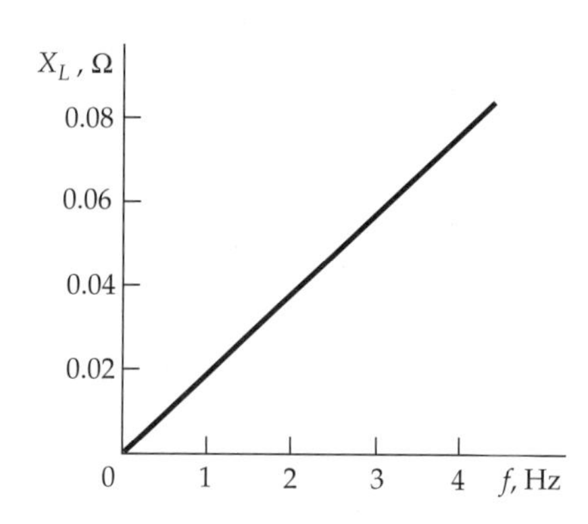

115. (*a*) 6.08 A; (*b*) 444 W; (*c*) 73 V

117. (*a*) 4.74 A; (*b*) 150 W

119. (*a*) 2.00 A, 2.31 A; (*b*) 2.00 A, 2.83 A

121. 45.2 mA, −45.2 mA, 0, 32.0 mA

Chapter 31

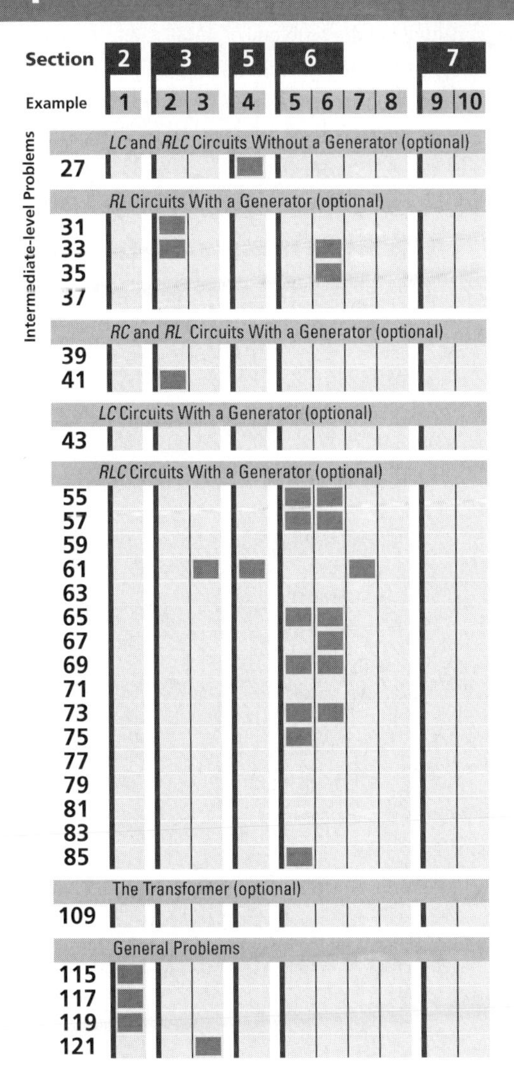

1. (*a*) 39.8 Hz; (*b*) 15.1 V

3. (*a*) 13.6 V; (*b*) 486 Hz

5. (*b*)

7. (*a*) 0.833 A; (*b*) 1.18 A; (*c*) 200 W

9. (*a*) 21.2 A; (*b*) 1.8 kW

11. (*a*)

13. Yes, yes

15. (*a*) 0.377 Ω; (*b*) 3.77 Ω; (*c*) 37.7 Ω

17. 1.59 kHz

19. (*a*) 0.025 A; (*b*) 0.0178 A

21.

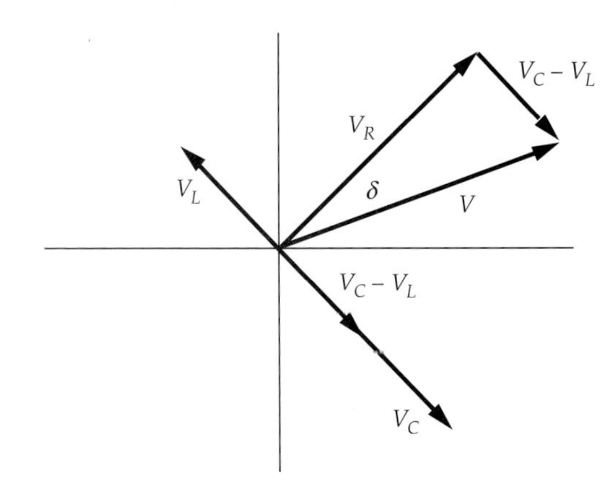

The net emf V lags the current I, which is in the same direction as V_R, by the angle δ. In the diagram, it is clear that $\tan\delta = (V_C - V_L)/V$.

23. (*a*)

25. Answer given in the problem.

27. Answer given in the problem. (*b*) third circuit

29. (*a*) 0.553; (*b*) 0.663 A; (*c*) 44 W

31. 29.2 mH

33. (*a*) 0.333; (*b*) 26.7 Ω; (*c*) 0.2 H; (*d*) lags, 70.5°

35. 0.397

37. (*a*) 8.81 A, 7.87 A, 3.95 A; (*b*) 4.64 A, 2.07 A, 4.15 A; (*c*) 0.502, 0.800

39. 60 V

41. (*a*) 6 Ω; (*b*) 35.5 mH

43. (*a*) $I_C = \omega CV$ leads voltage by 90°, $I_L = V/\omega L$ lags voltage by 90°; (*b*) 100 rad/s; (*c*) 0.25 A, 0.25 A;
(*d*)

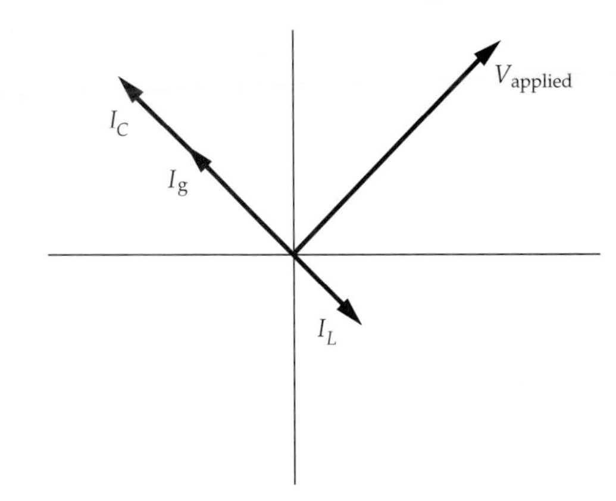

45. 1.24×10^9 N/m^2

47. (*a*) True; (*b*) True; (*c*) True

49. Yes, the freqency needs to be very precisely tuned to get a good signal.

51. 9.89×10^{-9} F to 1.01×10^{-7} F

53. (*a*) 0.539; (*b*) 95.3 mA; (*c*) 0.727 W

55. (*a*) 7.07×10^3 rad/s; (*b*) 14.1 A; (*c*) 62.5 Ω, 80 Ω; (*d*) 18.2 Ω, 3.86 A; (*e*) 74.1°

57. (*a*) 0.155, −81.1°; (*b*) 0.839, −33.0°; (*c*) 0.237, 76.3°

59. 2002

61. (*a*) 18.8 μF, 0.531 A; (*b*) 25 V

27. (a)

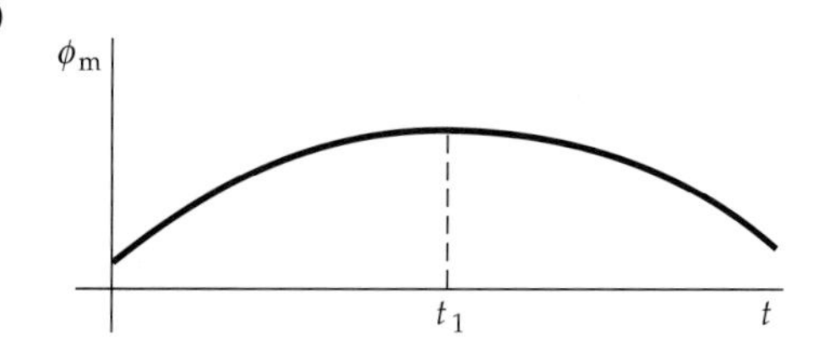

(b)

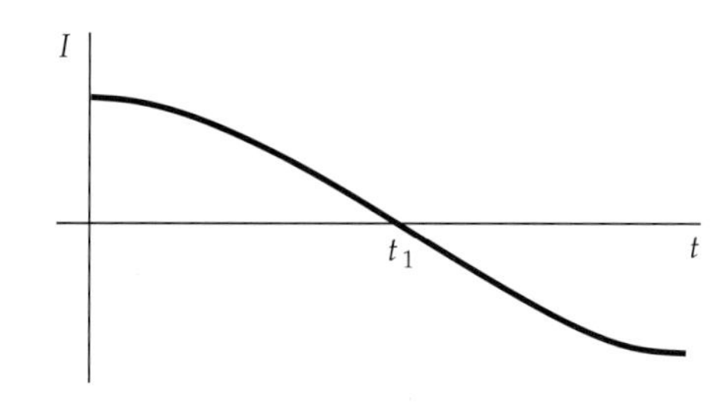

29. (a) 6.4×10^{-20} N; (b) 0.4 V/m; (c) 0.12 V

31. (a) 1.6 V; (b) 0.8 A; (c) 0.128 N; (d) 1.28 W; (e) 1.28 W

33. (a)

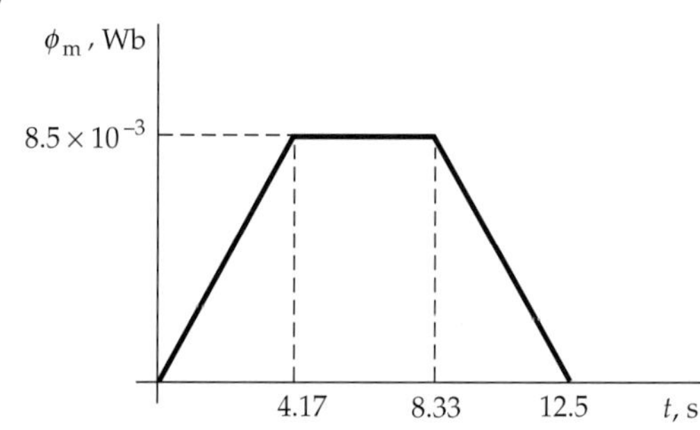

(b)

$\mathcal{E}$, V

2.04×10^{-3}

4.17

8.33

12.5

t, s

-2.04×10^{-3}

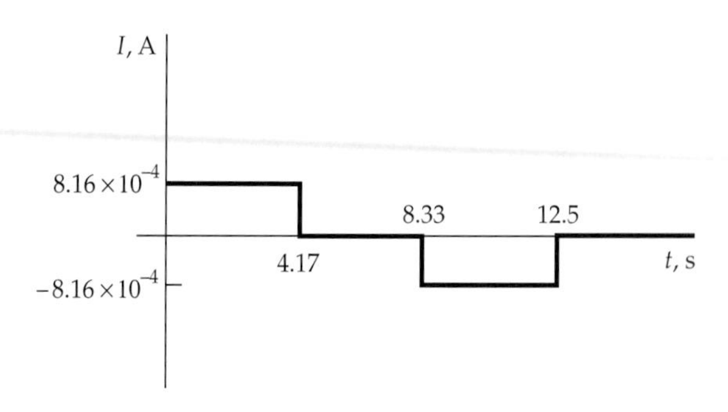

35. (a) $[(\mathcal{E} - B\ell v)\ell B]/R = ma$; (b) $|v| = \mathcal{E}/B\ell$; (c) 0

37. $v_0 mR/B^2\ell^2$

39. Answer given in the problem.

41. (a) 1.38×10^{-4} V/m; (b) 5.51×10^{-6} V

43. $m^2g^2R \sin^2\theta/B^2\ell^2 \cos^2\theta$

45. Answers given in the problem.

47 Answers given in the problem.

49. (a) 24 Wb; (b) -1600 V

51. (a) 6.03×10^{-3} T; (b) 7.58×10^{-4} Wb; (c) 253 μH; (d) $|\mathcal{E}| = 37.9$ mV

53. 162 Ω, 0

55. 12 mH

57. (c)

59. (a) 4.43×10^{-4} J; (b) 9.95×10^4 J; (c) 9.95×10^4 J

61. (a) 0.0536 J; (b) 447 J/m^3; (c) 0.0335 T; (d) 447 J/m^3

63. (a) 7958; (b) 2.55 T; (c) 2.58×10^6 J/m^3, 5093 J

65. (a) 13.5 mA; (b) 7.44×10^{-44} A

67. (a) 47.7 W, 48.0 W; (b) 47.4 W, 48.0 W; (c) 0.321 W, 0

69. (a) 2.30; (b) 4.61; (c) 6.91

71. (a) 2.41 s; (b) 20.1 s

73. Answer given in the problem.

75. 20 Ω, 4 H

77. 2.31×10^{-4} s

79. (a)

81. In the metal tube, the moving magnet creates eddy currents, which produce a retarding force.

83. (a) 7.07×10^{-3} V; (b) 6.64×10^{-3} V

85. (a) Answer given in the problem; (b) 275 rad/s

87. Answer given in the problem.

89. $(0.35 \sin 2t)$ A

91. Answer given in the problem.

93. (a) $-\frac{1}{2}\mu_0 nrI_0\omega \cos \omega t$; (b) $-\frac{1}{2}\mu_0 n(R^2/r)I_0\omega \cos \omega t$. In each case, the electric field is tangent to the circle of radius r. The minus sign indicates the direction of $\vec{E}$ relative to the current.

95. (a) $\mu_0 I/2\pi r$, no current is enclosed for $r < r_1$ and $r > r_2$; (b)–(d) Answer given in the problem

97. (a) $0.600v$ A, counterclockwise; (b) $-0.180v$ N, up ($+y$ is down); (c) $(4.90 - 0.180v)$ N; (d) $a_y = \frac{-1}{2.78}(v - 27.2)$ m/s^2; (e) $v = 27.2(1 - e^{-t/2.78})$ m/s; (f) $y = 27.2[t - 2.78(1 - e^{-t/2.78})]$m; (g) 0.553 s; (h) 4.91 m/s; (i) 5.24 m/s

99. Answer given in the problem.

tablished. The time for equilibrium to be established depends on the conductivity of the elements in the circuit, but is practically instantaneous for most purposes. In equilibrium, charge no longer accumulates at points along the circuit and the current is steady. (For circuits containing capacitors, the current may increase or decrease slowly, but appreciable changes occur only over times much longer than the time needed to reach the steady state.)

26-1 Current and the Motion of Charges

Electric **current** is defined as the rate of flow of electric charge through a cross-sectional area. Figure 26-1 shows a segment of a current-carrying wire in which charge carriers are moving. If ΔQ is the charge that flows through the cross-sectional area A in time Δt, the current I is

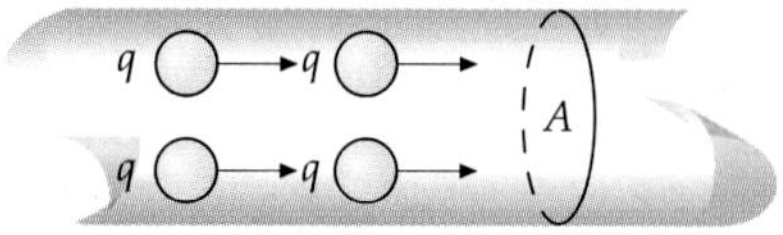

$$I = \frac{\Delta Q}{\Delta t} \qquad \text{26-1}$$

Definition—Electric current

Figure 26-1 A segment of a current-carrying wire. If ΔQ is the amount of charge that flows through the cross-sectional area A in time Δt, the current is $I = \Delta Q/\Delta t$.

The SI unit of current is the **ampere** (A):

$$1\text{ A} = 1\text{ C/s} \qquad \text{26-2}$$

By convention, the direction of current is considered to be the direction of flow of positive charge. This convention was established before it was known that free electrons are the particles that actually move in a conducting wire. Thus, electrons move in the direction *opposite* to the direction of the current. (In an accelerator that produces a proton beam, the direction of motion of the positively charged protons is in the direction of the current.)

In a conducting wire, the motion of negatively charged free electrons is quite complex. When there is no electric field in the wire, the free electrons move in random directions with relatively large speeds of the order of 10^6 m/s.* Since the velocity vectors of the electrons are randomly oriented, the *average* velocity is zero. When an electric field is applied, a free electron experiences an acceleration due to the force $-e\vec{E}$, and acquires an additional velocity in the direction opposite the field. However, the kinetic energy acquired is quickly dissipated by collisions with the lattice ions in the wire. The electron is then accelerated again by the field. The net result of this repeated acceleration and dissipation of energy is that the electron has a small average velocity called its **drift velocity** opposite to the electric field.

The motion of the free electrons in a metal is similar to that of the molecules of a gas, such as air. In still air, the molecules move with large instantaneous velocities (due to their thermal energy) between collisions, but the average velocity is zero. When there is a breeze, the air molecules have a small drift velocity in the direction of the breeze superimposed on their much larger instantaneous velocities. Similarly, when there is no applied electric field, the "electron gas" in a metal has a zero average velocity, but when there is an applied electric field, the electron gas acquires a small drift velocity.

* The average energy of the free electrons in a metal is quite large even at very low temperatures. These electrons do not have the classical Maxwell–Boltzmann energy distribution and do not obey the classical equipartition theorem. We discuss the energy distribution of these electrons and calculate their average speed in Chapter 27.

Let n be the number of free charge-carrying particles per unit volume in a conducting wire of cross-sectional area A. We call n the **number density** of charge carriers. Assume that each particle carries a charge q and moves with a drift velocity v_d. In a time Δt, all the particles in the volume $Av_d\,\Delta t$, shown in Figure 26-2 as a shaded area, pass through the area element. The number of particles in this volume is $nAv_d\,\Delta t$, and the total charge is

$$\Delta Q = qnAv_d\,\Delta t$$

The current is thus

$$I = \frac{\Delta Q}{\Delta t} = qnAv_d \qquad \text{26-3}$$

Relation between current and drift velocity

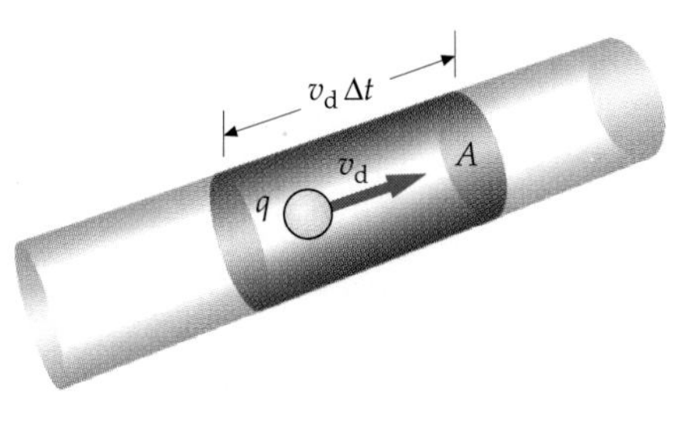

Figure 26-2 In time Δt, all the charges in the shaded volume pass through A. If there are n charge carriers per unit volume, each with charge q, the total charge in this volume is $\Delta Q = qnAv_d\,\Delta t$, where v_d is the drift velocity of the charge carriers.

Equation 26-3 can be used to find the current due to the flow of any type of charged particle, simply by substituting the velocity of the particle for the drift velocity v_d.

The number of charge carriers in a conductor can be measured by the Hall effect, which is discussed in Chapter 28. The result is that, in most metals, there is about one free electron per atom.

Example 26-1

A typical wire for laboratory experiments is made of copper and has a radius 0.815 mm. Calculate the drift velocity of electrons in such a wire carrying a current of 1 A, assuming one free electron per atom.

Picture the Problem Equation 26-3 relates the drift velocity to the number density of charge carriers, which equals the number density of copper atoms n_a. We can find n_a from the mass density of copper, its molecular mass, and Avogadro's number.

1. The drift velocity is related to the current and number density of charge carriers:
$$v_d = \frac{I}{nqA}$$

2. If there is one free electron per atom, the number density of free electrons equals the number density of atoms n_a:
$$n = n_a$$

3. The number density of atoms n_a is related to the mass density ρ_m. Avogadro's number N_A, and the molar mass M. For copper, $\rho = 8.93\ \text{g/cm}^3$ and $M = 63.5\ \text{g/mol}$:
$$n_a = \frac{\rho_m N_A}{M} = \frac{(8.93\ \text{g/cm}^3)(6.02 \times 10^{23}\ \text{atoms/mol})}{63.5\ \text{g/mol}}$$
$$= 8.47 \times 10^{22}\ \text{atoms/cm}^3 = 8.47 \times 10^{28}\ \text{atoms/m}^{-3}$$

4. The magnitude of the charge is e, and the area is related to the radius r of the wire:
$$q = e$$
$$A = \pi r^2$$

5. Substituting numerical values yields v_d:
$$v_d = \frac{I}{nqA} = \frac{I}{n_a e \pi r^2}$$
$$= \frac{1\ \text{C/s}}{(8.47 \times 10^{28}\ \text{m}^{-3})(1.6 \times 10^{-19}\ \text{C})\pi(0.000815\ \text{m})^2}$$
$$= 3.54 \times 10^{-5}\ \text{m/s}$$

Remark Typical drift velocities are of the order of a few hundredths of a millimeter per second, quite small by macroscopic standards.

Exercise How long would it take for an electron to drift from your car battery to the starter motor, a distance of about 1 m, if its drift velocity is 3.5×10^{-5} m/s? (*Answer* 7.9 h)

If electrons drift down a wire at such low speeds, why does an electric light come on instantly when the switch is thrown? An analogy with water in a hose may prove useful. When you turn on a faucet and water rushes into a long, empty hose, it takes quite a few seconds for the water to travel from the faucet to the nozzle. However, if the hose is already full of water, the water emerges from the nozzle almost instantaneously. Because of the water pressure at the faucet, the segment of water near the faucet pushes on the water immediately next to it, which pushes on the next segment of water and so on, until the last segment of water is pushed out the nozzle. This pressure wave moves down the hose at the speed of sound in water, and the water quickly reaches a steady flow rate.

Similarly, when a light is switched on, an electric field propagates down the wire with nearly the speed of light, and the free electrons throughout the wire acquire their drift velocity almost immediately. The charge that flows out of a segment of wire is replaced by an equal amount of charge that flows into the segment at the other end. Thus, charge starts moving through the filament almost immediately after the light switch is thrown. The transport of a significant amount of charge in a wire is accomplished not by a few charges moving rapidly down the wire, but by a very large number of charges slowly drifting down the wire.

Example 26-2

In a certain particle accelerator, a current of 0.5 mA is carried by a 5-MeV proton beam that has a radius of 1.5 mm. (*a*) Find the number density of protons in the beam. (*b*) If the beam hits a target, how many protons hit the target in 1 s?

(*a*)1. The number density is related to the current, charge, cross-sectional area, and speed:

$$n = \frac{I}{qAv}$$

2. We find the speed of the protons from their kinetic energy:

$$K = \tfrac{1}{2}mv^2 = 5\ \text{MeV}$$

$$= 5 \times 10^6\ \text{eV} \times \frac{1.6 \times 10^{-19}\ \text{J}}{1\ \text{eV}}$$

$$= 8 \times 10^{-13}\ \text{J}$$

3. Use $m = 1.67 \times 10^{-27}$ kg for the mass of a proton, and solve for the speed:

$$v = \sqrt{\frac{2K}{m}} = \sqrt{\frac{(2)(8 \times 10^{-13}\ \text{J})}{1.67 \times 10^{-27}\ \text{kg}}} = 3.10 \times 10^7\ \text{m/s}$$

4. Substitute to calculate n:

$$n = \frac{I}{qAv}$$

$$= \frac{0.5 \times 10^{-3}\ \text{A}}{(1.6 \times 10^{-19}\ \text{C/proton})\pi(1.5 \times 10^{-3}\ \text{m})^2(3.10 \times 10^7\ \text{m/s})}$$

$$= 1.43 \times 10^{13}\ \text{proton/m}^3$$

(b)1. The number of protons that hit the target in 1 s is related to the total charge ΔQ that hits in 1 s and the proton charge q:

$$N = \frac{\Delta Q}{q}$$

2. The charge ΔQ that strikes the target in some time Δt is the current times the time:

$$\Delta Q = I\,\Delta t = (0.5\text{ mA})(1\text{ s}) = 0.5\text{ mC}$$

3. The number of protons is then:

$$N = \frac{\Delta Q}{q} = \frac{0.5 \times 10^{-3}\text{ C}}{1.6 \times 10^{-19}\text{ C/proton}} = 3.13 \times 10^{15}\text{ protons}$$

Check the Result The number of protons hitting the target in time Δt is also the number in the volume $Av\,\Delta t$. Then $N = nAv\,\Delta t$. Substituting $n = (I/qAv)$ then gives $N = nAv\,\Delta t = (I/qAv)(Av)\Delta t = I\,\Delta t/q = \Delta Q/q$, which is what we used in part (b).

Remarks We were able to use the classical expression for kinetic energy in step 2 without taking relativity into consideration because the proton kinetic energy of 5 MeV is much less than the proton rest energy (about 931 MeV). The speed found, 3.1×10^7 m/s, is about one-tenth the speed of light.

26-2 Resistance and Ohm's Law

Current in a conductor is driven by an electric field $\vec{E}$ inside the conductor that exerts a force $q\vec{E}$ on the free charges. (In electrostatic equilibrium, the electric field must be zero inside a conductor, but when a conductor carries a current, it is no longer in electrostatic equilibrium and the free charge drifts down the conductor, driven by the electric field.) Since $\vec{E}$ is the direction of the force on a positive charge, it is in the direction of the current.

Figure 26-3 shows a wire segment of length ΔL and cross-sectional area A carrying a current I. Since the electric field points in the direction of decreasing potential, the potential at point a is greater than that at point b. Assuming that ΔL is small enough so that we may consider the electric field E to be constant across the segment, the potential difference V between points a and b is

$$V = V_a - V_b = E\,\Delta L \qquad \text{26-4}$$

Again we use V rather than ΔV for the potential difference (which in this case is a potential *decrease*) to simplify the notation. The ratio of the potential drop to the current is called the **resistance** of the segment.

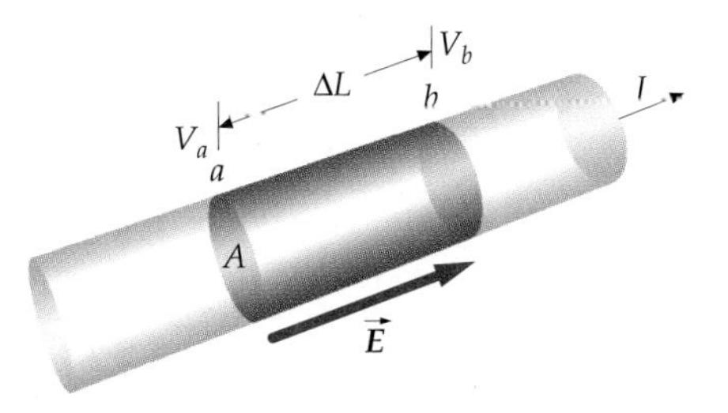

Figure 26-3 A segment of wire carrying a current I. The potential difference is related to the electric field by $V_a - V_b = E\,\Delta L$.

$$R = \frac{V}{I} \qquad \text{26-5}$$

Definition—Resistance

The SI unit of resistance, the volt per ampere, is called an ohm (Ω):

$$1\,\Omega = 1\text{ V/A} \qquad \text{26-6}$$

For many materials, the resistance does not depend on the voltage drop or the current. Such materials, which include most metals, are called **ohmic materials.** For ohmic materials, the potential drop across a segment is proportional to the current:

$$V = IR, \qquad R\text{ constant} \qquad \text{26-7}$$

Ohm's law

For **nonohmic materials,** the resistance depends on the current I, so V is not proportional to I. Figure 26-4 shows the potential difference V versus the current I for ohmic and nonohmic materials. For ohmic materials (Figure 26-4*a*), the relation is linear, but for nonohmic materials (Figure 26-4*b*), the relation is not linear. Ohm's law is not a fundamental law of nature, like Newton's laws or the laws of thermodynamics, but rather is an empirical description of a property shared by many materials.

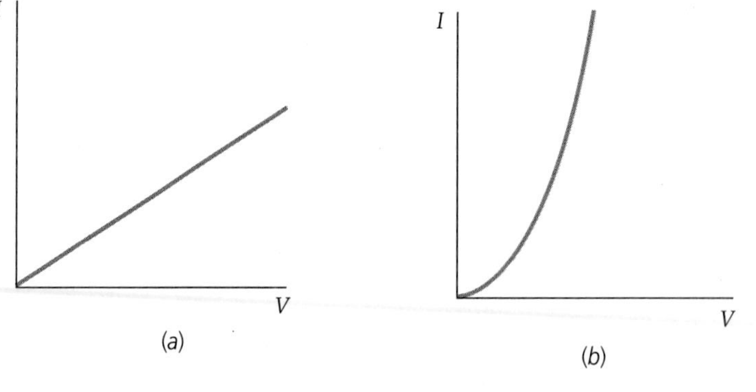

Figure 26-4 Plots of V versus I for (*a*) ohmic and (*b*) nonohmic materials. The resistance $R = V/I$ is independent of I for ohmic materials, as is indicated by the constant slope of the line in (*a*).

Exercise A wire of resistance 3 Ω carries a current of 1.5 A. What is the potential drop across the wire? (*Answer* 4.5 V)

The resistance of a conducting wire is found to be proportional to the length of the wire and inversely proportional to its cross-sectional area:

$$R = \rho\frac{L}{A} \qquad \text{26-8}$$

where the proportionality constant ρ is called the **resistivity** of the conducting material.* The unit of resistivity is the ohm-meter (Ω·m). Note that Equations 26-7 and 26-8 for electrical conduction and electrical resistance are of the same form as Equations 21-9 ($\Delta T = IR$) and 21-10 ($R = \Delta x/kA$) for thermal conduction and thermal resistance. For electrical equations, the potential difference V replaces the temperature difference ΔT and $1/\rho$ replaces the thermal conductivity k. Ohm was, in fact, led to his law by the similarity between the conduction of electricity and the conduction of heat.

Example 26-3

A Nichrome wire ($\rho = 10^{-6}$ Ω·m) has a radius of 0.65 mm. What length of wire is needed to obtain a resistance of 2.0 Ω?

Solve $R = \rho L/A$ (Equation 26-8) for L:

$$L = \frac{RA}{\rho} = \frac{(2\ \Omega)\pi(0.00065\ \text{m})^2}{10^{-6}\ \Omega\cdot\text{m}} = 2.65\ \text{m}$$

The resistivity of any given metal depends on the temperature. Figure 26-5 shows the temperature dependence of the resistivity of copper. This graph is nearly a straight line, which means that the resistivity varies nearly linearly with temperature.† In tables, the resistivity is usually given in terms of its value at 20°C, ρ_{20}, along with the **temperature coefficient of resistivity**, α, which is defined by

$$\rho = \rho_{20}[1 + \alpha\,(t_c - 20\ \text{C}°)] \qquad \text{26-9}$$

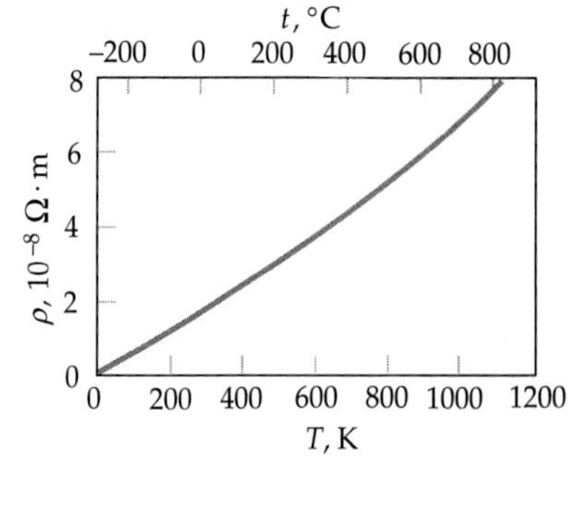

Figure 26-5 Plot of resistivity ρ versus temperature for copper. Since the Celsius and absolute temperatures differ only in the choice of zero, the resistivity has the same slope whether it is plotted against t or T.

* The symbol ρ used here for the resistivity was used in previous chapters for volume charge density. Care must be taken to distinguish which quantity ρ refers to. Usually this will be clear from the context.

† There is a breakdown in this linearity for all metals at very low temperatures that is not shown in Figure 26-5.

Table 26-1 gives the resistivity at 20°C and the temperature coefficient α for various materials. Note the tremendous range of values for ρ.

Electrical wires are manufactured in standard sizes. The diameter of the circular cross section is indicated by a *gauge number,* with higher numbers corresponding to smaller diameters, as can be seen from Table 26-2. Handbooks typically give R/L in ohms per centimeter or ohms per foot.

Table 26-1

Resistivities and Temperature Coefficients

Material	Resistivity ρ at 20°C, $\Omega\cdot$m	Temperature Coefficient α at 20°C, K^{-1}
Silver	1.6×10^{-8}	3.8×10^{-3}
Copper	1.7×10^{-8}	3.9×10^{-3}
Aluminum	2.8×10^{-8}	3.9×10^{-3}
Tungsten	5.5×10^{-8}	4.5×10^{-3}
Iron	10×10^{-8}	5.0×10^{-3}
Lead	22×10^{-8}	4.3×10^{-3}
Mercury	96×10^{-8}	0.9×10^{-3}
Nichrome	100×10^{-8}	0.4×10^{-3}
Carbon	3500×10^{-8}	-0.5×10^{-3}
Germanium	0.45	-4.8×10^{-2}
Silicon	640	-7.5×10^{-2}
Wood	10^{8}–10^{14}	
Glass	10^{10}–10^{14}	
Hard rubber	10^{13}–10^{16}	
Amber	5×10^{14}	
Sulfur	1×10^{15}	

Table 26-2

Wire Diameters and Cross-Sectional Areas for Commonly Used Copper Wires

Gauge Number	Diameter at 20°C, mm	Area, mm^2
4	5.189	21.15
6	4.115	13.30
8	3.264	8.366
10	2.588	5.261
12	2.053	3.309
14	1.628	2.081
16	1.291	1.309
18	1.024	0.8235
20	0.8118	0.5176
22	0.6438	0.3255

Example 26-4

Calculate the resistance per unit length of a 14-gauge copper wire.

1. From Equation 26-8, the resistance per unit length equals the resistivity per unit area:
$$\frac{R}{L} = \frac{\rho}{A}$$

2. Find the resistivity of copper from Table 26-1 and the area from Table 26-2:
$$\rho = 1.7 \times 10^{-8}\,\Omega\cdot\text{m}$$
$$A = 2.08\ \text{mm}^2$$

3. Use these values to find R/L:
$$\frac{R}{L} = \frac{\rho}{A} = \frac{1.7 \times 10^{-8}\,\Omega\cdot\text{m}}{2.08 \times 10^{-6}\,\text{m}^2} = 8.17 \times 10^{-3}\,\Omega/\text{m}$$

Remark 14-gauge copper wire is commonly used for low-current circuits. As this example shows, it has a very small resistance.

Carbon, which has a relatively high resistivity, is used in resistors found in electronic equipment. Resistors are often marked with colored stripes that indicate their resistance value.

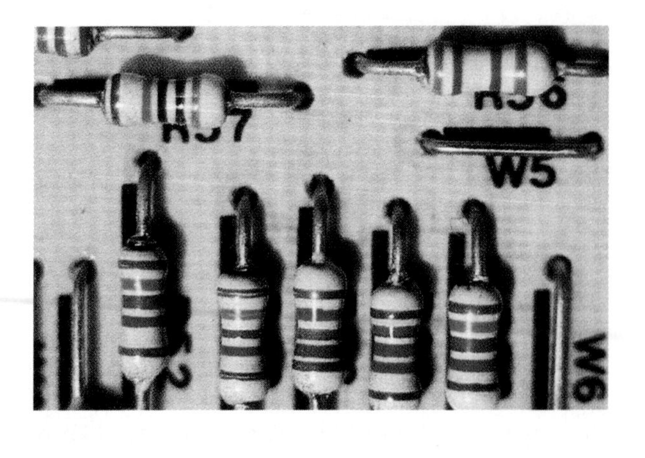

Color-coded carbon resistors on a circuit board.

Example 26-5

Assuming that an electric field E is uniform, find its magnitude in the 14-gauge copper wire of Example 26-4 when the wire is carrying a current of 1.3 A.

Picture the Problem We find the electric field as the voltage drop for a given length of wire, $E = V/L$. The voltage drop is found using Ohm's law, $V = IR$, and the resistance per length is given in Example 26-4.

1. The electric field equals the voltage drop per unit length: $E = \frac{V}{L}$

2. Write Ohm's law for the voltage drop: $V = IR$

3. Substitute this expression into the equation for E: $E = \frac{V}{L} = \frac{IR}{L} = I\frac{R}{L}$

4. Substitute the value of R/L found in Example 26-4 to calculate E: $E = I\frac{R}{L} = (1.3\text{ A})(8.17 \times 10^{-3}\ \Omega/\text{m}) = 1.06 \times 10^{-2}\text{ V/m}$

26-3 Energy in Electric Circuits

When there is an electric field in a conductor, the free electrons are accelerated for a short time, giving the "electron gas" increased kinetic energy; but this additional energy is quickly converted into thermal energy of the conductor by collisions between the electrons and the lattice ions of the conductor. The increase in thermal energy in a conductor is called **Joule heat.**

Consider the segment of wire of length ΔL and cross-sectional area A shown in Figure 26-6. In time Δt, charge ΔQ enters from the left at potential V_1 and the same amount of charge exits from the right at potential V_2. The effect is just as if the same charge ΔQ entered the segment at a high potential V_1 and left it at a low potential V_2. The change in potential energy of the charge passing though the segment is

$$\Delta U = \Delta Q(V_2 - V_1) = \Delta Q(-V) = -(\Delta Q)V$$

where $V = V_1 - V_2$ is the potential drop across the segment. The potential energy lost in this segment of the wire is thus

$$-\Delta U = (\Delta Q)V$$

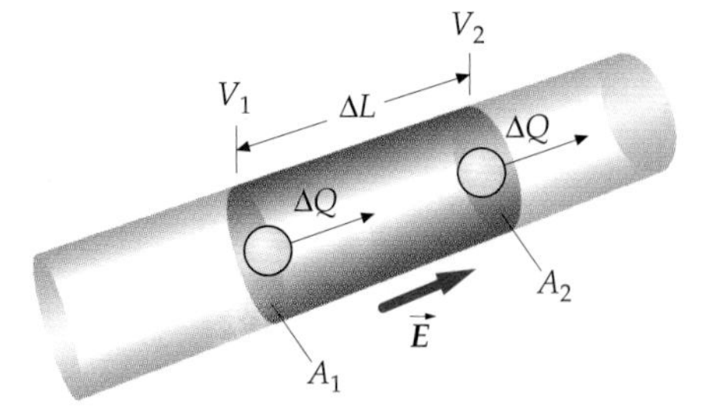

Figure 26-6 During a time Δt, an amount of charge ΔQ passes through area A_1, where the potential is V_1. During the same time interval, an equal amount of charge leaves the segment, passing through area A_2, where the potential is V_2. The effect is just as if the same charge ΔQ entered the segment at a high potential V_1 and left it at a low potential V_2, thereby losing potential energy in the segment.

The rate of energy loss is

$$-\frac{\Delta U}{\Delta t} = \frac{\Delta Q}{\Delta t}V = IV$$

where $I = \Delta Q/\Delta t$ is the current. The energy loss per unit time is the power P dissipated in the conducting segment:

$$P = VI \qquad \text{26-10}$$

Power dissipated in a conductor

If V is in volts and I is in amperes, the power is in watts. The power loss is the product of the decrease in potential energy per unit charge, V, and the charge flowing per unit time, I. Equation 26-10 applies to any device in a circuit. The power delivered to the device is the product of the potential drop and the current. In a conductor, this power goes into thermal energy in the conductor. Using $V = IR$, or $I = V/R$, we can write Equation 26-10 in other useful forms

$$P = VI = I^2R = \frac{V^2}{R} \qquad \text{26-11}$$

Power dissipated in a resistor

Example 26-6

A 12-Ω resistor carries a current of 3 A. Find the power dissipated in this resistor.

Picture the Problem Since we are given the current and the resistance, but not the potential drop, $P = I^2R$ is the most convenient equation to use. Alternatively, we could find the potential drop from $V = IR$, then use $P = IV$.

1. Compute I^2R: $P = I^2R = (3\text{ A})^2(12\ \Omega) = 108\text{ W}$

Check the Result The potential drop across the resistor is $V = IR = (3\text{ A})(12\ \Omega) = 36\text{ V}$. We can use this to find the power from $P = IV = (3\text{ A})(36\text{ V}) = 108\text{ W}$.

Exercise A wire of resistance 5 Ω carries a current of 3 A for 6 s. (*a*) How much power is put into the wire? (*b*) How much thermal energy is produced? (*Answers* (*a*) 45 W, (*b*) 270 J)

EMF and Batteries

To maintain a steady current in a conductor, we need a constant supply of electrical energy. A device that supplies electrical energy is called a **source of emf.** (The letters *emf* stand for *electromotive force*, a term that is now rarely used.) Examples of emf sources are a battery, which converts chemical energy into electrical energy, and a generator, which converts mechanical energy into electrical energy. A source of emf does work on the charge passing through it, raising the potential energy of the charge. The work per unit charge is called the **emf,** $\mathcal{E}$, of the source. The unit of emf is the volt, the same as the unit of potential difference. An **ideal battery** is a source of emf that maintains a constant potential difference between its two terminals, indepen-

CHAPTER 27

optional

The Microscopic Theory of Electrical Conduction

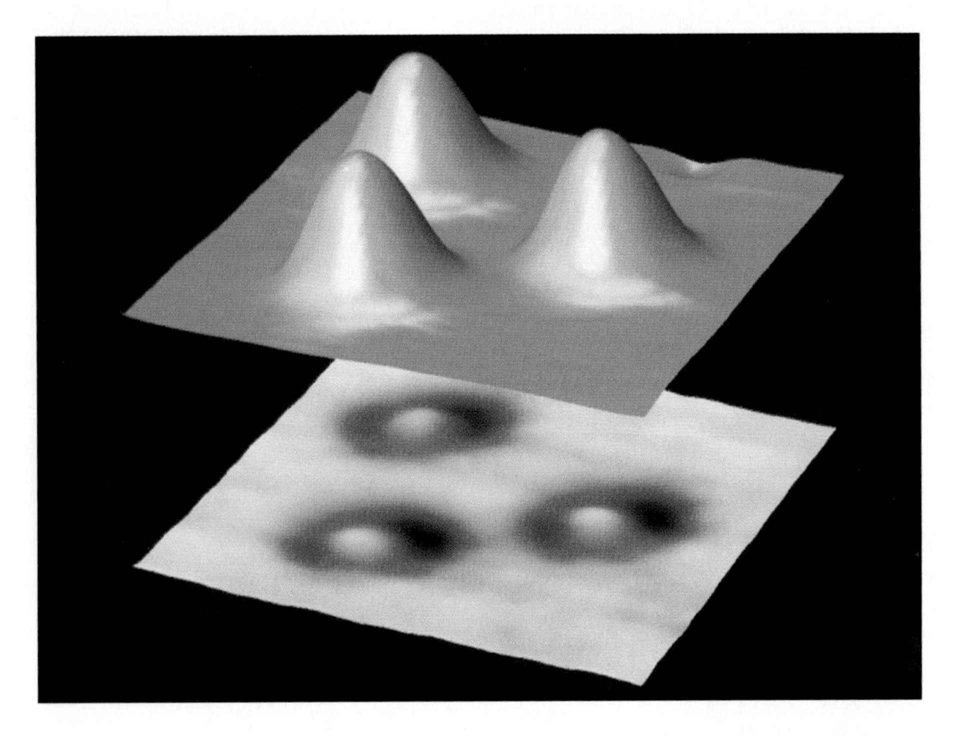

Simultaneously acquired topographic (top) and spectroscopic (bottom) images of three gadolinium atoms on top of a superconducting niobium surface. In the region near the gadolinium atoms the magnetic properties of these individual atoms break up Cooper electron pairs, thereby modifying the superconductivity of the niobium.

In earlier chapters we used Ohm's law to find the currents in various circuits while making no attempt to relate Ohm's law to the properties of metals. Here we introduce a microscopic model of electrical conduction to relate the resistivity of a metal to the properties of electrons and the lattice ions of the metal. This model is similar to the model of a gas that we discussed in Chapter 18 to relate the pressure exerted by a gas on the walls of its container to the average energy of the gas molecules, which in turn we could relate to the absolute temperature of the gas.

The first microscopic model of electric conduction was proposed by P. Drude in 1900 and developed by Hendrik A. Lorentz about 1909. This model successfully predicts Ohm's law and relates the resistivity of conductors to the mean speed v_{av} and the mean free path λ of the free electrons within the conductor. However, when v_{av} and λ are interpreted classically, there is a disagreement between the calculated and measured values of the resistivity, and a similar disagreement between the predicted and observed temperature dependence. Thus, the classical theory fails to adequately describe the resistivity of metals. Furthermore, the classical theory says nothing about the most striking property of solids, namely that some materials are conductors, others are insulators, and still others are semiconductors, materials whose resistivity falls between that of conductors and insulators.

When v_{av} and λ are interpreted using quantum theory, the magnitude and temperature dependence of the resistivity are correctly predicted. In addition, quantum theory allows us to determine if a material will be a conductor, insulator, or semiconductor.

27-1 A Microscopic Picture of Conduction

We consider a metal as a regular three-dimensional lattice of ions filling some volume V and containing a large number N of electrons that are free to move throughout the whole metal. Experimentally the number of free electrons in a metal is about 1 to 4 electrons per atom. In the absence of an electric field, the free electrons move about the metal randomly, much the way gas molecules move about in a container. We will often refer to these free electrons in a metal as an electron gas.

The current in a conducting wire segment is proportional to the voltage drop across the segment:

$$I = \frac{V}{R}$$

or

$$V = IR$$

The resistance R is proportional to the length L of the wire segment and inversely proportional to the cross-sectional area A:

$$R = \rho\frac{L}{A}$$

where ρ is the resistivity. According to Ohm's law, the resistance of a given wire segment is independent of the current in the segment and therefore independent of the voltage drop across the segment. For a uniform electric field E, the voltage across a segment of length L is $V = EL$. Then Ohm's law implies that the resistance, and therefore the resistivity, is independent of the applied electric field E.

Substituting $\rho L/A$ for R, and EL for V, we can write the current in terms of the electric field E and the resistivity. We have

$$I = \frac{V}{R} = \frac{EL}{\rho L/A} = \frac{1}{\rho}EA \qquad 27\text{-}1$$

The current in a wire is related by Equation 26-3 to the number of electrons per unit volume $n_e = N/V$, the drift velocity v_d, the magnitude of the charge of an electron $q = e$, and the cross-sectional area A:

$$I = nqAv_d = n_e eAv_d \qquad 27\text{-}2$$

Combining Equations 27-1 and 27-2 for the current I gives

$$n_e eAv_d = \frac{1}{\rho}EA$$

or

$$\rho = \frac{E}{n_e ev_d} \qquad 27\text{-}3$$

If Ohm's law holds, ρ must be independent of E, so according to Equation 27-3, the drift velocity v_d must be proportional to E.

In the presence of an electric field, a free electron experiences a force of magnitude eE. If this were the only force acting, the electron would have an acceleration eE/m_e and its velocity would steadily increase. However, Equation 27-3 implies a steady-state situation with a constant drift velocity that is proportional to the field E. In the microscopic model, it is assumed that a free electron is accelerated for a short time and then makes a collision with a lattice ion. The velocity of the electron after the collision is completely unrelated to the drift velocity. The justification for this assumption is that the drift velocity is very small compared with the random thermal velocity.

Let τ be the average time since the last collision for an electron picked at random. Since the acceleration of each electron is eE/m_e, the drift velocity of the electrons is

$$v_d = \frac{eE}{m_e}\tau \qquad \text{27-4}$$

Using this result in Equation 27-3, we obtain

$$\rho = \frac{E}{n_e e(eE\tau/m_e)} = \frac{m_e}{n_e e^2 \tau} \qquad \text{27-5}$$

The time τ, called the **collision time**, is also the average time between collisions.* The average distance the electron travels between collisions is $v_{av}\tau$, which is called the **mean free path** λ:

$$\lambda = v_{av}\tau \qquad \text{27-6}$$

In terms of the mean free path and the mean speed, the resistivity is

$$\rho = \frac{m_e v_{av}}{n_e e^2 \lambda} \qquad \text{27-7}$$

Resistivity in terms of v_{av} and λ

According to Ohm's law, the resistivity is independent of the electric field E. Since m_e, n_e, and e^2 are constants, the only quantities that could possibly depend on E are the mean speed v_{av} and the mean free path λ. Let us examine these quantities to see if they can possibly depend on the applied field E.

Classical Interpretation of λ and v_{av}

Classically, at $T = 0$ all the free electrons in a conductor should have zero kinetic energy. As the conductor is heated, the lattice ions acquire an average kinetic energy of $\frac{3}{2}kT$, which is imparted to the electron gas by the collisions between the electrons and the ions. (This is a result of the equipartition theorem studied in Chapters 18 and 19.) The electron gas would then have a Maxwell–Boltzmann distribution just like a gas of molecules. In equilibrium, the electrons would be expected to have a mean kinetic energy of $\frac{3}{2}kT$, which at ordinary temperatures (~300 K) is about 0.04 eV. At $T = 300$ K, their root mean square (rms) speed,† which is slightly greater than the mean speed, is

$$v_{av} \approx v_{rms} = \sqrt{\frac{3kT}{m_e}} = \sqrt{\frac{3(1.38 \times 10^{-23}\,\text{J/K})(300\,\text{K})}{9.11 \times 10^{-31}\,\text{kg}}}$$

$$= 1.17 \times 10^5\ \text{m/s} \qquad \text{27-8}$$

---

* It is tempting but incorrect to think that if τ is the average time between collisions, the average time since its last collision is $\frac{1}{2}\tau$ rather than τ. If you find this confusing, you may take comfort in the fact that Drude used the incorrect result $\frac{1}{2}\tau$ in his original work.

† See Equation 18-23.

optional

Note that this is about three billion times greater than the typical drift speed of 3.5×10^{-5} m/s calculated in Example 26-1. The very small drift velocity caused by the electric field therefore has essentially no effect on the very large mean speed of the electrons, so v_{av} in Equation 27-7 cannot depend on the electric field E.

The mean free path is related classically to the size of the lattice ions in the conductor and to the number of ions per unit volume. Consider one electron moving with speed v through a region of stationary ions, assumed to be hard spheres (Figure 27-1). The size of the electron is assumed to be negligible. The electron will collide with an ion if it comes within a distance r from the center of the ion, where r is the radius of the ion. In some time t_1, the electron moves a distance vt_1. If there is an ion whose center is in the cylindrical volume $\pi r^2 vt_1$, the electron will collide with it. The electron will then change directions and collide with another ion in time t_2 if the center of the ion is in the volume $\pi r^2 vt_2$. Thus, in the total time $t = t_1 + t_2 + \cdots$, the electron will collide with the number of ions whose centers are in the volume $\pi r^2 vt$. The number of ions in this volume is $n_{ion}\pi r^2 vt$, where n_{ion} is the number of ions per unit volume. The total path length divided by the number of collisions is the mean free path:

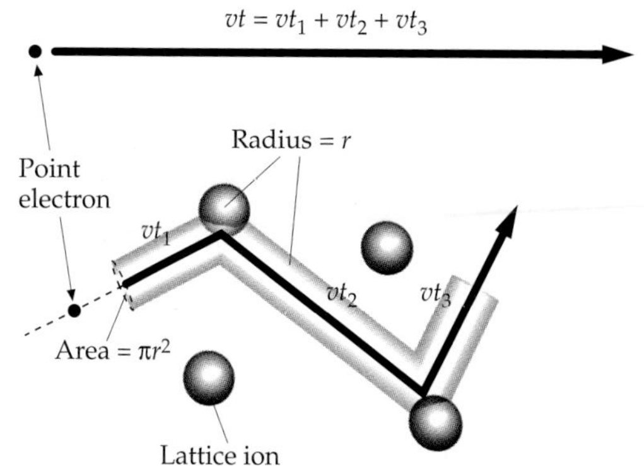

Figure 27-1 Model of an electron moving through the lattice ions of a conductor. The electron, which is considered to be a point, collides with an ion if it comes within a distance r of the center of the ion, where r is the radius of the ion. If the electron speed is v, it collides in time t with all the ions whose centers are in the volume $\pi r^2 vt$.

$$\lambda = \frac{vt}{n_{ion}\pi r^2 vt} = \frac{1}{n_{ion}\pi r^2} = \frac{1}{n_{ion}A} \tag{27-9}$$

where $A = \pi r^2$ is the cross-sectional area of a lattice ion.

Neither n_{ion} nor r depends on the electric field E, so λ also does not depend on E. Thus, according to the classical interpretation of v_{av} and λ, neither depend on E, so the resistivity ρ does not depend on E, in accordance with Ohm's law. However, the classical theory gives an incorrect temperature dependence for the resistivity. Since λ depends only on the number density of lattice ions and their radius, the only quantity in Equation 27-7 that depends on temperature in the classical theory is v_{av}, which is proportional to $\sqrt{T}$. But experimentally, ρ varies linearly with temperature. Furthermore, when ρ is calculated at $T = 300$ K using the Maxwell–Boltzmann distribution for v_{av} and Equation 27-9 for λ, the numerical result is about six times greater than the measured value.

The classical theory of conduction fails because electrons are not classical particles. The wave nature of the electrons must be considered. Because of the wave properties of electrons and the exclusion principle (to be discussed below), the energy distribution of the free electrons in a metal is not even approximately given by the Maxwell–Boltzmann distribution. Furthermore, the collision of an electron with a lattice ion is not similar to the collision of a baseball with a tree. Instead, it involves the scattering of electron waves by the lattice. To understand the quantum theory of conduction, we need a qualitative understanding of the energy distribution of free electrons in a metal. This will also help us understand the origin of contact potentials between two dissimilar metals in contact, and the contribution of free electrons to the heat capacity of metals.

27-2 The Fermi Electron Gas

We have used the term *electron gas* to describe the free electrons in a metal. Whereas the molecules in an ordinary gas such as air obey the classical Maxwell–Boltzmann energy distribution, the free electrons in a metal do not. Instead they obey a quantum energy distribution called the Fermi–Dirac dis-

27-4 Band Theory of Solids

Resistivities vary enormously between insulators and conductors. For a typical insulator, such as quartz, $\rho \sim 10^{16}\ \Omega\cdot\text{m}$, whereas for a typical conductor, $\rho \sim 10^{-8}\ \Omega\cdot\text{m}$. The reason for this enormous variation is the variation in the number density of free electrons n_e. To understand this variation, we consider the effect of the lattice on the electron energy levels.

We begin by considering the energy levels of the individual atoms as they are brought together. The allowed energy levels in an isolated atom are often far apart. For example, in hydrogen, the lowest allowed energy ($E_1 = -13.6$ eV) is 10.2 eV below the next lowest ($E_2 = (-13.6\text{ eV})/4 = -3.4$ eV).* Let us consider two identical atoms and focus our attention on one particular energy level. When the atoms are far apart, the energy of a particular level is the same for each atom. As the atoms are brought closer together, the energy level for each atom changes because of the influence of the other atom. As a result, the level splits into two levels of slightly different energies for the two-atom system. If we bring three atoms close together, a particular energy level splits into three separate levels of slightly different energies. Figure 27-10 shows the energy splitting of two energy levels for six atoms as a function of the separation of the atoms.

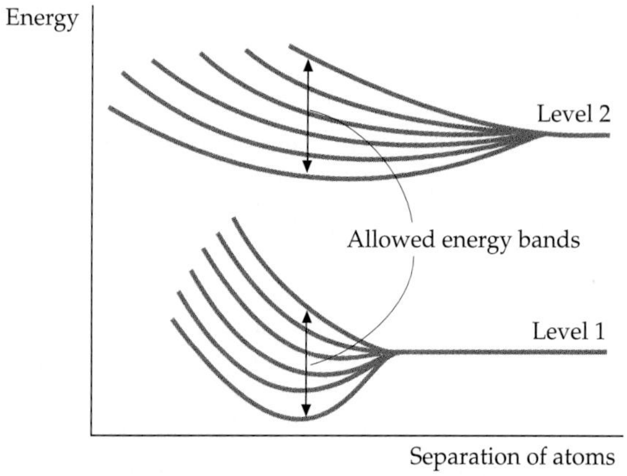

Figure 27-10 Energy splitting of two energy levels for six atoms as a function of the separation of the atoms. When there are many atoms, each level splits into a near-continuum of levels called a band.

If we have N identical atoms, a particular energy level in the isolated atom splits into N different, closely spaced energy levels when the atoms are close together. In a macroscopic solid, N is very large—of the order of 10^{23}—so each energy level splits into a very large number of levels called a **band**. The levels are spaced almost continuously within the band. There is a separate band of levels for each particular energy level of the isolated atom. The bands may be widely separated in energy, they may be close together, or they may even overlap, depending on the kind of atom and the type of bonding in the solid.

The lowest-energy bands, corresponding to the lowest energy levels of the atom in the lattice, are filled with electrons that are bound to the atom. The electrons that can take part in conduction occupy the higher energy bands. The highest energy band that contains electrons is called the **valence band**. The valence band may be completely filled with electrons or only partially filled, depending on the kind of atom and type of bonding in the solid.

We can now understand why some solids are conductors and others are insulators. If the valence band is only partially full, there are many available empty energy states in the band, and the electrons in the band can easily be raised to a higher energy state by an electric field. Accordingly, this material is a good conductor. If the valence band is full and there is a large energy gap between it and the next available band, a typical applied electric field will be too weak to excite an electron from the upper energy levels of the filled band across the large gap into the energy levels of the empty band, so the material is an insulator. The lowest band in which there are unoccupied states is called the **conduction band**. In a conductor, the valence band is only partially filled, so the valence band is also the conduction band. An energy gap between allowed bands is called a **forbidden energy band**.

The band structure for a conductor such as copper is shown in Figure

* The energy levels in hydrogen are discussed in Chapter 37.

optional

27-11*a*. The lower bands (not shown) are filled with the inner electrons of the atoms. The valence band is only about half full. When an electric field is established in the conductor, the electrons in the conduction band are accelerated, which means that their energy is increased. This is consistent with the Pauli exclusion principle because there are many empty energy states just above those occupied by electrons in this band. These electrons are thus the conduction electrons.

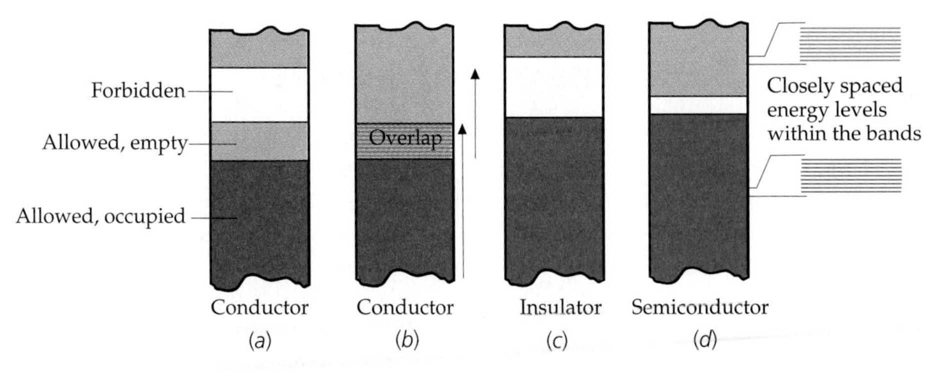

Figure 27-11 Four possible band structures for a solid. (*a*) A typical conductor. The valence band is only partially full, so electrons can be easily excited to nearby energy states. (*b*) A conductor in which the allowed energy bands overlap. (*c*) A typical insulator. There is a forbidden band with a large energy gap between the filled valence band and the conduction band. (*d*) A semiconductor. The energy gap between the filled valence band and the conduction band is very small, so some electrons are excited to the conduction band at normal temperatures, leaving holes in the valence band.

Figure 27-11*b* shows the band structure for magnesium, which is also a conductor. In this case, the highest occupied band is full, but there is an empty band above it that overlaps it. The two bands thus form a combined valence–conduction band that is only partially filled.

Figure 27-11*c* shows the band structure for a typical insulator. At $T = 0$ K, the valence band is completely full. The next energy band containing empty energy states, the conduction band, is separated from the valence band by a large energy gap. At $T = 0$, the conduction band is empty. At ordinary temperatures, a few electrons can be excited to states in this band, but most cannot be because the energy gap is large compared with the energy an electron might obtain by thermal excitation. Very few electrons can be thermally excited to the nearly empty conduction band, even at fairly high temperatures. When an electric field of ordinary magnitude is established in the solid, electrons cannot be accelerated because there are no empty energy states at nearby energies. We describe this by saying that there are no free electrons. The small conductivity that is observed is due to the very few electrons that are thermally excited into the nearly empty conduction band. When an electric field applied to an insulator is sufficiently strong to cause an electron to be excited across the energy gap to the empty band, dielectric breakdown occurs.

In some materials, the energy gap between the filled valence band and the empty conduction band is very small, as shown in Figure 27-11*d*. At $T = 0$, there are no electrons in the conduction band and the material is an insulator. However, at ordinary temperatures, there are an appreciable number of electrons in the conduction band due to thermal excitation. Such a material is called an **intrinsic semiconductor**. For typical semiconductors such as silicon and germanium, the energy gap is only about 1 eV. In the presence of an electric field, the electrons in the conduction band can be accelerated because there are empty states nearby. Also, for each electron in the conduction band there is a vacancy, or hole, in the nearly filled valence band. In the presence of an electric field, electrons in this band can also be excited to a vacant energy level. This contributes to the electric current and is most easily described as the motion of a hole in the direction of the field and opposite to the motion of the electrons. The hole thus acts like a positive charge. To visualize the conduction of holes, think of a two-lane, one-way road with one lane full of parked cars and the other empty. If a car moves out of the filled lane into the empty lane, it can move ahead freely. As the other cars move up to occupy the space left, the empty space propagates backward in the direction opposite the motion of the cars. Both the forward motion of the car in the nearly empty lane and the backward propagation of the empty space contribute to a net forward propagation of the cars.

An interesting characteristic of semiconductors is that the resistivity of the material decreases as the temperature increases, which is contrary to the case for normal conductors. The reason is that as the temperature increases, the number of free electrons increases because there are more electrons in the

conduction band. The number of holes in the valence band also increases, of course. In semiconductors, the effect of the increase in the number of charge carriers, both electrons and holes, exceeds the effect of the increase in resistivity due to the increased scattering of the electrons by the lattice ions due to thermal vibrations. Semiconductors therefore have a negative temperature coefficient of resistivity.

27-5 Superconductivity

There are some materials for which the resistivity suddenly drops to zero below a certain temperature T_c, called the **critical temperature**. This amazing phenomenon, called **superconductivity**, was discovered in 1911 by the Dutch physicist H. Kamerlingh Onnes, who had developed a technique for liquefying helium (boiling point 4.2 K) and was putting his technique to work exploring the properties of materials at extremely low temperatures. Figure 27-12 shows his plot of the resistance of mercury versus temperature. The critical temperature for mercury is the same as the boiling point of helium, 4.2 K. Critical temperatures for other superconducting elements range from less than 0.1 K for hafnium and iridium to 9.2 K for niobium. The temperature range for superconductors goes much higher for a number of metallic compounds. For example, the superconducting alloy Nb_3Ge, discovered in 1973, has a critical temperature of 23.2 K, which was the highest known until 1986, when the discoveries of Bednorz and Müller launched the era of high-temperature superconductors, now defined as materials that exhibit superconductivity at temperatures above 77 K. To date (mid 1998), the highest temperature at which superconductivity has been demonstrated, using compounds containing mercury, is 125 K at atmospheric pressure and 164 K at high pressure.

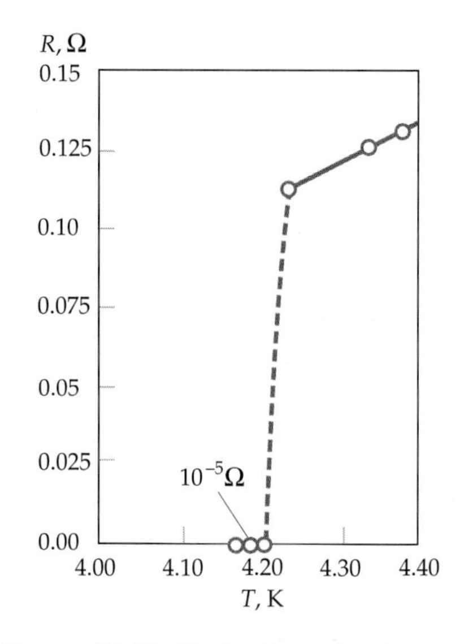

Figure 27-12 Plot by Kamerlingh Onnes of the resistance of mercury versus temperature, showing the sudden decrease at the critical temperature of $T = 4.2$ K.

The resistivity of a superconductor is zero. There can be a current in a superconductor even when the electric field in the superconductor is zero. Indeed, in superconducting rings in which there was no electric field, steady currents have been observed to persist for years without apparent loss. Despite the cost and inconvenience of refrigeration with expensive liquid helium, many superconducting magnets have been built using superconducting materials, because such magnets require no power expenditure to maintain the large current needed to produce a large magnetic field.

The discovery of high-temperature superconductors has revolutionized the study of superconductivity because relatively inexpensive liquid nitrogen, which boils at 77 K, can be used for a coolant. However, many problems, such as brittleness and the toxicity of the materials, make these new superconductors difficult to use. The search continues for new materials that will superconduct at even higher temperatures.

The BCS Theory

It had been recognized for some time that superconductivity is due to a collective action of the conducting electrons. In 1957, John Bardeen, Leon Cooper, and Robert Schrieffer published a successful theory of superconductivity now known by the initials of the inventors as the **BCS theory**. According to this theory, the electrons in a superconductor are coupled in pairs at low temperatures. The coupling comes about because of the interaction between electrons and the crystal lattice. One electron interacts with the lattice and perturbs it. The perturbed lattice interacts with another electron in such a way that there is an attraction between the two electrons that at low tem-

peratures can exceed the Coulomb repulsion between them. The electrons form a bound state called a **Cooper pair**. The electrons in a Cooper pair have equal and opposite spins, so they form a system with zero spin. Each Cooper pair acts as a *single particle* with zero spin—in other words, as a boson. Bosons do not obey the exclusion principle. Any number of Cooper pairs may be in the same quantum state with the same energy. In the ground state of a superconductor (at $T = 0$), all the electrons are in Cooper pairs and all the Cooper pairs are in the same energy state. In the superconducting state, the Cooper pairs are correlated so that they all act together. An electric current can be produced in a superconductor because all of the electrons in this collective state move together. But energy cannot be dissipated by individual collisions of electron and lattice ions unless the temperature is high enough to break the binding of the Cooper pairs. The energy needed to break up a Cooper pair is similar to that needed to break up a molecule into its constituent atoms. This energy is called the **superconducting energy gap** E_g. In the BCS theory, this energy at absolute zero is related to the critical temperature by

$$E_g = 3.5kT_c \tag{27-24}$$

The energy gap can be determined by measuring the current across a junction between a normal metal and a superconductor as a function of voltage. Consider two metals separated by a layer of insulating material, such as aluminum oxide, that is only a few nanometers thick. The insulating material between the metals forms a barrier that prevents most electrons from traversing the junction. But, as mentioned in Chapter 15, waves can tunnel through a barrier if the barrier is not too thick even if the energy of the wave is less than that of the barrier (see Figures 15-21 and 15-22).

When the materials on either side of the gap are normal nonsuperconducting metals, the current resulting from the tunneling of electrons through the insulating layer obeys Ohm's law for low applied voltages (Figure 27-13*a*). When one of the metals is a normal metal and the other is a superconductor, there is no current (at absolute zero) unless the applied voltage V is greater than a critical voltage $V_c = E_g/2e$, where E_g is the superconductor energy gap. Figure 27-13*b* shows the plot of current versus voltage for this situation. The current jumps abruptly when the energy $2eV_c$ absorbed by a Cooper pair is great enough to break up the pair. (The small current visible in Figure 27-13*b* before the critical voltage is reached is present because at any temperature above absolute zero some of the electrons in the superconductor are thermally excited above the energy gap and are therefore not paired.) The superconducting energy gap can thus be accurately measured by measuring the critical voltage V_c.

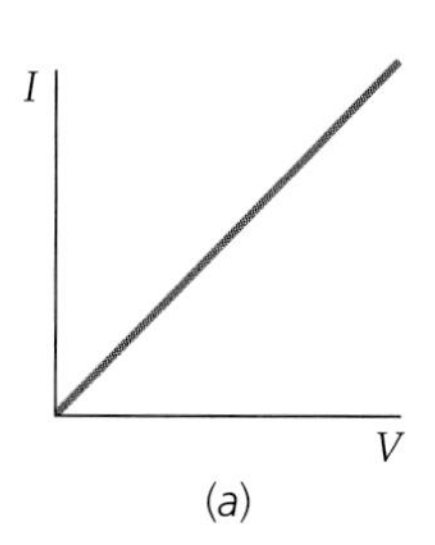

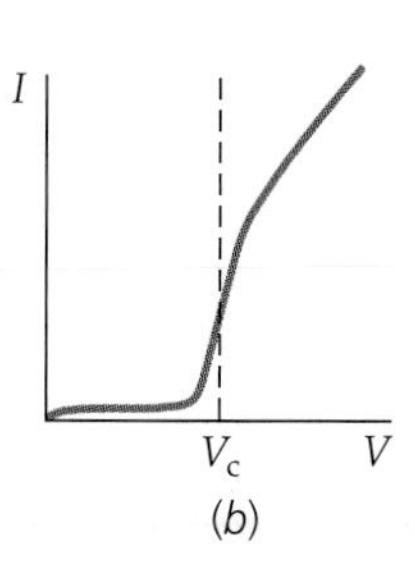

Figure 27-13 Tunneling current versus voltage for a junction of two metals separated by a thin oxide layer. (*a*) When both metals are normal metals, the current is proportional to the voltage, as predicted by Ohm's law. (*b*) When one metal is a normal metal and one is a superconductor, the current is approximately zero until the applied voltage exceeds the critical voltage $V_c = E_g/2e$.

Example 27-6

Calculate the superconducting energy gap for mercury (T_c = 4.2 K) predicted by the BCS theory.

1. The BCS prediction for the energy gap is: $E_g = 3.5kT_c$

2. Substitute $T_c = 4.2$ K:

$$E_g = 3.5kT_c$$
$$= 3.5(1.38 \times 10^{-23}\ \text{J/K})(4.2\ \text{K})\frac{1\ \text{eV}}{1.6 \times 10^{-19}\ \text{J}}$$
$$= 1.27 \times 10^{-3}\ \text{eV}$$

5. The initial flux through the coil is $\phi_{mi} = NBA$. When the coil is flipped, the flux reverses. Thus, $\phi_{mf} = -NBA$. This gives us $\Delta\phi_m$:

$$\Delta\phi_m = \phi_{mf} - \phi_{mi} = (NBA) - (-NBA) = 2NBA$$

6. Combining the previous two results yields Q:

$$Q = \frac{2NBA}{R}$$

Remarks Note that the charge Q is independent of the time involved in rotating the coil—all that matters is the change in magnetic flux. A coil used in this way is called a *flip coil*. It is used to measure magnetic fields. For example, if the current integrator Ⓒ measures a total charge Q passing through the coil when it is flipped, the magnetic field can be found from $B = RQ/2NA$.

Exercise A flip coil of 40 turns has a radius of 3 cm and a resistance of 16 Ω. If the coil is turned through 180° in a magnetic field of 5000 G, how much charge passes through it? (*Answer* 7.07 mC)

30-3 Lenz's Law

The negative sign in Faraday's law has to do with the direction of the induced emf, which can be found from a general physical principle known as **Lenz's law:**

> The induced emf and induced current are in such a direction as to oppose the change that produces them.

Lenz's law

Note that we didn't specify just what kind of change causes the induced emf and current. We purposefully left the statement vague to cover a variety of conditions, which we will now illustrate.

Figure 30-6 shows a bar magnet moving toward a loop that has a resistance R. Since $\vec{B}$ from the bar magnet is to the right, out of the north pole of the magnet, the movement of the magnet toward the loop tends to increase the flux through the loop to the right. (The magnetic field at the loop is stronger when the magnet is closer.) The induced current in the loop produces a magnetic field of its own. This induced current is in the direction shown, so the magnetic flux it produces is opposite that of the magnet. The induced magnetic field tends to *decrease* the flux through the loop. If the magnet were moved away from the loop, which would decrease the flux through the loop due to the magnet, the induced current would be in the opposite direction from that in Figure 30-6. In that case, the current would produce a magnetic field to the right, which would tend to increase the flux through the loop. As we might expect, moving the loop toward or away from the magnet has the same effect as moving the magnet. Only the relative motion is important.

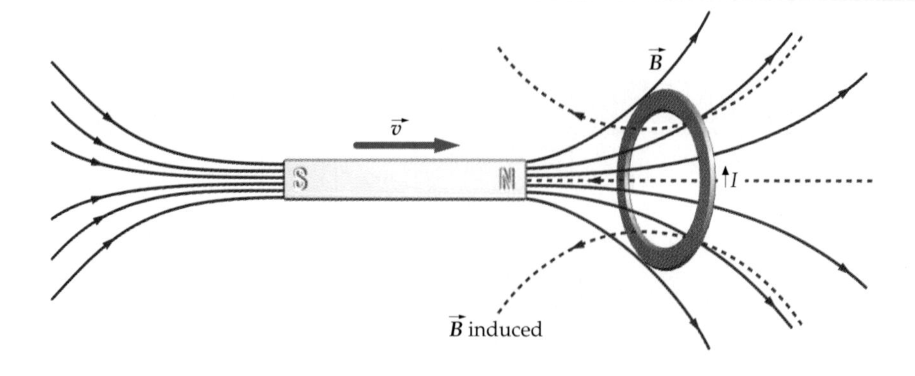

Figure 30-6 When the bar magnet is moving toward the loop, the emf induced in the loop produces a current in the direction shown. The magnetic field due to the induced current in the loop (indicated by the dashed lines) produces a flux that opposes the increase in flux through the loop due to the motion of the magnet.

Figure 30-7 shows the induced magnetic moment of the current loop when the magnet is moving toward it as in Figure 30-6. The loop acts like a small magnet with its north pole to the left and its south pole to the right. Since opposite poles attract and like poles repel, the induced magnetic moment of the loop repels the bar magnet, that is, it opposes its motion toward the loop. Thus, we can express Lenz's law in terms of forces rather than flux. If the bar magnet is moved toward the loop, the induced current must produce a magnetic moment to oppose this change.

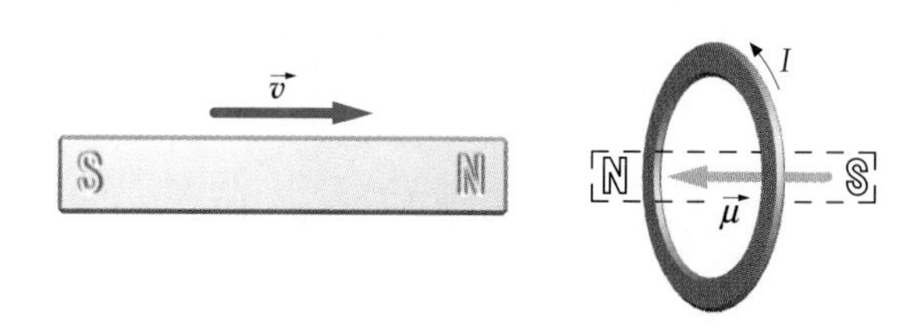

Figure 30-7 The magnetic moment of the loop $\vec{\mu}$ (shown in outline as if it were a magnet) due to the induced current is such as to oppose the motion of the bar magnet. Here the bar magnet is moving toward the loop so the induced magnetic moment repels the bar magnet.

Lenz's law is required by the law of conservation of energy. If the current in the loop in Figure 30-7 were opposite the direction shown, the induced magnetic moment of the loop would attract the magnet when it is moving toward the loop and cause it to accelerate toward the loop. If we begin with the magnet a great distance from the loop and give it a very slight push toward the loop, the force due to the induced current would be toward the loop, which would increase the velocity of the magnet. As the speed of the magnet increases, the rate of change of the flux would increase, thereby increasing the induced current. This would further increase the force on the magnet. Hence, the kinetic energy of the magnet and the rate at which Joule heat is produced in the loop (I^2R) would both increase with no source of energy. This would violate the law of conservation of energy.

In Figure 30-8, the bar magnet is at rest, and the loop is moving away from it. The induced current and magnetic moment are shown in the figure. In this case, the magnetic moment of the loop attracts the bar magnet, thus opposing the motion of the loop as required by Lenz's law.

Figure 30-8 When the loop is moving away from the stationary bar magnet, the induced magnetic moment in the loop attracts the bar magnet, again opposing the relative motion.

In Figure 30-9, when the current in circuit 1 is changing, there is a changing flux through circuit 2. Suppose that the switch S in circuit 1 is initially open so that there is no current in the circuit (Figure 30-9*a*). When we close the switch (Figure 30-9*b*), the current in circuit 1 does not reach its steady value $\mathcal{E}_1/R_1$ instantaneously but takes some time to change from zero to this value. During this time, while the current is increasing, the flux through circuit 2 is changing, and there is an induced current in that circuit in the direction shown. When the current in circuit 1 reaches its steady value, the flux through circuit 2 is no longer changing, so there is no induced current in circuit 2. An induced current in circuit 2 in the opposite direction appears momentarily when the switch in circuit 1 is opened (Figure 30-9*c*) and the current is decreasing to zero. It is important to understand that there is an induced emf *only while the flux is changing*. The emf does not depend on the magnitude of the flux, only on its rate of change. If there is a large, steady flux through a circuit, there is no induced emf.

Figure 30-9 (*a*) Two adjacent circuits. (*b*) Just after the switch is closed, I_1 is increasing in the direction shown. The changing flux in circuit 2 induces the current I_2. The flux due to I_2 opposes the increase in flux due to I_1. (*c*) As the switch is opened, I_1 decreases and B decreases. The induced current I_2 then tends to maintain the flux in the circuit, opposing the change.

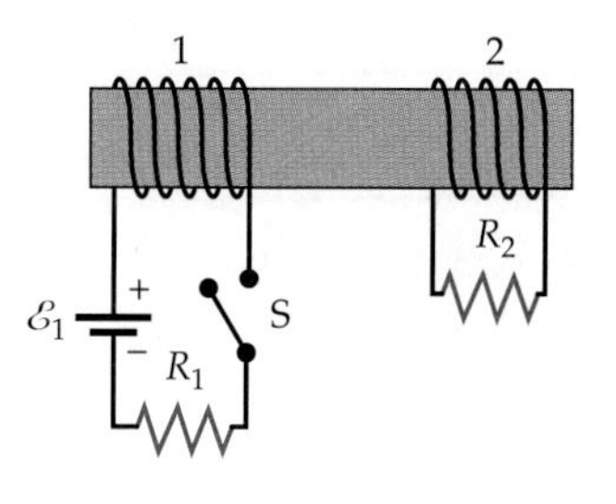

(*a*)

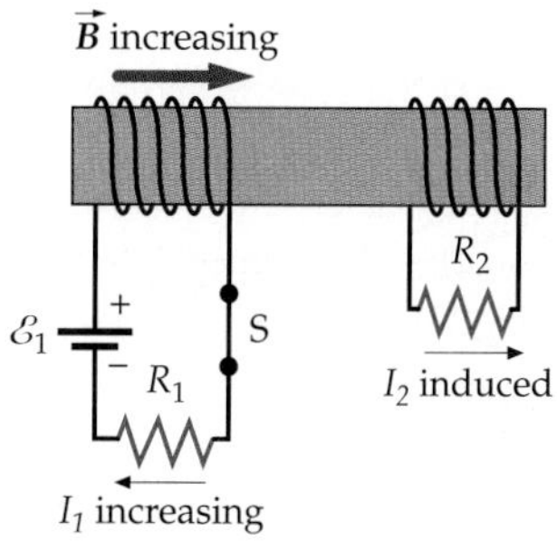

(*b*)

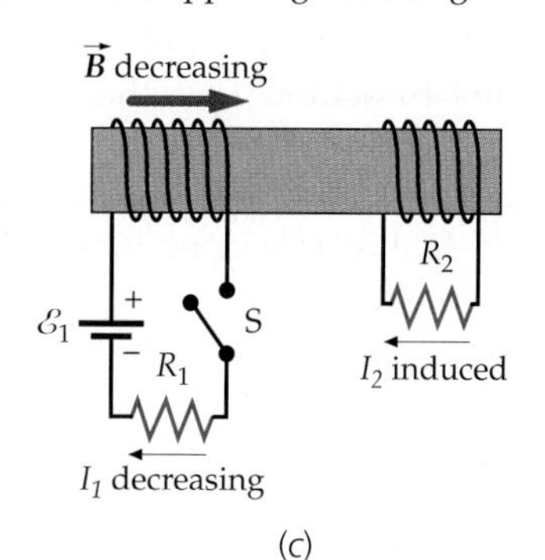

(*c*)

For our next example, we consider the single, isolated circuit shown in Figure 30-10. When there is a current in the circuit, there is a magnetic flux through the coil due to its own current. When the current is changing, the flux in the coil is changing and there is an induced emf in the circuit. This *self-induced* emf opposes the change in the current. It is therefore called a **back emf**. Because of this self-induced emf, the current in a circuit cannot jump instantaneously from zero to some finite value or from some finite value to zero. Henry first noticed this effect when he was experimenting with a circuit consisting of many turns of a wire like that in Figure 30-10. This arrangement gives a large flux through the circuit for even a small current. Henry noticed a spark across the switch when he tried to break the circuit. Such a spark is due to the large induced emf that occurs when the current varies rapidly, as during the opening of the switch. In this case, the induced emf is directed so as to maintain the original current. The large induced emf produces a large voltage drop across the switch as it is opened. The electric field between the contacts of the switch is large enough to tear electrons from surrounding air molecules, causing dielectric breakdown. When the molecules in the air dielectric are ionized, the air conducts electric current in the form of a spark.

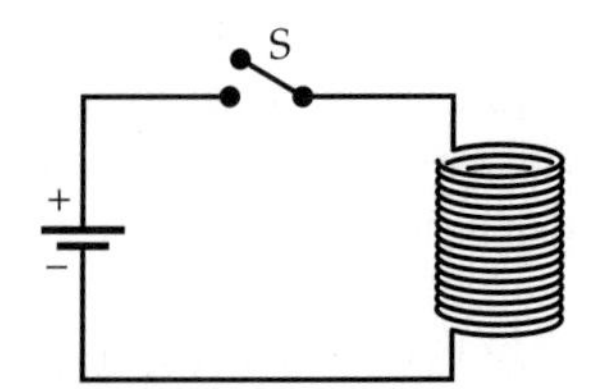

Figure 30-10 The coil with many turns of wire gives a large flux for a given current in the circuit. When the current changes, there is a large emf induced in the coil opposing the change.

Example 30-6

A rectangular coil of 80 turns, 20 cm wide and 30 cm long, is located in a magnetic field $B = 0.8$ T directed into the page (Figure 30-11), with only half of the coil in the region of the magnetic field. The resistance of the coil is 30 Ω. Find the magnitude and direction of the induced current if the coil is moved with a speed of 2 m/s (*a*) to the right, (*b*) up, and (*c*) down.

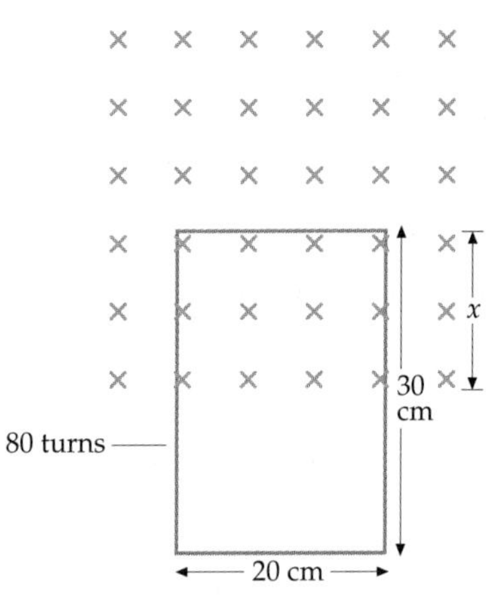

Figure 30-11

Picture the Problem The current equals the induced emf divided by the resistance. We can calculate the emf induced in the circuit as the coil moves by calculating the rate of change of the flux through the coil. The flux is proportional to the distance x. The direction of the current is found from Lenz's law.

(*a*) 1. The magnitude of the induced current equals the emf divided by the resistance: $I = \dfrac{|\mathcal{E}|}{R}$

2. The magnitude of the induced emf is given by Faraday's law: $|\mathcal{E}| = \dfrac{d\phi_m}{dt}$

3. When the coil is moving to the right (or to the left), the flux does not change (until the coil leaves the region of magnetic field). The current is therefore zero: $|\mathcal{E}| = \dfrac{d\phi_m}{dt} = 0$

$I = 0$

(*b*) 1. The flux is the product of B and the area, which is given by (20 cm)x: $\phi_m = NB(20\text{ cm})x$

2. Compute the rate of change of the flux when the coil is moving up: $\dfrac{d\phi_m}{dt} = NB(20\text{ cm})\dfrac{dx}{dt} = (80)(0.8\text{ T})(0.20\text{ m})(2\text{ m/s}) = 25.6\text{ V}$

3. Calculate the magnitude of the current: $I = |\mathcal{E}|/R = \dfrac{25.6\text{ V}}{30\ \Omega} = 0.853\text{ A}$

4. Since the inward flux is increasing, the induced current will be in the sense as to produce outward flux: The current is counterclockwise.

(*c*) When the coil moves downward at 2 m/s, the current has the same magnitude as when it moves upward, but is oppositely directed: $I = 0.853$ A clockwise

Remarks As the coil moves downward, the inward flux decreases so the induced current is clockwise so as to produce a flux inward.

optional

30-4 Motional emf

Figure 30-12 shows a conducting rod sliding to the right along conducting rails that are connected by a resistor. A uniform magnetic field $\vec{B}$ is directed into the paper. Since the area of the circuit increases as the rod moves to the right, the magnetic flux through the circuit is increasing. An emf is therefore induced in the circuit. Let ℓ be the separation of the rails and x be the distance from the left end of the rails to the rod at some time. The area enclosed by the circuit is then ℓx, and the magnetic flux through the circuit at this time is

$$\phi_m = BA = B\ell x$$

When the rod moves through a distance dx, the area enclosed by the circuit changes by $dA = \ell\, dx$ and the flux changes by $d\phi_m = B\ell\, dx$. The rate of change of the flux is

$$\frac{d\phi_m}{dt} = B\ell \frac{dx}{dt} = B\ell v$$

where $v = dx/dt$ is the speed of the rod. The magnitude of the emf induced in this circuit is therefore

$$|\mathcal{E}| = \frac{d\phi_m}{dt} = B\ell v$$

The direction of the emf in this case is such as to produce a current in the counterclockwise sense. The flux produced by this induced current is out of the paper, opposing the increase in flux due to the motion of the rod. Because of the induced current, which is upward in the rod, there is a magnetic force on the rod of magnitude $I\ell B$. The direction of this force, obtained from the right-hand rule, is to the left, opposing the motion of the rod. If the rod is given some initial velocity $\vec{v}$ to the right and is then released, the force due to the induced current slows the rod until it stops. To maintain the motion of the rod, an external force must be exerted on it to the right.

The emf induced in a conductor moving through a magnetic field is called **motional emf.** More generally,

> Motional emf is any emf induced by the relative motion of a magnetic field and a current path.

Motional emf defined

Motional emf is induced in a conducting rod or wire moving in a magnetic field even when there is no complete circuit and thus no current.

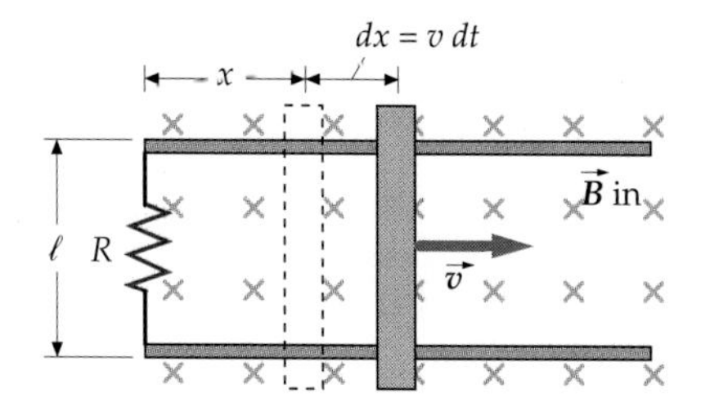

Figure 30-12 A conducting rod sliding on conducting rails in a magnetic field. As the rod moves to the right, the area of the circuit increases so the magnetic flux through the circuit into the paper increases. An emf of magnitude $B\ell v$ is induced in the circuit, producing a counterclockwise current that produces flux out of the paper opposing the charge.

Figure 30-13 shows an electron in a conducting rod that is moving through a uniform magnetic field directed into the paper. Because the electron is moving horizontally with the rod, there is a magnetic force on the electron that has a downward component of magnitude qvB. Because of this magnetic force, free electrons in the rod move downward, producing a net negative charge at the bottom and leaving a net positive charge at the top. The electrons continue to move down until the electric field produced by the separated charges exerts an upward force of magnitude qE on the electrons that balances the magnetic force qvB. In equilibrium, the electric field in the rod is thus

$$E = vB$$

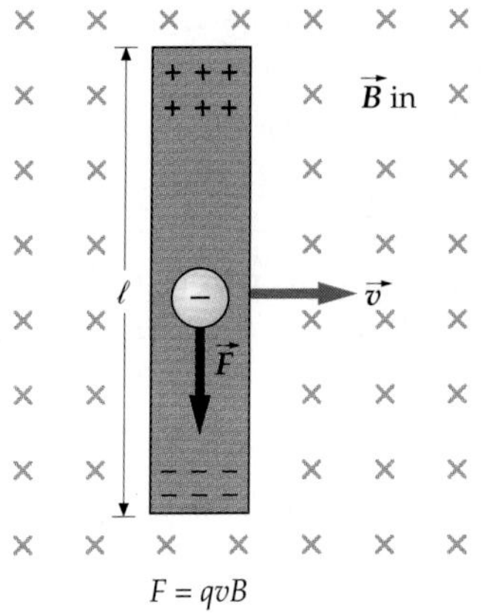

Figure 30-13 An electron in a conducting rod that is moving through a magnetic field experiences a magnetic force that has a downward component. Electrons move to the bottom of the rod, leaving the top of the rod positive. The charge separation produces an electric field of magnitude $E = vB$. The potential at the top of the rod is greater than that at the bottom by $E\ell = vB\ell$.

The potential difference across the rod is

$$\Delta V = E\ell = vB\ell$$

This potential difference equals the magnitude of the induced emf, that is, the motional emf:

$$|\mathcal{E}| = vB\ell \qquad 30\text{-}6$$

Motional emf

Exercise A rod 40 cm long moves at 12 m/s in a plane perpendicular to a magnetic field of 3000 G. Its velocity is perpendicular to its length. Find the emf induced in the rod. (*Answer* 1.44 V)

Example 30-7 *try it yourself*

In Figure 30-12, let $B = 0.6$ T, $v = 8$ m/s, $\ell = 15$ cm, and $R = 25\ \Omega$, and assume that the resistance of the rods and rails is negligible. Find (*a*) the induced emf in the circuit, (*b*) the current in the circuit, (*c*) the force needed to move the rod with constant velocity, and (*d*) the power dissipated in the resistor.

Cover the column to the right and try these on your own before looking at the answers.

Steps	*Answers*
(*a*) Calculate the induced emf from Equation 30-6.	$\mathcal{E} = Bv\ell = 0.72$ V
(*b*) Find the current from Ohm's law.	$I = \dfrac{\mathcal{E}}{R} = 28.8$ mA
(*c*) The force needed to move the rod with constant velocity is equal and opposite to the force exerted by the magnetic field on the rod, which has the magnitude $IB\ell$. Calculate the magnitude of this force.	$F = IB\ell = 2.59$ mN
(*d*)1. Find the power dissipated in the resistor.	$P = I^2R = 20.7$ mW
2. Check your answer for the previous step by computing the power input of the force from $P = Fv$.	$P = Fv = 20.7$ mW

optional

Example 30-8

A rod of mass m slides on frictionless, conducting rails in a region of constant magnetic field B. At time $t = 0$, the rod is moving with an initial speed v_0 and the external force acting on it is removed. Find the speed of the rod as a function of time.

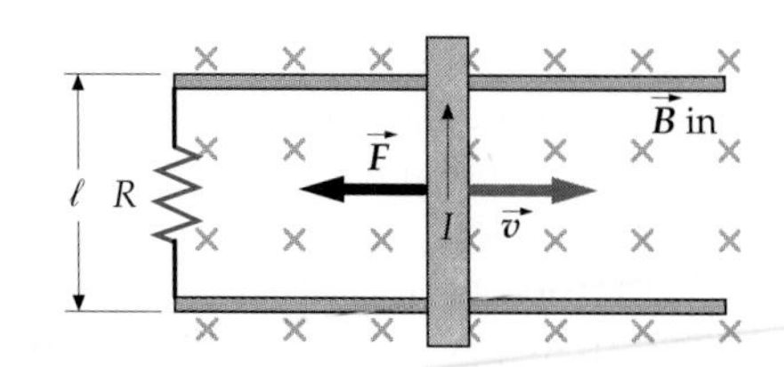

Figure 30-14

Picture the Problem The speed of the rod changes because a magnetic force acts on the induced current. The motion of the rod through a magnetic field induces an emf $\mathcal{E} = B\ell v$ and therefore a current in the rod, $I = \mathcal{E}/R$. This causes a magnetic force to act on the rod, $F = IB\ell$. With the force known, we apply Newton's second law to find the speed as a function of time. Take the direction of the initial velocity to be positive, as shown in Figure 30-14.

1. Apply Newton's second law to the rod:
$$F = ma = m\frac{dv}{dt}$$

2. The force exerted on the rod is the magnetic force, which is proportional to the current:
$$F = IB\ell$$

3. The current equals the motional emf divided by the resistance of the rod:
$$I = \frac{\mathcal{E}}{R} = \frac{B\ell v}{R}$$

4. With these results we find the magnitude of the magnetic force exerted on the rod:
$$F = IB\ell = \frac{B\ell v}{R}B\ell = \frac{B^2\ell^2 v}{R}$$

5. The force is opposite to the direction of motion, as shown in Figure 30-14, hence it is negative. Newton's second law then gives:
$$-\frac{B^2\ell^2 v}{R} = m\frac{dv}{dt}$$

6. Separate variables, then integrate the speed from v_0 to v and integrate the time from 0 to t:
$$\frac{dv}{v} = -\frac{B^2\ell^2}{mR}dt$$
$$\int_{v_0}^{v}\frac{dv}{v} = -\frac{B^2\ell^2}{mR}\int_0^t dt$$
$$\ln\frac{v}{v_0} = -\frac{B^2\ell^2}{mR}t$$

7. Solve for v:
$$v = v_0 e^{-(B^2\ell^2/mR)t}$$

Remarks If the force were constant, the rod's speed would decrease linearly with time. However, because the force is proportional to the rod's speed, as found in step 4, the force is large initially but decreases as the speed decreases. In principle, the rod never stops moving. Even so, it travels only a finite distance. (See Problem 37.)

30-5 Eddy Currents

In the examples we have discussed, the currents produced by a changing flux were set up in definite circuits. Often a changing flux sets up circulating currents, called *eddy currents*, in a piece of bulk metal like the core of a transformer. The heat produced by such current constitutes a power loss in the transformer. Consider a conducting slab between the pole faces of an electromagnet (Figure 30-15). If the magnetic field B between the pole faces is

changing with time (as it will if the current in the magnet windings is alternating current), the flux through any closed loop in the slab such as through the curve C indicated in the figure will be changing. Since path C is in a conductor, there will be a current along the path. At the right in Figure 30-15 we have indicated just one of the many closed paths that will contain currents if B varies.

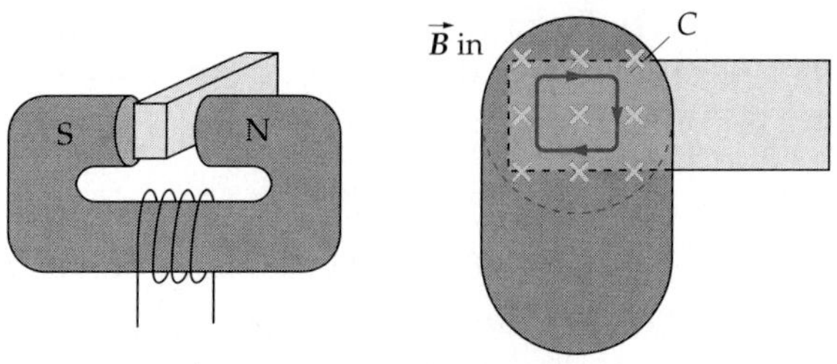

Figure 30-15 Eddy currents. When the magnetic field through a metal slab is changing, an emf is induced in any closed loop in the metal such as loop C. The induced emf causes a current in the loop.

The existence of eddy currents can be demonstrated by pulling a copper or aluminum sheet between the poles of a strong permanent magnet (Figure 30-16). Part of the area enclosed by curve C in this figure is in the magnetic field, and part is outside the field. As the sheet is pulled to the right, the flux through this curve decreases (assuming that the flux into the paper is positive). According to Faraday's law and Lenz's law, a clockwise current will be induced around this curve. Since this current is directed upward in the region between the pole faces, the magnetic field exerts a force on the current to the left, opposing motion of the sheet. You can feel this force on the sheet if you try to pull a conducting sheet suddenly through a strong magnetic field.

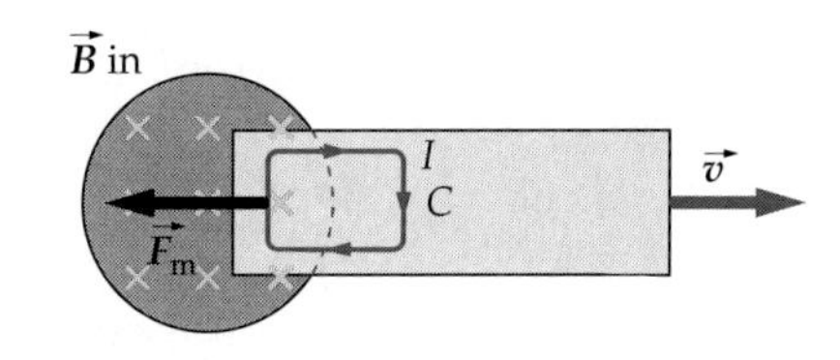

Figure 30-16 Demonstration of eddy currents. When the metal sheet is pulled to the right, there is a magnetic force to the left on the induced current opposing the motion.

Eddy currents are usually unwanted because power is lost in the form of heat generated by the current, and the heat itself must be dissipated. The power loss can be reduced by increasing the resistance of the possible paths for the eddy currents, as shown in Figure 30-17*a*. Here the conducting slab is laminated, that is, made up of small strips glued together. Because insulating glue separates the strips, the eddy currents are essentially confined to the strips. The large eddy-current loops are broken up, and the power loss is greatly reduced. Similarly if the sheet has cuts in it, as in Figure 30-17*b*, the eddy currents are lessened and the magnetic force is greatly reduced.

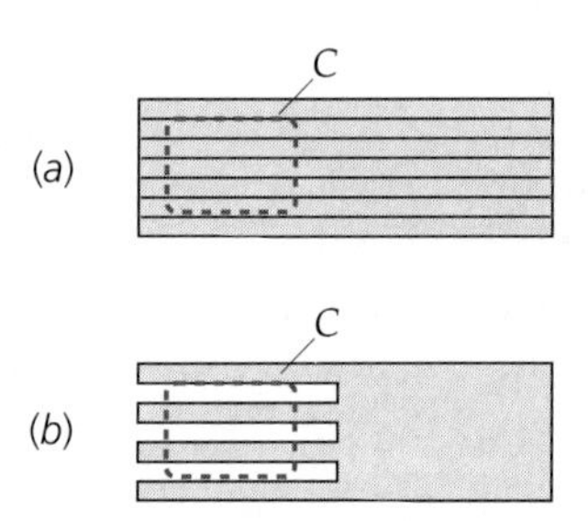

Figure 30-17 Eddy currents in a metal slab can be reduced by disrupting the conduction paths in the slab. (*a*) If the slab is constructed from strips of metal glued together, the insulating glue between the slabs increases the resistance of the closed loop C. (*b*) Slots cut into the metal slab also reduce the eddy current.

Eddy currents are not always undesirable. For example, they are often used to damp unwanted oscillations. With no damping present, sensitive mechanical balance scales used to weigh small masses might oscillate back and forth around their equilibrium reading many times. Such scales are usually designed so that a small piece of metal moves between the poles of a magnet as the scales oscillate. The resulting eddy currents dampen the oscillations so that equilibrium is quickly reached. Eddy currents also have a role in the magnetic braking systems of some rapid transit cars. A large electromagnet is positioned in the vehicle over the rails. When the magnet is energized by a current in its windings, eddy currents are induced in the rails by the motion of the magnet, and the magnetic forces provide a drag force on the car that stops it.

30-6 Inductance

Self-inductance

The magnetic flux through a circuit is related to the current in that circuit and the currents in other, nearby circuits.* Consider a coil carrying a current I. The current produces a magnetic field B that varies from point to point, but B is proportional to I at every point. The magnetic flux through the coil is therefore also proportional to I:

$$\phi_m = LI \qquad 30\text{-}7$$

Definition—Self-inductance

where L is a constant called the **self-inductance** of the coil. The self-inductance depends on the geometric shape of the coil. The SI unit of inductance is

*We will assume that there are no permanent magnets around.

the **henry** (H). From Equation 30-7, we can see that the unit of inductance equals the unit of flux divided by the unit of current:

$$1\,\text{H} = 1\,\frac{\text{Wb}}{\text{A}} = 1\,\frac{\text{T}\cdot\text{m}^2}{\text{A}}$$

In principle, the self-inductance of any coil or circuit can be calculated by assuming a current I, finding $\vec{B}$ at every point, calculating the flux ϕ_m, and using $L = \phi_m/I$. In actual practice, the calculation is very difficult. However, there is one case, that of the tightly wound solenoid, for which the self-inductance can be calculated directly. The magnetic flux through a solenoid of length ℓ and N turns carrying a current I was calculated in Example 30-1:

$$\phi_m = \frac{\mu_0 N^2 I A}{\ell} = \mu_0 n^2 I A \ell \qquad \text{30-8}$$

where $n = N/\ell$ is the number of turns per unit length. As expected, the flux is proportional to the current I. The proportionality constant is the self-inductance:

$$L = \frac{\phi_m}{I} = \mu_0 n^2 A \ell \qquad \text{30-9}$$

Self-inductance of a solenoid

The self-inductance of a solenoid is proportional to the square of the number of turns per unit length n and to the volume $A\ell$. Thus, like capacitance, self-inductance depends only on geometric factors. From the dimensions of Equation 30-9, we can see that μ_0 can be expressed in henrys per meter:

$$\mu_0 = 4\pi \times 10^{-7}\,\text{H/m}$$

Example 30-9

Find the self-inductance of a solenoid of length 10 cm, area 5 cm², and 100 turns.

Picture the Problem We can calculate the self-inductance in henrys from Equation 30-9 if we put all the quantities in SI units.

1. L is given by Equation 30-9: $L = \mu_0 n^2 A \ell$

2. Convert the given quantities to SI units:

$\ell = 10\text{ cm} = 0.1\text{ m}$

$A = 5\text{ cm}^2 = 5 \times 10^{-4}\text{ m}^2$

$n = N/\ell = (100\text{ turns})/(0.1\text{ m}) = 1000\text{ turns/m}$

$\mu_0 = 4\pi \times 10^{-7}\,\text{H/m}$

3. Substitute the given quantities:

$L = \mu_0 n^2 A \ell$

$= (4\pi \times 10^{-7}\,\text{H/m})(10^3\text{ turns/m})^2(5 \times 10^{-4}\text{ m}^2)(0.1\text{ m})$

$= 6.28 \times 10^{-5}\,\text{H}$

When the current in a circuit is changing, the magnetic flux due to the current is also changing, so an emf is induced in the circuit. Since the self-

(a)

(b)

(*a*) River-level view of Hoover Dam with the Nevada wing of its power plant on the left and the Arizona wing on the right. The mechanical energy of falling water drives turbines (*b*) for the generation of electricity.

(*c*) Schematic drawing of Hoover Dam showing the intake towers and pipes (penstocks) that carry the water to the generators below.

(c)

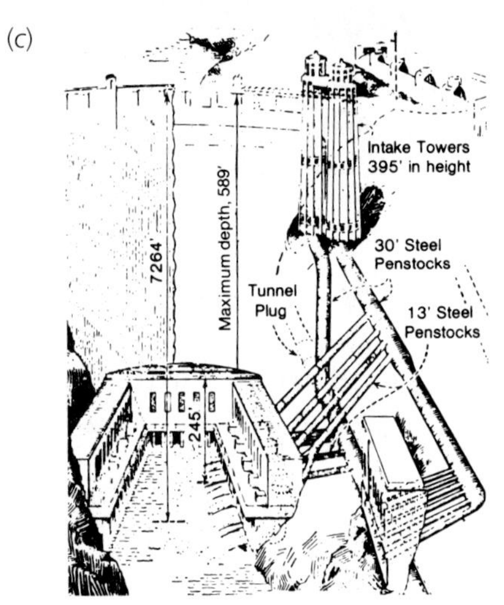

Exercise A 250-turn coil has an area of 3 cm^2. If it rotates in a magnetic field of 0.4 T at 60 Hz, what is $\mathcal{E}_{max}$? (*Answer* $\mathcal{E}_{max} = 11.3$ V)

31-2 Alternating Current in a Resistor

Figure 31-2 shows a simple ac circuit consisting of a generator and a resistor. The plus and minus signs around the generator symbol indicate the sides of higher and lower potential, and the same signs around the resistor indicate the assumed direction of the current. The voltage drop across the resistor V_R is given by

$$V_R = V_+ - V_- = IR \qquad \text{31-5}$$

If $\mathcal{E}$ is the emf supplied by the generator, applying Kirchhoff's loop rule to this circuit gives

$$\mathcal{E} - V_R = 0$$

If the generator produces an emf given by Equation 31-3, we have

$$\mathcal{E}_{max} \sin(\omega t + \delta) - IR = 0$$

In this equation, the phase constant δ is arbitrary. It is convenient to choose $\delta = \pi/2$ so that $\mathcal{E} = \mathcal{E}_{max} \sin(\omega t + \delta) = \mathcal{E}_{max} \cos \omega t$. Then

$$\mathcal{E}_{max} \cos \omega t - IR = 0 \qquad \text{31-6}$$

The current in the resistor is

$$I = \frac{\mathcal{E}_{max}}{R} \cos \omega t = I_{max} \cos \omega t \qquad \text{31-7}$$

where

$$I_{max} = \frac{\mathcal{E}_{max}}{R} \qquad \text{31-8}$$

Note that the current through the resistor is in phase with the voltage across the resistor.

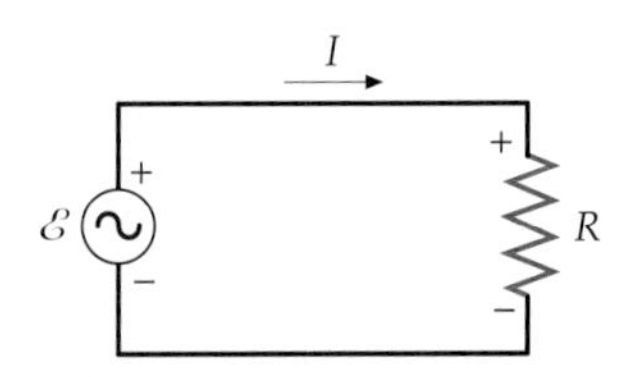

Figure 31-2 An ac generator in series with a resistor *R*.

The power dissipated in the resistor varies with time. Its instantaneous value is

$$P = I^2R = (I_{\text{max}} \cos \omega t)^2 R = I^2_{\text{max}} R \cos^2 \omega t \qquad \text{31-9}$$

Figure 31-3 shows the power as a function of time. It varies from zero to its maximum value $I^2_{\text{max}}R$ as shown. We are usually interested in the average power over one or more cycles:

$$P_{\text{av}} = (I^2R)_{\text{av}} = I^2_{\text{max}} R (\cos^2 \omega t)_{\text{av}}$$

The average value of $\cos^2 \omega t$ over one or more periods is $\frac{1}{2}$.* The average power dissipated in the resistor is thus

$$P_{\text{av}} = (I^2R)_{\text{av}} = \tfrac{1}{2} I^2_{\text{max}} R \qquad \text{31-10}$$

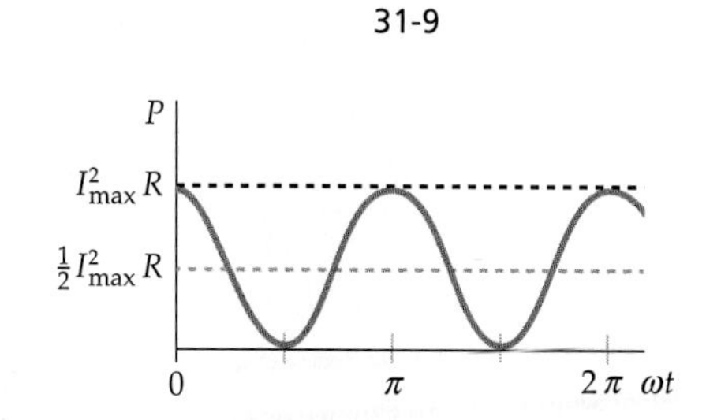

Figure 31-3 Plot of the power dissipated in the resistor in Figure 31-2 versus time. The power varies from zero to a maximum value $I^2_{\text{max}}R$. The average power is half the maximum power.

rms Values

Most ac ammeters and voltmeters are designed to measure **root-mean-square (rms) values** of current and voltage rather than the maximum or peak values. The **rms** value of a current I_{rms} is defined by

$$I_{\text{rms}} = \sqrt{(I^2)_{\text{av}}} \qquad \text{31-11}$$

Definition—rms current

For a sinusoidal current, the average value of I^2 is

$$(I^2)_{\text{av}} = [(I_{\text{max}} \cos \omega t)^2]_{\text{av}} = \tfrac{1}{2} I^2_{\text{max}}$$

Substituting $\frac{1}{2}I^2_{\text{max}}$ for $(I^2)_{\text{av}}$ in Equation 31-11, we obtain

$$I_{\text{rms}} = \frac{1}{\sqrt{2}} I_{\text{max}} \qquad \text{31-12}$$

rms value related to maximum value

The rms value of any quantity that varies sinusoidally equals the maximum value of that quantity divided by $\sqrt{2}$.

Substituting I^2_{rms} for $\frac{1}{2}I^2_{\text{max}}$ in Equation 31-10, we obtain for the average power dissipated in the resistor

$$P_{\text{av}} = I^2_{\text{rms}} R \qquad \text{31-13}$$

The rms current equals the steady dc current that would produce the same Joule heating as the actual ac current.

For the simple circuit in Figure 31-2, the average power delivered by the generator is equal to that dissipated in the resistor:

$$P_{\text{av}} = (\mathcal{E}I)_{\text{av}} = [(\mathcal{E}_{\text{max}} \cos \omega t)(I_{\text{max}} \cos \omega t)]_{\text{av}}$$

$$= \mathcal{E}_{\text{max}} I_{\text{max}} (\cos^2 \omega t)_{\text{av}}$$

or

$$P_{\text{av}} = \tfrac{1}{2} \mathcal{E}_{\text{max}} I_{\text{max}}$$

* This can be seen from the identity $\cos^2 \omega t + \sin^2 \omega t = 1$. A plot of $\sin^2 \omega t$ looks the same as one of $\cos^2 \omega t$ except that it is shifted by 90°. Both have the same average value over one or more periods, and since their sum is 1, the average value of each must be $\frac{1}{2}$.

Using $I_{rms} = I_{max}/\sqrt{2}$ and $\mathcal{E}_{rms} = \mathcal{E}_{max}/\sqrt{2}$, this can be written

$$P_{av} = \mathcal{E}_{rms} I_{rms} \tag{31-14}$$

Average power delivered by a generator

The rms current is related to the rms emf in the same way that the maximum current is related to the maximum emf. We can see this by dividing each side of Equation 31-8 by $\sqrt{2}$ and using $I_{rms} = I_{max}/\sqrt{2}$ and $\mathcal{E}_{rms} = \mathcal{E}_{max}/\sqrt{2}$:

$$I_{rms} = \frac{\mathcal{E}_{rms}}{R} \tag{31-15}$$

Equations 31-13, 31-14, and 31-15 are of the same form as the corresponding equations for direct-current circuits with I replaced by I_{rms} and $\mathcal{E}$ replaced by $\mathcal{E}_{rms}$. We can therefore calculate the power input and the heat generated using the same equations that we used for direct current if we use rms values for the current and emf.

In the circuit of Figure 31-2, which contains only a generator and a resistor, the voltage drop across the resistor equals the voltage of the generator. In more complicated circuits containing several elements, the voltage drop across a resistor is usually not equal to the generator voltage. It is useful, therefore, to write Equation 31-15 in terms of the voltage drop across the resistor $V_{R,rms}$:

$$I_{rms} = \frac{V_{R,rms}}{R} \tag{31-16}$$

Exercise A 12-Ω resistor is connected across a sinusoidal emf that has a peak value of 48 V. Find (*a*) the rms current, (*b*) the average power, and (*c*) the maximum power. (*Answers* (*a*) 2.83 A, (*b*) 96 W, (*c*) 192 W)

The ac power supplied by power companies to dwellings in the United States has a frequency of 60 Hz and a voltage of 120 V_{rms}.* If you plug in a 1600-W heater, it will draw a current of

$$I_{rms} = \frac{P_{av}}{\mathcal{E}_{rms}} = \frac{1600\text{ W}}{120\text{ V}} = 13.3\text{ A}$$

The voltage across the outlets is maintained at 120 V, independent of the current drawn. Thus, all appliances plugged into the outlets of a single circuit are essentially in parallel. If you plug a 500-W toaster into another outlet of the same circuit, it will draw a current of 500 W/120 V = 4.17 A and the total current through the circuit will be 17.5 A. Currents greater than about 20 A may overheat household wiring and create a fire hazard. Each circuit is therefore equipped with a circuit breaker (or a fuse in older houses) that trips (or blows) when the current exceeds 20 A. The maximum power load that can be handled by a circuit with a 20-A circuit breaker is

$$P_{av} = \mathcal{E}_{rms} I_{rms} = (120\text{ V})(20\text{ A}) = 2.4\text{ kW}$$

Since most modern houses require considerably more than 2.4 kW of power, several circuits are supplied, each with its own circuit breaker and each having several outlets.

* For some high-power appliances, such as an electric clothes dryer or an oven, separate lines carrying power at 240 V are often required. For a given power requirement, only half as much current is required at 240 V as at 120 V, but a shock at 240 V is much more likely to be fatal than one at 120 V.

Example 31-1

Find (*a*) the average current and (*b*) the rms current for the sawtooth waveform shown in Figure 31-4. In the region $0 < t < T$, the current is given by $I = (I_0/T)t$.

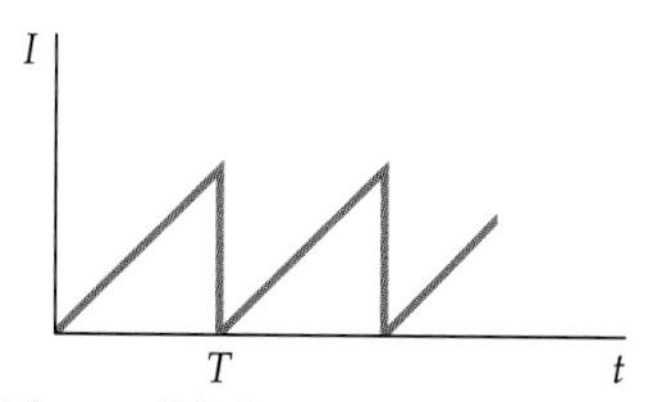

Figure 31-4

Picture the Problem The average of any quantity over a time interval T is the integral of the quantity over the interval divided by T. We use this to find both the average current, I_{av}, and the average of the current squared, $(I^2)_{av}$.

(*a*) Calculate I_{av} by integrating I from $t = 0$ to $t = T$ and dividing by T:

$$I_{av} = \frac{1}{T}\int_0^T I\,dt = \frac{1}{T}\int_0^T \frac{I_0}{T}t\,dt = \frac{I_0}{T^2}\frac{T^2}{2} = \frac{1}{2}I_0$$

(*b*)1. Find $(I^2)_{av}$ by integrating I^2:

$$(I^2)_{av} = \frac{1}{T}\int_0^T I^2\,dt = \frac{1}{T}\int_0^T \left(\frac{I_0}{T}\right)^2 t^2\,dt = \frac{I_0^2}{T^3}\frac{T^3}{3} = \frac{1}{3}I_0^2$$

2. The rms current is the square root of the above result:

$$I_{rms} = \sqrt{(I^2)_{av}} = \frac{I_0}{\sqrt{3}}$$

31-3 Alternating Current in Inductors and Capacitors

Alternating current behaves differently than direct current in inductors and capacitors. When a capacitor becomes fully charged in a dc circuit, it stops the current, that is, it acts like an open circuit. But if the current alternates, charge continually flows onto or off of the plates of the capacitor. We will see that at high frequencies a capacitor hardly impedes the current at all, that is, it acts like a short circuit. Conversely, an inductor coil usually has a very small resistance and is essentially a short circuit for direct current, but when the current is changing, a back emf is generated in an inductor that is proportional to dI/dt. At high frequencies the back emf is large and the inductor acts like an open circuit.

Inductors in ac Circuits

Figure 31-5 shows an inductor coil in series with an ac generator. When the current increases in the inductor, a back emf of magnitude $L\,dI/dt$ is generated due to the changing flux. Usually this back emf is much greater than the drop IR due to the resistance of the coil, so we normally neglect the resistance of the coil. The voltage drop across the inductor V_L is then given by

$$V_L = V_+ - V_- = L\frac{dI}{dt} \qquad \text{31-17}$$

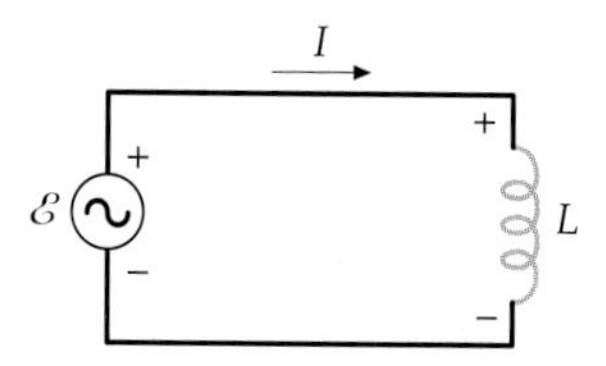

Figure 31-5 An ac generator in series with an inductor L. Plus and minus signs have been placed on the inductor to indicate the direction of the potential drop when dI/dt is positive for the assumed direction of the current. Note that for a positive value of dI/dt, the point at which the current enters the inductor is at a higher potential than the point at which the current leaves.

Applying Kirchhoff's loop rule to this circuit gives

$$\mathcal{E} - V_L = 0$$

or

$$V_L = \mathcal{E} = \mathcal{E}_{max}\cos\omega t$$

$$L\frac{dI}{dt} = \mathcal{E}_{max}\cos\omega t \qquad \text{31-18}$$

Summary

1. Reactance is a frequency-dependent property of capacitors and inductors that is analogous to the resistance of a resistor.
2. Impedance is a frequency-dependent property of an ac circuit or circuit loop that is analogous to the resistance in a dc circuit.
3. Phasors are two-dimensional vectors that allow us to picture the phase relations in a circuit.
4. Resonance occurs when the frequency of the generator equals the natural frequency of the oscillating circuit.

Topic	Remarks and Relevant Equations	
1. ac Generator	An ac generator is a device for transforming mechanical energy into electrical energy. This transformation is accomplished by using the mechanical energy to rotate a conducting coil in a magnetic field.	
emf generated	$\mathcal{E} = \mathcal{E}_{max} \sin(\omega t + \delta) = NBA\omega \sin(\omega t + \delta)$	**31-3, 31-4**
2. Current		
rms current	$I_{rms} = \sqrt{(I^2)_{av}}$	**31-11**
rms current and maximum current	$I_{rms} = \frac{I_{max}}{\sqrt{2}}$	**31-12**
In a resistor	$I_{rms} = \frac{V_{R,rms}}{R}$, voltage and current in phase	**31-16**
In an inductor	$I_{rms} = \frac{V_{L,rms}}{\omega L} = \frac{V_{L,rms}}{X_L}$, voltage leads current by 90°	**31-32**
In a capacitor	$I_{rms} = \frac{V_{C,rms}}{1/\omega C} = \frac{V_{C,rms}}{X_C}$, voltage lags current by 90°	**31-33**
3. Reactance		
Inductive reactance	$X_L = \omega L$	**31-25**
Capacitive reactance	$X_C = \frac{1}{\omega C}$	**31-31**
4. Average Power Dissipation		
In a resistor	$P_{av} = \frac{1}{2}\mathcal{E}_{max} I_{max} = \mathcal{E}_{rms} I_{rms} = I_{rms}^2 R$	**31-13, 31-14, 31-56**
In an inductor	$P_{av} = 0$	
In a capacitor	$P_{av} = 0$	
5. Phasors	Phasors are two-dimensional vectors that represent the current $\vec{I}$, the voltage across a resistor $\vec{V}_R$, the voltage across a capacitor $\vec{V}_C$, and the voltage across an inductor $\vec{V}_L$ in	

an ac circuit. These phasors rotate in the counterclockwise direction with an angular frequency ω that is equal to the angular frequency of the current. $\vec{V}_R$ is in phase with the current, $\vec{V}_L$ leads the current by 90°, and $\vec{V}_C$ lags the current by 90°. The x component of each phasor equals the magnitude of the current or the corresponding voltage drop at any instant.

6. *LC* and *RLC* Series Circuit

If a capacitor is discharged through an inductor, the charge and voltage on the capacitor oscillate with angular frequency

$$\omega = \frac{1}{\sqrt{LC}} \qquad \textbf{31-41}$$

The current in the inductor oscillates with the same frequency, but it is out of phase with the charge by 90°. The energy oscillates between electric energy in the capacitor and magnetic energy in the inductor. If the circuit also has resistance, the oscillations are damped because energy is dissipated in the resistor.

7. Series *RLC* Circuit Driven by a Generator of Frequency ω

Current

$$I = \frac{\mathcal{E}_{\max}}{Z}\cos(\omega t - \delta) \qquad \textbf{31-54}$$

Impedance Z

$$Z = \sqrt{R^2 + (X_L - X_C)^2} \qquad \textbf{31-53}$$

Phase angle δ

$$\tan\delta = \frac{X_L - X_C}{R} \qquad \textbf{31-51}$$

Average power

$$P_{\text{av}} = \tfrac{1}{2}I_{\max}^2 R = I_{\text{rms}}^2 R = \mathcal{E}_{\text{rms}} I_{\text{rms}} \cos\delta$$

$$= \frac{\mathcal{E}_{\text{rms}}^2 R\omega^2}{L^2(\omega^2 - \omega_0^2)^2 + \omega^2 R^2} \qquad \textbf{31-56, 31-57, 31-58}$$

Power factor

The quantity $\cos\delta$ in Equation 31-57 is called the power factor of the *RLC* circuit. At resonance, δ is zero, and the power factor is 1. Then

$$(P_{\text{av}})_{\max} = \mathcal{E}_{\text{rms}} I_{\text{rms}}$$

Resonance

When the current is maximum, the circuit is said to be at resonance. The conditions for resonance are

$$X_L = X_C, \quad Z = R$$

$$\omega = \omega_0 = \frac{1}{\sqrt{LC}}$$

$$\delta = 0$$

8. *Q* Factor

The sharpness of the resonance curve is described by the Q factor,

$$Q = \frac{\omega_0 L}{R} \qquad \textbf{31-59}$$

When the resonance curve is reasonably narrow, the Q factor can be approximated by

$$Q \approx \frac{\omega_0}{\Delta\omega} = \frac{f_0}{\Delta f} \qquad \textbf{31-60}$$

9. Transformers

A transformer is a device used to raise or lower the voltage in a circuit without an appreciable loss in power. For a transformer with N_1 turns in the primary and N_2 turns

in the secondary, the voltage across the secondary coil is related to the generator emf across the primary coil by

$$V_2 = \frac{N_2}{N_1} V_1 \tag{31-65}$$

If there are no losses,

$$V_{1,\text{rms}} I_{1,\text{rms}} = V_{2,\text{rms}} I_{2,\text{rms}} \tag{31-67}$$

Problem-Solving Guide

1. Draw a neat sketch of the appropriate circuit, showing all circuit elements—capacitors, resistors, inductors, and generators. Indicate a direction for the current I and label the appropriate high- and low-voltage sides of the elements.
2. When using phasor analysis, draw vectors for $\vec{V}_R$, $\vec{V}_L$, and $\vec{V}_C$, being careful to draw $\vec{V}_C$ at 90° clockwise from $\vec{V}_R$, and $\vec{V}_L$ at 90° counterclockwise from $\vec{V}_R$.

Summary of Worked Examples

Type of Calculation	Procedure and Relevant Examples
1. rms Current	
Find the average and rms current for a given time interval.	The average current is $I_{\text{av}} = \frac{1}{T}\int_0^T I\,dt$. The rms current is $I_{\text{rms}} = \sqrt{(I^2)_{\text{av}}}$, where $(I^2)_{\text{av}} = \frac{1}{T}\int_0^T I^2\,dt$ **Example 31-1**
2. Reactance	
Find the reactance of inductors and capacitors.	Use $X_L = \omega L$ and $X_C = 1/\omega C$. **Examples 31-2, 31-3**
3. *LC* and *RLC* Circuits	
Determine the frequency and maximum current in an *LC* circuit.	The frequency of oscillation is $\omega = 1/\sqrt{LC}$ and the maximum current is $I_{\max} = \omega Q_0$, where Q_0 is the initial charge on the capacitor. **Example 31-4**
Analyze an *RLC* circuit with a generator.	Use $\omega_0 = 1/\sqrt{LC}$, $I_{\max} = \mathcal{E}_{\max}/Z$, $Z = \sqrt{R^2 + (X_L - X_C)^2}$, $\tan\delta = (X_L - X_C)/R$, and $Q = \omega_0 L/R$. **Examples 31-5, 31-6, 31-7, 31-8, 31-9**
4. Transformers	
Relate the currents in the primary and secondary coils of a transformer.	The currents are related by $N_1 I_1 = -N_2 I_2$, where subscripts 1 and 2 refer to the primary and secondary coils, respectively. **Example 31-10**

Problems

Conceptual Problems

Problems from Optional and Exploring sections

In a few problems, you are given more data than you actually need; in a few other problems, you are required to supply data from your general knowledge, outside sources, or informed estimates.

• Single-concept, single-step, relatively easy
•• Intermediate-level, may require synthesis of concepts
••• Challenging, for advanced students

Generators

1 • A 200-turn coil has an area of 4 cm^2 and rotates in a magnetic field of 0.5 T. (*a*) What frequency will generate a maximum emf of 10 V? (*b*) If the coil rotates at 60 Hz, what is the maximum emf?

2 • In what magnetic field must the coil of Problem 1 be rotating to generate a maximum emf of 10 V at 60 Hz?

3 • A 2-cm by 1.5-cm rectangular coil has 300 turns and rotates in a magnetic field of 4000 G. (*a*) What is the maximum emf generated when the coil rotates at 60 Hz? (*b*) What must its frequency be to generate a maximum emf of 110 V?

4 • The coil of Problem 3 rotates at 60 Hz in a magnetic field B. What value of B will generate a maximum emf of 24 V?

Alternating Current in a Resistor

5 • As the frequency in the simple ac circuit in Figure 31-26 increases, the rms current through the resistor

(*a*) increases.
(*b*) does not change.
(*c*) may increase or decrease depending on the magnitude of the original frequency.
(*d*) may increase or decrease depending on the magnitude of the resistance.
(*e*) decreases.

Figure 31-26 Problem 5

6 • If the rms voltage in an ac circuit is doubled, the peak voltage is

(*a*) increased by a factor of 2.
(*b*) decreased by a factor of 2.
(*c*) increased by a factor of $\sqrt{2}$.
(*d*) decreased by a factor of $\sqrt{2}$.
(*e*) not changed.

7 • A 100-W light bulb is plugged into a standard 120-V (rms) outlet. Find (*a*) I_{rms}, (*b*) I_{max}, and (*c*) the maximum power.

8 • A 3-Ω resistor is placed across a generator having a frequency of 60 Hz and a maximum emf of 12.0 V. (*a*) What is the angular frequency ω of the current? (*b*) Find I_{max} and I_{rms}. What is (*c*) the maximum power into the resistor, (*d*) the minimum power, and (*e*) the average power?

9 • A circuit breaker is rated for a current of 15 A rms at a voltage of 120 V rms. (*a*) What is the largest value of I_{max} that the breaker can carry? (*b*) What average power can be supplied by this circuit?

Alternating Current in Inductors and Capacitors

10 • If the frequency in the circuit shown in Figure 31-27 is doubled, the inductance of the inductor will

(*a*) increase by a factor of 2.
(*b*) not change.
(*c*) decrease by a factor of 2.
(*d*) increase by a factor of 4.
(*e*) decrease by a factor of 4.

Figure 31-27 Problems 10 and 11

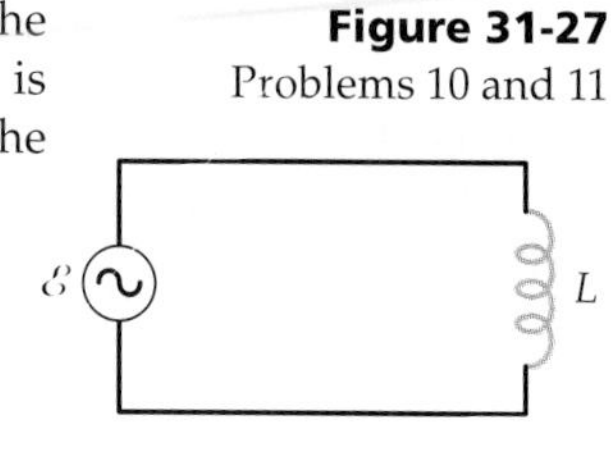

11 • If the frequency in the circuit shown in Figure 31-27 is doubled, the inductive reactance of the inductor will

(*a*) increase by a factor of 2.
(*b*) not change.
(*c*) decrease by a factor of 2.
(*d*) increase by a factor of 4.
(*e*) decrease by a factor of 4.

12 • If the frequency in the circuit in Figure 31-28 is doubled, the capacitative reactance of the circuit will

(*a*) increase by a factor of 2.
(*b*) not change.
(*c*) decrease by a factor of 2.
(*d*) increase by a factor of 4.
(*e*) decrease by a factor of 4.

Figure 31-28 Problem 12

13 • In a circuit consisting of a generator and an inductor, are there any times when the inductor absorbs power from the generator? Are there any times when the inductor supplies power to the generator?

14 • In a circuit consisting of a generator and a capacitor, are there any times when the capacitor absorbs power from the generator? Are there any times when the capacitor supplies power to the generator?

15 • What is the reactance of a 1.0-mH inductor at (*a*) 60 Hz, (*b*) 600 Hz, and (*c*) 6 kHz?

16 • An inductor has a reactance of 100 Ω at 80 Hz. (*a*) What is its inductance? (*b*) What is its reactance at 160 Hz?

17 • At what frequency would the reactance of a 10.0-μF capacitor equal that of a 1.0-mH inductor?

18 • What is the reactance of a 1.0-nF capacitor at (*a*) 60 Hz, (*b*) 6 kHz, and (*c*) 6 MHz?

19 • An emf of 10.0 V maximum and frequency 20 Hz is applied to a 20-μF capacitor. Find (*a*) I_{max} and (*b*) I_{rms}.

20 • At what frequency is the reactance of a 10-μF capacitor (*a*) 1 Ω, (*b*) 100 Ω, and (*c*) 0.01 Ω?

crowaves are therefore readily absorbed by the water molecules in foods, which is the mechanism by which food is heated in microwave ovens.

Production of Electromagnetic Waves

Electromagnetic waves are produced when free electric charges accelerate, or when electrons bound to atoms and molecules make transitions to lower energy states. Radio waves, which have frequencies from about 550 to 1600 kHz for AM and from about 88 to 108 MHz for FM, are produced by macroscopic electric currents oscillating in radio antennas. The frequency of the emitted waves equals the frequency of oscillation of the charges.

A continuous spectrum of X rays is produced by the deceleration of electrons when they crash into a metal target. The radiation produced is called **bremsstrahlung** (German for braking radiation). Accompanying the broad, continuous bremsstrahlung spectrum is a discrete spectrum of X-ray lines produced by transitions of inner electrons in the atoms of the target material.

Synchrotron radiation arises from the circular orbital motion of charged particles (usually electrons or positrons) in nuclear accelerators called synchrotrons. Originally considered a nuisance by accelerator scientists, synchrotron radiation X rays are now produced and used as a medical diagnostic tool because of the ease of manipulating the beams with reflection and diffraction optics. Synchrotron radiation is also emitted by charged particles trapped in magnetic fields associated with stars and galaxies. It is believed that most low-frequency radio waves reaching the earth from outer space originate as synchrotron radiation.

Heat is radiated by the thermally excited molecular charges. The spectrum of heat radiation is the blackbody radiation spectrum discussed in Section 21-4.

Light waves, which have frequencies of the order of 10^{14} Hz, are generally produced by transitions of bound atomic charges. We discuss sources of light waves in Chapter 33.

Electric-Dipole Radiation

Figure 32-4 is a schematic drawing of an electric-dipole radio antenna consisting of two conducting rods along a line fed by an alternating-current generator. At time $t = 0$ (Figure 32-4*a*), the ends of the rods are charged, and there is an electric field near the rod parallel to the rod. There is also a magnetic field, not shown, encircling the rods due to the current in the rods. These fields move out away from the rods with the speed of light. After one-fourth period, at $t = T/4$ (Figure 32-4*b*), the rods are uncharged, and the electric field near the rod is zero. At $t = T/2$ (Figure 32-4*c*), the rods are again charged, but the charges are opposite those at $t = 0$. The electric and magnetic fields at a great distance from the antenna are quite different from the fields near the antenna. Far from the antenna, the electric and magnetic fields oscillate in phase with simple harmonic

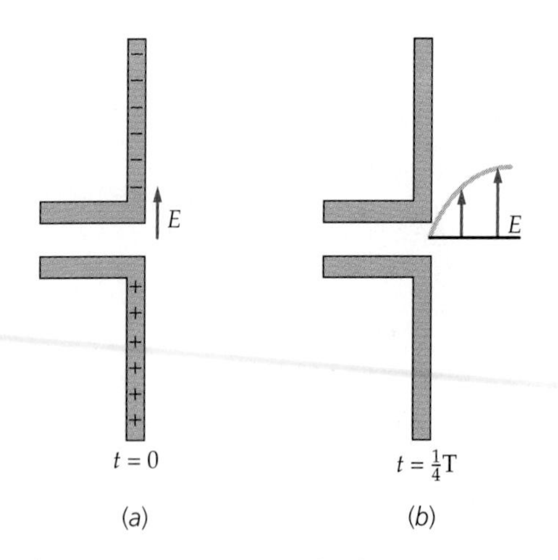

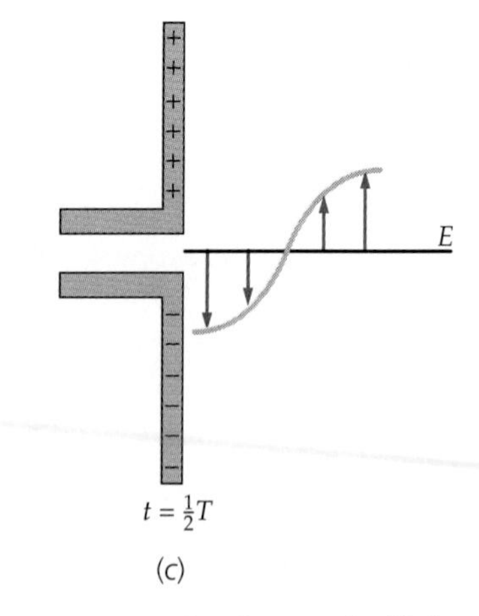

Figure 32-4 An electric-dipole antenna for radiating electromagnetic waves. Alternating current is supplied to the antenna by a generator (not shown). The electric field due to the charges in the antenna propagates outward at the speed of light. There is also a propagating magnetic field (not shown) perpendicular to the paper due to the current in the antenna.

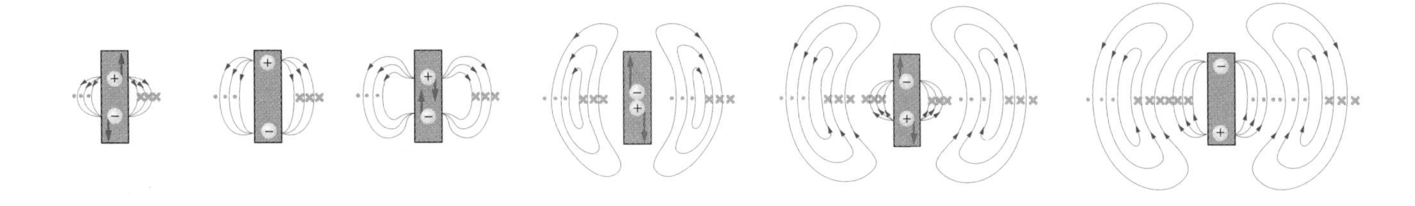

motion, perpendicular to each other and to the direction of propagation of the wave. Figure 32-5 shows the electric and magnetic fields far from an electric-dipole antenna.

Electromagnetic waves of radio or television frequencies can be detected by a dipole antenna placed parallel to the electric field so that it induces an alternating current in the antenna (Figure 32-6). They can also be detected by a loop antenna placed perpendicular to the magnetic field so that the changing magnetic flux through the loop induces a current in the loop (Figure 32-7). Electromagnetic waves of frequency in the visible light range are detected by the eye or by photographic film, both of which are mainly sensitive to the electric field.

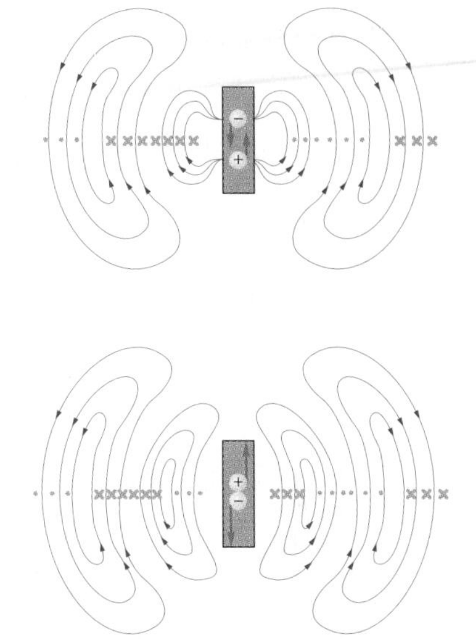

Figure 32-5 Electric and magnetic field lines produced by an oscillating electric dipole.

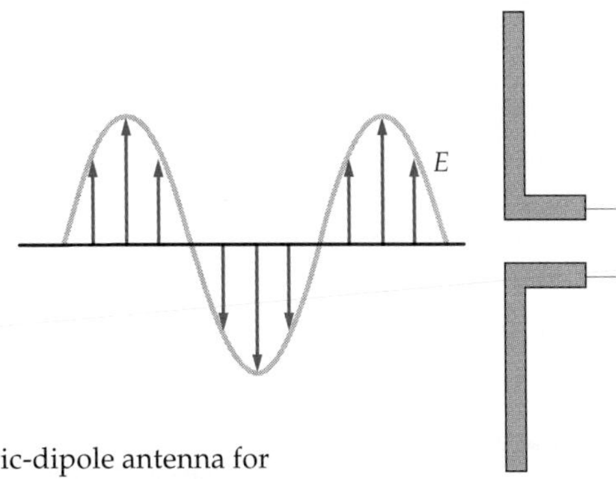

Figure 32-6 Electric-dipole antenna for detecting electromagnetic waves. The alternating electric field of the wave produces an alternating current in the antenna.

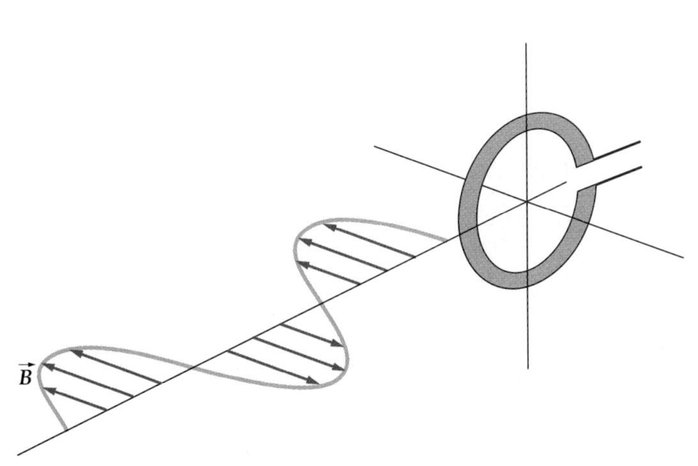

Figure 32-7 A loop antenna for detecting electromagnetic radiation. The alternating magnetic flux in the loop due to the magnetic field of the radiation induces an alternating current in the loop.

The radiation from a dipole antenna such as that in Figure 32-4 is called electric-dipole radiation. Many electromagnetic waves exhibit the characteristics of electric-dipole radiation. An important feature of this type of radiation is that the intensity of the electromagnetic waves radiated by a dipole antenna is zero along the axis of the antenna and maximum in the directions perpendicular to the axis. If the dipole is in the y direction with its center at the origin as in Figure 32-8, the intensity is zero along the y axis and maximum in the xz plane. In the direction of a line making an angle θ with the y axis, the intensity is proportional to $\sin^2 \theta$.

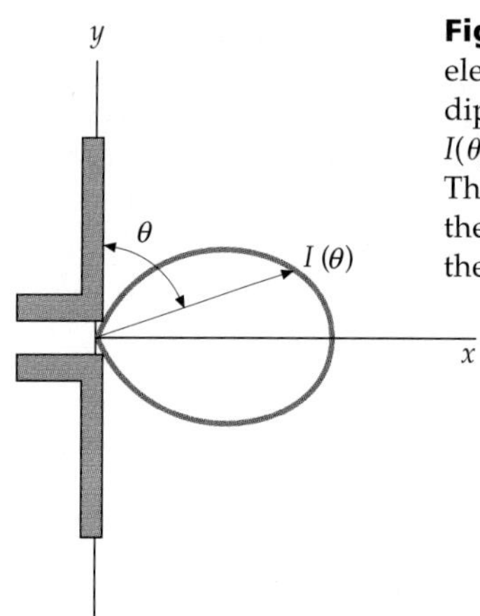

Figure 32-8 Polar plot of the intensity of electromagnetic radiation from an electric-dipole antenna versus angle. The intensity $I(\theta)$ is proportional to the length of the arrow. The intensity is maximum perpendicular to the antenna at $\theta = 90°$ and minimum along the antenna at $\theta = 0°$ or $\theta = 180°$.

Example 32-3

An antenna consisting of a single loop of wire of radius 10 cm is used to detect electromagnetic waves for which $E_{rms} = 0.15$ V/m. Find the rms emf induced in the loop if the wave frequency is (*a*) 600 kHz and (*b*) 600 MHz.

Picture the Problem The induced emf in the coil is related to the rate of change of the magnetic flux by Faraday's law.

(*a*)1. Faraday's law relates the magnitude of the emf to the rate of change of the magnetic flux:

$$|\mathcal{E}| = \frac{d\phi_m}{dt} = \pi r^2 \frac{dB}{dt}$$

$$\mathcal{E}_{rms} = \pi r^2 \left(\frac{dB}{dt}\right)_{rms}$$

2. Compute dB_{rms}/dt from a sinusoidal B:

$$B = B_0 \sin(kx - \omega t)$$

$$\frac{dB}{dt} = -\omega B_0 \cos(kx - \omega t)$$

$$\left(\frac{dB}{dt}\right)_{rms} = \omega B_{rms}$$

3. Relate B_{rms} to E_{rms}:

$$B_{rms} = \frac{E_{rms}}{c}$$

$$\left(\frac{dB}{dt}\right)_{rms} = \frac{\omega E_{rms}}{c} = \frac{2\pi f}{c} E_{rms}$$

4. Calculate $\mathcal{E}_{rms}$ at $f = 600$ Hz:

$$\mathcal{E}_{rms} = \pi r^2 \left(\frac{dB}{dt}\right)_{rms} = \pi r^2 \frac{2\pi f}{c} E_{rms}$$

$$= \pi(0.1\text{ m})^2\, 2\pi(6 \times 10^5\text{ Hz})(0.15\text{ V/m})/(3 \times 10^8\text{ m/s})$$

$$= 5.92 \times 10^{-5}\text{ V}$$

(*b*) The induced emf is proportional to the frequency, so at 600 MHz it will be 1000 times greater than at 600 kHz:

$$\mathcal{E}_{rms} = (10^3)(5.92 \times 10^{-5}\text{ V}) = 0.0592\text{ V}$$

Energy and Momentum in an Electromagnetic Wave

Like all waves, electromagnetic waves carry energy and momentum. The energy carried is described by the intensity, the average power per unit area incident on a surface perpendicular to the direction of propagation. The momentum per unit time per unit area carried by an electromagnetic wave is called the **radiation pressure**.

Intensity The intensity of a wave equals the product of the wave speed and the average energy density u_{av} (Section 15-3).

$$I = u_{av}\, c$$

The total energy density in the wave u is the sum of the electric and magnetic energy densities. The electric energy density u_e (Equation 25-13) and magnetic energy density u_m (Equation 30-19) are given by

$$u_e = \frac{1}{2}\epsilon_0 E^2 \quad \text{and} \quad u_m = \frac{B^2}{2\mu_0}$$

In an electromagnetic wave in free space, E equals cB, so we can express the magnetic energy density in terms of the electric field:

$$u_m = \frac{B^2}{2\mu_0} = \frac{(E/c)^2}{2\mu_0} = \frac{E^2}{2\mu_0 c^2} = \frac{1}{2}\epsilon_0 E^2$$

where we have used $c^2 = 1/\epsilon_0\mu_0$. Thus, the electric and magnetic energy densities are equal. Using $E = cB$, we may express the total energy density in several useful ways:

$$u = u_e + u_m = \epsilon_0 E^2 = \frac{B^2}{\mu_0} = \frac{EB}{\mu_0 c} \qquad \text{32-8}$$

Energy density in an electromagnetic wave

To compute the *average* energy density, we replace the instantaneous fields E and B by their rms values $E_{rms} = E_0/\sqrt{2}$ and $B_{rms} = B_0/\sqrt{2}$, where E_0 and B_0 are the maximum values of the fields. The intensity is then

$$I = u_{av}c = \frac{E_{rms}B_{rms}}{\mu_0} = \frac{1}{2}\frac{E_0B_0}{\mu_0} = |\vec{S}|_{av} \qquad \text{32-9}$$

Intensity of an electromagnetic wave

where the vector

$$\vec{S} = \frac{\vec{E} \times \vec{B}}{\mu_0} \qquad \text{32-10}$$

Definition—Poynting vector

is called the **Poynting vector** after its discoverer, Sir John Poynting. The average magnitude of $\vec{S}$ is the intensity of the wave, and the direction of $\vec{S}$ is the direction of propagation of the wave.

Radiation Pressure We now show by a simple example that an electromagnetic wave carries momentum. Consider a wave moving along the x axis that is incident on a stationary charge as shown in Figure 32-9. For simplicity, we assume that $\vec{E}$ is in the y direction and $\vec{B}$ is in the z direction, and we neglect

Physical Mechanisms for Reflection and Refraction

The physical mechanism of the reflection and refraction of light can be understood in terms of the absorption and reradiation of the light by the atoms in the reflecting or refracting medium. When light traveling in air strikes a glass surface, the atoms in the glass absorb the light and reradiate it at the same frequency in all directions. The waves radiated backward by the glass atoms interfere constructively at an angle equal to the angle of incidence to produce the reflected wave.

The transmitted wave is the result of the interference of the incident wave and the wave produced by the absorption and reradiation of light energy by the atoms in the medium. For light entering glass from air, there is a phase lag between the reradiated wave and the incident wave. There is therefore also a phase lag between the resultant wave and the incident wave. This phase lag means that the position of a wave crest of the transmitted wave is retarded relative to the position of a wave crest of the incident wave in the medium. As a result, the transmitted wave does not travel as far in a given time as the original incident wave; that is, the velocity of the transmitted wave is less than that of the incident wave. The index of refraction is therefore greater than 1. The frequency of the light in the second medium is the same as that of the incident light—the atoms absorb and reradiate the light at the same frequency—but the wave speed is different, so the wavelength of the transmitted light is different from that of the incident light. If λ is the wavelength of light in a vacuum, the wavelength λ' in a medium of index of refraction n is

$$\lambda' = \frac{v}{f} = \frac{c/n}{f} = \frac{\lambda}{n} \qquad 33\text{-}10$$

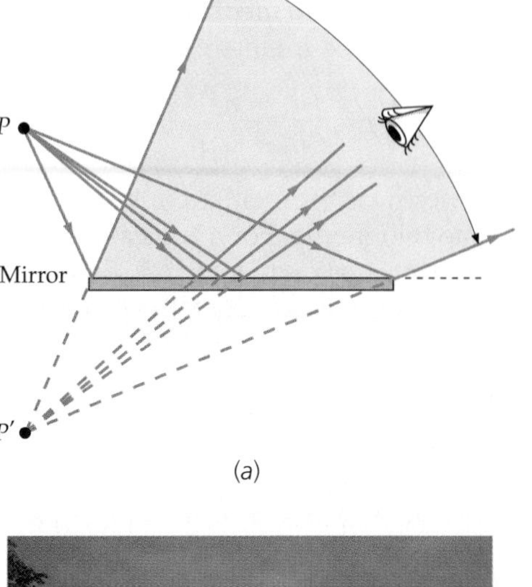

(a)

(b)

Figure 33-19 (*a*) Specular reflection from a smooth surface. (*b*) Specular reflection of trees from water.

Specular and Diffuse Reflection

Figure 33-19*a* shows a bundle of light rays from a point source P that are reflected from a flat surface. After reflection, the rays diverge exactly as if they came from a point P' behind the surface. (This point is called the *image point.* We will study the formation of images by reflecting and refracting surfaces in the next chapter.) When these rays enter the eye, they cannot be distinguished from rays actually diverging from a source at P'.

Reflection from a smooth surface is called **specular reflection.** It differs from **diffuse reflection,** which is illustrated in Figure 33-20. Here, because the surface is rough, the rays from a point reflect in random directions and do not diverge from any point, so there is no image. The reflection of light from the page of this book is diffuse reflection. The glass used in picture frames is sometimes ground slightly to give diffuse reflection and thereby cut down on glare from the light used to illuminate the picture. Diffuse reflection from the surface of the road allows you to see the road when you are driving at night because some of the light from your headlights reflects back toward you.

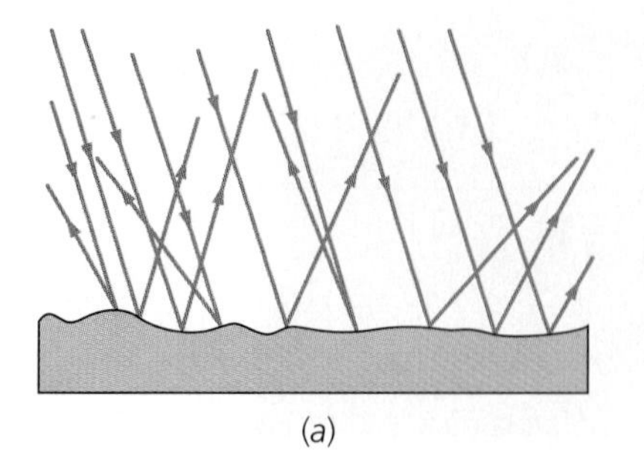
(a)

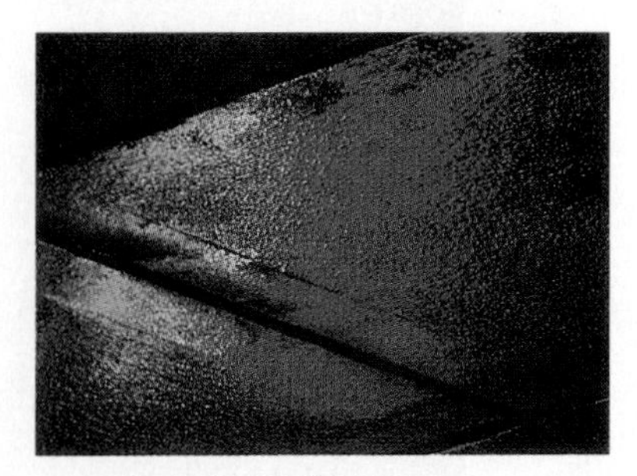
(b)

Figure 33-20 (*a*) Diffuse reflection from a rough surface. (*b*) Diffuse reflection of colored lights from a sidewalk.

Relative Intensity of Reflected and Transmitted Light

The fraction of light energy reflected at a boundary such as an air–glass interface depends in a complicated way on the angle of incidence, the orientation of the electric field vector associated with the wave, and the indices of refraction of the two media. For the special case of normal incidence ($\theta_1 = \theta_1' = 0$), the reflected intensity can be shown to be

$$I = \left(\frac{n_1 - n_2}{n_1 + n_2}\right)^2 I_0 \qquad \text{33-11}$$

where I_0 is the incident intensity and n_1 and n_2 are the indexes of refraction of the two media. For a typical case of reflection from an air–glass interface for which $n_1 = 1$ and $n_2 = 1.5$, Equation 33-11 gives $I = I_0/25$. Only about 4% of the energy is reflected; the rest is transmitted.

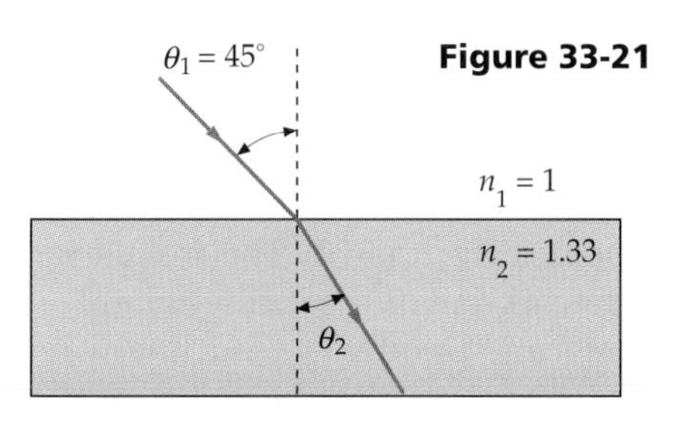

Figure 33-21

Example 33-4

Light traveling in air enters water with an angle of incidence of 45°. If the index of refraction of water is 1.33, what is the angle of refraction?

Picture the Problem The angle of refraction is found using Snell's law. Let subscripts 1 and 2 refer to the air and water, respectively. Then $n_1 = 1$, $\theta_1 = 45°$, $n_2 = 1.33$, and θ_2 is the angle of refraction (Figure 33-21).

1. Use Snell's law to solve for $\sin\theta_2$, the sine of the angle of refraction:

$$n_1 \sin\theta_1 = n_2 \sin\theta_2$$

$$\sin\theta_2 = \left(\frac{n_1}{n_2}\right)\sin\theta_1 = \left(\frac{1}{1.33}\right)\sin 45° = \frac{0.707}{1.33} = 0.532$$

2. Find the angle whose sine is 0.532:

$$\theta_2 = \sin^{-1} 0.532 = 32.1°$$

Remark Note that light is bent closer to the normal in the medium with the larger index of refraction.

Total Internal Reflection

Figure 33-22 shows a point source in glass with rays striking the glass–air interface at various angles. All the rays not perpendicular to the interface are bent away from the normal. As the angle of incidence is increased, the angle of refraction increases until a critical angle of incidence θ_c is reached for

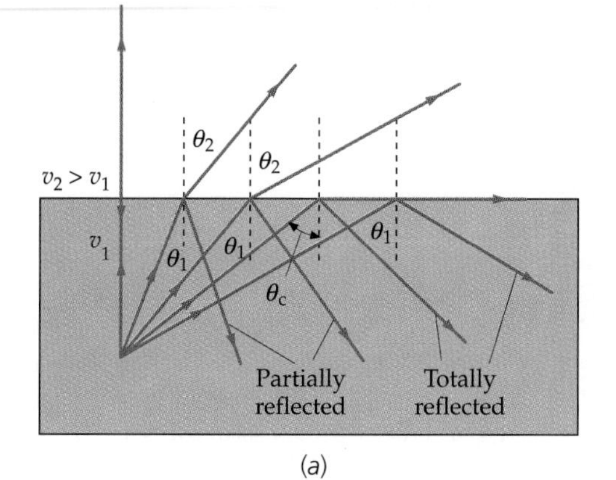

(a)

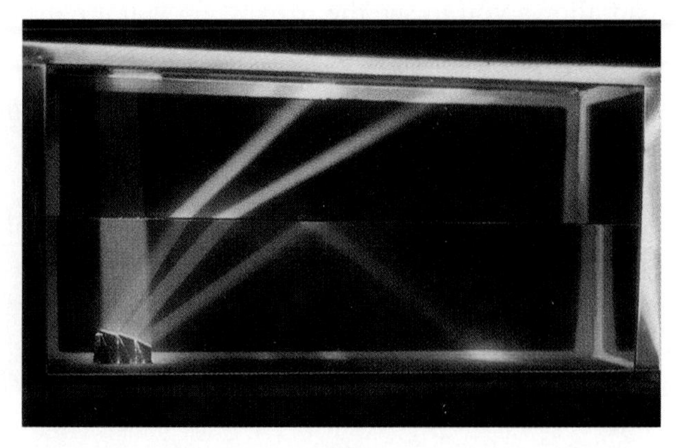
(b)

Figure 33-22 (*a*) Total internal reflection. As the angle of incidence is increased, the angle of refraction is increased until, at a critical angle of incidence θ_c, the angle of refraction is 90°. For angles of incidence greater than the critical angle, there is no refracted ray. (*b*) Photograph of refraction and total internal reflection from a water–air interface.

which the angle of refraction is 90°. For incident angles greater than this critical angle, there is no refracted ray. All the energy is reflected. This phenomenon is called **total internal reflection.** The critical angle can be found in terms of the indexes of refraction of the two media by solving Equation 33-9*b* for sin θ_1 and setting θ_2 equal to 90°:

$$\sin \theta_c = \frac{n_2}{n_1} \sin 90° = \frac{n_2}{n_1} \qquad 33\text{-}12$$

Critical angle for total internal reflection

Note that total internal reflection occurs only when the light is originally in the medium with the higher index of refraction. Mathematically, if n_2 is greater than n_1, Snell's law cannot be satisfied because there is no real angle whose sine is greater than 1.

Example 33-5 *try it yourself*

A particular glass has an index of refraction of $n = 1.50$. What is the critical angle for total internal reflection for light leaving this glass and entering air, for which $n = 1.00$?

Cover the column to the right and try these on your own before looking at the answers.

Steps	***Answers***
1. Calculate the sine of the critical angle from Equation 33-12.	$\sin \theta_c = 0.667$
2. Use your result to find the angle.	$\theta_c = 41.8°$

Figure 33-23*a* shows light incident normally on one of the short sides of a 45–45–90° glass prism. If the index of refraction of the prism is 1.5, the critical angle for total internal reflection is 41.8°, as you found in Example 33-5. Since the angle of incidence of the ray on the glass–air interface is 45°, the light will be totally reflected and will exit perpendicular to the other face of the prism as shown. In Figure 33-23*b*, the light is incident perpendicular to the hypotenuse of the prism and is totally reflected twice such that it emerges at 180° to its original direction. Prisms are used to change the directions of light rays. In binoculars, four prisms are used to reinvert the image that was inverted by the binocular lens. Diamonds have a very high index of refraction ($n \approx 2.4$), so nearly all the light that enters a diamond is eventually reflected back out, giving the diamond its sparkle.

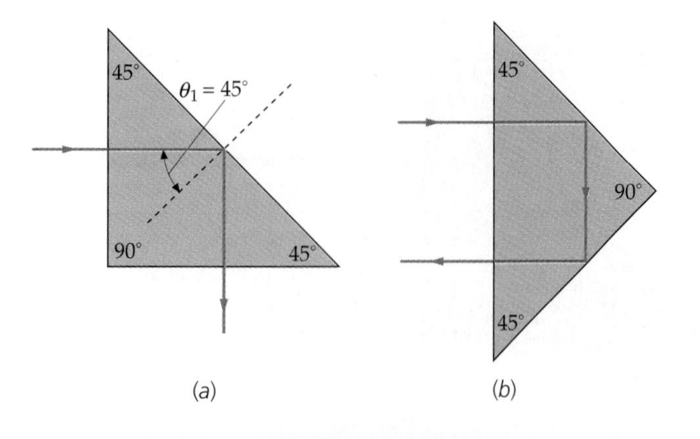

Figure 33-23 (*a*) Light entering through one of the short sides of a 45–45–90° glass prism is totally reflected. (*b*) Light entering through the long side of the prism is totally reflected twice.

Fiber Optics An interesting application of total internal reflection is the transmission of a beam of light down a long, narrow, transparent glass fiber (Figure 33-24*a*). If the beam begins approximately parallel to the axis of the fiber, it will strike the walls of the fiber at angles greater than the critical angle (if the bends in the fiber are not too sharp) and no light energy will be lost through the walls of the fiber. A bundle of such fibers can be used for imaging, as illustrated in Figure 33-24*b*. Fiber optics has many applications in medicine and in communications. In medicine, light is transmitted along tiny fibers to visually probe various internal organs without surgery. In communications, the rate at which information can be transmitted is related to the signal frequency. A transmission system using light of frequencies of the order of 10^{14} Hz can transmit information at a much greater rate than one using radio waves, which have frequencies of the order of 10^{6} Hz. In telecommunication systems, a single glass fiber the thickness of a human hair can transmit audio or video information equivalent to 32,000 voices speaking simultaneously.

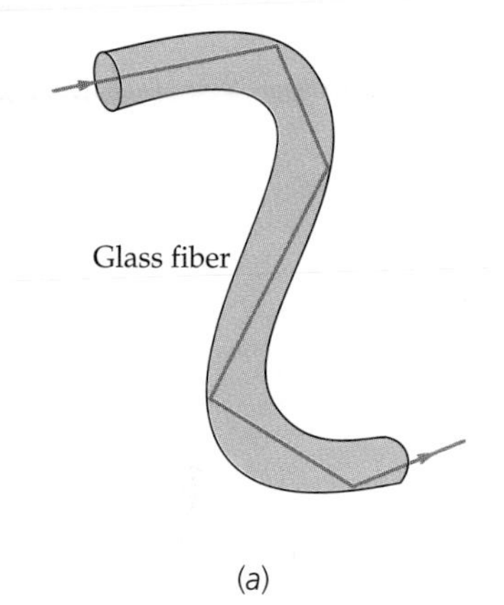

(*a*)

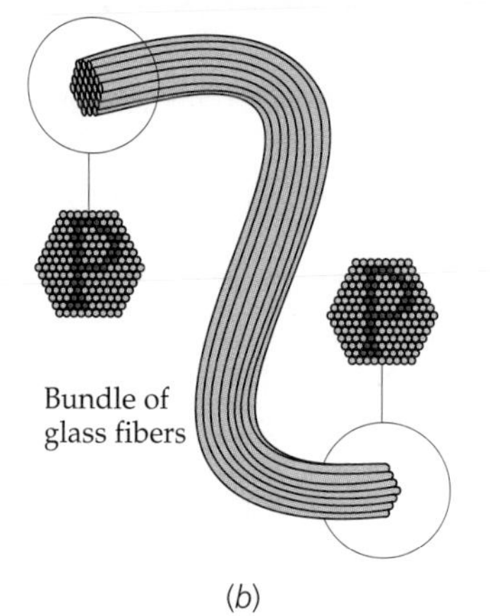

(*b*)

(*c*)

Figure 33-24 (*a*) A light pipe. Light inside the pipe is always incident at an angle greater than the critical angle, so no light escapes the pipe by refraction. (*b*) Light from the object is transported by a bundle of glass fibers to form an image of the object at the other end of the pipe. (*c*) Light emerging from a bundle of glass fibers.

(*a*)

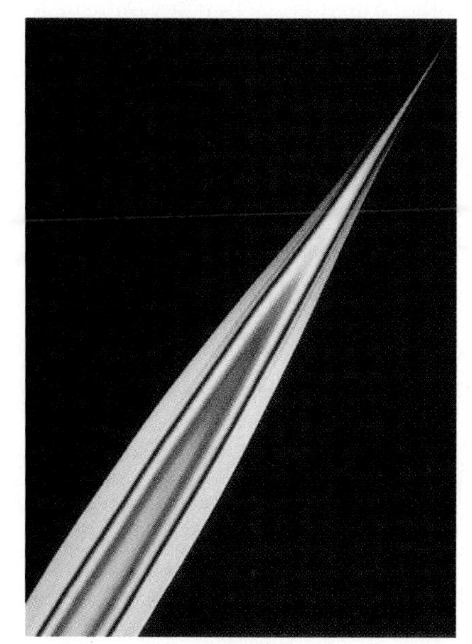

(*b*)

(*a*) In this demonstration at the Naval Research Laboratory, a combination of laser sources generates different colors that excite adjacent fiber sensor elements, leading to a separation of the information as indicated by the separation of the colors. (*b*) The tip of a light guide preform is softened by heat and drawn into a long, tiny fiber. The colors in the preform indicate a layered structure of differing compositions, which is retained in the fiber.

Example 34-3 *try it yourself*

Derive an expression for the magnification $m = y'/y$ of an image formed by a refracting surface.

Picture the Problem The magnification is the ratio of y' to y. These heights are related to the tangents of the angles θ_1 and θ_2, as shown in Figure 34-22. The angles are related by Snell's law. For paraxial rays, you can approximate $\tan \theta \approx \sin \theta \approx \theta$.

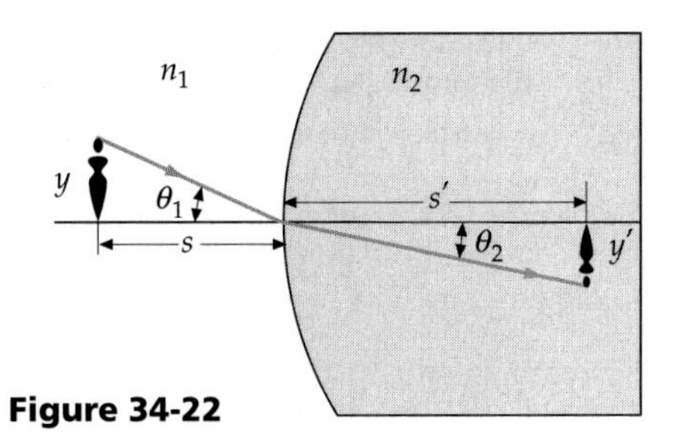

Figure 34-22

Cover the column to the right and try these on your own before looking at the answers.

Steps	*Answers*
1. Write expressions for $\tan \theta_1$ and $\tan \theta_2$ in terms of the heights y and $-\,y'$, and the object and image distances s and s'. (Since y' is negative, use $-y'$ so that $\tan \theta_2$ is positive.)	$\tan \theta_1 = \dfrac{y}{s};\quad \tan \theta_2 = \dfrac{-y'}{s'}$
2. Apply the small angle approximation $\tan \theta \approx \theta$ to your expressions.	$\theta_1 = \dfrac{y}{s};\quad \theta_2 = \dfrac{-y'}{s'}$
3. Write Snell's law relating the angles θ_1 and θ_2 using the small-angle approximation $\sin \theta \approx \theta$.	$n_1\theta_1 = n_2\theta_2$
4. Substitute the expressions for θ_1 and θ_2 found in step 2.	$n_1\left(\dfrac{y}{s}\right) = n_2\left(\dfrac{-y'}{s'}\right)$
5. Solve for the magnification $m = y'/y$.	$m = \dfrac{y'}{y} = -\dfrac{n_1 s'}{n_2 s}$

We see from Example 34-3 that the magnification due to refraction at a spherical surface is

$$m = \frac{y'}{y} = -\frac{n_1 s'}{n_2 s} \qquad \text{34-6}$$

Example 34-4

A fish is in a spherical bowl of water with an index of refraction of 1.33. The radius of the bowl is 15 cm. The fish looks through the bowl and sees a cat sitting on the table with its nose 10 cm from the bowl (Figure 34-23). Find (*a*) the image distance for the cat's nose, and (*b*) the magnification of the nose. Neglect any effect of the bowl's thin glass wall.

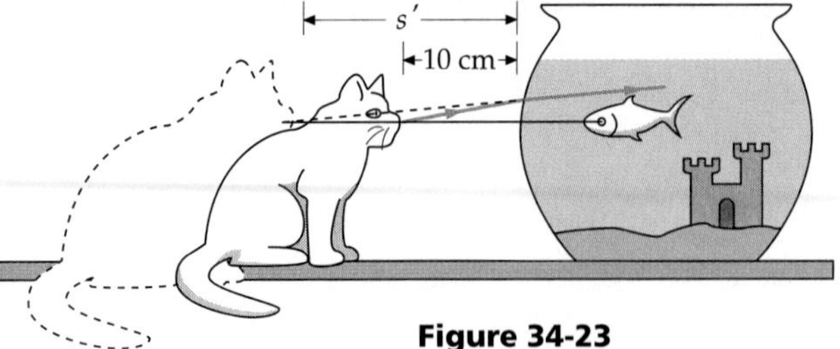

Figure 34-23

Picture the Problem We find the image distance s' using Equation 34-5 and the magnification using Equation 34-6. Since we are interested in light that goes from the cat's nose to the fish, it follows that air is the incident side and water is the transmission side. With these identifications, we have $n_1 = 1$, $n_2 = 1.33$, $s = +10$ cm (real object), and $r = +15$ cm (center of curvature on the transmission side).

(*a*)1. Substitute numerical values into Equation 34-5 to find a relation for the image distance s':

$$\frac{n_1}{s} + \frac{n_2}{s'} = \frac{n_2 - n_1}{r}$$

$$\frac{1}{10\text{ cm}} + \frac{1.33}{s'} = \frac{1.33 - 1}{15\text{ cm}}$$

$$\frac{1.33}{s'} = \frac{0.33}{15\text{ cm}} - \frac{1}{10\text{ cm}}$$

2. Solve for s':

$$s' = -17.1\text{ cm}$$

(*b*) Substitute numerical values into Equation 34-6 to find the magnification m:

$$m = -\frac{n_1 s'}{n_2 s} = -\frac{(1)(-17.1\text{ cm})}{(1.33)(10\text{ cm})} = 1.29$$

Remarks Since s' is negative, the image is virtual; that is, the image is on the incident side of the bowl, as shown in Figure 34-23. The cat appears to be slightly farther away ($|s'| > s$), but larger ($|m| > 1$). The fact that m is positive indicates that the image is upright.

Exercise If the fish is 7.5 cm from the surface of the bowl, find (*a*) the location of the fish and (*b*) its magnification, as seen by the cat. (*Answers* $n_1 = 1.33$, $n_2 = 1$, $s = 7.5$ cm, $r = -15$ cm, thus (*a*) $s' = -6.44$ cm, and (*b*) $m = 1.14$. The fish appears slightly closer and larger.)

Because of refraction, the apparent depth of the submerged portion of the straw is less than the real depth. Consequently, the straw appears to be bent. A reflected image of the straw is also seen.

We can use Equation 34-5 to find the **apparent depth** of an object under water when it is viewed from directly overhead. For this case, the surface is a plane surface, so the radius of curvature is infinite. The image and object distances are related by

$$\frac{n_1}{s} + \frac{n_2}{s'} = 0$$

where n_1 is the index of refraction of the first medium (water) and n_2 is that of the second medium (air). The apparent depth is therefore

$$s' = -\frac{n_2}{n_1}s \qquad 34\text{-}7$$

The negative sign indicates that the image is virtual and on the same side of the refracting surface as the object, as shown in the ray diagram in Figure 34-24. The magnification is

$$m = -\frac{n_1 s'}{n_2 s} = +1$$

Since $n_2 = 1$ for air, we see from Equation 34-7 that the apparent depth equals the real depth divided by the index of refraction of water.

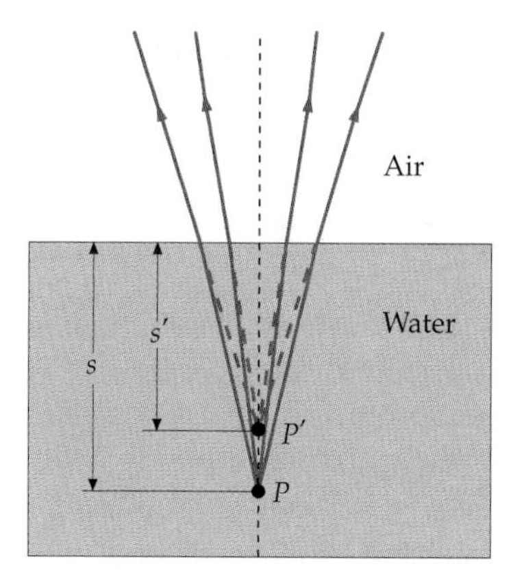

Figure 34-24 Ray diagram for the image of an object in water as viewed from directly overhead. The depth of the image is less than the depth of the object.

Example 34-5

Find the apparent depth of a fish resting 1 m below the surface of water ($n = \frac{4}{3}$).

Picture the Problem Since the light originates in the water, use $n_1 = \frac{4}{3}$, and $n_2 = 1$.

Calculate s' from Equation 34-7 using the given values for n_1, n_2, and s:

$$s' = \frac{1}{4/3}(1\text{ m}) = \frac{3}{4}(1\text{ m}) = 0.75\text{ m}$$

Remarks The apparent depth is three-fourths the actual depth, so the fish appears to be 75 cm below the surface. Note that this result holds only when the object is viewed from directly overhead so that the rays are paraxial.

Thin Lenses

The most important application of Equation 34-5 for refraction at a single surface is finding the position of the image formed by a lens. This is done by considering the refraction at each surface of the lens separately to derive an equation relating the image distance to the object distance, the radius of curvature of each surface of the lens, and the index of refraction of the lens.

We will consider a very thin lens of index of refraction n with air on both sides. Let the radii of curvature of the surfaces of the lens be r_1 and r_2. If an object is at a distance s from the first surface (and therefore from the lens), the distance s_1' of the image due to refraction at the first surface can be found using Equation 34-5:

$$\frac{1}{s} + \frac{n}{s_1'} = \frac{n-1}{r_1} \qquad \text{34-8}$$

This image is not formed because the light is again refracted at the second surface. Figure 34-25 shows the case when the image distance s_1' for the first surface is negative, indicating a virtual image to the left of the surface. Rays in the glass refracted from the first surface diverge as if they came from the image point P_1'. They strike the second surface at the same angles as if there were an object at this image point. The image for the first surface therefore becomes the object for the second surface. Since the lens is of negligible thickness, the object distance is equal in magnitude to s_1', but since object distances in front of the surface are positive whereas image distances are negative there, the object distance for the second surface is $s_2' = -s_1'$.* We now write Equation 34-5 for the second surface with $n_1 = n$, $n_2 = 1$, and $s = -s_1'$. The image distance for the second surface is the final image distance s' for the lens:

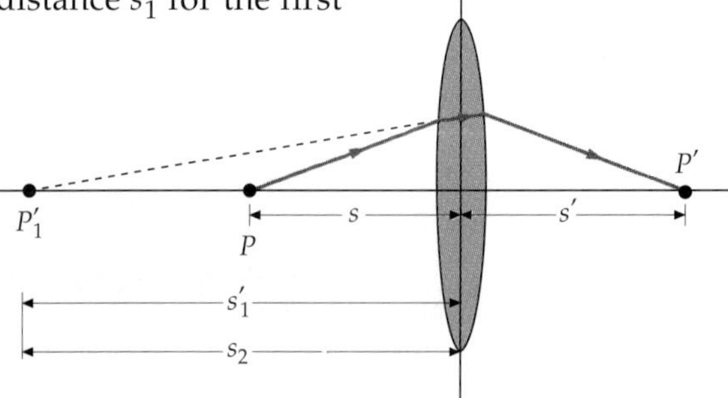

Figure 34-25 Refraction occurs at both surfaces of a lens. Here, the refraction at the first surface leads to a virtual image at P_1'. The rays strike the second surface as if they came from P_1'. Since image distances are negative when the image is on the incident side of the surface whereas object distances are positive for objects there, $s_2 = -s_1'$ is the object distance for the second surface of the lens.

$$\frac{n}{-s_1'} + \frac{1}{s'} = \frac{1-n}{r_2} \qquad \text{34-9}$$

* If s_1' were positive, the rays would converge as they strike the second surface. The object for the second surface would then be to the right of the surface. This object would be a virtual object. Again, $s_2 = -s_1'$.

We can eliminate the image distance for the first surface s'_1 by adding Equations 34-8 and 34-9. We obtain

$$\frac{1}{s} + \frac{1}{s'} = (n - 1)\left(\frac{1}{r_1} - \frac{1}{r_2}\right) \qquad 34\text{-}10$$

Equation 34-10 gives the image distance s' in terms of the object distance s and the properties of the thin lens—r_1, r_2, and the index of refraction n. As with mirrors, the focal length f of a thin lens is defined as the image distance when the object distance is infinite. Setting s equal to infinity and writing f for the image distance s', we obtain

$$\frac{1}{f} = (n - 1)\left(\frac{1}{r_1} - \frac{1}{r_2}\right) \qquad 34\text{-}11$$

Lens-maker's equation

Equation 34-11 is called the **lens-maker's equation;** it gives the focal length of a thin lens in terms of the properties of the lens. Substituting $1/f$ for the right side of Equation 34-10, we obtain

$$\frac{1}{s} + \frac{1}{s'} = \frac{1}{f} \qquad 34\text{-}12$$

Thin-lens equation

This **thin-lens equation** is the same as the mirror equation (Equation 34-3). Recall, however, that the sign conventions for refraction are somewhat different from those for reflection. For lenses, the image distance s' is positive when the image is on the transmission side of the lens, that is, when it is on the side opposite the side upon which light is incident. The sign convention for r in Equation 34-11 is the same as that for refraction at a single surface. The radius is positive if the center of curvature is on the transmission side of the lens and negative if it is on the incident side.

Figure 34-26*a* shows the wavefronts of plane waves incident on a double convex lens. The central part of the wavefront strikes the lens first. Since the wave speed in the lens is less than that in air (assuming $n > 1$), the central part of the wavefront lags behind the outer parts, resulting in a spherical wave that converges at the focal point F'. The rays for this situation are shown in Figure 34-26*b*. Such a lens is called a **converging lens.** Since its focal length as calculated from Equation 34-11 is positive, it is also called a **positive lens.** Any lens that is thicker in the middle than at the edges is a converging lens (providing that the index of refraction of the lens is greater than that of the surrounding medium). Figure 34-27 shows the

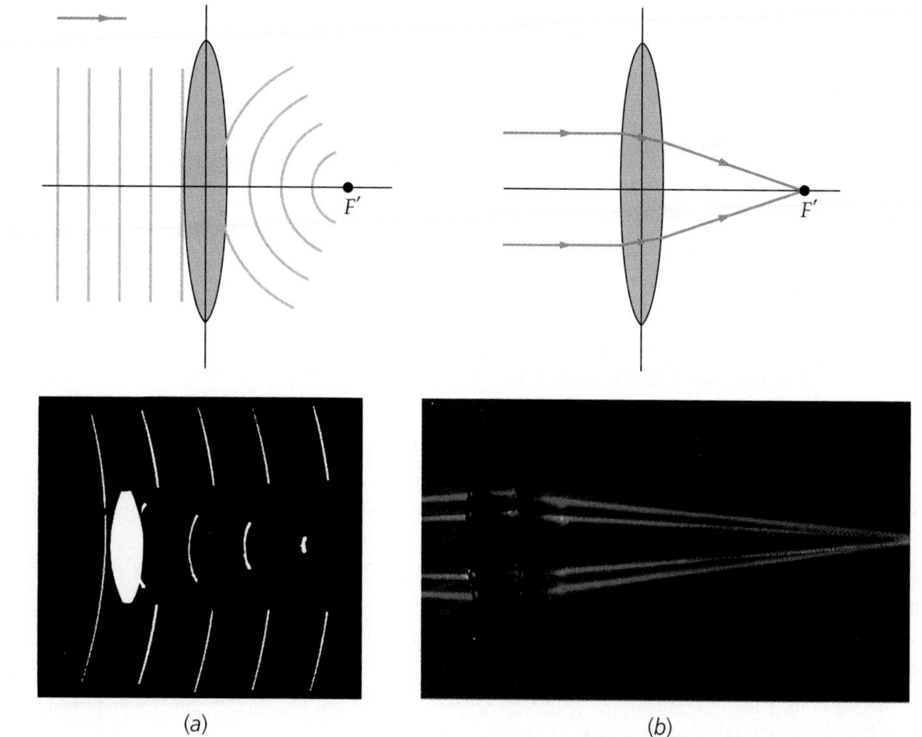

Figure 34-26 (*a*) *Top:* Wavefronts for plane waves striking a converging lens. The central part of the wavefront is retarded more by the lens than the outer part, resulting in a spherical wave that converges at the focal point F'. *Bottom:* Wavefronts passing through a lens, shown by a photographic technique called *light-in-flight-recording* that uses a pulsed laser to make a hologram of the wavefronts of light. (*b*) *Top:* Rays for plane waves striking a converging lens. The rays are bent at each surface and converge at the focal point. *Bottom:* Photograph of rays focused by a converging lens.